BINARY FLUORIDES
Free Molecular Structures and Force Fields
A Bibliography (1957-1975)

NSRDS Bibliographic Series

Office of Standard Reference Data
National Bureau of Standards
Washington, D.C.

EQUILIBRIUM PROPERTIES OF FLUID MIXTURES • 1975
M. J. Hiza, A. J. Kidnay, and R. C. Miller

BINARY FLUORIDES • 1976
Donald T. Hawkins, Lawrence S. Bernstein,
Warren E. Falconer, and William Klemperer

BINARY FLUORIDES
Free Molecular Structures and Force Fields
A Bibliography (1957-1975)

Compiled by

Donald T. Hawkins,
Lawrence S. Bernstein,
Warren E. Falconer,

Bell Telephone Laboratories
Murray Hill, New Jersey

and

William Klemperer

Harvard University
Cambridge, Massachusetts

IFI/PLENUM • NEW YORK-WASHINGTON-LONDON

Library of Congress Cataloging in Publication Data

Main entry under title:

Binary fluorides.

(NSRDS bibliographic series)
1. Fluorides—Bibliography. I. Hawkins, Donald T. II. Series.
Z5524.F6B54 [QD181.F1]   016.5304'1'08s   [016.546'731'2]
ISBN 978-1-4684-6149-7   ISBN 978-1-4684-6147-3 (eBook)   76-40174
DOI 10.1007/978-1-4684-6147-3

© 1976 IFI/Plenum Data Company
Softcover reprint of the hardcover 1st edition 1976

A Division of Plenum Publishing Corporation
227 West 17th Street, New York, N.Y. 10011

ACKNOWLEDGMENT

We are heavily indebted to Mrs. Marion A. Marasco for her substantial assistance in verifying references and performing a multitude of other editorial tasks. The layout and appearance of this volume are largely a result of her care and diligence.

INTRODUCTION

Coverage

For some time, we have contemplated a comprehensive review of the structures and force fields of the binary fluorides. This bibliography of 1498 references marks the first step of that effort. We are publishing this material now rather than waiting until the review is complete some two years hence because we believe that the information already accumulated will be of immediate use to a broad spectrum of researchers.

Anyone ambitious enough to read through all the articles on binary fluorides will find that the structures and force fields of many of these molecules are at present unknown. For example, it has not been clearly established to which point group(s) the lanthanide trifluorides should be assigned. There remain interesting problems relating to the role of Jahn–Teller and pseudo-Jahn–Teller distortions in some of the transition metal fluorides such as VF_5, MoF_5, ReF_6, and ReF_7, to name only a few. One also finds fascinating examples of large-amplitude motions, or pseudorotations, as they are often called, in such molecules as XeF_6, IF_7, and PF_5. For those binary fluorides whose equilibrium geometries are precisely known, there still exists the problem of accurately determining the harmonic force field. In a few cases, most notably the Group VA trifluorides, there has been some attempt made at extracting the cubic and quartic contributions to the force field.

Even those not directly interested in binary fluorides will find a wealth of useful information on the application of various techniques to the study of structure and force fields. There are numerous examples here of the use of electron diffraction, infrared, Raman, and microwave spectroscopy, NMR, ESR, mass spectrometry, electric field deflection, thermodynamic measurements (vapor pressure and heat capacity), and *ab initio* and semiempirical calculations, as applied to the solution of structural problems. We have included a representative collection of material (not necessarily exclusively relating to fluorides) on vibrational analysis, which can be found in the sections labeled MX_2 to MX_9 and also in a section labeled Vibrational Analysis. These sections contain examples of a variety of different force fields. One can find formulas for many of the standard G matrices, F matrices, and Coriolis zeta coefficients employed in force field calculations. There is also information on the calculation and use of vibrational amplitudes, which are normally coupled with electron-diffraction-determined amplitudes as a means of refining force fields.

We have limited the scope of this bibliography to include information directly related to the structure and force field of an *isolated* binary fluoride molecule in its ground electronic state. Free molecular structures are therefore of greatest interest; the most commonly studied diatomic fluorides such as the alkali fluorides, HF, or molecules with fluorine-bridged chains are omitted. Likewise, we have not included molecules with more than one nonfluorine atom, even though many of these are true binary fluorides. Our interest focuses on the stereochemical understanding of the simpler fluorides with three or more atoms having only M–F and not M–M or M–X bonds. Properties of the solids, such as magnetic properties, have generally been omitted. Crystallographic data have been included only for molecular solids or near-molecular solids in which intermolecular distortions are minimal. Thermodynamic properties are covered only to the extent that structural information can be derived from them.

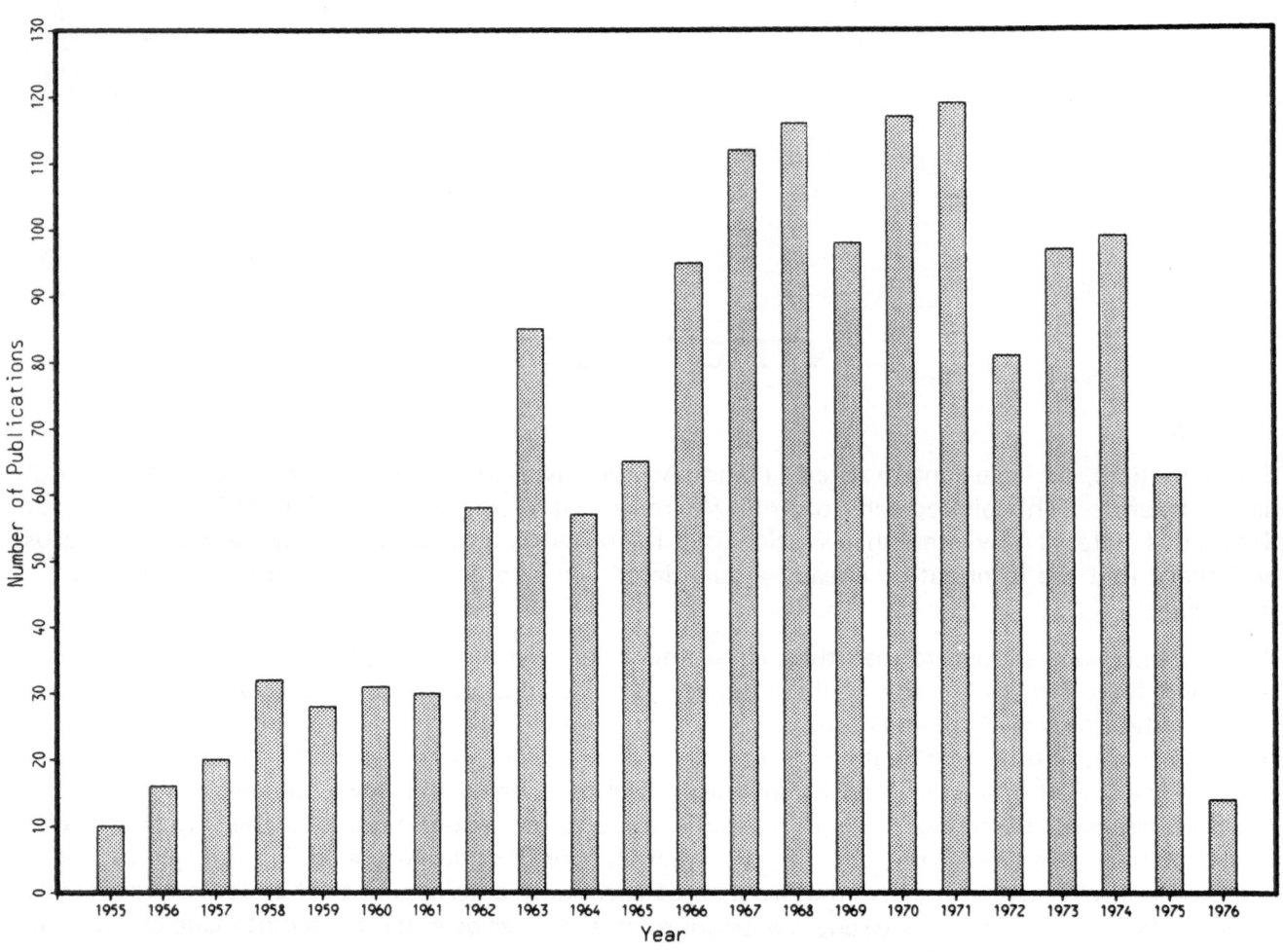

Figure 1. Growth of the literature on binary fluorides.

In any project of this magnitude, it is unlikely that one unearths every article of importance. Consequently we appeal to the users of this compilation to inform us of any omissions. As the total undertaking will be evolving in time, we would also appreciate receiving information about current work in this area.

Organization

The bibliography is organized into 240 sections, according to substance. The sections are ordered alphabetically by element name and, for a given element, in order of increasing number of fluorine atoms. Within each section, entries are listed alphabetically by first author. Each unique reference bears a serial number In order that each section contain a complete listing of the references on a given binary fluoride, references dealing with more than one fluoride are listed in all sections to which they are pertinent, but only one occurrence of the reference bears a serial number or generates an entry in the indexes. Relevant reviews, books, and miscellaneous papers are listed in the first section of the bibliography. Following the sections dealing with the specific fluorides, sections containing general references on groups of fluorides, MX_n molecules (n = 2 to 9), and vibrational analysis in general appear. In order to prevent undue proliferation of entries, a few papers dealing with a large number of fluorides have been listed either in the review section or, when appropriate, in one of the sections covering groups of fluorides.

In the various bibliography sections, the title, author(s), and source are listed for each reference. Most of the references were obtained from *Chemical Abstracts,* which was searched from Volume 51 through Volume 84, issue 22. Although most of the references fall in the time period 1955–1975, a few earlier references are included.

Permuted title and author indexes follow the bibliographic listing. The permuted title index serves as a subject index and provides many additional entry points into the bibliography. Terms beginning

Table 1. Authorship of Papers on Binary Fluorides

Number of papers, P	Number of authors' names appearing on P papers
1	1146
2	283
3	101
4	62
5	34
6	13
7	20
8	7
9	6
10	7
11	2
12	2
13	2
14	1
15	0
16	8
17	3
21	1
22	1
24	2
27	1
29	2
40	1
63	1
Total	1706

Table 2. Sources of Literature on Binary Fluorides

Source	Number of papers
Books, conference proceedings	63
Reports	23
Theses	24
Journals:	
1 paper each (76 journals)	76
2 papers each (25 journals)	50
3 papers each (7 journals)	21
4 papers each (5 journals)	20
5 papers each (2 journals)	10
More than 5 papers each (48 journals) (see Table 3)	1211
Total	1498

Table 3. Journals Publishing More Than 5 Papers on Binary Fluorides

Journal	Number of papers
J. Chem. Phys.	313
J. Mol. Spectrosc.	86
J. Phys. Chem.	60
J. Amer. Chem. Soc.	53
Indian J. Pure Appl. Phys.	41
J. Chem. Soc.	37
Trans. Faraday Soc.	31
Spectrochim. Acta	30
Inorg. Chem.	28
Chem. Phys. Lett.	28
J. Mol. Struct.	26
Can. J. Phys.	26
Z. Naturforsch.	23
Russ. J. Phys. Chem.	23
Opt. Spectrosc.	21
Mol. Phys.	20
Z. Phys. Chem.	18
Russ. J. Inorg. Chem.	18
J. Phys. B	17
Can. J. Chem.	17
Acta Chem. Scand.	17
J. Inorg. Nucl. Chem.	16
Vest. Mosk. Univ.	15
Proc. Phys. Soc.	15
Chem. Commun.	15
High Temp. Sci.	14
J. Struct. Chem.	13
Theor. Chim. Acta	12
Spectrosc. Lett.	12
J. Fluorine Chem.	12
Compt. Rend.	12
Bull. Soc. Chim. Belg.	11
Z. Anorg. Allgem. Chem.	10
Phys. Rev.	10
J. Appl. Spectrosc.	10
Inorg. Nucl. Chem. Lett.	10
Curr. Sci.	10
Science	9
J. Chem. Thermodyn.	9
Indian J. Phys.	9
Bull. Chem. Soc. Jap.	9
Aust. J. Chem.	8
High Temp.	7
J. Res. NBS	6
J. Less Common Metals	6
J. Chem. Eng. Data.	6
Izv. Sib. Otd. Akad. Nauk SSSR	6
Acta Crystallogr.	6
Total	1211

with common prefixes such as DI . . . , TRI . . . , NON . . . , etc. are indexed under the parent word, except when the parent word is "fluoride." This treatment allows the numerous entries under difluoride, trifluoride, etc. to file separately.

This bibliography is computer-produced and formated by the Bell Laboratories proprietary system BELDEX.* The BELDEX system formats and indexes bibliographic and other types of material.

Statistical Information

Figure 1 shows the yearly growth of the literature on binary fluorides. The number of unique references (excluding cross-references) appearing in each year is plotted against the year. Growth was slow from 1955 to 1961, leveling off around 30 papers per year in 1958 to 1961. In 1962 there was a large increase, with nearly twice as many papers published. We ascribe this discontinuity to the upsurge of interest in inorganic fluorine chemistry, including the properties of inorganic fluorides, following the landmark discovery in 1961 by Neil Bartlett that fluorides of the noble gas xenon (which was previously thought to be chemically inert) could be synthesized. From 1962 to the peak year of 1971, the yearly number of publications grew steadily to a maximum of 119. Since 1971, there has been a slight decline, with an average of 85 to 90 in recent years. (Data for 1975 and 1976 are incomplete due to the time lag in abstracting services.)

A total of 1706 authors' names appear on the 1498 papers in this bibliography. Table 1 shows the numbers of authors whose names appear on one paper, two papers, and so on. 67% of the names are associated with only a single paper. J. L. Margrave and G. Nagarajan are the most prolific authors in this collection of references, with 63 and 40 papers, respectively.

The sources of the literature on fluorides were analyzed and are listed in Table 2. For each type of literature — books and conference proceedings, reports, theses, and journal articles — the number of entries is listed. The journal literature is further subdivided into those journals that contributed one, two, three, four, five, or more than five articles. The 48 journals contributing more than five articles each are listed in Table 3. These 48 journals account for 1211 papers, 81% of the total. The *Journal of Chemical Physics* heads the list with 313 papers, more than the next six journals combined. Approximately half of the references in this bibliography appear in 12 journals.

There are 240 sections in this bibliography, containing 1498 unique references. Including the 686 cross references, the bibliography has 2184 entries. There are 9651 entries in the permuted title index, and 3380 entries in the author index.

*The BELDEX system has been copyrighted by Bell Laboratories. For further information see W. K. Lowry, Use of computers in information systems, *Science* 175:841–846 (1972).

CONTENTS

BIBLIOGRAPHY

REVIEWS, MISCELLANEOUS   ☐ ☐

0001 HALIDES OF THE TRANSITION ELEMENTS. VOLUME-1: FIRST ROW TRANSITION METALS. VOLUME-2: SECOND AND THIRD ROW TRANSITION METALS. VOLUME-3: LANTHANIDES AND ACTINIDES.
CANTERFORD JH + COLTON R + BROWN D
WILEY-INTERSCIENCE, 3 VOLUMES, 1968.

0002 FLUORIDES OF THE HEAVY TRANSITION METALS. (REVIEW)
CANTERFORD JH + COLTON R + O'DONNELL TA
REV PURE APPL CHEM 17: 123-32 (1967)

0003 GAS PHASE STRUCTURES AND MASS SPECTRA OF BINARY PENTAFLUORIDES.
FALCONER WE + JONES GR + SUNDER WA + VASILE MJ + MUENTER AA + DYKE TR + KLEMPERER W
J FLUORINE CHEM 4: 213-34 (1974)

0004 REACTIVITY AND THERMAL STABILITY OF HEXAFLUORIDES.
GALKIN NP + TUMANOV YN
RUSS CHEM REV 40(2): 154-63 (1971)

0005 HALOGEN CHEMISTRY. (VOLUMES 1, 2, 3)
GUTMANN V (ED)
ACADEMIC PRESS, NEW YORK, 1967.

0006 FLUORIDES OF THE ACTINIDE ELEMENTS.
HODGE N
P138-82 OF ADVAN FLUORINE CHEM, VOL 2, STACEY M + TATLOW JC + SHARPE AG(EDS), BUTTERWORTHS, 1961.

0007 FLUORIDES OF THE MAIN GROUP ELEMENTS.
KEMMITT RDW + SHARP DWA
P142-252 OF ADVAN FLUORINE CHEM, VOL 4, STACEY M + TATLOW JC + SHARPE AG(EDS), BUTTERWORTHS, 1965.

0008 MEAN SQUARE AMPLITUDES OF INTERATOMIC DISTANCES IN HEXAFLUORIDE MOLECULES.
KIMURA M + KIMURA K
J MOL SPECTROSC 11: 368-77 (1963)

0009 CHARACTERISTICS OF THE ELECTRONIC STRUCTURE OF FLUORIDES OF NONTRANSITION ELEMENTS FROM THE DATA OF LCAO MO SCF NONEMPIRICAL CALCULATIONS.
KLIMENKO NM + CHARKIN OP + SMOLYAR AE + RYABOV MA + ZYUBIN AS + ZYUBINA TS
P10-11 OF MENDELEEVSK S'EZD PO OBSHCH I PRIKL KHIMMI REF DOKL I SOOBSHCH, 1975, ISSUE 1. (IN RUSSIAN)

0010 CHEMICAL COMPOUNDS OF XENON AND OTHER NOBLE GASES.
MALM JG + APPELMAN EH
AT ENERGY REV 7(3): 3-48 (1969)

0011 MATRIX ISOLATION INFRARED SPECTROSCOPY OF HIGH TEMPERATURE SPECIES. (FLUORIDES, OXIDES, CHLORIDES)
MARGRAVE JL + HASTIE JW + HAUGE RH
AMER CHEM SOC, DIV FUEL CHEM PREPR 14(2): E11-13 (1969)

0012 THERMODYNAMIC PROPERTIES OF SOME CHALCOGEN FLUORIDES.
O'HARE PAG
US AT ENERGY COMM, 1968, 95P. ANL-7315

0013 TRANSITION METAL PENTAFLUORIDES AND RELATED COMPOUNDS.
PEACOCK RD
P113-46 OF ADVAN FLUORINE CHEM, VOL 7, TATLOW JC + PEACOCK RD + HYMAN HH(EDS), BUTTERWORTHS, 1973.

0014 VIBRATIONAL SPECTRA OF INORGANIC FLUORIDES.
REYNOLDS DJ
P1-68 OF ADVAN FLUORINE CHEM, VOL 7, TATLOW JC + PEACOCK RD + HYMAN HH(EDS), BUTTERWORTHS, 1973.

0015 TRANSITION METAL FLUORIDES AND THEIR COMPLEXES.
SHARPE AG
P29-67 OF ADVAN FLUORINE CHEM, VOL 1, STACEY M + TATLOW JC + SHARPE AG(EDS), BUTTERWORTHS, 1960.

0016 CALCULATIONS OF ELECTRON STRUCTURES OF COMPOUNDS OF HALOGENS AND INERT GASES BY THE NONEMPIRICAL NDDO-2 (ALPHA, BETA) METHOD. PART-3: ENERGY LEVELS, WAVE FUNCTIONS AND AO POPULATIONS OF POLYATOMIC FLUORIDES.
SMOLYAR AE + CHARKIN OP + KLIMENKO NM
J STRUCT CHEM 15(6): 885-93 (1974)

0017 PHYSICAL AND CHEMICAL PROPERTIES OF HALOGEN FLUORIDES.
STEIN L
P133-224 OF HALOGEN CHEMISTRY, VOL-1, GUTMANN V(ED), ACADEMIC PRESS, 1967.

0018 FORCE CONSTANTS OF METAL HEXAFLUORIDES.
THAKUR SN + RAI DK
J MOL SPECTROSC 19: 341-8 (1966)

0019 APPLICATION OF A MOLECULAR BEAM SOURCE MASS SPECTROMETER TO THE STUDY OF REACTIVE FLUORIDES.
VASILE MJ + JONES GR + FALCONER WE
P557-63 OF INT MASS SPECTROM CONF, 6TH PROC, EDINBURGH, SCOTLAND, 1973.

0020 SOME PROPERTIES OF THE HEXAFLUORIDE MOLECULES. (REVIEW)
WEINSTOCK B
REC CHEM PROGR 23(1): 23-50 (1962)

0021 VIBRATIONAL PROPERTIES OF HEXAFLUORIDE MOLECULES.
WEINSTOCK B + GOODMAN GL
P169-316 OF ADVAN CHEM PHYS, VOL 9, PRIGOGENE I(ED), INTERSCIENCE, 1965.

ALUMINUM MONOFLUORIDE   AlF

DISSOCIATION ENERGIES OF THE GASEOUS MONOHALIDES OF BORON, ALUMINUM, GALLIUM, INDIUM, AND THALLIUM.
BARROW RF
TRANS FARADAY SOC 56: 952-58 (1960)

0022 SPECTROSCOPIC AND THERMODYNAMIC PROPERTIES OF ALUMINUM MONOFLUORIDE.
BARROW RF + JOHNS JWC + SMITH FJ
TRANS FARADAY SOC 52: 913-16 (1956)

0023 SINGLET ELECTRONIC STATES OF ALUMINUM MONOFLUORIDE.
BARROW RF + KOPP I + SCULLMAN R
PROC PHYS SOC 82: 635-36 (1963)

0024 ALUMINUM AND ALKALI HALIDE MONOMERS MEAN AMPLITUDES OF VIBRATION WITH LOW TEMPERATURE ANOMALIES.
CYVIN SJ
J MOL STRUCT 8: 43-8 (1971)

INFRARED SPECTRA OF MATRIX ISOLATED SPECIES IN THE GALLIUM- FLUORINE SYSTEM.
HASTIE JW + HAUGE RH + MARGRAVE JL
J FLUORINE CHEM 3(3-4): 285-91 (1973)

0025 MICROWAVE ABSORPTION SPECTRA OF ALUMINUM MONOFLUORIDE, GALLIUM MONOFLUORIDE, INDIUM MONOFLUORIDE, AND THALLIUM MONOFLUORIDE.
HOEFT J + LOVAS FJ + TIEMANN E + TORRING T
Z NATURFORSCH 25A: 1029-35 (1970)

0026 ZEEMAN EFFECT IN THE MICROWAVE ROTATIONAL SPECTRUM OF THE ALUMINUM FLUORIDE MOLECULE.
HONERJAEGER R + TISCHER R
Z NATURFORSCH 29A(2): 342-5 (1974) (IN GERMAN)

0027 ELECTRONIC STATES OF GASEOUS ALUMINUM MONOFLUORIDE.
KOPP I + BARROW RF
J PHYS B 3(10): L118-20 (1970)

1

0028 HIGH TEMPERATURE MICROWAVE SPECTROSCOPY OF
      ALUMINUM MONOFLUORIDE AND ALUMINUM CHLORIDE.
      LIDE DR
      J CHEM PHYS 42: 1013-18 (1965)

0029 MICROWAVE SPECTRUM OF ALUMINUM MONOFLUORIDE.
      LIDE DR
      J CHEM PHYS 38: 2027 (1963)

0030 EMISSION SPECTRUM OF ALUMINUM MONOFLUORIDE IN
      THE VACUUM ULTRAVIOLET.
      NAUDE SM + HUGO TJ
      CAN J PHYS 35: 65-70 (1957)

0031 ALUMINUM MONOFLUORIDE. (VAPOR PRESSURE)
      SEMENKOVICH SA
      TR VSES ALYUMIN MAGNIEVYI INST NO 44: 113-19
      (1960) (IN RUSSIAN)

      POTENTIAL ENERGY CURVES AND DISSOCIATION
      ENERGIES OF DIATOMIC BORON AND ALUMINUM HALIDES.
      SINGH J + NAIR KPR + RAI DK
      J MOL STRUCT 6: 328-32 (1970)

0032 POTENTIAL ENERGY CURVES AND NATURE OF BINDING IN
      GROUP-3A MONOHALIDES.
      THAKUR SN + SINGH RB + RAI DK
      J SCI IND RES 27(9): 339-47 (1968)

0033 MILLIMETER AND SUBMILLETER WAVE SPECTRUM AND
      MOLECULAR CONSTANTS OF ALUMINUM MONOFLUORIDE.
      WUSE FC + GORDY W + PEARSON EF
      J CHEM PHYS 52: 3887-9 (1970)

ALUMINUM DIFLUORIDE AlF_2

0034 HEAT OF ATOMIZATION OF ALUMINUM DIFLUORIDE.
      EHLERT TC + MARGRAVE JL
      J AMER CHEM SOC 86: 3901 (1964)

      MEAN AMPLITUDES OF VIBRATION FOR BORON
      DIFLUORIDE CARBON DIFLUORIDE, NITROGEN
      DIFLUORIDE, ALUMINUM DIFLUORIDE, PHOSPHORUS
      DIFLUORIDE, AND ZIRCONIUM DIFLUORIDE.
      NAGARAJAN G
      INDIAN J PURE APPL PHYS 2: 341-3 (1964)

0035 MASS SPECTROMETRIC DETERMINATION OF THE HEATS OF
      FORMATION OF THE GASEOUS MOLECULES ALUMINUM
      OXYDIFLUORIDE AND ALUMINUM DIFLUORIDE.
      UY OM + SRIVASTAVA RD + FARBER M
      HIGH TEMP SCI 4(3): 227-30 (1972)

ALUMINUM TRIFLUORIDE AlF_3

0036 ELECTRON DIFFRACTION INVESTIGATION OF THE
      MOLECULAR STRUCTURES OF THE ALUMINUM HALIDES.
      AKISHIN PA + RAMBIDI NG + ZASORIN EZ
      SOV PHYS CRYSTALLOGR 4(2): 167-73 (1960)

0037 SUBLIMATION OF ALUMINUM TRIFLUORIDE AND THE
      INFRARED SPECTRUM OF ALUMINUM TRIFLUORIDE.
      BUCHLER A + MARRAM EP + STAUFFER JL
      J PHYS CHEM 71: 4139-40 (1967)

0038 HARMONIC FORCE FIELDS AND MEAN AMPLITUDES FOR
      ALUMINUM TRIFLUORIDE MONOMER AND DIMER.
      CYVIN SJ
      SPECTROSC LETT 7(6): 255-61 (1974)

0039 HEAT OF FORMATION OF ALUMINUM TRIFLUORIDE BY
      DIRECT COMBINATION OF THE ELEMENTS.
      DOMALSKI ES + ARMSTRONG GT
      J RES NBS A 69: 137-47 (1965)

0040 MASS SPECTROMETRIC INVESTIGATION OF THE
      THERMODYNAMIC PROPERTIES OF ALUMINUM
      TRIFLUORIDE.
      EROKHIN EV + ZHEGUL'SKAYA NA + SIDOROV LN
      + AKISHIN NA
      INORG MATER 3(5): 779-80 (MAY 1967)

0041 VAPOR PRESSURE OF ALUMINUM TRIFLUORIDE.
      EVSEEV EM + POZHARSKAYA GV + NESMEYANOV AN
      + GERASIMOV YI
      RUSS J INORG CHEM 4: 1000-1 (1959)

0042 RECENT DEVELOPMENTS IN MULTIPLE BEAM
      INTERFEROMETRY IN THE ULTRAVIOLET REGION
      1700-2500 ANGSTROMS. (ALUMINUM- MAGNESIUM
      FLUORIDE)
      GUERN Y + BIDEAU-MEHU A + ABJEAN R
      + JOHANNIN-GILLES A
      OPT COMMUN 12(1): 66-70 (1974)

      INFRARED SPECTRA OF MATRIX ISOLATED SPECIES IN
      THE GALLIUM- FLUORINE SYSTEM.
      HASTIE JW + HAUGE RH + MARGRAVE JL
      J FLUORINE CHEM 3(3-4): 285-91 (1973)

      FREE ENERGIES OF FORMATION OF SOME INORGANIC
      FLUORIDES BY SOLID STATE EMF MEASUREMENTS.
      (THORIUM TETRAFLUORIDE, ALUMINUM TRIFLUORIDE,
      NICKEL DIFLUORIDE, LEAD DIFLUORIDE, COLBALT
      DIFLUORIDE, AND URANIUM TRIFLUORIDE)
      HEUS RJ + EGAN JJ
      Z PHYS CHEM NEUE FOLGE 49: 38-43 (1966)

0043 STANDARD ENTHALPY OF FORMATION OF ALUMINUM
      TRIFLUORIDE.
      KOLESOV VP + MARTYNOV AM + SKURATOV SM
      RUSS J INORG CHEM 6: 1326 (1961)

0044 BOND ENERGY IN THE ALUMINUM TRIFLUORIDE
      MOLECULE.
      KRASNOV KS
      IZV VYSSH UCHEB ZAVED, FIZ 7(2): 344-5 (1964)
      (IN RUSSIAN)

0045 INFRARED SPECTRA OF GASEOUS ALUMINUM
      TRIFLUORIDE, LITHIUM ALUMINUM TETRAFLUORIDE, AND
      SODIUM ALUMINUM TETRAFLUORIDE.
      MCCORY LD + PAULE RC + MARGRAVE JL
      J PHYS CHEM 67: 1086-7 (1963)

0046 SPECTROSCOPIC STUDIES OF MOLECULAR CONSTANTS IN
      SOME POLYATOMIC MOLECULES.
      NAGARAJAN G
      Z NATURFORSCH 21A: 244-51 (1966)

0047 HIGH TEMPERATURE NEGATIVE IONS. GASEOUS GROUP-3
      FLUORIDES.
      PETTY F + WANG JLF + STEIGER RP + HARLAND PW
      + FRANKLIN JL + MARGRAVE JL
      HIGH TEMP SCI 5(1): 25-33 (1973)

0048 FLUORINE BOMB CALORIMETRY. PART-22: THE ENTHALPY
      OF FORMATION OF ALUMINUM TRIFLUORIDE.
      RUDZITIS E + FEDER HM + HUBBARD WN
      INORG CHEM 9: 1716-7 (1967)

0049 VAPOR PRESSURE OF ZINC DIFLUORIDE, CADMIUM
      DIFLUORIDE, MAGNESIUM DIFLUORIDE, CALCIUM
      DIFLUORIDE, STRONTIUM DIFLUORIDE, BARIUM
      DIFLUORIDE, AND ALUMINUM TRIFLUORIDE.
      RUFF O + LE BOUCHER L
      Z ANORG ALLGEM CHEM 219: 376-81 (1934) (IN
      GERMAN)

0050 MEAN AMPLITUDES OF VIBRATION, CORIOLIS COUPLING
      CONSTANTS, SHRINKAGE EFFECT, AND THERMODYNAMIC
      FUNCTIONS OF ALUMINUM TRIFLUORIDE.
      SHANMUGASUNDARAM G + NAGARAJAN G
      Z PHYS CHEM 240(5-6): 363-70 (1969)

0051 INFRARED SPECTRUM OF ALUMINUM TRIFLUORIDE,
      DIALUMINUM HEXAFLUORIDE, AND ALUMINUM
      MONOFLUORIDE BY MATRIX ISOLATION.
      SNELSON A
      J PHYS CHEM 71: 3202-7 (1967)

0052 THERMODYNAMICS OF THE VAPORIZATION OF ALUMINUM
      FLUORIDE AND CESIUM FLUORIDE.
      SONIN VI + POLYACHENOK OG
      RUSS J PHYS CHEM 47(6): 1612-13 (1973)

0053 HEAT OF SUBLIMATION OF ALUMINUM TRIFLUORIDE AND
HEAT OF FORMATION OF ALUMINUM MONOFLUORIDE.
WITT WP + BARROW RF
TRANS FARADAY SOC 55: 730-5 (1959)

0054 INFRARED SPECTRA OF ALUMINUM FLUORIDE IN SOLID
ARGON.
YANG YS + SHIRK JS
J MOL SPECTROSC 54(1): 39-42 (1975)

AMERICIUM TRIFLUORIDE AmF_3

MELTING POINTS OF CURIUM TRIFLUORIDE AND
AMERICIUM TRIFLUORIDE.
BURNETT JL
J INORG NUCL CHEM 28: 2454-5 (1966)

0055 VAPOR PRESSURES OF AMERICIUM TRIFLUORIDE AND
PLUTONIUM TRIFLUORIDE, HEATS AND FREE ENERGIES
OF SUBLIMATION.
CARNIGLIA SC + CUNNINGHAM BB
J AMER CHEM SOC 77: 1451-3 (1955)

AMERICIUM TETRAFLUORIDE AmF_4

0056 INVESTIGATION OF THE CHEMISTRY OF AMERICIUM.
(AMERICIUM TETRAFLUORIDE)
YAKOVLEV GN + KOSYAKOV VN
P373-84 OF CONF PEACEFUL USES AT ENERGY, 2ND,
1958, GENEVA, VOL 28, 686P.

ANTIMONY MONOFLUORIDE SbF

0057 ROTATIONAL ANALYSIS OF C3 SYSTEM OF THE ANTIMONY
MONOFLUORIDE MOLECULE.
ABRAHAM KC + PATEL MM
J PHYS B 3: 1183-5 (1970)

0058 A NEW BAND SYSTEM OF THE ANTIMONY MONOFLUORIDE
MOLECULE IN 4050-5450 ANGSTROM REGION.
ABRAHAM KC + PATEL MM
J PHYS B 3: 882-3 (1970)

0059 FINE STRUCTURE ANALYSIS OF THE C1 SYSTEM OF
ANTIMONY MONOFLUORIDE.
ABRAHAM KC + PATEL MM
J PHYS B 4: 1398-1402 (1971)

0060 EMISSION OF MOLECULAR ANTIMONY MONOFLUORIDE.
CHAKRAVORTY M + ABRAHAM KC + PATEL MM
J PHYS B 6: 757-60 (1973)

0061 DETERMINATION OF THE EFFECTIVE VIBRATIONAL
TEMPERATURE IN THE A-2 SYSTEM OF ANTIMONY
MONOFLUORIDE.
DUBE PS + RAI DK + SINGH NI
INDIAN J PURE APPL PHYS 9: 484-5 (1971)

0062 VIBRATIONAL ANALYSIS OF ULTRAVIOLET BANDS OF
ANTIMONY MONOFLUORIDE.
PATEL MM + ABRAHAM KC
INDIAN J PURE APPL PHYS 7: 641-3 (1969)

0063 ROTATIONAL ANALYSIS OF THE ULTRAVIOLET BANDS OF
THE ANTIMONY MONOFLUORIDE MOLECULE.
PREVOT F + COLIN R + JONES WE
J MOL SPECTROSC 56(3): 432-40 (1975)

0064 EMISSION SPECTRA OF ANTIMONY MONOFLUORIDE AND
BISMUTH MONOFLUORIDE.
RAO TAP + RAO PT
INDIAN J PHYS 36: 85-92 (1962)

0065 BAND SPECTRA OF BISMUTH MONOFLUORIDE AND
ANTIMONY MONOFLUORIDE.
ROCHESTER GD
PROC PHYS SOC 78: 614 (1961)

0066 ROTATIONAL STRUCTURE OF THE C3 SYSTEM OF
ANTIMONY FLUORIDE MOLECULE.
SHANKER R + SINGH IS
INDIAN J PURE APPL PHYS 10(5): 395-7 (1972)

0067 ROTATIONAL ANALYSIS OF THE 2550-2750 ANGSTROM
BAND SYSTEM AT ANTIMONY FLUORIDE MOLECULE.
SHANKER R + SRIVASTAVA SC + SINGH IS
INDIAN J PURE APPL PHYS 10(7): 541-4 (1972)

0068 ROTATIONAL ANALYSIS OF THE C2 SYSTEM OF ANTIMONY
MONOFLUORIDE.
SHANKER R + SINGH IS
CURR SCI 40: 341-2 (1971)

0069 ROTATIONAL ANALYSIS OF THE C1 SYSTEM OF ANTIMONY
MONOFLUORIDE.
SIVAJI C + RAO PT
CURR SCI 38: 432-33 (1969)

0070 NEAR ULTRAVIOLET AND VISIBLE BANDS OF ANTIMONY
MONOFLUORIDE.
VASUDEV R + JONES WE
J MOL SPECTROSC 59(3): 442-58 (1976)

0071 BO(+)-X(21) AND BO(+)-X(10+) BANDS OF ANTIMONY
MONOFLUORIDE.
WANG DKW + JONES WE + PREVOT F + COLIN R
J MOL SPECTROSC 49(3): 377-8 (1974)

ANTIMONY TRIFLUORIDE SbF_3

VAPOR PHASE RAMAN SPECTRA, RAMAN BAND CONTOUR
ANALYSES, CORIOLIS CONSTANTS, FORCE CONSTANTS,
AND VALUES FOR THERMODYNAMIC FUNCTIONS OF THE
TRIHALIDES OF GROUP-5.
CLARK RJH + RIPPON DM
J MOL SPECTROSC 52(1): 58-71 (1974)

0072 THERMODYNAMIC PROPERTIES OF ANTIMONY
TRIFLUORIDE.
CUBICCIOTTI D
HIGH TEMP SCI 1: 268-76 (1969)

0073 RAMAN SPECTRUM OF ANTIMONY TRIFLUORIDE.
FOMICHEV VV + PETROV KI + SADOKHINA LA
RUSS J INORG CHEM 17(9): 1348-9 (1972)

CALCULATION OF UREY-BRADLEY POTENTIAL CONSTANTS.
(PHOSPHORUS TRIFLUORIDE, ARSENIC TRIFLUORIDE,
AND ANTIMONY TRIFLUORIDE)
MEISINGSETH E
ACTA CHEM SCAND 17: 509-12 (1963)

0074 CALCULATION OF THE FREQUENCIES OF THE NORMAL
VIBRATIONS OF ANTIMONY TRIFLUORIDE.
TIMOSHININ VS + KHANDOZHKO SV
J APPL SPECTROSC 7(6): 638-40 (1967)

0075 THERMODYNAMIC FUNCTIONS OF ANTIMONY FLUORIDE,
NIOBIUM BROMIDE, TANTALUM CHLORIDE, TANTALUM
BROMIDE, AND MOLYBDENUM PENTACHLORIDE IN THE
GASEOUS STATE.
YUSHIN AS + OSIPOVA LI + SLEGINA VI
RUSS J PHYS CHEM 47(1): 161 (1973)

BINARY FLUORIDES: FREE MOLECULAR STRUCTURES AND FORCE FIELDS

ANTIMONY PENTAFLUORIDE SbF_5

0076 POLYMER MONOMER EQUILIBRIA IN NIOBIUM
PENTAFLUORIDE, TANTALUM PENTAFLUORIDE, AND
ANTIMONY PENTAFLUORIDE. THE SHAPE OF ANTIMONY
PENTAFLUORIDE.
ALEXANDER LE
INORG NUCL CHEM LETT 7: 1053-6 (1970)

0077 RAMAN SPECTRUM OF ANTIMONY PENTAFLUORIDE IN THE
GAS PHASE AS A FUNCTION OF TEMPERATURE. EVIDENCE
FOR A TRIGONAL BIPYRAMIDAL SHAPE FOR THE
MONOMERIC SPECIES.
ALEXANDER LE + BEATTIE IR
J CHEM PHYS 56: 5829-31 (1972)

INFRARED SPECTRA OF MATRIX ISOLATED ANTIMONY
PENTAFLUORIDE AND ARSENIC PENTAFLUORIDE.
ALJIBURY ALK + REDINGTON RL
J CHEM PHYS 52: 453-9 (1970)

0078 FLUORINE NMR OF ANTIMONY PENTAFLUORIDE.
BACON J + DEAN PAW + GILLESPIE RJ
CAN J PHYS 48: 3414-23 (1970)

0079 A FLUORINE NMR AND RAMAN STUDY OF COMPLEX
FORMATION BETWEEN ANTIMONY PENTAFLUORIDE AND
NIOBIUM PENTAFLUORIDE AND TANTALUM
PENTAFLUORIDE.
DEAN PAW + GILLESPIE RJ
CAN J CHEM 49: 1736-46 (1971)

0080 MOLECULAR STRUCTURE OF ANTIMONY PENTAFLUORIDE.
GAUNT J + AINSCOUGH JB
SPECTROCHIM ACTA 10: 57-60 (1957)

THERMODYNAMIC PROPERTIES OF PENTAFLUORIDES AT
HIGH TEMPERATURES. PART-1: CALCULATION METHODS.
(CHLORINE PENTAFLUORIDE, BROMINE PENTAFLUORIDE,
IODINE PENTAFLUORIDE, PHOSPHORUS PENTAFLUORIDE,
VANADIUM PENTAFLUORIDE, ANTIMONY PENTAFLUORIDE)
GALKIN NP + TUMANOV YN + BUTYLKIN YP
KHIM VYS ENERG 4(6): 512-8 (1970) (IN RUSSIAN)

POTENTIAL FIELD AND FORCE CONSTANTS OF SOME
TRIGONAL BIPYRAMIDAL PENTAHALIDES. (PHOSPHORUS
PENTAFLUORIDE AND ANTIMONY PENTAFLUORIDE)
HAARHOFF PC + PISTORIUS CWFT
Z NATURFORSCH 14A: 972-74 (1959)

0081 STRUCTURE OF LIQUID ANTIMONY PENTAFLUORIDE.
HOFFMAN CJ + HOLDER BE + JOLLY WL
J PHYS CHEM 62: 364-6 (1958)

0082 DENSITY, MELTING POINT AND VAPOR PRESSURE OF
ANTIMONY PENTAFLUORIDE.
HOFFMAN CJ + JOLLY WL
J PHYS CHEM 61: 1574-5 (1957)

0083 CONSTRUCTION OF A FAR INFRARED SPECTROMETER AND
ITS USE IN THE STUDY OF THE LOW FREQUENCY
VIBRATIONS OF ANTIMONY PENTAFLUORIDE AND THE
PURE ROTATIONAL SPECTRUM OF HYDROGEN CYANIDE.
JONES WD
VANDERBILT UNIV, PHD THESIS, 1963, 192P.

0084 MASS SPECTROMETRIC EVIDENCE OF DIMERS IN BISMUTH
PENTAFLUORIDE AND ANTIMONY PENTAFLUORIDE.
LAWLESS EW
INORG CHEM 10: 2084-6 (1971)

0085 MASS SPECTRUM OF ANTIMONY PENTAFLUORIDE.
MUELLER A + ROESKY HW + BOEHLER D
Z CHEM 7(12): 469-70 (1967) (IN GERMAN)

0086 POTENTIAL FIELD AND FORCE CONSTANTS OF ANTIMONY
PENTAFLUORIDE AND NIOBIUM PENTAFLUORIDE.
NAGARAJAN G
BULL SOC CHIM BELG 72: 563-71 (1963)

0087 VIBRATIONAL ASSIGNMENTS AND NORMAL COORDINATE
ANALYSIS FOR ANTIMONY PENTAFLUORIDE.
PILLAI MGK + RAMASWAMY K + PERUMAL A
INDIAN J PURE APPL PHYS 3(5): 180-2 (1965)

INFRARED MATRIX ISOLATION STUDIES. (BORON
TRIFLUORIDE, SULFUR TETRAFLUORIDE, ANTIMONY
PENTAFLUORIDE, ARSENIC PENTAFLUORIDE, CHLORINE
TRIFLUORIDE, BROMINE TRIFLUORIDE, AND BROMINE
PENTAFLUORIDE)
REDINGTON RL
TEXAS TECH UNIV, 1969, 89P. AD701120

ASSOCIATION OF GROUP-5 PENTAFLUORIDES IN THE GAS
PHASE. (PHOSPHORUS PENTAFLUORIDE, ARSENIC
PENTAFLUORIDE, BISMUTH PENTAFLUORIDE AND
ANTIMONY PENTAFLUORIDE)
VASILE MJ + FALCONER WE
INORG CHEM 11: 2282-3 (1972)

APPLICATION OF A MOLECULAR BEAM SOURCE MASS
SPECTROMETER TO THE STUDY OF REACTIVE FLUORIDES.
(SECOND ROW, THIRD ROW, ANTIMONY PENTAFLUORIDE,
AND BISMUTH PENTAFLUORIDE)
VASILE MJ + JONES GR + FALCONER WE
INT J MASS SPECTROM ION PHYS 10: 457-9 (1972)

A MOLECULAR BEAM MASS SPECTROMETRIC STUDY OF
BINARY PENTAFLUORIDES. (ANTIMONY PENTAFLUORIDE)
VASILE MJ + JONES GR + FALCONER WE
CHEM COMMUN 1971: 1355-6

GENERALIZED VALENCE FORCE FIELD FORCE CONSTANTS
AND MEAN-SQUARE AMPLITUDES OF VIBRATION OF SOME
TRIGONAL BIPYRAMIDAL MOLECULES BY THE
LOGARITHMIC STEPS METHOD. (PHOSPHORUS
PENTAFLUORIDE, ARSENIC PENTAFLUORIDE, ANTIMONY
PENTAFLUORIDE, VANADIUM PENTAFLUORIDE, AND
MOLYBDENUM PENTAFLUORIDE)
WENDLING EJL + MAHMOUDI S + MACCORDICK HJ
J CHEM SOC A 1971: 1747-54

0088 ASSOCIATION OF ANTIMONY PENTAFLUORIDE IN VAPOR.
ZEMSKII GI + SUDARIKOV BN + GROMOV BV
TR MOSK TEKHNOL INST 62: 23-5 (1969) (IN
RUSSIAN)

ARGON DIFLUORIDE ArF_2

D ORBITALS IN THE NOBLE GAS DIHALIDES. (ARGON
DIFLUORIDE, KRYPTON DIFLUORIDE, AND XENON
DIFLUORIDE)
CATTON RC + MITCHELL KAR
CAN J CHEM 48: 2695-2701 (1970)

ARSENIC MONOFLUORIDE AsF

0089 ULTRAVIOLET SPECTRUM OF ARSENIC MONOFLUORIDE.
CHATALIC A + DANON N + PANNETIER G
+ DESCHAMPS P + GUILLAUME J
COMPT REND 261: 3396-97 (1965); COMPT REND C
265: 710-11 (1967) (IN FRENCH)

0090 ANALYSIS OF THE ABSORPTION SPECTRUM AND OF TWO
NEW SINGLET SYSTEMS OF ARSENIC MONOFLUORIDE.
CHATALIC A + DANON N + PANNETIER G
COMPT REND C 273(15): 874-7 (1971) (IN FRENCH)

0091 SPECTRUM OF ARSENIC MONOFLUORIDE IN THE NEAR
VACUUM ULTRAVIOLET AND NEAR INFRARED REGIONS.
LIU DS + JONES WE
CAN J PHYS 50(12): 1230-51 (1972)

0092 ROTATIONAL ANALYSIS OF FOUR SINGLET SYSTEMS OF
ARSENIC MONOFLUORIDE.
LIU DS + YEE KK + JONES WE
J MOL SPECTROSC 38: 512-23 (1971)

4

0093 THERMODYNAMIC PROPERTIES OF SOME ARSENIC
FLUORIDES. (ARSENIC MONOFLUORIDE, ARSENIC
DIFLUORIDE, ARSENIC TRIFLUORIDE, ARSENIC
TETRAFLUORIDE, AND ARSENIC PENTAFLUORIDE)
O'HARE PAG
USAEC, 1968, 20P. ANL-7456

0094 ARSENIC MONOFLUORIDE, 3 SIGMA. DISSOCIATION
ENTHALPY, IONIZATION POTENTIAL, ELECTRON
AFFINITY, DIPOLE MOMENT, SPECTROSCOPIC
CONSTANTS, AND IDEAL GAS THERMODYNAMIC FUNCTIONS
FROM A HARTREE-FOCK MOLECULAR ORBITAL
INVESTIGATION.
O'HARE PAG + BATANA A + WAHL AC
J CHEM PHYS 59: 6495-6501 (1973)

0095 VIBRATIONAL ANALYSIS OF THE SINGLET SYSTEMS OF
ARSENIC MONOFLUORIDE.
YEE KK + LIU DS + JONES WE
J MOL SPECTROSC 35: 153-7 (1970)

ARSENIC TRIFLUORIDE AsF_3

0096 THE INFRARED SPECTRA OF SOME VOLATILE INORGANIC
FLUORIDES IN THE SOLID STATE. (ARSENIC
TRIFLUORIDE, SULFUR TETRAFLUORIDE, AND SELENIUM
TETRAFLUORIDE)
AYNSLEY EE + DODD RE + LITTLE R
SPECTROCHIM ACTA 18: 1005-8 (1962)

0097 MICROWAVE SPECTRUM OF ARSENIC TRIFLUORIDE IN THE
EXCITED VIBRATIONAL STATE.
CHIKARAISHI T + HIROTA E
BULL CHEM SOC JAP 46(8): 2314-16 (1973)

VAPOR PHASE RAMAN SPECTRA, RAMAN BAND CONTOUR
ANALYSES, CORIOLIS CONSTANTS, FORCE CONSTANTS,
AND VALUES FOR THERMODYNAMIC FUNCTIONS OF THE
TRIHALIDES OF GROUP-5.
CLARK RJH + RIPPON DM
J MOL SPECTROSC 52(1): 58-71 (1974)

0098 AN ELECTRON DIFFRACTION INVESTIGATION OF ARSENIC
TRIFLUORIDE, ARSENIC PENTAFLUORIDE, AND OTHER
MOLECULES.
CLIPPARD FB
UNIV MICHIGAN, PHD THESIS, 1969, 90P.

0099 MOLECULAR STRUCTURES OF ARSENIC TRIFLUORIDE AND
ARSENIC PENTAFLUORIDE AS DETERMINED BY ELECTRON
DIFFRACTION.
CLIPPARD FB + BARTELL LS
INORG CHEM 9: 805-11 (1970)

A BENT BOND MODEL FOR THE VIBRATIONAL FORCE
CONSTANTS OF NONPLANAR XY3 MOLECULES. (NITROGEN
TRIFLUORIDE)
CURTIS EC
ROCKETDYNE, CANOGA PARK, CALIF, 1968, 21P.
AD666456

P-R SEPARATIONS AND CORIOLIS CONSTANTS FOR
SYMMETRIC TOP MOLECULES AND FORCE CONSTANTS FOR
NITROGEN TRIFLUORIDE, PHOSPHORUS TRIFLUORIDE,
AND ARSENIC TRIFLUORIDE.
HOSKINS LC
J CHEM PHYS 45: 4594-4600 (1966)

0100 INFRARED SPECTRUM AND VIBRATIONAL POTENTIAL
FUNCTION OF ARSENIC TRIFLUORIDE.
HOSKINS LC + LORD RC
J CHEM PHYS 43(11): 155-8 (1965)

AMBIGUITIES IN THE HARMONIC FORCE FIELDS OF XY3
MOLECULES. (NITROGEN TRIFLUORIDE, BORON
TRIFLUORIDE, PHOSPHORUS TRIFLUORIDE, AND ARSENIC
TRIFLUORIDE)
HOY AR + STONE JMR + WATSON JKG
J MOL SPECTROSC 42: 393-99 (1972)

0101 CENTRIFUGAL STRETCHING OF TRIFLUORO CHLORO
METHANE AND SOME OTHER SYMMETRIC TOPS. (ARSENIC
TRIFLUORIDE)
JOHNSON RC + WILLIAMS Q + WEATHERLY TL
J CHEM PHYS 36: 1588-90 (1962)

0102 MICROWAVE SPECTRUM OF ARSENIC TRIFLUORIDE.
KISLIUK P + GESCHWIND S
J CHEM PHYS 21(5): 828-9 (1953)

0103 FORCE CONSTANTS AND AVERAGE STRUCTURES OF
ARSENIC TRIFLUORIDE AND ARSENIC TRICHLORIDE
DETERMINED BY ELECTRON DIFFRACTION AND
SPECTROSCOPY.
KONAKA S
BULL CHEM SOC JAP 43: 3107-3115 (1970)

0104 DETERMINATION OF THE MOLECULAR STRUCTURES OF
ARSENIC TRIFLUORIDE AND ARSENIC TRICHLORIDE BY
GAS ELECTRON DIFFRACTION.
KONAKA S + KIMURA M
BULL CHEM SOC JAP 43: 1693-1703 (1970)

0105 THE INFRARED SPECTRUM AND FORCE CONSTANTS FOR
ARSENIC TRIFLUORIDE.
MANDIROLA OBD
J MOL STRUCT 1: 203-10 (1968)

0106 LIQUID GAS ARSENIC INFRARED FREQUENCY SHIFTS.
MANDIROLA OBD
J MOL STRUCT 3: 465-72 (1969)

CALCULATION OF UREY-BRADLEY POTENTIAL CONSTANTS.
(PHOSPHORUS TRIFLUORIDE, ARSENIC TRIFLUORIDE,
AND ANTIMONY TRIFLUORIDE)
MEISINGSETH E
ACTA CHEM SCAND 17: 509-12 (1963)

GENERAL FORCE FIELD OF NITROGEN TRIFLUORIDE,
PHOSPHORUS TRIFLUORIDE, AND ARSENIC TRIFLUORIDE
BY THE COMBINED USE OF VIBRATIONAL FREQUENCIES,
CENTRIFUGAL STRETCHING, AND CORIOLIS COUPLING
CONSTANTS.
MIRRI AM
J CHEM PHYS 47: 2823-8 (1967)

MEAN AMPLITUDES OF VIBRATION AND THERMODYNAMIC
FUNCTIONS OF SOME GROUP-5 TRIHALIDES. (NITROGEN
TRIFLUORIDE, ARSENIC TRIFLUORIDE, AND PHOSPHORUS
TRIFLUORIDE)
NAGARAJAN G
INDIAN J PURE APPL PHYS 2: 237-41 (1964)

THERMODYNAMIC PROPERTIES OF SOME ARSENIC
FLUORIDES. (ARSENIC MONOFLUORIDE, ARSENIC
DIFLUORIDE, ARSENIC TRIFLUORIDE, ARSENIC
TETRAFLUORIDE, AND ARSENIC PENTAFLUORIDE)
O'HARE PAG
USAEC, 1968, 20P. ANL-7456

MOLECULAR FORCE FIELD FOR SOME XY3 TYPE
MOLECULES. (PHOSPHORUS TRIFLUORIDE AND ARSENIC
TRIFLUORIDE)
PILLAI MGK + PILLAI PP
INDIAN J PURE APPL PHYS 6: 404-7 (1968)

RAMAN SPECTRUM OF GASEOUS NITROGEN TRIFLUORIDE.
SHAMIR J + HYMAN HH
SPECTROCHIM ACTA 23A: 1899-1901 (1967)

0107 CALCULATIONS ON IONIC CHARACTER AND
HYBRIDIZATION FROM DIPOLE MOMENTS FOR ARSENIC
TRIFLUORIDE AND ARSENIC TRICHLORIDE.
SOBHANADRI J
PROC PHYS SOC 76: 267-72 (1960)

0108 ESR STUDY OF RADICALS IN GAMMA IRRADIATED
POLYCRYSTALLINE ARSENIC TRIFLUORIDE AND ARSENIC
TRICHLORIDE.
SUBRAMANIAN S + ROGERS MT
J CHEM PHYS 57(11): 4582-9 (1972)

0109 MEAN AMPLITUDES AND THERMODYNAMIC FUNCTIONS OF
SOME TRIHALIDES. (ARSENIC TRIFLUORIDE)
SUNDARAM S
Z PHYS CHEM NEUE FOLGE 34: 233-7 (1962)

0110 STANDARD HEAT OF FORMATION OF ARSENIC
TRIFLUORIDE.
WOOLF AA
J FLUORINE CHEM 5(2): 172-4 (1975)

0111 RAMAN SPECTRUM OF ARSENIC TRIFLUORIDE AND THE
MOLECULAR CONSTANTS OF ARSENIC TRIFLUORIDE,
ARSENIC TRICHLORIDE, AND PHOSPHORUS TRICHLORIDE.
YOST DM + SHERBORNE JE
J CHEM PHYS 2: 125-7 (1934)

ARSENIC TETRAFLUORIDE AsF_4

0112 ESR SPECTRUM OF ARSENIC TETRAFLUORIDE.
COLUSSI AJ + MORTON JR + PRESTON KF
CHEM PHYS LETT 30: 317-19 (1975)

THERMODYNAMIC PROPERTIES OF SOME ARSENIC
FLUORIDES. (ARSENIC MONOFLUORIDE, ARSENIC
DIFLUORIDE, ARSENIC TRIFLUORIDE, ARSENIC
TETRAFLUORIDE, AND ARSENIC PENTAFLUORIDE)
O'HARE PAG
USAEC, 1968, 20P. ANL-7456

ARSENIC PENTAFLUORIDE AsF_5

0113 INFRARED SPECTRA OF MATRIX ISOLATED ANTIMONY
PENTAFLUORIDE AND ARSENIC PENTAFLUORIDE.
ALJIBURY ALK + REDINGTON RL
J CHEM PHYS 52: 453-9 (1970)

ORBITALS AND STRUCTURES OF PENTAFLUORIDES.
(PHOSPHORUS PENTAFLUORIDE, ARSENIC
PENTAFLUORIDE, AND BROMINE PENTAFLUORIDE)
BERRY RS + TAMRES M + BALLHAUSEN CH + HOHANSEN H
ACTA CHEM SCAND 22: 231-46 (1968)

0114 INFRARED SPECTRUM OF ARSENIC PENTAFLUORIDE.
BLANCHARD S
COMM ENERG AT, 1967, 10P. CEA-R-3195 (IN FRENCH)

MOLECULAR STRUCTURES OF ARSENIC TRIFLUORIDE AND
ARSENIC PENTAFLUORIDE AS DETERMINED BY ELECTRON
DIFFRACTION.
CLIPPARD FB + BARTELL LS
INORG CHEM 9: 805-11 (1970)

AN ELECTRON DIFFFRACTION INVESTIGATION OF
ARSENIC TRIFLUORIDE, ARSENIC PENTAFLUORIDE, AND
OTHER MOLECULES.
CLIPPARD FB
UNIV MICHIGAN, PHD THESIS, 1969, 90P.

MEAN AMPLITUDES OF VIBRATION AND CORIOLIS
CONSTANTS FOR PHOSPHORUS PENTAFLUORIDE AND
ARSENIC PENTAFLUORIDE.
CYVIN SJ + BRUNVOLL J
J MOL STRUCT 3: 151-4 (1969)

0115 PENTACOORDINATED MOLECULES. PART-12: CORRELATION
OF EXCHANGE RATES FOR SOME GROUP-5 PENTAHALIDE
MOLECULES.
HOLMES RR
INORG CHEM 7(11): 2229-35 (1968)

VIBRATIONAL SPECTRA OF PHOSPHORUS PENTAFLUORIDE
AND ARSENIC PENTAFLUORIDE AND THE HEIGHT OF THE
BARRIER TO INTERNAL EXCHANGE OF FLUORINE NUCLEI.
HOSKINS LC + LORD RC
J CHEM PHYS 46: 2402-12 (1967)

0116 PERPENDICULAR BAND CONTOURS AND VIBRATIONAL
POTENTIAL FUNCTION OF THE E' VIBRATIONS OF
ARSENIC PENTAFLUORIDE.
HOSKINS LC + PERNG CN
J CHEM PHYS 55: 5063-5 (1971)

MEAN AMPLITUDES OF VIBRATION FOR A TRIGONAL
BIPYRAMIDAL MOLECULE WITH APPLICATION TO
PHOSPHORUS PENTAFLUORIDE AND ARSENIC
PENTAFLUORIDE.
NAGARAJAN G + DURIG JR
BULL SOC ROY SCI LIEGE 5: 334-46 (1967)

THERMODYNAMIC PROPERTIES OF SOME ARSENIC
FLUORIDES. (ARSENIC MONOFLUORIDE, ARSENIC
DIFLUORIDE, ARSENIC TRIFLUORIDE, ARSENIC
TETRAFLUORIDE, AND ARSENIC PENTAFLUORIDE)
O'HARE PAG
USAEC, 1968, 20P. ANL-7456

0117 ENTHALPY OF FORMATION OF ARSENIC PENTAFLUORIDE.
O'HARE PAG + HUBBARD WN
J PHYS CHEM 68: 4358-60 (1965)

INFRARED MATRIX ISOLATION STUDIES. (BORON
TRIFLUORIDE, SULFUR TETRAFLUORIDE, ANTIMONY
PENTAFLUORIDE, ARSENIC PENTAFLUORIDE, CHLORINE
TRIFLUORIDE, BROMINE TRIFLUORIDE, AND BROMINE
PENTAFLUORIDE)
REDINGTON RL
TEXAS TECH UNIV, 1969, 89P. AD701120

SPECTROSCOPIC STUDIES IN MEAN AMPLITUDES OF
VIBRATION OF PENTAFLUORIDES OF ARSENIC,
PHOSPHORUS, AND VANADIUM.
SANYAL NK + DIXIT L
INDIAN J PURE APPL PHYS 12(8): 550-3 (1974)

RAMAN SPECTRA OF ARSENIC PENTAFLUORIDE AND
VANADIUM PENTAFLUORIDE AND THEIR FORCE CONSTANTS
INCLUDING THOSE OF PHOSPHORUS PENTAFLUORIDE.
SELIG H + HOLLOWAY JH + TYSON J + CLAASSEN HH
J CHEM PHYS 53: 2559-64 (1970)

ASSOCIATION OF GROUP-5 PENTAFLUORIDES IN THE GAS
PHASE. (PHOSPHORUS PENTAFLUORIDE, ARSENIC
PENTAFLUORIDE, BISMUTH PENTAFLUORIDE, AND
ANTIMONY PENTAFLUORIDE)
VASILE MJ + FALCONER WE
INORG CHEM 11: 2282-3 (1972)

BARIUM MONOFLUORIDE BaF

0118 ROTATIONAL ANALYSIS OF SOME BANDS OF GASEOUS
BARIUM MONOFLUORIDE.
BARROW RF + BASTIN MW + LONGBOROUGH B
PROC PHYS SOC 92(2): 518-19 (1967)

0119 DISSOCIATION ENERGIES OF THE ALKALINE EARTH
MONOFLUORIDES.
EHLERT TC + BLUE GD + GREEN JW + MARGRAVE JL
J CHEM PHYS 41: 2250-5 (1964)

0120 THE C DOUBLET PI AND THE X DOUBLET SIGMA STATES
OF BARIUM MONOFLUORIDE.
KUSHAWAHA VS
SPECTROSC LETT 6(10): 633-45 (1973)

0121 GREEN BAND SYSTEM OF BARIUM MONOFLUORIDE.
KUSHAWAHA VS + ASTHANA BP + SHANKER R
+ PATHAK CM
SPECTROSC LETT 5(11): 407-14 (1972)

0122 VIOLET AND ULTRAVIOLET SPECTRUM OF BARIUM
MONOFLUORIDE MOLECULE.
SINGH J + MOHAN H
INDIAN J PURE APPL PHYS 11(12): 918-22 (1973)

0123 THERMALLY EXCITED EMISSION SPECTRUM OF BARIUM
MONOFLUORIDE.
SINGH J + MOHAN H
J PHYS B 4(10): 1395-7 (1971)

BARIUM DIFLUORIDE BaF_2

BERYLLIUM MONOFLUORIDE BeF

ELECTRON DIFFRACTION ANALYSIS OF THE MOLECULAR STRUCTURES OF GROUP-2 ELEMENTS. (BERYLLIUM DIFLUORIDE, MAGNESIUM DIFLUORIDE, CALCIUM DIFLUORIDE, STRONTIUM DIFLUORIDE, BARIUM DIFLUORIDE, ZINC DIFLUORIDE, CADMIUM DIFLUORIDE)
AKISHIN PA + SPIRIDONOV VP
SOV PHYS CRYSTALLOGR 2(4): 472-9 (1957)

SPECTRA OF ELECTRONIC EXITATIONS IN CALCIUM DIFLUORIDE, STRONTIUM DIFLUORIDE, AND BARIUM DIFLUORIDE IN THE 8 TO 150 EV RANGE.
FRANDON J + LAHAYE B + PRADAL F
PHYS STAT SOL B 53: 565-75 (1972)

MELTING POINTS OF INORGANIC FLUORIDES. (CADMIUM DIFLUORIDE, MAGNESIUM DIFLUORIDE, STRONTIUM DIFLUORIDE, AND BARIUM DIFLUORIDE)
KOJIMA H + WHITEWAY SG + MASSON CR
CAN J CHEM 46: 2968-71 (1968)

0124 MEAN AMPLITUDES OF VIBRATION, BASTIANSEN-MORINO SHRINKAGE EFFECT, AND MOLECULAR POLARIZABILITIES FOR DIHALIDES OF SOME GROUP-2A ELEMENTS.
NAGARAJAN G
INDIAN J PHYS 39(9): 405-20 (1965)

FAR ULTRAVIOLET ELECTRONIC SPECTRA OF STRONTIUM DIFLUORIDE AND BARIUM DIFLUORIDE.
NISAR M + ROBIN S
PAK J SCI IND RES 17(2-3): 49-54 (1974)

VAPOR PRESSURE OF ZINC DIFLUORIDE, CADMIUM DIFLUORIDE, MAGNESIUM DIFLUORIDE, CALCIUM DIFLUORIDE, STRONTIUM DIFLUORIDE, BARIUM DIFLUORIDE, AND ALUMINUM TRIFLUORIDE.
RUFF O + LE BOUCHER L
Z ANORG ALLGEM CHEM 219: 376-81 (1934) (IN GERMAN)

0125 INFRARED SPECTRA OF SOME ALKALINE EARTH HALIDES BY THE MATRIX ISOLATION TECHNIQUE.
SNELSON A
J PHYS CHEM 70: 3208-16 (1966)

X-RAY STUDY OF ANHARMONIC VIBRATIONS IN CALCIUM DIFLUORIDE AND BARIUM DIFLUORIDE.
STROCK HB
CORNELL UNIV, PHD THESIS, 1971, 111P.

0126 GEOMETRY OF THE ALKALINE EARTH DIHALIDES. (BARIUM DIFLUORIDE, STRONTIUM DIFLUORIDE, CALCIUM DIFLUORIDE)
WHARTON L + BERG RA + KLEMPERER W
J CHEM PHYS 39: 2023-31 (1963)

BERKELIUM TETRAFLUORIDE BkF_4

LATTICE CONSTANTS OF ACTINIDE TETRAFLUORIDES INCLUDING BERKELIUM.
KEENAN TK + ASPREY LB
INORG CHEM 8: 235-8 (1969)

0127 RYDBERG-KLEIN-REES FRANCK-CONDON FACTORS AND R CENTROIDS OF THE A-X BAND SYSTEM OF BERYLLIUM FLUORIDE.
GOHEL VB + SHAH NP
INDIAN J PHYS 48(10): 932-40 (1974)

0128 DISSOCIATION ENERGIES OF BERYLLIUM FLUORIDE AND BERYLLIUM CHLORIDE AND HEAT OF FORMATION OF BERYLLIUM CHLOROFLUORIDE.
FARBER M + SRIVASTAVA RD
J CHEM SOC, FARADAY TRANS 70(9): 1581-9 (1970)

0129 MASS SPECTROMETRIC DETERMINATION OF THE DISSOCIATION ENERGY OF BERYLLIUM MONOFLUORIDE.
HILDENBRAND DL + MURAD E
J CHEM PHYS 44(4): 1524-9 (1966)

0130 POTENTIAL CURVES FOR SOME DIATOMIC MOLECULES. (BERYLLIUM MONOFLUORIDE, SILICON MONOFLUORIDE, AND TIN MONOFLUORIDE)
SINGH RB + RAI DK
INDIAN J PURE APPL PHYS 4: 102-5 (1966)

BERYLLIUM DIFLUORIDE BeF_2

0131 ELECTRON DIFFRACTION ANALYSIS OF THE MOLECULAR STRUCTURES OF GROUP-2 ELEMENTS. (BERYLLIUM DIFLUORIDE, MAGNESIUM DIFLUORIDE, CALCIUM DIFLUORIDE, STRONTIUM DIFLUORIDE, BARIUM DIFLUORIDE, ZINC DIFLUORIDE, CADMIUM DIFLUORIDE)
AKISHIN PA + SPIRIDONOV VP
SOV PHYS CRYSTALLOGR 2(4): 472-9 (1957)

0132 HEAT OF FORMATION OF BERYLLIUM DIFLUORIDE.
ARMSTRONG GT
P73-8 OF WORKING GROUP ON THERMOCHEMISTRY, 2ND PROC, 1964, CHEM PROPULSION INFO AGENCY, AD451711.

0133 INFRARED SPECTRA OF THE ALKALINE EARTH HALIDES. PART-1: BERYLLIUM DIFLUORIDE, BERYLLIUM DICHLORIDE, MAGNESIUM DICHLORIDE.
BUECHLER A + KLEMPERER W
J CHEM PHYS 29: 121-3 (1958)

0134 DETERMINATION OF THE GEOMETRY OF HIGH TEMPERATURE SPECIES BY ELECTRIC DEFLECTION AND MASS SPECTROMETRIC DETECTION. (FIRST ROW FLUORIDES, BERYLLIUM, MAGNESIUM, AND LEAD DIFLUORIDES)
BUECHLER A + STAUFFER JL + KLEMPERER W
J AMER CHEM SOC 86: 4544-50 (1964)

0135 VAPOR PRESSURES OF BERYLLIUM DIFLUORIDE AND NICKEL DIFLUORIDE.
CANTOR S
J CHEM ENG DATA 10: 237-8 (1965)

MEAN AMPLITUDES AND SHRINKAGE EFFECTS OF VIBRATION FOR SOME LINEAR SYMMETRICAL DIFLUORIDES. (BERYLLIUM DIFLUORIDE AND MAGNESIUM DIFLUORIDE)
CYVIN SJ + VIZI B
VESZPREMI VEGYIPARI EGYETEM KOZLEMENYEI 11: 83-9 (1968)

0136 THERMODYNAMIC DATA FOR BERYLLIUM DIFLUORIDE.
DOUGLAS TB
P67-72 OF WORKING GROUP ON THERMOCHEMISTRY, 2ND PROC, 1964, CHEM PROPULSION INFO AGENCY, AD451711.

0137 NONEMPIRICAL LCAO MO SCF STUDIES OF THE LOW LYING STATES OF BERYLLIUM DIFLUORIDE.
GOLE JL
J CHEM PHYS 58: 869-75 (1973)

NONEMPERICAL LCAO MO SCF STUDIES OF THE GROUP-2A
DIHALIDES: BERYLLIUM DIFLUORIDE, MAGNESIUM
DIFLUORIDE, AND CALCIUM DIFLUORIDE.
GOLE JL + SIU AKQ + HAYES EF
J CHEM PHYS 58: 857-68 (1973)

0138 THERMODYNAMIC AND PHYSICAL PROPERTIES OF
BERYLLIUM COMPOUNDS. PART-1: ENTHALPY AND
ENTROPY OF VAPORIZATION OF BERYLLIUM DIFLUORIDE.
GREENBAUM MA + FOSTER JN + ARIN ML + FARBER M
J PHYS CHEM 67: 36-40 (1963)

0139 ENTHALPY OF FORMATION OF BERYLLIUM DIFLUORIDE.
GROSS P + HAYMAN C + BINGHAM JT
FULMER RES INST, RES REP, 1971, 3P.

0140 EFFUSION STUDIES, MASS SPECTRA, AND
THERMODYNAMICS OF BERYLLIUM FLUORIDE VAPOR.
HILDENBRAND DL + THEARD LP
J CHEM PHYS 42(9): 3230-6 (1965)

MEAN AMPLITUDES OF VIBRATION FOR THE DIHALIDES
OF MERCURY, BERYLLIUM, AND MAGNESIUM.
NAGARAJAN G
ACTA PHYS AUSTRIACA 17: 246-53 (1973)

MEAN AMPLITUDES OF VIBRATION AND
BASTIANSEN-MORINC SHRINKAGE EFFECT IN SOME
LINEAR SYMMETRICAL DIHALIDES. (BERYLLIUM
DIFLUORIDE AND MAGNESIUM DIFLUORIDE)
NAGARAJAN G
J MOL SPECTROSC 13: 361-92 (1964)

0141 MOLECULAR PROPERTIES OF THE TRIATOMIC
DIFLUORIDES BERYLLIUM DIFLUORIDE, BORON
DIFLUORIDE, CARBON DIFLUORIDE, NITROGEN
DIFLUORIDE, AND OXYGEN DIFLUORIDE. (SCF)
ROTHENBERG S + SCHAEFER HF
J AMER CHEM SOC 95: 2095-100 (1973)

INFRARED SPECTRA OF SOME ALKALINE EARTH HALIDES
BY THE MATRIX ISOLATION TECHNIQUE.
SNELSON A
J PHYS CHEM 70: 3208-16 (1966)

0142 ENTHALPY OF BERYLLIUM FLUORIDE FROM 456 TO 1083
DEGREES K BY TRANSPOSED TEMPERATURE DROP
CALORIMETRY.
TAMURA S + YOKOKAWA T + NIWA K
J CHEM THERMODYN 7(7): 633-43 (1975)

BISMUTH MONOFLUORIDE BiF

0143 ABSORPTION SPECTRUM OF THE BISMUTH MONOFLUORIDE
MOLECULE IN THE ULTRAVIOLET REGION.
JOSHI KC
PROC PHYS SOC 78: 610-13 (1961)

FRANCK-CONDON FACTORS AND R CENTROIDS OF SOME
BAND SYSTEMS OF SILICON MONOFLUORIDE, CALCIUM
MONOFLUORIDE, AND BISMUTH MONOFLUORIDE.
MOHANTY BS + SINGH ON
INDIAN J PURE APPL PHYS 7: 109-11 (1969)

0144 NEW BAND SYSTEM OF BISMUTH FLUORIDE MOLECULE IN
THE REGION 6200-7000 ANGSTROMS.
MURTY PS + RAO DVK + REDDY YP + RAO PT
SPECTROSC LETT 8(4): 217-21 (1975)

0145 VISIBLE EMISSION SPECTRUM OF BISMUTH
MONOFLUORIDE.
RAO KM + RAO PT
INDIAN J PHYS 39: 572-9 (1965)

0146 ROTATIONAL ANALYSIS OF THE VISIBLE BAND SYSTEM
OF THE BISMUTH MONOFLUORIDE MOLECULE.
RAO TAP + RAO PT
CAN J PHYS 40: 1077-84 (1962)

EMISSION SPECTRA OF ANTIMONY MONOFLUORIDE AND
BISMUTH MONOFLUORIDE.
RAO TAP + RAO PT
INDIAN J PURE APPL PHYS 7: 641-3 (1969)

BAND SPECTRA OF BISMUTH MONOFLUORIDE AND
ANTIMONY MONOFLUORIDE.
ROCHESTER GD
PROC PHYS SOC 78: 614 (1961)

0147 EMISSION SPECTRUM OF BISMUTH MONOFLUORIDE.
SANKARANARAYANAN S + NARAYANAN PS + PATEL MM
PROC INDIAN ACAD SCI A 59(6): 378-84 (1964)

DISSOCIATION OF LEAD MONOFLUORIDE AND BISMUTH
MONOFLUORIDE.
SINGH RB + RAI DK
CAN J PHYS 43: 829-35 (1965)

BISMUTH TRIFLUORIDE BiF_3

STRUCTURES OF YTTRIUM AND BISMUTH TRIFLUORIDES
BY NEUTRON DIFFRACTION.
CHEETHAM AK + NORMAN N
ACTA CHEM SCAND 28(1): 55-60 (1974)

0148 THERMODYNAMIC PROPERTIES OF BISMUTH TRIFLUORIDE.
CUBICCIOTTI D
J ELECTROCHEM SOC 115(11): 1138-43 (1968)

0149 THERMODYNAMIC FUNCTIONS OF BISMUTH TRIHALIDE
VAPORS.
ZAVALISHIN NI + MAL'TSEV AA
VEST MOSK UNIV KHIM 16(3): 380 (IN RUSSIAN)

BISMUTH PENTAFLUORIDE BiF_5

MASS SPECTROMETRIC EVIDENCE OF DIMERS IN BISMUTH
PENTAFLUORIDE AND ANTIMONY PENTAFLUORIDE.
LAWLESS EW
INORG CHEM 10: 2084-6 (1971)

ASSOCIATION OF GROUP-5 PENTAFLUORIDES IN THE GAS
PHASE. (PHOSPHORUS PENTAFLUORIDE, ARSENIC
PENTAFLUORIDE, BISMUTH PENTAFLUORIDE AND
ANTIMONY PENTAFLUORIDE)
VASILE MJ + FALCONER WE
INORG CHEM 11: 2282-3 (1972)

APPLICATION OF A MOLECULAR BEAM SOURCE MASS
SPECTROMETER TO THE STUDY OF REACTIVE FLUORIDES.
(SECOND ROW, THIRD ROW, ANTIMONY PENTAFLUORIDE,
AND BISMUTH PENTAFLUORIDE)
VASILE MJ + JONES GR + FALCONER WE
INT J MASS SPECTROM ION PHYS 10: 457-9 (1972)

BORON MONOFLUORIDE BF

0150 MOLECULAR CHARGE DISTRIBUTIONS AND CHEMICAL
BINDING. PART-3: THE ISOELECTRONIC NITROGEN,
CARBON MONOXIDE, BORON FLUORIDE, AND CARBON,
BERYLLIUM OXIDE, LITHIUM FLUORIDE.
BADER RFW + BANDRAUK AD
J CHEM PHYS 49: 1653-65 (1968)

0151 RELAXATION OF THE MOLECULAR CHARGE DISTRIBUTION
AND THE VIBRATIONAL FORCE CONSTANT.
BADER RFW + BANDRAUK AD
J CHEM PHYS 49: 1666-75 (1968)

0152 DISSOCIATION ENERGIES OF THE GASEOUS MONOHALIDES
OF BORON, ALUMINUM, GALLIUM, INDIUM, AND
THALLIUM.
BARROW RF
TRANS FARADAY SOC 56: 952-58 (1960)

0153 ROTATIONAL ANALYSIS OF BANDS OF THE C TRIPLET
SIGMA, B TRIPLET SIGMA: A TRIPLET PI SYSTEM OF
BORON MONOFLUORIDE.
BARROW RF + PREMASWARUP D + WINTERNITZ J
+ ZEEMAN PB
PROC PHYS SOC 71: 61-4 (1958)

0154 HEAT AND ENTROPY OF FORMATION OF BORON FLUORIDE.
BLAUER JA + GREENBAUM MA + FARBER M
J PHYS CHEM 68: 2332-34 (1964)

0155 ELECTRONIC SPECTRUM OF THE BORON MONOFLUORIDE
MOLECULE.
CATON RB + DOUGLAS AE
CAN J PHYS 48: 431-52 (1969)

0156 ROTATIONAL ANALYSIS OF THE BAND AT 14900 CM-1 OF
A 3 SIGMA TO 3 SIGMA SYSTEM OF THE BORON
MONOFLUORIDE MOLECULE.
CZARNY J + FELENBOK P
CHEM PHYS LETT 2: 533-5 (1968)

0157 ELECTRONIC POPULATION ANALYSIS OF MOLECULAR WAVE
FUNCTIONS.
DAVIDSON ER
J CHEM PHYS 46: 3320-24 (1967)

0158 LOCALIZED ORBITALS IN BORON FLUORIDES. HIGHLY
POLARIZED BORON- FLUORINE DOUBLE AND TRIPLE
BONDS.
HALL JH + HALGREN TA + KLEIER DA + LIPSCOMB WN
INORG CHEM 13(10): 2520-1 (1974)

0159 MAGNETIC PROPERTIES OF THE BORON FLUORIDE AND
BORON HYDRIDE MOLECULES.
HEGSTROM RA + LIPSCOMB WN
J CHEM PHYS 48: 809-11 (1968)

0160 FIRST IONIZATION POTENTIALS OF THE MOLECULES
BORON FLUORIDE, SILICON OXIDE AND GERMANIUM
OXIDE.
HILDENBRAND DL
INT J MASS SPECTROM ION PHYS 7: 255-60 (1971)

0161 DISSOCIATION ENERGY OF BORON MONOFLUORIDE FROM
MASS SPECTROMETRIC STUDIES.
HILDENBRAND DL + MURAD E
J CHEM PHYS 43: 1400-3 (1965)

0162 ELECTRONIC STRUCTURE OF CARBON MONOXIDE AND
BORON MONOFLUORIDE.
HUO WM
J CHEM PHYS 43: 624-47 (1965)

0163 CAMERON SYSTEM OF BORON MONOFLUORIDE.
LEBRETON J + FERRAN J + MARSIGNY L
J PHYS B 8(17): L465-6 (1975)

0164 CALCULATION OF RYDBERG LEVELS IN NITROGEN OXIDE
AND BORON FLUORIDE.
LEFEBVRE-BRION H + MOSER CM
J MOL SPECTROSC 15: 211-19 (1965)

0165 MICROWAVE SPECTRUM OF BORON FLUORIDE.
LOVAS FJ + JOHNSON DR
J CHEM PHYS 55: 41-44 (1971)

0166 ISOTOPE EFFECT IN THE SINGLET SPECTRAL BANDS OF
THE BORON FLUORIDE MOLECULE.
MAL'TSEV AA
OPT SPECTROSC 9(4): 225-6 (OCT 1960)

0167 FLUORESCENCE SPECTRA OF PHOSPHORUS NITRIDE AND
BORON FLUORIDE.
MOELLER MB + SILVERS SJ
CHEM PHYS LETT 19(1): 78-81 (1973)

0168 VIBRATIONAL TRANSITION PROBABILITIES FOR THE A-X
SYSTEM OF BORON MONOFLUORIDE.
MISHIRA RK + KHANNA BN
J QUANT SPECT RAD TRANSFER 10: 703-4 (1970)

0169 VALENCE EXCITED STATES OF NITROGEN, CARBON
MONOXIDE, AND BORON FLUORIDE.
NESBET RK
J CHEM PHYS 43: 4403-9 (1965)

0170 STUDY OF THE A SINGLET PI- X SINGLET SIGMA(+)
BANDS OF BORON-11 MONOFLUORIDE WITH A VACUUM
ECHELLE SPECTROGRAPH.
ONAKA R
J CHEM PHYS 27: 374-77 (1957)

0171 FRANCK-CONDON FACTORS AND R CENTROIDS FOR THE B
TRIPLET SIGMA(+) TO A TRIPLET PI SYSTEM OF BORON
MONOFLUORIDE.
PATHAK AN + MAHESHWARI RC
INDIAN J PURE APPL PHYS 5: 138-139 (1967)

0172 THE ELECTRONIC SPECTRUM OF BORON FLUORIDE.
ROBINSON DW
J MOL SPECTROSC 11: 275-3000 (1963)

0173 POTENTIAL ENERGY CURVES AND DISSOCIATION
ENERGIES OF DIATOMIC BORON AND ALUMINUM HALIDES.
SINGH J + NAIR KPR + RAI DK
J MOL STRUCT 6: 328-32 (1970)

0174 METHODS FOR CORRELATING MOLECULES AND SOME
OPTIMIZED VALENCE CONFIGURATION RESULTS ON THE
DIATOMIC MOLECULES LITHIUM, BERYLLIUM, BORON,
CARBON, NITROGEN, FLUORINE, BORON NITRIDE,
BERYLLIUM OXIDE, LITHIUM FLUORIDE, HELIUM- NEON,
CARBON MONOXIDE, AND BORON MONOFLUORIDE.
SUTTON P + BERTONICINI P + DAS G + GILBERT TL
+ WAHL AC + SINANOGLU O
INT J QUANT CHEM 3S: 479-97 (1970)

POTENTIAL ENERGY CURVES AND NATURE OF BINDING IN
GROUP-3A MONOHALIDES.
THAKUR SN + SINGH RB + RAI DK
J SCI IND RES 27(9): 339-47 (1968)

0175 DENSITY LOCALIZATION OF ATOMIC AND MOLECULAR
ORBITALS. PART-3: HETERONUCLEAR DIATOMIC AND
POLYATOMIC MOLECULES.
VAN NIESSEN W
THEOR CHIM ACTA 29(1): 29-48 (1973)

BORON DIFLUORIDE BF_2

0176 AB INITIO CALCULATIONS ON SOME SMALL RADICALS BY
THE UNRESTRICTED HARTREE-FOCK METHOD. (BORON
DIFLUORIDE, NITROGEN DIFLUORIDE, OXYGEN
DIFLUORIDE, CARBON DIFLUORIDE)
BROWN RD + WILLIAMS GR
CHEM PHYS 3: 19-34 (1974)

LOCALIZED ORBITALS IN BORON FLUORIDES. HIGHLY
POLARIZED BORON- FLUORINE DOUBLE AND TRIPLE
BONDS.
HALL JH + HALGREN TA + KLEIER DA + LIPSCOMB WN
INORG CHEM 13(10): 2520-1 (1974)

EMISSION SPECTRA OF CERTAIN POLYATOMIC FREE
RADICALS OBTAINED IN ELECTRICAL DISCHARGES
THROUGH BORON TRIFLUORIDE. (BORON DIFLUORIDE)
KRISHNAMACHARI SLNG + NARASIMHAM NA + SINGH M
P181-4 OF INT CONF ON SPECTROSC, 1ST PROC, 1967,
BOMBAY, INDIA.

0177 HEAT OF FORMATION OF BORON DIFLUORIDE.
MARGRAVE JL
J PHYS CHEM 66: 1209-10 (1962)

MEAN AMPLITUDES OF VIBRATION FOR BORON
DIFLUORIDE, CARBON DIFLUORIDE, NITROGEN
DIFLUORIDE, ALUMINUM DIFLUORIDE, PHOSPHORUS
DIFLUORIDE, AND ZIRCONIUM DIFLUORIDE.
NAGARAJAN G
INDIAN J PURE APPL PHYS 2: 341-3 (1964)

0178 ELECTRON SPIN RESONANCE AND STRUCTURE OF THE
BORON DIFLUORIDE RADICAL.
NELSON W + GORDY W
J CHEM PHYS 51: 4710-13 (1969)

MOLECULAR PROPERTIES OF THE TRIATOMIC DIFLUORIDES BERYLLIUM DIFLUORIDE, BORON DIFLUORIDE, ARGON DIFLUORIDE, NITROGEN DIFLUORIDE, AND OXYGEN DIFLUORIDE. (SCF)
ROTHENBERG S + SCHAEFER HF
J AMER CHEM SOC 95: 2095-100 (1973)

0179 THERMODYNAMIC PROPERTIES OF THE BORON CHLORIDE FLUORIDE SYSTEM FROM MASS SPECTROMETER INVESTIGATIONS. (BORON DIFLUORIDE)
SRIVASTAVA RD + FARBER M
TRANS FARADAY SOC 67(8): 2298-302 (1971)

0180 AB INITIO INVESTIGATION OF THE GEOMETRY, BONDING, AND COUPLING CONSTANTS OF BORON DIFLUORIDE.
THOMSON C + BROTCHIE DA
THEOR CHIM ACTA 32(2): 101-9 (1973)

0181 ESR PARAMETERS OF THE BORON DIFLUORIDE RADICAL.
ZAUCER M + AZMAN A
CROAT CHEM ACTA 43: 139-140 (1971)

BORON TRIFLUORIDE BF_3

0182 ROTATIONAL SPECTRA OF NONPOLAR MOLECULES.
ALIEV MR + MIKHAILOV VM
OPT SPECTROSC 35(2): 147-51 (AUG 1973)

0183 NUCLEAR SPIN RELAXATION STUDY OF THE SPIN ROTATION INTERACTION IN SYMMETRIC TOP MOLECULES.
ARMSTRONG RL + COURTNEY JA
CAN J PHYS 50(12): 1262-72 (1972)

0184 ELECTRONIC STRUCTURE OF BORON TRIFLUORIDE.
ARMSTRONG DR + PERKINS PG
THEOR CHIM ACTA 15: 413-22 (1969) ; CHEM COMMUN 1969: 856

0185 GROUND STATE PROPERTIES OF THE GROUP-2 TRIHALIDES. (BORON TRIFLUORIDE)
ARMSTRONG DR + PERKINS PG
J CHEM SOC A 1967: 1218-22

0186 INFRARED ABSORPTION SPECTRA OF SOME POLYATOMIC FLUORIDES. (SILICON TETRAFLUORIDE, CARBON TETRAFLUORIDE, BORON TRIFLUORIDE, NITROGEN TRIFLUORIDE)
BAILEY CR + HALE JB + THOMPSON JW
J CHEM PHYS 5: 274-5 (1937)

0187 MOLECULAR STRUCTURE OF BORON TRIFLUORIDE.
BAILEY CR + HALE JB + THOMPSON JW
PROC ROY SOC A 161: 107-14 (1937)

0188 PHOTOELECTRON SPECTRA OF HALIDES. PART-2: HIGH RESOLUTION SPECTRA OF THE BORON TRIHALIDES.
BASSETT PJ + LLOYD DR
J CHEM SOC A 1971: 1551-9

0189 PHOTOELECTRON SPECTRUM OF BORON TRIFLUORIDE.
BASSETT PJ + LLOYD DR
CHEM COMMUN 1970: 36-7

0190 FORCE CONSTANTS FOR MOLECULES WITH D3H SYMMETRY.
BECKMANN L + GUTJAHR L + MECKE R
SPECTROCHIM ACTA 21: 141-53 (1965)

0191 VIBRATION- ROTATION BANDS OF BORON TRIFLUORIDE.
BROWN CW
UNIV MINNESOTA, PHD THESIS, 1966, 243P.

0192 BORON TRIFLUORIDE. THE ANALYSIS OF 2NU-3 AND THE DETERMINATION OF THE 33 CORIOLIS CONSTANT.
BROWN CW + OVEREND J
CAN J PHYS 46: 977-85 (1968)

0193 ROTATION VIBRATION BAND CONTOURS IN THE INFRARED SPECTRUM OF BORON TRIFLUORIDE.
BROWN CW + OVEREND J
SPECTROCHIM ACTA 26A: 1535-46 (1969)

0194 INTERMOLECULAR FLUORINE EXCHANGE BETWEEN TETRABUTYL AMMONIUM HEXAFLUORO PHOSPHATE AND BORON TRIFLUORIDE OR PHOSPHORUS PENTAFLUORIDE.
BROWNSTEIN S + BORNAIS J
CAN J CHEM 46: 225-28 (1968)

0195 CNDO CALCULATION OF DIPOLE MOMENT DERIVATIVES AND INFRARED INTENSITIES OF BORON TRIFLUORIDE.
BRUNS R + PERSON WB
J CHEM PHYS 55: 5401-5404 (1971)

0196 BARRIER TO ELECTRON PASSAGE THROUGH ELECTRONEGATIVE ATOMS IN BORON TRIFLUORIDE.
CADIOLI B + PINCELLI U + TOSATTI E + FANO U + DEHMER JL
CHEM PHYS LETT 17: 15-18 (1972)

0197 CALCULATED MEAN AMPLITUDES OF VIBRATION IN BORON TRIHALIDES.
CYVIN SJ
ACTA CHEM SCAND 13: 334-6 (1959)

0198 HEAT FORMATION BORON TRIFLUORIDE BY DIRECT COMBINATION OF THE ELEMENTS.
DOMALSKI ES + ARMSTRONG GT
J RES NBS A 71: 195-202 (1967)

0199 VIBRATION ROTATION BANDS OF BORON TRIFLUORIDE.
DRESKA M
OHIO STATE UNIV, PHD THESIS, 1964, 151P.

0200 NU-3 OF BORON TRIFLUORIDE.
DRESKA SN + RAO KN
J MOL SPECTROSC 18: 404-11 (1965)

0201 FORCE CONSTANTS OF THE BORON TRIHALIDES - A SURVEY.
DUNCAN JL
J MOL SPECTROSC 13: 338-43 (1964)

0202 PERPENDICULAR VIBRATIONS OF BORON TRIFLUORIDE.
DUNCAN JL
J MOL SPECTROSC 22: 247-61 (1967)

0203 COMPARATIVE AB INITIO CALCULATIONS ON BORON FLUORIDES.
FITZPATRICK NJ
INORG NUCL CHEM LETT 9(9): 965-70 (1973)

0204 ULTRASOFT X-RAY ABSORPTION SPECTRA OF BORON TRIFLUORIDE.
FOMICHEV VA + BARINSKII RL
J STRUCT CHEM 11(5): 810-13 (1970)

0205 ANALYSIS OF SYMMETRIC ROTOR RAMAN SPECTRA. PURE ROTATIONAL SPECTRUM OF BORON TRIFLUORIDE.
FREEDMAN PA + JONES WJ
J MOL SPECTROSC 54: 182-90 (1975)

0206 INFRARED ABSORPTION SPECTRUM OF BORON TRIFLUORIDE.
GAGE DM + BARKER EF
J CHEM PHYS 7: 455-9 (1939)

0207 NU-3 BANDS OF BORON TRIFLUORIDE.
GINN SGW + BROWN CW + KENNEY JK + OVEREND J
J MOL SPECTROSC 28: 509-25 (1968)

0208 NU-4 BANDS OF THE TEN AND ELEVEN ISOTOPES OF BORON TRIFLUORIDE AT HIGH RESOLUTION.
GINN SGW + JOHANSEN D + OVEREND J
J MOL SPECTROSC 36: 448-63 (1970)

0209 BORON TRIFLUORIDE. THE ANALYSIS OF NU-2 AND THE DETERMINATION OF THE MOLECULAR GEOMETRY.
GINN SGW + KENNEY JK + OVEREND J
J CHEM PHYS 48: 1571-79 (1968)

0210 VIBRATIONAL ANHARMONICITY IN BORON TRIFLUORIDE.
GINN SGW + REICHMAN S + OVEREND J
SPECTROCHIM ACTA 26A: 291-96 (1970)

0211 APPLICATION OF THE PROGRESSIVE RIGIDITY METHOD TO SOME SIMPLE XYN TYPE MOLECULES. (BORON TRIFLUORIDE)
GODNEV IN + ALEKSANDROVSKAYA AM
J APPL SPECTROSC 4(4): 265-7 (APR 1966)

0212 FORCE CONSTANTS OF CARBON TETRAFLUORIDE AND BORON TRIFLUORIDE AND COMPARISON WITH FORCE CONSTANTS OF THE EIGHTH PERIOD.
GOUBEAU J + BUES W + KAMPMANN FW
Z ANORG ALLGEM CHEM 283: 123-37 (1956) (IN GERMAN)

0213 AB INITIO CALCULATION ON BORON TRIFLUORIDE AND BORON TRICHLORIDE.
GOUTIER D + BURNELLE LA
CHEM PHYS LETT 18: 460-64 (1973)

0214 HEATS OF FORMATION OF INORGANIC FLUORIDES ESPECIALLY THE ELEMENTS OF ATOMIC NUMBER BELOW 20.
GROSS P + HAYMAN C + LEVI DL + STUART MC
FULMER RES INST, 1960, 32P. PB153445

0215 ABSORPTION OF BORON TRIFLUORIDE GAS IN THE EXTREME ULTRAVIOLET.
HAYES W + BROWN FC
J PHYS B 4(10): L85-8 (1971)

0216 RADIATIVE LIFETIMES OF ULTRAVIOLET EMISSION SYSTEMS EXCITED IN BORON TRIFLUORIDE, CARBON TETRAFLUORIDE, AND SILICON TETRAFLUORIDE.
HESSER JE + DRESSLER K
J CHEM PHYS 47: 3443-50 (1967)

0217 NUCLEAR MAGNETIC RELAXATION IN BORON TRIFLUORIDE GAS.
HINSHAW WS
UNIV NORTH CAROLINA, PHD THESIS, 1970, 80P.

0218 ELECTRON DENSITY DISTRIBUTION IN SOME CHLORIDES.
IONOV SP + IONOVA GV
RUSS J INORG CHEM 13(1): 157-8 (1968)

0219 FORCE CONSTANTS AND THERMODYNAMIC CONSTANTS FOR THE BORON HALIDES.
JAKES J + PAPOUSEK D
COLLECT CZECH CHEM COMMUN 26: 2110-23 (1961)

0220 FLUORINE BOMB CALORIMETRY. PART-15: THE ENTHALPY OF FORMATION OF BORON TRIFLUORIDE.
JOHNSON GK + FEDER HM + HUBBARD WN
J PHYS CHEM 70: 1-6 (1966)

0221 MOLECULAR STRUCTURE AND FORCE CONSTANTS OF BORON TRIFLUORIDE AND BORON TRICHLORIDE.
KONAKA S + MURATA Y + KUCHITSU K + MORINO Y
BULL CHEM SOC JAP 39: 1134-46 (1966)

0222 EMISSION SPECTRA OF CERTAIN POLYATOMIC FREE RADICALS OBTAINED IN ELECTRICAL DISCHARGES THROUGH BORON TRIFLUORIDE. (BORON DIFLUORIDE)
KRISHNAMACHARI SLNG + NARASIMHAM NA + SINGH M
P181-4 OF INT CONF SPECTROSC, 1ST PROC, VOL 1, BOMBAY, 1967.

0223 BORON FLUORINE BOND DISTANCE OF BORON TRIFLUORIDE DETERMINED BY GAS ELECTRON DIFFRACTION.
KUCHITSU K + KINAKA S
J CHEM PHYS 45: 4342-47 (1966)

0224 PHOTOELECTRON SPECTRA OF COMPLEXES OF BORON TRIFLUORIDE AND AMINES.
LAKE RF
SPECTROCHIM ACTA 27A: 1220-21 (1971)

0225 A THEORETICAL COMPARISON OF THE ELECTRONIC STRUCTURES OF BORON TRIHALIDES AND BORON TETRAHALIDE ANIONS.
LEIBOVICI C
J MOL STRUCT 14: 459-60 (1972)

0226 SEMIEMPIRICAL STUDY OF THE ELECTRONIC STRUCTURE OF BORON HYDRIDE FLUORIDES. (BORON TRIFLUORIDE)
LEIBOVICI C + LABARRE JF
J CHIM PHYS PHYSICOCHIM BIOL 68(5): 726-9 (1971) (IN FRENCH)

0227 FORCE FIELDS FOR THE BORON TRIHALIDES.
LEVIN IW + ABRAMOWITZ S
J CHEM PHYS 43: 4213-22 (1965)

0228 ISOTOPIC SPLITTING IN MATRIX ISOLATED BORON TRIFLUORIDE.
LEVIN IW + ABRAMOWITZ S
CHEM PHYS LETT 10: 247-8 (1971)

0229 ELECTRON STRUCTURE AND NMR SPECTRA OF BORON COMPOUNDS.
LYUBIMOV VS + IONOV SP
BULL ACAD SCI USSR, DIV CHEM SCI 20(11): 2455-6 (NOV 1971)

0230 STUDY OF THE ISOTOPE EFFECT AND A MORE PRECISE DETERMINATION OF THE FAR INFRARED SPECTRUM OF BORON TRIFLUORIDE.
MAL'TSEV AA + MOSKVITINA EN + TATAEVSKII VM
FIZ SB LVOV GOS UNIV NO 3: 465-73 (1957) (IN RUSSIAN)

0231 CUBIC POTENTIAL CONSTANTS AND 1-TYPE RESONANCE IN BORON TRIFLUORIDE.
MASRI FN
J MOL SPECTROSC 43(1): 168-70 (1972)

0232 ACCURATE FORCE CONSTANTS FROM HEAVY ISOTOPIC SUBSTITUTION. PART-2: BORON TRIFLUORIDE, SILICON TETRAFLUORIDE, AND NITROGEN DIOXIDE.
MCKEAN DC
SPECTROCHIM ACTA 22: 269-79 (1966)

0233 MEAN AMPLITUDES OF VIBRATION AND SHRINKAGE EFFECT OF BORON TRIHALIDES, CALCULATED FROM UREY-BRADLEY FORCE CONSTANTS.
MEISINGSETH E
ACTA CHEM SCAND 16: 778 (1962)

0234 UTILITY OF HEAVY ATOM ISOTOPIC SUBSTITUION DATA FOR THE CALCULATION OF FORCE CONSTANTS. ANHARMONICITY EFFECTS ON THE FORCE CONSTANTS OF BORON TRIFLUORIDE.
MOHAN N + MUELLER A
CAN J SPECTROSC 17(5): 132-4 (1972)

0235 CHEMICAL SHIFTS OF BORON, PHOSPHORUS, PHOSPHORYL, AND THIOPHOSPHORYL HALIDES.
MULLER A + NIECKE E + KREBS B
MOL PHYS 14: 591-94 (1968)

0236 THERMODYNAMIC PROPERTIES OF SOME BORON TRIHALIDES.
NAGARAJAN G
Z PHYS CHEM 31: 347-9 (1962)

0237 VACUUM ULTRAVIOLET ABSORPTION SPECTRA OF BORON TRIHALIDES.
PLANCKAERT AA + SAUVAGEAU P + SANDORFY C
CHEM PHYS LETT 20(2): 170-3 (1973)

CONSTRUCTION OF A HIGH RESOLUTION PHOTOELECTRON SPECTROMETER. (CARBON TETRAFLUORIDE, BORON TRIFLUORIDE, SULFUR HEXAFLUORIDE, NITROGEN TRIFLUORIDE, AND SILICON TETRAFLUORIDE)
PULLEN BP
UNIV TENNESSEE, PHD THESIS, 1970, 117P.

0238 UNIQUE FORCE FIELD FOR BORON TRIHALIDES.
RAMASWAMY K + JAYARAMAN L
INDIAN J PHYS 49(1): 1-8 (1975)

0239 INFRARED MATRIX ISOLATION STUDIES. (BORON TRIFLUORIDE, SULFUR TETRAFLUORIDE, ANTIMONY PENTAFLUORIDE, ARSENIC PENTAFLUORIDE, CHLORINE TRIFLUORIDE, BROMINE TRIFLUORIDE, AND BROMINE PENTAFLUORIDE)
REDINGTON RL
TEXAS TECH UNIV, 1969, 89P. AD701120

0240 THERMODYNAMIC PROPERTIES OF BORON HALIDES.
ROMASHKO BV + SOROKIN YV + FEDOROVA TA + CHICHIKALYUK EM + MASLOV PG
J APPL CHEM USSR 43(12): 2692-5 (DEC 1970)

0241 AB INITIO STUDIES OF THE ELECTRONIC STRUCTURES OF BORON TRIHYDRIDE, BORON DIHYDRIDE FLUORIDE, BORON HYDRIDE DIFLUORIDE, AND BORON TRIFLUORIDE.
SCHWARTZ ME + ALLEN LC
J AMER CHEM SOC 92: 1466-71 (1970)

0242 FERMI RESONANCE IN BORON TRIFLUORIDE.
SHELDRICK GM
J MOL SPECTROSC 20(3): 295-6 (1966)

0243 FORCE CONSTANTS OF CARBON TETRAFLUORIDE, SILICON
      TETRAFLUORIDE, BORON TRIFLUORIDE, ETHANE,
      SILANE, AMMONIA, AND PHOSPHINE.
      SHIMANOUCHI T + NAKAGAWA I + HIRAISHI J
      + ISHII M
      J MOL SPECTROSC 19: 78-107 (1966)

0244 EFFECTS OF THE LENNARD-JONES POTENTIALS ON THE
      UREY-BRADLEY FORCE CONSTANTS OF BORON
      TRIHALIDES.
      SHIMIZU K
      KAGAKU TO KOGYO 40(10): 477-80 (1966) (IN
      JAPANESE)

0245 INFRARED ABSORPTION SPECTRUM OF LIQUID BORON
      TRIFLUORIDE.
      STEINHARDT RG + FETSCH GES + JORDAN MW
      J CHEM PHYS 43(12): 4528-30 (1965)

0246 UREY-BRADLEY FORCE CONSTANTS OF BORON HALIDES
      AND THEIR TRANSFERABILITY. (BORON TRIFLUORIDE)
      TOYUKI H + SHIMIZU K + HYAMA H
      BULL CHEM SOC JAP 36(7): 783-5 (1963)

0247 NORMAL VIBRATIONS OF BORON TRIHALIDES. (BORON
      TRIFLUORIDE)
      TOYUKI H + SHIMIZU K + HIYAMA H
      KAGAKU TO KOGYO 37: 288-91 (1963) (IN JAPANESE)

0248 KINEMATICAL EVALUATION OF FORCE CONSTANTS.
      APPLICATION TO BORON TRIHALIDES.
      TRIPATHI DN + THAKUR SN
      J MOL STRUCT 5: 345-49 (1970)

0249 INFRARED SPECTRUM OF BORON TRIFLUORIDE.
      VANDERRYN J
      J CHEM PHYS 30: 331-2 (1959)

0250 FLUORINE BOMB CALORIMETRY. PART-3: THE HEAT OF
      FORMATION OF BORON TRIFLUORIDE.
      WISE SS + MARGRAVE JL + FEDER HM + HUBBARD WN
      J PHYS CHEM 65: 2157-9 (1961)

0251 RAMAN SPECTRUM OF BORON TRIFLUORIDE GAS.
      YOST DM + DEVAULT D + ANDERSON TF + LASSETTRE EN
      J CHEM PHYS 6: 424-5 (1938)

BROMINE MONOFLUORIDE   **BrF**

      PHOTOELECTRON SPECTRA OF HALIDES. (CHLORINE
      MONOFLUORIDE, CHLORINE TRIFLUORIDE, BROMINE
      MONOFLUORIDE, AND BROMINE TRIFLUORIDE)
      DEKOCK RL + HIGGINSON BR + LLOYD DR + BREEZE A +
      ARMSTRONG DR
      MOL PHYS 24: 1059-72 (1972)

      ROTATIONAL ANALYSIS OF VISIBLE BROMINE FLUORIDE
      BANDS AND THE POTENTIAL CURVE OF THE ZERO(+)
      TERMS.
      BODERSON PH + CUDMANI LYC
      BER DEUT VERSUCHT LUFT RAUMFAHRT 250: 3-18
      (1963) (IN GERMAN)

0252 SPECTRUM OF BROMINE MONOFLUORIDE AND ITS
      DISSOCIATION ENERGY.
      BODERSON PH + SICRE JE
      Z PHYS 141(4): 515-24 (1955)

      ELECTRONEGATIVITY, NCN-BONDED INTERACTIONS AND
      POLARIZABILITY IN HYDROGEN HALIDES AND THE
      INTERHALOGEN COMPOUNDS. (CHLORINE MONOFLUORIDE,
      BROMINE MONOFLUORIDE, AND IODINE MONOFLUORIDE)
      BROWN RF
      J AMER CHEM SOC 83: 36-42 (1961)

0253 THERMODYNAMIC PROPERTIES OF THE DIATOMIC
      INTERHALOGENS FRCM SPECTROSCOPIC DATA.
      (CHLORINE, BROMINE, AND IODINE MONOFLUORIDES)
      COLE LG + ELVERUM GW
      J CHEM PHYS 20(10): 1543-51 (1952)

      DISSOCIATION ENERGIES OF DIATOMIC HALOGEN
      FLUORIDES. (IODINE FLUORIDE, BROMINE FLUORIDE)
      COXON JA
      CHEM PHYS LETT 33(1): 136-40 (1975)

0254 MOLECULAR ZEEMAN EFFECT, MAGNETIC PROPERTIES,
      AND ELECTRIC QUADRUPOLE MOMENTS IN CHLORINE
      MONOFLUORIDE, BROMINE MONOFLUORIDE, BROMINE
      CYANIDE, AND IODINE CYANIDE.
      EWING JJ + TIGELAAR HL + FLYGARE WH
      J CHEM PHYS 56: 1957-66 (1972)

      SOME CORRECTIONS TO THE SECOND ORDER STARK
      EFFECT OF LINEAR MOLECULES. (CHLORINE
      MONOFLUORIDE AND BROMINE MONOFLUORIDE)
      MIZUSHIMA M
      P1167-82 OF INT MEETING MOL SPECSTOSC, 4TH,
      1959, BOLOGNA, VOL 3, MANGINI A (ED), MACMILLAN,
      1962.

0255 MATRIX ISOLATION OF BROMINE MONOFLUORIDE.
      SMARDZEWSKI RR + FOX WB
      J FLUORINE CHEM 7(4): 453-5 (1976)

0256 INFRARED STUDIES OF THE BROMINE FLUORIDES.
      (BROMINE MONOFLUORIDE, BROMINE TRIFLUORIDE, AND
      BROMINE PENTAFLUORIDE)
      STEIN L
      J AMER CHEM SOC 81: 1273-6 (1959)

BROMINE TRIFLUORIDE   **BrF₃**

0257 VIBRATIONAL SPECTRUM OF BROMINE TRIFLUORIDE.
      CHRISTE KO + CURTIS EC + PILIPOVICH D
      SPECTROCHIM ACTA 27A: 931-6 (1971)

      VIBRATIONAL SPECTRA AND THERMODYNAMIC PROPERTIES
      OF CHLORINE TRIFLUORIDE AND BROMINE TRIFLUORIDE.
      CLAASSEN HH + WEINSTOCK B + MALM JG
      J CHEM PHYS 28: 285-9 (1958)

      CNDO/2 AND INDO ALL VALENCE ELECTRON
      CALCULATIONS ON THE GEOMETRY AND PROPERTIES OF
      SOME INTERHALOGENS. (CHLORINE TRIFLUORIDE,
      BROMINE TRIFLUORIDE, IODINE TRIFLUORIDE,
      CHLORINE PENTAFLUORIDE, BROMINE PENTAFLUORIDE,
      IODINE PENTAFLUORIDE, CHLORINE HEPTAFLUORIDE,
      BROMINE HEPTAFLUORIDE, AND IODINE HEPTAFLUORIDE)
      DEB BM + COULSON CA
      J CHEM SOC A 1971: 958-70

      RAMAN SPECTRA AND THE STRUCTURE OF CHLORINE
      TRIFLUORIDE AND BROMINE TRIFLUORIDE.
      DRIFFORD M + MARTIN D + BOUGON R
      REV CHIM MINERALE 7: 1069-86 (1970)

      INFRARED SPECTRA OF MATRIX ISOLATED CHLORINE
      TRIFLUORIDE, BROMINE TRIFLUORIDE, AND BROMINE
      PENTAFLUORIDE. FLOURINE EXCHANGE MECHANISM OF
      LIQUID CHLORINE TRIFLUORIDE, BROMINE
      TRIFLUORIDE, AND SULFUR TETRAFLUORIDE.
      FREY RA + REDINGTON RL + ALJIBURY ALK
      J CHEM PHYS 54: 344-55 (1971)

      THERMODYNAMIC PROPERTIES OF PENTAFLUORIDES AT
      HIGH TEMPERATURES. PART-1: CALCULATION METHODS.
      (CHLORINE PENTAFLUORIDE, BROMINE PENTAFLUORIDE,
      IODINE PENTAFLUORIDE, PHOSPHORUS PENTAFLUORIDE,
      VANADIUM PENTAFLUORIDE, ANTIMONY PENTAFLUORIDE)
      GALKIN NP + TUMANOV YN + BUTYLKIN YP
      KHIM VYS ENERG 4(6): 512-8 (1970) (IN RUSSIAN)

      QUADRUPOLE COUPLING CONSTANTS IN CHLORINE
      TRIFLUORIDE AND BROMINE TRIFLUORIDE.
      GUPTA LC
      INDIAN J PURE APPL PHYS 5: 437-8 (1967)

0258 MICROWAVE SPECTRUM AND MOLECULAR STRUCTURE OF
      BROMINE TRIFLUORIDE.
      MAGNUSON LW
      J CHEM PHYS 27: 223-6 (1957)

MOLECULAR FORCE FIELD FOR INTERHALOGEN
COMPOUNDS. (CHLORINE TRIFLUORIDE, BROMINE
TRIFLUORIDE)
RAMASWAMY K + MUTHUSUBRAMANIAN P
J MOL STRUCT 9: 193-6 (1971)

INFRARED MATRIX ISOLATION STUDIES. (BORON
TRIFLUORIDE, SULFUR TETRAFLUORIDE, ANTIMONY
PENTAFLUORIDE, ARSENIC PENTAFLUORIDE, CHLORINE
TRIFLUORIDE, BROMINE TRIFLUORIDE, AND BROMINE
PENTAFLUORIDE)
REDINGTON RL
TEXAS TECH UNIV, 1969, 89P. AD701120

INFRARED AND RAMAN SPECTRA OF CHLORINE
TRIFLUORIDE AND BROMINE TRIFLUORIDE.
SELIG H + CLAASSEN HH + HOLLOWAY JH
J CHEM PHYS 52: 3517-21 (1970)

INFRARED STUDIES OF THE BROMINE FLUORIDES.
(BROMINE MONOFLUORIDE, BROMINE TRIFLUORIDE, AND
BROMINE PENTAFLUORIDE)
STEIN L
J AMER CHEM SOC 81: 1273-6 (1959)

0259 ELECTRICAL CONDUCTIVITY OF SOLID CHLORINE
TRIFLUORIDE AND BROMINE TRIFLUORIDE.
TOY MS + CANNON WA
P237-45 OF ADVANCED PROPELLANT CHEMISTRY, AMER
CHEM SOC, 1966, 290P. (ADVANCES IN CHEMISTRY
SERIES 54)

BROMINE PENTAFLUORIDE   BrF$_5$

VIBRATIONAL SPECTRA AND VALENCE FORCE CONSTANTS
OF SQUARE PYRAMIDAL MOLECULES. XENON OXIDE
TETRAFLUORIDE, IODINE PENTAFLUORIDE, BROMINE
PENTAFLUORIDE, AND CHLORINE PENTAFLUORIDE.
BEGUN GM + FLETCHER WH + SMITH DF
J CHEM PHYS 42: 2236-42 (1965)

ORBITALS AND STRUCTURES OF PENTAFLUORIDES.
(PHOSPHORUS PENTAFLUORIDE, ARSENIC
PENTAFLUORIDE, AND BROMINE PENTAFLUORIDE)
BERRY RS + TAMRES M + BALLHAUSEN CH + HOHANSEN H
ACTA CHEM SCAND 22: 231-46 (1968)

0260 K-TYPE DOUBLING IN THE MILLIMETER WAVE SPECTRUM
OF BROMINE PENTAFLUORIDE.
BRADLEY RH + BRIER PN + WHITTLE MJ
J MOL SPECTROSC 44: 536-48 (1972)

REPARAMETERIZED VIBRATIONAL FORCE CONSTANTS.
PART-1: METHOD AND APPLICATION TO SOME SQUARE
PYRAMIDAL MOLECULES. (IODINE PENTAFLUORIDE,
BROMINE PENTAFLUORIDE, AND CHLORINE BROMINE
PENTAFLUORIDE)
CURTIS EC
SPECTROCHIM ACTA 27A: 1989-97 (1971)

MEAN AMPLITUDES OF VIBRATION FOR SOME PYRAMIDAL
XY4Z MOLECULES. (IODINE PENTAFLUORIDE, BROMINE
PENTAFLUORIDE, AND CHLORINE PENTAFLUORIDE)
CYVIN SJ + BRUNVOLL J + ROBIETTE AG
J MOL STRUCT 3: 259-61 (1969)

CNDO/2 AND INDO ALL VALENCE ELECTRON
CALCULATIONS ON THE GEOMETRY AND PROPERTIES OF
SOME INTERHALOGENS. (CHLORINE TRIFLUORIDE,
BROMINE TRIFLUORIDE, IODINE TRIFLUORIDE,
CHLORINE PENTAFLUORIDE, BROMINE PENTAFLUORIDE,
IODINE PENTAFLUORIDE, CHLORINE HEPTAFLUORIDE,
BROMINE HEPTAFLUORIDE, AND IODINE HEPTAFLUORIDE)
DEB BM + COULSON CA
J CHEM SOC A 1971: 958-70

INFRARED SPECTRA OF MATRIX ISOLATED CHLORINE
TRIFLUORIDE, BROMINE TRIFLUORIDE, AND BROMINE
PENTAFLUORIDE. FLOURINE EXCHANGE MECHANISM OF
LIQUID CHLORINE TRIFLUORIDE, BROMINE
TRIFLUORIDE, AND SULFUR TETRAFLUORIDE.
FREY RA + REDINGTON RL + ALJIBURY ALK
J CHEM PHYS 54: 344-55 (1971)

0261 MICROWAVE MEASUREMENTS OF THE J-8 FAR-9
TRANSITION OF BROMINE PENTAFLUORIDE.
JONES SR + BRIER PN + BROOKBANKS DM + BAKER JG
J MOL SPECTROSC 47(2): 351-2 (1973)

MOLECULAR STRUCTURE FORCE CONSTANTS AND
THERMODYNAMIC FUNCTIONS OF IODINE PENTAFLUOIRDE
AND BROMINE PENTAFLUORIDE.
KHANNA RK
J SCI IND RES (INDIA) 21B: 352-6 (1962)

IDEAL GAS THERMODYNAMIC PROPERTIES OF CHLORINE
PENTAFLUORIDE, BROMINE PENTAFLUORIDE, AND IODINE
PENTAFLUORIDE.
KUDCHADKER AP + KUDCHADKER SA + AGARWAL PM
INDIAN J CHEM 9: 722-24 (1971)

0262 INFRARED SPECTRUM OF BROMINE PENTAFLUORIDE.
MCDOWELL RS + ASPREY LB
J CHEM PHYS 37: 165-7 (1962)

0263 SELF IONIZATION OF HALOGEN FLUORIDES. (BROMINE
TRIFLUORIDE)
MEINERT H + GROSS U
J FLUORINE CHEM 2: 381-6 (1972)

0264 MOLECULAR FORCE FIELD FOR BROMINE PENTAFLUORIDE.
PILLAI MGK + PILLAI PP
CAN J CHEM 46: 2393-97 (1968)

MOLECULAR FORCE FIELDS FOR INTERHALOGEN
COMPOUNDS. (CHLORINE PENTAFLUORIDE AND BROMINE
PENTAFLUORIDE)
RAMASWAMY K + MUTHUSUBRAMANIAN P
J MOL STRUCT 7: 45-50 (1971)

INFRARED MATRIX ISOLATION STUDIES. (BORON
TRIFLUORIDE, SULFUR TETRAFLUORIDE, ANTIMONY
PENTAFLUORIDE, ARSENIC PENTAFLUORIDE, CHLORINE
TRIFLUORIDE, BROMINE TRIFLUORIDE, AND BROMINE
PENTAFLUORIDE)
REDINGTON RL
TEXAS TECH UNIV, 1969, 89P. AD701120

0265 GAS PHASE MOLECULAR STRUCTURES OF BROMINE
PENTAFLUORIDE AND IODINE PENTAFLUORIDE FROM
ELECTRON DIFFRACTION AND ROTATIONAL CONSTANT
DATA.
ROBIETTE AG + BRADLEY RH + BRIER PN
CHEM COMMUN 1971: 1567-8

0266 INFRARED AND RAMAN SPECTRA OF BROMINE
PENTAFLUORIDE AND CHLORINE TRIFLUORIDE IN THE
CONDENSED PHASE. EVIDENCE OF AN ASSOCIATED STATE
IN LIQUID CHLORINE TRIFLUORIDE.
ROUSSON R + DRIFFORD M
J CHEM PHYS 62(5): 1806-11 (1975)

INFRARED STUDIES OF THE BROMINE FLUORIDES.
(BROMINE MONOFLUORIDE, BROMINE TRIFLUORIDE, AND
BROMINE PENTAFLUORIDE)
STEIN L
J AMER CHEM SOC 81: 1273-6 (1959)

0267 RAMAN SPECTRUM, STRUCTURE, FORCE CONSTANTS, AND
THERMODYNAMIC PROPERTIES OF BROMINE
PENTAFLUORIDE.
STEPHENSON CV + JONES EA
J CHEM PHYS 20(12): 1830-6 (1952)

0268 ROTATIONAL SPECTRA OF BROMINE PENTAFLUORIDE IN
THE MILLIMETER WAVE LENGTH REGION.
SUZEAU P + JUREK R + CHANUSSOT J
COMPT REND B 276(19): 777-9 (1973) (IN FRENCH)

CORIOLIS COUPLING COEFFICIENTS OF BROMINE
PENTAFLUORIDE AND CHLORINE PENTAFLUORIDE.
VENKATESWARLU K + MATHEW MP
CURR SCI 37: 252-3 (1968)

0269 MICROWAVE SPECTRUM OF BROMINE PENTAFLUORIDE.
WHITTLE MJ + BRADLEY RH + BRIER PN
TRANS FARADAY SOC 67(9): 2505-9 (1971)

BROMINE HEXAFLUORIDE $\quad$ BrF_6

0270   ESR SPECTRUM AND STRUCTURE OF BROMINE
HEXAFLUORIDE.
NISHIKIDA K + WILLIAMS F + MAMANTOV G + SMYRL N
J CHEM PHYS 63(4): 1693-4 (1975)

0271   ELECTRON SPIN RESONANCE SPECTRA OF HALOGEN
HEXAFLUORIDES. (CHLORINE HEXAFLUORIDE, IODINE
HEXAFLUORIDE, BROMINE HEXAFLUORIDE)
BOATE AR + MORTON JR + PRESTON KF
INORG CHEM 14(12): 3127-8 (1975)

CADMIUM DIFLUORIDE $\quad$ CdF_2

ELECTRON DIFFRACTION ANALYSIS OF THE MOLECULAR
STRUCTURES OF GROUP-2 ELEMENTS. (BERYLLIUM
DIFLUORIDE, MAGNESIUM DIFLUORIDE, CALCIUM
DIFLUORIDE, STRONTIUM DIFLUORIDE, BARIUM
DIFLUORIDE, ZINC DIFLUORIDE, CADMIUM DIFLUORIDE)
AKISHIN PA + SPIRIDONOV VP
SOV PHYS CRYSTALLOGR 2(4): 472-9 (1957)

0272   KNUDSEN AND LANGMUIR MEASUREMENTS OF THE
SUBLIMATION PRESSURE OF CADMIUM DIFLUORIDE.
BESENBRUCH G + KANA'AN AS + MARGRAVE JL
J PHYS CHEM 68: 3174-6 (1965)

POLARIZED ION MODEL AND THE BENDING FORCE
CONSTANTS OF THE GROUP-2B HALIDES. (ZINC
DIFLUORIDE, CADMIUM DIFLUORIDE, AND MERCURY
DIFLUORIDE)
ELIEZER I
THEOR CHIM ACTA 18: 77-85 (1970)

0273   MELTING POINTS OF INORGANIC FLUORIDES. (CADMIUM
DIFLUORIDE, MAGNESIUM DIFLUORIDE, STRONTIUM
DIFLUORIDE, AND BARIUM DIFLUORIDE)
KOJIMA H + WHITEWAY SG + MASSON CR
CAN J CHEM 46: 2968-71 (1968)

0274   RAMAN SPECTRA OF CADMIUM DIFLUORIDE AND LEAD
DIFLUORIDE.
KRISHNAMURTHY N + SOOTS V
CAN J PHYS 48: 1104-7 (1970)

0275   VIBRATIONAL SPECTRA OF THE DIHALIDES OF MERCURY
AND CADMIUM. (CADMIUM DIFLUORIDE AND MERCURY
DIFLUORIDE)
LOEWENSCHUSS A + RON A + SCHNEPP O
J CHEM PHYS 50: 2502-12 (1969)

VAPOR PRESSURE OF ZINC DIFLUORIDE, CADMIUM
DIFLUORIDE, MAGNESIUM DIFLUORIDE, CALCIUM
DIFLUORIDE, STRONTIUM DIFLUORIDE, BARIUM
DIFLUORIDE, AND ALUMINUM TRIFLUORIDE.
RUFF O + LE BOUCHER L
Z ANORG ALLGEM CHEM 219: 376-81 (1934) (IN
GERMAN)

CALCIUM MONOFLUORIDE $\quad$ CaF

0276   SUBLIMATION PRESSURE OF CALCIUM DIFLUORIDE AND
THE DISSOCIATION ENERGY OF CALCIUM MONOFLUORIDE.
BLUE GD + GREEN JW + BAUTISTA RG + MARGRAVE JL
J CHEM PHYS 67(4): 877-82 (1963)

0277   REEVALUATION OF THE DISSOCIATION ENERGY OF
CALCIUM FLUORIDE. (ALSO SILICON AND GERMANIUM
FLUORIDES)
HASTIE JW + MARGRAVE JL
J CHEM ENG DATA 13(3): 428-9 (1968)

0278   DISSOCIATION ENERGY OF MONOFLUORIDE COMPOUNDS OF
MAGNESIUM AND CALCIUM AND THE IONIC MODEL OF A
MOLECULE.
KRASNOV KS
RUSS J PHYS CHEM 39(7): 842-4 (1965)

CALCIUM DIFLUORIDE $\quad$ CaF_2

ELECTRON DIFFRACTION ANALYSIS OF THE MOLECULAR
STRUCTURES OF GROUP-2 ELEMENTS. (BERYLLIUM
DIFLUORIDE, MAGNESIUM DIFLUORIDE, CALCIUM
DIFLUORIDE, STRONTIUM DIFLUORIDE, BARIUM
DIFLUORIDE, ZINC DIFLUORIDE, CADMIUM DIFLUORIDE)
AKISHIN PA + SPIRIDONOV VP
SOV PHYS CRYSTALLOGR 2(4): 472-9 (1957)

SUBLIMATION PRESSURE OF CALCIUM DIFLUORIDE AND
THE DISSOCIATION ENERGY OF CALCIUM MONOFLUORIDE.
BLUE GD + GREEN JW + BAUTISTA RG + MARGRAVE JL
J CHEM PHYS 67(4): 877-82 (1963)

0279   SPECTRA OF ELECTRONIC EXCITATIONS IN CALCIUM
DIFLUORIDE, STRONTIUM DIFLUORIDE, AND BARIUM
DIFLUORIDE IN THE 8 TO 150 EV RANGE.
FRANDON J + LAHAYE B + PRADAL F
PHYS STAT SOL B 53: 565-75 (1972)

NONEMPERICAL LCAO MO SCF STUDIES OF THE GROUP-2A
DIHALIDES: BERYLLIUM DIFLUORIDE, MAGNESIUM
DIFLUORIDE, AND CALCIUM DIFLUORIDE.
GOLE JL + SIU AKQ + HAYES EF
J CHEM PHYS 58: 857-68 (1973)

0280   ANALYSIS OF THE BANDS OF B DOUBLET SIGMA- X
DOUBLET SIGMA. SYSTEM OF CALCIUM FLUORIDE
MOLECULE IN EMISSION SPECTRUM.
KHANNA LK + DUBEY VS
INDIAN J PURE APPL PHYS 11(6): 444-5 (1973)

0281   ENTROPIES AND ENTHALPIES OF SUBLIMATION OF
CALCIUM AND CERIUM FLUORIDES. CORRELATION OF
ENTROPY AND ENTHALPY IN ERRORS.
MCCREARY JR + THORN RJ
HIGH TEMP SCI 5(5): 365-82 (1973)

MEAN AMPLITUDES OF VARIATION, BASTIANSEN-MORINO
SKRINKAGE EFFECT, AND MOLECULAR POLARIZABILITIES
FOR DIHALIDES OF SOME GROUP-2A ELEMENTS.
NAGARAJAN G
INDIAN J PHYS 39(9): 405-20 (1965)

VAPOR PRESSURE OF ZINC DIFLUORIDE, CADMIUM
DIFLUORIDE, MAGNESIUM DIFLUORIDE, CALCIUM
DIFLUORIDE, STRONTIUM DIFLUORIDE, BARIUM
DIFLUORIDE, AND ALUMINUM TRIFLUORIDE.
RUFF O + LE BOUCHER L
Z ANORG ALLGEM CHEM 219: 376-81 (1934) (IN
GERMAN)

0282   VAPOR PRESSURE AND HEAT OF SUBLIMATION OF
CALCIUM FLUORIDE.
SCHULZ DA + SEARCY AW
J PHYS CHEM 67(1): 103-6 (1963)

INFRARED SPECTRA OF SOME ALKALINE EARTH HALIDES
BY THE MATRIX ISOLATION TECHNIQUE.
SNELSON A
J PHYS CHEM 70: 3208-16 (1966)

0283   X-RAY STUDY OF ANHARMONIC VIBRATIONS IN CALCIUM
DIFLUORIDE.
STROCK HB + BATTERMAN BW
PHYS REV B 5: 2337-43 (1972)

INTERACTION OF MATRIX ISOLATED NICKEL FLUORIDE
AND NICKEL CHLORIDE WITH CARBON MONOXIDE,
MOLECULAR NITROGEN, NITRIC OXIDE, AND MOLECULAR
OXYGEN AND OF CALCIUM FLUORIDE, CHROMIUM(II)
FLUORIDE, MANGANESE(II) FLUORIDE, COPPER(II)
FLUORIDE, AND ZINC(II) FLUORIDE WITH CARBON
MONOXIDE IN ARGON MATRICES.
VAN LEIRSBURG DA + DEKOCK CW
J PHYS CHEM 78(2): 134-42 (1974)

GEOMETRY OF THE ALKALINE EARTH DIHALIDES.
(BARIUM DIFLUORIDE, STRONTIUM DIFLUORIDE,
CALCIUM DIFLUORIDE)
WHARTON L + BERG RA + KLEMPERER W
J CHEM PHYS 39: 2023-31 (1963)

CARBON DIFLUORIDE CF_2

CARBON MONOFLUORIDE CF

0297   ABSORPTION SPECTRUM OF CARBON DIFLUORIDE TRAPPED
       IN AN ARGON MATRIX.
       BASS AM + MANN DE
       J CHEM PHYS 36: 3501-2 (1962)

0284   GAS PHASE ELECTRON RESONANCE SPECTRUM AND DIPOLE
       MOMENT OF CARBON MONOFLUORIDE.
       CARRINGTON A + HOWARD BJ
       MOL PHYS 18: 225-31 (1970)

0298   HEAT OF FORMATION OF CARBON DIFLUORIDE.
       BREWER L + MARGRAVE JL + PORTER RF + WIELAND K
       J PHYS CHEM 65: 1913 (1961)

0285   THE B-X SYSTEM OF CARBON MONOFLUORIDE.
       CARROLL PK + GRENNAN TP
       J PHYS B 3: 865-77 (1970)

       AB INITIO CALCULATIONS ON SOME SMALL RADICALS BY
       THE UNRESTRICTED HARTREE-FOCK METHOD. (BORON
       DIFLUORIDE, NITROGEN DIFLUORIDE, OXYGEN
       DIFLUORIDE, CARBON DIFLUORIDE)
       BROWN RD + WILLIAMS GR
       CHEM PHYS 3: 19-34 (1974)

0286   THERMODYNAMIC PROPERTIES OF CARBON MONOFLUORIDE
       AND CARBON DIFLUORIDE FROM MOLECULAR FLOW
       EFFUSION AND MASS SPECTROMETER INVESTIGATIONS.
       FARBER M + FRISCH MA + KO HC
       TRANS FARADAY SOC 65(12): 3202-9 (1969)

0299   A THEORETICAL STUDY OF THE BOND INTERACTION
       FORCE CONSTANTS IN DIFLUORIDE MOLECULES. (OXYGEN
       DIFLUORIDE, NITROGEN DIFLUORIDE, CARBON
       DIFLUORIDE)
       BRUNS R + RAFF L + DEVLIN JP
       THEOR CHIM ACTA 14: 232-41 (1969)

0287   THEORETICAL STUDY OF THE SPECTROSCOPIC STATES OF
       THE CARBON MONOFLUORIDE MOLECULE.
       HALL JA + RICHARDS WG
       MOL PHYS 23: 331-43 (1972)

0300   BONDING IN CARBON(1) AND CARBON(2) FLUORIDES.
       EHLERT TC
       J PHYS CHEM 73(4): 949-53 (1969)

0288   SHOCK TUBE DETERMINATION OF THE C2 (A TRIPLET PI
       TO X TRIPLET PI) AND CARBON MONOFLUORIDE A
       DOUBLET SIGMA(+) TO X DOUBLET PI BAND SYSTEM
       OSCILLATOR STRENGTHS.
       HARRINGTON JA + MODICA AP + LIBBY DR
       J CHEM PHYS 44: 3380-87 (1966)

       THERMODYNAMIC PROPERTIES OF CARBON MONOFLUORIDE
       AND CARBON DIFLUORIDE FROM MOLECULAR FLOW
       EFFUSION AND MASS SPECTROMETER INVESTIGATIONS.
       FARBER M + FRISCH MA + KO HC
       TRANS FARADAY SOC 65(12). 3202-9 (1969)

0289   DETERMINATION OF THE DISSOCIATION ENERGIES OF
       CARBON MONOHALIDES.
       KUZYAKOV YY
       VEST MOSK UNIV KHIM 23: 21-4 (1968) (IN RUSSIAN)

0301   WAVE FUNCTIONS FOR THE FOUR ELECTRON THREE
       CENTER BONDING OF FOUR AND EIGHT PI ELECTRON
       SYSTEMS.
       HARCOURT RD + SILLITOE JF
       AUST J CHEM 27(4): 691-711 (1974)

0290   MOLECULAR ORBITAL INVESTIGATION OF CARBON
       MONOFLUORIDE AND SILICON MONOFLUORIDE AND THEIR
       POSITIVE AND NEGATIVE IONS.
       O'HARE PAG + WAHL AC
       J CHEM PHYS 55: 666-76 (1971)

0302   ABSORPTION SPECTRUM OF CARBON DIFLUORIDE.
       LAIRD RK + ANDREWS EB + BARROW RF
       TRANS FARADAY SOC 46: 803-5 (1950)

0291   EMISSION SPECTRUM OF CARBON MONOFLUORIDE.
       PORTER TL + MANN DE + ACQUISTA N
       J MOL SPECTROSC 16: 228-63 (1965)

0303   ABSORPTION SPECTRUM OF CARBON DIFLUORIDE AND ITS
       VIBRATIONAL ANALYSIS.
       MANN DE + THRUSH BA
       J CHEM PHYS 33: 1732-34 (1960)

0292   NEW BANDS OF THE CARBON MONOFLUORIDE MOLECULE.
       TATAEVSKII VM + KUZYAKOV YY
       OPT SPEKTROSK 5: 98 (1958) (IN RUSSIAN)

0304   HEAT OF FORMATION OF THE CARBON DIFLUORIDE
       RADICAL.
       MARGRAVE JL
       NATURE 197: 376 (1963)

0293   PREDISSOCIATION IN THE ABSORPTION SPECTRA OF
       CARBON MONOFLUORIDE AND CARBON DIFLUORIDE.
       THRUSH BA + ZWOLENIK JJ
       TRANS FARADAY SOC 59: 582-7 (1963)

0305   IONIZATION POTENTIALS FOR TETRAFLUORO ETHYLENE,
       TRIFLUORO CHLORO ETHYLENE, DIFLUORO DICHLORO
       ETHYLENE, AND THE APPEARANCE POTENTIAL OF CARBON
       DIFLUORIDE(X) FROM TETRAFLUORO ETHYLENE. (CARBON
       DIFLUORIDE BOND ENERGY)
       MARGRAVE JL
       J CHEM PHYS 31: 1432 (1959)

0294   ROTATIONAL CONSTANTS OF THE B STATE OF THE
       CARBON MONOFLUORIDE MOLECULE .
       UZIKOV AN + KUZYAKOV YY
       VEST MOSK UNIV KHIM 22: 22-5 (1967) (IN RUSSIAN)

0295   OSCILLATOR STRENGTHS OF CARBON MONOFLUORIDE AND
       COMMENTS ON HEATS OF FORMATION OF CARBON
       MONOFLUORIDE AND CARBON DIFLUORIDE.
       WENTINK T + ISAACSON L
       J CHEM PHYS 46: 603-5 (1967)

0306   2500 ANGSTROM ABSORPTION SPECTRUM OF CARBON
       DIFLUORIDE.
       MATHEWS WC
       J CHEM PHYS 45: 1068-9 (1966)

       MEAN AMPLITUDES OF VIBRATION FOR THE BORON
       DIFLUORIDE, CARBON DIFLUORIDE, NITROGEN
       DIFLUORIDE, ALUMINUM DIFLUORIDE, PHOSPHORUS
       DIFLUORIDE, AND ZIRCONIUM DIFLUORIDE.
       NAGARAJAN G
       INDIAN J PURE APPL PHYS 2: 341-3 (1964)

0296   THERMODYNAMIC, ELECTROCHEMICAL AND SYNTHETIC
       STUDIES OF THE GRAPHITE FLUORINE COMPOUNDS OF
       CARBON MONOFLUORIDE AND TETRACARBON FLUORIDE.
       WOOD JL+ VALERGA AJ + BADACHHAPE RB
       + MARGRAVE JL
       RICE UNIV, 1972, 30P. AD755934

0307   MICROWAVE SPECTRUM OF CARBON DIFLUORIDE.
       POWELL FX + LIDE DR
       J CHEM PHYS 45: 1067-68 (1966)

0308   ELECTRONIC SPECTRA OF POLYATOMIC FREE RADICALS.
       (CARBON DIFLUORIDE)
       RAMSAY DA
       ANN NY ACAD SCI 67: 485-98 (1957)

15

MOLECULAR PROPERTIES OF THE TRIATOMIC
DIFLUORIDES BERYLLIUM DIFLUORIDE: BORON
DIFLUORIDE, CARBON DIFLUORIDE, NITROGEN
DIFLUORIDE, AND OXYGEN DIFLUORIDE. (SCF)
ROTHENBERG S + SCHAEFER HF
J AMER CHEM SOC 95: 2095-100 (1973)

PREDISSOCIATION IN THE ABSORPTION SPECTRA OF
CARBON MONOFLUORIDE AND CARBON DIFLUORIDE.
THRUSH BA + ZWOLENIK JJ
TRANS FARADAY SOC 59: 582-7 (1963)

0309 EMISSION BANDS OF CARBON DIFLUORIDE.
VENKATESWARLU P
PHYS REV 77(5): 676-80 (1950)

OSCILLATOR STRENGTHS OF CARBON MONOFLUORIDE AND
COMMENTS ON HEATS OF FORMATION OF CARBON
MONOFLUORIDE AND CARBON DIFLUORIDE.
WENTINK T + ISAACSON L
J CHEM PHYS 46: 603-5 (1967)

CARBON TRIFLUORIDE CF_3

0310 INFRARED DETECTION OF GASEOUS CARBON TRIFLUORIDE
RADICAL.
CARLSON GA + PIMENTEL GC
J CHEM PHYS 44(10): 4053-4 (1966)

0311 INFRARED SPECTRUM OF THE FREE RADICAL CARBON
TRIFLUORIDE ISOLATED IN INERT MATRICES.
PHOTOCHLORINATION OF METHANE AND FLUOROFORM.
DISSOCIATION ENERGY AND ENTROPY OF CARBON
TRIFLUORIDE.
COOMBER JW + WHITTLE E
TRANS FARADAY SOC 62(8): 2183-90 (1966)

BONDING IN CARBON(1) AND CARBON(2) FLUORIDES.
EHLERT TC
J PHYS CHEM 73(4): 949-53 (1969)

0312 ELECTRON IMPACT STUDIES OF SOME TRIHALO
METHANES. (CARBON TRIFLUORIDE)
HOBROCK DL + KISER RW
J PHYS CHEM 68: 575-9 (1964)

CARBON TETRAFLUORIDE CF_4

0313 INFRARED SPECTRA OF A CRYOSYSTEM. CARBON
TETRAFLUORIDE.
AKHMEDZHANOV R + BERTSEV VV + BULANIN MO
+ ZHIGULA LA
OPT SPECTROSC 36(6): 709-10 (1974)

0314 VAPOR PHASE RAMAN SPECTRA OF MH4 (M= CARBON,
SILICON, GERMANIUM, AND TIN) AND MF4 (M= CARBON,
SILICON, AND GERMANIUM).
ARMSTRONG RS + CLARK RJH
J CHEM SOC, FARADAY TRANS 2 72(PT 1): 11-21
(1976)

0315 EXPLOSION METHOD AND THE HEATS OF FORMATION OF
CARBON TETRAFLUORIDE AND OTHER CARBON HALIDES.
BAIBUZ VF
DOKL AKAD NAUK SSSR 140: 1358-60 (1961) (IN
RUSSIAN)

INFRARED ABSORPTION SPECTRA OF SOME POLYATOMIC
FLUORIDES. (SILICON TETRAFLUORIDE, CARBON
TETRAFLUORIDE, BORON TRIFLUORIDE, NITROGEN
TRIFLUORIDE)
BAILEY CR + HALE JB + THOMPSON JW
J CHEM PHYS 5: 274-5 (1937)

0316 PHOTOELECTRON SPECTRA OF HALIDES. PART-1.
TETRAFLUORIDES AND TETRACHLORIDES OF GROUP-5B.
(CARBON TETRAFLUORIDE, SILICON TETRAFLUORIDE,
GERMANIUM TETRAFLUORIDE, TIN TETRAFLUORIDE)
BASSETT PJ + LLOYD DR
J CHEM SOC A 1971: 641-54

0317 HELIUM-1 RESONANCE PHOTOELECTRON SPECTRA OF
GROUP-4 TETRAFLUORIDES.
BASSETT PJ + LLOYD DR
CHEM PHYS LETT 3(1): 22-4 (1969)

0318 STRUCTURES OF TRIFLUORO METHYL HALIDES. (CARBON
TETRAFLUORIDE)
BOWEN HJM
TRANS FARADAY SOC 50: 444-51 (1954)

0319 HIGH RESOLUTION PHOTOELECTRON SPECTROSCOPY OF
CARBON TETRAFLUORIDE AND SILICON TETRAFLUORIDE.
BULL WE + PULLEN BP + GRIMM FA + MODDEMAN WE
+ SCHWEITZER GK + CARLSON TA
INORG CHEM 9: 2474-78 (1970)

0320 ACCURATE FORCE CONSTANTS FROM HEAVY ISOTOPIC
SUBSTITUTION. PART-1. CARBON TETRAFLUORIDE.
CHALMERS AA + MCKEAN DC
SPECTROCHIM ACTA 22: 251-67 (1966)

0321 THERMODYNAMIC PROPERTIES OF CARBON
TETRAFLUORIDE.
CHARI N
UNIV MICHIGAN, PHD THESIS, 1960, 177P.

0322 RAMAN SPECTRA OF CHLOROFLUORO METHANES IN THE
GASEOUS STATE. (CARBON TETRAFLUORIDE)
CLAASSEN HH
J CHEM PHYS 22(1): 50-2 (1954)

0323 NUCLEAR SPIN RELAXATION STUDY OF THE SPIN
ROTATION INTERACTION IN SPHERICAL TOP MOLECULES.
COURTNEY JA + ARMSTRONG RL
CAN J PHYS 50(12): 1252-61 (1972)

0324 FORCE FIELDS OF CARBON TETRAFLUORIDE AND SILICON
TETRAFLUORIDE.
DUNCAN JL + MILLS IM
SPECTROCHIM ACTA 20(6): 1089-92 (1964)

0325 INFRARED BAND CONTOURS. PART-1: SPHERICAL TOP
MOLECULES. (CARBON TETRAFLUORIDE)
EDGELL WF + MOYNIHAN RE
J CHEM PHYS 27: 155-9 (1957)

BONDING IN CARBON(1) AND CARBON(2) FLUORIDES.
EHLERT TC
J PHYS CHEM 73(4): 949-53 (1969)

0326 THERMODYNAMIC PROPERTIES OF CARBON TETRAFLUORIDE
FROM 4 DEGREES K TO ITS MELTING POINT.
ENOKIDO H + SHIMODA T + MASHIKO Y
BULL CHEM SOC JAP 42: 3415-21 (1969)

0327 VIBRATIONAL SPECTRA OF LIQUID AND CRYSTALLINE
CARBON TETRAFLUORIDE.
FOURNIER RP + SAVOIE R + CABANA A + BESSETTE F
J CHEM PHYS 49: 1159-64 (1968)

0328 MOLECULAR SPECTROSCOPY BY MEANS OF ESCA. (CARBON
TETRAFLUORIDE)
GELIUS U + HEDEN PF + HEDMAN J + LINDBERG BJ
+ MANNE R + NORDBERG R + NORDLING C + SIEGBAHN K
PHYS SCRIPTA 2: 70-80 (1970)

FORCE CONSTANTS OF CARBON TETRAFLUORIDE AND
BORON TRIFLUORIDE AND COMPARISON WITH FORCE
CONSTANTS OF THE EIGHTH PERIOD.
GOUBEAU J + BUES W + KAMPMANN FW
Z ANORG ALLGEM CHEM 283: 123-37 (1956) (IN
GERMAN)

0329 ENTHALPY OF FORMATION OF CARBON TETRAFLUORIDE.
GREENBERG E + HUBBARD WN
J PHYS CHEM 72: 222-7 (1968)

0330 ELECTRONIC STRUCTURE OF THE CARBON TETRAFLUORIDE
MOLECULE. PART-1. THE GROUND STATE.
GUILLEMIN YGS + MASIA AP
ANALES FISICA QUIMICA 60: 177-88 (1964)

0331 MOLECULAR CONSTANTS FROM INFRARED BAND SHAPES. THE ATOMIC DISPLACMENTS AND INTRAMOLECULAR FORCES IN CARBON TETRAFLUORIDE.
HANNAH RW
PURDUE UNIV, PHD THESIS, 1957, 108P.

0332 CARBON TETRAFLUORIDE. THERMODYNAMIC PROPERTIES OF THE REAL GAS.
HARRISON RH + DOUSLIN DR
J CHEM ENG DATA 11: 383-88 (1966)

0333 ELECTRON IMPACT SPECTRA OF METHANE AND CARBON TETRAFLUORIDE.
HARSHBARGER WR + LASSETTRE EN
J CHEM PHYS 58(4): 1505-13 (1973)

RADIATIVE LIFETIMES OF ULTRAVIOLET EMISSION SYSTEMS EXCITED IN BORON TRIFLUORIDE, CARBON TETRAFLUORIDE, AND SILICON TETRAFLUORIDE.
HESSER JE + DRESSLER K
J CHEM PHYS 47: 3443-50 (1967)

0334 MOLECULAR STRUCTURE OF CARBON TETRAFLUORIDE.
HOFFMAN CWW + LIVINGSTON RL
J CHEM PHYS 21: 565 (1953)

0335 CONSTANT VOLUME HEAT CAPACITIES OF GASEOUS CARBON TETRAFLUORIDE AND OTHER MOLECULES.
HWANG YT
UNIV MICHIGAN, PHD THESIS, 1961, 159P.

0336 CONSTANT VOLUME HEAT CAPACITY OF GASEOUS CARBON TETRAFLUORIDE.
HWANG YT + MARTIN JJ
AICHE J 10: 89-91 (1964)

0337 ELECTRON DENSITY DISTRIBUTIONS IN SOME FLUORIDES AND OXYCHLORIDES. (CARBON TETRAFLUORIDE AND SILICON TETRAFLUORIDE)
IONOV SP + IONOVA GV
RUSS J INORG CHEM 14(6): 886-7 (1969)

0338 INFRARED SPECTRA OF SIMPLE MOLECULES IN LIQUID ARGON SOLUTION. CARBON TETRAFLUORIDE.
JEANNOTTE AC + LEGLER D + OVEREND J
SPECTROCHIM ACTA 29A: 1915-21 (1973)

0339 IMPLICATIONS OF PHOTOELECTRON SPECTROSCOPIC MEASUREMENTS FOR COMPOUNDS WHICH PRODUCE NO PARENT ION. (CARBON TETRAFLUORIDE)
KAUFMAN JJ + KERMAN E + KOSKI WS
INT J QUANT CHEM 4: 391-4 (1971)

0340 ABSOLUTE INFRARED INTENSITIES OF CARBON TETRAFLUORIDE.
LEVIN IW + LEWIS TP
J CHEM PHYS 52: 1608-09 (1970)

0341 PHOTOELECTRON SPECTRA OF HALIDES. PART-7: VARIABLE TEMPERATURE HELIUM-1 AND HELIUM-2 STUDIES OF CARBON TETRAFLUORIDE, SILICON TETRAFLUORIDE, AND GERMANIUM TETRAFLUORIDE.
LLOYD DR + ROBERTS PJ
J ELECTRON SPECTROSC RELAT PHENOM 7(4): 325-30 (1975)

0342 ROOT MEAN SQUARE AMPLITUDES OF VIBRATION IN SOME GROUP-4 TETRAHALIDES. (CARBON TETRAFLUORIDE)
LONG DA + CHAU JYH
TRANS FARADAY SOC 58: 2328-35 (1962)

ROOT MEAN SQUARE AMPLITUDES OF VIBRATION IN SOME GROUP-4 TETRAHALIDES. (CARBON TETRAFLUORIDE AND SILICON MONOFLUORIDE)
LONG DA + SEIBOLD EA
TRANS FARADAY SOC 56: 1105-9 (1960)

0343 INFRARED SPECTRUM OF CARBON TETRAFLUORIDE.
MAKI A + PLYLER EK + THIBAULT R
J CHEM PHYS 37: 1899-1900 (1962)

INTENSITIES OF THE INFRARED FUNDAMENTALS OF CARBON TETRAFLUORIDE, SILICON TETRAFLUORIDE, CARBON TETRACHLORIDE, SILICON TETRACHLORIDE, AND GERMANIUM TETRACHLORIDE.
MEDINA A + MORCILLO J
ANALES FISICA 64(7-8): 251-61 (1968) (IN SPANISH)

0344 CALCULATED IONIZATION POTENTIAL OF CHLORO- AND FLUOROMETHANES, TETRAFLUORO METHANE XENON TETRAFLUORIDE, AND XENON TETRACHLORIDE.
MELTON CE + JOY HW
J CHEM PHYS 42: 2982 (1965)

0345 VAPOR PRESSURE OF CARBON TETRAFLUORIDE AND NITROGEN TRIFLUORIDE AND THE TRIPLE POINT OF CARBON TETRAFLUORIDE.
MENZEL W + MOHRY F
Z ANORG ALLGEM CHEM 210: 257-63 (1933) (IN GERMAN)

0346 FLUORINE RELAXATION BY NMR ABSORPTION IN GASEOUS CARBON TETRAFLUORIDE, SILICON TETRAFLUORIDE, AND SULFUR HEXAFLUORIDE.
MOHANTY BS + BERNSTEIN HJ
J CHEM PHYS 53: 461-62 (1970)

0347 RAMAN SPECTRUM OF GASEOUS CARBON TETRAFLUORIDE.
MONOSTORI B + WEVER A
J CHEM PHYS 33: 1867-68 (1960)

0348 COMPARISON OF CALCULATIONS OF FORCE CONSTANTS FOR BORON TETRAFLUORIDE(-), CARBON TETRAFLUORIDE, SILICON TETRAFLUORIDE, AND SILICON TETRACHLORIDE.
MUELLER A + FADINI A
Z CHEM 7(3): 115-16 (1967) (IN GERMAN)

0349 MOLECULAR BEAM ELECTRIC DEFLECTION OF THE TETRAHALIDES CARBON TETRAFLUORIDE, CARBON TETRACHLORIDE, SILICON TETRAFLUORIDE, SILICON TETRACHLORIDE, GERMANIUM TETRACHLORIDE, TITANIUM TETRAFLUORIDE, TITANIUM TETRACHLORIDE, VANADIUM TETRAFLUORIDE, AND VANADIUM TETRACHLORIDE.
MUENTER AA + DYKE TR + FALCONER WE + KLEMPERER W
J CHEM PHYS 63: 1231-6 (1975)

0350 POTENTIAL CONSTANTS AND THERMODYNAMIC FUNCTIONS OF CARBON TETRAFLUORIDE.
NAGARAJAN G
AUST J CHEM 15: 566-8 (1962)

0351 A TRANSFERABLE UREY-BRADLEY FORCE FIELD AND THE ASSIGNMENTS OF SOME MIXED HALOMETHANES. (CARBON TETRAFLUORIDE)
NGAI LH + MANN RH
J MOL SPECTROSC 38: 322-45 (1971)

0352 A GEOMETRIC VISUALIZATION OF NORMAL COORDINATE TRANSFORMATIONS. APPLICATION TO THE CALCULATION BOND MOMENT PARAMETERS AND FORCE CONSTANTS. (CARBON TETRAFLUORIDE)
PERSON WB + CRAWFORD B
J CHEM PHYS 26: 1295-1301 (1957)

0353 HARMONIC POTENTIAL CONSTANTS OF OSMIUM TETROXIDE AND SOME TETRAHALIDE (CARBON TETRAFLUORIDE)
PISTORIUS CWFT + HAARHOFF PC
Z PHYS CHEM NEUE FOLGE 19: 202-5 (1959)

0354 FORCE CONSTANTS OF THE CARBON TETRAHALIDES. (CARBON TETRAFLUORIDE)
PISTORIUS CWFT + PISTORIUS MC
Z PHYS CHEM NEUE FOLGE 35: 196-98 (1962)

0355 INFRARED SPECTRA OF HALOGEN-SUBSTITUTED METHANES. (CARBON TETRAFLUORIDE)
PLYLER EK + BENEDICT WS
J RES NBS 47: 202-20 (1951)

CONSTRUCTION OF A HIGH RESOLUTION PHOTOELECTRON SPECTROMETER. (CARBON TETRAFLUORIDE, BORON TRIFLUORIDE, SULFUR HEXAFLUORIDE, NITROGEN TRIFLUORIDE, AND SILICON TETRAFLUORIDE)
PULLEN BP
UNIV TENNESSEE, PHD THESIS, 1970, 117P.

0356 FAR INFRARED ABSORPTION IN GASEOUS CARBON TETRAFLUORIDE.
ROSENBERG A + BIRNBAUM G
J CHEM PHYS 48: 1396-97 (1968)

0357 ELECTRONIC STRUCTURE OF THE CARBON TETRAFLUORIDE MOLECULE.
ROZENBERG EL + DYATKINA ME
J STRUCT CHEM 12(2): 270-5 (1971)

0358 THE FORCE CONSTANTS OF VARIOUS ISOTOPES OF
CARBON TETRAFLUORIDE, SILICON TETRAFLUORIDE,
GERMANIUM TETRAFLUORIDE AND SULFUR TRIOXIDE.
RUOFF VA
SPECTROCHIM ACTA 23A: 2421-31 (1967)

0359 EVALUATION OF THE FORCE CONSTANTS OF SOME
TETRAHEDRAL MOLECULES. (CARBON TETRAFLUORIDE)
SAHINI VE + FULEA AO
REV ROUM CHEM 11: 1045-50 (1966)

0360 UREY-BRADLEY FORCE FIELD FOR CARBON
TETRAFLUORIDE.
SHIMANOUCHI T
NIPPON KAGAKU ZAISSHI 86(8): 768-69 (1965) (IN
JAPANESE)

FORCE CONSTANTS OF CARBON TETRAFLUORIDE, SILICON
TETRAFLUORIDE, BORON TRIFLUORIDE, ETHANE,
SILANE, AMMONIA, AND PHOSPHINE.
SHIMANOUCHI T + NAKAGAWA I + HIRAISHI J + ISHII
J MOL SPECTROSC 19: 78-107 (1966)

0361 THERMODYNAMIC PROPERTIES OF CARBON TETRAFLUORIDE
FROM 12 DEGREES K TO ITS BOILING POINT. THE
SIGNIFICANCE OF THE PARAMETER NU.
SMITH JH + PACE EL
J PHYS CHEM 73: 4232-36 (1969)

0362 THEORY OF INTENSITIES IN THE INFRARED SPECTRA OF
THE MOLECULES CARBON TETRAFLUORIDE AND DICARBON
HEXAFLUORIDE.
SVERDLOV LM
OPT SPECTROSC 7(5): 368-71 (NOV 1959)

0363 INFRARED SPECTRA OF CARBON TETRAFLUORIDE AND
GERMANIUM TETRAFLUORIDE.
WOLTZ PJH + NIELSEN AH
J CHEM PHYS 20(2): 307-12 (1952)

0364 RAMAN SPECTRA OF CARBON AND SILICON
TETRAFLUORIDES.
YOST DM + LASSETTRE EN + GROSS ST
J CHEM PHYS 4: 325 (1936)

0365 VACUUM ULTRAVIOLET ABSORPTION SPECTRA OF SOME
HALOGEN DERIVATIVES OF METHANE. (CARBON
TETRAFLUORIDE) CORRELATION OF THE SPECTRA.
ZOBEL CR + DUNCAN ABF
J AMER CHEM SOC 77: 2611-15 (1955)

CERIUM DIFLUORIDE CeF_2

ENTROPIES AND ENTHALPIES OF SUBLIMATION OF
CALCIUM AND CERIUM FLUORIDES. CORRELATION OF
ENTROPY AND ENTHALPY IN ERRORS.
MCCREARY JR + THORN RJ
HIGH TEMP SCI 5(5): 365-82 (1973)

CERIUM TRIFLUORIDE CeF_3

GEOMETRIES AND ENTROPIES OF METAL TRIFLUORIDES
FROM INFRARED SPECTRA. (SCANDIUM, YTTRIUM,
LANTHANUM, CERIUM, NEODYMIUM, EUROPIUM, AND
GADOLINIUM TRIFLUORIDES)
HASTIE JW + HAUGE RH + MARGRAVE JL
J LESS COMMON METALS 39(2): 309-34 (1975)

0366 VAPOR PRESSURE AND HEAT OF SUBLIMATION OF CERIUM
TRIFLUORIDE.
LIM M + SEARCY AW
J PHYS CHEM 70: 1762-5 (1966)

CERIUM TETRAFLUORIDE CeF_4

0367 ELECTRIC DEFLECTION AND THERMAL DECOMPOSITION
STUDIES ON CERIUM TETRAFLUORIDE, TERBIUM
TETRAFLUORIDE, AND PRASEODYMIUM TETRAFLUORIDE.
KAISER EW + SUNDER WA + FALCONER WE
J LESS COMMON METALS 27(3): 383-7 (1972)

CHLORINE MONOFLUORIDE ClF

0368 NMR SPECTRA OF CHLORINE TRIFLUORIDE AND CHLORINE
MONOFLUORIDE. GASEOUS SPECTRA AND GAS TO LIQUID
SHIFTS.
ALEXAKOS LG + CORNWELL CD
J CHEM PHYS 41: 2098-2107 (1964)

0369 PHOTOELECTRON SPECTRUM OF CHLORINE MONOFLUORIDE.
ANDERSON CP + MAMANTOV G + BULL WE + GRIMM FA
CHEM PHYS LETT 12: 137-9 (1971)

0370 HEATS OF FORMATION OF CHLORINE FLUORIDES.
(CHLORINE MONOFLUORIDE, CHLORINE DIFLUORIDE,
AND CHLORINE TRIFLUORIDE)
BARBERI P + CATON J + GUILLIN J + HARTMANSHENN O
COMM ENERG AT, 1964, 38P. CEA-R-3761 (IN FRENCH)

0371 SCF- MO CALCULATIONS OF SOME MOLECULAR
PROPERTIES OF THE ISOELECTRONIC SERIES CHLORINE
MONOFLUORIDE, HYDROGEN OXYCHLORIDE, CHLORO
AMMONIA, AND CHLORO METHANE.
BENDAZZOLI GL + LISTE DG + PALMIERI P
J CHEM SOC, FARADAY TRANS 69(PT 6): 791-7 (1973)

0372 ELECTRONIC STRUCTURES OF CHLORINE MONOFLUORIDE
AND CHLORINE TRIFLUORIDE.
BREEZE A + CRUICKSHANK DWJ + ARMSTRONG DR
J CHEM SOC, FARADAY TRANS 68(12): 2144-9 (1972)

0373 ELECTRONEGATIVITY, NONBONDED INTERACTIONS AND
POLARIZABILITY IN HYDROGEN HALIDES AND THE
INTERHALOGEN COMPOUNDS. (CHLORINE MONOFLUORIDE,
BROMINE MONOFLUORIDE, AND IODINE MONOFLUORIDE)
BROWN RF
J AMER CHEM SOC 83: 36-42 (1961)

0374 CHARGE DISTRIBUTION IN CHLORINE MONOFLUORIDE
FROM CORE ELECTRON BINDING ENERGIES.
CARROLL TX + THOMAS TD
J CHEM PHYS 60: 2186-7 (1974)

THERMODYNAMIC PROPERTIES OF THE DIATOMIC
INTERHALOGENS FROM SPECTROSCOPIC DATA.
(CHLORINE, BROMINE, AND IODINE MONOFLUORIDES)
COLE LG + ELVERUM GW
J CHEM PHYS 20(10): 1543-51 (1952)

0375 HYPERFINE STRUCTURE CONSTANTS OF CHLORINE
MONOFLUORIDE.
DAVIS RE + MUENTER JS
J CHEM PHYS 57: 2836-8 (1972)

0376 PHOTOELECTRON SPECTRA OF HALIDES. (CHLORINE
MONOFLUORIDE, CHLORINE TRIFLUORIDE, BROMINE
MONOFLUORIDE, AND BROMINE TRIFLUORIDE)
DEKOCK RL + HIGGINSON BR + LLOYD DR + BREEZE A
+ CRUICKSHANK DWJ + ARMSTRONG DR
MOL PHYS 24: 1059-72 (1972)

0377 PHOTOIONIZATION STUDIES AND THE THERMODYNAMIC
PROPERTIES OF SOME HALOGEN MOLECULES. (CHLORINE
FLUORIDE)
DIBELER VH
P781-90 OF INT CONF MASS SPECTROSC, RECENT
DEVELOP MASS SPECTROSC, 1969, KYOTO, 1324P.

0378 PHOTOIONIZATION STUDY OF CHLORINE MONOFLUORIDE
AND THE DISSOCIATION ENERGY OF FLUORINE.
DIBELER VH + WALTER JA + MCCULLOH KE
J CHEM PHYS 53: 4414-7 (1970)

MOLECULAR ZEEMAN EFFECT, MAGNETIC PROPERTIES,
AND ELECTRIC QUADRUPOLE MOMENTS IN CHLORINE
MONOFLUORIDE, BROMINE MONOFLUORIDE, BROMINE
CYANIDE, AND IODINE CYANIDE.
EWING JJ + TIGELAAR HL + FLYGARE WH
J CHEM PHYS 56: 1957-66 (1972)

0379 AB INITIO MOLECULAR ORBITAL STUDY OF THE
GEOMETRY OF INTERHALOGENS. (CHLORINE
MONOFLUORIDE, CHLORINE TRIFLUORIDE, AND CHLORINE
PENTAFLUORIDE)
GUEST MF + HALL MB + HILLIER IH
J CHEM SOC, FARADAY TRANS 69(12): 1829-34 (1973)

0380 SCF CALCULATIONS ON CHLORATE ION, HYDROGEN
CHLORIDE AND CHLORINE MONOFLUORIDE.
JOHANSEN H
CHEM PHYS LETT 11: 466-7 (1971)

0381 INFRARED AND RAMAN SPECTRA OF CHLORINE
MONOFLUORIDE.
JONES EA + PARKINSON TF + BURKE TG
J CHEM PHYS 18: 235-6 (1950)

0382 GEOMETRY OF MOLECULES. PART-2: GEOMETRICAL AND
ELECTRICAL STRUCTURE OF THE MOLECULE CHLORINE
DIFLUORIDE ON THE BASIS OF A CALCULATION BY THE
MO LCAO SCF IN THE NDDO APPROXIMATION.
KLYAGINA AP + KLIMENKO NM + DYATKINA ME
J STRUCT CHEM 14(5): 838-44 (1973)

0383 PURE QUADRUPOLE SPECTRA OF SOLID CHLORINE
COMPOUNDS.
LIVINGSTON RL
J PHYS CHEM 57: 496-501 (1953)

0384 MOLECULAR MAGNETIC PROPERTIES OF CHLORINE
MONOFLUORIDE.
MCGURK J + NORRIS CL + TIGELAAR HL + FLYGARE WH
J CHEM PHYS 58: 3118-20 (1973)

0385 SOME CORRECTIONS TO THE SECOND ORDER STARK
EFFECT OF LINEAR MOLECULES. (CHLORINE
MONOFLUORIDE AND BROMINE MONOFLUORIDE)
MIZUSHIMA M
P1167-82 OF INT MEETING MOL SPECTROSC, 1959,
4TH, BOLOGNA, ITALY, VOL 3, MANGINI A (ED),
MACMILLAN, 1962.

0386 ANALYSIS OF THE INFRARED SPECTRUM OF CHLORINE
MONOFLUORIDE.
NIELSEN AH + JONES EA
J CHEM PHYS 19(9): 1117-21 (1951)

0387 DISSOCIATION ENERGY OF CHLORINE MONOFLUORIDE.
NORDINE PC
J CHEM PHYS 61(1): 224-6 (1974)

0388 ABSORPTION SPECTRA OF CHLORINE FLUORIDE,
CHLORINE TRIFLUORIDE, AND CHLORINE PENTAFLUORIDE
MOLECULES.
RYMARCHUK YUA + IVANOV VS
ZH PRIKL SPEKTROSK 22(5): 950 (1975) (IN
RUSSIAN)

0389 HIGH RESOLUTION RAMAN SPECTROSCOPY IN THE RED
AND NEAR INFRARED. (CHLORINE MONOFLUORIDE)
STAMMREICH H + FORNERIS R + TAVARES Y
SPECTROCHIM ACTA 17: 1173-84 (1961)

0390 O(+) TO SINGLET SIGMA(+) SYSTEMS OF BOTH
ISOTOPES OF CHLORINE MONOFLUORIDE.
STRIKER W + KRAUSS L
Z NATURFORSCH 23A: 1116-21 (1968)

0391 METHOD OF DETERMINING SATURATED LIQUID AND
SATURATED VAPOR ENTROPY. (CHLORINE MONOFLUORIDE,
OXYGEN MONOFLUORIDE, PHOSPHORUS TRIFLUORIDE)
WALKER MA
AIAA J 1: 2636-8 (1963)

CHLORINE DIFLUORIDE CIF_2

HEATS OF FORMATION OF CHLORINE FLUORIDES.
(CHLORINE MONOFLUORIDE, CHLORINE DIFLUORIDE, AND
CHLORINE TRIFLUORIDE)
BARBERI P + CATON J + GUILLIN J + HARTMANSHENN O
COMM ENERG AT, 1964, 38P. CEA-R-3761 (IN FRENCH)

0392 CHLORINE FLUORINE SYSTEM AT LOW TEMPERATURE.
CHARACTERIZATION OF THE CHLORINE DIFLUORIDE FREE
RADICAL.
MAMANTOV G + VASINI EJ + MOULTON MC
+ VICKROY DG + MAEKAWA T
J CHEM PHYS 54: 3419-21 (1971)

0393 THE CHLORINE DIFLUORIDE FREE RADICAL.
MAMANTOV G + VICKROY DG + VASINI EJ + MAEKAWA T
+ MOULTON MC
INORG NUCL CHEM LETT 6: 701-2 (1970)

CHLORINE TRIFLUORIDE CIF_3

NMR SPECTRA OF CHLORINE TRIFLUORIDE AND CHLORINE
MONOFLUORIDE. GASEOUS SPECTRA AND GAS TO LIQUID
SHIFTS.
ALEXAKOS LG + CORNWELL CD
J CHEM PHYS 41: 2098-2107 (1964)

0394 THERMAL DISSOCIATION OF CHLORINE TRIFLUORIDE
BEHIND INCIDENT SHOCK WAVES.
BLAUER JA + MCMATH HG + JAYE FC
J PHYS CHEM 73: 2683-8 (1969)

PHOTOELECTRON SPECTRA OF HALIDES. (CHLORINE
MONOFLUORIDE, CHLORINE TRIFLUORIDE, BROMINE
MONOFLUORIDE, AND BROMINE TRIFLUORIDE)
DEKOCK RL + HIGGINSON BR + LLOYD DR + BREEZE A +
ARMSTRONG DR
MOL PHYS 24: 1059-72 (1972)

HEATS OF FORMATION OF CHLORINE FLUORIDES.
(CHLORINE MONOFLUORIDE, CHLORINE DIFLUORIDE,
AND CHLORINE TRIFLUORIDE)
BARBERI P + CATON J + GUILLIN J + HARTMANSHENN O
COMM ENERG AT, 1964, 38P. CEA-R-3761 (IN FRENCH)

ELECTRONIC STRUCTURES OF CHLORINE MONOFLUORIDE
AND CHLORINE TRIFLUORIDE.
BREEZE A + CRUICKSHANK DWJ + ARMSTRONG DR
J CHEM SOC, FARADAY TRANS 68(12): 2144-9 (1972)

VESCF-MO STUDIES OF PHOSPHORUS PENTAFLUORIDE,
SULFUR TETRAFLUORIDE, AND CHLORINE TRIFLUORIDE.
BROWN RD + PEEL JB
AUST J CHEM 21: 2617-29 (1968)

0395 STRUCTURES OF THE INTERHALOGEN COMPOUNDS.
PART-1: CHLORINE TRIFLUORIDE AT -120 DEGREES C.
BURBANK RD + BENSEY FN
J CHEM PHYS 21(4): 602-8 (1953)

0396 VIBRATIONAL SPECTRA AND THERMODYNAMIC PROPERTIES
OF CHLORINE TRIFLUORIDE AND BROMINE TRIFLUORIDE.
CLAASSEN HH + WEINSTOCK B + MALM JG
J CHEM PHYS 28: 285-9 (1958)

0397 CNDO/2 AND INDO ALL VALENCE ELECTRON
CALCULATIONS ON THE GEOMETRY AND PROPERTIES OF
SOME INTERHALOGENS. (CHLORINE TRIFLUORIDE,
BROMINE TRIFLUORIDE, IODINE TRIFLUORIDE,
CHLORINE PENTAFLUORIDE, BROMINE PENTAFLUORIDE,
IODINE PENTAFLUORIDE, CHLORINE HEPTAFLUORIDE,
BROMINE HEPTAFLUORIDE, AND IODINE HEPTAFLUORIDE)
DEB BM + COULSON CA
J CHEM SOC A 1971: 958-70

0398 RAMAN SPECTRA AND THE STRUCTURE OF CHLORINE
TRIFLUORIDE AND BROMINE TRIFLUORIDE.
DRIFFORD M + MARTIN D + BOUGON R
REV CHIM MINERALE 7: 1069-86 (1970)

0399 INFRARED SPECTRA OF MATRIX ISOLATED CHLORINE
TRIFLUORIDE, BROMINE TRIFLUORIDE, AND BROMINE
PENTAFLUORIDE. FLUORINE EXCHANGE MECHANISM OF
LIQUID CHLORINE TRIFLUORIDE, BROMINE
TRIFLUORIDE, AND SULFUR TETRAFLUORIDE.
FREY RA + REDINGTON RL + ALJIBURY ALK
J CHEM PHYS 54: 344-55 (1971)

AB INITIO MOLECULAR ORBITAL STUDY OF THE
GEOMETRY OF INTERHALOGENS. (CHLORINE
MONOFLUORIDE, CHLORINE TRIFLUORIDE, AND CHLORINE
PENTAFLUORIDE)
GUEST MF + HALL MB + HILLIER IH
J CHEM SOC, FARADAY TRANS 69(12): 1829-34 (1973)

0400 QUADRUPOLE COUPLING CONSTANTS IN CHLORINE
TRIFLUORIDE AND BROMINE TRIFLUORIDE.
GUPTA LC
INDIAN J PURE APPL PHYS 5: 437-8 (1967)

0401 HEAT OF FORMATION OF CHLORINE TRIFLUORIDE AT
298.14 DEGREES K.
KING RC + ARMSTRONG GT
J RES NBS A 74: 769-79 (1970)

0402 MOLECULAR FORCE FIELD FOR CHLORINE TRIFLUORIDE.
KRISHNA MG + PILLAI MGK
AUST J CHEM 18(3): 261-70 (1965)

PURE QUADRUPOLE SPECTRA OF SOLID CHLORINE
COMPOUNDS.
LIVINGSTON RL
J PHYS CHEM 57: 496-501 (1953)

0403 FORCE CONSTANT CALCULATION FOR CHLORINE
TRIFLUORIDE.
LONG DA + JONES DTL
TRANS FARADAY SOC 59: 273-5 (1963)

ORBITAL MODIFICATION BY THE COULOMB FIELD OF
LIGAND ATOMS OF LATER SECOND ROW ELEMENTS IN
PERFECT PAIRING VALENCE STATES. (PHOSPHORUS
PENTAFLUORIDE, SULFUR HEXAFLUORIDE, AND CHLORINE
TRIFLUORIDE)
MACLAGAN RGAR
J CHEM SOC A 1971: 222-6

0404 ELECTRONIC STRUCTURE OF CHLORINE TRIFLUORIDE. AN
APPROXIMATE SCF CALCULATION.
MANNE R
THEOR CHIM ACTA 6: 312-19 (1966)

0405 CONSTRUCTION OF HYBRID ORBITALS. (CHLORINE
TRIFLUORIDE)
MURRELL JN
J CHEM PHYS 32: 767-8 (1960)

0406 MEAN AMPLITUDES OF VIBRATION OF SELENIUM
TRIOXIDE AND CHLORINE TRIFLUORIDE.
PURUSHOTHAMAN C
PROC INDIAN ACAD SCI A 60(6): 431-7 (1964)

0407 MOLECULAR FORCE FIELD FOR INTERHALOGEN
COMPOUNDS. (CHLORINE TRIFLUORIDE AND BROMINE
TRIFLUORIDE)
RAMASWAMY K + MUTHUSUBRAMANIAN P
J MOL STRUCT 9: 193-6 (1971)

INFRARED MATRIX ISOLATION STUDIES. (BORON
TRIFLUORIDE, SULFUR TETRAFLUORIDE, ANTIMONY
PENTAFLUORIDE, ARSENIC PENTAFLUORIDE, CHLORINE
TRIFLUORIDE, BROMINE TRIFLUORIDE, AND BROMINE
PENTAFLUORIDE)
REDINGTON RL
TEXAS TECH UNIV, 1969, 89P. AD701120

ABSORPTION SPECTRA OF CHLORINE FLUORIDE,
CHLORINE TRIFLUORIDE, AND CHLORINE PENTAFLUORIDE
MOLECULES.
RYMARCHUK YUA + IVANOV VS
ZH PRIKL SPEKTROSK 22(5): 950 (1975) (IN
RUSSIAN)

0408 INFRARED AND RAMAN SPECTRA OF CHLORINE
TRIFLUORIDE AND BROMINE TRIFLUORIDE.
SELIG H + CLAASSEN HH + HOLLOWAY JH
J CHEM PHYS 52: 3517-21 (1970)

0409 X-RAY PHOTOELECTRON SPECTROSCOPY OF CHLORINE
TRIFLUORIDE, SULFUR TETRAFLUORIDE, AND
PHOSPHORUS PENTAFLUORIDE.
SHAW RW + CARROLL TX + THOMAS TD
J AMER CHEM SOC 95(18): 5870-5 (1973)

OBSERVATION BY ESCA OF INEQUIVALENT FLUORINES IN
CHLORINE TRIFLUORIDE, SULFUR TETRAFLUORIDE, AND
PHOSPHORUS PENTAFLUORIDE. INTERMOLECULAR
FLUORINE EXCHANGE BETWEEN TETRABUTYLAMMONIUM
HEXAFLUOROPHOSPHATE AND BORON TRIFLUORIDE OR
PHOSPHORUS PENTAFLUORIDE.
SHAW RW + CARROLL TX + THOMAS TD + BROWNSTEIN S
CAN J CHEM 46: 225-8 (1968)

0410 MICROWAVE SPECTRUM AND STRUCTURE OF CHLORINE
TRIFLUORIDE.
SMITH DF
J CHEM PHYS 21(4): 609-14 (1953)

ELECTRICAL CONDUCTIVITY OF SOLID CHLORINE
TRIFLUORIDE AND BROMINE TRIFLUORIDE.
TOY MS + CANNON WA
P237-45 OF ADVANCED PROPELLANT CHEMISTRY, AMER
CHEM SOC, 1966, 290P. (ADVANCES IN CHEMISTRY
SERIES 54)

0411 GENERALIZED MEAN AMPLITUDES AND CORIOLIS
CONSTANTS IN DICHLORO BORANE, DIBROMO BORANE AND
CHLORINE TRIFLUORIDE.
VENKATESWARLU K + PURUSHOTHAMAN C
ACTA PHYS POLON 30: 801-6 (1966)

CHLORINE TETRAFLUORIDE ClF_4

0412 IS THE RADICAL CHLORINE TETRAFLUORIDE PLANAR OR
NOT.
GREGORY AR
J CHEM PHYS 60: 3713-14 (1974)

0413 EPR SPECTRUM OF THE RADICAL CHLORINE
TETRAFLUORIDE.
MORTON JR + PRESTON KF
J CHEM PHYS 58(7): 3112-13 (1973)

CHLORINE PENTAFLUORIDE ClF_5

0414 STUDY OF LIQUID CHLORINE FLUORIDES AND
OXYFLUORIDES, CHLORINE PENTAFLUORIDE AND
CHLORINE OXYFLUORIDE BY NUCLEAR MAGNETIC
RESONANCE.
ALEXANDRE M + RIGNY P
CAN J CHEM 52(21): 3676-81 (1974) (IN FRENCH)

0415 CALCULATION OF THE ELECTRONIC STRUCTURE OF A
CHLORINE PENTAFLUORIDE MOLECULE BY THE MO LCAO
SELF CONSISTENT FIELD METHOD IN AN NDDO
APPROXIMATION.
BAGATURYANTS AA + KLIMENKO NM + DYATKINA ME
J STRUCT CHEM 14(4): 710-14 (1973)

0416 FLUORINE NMR OF CHLORINE PENTAFLUORIDE.
BANTOV DV + DZEVITSKII BE + KONSTANTINOV YS
+ SUKHOVERKHOV VF
IZV SIB OTDEL AKAD NAUK SSSR, SER KHIM NAUK 1:
81-3 (1968) (IN RUSSIAN)

0417 EQUILIBRIUM STUDIES OF CHLORINE PENTAFLUORIDE.
BAUER HF + SHEEHAN DF
INORG CHEM 9: 1736-7 (1967)

0418 VIBRATIONAL SPECTRA AND VALENCE FORCE CONSTANTS
OF SQUARE PYRAMIDAL MOLECULES. XENON OXIDE
TETRAFLUORIDE, IODINE PENTAFLUORIDE, BROMINE
PENTAFLUORIDE, AND CHLORINE PENTAFLUORIDE.
BEGUN GM + FLETCHER WH + SMITH DF
J CHEM PHYS 42: 2236-42 (1965)

0419 MATRIX ISOLATION STUDY OF CHLORINE
PENTAFLUORIDE.
CHRISTE KO
SPECTROCHIM ACTA 27A: 631-35 (1971)

0420 REPARAMETERIZED VIBRATIONAL FORCE CONSTANTS.
PART-1: METHOD AND APPLICATION TO SOME SQUARE
PYRAMIDAL MOLECULES. (IODINE PENTAFLUORIDE,
BROMINE PENTAFLUORIDE, AND CHLORINE
PENTAFLUORIDE)
CURTIS EC
SPECTROCHIM ACTA 27A: 1989-97 (1971)

0421 MEAN AMPLITUDES OF VIBRATION FOR SOME PYRAMIDAL
XY4Z MOLECULES. (IODINE PENTAFLUORIDE, BROMINE
PENTAFLUORIDE, AND CHLORINE PENTAFLUORIDE)
CYVIN SJ + BRUNVOLL J + ROBIETTE AG
J MOL STRUCT 3: 259-61 (1969)

CNDO/2 AND INDO ALL VALENCE ELECTRON
CALCULATIONS ON THE GEOMETRY AND PROPERTIES OF
SOME INTERHALOGENS. (CHLORINE TRIFLUORIDE,
BROMINE TRIFLUORIDE, IODINE TRIFLUORIDE,
CHLORINE PENTAFLUORIDE, BROMINE PENTAFLUORIDE,
IODINE PENTAFLUORIDE, CHLORINE HEPTAFLUORIDE,
BROMINE HEPTAFLUORIDE, AND IODINE HEPTAFLUORIDE)
DEB BM + COULSON CA
J CHEM SOC A 1971: 958-70

0422 THERMODYNAMIC PROPERTIES OF PENTAFLUORIDES AT
HIGH TEMPERATURES. PART-1: CALCULATION METHODS.
(CHLORINE PENTAFLUORIDE, BROMINE PENTAFLUORIDE,
IODINE PENTAFLUORIDE, PHOSPHORUS PENTAFLUORIDE,
VANADIUM PENTAFLUORIDE, ANTIMONY PENTAFLUORIDE)
GALKIN NP + TUMANOV YN + BUTYLKIN YP
KHIM VYS ENERG 4(6): 512-8 (1970) (IN RUSSIAN)

AB INITIO MOLECULAR ORBITAL STUDY OF THE
GEOMETRY OF INTERHALOGENS. (CHLORINE
MONOFLUORIDE, CHLORINE TRIFLUORIDE, AND CHLORINE
PENTAFLUORIDE)
GUEST MF + HALL MB + HILLIER IH
J CHEM SOC, FARADAY TRANS 69(12): 1829-34 (1973)

0423 MICROWAVE SPECTROSCOPY OF CHLORINE PENTAFLUORIDE
AT 70 AND 140 GHZ
JUREK R + SUZEAU P + CHANUSSOT J + CHAMPION JP
J PHYS (PARIS) 35(7/8): 533-40 (1974) (IN
FRENCH)

0424 IDEAL GAS THERMODYNAMIC PROPERTIES OF CHLORINE
PENTAFLUORIDE, BROMINE PENTAFLUORIDE, AND IODINE
PENTAFLUORIDE.
KUDCHADKER AP + KUDCHADKER SA + AGARWAL PM
INDIAN J CHEM 9: 722-24 (1971)

0425 MOLECULAR FORCE FIELDS FOR INTERHALOGEN
COMPOUNDS. (CHLORINE PENTAFLUORIDE AND BROMINE
PENTAFLUORIDE)
RAMASWAMY K + MUTHUSUBRAMANIAN P
J MOL STRUCT 7: 45-50 (1971)

0426 DENSITY, VAPOR PRESSURE, CRITICAL PROPERTIES,
DIELECTRIC CONSTANT, AND SPECIFIC CONDUCTIVITY
OF CHLORINE PENTAFLUORIDE.
ROGERS HH + CONSTANTINE MT + QUAGLINO J
+ DUBB JE + OGIMACHI NN
J CHEM ENG DATA 13: 307-12 (1968)

ABSORPTION SPECTRA OF CHLORINE FLUORIDE,
CHLORINE TRIFLUORIDE, AND CHLORINE PENTAFLUORIDE
MOLECULES.
RYMARCHUK YUA + IVANOV VS
ZH PRIKL SPEKTROSK 22(5): 950 (1975) (IN
RUSSIAN)

0427 CORIOLIS COUPLING COEFFICIENTS OF BROMINE
PENTAFLUORIDE AND CHLORINE PENTAFLUORIDE.
VENKATESWARLU K + MATHEW MP
CURR SCI 37: 252-3 (1968)

CHLORINE HEXAFLUORIDE ClF_6

ELECTRON SPIN RESONANCE SPECTRA OF HALOGEN
HEXAFLUORIDES. (CHLORINE HEXAFLUORIDE, IODINE
HEXAFLUORIDE, BROMINE HEXAFLUORIDE)
BOATE AR + MORTON JR + PRESTON KF
INORG CHEM 14(12): 3127-8 (1975)

0428 CHLORINE HEXAFLUORIDE RADICAL. PREPARATION,
ELECTRON SPIN RESONANCE SPECTRUM, AND STRUCTURE.
NISHIKIDA K + WILLIAMS F + MAMANTOV G + SMYRL N
J AMER CHEM SOC 97(12): 3526-7 (1975)

CHROMIUM MONOFLUORIDE CrF

0429 BAND SPECTRUM OF CHROMIUM MONOFLUORIDE.
DURGAVATH BK + RAO VRA
INDIAN J PHYS 28: 525-32 (1954)

SUBLIMATION PRESSURE OF CHROMIUM DIFLUORIDE AND
THE DISSOCIATION ENERGY OF CHROMIUM
MONOFLUORIDE.
KENT RA + MARGRAVE JL
J AMER CHEM SOC 87: 3582-5 (1965)

CHROMIUM DIFLUORIDE CrF_2

0430 SUBLIMATION PRESSURE OF CHROMIUM DIFLUORIDE AND
THE DISSOCIATION ENERGY OF CHROMIUM
MONOFLUORIDE.
KENT RA + MARGRAVE JL
J AMER CHEM SOC 87: 3582-5 (1965)

0431 SPECTROSCOPIC STUDIES OF THE VAPORIZATION OF
HIGH TEMPERATURE MATERIALS. (CHROMIUM
DIFLUORIDE, CHROMIUM TRIFLUORIDE, AND IRON
DIFLUORIDE)
LINEVSKY MJ
GENERAL ELECTRIC CORP, 1968, 45P. AD-670626

INTERACTION OF MATRIX ISOLATED NICKEL FLUORIDE
AND NICKEL CHLORIDE WITH CARBON MONOXIDE,
MOLECULAR NITROGEN, NITRIC OXIDE, AND MOLECULAR
OXYGEN AND OF CALCIUM FLUORIDE, CHROMIUM(II)
FLUORIDE, MANGANESE(II) FLUORIDE, COPPER(II)
FLUORIDE, AND ZINC(II) FLUORIDE WITH CARBON
MONOXIDE IN ARGON MATRICES.
VAN LEIRSBURG DA + DEKOCK CW
J PHYS CHEM 78(2): 134-42 (1974)

0432 THERMODYNAMIC PROPERTIES OF CHROMIUM DIFLUORIDE.
VECHER RA + VECHER AA + ZILBERMAN TB
NEORG MATER 11(8): 1520-1 (1975) (IN RUSSIAN)

CHROMIUM TRIFLUORIDE CrF_3

0433 STRUCTURE OF CRYSTALLINE CHROMIUM TRIFLUORIDE.
KNOX K
ACTA CRYSTALLOGR 13: 507-8 (1960)

SPECTROSCOPIC STUDIES OF THE VAPORIZATION OF
HIGH TEMPERATURE MATERIALS. (CHROMIUM
DIFLUORIDE, CHROMIUM TRIFLUORIDE, AND IRON
DIFLUORIDE)
LINEVSKY MJ
GENERAL ELECTRIC CORP, 1968, 45P. AD-670626

0434 SUBLIMATION PRESSURES OF CHROMIUM TRIFLUORIDE,
MANGANESE TRIFLUORIDE, AND IRON TRIFLUORIDE.
ZMBOV KF + MARGRAVE JL
J INORG NUCL CHEM 29: 673-80 (1967)

COPPER DIFLUORIDE CuF_2

CHROMIUM PENTAFLUORIDE CrF_5

0442 MASS SPECTROMETRIC STUDIES AT HIGH TEMPERATURES.
PART-9: SUBLIMATION PRESSURE OF COPPER
DIFLUORIDE.
KENT RA + MCDONALD JD + MARGRAVE JL
J PHYS CHEM 70: 874-6 (1966)

0435 VIBRATIONAL SPECTRUM OF LIQUID CHROMIUM
PENTAFLUORIDE.
BROWN SD + LOEHR TM + GARD GL
J CHEM PHYS 64(1): 260-2 (1976)

INFRARED SPECTRA AND GEOMETRY OF MATRIX ISOLATED
COBALT DIFLUORIDE, NICKEL DIFLUORIDE, COPPER
DIFLUORIDE, AND ZINC DIFLUORIDE.
HASTIE JW + HAUGE RH + MARGRAVE JL
HIGH TEMP SCI 1: 76-85 (1969)

CHROMIUM HEXAFLUORIDE CrF_6

MASS SPECTROMETRIC STUDIES AT HIGH TEMPERATURES.
PART-9: SUBLIMATION PRESSURE OF COPPER
DIFLUORIDE. (AND COPPER MONOFLUORIDE)
KENT RA + MCDONALD JD + MARGRAVE JL
J PHYS CHEM 70(3): 874-7 (1966)

0436 INFRARED SPECTRUM OF CHROMIUM HEXAFLUORIDE,
MOLYBDENUM HEXAFLUORIDE, AND OSMIUM
HEXAFLUORIDE.
HELLBERG KH + MULLER A + GLEMSER O
Z NATURFORSCH 21B: 118-21 (1966)

0443 STRUCTURES OF FLUORIDES. PART-6: PRECISE
STRUCTURAL PARAMETERS IN COPPER DIFLUORIDE BY
NEUTRON DIFFRACTION.
TAYLOR JC + WILSON PW
J LESS COMMON METALS 34(2): 257-9 (1974)

INTERACTION OF MATRIX ISOLATED NICKEL FLUORIDE
AND NICKEL CHLORIDE WITH CARBON MONOXIDE,
MOLECULAR NITROGEN, NITRIC OXIDE, AND MOLECULAR
OXYGEN AND OF CALCIUM FLUORIDE, CHROMIUM(II)
FLUORIDE, MANGANESE(II) FLUORIDE, COPPER(II)
FLUORIDE, AND ZINC(II) FLUORIDE WITH CARBON
MONOXIDE IN ARGON MATRICES.
VAN LEIRSBURG DA + DEKOCK CW
J PHYS CHEM 78(2): 134-42 (1974)

COBALT DIFLUORIDE CoF_2

0437 INFRARED SPECTRA AND GEOMETRY OF MATRIX ISOLATED
COBALT DIFLUORIDE, NICKEL DIFLUORIDE, COPPER
DIFLUORIDE, AND ZINC DIFLUORIDE.
HASTIE JW + HAUGE RH + MARGRAVE JL
HIGH TEMP SCI 1: 76-85 (1969)

FREE ENERGIES OF FORMATION OF SOME INORGANIC
FLUORIDES BY SOLID STATE EMF MEASUREMENTS.
(THORIUM TETRAFLUORIDE, ALUMINUM TRIFLUORIDE,
NICKEL DIFLUORIDE, LEAD DIFLUORIDE, COLBALT
DIFLUORIDE, AND URANIUM TRIFLUORIDE)
HEUS RJ + EGAN JJ
Z PHYS CHEM NEUE FOLGE 49: 38-43 (1966)

COPPER TRIFLUORIDE CuF_3

0444 NEW FLUORIDES OF TRIVALENT COPPER.
GRANNEC J + SORBE P + PORTIER J + HAGENMULLER P
COMPT REND C 280(2): 45-7 (1975) (IN FRENCH)

0438 VAPOR PRESSURE AND HEAT OF SUBLIMATION OF COBALT
DIHALIDES. (COBALT DIFLUORIDE)
HILL SD + CLELAND CA + ADAMS A + LANDSBERG A
+ BLOCK FE
J CHEM ENG DATA 14: 84-9 (1969)

CURIUM TRIFLUORIDE CmF_3

0439 KNUDSEN AND LANGMUIR MEASUREMENTS OF THE
SUBLIMATION PRESSURE OF COBALT DIFLUORIDE.
KANA'AN AS + BESENBRUCH G + MARGRAVE JL
J INORG NUCL CHEM 28: 1035-7 (1966)

0445 ABSORPTION SPECTRA OF SOLID AMERICIUM
TRIFLUORIDE, AMERICIUM TETRAFLUORIDE, CURIUM
TRIFLUORIDE AND CURIUM TETRAFLUORIDE.
ASPREY LB + KEENAN TK
J INORG NUCL CHEM 7: 27-31 (1958)

COPPER MONOFLUORIDE CuF

0446 MELTING POINTS OF CURIUM TRIFLUORIDE AND
AMERICIUM TRIFLUORIDE.
BURNETT JL
J INORG NUCL CHEM 28: 2454-5 (1966)

0440 DISSOCIATION ENERGY OF COPPER MONOFLUORIDE.
HILDENBRAND DL
J CHEM PHYS 48: 2457-9 (1968)

CURIUM TETRAFLUORIDE CmF_4

0441 MASS SPECTROMETRIC STUDIES AT HIGH TEMPERATURES.
PART-9: SUBLIMATION PRESSURE OF COPPER
DIFLUORIDE. (AND COPPER MONOFLUORIDE)
KENT RA + MCDONALD JD + MARGRAVE JL
J PHYS CHEM 70(3): 874-7 (1966)

0447 EVIDENCE FOR CURIUM TETRAFLUORIDE.
ASPREY LB + ELLINGER FH + FRIED SM
+ ZACHARIASEN WH
J AMER CHEM SOC 79: 5825 (1957)

ABSORPTION SPECTRA OF SOLID AMERICIUM
TRIFLUORIDE, AMERICIUM TETRAFLUORIDE, CURIUM
TRIFLUORIDE, AND CURIUM TETRAFLUORIDE.
ASPREY LB + KEENAN TK
J INORG NUCL CHEM 7: 27-31 (1958)

DYSPROSIUM TRIFLUORIDE DyF_3

0448 MASS SPECTROMETRIC STUDIES AT HIGH TEMPERATURES.
PART-17: SUBLIMATION AND VAPOR PRESSURES OF
DYSPROSIUM, HOLMIUM, AND ERBIUM TRIFLUORIDES.
BESENBRUCH G + CHARLU TV + ZMBOV KF
+ MARGRAVE JL
J LESS COMMON METALS 12: 375-81 (1967)

0449 STABILITIES OF DYSPROSIUM TRIFLUORIDE, HOLMIUM
TRIFLUORIDE, AND ERBIUM TRIFLUORIDE.
ZMBOV KF + MARGRAVE JL
J PHYS CHEM 70: 3379-81 (1966)

EUROPIUM MONOFLUORIDE EuF

MASS SPECTROMETRIC STUDIES AT HIGH TEMPERATURES.
PART-17: SUBLIMATION AND VAPOR PRESSURES OF
DYSPROSIUM, HOLMIUM, AND EUROPIUM
MONOFLUORIDES.
BESENBRUCH G + CHARLU TV + ZMBOV KF + MARGRAVE J
J LESS COMMON METALS 12(5): 375-81 (1967)

0450 MASS SPECTROMETRIC STUDIES AT HIGH TEMPERATURES.
PART-13: STABILITIES OF SAMARIUM, EUROPIUM, AND
GADOLINIUM MONO AND DIFLUORIDES.
ZMBOV KF + MARGRAVE JL
J INORG NUCL CHEM 29(1): 59-63 (1967)

EUROPIUM DIFLUORIDE EuF_2

0451 INFRARED SPECTRUM OF HEAVY METAL DIHALIDES.
(EUROPIUM DIFLUORIDE)
HASTIE JW + HAUGE RH + MARGRAVE JL
HIGH TEMP SCI 3: 56-72 (1971)

0452 PHASE INVESTIGATIONS AND VAPOR PRESSURE
MEASUREMENTS ON EUROPIUM DIFLUORIDE.
PETZEL VT + GREIS O
Z ANORG ALLGEM CHEM 388(2): 137-57 (1972) (IN
GERMAN)

EUROPIUM TRIFLUORIDE EuF_3

0453 OPTICAL ABSORPTION AND FLUORESCENCE SPECTRA OF
EUROPIUM TRIFLUORIDE.
CASPERS HH + RAST HE + FRY JL
J CHEM PHYS 47: 4505-13 (1967)

GEOMETRIES AND ENTROPIES OF METAL TRIFLUORIDES
FROM INFRARED SPECTRA. (SCANDIUM, YTTRIUM,
LANTHANUM, CERIUM, NEODYMIUM, EUROPIUM, AND
GADOLINIUM TRIFLUORIDES)
HASTIE JW + HAUGE RH + MARGRAVE JL
J LESS COMMON METALS 39(2): 309-34 (1975)

0454 MAGNETIC SUSCEPTIBILITY OF EUROPIUM TRIFLUORIDE.
KERN S + RACCAH PM + TVETEN A
J PHYS CHEM SOLIDS 31: 2639-42 (1970)

GADOLINIUM MONOFLUORIDE GdF

MASS SPECTROMETRIC STUDIES AT HIGH TEMPERATURES.
PART-13: STABILITIES OF SAMARIUM, EUROPIUM, AND
GADOLINIUM MONO AND DIFLUORIDES.
ZMBOV KF + MARGRAVE JL
J INORG NUCL CHEM 29(1): 59-63 (1967)

GADOLINIUM TRIFLUORIDE GdF_3

0455 REFLECTION SPECTRA OF GADOLINIUM OXIDE AND
GADOLINIUM TRIFLUORIDE POWDERS.
GIL'FANOV FZ + STOLOV AL
J APPL SPECTROSC 6(4): 334-5 (1967)

GEOMETRIES AND ENTROPIES OF METAL TRIFLUORIDES
FROM INFRARED SPECTRA. (SCANDIUM, YTTRIUM,
LANTHANUM, CERIUM, NEODYMIUM, EUROPIUM, AND
GADOLINIUM TRIFLUORIDES)
HASTIE JW + HAUGE RH + MARGRAVE JL
J LESS COMMON METALS 39(2): 309-34 (1975)

0456 ENTROPY AND ENTHALPY OF SUBLIMATION OF
GADOLINIUM TRIFLUORIDE. ROLE OF CORRELATION OF
ENTROPY AND ENTHALPY IN ERRORS.
MCCREARY JR + THORN RJ
HIGH TEMP SCI 5(2): 97-112 (1973)

GALLIUM MONOFLUORIDE GaF

MICROWAVE SPECTRA OF THE THALLIUM, INDIUM AND
GALLIUM MONOHALIDES.
BARRETT AH + MANDEL M
PHYS REV 109: 1572-89 (1958)

DISSOCIATION ENERGIES OF THE GASEOUS MONOHALIDES
OF BORON, ALUMINUM, GALLIUM, INDIUM, AND
THALLIUM.
BARROW RF
TRANS FARADAY SOC 56: 952-58 (1960)

0457 ROTATIONAL ANALYSIS OF BANDS OF THE A TRIPLET PI
ZERO(+) B TRIPLET PI(1) TO X SINGLET SIGMA(+)
SYSTEMS OF GALLIUM MONOFLUORIDE.
BARROW RF + DODSWORTH PG + ZEEMAN PB
PROC PHYS SOC 70A: 34-40 (1957)

INFRARED SPECTRA OF MATRIX ISOLATED SPECIES IN
THE GALLIUM- FLUORINE SYSTEM.
HASTIE JW + HAUGE RH + MARGRAVE JL
J FLUORINE CHEM 3(3-4): 285-91 (1973)

0458 POTENTIAL ENERGY CURVES AND DISSOCIATION
ENERGIES OF DIATOMIC FLUORIDES AND CHLORIDES OF
GALLIUM, INDIUM, AND THALLIUM. (MONOFLUORIDES)
SINGH J + NAIR KPR + RAI DK
J QUANT SPECT RAD TRANSFER 11: 1577-81 (1971)

0459 MOLECULAR SCF CALCULATIONS FOR GALLIUM
MONOFLUORIDE, GALLIUM TRIHYDRIDE, GERMANIUM
TETRAHYDRIDE, ARSENIC TRIHYDRIDE, AND SELENIUM
DIHYDRIDE.
STEVENSON PE + LIPSCOMB WN
J CHEM PHYS 52: 5343-53 (1970)

POTENTIAL ENERGY CURVES AND NATURE OF BINDING IN
GROUP-3A MONOHALIDES.
THAKUR SN + SINGH RB + RAI DK
J SCI IND RES 27(9): 339-47 (1968)

GALLIUM TRIFLUORIDE   **GaF₃**

0460   ELECTRON DIFFRACTION ANALYSIS OF GALLIUM
       HALIDES. (GALLIUM TRIFLUORIDE)
       AKISHIN PA + NAUMOV VA + TATAEVSKII VM
       NAUCH DOKL VYSSH SHKOL, KHIM KHIM TEKHNOL 2:
       205-9 (1958) (IN RUSSIAN)

       ELECTRON DIFFRACTION STUDY OF THE MOLECULAR
       STRUCTURE OF THE VAPOR PHASE HALIDES OF GALLIUM,
       YTTRIUM, LANTHANUM, AND NEODYMIUM.
       AKISHIN PA + NAUMOV VA + TATAEVSKII VM
       VEST MOSK UNIV 1959(1): 229-36 (IN RUSSIAN)

0461   PREPARATION AND CRYSTAL STRUCTURE OF GALLIUM
       TRIFLUORIDE.
       BREWER FM + GARTON G + GOODGAME DML
       J INORG CHEM 9: 56-64 (1959)

0462   INFRARED SPECTRA OF MATRIX ISOLATED SPECIES IN
       THE GALLIUM- FLUORINE SYSTEM.
       HASTIE JW + HAUGE RH + MARGRAVE JL
       J FLUORINE CHEM 3(3-4): 285-91 (1973)

       HIGH TEMPERATURE NEGATIVE IONS. GASEOUS GROUP-3
       FLUORIDES.
       PETTY F + LING-FAI WANG J + STEIGER RP + HARLAND
       FRANKLIN JL + MARGRAVE JL
       HIGH TEMP SCI 5(1): 25-33 (1973)

GERMANIUM MONOFLUORIDE   **GeF**

       SPECTRA OF SILICON MONOFLUORIDE, GERMANIUM
       MONOFLUORIDE, TIN MONOFLUORIDE, AND LEAD
       MONOFLUORIDE.
       BARROW RF + BUTLER D + JOHNS JWC + POWELL JL
       PROC PHYS SOC 73: 317-20 (1959)

0463   MASS SPECTROMETRIC STUDIES AT HIGH TEMPERATURES.
       PART-2: THE DISSOCIATION ENERGIES OF THE
       MONOFLUORIDES AND DIFLUORIDES OF SILICON AND
       GERMANIUM.
       EHLERT TC + MARGRAVE JL
       J CHEM PHYS 41(4): 1066-72 (1964)

0464   BOND DISSOCIATION ENERGIES, IONIZATION
       POTENTIALS AND ELECTRON AFFINITIES OF SOME
       GERMANIUM FLUORIDE SPECIES.
       HARLAND PW + CRADOCK S + THYNNE JCJ
       INORG NUCL CHEM LETT 9: 53-8 (1973)

       REEVALUATION OF THE DISSOCIATION ENERGY OF
       CALCIUM FLUORIDE. (ALSO SILICON AND GERMANIUM
       FLUORIDES)
       HASTIE JW + MARGRAVE JL
       J CHEM ENG DATA 13(3): 428-9 (1968)

0465   ANALYSIS OF THE ROTATIONAL STRUCTURE OF THE
       BANDS 0-0 AND 0-1 IN GERMANIUM MONOFLUORIDE.
       KUZYAKOV YY + UZIKOV AN + OSMININ EN
       VEST MOSK UNIV KHIM 12: 110-11 (1971) (IN
       RUSSIAN)

0466   ROTATIONAL STRUCTURE OF THE A DOUBLET SIGMA(+),
       B DOUBLET SIGMA(+), AND A QUARTET SIGMA(-) TO X
       DOUBLET PI TRANSITIONS OF GERMANIUM
       MONOFLUORIDE.
       MARTIN RW + MERER AJ
       CAN J PHYS 51: 125-43 (1973)

0467   ROTATIONAL STRUCTURE IN SOME HIGHER EXCITED
       STATES OF THE GERMANIUM(I) FLUORIDE MOLECULE.
       MARTIN RW + MERER AJ
       CAN J PHYS 52(15): 1458-75 (1974)

0468   THERMODYNAMIC PROPERTIES OF GASEOUS GERMANIUM,
       GERMANIUM TETRAFLUORIDE, GERMANIUM DIFLUORIDE,
       AND GERMANIUM MONOFLUORIDE.
       O'HARE PAG
       USAEC, 1968, 18P. ANL-7523

0469   ENTHALPY OF FORMATION OF GERMANIUM
       TETRAFLUORIDE.
       O'HARE PAG + JOHNSON J + KLAMECKI B
       + MULVIHILL M + HUBBARD WN
       J CHEM THERMODYN 1(2): 177-81 (1969)

       CALCULATION OF THE FRANCK-CONDON FACTOR FOR
       MOLECULES OF SILICON FLUORIDE, GERMANIUM
       FLUORIDE, AND GERMANIUM CHLORIDE.
       SMIRNOV AD + KUZ'MENKO NE + BOLOTIN AB
       LIET FIZ RINKINYS 14(4): 573-6 (1974) (IN
       RUSSIAN)

0470   VIBRATIONAL ANALYSIS OF C-X AND C'-X' BAND
       SYSTEMS AND ENERGY OF DISSOCIATION IN GERMANIUM
       MONOFLUORIDE.
       UZIKOV AN + KUZYAKOV YY
       VEST MOSK UNIV KHIM 24: 22-5 (1969) (IN RUSSIAN)

GERMANIUM DIFLUORIDE   **GeF₂**

0471   ENTHALPY OF SUBLIMATION OF GERMANIUM DIFLUORIDE
       AND THE THERMODYNAMICS OF SUBLIMATION OF THE
       GROUP-4A DIFLUORIDES.
       ADAMS GP + MARGRAVE JL + STEIGER RP
       J CHEM THERMODYN 3: 297-305 (1971)

0472   ENTHALPY OF FORMATION OF GERMANIUM DIFLUORIDE.
       ADAMS GP + MARGRAVE JL + WILSON PW
       J CHEM THERMODYN 2: 741-4 (1970)

       MASS SPECTROMETRIC STUDIES AT HIGH TEMPERATURES.
       PART-2: THE DISSOCIATION ENERGIES OF THE
       MONOFLUORIDES AND DIFLUORIDES OF SILICON AND
       GERMANIUM.
       EHLERT TC + MARGRAVE JL
       J CHEM PHYS 41(4): 1066-72 (1964)

       VACUUM ULTRAVIOLET SPECTRA OF SILICON DIFLUORIDE
       AND GERMANIUM DIFLUORIDE.
       GOLE JL + HAUGE RH + MARGRAVE JL + HASTIE JW
       J MOL SPECTROSC 43(3): 441-51 (1972)

0473   INFRARED VIBRATIONAL PROPERTIES OF GERMANIUM
       DIFLUORIDE.
       HASTIE JW + HAUGE RH + MARGRAVE JL
       J PHYS CHEM 72: 4492-6 (1968)

0474   ULTRAVIOLET ABSORPTION SPECTRUM OF GERMANIUM
       DIFLUORIDE.
       HAUGE RH + KHANNA VM + MARGRAVE JL
       J MOL SPECTROSC 27: 143-7 (1968)

0475   MATRIX INFRARED AND LASER RAMAN SPECTRA,
       MOLECULAR STRUCTURES AND NORMAL COORDINATE
       ANALYSES OF GERMANIUM DIFLUORIDE, MONOMERIC
       GERMANIUM DIFLUORIDE, AND DIMERIC GERMANIUM
       DIFLUORIDE.
       HUBER H + KUENDIG EP + OZIN GA + VANDERVOET A
       CAN J CHEM 52(1): 95-9 (1974)

0476   NEW ELECTRONIC EMISSION SYSTEM OF GERMANIUM
       DIFLUORIDE.
       MARTIN RW + MERER AJ
       CAN J PHYS 51(7): 727-30 (1973)

       THERMODYNAMIC PROPERTIES OF GASEOUS GERMANIUM,
       GERMANIUM TETRAFLUORIDE, GERMANIUM DIFLUORIDE,
       AND GERMANIUM MONOFLUORIDE.
       O'HARE PAG
       USAEC, 1968, 18P. ANL-7523

       VIRIAL THEOREM DECOMPOSITION OF MOLECULAR FORCE
       FIELDS.
       SIMONS G + NOVICK JL
       J PHYS CHEM 78(10): 989-93 (1974)

VIBRATIONAL TRANSITION PROBABILITIES AND R
CENTROIDS FOR DIATOMIC FLUORIDES OF SILICON AND
GERMANIUM.
SINGH J
INDIAN J PURE APPL PHYS 13(3): 204-6 (1975)

0477   MICROWAVE SPECTRUM OF GERMANIUM DIFLUORIDE.
QUADRUPOLE COUPLING AND CENTRIFUGAL DISTORTION.
TAKEO H + CURL RF
J MOL SPECTROSC 43: 21-30 (1972)

0478   MICROWAVE SPECTRUM OF GERMANIUM DIFLUORIDE.
TAKEO H + CURL RF + WILSON PW
J MOL SPECTROSC 38: 464-75 (1971)

0479   FREQUENCIES OF THE DEFORMATION VIBRATIONS OF THE
DIHALIDES OF GERMANIUM, TIN, AND LEAD.
(GERMANIUM DIFLUORIDE, TIN DIFLUORIDE, LEAD
DIFLUORIDE)
TIMOSHININ VS + DANILOVA TG
RUSS J PHYS CHEM 42: 1596 (1968)

GERMANIUM TRIFLUORIDE GeF_3

0480   ENTHALPY OF FORMATION OF GERMANIUM TRIFLUORIDE.
WANG JL + MARGRAVE JL + FRANKLIN JL
J CHEM PHYS 60(5): 2158-62 (1974)

GERMANIUM TETRAFLUORIDE GeF_4

0481   ORBITAL VALENCE FORCE CONSTANTS OF TETRAHEDRAL
XY4 MOLECULES. (GERMANIUM TETRAFLUORIDES)
CURR SCI 37: 374-5 (1962)

VAPOR PHASE RAMAN SPECTRA OF MH4 (M= CARBON,
SILICON, GERMANIUM, AND TIN) AND MF4 (M= CARBON,
SILICON, AND GERMANIUM)
ARMSTRONG RS + CLARK RJH
J CHEM SOC, FARADAY TRANS 2 72: 11-21 (1976)

PHOTOELECTRON SPECTRA OF HALIDES. PART-1:
TETRAFLUORIDES AND TETRACHLORIDES OF GROUP-5B.
(CARBON TETRAFLUORIDE, SILICON TETRAFLUORIDE,
GERMANIUM TETRAFLUORIDE, TIN TETRAFLUORIDE)
BASSETT PJ + LLOYD DR
J CHEM SOC A 1971: 641-54

HELIUM-1 RESONANCE PHOTOELECTRON SPECTRA OF
GROUP-4 TETRAFLUORIDES.
BASSETT PJ + LLOYD DR
CHEM PHYS LETT 3(1): 22-4 (1969)

0482   RAMAN AND INFRARED SPECTRA OF GERMANIUM
TETRAFLUORIDE.
CAUNT AD + SHORT LN + WOODWARD LA
TRANS FARADAY SOC 48: 873-7 (1952)

NUCLEAR SPIN RELAXATION STUDY OF THE SPIN
ROTATION INTERACTION IN SPHERICAL TOP MOLECULES.
COURTNEY JA + ARMSTRONG RL
CAN J PHYS 50(12): 1252-61 (1972)

0483   HEATS OF FORMATION OF GERMANIUM TETRAFLUORIDE
AND OF THE GERMANIUM DIOXIDES.
GROSS P + HAYMAN C + BINGHAM JT
TRANS FARADAY SOC 62(9): 2388-94 (1966)

PHOTOELECTRON SPECTRA OF HALIDES. PART-7:
VARIABLE TEMPERATURE HELIUM-1 AND HELIUM-2
STUDIES OF CARBON TETRAFLUORIDE, SILICON
TETRAFLUORIDE, AND GERMANIUM TETRAFLUORIDE.
LLOYD DR + ROBERTS PJ
J ELECTRON SPECTROSC RELAT PHENOM 7(4): 325-30
(1975)

0484   A STUDY OF THE RAMAN AND BROAD LINE NMR SPECTRA
OF GERMANIUM TETRAFLUORIDE.
MARGRAVE JL + WILSON PW
SPECTROSC LETT 6: 191-5 (1973)

THERMODYNAMIC PROPERTIES OF GASEOUS GERMANIUM,
GERMANIUM TETRAFLUORIDE, GERMANIUM DIFLUORIDE,
AND GERMANIUM MONOFLUORIDE.
O'HARE PAG
USAEC, 1968, 18P. ANL-7523

FORCE CONSTANTS OF VARIOUS ISOTOPES OF CARBON
TETRAFLUORIDE, SILICON TETRAFLUORIDE, GERMANIUM
TETRAFLUORIDE AND SULFUR TRIOXIDE.
RUOFF A
SPECTROCHIM ACTA 23A: 2421-31 (1967)

INFRARED SPECTRA OF CARBON TETRAFLUORIDE AND
GERMANIUM TETRAFLUORIDE.
WOLTZ PJH + NIELSEN AH
J CHEM PHYS 20(2): 307-12 (1952)

GOLD PENTAFLUORIDE AuF_5

0485   PREPARATION AND CHARACTERIZATION OF GOLD
PENTAFLUORIDE.
VASILE MJ + RICHARDSON TJ + STEVIE FA
+ FALCONER WE
J CHEM SOC, DALTON TRANS 4: 351-3 (1976)

HAFNIUM TETRAFLUORIDE HfF_4

INFRARED SPECTRA OF SOME GROUP-4 HALIDES.
(ZIRCONIUM TETRAFLUORIDE, HAFNIUM TETRAFLUORIDE,
AND THORIUM TETRAFLUORIDE)
BUCHLER A + MATTUCK JBB + DUGRE DH
J CHEM PHYS 34: 2202-3 (1961)

0486   VAPOR PRESSURE AND HEAT OF SUBLIMATION OF
ZIRCONIUM AND HAFNIUM TETRACHLORIDES.
DENISOVA DN + SAFRONOV EK + BYSTROVA ON
RUSS J INORG CHEM 11(9): 1171-3 (OCT 1966)

VAPOR PRESSURES OF ZIRCONIUM TETRAFLUORIDE,
HAFNIUM TETRAFLUORIDE, AND TIN DIFLUORIDE.
FISCHER VW + PETEZEL T + LAUTER S
Z ANORG ALLGEM CHEM 333: 226-34 (1964)

0487   MEAN AMPLITUDES OF VIBRATION OF SOME TETRAHEDRAL
XY4 TYPE MOLECULES. PART-5: TETRAHALIDES OF
HAFNIUM AND LEAD.
NAGARAJAN G
INDIAN J PURE APPL PHYS 2(5): 145-9 (1964)

HEATS OF FORMATION OF TITANIUM TETRAFLUORIDE AND
HAFNIUM TETRAFLUORIDE.
GREENBERG E + SETTLE JL + HUBBARD WN
J PHYS CHEM 66: 1345-8 (1962)

VAPOR PRESSURES OF ZIRCONIUM TETRAFLUORIDE,
HAFNIUM TETRAFLUORIDE, AND TIN DIFLUORIDE.
JUZA R
Z ANORG ALLGEM CHEM 333: 226-34 (1964)

IONICITY OF THE MX BOND IN THE TETRAHALIDES OF
GROUP-4 ELEMENTS. (TIN TETRAFLUORIDE AND HAFNIUM
TETRAFLUORIDE)
RAI SN + THAKUR SN + RAI DK
J MOL STRUCT 8: 55-61 (1971)

HOLMIUM MONOFLUORIDE **HoF**

MASS SPECTROMETRIC STUDIES AT HIGH TEMPERATURES. PART-17: SUBLIMATION AND VAPOR PRESSURES OF DYSPROSIUM, HOLMIUM, AND EUROPIUM MONOFLUORIDES.
BESENBRUCH G + CHARLU TV + ZMBOV KF + MARGRAVE J
J LESS COMMON METALS 12(5): 375-81 (1967)

0488 ABSORPTION SPECTRUM OF GASEOUS HOLMIUM MONOFLUORIDE.
ROBBINS DJW + BARROW RF
J PHYS B 7(7): L234-L235 (1974)

HOLMIUM TRIFLUORIDE **HoF₃**

STABILITIES OF DYSPROSIUM TRIFLUORIDE, HOLMIUM TRIFLUORIDE AND ERBIUM TRIFLUORIDE.
ZMBOV KF + MARGRAVE JL
J PHYS CHEM 70: 3379-81 (1966)

INDIUM MONOFLUORIDE **InF**

MICROWAVE SPECTRA OF THE THALLIUM, INDIUM AND GALLIUM MONOHALIDES.
BARRETT AH + MANDEL M
PHYS REV 109: 1572-89 (1958)

DISSOCIATION ENERGIES OF THE GASEOUS MONOHALIDES OF BORON, ALUMINUM, GALLIUM, INDIUM, AND THALLIUM.
BARROW RF
TRANS FARADAY SOC 56: 952-8 (1960)

0489 HYPERFINE STRUCTURE OF INDIUM MONOFLUORIDE.
HAMMERLE RH + AUSDAL RV + ZORN JC
J CHEM PHYS 57: 4068-69 (1972)

0490 MICROWAVE ROTATIONAL SPECTRUM OF INDIUM MONOFLUORIDE.
LOVAS FJ + TORRING T
Z NATURFORSCH 24A: 634-36 (1969)

0491 ROTATIONAL ANALYSIS OF THE C SINGLET PI- X SINGLET SIGMA(+) SYSTEM OF INDIUM MONOFLUORIDE.
NAMPOORI VPN + KAMALASANAN MN + PATEL MM
J PHYS B 8(17): 2841-5 (1975)

POTENTIAL ENERGY CURVES AND DISSOCIATION ENERGIES OF DIATOMIC FLUORIDES AND CHLORIDES OF GALLIUM, INDIUM, AND THALLIUM. (MONOFLUORIDES)
SINGH J + NAIR KPR + RAI DK
J QUANT SPECT RAD TRANSFER 11: 1577-81 (1971)

POTENTIAL ENERGY CURVES AND NATURE OF BINDING IN GROUP-3A MONOHALIDES.
THAKUR SN + SINGH RB + RAI DK
J SCI IND RES 27(9): 339-47 (1968)

INDIUM TRIFLUORIDE **InF₃**

HIGH TEMPERATURE NEGATIVE IONS. GASEOUS GROUP-3 FLUORIDES.
PETTY F + LING-FAI WANG J + STEIGER RP + HARLAND FRANKLIN JL + MARGRAVE JL
HIGH TEMP SCI 5(1): 25-33 (1973)

0492 VAPORIZATION THERMODYNAMICS OF INDIUM TRIFLUORIDE.
STEIGER RP + MARGRAVE JL
HIGH TEMP HIGH PRESS 73(5): 471-4 (1974)

IODINE MONOFLUORIDE **IF**

0493 THEORY OF THE DISSOCIATION OF DIATOMIC MOLECULES AND A STUDY OF THE EMISSION OF IODINE MONOFLUORIDE.
BIRKS JW
LAWRENCE BERKELEY LAB, UNIV CALIF, BERKELEY, 1974, 155P. LBL-2743

ELECTRONEGATIVITY, NONBONDED INTERACTIONS AND POLARIZABILITY IN HYDROGEN HALIDES AND THE INTERHALOGEN COMPOUNDS. (CHLORINE MONOFLUORIDE, BROMINE MONOFLUORIDE, AND IODINE MONOFLUORIDE)
BROWN RF
J AMER CHEM SOC 83: 36-42 (1961)

THERMODYNAMIC PROPERTIES OF THE DIATOMIC INTERHALOGENS FROM SPECTROSCOPIC DATA. (CHLORINE, BROMINE, AND IODINE MONOFLUORIDES)
COLE LG + ELVERUM GW
J CHEM PHYS 20(10): 1543-51 (1952)

0494 DISSOCIATION ENERGIES OF DIATOMIC HALOGEN FLUORIDES. (IODINE FLUORIDE, BROMINE FLUORIDE)
COXON JA
CHEM PHYS LETT 33(1): 136-40 (1975)

CNDO/2 AND INDO ALL VALENCE ELECTRON CALCULATIONS ON THE GEOMETRY AND PROPERTIES OF SOME INTERHALOGENS. (CHLORINE TRIFLUORIDE, BROMINE TRIFLUORIDE, IODINE TRIFLUORIDE, CHLORINE PENTAFLUORIDE, BROMINE PENTAFLUORIDE, IODINE PENTAFLUORIDE, CHLORINE HEPTAFLUORIDE, BROMINE HEPTAFLUORIDE, AND IODINE HEPTAFLUORIDE)
DEB BM + COULSON CA
J CHEM SOC A 1971: 958-70

0495 A SMALL GAUSSIAN BASIS SET FOR NONEMPIRICAL ALL-ELECTRON SCF CALCULATIONS IN IODINE COMPOUNDS.
RODE BM
CHEM PHYS LETT 27(2): 264-8 (1974)

ELECTRONIC STRUCTURE AND CHEMISTRY OF IODINE COMPOUNDS. (FLUORIDES)
RODE BM
J CHEM SOC, FARADAY TRANS 71(3,PART-2): 481-95 (1975)

0496 ELECTRONIC EMISSION SPECTRUM AND MOLECULAR CONSTANTS OF IODINE MONOFLUORIDE.
DURIE RA
CAN J PHYS 44: 337-52 (1966)

0497 ROTATIONAL SPECTRUM OF IODINE FLUORIDE.
TIEMANN E + HOEFT J + TOERRING T
Z NATURFORSCH 28A(9): 1405-7 (1973) (IN GERMAN)

0498 SPIN ORBITAL INTERACTION AND RELATIVE INTENSITIES OF FORBIDDEN TRANSITIONS IN IODINE, OR IODINE BROMIDE (CHLORIDE, FLUORIDE) MOLECULES.
YARUNIN VS + GANIN VA
OPT SPECTROSC 37(5): 495-7 (NOV 1974)

IODINE TRIFLUORIDE   IF$_3$

0499   ELECTRONIC STRUCTURE AND CHEMISTRY OF IODINE
       COMPOUNDS. (FLUORIDES)
       RODE BM
       J CHEM SOC, FARADAY TRANS 71(3,PART-2): 481-95
       (1975)

0500   VIBRATIONAL SPECTRUM OF IODINE TRIFLUORIDE.
       SCHMEISSER M + NAUMANN D + LEHMANN E
       J FLUORINE CHEM 3(3-4): 441-4 (1973)

IODINE PENTAFLUORIDE   IF$_5$

0501   VIBRATIONAL SPECTRA OF THE ISOELECTRONIC SPECIES
       IODINE PENTAFLUORIDE, TELLURIUM PENTAFLUORIDE(-)
       AND ANTIMONY PENTAFLUORIDE(-2).
       ALEXANDER LE + BEATTIE IR
       J CHEM SOC A 1971: 3091-5

       VIBRATIONAL SPECTRA AND VALENCE FORCE CONSTANTS
       OF SQUARE PYRAMIDAL MOLECULES. XENON OXIDE
       TETRAFLUORIDE, IODINE PENTAFLUORIDE, BROMINE
       PENTAFLUORIDE, AND CHLORINE PENTAFLUORIDE.
       BEGUN GM + FLETCHER WH + SMITH DF
       J CHEM PHYS 42: 2236-42 (1965)

0502   MICROWAVE SPECTRUM OF IODINE PENTAFLUORIDE.
       BRADLEY RH + BRIER PN + WHITTLE MJ
       CHEM PHYS LETT 11: 192-3 (1971)

       STRUCTURE OF THE INTERHALOGEN COMPOUNDS. PART-2:
       IODINE HEPTAFLUORIDE AT -110 DEGREES C AND AT
       -145 DEGREES C. PART-3: BROMINE TRIFLUORIDE,
       BROMINE PENTAFLUORIDE, AND IODINE PENTAFLUORIDE.
       BURBANK RD + BENSEY FN
       J CHEM PHYS 27: 981-3 (1957)

0503   MOSSBAUER STUDY OF IODINE PENTAFLUORIDE AND
       IODINE HEPTAFLUORIDE.
       BUKSHPAN S + GOLDSTEIN C + SORIANO J + SHAMIR J
       J CHEM PHYS 51: 3976-8 (1969)

       REPARAMETERIZED VIBRATIONAL FORCE CONSTANTS.
       PART-1: METHOD AND APPLICATION TO SOME SQUARE
       PYRAMIDAL MOLECULES. (IODINE PENTAFLUORIDE,
       BROMINE PENTAFLUORIDE, AND CHLORINE
       PENTAFLUORIDE)
       CURTIS EC
       SPECTROCHIM ACTA 27A: 1989-97 (1971)

       MEAN AMPLITUDES OF VIBRATION FOR SOME PYRAMIDAL
       XY4Z MOLECULES. (IODINE PENTAFLUORIDE, BROMINE
       PENTAFLUORIDE, AND CHLORINE PENTAFLUORIDE)
       CYVIN SJ + BRUNVOLL J + ROBIETTE AG
       J MOL STRUCT 3: 259-61 (1969)

       CNDO/2 AND INDO ALL VALENCE ELECTRON
       CALCULATIONS ON THE GEOMETRY AND PROPERTIES OF
       SOME INTERHALOGENS. (CHLORINE TRIFLUORIDE,
       BROMINE TRIFLUORIDE, IODINE TRIFLUORIDE,
       CHLORINE PENTAFLUORIDE, BROMINE PENTAFLUORIDE,
       IODINE PENTAFLUORIDE, CHLORINE HEPTAFLUORIDE,
       BROMINE HEPTAFLUORIDE, AND IODINE HEPTAFLUORIDE)
       DEB BM + COULSON CA
       J CHEM SOC A 1971: 958-70

       THERMODYNAMIC PROPERTIES OF PENTAFLUORIDES AT
       HIGH TEMPERATURES. PART-1: CALCULATION METHODS.
       (CHLORINE PENTAFLUORIDE, BROMINE PENTAFLUORIDE,
       IODINE PENTAFLUORIDE, PHOSPHORUS PENTAFLUORIDE,
       VANADIUM PENTAFLUORIDE, ANTIMONY PENTAFLUORIDE)
       GALKIN NP + TUMANOV YN + BUTYLKIN YP
       KHIM VYS ENERG 4(6): 512-8 (1970) (IN RUSSIAN)

0504   VIBRATIONAL SPECTRA AND ASSIGNMENT OF IODINE
       PENTAFLUORIDE. FERMI RESONANCE BETWEEN A
       FUNDAMENTAL AND A TERNARY COMBINATION.
       GILLESPIE RJ + CLASE HJ
       J CHEM PHYS 47: 1071-3 (1967)

0505   MOLECULAR STRUCTURE, FORCE CONSTANTS, AND
       THERMODYNAMIC FUNCTIONS OF IODINE PENTAFLUORIDE
       AND BROMINE PENTAFLUORIDE.
       KHANNA RK
       J SCI IND RES (INDIA) 21B: 352-6 (1962)

       IDEAL GAS THERMODYNAMIC PROPERTIES OF CHLORINE
       PENTAFLUORIDE, BROMINE PENTAFLUORIDE, AND IODINE
       PENTAFLUORIDE.
       KUDCHADKER AP + KUDCHADKER SA + AGARWAL PM
       INDIAN J CHEM 9: 722-24 (1971)

0506   VIBRATIONAL SPECTRA AND STRUCTURES OF IODINE
       PENTAFLUORIDE AND HEPTAFLUORIDE.
       LORD RC + LYNCH MA + SCHUMB WC + SLOWINSKI EJ
       J AMER CHEM SOC 72: 522-7 (1950)

0507   POTENTIAL CONSTANTS AND THERMODYNAMIC FUNCTIONS
       OF IODINE PENTAFLUORIDE.
       NAGARAJAN G
       BULL SOC CHIM BELG 72: 5-15 (1962)

0508   POTENTIAL CONSTANTS AND THERMODYNAMIC PROPERTIES
       OF IODINE PENTAFLUORIDE.
       NAGARAJAN G
       Z NATURFORSCH 17A: 871-4 (1962)

0509   POTENTIAL CONSTANTS AND THERMODYNAMIC FUNCTIONS
       FOR IODINE PENTAFLUORIDE AND IODINE
       HEPTAFLUORIDE.
       NAGARAJAN G
       CURR SCI 30: 413-14 (1961)

0510   CALORIMETRIC STUDY OF IODINE PENTAFLUORIDE. HEAT
       CAPACITY BETWEEN 5 AND 350 DEGREES K, ENTHALPY
       OF FUSION AND VAPORIZATION, STANDARD ENTROPY OF
       THE VAPOR, AND OTHER THERMODYNAMIC PROPERTIES.
       OSBORNE DW + SCHREINER F + SELIG H
       J CHEM PHYS 54: 3790-7 (1971)

0511   MOLECULAR FORCE FIELD FOR INTERHALOGEN
       COMPOUNDS. IODINE PENTAFLUORIDE, IODINE OXYGEN
       PENTAFLUORIDE, AND IODINE HEPTAFLUORIDE.
       RAMASWAMY K + MUTHUSUBRAMANIAN P
       J MOL STRUCT 6: 205-14 (1970)

       GAS PHASE MOLECULAR STRUCTURES OF BROMINE
       PENTAFLUORIDE AND IODINE PENTAFLUORIDE FROM
       ELECTRON DIFFRACTION AND ROTATIONAL CONSTANT
       DATA.
       ROBIETTE AG + BRADLEY RH + BRIER PN
       CHEM COMMUN 1971: 1567-8

       ELECTRONIC STRUCTURE AND CHEMISTRY OF IODINE
       COMPOUNDS. (FLUORIDES)
       RODE BM
       J CHEM SOC, FARADAY TRANS 71(3,PART-2): 481-95
       (1975)

0512   RAMAN SPECTRUM OF IODINE PENTAFLUORIDE. EVIDENCE
       FOR POLYMERIZATION IN THE LIQUID STATE.
       SELIG H + HOLZMAN H
       ISRAEL J CHEM 7: 417-20 (1969)

0513   UREY-BRADLEY FORCE CONSTANTS, MEAN AMPLITUDES OF
       VIBRATION, SHRINKAGE EFFECT AND CORIOLIS
       CONSTANTS IN IODINE PENTAFLUORIDE AND IODINE
       OXYGEN PENTAFLUORIDE.
       VENKATESWARLU K + PURUSHOTHAMAN C
       ACTA PHYS ACAD SCI HUNG 25: 133-40 (1968)

IODINE HEXAFLUORIDE IF_6

ELECTRON SPIN RESONANCE SPECTRA OF HALOGEN
HEXAFLUORIDES. (CHLORINE HEXAFLUORIDE, IODINE
HEXAFLUORIDE, BROMINE HEXAFLUORIDE)
BOATE AR + MORTON JR + PRESTON KF
INORG CHEM 14(12): 3127-8 (1975)

IODINE HEPTAFLUORIDE IF_7

0514 ACCEPTOR PROPERTIES OF IODINE HEPTAFLUORIDE.
OCTAFLUORO PERIODATES(VII).
ADAMS CJ
INORG NUCL CHEM LETT 10(10): 831-5 (1974)

0515 STRUCTURE, PSEUDOROTATION, AND VIBRATIONAL MODE
COUPLING IN IODINE HEPTAFLUORIDE. AN ELECTRON
DIFFRACTION STUDY.
ADAMS WJ + THOMPSON HB + BARTELL LS
J CHEM PHYS 53: 404-6 (1970)

0516 NMR SPECTRA OF IODINE HEPTAFLUORIDE AND IODINE
OXYGEN PENTAFLUORIDE.
ALEXAKOS LG + CORNWELL CD + PIERCE SB
PROC CHEM SOC 1963(11): 341-2

0517 FLUORINE NMR OF IODINE HEPTAFLUORIDE, RHENIUM
HEPTAFLUORIDE, AND OXIDE PENTAFLUORIDES.
BARTLETT N + BEATON S + REEVES LW + WELLS EJ
CAN J CHEM 42: 2531-9 (1964)

ELECTRIC FIELD DEFLECTION OF MOLECULES WITH
LARGE AMPLITUDE MOTION. (XENON HEXAFLUORIDE,
IODINE HEPTAFLUORIDE, RHENIUM HEPTAFLUORIDE)
BERNSTEIN LS + PITZER KS
J CHEM PHYS 62: 2530-4 (1975)

0518 SPECTROSCOPIC STUDY OF IODINE HEPTAFLUORIDE AND
IODINE HEXAFLUORIDE(X) IN ANHYDROUS HYDROGEN
FLUORIDE.
BROWNSTEIN M + SELIG H
INORG CHEM 11(3): 656-8 (1972)

MOSSBAUER STUDY OF IODINE PENTAFLUORIDE AND
IODINE HEPTAFLUORIDE.
BUKSHPAN S + GOLDSTEIN C + SORIANO J + SHAMIR J
J CHEM PHYS 51: 3976-8 (1969)

0519 A REDETERMINATION OF THE ORTHORHOMBIC IODINE
HEPTAFLUORIDE STRUCTURE. A CORRECTION.
BURBANK RD
ACTA CRYSTALLOGR 15: 1207-14 (1962); 16: 700
(1965)

0520 INTRAMOLECULAR REARRANGEMENT IN IODINE
HEPTAFLUORIDE AND XENON HEXAFLUORIDE.
BURBANK RD + BARTLETT N
CHEM COMMUN 1968: 645-7

0521 STRUCTURE OF INTERHALOGEN COMPOUNDS. PART-2:
IODINE HEPTAFLUORIDE AT -110 DEGREES C AND AT
-145 DEGREES C.
BURBANK RD + BENSEY FN
J CHEM PHYS 27: 981-2 (1957)

0522 VIBRATIONAL SPECTRA OF IODINE HEPTAFLUORIDE AND
RHENIUM HEPTAFLUORIDE.
CLAASSEN HH + GASNER EL + SELIG H
J CHEM PHYS 49: 1803-7 (1968)

CNDO/2 AND INDO ALL VALENCE ELECTRON
CALCULATIONS ON THE GEOMETRY AND PROPERTIES OF
SOME INTERHALOGENS. (CHLORINE TRIFLUORIDE,
BROMINE TRIFLUORIDE, IODINE TRIFLUORIDE,
CHLORINE PENTAFLUORIDE, BROMINE PENTAFLUORIDE,
IODINE PENTAFLUORIDE, CHLORINE HEPTAFLUORIDE,
BROMINE HEPTAFLUORIDE, AND IODINE HEPTAFLUORIDE)
DEB BM + COULSON CA
J CHEM SOC A 1971: 958-70

CORRELATION OF ELECTRONIC STATES OF THE POSITIVE
IONS OF XENON TETRAFLUORIDE, XENON OXIDE
TETRAFLUORIDE, AND IODINE PENTAFLUORIDE.
DEKOCK RL
J ELECTRON SPECTROSC RELAT PHENOM 4: 155-61
(1974)

0523 EVIDENCE FOR THE MOLECULAR SYMMETRY OF IODINE
HEPTAFLUORIDE.
DONOHUE J
ACTA CRYSTALLOGR 18: 1018-21 (1965)

0524 MOLECULAR SYMMETRY OF IODINE HEPTAFLUORIDE.
DONOHUE J
J CHEM PHYS 30: 1618-19 (1959)

0525 VIBRATIONAL ASSIGNMENT FOR IODINE PENTAFLUORIDE,
A NONRIGID D5H MOLECULE.
EYSEL HH + SEPPELT K
J CHEM PHYS 56(10): 5081-6 (1972)

0526 MOLECULAR STRUCTURE OF XENON HEXAFLUORIDE AND
IODINE HEPTAFLUORIDE.
FALCONER WE + BUCHLER A + STAUFFER JL
+ KLEMPERER W
J CHEM PHYS 48: 312-19 (1968)

PREPARATION OF SINGLE CRYSTALS OF MANGANOUS
FLUORIDE. CRYSTAL STRUCTURE FROM X-RAY
DIFFRACTION. MELTING POINT AND DENSITY.
GRIFFEL M + STOUT JW
J AMER CHEM SOC 72: 4351-3 (1950)

0527 STRUCTURE DETERMINATION OF TRIS (TRIMETHYL
SILYL) SILANE, INFRARED INVESTIGATION OF IODINE
HEPTAFLUORIDE AND AN APPLICATION OF THREE ATOM
SCATTERING THEORY TO IODINE HEPTAFLUORIDE AND
RHENIUM HEPTAFLUORIDE.
KARRENBROCK AH
UNIV MICHIGAN, PHD THESIS, 1975, 133P.

0528 FORCE CONSTANTS AND THERMODYNAMIC FUNCTIONS OF
IODINE HEPTAFLUORIDE.
KHANNA RK
J MOL SPECTROSC 8: 134-41 (1962)

0529 POLAR DISTORTIONS IN RHENIUM HEPTAFLUORIDE AND
IODINE HEPTAFLUORIDE.
KAISER EW + MUENTER JS + KLEMPERER W
+ FALCONER WE
J CHEM PHYS 53: 53-55 (1970)

0530 MOLECULAR STRUCTURE OF IODINE HEPTAFLUORIDE.
LAVILLA RE + BAUER SH
J CHEM PHYS 33: 182-6 (1960)

0531 MOLECULAR SYMMETRY OF IODINE HEPTAFLUORIDE.
LOHR LL + LIPSCOMB WN
J CHEM PHYS 36: 2225-6 (1962)

VIBRATIONAL SPECTRA AND STRUCTURES OF IODINE
PENTAFLUORIDE AND HEPTAFLUORIDE.
LORD RC + LYNCH MA + SCHUMB WC + SLOWINSKI EJ
J AMER CHEM SOC 72: 522-7 (1950)

INTRAMOLECULAR LIGAND EXCHANGE IN SEVEN
COORDINATE STRUCTURES. (RHENIUM HEPTAFLUORIDE
AND IODINE HEPTAFLUORIDE)
MUETTERTIES EL + PACKER KH
J AMER CHEM SOC 86: 293-4 (1964)

POTENTIAL CONSTANTS AND THERMODYNAMIC FUNCTIONS
FOR IODINE PENTAFLUORIDE AND IODINE
HEPTAFLUORIDE.
NAGARAJAN G
CURR SCI 30: 413-14 (1961)

0532 POTENTIAL CONSTANTS OF IODINE HEPTAFLUORIDE.
NAGARAJAN G
BULL SOC CHIM BELG 71: 82-99 (1962)

0533 MOLECULAR ORBITAL TREATMENT OF IODINE
HEPTAFLUORIDE.
OAKLAND RL + DUFFEY GH
J CHEM PHYS 46(1): 19-22 (1967)

MOLECULAR FORCE FIELD FOR INTERHALOGEN
COMPOUNDS. IODINE PENTAFLUORIDE, IODINE OXIDE
PENTAFLUORIDE, AND IODINE HEPTAFLUORIDE.
RAMASWAMY K + MUTHUSUBRAMANIAN P
J MOL STRUCT 6: 205-14 (1970)

0534 MASS SPECTRA AND SUBLIMATION PRESSURES OF IODINE
HEPTAFLUORIDE AND IODINE OXYGEN PENTAFLUORIDE.
SCHACK CJ + PILIPOVICH D + COHZ SN + SHEEHAN DF
J PHYS CHEM 72: 4697-8 (1968)

0535 FORCE CONSTANTS AND MEAN VIBRATION AMPLITUDES OF
IODINE HEPTAFLUORIDE CALCULATED BY THE
LOGARITHMIC STEP METHOD IN THE RANGE OF THE
GENERALIZED VALENCE FORCE FIELD.
WENDLING EJL + MAHMOUDI S
BULL SOC CHIM FR 1972(1): 33-9 (IN FRENCH)

IRIDIUM PENTAFLUORIDE IrF_5

CONFIGURATION OF THE PENTAFLUORIDES OF RHODIUM
AND IRIDIUM BY MAGNETIC RESONANCE.
CYR T
CAN J SPECTROSC 19(5): 136-40 (1974)

0536 THERMODYNAMIC PROPERTIES OF PENTAFLUORIDES AT
HIGH TEMPERATURES. PART-3: PENTAFLUORIDES OF
SOME 5D-ELEMENTS.
GALKIN NP + TUMANOV YN + BUTYLKIN YP
RUSS J PHYS CHEM 44(12): 1724-6 (1970)

IRIDIUM HEXAFLUORIDE IrF_6

0537 VIBRONICALLY INDUCED JAHN-TELLER PROGRESSIONS IN
THE ELECTRONIC SPECTRUM OF IRIDIUM HEXAFLUORIDE.
BRAND JCD + GOODMAN GL
CAN J PHYS 46: 1721-4 (1968)

VAPOR PRESSURES CF SOME HEAVY TRANSITON METAL
HEXAFLUORIDES. (TUNGSTEN HEXAFLUORIDE,
MOLYBDENUM HEXAFLUORIDE, RHENIUM HEXAFLUORIDE,
OSMIUM HEXAFLUORIDE, AND IRIDIUM HEXAFLUORIDE)
CADY GH + HARGREAVES GB
J CHEM SOC A 1961: 1563-74

0538 VIBRATIONAL ELECTRONIC COUPLING IN IRIDIUM
HEXAFLUORIDE, OSMIUM HEXACHLORIDE(2-), AND
IRIDIUM HEXACHLORIDE(2-).
CHILD MS
MOL PHYS 3: 605-7 (1960)

0539 SEARCH FOR A JAHN-TELLER EFFECT IN IRIDIUM
HEXAFLUORIDE.
CLAASSEN HH + WEINSTOCK B
J CHEM PHYS 33: 436-7 (1960)

VIBRONIC INTERACTIONS IN METAL HEXAFLUORIDE
MOLECULES.
GOODMAN GL + FRED M
PAPER B102 OF INT SYMP MOL STRUCT SPECTROSC,
PROC, 1962, TOKYC, BUTTERWORTHS, 1963

ELECTRIC DEFLECTION OF BINARY HEXAFLUORIDES.
(SULFUR HEXAFLUORIDE, SELENIUM HEXAFLUORIDE,
TELLURIUM HEXAFLUORIDE, MOLYBDENUM HEXAFLUORIDE,
TUNGSTEN HEXAFLUORIDE, URANIUM HEXAFLUORIDE,
RUTHENIUM HEXAFLUORIDE, RHENIUM HEXAFLUORIDE,
RHODIUM HEXAFLUORIDE, OSMIUM HEXAFLUORIDE,
IRIDIUM HEXAFLUORIDE, AND PLATINUM HEXAFLUORIDE)
KAISER EW + MUENTER JS + KLEMPERER W +
FALCONER WE + SUNDER WA
J CHEM PHYS 53: 1411-12 (1970)

ELECTRON DIFFRACTION INVESTIGATION OF TUNGSTEN
HEXAFLUORIDE, OSMIUM HEXAFLUORIDE, IRIDIUM
HEXAFLUORIDE, URANIUM HEXAFLUORIDE, NEPTUNIUM
HEXAFLUORIDE, AND PLUTONIUM HEXAFLUORIDE.
KIMURA M + SCHOMAKER V + SMITH DW + WEINSTOCK B
J CHEM PHYS 48: 4001-12 (1968)

0540 INFRARED SPECTRUM OF IRIDIUM HEXAFLUORIDE.
MATTRAW HC + HAWKINS NJ + CARPENTER DR
+ SABOL WW
J CHEM PHYS 23: 985-6 (1955)

COLORS OF TRANSITION METAL HEXAFLUORIDES.
(RHENIUM HEXAFLUORIDE OSMIUM HEXAFLUORIDE,
IRIDIUM HEXAFLUORIDE, AND PLATINUM HEXAFLUORIDE)
MOFFITT W + GOODMAN GL + FRED M + WEINSTOCK B
MOL PHYS 2: 109-22 (1959)

HEAT CAPACITIES AND ELECTRONIC SPECTRA OF THE
PLATINUM GROUP METAL HEXAFLUORIDE MOLECULES DOWN
TO HELIUM TEMPERATURES.
WEINSTOCK B + WESTRUM EF + GOODMAN GL
P405-6 OF INT CONF LOW TEMP PHYS, 8TH PROC,
LONDON, 1962.

IRON DIFLUORIDE FeF_2

STRUCTURE OF SEVERAL RUTILE-TYPE FLUORIDES:
MANGANESE DIFLUORIDE, IRON DIFLUORIDE, COBALT
DIFLUORIDE, NICKEL DIFLUORIDE, ZINC DIFLUORIDE.
BAUR WH
NATURWISSENSCHAFTEN 44: 349-50 (1957)

FORCE FIELDS OF THE RUTILE COUNTERPARTS:
TITANIUM DIOXIDE, MAGNESIUM FLUORIDE, ZINC
FLUORIDE, AND FERROUS FLUORIDE.
IIISHI K + TOMISAKA T + UMEGAKI Y
MINERALOG J 6(1-2): 77-84 (1969)

0541 SUBLIMATION PRESSURE OF IRON DIFLUORIDE.
KENT RA + MARGRAVE JL
J AMER CHEM SOC 87: 4754-6 (1965)

SPECTROSCOPIC STUDIES OF THE VAPORIZATION OF
HIGH TEMPERATURE MATERIALS. (CHROMIUM
DIFLUORIDE, CHROMIUM TRIFLUORIDE, AND IRON
DIFLUORIDE)
LINEVSKY MJ
GENERAL ELECTRIC CORP, 1968, 45P. AD-670626

RAMAN SPECTRA OF TITANIUM DIOXIDE, MAGNESIUM
DIFLUORIDE, ZINC DIFLUORIDE, IRON DIFLUORIDE,
AND MANGANESE DIFLUORIDE.
PORTO SPS + FLEURY PA + DAMEN TC
PHYS REV 154: 522-26 (1967)

CRYSTAL STRUCTURE OF MANGANESE DIFLUORIDE, IRON
DIFLUORIDE, COBALT DIFLUORIDE, NICKEL
DIFLUORIDE, AND ZINC DIFLUORIDE.
STOUT JW + REED SA
J AMER CHEM SOC 76: 5279-81 (1954)

IRON TRIFLUORIDE FeF_3

0542 ELECTRONIC STRUCTURE OF IRON TRIFLUORIDE.
HAND RW + HUNT WJ + SCHAEFER HF
J AMER CHEM SOC 95: 4517-22 (1973)

0543 ELECTRONIC STRUCTURE AND ANOMALOUS THERMAL
EXPANSION ON IRON DIFLUORIDE AND VANADIUM
DIOXIDE.
HAZONY Y + PERKINS HK
J APPL PHYS 41: 5130-1 (1970)

SUBLIMATION PRESSURES OF CHROMIUM TRIFLUORIDE,
MANGANESE TRIFLUORIDE, AND IRON TRIFLUORIDE.
ZMBOV KF + MARGRAVE JL
J INORG NUCL CHEM 29: 673-80 (1967)

KRYPTON MONOFLUORIDE

KrF

0544 ABSORPTION AND EMISSION SPECTRA OF ARGON MATRIX
ISOLATED XENON MONOFLUORIDE AND KRYPTON
MONOFLUORIDE.
AULT BS + ANDREWS L
J CHEM PHYS 64(7): 3075-6 (1976)

0545 EMISSION SPECTRA OF XENON MONOBROMIDE, XENON
MONOCHLORIDE, XENON MONOFLUORIDE, AND KRYPTON
MONOFLUORIDE.
BRAU CA + EWING JJ
J CHEM PHYS 63(11): 4640-7 (1975)

0546 ELECTRON SPIN RESONANCE SPECTRUM OF KRYPTON
MONOFLUORIDE.
FALCONER WE + MORTON JR + STRENG AG
J CHEM PHYS 41(3): 902-3 (1964)

0547 KRYPTON MONOFLUORIDE AND ITS POSITIVE ION. (SCF
CALCULATIONS)
LIU B + SCHAEFER HF
J CHEM PHYS 55: 2369-74 (1971)

KRYPTON DIFLUORIDE

KrF$_2$

0548 ELECTRONIC STRUCTURE AND PROPERTIES OF KRYPTON
DIFLUORIDE.
BAGUS PS + LIU B + SCHAEFER HF
J AMER CHEM SOC 94: 6635-41 (1972)

0549 MOLECULAR ORBITAL ENERGY LEVELS AND BONDING IN
KRYPTON DIFLUORIDE.
BRUNDLE CR + JONES GR
CHEM COMMUN 1971: 1198-9

0550 ELECTRONIC STRUCTURE OF KRYPTON DIFLUORIDE
STUDIED BY PHOTOELECTRON SPECTROSCOPY.
BRUNDLE CR + JONES GR
J CHEM SOC, FARADAY TRANS 68 (PT-2): 959-66
(1972)

0551 CRYSTAL STRUCTURE OF KRYPTON DIFLUORIDE AT -80C.
BURBANK RD + FALCONER WE + SUNDER WA
SCIENCE 178: 1285-6 (1972)

D ORBITALS IN THE NOBLE GAS DIHALIDES. (ARGON
DIFLUORIDE, KRYPTON DIFLUORIDE, AND XENON
DIFLUORIDE)
CATTON RC + MITCHELL KAR
CAN J CHEM 48: 2695-2701 (1970)

0552 INFRARED AND RAMAN SPECTRA OF KRYPTON
DIFLUORIDE.
CLAASSEN HH + GOODMAN GL + MALM JG + SCHREINER F
J CHEM PHYS 42: 1229-32 (1965)

0553 AB INITIO CALCULATIONS OF THE BONDING IN KRYPTON
DIFLUORIDE.
COLLINS GAD + CRUICKSHANK DWJ + BREEZE A
CHEM COMMUN 1970: 884-5

0554 BONDING IN KRYPTON DIFLUORIDE.
COLLINS GAD + CRUICKSHANK DWJ + BREEZE A
J CHEM SOC, FARADAY TRANS 70(2): 393-7 (1974)

0555 BONDING IN KRYPTON DIFLUORIDE.
COLLINS GAD + CRUICKSHANK DWJ + BREEZE A
J CHEM SOC, FARADAY TRANS 68(PT 7): 1189-95
(1972)

0556 FORCE FIELDS IN KRYPTON DIFLUORIDE AND XENON
DIFLUORIDE.
COULSON CA
J CHEM PHYS 44: 468-9 (1966)

HELIUM-2 PHOTOELECTRON SPECTRA OF XENON
DIFLUORIDE AND KRYPTON DIFLUORIDE.
DEKOCK RL
J CHEM PHYS 58: 1267-8 (1973)

0557 HEAT FORMATION OF KRYPTON DIFLUORIDE.
GUNN SR
J AMER CHEM SOC 88: 5924 (1966); J PHYS CHEM 71:
2934-7 (1967)

0558 STRUCTURE OF KRYPTON DIFLUORIDE AS INVESTIGATED
BY ELECTRON DIFFRACTION.
HARSHBARGER WR + BOHN RK + BAUER SH
J AMER CHEM SOC 89: 6466-9 (1967)

0559 ULTRAVIOLET EMISSION SPECTRA. (XENON DIFLUORIDE
AND KRYPTON DIFLUORIDE)
KRISHNAMACHARI SLNG + NARASIMHAM NA + SINGH M
CURR SCI 34: 75-7 (1965)

0560 ABSORPTION SPECTRA OF CERTAIN FLUORIDES IN THE
NEAR ULTRAVIOLET REGION. (KRYPTON DIFLUORIDE,
NITROGEN TRIFLUORIDE)
MAKEEV GN + SINYANSKII VF + SMIRNOV BM
DOKL AKAD NAUK SSSR 222(1): 151-4 (1975) (IN
RUSSIAN)

0561 STRUCTURE OF KRYPTON DIFLUORIDE.
MURCHISON C + REICHMAN S + ANDERSON D
+ OVEREND J + SCHREINER F
J AMER CHEM SOC 90: 5690-3 (1968)

0562 MEAN AMPLITUDES OF VIBRATION, BASTIANSEN-MORINO
SHRINKAGE EFFECT AND THERMODYNAMIC FUNCTIONS OF
KRYPTON DIFLUORIDE.
NAGARAJAN G
ACTA PHYS POLON 29: 831-6 (1966)

0563 KRYPTON DIFLUORIDE. (PHYSICAL AND CHEMICAL
PROPERTIES)
PRUSAKOV VN + SOKOLOV VB
AT ENERG 31(3): 259-68 (1971)

0564 ABSENCE OF FERMI RESONANCE IN KRYPTON
DIFLUORIDE.
REICHMAN S + OVEREND J
J CHEM PHYS 47: 3690 (1967)

0565 MASS SPECTRUM AND MOLECULAR ENERGETICS OF
KRYPTON DIFLUORIDE.
SESSA PA + MCGEE JA
J PHYS CHEM 73: 2078-9 (1969)

0566 PREPARATION OF INERT GAS COMPOUNDS BY MATRIX
ISOLATION: KRYPTON DIFLUORIDE.
TURNER JJ + PIMENTEL GC
P101-5 OF NOBLE GAS COMPOUNDS, HYMAN HH(ED),
UNIV OF CHICAGO PRESS, 1963.

0567 KRYPTON DIFLUORIDE. PREPARATION BY THE MATRIX
ISOLATION TECHNIQUE.
TURNER JJ + PIMENTEL GC
SCIENCE 140: 974-5 (1963)

BOND ENERGIES AND IONIC CHARACTER OF INERT GAS
HALIDES.
WATERS JH + GRAY HB
J AMER CHEM SOC 85: 825-6 (1963)

LANTHANUM MONOFLUORIDE

LaF

ELECTRONIC STATES OF GASEOUS SCANDIUM
MONOFLUORIDE, YTTRIUM MONOFLUORIDE, AND
LANTHANUM MONOFLUORIDE.
BARROW RF + BASTIN MW + MOORE DLG + POTT CJ
NATURE 215: 1072-3 (1967)

EVALUATION OF MOLECULAR VIBRATION FREQUENCIES OF
SCADIUM MONOFLUORIDE, YTTRIUM MONOFLUORIDE, AND
LANTHANUM MONOFLUORIDE.
KRASNOV KS
IZV VYSSH UCHEB ZAVED, KHIM 1: 594-6 (1967) (IN
RUSSIAN)

DISSOCIATION ENERGIES AND STABILITIES OF
SCANDIUM MONOFLUORIDE, YTTRIUM MONOFLUORIDE, AND
LANTHANUM MONOFLUORIDE.
KRASNOV KS
IZV VYSSH UCHEB ZAVED, KHIM 12: 578-82 (1969)
(IN RUSSIAN)

THERMODYNAMIC FUNCTIONS OF GASEOUS SCANDIUM
MONOFLUORIDE, SCANDIUM DIFLUORIDE, SCANDIUM
TRIFLUORIDE, YTTRIUM MONOFLUORIDE, YTTRIUM
DIFLUORIDE, YTTRIUM TRIFLUORIDE, LANTHANUM
MONOFLUORIDE, LANTHANUM DIFLUORIDE, AND
LANTHANUM TRIFLUORIDE.
KRASNOV KS + DANILOVA TG
HIGH TEMP 7(6): 1131-3 (1969)

MOLECULAR CONSTANTS OF SCANDIUM, YTTRIUM, AND
LANTHANUM HALIDES. (SCANDIUM DIFLUORIDE,
SCANDIUM MONOFLUORIDE, YTTRIUM DIFLUORIDE,
YTTRIUM MONOFLUORIDE, LANTHANUM DIFLUORIDE, AND
LANTHANUM MONOFLUORIDE)
KRASNOV KS + TIMOSHININ VS
HIGH TEMP 7(2): 333-4 (1969)

0568 THERMODYNAMIC FUNCTIONS OF GASEOUS SCANDIUM,
YTTRIUM, AND LANTHANUM HALIDES IN THE 293-3000
DEGREES K RANGE.
KRASNOV KS + DANILOVA TG
HIGH TEMP 7(6): 1131-3 (1969)

VIBRONIC SPECTRUM OF SCANDIUM MONOFLUORIDE,
YTTRIUM MONOFLUORIDE, AND LANTHANUM
MONOFLUORIDE.
SHENYAVSKAYA EA + MAL'TSEV AA
VEST MOSK UNIV KHIM 22: 104-5 (1967) (IN
RUSSIAN)

0569 ELECTRONIC SPECTRUM OF LANTHANUM MONOFLUORIDE.
SHENYAVSKAYA EA + GURVICH LV + MAL'TSEV AA
VEST MOSK UNIV KHIM 20: 10-13 (1965) (IN
RUSSIAN)

0570 ELECTRONIC SPECTRA OF LANTHANUM FLUORIDE IN THE
EXTREME ULTRAVIOLET.
STEPHAN G + NISAR M + ROTH A
COMPT REND B 274(12): 807-10 (1972) (IN FRENCH)

LANTHANUM DIFLUORIDE LaF_2

THERMODYNAMIC FUNCTIONS OF GASEOUS SCANDIUM
MONOFLUORIDE, SCANDIUM DIFLUORIDE, SCANDIUM
TRIFLUORIDE, YTTRIUM MONOFLUORIDE, YTTRIUM
DIFLUORIDE, YTTRIUM TRIFLUORIDE, LANTHANUM
MONOFLUORIDE, LANTHANUM DIFLUORIDE, AND
LANTHANUM TRIFLUORIDE.
KRASNOV KS + DANILOVA TG
HIGH TEMP 7(6): 1131-3 (1969)

MOLECULAR CONSTANTS OF SCANDIUM, YTTRIUM, AND
LANTHANUM HALIDES. (SCANDIUM DIFLUORIDE,
SCANDIUM MONOFLUORIDE, YTTRIUM DIFLUORIDE,
YTTRIUM MONOFLUORIDE, LANTHANUM DIFLUORIDE, AND
LANTHANUM MONOFLUORIDE)
KRASNOV KS + TIMOSHININ VS
HIGH TEMP 7(2): 333-4 (1969)

LANTHANUM TRIFLUORIDE LaF_3

0571 SYMMETRY AND CRYSTAL STRUCTURE OF RARE EARTH
TRIFLUORIDES. (LANTHANUM TRIFLUORIDE)
AFANASHEV ML + HABUDA SP + LUNDIN AG
ACTA CRYSTALLOGR 28: 2903-5 (1972)

0572 LATTICE VIBRATIONS AND STRUCTURE OF RARE EARTH
HALIDES. (LANTHANUM TRIFLUORIDE)
BAUMAN RP + PORTO SPS
PHYS REV 161: 842-8 (1967)

STANDARD ENTHALPIES OF FORMATION OF SCANDIUM
TRIFLUORIDE, YTTRIUM TRIFLUORIDE, AND LANTHANUM
TRIFLUORIDE.
FINOGENOV AD
RUSS J PHYS CHEM 45: 900-1 (1971)

GEOMETRIES AND ENTROPIES OF METAL TRIFLUORIDES
FROM INFRARED SPECTRA. (SCANDIUM, YTTRIUM,
LANTHANUM, CERIUM, NEODYMIUM, EUROPIUM, AND
GADOLINIUM TRIFLUORIDES)
HASTIE JW + HAUGE RH + MARGRAVE JL
J LESS COMMON METALS 39(2): 309-34 (1975)

FORCE CONSTANTS AND GEOMETRIES OF MATRIX
ISOLATED RARE EARTH TRIFLUORIDES. (SCANDIUM
TRIFLUORIDE, YTTRIUM TRIFLUORIDE, AND LANTHANUM
TRIFLUORIDE)
HAUGE RH + HASTIE JW + MARGRAVE JL
J LESS COMMON METALS 23: 359-65 (1971)

ELECTRIC DEFLECTION OF MOLECULAR BEAMS OF THE
RARE EARTH DIFLUORIDES AND TRIFLUORIDES.
(LANTHANIDE TRIFLUORIDES, LANTHANIDE
DIFLUORIDES, SCANDIUM TRIFLUORIDE, AND YTTRIUM
TRIFLUORIDE)
KAISER EW + FALCONER WE + KLEMPERER W
J CHEM PHYS 56: 5392-8 (1972)

SUBLIMATION PRESSURES OF SCANDIUM TRIFLUORIDE,
YTTRIUM TRIFLUORIDE, AND LANTHANUM TRIFLUORIDE.
KENT RA + ZMBOV KF + KANA'AN AS + BESENBRUCH G
MCDONALD JD + MARGRAVE JL
J INORG NUCL CHEM 28: 1419-27 (1966)

CALCULATION OF DISSOCIATION ENERGIES FOR THE
HALIDES OF THE SCANDIUM SUBGROUP. (SCANDIUM
TRIFLUORIDE, YTTRIUM TRIFLUORIDE, AND LANTHANUM
TRIFLUORIDE)
KRASNOV KS
HIGH TEMP 4(1): 128-30 (1966)

CALCULATION OF THE VIBRATIONAL FREQUENCIES OF
SCANDIUM SUBGROUP HALIDES.
KRASNOV KS
HIGH TEMP 5(4): 639-40 (1967)

0573 FORCE CONSTANT OF THE NONPLANAR VIBRATION OF AN
XY3 MOLECULE AND A MODEL WITH POLARIZABLE IONS.
KRASNOV KS
IZV VYSSH UCHEB ZAVED KHIM KHIM TEKHNOL 10(9):
997-1000 (1967) (IN RUSSIAN

THERMODYNAMIC FUNCTIONS OF GASEOUS SCANDIUM
MONOFLUORIDE, SCANDIUM DIFLUORIDE, SCANDIUM
TRIFLUORIDE, YTTRIUM MONOFLUORIDE, YTTRIUM
DIFLUORIDE, YTTRIUM TRIFLUORIDE, LANTHANUM
MONOFLUORIDE, LANTHANUM DIFLUORIDE, AND
LANTHANUM TRIFLUORIDE.
KRASNOV KS + DANILOVA TG
HIGH TEMP 7(6): 1131-3 (1969)

0574 VALENCE VIBRATION FREQUENCIES OF TRIHALIDE
MOLECULES OF LANTHANUM HALIDES.
KRASNOV KS + ZAITSEV AA
RUSS J PHYS CHEM 39(10): 1322-4 (1965)

0575 VAPORIZATION STUDIES OF LANTHANUM TRIFLUORIDE.
MAR RW
UNIV CALIF, BERKELEY, MS THESIS, 1966, 21P.
UCRL-16649

0576 VAPOR PRESSURE, HEAT OF SUBLIMATION, AND
EVAPORATION COEFFICIENT OF LANTHANUM
TRIFLUORIDE.
MAR RW + SEARCY AW
J PHYS CHEM 71: 888-94 (1967)

0577 ENTHALPY OF FORMATION OF LANTHANUM TRIFLUORIDE
AND PRASEODYMIUM TRIFLUORIDE.
POLYACHENOK OG
RUSS J INORG CHEM 10: 1057 (1965)

0578 EXPERIMENTAL DETERMINATION OF THE ENTHALPIES OF
FORMATION OF RARE EARTH FLUORIDES. (LANTHANUM
TRIFLUORIDE)
POLYACHENOK OG
RUSS J INORG CHEM 12: 449-52 (1967)

0579 NUCLEAR RELAXATION TIMES OF LANTHANUM
TRIFLUORIDE.
SHEN L
UNIV WASHINGTON, PHD THESIS, 1967, 76P.

0580 BAND SYSTEM IN THE DISCHARGE SPECTRUM OF
LANTHANUM TRIFLUORIDE VAPORS.
SHENYAVSKAYA EA + GURVICH LV + MAL'TSEV AA
OPT SPECTROSC 24(6): 556 (1968)

0581 DEMONSTRATION OF THE EXISTENCE OF DILANTHANUM
HEXAFLUORIDE GAS AND DETERMINATION OF ITS
STABILITY. (LANTHANUM TRIFLUORIDE)
SKINNER HB + SEARCY AW
J PHYS CHEM 75: 108-11 (1971)

0582 ELECTRONIC SPECTRUM OF LANTHANUM TRIFLUORIDE.
STEPHAN G + NISAR M + ROTH A + KASTLER A
COMPT REND C 274: 807-10 (1972) (IN FRENCH)

SATURATED VAPOR PRESSURE OF SCANDIUM
TRIFLUORIDE, YTTRIUM TRIFLUORIDE, AND LANTHANUM
TRIFLUORIDE.
SUVOROV AL + NOVIKOV GI
VEST LENINGRAD UNIV, FIZ KHIM 23: 83-8 (1968)
(IN RUSSIAN)

0583 GEOMETRY AND INFRARED SPECTRA OF MATRIX ISOLATED
RARE EARTH HALIDES. (ALSO LANTHANUM TRIFLUORIDE)
WESLEY RD + DEKOCK CW
J CHEM PHYS 55: 3866-77 (1971)

0584 NEAR ULTRAVIOLET OPTICAL CONSTANTS OF LANTHANUM
TRIFLUORIDE.
WIRICK MP
APPL OPT 5(12): 1966-7 (1966)

MASS SPECTROMETRIC STUDIES OF SCANDIUM
TRIFLUORIDE, YTTRIUM TRIFLUORIDE, LANTHANUM
TRIFLUORIDE AND THE THE RARE EARTH TRIFLUORIDES.
ZMBOV KF + MARGRAVE JL
P267 OF MASS SPECTROMETRY IN INORGANIC
CHEMISTRY, MARGRAVE JL (ED), AMER CHEM SOC,
1968, 329P. (ADVANCES IN CHEMISTRY SERIES NO 72)

LEAD MONOFLUORIDE   PbF

SPECTRA OF SILICON MONOFLUORIDE, GERMANIUM
MONOFLUORIDE, TIN MONOFLUORIDE, AND LEAD
MONOFLUORIDE.
BARROW RF + BUTLER D + JOHNS JWC + POWELL JL
PROC PHYS SOC 73: 317-20 (1959)

0585 B SINGLET SIGMA(+) TO X TRIPLET SIGMA(-) BAND
SYSTEM IN LEAD MONOFLUORIDE.
COLIN R + DEVILLERS J + PREVOT F
J MOL SPECTROSC 44: 230-35 (1972)

0586 ROTATIONAL ANALYSIS OF THE B-X SYSTEM OF LEAD
MONOFLUORIDE IN ULTRAVIOLET.
REDDY YP + RAO PT
P129-32 OF INT CONF SPECTROSCOPY, 1ST PROC, VOL
1, 1967, BOMBAY, INDIA.

0587 ROTATIONAL ANALYSIS OF THE VISIBLE EMISSION
BANDS OF LEAD MONOFLUORIDE.
RAO KM + RAO PT
CAN J PHYS 42: 690-95 (1964)

0588 EMISSION SPECTRUM OF LEAD MONOFLUORIDE.
SINGH SP
INDIAN J PURE APPL PHYS 5: 292-4 (1967)

0589 ON THE DISSOCIATION OF LEAD MONOFLUORIDE AND
BISMUTH MONOFLUORIDE.
SINGH RB + RAI DK
CAN J PHYS 43: 829-35 (1965)

0590 ROTATIONAL ANALYSIS OF B-X2 SYSTEM OF LEAD-208
MONOFLUORIDE.
SINGH ON + SINGH IS + SINGH ON
CAN J PHYS 50: 2206-2210 (1972)

0591 ROTATIONAL ANALYSIS OF A-X1 BANDS OF LEAD
MONOFLUORIDE.
SINGH ON + SRIVASTAVA MP + SINGH IS
CAN J PHYS 47: 1639-41 (1969)

0592 MASS SPECTROMETRIC STUDIES AT HIGH TEMPERATURES.
(TIN DIFLUORIDE, LEAD DIFLUORIDE, TIN
MONOFLUORIDE, AND LEAD MONOFLUORIDE)
ZMBOV KF + HASTIE JW + MARGRAVE JL
TRANS FARADAY SOC 64(4): 861-7 (1968)

LEAD DIFLUORIDE   PbF$_2$

0593 NEUTRON DIFFRACTION INVESTIGATION OF
ORTHORHOMBIC LEAD DIFLUORIDE.
BOLDRINI P + LOOPSTRA BO
ACTA CRYSTALLOGR 22: 744-5 (1967)

DETERMINATION OF THE GEOMETRY OF HIGH
TEMPERATURE SPECIES BY ELECTRIC DEFLECTION AND
MASS SPECTROMETRIC DETECTION. (FIRST ROW
FLUORIDES, BERYLLIUM, MAGNESIUM, AND LEAD
DIFLUORIDES)
BUECHLER A + STAUFFER JL + KLEMPERER W
J AMER CHEM SOC 86: 4544-50 (1964)

0594 INFRARED SPECTRA OF MATRIX ISOLATED TIN
DIFLUORIDE AND LEAD DIFLUORIDE.
HAUGE RH + HASTIE JW + MARGRAVE JL
J MOL SPECTROSC 45: 420-7 (1973)

0595 ULTRAVIOLET ABSORPTION SPECTRA OF GASEOUS TIN
DIFLUORIDE AND LEAD DIFLUORIDE.
HAUGE RH + HASTIE JW + MARGRAVE JL
J PHYS CHEM 72: 3510-11 (1968)

FREE ENERGIES OF FORMATION OF SOME INORGANIC
FLUORIDES BY SOLID STATE EMF MEASUREMENTS.
(THORIUM TETRAFLUORIDE, ALUMINUM TRIFLUORIDE,
NICKEL DIFLUORIDE, LEAD DIFLUORIDE, COLBALT
DIFLUORIDE, AND URANIUM TRIFLUORIDE)
HEUS RJ + EGAN JJ
Z PHYS CHEM NEUE FOLGE 49: 38-43 (1966)

RAMAN SPECTRA OF CADMIUM DIFLUORIDE AND LEAD
DIFLUORIDE.
KRISHNAMURTHY N + SOOTS V
CAN J PHYS 48: 1104-7 (1970)

FREQUENCIES OF THE DEFORMATION VIBRATIONS OF THE
DIHALIDES OF GERMANIUM, TIN, AND LEAD.
(GERMANIUM DIFLUORIDE, TIN DIFLUORIDE, LEAD
DIFLUORIDE)
TIMOSHININ VS + DANILOVA TG
RUSS J PHYS CHEM 42: 1596 (1968)

MASS SPECTROMETRIC STUDIES AT HIGH TEMPERATURES.
(TIN DIFLUORIDE, LEAD DIFLUORIDE, TIN
MONOFLUORIDE, AND LEAD MONOFLUORIDE)
ZMBOV KF + HASTIE JW + MARGRAVE JL
TRANS FARADAY SOC 64(4): 861-7 (1968)

LEAD TETRAFLUORIDE   PbF$_4$

MEAN AMPLITUDES OF VIBRATION OF SOME TETRAHEDRAL
XY4 TYPE MOLECULES. PART-5: TETRAHALIDES OF
HAFNIUM AND LEAD.
NAGARAJAN G
INDIAN J PURE APPL PHYS 2(5): 145-9 (1964)

0596 UREY-BRADLEY FORCE FIELD. (LEAD TETRAFLUORIDE)
RADHAKRISHNAN M
INDIAN J PURE APPL PHYS 1: 402-3 (1963)

0597 THERMODYNAMIC FUNCTIONS OF LEAD TETRAHALIDES.
SINGH SP
LABDEV J SCI TECH 7A(4): 185-6 (1969)

LUTETIUM MONOFLUORIDE   **LuF**

0598 ELECTRONIC SPECTRUM OF LUTETIUM MONOFLUORIDE.
      D'INCAN J + EFFANTIN C + BACIS R
      J PHYS B 5: L189-90 (1972)

0599 ROTATIONAL ANALYSIS OF SELECTED BANDS FROM THE
      ELECTRONIC SPECTRUM OF THE LUTETIUM MONOFLUORIDE
      MOLECULE.
      EFFANTIN C + WANNOUS G + D'INCAN J + ATHENOUR C
      CAN J PHYS 54(3): 279-94 (1976)

LUTETIUM TRIFLUORIDE   **LuF$_3$**

SUBLIMATION PRESSURES AND HEATS OF SUBLIMATION
OF THULIUM TRIFLUORIDE, YTTERBIUM TRIFLUORIDE,
AND LUTETIUM TRIFLUORIDE.
ZMBOV KF + MARGRAVE JL
J LESS COMMON METALS 12: 494-6 (1967)

MAGNESIUM MONOFLUORIDE   **MgF**

0600 ROTATIONAL ANALYSIS OF ELECTRONIC BANDS OF
      GASEOUS MAGNESIUM MONOFLUORIDE.
      BARROW RF + BEALE JR
      PROC PHYS SOC 91(2): 483-8 (1967)

      DISSOCIATION ENERGIES OF THE ALKALINE EARTH
      MONOFLUORIDES.
      EHLERT TC + BLUE GD + GREEN JW + MARGRAVE JL
      J CHEM PHYS 41: 2250-5 (1964)

      DISSOCIATION ENERGY OF MONOFLUORIDE COMPOUNDS OF
      MAGNESIUM AND CALCIUM AND THE IONIC MODEL OF A
      MOLECULE.
      KRASNOV KS
      RUSS J PHYS CHEM 39(7): 842-4 (1965)

0601 DISSOCIATION ENERGY OF MAGNESIUM MONOFLUORIDE.
      SINGH ON + ASTHANA BP + SINGH ON
      SPECTROSC LETT 7(3): 175-86 (1974)

MAGNESIUM DIFLUORIDE   **MgF$_2$**

ELECTRON DIFFRACTION ANALYSIS OF THE MOLECULAR
STRUCTURES OF GROUP-2 ELEMENTS. (BERYLLIUM
DIFLUORIDE, MAGNESIUM DIFLUORIDE, CALCIUM
DIFLUORIDE, STRONTIUM DIFLUORIDE, BARIUM
DIFLUORIDE, ZINC DIFLUORIDE, CADMIUM DIFLUORIDE)
AKISHIN PA + SPIRIDONOV VP
SOV PHYS CRYSTALLOGR 2(4): 472-9 (1957)

0602 STRUCTURE OF THE MAGNESIUM DIFLUORIDE MOLECULE.
      ASTIER M + BERTHIER G + MILLE P
      J CHEM PHYS 57: 5008-9 (1972)

0603 MASS SPECTROMETRIC STUDY OF MAGNESIUM HALIDES.
      (MAGNESIUM DIFLUORIDE)
      BERKOWITZ J + MARQUART JR
      J CHEM PHYS 37: 1853-65 (1962)

DETERMINATION OF THE GEOMETRY OF HIGH
TEMPERATURE SPECIES BY ELECTRIC DEFLECTION AND
MASS SPECTROMETRIC DETECTION. (FIRST ROW
FLUORIDES, BERYLLIUM, MAGNESIUM, AND LEAD
DIFLUORIDES)
BUECHLER A + STAUFFER JL + KLEMPERER W
J AMER CHEM SOC 86: 4544-50 (1964)

MEAN AMPLITUDES AND SHRINKAGE EFFECTS OF
VIBRATION FOR SOME LINEAR SYMMETRICAL
DIFLUORIDES. (BERYLLIUM DIFLUORIDE AND MAGNESIUM
DIFLUORIDE)
CYVIN SJ + VIZI B
VESZPREMI VEGYIPARI EGYETEM KOZLEMENYEI 11: 83-9
(1968)

NONEMPERICAL LCAO MO SCF STUDIES OF THE GROUP-2A
DIHALIDES: BERYLLIUM DIFLUORIDE, MAGNESIUM
DIFLUORIDE, AND CALCIUM DIFLUORIDE.
GOLE JL + SIU AKQ + HAYES EF
J CHEM PHYS 58: 857-68 (1973)

RECENT DEVELOPMENTS IN MULTIPLE BEAM
INTERFEROMETRY IN THE ULTRAVIOLET REGION
1700-2500 ANGSTROMS. (ALUMINUM- MAGNESIUM
FLUORIDE)
GUERN Y + BIDEAU-MEHU A + ABJEAN R + JOHANNIN-GI
OPT COMMUN 12(1): 66-70 (1974)

0604 INFRARED SPECTRA OF MATRIX ISOLATED MAGNESIUM
      DIFLUORIDE.
      HAUGE RH + MARGRAVE JL + KANA'AN AD
      J CHEM SOC, FARADAY TRANS 71(5,PART-2): 1082-90
      (1975)

0605 THEORETICAL STUDIES OF THE INTERACTION OF
      MAGNESIUM DIFLUORIDE WITH RARE GAS ATOMS.
      HAYES EF + SIU AKQ + KISKER DW
      J CHEM PHYS 59: 4587-9 (1973)

0606 FORCE FIELDS OF THE RUTILE COUNTERPARTS:
      TITANIUM DIOXIDE, MAGNESIUM FLUORIDE, ZINC
      FLUORIDE, AND FERROUS FLUORIDE.
      IIISHI K + TOMISAKA T + UMEGAKI Y
      MINERALOG J 6(1-2): 77-84 (1969)

      MELTING POINTS OF INORGANIC FLUORIDES. (CADMIUM
      DIFLUORIDE, MAGNESIUM DIFLUORIDE, STRONTIUM
      DIFLUORIDE, AND BARIUM DIFLUORIDE)
      KOJIMA H + WHITEWAY SG + MASSON CR
      CAN J CHEM 46: 2968-71 (1968)

0607 INFRARED AND RAMAN SPECTRA AND STRUCTURES OF
      MATRIX ISOLATED MAGNESIUM DIHALIDES: MAGNESIUM
      DIFLUORIDE, MAGNESIUM DICHLORIDE, MAGNESIUM
      DIBROMIDE, AND MAGNESIUM DIIODIDE.
      LESIECKI ML + NIBLER JW
      J CHEM PHYS 64(2): 871-84 (1976)

0608 INFRARED SPECTRA OF MAGNESIUM AND ALUMINUM
      FLUORIDES BY MATRIX ISOLATION.
      LINEVSKY MJ
      P87-96 OF WORKING GROUP ON THERMOCHEMISTRY, 2ND
      PROC, 1964, CHEM PROPULSION INFO AGENCY,
      AD451711.

0609 GEOMETRY AND VIBRATIONAL SPECTRA OF THE ALKALINE
      EARTH DIHALIDES. PART-1: MAGNESIUM DIFLUORIDE.
      MANN DE + CALDER GV + SESHADRI KS + WHITE D
      + LINEVSKY MJ
      J CHEM PHYS 46(3): 1138-43 (1967)

      MEAN AMPLITUDES OF VIBRATION FOR THE DIHALIDES
      OF MERCURY, BERYLLIUM, AND MAGNESIUM.
      NAGARAJAN G
      ACTA PHYS AUSTRIACA 17: 246-53 (1973)

      MEAN AMPLITUDES OF VIBRATION AND
      BASTIANSEN-MORINO SHRINKAGE EFFECT IN SOME
      LINEAR SYMMETRICAL DIHALIDES. (BERYLLIUM
      DIFLUORIDE AND MAGNESIUM DIFLUORIDE)
      NAGARAJAN G
      J MOL SPECTROSC 13: 361-92 (1964)

VAPOR PRESSURE OF ZINC DIFLUORIDE, CADMIUM
DIFLUORIDE, MAGNESIUM DIFLUORIDE, CALCIUM
DIFLUORIDE, STRONTIUM DIFLUORIDE, BARIUM
DIFLUORIDE, AND ALUMINUM TRIFLUORIDE.
RUFF O + LE BOUCHER L
Z ANORG ALLGEM CHEM 219: 376-81 (1934) (IN
GERMAN)

INFRARED SPECTRA OF SOME ALKALINE EARTH HALIDES
BY THE MATRIX ISOLATION TECHNIQUE.
SNELSON A
J PHYS CHEM 70: 3208-16 (1966)

MANGANESE MONOFLUORIDE $\qquad$ **MnF**

SUBLIMATION PRESSURE OF MANGANESE DIFLUORIDE AND
MANGANESE MONOFLUORIDE.
KENT RA + EHLERT TC + MARGRAVE JL
J AMER CHEM SOC 86(23): 5093-5 (1964)

0610 COMPLEX BAND SYSTEM OF MANGANESE MONOFLUORIDE IN
THE NEAR ULTRAVIOLET.
RAO VK + REDDY SP + RAO PT
PROC PHYS SOC 79: 741-4 (1962)

MANGANESE DIFLUORIDE $\qquad$ **MnF$_2$**

0611 A LANGMUIR MEASUREMENT OF THE SUBLIMATION
PRESSURE OF MANGANESE DIFLUORIDE.
BAUTISTA RG + MARGRAVE JL
J PHYS CHEM 67: 1564-5 (1963)

0612 MASS SPECTROMETRIC AND THERMOCHEMICAL STUDIES OF
THE MANGANESE FLUORIDES.
EHLERT TC + HSIA M
J FLUORINE CHEM 2(1): 33-51 (1972)

0613 KNUDSEN MEASUREMENT OF THE VAPOR PRESSURE OF
MANGANESE DIFLUORIDE.
HITCHINGHAM WC + KANA'AN AS
HIGH TEMP SCI 1: 216-221 (1969)

0614 SUBLIMATION PRESSURE OF MANGANESE DIFLUORIDE AND
MANGANESE MONOFLUORIDE.
KENT RA + EHLERT TC + MARGRAVE JL
J AMER CHEM SOC 86(23): 5093-5 (1964)

0615 RAMAN SPECTRA OF TITANIUM DIOXIDE, MAGNESIUM
DIFLUORIDE, ZINC DIFLUORIDE, DIFLUORIDE, AND
MANGANESE DIFLUORIDE.
PORTO SPS + FLEURY PA + DAMEN TC
PHYS REV 154: 522-6 (1967)

CRYSTAL STRUCTURE OF MANGANESE DIFLUORIDE, IRON
DIFLUORIDE, COBALT DIFLUORIDE, NICKEL
DIFLUORIDE, AND ZINC DIFLUORIDE.
STOUT JW + REED SA
J AMER CHEM SOC 76: 5279-81 (1954)

INTERACTION OF MATRIX ISOLATED NICKEL FLUORIDE
AND NICKEL CHLORIDE WITH CARBON MONOXIDE,
MOLECULAR NITROGEN, NITRIC OXIDE, AND MOLECULAR
OXYGEN AND OF CALCIUM FLUORIDE, CHROMIUM(II)
FLUORIDE, MANGANESE(II) FLUORIDE, COPPER(II)
FLUORIDE, AND ZINC(II) FLUORIDE WITH CARBON
MONOXIDE IN ARGON MATRICES.
VAN LEIRSBURG DA + DEKOCK CW
J PHYS CHEM 78(2): 134-42 (1974)

MANGANESE TRIFLUORIDE $\qquad$ **MnF$_3$**

0616 CRYSTAL STRUCTURE OF MANGANESE TRIFLUORIDE.
HEPWORTH MA + JACK KH
ACTA CRYSTALLOGR 10: 345-51 (1957)

0617 INTERATOMIC BONDING IN CRYSTALLINE MANGANESE
TRIFLUORIDE.
HEPWORTH MA + JACK KH + NYHOLM RS
NATURE 178: 211-12 (1957)

SUBLIMATION PRESSURES OF CHROMIUM TRIFLUORIDE,
MANGANESE TRIFLUORIDE, AND IRON TRIFLUORIDE.
ZMBOV KF + MARGRAVE JL
J INORG NUCL CHEM 29: 673-80 (1967)

MANGANESE TETRAFLUORIDE $\qquad$ **MnF$_4$**

MASS SPECTROMETRIC AND THERMOCHEMICAL STUDIES OF
THE MANGANESE FLUORIDES.
EHLERT TC + HSIA M
J FLUORINE CHEM 2(1): 33-51 (1972)

MERCURY MONOFLUORIDE $\qquad$ **HgF**

0618 A NEW ELECTRONIC TRANSITION OF THE MERCURY
MONOFLUORIDE MOLECULE.
BABU YKSC + RAO PT + REDDY BR
INDIAN J PURE APPL PHYS 4: 467-9 (1966)

MERCURY DIFLUORIDE $\qquad$ **HgF$_2$**

POLARIZED ION MODEL AND THE BENDING FORCE
CONSTANTS OF THE GROUP-2B HALIDES. (ZINC
DIFLUORIDE, CADMIUM DIFLUORIDE, AND MERCURY
DIFLUORIDE)
ELIEZER I
THEOR CHIM ACTA 18: 77-85 (1970)

VIBRATIONAL SPECTRA OF THE DIHALIDES OF MERCURY
AND CADMIUM. (CADMIUM DIFLUORIDE AND MERCURY
DIFLUORIDE)
LOEWENSCHUSS A + RON A + SCHNEPP O
J PHYS CHEM 50: 2502-12 (1969)

0619 MEAN AMPLITUDES OF VIBRATION FOR THE DIHALIDES
OF MERCURY, BERYLLIUM, AND MAGNESIUM.
NAGARAJAN G
ACTA PHYS AUSTRIACA 17: 246-53 (1973)

MOLYBDENUM TRIFLUORIDE $\qquad$ **MoF$_3$**

0620 PREPARATION AND CRYSTAL STRUCTURE OF MOLYBDENUM
TRIFLUORIDE.
LAVALLE DE + STEELE RM + WILKINSON MK + YAKEL HL
J AMER CHEM SOC 82: 2433-4 (1960)

MOLYBDENUM TETRAFLUORIDE MoF_4

0621 RAMAN SPECTRUM OF SOLID MOLYBDENUM
TETRAFLUORIDE.
BATES JB
INORG NUCL CHEM LETT 7: 957-60 (1971)

0622 EMPIRICAL METHOD FOR DETERMINING EFFECTIVE
VIBRATIONAL AND ROTATIONAL CHARACTERISTICS OF
MOLECULES OF SOME TETRAFLUCRIDES FOR CALCULATING
THEIR THERMODYNAMIC FUNCTIONS.
TUMANOV YN + GALKIN NP
RUSS J PHYS CHEM 43(4): 464-7 (1969)

MOLYBDENUM PENTAFLUORIDE MoF_5

0623 RAMAN SPECTRUM OF CRYSTALLINE MOLYBDENUM
PENTAFLUORIDE.
BATES JB
SPECTROCHIM ACTA 27A: 1255-68 (1971)

0624 VAPOR PRESSURES OF MOLYBDENUM PENTAFLUORIDE,
TUNGSTEN PENTAFLUORIDE, RHENIUM PENTAFLUORIDE,
AND OSMIUM PENTAFLUORIDE.
CADY GH + HARGREAVES GB
J CHEM SOC A 1961: 1568-74

0625 VIBRATIONAL SPECTRA OF MOLYBDENUM AND TUNGSTEN
PENTAFLUORIDES.
OUELLETTE TJ + RATCLIFFE CT + SHARP DWA
J CHEM SOC A 1969: 2351-4

THERMODYNAMIC PROPERTIES OF NIOBIUM(V),
TANTALUM(V), AND MOLYBDENUM(V) FLUORIDES.
OREKHOV VT + RYBAKOV AG
RUSS J PHYS CHEM 47(6): 1612 (1973)

0626 VIBRATIONAL SPECTRA OF MOLYBDENUM PENTAFLUORIDE
AND TUNGSTEN PENTAFLUORIDE.
OUELLETTE TH + RATCLIFFE CT + SHARP DWA
J CHEM SOC A 1969: 2351-4

ELECTRONIC SPECTRA OF LIQUID RUTHENIUM AND
MOLYBDENUM PENTAFLUORIDES.
PEACOCK RD + SLEIGHT TP
J FLUORINE CHEM 1(2): 243-5 (1971)

GENERALIZED VALENCE FORCE FIELD FORCE CONSTANTS
AND MEAN-SQUARE AMPLITUDES OF VIBRATION OF SOME
TRIGONAL BIPYRAMIDAL MOLECULES BY THE
LOGARITHIMIC STEPS METHOD. (PHOSPHORUS
PENTAFLUORIDE, ARSENIC PENTAFLUORIDE, ANTIMONY
PENTAFLUORIDE, VANADIUM PENTAFLUORIDE, AND
MOLYBDENUM PENTAFLUORIDE)
WENDLING EJL + MAHMOUDI S + MACCORDICK HJ
J CHEM SOC A 1971: 1747-54

MOLYBDENUM HEXAFLUORIDE MoF_6

VAPOR PHASE RAMAN SPECTRA, RAMAN BAND CONTOUR
ANALYSES, AND CORIOLIS CONSTANTS OF THE
SPHERICAL TOP MOLECULES: SULFUR HEXAFLUORIDE,
SELENIUM HEXAFLUCRIDE, TELLURIUM HEXAFLUORIDE,
MOLYBDENUM HEXAFLUCRIDE, TUNGSTEN HEXAFLUORIDE,
AND URANIUM HEXAFLUORIDE.
BOSWORTH YM + CLARK RJH + RIPPON DM
J MOL SPECTROSC 46: 240-55 (1973)

HEAT CAPACITY, ENTROPY, AND HEATS OF TRANSITION
OF MOLYBDENUM HEXAFLUORIDE AND NIOBIUM
PENTAFLUORIDE.
BRADY AP + MYERS OW + CLAUSS JK
J PHYS CHEM 64: 588-91 (1960)

0627 MOLECULAR STRUCTURE OF MOLYBDENUM, TUNGSTEN, AND
URANIUM HEXAFLUORIDES FROM INFRARED AND RAMAN
SPECTRA.
BURKE TG + SMITH DF + NIELSEN AH
J CHEM PHYS 20(3): 447-54 (1953)

VAPOR PRESSURES OF SOME HEAVY TRANSITON METAL
HEXAFLUORIDES. (TUNGSTEN HEXAFLUORIDE,
MOLYBDENUM HEXAFLUORIDE, RHENIUM HEXAFLUORIDE,
OSMIUM HEXAFLUORIDE, AND IRIDIUM HEXAFLUORIDE)
CADY GH + HARGREAVES GB
J CHEM SOC A 1961: 1563-74

RAMAN SPECTRA OF MOLYBDENUM HEXAFLUORIDE,
TECHNETIUM HEXAFLUORIDE, RHENIUM HEXAFLUORIDE,
URANIUM HEXAFLUORIDE, SULFUR HEXAFLUORIDE,
SELENIUM HEXAFLUORIDE, AND TELLURIUM
HEXAFLUORIDE IN THE VAPOR STATE.
CLAASSEN HH + GOODMAN GL + HOLLOWAY JH + SELIG H
J CHEM PHYS 53: 341-8 (1970)

0628 VIBRATIONAL SPECTRA OF MOLYBDENUM HEXAFLUORIDE
AND TECHNETIUM HEXAFLUORIDE.
CLAASSEN HH + SELIG H + MALM JG
J CHEM PHYS 36: 2888-90 (1962)

SPECTROSCOPIC STUDIES IN CONNECTION WITH
ELECTRON DIFFRACTION INVESTIGATION OF SOME
SIMPLE MOLECULES. (TUNGSTEN HEXAFLUORIDE,
TELLURIUM HEXAFLUORIDE, AND MOLYBDENUM
HEXAFLUORIDE)
CYVIN SJ + CYVIN BN + BRUNVOLL J
P69-89 OF SELECTED TOPICS IN STRUCTURE
CHEMISTRY, ANDERSON P + BASTIANSEN O + FURBERG S
(EDS), UNIVERSITETSFORLAGET, OSLO, NORWAY, 1967.

0629 THERMODYNAMIC STABILITY OF HEXAFLUORIDES AT HIGH
TEMPERATURES. PART-2: PLUTONIUM HEXAFLUORIDE.
PART-3: URANIUM HEXAFLUORIDE. PART-4: MOLYBDENUM
AND TUNGSTEN HEXAFLUORIDES.
GALKIN NP + TUMANOV YN
P188-91, 191-5, 195-9 OF TERMODIN TERMOKHIM
KONSTANTY, ASTAKHOV KV(ED), NAUKA, MOSCOW, 1970.
(IN RUSSIAN)

0630 THERMAL STABILITY AND REACTIVITY OF D- AND F
ELEMENT HEXAFLUORIDES.
GALKIN NP + TUMANOV YN + BUTYLKIN YP
IZV SIB OTDEL AKAD NAUK SSSR SER KHIM NAUK (2):
12-21 (1968) (IN RUSSIAN)

INFRARED SPECTRA AND MOLECULAR STRUCTURE OF SOME
GROUP-6 HEXAFLUORIDES. (OF SULFUR, SELENIUM,
TELLURIUM, MOLYBDENUM, TUNGSTEN, AND URANIUM)
GAUNT J
TRANS FARADAY SOC 49: 1122-31 (1953)

0631 FORCE CONSTANTS AND BOND LENGTHS OF SOME
INORGANIC HEXAFLUORIDES. (OF SULFUR, SELENIUM,
TELLURIUM, MOLYBDENUM, TUNGSTEN, URANIUM, AND
RHENIUM)
GAUNT J
TRANS FARADAY SOC 50: 546-51 (1954)

INFRARED SPECTRUM OF CHROMIUM HEXAFLUORIDE,
MOLYBDENUM HEXAFLUORIDE, AND OSMIUM
HEXAFLUORIDE.
HELLBERG KH + MULLER A + GLEMSER O
Z NATURFORSCH 21B: 118-21 (1966)

ELECTRIC DEFLECTION OF BINARY HEXAFLUORIDES.
(SULFUR HEXAFLUORIDE, SELENIUM HEXAFLUORIDE,
TELLURIUM HEXAFLUORIDE, MOLYBDENUM HEXAFLUORIDE,
TUNGSTEN HEXAFLUORIDE, URANIUM HEXAFLUORIDE,
RUTHENIUM HEXAFLUORIDE, RHENIUM HEXAFLUORIDE,
RHODIUM HEXAFLUORIDE, OSMIUM HEXAFLUORIDE,
IRIDIUM HEXAFLUORIDE, AND PLATINUM HEXAFLUORIDE)
KAISER EW + MUENTER JS + KLEMPERER W +
FALCONER WE + SUNDER WA
J CHEM PHYS 53: 1411-12 (1970)

0632 THERMODYNAMIC SIMILARITY AND UNIVERSAL EQUATIONS
OF STATE OF HEXAFLUORIDES. (OF SULFUR,
MOLYBDENUM TUNGSTEN, AND URANIUM)
MALYSHEV VV
TEPLOFIZ VYS TEMP 14(1): 47-55 (1976) (IN
RUSSIAN)

0633 ASSIGNMENTS IN THE ULTRAVIOLET SPECTRA OF
MOLYBDENUM HEXAFLUORIDE AND TUNGSTEN
HEXAFLUORIDE.
MCDIARMID R
J CHEM PHYS 61: 3333-9 (1974)

0634 VIBRATIONAL SPECTRUM AND FORCE FIELD OF
MOLYBDENUM HEXAFLUORIDE.
MCDOWELL RS + SHERMAN RJ + ASPREY LB
+ KENNEDY RC
J CHEM PHYS 62: 3974-8 (1975)

0635 POTENTIAL CONSTANTS FOR MOLYBDENUM HEXAFLUORIDE
AND RHENIUM HEXAFLUORIDE.
NAGARAJAN G
AUST J CHEM 16: 906-7 (1963)

0636 HEAT CAPACITY AND OTHER THERMODYNAMIC PROPERTIES
OF MOLYBDENUM HEXAFLUORIDE BETWEEN 4 AND 350
DEGREES K.
OSBORNE DW + SCHREINER F + MALM JG + SELIG H
+ ROCHESTER L
J CHEM PHYS 44: 2802-9 (1966)

0637 HIGH RESOLUTION NMR SPECTRUM OF LIQUID
MOLYBDENUM HEXAFLUORIDE, TUNGSTEN HEXAFLUORIDE,
AND URANIUM HEXAFLUORIDE.
RIGNY P + DEMORTIER A + PERRIN F
COMPT REND 263: 1408-10 (1966) (IN FRENCH)

0638 MOLECULAR MOTION AND FLUORINE RELAXATION IN THE
LIQUIDS OF MOLYBDENUM HEXAFLUORIDE, TUNGSTEN
HEXAFLUORIDE, AND URANIUM HEXAFLUORIDE.
RIGNY P + VIRLET J
J CHEM PHYS 47: 4645-52 (1967)

0639 PHYSICAL CONSTANTS (VAPOR PRESSURE) OF SILICON
TETRAFLUORIDE, TUNGSTEN HEXAFLUORIDE, AND
MOLYBDENUM HEXAFLUORIDE.
RUFF O + ASCHER E
Z ANORG ALLGEM CHEM 196: 413-20 (1931) (IN
GERMAN)

FAILURE OF THE FIRST BORN APPROXIMATION IN
ELECTRON DIFFRACTION. (TUNGSTEN HEXAFLUORIDE,
URANIUM HEXAFLUORIDE, TELLURIUM HEXAFLUORIDE,
MOLYBDENUM HEXAFLUORIDE)
SEIP HM
P26-68 OF SELECTED TOPICS IN STRUCTURE
CHEMISTRY, ANDERSON P + BASTIANSEN O + FURBERG S
(EDS), UNIVERSITETSFORLAGET, OSLO, NORWAY, 1967.

0640 STUDIES ON THE FAILURE OF THE FIRST BORN
APPROXIMATION IN ELECTRON DIFFRACTION.
(MOLYBDENUM HEXAFLUORIDE AND TUNGSTEN
HEXAFLUORIDE)
SEIP HM + SEIP R
ACTA CHEM SCAND 20: 2698-2710 (1966)

0641 VALENCE FORCE CONSTANTS OF THE HEXAFLUORIDES OF
MOLYBDENUM, TECHNETIUM, AND RHENIUM.
SINGH RB + RAI DK
CAN J PHYS 43: 167-9 (1965)

0642 RAMAN EFFECT AND ULTRAVIOLET ABSORPTION SPECTRA
OF MOLYBDENUM AND TUNGSTEN HEXAFLUORIDES.
TANNER KN + DUNCAN ABF
J AMER CHEM SOC 73: 1164-7 (1951)

NEODYMIUM MONOFLUORIDE NdF

SUBLIMATION PRESSURE OF NEODYMIUM TRIFLUORIDE
AND THE STABILITIES OF GASEOUS NEODYMIUM
DIFLUORIDE AND NEODYMIUM MONOFLUORIDE.
ZMBOV KF + MARGRAVE JL
J CHEM PHYS 45: 3167-70 (1966)

NEODYMIUM DIFLUORIDE NdF_2

0643 SUBLIMATION PRESSURE OF NEODYMIUM TRIFLUORIDE
AND THE STABILITIES OF GASEOUS NEODYMIUM
DIFLUORIDE AND NEODYMIUM MONOFLUORIDE.
ZMBOV KF + MARGRAVE JL
J CHEM PHYS 45: 3167-70 (1966)

NEODYMIUM TRIFLUORIDE NdF_3

0644 ELECTRONOGRAPHIC INVESTIGATION OF THE STRUCTURE
OF THE MOLECULES OF NEODYMIUM HALIDES.
AKISHIN PA + NAUMOV VA + TATAEVSKII VM
NAUCH DOKL VYSSH SHKOL, KHIM KHIM TEKHNOL 3(2):
229-32 (1959) (IN RUSSIAN)

GEOMETRIES AND ENTROPIES OF METAL TRIFLUORIDES
FROM INFRARED SPECTRA. (SCANDIUM, YTTRIUM,
LANTHANUM, CERIUM, NEODYMIUM, EUROPIUM, AND
GADOLINIUM TRIFLUORIDES)
HASTIE JW + HAUGE RH + MARGRAVE JL
J LESS COMMON METALS 39(2): 309-34 (1975)

0645 ENTROPIES AND ENTHALPIES OF SUBLIMATION OF
NEODYMIUM AND TERBIUM TRIFLUORIDES.
MCCREARY JR + THORN RJ
HIGH TEMP SCI 6(3): 205-14 (1974)

0646 ENERGY LEVELS AND SPECTROSCOPIC PARAMETERS OF
NEODYMIUM TRIFLUORIDES.
VAISHNAVA PP + TANDON SP + BHUTRA MP
SPECTROSC LETT 7(10): 515-21 (1974)

SUBLIMATION PRESSURE OF NEODYMIUM TRIFLUORIDE
AND THE STABILITIES OF GASEOUS NEODYMIUM
DIFLUORIDE AND NEODYMIUM MONOFLUORIDE.
ZMBOV KF + MARGRAVE JL
J CHEM PHYS 45: 3167-70 (1966)

NEPTUNIUM PENTAFLUORIDE NpF_5

0647 THERMODYNAMIC PROPERTIES OF PENTAFLUORIDES AT
HIGH TEMPERATURES. PART-4: PENTAFLUORIDES OF 5F
ELEMENTS.
GALKIN NP + TUMANOV YN + BUTYLKIN YP
RUSS J PHYS CHEM 44(12): 1726-7 (1970)

CRYSTAL STRUCTURE OF MANGANESE DIFLUORIDE, IRON
DIFLUORIDE, COBALT DIFLUORIDE, NICKEL
DIFLUORIDE, AND ZINC DIFLUORIDE.
STOUT JW + REED SA
J AMER CHEM SOC 76: 5279-81 (1954)

NEPTUNIUM HEXAFLUORIDE NpF_6

NIOBIUM MONOFLUORIDE NbF

0648 ELECTRONIC STRUCTURE OF NEPTUNIUM HEXAFLUORIDE.
      GOODMAN GL + FRED M
      J CHEM PHYS 30: 849-50 (1959)

      THERMAL STABILITY AND REACTIVITY OF D- AND F
      ELEMENT HEXAFLUORIDES.
      GALKIN NP + TUMANOV YN + BUTYLKIN YP
      IZV SIB OTD AKAD NAUK SSSR SER KHIM NAUK (2):
      12-21 (1968) (IN RUSSIAN)

0649 INFRARED SPECTRA OF NEPTUNIUM HEXAFLUORIDE AND
      PLUTONIUM HEXAFLUORIDE.
      MALM JG + WEINSTOCK B + CLAASSEN HH
      J CHEM PHYS 23: 2192-3 (1955)

0650 HEAT CAPACITY, ENTHALPY OF FUSION, AND
      THERMODYNAMIC PROPERTIES OF NEPTUNIUM
      HEXAFLUORIDE FROM 7 TO 350 DEGREES K.
      OSBORNE DW + WEINSTOCK B + BURNS JH
      J CHEM PHYS 52(4): 1803-10 (1970)

      THERMODYNAMIC PROPERTIES OF NIOBIUM
      MONOFLUORIDE, NIOBIUM DIFLUORIDE, TANTALUM
      MONOFLUORIDE, AND TANTALUM DIFLUORIDE AT HIGH
      TEMPERATURES.
      GALKIN NP + TUMANOV YN + KOROBSTEV VP + BATAREV
      + PAVLOV AA
      RUSS J PHYS CHEM 45: 1532 (1971)

0654 THERMODYNAMIC PROPERTIES OF NIOBIUM AND TANTALUM
      FLUORIDES AT HIGH TEMPERATURES. PART-3:
      DIFLUORIDES AND MONOFLUORIDES.
      GALKIN NP + TUMANOV YN + KOROBTSEV VP
      + BATAREV GA + KHOKHLOV VA + PAVLOV AA
      RUSS J PHYS CHEM 45(10): 1532 (1971)

NICKEL MONOFLUORIDE NiF

NIOBIUM DIFLUORIDE NbF_2

0651 BAND SPECTRUM OF NICKEL MONOFLUORIDE.
      KRISHNAMURTY VG
      INDIAN J PHYS 27: 354-8 (1953)

      INTERACTION OF MATRIX ISOLATED NICKEL FLUORIDE
      AND NICKEL CHLORIDE WITH CARBON MONOXIDE,
      MOLECULAR NITROGEN, NITRIC OXIDE, AND MOLECULAR
      OXYGEN AND OF CALCIUM FLUORIDE, CHROMIUM(II)
      FLUORIDE, MANGANESE(II) FLUORIDE, COPPER(II)
      FLUORIDE, AND ZINC(II) FLUORIDE WITH CARBON
      MONOXIDE IN ARGON MATRICES.
      VAN LEIRSBURG DA + DEKOCK CW
      J PHYS CHEM 78(2): 134-42 (1974)

0655 THERMODYNAMIC PROPERTIES OF NIOBIUM
      MONOFLUORIDE, NIOBIUM DIFLUORIDE, TANTALUM
      MONOFLUORIDE, AND TANTALUM DIFLUORIDE AT HIGH
      TEMPERATURES.
      GALKIN NP + TUMANOV YN + KOROBTSEV VP
      + BATAREV GA + KHOKHLOV VA + PAVLOV AA
      RUSS J PHYS CHEM 45: 1532 (1971)

      THERMODYNAMIC PROPERTIES OF NIOBIUM AND TANTALUM
      FLUORIDES AT HIGH TEMPERATURES. PART-3:
      DIFLUORIDES AND MONOFLUORIDES.
      GALKIN NP + TUMANOV YN + KOROBTSEV VP + BATAREV
      + KHOKHLOV VA + PAVLOV AA
      RUSS J PHYS CHEM 45(10): 1532 (1971)

NICKEL DIFLUORIDE NiF_2

NIOBIUM TRIFLUORIDE NbF_3

      VAPOR PRESSURES OF BERYLLIUM DIFLUORIDE AND
      NICKEL DIFLUORIDE.
      CANTOR S
      J CHEM ENG DATA 10: 237-8 (1965)

0652 VAPOR PRESSURE OF NICKEL DIFLUORIDE.
      FARBER M + MEYER RT + MARGRAVE JL
      J PHYS CHEM 62: 883-4 (1958)

      INFRARED SPECTRA AND GEOMETRY OF MATRIX ISOLATED
      COBALT DIFLUORIDE, NICKEL DIFLUORIDE, COPPER
      DIFLUORIDE, AND ZINC DIFLUORIDE.
      HASTIE JW + HAUGE RH + MARGRAVE JL
      HIGH TEMP SCI 1: 76-85 (1969)

      FREE ENERGIES OF FORMATION OF SOME INORGANIC
      FLUORIDES BY SOLID STATE EMF MEASUREMENTS.
      (THORIUM TETRAFLUORIDE, ALUMINUM TRIFLUORIDE,
      NICKEL DIFLUORIDE, LEAD DIFLUORIDE, COLBALT
      DIFLUORIDE, AND URANIUM TRIFLUORIDE)
      HEUS RJ + EGAN JJ
      Z PHYS CHEM NEUE FOLGE 49: 38-43 (1966)

0653 SPECTRA OF MATRIX ISOLATED NICKEL DIFLUORIDE AND
      NICKEL DICHLORIDE.
      MILLIGAN DE + JACOX ME + MCKINLEY JD
      J CHEM PHYS 42: 902-5 (1965)

0656 THERMODYNAMIC PROPERTIES OF NIOBIUM TRIFLUORIDE,
      NIOBIUM TETRAFLUORIDE, TANTALUM TRIFLUORIDE, AND
      TANTALUM TETRAFLUORIDE AT HIGH TEMPERATURES.
      GALKIN NP + TUMANOV YN + KOROBTSEV VP
      + BATAREV GA + KHOKHLOV VA + PAVLOV AA
      RUSS J PHYS CHEM 45: 1531 (1971)

0657 THERMODYNAMIC PROPERTIES OF NIOBIUM AND TANTALUM
      FLUORIDES AT HIGH TEMPERATURES. PART-2:
      TETRAFLUORIDES AND TRIFLUORIDES.
      GALKIN NP + TUMANOV YN + KOROBTSEV VP
      + BATAREV GA + KHOKHLOV VA + PAVLOV AA
      RUSS J PHYS CHEM 45(10): 1532 (1971)

NIOBIUM TETRAFLUORIDE NbF_4

0658 INFRARED SPECTRA OF NIOBIUM TETRAFLUORIDE.
      DICKSON FE
      J INORG NUCL CHEM 31: 2636-8 (1969)

THERMODYNAMIC PROPERTIES OF NIOBIUM TRIFLUORIDE,
NIOBIUM TETRAFLUORIDE, TANTALUM TRIFLUORIDE, AND
TANTALUM TETRAFLUORIDE AT HIGH TEMPERATURES.
GALKIN NP + TUMANOV YN + KOROBTSEV VP + BATAREV
+ PAVLOV AA
RUSS J PHYS CHEM 45: 1531 (1971)

THERMODYNAMIC PROPERTIES OF NIOBIUM AND TANTALUM
FLUORIDES AT HIGH TEMPERATURES. PART-2:
TETRAFLUORIDES AND TRIFLUORIDES.
GALKIN NP + TUMANOV YUN + KOROBTSEV VP + BATAREV
+ PAVLOV AA
ZH FIZ KHIM 45(10): 2695 (1971) (IN RUSSIAN)

NIOBIUM PENTAFLUORIDE NbF_5

0659 INFRARED SPECTRUM OF MATRIX ISOLATED NIOBIUM
PENTAFLUORIDE.
ACQUISTA N + ABRAMOWITZ S
J CHEM PHYS 56(11): 4609-17 (1972)

POLYMER MONOMER EQUILIBRIA IN NIOBIUM
PENTAFLUORIDE, TANTALUM PENTAFLUORIDE, AND
ANTIMONY PENTAFLUORIDE. THE SHAPE OF ANTIMONY
PENTAFLUORIDE.
ALEXANDER LE
INORG NUCL CHEM LETT 7: 1053-6 (1970)

0660 VIBRATIONAL SPECTRA OF NIOBIUM AND TANTALUM
PENTAFLUORIDES IN THE GAS PHASE. VAPOR DENSITY
OF NIOBIUM PENTAFLUORIDE.
ALEXANDER LE + BEATTIE IR + JONES PJ
J CHEM SOC, DALTON TRANS 1972(2): 210-12

0661 INFRARED SPECTRUM OF NIOBIUM PENTAFLUORIDE.
BLANCHARD S
J CHIM PHYS 62: 919-20 (1965)

0662 HEAT CAPACITY, ENTROPY, AND HEATS OF TRANSITION
OF MOLYBDENUM HEXAFLUORIDE AND NIOBIUM
PENTAFLUORIDE.
BRADY AP + MYERS OE + CLAUSS JK
J PHYS CHEM 64: 588-91 (1960)

FLUORINE NMR AND RAMAN STUDY OF COMPLEX
FORMATION BETWEEN ANTIMONY PENTAFLUORIDE AND
NIOBIUM PENTAFLUORIDE AND TANTALUM
PENTAFLUORIDE.
DEAN PAW + GILLESPIE RJ
CAN J CHEM 49: 1736-46 (1971)

0663 HALIDES OF NIOBIUM AND TANTALUM. PART-3: VAPOR
PRESSURES OF NIOBIUM AND TANTALUM
PENTAFLUORIDES.
FAIRBROTHER F + FRITH WC
J CHEM SOC 1951: 3051-6

0664 NIOBIUM-93 NUCLEAR QUADRUPOLE RESONANCE STUDIES
OF NIOBIUM PENTAFLUORIDE AND ITS COMPLEXES WITH
XENON DIFLUORIDE AND ORGANIC BASES.
FUGGLE JC + TONG DA + SHARP DWA + WINFIELD JM
+ HOLLOWAY JH
J CHEM SOC, DALTON TRANS 1974(2): 205-10

0665 THERMODYNAMIC PROPERTIES OF NIOBIUM AND TANTALUM
FLUORIDES AT HIGH TEMPERATURES. PART-1:
PENTAFLUORIDES.
GALKIN NP + TUMANOV YUN + KOROBTSEV VP
+ BATAREV GA + KHOKHLOV VA + PAVLOV AA
RUSS J PHYS CHEM 45(10): 1531 (1971)

0666 MASS SPECTRUM AND COMPOSITION OF NIOBIUM
PENTAFLUORIDE VAPOR.
GOTKIS IS + GUSAROV AV + GOROKHOV LN
ZH NEORG KHIM 20(5): 1250-3 (1975) (IN RUSSIAN)

0667 ENTHALPIES OF FORMATION OF NIOBIUM PENTAFLUORIDE
AND TANTALUM PENTAFLUORIDE.
GREENBERG E + NATKE CA + HUBBARD WN
J PHYS CHEM 69: 2089-93 (1965)

POTENTIAL FIELD AND FORCE CONSTANTS OF ANTIMONY
PENTAFLUORIDE AND NIOBIUM PENTAFLUORIDE.
NAGARAJAN G
BULL SOC CHIM BELG 72: 563-71 (1963)

0668 THERMODYNAMIC PROPERTIES OF NIOBIUM(V),
TANTALUM(V), AND MOLYBDENUM(V) FLUORIDES.
OREKHOV VT + RYBAKOV AG
RUSS J PHYS CHEM 47(6): 1612 (1973)

ELECTRON DIFFRACTION STUDY OF THE MOLECULAR
STRUCTURES OF VANADIUM PENTAFLUORIDE, NIOBIUM
PENTAFLUORIDE, AND TANTALUM PENTAFLUORIDE.
ROMANOV GV + SPIRIDONOV VP
IZV SIB OTDEL AKAD NAUK SSSR, SER KHIM NAUK 1:
126-31 (1968) (IN RUSSIAN)

0669 THERMODYNAMIC PROPERTIES OF NIOBIUM AND TANTALUM
PENTAFLUORIDES.
SELEZNEV VP + RAKOV EG + MIKULENOK VV
RUSS J PHYS CHEM 45(11): 1667 (1971)

0670 RAMAN SPECTRA OF LIQUID NIOBIUM PENTAFLUORIDE
AND TANTALUM PENTAFLUORIDE.
SELIG H + REIS A + GASNER EL
J INORG NUCL CHEM 30: 2087-90 (1968)

0671 ELECTRON DIFFRACTION INVESTIGATION OF NIOBIUM
PENTAFLUORIDE AND TANTALUM PENTAFLUORIDE.
SPIRIDONOV VP
VEST MOSK UNIV KHIM 23: 7-11 (1968) (IN RUSSIAN)

NITROGEN MONOFLUORIDE NF

0672 CONFIGURATION INTERACTION STUDIES OF NITROGEN
MONOFLUORIDE AND ITS POSITIVE ION.
ANDERSEN A + OHRN Y
J MOL SPECTROSC 45: 358-65 (1973)

0673 GAS PHASE ELECTRON PARAMAGNETIC RESONANCE
SPECTRUM AND DIPOLE MOMENT OF NITROGEN
MONOFLUORIDE IN THE SINGLET DELTA STATE.
CURRAN AH + MACDONALD RG + STONE AJ + THRUSH BA
PROC ROY SOC A 332: 355-63 (1973)

0674 NEW SPECTRA OF OXYGEN LIKE MOLECULES. (NITROGEN
MONOFLUORIDE)
DOUGLAS AE
P81-4 OF INT CONF SPECT, 1ST PROC, VOL 1, 1967,
BOMBAY, INDIA.

0675 B SINGLET SIGMA(+) TO X TRIPLET SIGMA(-) BAND
SYSTEM OF NITROGEN MONOFLUORIDE.
DOUGLAS AE + JONES WE
CAN J PHYS 44: 2251-58 (1966)

0676 CALCULATION OF THE SPECTROSCOPIC CONSTANTS FOR
SEVERAL STATES OF NITROGEN MONOFLUORIDE.
ELLIS DJ + BANYARD KE
J PHYS B 7: 2021-4 (1974)

0677 FRANCK-CONDON FACTORS AND R CENTROIDS FOR THE 6
SINGLET SIGMA- X TRIPLET SIGMA BAND SYSTEM OF
NITROGEN MONOFLUORIDE.
MOHAMED KA + KHANNA BN + LAL KM
INDIAN J PURE APPL PHYS 12(3): 243-4 (1974)

0678 QUANTUM CHEMICAL STUDY OF SOME PNICOGEN
MONOFLUORIDES. (NITROGEN MONOFLUORIDE AND
PHOSPHORUS MONOFLUORIDE)
O'HARE PAG + WAHL AC
J CHEM PHYS 54: 4563-77 (1971)

NITROGEN DIFLUORIDE NF_2

0679 ELECTRON DIFFRACTION STUDIES AT ELEVATED
TEMPERATURES. (NITROGEN DIFLUORIDE)
BAUER SH
CORNELL UNIV, 1967, 38P. AD665316; US GOVT RES
DEVELOP REP 68(8): 51 (1968)

0680 DIFFRACTION STUDY OF THE STRUCTURES OF NITROGEN
DIFLUORIDE AND DINITROGEN TETRAFLUORIDE.
BOHN RK + BAUER SH
INORG CHEM 6: 304-9 (1967)

0681 MICROWAVE SPECTRUM OF NITROGEN DIFLUORIDE
RADICAL.
BROWN RD + BURDEN FR + GODFREY PD + GILLARD IR
J MOL SPECTROSC 52(2): 301-21 (1974)

0682 ELECTRONIC STRUCTURE OF THE NITROGEN DIFLUORIDE
RADICAL.
BROWN RD + BURDEN FR + HART BT + WILLIAMS GR
THEOR CHIM ACTA 28: 339-53 (1973)

AB INITIO CALCULATIONS ON SOME SMALL RADICALS BY
THE UNRESTRICTED HARTREE-FOCK METHOD. (BORON
DIFLUORIDE, NITROGEN DIFLUORIDE, OXYGEN
DIFLUORIDE, CARBON DIFLUORIDE)
BROWN RD + WILLIAMS GR
CHEM PHYS 3: 19-34 (1974)

THEORETICAL STUDY OF THE BOND-BOND INTERACTION
FORCE CONSTANTS IN DIFLUORIDE MOLECULES. (OXYGEN
DIFLUORIDE, NITROGEN DIFLUORIDE, CARBON
DIFLUORIDE)
BRUNS R + RAFF L + DEVLIN JP
THEOR CHIM ACTA 14: 232-41 (1969)

0683 MULTIPLET SPLITTING IN 1S HOLE STATES OF
MOLECULES.
DAVIS DW + MARTIN RL + BANNA MS + SHIRLEY DA
J CHEM PHYS 59(8): 4235-45 (1973)

0684 MOLECULAR ORBITAL STUDY OF NITROGEN DIHYDRIDE,
NITROGEN DIOXIDE, AND NITROGEN DIFLUORIDE.
DELBENE JE
J CHEM PHYS 54: 3487-90 (1971)

0685 ABSORPTION SPECTRUM OF NITROGEN DIFLUORIDE.
GOODFRIEND PL + WOODS HP
J MOL SPECTROSC 13: 63-66 (1964)

0686 INFRARED SPECTRUM AND THERMODYNAMIC FUNCTIONS OF
THE NITROGEN DIFLUORIDE RADICAL.
HARMONY MD + MYERS RJ
J CHEM PHYS 37: 636-41 (1962)

0687 INFRARED SPECTRUM AND STRUCTURE OF THE NITROGEN
DIFLUORIDE RADICAL.
HARMONY MD + MYERS RJ + SCHOEN LH + LIDE DR
+ MANN DE
J CHEM PHYS 35: 1129-30 (1961)

0688 AB INITIO MOLECULAR ORBITAL STUDY OF THE
NITROGEN DIFLUORIDE AND DINITROGEN TETRAFLUORIDE
MOLECULES, AND ESR HYPERFINE COUPLING CONSTANTS
IN NITROGEN DIFLUORIDE.
HINCHLIFFE A + COBB JC
CHEM PHYS 3: 271-76 (1974)

0689 A CAVITY SEARCH SPECTROMETER FOR FREE RADICAL
MICROWAVE ROTATIONAL ABSORPTION STUDIES.
(NITROGEN DIFLUORIDE)
HRUBESH LW + ANDERSON RE + RINEHART EA
REV SCI INSTR 42: 789-96 (1971)

0690 MICROWAVE ROTATIONAL SPECTRUM OF THE NITROGEN
DIFLUORIDE FREE RADICAL.
HRUBESH LW + RINEHART EA + ANDERSON RE
J MOL SPECTROSC 36: 354-6 (1970)

0691 MATRIX ISOLATION STUDY OF THE VACUUM ULTRAVIOLET
PHOTOLYSIS OF NITROGEN TRIFLUORIDE. ELECTRONIC
SPECTRUM OF THE NITROGEN DIFLUORIDE FREE
RADICAL.
JACOX ME + MILLIGAN DE + GUILLORY WA + SMITH JJ
J MOL SPECTROSC 52(2): 322-7 (1974)

0692 INFRARED SPECTRUM OF NITROGEN DIFLUORIDE.
JOHNSON FA + COLBURN CB
INORG CHEM 2: 431-2 (1962)

0693 CALCULATION OF FORCE CONSTANTS FOR NITROGEN
DIFLUORIDE AND NITROGEN TRIFLUORIDE.
KOTOV YI + TATAEVSKII VM
VEST MOSK UNIV KHIM 18: 3-5 (1963) (IN RUSSIAN)

0694 ELECTRONIC ABSORPTION SPECTRUM OF THE NITROGEN
DIFLUORIDE RADICAL.
KUZNETSOVA LA + KUZYAKOV YY + TATAEVSKII VM
OPT SPECTROSC 16(3): 295 (MAR 1964)

MEAN AMPLITUDES OF VIBRATION FOR BORON
DIFLUORIDE, CARBON DIFLUORIDE, NITROGEN
DIFLUORIDE, ALUMINUM DIFLUORIDE, PHOSPHORUS
DIFLUORIDE, AND ZIRCONIUM DIFLUORIDE.
NAGARAJAN G
INDIAN J PURE APPL PHYS 2: 341-3 (1964)

MOLECULAR ORBITAL STUDY OF NITROGEN TRIFLUORIDE,
PHOSPHORUS TRIFLUORIDE, AND NITROGEN DIFLUORIDE.
OLMSTEAD M
UNIV WISCONSIN, PHD THESIS, 1969, 111P.

MOLECULAR PROPERTIES OF THE TRIATOMIC
DIFLUORIDES BERYLLIUM DIFLUORIDE, BORON
DIFLUORIDE, CARBON DIFLUORIDE, NITROGEN
DIFLUORIDE, AND OXYGEN DIFLUORIDE. (SCF)
ROTHENBERG S + SCHAEFER HF
J AMER CHEM SOC 95: 2095-100 (1973)

0695 RAMAN SPECTRUM OF THE NITROGEN DIFLUORIDE
RADICAL.
SELIG H + HOLLOWAY JH
J INORG NUCL CHEM 33: 3169-71 (1971)

ELECTRONIC STRUCTURE AND GEOMETRY OF BORON
DIFLUORIDE AND NITROGEN DIFLUORIDE.
THOMSON C + BROTCHIE DA
CHEM PHYS LETT 16: 573-75 (1972)

NITROGEN TRIFLUORIDE NF_3

0696 THERMOCHEMICAL AND THERMODYNAMIC PROPERTIES OF
INORGANIC NITROGEN FLUORIDES. (NITROGEN
TRIFLUORIDE)
AL ZERCHENIOV AN + CHESNOKOV VI + PANKTRATOV AV
IZV SIB OTDEL AKAD NAUK SSSR, SER KHIM NAUK 1:
74-77 (1968) (IN RUSSIAN)

0697 INFRARED SPECTRUM OF NITROGEN TRIFLUORIDE AND
FORCE FIELD.
ALLAN A + DUNCAN JL + HOLLOWAY JH + MCKEAN DC
J MOL SPECTROSC 31: 368-77 (1960)

0698 ABSORPTION SPECTRA OF MATRIX ISOLATED NITROGEN
TRIFLUORIDE FROM 18 TO 50 GC.
ARMSTRONG RL + NEWMAN JB + WHALEN JJ
JOHNS HOPKINS UNIV, 1964, 33P. AD608739

INFRARED ABSORPTION SPECTRA OF SOME POLYATOMIC
FLUORIDES. (SILICON TETRAFLUORIDE, CARBON
TETRAFLUORIDE, BORON TRIFLUORIDE, NITROGEN
TRIFLUORIDE)
BAILEY CR + HALE JB + THOMPSON JW
J CHEM PHYS 5: 274-5 (1937)

0699 INFRARED BAND SHAPES AND VIBRATIONS OF NITROGEN
TRIFLUORIDE.
BAKERMAN S
PURDUE UNIV, PHD THESIS, 1957, 54P.

0700 PHOTOELECTRON SPECTRA OF NITROGEN TRIFLUORIDE
AND NITROGEN OXIDE TRIFLUORIDE, AND A
REASSIGNMENT OF THE SPECTRA OF TETRAFLUORIDES OF
GROUP-4.
BASSETT PJ + LLOYD DR
CHEM PHYS LETT 6: 166-8 (1970)

0701 PHOTOELECTRON SPECTRA OF HALIDES. PART-3:
TRIFLUORIDES AND OXIDE TRIFLUORIDES OF NITROGEN
AND PHOSPHORUS, AND PHOSPHORUS OXIDE
TRICHLORIDE.
BASSETT PJ + LLOYD DR
J CHEM SOC, DALTON TRANS 1972(3): 248-54

0702 FAR INFRARED SPECTRUM OF NITROGEN TRIFLUORIDE.
CHANTRY GW + GEBBIE HA
PROC ROY SOC A 304: 45-51 (1968)

0703 NITROGEN FLUORIDES AND THEIR INORGANIC
DERIVATIVES.
COLBURN CB
P92-116 OF ADVAN FLUORINE CHEM, VOL 3, STACEY M
+ TATLOW JC + SHARPE AG(EDS), BUTTERWORTHS,
1963.

0704 A BENT BOND MODEL FOR THE VIBRATIONAL FORCE
CONSTANTS OF NONPLANAR XY3 MOLECULES. (NITROGEN
TRIFLUORIDE)
CURTIS EC
ROCKETDYNE, CANOGA PARK, CALIF, 1968, 21P.
AD666456

0705 LONE PAIR MODEL AND THE VIBRATIONAL FORCE
CONSTANTS OF NITROGEN TRIFLUORIDE.
CURTIS EC + MUIRHEAD JS
J PHYS CHEM 70: 3330-6 (1966)

0706 MEAN AMPLITUDES OF VIBRATION FOR SOME PYRAMIDAL
XY3 MOLECULES. (NITROGEN TRIFLUORIDE,
PHOSPHORUS TRIFLUORIDE, AND ARSENIC TRIFLUORIDE)
CYVIN SJ + CYVIN BN
J MOL STRUCT 4: 341-9 (1969)

0707 MOLECULAR ORBITAL THEORY OF THE ELECTRONIC
STRUCTURE OF ORGANIC COMPOUNDS. PART-3: AB
INITIO STUDIES OF CHARGE DISTRIBUTION USING A
MINIMAL SLATER-TYPE BASIS. (NITROGEN
TRIFLUORIDE)
HEHRE WJ + POPLE JA
J AMER CHEM SOC 92: 2191-7 (1970)

0708 P-R SEPARATIONS AND CORIOLIS CONSTANTS FOR
SYMMETRIC TOP MOLECULES AND FORCE CONSTANTS FOR
NITROGEN TRIFLUORIDE, PHOSPHORUS TRIFLUORIDE,
AND ARSENIC TRIFLUORIDE.
HOSKINS LC
J CHEM PHYS 45: 4594-4600 (1966)

0709 AMBIGUITIES IN THE HARMONIC FORCE FIELDS OF XY3
MOLECULES. (NITROGEN TRIFLUORIDE, BORON
TRIFLUORIDE, PHOSPHORUS TRIFLUORIDE, AND ARSENIC
TRIFLUORIDE)
HOY AR + STONE JMR + WATSON JKG
J MOL SPECTROSC 42: 393-99 (1972)

0710 LIQUID DENSITY, VAPOR PRESSURE AND CRITICAL
TEMPERATURE AND PRESSURE OF NITROGEN
TRIFLUORIDE.
JARRY RL + MILLER HC
J PHYS CHEM 60: 1412-13 (1956)

0711 MICROWAVE SPECTROSCOPY IN THE REGION FROM TWO TO
THREE MILLIMETERS, PART-2: (NITROGEN TRIFLUORIDE
AND PHOSPHORUS TRIFLUORIDE)
JOHNSON CM + TRAMBARULO R + GORDY W
PHYS REV 84: 117880 (1951)

0712 STRENGTH OF THE NITROGEN FLUORINE BOND IN
NITROGEN TRIFLUORIDE AND DINITROGEN
TETRAFLUORIDE.
KENNEDY A + COLBURN CB
J CHEM PHYS 35: 1892-3 (1961)

CALCULATION OF FORCE CONSTANTS FOR NITROGEN
DIFLUORIDE AND NITROGEN TRIFLUORIDE.
KOTOV YI + TATAEVSKII VM
VEST MOSK UNIV KHIM 18: 3-5 (1963) (IN RUSSIAN)

0713 PHYSICAL CHEMICAL PROPERTIES OF INORGANIC
NITROGEN FLUORIDES. (NITROGEN TRIFLUORIDE)
KUZNETSOVA TV + EGOROVA LF + RIPD SM
IZV SIB OTDEL AKAD NAUK SSSR, SER KHIM NAUK 1:
68-73 (1968) (IN RUSSIAN)

0714 INFRARED INTENSITIES AND FORCE FIELD FOR
NITROGEN TRIFLUORIDE.
LEVIN IW
J CHEM PHYS 52: 2783-4 (1970)

0715 FORCE FIELDS FOR GROUP-4 TETRAFLUORIDES AND
GROUP-5 TRIFLUORIDES.
LEVIN IW + ABRAMOWITZ S
J PHYS CHEM 44: 2562-7 (1966)

ABSORPTION SPECTRA OF CERTAIN FLUORIDES IN THE
NEAR ULTRAVIOLET REGION. (KRYPTON DIFLUORIDE,
NITROGEN TRIFLUORIDE)
MAKEEV GN + SINYANSKII VF + SMIRNOV BM
DOKL AKAD NAUK SSSR 222(1): 151-4 (1975) (IN
RUSSIAN)

0716 PSEUDO-PARALLEL INFRARED BANDS OF OBLATE
SYMMETRIC TOPS NITROGEN TRIFLUORIDE NU-4 AND
NU-4 + NU-2 AND TRIFLUORO METHANE NU-6.
MASRI FN + BLASS WE
J MOL SPECTROSC 39: 98-114 (1971)

VAPOR PRESSURE OF CARBON TETRAFLUORIDE AND
NITROGEN TRIFLUORIDE AND THE TRIPLE POINT OF
CARBON TETRAFLUORIDE.
MENZEL W + MOHRY F
Z ANORG ALLGEM CHEM 210: 257-63 (1933) (IN
GERMAN)

0717 GENERAL FORCE FIELD OF NITROGEN TRIFLUORIDE,
PHOSPHORUS TRIFLUORIDE, AND ARSENIC TRIFLUORIDE
BY THE COMBINED USE OF VIBRATIONAL FREQUENCIES,
CENTRIFUGAL STRETCHING, AND CORIOLIS COUPLING
CONSTANTS.
MIRRI AM
J CHEM PHYS 47: 2823-8 (1967)

0718 MILLIMETER WAVE SPECTRUM AND CENTRIFUGAL
DISTORTION CONSTANTS OF NITROGEN TRIFLUORIDE.
MIRRI AM + CAZZOLI G
J CHEM PHYS 47: 1197-8 (1967)

0719 MEAN AMPLITUDES OF VIBRATION AND THERMODYNAMIC
FUNCTIONS OF SOME GROUP-5 TRIHALIDES. (NITROGEN
TRIFLUORIDE, ARSENIC TRIFLUORIDE, AND PHOSPHORUS
TRIFLUORIDE)
NAGARAJAN G
INDIAN J PURE APPL PHYS 2: 237-41 (1964)

0720 SELF CONSISTENT MOLECULAR ORBITAL METHODS.
PART-5: AB INITIO CALCULATION OF EQUILIBRIUM
GEOMETRIES AND QUADRATIC FORCE CONSTANTS.
(NITROGEN TRIFLUORIDE)
NEWTON MD + LATHAN WA + HEHRE WJ + POPLE JA
J CHEM PHYS 52: 4064-72 (1970)

0721 MOLECULAR ORBITAL STUDY OF NITROGEN TRIFLUORIDE,
PHOSPHORUS TRIFLUORIDE, AND NITROGEN DIFLUORIDE.
OLMSTEAD M
UNIV WISCONSIN, PHD THESIS, 1969, 111P.

0722 MICROWAVE SPECTRA OF NITROGEN TRIFLUORIDE IN THE
EXCITED VIBRATIONAL STATES. (EQUILIBRIUM
STRUCTURE)
OTAKE M + HIROTA E + MORINO Y
J MOL SPECTROSC 28: 325-40 (1968)

0723 MICROWAVE SPECTRA OF NITROGEN TRIFLUORIDE IN THE
EXCITED VIBRATIONAL STATES. (ZETA TYPE DOUBLING
AND RESONANCE IN THE V3 AND V4 STATES)
OTAKE M + MATSUMURA C + MORINO Y
J MOL SPECTROSC 28: 316-24 (1968)

0724 INFRARED AND RAMAN SPECTRUM OF NITROGEN
TRIFLUORIDE.
PACE EL + PIERCE L
J CHEM PHYS 23: 1248-50 (1955)

0725 VACUUM ULTRAVIOLET ABSORPTION SPECTRUM AND
DIPOLE MOMENT OF NITROGEN TRIFLUORIDE.
PAGLIA SRL + DUNCAN ABF
J CHEM PHYS 34: 1003-7 (1961)

0726 HARMONIC FREQUENCIES, POTENTIAL FUNCTION, CORIOLIS COUPLING CONSTANTS, CENTRIFUGAL STRETCHING CONSTANTS, AND MEAN SQUARE AMPLITUDES OF THE VIBRATIONS OF THE NITROGEN TRIFLUORIDE MOLECULE.
PONOMAREV YI + KHOVRIN GV
OPT SPECTROSC 26(6): 580-1 (JUN 1969)

0727 DETERMINATION OF THE STRUCTURE OF PYRAMIDAL MOLECULES OF THE XY3 TYPE BY SOLUTION OF THE VIBRATIONAL PROBLEM WITH THE USE OF CENTRIFUGAL STRETCHING CONSTANTS. (NITROGEN TRIFLUORIDE, PHOSPHORUS TRIFLUORIDE, ARSENIC TRIFLUORIDE)
PONOMAREV YI + KHOVRIN GV
OPT SPECTROSC 30(2): 122-4 (1971)

0728 INFRARED SPECTRUM AND MOLECULAR CONSTANTS OF NITROGEN TRIFLUORIDE.
POPPLEWELL RJL + MASRI FN + THOMPSON HW
SPECTROCHIM ACTA 23A: 2797-2807 (1967)

CONSTRUCTION OF A HIGH RESOLUTION PHOTOELECTRON SPECTROMETER. (CARBON TETRAFLUORIDE, BORON TRIFLUORIDE, SULFUR HEXAFLUORIDE, NITROGEN TRIFLUORIDE, AND SILICON TETRAFLUORIDE)
PULLEN BP
UNIV TENNESSEE, PHD THESIS, 1970, 117P.

0729 GREEN'S FUNCTION ANALYSIS OF NITROGEN TRIFLUORIDE AND NITROGEN TRICHLORIDE.
RAMASWAMY K + MOHAN N
INDIAN J PHYS 43: 693-99 (1969)

0730 NORMAL COORDINATE ANALYSIS OF SOME NITROGEN HALIDES. (NITROGEN TRIFLUORIDE)
RAMASWAMY K + MOHAN N
Z NATURFORSCH 25B: 169-73 (1970)

0731 HIGH RESOLUTION STUDY OF L-TYPE RESONANCE IN THE NU-4 BAND OF NITROGEN TRIFLUORIDE.
REICHMAN S
J MOL SPECTROSC 40: 27-32 (1971)

0732 1-TYPE RESONANCE IN AN OVERTONE BAND. THE 2NU-4 SPECTRUM OF NITROGEN TRIFLUORIDE.
REICHMAN S + SCHATZ J
J MOL SPECTROSC 48(2): 277-82 (1973)

0733 ANHARMONIC POTENTIAL CONSTANTS FROM THE BAND CONTOURS OF HOT BANDS FOR A SYMMETRIC TOP. (NITROGEN TRIFLUORIDE)
RUOFF VA
MOL PHYS 19: 23-31 (1970)

0734 AB INITIO STUDY OF THE FORCE CONSTANTS OF INORGANIC MOLECULES NITROGEN OXYFLUORIDE AND NITROGEN TRIFLUORIDE.
SAWODNY W
J MOL SPECTROSC 51: 135-41 (1974)

0735 FORCE FIELD AND MOLECULAR CONSTANTS OF NITROGEN TRIFLUORIDE.
SAWODNY W + RUOFF VA + PEACOCK CJ + MULLER A
MOL PHYS 14: 433-40 (1968)

0736 ABSOLUTE INFRARED INTENSITIES OF THE FUNDAMENTAL VIBRATIONS OF NITROGEN TRIFLUORIDE.
SCHATZ PN + LEVIN IW
J CHEM PHYS 29: 475-80 (1958)

0737 POTENTIAL FUNCTION OF NITROGEN TRIFLUORIDE.
SCHATZ PN
J CHEM PHYS 29: 481-3 (1958)

0738 AN ELECTRON DIFFRACTION INVESTIGATION OF NITROGEN TRIFLUORIDE.
SCHOMAKER V + LU CS
J AMER CHEM SOC 72: 1182-5 (1950)

0739 THERMODYNAMIC PROPERTIES OF NITROGEN TRIFLUORIDE.
SESHADRI DN + VISWANATH DS + KULOOR NS
INDIAN J TECHNOL 8(5): 153-60 (1970)

0740 RAMAN SPECTRUM OF GASEOUS NITROGEN TRIFLUORIDE.
SHAMIR J + HYMAN HH
SPECTROCHIM ACTA 23A: 1899-1901 (1967)

0741 NUCLEAR QUADRUPOLE MOMENT OF NITROGEN-14 AND THE STRUCTURE OF NITROGEN TRIFLUORIDE FROM MICROWAVE SPECTRA.
SHERIDAN J + GORDY W
PHYS REV 79: 513-15 (1950)

0742 ENTHALPY OF DISSOCIATION OF NITROGEN TRIFLUORIDE.
SINKE GC
J PHYS CHEM 71: 359-61 (1967)

0743 SIGNS OF L DOUBLING CONSTANTS IN NITROGEN TRIFLUORIDE.
STONE JMR + MILLS IM
J MOL SPECTROSC 35: 354-8 (1970)

0744 ZEEMAN STUDIES INCLUDING THE MOLECULAR G VALUES, MAGNETIC SUSCEPTIBILITY, AND MOLECULAR QUADRUPOLE MOMENTS IN PHOSPHORUS TRIFLUORIDE, NITROGEN TRIFLUORIDE, AND PHOSPHORYL, THIONYL AND SULFURYL FLUORIDES.
STONE RG + POCHAN JM + FLYGARE WH
INORG CHEM 8: 2647-55 (1969)

0745 VIBRATIONAL ANHARMONICITY IN THE PYRAMIDAL TRIFLUORIDES.
TEICHMAN S + SMITH DF + OVEREND J
SPECTROCHIM ACTA 26A: 927-35 (1970)

0746 CHARACTERIZATION OF GROUND STATE WAVE FUNCTIONS BY MEASURED ELECTRONIC PROPERTIES. PART-2: DIPOLE MOMENT AND FIELD GRADIENT OF NITROGEN TRIFLUORIDE. PART-3: A GAUSSIAN BASIS SELF CONSISTENT FIELD CALCULATION FOR NITROGEN TRIFLUORIDE.
UNLAND ML + DUNNING TH + WAZER JRV + LETCHER JH
J CHEM PHYS 50: 3208-13 (1969); 3214-19 (1969)

0747 CALCULATION OF PERPENDICULAR INFRARED BAND CONTOURS: APPLICATION TO THE NU-4 BAND OF NITROGEN TRIFLUORIDE.
WEINER RS
PURDUE UNIV, PHD THESIS, 1970, 247P.

0748 INFRARED SPECTRA OF NITROGEN TRIFLUORIDE AND PHOSPHORUS TRIFLUORIDE.
WILSON MK + POLO SR
J CHEM PHYS 20(11): 1716-19 (1952)

OSMIUM TETRAFLUORIDE OsF_4

PREPARATION AND CRYSTAL STRUCTURE OF OSMIUM PENTAFLUORIDE. TWO FLUORIDES OF OSMIUM. (OSMIUM PENTAFLUORIDE AND OSMIUM TETRAFLUORIDE)
HARGREAVES GB + PEACOCK RD
J CHEM SOC A 1960: 2618-20

OSMIUM PENTAFLUORIDE OsF_5

VAPOR PRESSURES OF MOLYBDENUM PENTAFLUORIDE, TUNGSTEN PENTAFLUORIDE, RHENIUM PENTAFLUORIDE, AND OSMIUM PENTAFLUORIDE.
CADY GH + HARGREAVES GB
J CHEM SOC A 1961: 1568-74

THERMODYNAMIC PROPERTIES OF PENTAFLUORIDES AT HIGH TEMPERATURES. PART-3: PENTAFLUORIDES OF SOME 5D-ELEMENTS.
GALKIN NP + TUMANOV YN + BUTYLKIN YP
RUSS J PHYS CHEM 44(12): 1724-6 (1970)

0749 PREPARATION AND CRYSTAL STRUCTURE OF OSMIUM PENTAFLUORIDE. TWO FLUORIDES OF OSMIUM. (OSMIUM PENTAFLUORIDE, AND OSMIUM TETRAFLUORIDE)
HARGREAVES GB + PEACOCK RD
J CHEM SOC A 1960: 2618-20

0750 PREPARATION AND CRYSTAL STRUCTURE OF OSMIUM
PENTAFLUORIDE.
MITCHELL SJ + HOLLOWAY JH
J CHEM SOC A 1971: 2789-94

OSMIUM HEXAFLUORIDE OsF_6

VAPOR PRESSURES OF SOME HEAVY TRANSITON METAL
HEXAFLUORIDES. (TUNGSTEN HEXAFLUORIDE,
MOLYBDENUM HEXAFLUORIDE, RHENIUM HEXAFLUORIDE,
OSMIUM HEXAFLUORIDE, AND IRIDIUM HEXAFLUORIDE)
CADY GH + HARGREAVES GB
J CHEM SOC A 1961: 1563-74

0751 ABSORPTION SPECTRUM AND MAGNETIC PROPERTIES OF
OSMIUM HEXAFLUORIDE.
EISENSTEIN JC
J CHEM PHYS 34: 310-18 (1961)

INFRARED SPECTRUM OF CHROMIUM HEXAFLUORIDE,
MOLYBDENUM HEXAFLUORIDE, AND OSMIUM
HEXAFLUORIDE.
HELLBERG KH + MULLER A + GLEMSER O
Z NATURFORSCH 21B: 118-21 (1966)

ELECTRIC DEFLECTION OF BINARY HEXAFLUORIDES.
(SULFUR HEXAFLUORIDE, SELENIUM HEXAFLUORIDE,
TELLURIUM HEXAFLUORIDE, MOLYBDENUM HEXAFLUORIDE,
TUNGSTEN HEXAFLUORIDE, URANIUM HEXAFLUORIDE,
RUTHENIUM HEXAFLUORIDE, RHENIUM HEXAFLUORIDE,
RHODIUM HEXAFLUORIDE, OSMIUM HEXAFLUORIDE,
IRIDIUM HEXAFLUORIDE, AND PLATINUM HEXAFLUORIDE)
KAISER EW + MUENTER JS + KLEMPERER W +
FALCONER WE + SUNDER WA
J CHEM PHYS 53: 1411-12 (1970)

ELECTRON DIFFRACTION INVESTIGATION OF TUNGSTEN
HEXAFLUORIDE, OSMIUM HEXAFLUORIDE, IRIDIUM
HEXAFLUORIDE, URANIUM HEXAFLUORIDE, NEPTUNIUM
HEXAFLUORIDE, AND PLUTONIUM HEXAFLUORIDE.
KIMURA M + SCHOMAKER V + SMITH DW + WEINSTOCK B
J CHEM PHYS 48: 4001-12 (1968)

COLORS OF TRANSITION METAL HEXAFLUORIDES.
(RHENIUM HEXAFLUORIDE, OSMIUM HEXAFLUORIDE,
IRIDIUM HEXAFLUORIDE, AND PLATINUM HEXAFLUORIDE)
MOFFITT W + GOODMAN GL + FRED M + WEINSTOCK B
MOL PHYS 2: 109-22 (1959)

0752 VIBRATIONAL SPECTRA OF OSMIUM HEXAFLUORIDE AND
PLATINUM HEXAFLUORIDE.
WEINSTOCK B + CLAASSEN HH + MALM JG
J CHEM PHYS 32: 181-5 (1960)

0753 OSMIUM HEXAFLUORIDE AND ITS IDENTITY WITH THE
PREVIOUSLY REPORTED OCTAFLUORIDE.
WEINSTOCK B + MALM JG
J AMER CHEM SOC 80: 4466-8 (1958)

HEAT CAPACITIES AND ELECTRONIC SPECTRA OF THE
PLATINUM GROUP METAL HEXAFLUORIDE MOLECULES DOWN
TO HELIUM TEMPERATURES.
WEINSTOCK B + WESTRUM EF + GOODMAN GL
P405-6 OF INT CONF LOW TEMP PHYS, 8TH PROC,
LONDON, 1962

OSMIUM HEPTAFLUORIDE OsF_7

0754 PREPARATION AND PROPERTIES (INFRARED SPECTRUM)
OF OSMIUM HEPTAFLUORIDE.
GLEMSER O + ROESKY HW + HELLBERG KH + WERTHER HU
CHEM BER 99: 2652-62 (1966) (IN GERMAN)

OXYGEN MONOFLUORIDE OF

ARGON MATRIX RAMAN SPECTRA OF OXYGEN DIFLUORIDE
AND THE OXYGEN FLUORIDE FREE RADICAL.
ANDREWS L
J CHEM PHYS 57(1): 51-5 (1972)

0755 MATRIX INFRARED SPECTRUM OF OXYGEN MONOFLUORIDE
AND DETECTION OF LITHIUM OXYFLUORIDE.
ANDREWS L + RAYMOND JI
J CHEM PHYS 55: 3078-86 (1971)

0756 OXYGEN MONOFLUORIDE. HARTREE-FOCK WAVE FUNCTION,
BINDING ENERGY, IONIZATION POTENTIAL, ELECTRON
AFFINITY, DIPOLE AND QUADRUPOLE MOMENTS, AND
SPECTROSCOPIC CONSTANTS. COMPARISON OF
THEORETICAL AND EXPERIMENTAL RESULTS. COMMENTS.
LIEBMAN JF
J CHEM PHYS 56(8): 4242-3 (1972)

0757 OXYGEN MONOFLUORIDE, DOUBLET PI HARTREE-FOCK
WAVE FUNCTION, BINDING ENERGY, IONIZATION
POTENTIAL, ELECTRON AFFINITY, DIPOLE MOMENT,
QUADRUPOLE MOMENT, AND SPECTROSCOPIC CONSTANTS.
A COMPARISON OF THEORETICAL AND EXPERIMENTAL
RESULTS.
O'HARE PAG + WAHL AC
J CHEM PHYS 53: 2469-78 (1970)

HARTREE-FOCK WAVEFUNCTIONS AND COMPUTED
PROPERTIES FOR THE DOUBLET PI GROUND STATES OF
SULFUR MONOFLUORIDE AND SELENIUM MONOFLUORIDE
AND THEIR POSITIVE AND NEGATIVE IONS. A
COMPARISON OF THE THEORETICAL AND EXPERIMENTAL
RESULTS. (OXYGEN MONOFLUORIDE)
O'HARE PAG + WAHL AC
J CHEM PHYS 53: 2834-46 (1970)

METHOD OF DETERMINING SATURATED LIQUID AND
SATURATED VAPOR ENTROPY. (CHLORINE MONOFLUORIDE,
OXYGEN MONOFLUORIDE, PHOSPHORUS TRIFLUORIDE)
WALKER MA
AIAA J 1: 2636-8 (1963)

OXYGEN DIFLUORIDE OF_2

0758 EXTENDED HUCKEL THEORY AND THE SHAPE OF
MOLECULES. (OXYGEN DIFLUORIDE)
ALLEN LC + RUSSELL JD
J CHEM PHYS 46: 1029-37 (1967)

0759 ARGON MATRIX RAMAN SPECTRA OF OXYGEN DIFLUORIDE
AND THE OXYGEN FLUORIDE FREE RADICAL.
ANDREWS L
J CHEM PHYS 57(1): 51-5 (1972)

0760 PHOTOIONIZATION MASS SPECTROMETRY OF OXYGEN
DIFLUORIDE.
BERKOWITZ J + DEHMER PM + CHUPKA WA
J CHEM PHYS 59: 925-28 (1973)

0761 VIBRATIONAL SPECTRA AND STRUCTURE OF INORGANIC
MOLECULES. PART-1: INFRARED SPECTRUM OF OXYGEN
DIFLUORIDE FROM 2.5 TO 25 MICRONS.
BERNSTEIN HJ + POWLING J
J CHEM PHYS 18(5): 685-8 (1950)

0762 DOUBLE ZETA SCF CALCULATIONS FOR NITRITE ION AND
OXYGEN DIFLUORIDE.
BONACCORSI R + PETRONGOLO C + SCROCCO E
+ TOMASI J
J CHEM PHYS 48: 1497-99 (1968)

0763 CONFIGURATION INTERACTION CALCULATION FOR THE
GROUND STATE OF OXYGEN DIFLUORIDE, NITRITE ION
AND CYANIDE ION.
BONACCORSI R + PETRONGOLO C + SCROCCO E
+ TOMASI J
THEOR CHIM ACTA 15: 332-43 (1969)

AB INITIO CALCULATIONS ON SOME SMALL RADICALS BY
THE UNRESTRICTED HARTREE-FOCK METHOD. (BORON
DIFLUORIDE, NITROGEN DIFLUORIDE, OXYGEN
DIFLUORIDE, CARBON DIFLUORIDE)
BROWN RD + WILLIAMS GR
CHEM PHYS 3: 19-34 (1974)

A THEORETICAL STUDY OF THE BOND-BOND INTERACTION
FORCE CONSTANTS IN DIFLUORIDE MOLECULES. (OXYGEN
DIFLUORIDE, NITROGEN DIFLUORIDE, CARBON
DIFLUORIDE)
BRUNS R + RAFF L + DEVLIN JP
THEOR CHIM ACTA 14: 232-41 (1969)

0764 GEOMETRY OF MOLECULES. (OXYGEN DIFLUORIDE)
BUENKER RJ + PEYERIMHOFF SD
J CHEM PHYS 45: 3682-3700 (1966)

0765 APPROXIMATE MOLECULAR ORBITAL THEORY. ESSENTIAL
STRUCTURAL ELEMENTS- MO FORMALISM. PART-2:
DIRECT COMPARISON WITH AB INITIO RESULTS.
BURTON PG
CHEM PHYS 4(2): 226-35 (1974)

0766 PHOTOELECTRON SPECTRA OF OXYGEN DIFLUORIDE AND
OXYGEN DICHLORIDE.
CORNFORD AB + FROST DC + HERRING FG
+ MCDOWELL CA
J CHEM PHYS 55: 2820-22 (1971)

0767 RAMAN SPECTRUM OF LIQUID OXYGEN DIFLUORIDE.
GARDINER DJ + TURNER JJ
J MOL SPECTROSC 38: 428-30 (1971)

0768 MILLIMETER WAVE SPECTRUM AND STRUCTURE OF OXYGEN
DIFLUORIDE.
HILTON AR + JACHE AW + BEAL JB + HENDERSON WD
+ ROBINSON RJ
J CHEM PHYS 34: 1137-41 (1961)

0769 HEAT OF FORMATION OF OXYGEN DIFLUORIDE.
KING RC + ARMSTRONG GT
J RES NBS A 72: 113-31 (1968)

0770 THIRD ORDER POTENTIAL CONSTANTS OF BENT XY2
MOLECULES. (OXYGEN DIFLUORIDE)
KUCHITSU K + MORINO Y
SPECTROCHIM ACTA 22: 33-46 (1966)

0771 MICROWAVE SPECTRUM OF OXYGEN DIFLUORIDE IN
VIBRATIONALLY EXCITED STATES, NU-1- 2 NU-2 FERMI
RESONANCE AND EQUILIBRIUM STRUCTURE.
MORINO Y + SAITO S
J MOL SPECTROSC 19: 435-53 (1966)

0772 FERMI DIAD OF NU-1 AND 2NU-2 IN OXYGEN
DIFLUORIDE.
MORINO Y + YAMAMOTO S
J MOL SPECTROSC 23(2): 235-7 (1967)

0773 MEAN AMPLITUDES OF VIBRATION AND THERMODYNAMIC
FUNCTIONS FOR SILICON DIFLUORIDE AND OXYGEN
DIFLUORIDE.
NAGARAJAN G
BULL SOC CHIM BELG 71: 337-46 (1962)

0774 A REINVESTIGATION OF THE INFRARED SPECTRUM OF
OXYGEN DIFLUORIDE.
NEBGEN JW + METZ FI + ROSE WB
J MOL SPECTROSC 21: 99-103 (1966)

0775 FLUORINE NMR SPECTRA OF OXYGEN FLUORIDES.
(OXYGEN DIFLUORIDE)
NEBGEN JW + METZ FI + ROSE WB
J AMER CHEM SOC 89: 3118-21 (1967)

0776 MINIMAL BASIS SET SCF CALCULATIONS FOR THE
GROUND STATE OF OZONE, NITRITE ION, NITROGEN
OXYFLUORIDE, AND OXYGEN DIFLUORIDE.
PETRONGOLO C + SCROCCO E + TOMASI J
J CHEM PHYS 48: 407-11 (1968)

0777 SPIN ROTATIONAL HYPERFINE STRUCTURE IN THE
MICROWAVE SPECTRUM OF OXYGEN DIFLUORIDE.
PIERCE L + DICIANNI N
J CHEM PHYS 38: 2029-30 (1963)

0778 CENTRIFUGAL DISTORTION EFFECTS IN ASYMMETRIC
ROTOR MOLECULES. PART-1. QUADRATIC POTENTIAL
CONSTANTS AND AVERAGE STRUCTURE OF OXYGEN
DIFLUORIDE FROM THE GROUND STATE ROTATIONAL
SPECTRUM.
PIERCE L + DICIANNI N
J CHEM PHYS 38: 739-9 (1963)

0779 MICROWAVE SPECTRUM, STRUCTURE, AND DIPOLE MOMENT
OF OXYGEN DIFLUORIDE.
PIERCE L + JACKSON R + DICIANNI N
J CHEM PHYS 35: 2240-1 (1961)

0780 MOLECULAR G VALUES, MAGNETIC SUSCEPTIBILITIES,
MOLECULAR QUADRUPOLE MOMENTS AND SECOND MOMENTS
OF THE ELECTRONIC CHARGE DISTRIBUTION IN OXYGEN
DIFLUORIDE, OZONE, AND SULFUR DIOXIDE.
POCHAN JM + STONE RG + FLYGARE WH
J CHEM PHYS 51: 4278-85 (1969)

MOLECULAR PROPERTIES OF THE TRIATOMIC
DIFLUORIDES BERYLLIUM DIFLUORIDE, BORON
DIFLUORIDE, ARGON DIFLUORIDE, NITROGEN
DIFLUORIDE, AND OXYGEN DIFLUORIDE. (SCF)
ROTHENBERG S + SCHAEFER HF
J AMER CHEM SOC 95: 2095-100 (1973)

0781 PREPARATION AND PURIFICATION OF OXYGEN
DIFLUORIDE AND DETERMINATION OF ITS VAPOR
PRESSURE.
SCHNIZLEIN JG + SHEARD JL + TOOLE RC
+ O'BRIEN TD
J PHYS CHEM 56: 233-4 (1952)

0782 CNDO STUDY OF THE OZONIDE ION AND RELATED
SPECIES. (OXYGEN DIFLUORIDE)
SICHEL JM
CAN J CHEM 51: 2124-8 (1973)

0783 VIRIAL THEOREM DECOMPOSITION OF MOLECULAR FORCE
FIELDS.
SIMONS G + NOVICK JL
J PHYS CHEM 78(10): 989-93 (1974)

0784 SPECTROSCOPIC STUDIES OF VIBRATIONAL CONSTANTS,
STATISTCIAL THERMODYNAMICS AND QUANTUM
MECHANICAL STUDIES OF POLARIZABILITIES FOR THE
DIFLUORIDES OF OXYGEN AND SULFUR.
SINGH Z + NAGARAJAN G
ACTA PHYS ACAD SCI HUNG 36(4): 415-30 (1974)

0785 OXYGEN FLUORIDES. (OXYGEN DIFLUORIDE)
STRENG AG
CHEM REV 63(6): 607-24 (1960)

PHOSPHORUS MONOFLUORIDE   PF

0786 ELECTRONIC SPECTRA OF PHOSPHORUS MONOFLUORIDE
AND ITS POSITIVE ION.
DOUGLAS AE + FRACKOWIAK M
CAN J PHYS 40: 832-49 (1962)

0787 ROTATIONAL FINE STRUCTURE OF THE TRIPLET PI
STATE OF PHOSPHORUS MONOFLUORIDE.
KOVACS I
CAN J PHYS 42: 2180-4 (1964)

0788 THERMODYNAMIC PROPERTIES OF PHOSPHORUS(2),
PHOSPHORUS(4), AND SOME PHOSPHORUS FLUORIDES.
(PHOSPHORUS MONOFLUORIDE, PHOSPHORUS DIFLUORIDE,
PHOSPHORUS TRIFLUORIDE, PHOSPHORUS
TETRAFLUORIDE, AND PHOSPHORUS PENTAFLUORIDE)
O'HARE PAG
USAEC, 1968, 29P. ANL-7459

QUANTUM CHEMICAL STUDY OF SOME PNICOGEN
MONOFLUORIDES. (NITROGEN MONOFLUORIDE AND
PHOSPHORUS MONOFLUORIDE)
O'HARE PAG + WAHL AC
J CHEM PHYS 54: 4563-77 (1971)

0789 THERMODYNAMIC FUNCTIONS OF DIATOMIC GASES WITH
MOLECULES IN TRIPLET SIGMA STATES. (PHOSPHORUS
MONOFLUORIDE)
YURKOV GN
HIGH TEMP 4(6): 738-42 (1966)

PHOSPHORUS DIFLUORIDE PF_2

0790 AB INITIO MOLECULAR ORBITAL STUDY OF THE
PHOSPHORUS DIFLUORIDE RADICAL.
COBB JC + HINCHLIFFE A
CHEM PHYS LETT 24(1): 75-6 (1974)

0791 ESR SPECTRA OF PHOSPHORUS DIFLUORIDE AND SULFUR
TRIFLUORIDE RADICALS.
COLUSSI AJ + MORTON JR + PRESTON KF
+ FESSENDEN RW
J CHEM PHYS 61(3): 1247-8 (1974)

MEAN AMPLITUDES OF VIBRATION FOR BORON
DIFLUORIDE, CARBON DIFLUORIDE, NITROGEN
DIFLUORIDE, ALUMINUM DIFLUCRIDE, PHOSPHCRUS
DIFLUORIDE, AND ZIRCONIUM DIFLUORIDE.
NAGARAJAN G
INDIAN J PURE APPL PHYS 2: 341-3 (1964)

0792 ELECTRON SPIN RESONANCE OF TRAPPED PHOSPHORUS
DIFLUORIDE AND PHOSPHORUS TETRAFLUORIDE
RADICALS.
NELSON W + JACKEL G + GORDY W
J CHEM PHYS 52: 4572-8 (1970)

THERMODYNAMIC PROPERTIES OF PHOSPHOROUS(2),
PHOSPHORUS(4), AND SOME PHOSPHORUS FLUORIDES.
(PHOSPHORUS MONOFLUORIDE, PHOSPHORUS DIFLUORIDE,
PHOSPHORUS TRIFLUORIDE, PHCSPHORUS
TETRAFLUORIDE, AND PHOSPHORUS PENTAFLUORIDE)
O'HARE PAG
USAEC, 1968, 29P. ANL-7459

0793 ELECTRON SPIN RESONANCE STUDIES OF PHOSPHORUS
DIFLUORIDE, PHOSPHORUS DICHLORIDE, AND NITROGEN
DICHLORIDE IN LOW TEMPERATURE MATRICES.
WEI MSS
UNIV MICHIGAN, PHD THESIS, 1970, 155P.

PHOSPHORUS TRIFLUORIDE PF_3

0794 THEORETICAL STUDY OF THE GEOMETRY OF PHOSPHINE
AND PHOSPHORUS TRIFLUORIDE AND THEIR GROUND
IONIC STATES.
AARONS LJ + GUEST MF + HALL MB + HILLIER IH
J CHEM SOC, FARADAY TRANS 69 (PT 5): 643-7
(1973)

PHOTOELECTRON SPECTRA OF HALIDES. PART-3:
TRIFLUORIDES AND OXIDE TRIFLUORIDES OF NITROGEN
AND PHOSPHORUS, AND PHOSPHORUS OXIDE
TRICHLORIDE.
BASSETT PJ + LLOYD DR
J CHEM SOC, DALTON TRANS 1972(3): 248-54

0795 VAPOR PHASE RAMAN SPECTRA, RAMAN BAND CONTOUR
ANALYSES, CORIOLIS CONSTANTS, FORCE CONSTANTS,
AND VALUES FOR THERMODYNAMIC FUNCTIONS OF THE
TRIHALIDES OF GROUP-5.
CLARK RJH + RIPPON DM
J MOL SPECTROSC 52(1): 58-71 (1974)

0796 ZEEMAN EFFECT OF SOME LINEAR AND SYMMETRIC TOP
MOLECULES. (PHOSPHORUS TRIFLUORIDE)
COX JT + GORDY W
PHYS REV 101(4): 1298-1300 (1956)

BENT BOND MODEL FOR THE VIBRATIONAL FORCE
CONSTANTS OF NON-PLANAR XY3 MOLECULES. (NITROGEN
TRIFLUORIDE)
CURTIS EC
ROCKETDYNE, CANOGA PARK, CALIF, 1968, 21P.
AD666456

0797 INFRARED INTENSITIES, BOND MOMENTS AND BOND
MOMENT DERIVATIVES FOR PHOSPHORUS TRIFLUORIDE
AND PHOSPHORUS OXYGEN TRIFLUORIDE.
DUNLAP JL
VANDERBILT UNIV, PHD THESIS, 1959, 135P.

0798 MICROWAVE INVESTIGATIONS OF METHYL FLUORIDE,
FLUOROFORM, AND PHOSPHORUS TRIFLUORIDE.
GILLIAM OR + EDWARDS HD + GORDY W
PHYS REV 75(7): 1014-16 (1949)

0799 INFRARED SPECTRA OF PHOSPHORUS TRIFLUORIDE,
PHOSPHORUS OXIDE TRIFLUORIDE, AND PHOSPHORUS
PENTAFLUORIDE.
GUTOWSKY HS + LIEHR AD
J CHEM PHYS 20: 1652-3 (1952)

0800 IONIZATION BY ELECTRON IMPACT OF PHOSPHORUS
TRIFLUORIDE AND DIFLUORO CYANO PHOSPHINE.
HARLAND PW + RANKIN WH + THYNNE JCJ
INT J MASS SPECTROM ION PHYS 13: 395-410 (1974)

0801 ELECTRON DIFFRACTION BY GASES AND THE STRUCTURES
OF PHOSPHORUS TRIFLUORIDE, PHOSPHORUS
PENTAFLUORIDE, DIFLUORO AMINE, AND TETRAFLUORO
HYDRAZINE.
HERSH OL
UNIV MICHIGAN, PHD THESIS, 1963, 74P.

0802 AB INITIO CALCULATIONS OF THE BONDING IN
PHOSPHINE, PHOSPHORUS TRIFLUORIDE AND TRIMETHYL
PHOSPHINE.
HILLIER IH + SAUNDERS VR
CHEM COMMUN 1970: 316-18

0803 L-TYPE DOUBLING TRANSITIONS OF PHOSPHORUS
TRIFLUORIDE IN THE NU-4 EQUALS 1 STATE.
HIROTA E
J MOL SPECTROSC 37: 20-32 (1971)

0804 EFFECTS OF THE THIRD ORDER CONSTANTS ON THE
L-TYPE DOUBLING TRANSITIONS OF PHOSPHORUS
TRIFLUORIDE IN THE NU-4 EQUALS 1 STATE.
HIROTA E
J MOL SPECTROSC 38: 195-6 (1971)

0805 MICROWAVE SPECTRUM OF PHOSPHORUS TRIFLUORIDE.
HIROTA E + MORINO Y
J MOL SPECTROSC 33: 460-473 (1970)

P-R SEPARATIONS AND CORIOLIS CONSTANTS FOR
SYMMETRIC TOP MOLECULES AND FORCE CONSTANTS FOR
NITROGEN TRIFLUORIDE, PHOSPHORUS TRIFLUORIDE,
AND ARSENIC TRIFLUORIDE.
HOSKINS LC
J CHEM PHYS 45: 4594-4600 (1966)

AMBIGUITIES IN THE HARMONIC FORCE FIELDS OF XY3
MOLECULES. (NITROGEN TRIFLUORIDE, BORON
TRIFLUORIDE, PHOSPHORUS TRIFLUORIDE, AND ARSENIC
TRIFLUORIDE)
HOY AR + STONE JMR + WATSON JKG
J MOL SPECTROSC 42: 393-99 (1972)

0806 ABSORPTION SPECTRA OF THE HYDRIDES, DEUTERIDES,
AND HALIDES OF GROUP-5 ELEMENTS. (PHOSPHORUS
TRIFLUORIDE)
HUMPHRIES CM + WALSH AD + WARSOP PA
DISC FARADAY SOC 35: 148-57 (1963)

NONEMPIRICAL VALENCE ELECTRON CALCULATIONS ON
SMALL MOLECULES CONTAINING PHOSPHORUS OR SULFUR.
HYDE RG + PEEL JB + TERAUDS K
J CHEM SOC, FARADAY TRANS 73(69): 1563-8 (1973)

MICROWAVE SPECTROSCOPY IN THE REGION FROM TWO
THREE MILLIMETERS, PART-2: (NITROGEN TRIFLUORIDE
AND PHOSPHORUS TRIFLUORIDE)
JOHNSON CM + TRAMBARULO R + GORDY W
PHYS REV 84: 1178-80 (1951)

0807 ENTHALPIES OF FORMATION OF XENON HEXAFLUORIDE,
XENON TETRAFLUORIDE, XENON DIFLUORIDE, AND
PHOSPHORUS TRIFLUORIDE.
JOHNSON GK + MALM JG + HUBBARD WN
J CHEM THERMODYN 4: 879-91 (1972)

0808 MEASUREMENT AND CALCULATION OF THE ABSOLUTE
INFRARED INTENSITIES OF PHOSPHORUS TRIFLUORIDE.
LEVIN IW + ADAMS OW
J MOL SPECTROSC 39: 380-91 (1971)

0809 CALCULATION OF UREY-BRADLEY POTENTIAL CONSTANTS.
(PHOSPHORUS TRIFLUORIDE, ARSENIC TRIFLUORIDE,
AND ANTIMONY TRIFLUORIDE)
MEISINGSETH E
ACTA CHEM SCAND 17: 509-12 (1963)

GENERAL FORCE FIELD OF NITROGEN TRIFLUORIDE,
PHOSPHORUS TRIFLUORIDE, AND ARSENIC TRIFLUORIDE
BY THE COMBINED USE OF VIBRATIONAL FREQUENCIES,
CENTRIFUGAL STRETCHING, AND CORIOLIS COUPLING
CONSTANTS.
MIRRI AM
J CHEM PHYS 47: 2823-8 (1967)

0810 MOLECULAR STRUCTURE OF PHOSPHORUS TRIFLUORIDE
STUDIED BY GAS ELECTRON DIFFRACTION.
MORINO Y + KUCHITSU K + MORITANI T
INORG CHEM 8: 867-71 (1969)

0811 RAMAN AND INFRARED SPECTRA AND THERMODYNAMIC
FUNCTIONS FOR VARIOUS PHOSPHORUS AND ARSENIC
HALIDES. (PHOSPHORUS TRIFLUORIDE)
MULLER A + NIECKE E + KREBS B + GLEMSER O
Z NATURFORSCH 23B: 588-94 (1968)

MEAN AMPLITUDES OF VIBRATION AND THERMODYNAMIC
FUNCTIONS OF SOME GROUP-5 TRIHALIDES. (NITROGEN
TRIFLUORIDE, ARSENIC TRIFLUORIDE, AND PHOSPHORUS
TRIFLUORIDE)
NAGARAJAN G
INDIAN J PURE APPL PHYS 2: 237-41 (1964)

THERMODYNAMIC PROPERTIES OF PHOSPHORUS(2),
PHOSPHORUS(4), AND SOME PHOSPHORUS FLUORIDES.
(PHOSPHORUS MONOFLUORIDE, PHOSPHORUS DIFLUORIDE,
PHOSPHORUS TRIFLUORIDE, PHOSPHORUS
TETRAFLUORIDE, AND PHOSPHORUS PENTAFLUORIDE)
O'HARE PAG
USAEC, 1968, 29P. ANL-7459

MOLECULAR ORBITAL STUDY OF NITROGEN TRIFLUORIDE,
PHOSPHORUS TRIFLUORIDE, AND NITROGEN DIFLUORIDE.
OLMSTEAD M
UNIV WISCONSIN, PHD THESIS, 1969, 111P.

MOLECULAR FORCE FIELD FOR SOME XY3 TYPE
MOLECULES. (PHOSPHORUS TRIFLUORIDE AND ARSENIC
TRIFLUORIDE)
PILLAI MGK + PILLAI PP
INDIAN J PURE APPL PHYS 6: 404-7 (1968)

0812 THERMODYNAMIC FUNCTIONS OF SOME PHOSPHORUS
COMPOUNDS. (PHOSPHORUS TRIFLUORIDE)
POTTER RL + DISTEFANO VN
J PHYS CHEM 65: 849-55 (1961)

0813 MOLECULAR CONSTANTS OF PHOSPHORUS HALOGEN
COMPOUNDS. (PHOSPHORUS TRIFLUORIDE)
RAO BK
Z PHYS CHEM 242: 155-60 (1969)

0814 CENTRIFUGAL DISTORTION CONSTANTS OF PHOSPHORUS
TRIFLUORIDE.
RAO CGR + MURTY AAN
SCI CULT 40: 315-16 (1974)

0815 NU-3 BAND OF PHOSPHORUS TRIFLUORIDE.
REICHMAN S
J MOL SPECTROSC 35: 329-31 (1970)

0816 HIGH RESOLUTION SPECTRA OF ARSENIC TRIFLUORIDE
AND PHOSPHORUS TRIFLUORIDE.
REICHMAN S + OVEREND J
SPECTROCHIM ACTA 26A: 379-89 (1970)

0817 2NU-4 SPECTRUM OF PHOSPHORUS TRIFLUORIDE.
REICHMAN S + SMITH DF
J MOL SPECTROSC 59(3): 502-4 (1976)

0818 ENTHALPY OF FORMATION PHOSPHORUS TRIFLUORIDE.
RUDZITIS E + DEVENTER EHV + HUBBARD WN
J CHEM THERMODYN 2: 221-5 (1970)

0819 FLUORIDES OF PHOSPHORUS.
SCHMULTZER R
P31-285 OF ADVAN FLUORINE CHEM, VOL 5, STACEY M
+ TATLOW JC + SHARPE AG(EDS), BUTTERWORTHS,
1965.

0820 BONDING IN PHOSPHORUS TRIFLUORIDE AND PHOSPHORUS
OXYFLUORIDE. AN AB INITIO SCF STUDY.
SERAFINI A + LABARRE JF
CHEM COMMUN 1971: 996-8

RAMAN SPECTRUM OF GASEOUS NITROGEN TRIFLUORIDE.
SHAMIR J + HYMAN HH
SPECTROCHIM ACTA 23A: 1899-1901 (1967)

ZEEMAN STUDIES INCLUDING THE MOLECULAR G VALUES,
MAGNETIC SUSCEPTIBILITY, AND MOLECULAR
QUADRUPOLE MOMENTS IN PHOSPHORUS TRIFLUORIDE,
NITROGEN TRIFLUORIDE, AND PHOSPHORYL, THIONYL
AND SULFURYL FLUORIDES.
STONE RG + POCHAN JM + FLYGARE WH
INORG CHEM 8: 2647-55 (1969)

VIBRATIONAL ANHARMONICITY IN THE PYRAMIDAL
TRIFLUORIDES.
TEICHMAN S + SMITH DF + OVEREND J
SPECTROCHIM ACTA 26A: 927-35 (1970)

METHOD OF DETERMINING SATURATED LIQUID AND
SATURATED VAPOR ENTROPY. (CHLORINE MONOFLUORIDE,
OXYGEN MONOFLUORIDE, PHOSPHORUS TRIFLUORIDE)
WALKER MA
AIAA J 1: 2636-8 (1963)

0821 SELF CONSISTENT CHARGE AND CONFIGURATION
MOLECULAR ORBITAL CALCULATIONS ON PHOSPHORUS
TRIFLUORIDE.
WENSKY DA
J CHEM PHYS 60: 1-11 (1974)

INFRARED SPECTRA OF NITROGEN TRIFLUORIDE AND
PHOSPHORUS TRIFLUORIDE.
WILSON MK + POLO SR
J CHEM PHYS 20(11): 1716-19 (1952)

0822 RAMAN SPECTRA AND MOLECULAR CONSTANTS OF
PHOSPHORUS TRIFLUORIDE AND PHOSPHINE.
YOST DM + ANDERSON TF
J CHEM PHYS 2: 624-7 (1934)

PHOSPHORUS TETRAFLUORIDE PF_4

0823 ESR SPECTRA OF PHOSPHORUS TETRAFLUORIDE RADICALS
PRODUCED IN A SINGLE CRYSTAL OF PHOSPHORUS
TRIFLUORIDE.
HASEGAWA A + OHNISHI K + SOGABE K + MIURA M
MOL PHYS 30(5): 1367-75 (1975)

0824 ISOTROPIC HYPERFINE SPLITTINGS AND MOLECULAR
STRUCTURE IN THE PHOSPHORUS TETRAFLUORIDE
RADICAL.
JIGUCHI J
J CHEM PHYS 50: 1001-9 (1969)

ELECTRON SPIN RESONANCE OF TRAPPED PHOSPHORUS
DIFLUORIDE AND PHOSPHORUS TETRAFLUORIDE
RADICALS.
NELSON W + JACKEL G + GORDY W
J CHEM PHYS 52: 4572-8 (1970)

THERMODYNAMIC PROPERTIES CF PHOSPHORUS(2),
PHOSPHORUS(4), AND SOME PHOSPHORUS FLUORIDES.
(PHOSPHORUS MONOFLUORIDE, PHOSPHORUS DIFLUORIDE,
PHOSPHORUS TRIFLUORIDE, PHOSPHORUS
TETRAFLUORIDE, AND PHOSPHORUS PENTAFLUORIDE)
O'HARE PAG
USAEC, 1968, 29P. ANL-7459

0825 INFRARED AND RAMAN SPECTRA AND STRUCTURE OF
PHOSPHORUS TETRAFLUORIDE.
RHEE KH + SNIDER AM + MILLER FA
SPECTROCHIM ACTA 29A(6): 1029-35 (1973)

PHOSPHORUS PENTAFLUORIDE PF_5

0826 CALCULATION OF FORCE CONSTANTS AND ROOT MEAN
SQUARE VIBRATION AMPLITUDES OF PHOSPHORUS
PENTAFLUORIDE, PHOSPHORUS PENTACHLORIDE, AND
PHOSPHORUS FLUORO CHLORIDE MOLECULES.
ALESHONKOVA YA + PLOTNIKOVA AD + SHARKOV VI
ZH PRIKL SPEKTROSK 22(6): 1131 (1975) (IN
RUSSIAN)

0827 PARAMETER METHOD IN THE CALCULATION OF FORCE
CONSTANTS. APPLICATION TO E' SPECIES FORCE FIELD
OF PHOSPHORUS PENTAFLUORIDE.
ANANTHAKRISHNAN TR + ARULDHAS G
J MOL STRUCT 26: 1-15 (1975)

0828 GILLESPIE-NYHOLM ASPECTS OF FORCE FIELDS.
PART-1. POINTS-ON-A-SPHERE AND EXTENDED HUCKEL
MOLECULAR ORBITAL ANALYSES OF TRIGONAL
BIPYRAMIDS. (PHOSPHORUS PENTAFLUORIDE)
BARTELL LS + PLATO V
J AMER CHEM SOC 95: 3097-3104 (1973)

0829 VIBRATIONAL SPECTRA OF MIXED PHOSPHORUS HALIDES.
(PHOSPHORUS PENTAFLUORIDE)
BEATTIE IR + LIVINGSTON KMS + REYNOLDS DJ
J CHEM PHYS 51: 4269-71 (1969)

0830 POTENTIAL FUNCTION FOR THE NU-7 VIBRATION OF
PHOSPHORUS PENTAFLUORIDE.
BERNSTEIN LS + KIM JJ + PITZER KS
+ ABRAMOWITZ S + LEVIN IW
J CHEM PHYS 62: 3671-5 (1975)

0831 ORBITALS AND STRUCTURES OF PENTAFLUORIDES.
(PHOSPHORUS PENTAFLUORIDE, ARSENIC
PENTAFLUORIDE, AND BROMINE PENTAFLUORIDE)
BERRY RS + TAMRES M + BALLHAUSEN CH + HOHANSEN H
ACTA CHEM SCAND 22: 231-46 (1968)

0832 LASER SPECTROSCOPY. (PHOSPHORUS PENTAFLUORIDE)
BORDE C + KASTLER A
COMPT REND C 271: 371-4 (1970) (IN FRENCH)

0833 PSEUDOROTATION IN TRIGONAL BIPYRAMIDAL
MOLECULES. (PHOSPHORUS PENTAFLUORIDE)
BRICKMANN V
BER BUNSENGES PHYS CHEM 75(8): 747-51 (1971) (IN
GERMAN)

0834 SPECTROSCOPIC CONSEQUENCES OF DIFFERENT
TUNNELING MECHANISMS IN NONRIGID PHOSPHORUS
PENTAFLUORIDE.
BROCAS J + FASTENAKEL D
MOL PHYS 30(1): 193-8 (1975)

0835 VESCF-MO STUDIES OF PHOSPHORUS PENTAFLUORIDE,
SULFUR TETRAFLUORIDE, AND CHLORINE TRIFLUORIDE.
BROWN RD + PEEL JB
AUST J CHEM 21: 2617-29 (1968)

0836 EXCHANGE OF FLUORINE IN PHOSPHORUS
PENTAFLUORIDE.
BROWNSTEIN S
CAN J CHEM 45: 1711-13 (1967)

INTERMOLECULAR FLUORINE EXCHANGE BETWEEN
TETRABUTYL AMMONIUM HEXAFLUORO PHOSPHATE AND
BORON TRIFLUORIDE OR PHOSPHORUS PENTAFLUORIDE.
BROWNSTEIN S + BORNAIS J
CAN J CHEM 46: 225-8 (1968)

0837 SYMMETRY GROUP OF PHOSPHORUS PENTAFLUORIDE.
CHERON M + BORDE J
J PHYS (PARIS) 35: 641-6 (1974) (IN FRENCH)

0838 UREY-BRADLEY FORCE CONSTANTS OF TRIGONAL
BIPYRAMIDAL XY5 MOLECULES. (PHOSPHORUS
PENTAFLUORIDE)
CONDRATE RA + NAKAMOTO K
DEVEL APPL SPECT 3: 169-77 (1963)

0839 RELATIVE BOND STRENGTHS IN TRIGONAL BIPYRAMIDAL
MOLECULES. (PHOSPHORUS PENTAFLUORIDE)
COTTON FA
J CHEM PHYS 35: 228-31 (1961)

0840 MEAN AMPLITUDES OF VIBRATION AND CORIOLIS
CONSTANTS FOR PHOSPHORUS PENTAFLUORIDE AND
ARSENIC PENTAFLUORIDE.
CYVIN SJ + BRUNVOLL J
J MOL STRUCT 3: 151-4 (1969)

0841 NONRIGID MOLECULE EFFECTS ON THE ROVIBRONIC
ENERGY LEVELS AND SPECTRA OF PHOSPHORUS
PENTAFLUORIDE.
DALTON BJ
J CHEM PHYS 54: 4745-62 (1971)

0842 FAR INFRARED SPECTRUM OF PHOSPHORUS
PENTAFLUORIDE.
DEITERS RM + HOLMES RR
J CHEM PHYS 48: 4796-9 (1968)

0843 SEMIEMPIRICAL MOLECULAR ORBITAL CALCULATIONS.
PSEUDOROTATION IN PHOSPHORUS PENTAFLUORIDE.
FLOREY JB + CUSACHS LC
J AMER CHEM SOC 94(9): 3040-7 (1972)

THERMODYNAMIC PROPERTIES OF PENTAFLUORIDES AT
HIGH TEMPERATURES. PART-1: CALCULATION METHODS.
(CHLORINE PENTAFLUORIDE, BROMINE PENTAFLUORIDE,
IODINE PENTAFLUORIDE, PHOSPHORUS PENTAFLUORIDE,
VANADIUM PENTAFLUORIDE, ANTIMONY PENTAFLUORIDE)
GALKIN NP + TUMANOV YN + BUTYLKIN YP
KHIM VYS ENERG 4(6): 512-8 (1970) (IN RUSSIAN)

0844 RAMAN INTENSITIES AND THE FORCE FIELD OF
PHOSPHORUS PENTAFLUORIDE.
GAY RS + FONTAL B + SPIRO TG
INORG CHEM 12(8): 1881-3 (1973)

FLUORINE NMR SPECTRA OF SULFUR, PHOSPHORUS, AND
SILICON FLUORIDES. HYDROLYSIS AND INTRA-
INTERMOLECULAR MECHANISM OF FLUORINE EXCHANGE.
GIBSON JA + IBBOTT DG + JANZEN AF
CAN J CHEM 51(19): 3203-10 (1973)

0845 STRUCTURES OF PHOSPHORUS PENTAFLUORIDE, METHYL
PHOSPHORUS TETRAFLUORIDE AND DIMETHYL PHOSPHORUS
TRIFLUORIDE.
GILLESPIE RJ
INORG CHEM 5(9): 1634-5 (1966)

0846 PHOTOELECTRON SPECTRUM OF PHOSPHORUS
PENTAFLUORIDE.
GOODMAN DW + DEWAR MJS + SCHWEIGER JR
+ COWLEY AH
CHEM PHYS LETT 21(3): 474-5 (1973)

0847 RAMAN AND INFRARED SPECTRUM OF TRIFLUORO
PHOSPHORANE. (PHOSPHORUS PENTAFLUORIDE)
GOUBEAU VJ + BAUMGARTNER R + WEISS H
Z ANORG ALLGEM CHEM 348: 286-97 (1966)

0848 NU-7 FUNDAMENTAL OF PHOSPHORUS PENTAFLUORIDE.
GRIFFITHS JE
J CHEM PHYS 42(7): 2632-3 (1965)

0849 MOLECULAR STRUCTURES OF VARIOUS PHOSPHORUS
HALIDES AND PHOSPHORUS PENTAFLUORIDE. INFRARED
AND LOW TEMPERATURE RAMAN VIBRATIONAL SPECTRA.
GRIFFITHS JE + CARTER RP + HOLMES RR
J CHEM PHYS 41: 863-76 (1964)

HEATS OF FORMATION OF INORGANIC FLUORIDES
ESPECIALLY THE ELEMENTS OF ATOMIC NUMBER BELOW
20.
GROSS P + HAYMAN C + LEVI DL + STUART MC
FULMER RES INST, 1960, 32P. PB153445

INFRARED SPECTRA OF PHOSPHORUS TRIFLUORIDE,
PHOSPHORUS OXIDE TRIFLUORIDE, AND PHOSPHORUS
PENTAFLUORIDE.
GUTOWSKY HS + LIEHR AD
J CHEM PHYS 20: 1652-3 (1952)

0850 POTENTIAL FIELD AND FORCE CONSTANTS OF SOME
TRIGONAL BIPYRAMIDAL PENTAHALIDES. (PHOSPHORUS
PENTAFLUORIDE AND ANTIMONY PENTAFLUORIDE)
HAARHOFF PC + PISTORIUS CWFT
Z NATURFORSCH 14A: 972-74 (1959)

0851 ELECTRON DIFFRACTION STUDY OF THE STRUCTURE OF
PHOSPHORUS PENTAFLUORIDE.
HANSEN KW + BARTELL LS
INORG CHEM 4: 1755-6 (1965)

ELECTRON DIFFRACTION BY GASES AND THE STRUCTURES
OF PHOSPHORUS TRIFLUORIDE, PHOSPHORUS
PENTAFLUORIDE, DIFLUORO AMINE, AND TETRAFLUORO
HYDRAZINE.
HERSH OL
UNIV MICHIGAN, PHD THESIS, 1963, 74P.

PENTACOORDINATED MOLECULES. PART-12: CORRELATION
OF EXCHANGE RATES FOR SOME GROUP-5 PENTAHALIDE
MOLECULES.
HOLMES RR
INORG CHEM 7(11): 2229-35 (1968)

0852 INTRAMOLECULAR EXCHANGE IN PHOSPHORUS
PENTAHALIDE MOLECULES.
HOLMES RR
J AMER CHEM SOC 90(18): 5021-3 (1968)

0853 ANISOTROPIC THERMAL MOTION OF TRIGONAL
BIPYRAMIDAL MOLECULES FROM SPECTROSCOPIC DATA.
PENTACOORDINATED MOLECULES. PART-13: (PHOSPHORUS
PENTAFLUORIDE)
HOLMES RR + DEITERS RM
J CHEM PHYS 51: 4043-54 (1969)

0854 PENTACOORDINATED MOLECULES. PART-14: MOLECULAR
VIBRATIONS AND STEREOCHEMICAL NONRIGIDITY OF THE
TRIGONAL BIPYRAMIDAL MODEL MX3Y2.
HOLMES RR + DEITERS RM + GOLEN JA
INORG CHEM 12: 2612-20 (1969)

0855 FAR INFRARED SPECTRUM OF PHOSPHORUS
PENTAFLUORIDE.
HOSKINS LC
J CHEM PHYS 42(7): 2631-2 (1965)

0856 VIBRATIONAL SPECTRA OF PHOSPHORUS PENTAFLUORIDE
AND ARSENIC PENTAFLUORIDE AND THE HEIGHT OF THE
BARRIER TO INTERNAL EXCHANGE OF FLUORINE NUCLEI.
HOSKINS LC + LORD RC
J CHEM PHYS 46: 2402-12 (1967)

0857 ELECTRONIC STRUCTURES OF PHOSPHORUS
PENTAFLUORIDE AND TETRAFLUORO PHOSPHORANE.
HOWELL JM + VANWAZER JR + ROSSI AR
INORG CHEM 13(7): 1747-52 (1974)

NONEMPIRICAL VALENCE ELECTRON CALCULATIONS ON
SMALL MOLECULES CONTAINING PHOSPHORUS OR SULFUR.
HYDE RG + PEEL JB + TERAUDS K
J CHEM SOC, FARADAY TRANS 73(69): 1563-8 (1973)

0858 RAMAN SPECTRUM OF SOLID PHOSPHORUS PENTAFLUORIDE
LEVIN IW
J CHEM PHYS 50: 1031 (1969)

0859 POTENTIAL FUNCTION FOR PHOSPHORUS PENTAFLUORIDE.
LEVIN IW
J MOL SPECTROSC 33: 61-71 (1970)

0860 DIRECTED VALENCY IN CERTAIN MOLECULES AND
COMPLEX IONS. (PHOSPHORUS PENTAFLUORIDE, SULFUR
HEXAFLUORIDE)
LINNETT JW + MELLISH CE
TRANS FARADAY SOC 50: 665-70 (1954)

0861 E´ FORCE FIELD OF PHOSPHORUS PENTAFLUORIDE.
LOCKETT P + FOWLER W + WILT PM
J CHEM PHYS 53: 452-3 (1970)

0862 ORBITAL MODIFICATION BY THE COULOMB FIELD OF
LIGAND ATOMS OF LATER SECOND ROW ELEMENTS IN
PERFECT PAIRING VALENCE STATES. (PHOSPHORUS
PENTAFLUORIDE, SULFUR HEXAFLUORIDE, AND CHLORINE
TRIFLUORIDE)
MACLAGAN RGAR
J CHEM SOC A 1971: 222-6

0863 POTENTIAL FIELD AND MOLECULAR VIBRATIONS OF THE
TRIGONAL BIPYRAMIDAL THE RAMAN SPECTRUM OF
PHOSPHORUS PENTAFLUORIDE GAS.
MILLER FA + CAPWELL RJ
SPECTROCHIM ACTA 27A: 125-9 (1971)

0864 MEAN AMPLITUDES OF VIBRATION FOR A TRIGONAL
BIPYRAMIDAL MOLECULE WITH APPLICATION TO
PHOSPHORUS PENTAFLUORIDE AND ARSENIC
PENTAFLUORIDE.
NAGARAJAN G + DURIG JR
BULL SOC ROY SCI LIEGE 5: 334-46 (1967)

THERMODYNAMIC PROPERTIES OF PHOSPHORUS(2),
PHOSPHORUS(4), AND SOME PHOSPHORUS FLUORIDES.
(PHOSPHORUS MONOFLUORIDE, PHOSPHORUS DIFLUORIDE,
PHOSPHORUS TRIFLUORIDE, PHOSPHORUS
TETRAFLUORIDE, AND PHOSPHORUS PENTAFLUORIDE)
O´HARE PAG
USAEC, 1968, 29P. ANL-7459

0865 FLUORINE BOMB CALORIMETRY. PART-18: STANDARD
ENTHALPY OF FORMATION OF PHOSPHORUS
PENTAFLUORIDE AND ENTHALPIES OF TRANSITION
BETWEEN VARIOUS FORMS OF PHOSPHORUS.
THERMODYNAMIC FUNCTIONS OF PHOSPHORUS
PENTAFLUORIDE BETWEEN 0 AND 1500 DEGREES K.
O´HARE PAG + HUBBARD WN
TRANS FARADAY SOC 62(10): 2709-15 (1966)

0866 INFRARED SPECTRUM OF PHOSPHORUS PENTAFLUORIDE.
PEMSLER JP + PLANET WG
J CHEM PHYS 24: 920-1 (1956)

0867 MOLECULAR CONSTANTS OF PHOSPHORUS HALIDES.
PART-4. (PHOSPHORUS PENTAFLUORIDE)
RAMASWAMY K + RAO BK
Z PHYS CHEM 242(3-4): 215-19 (1969)

0868 MOLECULAR CONSTANTS OF SOME PHOSPHORUS HALOGEN
COMPOUNDS. (PHOSPHORUS PENTAFLUORIDE)
RAMASWAMY K + RAO BK
Z PHYS CHEM 240(1-2): 127-34 (1969)

0869 ELECTRON DIFFRACTION INVESTIGATIONS OF
PHOSPHORUS PENTAFLUORIDE AND PHOSPHORUS
PENTACHLORIDE.
ROMANOV GV + SPIRIDONOV VP
J STRUCT CHEM 8(1): 131-2 (1967)

0870 PSEUDOROTATION OF TRIGONAL BIPYRAMIDAL
MOLECULES. BERRY ROTATION CONTRA TURNSTILE
ROTATION IN PHOSPHORUS PENTAFLUORIDE.
RUSSEGGER P + BRICKMANN J
CHEM PHYS LETT 30(2): 276-8 (1975)

0871 SPECTROSCOPIC STUDIES IN MEAN AMPLITUDES OF
VIBRATION OF PENTAFLUORIDES OF ARSENIC,
PHOSPHORUS, AND VANADIUM.
SANYAL NK + DIXIT L
INDIAN J PURE APPL PHYS 12(8): 550-3 (1974)

0872 HIGH RESOLUTION INFRARED SPECTRA OF THE PARALLEL
BANDS OF PHOSPHORUS PENTAFLUORIDE.
SCHATZ J + REICHMAN S
J CHEM PHYS 57: 4571-75 (1972)

FLUORIDES OF PHOSPHORUS.
SCHMULTZER R
P31-285 OF ADVAN FLUORINE CHEM, VOL 5, STACEY M
+ TATLOW JC + SHARPE AG (EDS), BUTTERWORTHS,
1965.

0873 RAMAN SPECTRA OF ARSENIC PENTAFLUORIDE AND
VANADIUM PENTAFLUORIDE AND THEIR FORCE
CONSTANTS INCLUDING THOSE OF PHOSPHORUS
PENTAFLUORIDE.
SELIG H + HOLLOWAY JH + TYSON J + CLAASSEN HH
J CHEM PHYS 53: 2559-64 (1970)

0874 OBSERVATION BY ESCA OF INEQUIVALENT FLUORINES IN CHLORINE TRIFLUORIDE, SULFUR TETRAFLUORIDE, AND PHOSPHORUS PENTAFLUORIDE.
SHAW RW + CARROLL TX + THOMAS TD
J AMER CHEM SOC 95: 2033-4 (1973)

X-RAY PHOTOELECTRON SPECTROSCOPY OF CHLORINE TRIFLUORIDE, SULFUR TETRAFLUORIDE, AND PHOSPHORUS PENTAFLUORIDE.
SHAW RW + CARROLL TX + THOMAS TD
J AMER CHEM SOC 95(18): 5870-5 (1973)

0875 ELECTRONIC STRUCTURE OF PHOSPHORUS PENTAFLUORIDE AND POLYTOPAL REARRANGEMENT IN PHOSPHORANES.
STRICH A + VEILLARD A
J AMER CHEM SOC 95: 5574-81 (1973)

0876 ASSOCIATION OF GROUP-5 PENTAFLUORIDES IN THE GAS PHASE. (PHOSPHORUS PENTAFLUORIDE, ARSENIC PENTAFLUORIDE, BISMUTH PENTAFLUORIDE AND ANTIMONY PENTAFLUORIDE)
VASILE MJ + FALCONER WE
INORG CHEM 11: 2282-3 (1972)

0877 NORMAL COORDINATE ANALYSIS OF PHOSPHORUS PENTAFLUORIDE, PHOSPHORUS DIFLUORIDE DICHLORIDE, AND PHOSPHORUS PENTACHLORIDE.
VOORN PCVD + PURCELL KF + DRAGO RS
J CHEM PHYS 43: 3457-62 (1965)

0878 GENERALIZED VALENCE FORCE FIELD FORCE CONSTANTS AND MEAN SQUARE AMPLITUDES OF VIBRATION OF SOME TRIGONAL BIPYRAMIDAL MOLECULES BY THE LOGARITHIMIC STEPS METHOD. (PHOSPHORUS PENTAFLUORIDE, ARSENIC PENTAFLUORIDE, ANTIMONY PENTAFLUORIDE, VANADIUM PENTAFLUORIDE, AND MOLYBDENUM PENTAFLUORIDE)
WENDLING EJL + MAHMOUDI S + MACCORDICK HJ
J CHEM SOC A 1971: 1747-54

0879 RAMAN SPECTRA OF GASES. PART-10: RAMAN BAND CONTOUR OF THE NU-7 (E') FUNDAMENTAL OF PHOSPHORUS PENTAFLUORIDE.
WITT JD + CARREIRA LA + DURIG JR
J MOL STRUCT 18(2): 157-62 (1973)

0880 ANALYSES OF THE VIBRATION ROTATION BANDS OF TRIFLUORO METHANE, PHOSPHORUS PENTAFLUORIDE, AND CYCLOPROPANE BY COMPUTER TECHNIQUES.
WYATT R + ROBERTS JT + WENTZ RE + WILT PM
J CHEM PHYS 50: 2552-58 (1969)

PLATINUM PENTAFLUORIDE PtF_5

THERMODYNAMIC PROPERTIES OF PENTAFLUORIDES AT HIGH TEMPERATURES. PART-3: PENTAFLUORIDES OF SOME 5D-ELEMENTS.
GALKIN NP + TUMANOV YN + BUTYLKIN YP
RUSS J PHYS CHEM 44(12): 1724-6 (1970)

PLATINUM HEXAFLUORIDE PtF_6

ELECTRIC DEFLECTION OF BINARY HEXAFLUORIDES. (SULFUR HEXAFLUORIDE, SELENIUM HEXAFLUORIDE, TELLURIUM HEXAFLUORIDE, MOLYBDENUM HEXAFLUORIDE, TUNGSTEN HEXAFLUORIDE, URANIUM HEXAFLUORIDE, RUTHENIUM HEXAFLUORIDE, RHENIUM HEXAFLUORIDE, RHODIUM HEXAFLUORIDE, OSMIUM HEXAFLUORIDE, IRIDIUM HEXAFLUORIDE, AND PLATINUM HEXAFLUORIDE)
KAISER EW + MUENTER JS + KLEMPERER W + FALCONER WE + SUNDER WA
J CHEM PHYS 53: 1411-12 (1970)

COLORS OF TRANSITION METAL HEXAFLUORIDES. (RHENIUM HEXAFLUORIDE OSMIUM HEXAFLUORIDE, IRIDIUM HEXAFLUORIDE, AND PLATINUM HEXAFLUORIDE)
MOFFITT W + GOODMAN GL + FRED M + WEINSTOCK B
MOL PHYS 2: 109-22 (1959)

0881 MOLECULAR ORBITALS OF PLATINUM HEXAFLUORIDE AND ELEMENT 110 HEXAFLUORIDE CALCULATED BY THE SELF CONSISTENT MULTIPLE SCATTERING X-ALPHA METHOD.
WABER JT + AVERILL FW
J CHEM PHYS 60(11): 4466-70 (1974)

VIBRATIONAL SPECTRA OF OSMIUM HEXAFLUORIDE AND PLATINUM HEXAFLUORIDE.
WEINSTOCK B + CLAASSEN HH + MALM JG
J CHEM PHYS 32: 181-5 (1960)

0882 PREPARATION AND SOME PROPERTIES OF PLATINUM HEXAFLUORIDE.
WEINSTOCK B + MALM JG + WEAVER EE
J AMER CHEM SOC 83: 4310-17 (1961)

0883 SUSCEPTIBILITY MEASUREMENTS AND FLUORINE NMR OF PLATINUM HEXAFLUORIDE.
ZUPAN J + PIRKMAJER E + SLIVNIK J
CROAT CHEM ACTA 39: 135-6 (1967)

PLUTONIUM MONOFLUORIDE PuF

0884 MASS SPECTROMETRIC STUDIES OF PLUTONIUM COMPOUNDS AT HIGH TEMPERATURES. PART-2: ENTHALPY OF SUBLIMATION OF PLUTONIUM TRIFLUORIDE AND THE DISSOCIATION ENERGY OF PLUTONIUM MONOFLUORIDE.
KENT RA
J AMER CHEM SOC 90(21): 5657-9 (1968)

PLUTONIUM DIFLUORIDE PuF_2

MASS SPECTROMETRIC STUDIES OF PLUTONIUM COMPOUNDS AT HIGH TEMPERATURES. PART-2: ENTHALPY OF SUBLIMATION OF PLUTONIUM TRIFLUORIDE AND THE DISSOCIATION ENERGY OF PLUTONIUM MONOFLUORIDE.
KENT RA
J AMER CHEM SOC 90(21): 5657-9 (1968)

PLUTONIUM TRIFLUORIDE PuF_3

VAPOR PRESSURES OF AMERICIUM TRIFLUORIDE AND PLUTONIUM TRIFLUORIDE, HEATS AND FREE ENERGIES OF SUBLIMATION.
CARNIGLIA SC + CUNNINGHAM BB
J AMER CHEM SOC 77: 1451-3 (1955)

0885 EXPERIMENTAL HEAT CAPACITIES OF PLUTONIUM-242 TRIFLUORIDE AND PLUTONIUM-242 TETRAFLUORIDE FROM 10 TO 350 DEGREES K, AND OF PLUTONIUM-244 DIOXIDE FROM 4 TO 25 DEGREES K. DERIVED ENTROPIES AND OTHER THERMODYNAMIC PROPERTIES AT 298.15 DEGREES K.
FLOTOW HE + OSBORNE DW + FRIED SM + MALM JG
P477-88 OF INT SYMP THERMODYNAMICS NUCLEAR MATER, 4TH PROC, 1974, IAEA, VIENNA, 1975.

MASS SPECTROMETRIC STUDIES OF PLUTONIUM COMPOUNDS AT HIGH TEMPERATURES. PART-2: ENTHALPY OF SUBLIMATION OF PLUTONIUM TRIFLUORIDE AND THE DISSOCIATION ENERGY OF PLUTONIUM MONOFLUORIDE.
KENT RA
J AMER CHEM SOC 90(21): 5657-9 (1968)

0886 HEAT CAPACITY, ENTROPY, AND ENTHALPY OF PLUTONIUM-242 TRIFLUORIDE FROM 10 TO 350 DEGREES K.
OSBORNE DW + FLOTOW HE + FRIED SM + MALM JG
J CHEM PHYS 61: 1463-8 (1974)

0887 VAPOR PRESSURE OF PLUTONIUM HALIDES. (PLUTONIUM
      TRIFLUORIDE)
      PHIPPS TE + SEARS GW + SEIFERT RL + SIMPSON OC
      J CHEM PHYS 18(5): 713-23 (1950)

PLUTONIUM TETRAFLUORIDE   PuF$_4$

0888 SUBLIMATION OF PLUTONIUM TETRAFLUORIDE.
      BERGER R + GAEUMANN T T
      HELV CHIM ACTA 44: 1084-8 (1961) (IN GERMAN)

      EXPERIMENTAL HEAT CAPACITIES OF PLUTONIUM-242
      TRIFLUORIDE AND PLUTONIUM-242 TETRAFLUORIDE FROM
      10 TO 350 DEGREES K, AND OF PLUTONIUM-244
      DIOXIDE FROM 4 TO 25 DEGREES K. DERIVED
      ENTROPIES AND OTHER THERMODYNAMIC PROPERTIES AT
      298.15 DEGREES K.
      FLOTOW HE + OSBORNE DW + FRIED SM + MALM JG
      P477-88 OF INT SYMP THERMODYNAMICS NUCLEAR
      MATER, 4TH PROC, 1974, IAEA, VIENNA, 1975.

0889 HEAT CAPACITY, ENTROPY, ENTHALPY, AND GIBBS
      ENERGY OF PLUTONIUM-242 TETRAFLUORIDE FROM 10 TO
      350 DEGREES K.
      OSBORNE DW + FLOTOW HE + FRIED SM + MALM JG
      J CHEM PHYS 63: 4613-17 (1975)

      EMPIRICAL METHOD FOR DETERMINING EFFECTIVE
      VIBRATIONAL AND ROTATIONAL CHARACTERISTICS OF
      MOLECULES OF SOME TETRAFLUORIDES FOR CALCULATING
      THEIR THERMODYNAMIC FUNCTIONS.
      TUMANOV YN + GALKIN NP
      RUSS J PHYS CHEM 43(4): 464-7 (1969)

PLUTONIUM PENTAFLUORIDE   PuF$_5$

      THERMODYNAMIC PROPERTIES OF URANIUM
      PENTAFLUORIDE NEPTUNIUM PENTAFLUORIDE, AND
      PLUTONIUM PENTAFLUORIDE.
      GALKIN NP + TUMANOV YN + BUTYLKIN YP
      RUSS J PHYS CHEM 44: 1726 (1970)

      THERMODYNAMIC PROPERTIES OF PENTAFLUORIDES AT
      HIGH TEMPERATURES. PART-4: PENTAFLUORIDES OF 5F
      ELEMENTS.
      GALKIN NP + TUMANOV YN + BUTYLKIN YP
      RUSS J PHYS CHEM 44(12): 1726-7 (1970)

PLUTONIUM HEXAFLUORIDE   PuF$_6$

0890 PREPARATION AND PROPERTIES OF (VAPOR PRESSURE)
      OF PLUTONIUM HEXAFLUORIDE AND IDENTIFICATION OF
      PLUTONIUM(VI) OXYFLUORIDE.
      FLORIN AE + TANNENBAUM IR + LEMONS JF
      J INORG NUCL CHEM 2: 368-79 (1956)

      LONG WAVELENGTH INFRARED ACTIVE FUNDAMENTAL FOR
      URANIUM HEXAFLUORIDE, NEPTUNIUM HEXAFLUORIDE,
      AND PLUTONIUM HEXAFLUORIDE.
      FRIEC B + CLAASSEN HH
      J CHEM PHYS 46: 4603-4 (1967)

      THERMODYNAMIC STABILITY OF HEXAFLUORIDES AT HIGH
      TEMPERATURES. PART-2: PLUTONIUM HEXAFLUORIDE.
      PART-3: URANIUM HEXAFLUORIDE. PART-4: MOLYBDENUM
      AND TUNGSTEN HEXAFLUORIDES.
      GALKIN NP + TUMANOV YN
      P188-91, 191-5, 195-9 OF TERMODIN TERMOKHIM
      KONSTANTY, ASTAKHOV KV(ED), NAUKA, MOSCOW, 1970.
      (IN RUSSIAN)

      THERMAL STABILITY AND REACTIVITY OF D- AND F
      ELEMENT HEXAFLUORIDES.
      GALKIN NP + TUMANOV YN + BUTYLKIN YP
      IZV SIB OTD AKAD NAUK SSSR SER KHIM NAUK (2):
      12-21 (1968) (IN RUSSIAN)

0891 INFRARED SPECTRUM OF PLUTONIUM HEXAFLUORIDE.
      HAWKINS NJ + MATTRAW HC + SABOL WW
      J CHEM PHYS 23: 2191-2 (1955)

      ELECTRON DIFFRACTION INVESTIGATION OF TUNGSTEN
      HEXAFLUORIDE, OSMIUM HEXAFLUORIDE, IRIDIUM
      HEXAFLUORIDE, URANIUM HEXAFLUORIDE, NEPTUNIUM
      HEXAFLUORIDE, AND PLUTONIUM HEXAFLUORIDE.
      KIMURA M + SCHOMAKER V + SMITH DW + WEINSTOCK B
      J CHEM PHYS 48: 4001-12 (1968)

      INFRARED SPECTRA OF NEPTUNIUM HEXAFLUORIDE AND
      PLUTONIUM HEXAFLUORIDE.
      MALM JG + WEINSTOCK B + CLAASSEN HH
      J CHEM PHYS 23: 2192-3 (1955)

0892 ABSORPTION SPECTRUM OF PLUTONIUM HEXAFLUORIDE.
      STEINDLER MJ + GUNTHER WH
      SPECTROCHIM ACTA 20: 1319-22 (1964)

0893 PLUTONIUM HEXAFLUORIDE: ITS PREPARATION AND
      PROPERTIES.
      VANKA M
      JADERNA ENERG 17(2): 46-54 (1971) (IN CZECH)

      SOME RECENT STUDIES WITH HEXAFLUORIDES. (URANIUM
      HEXAFLUORIDE, PLUTONIUM HEXAFLUORIDE, AND
      NEPTUNIUM HEXAFLUORIDE)
      WEINSTOCK B + MALM JG
      P125-9 OF INT CONF ON PEACEFUL USES OF ATOMIC
      ENERGY, 2ND, 1958, GENEVA, VOL 28, 686P.

0894 PROPERTIES OF PLUTONIUM HEXAFLUORIDE. (MOLECULAR
      STRUCTURE, ABSORPTION SPECTRUM)
      WEINSTOCK B + MALM JG
      J INORG NUCL CHEM 2: 380-94 (1956)

PRASEODYMIUM TRIFLUORIDE   PrF$_3$

0895 LASER EXCITED RAMAN SPECTRA OF MATRIX ISOLATED
      PRASEODYMIUM TRIFLUORIDE.
      LESIECKI ML + NIBLER JW + DEKOCK CW
      J CHEM PHYS 57(3): 1352-3 (1972)

      ENTHALPY OF FORMATION OF LANTHANUM TRIFLUORIDE
      AND PRASEODYMIUM TRIFLUORIDE.
      POLYACHENOK OG
      RUSS J INORG CHEM 10: 1057 (1965)

0896 VAPOR PRESSURE, HEAT OF SUBLIMATION, AND THE
      EVAPORATION COEFFICIENT OF PRASEODYMIUM
      TRIFLUORIDE.
      SKINNER HB + SEARCY AW
      J PHYS CHEM 72: 3375-81 (1968)

PRASEODYMIUM TETRAFLUORIDE   PrF$_4$

0897 PREPARATION, STRUCTURE AND SPECTRA OF
      PRASEODYMIUM TETRAFLUORIDE.
      ASPREY LB + COLEMAN JS + REISFELD MJ
      P122-6 OF LANTHANIDE-ACTINIDE CHEMISTRY, FIELDS
      PR + MOLLER T (EDS), AMER CHEM SOC, 1967, 359P.
      (ADVANCES IN CHEMISTRY SERIES NO 71)

      ELECTRIC DEFLECTION AND THERMAL DECOMPOSITION
      STUDIES ON CERIUM TETRAFLUORIDE, TERBIUM
      TETRAFLUORIDE, AND PRASEODYMIUM TETRAFLUORIDE.
      KAISER EW + SUNDER WA + FALCONER WE
      J LESS COMMON METALS 27(3): 383-7 (1972)

PROMETHIUM TRIFLUORIDE PmF_3

0898 PROPERTIES OF PROMETHIUM COMPOUNDS. (PROMETHIUM
TRIFLUORIDE)
WEIGEL F + SCHERER V
RADIOCHIM ACTA 7(1): 40-6 (1967) (IN GERMAN)

PROTACTINIUM TETRAFLUORIDE PaF_4

0899 SOME PHYSICAL PROPERTIES OF CURIUM AND
PROTACTINIUM METALS AND PROTACTINIUM
TETRAFLUORIDE.
BANSAL BM
UNIV CALIF, BERKELEY, PHD THESIS, 1966, 144P.

0900 ESTIMATED FREE ENERGIES OF FORMATION OF
PROTACTINIUM HALIDES AND ACTIVITY COEFFICIENTS
FOR PROTACTINIUM IN LIQUID BISMUTH SOLUTIONS.
(PROTACTINIUM TETRAFLUORIDE)
FERRIS LM
INORG NUCL CHEM LETT 7: 791-9 (1971)

RHENIUM PENTAFLUORIDE ReF_5

THERMODYNAMIC PROPERTIES OF PENTAFLUORIDES AT
HIGH TEMPERATURES. PART-3: PENTAFLUORIDES OF
SOME 5D-ELEMENTS.
GALKIN NP + TUMANOV YN + BUTYLKIN YP
RUSS J PHYS CHEM 44(12): 1724-6 (1970)

RHENIUM HEXAFLUORIDE ReF_6

0901 POSSIBLE ROTATIONAL RAMAN SPECTRUM AND THE
DEPOLARIZATION OF RAYLEIGH SCATTERING IN CERTAIN
HEXAFLUORIDES. (RHENIUM HEXAFLUORIDE)
BERSUKER IB
OPT SPECTROSC 12(4): 294 (APR 1962)

0902 NEAR INFRARED SYSTEM OF RHENIUM HEXAFLUORIDE.
BRAND JCD + GOODMAN GL + WEINSTOCK B
J MOL SPECTROSC 38: 449-63 (1971)

0903 VAPOR PRESSURES OF SOME HEAVY TRANSITION METAL
HEXAFLUORIDES. (TUNGSTEN HEXAFLUORIDE,
MOLYBDENUM HEXAFLUORIDE, RHENIUM HEXAFLUORIDE,
OSMIUM HEXAFLUORIDE, AND IRIDIUM HEXAFLUORIDE)
CADY GH + HARGREAVES GB
J CHEM SOC A 1961: 1563-74

VAPOR PRESSURES OF MOLYBDENUM PENTAFLUORIDE,
TUNGSTEN PENTAFLUORIDE, RHENIUM PENTAFLUORIDE,
AND OSMIUM PENTAFLUORIDE.
CADY GH + HARGREAVES GB
J CHEM SOC A 1961: 1568-74

0904 JAHN-TELLER EFFECT IN RHENIUM HEXAFLUORIDE.
CHILD MS + ROACH AC
MOL PHYS 9(3): 281-5 (1965)

RAMAN SPECTRA OF MOLYBDENUM HEXAFLUORIDE,
TECHNETIUM HEXAFLUORIDE, RHENIUM HEXAFLUORIDE,
URANIUM HEXAFLUORIDE, SULFUR HEXAFLUORIDE,
SELENIUM HEXAFLUORIDE, AND TELLURIUM
HEXAFLUORIDE IN THE VAPOR STATE.
CLAASSEN HH + GOODMAN GL + HOLLOWAY JH + SELIG H
J CHEM PHYS 53: 341-8 (1970)

0905 VIBRATIONAL SPECTRA OF RHENIUM HEXAFLUORIDE.
CLAASSEN HH + MALM JG + SELIG H
J CHEM PHYS 36: 2890-92 (1962)

0906 SPECTRUM OF RHENIUM HEXAFLUORIDE.
EISENSTEIN JC
J CHEM PHYS 33: 1530-1 (1960)

0907 VIBRATIONAL STRUCTURE AND MOLECULAR STRUCTURE OF
RHENIUM HEXAFLUORIDE.
GAUNT J
TRANS FARADAY SOC 50: 209-12 (1954)

0908 VIBRONIC INTERACTIONS IN METAL HEXAFLUORIDE
MOLECULES.
GOODMAN GL + FRED M
PAPER B102 OF INT SYMP MOL STRUCT SPECTROSC,
PROC, 1962, TOKYO, BUTTERWORTHS, 1963

0909 ELECTRON DIFFRACTION STUDY OF RHENIUM
HEXAFLUORIDE.
JACOB EJ + BARTELL LS
J CHEM PHYS 53: 2231-35 (1970)

ELECTRIC DEFLECTION OF BINARY HEXAFLUORIDES.
(SULFUR HEXAFLUORIDE, SELENIUM HEXAFLUORIDE,
TELLURIUM HEXAFLUORIDE, MOLYBDENUM HEXAFLUORIDE,
TUNGSTEN HEXAFLUORIDE, URANIUM HEXAFLUORIDE,
RUTHENIUM HEXAFLUORIDE, RHENIUM HEXAFLUORIDE,
RHODIUM HEXAFLUORIDE, OSMIUM HEXAFLUORIDE,
IRIDIUM HEXAFLUORIDE, AND PLATINUM HEXAFLUORIDE)
KAISER EW + MUENTER JS + KLEMPERER W +
FALCONER WE + SUNDER WA
J CHEM PHYS 53: 1411-12 (1970)

0910 JAHN-TELLER VIBRATIONS OF RHENIUM HEXAFLUORIDE.
LEVIN IW + ABRAMOWITZ S + MULLER A
J MOL SPECTROSC 41: 415-19 (1972)

0911 MULTIPLE INTRAMOLECULAR SCATTERING EFFECTS ON
ELECTRON DIFFRACTION PATTERNS FOR THE RHENIUM
HEXAFLUORIDE MOLECULE.
LIU JW + BONHAM RA
J MOL STRUCT 11(2): 297-304 (1972)

0912 VAPOR PRESSURES AND OTHER PROPERTIES OF RHENIUM
HEXAFLUORIDE AND RHENIUM HEPTAFLUORIDE.
MALM JG + SELIG H
J INORG CHEM 20: 189-97 (1961)

0913 JAHN-TELLER EFFECTS IN THE 2E5/2G TO 2G3/2G
TRANSITION OF RHENIUM HEXAFLUORIDE.
MCDIARMID R
J MOL SPECTROSC 38: 495-502 (1971)

0914 HIGHER ELECTRONIC STATES OF RHENIUM
HEXAFLUORIDE.
MCDIARMID R
J MOL SPECTROSC 39: 332-9 (1971)

0915 INFRARED BAND CONTOURS OF RHENIUM HEXAFLUORIDE.
MCDOWELL RS + ASPREY LB
J MOL SPECTROSC 45: 491-3 (1973)

0916 COLORS OF TRANSITION METAL HEXAFLUORIDES.
(RHENIUM HEXAFLUORIDE, OSMIUM HEXAFLUORIDE,
IRIDIUM HEXAFLUORIDE, AND PLATINUM HEXAFLUORIDE)
MOFFITT W + GOODMAN GL + FRED M + WEINSTOCK B
MOL PHYS 2: 109-22 (1959)

POTENTIAL CONSTANTS FOR MOLYBDENUM HEXAFLUORIDE
AND RHENIUM HEXAFLUORIDE.
NAGARAJAN G
AUST J CHEM 16: 906-7 (1963)

0917 MAGNETIC SUSCEPTIBILITY OF RHENIUM HEXAFLUORIDE.
SELIG H + CAFASSO FA + GRUEN DM + MALM JG
J CHEM PHYS 36: 3440-4 (1962)

VALENCE FORCE CONSTANTS OF THE HEXAFLUORIDES OF
MOLYBDENUM, TECHNETIUM, AND RHENIUM.
SINGH RB + RAI DK
CAN J PHYS 43: 167-9 (1965)

HEAT CAPACITIES AND ELECTRONIC SPECTRA OF THE
PLATINUM GROUP METAL HEXAFLUORIDE MOLECULES DOWN
TO HELIUM TEMPERATURES.
WEINSTOCK B + WESTRUM EF + GOODMAN GL
P405-6 OF INT CONF LOW TEMP PHYS, 8TH PROC,
LONDON, 1962

RHENIUM HEPTAFLUORIDE ReF_7

FLUORINE NMR OF IODINE HEPTAFLUORIDE RHENIUM
HEPTAFLUORIDE AND OXIDE PENTAFLUORIDES.
BARTLETT N + BEATON S + REEVES LW + WELLS EJ
CAN J CHEM 42: 2531-9 (1964)

ELECTRIC FIELD DEFLECTION OF MOLECULES WITH
LARGE AMPLITUDE MOTION. (XENON HEXAFLUORIDE,
IODINE HEPTAFLUORIDE, RHENIUM HEPTAFLUORIDE)
BERNSTEIN LS + PITZER KS
J CHEM PHYS 62: 2530-4 (1975)

0918 MOLECULAR CONSTANTS OF RHENIUM HEPTAFLUORIDE
MOLECULE.
BHAVSAR GP + SATHIANANDAN K
CURR SCI 41(5): 173-4 (1972)

VIBRATIONAL SPECTRA OF IODINE HEPTAFLUORIDE AND
RHENIUM HEPTAFLUORIDE.
CLAASSEN HH + GASNER EL + SELIG H
J CHEM PHYS 49: 1803-7 (1968)

0919 VIBRATIONAL SPECTRA OF RHENIUM HEPTAFLUORIDE.
CLAASSEN HH + SELIG H
J CHEM PHYS 43: 103-5 (1965)

0920 ELECTRON DIFFRACTION STUDY OF RHENIUM
HEPTAFLUORIDE. STRUCTURE, PSEUDOROTATION, AND
ANHARMONIC COUPLING OF MODES.
JACOB EJ + BARTELL LS
J CHEM PHYS 53: 2235-41 (1970)

STRUCTURE DETERMINATION OF TRIS (TRIMETHYL
SILYL) SILANE. INFRARED INVESTIGATION OF IODINE
HEPTAFLUORIDE AND AN APPLICATION OF THREE ATOM
SCATTERING THEORY TO IODINE HEPTAFLUORIDE AND
RHENIUM HEPTAFLUORIDE.
KARRENBROCK AH
UNIV MICHIGAN, PHD THESIS, 1975, 133P.

POLAR DISTORTIONS IN RHENIUM HEPTAFLUORIDE AND
IODINE HEPTAFLUORIDE.
KAISER EW + MUENTER JS + KLEMPERER W + FALCONER
J CHEM PHYS 53: 53-55 (1970)

VAPOR PRESSURES AND OTHER PROPERTIES OF RHENIUM
HEXAFLUORIDE AND RHENIUM HEPTAFLUORIDE.
MALM JG + SELIG H
J INORG CHEM 20: 189-97 (1961)

0921 PREPARATION AND PROPERTIES OF RHENIUM
HEPTAFLUORIDE.
MALM JG + SELIG H + FRIED SM
J AMER CHEM SOC 82: 1510 (1960)

0922 INTRAMOLECULAR LIGAND EXCHANGE IN SEVEN
COORDINATE STRUCTURES. (RHENIUM HEPTAFLUORIDE
AND IODINE HEPTAFLUORIDE)
MUETTERTIES EL + PACKER KH
J AMER CHEM SOC 86: 293-4 (1964)

X-RAY DIFFRACTION STUDIES OF SOME TRANSITION
METAL HEXAFLUORIDES. (SECOND ROW, THIRD ROW, AND
RHENIUM HEPTAFLUORIDE)
SIEGEL S + NORTHROP DA
INORG CHEM 5: 2187-8 (1966)

RHODIUM PENTAFLUORIDE RhF_5

0923 CONFIGURATION OF THE PENTAFLUORIDES OF RHODIUM
AND IRIDIUM BY MAGNETIC RESONANCE.
CYR T
CAN J SPECTROSC 19(5): 136-40 (1974)

RHODIUM HEXAFLUORIDE RhF_6

ELECTRIC DEFLECTION OF BINARY HEXAFLUORIDES.
(SULFUR HEXAFLUORIDE, SELENIUM HEXAFLUORIDE,
TELLURIUM HEXAFLUORIDE, MOLYBDENUM HEXAFLUORIDE,
TUNGSTEN HEXAFLUORIDE, URANIUM HEXAFLUORIDE,
RUTHENIUM HEXAFLUORIDE, RHENIUM HEXAFLUORIDE,
RHODIUM HEXAFLUORIDE, OSMIUM HEXAFLUORIDE,
IRIDIUM HEXAFLUORIDE, AND PLATINUM HEXAFLUORIDE)
KAISER EW + MUENTER JS + KLEMPERER W +
FALCONER WE + SUNDER WA
J CHEM PHYS 53: 1411-12 (1970)

JAHN-TELLER EFFECT IN THE E(8) VIBRATIONAL MODE
OF HEXAFLUORIDE MOLECULES. THE INFRARED SPECTRA
OF RUTHENIUM HEXAFLUORIDE AND RHODIUM
HEXAFLUORIDE.
WEINSTOCK B + CLAASSEN HH + CHERNICK CL
J CHEM PHYS 38: 1470-5 (1963)

RUTHENIUM PENTAFLUORIDE RuF_5

0924 RUTHENIUM PENTAFLUORIDE AND RUTHENIUM
OXYTETRAFLUORIDE. (VAPOR PRESSURES)
HOLLOWAY JH + PEACOCK RD
J CHEM SOC 1963: 527-30

0925 ELECTRONIC SPECTRA OF LIQUID RUTHENIUM AND
MOLYBDENUM PENTAFLUORIDES.
PEACOCK RD + SLEIGHT TP
J FLUORINE CHEM 1(2): 243-5 (1971)

RUTHENIUM HEXAFLUORIDE RuF_6

ELECTRIC DEFLECTION OF BINARY HEXAFLUORIDES.
(SULFUR HEXAFLUORIDE, SELENIUM HEXAFLUORIDE,
TELLURIUM HEXAFLUORIDE, MOLYBDENUM HEXAFLUORIDE,
TUNGSTEN HEXAFLUORIDE, URANIUM HEXAFLUORIDE,
RUTHENIUM HEXAFLUORIDE, RHENIUM HEXAFLUORIDE,
RHODIUM HEXAFLUORIDE, OSMIUM HEXAFLUORIDE,
IRIDIUM HEXAFLUORIDE, AND PLATINUM HEXAFLUORIDE)
KAISER EW + MUENTER JS + KLEMPERER W +
FALCONER WE + SUNDER WA
J CHEM PHYS 53: 1411-12 (1970)

0926 JAHN-TELLER EFFECT IN THE E(8) VIBRATIONAL MODE
OF HEXAFLUORIDE MOLECULES. THE INFRARED SPECTRA
OF RUTHENIUM HEXAFLUORIDE AND RHODIUM
HEXAFLUORIDE.
WEINSTOCK B + CLAASSEN HH + CHERNICK CL
J CHEM PHYS 38: 1470-5 (1963)

SAMARIUM MONOFLUORIDE SmF

MASS SPECTROMETRIC STUDIES AT HIGH TEMPERATURES.
PART-13: STABILITIES OF SAMARIUM, EUROPIUM, AND
GADOLINIUM MONO AND DIFLUORIDES.
ZMBOV KF + MARGRAVE JL
J INORG NUCL CHEM 29(1): 59-63 (1967)

SAMARIUM TRIFLUORIDE SmF_3

0927 ANOMALOUS NMR SHIFT IN SAMARIUM TRIFLUORIDE.
MALIK SK + VIJAYARAGHAVAN R
J MAG RESON 8: 161-3 (1972)

SCANDIUM MONOFLUORIDE ScF

0928 ELECTRONIC STATES OF GASEOUS SCANDIUM
MONOFLUORIDE, YTTRIUM MONOFLUORIDE, AND
LANTHANUM MONOFLUORIDE.
BARROW RF + BASTIN MW + MOORE DLG + POTT CJ
NATURE 215: 1072-3 (1967)

0929 GROUND STATES OF SCANDIUM MONOFLUORIDE AND
YTTRIUM MONOFLUORIDE.
BARROW RF + GISSANE WJM
PROC PHYS SOC 84: 615-6 (1964)

0930 ROTATIONAL ANALYSIS OF SOME SINGLET TRANSITIONS
IN THE SPECTRUM OF GASEOUS SCANDIUM
MONOFLUORIDE.
BARROW RF + GISSANE WJM + LEBARGY RC + ROSE GVM
+ ROSS PA
PROC PHYS SOC 83: 889-90 (1964)

0931 ROTATIONAL ANALYSIS OF THE SYSTEM B SINGLET PI
TO SINGLET SIGMA(+) OF GASEOUS SCANDIUM
MONOFLUORIDE.
BARROW RF + PEDERSEN L
J PHYS B 4: 11-13 (1971)

0932 ELECTRONIC STRUCTURE AND GROUND STATE OF
SCANDIUM MONOFLUCRIDE.
CARLLON KD + MOSER CM
J CHEM PHYS 46: 35-45 (1967)

0933 SPECTROSCOPIC STUDIES OF SCANDIUM MONOFLUORIDE
AND MAGNESIUM OXIDE.
GREEN DW
UNIV CALIF, BERKELEY, PHD THESIS, 1968, 139P.

0934 ELECTRONIC SPECTRUM OF SCANDIUM MONOFLUORIDE.
GURVICH LV + SHENYAVSKAYA EA
OPT SPECTROSC 14: 161-2 (1963)

THERMODYNAMIC FUNCTIONS OF GASEOUS SCANDIUM
MONOFLUORIDE, SCANDIUM DIFLUORIDE, SCANDIUM
TRIFLUORIDE, YTTRIUM MONOFLUORIDE, YTTRIUM
DIFLUORIDE, YTTRIUM TRIFLUORIDE, LANTHANUM
MONOFLUORIDE, LANTHANUM DIFLUORIDE, AND
LANTHANUM TRIFLUCRIDE.
KRASNOV KS + DANILOVA TG
HIGH TEMP 7(6): 1131-3 (1969)

MOLECULAR CONSTANTS OF SCANDIUM, YTTRIUM, AND
LANTHANUM HALIDES. (SCANDIUM DIFLUORIDE,
SCANDIUM MONOFLUORIDE, YTTRIUM DIFLUORIDE,
YTTRIUM MONOFLUORIDE, LANTHANUM DIFLUORIDE, AND
LANTHANUM MONOFLUORIDE)
KRASNOV KS + TIMOSHININ VS
HIGH TEMP 7(2): 333-4 (1969)

0935 DISSOCIATION ENERGIES AND STABILITIES OF
SCANDIUM MONOFLUORIDE, YTTRIUM MONOFLUORIDE, AND
LANTHANUM MONOFLUORIDE.
KRASNOV KS
IZV VYSSH UCHEB ZAVED, KHIM 12: 578-82 (1969)

0936 EVALUATION OF MOLECULAR VIBRATION FREQUENCIES OF
SCANDIUM MONOFLUCRIDE, YTTRIUM MONOFLUORIDE, AND
LANTHANUM MONOFLUORIDE.
KRASNOV KS
IZV VYSSH UCHEB ZAVED, KHIM 1: 594-6 (1967)

0937 REASSIGNMENT OF MOLECULAR ORBITAL CONFIGURATIONS
OF THE ELECTRONIC STATES OF SCANDIUM
MONOFLUORIDE.
SCOTT PR + RICHARDS WG
CHEM PHYS LETT 28(1): 101-3 (1974)

0938 VIBRONIC SPECTRUM OF SCANDIUM MONOFLUORIDE,
YTTRIUM MONOFLUORIDE, AND LANTHANUM
MONOFLUORIDE.
SHENYAVSKAYA EA + MAL'TSEV AA
VEST MOSK UNIV KHIM 22: 104-5 (1967) (IN
RUSSIAN)

0939 MASS SPECTROMETRIC STUDIES AT HIGH TEMPERATURES.
PART-18: THE STABILITIES OF THE MONO- AND
DIFLUORIDES OF SCANDIUM AND YTTRIUM.
ZMBOV KF + MARGRAVE JL
J CHEM PHYS 47(9): 3122-5 (1967)

SCANDIUM DIFLUORIDE ScF_2

0940 THERMODYNAMIC CHARACTERISTICS OF SCANDIUM
DIHALIDES AND THE THE POSSIBILITY OF THEIR
SYNTHESIS BY THE REDUCTION OF THE TRIHALIDES.
(SCANDIUM DIFLUORIDE)
KOMISSAROVA LN + TARASOV LK
RUSS J INORG CHEM 14: 324-5 (1969)

THERMODYNAMIC FUNCTIONS OF GASEOUS SCANDIUM
MONOFLUORIDE, SCANDIUM DIFLUORIDE, SCANDIUM
TRIFLUORIDE, YTTRIUM MONOFLUORIDE, YTTRIUM
DIFLUORIDE, YTTRIUM TRIFLUORIDE, LANTHANUM
MONOFLUORIDE, LANTHANUM DIFLUORIDE, AND
LANTHANUM TRIFLUORIDE.
KRASNOV KS + DANILOVA TG
HIGH TEMP 7(6): 1131-3 (1969)

0941 MOLECULAR CONSTANTS OF SCANDIUM, YTTRIUM, AND
LANTHANUM HALIDES. (SCANDIUM DIFLUORIDE,
SCANDIUM MONOFLUORIDE, YTTRIUM DIFLUORIDE,
YTTRIUM MONOFLUORIDE, LANTHANUM DIFLUORIDE, AND
LANTHANUM MONOFLUORIDE)
KRASNOV KS + TIMOSHININ VS
HIGH TEMP 7(2): 333-4 (1969)

MASS SPECTROMETRIC STUDIES AT HIGH TEMPERATURES.
PART-18: THE STABILITIES OF THE MONO- AND
DIFLUORIDES OF SCANDIUM AND YTTRIUM.
ZMBOV KF + MARGRAVE JL
J CHEM PHYS 47(9): 3122-5 (1967)

SCANDIUM TRIFLUORIDE ScF_3

0942 STANDARD ENTHALPIES OF FORMATION OF SCANDIUM
TRIFLUORIDE, YTTRIUM TRIFLUORIDE, AND LANTHANUM
TRIFLUORIDE.
FINOGENOV AD
RUSS J PHYS CHEM 45: 900-1 (1971)

0943 GEOMETRIES AND ENTROPIES OF METAL TRIFLUORIDES
FROM INFRARED SPECTRA. (SCANDIUM, YTTRIUM,
LANTHANUM, CERIUM, NEODYMIUM, EUROPIUM, AND
GADOLINIUM TRIFLUORIDES)
HASTIE JW + HAUGE RH + MARGRAVE JL
J LESS COMMON METALS 39(2): 309-34 (1975)

0944 FORCE CONSTANTS AND GEOMETRIES OF MATRIX
ISOLATED RARE EARTH TRIFLUORIDES. (SCANDIUM
TRIFLUORIDE, YTTRIUM TRIFLUORIDE, AND LANTHANUM
TRIFLUORIDE)
HAUGE RH + HASTIE JW + MARGRAVE JL
J LESS COMMON METALS 23: 359-65 (1971)

0945 STRUCTURE AND SOME PROPERTIES OF SCANDIUM
TRIFLUORIDE.
IPPOLITOV YEG + MAKLACHKOV AG
RUSS J INORG CHEM 15: 753-5 (1970)

0946 ELECTRIC DEFLECTION OF MOLECULAR BEAMS OF THE
RARE EARTH DIFLUORIDES AND TRIFLUORIDES.
(LANTHANIDE TRIFLUORIDES, LANTHANIDE
DIFLUORIDES, SCANDIUM TRIFLUORIDE, AND YTTRIUM
TRIFLUORIDE)
KAISER EW + FALCONER WE + KLEMPERER W
J CHEM PHYS 56: 5392-8 (1972)

0947 SUBLIMATION PRESSURES OF SCANDIUM TRIFLUORIDE,
YTTRIUM TRIFLUORIDE, AND LANTHANUM TRIFLUORIDE.
KENT RA + ZMBOV KF + KANA'AN AS + BESENBRUCH G
+ MCDONALD JD + MARGRAVE JL
J INORG NUCL CHEM 28: 1419-27 (1966)

0948 CALCULATION OF DISSOCIATION ENERGIES FOR THE
HALIDES OF THE SCANDIUM SUBGROUP. (SCANDIUM
TRIFLUORIDE, YTTRIUM TRIFLUORIDE, AND LANTHANUM
TRIFLUORIDE)
KRASNOV KS
HIGH TEMP 4(1): 128-30 (1966)

0949 CALCULATION OF THE VIBRATIONAL FREQUENCIES OF
SCANDIUM SUBGROUP HALIDES.
KRASNOV KS
HIGH TEMP 5(4): 639-40 (1967)

FORCE CONSTANT OF THE NONPLANAR VIBRATION OF AN
XY3 MOLECULE AND A MODEL WITH POLARIZABLE IONS.
KRASNOV KS
IZV VYSSH UCHEB ZAVED KHIM KHIM TEKHNOL 10(9):
997-1000 (1967) (IN RUSSIAN)

THERMODYNAMIC FUNCTIONS OF GASEOUS SCANDIUM
MONOFLUORIDE, SCANDIUM DIFLUORIDE, SCANDIUM
TRIFLUORIDE, YTTRIUM MONOFLUORIDE, YTTRIUM
DIFLUORIDE, YTTRIUM TRIFLUORIDE, LANTHANUM
MONOFLUORIDE, LANTHANUM DIFLUORIDE, AND
LANTHANUM TRIFLUORIDE.
KRASNOV KS + DANILOVA TG
HIGH TEMP 7(6): 1131-3 (1969)

0950 THERMODYNAMIC PROPERTIES OF SCANDIUM
TRIFLUORIDE.
PETZEL VT
Z ANORG ALLGEM CHEM 395: 1-18 (1973)

0951 ENERGETICS AND STABILITY OF GAS PHASE SCANDIUM
FLUORIDES.
SONIN VI + POLYACHENOK OG + IPPOLITOV YEG
RUSS J INORG CHEM 18(11): 1552-4 (1973)

0952 SATURATED VAPOR PRESSURE OF SCANDIUM
TRIFLUORIDE, YTTRIUM TRIFLUORIDE, AND LANTHANUM
TRIFLUORIDE.
SUVOROV AL + NOVIKOV GI
VEST LENINGRAD UNIV, FIZ KHIM 23: 83-8 (1968)
(IN RUSSIAN)

0953 MASS SPECTROMETRIC STUDIES OF SCANDIUM
TRIFLUORIDE, YTTRIUM TRIFLUORIDE, LANTHANUM
TRIFLUORIDE AND THE RARE EARTH TRIFLUORIDES.
ZMBOV KF + MARGRAVE JL
P267-90 OF MASS SPECTROMETRY IN INORGANIC
CHEMISTRY, MARGRAVE JL (ED), AMER CHEM SOC,
1968, 329P. (ADVANCES IN CHEMISTRY SERIES NO 72)

SELENIUM MONOFLUORIDE   SeF

0954 ESR SPECTRUM OF BROMINE OXIDE, IODINE OXIDE AND
SELENIUM MONOFLUORIDE IN J-5/2 ROTATIONAL
LEVELS.
BROWN JM + BYFLEET CR + HOWARD BJ + RUSSELL DK
MOL PHYS 23: 457-68 (1972)

ELECTRIC DIPOLE MOMENTS OF OPEN SHELL DIATOMIC
MOLECULES. (SULFUR MONOFLUORIDE AND SELENIUM
MONOFLUORIDE)
BYFLEET CR + CARRINGTON A + RUSSELL DK
MOL PHYS 20: 271-77 (1971)

GAS PHASE ELECTRON RESONACE SPECTRA OF SULFUR
MONOFLUORIDE, SELENIUM MONOFLUORIDE, SELENIUM
OXIDE AND IODINE OXIDE.
CARRINGTON A + CURRIE GN + DYER PN + LEVY DH +
MILLER TA
CHEM COMMUN 1967: 641-2

GAS PHASE ELECTRON RESONANCE SPECTRA OF SULFUR
MONOFLUORIDE AND SELENIUM MONOFLUORIDE.
CARRINGTON A + CURRIE GN + MILLER T
J CHEM PHYS 50: 2726-32 (1969)

0955 DISSOCIATION ENTHALPY, SPECTROSCOPIC CONSTANTS,
AND OTHER PROPERTIES OF SELENIUM MONOFLUORIDE
FROM A MOLECULAR ORBITAL STUDY.
O'HARE PAG
J CHEM PHYS 60(10): 4084-5 (1974)

HARTREE-FOCK WAVEFUNCTIONS AND COMPUTED
PROPERTIES FOR THE DOUBLET PI GROUND STATES OF
SULFUR MONOFLUORIDE AND SELENIUM MONOFLUORIDE
AND THEIR POSITIVE AND NEGATIVE IONS. A
COMPARISON OF THE THEORETICAL AND EXPERIMENTAL
RESULTS. (OXYGEN MONOFLUORIDE)
O'HARE PAG + WAHL AC
J CHEM PHYS 53: 2834-46 (1970)

SELENIUM TETRAFLUORIDE   SeF$_4$

0956 VIBRATIONAL SPECTRA AND STRUCTURES OF SELENIUM
TETRAFLUORIDE AND TELLURIUM TETRAFLUORIDE,
INCLUDING A MATRIX ISOLATION STUDY.
ADAMS CJ + DOWNS AJ
SPECTROCHIM ACTA 28A: 1841-1854 (1972)

INFRARED SPECTRA OF SOME VOLATILE INORGANIC
FLUORIDES IN THE SOLID STATE. (ARSENIC
TRIFLUORIDE, SULFUR TETRAFLUORIDE, AND SELENIUM
TETRAFLUORIDE)
AYNSLEY EE + DODD RE + LITTLE R
SPECTROCHIM ACTA 18: 1005-8 (1962)

0957 MICROWAVE SPECTRUM, DIPOLE MOMENT, AND STRUCTURE
ANALYSIS OF SELENIUM TETRAFLUORIDE.
BOWATER IC + BROWN RD + BURDEN FR
J MOL SPECTROSC 28: 454-60 (1968)

INVESTIGATION BY ELECTRON DIFFRACTION OF THE
MOLECULAR STRUCTURES OF SULFUR HEXAFLUORIDE,
SULFUR TETRAFLUORIDE, SELENIUM HEXAFLUORIDE, AND
SELENIUM TETRAFLUORIDE.
EWING VC + SUTTON LE
TRANS FARADAY SOC 59: 1242-7 (1963)

STRUCTURE OF EXCHANGE PROCESSES IN SOME
INORGANIC FLUORIDES BY NMR. (SULFUR
TETRAFLUORIDE AND SELENIUM TETRAFLUORIDE)
MUETTERTIES EL + PHILLIPS ED
J AMER CHEM SOC 81: 1084-8 (1959)

VIBRATIONAL SPECTRA OF MOLYBDENUM AND TUNGSTEN
PENTAFLUORIDES.
OUELLETTE TJ + RATCLIFFE CT + SHARP DWA
J CHEM SOC A 1969: 2351-4

0958 VIBRATIONAL ASSIGNMENT AND NORMAL COORDINATE
ANALYSIS OF SELENIUM TETRAFLUORIDE.
RAMASWAMY K + JAYARAMAN S
INDIAN J PURE APPL PHYS 8: 625-8 (1970)

SELENIUM PENTAFLUORIDE SeF_5

0959 PARAMAGNETIC HEXAFLUORIDE ANIONS OF GROUP-6.
(SELENIUM PENTAFLUORIDE)
MORTON JR + PREATON KF + TAIT JC
J CHEM PHYS 62: 2029-1 (1975)

SELENIUM HEXAFLUORIDE SeF_6

0960 VIBRATIONAL ANALYSIS OF SELENIUM HEXAFLUORIDE
AND TUNGSTEN HEXAFLUORIDE.
ABRAMOWITZ S + LEVIN IW
INORG CHEM 3: 538-41 (1967)

VAPOR PHASE RAMAN SPECTRA, RAMAN BAND CONTOUR
ANALYSES, AND CORIOLIS CONSTANTS OF THE
SPHERICAL TOP MOLECULES, SULFUR HEXAFLUCRIDE,
SELENIUM HEXAFLUCRIDE, TELLURIUM HEXAFLUORIDE,
MOLYBDENUM HEXAFLUORIDE, TUNGSTEN HEXAFLUORIDE,
AND URANIUM HEXAFLUORIDE.
BOSWORTH YM + CLARK RJH + RIPPON DM
J MOL SPECTROSC 46: 240-55 (1973)

RAMAN SPECTRA OF MOLYBDENUM HEXAFLUORIDE,
TECHNETIUM HEXAFLUORIDE, RHENIUM HEXAFLUORIDE,
URANIUM HEXAFLUORIDE, SULFUR HEXAFLUORIDE,
SELENIUM HEXAFLUORIDE, AND TELLURIUM
HEXAFLUORIDE IN THE VAPOR STATE.
CLAASSEN HH + GOODMAN GL + HOLLOWAY JH + SELIG H
J CHEM PHYS 53: 341-8 (1970)

INVESTIGATION BY ELECTRON DIFFRACTION OF THE
MOLECULAR STRUCTURES OF SULFUR HEXAFLUORIDE,
SULFUR TETRAFLUORIDE, SELENIUM HEXAFLUORIDE, AND
SELENIUM TETRAFLUORIDE.
EWING VC + SUTTON LE
TRANS FARADAY SOC 59: 1242-7 (1963)

0961 INFRARED SPECTRA AND MOLECULAR STRUCTURE OF SOME
GROUP-6 HEXAFLUORIDES. (OF SULFUR, SELENIUM,
TELLURIUM, MOLYBDENUM, TUNGSTEN, AND URANIUM)
GAUNT J
TRANS FARADAY SOC 49: 1122-31 (1953)

MOLECULAR FORCE FIELDS. PART-8: VIBRATION
FREQUENCIES OF SOME OCTAHEDRAL XY6 MOLECULES.
(SULFUR HEXAFLUORIDE, SELENIUM HEXAFLUORIDE,
TELLURIUM HEXAFLUORIDE)
HEATH DF + LINNETT JW
TRANS FARADAY SOC 45: 264-71 (1949)

ELECTRIC DEFLECTION OF BINARY HEXAFLUORIDES.
(SULFUR HEXAFLUORIDE, SELENIUM HEXAFLUORIDE,
TELLURIUM HEXAFLUORIDE, MOLYBDENUM HEXAFLUORIDE,
TUNGSTEN HEXAFLUORIDE, URANIUM HEXAFLUORIDE,
RUTHENIUM HEXAFLUORIDE, RHENIUM HEXAFLUORIDE,
RHODIUM HEXAFLUORIDE, OSMIUM HEXAFLUORIDE,
IRIDIUM HEXAFLUORIDE, AND PLATINUM HEXAFLUORIDE)
KAISER EW + MUENTER JS + KLEMPERER W +
FALCONER WE + SUNDER WA
J CHEM PHYS 53: 1411-12 (1970)

ENTHALPIES OF FORMATION OF SULFUR HEXAFLUORIDE,
SELENIUM HEXAFLUORIDE, TELLURIUM HEXAFLUORIDE
AND THEIR THERMODYNAMIC PROPERTIES.
O'HARE PAG + SETTLE JL + HUBBARD WN
TRANS FARADAY SOC 62(3): 558-65 (1966)

0962 ROOT MEAN SQUARE AMPLITUDES IN SOME
HEXAFLUORIDES OF OCTAHEDRAL SYMMETRY. (SELENIUM
AND TELLURIUM FLUORIDES)
NAGARAJAN G + ADAMS TS
Z PHYS CHEM (LEIPZIG) 255: 869-88 (1974)

GREEN'S FUNCTION ANALYSIS OF SULFUR HEXAFLUORIDE
AND RELATED MOLECULES. (SELENIUM HEXAFLUORIDE,
TELLURIUM HEXAFLUORIDE)
RAMASWAMY K + MOHAN N
J MOL STRUCT 7: 51-8 (1971)

0963 RAMAN SPECTRA AND MOLECULAR CONSTANTS OF THE
HEXAFLUORIDES OF SULFUR, SELENIUM, AND
TELLURIUM.
YOST DM + STEFFENS CC + GROSS ST
J CHEM PHYS 2: 311-16 (1934)

SILICON MONOFLUORIDE SiF

0964 SPECTRA OF SILICON MONOFLUORIDE, GERMANIUM
MONOFLUORIDE, TIN MONOFLUORIDE, AND LEAD
MONOFLUORIDE.
BARROW RF + BUTLER D + JOHNS JWC + POWELL JL
PROC PHYS SOC 73: 317-20 (1959)

MASS SPECTROMETRIC STUDIES AT HIGH TEMPERATURES.
PART-2: THE DISSOCIATION ENERGIES OF THE
MONOFLUORIDES AND DIFLUORIDES OF SILICON AND
GERMANIUM.
EHLERT TC + MARGRAVE JL
J CHEM PHYS 41(4): 1066-72 (1964)

REEVALUATION OF THE DISSOCIATION ENERGY OF
CALCIUM FLUORIDE. (ALSO SILICON AND GERMANIUM
FLUORIDES)
HASTIE JW + MARGRAVE JL
J CHEM ENG DATA 13(3): 428-9 (1968)

0965 BAND SPECTRUM OF SILICON MONOFLUORIDE.
JOHNS JWC + BARROW RF
PROC PHYS SOC 71: 476-84 (1958)

0966 FRANCK-CONDON FACTORS AND R CENTROIDS OF A
SINGLET SIGMA- X SINGLET SIGMA PLATINUM OXIDE
AND SILICON MONOFLUORIDE (D DOUBLET SIGMA(+)- X
DOUBLET PI) MOLECULES.
KATTI PH + KORWAR VM
CURR SCI 44(21): 768-70 (1975)

0967 ELECTRONIC TRANSITION MOMENT AS A FUNCTION OF
THE INTERNUCLEAR SEPARATION FOR THE B-X BAND
SYSTEM OF MOLECULAR SILICON MONOFLUORIDE.
KUZMENKO NE + SMIRNOV AD + KUZ'YAKOV YY
VEST MOSK UNIV KHIM 11(4): 478-80 (1970) (IN
RUSSIAN)

0968 DISTRIBUTION OF INTENSITY IN THE ELECTRON
EMISSION SPECTRUM OF SILICON MONOFLUORIDE.
KUZ'MENKO NE + KUZYAKOV YY + SMIRNOV AD
J APPL SPECTROSC 13(4): 1304-6 (1970)

0969 CORRELATION OF THE ELECTRONIC TRANSITION MOMENT
WITH INTERNUCLEAR DISTANCE FOR BANDS OF THE X
SYSTEM.
KUZ'MENKO NE + SMIRNOV AD + KUZYAKOV YY
VEST MOSK UNIV KHIM 11: 357-59 (1970) (IN
RUSSIAN)

0970 DETERMINATION OF THE MATRIX ELEMENT OF THE
DIPOLE MOMENT OF AN A DOUBLET SIGMA TO X DOUBLET
PI ELECTRON TRANSITON IN A SILICON MONOFLUORIDE
MOLECULE.
KUZYAKOV YY + OVCHARENKO IE + KUZ'MENKO NE
+ KURDYUMOVA IN
J APPL SPECTROSC 12(3): 425-7 (1970)

ROOT MEAN SQUARE AMPLITUDES OF VIBRATION IN SOME
GROUP-4 TETRAHALIDES. (CARBON TETRAFLUORIDE AND
SILICON MONOFLUORIDE)
LONG DA + SEIBOLD EA
TRANS FARADAY SOC 56: 1105-9 (1960)

0971 A QUARTET SIGMA TO X DOUBLET PI ELECTRONIC
TRANSITION OF SILICON MONOFLUORIDE.
MARTIN RW + MERER AJ
CAN J PHYS 51: 634-43 (1973)

0972 FRANCK-CONDON FACTORS AND R CENTROIDS OF SOME
BAND SYSTEMS OF SILICON MONOFLUORIDE, CALCIUM
MONOFLUORIDE, AND BISMUTH MONOFLUORIDE.
MOHANTY BS + SINGH ON
INDIAN J PURE APPL PHYS 7: 109-11 (1969)

0973 PROPERTIES OF NITROGEN MONOFLUORIDE, SILICON
MONOFLUORIDE, PHOSPHORUS MONOFLUORIDE, AND
SULFUR MONOFLUORIDE FROM A MOLECULAR ORBITAL
STUDY.
O'HARE PAG
J CHEM PHYS 59: 3842-47 (1973)

MOLECULAR ORBITAL INVESTIGATION OF CARBON
MONOFLUORIDE AND SILICON MONOFLUORIDE AND THEIR
POSITIVE AND NEGATIVE IONS.
O'HARE PAG + WAHL AC
J CHEM PHYS 55: 666-76 (1971)

0974 EMISSION SPECTRUM OF THE GAMMA SYSTEM OF SILICON
MONOFLUORIDE.
SANKARANARAYANAN S + NARAYANAN PS
PROC NAT INST SCI INDIA A 32: 56-62 (1965)

0975 FRANCK-CONDON FACTORS AND R CENTROIDS FOR THE
C-X SYSTEM OF SILICON MONOFLUORIDE MOLECULE.
SINGH ID + MAHESHWARI RC
INDIAN J PURE APPL PHYS 7: 708-9 (1969)

POTENTIAL CURVES FOR SOME DIATOMIC MOLECULES.
(BERYLLIUM MONOFLUORIDE, SILICON MONOFLUORIDE,
AND TIN MONOFLUORIDE)
SINGH RB + RAI DK
INDIAN J PURE APPL PHYS 4: 102-5 (1966)

0976 ROTATIONAL ANALYSIS OF THE BETA BAND SYSTEM OF
SILICON MONOFLUORIDE MOLECULE.
SINGH ON + SINGH IS
CURR SCI 37: 9-10 (1968)

0977 CALCULATION OF THE FRANCK-CONDON FACTOR FOR
MOLECULES OF SILICON FLUORIDE, GERMANIUM
FLUORIDE, AND GERMANIUM CHLORIDE.
SMIRNOV AD + KUZ'MENKO NE + BOLOTIN AB
LIET FIZ RINKINYS 14(4): 573-6 (1974) (IN
RUSSIAN)

0978 A QUARTET SIGMA TO DOUBLET PI TRANSITION OF THE
SILICON MONOFLUORIDE MOLECULE.
VERMA RD
CAN J PHYS 40: 586-97 (1962)

SILICON DIFLUORIDE SiF_2

0979 ROTATIONAL ANALYSIS OF ABSORPTION BANDS IN THE
2266 ANGSTROM SYSTEM OF SILICON DIFLUORIDE.
DIXON RN + HALLE M
J MOL SPECTROSC 36: 192-203 (1970)

MASS SPECTROMETRIC STUDIES AT HIGH TEMPERATURES.
PART-2: THE DISSOCIATION ENERGIES OF THE
MONOFLUORIDES AND DIFLUORIDES OF SILICON AND
GERMANIUM.
EHLERT TC + MARGRAVE JL
J CHEM PHYS 41(4): 1066-72 (1964)

0980 PHOTOELECTRON SPECTRUM OF SILICON DIFLUORIDE.
FEHLNER TP + TURNER DW
INORG CHEM 13(3): 754-5 (1974)

0981 VACUUM ULTRAVIOLET SPECTRA OF SILICON DIFLUORIDE
AND GERMANIUM DIFLUORIDE.
GOLE JL + HAUGE RH + MARGRAVE JL + HASTIE JW
J MOL SPECTROSC 43(3): 441-51 (1972)

0982 INFRARED SPECTRA OF SILICON DIFLUORIDE IN INERT
GAS MATRICES.
HASTIE JW + HAUGE RH + MARGRAVE JL
J AMER CHEM SOC 91: 2536-8 (1969)

REEVALUATION OF THE DISSOCIATION ENERGY OF
CALCIUM FLUORIDE. (ALSO SILICON AND GERMANIUM
FLUORIDES)
HASTIE JW + MARGRAVE JL
J CHEM ENG DATA 13(3): 428-9 (1968)

0983 ULTRAVIOLET SPECTRUM OF SILICON DIFLUORIDE.
JOHNS JWC + CHANTRY GW + BARROW RF
TRANS FARADAY SOC 54: 1589-91 (1958)

0984 ULTRAVIOLET ABSORPTION SPECTRUM OF SILICON
DIFLUORIDE.
KHANNA VM + BESENBRUCH G + MARGRAVE JL
J CHEM PHYS 46: 2310-14 (1967)

0985 INFRARED SPECTRUM, FORCE CONSTANTS, AND
THERMODYNAMIC FUNCTIONS OF SILICON DIFLUORIDE.
KHANNA VM + HAUGE RH + CURL RF + MARGRAVE JL
J CHEM PHYS 47: 5031-36 (1967)

0986 MATRIX ISOLATION STUDY OF THE VACUUM ULTRAVIOLET
PHOTOLYSIS OF SILICON DIFLUORIDE DIHYDRIDE. THE
INFRARED AND ULTRAVIOLET SPECTRA OF THE FREE
RADICAL SILICON DIFLUORIDE.
MILLIGAN DE + JACOX ME
J CHEM PHYS 49: 4269-75 (1968)

MEAN AMPLITUDES OF VIBRATION AND THERMODYNAMIC
FUNCTIONS FOR SILICON DIFLUORIDE AND OXYGEN
DIFLUORIDE.
NAGARAJAN G
BULL SOC CHIM BELG 71: 337-46 (1962)

0987 NEW ELECTRONIC EMISSION FROM SILICON DIFLUORIDE.
RAO DR
J MOL SPECTROSC 34: 284-87 (1970)

0988 EMISSION SPECTRUM OF SILICON DIFLUORIDE. PART-1:
THE BAND SYSTEM IN THE REGION 2755-2179
ANGSTROMS.
RAO DR + VENKATESWARLU P
J MOL SPECTROSC 7: 287-303 (1961)

0989 MICROWAVE SPECTRUM AND FORCE CONSTANTS OF
SILICON DIFLUORIDE. CENTRIFUGAL DISTORTION.
RAO VM + CURL RF
TRANS AMER CRYSTALLOGR ASSOC 2: 183-9 (1966)

0990 MICROWAVE SPECTRUM OF SILICON DIFLUORIDE.
RAO VM + CURL RF + TIMMS PL + MARGRAVE JL
J CHEM PHYS 43(7): 2557-8 (1965)

0991 ULTRAVIOLET BAND SPECTRUM OF SILICON DIFLUORIDE.
SANKARANARAYANAN S
PROC NAT INST SCI INDIA A 28: 311-16 (1962)

0992 THERMOCHEMISTRY OF SILICON DIFLUORIDE.
SCHAFER VH + BRUDERRECK H + MORCHER B
Z ANORG ALLGEM CHEM 352: 122-37 (1967)

0993 MICROWAVE SPECTRUM OF SILICON DIFLUORIDE IN THE
EXCITED VIBRATIONAL STATES, EQUILIBRIUM
STRUCTURE, ANHARMONIC POTENTIAL FUNCTION, AND
NU-1- NU-3 CORIOLIS RESONANCE.
SHOJI H + TANAKA T + HIROTA E
J MOL SPECTROSC 47(2): 268-74 (1973)

0994 VIBRATIONAL TRANSITION PROBABILITIES AND R
CENTROIDS FOR DIATOMIC FLUORIDES OF SILICON AND
GERMANIUM.
SINGH J
INDIAN J PURE APPL PHYS 13(3): 204-6 (1975)

0995 NONEMPIRICAL LCAO MO SCF CALCULATIONS OF THE
ELECTRONIC STRUCTURE OF SILICON DIFLUORIDE.
THOMSON C
THEOR CHIM ACTA 32(2): 93-100 (1973)

0996 PHOTOELECTRON SPECTRUM OF SILICON DIFLUORIDE.
WESTWOOD NPC
CHEM PHYS LETT 25(4): 558-61 (1974)

0997 ELECTRONIC SPECTRUM OF SILICON DIFLUORIDE.
THEORETICAL STUDY.
WIRSAM B
CHEM PHYS LETT 22(2): 360-3 (1973)

SILICON TRIFLUORIDE SiF_3

0998   EMISSION SPECTRUM OF SILICON TRIFLUORIDE.
        WANG JL + KRISHNAN CN + MARGRAVE JL
        J MOL SPECTROSC 48(2): 346-53 (1973)

SILICON TETRAFLUORIDE SiF_4

0999   ROTATIONAL ANALYSIS OF THE GAMMA BANDS OF
        SILICON MONOFLUORIDE.
        APPELBLAD O + BARROW RF + VERMA RD
        J PHYS B (PROC PHYS SOC) 1: 274-81 (1968)

        VAPOR PHASE RAMAN SPECTRA OF MH4 (M= CARBON,
        SILICON, GERMANIUM, AND TIN) AND MF4 (M= CARBON,
        SILICON, AND GERMANIUM).
        ARMSTRONG RS + CLARK RJH
        J CHEM SOC, FARADAY TRANS 2 72: 11-21 (1976)

        INFRARED ABSORPTION SPECTRA OF SOME POLYATOMIC
        FLUORIDES. (SILICON TETRAFLUORIDE, CARBON
        TETRAFLUORIDE, BORON TRIFLUORIDE, NITROGEN
        TRIFLUORIDE)
        BAILEY CR + HALE JB + THOMPSON JW
        J CHEM PHYS 5: 274-5 (1937)

        HELIUM-1 RESONANCE PHOTOELECTRON SPECTRA OF
        GROUP-4 TETRAFLUORIDES.
        BASSETT PJ + LLOYD DR
        CHEM PHYS LETT 3(1): 22-4 (1969)

        PHOTOELECTRON SPECTRA OF HALIDES. PART-1.
        TETRAFLUORIDES AND TETRACHLORIDES OF GROUP-5B.
        (CARBON TETRAFLUORIDE, SILICON TETRAFLUORIDE,
        GERMANIUM TETRAFLUORIDE, TIN TETRAFLUORIDE)
        BASSETT PJ + LLOYD DR
        J CHEM SOC A 1971: 641-54

1000   SILICON- FLUORINE LENGTH IN SILICON
        TETRAFLUORIDE. NEW ELECTRON DIFFRACTION STUDY.
        BEAGLEY B + BROWN DP + FREEMAN JM
        J MOL STRUCT 18(2): 337-8 (1973)

1001   INFRARED AND RAMAN SPECTRA OF LIQUID AND
        CRYSTALLINE SILICON TETRAFLUORIDE.
        BESSETTE F + CABANA A + FOURNIER RP + SAVOIE R
        CAN J CHEM 48: 410-16 (1970)

        NUCLEAR SPIN RELAXATION STUDY OF THE SPIN
        ROTATION INTERACTION IN SPHERICAL TOP MOLECULES.
        COURTNEY JA + ARMSTRONG RL
        CAN J PHYS 50(12): 1252-61 (1972)

        FORCE FIELDS OF CARBON TETRAFLUORIDE AND SILICON
        TETRAFLUORIDE.
        DUNCAN JL + MILLS IM
        SPECTROCHIM ACTA 20(6): 1089-92 (1964)

1002   INTERATOMIC DISTANCES AND ROOT MEAN SQUARE
        AMPLITUDES OF VIBRATION OF GASEOUS SILICON
        TETRAFLUORIDE FROM ELECTRON DIFFRACTION.
        HAGEN K + HEDBERG K
        J CHEM PHYS 59(3): 1549-50 (1973)

1003   INFRARED SPECTRUM OF SILICON TETRAFLUORIDE.
        HEICKLEN J + KNIGHT V
        SPECTROCHIM ACTA 20: 295-8 (1964)

        RADIATIVE LIFETIMES OF ULTRAVIOLET EMISSION
        SYSTEMS EXCITED IN BORON TRIFLUORIDE, CARBON
        TETRAFLUORIDE, AND SILICON TETRAFLUORIDE.
        HESSER JE + DRESSLER K
        J CHEM PHYS 47: 3443-50 (1967)

1004   FORCE CONSTANTS OF THE SILICON TETRAHALIDES AND
        FREQUENCIES OF CHLORO BROMO SILICON.
        HOFLER F
        Z NATURFORSCH 26A: 547-50 (1971)

ELECTRON DENSITY DISTRIBUTIONS IN SOME FLUORIDES
AND OXYCHLORIDES. (CARBON TETRAFLUORIDE AND
SILICON TETRAFLUORIDE)
IONOV SP + IONOVA GV
RUSS J INORG CHEM 13(1): 157-8 (1968)

1005   INFRARED AND RAMAN SPECTRA OF SILICON
        TETRAFLUORIDE.
        JONES EA + KIRBY-SMITH JS + WOLTZ PJH
        + NIELSEN AH
        J CHEM PHYS 19(2): 242-5 (1951)

1006   FORCE FIELD FOR SILICON TETRAFLUORIDE.
        LEVIN IW + ABRAMOWITZ S
        J RES NBS A 72: 247-49 (1968)

        PHOTOELECTRON SPECTRA OF HALIDES. PART-7:
        VARIABLE TEMPERATURE HELIUM-1 AND HELIUM-2
        STUDIES OF CARBON TETRAFLUORIDE, SILICON
        TETRAFLUORIDE, AND GERMANIUM TETRAFLUORIDE.
        LLOYD DR + ROBERTS PJ
        J ELECTRON SPECTROSC RELAT PHENOM 7(4): 325-30
        (1975)

        ACCURATE FORCE CONSTANTS FROM HEAVY ISOTOPIC
        SUBSTITIUTION. PART-2: BORON TRIFLUORIDE,
        SILICON TETRAFLUORIDE, AND NITROGEN DIOXIDE.
        MCKEAN DC
        SPECTROCHIM ACTA 22: 269-79 (1966)

1007   INTENSITIES OF THE INFRARED FUNDAMENTALS OF
        CARBON TETRAFLUORIDE, SILICON TETRAFLUORIDE,
        CARBON TETRACHLORIDE, SILICON TETRACHLORIDE, AND
        GERMANIUM TETRACHLORIDE.
        MEDINA A + MORCILLO J
        ANALES FISICA 64(7-8): 251-61 (1968) (IN
        SPANISH)

        FLUORINE RELAXATION BY NMR ABSORPTION IN GASEOUS
        CARBON TETRAFLUORIDE, SILICON TETRAFLUORIDE, AND
        SULFUR HEXAFLUORIDE.
        MOHANTY S + BERNSTEIN HJ
        J CHEM PHYS 53: 461-2 (1970)

        COMPARISON OF CALCULATIONS OF FORCE CONSTANTS
        FOR BORON TETRAFLORIDE(-), CARBON TETRAFLUORIDE,
        SILICON TETRAFLUORIDE, AND SILICON
        TETRACHLORIDE.
        MUELLER A + FADINI A
        Z CHEM 7(3): 115-16 (1967) (IN GERMAN)

        MOLECULAR BEAM ELECTRIC DEFLECTION OF THE
        TETRAHALIDES CARBON TETRAFLUORIDE, CARBON
        TETRACHLORIDE, SILICON TETRAFLUORIDE, SILICON
        TETRACHLORIDE, GERMANIUM TETRACHLORIDE, TITANIUM
        TETRAFLUORIDE, TITANIUM TETRACHLORIDE, VANADIUM
        TETRAFLUORIDE, AND VANADIUM TETRACHLORIDE.
        MUENTER AA + DYKE TR + FALCONER WE + KLEMPERER W
        J CHEM PHYS 63: 1231-6 (1975)

1008   MEAN AMPLITUDES OF VIBRATION OF SOME XY4
        TETRAHEDRAL MOLECULES. (SILICON TETRAFLUORIDE)
        NAGARAJAN G
        BULL SOC CHIM BELG 73: 768-81 (1964)

        CONSTRUCTION OF A HIGH RESOLUTION PHOTOELECTRON
        SPECTROMETER. (CARBON TETRAFLUORIDE, BORON
        TRIFLUORIDE, SULFUR HEXAFLUORIDE, NITROGEN
        TRIFLUORIDE, AND SILICON TETRAFLUORIDE)
        PULLEN BP
        UNIV TENNESSEE, PHD THESIS, 1970, 117P.

1009   POTENTIAL CONSTANTS OF SOME XY4 TYPE OF SILANE
        COMPOUNDS. (SILICON TETRAFLUORIDE)
        RADHAKRISHNAN M
        Z PHYS CHEM NEUE FOLGE 35: 247-252 (1962)

        FORCE CONSTANTS OF VARIOUS ISOTOPES OF CARBON
        TETRAFLUORIDE, SILICON TETRAFLUORIDE, GERMANIUM
        TETRAFLUORIDE AND SULFUR TRIOXIDE.
        RUOFF A
        SPECTROCHIM ACTA 23A: 2421-31 (1967)

        PHYSICAL CONSTANTS (VAPOR PRESSURE) OF SILICON
        TETRAFLUORIDE, TUNGSTEN HEXAFLUORIDE, AND
        MOLYBDENUM HEXAFLUORIDE.
        RUFF O + ASCHER E
        Z ANORG ALLGEM CHEM 196: 413-20 (1931) (IN
        GERMAN)

HIGH RESOLUTION PHOTOELECTRON SPECTROSCOPY OF
CARBON TETRAFLUORIDE AND SILICON TETRAFLUORIDE.
BULL WE + PULLEN BP + GRIMM FA + MODDEMAN WE
SCHWEITZER GK + CARLSON TA
INORG CHEM 9: 2474-78 (1970)

FORCE CONSTANTS OF CARBON TETRAFLUORIDE, SILICON
TETRAFLUORIDE, BORON TRIFLUORIDE, ETHANE,
SILANE, AMMONIA, AND PHOSPHINE.
SHIMANOUCHI T + NAKAGAWA I + HIRAISHI J + ISHII
J MOL SPECTROSC 19: 78-107 (1966)

1010 INTERMOLECULAR POTENTIAL FOR SILICON
TETRAFLUORIDE.
SHINODA T
J PHYS SOC JAP 38(1): 224-30 (1975)

RAMAN SPECTRA OF CARBON AND SILICON
TETRAFLUORIDES.
YOST DM + LASSETTRE EN + GROSS ST
J CHEM PHYS 4: 325 (1936)

SILICON PENTAFLUORIDE   SiF$_5$

FLUORINE NMR SPECTRA OF SULFUR, PHOSPHORUS, AND
SILICON FLUORIDES. HYDROLYSIS AND INTRA-
INTERMOLECULAR MECHANISM OF FLUORINE EXCHANGE.
GIBSON JA + IBBOTT DG + JANZEN AF
CAN J CHEM 51(19): 3203-10 (1973)

SILVER MONOFLUORIDE   AgF

1011 ROTATIONAL ANALYSIS OF THE AO(+), BO(+), TO X
SINGLET SIGMA(+) SYSTEMS OF GASEOUS SILVER
MONOFLUORIDE.
BARROW RF + CLEMENTS RM
PROC ROY SOC A 322: 243-9 (1971)

1012 ABSORPTION SPECTRUM OF GASEOUS SILVER
MONOFLUORIDE.
CLEMENTS RM + BARROW RF
CHEM COMMUN 1968: 27-8

1013 NEAR ULTRAVIOLET ABSORPTION SPECTRUM OF SILVER
MONOFLUORIDE.
JOSHI MM + SHARMA D
INDIAN J PURE APPL PHYS 1: 86 (1963)

1014 MASS SPECTROMETRIC STUDIES AT HIGH TEMPERATURES.
PART-14: VAPOR PRESSURE AND DISSOCIATION ENERGY
OF SILVER MONOFLUORIDE.
ZMBOV KF + MARGRAVE JL
J PHYS CHEM 71(2): 446-8 (1967)

SILVER DIFLUORIDE   AgF$_2$

1015 MAGNETIC SUSCEPTIBILITY OF SILVER DIFLUORIDE.
CHARPIN P + DIANOUX AJ + ELLIS HM + NGU N
+ PERRIN F
COMPT REND 264: 1108-10 (1967) (IN FRENCH)

STRONTIUM MONOFLUORIDE   SrF

1016 INTERNUCLEAR DISTANCE IN GASEOUS STRONTIUM
MONOFLUORIDE.
BARROW RF + BEALE JR
CHEM COMMUN 1967(12): 606

DISSOCIATION ENERGIES OF THE ALKALINE EARTH
MONOFLUORIDES.
EHLERT TC + BLUE GD + GREEN JW + MARGRAVE JL
J CHEM PHYS 41: 2250-5 (1964)

STRONTIUM DIFLUORIDE   SrF$_2$

ELECTRON DIFFRACTION ANALYSIS OF THE MOLECULAR
STRUCTURES OF GROUP-2 ELEMENTS. (BERYLLIUM
DIFLUORIDE, MAGNESIUM DIFLUORIDE, CALCIUM
DIFLUORIDE, STRONTIUM DIFLUORIDE, BARIUM
DIFLUORIDE, ZINC DIFLUORIDE, CADMIUM DIFLUORIDE)
AKISHIN PA + SPIRIDONOV VP
SOV PHYS CRYSTALLOGR 2(4): 472-9 (1957)

SPECTRA OF ELECTRONIC EXITATIONS IN CALCIUM
DIFLUORIDE, STRONTIUM DIFLUORIDE, AND BARIUM
DIFLUORIDE IN THE 8 TO 150 EV RANGE.
FRANDON J + LAHAYE B + PRADAL F
PHYS STAT SOL B 53: 565-75 (1972)

MELTING POINTS OF INORGANIC FLUORIDES. (CADMIUM
DIFLUORIDE, MAGNESIUM DIFLUORIDE, STRONTIUM
DIFLUORIDE, AND BARIUM DIFLUORIDE)
KOJIMA H + WHITEWAY SG + MASSON CR
CAN J CHEM 46: 2968-71 (1968)

MEAN AMPLITUDES OF VARIATION, BASTIANSEN-MORINO
SKRINKAGE EFFECT, AND MOLECULAR POLARIZABILITIES
FOR DIHALIDES OF SOME GROUP-2A ELEMENTS.
NAGARAJAN G
INDIAN J PHYS 39(9): 405-20 (1965)

1017 FAR ULTRAVIOLET ELECTRONIC SPECTRA OF STRONTIUM
DIFLUORIDE AND BARIUM DIFLUORIDE.
NISAR M + ROBIN S
PAK J SCI IND RES 17(2-3): 49-54 (1974)

VAPOR PRESSURE OF ZINC DIFLUORIDE, CADMIUM
DIFLUORIDE, MAGNESIUM DIFLUORIDE, CALCIUM
DIFLUORIDE, STRONTIUM DIFLUORIDE, BARIUM
DIFLUORIDE, AND ALUMINUM TRIFLUORIDE.
RUFF O + LE BOUCHER L
Z ANORG ALLGEM CHEM 219: 376-81 (1934) (IN
GERMAN)

INFRARED SPECTRA OF SOME ALKALINE EARTH HALIDES
BY THE MATRIX ISOLATION TECHNIQUE.
SNELSON A
J PHYS CHEM 70: 3208-16 (1966)

GEOMETRY OF THE ALKALINE EARTH DIHALIDES.
(BARIUM DIFLUORIDE, CALCIUM DIFLUORIDE,
STRONTIUM DIFLUORIDE)
WHARTON L + BERG RA + KLEMPERER W
J CHEM PHYS 39: 2023-31 (1963)

SULFUR MONOFLUORIDE   **SF**

1018   MICROWAVE SPECTRUM OF THE SULFUR MONOFLUORIDE
       RADICAL.
       AMANO T + HIROTA E
       J MOL SPECTROSC 45: 417-9 (1973)

1019   ELECTRIC DIPOLE MOMENTS OF OPEN SHELL DIATOMIC
       MOLECULES. (SULFUR MONOFLUORIDE AND SELENIUM
       MONOFLUORIDE)
       BYFLEET CR + CARRINGTON A + RUSSELL DK
       MOL PHYS 20: 271-77 (1971)

1020   GAS PHASE ELECTRON RESONANCE SPECTRA OF SULFUR
       MONOFLUORIDE, SELENIUM MONOFLUORIDE, SELENIUM
       OXIDE AND IODINE OXIDE.
       CARRINGTON A + CURRIE GN + DYER PN + LEVY DH
       + MILLER TA
       CHEM COMMUN 1967: 641-2

1021   GAS PHASE ELECTRON RESONANCE SPECTRA OF SULFUR
       MONOFLUORIDE AND SELENIUM MONOFLUORIDE.
       CARRINGTON A + CURRIE GN + MILLER TA
       J CHEM PHYS 50: 2726-32 (1969)

1022   SPECTRUM OF SULFUR MONOFLUORIDE.
       DILONARDO G + TROMBETTI A
       TRANS FARADAY SOC 66(PT 11): 2694-8 (1970)

1023   MASS SPECTROMETRIC STUDIES OF SOME GASEOUS
       SULFUR FLUORIDES.
       HILDENBRAND DL
       J PHYS CHEM 77(7): 897-902 (1973)

1024   MICROWAVE AND MASS SPECTRA OF SULFUR
       MONOFLUORIDE.
       KUCZKOWSKI RL + WILSON EB
       J AMER CHEM SOC 85: 2028-9 (1963)

1025   HARTREE-FOCK WAVE FUNCTIONS AND COMPUTED
       PROPERTIES FOR THE DOUBLET PI GROUND STATES OF
       SULFUR MONOFLUORIDE AND SELENIUM MONOFLUORIDE
       AND THEIR POSITIVE AND NEGATIVE IONS. A
       COMPARISON OF THE THEORETICAL AND EXPERIMENTAL
       RESULTS. (OXYGEN MONOFLUORIDE)
       O'HARE PAG + WAHL AC
       J CHEM PHYS 53: 2834-46 (1970)

SULFUR DIFLUORIDE   **SF$_2$**

1026   ESTIMATION OF INFRARED FREQUENCIES OF SULFUR
       DIFLUORIDE.
       BLIEFERT C + WANCZEK KP
       Z NATURFORSCH 25A: 1770-1 (1970)

1027   MOLECULAR ORBITAL DESCRIPTION OF SULFUR
       DIFLUORIDE, SULFUR TETRAFLUORIDE AND SULFUR
       HEXAFLUORIDE.
       BONACIC-KOUTECKY V + MUSHER JI
       BELFER GRADUATE SCI SCHOOL, NY, 1973, 30P.
       AD770147/7GA

1028   MASS SPECTROMETRIC STUDIES OF SOME GASEOUS
       SULFUR FLUORIDES.
       HILDENBRAND DL
       J PHYS CHEM 77(7): 897-902 (1973)

1029   NONEMPIRICAL VALENCE ELECTRON CALCULATIONS ON
       SMALL MOLECULES CONTAINING PHOSPHORUS OR SULFUR.
       HYDE RG + PEEL JB + TERAUDS K
       J CHEM SOC, FARADAY TRANS 73(69): 1563-8 (1973)

1030   MICROWAVE SPECTRUM AND STRUCTURE OF SULFUR
       DIFLUORIDE.
       JOHNSON DR + POWELL FX
       SCIENCE 164: 950-1 (1969)

1031   CENTRIFUGAL DISTORTION EFFECTS IN SULFUR
       DIFLUORIDE. CALCULATION OF THE FORCE FIELD AND
       INFRARED SPECTRUM.
       KIRCHHOFF WH + JOHNSON DR + POWELL FX
       J MOL SPECTROSC 48(1): 157-64 (1973)

       MOLECULAR ORBITAL DESCRIPTION OF SULFUR
       COMPOUNDS OF VALENCES 2, 4, AND 6.
       KOUTECKY VB + MUSHER JI
       THEOR CHIM ACTA 33(3): 227-38 (1974)

1032   STRUCTURE OF THE SULFUR TRIFLUORIDE POSITIVE
       ION. (SULFUR DIFLUORIDE)
       LEIBOVICI C
       J FLUORINE CHEM 3: 437-40 (1973)

       SPECTROSCOPIC STUDIES OF VIBRATIONAL CONSTANTS,
       STATISTCIAL THERMODYNAMICS AND QUANTUM
       MECHANICAL STUDIES OF POLARIZABILITIES FOR THE
       DIFLUORIDES OF OXYGEN AND SULFUR.
       SINGH Z + NAGARAJAN G
       ACTA PHYS ACAD SCI HUNG 36(4): 415-30 (1974)

SULFUR TRIFLUORIDE   **SF$_3$**

       ESR SPECTRA OF PHOSPHORUS DIFLUORIDE AND SULFUR
       TRIFLUORIDE RADICALS.
       COLUSSI AJ + MORTON JR + PRESTON KF + FESSENDEN
       J CHEM PHYS 61(3): 1247-8 (1974)

SULFUR TETRAFLUORIDE   **SF$_4$**

       INFRARED SPECTRA OF SOME VOLATILE INORGANIC
       FLUORIDES IN THE SOLID STATE. (ARSENIC
       TRIFLUORIDE, SULFUR TETRAFLUORIDE, AND SELENIUM
       TETRAFLUORIDE)
       AYNSLEY EE + DODD RE + LITTLE R
       SPECTROCHIM ACTA 18: 1005-8 (1962)

1033   SOFT X-RAY SPECTROSCOPY ON SULFUR TETRAFLUORIDE
       AND SULFUR HEXAFLUORIDE.
       BARANOVSKII VI + ZIMKINA TM + FOMICHEV VA
       + DZEVITSKII BE
       THEOR EXPER KHIM 3: 354-62 (1967)

       MOLECULAR ORBITAL DESCRIPTION OF SULFUR
       DIFLUORIDE, SULFUR TETRAFLUORIDE AND SULFUR
       HEXAFLUORIDE.
       BONACIC-KOUTECKY V + MUSHER JI
       BELFER GRADUATE SCI SCHOOL, NY, 1973, 30P.
       AD770147/7GA

       VESCF-MO STUDIES OF PHOSPHORUS PENTAFLUORIDE,
       SULFUR TETRAFLUORIDE, AND CHLORINE TRIFLUORIDE.
       BROWN RD + PEEL JB
       AUST J CHEM 21: 2617-29 (1968)

1034   MOLECULAR VIBRATIONS OF SULFUR HEXAFLUORIDE AND
       SULFUR TETRAFLUORIDE.
       CHANTRY GW + EWING VC
       MOL PHYS 5: 209-15 (1962)

1035   RAMAN SPECTRUM AND FORCE CONSTANTS FOR SULFUR
       TETRAFLUORIDE.
       CHRISTE KO + SAWODNY W
       J CHEM PHYS 52: 6320-23 (1970)

1036   VIBRATIONAL ASSIGNMENT OF SULFUR TETRAFLUORIDE.
       CHRISTE KO + SAWODNY W + PULAY P
       J MOL STRUCT 21(1): 158-64 (1974)

1037   NMR SPECTRUM AND STRUCTURE OF SULFUR
       TETRAFLUORIDE.
       COTTON FA + GEORGE JW + WAUGH JS
       J CHEM PHYS 28: 994-5 (1958)

1038 MEAN AMPLITUDES OF VIBRATION FOR SULFUR
TETRAFLUORIDE.
CYVIN SJ
ACTA CHEM SCAND 23: 576-78 (1969)

1039 CLASSIFICATION OF THE STATES OF NONRIGID
MOLECULES. (SULFUR TETRAFLUORIDE)
DALTON BJ
MOL PHYS 11: 265-85 (1966)

1040 RAMAN AND INFRARED SPECTRUM OF SULFUR
TETRAFLUORIDE.
DODD RE + WOODWARD LA + ROBERTS HL
TRANS FARADAY SOC 52: 152-61 (1956)

1041 INVESTIGATION BY ELECTRON DIFFRACTION OF THE
MOLECULAR STRUCTURES OF SULFUR HEXAFLUORIDE,
SULFUR TETRAFLUORIDE, SELENIUM HEXAFLUORIDE, AND
SELENIUM TETRAFLUORIDE.
EWING VC + SUTTON LE
TRANS FARADAY SOC 59: 1242-7 (1963)

INFRARED SPECTRA OF MATRIX ISOLATED CHLORINE
TRIFLUORIDE, BROMINE TRIFLUORIDE, AND BROMINE
PENTAFLUORIDE. FLUORINE EXCHANGE MECHANISM OF
LIQUID CHLORINE TRIFLUORIDE, BROMINE
TRIFLUORIDE, AND SULFUR TETRAFLUORIDE.
FREY RA + REDINGTON RL + ALJIBURY ALK
J CHEM PHYS 54: 344-55 (1971)

1042 FLUORINE NMR SPECTRA OF SULFUR, PHOSPHORUS, AND
SILICON FLUORIDES. HYDROLYSIS AND INTRA-
INTERMOLECULAR MECHANISM OF FLUORINE EXCHANGE.
GIBSON JA + IBBOTT DG + JANZEN AF
CAN J CHEM 51(19): 3203-10 (1973)

1043 STRUCTURE OF SULFUR TETRAFLUORIDE AND RELATED
MOLECULES.
GILLESPIE RJ
J CHEM PHYS 37: 2498-9 (1962)

1044 FLUORINE-19 NMR INVESTIGATION OF THE ASSOCIATION
OF SULFUR TETRAFLUORIDE.
GOMBLER W + SEEL F
J FLUORINE CHEM 4(3): 333-9 (1974) (IN GERMAN)

NONEMPIRICAL VALENCE ELECTRON CALCULATIONS ON
SMALL MOLECULES CONTAINING PHOSPHORUS OR SULFUR.
HYDE RG + PEEL JB + TERAUDS K
J CHEM SOC, FARADAY TRANS 73(69): 1563-8 (1973)

MOLECULAR ORBITAL DESCRIPTION OF SULFUR
COMPOUNDS OF VALENCES 2, 4, AND 6.
KOUTECKY VB + MUSHER JI
THEOR CHIM ACTA 33(3): 227-38 (1974)

1045 INFRARED SPECTRUM AND NORMAL COORDINATE ANALYSIS
OF SULFUR TETRAFLUORIDE.
LEVIN IW + BERNEY CV
J CHEM PHYS 44: 2557-61 (1966)

1046 INTRAMOLECULAR EXCHANGE IN SULFUR TETRAFLUORIDE.
LEVIN IW + HARRIS WC
J CHEM PHYS 55: 3048-9 (1971)

1047 CALCULATION OF THE INFRARED SPECTRA OF SULFUR
TETRAFLUORIDE USING CNDO/2 TECHNIQUES.
LEVIN IW
J CHEM PHYS 55: 5393-5400 (1971)

1048 STRUCTURE OF EXCHANGE PROCESSES IN SOME
INORGANIC FLUORIDES BY NMR. (SULFUR
TETRAFLUORIDE AND SELENIUM TETRAFLUORIDE)
MUETTERTIES EL + PHILLIPS WD
J AMER CHEM SOC 81: 1084-8 (1959)

1049 FLUORINE EXCHANGE IN SULFUR TETRAFLUORIDE.
MUETTERTIES EL + PHILLIPS WD
J CHEM PHYS 46: 2861-2 (1967)

1050 MOLECULAR FORCE FIELD FOR SULFUR TETRAFLUORIDE.
PILLAI MGK + RAMASWAMY K + PICHAI R
CAN J CHEM 43: 463-9 (1965)

1051 THERMODYNAMIC PROPERTIES OF SULFUR TETRAFLUORIDE
AND SULFUR OXIDE TETRAFLUORIDE.
RADHAKRISHNAN M
Z NATURFORSCH 18A: 103-4 (1963)

1052 FORCE FIELD STUDY FOR SULFUR TETRAFLUORIDE.
RAMASWAMY K + SIVARAMAKRISHNAN TR
ACTA PHYS POLON A 46(2): 141-8 (1974)

INFRARED MATRIX ISOLATION STUDIES. (BORON
TRIFLUORIDE, SULFUR TETRAFLUORIDE, ANTIMONY
PENTAFLUORIDE, ARSENIC PENTAFLUORIDE, CHLORINE
TRIFLUORIDE, BROMINE TRIFLUORIDE, AND BROMINE
PENTAFLUORIDE)
REDINGTON RL
TEXAS TECH UNIV, 1969, 89P. AD701120

1053 FLUORINE EXCHANGE IN SULFUR TETRAFLUORIDE.
REDINGTON RL + BERNEY CV
J CHEM PHYS 46: 2862-4 (1967)

1054 SULFUR TETRAFLUORIDE AND SULFUR OXIDE
DIFLUORIDE. INFRARED EVIDENCE FOR DIMER
FORMATION AT LOW TEMPERATURE.
REDINGTON RL + BERNEY CV
J CHEM PHYS 43: 2020-6 (1965)

1055 PROBLEM OF FLUORINE EXCHANGE IN SULFUR
TETRAFLUORIDE.
SEEL F + GOMBLER W
J FLUORINE CHEM 4(3): 327-31 (1974) (IN GERMAN)

OBSERVATION BY ESCA OF INEQUIVALENT FLUORINES IN
CHLORINE TRIFLUORIDE, SULFUR TETRAFLUORIDE, AND
PHOSPHORUS PENTAFLUORIDE.
SHAW RW + CARROLL TX + THOMAS TD
J AMER CHEM SOC 95(18): 5870-5 (1973)

1056 STRUCTURE AND DIPOLE MOMENT OF SULFUR
TETRAFLUORIDE.
TOLLES WM + GWINN WD
J CHEM PHYS 36: 1119-21 (1962)

EMPIRICAL METHOD FOR DETERMINING EFFECTIVE
VIBRATIONAL AND ROTATIONAL CHARACTERISTICS OF
MOLECULES OF SOME TETRAFLUORIDES FOR CALCULATING
THEIR THERMODYNAMIC FUNCTIONS.
TUMANOV YN + GALKIN NP
RUSS J PHYS CHEM 43(4): 464-7 (1969)

1057 MEAN AMPLITUDES OF VIBRATION: SULFUR
TETRAFLUORIDE AND SULFUR PENTAFLUORO CHLORIDE
MOLECULES.
VENKATESWARLU K + MARIAM S
INDIAN J PURE APPL PHYS 3(12): 472-4 (1965)

1058 MOLECULAR VIBRATIONS OF SULFUR TETRAFLUORIDE.
VENKATESWARLU K + PILLAI MGK
OPT SPECTROSC 11(1): 26-7 (JUL 1961)

1059 UREY-BRADLEY FORCE FIELD FOR SULFUR
TETRAFLUORIDE.
VENKATESWARLU K + THANALAKSHMI R
PROC INDIAN ACAD SCI A 57(3): 181-5 (1963)

SULFUR PENTAFLUORIDE SF_5

1060 STRUCTURE OF THE RADICAL SULFUR PENTAFLUORIDE.
GREGORY AR
CHEM PHYS LETT 28: 552-4 (1974)

SULFUR HEXAFLUORIDE SF_6

1061 FORCE FIELDS FOR SOME GROUP-6 HEXAFLUORIDES.
(SULFUR HEXAFLUORIDE AND TELLURIUM HEXAFLUORIDE)
ABRAMOWITZ S + LEVIN IW
J CHEM PHYS 44: 3353-6 (1966)

1062 PHYSICAL AND THERMODYNAMIC PROPERTIES OF SULFUR
HEXAFLUORIDE. (REVIEW)
ADLER LS + YAWS CL
SOLID STATE TECHNOL 18(1): 35-8 (1975)

1063 OCTAHEDRAL FINE STRUCTURE SPLITTINGS IN NU-3 OF
SULFUR HEXAFLUORIDE.
ALDRIDGE JP + FILIP H + FLICKER H + HOLLAND RF
+ MCDOWELL RS + NERESON NG + FOX K
J MOL SPECTROSC 58(1): 165-8 (1975)

1064 SELECTIVE PHOTODISSOCIATION OF POLYATOMIC
MOLECULES (SULFUR HEXAFLUORIDE) BY INFRARED
RADIATION.
AMBARTSUMYAN RV + GOROKHOV YA + LETOKHOV VS
+ MAKAROV GN + PURETSKII AA
JETP LETT 22(7): 177-8 (1975)

1065 ELECTRONIC STRUCTURE OF SULFUR HEXAFLUORIDE
STUDIED BY ULTRALONG WAVE X-RAY SPECTROSCOPY.
BARANOVSKII VI + ZIMKINA TM + FOMICHEV VA
PROBLEMY SOVREMENNOI KHIMII KOORD SOED 2: 90-5
(1968)

1066 AB INITIO AND MZDO WAVE FUNCTIONS FOR SULFUR
HEXAFLUORIDE.
BENDAZZOLI GL + PALMIERI P + CADIOLI B
+ PINCELLI U
MOL PHYS 19: 865-70 (1970)

1067 X-RAY ABSORPTION IN SULFUR HEXAFLUORIDE.
BEST PE
J CHEM PHYS 47: 4002-6 (1967)

MOLECULAR ORBITAL DESCRIPTION OF SULFUR
DIFLUORIDE, SULFUR TETRAFLUORIDE, AND SULFUR
HEXAFLUORIDE.
BONACIC-KOUTECKY V + MUSHER JI
BELFER GRADUATE SCI SCHOOL, NY, 1973, 30P.
AD770147/7GA

1068 INFRARED SPECTRA OF CRYOSYSTEMS. PART-2: SULFUR
HEXAFLUORIDE.
BERTSEV VV + KOLOMIITSEVA TD + TSYANGENKO MM
OPT SPECTROSC 37(3): 263-4 (SEP 1974)

1069 VAPOR PHASE RAMAN SPECTRA, RAMAN BAND CONTOUR
ANALYSES, AND CORIOLIS CONSTANTS OF THE
SPHERICAL TOP MOLECULES, SULFUR HEXAFLUORIDE,
SELENIUM HEXAFLUORIDE, TELLURIUM HEXAFLUORIDE,
MOLYBDENUM HEXAFLUORIDE, TUNGSTEN HEXAFLUORIDE,
AND URANIUM HEXAFLUORIDE.
BOSWORTH YM + CLARK RJH + RIPPON DM
J MOL SPECTROSC 46: 240-55 (1973)

1070 NU-3 AND NU-4 BANDS OF SULFUR HEXAFLUORIDE.
BRUNET H + PEREZ M
J MOL SPECTROSC 29: 472-77 (1969)

1071 SPECTROSCOPIC CALCULATIONS ON SULFUR
HEXAFLUORIDE.
BYE BH + CYVIN SJ
ACTA CHEM SCAND 17: 1804 (1963)

MOLECULAR VIBRATIONS OF SULFUR HEXAFLUORIDE AND
SULFUR TETRAFLUORIDE.
CHANTRY GW + EWING VC
MOL PHYS 5: 209-15 (1962)

1072 RAMAN SPECTRA OF MOLYBDENUM HEXAFLUORIDE,
TECHNETIUM HEXAFLUORIDE, RHENIUM HEXAFLUORIDE,
URANIUM HEXAFLUORIDE, SULFUR HEXAFLUORIDE,
SELENIUM HEXAFLUORIDE, AND TELLURIUM
HEXAFLUORIDE IN THE VAPOR STATE.
CLAASSEN HH + GOODMAN GL + HOLLOWAY JH + SELIG H
J CHEM PHYS 53: 341-8 (1970)

NUCLEAR SPIN RELAXATION STUDY OF THE SPIN
ROTATION INTERACTION IN SPHERICAL TOP MOLECULES.
COURTNEY JA + ARMSTRONG RL
CAN J PHYS 50(12): 1252-61 (1972)

1073 EVIDENCE OF EFFECTIVE POTENTIAL BARRIERS IN THE
X-RAY ABSORPTION SPECTRA OF MOLECULES. (SULFUR
HEXAFLUORIDE)
DEHMER JL
J CHEM PHYS 56(9): 4496-504 (1972)

1074 OPTICAL SATURATION OF A SINGLE VIBRATION-
ROTATION TRANSITION IN THE NU-3 FUNDAMENTAL OF
SULFUR HEXAFLUORIDE.
DJEU N + WOLGA GJ
J CHEM PHYS 54: 774-8 (1971)

1075 INFRARED INTENSITIES IN CRYSTALLINE SULFUR
HEXAFLUORIDE.
DOWS DA + WIEDER GM
SPECTROCHIM ACTA 18: 1567-74 (1962)

SOFT X-RAY SPECTROSCOPY ON SULFUR TETRAFLUORIDE
AND SULFUR HEXAFLUORIDE.
BARANOVSKII VI + ZIMKINA TM + FOMICHEV VA
DZEVITSKII BE
THEOR EXPER KHIM 3: 354-62 (1967)

INVESTIGATION BY ELECTRON DIFFRACTION OF THE
MOLECULAR STRUCTURES OF SULFUR HEXAFLUORIDE,
SULFUR TETRAFLUORIDE, SELENIUM HEXAFLUORIDE, AND
SELENIUM TETRAFLUORIDE.
EWING VC + SUTTON LE
TRANS FARADAY SOC 59: 1242-7 (1963)

1076 PHOTOELECTRON SPECTRUM OF SULFUR HEXAFLUORIDE AT
584 ANGSTROMS.
FROST DC + MCDOWELL CA + SANDHU JS + VROOM DA
J CHEM PHYS 46: 2008-9 (1967)

INFRARED SPECTRA AND MOLECULAR STRUCTURE OF SOME
GROUP-6 HEXAFLUORIDES. (OF SULFUR, SELENIUM,
TELLURIUM, MOLYBDENUM, TUNGSTEN, AND URANIUM)
GAUNT J
TRANS FARADAY SOC 49: 1122-31 (1953)

1077 ELECTRONIC PROPERTIES OF SULFUR HEXAFLUORIDE. AB
INITIO CALCULATION FOR THE GROUND STATE.
GIANTURCO FA + GUIDOTTI C + LAMANNA U + MOCCIA R
CHEM PHYS LETT 10: 269-73 (1971)

1078 ELECTRONIC PROPERTIES OF SULFUR HEXAFLUORIDE.
PART-1: OPTICAL ABSORPTION AND X-RAY EMISSION
FROM SCF MO LCAO COMPUTATIONS. PART-2: MOLECULAR
ORBITAL INTERPRETATION OF ITS X-RAY ABSORPTION
SPECTRA.
GIANTURCO FA + GUIDOTTI C + LAMANNA U
CHEM PHYS LETT 17(1): 127-33 (1972); J CHEM PHYS
57(2): 840-6 (1972)

1079 RAMAN SPECTRUM OF SULFUR HEXAFLUORIDE IN THE
GASEOUS AND SOLID STATES.
GULLIKSON CW + NIELSEN JR + STAIR AT
J MOL SPECTROSC 1: 151-4 (1957)

1080 MOLECULAR FORCE FIELDS. PART-8: VIBRATION
FREQUENCIES OF SOME OCTAHEDRAL XY6 MOLECULES.
(SULFUR HEXAFLUORIDE, SELENIUM HEXAFLUORIDE,
TELLURIUM HEXAFLUORIDE)
HEATH DF + LINNETT JW
TRANS FARADAY SOC 45: 264-71 (1949)

MASS SPECTROMETRIC STUDIES OF SOME GASEOUS
SULFUR FLUORIDES.
HILDENBRAND DL
J PHYS CHEM 77(7): 897-902 (1973)

1081 FORBIDDEN RAMAN BANDS OF SULFUR HEXAFLUORIDE.
COLLISION-INDUCED RAMAN SCATTERING.
HOLZER W + OUILLON R
CHEM PHYS LETT 24(4): 589-93 (1974)

1082 LOW TEMPERATURE ABSORPTION CONTOUR OF THE NU-3
BAND OF SULFUR HEXAFLUORIDE.
HOUSTON PL + STEINFELD JI
J MOL SPECTROSC 54(2): 335-7 (1975)

1083 APPLICATION OF SYNCHROTON ORBITAL RADIATION TO
MOLECULAR SPECTROSCOPY. MOLECULAR NITROGEN AND
SULFUR HEXAFLUORIDE.
IWATA S
KAGAKU 44(10): 620-7 (1974) (IN JAPANESE)

1084 ELECTRIC DEFLECTION OF BINARY HEXAFLUORIDES.
(SULFUR HEXAFLUORIDE, SELENIUM HEXAFLUORIDE,
TELLURIUM HEXAFLUORIDE, MOLYBDENUM HEXAFLUORIDE,
TUNGSTEN HEXAFLUORIDE, URANIUM HEXAFLUORIDE,
RUTHENIUM HEXAFLUORIDE, RHENIUM HEXAFLUORIDE,
RHODIUM HEXAFLUORIDE, OSMIUM HEXAFLUORIDE,
IRIDIUM HEXAFLUORIDE, AND PLATINUM HEXAFLUORIDE)
KAISER EW + MUENTER JS + KLEMPERER W
+ FALCONER WE + SUNDER WA
J CHEM PHYS 53: 1411-12 (1970)

1085 MOLECULAR ORBITAL DESCRIPTION OF SULFUR
COMPOUNDS OF VALENCES 2, 4, AND 6.
KOUTECKY VB + MUSHER JI
THEOR CHIM ACTA 33(3): 227-38 (1974)

1086 INFRARED SPECTRUM OF SULFUR HEXAFLUORIDE.
LAGEMANN RT + JONES EA
J CHEM PHYS 19(5): 534-6 (1951)

1087 THERMODYNAMIC PROPERTIES OF SULFUR HEXAFLUORIDE.
LAGUTKIN OD + VERKHIVKER GP
KHOLODILNAYA TEKHNIKA 39: 24-9 (1962)

1088 SULFUR K AND L AND FLUORINE K X-RAY EMISSION AND
ABSORPTION SPECTRA OF GASEOUS SULFUR
HEXAFLUORIDE.
LAVILLA RE
J CHEM PHYS 57(2): 899-909 (1972)

DIRECTED VALENCY IN CERTAIN MOLECULES AND
COMPLEX IONS. (PHOSPHORUS PENTAFLUORIDE, SULFUR
HEXAFLUORIDE)
LINNETT JW + MELLISH CE
TRANS FARADAY SOC 50: 665-70 (1954)

ORBITAL MODIFICATION BY THE COULOMB FIELD OF
LIGAND ATOMS OF LATER SECOND ROW ELEMENTS IN
PERFECT PAIRING VALENCE STATES. (PHOSPHORUS
PENTAFLUORIDE, SULFUR HEXAFLUORIDE, AND CHLORINE
TRIFLUORIDE)
MACLAGAN RGAR
J CHEM SOC A 1971: 222-6

THERMODYNAMIC SIMILARITY AND UNIVERSAL EQUATIONS
OF STATE OF HEXAFLUORIDES. (OF SULFUR,
MOLYBDENUM, TUNGSTEN, AND URANIUM)
MALYSHEV VV
TEPLOFIZ VYS TEMP 14(1): 47-55 (1976) (IN
RUSSIAN)

FLUORINE RELAXATION BY NMR ABSORPTION IN GASEOUS
CARBON TETRAFLUORIDE, SILICON TETRAFLUORIDE, AND
SULFUR HEXAFLUORIDE.
MOHANTY S + BERNSTEIN HJ
J CHEM PHYS 53: 461-62 (1970)

1089 THERMODYNAMIC FUNCTIONS FOR SULFUR HEXAFLUORIDE.
MORSY TE
BER BUNSENGES PHYS CHEM 68(3): 277-80 (1964)

1090 ELECTRONIC STRUCTURE OF MOLECULES BY A MANY BODY
APPROACH. PART-6: THE ASSIGNMENT OF THE HELIUM-2
PHOTOELECTRON SPECTRUM OF SULFUR HEXAFLUORIDE.
VON NIESSEN W + CEDERBAUM LS + DIERCKSEN GHF
+ HOHLNEICHER G
CHEM PHYS 11(3): 399-407 (1975)

1091 TEMPERATURE DEPENDENT ABSORPTION SPECTRUM OF THE
NU-3 BAND OF SULFUR HEXAFLUORIDE AT 10.6
MICROMETERS.
NOWAK AV + LYMAN JL
J QUANT SPECT RAD TRANSFER 15(10): 945-61 (1975)

1092 ENTHALPIES OF FORMATION OF SULFUR HEXAFLUORIDE,
SELENIUM HEXAFLUORIDE, TELLURIUM HEXAFLUORIDE
AND THEIR THERMODYNAMIC PROPERTIES.
O'HARE PAG + SETTLE JL + HUBBARD WN
TRANS FARADAY SOC 62(3): 558-65 (1966)

CONSTRUCTION OF A HIGH RESOLUTION PHOTOELECTRON
SPECTROMETER. (CARBON TETRAFLUORIDE, BORON
TRIFLUORIDE, SULFUR HEXAFLUORIDE, NITROGEN
TRIFLUORIDE, AND SILICON TETRAFLUORIDE)
PULLEN BP
UNIV TENNESSEE, PHD THESIS, 1970, 117P.

1093 GREEN'S FUNCTION ANALYSIS OF SULFUR HEXAFLUORIDE
AND RELATED MOLECULES. (SELENIUM HEXAFLUORIDE,
TELLURIUM HEXAFLUORIDE)
RAMASWAMY K + MOHAN N
J MOL STRUCT 7: 51-8 (1971)

1094 LASER ISOTOPE SEPARATION. (SULFUR HEXAFLUORIDE
SPECTRUM)
ROBINSON CP
P275-95 OF INT CCNF LASER SPECTROSC, 2ND PROC,
1975. (LECT NOTES PHYSICS NO 43, SPRINGER,
1975.)

1095 FAR INFRARED SPECTRA OF GASEOUS AND LIQUID
SULFUR HEXAFLUORIDE.
ROSENBERG A + BIRNBAUM G
J CHEM PHYS 52: 683-6 (1970)

1096 FORCE CONSTANTS FOR SULFUR HEXAFLUORIDE.
RUOFF VA
J MOL STRUCT 4: 332-4 (1969)

1097 RAMAN SPECTRA OF MATRIX ISOLATED MOLECULES.
(SULFUR HEXAFLUORIDE)
SHIRK JS + CLAASSEN HH
J CHEM PHYS 54: 3237-8 (1971)

1098 RAMAN SPECTRUM OF SOLID SULFUR HEXAFLUORIDE.
SHURVELL HF + BERNSTEIN HJ
J MOL SPECTROSC 30(1): 153-7 (1969)

1099 INFRARED DOUBLE RESONANCE IN SULFUR
HEXAFLUORIDE.
STEINFELD JI + BURAK I + SUTTON DG + NOWAK AV
J CHEM PHYS 52: 5421-34 (1970)

1100 FORCE CONSTANTS OF SULFUR HEXAFLUORIDE FROM
ISOTOPIC SUBSTITUTION.
THAKUR SN
J MOL STRUCT 7: 315-22 (1971)

1101 X-RAY SPECTRA OF SULFUR AND FLUORINE IN SULFUR
HEXAFLUORIDE.
VINOGRADOV AS + ZIMKINA TM + FORICHEV VA
ZH STRUKT KHIM 12(5): 899-904 (1971) (IN
RUSSIAN)

RAMAN SPECTRA AND MOLECULAR CONSTANTS OF THE
HEXAFLUORIDES OF SULFUR, SELENIUM, AND
TELLURIUM.
YOST DM + STEFFENS CC + GROSS ST
J CHEM PHYS 2: 311-16 (1934)

1102 ULTRASOFT X-RAY ABSORPTION SPECTRUM OF SULFUR
HEXAFLUORIDE.
ZIMKINA TM + FOMICHEV VA
DOKL AKAD NAUK SSSR 169(6): 1304-6 (1966) (IN
RUSSIAN)

TANTALUM MONOFLUORIDE   **TaF**

THERMODYNAMIC PROPERTIES OF NIOBIUM AND TANTALUM
FLUORIDES AT HIGH TEMPERATURES. PART-3:
DIFLUORIDES AND MONOFLUORIDES.
GALKIN NP + TUMANOV YN + KOROBTSEV VP + BATAREV
+ KHOKHLOV VA + PAVLOV AA
RUSS J PHYS CHEM 45(10): 1531 (1971)

TANTALUM DIFLUORIDE   **TaF$_2$**

THERMODYNAMIC PROPERTIES OF NIOBIUM AND TANTALUM
FLUORIDES AT HIGH TEMPERATURES. PART-3:
DIFLUORIDES AND MONOFLUORIDES.
GALKIN NP + TUMANOV YN + KOROBTSEV VP + BATAREV
+ KHOKHLOV VA + PAVLOV AA
RUSS J PHYS CHEM 45(10): 1532 (1971)

TANTALUM TRIFLUORIDE TaF_3

THERMODYNAMIC PROPERTIES OF NIOBIUM AND TANTALUM
FLUORIDES AT HIGH TEMPERATURES. PART-2:
TETRAFLUORIDES AND TRIFLUORIDES.
GALKIN NP + TUMANOV YN + KOROBTSEV VP + BATAREV
+ KHOKHLOV VA + PAVLOV AA
RUSS J PHYS CHEM 45(10): 1531 (1971)

TANTALUM TETRAFLUORIDE TaF_4

THERMODYNAMIC PROPERTIES OF NIOBIUM AND TANTALUM
FLUORIDES AT HIGH TEMPERATURES. PART-2:
TETRAFLUORIDES AND TRIFLUORIDES.
GALKIN NP + TUMANOV YN + KOROBTSEV VP + BATAREV
+ KHOKHLOV VA + PAVLOV AA
RUSS J PHYS CHEM 45(10): 1531 (1971)

TANTALUM PENTAFLUORIDE TaF_5

POLYMER- MONOMER EQUILIBRIA IN NIOBIUM
PENTAFLUORIDE, TANTALUM PENTAFLUORIDE, AND
ANTIMONY PENTAFLUORIDE. THE SHAPE OF ANTIMONY
PENTAFLUORIDE.
ALEXANDER LE
INORG NUCL CHEM LETT 7: 1053-6 (1970)

VIBRATIONAL SPECTRA OF NIOBIUM AND TANTALUM
PENTAFLUORIDES IN THE GAS PHASE. VAPOR DENSITY
OF NIOBIUM PENTAFLUORIDE.
ALEXANDER LE + BEATTIE IR + JONES PJ
J CHEM SOC, DALTON TRANS 1972(2): 210-12

A FLUORINE NMR AND RAMAN STUDY OF COMPLEX
FORMATION BETWEEN ANTIMONY PENTAFLUORIDE,
NIOBIUM PENTAFLUORIDE, AND TANTALUM
PENTAFLUORIDE.
DEAN PAW + GILLESPIE RJ
CAN J CHEM 49: 1736-46 (1971)

HALIDES OF NIOBIUM AND TANTALUM. PART-3: VAPOR
PRESSURES OF NIOBIUM AND TANTALUM
PENTAFLUORIDES.
FAIRBROTHER F + FRITH WC
J CHEM SOC 1951: 3051-6

THERMODYNAMIC PROPERTIES OF NIOBIUM AND TANTALUM
FLUORIDES AT HIGH TEMPERATURES. PART-1:
PENTAFLUORIDES.
GALKIN NP + TUMANOV YN + KOROBTSEV VP + BATAREV
+ KHOKHLOV VA + PAVLOV AA
RUSS J PHYS CHEM 45(10): 1531 (1971)

ENTHALPIES OF FORMATION OF NIOBIUM PENTAFLUORIDE
AND TANTALUM PENTAFLUORIDE.
GREENBERG E + NATKE CA + HUBBARD WN
J PHYS CHEM 69: 2089-93 (1965)

THERMODYNAMIC PROPERTIES OF NIOBIUM, TANTALUM,
AND MOLYBDENUM PENTAFLUORIDES.
OREKHOV VT + RYBAKOV AG
RUSS J PHYS CHEM 47(6): 1612 (1973)

ELECTRON DIFFRACTION STUDY OF THE MOLECULAR
STRUCTURES OF VANADIUM PENTAFLUORIE, NIOBIUM
PENTAFLUORIDE, AND TANTALUM PENTAFLUORIDE.
ROMANOV GV + SPIRIDONOV VP
IZV SIB OTDEL AKAD NAUK SSSR, SER KHIM NAUK 1:
126-31 (1968) (IN RUSSIAN)

THERMODYNAMIC PROPERTIES OF NIOBIUM AND TANTALUM
PENTAFLUORIDES.
SELEZNEV VP + RAKOV EG + MIKULENOK VV
RUSS J PHYS CHEM 45(11): 1667 (1971)

RAMAN SPECTRA OF LIQUID NIOBIUM PENTAFLUORIDE
AND TANTALUM PENTAFLUORIDE.
SELIG H + REIS A + GASNER EL
J INORG NUCL CHEM 30: 2087-90 (1968)

ELECTRON DIFFRACTION INVESTIGATION OF NIOBIUM
PENTAFLUORIDE AND TANTALUM PENTAFLUORIDE.
SPIRIDONOV VP
VEST MOSK UNIV KHIM 23: 7-11 (1968) (IN RUSSIAN)

1103 STABILITIES OF TANTALUM PENTAFLUORIDE AND
TANTALUM OXIDE TRIFLUORIDE.
ZMBOV KF + MARGRAVE JL
J PHYS CHEM 72: 1099-1101 (1968)

TECHNETIUM HEXAFLUORIDE TcF_6

RAMAN SPECTRA OF MOLYBDENUM HEXAFLUORIDE,
TECHNETIUM HEXAFLUORIDE, RHENIUM HEXAFLUORIDE,
URANIUM HEXAFLUORIDE, SULFUR HEXAFLUORIDE,
SELENIUM HEXAFLUORIDE, AND TELLURIUM
HEXAFLUORIDE IN THE VAPOR STATE.
CLAASSEN HH + GOODMAN GL + HOLLOWAY JH + SELIG H
J CHEM PHYS 53: 341-8 (1970)

VIBRATIONAL SPECTRA OF MOLYBDENUM HEXAFLUORIDE
AND TECHNETIUM HEXAFLUORIDE.
CLAASSEN HH + SELIG H + MALM JG
J CHEM PHYS 36: 2888-90 (1962)

1104 MEAN AMPLITUDES OF VIBRATION AND THERMODYNAMIC
FUNCTIONS OF TECHNETIUM HEXAFLUORIDE.
NAGARAJAN G
INDIAN J PURE APPL PHYS 1: 232-4 (1963)

1105 VAPOR PRESSURE AND TRANSITION POINTS OF
TECHNETIUM HEXAFLUORIDE.
SELIG H + MALM JG
J INORG NUCL CHEM 24: 641-4 (1963)

VALENCE FORCE CONSTANTS OF THE HEXAFLUORIDES OF
MOLYBDENUM, TECHNETIUM, AND RHENIUM.
SINGH RB + RAI DK
CAN J PHYS 43: 167-9 (1965)

TELLURIUM TETRAFLUORIDE TeF_4

VIBRATIONAL SPECTRA AND STRUCTURES OF SELENIUM
TETRAFLUORIDE AND TELLURIUM TETRAFLUORIDE,
INCLUDING A MATRIX ISOLATION STUDY.
ADAMS CJ + DOWNS AJ
SPECTROCHIM ACTA 28A: 1841-1854 (1972)

1106 CRYSTAL STRUCTURE OF TELLURIUM TETRAFLUORIDE.
EDWARDS AJ + HEWAIDY FI
J CHEM SOC A 1968: 2977-80

TELLURIUM HEXAFLUORIDE TeF_6

FORCE FIELDS FOR SOME GROUP-6 HEXAFLUORIDES.
(SULFUR HEXAFLUORIDE AND TELLURIUM HEXAFLUORIDE)
ABRAMOWITZ S + LEVIN IW
J CHEM PHYS 44: 3353-6 (1966)

VAPOR PHASE RAMAN SPECTRA, RAMAN BAND CONTOUR ANALYSES, AND CORIOLIS CONSTANTS OF THE SPHERICAL TOP MOLECULES: SULFUR HEXAFLUORIDE, SELENIUM HEXAFLUORIDE, TELLURIUM HEXAFLUORIDE, MOLYBDENUM HEXAFLUORIDE, TUNGSTEN HEXAFLUORIDE, AND URANIUM HEXAFLUORIDE.
BOSWORTH YM + CLARK RJH + RIPPON DM
J MOL SPECTROSC 46: 240-55 (1973)

1107 INFRARED SPECTRA OF SELENIUM HEXAFLUORIDE AND TELLURIUM HEXAFLUORIDE.
BURKE TG
J CHEM PHYS 25: 791-2 (1956)

RAMAN SPECTRA OF MOLYBDENUM HEXAFLUORIDE, TECHNETIUM HEXAFLUORIDE, RHENIUM HEXAFLUORIDE, URANIUM HEXAFLUORIDE, SULFUR HEXAFLUORIDE, SELENIUM HEXAFLUORIDE, AND TELLURIUM HEXAFLUORIDE IN THE VAPOR STATE.
CLAASSEN HH + GOODMAN GL + HOLLOWAY JH + SELIG H
J CHEM PHYS 53: 341-8 (1970)

SPECTROSCOPIC STUDIES IN CONNECTION WITH ELECTRON DIFFRACTION INVESTIGATION OF SOME SIMPLE MOLECULES. (TUNGSTEN HEXAFLUORIDE, TELLURIUM HEXAFLUORIDE, AND MOLYBDENUM HEXAFLUORIDE)
CYVIN SJ + CYVIN BN + BRUNVOLL J
P69-89 OF SELECTED TOPICS IN STRUCTURE CHEMISTRY, ANDERSON P + BASTIANSEN O + FURBERG S (EDS), UNIVERSITETSFORLAGET, OSLO, NORWAY, 1967.

INFRARED SPECTRA AND MOLECULAR STRUCTURE OF SOME GROUP-6 HEXAFLUORIDES. (OF SULFUR, SELENIUM, TELLURIUM, MOLYBDENUM, TUNGSTEN, AND URANIUM)
GAUNT J
TRANS FARADAY SOC 49: 1122-31 (1953)

1108 INFRARED SPECTRUM OF TELLURIUM HEXAFLUORIDE FROM 25 TO 40 MICRONS.
GAUNT J
TRANS FARADAY SOC 51: 893-4 (1955)

MOLECULAR FORCE FIELDS. PART-8: VIBRATION FREQUENCIES OF SOME OCTAHEDRAL XY6 MOLECULES. (SULFUR HEXAFLUORIDE, SELENIUM HEXAFLUORIDE, TELLURIUM HEXAFLUORIDE)
HEATH DF + LINNETT JW
TRANS FARADAY SOC 45: 264-71 (1949)

ELECTRIC DEFLECTION OF BINARY HEXAFLUORIDES. (SULFUR HEXAFLUORIDE, SELENIUM HEXAFLUORIDE, TELLURIUM HEXAFLUORIDE, MOLYBDENUM HEXAFLUORIDE, TUNGSTEN HEXAFLUORIDE, URANIUM HEXAFLUORIDE, RUTHENIUM HEXAFLUORIDE, RHENIUM HEXAFLUORIDE, RHODIUM HEXAFLUORIDE, OSMIUM HEXAFLUORIDE, IRIDIUM HEXAFLUORIDE, AND PLATINUM HEXAFLUORIDE)
KAISER EW + MUENTER JS + KLEMPERER W + FALCONER WE + SUNDER WA
J CHEM PHYS 53: 1411-12 (1970)

ROOT MEAN SQUARE AMPLITUDES IN SOME HEXAFLUORIDES OF OCTAHEDRAL SYMMETRY. (SELENIUM AND TELLURIUM FLUORIDES)
NAGARAJAN G + ADAMS TS
Z PHYS CHEM (LEIPZIG) 255: 869-88 (1974)

ENTHALPIES OF FORMATION OF SULFUR HEXAFLUORIDE, SELENIUM HEXAFLUORIDE, TELLURIUM HEXAFLUORIDE AND THEIR THERMODYNAMIC PROPERTIES.
O'HARE PAG + SETTLE JL + HUBBARD WN
TRANS FARADAY SOC 62 (3): 558-65 (1966)

GREEN'S FUNCTION ANALYSIS OF SULFUR HEXAFLUORIDE AND RELATED MOLECULES. (SELENIUM HEXAFLUORIDE, TELLURIUM HEXAFLUORIDE)
RAMASWAMY K + MOHAN N
J MOL STRUCT 7: 51-8 (1971)

FAILURE OF THE FIRST BORN APPROXIMATION IN ELECTRON DIFFRACTION. (TUNGSTEN HEXAFLUORIDE, URANIUM HEXAFLUORIDE, TELLURIUM HEXAFLUORIDE, MOLYBDENUM HEXAFLUORIDE)
SEIP HM
P26-68 OF SELECTED TOPICS IN STRUCTURE CHEMISTRY, ANDERSON P + BASTIANSEN O + FURBERG S (EDS), UNIVERSITETSFORLAGET, OSLO, NORWAY, 1967.

1109 FAILURE OF THE FIRST BORN APPROXIMATION IN ELECTRON DIFFRACTION. (TELLURIUM HEXAFLUORIDE)
SEIP HM + STOELEVIK R
ACTA CHEM SCAND 20: 1535-45 (1966)

RAMAN SPECTRA AND MOLECULAR CONSTANTS OF THE HEXAFLUORIDES OF SULFUR, SELENIUM, AND TELLURIUM.
YOST DM + STEFFENS CC + GROSS ST
J CHEM PHYS 2: 311-16 (1934)

TERBIUM TRIFLUORIDE TbF_3

ENTROPIES AND ENTHALPIES OF SUBLIMATION OF NEODYMIUM AND TERBIUM TRIFLUORIDES.
MCCREARY JR + THORN RJ
HIGH TEMP SCI 6(3): 205-14 (1974)

TERBIUM TETRAFLUORIDE TbF_4

ELECTRIC DEFLECTION AND THERMAL DECOMPOSITION STUDIES ON CERIUM TETRAFLUORIDE, TERBIUM TETRAFLUORIDE, AND PRASEODYMIUM TETRAFLUORIDE.
KAISER EW + SUNDER WA + FALCONER WE
J LESS COMMON METALS 27(3): 383-7 (1972)

THALLIUM MONOFLUORIDE TlF

1110 MICROWAVE SPECTRA OF THE THALLIUM, INDIUM AND GALLIUM MONOHALIDES.
BARRETT AH + MANDEL M
PHYS REV 109: 1572-89 (1958)

DISSOCIATION ENERGIES OF THE GASEOUS MONOHALIDES OF BORON, ALUMINUM, GALLIUM, INDIUM, AND THALLIUM.
BARROW RF
TRANS FARADAY SOC 56: 952-8 (1960)

1111 ROTATIONAL ANALYSIS OF BANDS OF THE A TRIPLET PI ZERO(+), B TRIPLET PI(1) TO X SINGLET SIGMA(+) SYSTEMS OF THALLIUM MONOFLUORIDE.
BARROW RF + CHEALL HFK + THOMAS PM + ZEEMAN PB
PROC PHYS SOC 71: 128-30 (1958)

1112 PHOTOIONIZATION OF HIGH TEMPERATURE VAPORS. PART-4: THALLIUM MONOFLUORIDE, THALLIUM MONOCHLORIDE, AND THALLIUM MONOBROMIDE.
BERKOWITZ J + WALTER TA
J CHEM PHYS 49: 1184-89 (1968)

1113 INFRARED SPECTRA AND STRUCTURE OF MATRIX ISOLATED THALLIUM HALIDES.
BROM JM + FRANZEN HF
J CHEM PHYS 54: 2874-84 (1971)

1114 GASEOUS TRIMER AND TETRAMER OF THALLIUM MONOFLUORIDE.
CUBICCIOTTI D
HIGH TEMP SCI 2: 65-69 (1970)

1115 ENTHALPY OF FORMATION AND THE DISSOCIATION ENERGY OF THALLIUM MONOFLUORIDE
CUBICCIOTTI D + WITHERS GL
J PHYS CHEM 69(11): 4030-2 (1965)

1116 ENTHALPIES, ENTROPIES, AND FREE ENERGY FUNCTIONS OF THALLIUM MONOHALIDES ABOVE ROOM TEMPERATURE. (THALLIUM MONOFLUORIDE)
CUBICCIOTTI D + EDING H
J CHEM ENG DATA 10: 343-5 (1965)

1117 PHOTOELECTRON SPECTROSCOPY OF HIGH TEMPERATURE
VAPORS. PART-3: MONOMER AND DIMER SPECTRA OF
THALLOUS FLUORIDE.
DEHMER JL + BERKOWITZ J + CUSACHS LC
J CHEM PHYS 58(12): 5681-6 (1973)

1118 STARK EFFECT FOR THALLIUM MONOFLUORIDE AND
POTASSIUM FLUORIDE.
DIJKERMAN H + FLEGEL W + GRAFF G + MONTER B
Z NATURFORSCH 27A: 100-10 (1972)

1119 MICROWAVE ROTATIONAL SPECTRUM OF THALLIUM
MONOFLUORIDES.
FITZKY HG
Z PHYS 151: 351-64 (1958) (IN RUSSIAN)

1120 HIGH TEMPERATURE MICROWAVE SPECTROMETER FOR
MEASUREMENTS OF ZEEMAN EFFECT IN DIAMAGNETIC
MOLECULES. GJ FACTOR OF THALLIUM FLUORIDE,
CESIUM FLUORIDE, CESIUM CHLORIDE, CESIUM
BROMIDE, AND CESIUM IODIDE AND MAGNETIC
SUSCEPTIBILITY ANISOTROPY OF THALLIUM FLUORIDE,
CESIUM FLUORIDE, AND CESIUM CHLORIDE.
HONERJAEGER R + TISCHER R
Z NATURFORSCH 28A(3-4): 458-63 (1973) (IN
GERMAN)

1121 THERMODYNAMICS OF VAPORIZATION OF THALLIUM
MONOFLUORIDE.
KENESHEA FJ + CUBICCIOTTI D
J PHYS CHEM 71(6): 1958-60 (1967)

1122 NATURE OF BONDS IN THALLIUM MONOHALIDE
MOLECULES.
KRASNOV KS
RUSS J INORG CHEM 5: 805-7 (1960)

1123 INFRARED AND RAMAN SPECTRA AND STRUCTURE OF
MATRIX ISOLATED THALLOUS HALIDE DIMERS, THALLOUS
FLUORIDE, AND THALLOUS CHLORIDE.
LESIECKI ML + NIBLER JW
J CHEM PHYS 63(8): 3452-61 (1975)

1124 HEAT CAPACITY AND THERMODYNAMIC PROPERTIES OF
THALLIUM FLUORIDE FROM 5 TO 445 DEGREES K.
LYON WG + WESTRUM EF + CHAVRET M
J CHEM THERMODYN 3: 571-81 (1971)

1125 BAND SPECTRA OF THALLIUM MONOIODIDE AND
MONOFLUORIDE.
RAO JVR + RAO PT
INDIAN J PHYS 29: 20-6 (1955)

1126 INFRARED AND RAMAN SPECTRA AND THE STRUCTURE OF
CRYSTALLINE THALLIUM MONOFLUORIDE.
RUOFF VA + WEDLEIN J
Z ANORG ALLGEM CHEM 370: 113-19 (1969)

POTENTIAL ENERGY CURVES AND DISSOCIATION
ENERGIES OF DIATOMIC FLUORIDES AND CHLORIDES OF
GALLIUM, INDIUM, AND THALLIUM. (MONOFLUORIDES)
SINGH J + NAIR KPR + RAI DK
J QUANT SPECT RAD TRANSFER 11: 1577-81 (1971)

POTENTIAL ENERGY CURVES AND NATURE OF BINDING IN
GROUP-3A MONOHALIDES.
THAKUR SN + SINGH RB + RAI DK
J SCI IND RES 27(9): 339-47 (1968)

THALLIUM TRIFLUORIDE TlF_3

1127 CRYSTAL STRUCTURE OF THALLIUM TRIFLUORIDE.
HEBECKER VC
Z ANORG ALLGEM CHEM 393: 223-9 (1972)

HIGH TEMPERATURE NEGATIVE IONS. GASEOUS GROUP-3
FLUORIDES.
PETTY F + WANG JLF + STEIGER RP + HARLAND PW
+ FRANKLIN JL + MARGRAVE JL
HIGH TEMP SCI 5(1): 25-33 (1973)

THORIUM DIFLUORIDE ThF_2

1128 HEATS OF FORMATION GASEOUS THORIUM TRIFLUORIDE
AND THORIUM DIFLUORIDE FROM MASS SPECTROMETRIC
STUDIES.
ZMBOV KF
J INORG CHEM 32: 1378-81 (1970)

THORIUM TRIFLUORIDE ThF_3

HEATS OF FORMATION GASEOUS THORIUM TRIFLUORIDE
AND THORIUM DIFLUORIDE FROM MASS SPECTROMETRIC
STUDIES.
ZMBOV KF
J INORG CHEM 32: 1378-81 (1970)

THORIUM TETRAFLUORIDE ThF_4

INFRARED SPECTRA OF SOME GROUP-4 HALIDES.
(ZIRCONIUM TETRAFLUORIDE, HAFNIUM TETRAFLUORIDE,
AND THORIUM TETRAFLUORIDE)
BUCHLER A + MATTUCK JBB + DUGRE DH
J CHEM PHYS 34: 2202-3 (1961)

1129 VAPOR PRESSURE OF THORIUM TETRAFLUORIDE.
DARNELL AJ + KENESHEA FJ
J AMER CHEM SOC 62: 1143-5 (1958)

1130 ENTHALPY OF FORMATION OF THORIUM TETRAFLUORIDE
BY FLUORINE BOMB CALORIMETRY.
DEVENTER EHV + RUDZITIS E + HUBBARD WN
J INORG NUCL CHEM 32: 3233-6 (1970)

ELECTRON DIFFRACTION STUDY OF URANIUM AND
THORIUM TETRAHALIDES IN THE GAS PHASE. (URANIUM
TETRAFLUORIDE AND THORIUM TETRAFLUORIDE)
EZHOV YS + AKISHIN PA + RAMBIDI NG
J STRUCT CHEM 10(4): 483-6 (1969)

FREE ENERGIES OF FORMATION OF SOME INORGANIC
FLUORIDES BY SOLID STATE EMF MEASUREMENTS.
(THORIUM TETRAFLUORIDE, ALUMINUM TRIFLUORIDE,
NICKEL DIFLUORIDE, LEAD DIFLUORIDE, COLBALT
DIFLUORIDE, AND URANIUM TRIFLUORIDE)
HEUS RJ + EGAN JJ
Z PHYS CHEM NEUE FOLGE 49: 38-43 (1966)

THULIUM TRIFLUORIDE TmF_3

1131 VAPORIZATION REACTIONS IN THE THULIUM- FLUORINE
SYSTEM.
BIEFELD RM + EICK HA
J LESS COMMON METALS 45(1): 117-23 (1976)

1132 SUBLIMATION PRESSURES AND HEATS OF SUBLIMATION
OF THULIUM TRIFLUORIDE, YTTERBIUM TRIFLUORIDE,
AND LUTETIUM TRIFLUORIDE.
ZMBOV KF + MARGRAVE JL
J LESS COMMON METALS 12: 494-6 (1967)

TIN MONOFLUORIDE SnF

SPECTRA OF SILICON MONOFLUORIDE, GERMANIUM
MONOFLUORIDE, TIN MONOFLUORIDE, AND LEAD
MONOFLUORIDE.
BARROW RF + BUTLER D + JOHNS JWC + POWELL JL
PROC PHYS SOC 73: 317-20 (1959)

1133 ROTATIONAL ANALYSIS OF THE A DOUBLET SIGMA(+) TO
X DOUBLET PI, B DOUBLET SIGMA(+) TO X DOUBLET PI
SYSTEMS OF TIN MONOFLUORIDE.
BARROW RF + KOPP I + MERER AJ
PROC PHYS SOC 79: 749-52 (1962)

1134 ACCURACY OF THE VIBRATIONAL WAVE FUNCTIONS AND
DERIVED QUANTITIES. (TIN FLUORIDE)
GOHEL VB
SPECTROSC LETT 7(11): 575-80 (1974)

1135 ROTATIONAL STRUCTURE AND ISOTOPIC SHIFT IN THE
(1,0) BAND OF THE A DOUBLET SIGMA- X DOUBLET PI
SYSTEM OF THE TIN MONOFLUORIDE MOLECULE.
RAI SB + SINGH J
SPECTROSC LETT 5(5): 155-67 (1972)

1136 ROTATIONAL ANALYSIS OF THE C DOUBLET DELTA 5/2-
X DOUBLET PI 3/2 AND G DOUBLET DELTA 5/2- X
DOUBLET PI 3/2 SYSTEMS OF THE TIN MONOFLUORIDE
MOLECULE.
RAM RS + UPADHYA KN + RAI DK
J PHYS B 6(12): L372-4 (1973)

1137 VARIATION OF ELECTRONIC TRANSITION MOMENT WITH
THE INTERNUCLEAR SEPARATION IN SILICON CHLORIDE
AND TIN MONOFLUORIDE.
SINGH J + DUBE PS
INDIAN J PURE APPL PHYS 9: 164-5 (1971)

POTENTIAL CURVES FOR SOME DIATOMIC MOLECULES.
(BERYLLIUM MONOFLUORIDE, SILICON MONOFLUORIDE,
AND TIN MONOFLUORIDE)
SINGH RB + RAI DK
INDIAN J PURE APPL PHYS 4: 102-5 (1966)

1138 ANALYSIS OF VIBRATIONAL AND ROTATIONAL STRUCTURE
OF TIN MONOFLUORIDE IN THE 2500 ANGSTROM REGION.
UZIKOV AN + KUZYAKOV YY
VEST MOSK UNIV KHIM 23: 33-5 (1968) (IN RUSSIAN)

MASS SPECTROMETRIC STUDIES AT HIGH TEMPERATURES.
(TIN DIFLUORIDE, LEAD DIFLUORIDE, TIN
MONOFLUORIDE, AND LEAD MONOFLUORIDE)
ZMBOV KF + HASTIE JW + MARGRAVE JL
TRANS FARADAY SOC 64(4): 861-7 (1968)

TIN DIFLUORIDE SnF_2

1139 ENVIRONMENT OF THE TIN ATOM IN ORTHORHOMBIC TIN
DIFLUORIDE.
DONALDSON JD + OTENG R
INORG NUCL CHEM LETT 3: 163-4 (1967)

1140 MELTING POINT, VAPOR PRESSURE, AND HEAT OF
VAPORIZATION OF TIN DIFLUORIDE.
DUDASH JJ + SEARCY AW
HIGH TEMP SCI 1: 287-93 (1969)

VAPOR PRESSURES OF ZIRCONIUM TETRAFLUORIDE,
HAFNIUM TETRAFLUORIDE, AND TIN DIFLUORIDE.
FISCHER VW + PETEZEL T + LAUTER S
Z ANORG ALLGEM CHEM 333: 226-34 (1964)

INFRARED SPECTRA OF MATRIX ISOLATED TIN
DIFLUORIDE AND LEAD DIFLUORIDE.
HAUGE RH + HASTIE JW + MARGRAVE JL
J MOL SPECTROSC 45: 420-7 (1973)

ULTRAVIOLET ABSORPTION SPECTRA OF GASEOUS TIN
DIFLUORIDE AND LEAD DIFLUORIDE.
HAUGE RH + HASTIE JW + MARGRAVE JL
J PHYS CHEM 72: 3510-11 (1968)

FREQUENCIES OF THE DEFORMATION VIBRATIONS OF THE
DIHALIDES OF GERMANIUM, TIN, AND LEAD.
(GERMANIUM DIFLUORIDE, TIN DIFLUORIDE, LEAD
DIFLUORIDE)
TIMOSHININ VS + DANILOVA TG
RUSS J PHYS CHEM 42: 1596 (1968)

MASS SPECTROMETRIC STUDIES AT HIGH TEMPERATURES.
(TIN DIFLUORIDE, LEAD DIFLUORIDE, TIN
MONOFLUORIDE, AND LEAD MONOFLUORIDE)
ZMBOV KF + HASTIE JW + MARGRAVE JL
TRANS FARADAY SOC 64(4): 861-7 (1968)

TIN TETRAFLUORIDE SnF_4

1141 IONICITY OF THE MX BOND IN THE TETRAHALIDES OF
GROUP-4 ELEMENTS. (TIN TETRAFLUORIDE AND HAFNIUM
TETRAFLUORIDE)
RAI SN + THAKUR SN + RAI DK
J MOL STRUCT 8: 55-61 (1971)

TITANIUM MONOFLUORIDE TiF

1142 EMISSION SPECTRUM OF TITANIUM MONOFLUORIDE.
CHATALIC A + DESCHAMPS P + PANNETIER G
+ CHAMPETIER G
COMPT REND C 270: 146-9 (1970) (IN FRENCH)

1143 ABSORPTION SPECTRUM OF VAPORIZED TITANIUM
MONOFLUORIDE.
DIEBNER RL + KAY JF
J CHEM PHYS 51: 3547-54 (1969)

1144 SIMPLE THEORETICAL CALCULATIONS ON TITANIUM
MONOFLUORIDE.
ROBERTO C
GAZZ CHIM ITAL 105: 27-35 (1975)

1145 MASS SPECTROMETRIC STUDIES AT HIGH TEMPERATURES.
PART-16: SUBLIMATION PRESSURES FOR TITANIUM
TRIFLUORIDE, AND THE STABILITIES OF TITANIUM
DIFLUORIDE AND TITANIUM MONOFLUORIDE.
ZMBOV KF + MARGRAVE JL
J PHYS CHEM 71: 2893-5 (1967)

TITANIUM DIFLUORIDE TiF_2

1146 INFRARED SPECTRA AND GEOMETRY OF TITANIUM
DIFLUORIDE AND TITANIUM TRIFLUORIDE IN RARE GAS
MATRICES.
HASTIE JW + HAUGE RH + MARGRAVE JL
J CHEM PHYS 51: 2648-56 (1969)

MASS SPECTROMETRIC STUDIES AT HIGH TEMPERATURES.
PART-16: SUBLIMATION PRESSURES FOR TITANIUM
TRIFLUORIDE AND THE STABILITIES OF TITANIUM
DIFLUORIDE AND TITANIUM MONOFLUORIDE.
ZMBOV KF + MARGRAVE JL
J PHYS CHEM 71: 2893-5 (1967)

TITANIUM TETRAFLUORIDE TiF_4

1147 HEATS OF FORMATION OF TITANIUM TETRAFLUORIDE AND
HAFNIUM TETRAFLUORIDE.
GREENBERG E + SETTLE JL + HUBBARD WN
J PHYS CHEM 66: 1345-8 (1962)

1148 VAPOR PRESSURE OF TITANIUM TETRAFLUORIDE.
HALL EH + BLOCHER JM + CAMPBELL IE
J ELECTROCHEM SOC 105: 275-8 (1958)

INFRARED SPECTRA AND GEOMETRY OF TITANIUM
DIFLUORIDE AND TITANIUM TRIFLUORIDE IN RARE GAS
MATRICES.
HASTIE JW + HAUGE RH + MARGRAVE JL
J CHEM PHYS 51: 2648-56 (1969)

MOLECULAR BEAM ELECTRIC DEFLECTION OF THE
TETRAHALIDES CARBON TETRAFLUORIDE, CARBON
TETRACHLORIDE, SILICON TETRAFLUORIDE, SILICON
TETRACHLORIDE, GERMANIUM TETRACHLORIDE, TITANIUM
TETRAFLUORIDE, TITANIUM TETRACHLORIDE, VANADIUM
TETRAFLUORIDE, AND VANADIUM TETRACHLORIDE.
MUENTER AA + DYKE TR + FALCONER WE + KLEMPERER W
J CHEM PHYS 63: 1231-6 (1975)

TUNGSTEN TETRAFLUORIDE WF_4

EMPIRICAL METHOD FOR DETERMINING EFFECTIVE
VIBRATIONAL AND ROTATIONAL CHARACTERISTICS OF
MOLECULES OF SOME TETRAFLUORIDES FOR CALCULATING
THEIR THERMODYNAMIC FUNCTIONS.
TUMANOV YN + GALKIN NP
RUSS J PHYS CHEM 43(4): 464-7 (1969)

TUNGSTEN PENTAFLUORIDE WF_5

VAPOR PRESSURES OF MOLYBDENUM PENTAFLUORIDE,
TUNGSTEN PENTAFLUORIDE, RHENIUM PENTAFLUORIDE,
AND OSMIUM PENTAFLUORIDE.
CADY GH + HARGREAVES GB
J CHEM SOC A 1961: 1568-74

THERMODYNAMIC PROPERTIES OF PENTAFLUORIDES AT
HIGH TEMPERATURES. PART-3: PENTAFLUORIDES OF
SOME 5D-ELEMENTS.
GALKIN NP + TUMANOV YN + BUTYLKIN YP
RUSS J PHYS CHEM 44(12): 1724-6 (1970)

1149 THERMODYNAMIC PROPERTIES OF TUNGSTEN
PENTAFLUORIDE.
GUSAROV AV + PERVOV VS + GOTKIS IS + KLYUEV LI
+ BUTSKII VD
DOKL AKAD NAUK SSSR 216(6): 1296-9 (1974) (IN
RUSSIAN)

VIBRATIONAL SPECTRA OF MOLYBDENUM PENTAFLUORIDE
AND TUNGSTEN PENTAFLUORIDE.
OUELLETTE TH + RATCLIFFE CT + SHARP DWA
J CHEM SOC A 1969: 2351-4

1150 PREPARATION AND PROPERTIES (VAPOR PRESSURE) OF
TUNGSTEN PENTAFLUORIDE.
SCHRODER J + GREWE FJ
CHEM BER 103: 1536-46 (1970) (IN GERMAN)

1151 ENTHALPY OF FORMATION OF TUNGSTEN PENTAFLUORIDE
AND TUNGSTEN HEXAFLUORIDE.
SCHRODER J + SIEBEN FJ
CHEM BER 103: 76-81 (1970)

TUNGSTEN HEXAFLUORIDE WF_6

VIBRATIONAL ANALYSIS OF SELENIUM HEXAFLUORIDE
AND TUNGSTEN HEXAFLUORIDE.
ABRAMOWITZ S + LEVIN IW
INORG CHEM 3: 538-41 (1967)

1152 SOME PHYSICAL PROPERTIES OF TUNGSTEN
HEXAFLUORIDE. (VAPOR PRESSURE)
BARBER EJ + CADY GH
J PHYS CHEM 60: 505-7 (1956)

VAPOR PHASE RAMAN SPECTRA, RAMAN BAND CONTOUR
ANALYSES, AND CORIOLIS CONSTANTS OF THE
SPHERICAL TOP MOLECULES, SULFUR HEXAFLUORIDE,
SELENIUM HEXAFLUORIDE, TELLURIUM HEXAFLUORIDE,
MOLYBDENUM HEXAFLUORIDE, TUNGSTEN HEXAFLUORIDE,
AND URANIUM HEXAFLUORIDE.
BOSWORTH YM + CLARK RJH + RIPPON DM
J MOL SPECTROSC 46: 240-55 (1973)

MOLECULAR STRUCTURE OF MOLYBDENUM, TUNGSTEN, AND
URANIUM HEXAFLUORIDES FROM INFRARED AND RAMAN
SPECTRA.
BURKE TG + SMITH DF + NIELSEN AH
J CHEM PHYS 20(3): 447-54 (1953)

VAPOR PRESSURES OF SOME HEAVY TRANSITION METAL
HEXAFLUORIDES. (RHENIUM HEXAFLUORIDE, OSMIUM
HEXAFLUORIDE, IRIDIUM HEXAFLUORIDE, TUNGSTEN
HEXAFLUORIDE, MOLYBDENUM HEXAFLUORIDE)
CADY GH + HARGREAVES GB
J CHEM SOC 1961: 1563-74

1153 SPECTROSCOPIC STUDIES IN CONNECTION WITH
ELECTRON DIFFRACTION INVESTIGATION OF SOME
SIMPLE MOLECULES. (TUNGSTEN HEXAFLUORIDE,
TELLURIUM HEXAFLUORIDE, AND MOLYBDENUM
HEXAFLUORIDE)
CYVIN SJ + CYVIN BN + BRUNVOLL J + ANDERSEN B
+ STOELEVIK R
P69-89 OF SELECTED TOPICS IN STRUCTURE
CHEMISTRY, ANDERSON P + BASTIANSEN O + FURBERG S
(EDS), UNIVERSITETSFORLAGET, OSLO, NORWAY, 1967.

1154 CRYSTAL STRUCTURE OF TUNGSTEN PENTAFLUORIDE.
EDWARDS AJ
J CHEM SOC A 1969: 909

THERMODYNAMIC STABILITY OF HEXAFLUORIDES AT HIGH
TEMPERATURES. PART-2: PLUTONIUM HEXAFLUORIDE.
PART-3: URANIUM HEXAFLUORIDE. PART-4: MOLYBDENUM
AND TUNGSTEN HEXAFLUORIDES.
GALKIN NP + TUMANOV YN
P188-91, 191-5, 195-9 OF TERMODIN TERMOKHIM
KONSTANTY, ASTAKHOV KV(ED), NAUKA, MOSCOW, 1970.
(IN RUSSIAN)

THERMAL STABILITY AND REACTIVITY OF D- AND F
ELEMENT HEXAFLUORIDES.
GALKIN NP + TUMANOV YN + BUTYLKIN YP
IZV SIB OTD AKAD NAUK SSSR SER KHIM NAUK (2):
12-21 (1968) (IN RUSSIAN)

INFRARED SPECTRA AND MOLECULAR STRUCTURE OF SOME
GROUP-6 HEXAFLUORIDES. (OF SULFUR, SELENIUM,
TELLURIUM, MOLYBDENUM, TUNGSTEN, AND URANIUM)
GAUNT J
TRANS FARADAY SOC 49: 1122-31 (1953)

FORCE CONSTANTS AND BOND LENGTHS OF SOME
INORGANIC HEXAFLUORIDES. (OF SULFUR, SELENIUM,
TELLURIUM, MOLYBDENUM, TUNGSTEN, URANIUM, AND
RHENIUM)
GAUNT J
TRANS FARADAY SOC 50: 546-51 (1954)

ELECTRIC DEFLECTION OF BINARY HEXAFLUORIDES.
(SULFUR HEXAFLUORIDE, SELENIUM HEXAFLUORIDE,
TELLURIUM HEXAFLUORIDE, MOLYBDENUM HEXAFLUORIDE,
TUNGSTEN HEXAFLUORIDE, URANIUM HEXAFLUORIDE,
RUTHENIUM HEXAFLUORIDE, RHENIUM HEXAFLUORIDE,
RHODIUM HEXAFLUORIDE, OSMIUM HEXAFLUORIDE,
IRIDIUM HEXAFLUORIDE, AND PLATINUM HEXAFLUORIDE)
KAISER EW + MUENTER JS + KLEMPERER W +
FALCONER WE + SUNDER WA
J CHEM PHYS 53: 1411-12 (1970)

1155 ELECTRON DIFFRACTION INVESTIGATION OF TUNGSTEN
HEXAFLUORIDE, OSMIUM HEXAFLUORIDE, IRIDIUM
HEXAFLUORIDE, URANIUM HEXAFLUORIDE, NEPTUNIUM
HEXAFLUORIDE, AND PLUTONIUM HEXAFLUORIDE.
KIMURA M + SCHOMAKER V + SMITH DW + WEINSTOCK B
J CHEM PHYS 48: 4001-12 (1968)

THERMODYNAMIC SIMILARITY AND UNIVERSAL EQUATIONS
OF STATE OF HEXAFLUORIDES. (OF SULFUR,
MOLYBDENUM, TUNGSTEN, AND URANIUM)
MALYSHEV VV
TEPLOFIZ VYS TEMP 14(1): 47-55 (1976) (IN
RUSSIAN)

ASSIGNMENTS IN THE ULTRAVIOLET SPECTRA OF
MOLYBDENUM HEXAFLUORIDE AND TUNGSTEN
HEXAFLUORIDE.
MCDIARMID R
J CHEM PHYS 61: 3333-9 (1974)

1156 CORIOLIS CONSTANTS OF SPHERICAL TOP MOLECULES
FROM LOW TEMPERATURE INFRARED STUDIES OF VAPOR
BAND CONTOURS. APPLICATION TO THE FORCE FIELD OF
TUNGSTEN HEXAFLUORIDE.
MCDOWELL RS + ASPREY LB
J MOL SPECTROSC 48(2): 254-65 (1973)

1157 ENTHALPY OF FORMATION OF TUNGSTEN HEXAFLUORIDE.
O'HARE PAG + HUBBARD WN
J PHYS CHEM 70: 3353-5 (1966)

HIGH RESOLUTION NMR SPECTRUM OF LIQUID
MOLYBDENUM HEXAFLUORIDE, TUNGSTEN HEXAFLUORIDE,
AND URANIUM HEXAFLUORIDE.
RIGNY P + DEMORTIER A + PERRIN F
COMPT REND 263: 1408-10 (1966) (IN FRENCH)

MOLECULAR MOTION AND FLUORINE RELAXATION IN THE
LIQUIDS OF MOLYBDENUM HEXAFLUORIDE, TUNGSTEN
HEXAFLUORIDE, AND URANIUM HEXAFLUORIDE.
RIGNY P + VIRLET J
J CHEM PHYS 47: 4645-52 (1967)

PHYSICAL CONSTANTS (VAPOR PRESSURE) OF SILICON
TETRAFLUORIDE, TUNGSTEN HEXAFLUORIDE, AND
MOLYBDENUM HEXAFLUORIDE.
RUFF O + ASCHER E
Z ANORG ALLGEM CHEM 196: 413-20 (1931) (IN
GERMAN)

ENTHALPY OF FORMATION OF TUNGSTEN PENTAFLUORIDE
AND TUNGSTEN HEXAFLUORIDE.
SCHRODER J + SIEBEN FJ
CHEM BER 103: 76-81 (1970)

1158 FAILURE OF THE FIRST BORN APPROXIMATION IN
ELECTRON DIFFRACTION. (TUNGSTEN HEXAFLUORIDE,
URANIUM HEXAFLUORIDE, TELLURIUM HEXAFLUORIDE,
MOLYBDENUM HEXAFLUORIDE)
SEIP HM
P26-68 OF SELECTED TOPICS IN STRUCTURE
CHEMISTRY, ANDERSON P + BASTIANSEN O + FURBERG S
(EDS), UNIVERSITETSFORLAGET, OSLO, NORWAY, 1967.

FAILURE OF THE FIRST BORN APPROXIMATION IN
ELECTRON DIFFRACTION. (MOLYBDENUM HEXAFLUORIDE
AND TUNGSTEN HEXAFLUORIDE)
SEIP HM + SEIP R
ACTA CHEM SCAND 20: 2698-2710 (1966)

RAMAN EFFECT AND ULTRAVIOLET ABSORPTION SPECTRA
OF MOLYBDENUM AND TUNGSTEN HEXAFLUORIDES.
TANNER KN + DUNCAN ABF
J AMER CHEM SOC 73: 1164-7 (1951)

1159 HEAT CAPACITIES AND ELECTRONIC SPECTRA OF THE
PLATINUM GROUP METAL HEXAFLUORIDE MOLECULES DOWN
TO HELIUM TEMPERATURES.
WEINSTOCK B + WESTRUM EF + GOODMAN GL
P405-6 OF INT CONF LOW TEMP PHYS, 8TH PROC,
LONDON, 1962

URANIUM MONOFLUORIDE UF

1160 HEATS OF FORMATION OF GASEOUS URANIUM FLUORIDES.
(URANIUM MONOFLUORIDE, URANIUM DIFLUORIDE,
URANIUM TRIFLUORIDE, URANIUM TETRAFLUORIDE,
URANIUM PENTAFLUORIDE, AND URANIUM HEXAFLUORIDE)
GODNEV IN + SVERDLIN AS
IZV VYSSH UCHEB ZAVED, KHIM KHIM TEKN 9: 40-3
(1966) (IN RUSSIAN)

1161 CALCULATION OF THE DISSOCIATION ENERGY OF
CALCIUM MONOFLUORIDE AND URANIUM MONOFLUORIDE.
KUDRIN LP + MAZEEV MY
ATOMNAYA ENERGIYA 22: 85-9 (1967) (IN RUSSIAN)

URANIUM DIFLUORIDE UF_2

HEATS OF FORMATION OF GASEOUS URANIUM FLUORIDES.
(URANIUM MONOFLUORIDE, URANIUM DIFLUORIDE,
URANIUM TRIFLUORIDE, URANIUM TETRAFLUORIDE,
URANIUM PENTAFLUORIDE, AND URANIUM HEXAFLUORIDE)
GODNEV IN + SVERDLIN AS
IZV VYSSH UCHEB ZAVED, KHIM KHIM TEKN 9: 40-3
(1966) (IN RUSSIAN)

1162 STRUCTURE OF MOLECULAR URANIUM DIFLUORIDE AND
URANIUM TRIFLUORIDE.
VOLKOV VM + DYATKINA ME
J STRUCT CHEM 8(4): 607-11 (1967)

URANIUM TRIFLUORIDE UF_3

HEATS OF FORMATION OF GASEOUS URANIUM FLUORIDES.
(URANIUM MONOFLUORIDE, URANIUM DIFLUORIDE,
URANIUM TRIFLUORIDE, URANIUM TETRAFLUORIDE,
URANIUM PENTAFLUORIDE, AND URANIUM HEXAFLUORIDE)
GODNEV IN + SVERDLIN AS
IZV VYSSH UCHEB ZAVED, KHIM KHIM TEKN 9: 40-3
(1966) (IN RUSSIAN)

FREE ENERGIES OF FORMATION OF SOME INORGANIC
FLUORIDES BY SOLID STATE EMF MEASUREMENTS.
(THORIUM TETRAFLUORIDE, ALUMINUM TRIFLUORIDE,
NICKEL DIFLUORIDE, LEAD DIFLUORIDE, COBALT
DIFLUORIDE, AND URANIUM TRIFLUORIDE)
HEUS RJ + EGAN JJ
Z PHYS CHEM NEUE FOLGE 49: 38-43 (1966)

1163 NEUTRON DIFFRACTION STUDY OF THE CRYSTAL
STRUCTURE OF URANIUM TRIFLUORIDE.
LAVEISSIERE J
BULL SOC FR MINERAL CRISTALLOGR 90: 304-7 (1967)

STRUCTURE OF MOLECULAR URANIUM DIFLUORIDE AND
URANIUM TRIFLUORIDE.
VOLKOV VM + DYATKINA ME
J STRUCT CHEM 8(4): 607-11 (1967)

1164 PREPARATION AND PROPERTIES OF URANIUM
TRIFLUORIDE. (CRYSTAL STRUCTURE)
WARD JC
P81-90 OF CHEMISTRY OF URANIUM, COLLECTED
PAPERS, KATZ JJ + RABINOWITCH(EDS), USAEC, 1958,
769P. TID-5290

URANIUM TETRAFLUORIDE UF_4 URANIUM PENTAFLUORIDE UF_5

1165 DETERMINATION OF HEAT OF SUBLIMATION OF URANIUM
TETRAFLUORIDE BY THE MASS SPECTROMETRIC METHOD.
AKISHIN PA + KHODEEV YS
RUSS J PHYS CHEM 35(5): 574-6 (MAY 1961)

1166 LOW TEMPERATURE HEAT CAPACITIES, ENTHALPIES, AND
ENTROPIES OF URANIUM TETRAFLUORIDE AND URANIUM
HEXAFLUORIDE.
BRICKWEDDE FG + HOGE HJ + SCOTT RB
P368 OF CHEMISTRY OF URANIUM, COLLECTED PAPERS,
KATZ JJ + RABINOWITCH(EDS), USAEC, 1958, 769P.
TID-5290

1167 SATURATED VAPOR PRESSURE OF SOLID URANIUM
TETRAFLUORIDE.
CHUDINOV EG + CHOPOROV DY
RUSS J PHYS CHEM 44(8): 1106-9 (1970)

1168 ABSORPTION SPECTRUM OF URANIUM TETRAFLUORIDE AND
ENERGY LEVELS OF URANIUM(IV).
CONWAY JG
J CHEM PHYS 31: 1002-4 (1959)

1169 ENTHALPY OF URANIUM TETRAFLUORIDE FROM 298-1400
DEGREES K. ENTHALPY AND ENTROPY OF FUSION.
DWORKIN AS
J INORG NUCL CHEM 34(1): 135-8 (1972)

1170 ELECTRON DIFFRACTION STUDY OF URANIUM AND
THORIUM TETRAHALIDES IN THE GAS PHASE. (URANIUM
TETRAFLUORIDE AND THORIUM TETRAFLUORIDE)
EZHOV YS + AKISHIN PA + RAMBIDI NG
J STRUCT CHEM 10(4): 483-6 (1969)

HEATS OF FORMATION OF GASEOUS URANIUM FLUORIDES.
(URANIUM MONOFLUORIDE, URANIUM DIFLUORIDE,
URANIUM TRIFLUORIDE, URANIUM TETRAFLUORIDE,
URANIUM PENTAFLUORIDE, AND URANIUM HEXAFLUORIDE)
GODNEV IN + SVERDLIN AS
IZV VYSSH UCHEB ZAVED, KHIM KHIM TEKN 9: 40-3
(1966) (IN RUSSIAN)

1171 ENTHALPY OF FORMATION OF URANIUM TETRAFLUORIDE.
KHANAEV EI
IZV SIB OTDEL AKAD NAUK SSSR, SER KHIM NAUK
9(4): 123-5 (1968) (IN RUSSIAN)

1172 VIBRATIONAL SPECTRA AND FORCE CONSTANTS OF
URANIUM TETRAFLUORIDE.
KING EG + CHRISTENSEN AU
SPECTROCHIM ACTA 26A: 1059-62 (1970)

1173 HIGH TEMPERATURE HEAT CONTENT OF URANIUM
TETRAFLUORIDE.
KING EG + CHRISTENSEN AU
US BUR MINES REP 5709, 1961, 4P.

1174 VAPOR PRESSURE OF URANIUM TETRAFLUORIDE.
LANGER S + BLANKENSHIP FF
J INORG NUCL CHEM 14: 26-31 (1960)

1175 VAPOR PRESSURE OF URANIUM TETRAFLUORIDE.
POPOV M + KOSTYLEV FA + ZUBOVA NV
RUSS J INORG CHEM 4: 770-1 (1959)

EMPIRICAL METHOD FOR DETERMINING EFFECTIVE
VIBRATIONAL AND ROTATIONAL CHARACTERISTICS OF
MOLECULES OF SOME TETRAFLUORIDES FOR CALCULATING
THEIR THERMODYNAMIC FUNCTIONS.
TUMANOV YN + GALKIN NP
RUSS J PHYS CHEM 43(4): 464-7 (1969)

1176 THERMODYNAMICS OF INTERMEDIATE URANIUM FLUORIDES
FROM MEASUREMENTS OF THE DISPROPORTIONATION
PRESSURES. (URANIUM PENTAFLUORIDE, URANIUM
HEXAFLUORIDE)
AGRON PA
P610-26 OF CHEMISTRY OF URANIUM, COLLECTED
PAPERS, KATZ JJ + RABINOWITCH(EDS), USAEC, 1958,
769P. TID-5290

1177 THERMODYNAMIC PROPERTIES OF URANIUM
PENTAFLUORIDE, NEPTUNIUM PENTAFLUORIDE, AND
PLUTONIUM PENTAFLUORIDE.
GALKIN NP + TUMANOV YN + BUTYLKIN YP
RUSS J PHYS CHEM 44: 1726 (1970)

THERMODYNAMIC PROPERTIES OF PENTAFLUORIDES AT
HIGH TEMPERATURES. PART-4: PENTAFLUORIDES OF 5F
ELEMENTS.
GALKIN NP + TUMANOV YN + BUTYLKIN YP
RUSS J PHYS CHEM 44(12): 1726-7 (1970)

HEATS OF FORMATION OF GASEOUS URANIUM FLUORIDES.
(URANIUM MONOFLUORIDE, URANIUM DIFLUORIDE,
URANIUM TRIFLUORIDE, URANIUM TETRAFLUORIDE,
URANIUM PENTAFLUORIDE, AND URANIUM HEXAFLUORIDE)
GODNEV IN + SVERDLIN AS
IZV VYSSH UCHEB ZAVED, KHIM KHIM TEKN 9: 40-3
(1966) (IN RUSSIAN)

1178 URANIUM PENTAFLUORIDE, A NEW FLUORIDE OF
URANIUM.
GROSSE AV
P315-17 OF CHEMISTRY OF URANIUM, COLLECTED
PAPERS, KATZ JJ + RABINOWITCH(EDS), USAEC, 1958,
769P. TID-5290

1179 THEORETICAL CALCULATIONS OF THE ELECTRONIC
STRUCTURE AND OPTICAL TRANSITIONS OF URANIUM
HEXAFLUORIDE AND URANIUM PENTAFLUORIDE.
MAYLOTTE DH + ST PETERS R + MESSMER RP
CHEM PHYS LETT 38(1): 181-4 (1976)

1180 VAPOR PRESSURE OF URANIUM HEXAFLUORIDE IN
EQUILIBRIUM WITH URANIUM PENTAFLUORIDE.
PRIEST HF
P738-41 OF CHEMISTRY OF URANIUM, COLLECTED
PAPERS, KATZ JJ + RABINOWITCH(EDS), USAEC, 1958,
769P. TID-5290

URANIUM HEXAFLUORIDE UF_6

1181 MEASUREMENT OF THE GAS DENSITY OF URANIUM
HEXAFLUORIDE BY LASER RAMAN SCATTERING.
ADOLFSON WF + SCHLECHT RG + MORTON JB
UNIV VA, 1975, 64P. UVA-ER-001-75U

THERMODYNAMICS OF INTERMEDIATE URANIUM FLUORIDES
FROM MEASUREMENTS OF THE DISPROPORTIONATION
PRESSURES. (URANIUM PENTAFLUORIDE, URANIUM
HEXAFLUORIDE)
AGRON PA
P610-26 OF CHEMISTRY OF URANIUM, COLLECTED
PAPERS, KATZ JJ + RABINOWITCH(EDS), USAEC, 1958,
769P. TID-5290

1182 INFRARED SPECTRUM OF URANIUM HEXAFLUORIDE.
BAR-ZIV E + FREIBERG M + WEISS S
SPECTROCHIM ACTA 28A: 2025-8 (1972)

1183 STRUCTURE OF URANIUM HEXAFLUORIDE AS DETERMINED
BY DIFFRACTION OF ELECTRONS ON THE VAPOR.
BAUER SH
P321 OF CHEMISTRY OF URANIUM, COLLECTED PAPERS,
KATZ JJ + RABINOWITCH(EDS), USAEC, 1958, 769P.
TID-5290

1184 STRUCTURE OF URANIUM HEXAFLUORIDE AS DETERMINED
      BY THE DIFFRACTION OF ELECTRONS ON THE VAPOR.
      BAUER SH
      J CHEM PHYS 18(1): 27-41 (1950)

1185 PROPERTIES AND REACTIONS OF URANIUM HEXAFLUORIDE
      BY ION CYCLOTRON RESONANCE SPECTROSCOPY.
      BEAUCHAMP JL
      J CHEM PHYS 64(2): 718-23 (1976)

1186 VIBRATIONAL SPECTRUM AND THERMODYNAMIC
      PROPERTIES OF URANIUM HEXAFLUORIDE GAS.
      BIGELEISEN J + MEYER MG + STEVENSON PC
      + TURKEVICH J
      P748 OF CHEMISTRY OF URANIUM, COLLECTED PAPERS,
      KATZ JJ + RABINOWITCH(EDS), USAEC, 1958, 769P.
      TID-5290

1187 VIBRATIONAL SPECTRUM AND THERMODYNAMIC
      PROPERTIES OF URANIUM HEXAFLUORIDE GAS.
      BIGELEISEN J + MAYER MG + STEVENSON PC
      + TURKEVICH J
      J CHEM PHYS 16(5): 442-5 (1948)

1188 CALCULATED ELECTRONIC STRUCTURE OF URANIUM
      HEXAFLUORIDE.
      BORING M + MOSKOWITZ JW
      CHEM PHYS LETT 38(1): 185-9 (1976)

      VAPOR PHASE RAMAN SPECTRA, RAMAN BAND CONTOUR
      ANALYSES, AND CORIOLIS CONSTANTS OF THE
      SPHERICAL TOP MOLECULES, SULFUR HEXAFLUORIDE,
      SELENIUM HEXAFLUORIDE, TELLURIUM HEXAFLUORIDE,
      MOLYBDENUM HEXAFLUORIDE, TUNGSTEN HEXAFLUORIDE,
      AND URANIUM HEXAFLUORIDE.
      BOSWORTH YM + CLARK RJH + RIPPON DM
      J MOL SPECTROSC 46: 240-55 (1973)

1189 VIBRATIONAL SPECTRUM AND STRUCTURE OF SOLID
      URANIUM HEXAFLUORIDE.
      BOUGON R + RIGNY P + PERRIN F
      COMPT REND 263: 1321-4 (1966) (IN FRENCH)

      LOW TEMPERATURE HEAT CAPACITIES, ENTHALPIES, AND
      ENTROPIES OF URANIUM TETRAFLUORIDE AND URANIUM
      HEXAFLUORIDE.
      BRICKWEDDE FG + HOGE HJ + SCOTT RB
      P368 OF CHEMISTRY OF URANIUM, COLLECTED PAPERS,
      KATZ JJ + RABINOWITCH(EDS), USAEC, 1958, 769P.
      TID-5290

      MOLECULAR STRUCTURE OF MOLYBDENUM, TUNGSTEN, AND
      URANIUM HEXAFLUORIDES FROM INFRARED AND RAMAN
      SPECTRA.
      BURKE TG + SMITH DF + NIELSEN AH
      J CHEM PHYS 20(3): 447-54 (1953)

      RAMAN SPECTRA OF MOLYBDENUM HEXAFLUORIDE,
      TECHNETIUM HEXAFLUORIDE, RHENIUM HEXAFLUORIDE,
      URANIUM HEXAFLUORIDE, SULFUR HEXAFLUORIDE,
      SELENIUM HEXAFLUORIDE, AND TELLURIUM
      HEXAFLUORIDE IN THE VAPOR STATE.
      CLAASSEN HH + GOODMAN GL + HOLLOWAY JH + SELIG H
      J CHEM PHYS 53: 341-8 (1970)

1190 RAMAN SPECTRUM OF URANIUM HEXAFLUORIDE.
      CLAASSEN HH + WEINSTOCK B + MALM JG
      J CHEM PHYS 25: 426-7 (1956)

1191 ABSORPTION SPECTRUM OF URANIUM HEXAFLUORIDE FROM
      2000 TO 4200 ANGSTROMS.
      DEPOORTER GL + ROFER-DEPOORTER CK
      SPECTROSC LETT 8(8): 521-4 (1975)

1192 LONG WAVE LENGTH, INFRARED ACTIVE FUNDAMENTAL
      FOR URANIUM HEXAFLUORIDE, NEPTUNIUM
      HEXAFLUORIDE, AND PLUTONIUM HEXAFLUORIDE.
      FRIEC B + CLAASSEN HH
      J CHEM PHYS 46: 4603-4 (1967)

      THERMODYNAMIC STABILITY OF HEXAFLUORIDES AT HIGH
      TEMPERATURES. PART-2: PLUTONIUM HEXAFLUORIDE.
      PART-3: URANIUM HEXAFLUORIDE. PART-4: MOLYBDENUM
      AND TUNGSTEN HEXAFLUORIDES.
      GALKIN NP + TUMANOV YN
      P188-91, 191-5, 195-9 OF TERMODIN TERMOKHIM
      KONSTANTY, ASTAKHOV KV(ED), NAUKA, MOSCOW, 1970.
      (IN RUSSIAN)

      THERMAL STABILITY AND REACTIVITY OF D- AND F
      ELEMENT HEXAFLUORIDES.
      GALKIN NP + TUMANOV YN + BUTYLKIN YP
      IZV SIB OTD AKAD NAUK SSSR SER KHIM NAUK (2):
      12-21 (1968) (IN RUSSIAN)

      INFRARED SPECTRA AND MOLECULAR STRUCTURE OF SOME
      GROUP-6 HEXAFLUORIDES. (OF SULFUR, SELENIUM,
      TELLURIUM, MOLYBDENUM, TUNGSTEN, AND URANIUM)
      GAUNT J
      TRANS FARADAY SOC 49: 1122-31 (1953)

      FORCE CONSTANTS AND BOND LENGTHS OF SOME
      INORGANIC HEXAFLUORIDES. (OF SULFUR, SELENIUM,
      TELLURIUM, MOLYBDENUM, TUNGSTEN, URANIUM, AND
      RHENIUM)
      GAUNT J
      TRANS FARADAY SOC 50: 546-51 (1954)

      HEATS OF FORMATION OF GASEOUS URANIUM FLUORIDES.
      (URANIUM MONOFLUORIDE, URANIUM DIFLUORIDE,
      URANIUM TRIFLUORIDE, URANIUM TETRAFLUORIDE,
      URANIUM PENTAFLUORIDE, AND URANIUM HEXAFLUORIDE)
      GODNEV IN + SVERDLIN AS
      IZV VYSSH UCHEB ZAVED, KHIM KHIM TEKN 9: 40-3
      (1966) (IN RUSSIAN)

1193 THERMODYNAMIC PROPERTIES OF URANIUM HEXAFLUORIDE
      AT HIGH TEMPERATURES.
      HASSAN HA + DEESE JE
      N CAR STATE UNIV, 1974, 35P. NASA-CR-2373

1194 ELECTRONIC ABSORPTION SPECTRUM OF GASEOUS
      URANIUM HEXACHLORIDE AND URANIUM HEXAFLUORIDE.
      HURST HJ + WILSON PW
      SPECTROSC LETT 5: 275-9 (1972)

      ELECTRIC DEFLECTION OF BINARY HEXAFLUORIDES.
      (SULFUR HEXAFLUORIDE, SELENIUM HEXAFLUORIDE,
      TELLURIUM HEXAFLUORIDE, MOLYBDENUM HEXAFLUORIDE,
      TUNGSTEN HEXAFLUORIDE, URANIUM HEXAFLUORIDE,
      RUTHENIUM HEXAFLUORIDE, RHENIUM HEXAFLUORIDE,
      RHODIUM HEXAFLUORIDE, OSMIUM HEXAFLUORIDE,
      IRIDIUM HEXAFLUORIDE, AND PLATINUM HEXAFLUORIDE)
      KAISER EW + MUENTER JS + KLEMPERER W +
      FALCONER WE + SUNDER WA
      J CHEM PHYS 53: 1411-12 (1970)

      ELECTRON DIFFRACTION INVESTIGATION OF TUNGSTEN
      HEXAFLUORIDE, OSMIUM HEXAFLUORIDE, IRIDIUM
      HEXAFLUORIDE, URANIUM HEXAFLUORIDE, NEPTUNIUM
      HEXAFLUORIDE, AND PLUTONIUM HEXAFLUORIDE.
      KIMURA M + SCHOMAKER V + SMITH DW + WEINSTOCK B
      J CHEM PHYS 48: 4001-12 (1968)

1195 ALPHA SPECTROMETRIC INVESTIGATION OF URANIUM
      HEXAFLUORIDE. PART-1.
      KOSKE P + NENTWIG P + DREIZLER H + GAIDE W
      J BRIT NUCL ENERG SOC 14(2): 157-9 (1975)

      THERMODYNAMIC SIMILARITY AND UNIVERSAL EQUATIONS
      OF STATE OF HEXAFLUORIDES. (OF SULFUR,
      MOLYBDENUM, TUNGSTEN, AND URANIUM)
      MALYSHEV VV
      TEPLOFIZ VYS TEMP 14(1): 47-55 (1976) (IN
      RUSSIAN)

      THEORETICAL CALCULATIONS OF THE ELECTRONIC
      STRUCTURE AND OPTICAL TRANSITIONS OF URANIUM
      HEXAFLUORIDE AND URANIUM PENTAFLUORIDE.
      MAYLOTTE DH + ST PETERS RL + MESSMER RP
      CHEM PHYS LETT 38(1): 181-4 (1976)

1196 VIBRATIONAL SPECTRUM AND FORCE FIELD OF URANIUM
      HEXAFLUORIDE.
      MCDOWELL RS + ASPREY LB + PAINE RT
      J CHEM PHYS 61: 3571-80 (1974)

1197 VAPOR PRESSURE AND CRITICAL CONSTANTS OF URANIUM
      HEXAFLUORIDE.
      OLIVER GD + MILTON HT + GRISARD JW
      J AMER CHEM SOC 75: 2827-9 (1953)

1198 THERMODYNAMIC PROPERTIES OF GASEOUS URANIUM
      HEXAFLUORIDE.
      PARKS BH + BURTON DW
      USAEC, 1960, 43P. K-1458

VAPOR PRESSURE OF URANIUM HEXAFLUORIDE IN
EQUILIBRIUM WITH URANIUM PENTAFLUORIDE.
PRIEST HF
P738-41 OF CHEMISTRY OF URANIUM, COLLECTED
PAPERS, KATZ JJ + RABINOWITCH(EDS), USAEC, 1958,
769P. TID-5290

HIGH RESOLUTION NMR SPECTRUM OF LIQUID
MOLYBDENUM HEXAFLUORIDE, TUNGSTEN HEXAFLUORIDE,
AND URANIUM HEXAFLUORIDE.
RIGNY P + DEMORTIER A + PERRIN F
COMPT REND 263: 1408-10 (1966) (IN FRENCH)

MOLECULAR MOTION AND FLUORINE RELAXATION IN THE
LIQUIDS OF MOLYBDENUM HEXAFLUORIDE, TUNGSTEN
HEXAFLUORIDE, AND URANIUM HEXAFLUORIDE.
RIGNY P + VIRLET J
J CHEM PHYS 47: 4645-52 (1967)

1199 FAILURE OF THE FIRST BORN APPROXIMATION IN
ELECTRON DIFFRACTION. (URANIUM HEXAFLUORIDE)
SEIP HM
ACTA CHEM SCAND 19: 1955-68 (1965)

STUDIES ON THE FAILURE OF THE FIRST BORN
APPROXIMATION IN ELECTRON DIFFRACTION. (TUNGSTEN
HEXAFLUORIDE, URANIUM HEXAFLUORIDE, TELLURIUM
HEXAFLUORIDE, MOLYBDENUM HEXAFLUORIDE)
SEIP HM
P26-68 OF SELECTED TOPICS IN STRUCTURE
CHEMISTRY, ANDERSON P + BASTIANSEN O + FURBERG S
(EDS), UNIVERSITETSFORLAGET, OSLO, NORWAY, 1967.

1200 DIPOLE MOMENT AND MOLECULAR STRUCTURE OF URANIUM
HEXAFLUORIDE.
SMYTH CP + HANNAY NB
P437-40 OF CHEMISTRY OF URANIUM, COLLECTED
PAPERS, KATZ JJ + RABINOWITCH(EDS), USAEC, 1958,
769P. TID-5290

1201 SPECIFIC HEAT RATIO OF URANIUM HEXAFLUORIDE
MEASURED WITH A BALLISTIC PISTON COMPRESSOR.
STERRITT DE + LALOS GT + SCHNEIDER RT
NUCLEAR TECH 25: 150-65 (1975)

1202 STRUCTURES OF FLUORIDES. PART-10: NEUTRON POWDER
DIFFRACTION PROFILE STUDIES OF URANIUM
HEXAFLUORIDE AT 193 DEGREES AND 293 DEGREES K.
TAYLOR JC + WILSON PW
J SOLID STATE CHEM 14(4): 378-82 (1975)

1203 THERMODYNAMIC STABILITY OF URANIUM HEXAFLUORIDE.
TUMANOV YN
RUSS J INORG CHEM 13: 782-5 (1968)

1204 THERMODYNAMIC PROPERTIES OF URANIUM
HEXAFLUORIDE.
VERKHIVKER GP + TETL'BAUM SD + KONYAEVA GP
ATOMNAYA ENERGIYA 24: 158-62 (1968) (IN RUSSIAN)

1205 SOME RECENT STUDIES WITH HEXAFLUORIDES. (URANIUM
HEXAFLUORIDE, PLUTONIUM HEXAFLUORIDE, AND
NEPTUNIUM HEXAFLUORIDE)
WEINSTOCK B + MALM JG
P125-9 OF INT CONF ON PEACEFUL USES OF ATOMIC
ENERGY, 2ND, 1958, GENEVA, VOL 28, 686P.

VANADIUM DIFLUORIDE VF_2

1206 SYNTHESIS AND CHARACTERIZATION OF VANADIUM
DIFLUORIDE.
SHAFER MW
MATER RES BULL 4: 905-12 (1969)

VANADIUM TRIFLUORIDE VF_3

1207 VIBRATIONAL STUDY OF VANADIUM TRIFLUORIDE.
BURGOS A
VANDERBILT UNIV, PHD THESIS, 1960, 135P.

INFRARED SPECTRA OF VANADIUM PENTAFLUORIDE,
VANADIUM TETRAFLUORIDE, AND VANADIUM
TRIFLUORIDE.
CAVELL RG + CLARK HC
INORG CHEM 3: 1789-91 (1964)

1208 MASS SPECTROMETRIC DETERMINATION OF THE VAPOR
COMPOSITION AND VAPOR PRESSURE OF VANADIUM
TRIFLUORIDE.
SIDOROV LN + DENISOV MY + AKISHIN PA + SHOLTS VB
RUSS J PHYS CHEM 40: 620-22 (1966)

VANADIUM TETRAFLUORIDE VF_4

1209 PREPARATION AND PROPERTIES (CRYSTAL STRUCTURE)
OF VANADIUM TETRAFLUORIDE.
CAVELL RG + CLARK HC
J CHEM SOC 1962: 2692-8

INFRARED SPECTRA OF VANADIUM FLUORIDES.
(VANADIUM TRIFLUORIDE, TETRAFLUORIDE,
PENTAFLUORIDE)
CAVELL RG + CLARK HC
INORG CHEM 3(12): 1789-91 (1964)

MOLECULAR BEAM ELECTRIC DEFLECTION OF THE
TETRAHALIDES CARBON TETRAFLUORIDE, CARBON
TETRACHLORIDE, SILICON TETRAFLUORIDE, SILICON
TETRACHLORIDE, GERMANIUM TETRACHLORIDE, TITANIUM
TETRAFLUORIDE, TITANIUM TETRACHLORIDE, VANADIUM
TETRAFLUORIDE, AND VANADIUM TETRACHLORIDE.
MUENTER AA + DYKE TR + FALCONER WE + KLEMPERER W
J CHEM PHYS 63: 1231-6 (1975)

VANDIUM PENTAFLUORIDE VF_5

1210 INFRARED SPECTRA OF VANADIUM PENTAFLUORIDE,
VANADIUM TETRAFLUORIDE, AND VANADIUM
TRIFLUORIDE.
CAVELL RG + CLARK HC
INORG CHEM 3: 1789-91 (1964)

1211 VIBRATIONAL SPECTRA OF VANADIUM PENTAFLUORIDE.
CLAASSEN HH + SELIG H
J CHEM PHYS 44: 4039-43 (1966)

1212 SOME PHYSICAL AND CHEMICAL PROPERTIES OF
VANADIUM PENTAFLUORIDE.
CLARK HC + EMELEUS HJ
J CHEM SOC A 1957: 2119-22

1213 POLAR STATES IN VANADIUM PENTAFLUORIDE.
DYKE TR + MUENTER AA + KLEMPERER W + FALCONER WE
J CHEM PHYS 53: 3382-3 (1970)

THERMODYNAMIC PROPERTIES OF PENTAFLUORIDES AT
HIGH TEMPERATURES. PART-1: CALCULATION METHODS.
(CHLORINE PENTAFLUORIDE, BROMINE PENTAFLUORIDE,
IODINE PENTAFLUORIDE, PHOSPHORUS PENTAFLUORIDE,
VANADIUM PENTAFLUORIDE, ANTIMONY PENTAFLUORIDE)
GALKIN NP + TUMANOV YN + BUTYLKIN YP
KHIM VYS ENERG 4(6): 512-8 (1970) (IN RUSSIAN)

PENTACOORDINATED MOLECULES. PART-12: CORRELATION
OF EXCHANGE RATES FOR SOME GROUP-5 PENTAHALIDE
MOLECULES.
HOLMES RR
INORG CHEM 7(11): 2229-35 (1968)

1214 INTRAMOLECULAR EXCHANGE IN VANADIUM
PENTAFLUORIDE.
HOLMES RR + COUCH LS + HORA CJ
J CHEM SOC, CHEM COMMUN 1974(5): 175-7

1215 ENTHALPY OF FORMATION OF VANADIUM PENTAFLUORIDE
BY FLUORINE BOMB CALORIMETRY.
JOHNSON GK + HUBBARD WN
J CHEM THERMODYN 6(1): 59-63 (1974)

1216 MEAN AMPLITUDES OF VIBRATION AND THERMODYNAMIC
FUNCTIONS FOR VANADIUM PENTAFLUORIDE.
MULLER A + NAGARAJAN G
Z CHEM 7: 35 (1967)

1217 ELECTRON DIFFRACTION STUDY OF THE MOLECULAR
STRUCTURES OF VANADIUM PENTAFLUORIDE, NIOBIUM
PENTAFLUORIDE, AND TANTALUM PENTAFLUORIDE.
ROMANOV GV + SPIRIDONOV VP
IZV SIB OTDEL AKAD NAUK SSSR, SER KHIM NAUK 1:
126-31 (1968) (IN RUSSIAN)

1218 ELECTRON DIFFRACTION STUDY OF THE VANADIUM
PENTAFLUORIDE MOLECULE IN THE VAPOR FORM.
ROMANOV GV + SPIRIDONOV VP
J STRUCT CHEM 7(6): 816-17 (1966)

SPECTROSCOPIC STUDIES IN MEAN AMPLITUDES OF
VIBRATION OF PENTAFLUORIDES OF ARSENIC,
PHOSPHORUS, AND VANADIUM.
SANYAL NK + DIXIT L
INDIAN J PURE APPL PHYS 12(8): 550-3 (1974)

RAMAN SPECTRA OF ARSENIC PENTAFLUORIDE AND
VANADIUM PENTAFLUORIDE AND THEIR FORCE CONSTANTS
INCLUDING THOSE OF PHOSPHORUS PENTAFLUORIDE.
SELIG H + HOLLOWAY JH + TYSON J + CLAASSEN HH
J CHEM PHYS 53: 2559-64 (1970)

1219 FORCE CONSTANTS, MEAN AMPLITUDES OF VIBRATION,
SHRINKAGE EFFECTS, CORIOLIS CONSTANTS, AND
THERMODYNAMIC FUNCTIONS OF VANADIUM
PENTAFLUORIDE.
VIZI B + BRUNVOLL J
ACTA CHEM SCAND 22: 1279-82 (1968)

GENERALIZED VALENCE FORCE FIELD FORCE CONSTANTS
AND MEAN-SQUARE AMPLITUDES OF VIBRATION OF SOME
TRIGONAL BIPYRAMIDAL MOLECULES BY THE
LOGARITHIMIC STEPS METHOD. (PHOSPHORUS
PENTAFLUORIDE, ARSENIC PENTAFLUORIDE, ANTIMONY
PENTAFLUORIDE, VANADIUM PENTAFLUORIDE, AND
MOLYBDENUM PENTAFLUORIDE)
WENDLING EJL + MAHMOUDI S + MACCORDICK HJ
J CHEM SOC A 1971: 1747-54

XENON MONOFLUORIDE   XeF

ABSORPTION AND EMISSION SPECTRA OF ARGON MATRIX
ISOLATED XENON MONOFLUORIDE AND KRYPTON
MONOFLUORIDE.
AULT BS + ANDREWS L
J CHEM PHYS 64(7): 3075-6 (1976)

EMISSION SPECTRA OF XENON MONOBROMIDE, XENON
MONOCHLORIDE, XENON MONOFLUORIDE, AND KRYPTON
MONOFLUORIDE.
BRAU CA + EWING JJ
J CHEM PHYS 63(11): 4640-7 (1975)

1220 RADIATION DAMAGE IN XENON TETRAFLUORIDE.
ELECTRON SPIN RESONANCE OF THE TRAPPED RADICAL
XENON MONOFLUORIDE.
FALCONER WE + MORTON JR
P245-50 OF NOBLE GAS COMPOUNDS, HYMAN HH(ED),
CHICAGO UNIV PRESS, 1963.

1221 PROBABLE NONEXISTENCE OF XENON MONOFLUORIDE AS A
CHEMICALLY BOUND SPECIES IN THE GAS PHASE.
LISKOW DH + SCHAEFER HF + BAGUS PS + LIU B
J AMER CHEM SOC 95: 4056-7 (1973)

1222 ELECTRON SPIN RESONANCE SPECTRUM OF XENON
MONOFLUORIDE IN GAMMA IRRADIATED XENON
TETRAFLUORIDE.
MORTON JR + FALCONER WE
J CHEM PHYS 39: 427-31 (1963)

XENON DIFLUORIDE   XeF₂

1223 XENON DIFLUORIDE AND THE NATURE OF THE XENON-
FLUORINE BOND.
AGRON PA + BEGUN GM + LEVY HA
SCIENCE 139: 842-3 (1963)

1224 SELF CONSISTENT FIELD STUDY OF THE SERIES XENON
DIFLUORIDE, XENON TETRAFLUORIDE, XENON
HEXAFLUORIDE.
BASCH H + MOSKOWITZ JW + HOLISTER C + HANKIN D
J CHEM PHYS 55: 1923-33 (1971)

1225 PHOTOIONIZATION MASS SPECTROMETRIC STUDY OF
XENON DIFLUORIDE, XENON TETRAFLUORIDE, AND XENON
HEXAFLUORIDE.
BERKOWITZ J + CHUPKA WA + GUYON PM
+ HOLLOWAY JH + SPOHR R
J PHYS CHEM 75: 1461-71 (1971)

1226 ELECTRONIC STRUCTURE OF XENON DIFLUORIDE.
BILHAM J + LINNETT JW
NATURE 201: 1323 (1964)

1227 PHOTOELECTRON SPECTRUM OF XENON DIFLUORIDE.
BREHM B + MENZINGER M + ZORN C
CAN J CHEM 48: 3193-6 (1970)

1228 HELIUM-1 AND HELIUM-2 PHOTOELECTRON SPECTRA AND
THE ELECTRONIC STRUCTURES OF XENON DIFLUORIDE,
XENON TETRAFLUORIDE, AND XENON HEXAFLUORIDE.
BRUNDLE CR + JONES GR + BASCH H
J CHEM PHYS 55: 1098-1104 (1971)

1229 ELECTRON DISTRIBUTION IN THE XENON FLUORIDES AND
XENON OXIDE TETRAFLUORIDE BY ESCA AND EVIDENCE
FOR ORBITAL INDEPENDENCE IN THE XENON- FLUORINE
BONDING.
CARROLL TX + SHAE RW + THOMAS TD + KINDLE C
+ BARTLETT N
J AMER CHEM SOC 96(7): 1989-96 (1974)

1230 BONDING IN XENON DIFLUORIDE.
CATTON RC + MITCHELL KAR
CHEM COMMUN 1970: 457-8

D ORBITALS IN THE NOBLE GAS DIHALIDES. (ARGON
DIFLUORIDE KRYPTON DIFLUORIDE, AND XENON
DIFLUORIDE)
CATTON RC + MITCHELL KAR
CAN J CHEM 48: 2695-701 (1970)

1231 CALCULATIONS OF ELECTRONIC STRUCTURES OF
COMPOUNDS OF HALOGENS AND INERT GASES BY THE SCF
LCAO MO METHOD. PART-2: COMPARATIVE ANALYSIS OF
DIFFERENT SCHEMES OF NEGLECT OF DIFFERENTIAL
OVERLAP METHODS ILLUSTRATED BY THE XENON
FLUORIDES. MODIFIED NDDO-2 (ALPHA, BETA) METHOD
AS A USEFUL APPROXIMATION.
CHARKIN OP + SMOLYAR AE + ZYUBIN AS
+ KLIMENKO NM
J STRUCT CHEM 15(3): 461-7 (1974)

1232 VIBRATIONAL SPECTRA OF XENON COMPOUNDS.
CLAASSEN HH
P304-5 OF NOBLE GAS COMPOUNDS, HYMAN HH(ED),
CHICAGO UNIV PRESS, 1963.

1233 SPECTRA OF XENON DIFLUORIDE AND XENON
TETRAFLUORIDE IN THE FAR ULTRAVIOLET REGION.
COMES FJ + HAENSEL R + NIELSEN U + SCHWARZ WHE
J CHEM PHYS 58: 516-29 (1973)

1234 HELIUM-2 PHOTOELECTRON SPECTRA OF XENON
DIFLUORIDE AND KRYPTON DIFLUORIDE.
DEKOCK RL
J CHEM PHYS 58: 1267-8 (1973)

1235 STRUCTURAL CONSIDERATIONS IN THE CHEMISTRY OF
NOBLE GASES. (XENON DIFLUORIDE, XENON
TETRAFLUORIDE, XENON HEXAFLUORIDE)
FALCONER WE
P386-90 OF NOBLE GASES, STANLEY RE + MOGHISSI
AA(EDS), ENVIRONMENTAL PROTECTION AGENCY, 1973.
(CONF-730915)

1236 BOND LENGTHS IN XENON DIFLUORIDE.
GELLINGS PJ
Z PHYS CHEM NEUE FOLGE 43: 123-5 (1964)

ENTHALPIES OF FORMATION OF XENON HEXAFLUORIDE,
XENON TETRAFLUORIDE, XENON DIFLUORIDE, AND
PHOSPHORUS TRIFLUORIDE.
JOHNSON GK + MALM JG + HUBBARD WN
J CHEM THERMODYN 4: 879-91 (1972)

1237 HEATS OF SUBLIMATION OF XENON DIFLUORIDE AND
XENON TETRAFLUORIDE AND A CONJECTURE ON BONDING
IN THE SOLIDS.
JORTNER J + WILSON EG + RICE SA
J AMER CHEM SOC 85: 814-5 (1963)

1238 THEORETICAL AND EXPERIMENTAL STUDIES OF THE
ELECTRONIC STRUCTURE OF THE XENON FLUORIDES.
JORTNER J + WILSON EG + RICE SA
P358-88 OF NOBLE GAS COMPOUNDS, HYMAN HH(ED),
UNIV OF CHICAGO PRESS, 1963

ULTRAVIOLET EMISSION SPECTRA. (XENON DIFLUORIDE
AND KRYPTON DIFLUORIDE)
KRISHNAMACHARI SLNG + NARASIMHAM NA + SINGH M
CURR SCI 34: 75-7 (1965)

1239 CRYSTAL AND MOLECULAR STRUCTURE OF XENON
DIFLUORIDE BY NEUTRON DIFFRACTION.
LEVY HA + AGRON PA
J AMER CHEM SOC 85: 241-2 (1963)

1240 MOLECULAR SYMMETRY OF XENON DIFLUORIDE AND XENON
TETRAFLUORIDE.
LOHR LL + LIPSCOMB WN
J AMER CHEM SOC 85: 240-1 (1963)

ASSIGNMENTS IN THE ULTRAVIOLET SPECTRA OF
MOLYBDENUM HEXAFLUORIDE AND TUNGSTEN
HEXAFLUORIDE.
MCDIARMID R
J CHEM PHYS 61(8): 3333-9 (1974)

1241 INFRARED SPECTRA OF XENON DIFLUORIDE IN
DIFFERENT SOLVENTS.
MEINERT H + KAUSCHKA G
Z CHEM 9: 114-5 (1969)

1242 5D ORBITALS OF XENON IN ATOMIC VALENCE STATES
AND IN THE MOLECULES XENON DIFLUORIDE AND XENON
DICHLORIDE.
MITCHELL KAR
J CHEM SOC A 1969: 1637-44

1243 IONIZATION AND DISSOCIATION OF XENON DIFLUORIDE
INDUCED BY PHOTON IMPACT.
MORRISON JD + NICHOLSON AJC + O'DONNELL TA
J CHEM PHYS 49(2): 959-60 (1968)

1244 MOLECULAR CONSTANTS AND THERMODYNAMIC FUNCTIONS
OF SOME LINEAR SYMMETRICAL MOLECULES. (XENON
DIFLUORIDE)
NAGARAJAN G
ACTA PHYS AUSTRIACA 21(4): 355-65 (1966)

1245 SPECTROSCOPIC STUDIES OF MOLECULAR CONSTANTS IN
SOME POLYATOMIC MOLECULES. (XENON DIFLUORIDE)
NAGARAJAN G
Z NATURFORSCH 21A: 244-51 (1965)

1246 XENON DIFLUORIDE. HEAT CAPACITY FROM 5 TO 350
DEGREES K AND SOME DERIVED THERMODYNAMIC
PROPERTIES.
OSBORNE DW + FLOTOW HE + MALM JG
J CHEM PHYS 57: 4670-5 (1972)

1247 FORBIDDEN ELECTRONIC TRANSITIONS IN XENON
DIFLUORIDE AND XENON TETRAFLUORIDE.
PYSH ES + JORTNER J + RICE SA
J CHEM PHYS 40: 2018-32 (1964)

NOBLE GASES AND THE BINARY FLUORIDES OF XENON.
RAMALHO DE AZEVEDO MM
CIEN CULT 26(1): 3-18 (1974) (IN SPANISH)

1248 GAS PHASE STRUCTURE OF XENON DIFLUORIDE.
REICHMAN S + SCHREINER F
J CHEM PHYS 51: 2355-58 (1969)

1249 RELATIVISTIC MOLECULAR WAVE FUNCTIONS: XENON
DIFLUORIDE.
ROSEN A + ELLIS DE
CHEM PHYS LETT 27(4): 595-9 (1974)

1250 PROBABLE STRUCTURE OF XENON TETRAFLUORIDE AND
XENON DIFLUORIDE.
RUNDLE RE
J AMER CHEM SOC 85: 112-3 (1963)

1251 VAPOR PRESSURE AND MELTING POINTS OF XENON
DIFLUORIDE AND XENON TETRAFLUORIDE.
SCHREINER F + MCDONALD GN + CHERNICK CL
J PHYS CHEM 72: 1162-6 (1968)

1252 XENON DIFLUORIDE. (FORCE CONSTANTS, SPECTRA)
SMITH DF
J CHEM PHYS 38: 270-1 (1963)

1253 THERMOCHEMICAL PROPERTIES OF XENON DIFLUORIDE
AND XENON TETRAFLUORIDE FROM MASS SPECTRA.
SVEC HJ + FLESCH GD
SCIENCE 142: 954-5 (1963)

1254 RAMAN SPECTRA FOR XENON DIFLUORIDE, XENON
TETRAFLUORIDE, AND XENON OXIDE TETRAFLUORIDE
VAPORS, AND FORCE CONSTANT CALCULATIONS.
TSAO P + COBB CC + CLAASSEN HH
J CHEM PHYS 54: 5247-53 (1971)

BOND ENERGIES AND IONIC CHARACTER OF INERT GAS
HALIDES.
WATERS JH + GRAY HB
J AMER CHEM SOC 85: 825-6 (1963)

1255 THE XENON- FLUORINE SYSTEM.
WEINSTOCK B + WEAVER EE + KNOP CP
INORG CHEM 5: 2189-203 (1966)

1256 FAR ULTRAVIOLET SPECTROSCOPIC STUDY OF XENON
DIFLUORIDE.
WILSON EG + JORTNER J + RICE SA
J AMER CHEM SOC 85: 813-4 (1963)

1257 A THEORETICAL EVALUATION OF THE ROOT MEAN SQUARE
AMPLITUDES OF VIBRATION IN XENON DIFLUORIDE AND
XENON TETRAFLUORIDE.
YERANOS W
MOL PHYS 12: 529-32 (1967)

XENON TETRAFLUORIDE XeF_4

1258 RAMAN SPECTRA OF SOLID XENON TETRAFLUORIDE AND
ITS ADDUCT WITH XENON DIFLUORIDE.
ADAMS CJ
J RAMAN SPECTROSC 2: 391-7 (1974)

SELF-CONSISTENT FIELD STUDY OF THE SERIES XENON
DIFLUORIDE, XENON TETRAFLUORIDE, XENON
HEXAFLUORIDE.
BASCH H + MOSKOWITZ JW + HOLISTER C + HANKIN D
J CHEM PHYS 55: 1923-33 (1971)

PHOTOIONIZATION MASS SPECTROMETRIC STUDY OF
XENON DIFLUORIDE, XENON TETRAFLUORIDE, AND XENON
HEXAFLUORIDE.
BERKOWITZ J + CHUPKA WA + GUYON PM + HOLLOWAY JH
J PHYS CHEM 75: 1461-71 (1971)

1259 ELECTRON DIFFRACTION STUDIES OF XENON
TETRAFLUORIDE AND XENON HEXAFLUORIDE AND OTHER
COMPOUNDS.
BOHN RK
CORNELL UNIV, PHD THESIS, 1964, 107P.

HELIUM-1 AND HELIUM-2 PHOTOELECTRON SPECTRA AND
THE ELECTRONIC STRUCTURES OF XENON DIFLUORIDE,
XENON TETRAFLUORIDE, AND XENON HEXAFLUORIDE.
BRUNDLE CR + JONES GR + BASCH H
J CHEM PHYS 55: 1098-1104 (1971)

1260 CRYSTAL AND MOLECULAR STRUCTURE OF XENON
TETRAFLUORIDE BY NEUTRON DIFFRACTION.
BURNS JH + AGRON PA + LEVY HA
P211-20 OF NOBLE GAS COMPOUNDS, HYMAN HH(ED),
UNIV OF CHICAGO PRESS, 1963.

ELECTRON DISTRIBUTION IN THE XENON FLUORIDES AND
XENON OXIDE TETRAFLUORIDE BY ESCA AND EVIDENCE
FOR ORBITAL INDEPENDENCE IN THE XENON- FLUORINE
BONDING.
CARROLL TX + SHAE RW + THOMAS TD + KINDLE C +
BARTLETT N
J AMER CHEM SOC 96(7): 1989-96 (1974)

CALCULATIONS OF ELECTRONIC STRUCTURES OF
COMPOUNDS OF HALOGENS AND INERT GASES BY THE SCF
LCAO MO METHOD. PART-2: COMPARATIVE ANALYSIS OF
DIFFERENT SCHEMES OF NEGLECT OF DIFFERENTIAL
OVERLAP METHODS ILLUSTRATED BY THE XENON
FLUORIDES. MODIFIED NDDO-2 (ALPHA, BETA) METHOD
AS A USEFUL APPROXIMATION.
CHARKIN OP + SMOLYAR AE + ZYUBIN AS + KLIMENKO N
J STRUCT CHEM 15(3): 461-7 (1974)

VIBRATIONAL SPECTRA OF XENON COMPOUNDS.
CLAASSEN HH
P304-5 OF NOBLE GAS COMPOUNDS, HYMAN HH(ED),
CHICAGO UNIV PRESS, 1963.

1261 VIBRATIONAL SPECTRA AND STRUCTURES OF XENON
TETRAFLUORIDE AND XENON OXYTETRAFLUORIDE.
CLAASSEN HH + CHERNICK CL + MALM JG
P287-94 OF NOBLE GAS COMPOUNDS, HYMAN HH(ED),
CHICAGO UNIV PRESS, 1963.

1262 VIBRATIONAL SPECTRA AND STRUCTURE OF XENON
TETRAFLUORIDE.
CLAASSEN HH + CHERNICK CL + MALM JG
J AMER CHEM SOC 85(13): 1927-8 (1963)

1263 XENON TETRAFLUORIDE. (INFRARED SPECTRUM)
CLAASSEN HH + SELIG H + MALM JG
J AMER CHEM SOC 84: 3593 (1962)

SPECTRA OF XENON DIFLUORIDE AND XENON
TETRAFLUORIDE IN THE FAR ULTRAVIOLET REGION.
COMES FJ + HAENSEL R + NIELSEN U + SCHWARZ WHE
J CHEM PHYS 58: 516-29 (1973)

1264 CORRELATION OF ELECTRONIC STATES OF THE POSITIVE
IONS OF XENON TETRAFLUORIDE, XENON OXIDE
TETRAFLUORIDE, AND IODINE PENTAFLUORIDE.
DEKOCK RL
J ELECTRON SPECTROSC RELAT PHENOM 4: 155-61
(1974)

1265 GENERAL VALENCE FORCE FIELD FOR XENON
TETRAFLUORIDE.
FADINI A + MULLER A
MOL PHYS 12: 145-8 (1967)

STRUCTURAL CONSIDERATIONS IN THE CHEMISTRY OF
NOBLE GASES. (XENON DIFLUORIDE, XENON
TETRAFLUORIDE, XENON HEXAFLUORIDE)
FALCONER WE
P386-90 OF NOBLE GASES, STANLEY RE + MOGHISSI
AA(EDS), ENVIRONMENTAL PROTECTION AGENCY, 1973.
(CONF-730915)

1266 HEAT OF FORMATION OF XENON TETRAFLUORIDE.
GUNN SR + WILLIAMSON SM
P133-8 OF NOBLE GAS COMPOUNDS, HYMAN HH(ED),
UNIV OF CHICAGO PRESS, 1963

1267 XENON TETRAFLUORIDE: HEAT OF FORMATION.
GUNN SR + WILLIAMSON SM
SCIENCE 140: 177-8 (1963)

1268 VIBRATIONS OF PLANAR SYMMETRICAL XY4 MOLECULES
WITH APPLICATION XENON TETRAFLUORIDE.
HAGEN G
ACTA CHEM SCAND 21: 465-72 (1967)

1269 ENTHALPIES OF FORMATION OF XENON HEXAFLUORIDE,
XENON TETRAFLUORIDE, XENON TETRAFLUORIDE.
CRYSTAL STRUCTURE.
IBERS JA + HAMILTON WC
SCIENCE 139: 106-7 (1963)

1270 ULTRAVIOLET SPECTRUM OF XENON TETRAFLUORIDE.
ISRAELI YJ
BULL SOC CHIM FR 1964: 649 (IN FRENCH)

ENTHALPIES OF FORMATION OF XENON HEXAFLUORIDE,
XENON TETRAFLUORIDE, XENON DIFLUORIDE, AND
PHOSPHORUS TRIFLUORIDE.
JOHNSON GK + MALM JG + HUBBARD WN
J CHEM THERMODYN 4: 879-91 (1972)

1271 HEAT CAPACITY AND RELATED THERMODYNAMIC
FUNCTIONS OF XENON TETRAFLUORIDE.
JOHNSTON WV + PILIPOVICH D + SHEEHAN DE
P139-43 OF NOBLE GAS COMPOUNDS, HYMAN HH(ED),
UNIV OF CHICAGO PRESS, 1963.

HEATS OF SUBLIMATION OF XENON DIFLUORIDE AND
XENON TETRAFLUORIDE AND A CONJECTURE ON BONDING
IN THE SOLIDS.
JORTNER J + WILSON EG + RICE SA
J AMER CHEM SOC 85: 814-5 (1963)

THEORETICAL AND EXPERIMENTAL STUDIES OF THE
ELECTRONIC STRUCTURE OF THE XENON FLUORIDES.
JORTNER J + WILSON EG + RICE SA
P358-88 OF NOBLE GAS COMPOUNDS, HYMAN HH(ED),
UNIV OF CHICAGO PRESS, 1963

1272 FAR ULTRAVIOLET SPECTROSCOPIC STUDY OF XENON
TETRAFLUORIDE.
JORTNER J + WILSON EG + RICE SA
J AMER CHEM SOC 85: 815-6 (1963)

1273 IDEAL GAS THERMODYNAMIC PROPERTIES OF XENON
COMPOUNDS. (XENON TETRAFLUORIDE AND XENON
HEXAFLUORIDE)
KUDCHADKER SA + KUDCHADKER AP
PROC INDIAN ACAD SCI A 73(5): 261-7 (1971)

MOLECULAR SYMMETRY OF XENON DIFLUORIDE AND XENON
TETRAFLUORIDE.
LOHR LL + LIPSCOMB WN
J AMER CHEM SOC 85: 240-1 (1963)

CALCULATED IONIZATION POTENTIAL OF CHLORO- AND
FLUOROMETHANES, TETRAFLUORO METHANE XENON
TETRAFLUORIDE, AND XENON TETRACHLORIDE.
MELTON CE + JOY HW
J CHEM PHYS 42: 2982 (1965)

1274 MEAN AMPLITUDES OF VIBRATION AND THERMODYNAMIC
FUNCTIONS OF XENON TETRAFLUORIDE.
NAGARAJAN G
ACTA PHYS AUSTRIACA 18(1): 11-9 (1964)

1275 XENON TETRAFLUORIDE. HEAT CAPACITY AND
THERMODYNAMIC FUNCTIONS FROM 5 TO 350 DEGREES K.
RECONCILIATION OF THE ENTROPIES FROM MOLECULAR
AND THERMAL DATA.
OSBORNE DW + SCHREINER F + FLOTOW HE + MALM JG
J CHEM PHYS 57: 3401-7 (1972)

FORBIDDEN ELECTRONIC TRANSITIONS IN XENON
DIFLUORIDE AND XENON TETRAFLUORIDE.
PYSH ES + JORTNER J + RICE SA
J CHEM PHYS 40: 2018-32 (1964)

PROBABLE STRUCTURE OF XENON TETRAFLUORIDE AND
XENON DIFLUORIDE.
RUNDLE RE
J AMER CHEM SOC 85: 112-3 (1963)

VAPOR PRESSURE AND MELTING POINTS OF XENON
DIFLUORIDE AND XENON TETRAFLUORIDE.
SCHREINER F + MCDONALD GN + CHERNICK CL
J PHYS CHEM 72: 1162-6 (1968)

1276 THERMOCHEMICAL STUDIES OF XENON TETRAFLUORIDE
AND XENON HEXAFLUORIDE.
STEIN L + PLURIEN PL
P144-8 OF NOBLE GAS COMPOUNDS, HYMAN HH(ED),
UNIV OF CHICAGO PRESS, 1963

THERMOCHEMICAL PROPERTIES OF XENON DIFLUORIDE
AND XENON TETRAFLUORIDE FROM MASS SPECTRA.
SVEC HJ + FLESCH GD
SCIENCE 142: 954-5 (1963)

1277 CRYSTAL AND MOLECULAR STRUCTURE OF XENON
TETRAFLUORIDE.
TEMPLETON DH + ZALKIN A + FORRESTER JD
+ WILLIAMSON SM
J AMER CHEM SOC 85: 242 (1963)

RAMAN SPECTRA FOR XENON DIFLUORIDE, XENON
TETRAFLUORIDE AND XENON OXIDE TETRAFLUORIDE
VAPORS, AND FORCE CONSTANT CALCULATIONS.
TSAO P + COBB CC + CLAASSEN HH
J CHEM PHYS 54: 5247-53 (1971)

EMPIRICAL METHOD FOR DETERMINING EFFECTIVE
VIBRATIONAL AND ROTATIONAL CHARACTERISTICS OF
MOLECULES OF SOME TETRAFLUORIDES FOR CALCULATING
THEIR THERMODYNAMIC FUNCTIONS.
TUMANOV YN + GALKIN NP
RUSS J PHYS CHEM 43(4): 464-7 (1969)

1278 FORCE FIELD, CORIOLIS COUPLING COEFFICIENTS,
GENERALIZED MEAN SQUARE AMPLITUDES OF VIBRATION
AND SHRINKAGE CONSTANTS OF XENON TETRAFLUORIDE
AND XENON OXIDE TETRAFLUORIDE.
VENKATESWARLU K + JOSEPH KB
ACTA PHYS ACAD SCI HUNG 24: 139-45 (1968)

THE XENON- FLUORINE SYSTEM.
WEINSTOCK B + WEAVER EE + KNOP CP
INORG CHEM 5: 2189-203 (1966)

A THEORETICAL EVALUATION OF THE ROOT MEAN SQUARE
AMPLITUDES OF VIBRATION IN XENON DIFLUORIDE AND
XENON TETRAFLUORIDE.
YERANOS W
MOL PHYS 12: 529-32 (1967)

XENON HEXAFLUORIDE XeF_6

1279 EVIDENCE FOR PSEUDO JAHN-TELLER EFFECT IN XENON
HEXAFLUORIDE.
BARTELL LS
J CHEM PHYS 46: 4530-1 (1967)

1280 MOLECULAR STRUCTURE OF XENON HEXAFLUORIDE.
BARTELL LS + GAVIN RM + THOMPSON HB
J CHEM PHYS 43(7): 2547-8 (1965)

SELF-CONSISTENT FIELD STUDY OF THE SERIES XENON
DIFLUORIDE, XENON TETRAFLUORIDE, XENON
HEXAFLUORIDE.
BASCH H + MOSKOWITZ JW + HOLISTER C + HANKIN D
J CHEM PHYS 55: 1923-33 (1971)

PHOTOIONIZATION MASS SPECTROMETRIC STUDY OF
XENON DIFLUORIDE, XENON TETRAFLUORIDE, AND XENON
HEXAFLUORIDE.
BERKOWITZ J + CHUPKA WA + GUYON PM + HOLLOWAY JH
J PHYS CHEM 75: 1461-71 (1971)

1281 ELECTRIC FIELD DEFLECTION OF MOLECULES WITH
LARGE AMPLITUDE MOTION. (XENON HEXAFLUORIDE,
IODINE HEPTAFLUORIDE, RHENIUM HEPTAFLUORIDE)
BERNSTEIN LS + PITZER KS
J CHEM PHYS 62: 2530-4 (1975)

ELECTRON DIFFRACTION STUDIES OF XENON
TETRAFLUORIDE, XENON HEXAFLUORIDE, AND OTHER
COMPOUNDS.
BOHN RK
CORNELL UNIV, PHD THESIS, 1964, 107P.

1282 ELECTRON DIFFRACTION STUDY OF THE STRUCTURE OF
GASEOUS XENON TETRAFLUORIDE.
BOHN RK + KATADA K + MARTINEZ JV + BAUER SH
P238-42 OF NOBLE GAS COMPOUNDS, HYMAN HH(ED),
UNIV OF CHICAGO PRESS, 1963.

HELIUM-1 AND HELIUM-2 PHOTOELECTRON SPECTRA AND
THE ELECTRONIC STRUCTURES OF XENON DIFLUORIDE,
XENON TETRAFLUORIDE, AND XENON HEXAFLUORIDE.
BRUNDLE CR + JONES GR + BASCH H
J CHEM PHYS 55: 1098-1104 (1971)

INTRAMOLECULAR REARRANGEMENT IN IODINE
HEPTAFLUORIDE AND XENON HEXAFLUORIDE.
BURBANK RD + BARTLETT N
CHEM COMMUN 1968: 645-7

1283 XENON HEXAFLUORIDE. THE STRUCTURE OF A CUBIC
PHASE AT -30C.
BURBANK RD + JONES GR
SCIENCE 168: 248-9 (1970)

1284 XENON HEXAFLUORIDE. STRUCTURAL CRYSTALLOGRAPHY
OF TETRAMERIC PHASES.
BURBANK RD + JONES GR
SCIENCE 171: 485-7 (1971)

VIBRATIONAL SPECTRA OF XENON COMPOUNDS.
CLAASSEN HH
P304-5 OF NOBLE GAS COMPOUNDS, HYMAN HH(ED),
CHICAGO UNIV PRESS, 1963.

1285 SPECTRAL OBSERVATIONS ON MOLECULAR XENON
HEXAFLUORIDE. RAMAN SCATTERING AND INFRARED,
VISIBLE AND ULTRAVIOLET ABSORPTION IN THE VAPOR
AND IN MATRIX ISOLATION.
CLAASSEN HH + GOODMAN GL + KIM H
J CHEM PHYS 56: 5042-53 (1972)

1286 STEREOCHEMISTRY AND SEVEN COORDINATION. (XENON
HEXAFLUORIDE)
CLAXTON TA + BENSON GC
CAN J CHEM 44: 157-63, 1730-1 (1966)

1287 MAGNETIC FIELD DEFLECTION OF XENON HEXAFLUORIDE.
CODE RF + FALCONER WE + KLEMPERER W + OZIER I
J CHEM PHYS 47(12): 4955-8 (1967)

STRUCTURAL CONSIDERATIONS IN THE CHEMISTRY OF
NOBLE GASES. (XENON DIFLUORIDE, XENON
TETRAFLUORIDE, XENON HEXAFLUORIDE)
FALCONER WE
P386-90 OF NOBLE GASES, STANLEY RE + MOGHISSI
AA(EDS), ENVIRONMENTAL PROTECTION AGENCY, 1973.
(CONF-730915)

MOLECULAR STRUCTURE OF XENON HEXAFLUORIDE AND
IODINE HEPTAFLUORIDE.
FALCONER WE + BUCHLER A + STAUFFER JL
KLEMPERER W
J CHEM PHYS 48: 312-19 (1968)

1288 ELECTRON SPIN RESONANCE OF THE XENON FLUORIDE
RADICAL.
FALCONER WE + MORTON JR
PROC CHEM SOC 1963: 95-6

1289 RAMAN SPECTRUM OF XENON HEXAFLUORIDE.
GASNER EL + CLAASSEN HH
INORG CHEM 6: 1937-8 (1967)

1290 EFFECTS OF ELECTRON CORRELATION IN X-RAY
DIFFRACTION AND AN ELECTRON DIFFRACTION STUDY OF
XENON HEXAFLUORIDE.
GAVIN RM
IOWA STATE UNIV, PHD THESIS, 1966, 92P.

1291 MOLECULAR STRUCTURE OF XENON HEXAFLUORIDE.
PART-1: ANALYSIS OF ELECTRON DIFFRACTION
INTENSITIES. PART-2: INTERNAL MOTION AND MEAN
GEOMETRY DEDUCED BY ELECTRON DIFFRACTION.
GAVIN RM + BARTELL LS
J CHEM PHYS 48: 2460-5 (1968); 2466-73 (1968)

1292 STRUCTURES OF NOBLE GAS COMPOUNDS. (XENON
HEXAFLUORIDE)
GILLESPIE RJ
CHEMISTRY 39(4): 17-19 (1966)

1293 GAS PHASE STRUCTURE OF XENON HEXAFLUORIDE.
GLASS WK
CHEM COMMUN 1968: 455

1294 ELECTRONIC STATES AND MOLECULAR GEOMETRY OF
XENON HEXAFLUORIDE: A CASE OF ELECTRONIC
ISOMERISM.
GOODMAN GL
J CHEM PHYS 56: 5038-41 (1972)

1295 STRUCTURE OF GASEOUS XENON HEXAFLUORIDE.
HEDBERG K + PETERSON SH + RYAN RR
J CHEM PHYS 44: 1726 (1966)

ENTHALPIES OF FORMATION OF XENON HEXAFLUORIDE
XENON TETRAFLUCRIDE XENON TETRAFLUORIDE. CRYSTAL
STRUCTURE.
IBERS JA + HAMILTON WC
SCIENCE 139: 106-7 (1963)

1296 POLYMORPHISM IN XENON HEXAFLUORIDE.
JONES GR + BURBANK RD + FALCONER WE
J CHEM PHYS 52: 6450-1 (1970)

1297 CRYSTALLINE MODIFICATIONS OF XENON HEXAFLUORIDE.
JONES GR + BURBANK RD + FALCONER WE
J CHEM PHYS 53: 1605-6 (1970)

THEORETICAL AND EXPERIMENTAL STUDIES OF THE
ELECTRONIC STRUCTURE OF THE XENON FLUORIDES.
JORTNER J + WILSON EG + RICE SA
P358-88 OF NOBLE GAS COMPOUNDS, HYMAN HH(ED),
UNIV OF CHICAGO PRESS, 1963.

1298 BONDING IN XENON HEXAFLUORIDE.
KAUFMAN JJ
J CHEM EDUC 41: 183-4 (1964)

1299 FAR INFRARED AND MILLIMETER WAVE STUDIES OF
XENON HEXAFLUORIDE.
KIM H + CLAASSEN HH + PEARSON E
INORG CHEM 7: 616-7 (1968)

1300 STRUCTURE OF THE IODINE HEXAFLUORIDE ANION.
(XENON HEXAFLUORIDE)
KLANM H + MEINERT H + REICH P + WITKE K
Z CHEM 8: 469-79 (1968)

IDEAL GAS THERMODYNAMIC PROPERTIES OF XENON
COMPOUNDS. (XENON TETRAFLUORIDE AND XENON
HEXAFLUORIDE)
KUDCHADKER SA + KUDCHADKER AP
PROC INDIAN ACAD SCI A 73(5): 261-7 (1971)

1301 PREPARATION AND SOME THERMAL PROPERTIES OF PURE
XENON HEXAFLUORIDE.
MALM JG + SCHREINER F + OSBORNE DW
INORG NUCL CHEM LETT 1(3): 97-100 (1965)

1302 ELECTRONIC AND GEOMETRIC STRUCTURE OF THE FREE
XENON HEXAFLUORIDE MOLECULE.
NIELSEN U + HAENSEL R + SCHWARZ WHE
J CHEM PHYS 61(9): 3581-6 (1974)

1303 MOLECULAR STRUCTURE OF XENON HEXAFLUORIDE.
PITZER KS + BERNSTEIN LS
J CHEM PHYS 63(9): 3849-56 (1975)

1304 MULTIPLE SCATTERING X(ALPHA) CALCULATION OF THE
ENERGY LEVEL STRUCTURE OF XENON HEXAFLUORIDE.
PHILLIPS EW + CONNOLLY JWD + TRICKEY SB
CHEM PHYS LETT 17(2): 203-6 (1972)

1305 HEAT CAPACITY AND OTHER THERMODYNAMIC FUNCTIONS
OF XENON HEXAFLUORIDE FROM 5 TO 350 DEGREES K.
SCHREINER F + CSBORNE DW + MALM JG + MCDONALD GN
J CHEM PHYS 51: 4838-51 (1969)

1306 CHEMICAL APPLICATIONS OF CORE ELECTRON
EXCITATION SPECTROSCOPY. (XENON HEXAFLUORIDE)
SCHWARZ WHE
BER BUNSENGES PHYS CHEM 78(11): 1206-9 (1974)
(IN GERMAN)

1307 ELECTRICAL CONDUCTIVITY AND DIELECTRIC CONSTANT
OF LIQUID XENON HEXAFLUORIDE.
SELIG H + MOOTZ A
INORG NUCL CHEM LETT 3: 147-8 (1967)

1308 MAGNETIC SUSCEPTIBILITY OF XENON HEXAFLUORIDE.
SELIG H + SCHREINER F
J CHEM PHYS 45: 4755 (1966)

THERMOCHEMICAL STUDIES OF XENON TETRAFLUORIDE
AND XENON HEXAFLUORIDE.
STEIN L + PLURIEN PL
P144-8 OF NOBLE GAS COMPOUNDS, HYMAN HH(ED),
UNIV OF CHICAGO PRESS, 1963.

1309 THE STEREOCHEMICALLY INERT LONE PAIR. A
SPECULATION ON THE BONDING IN ANTIMONY
HEXACHLORIDE(3-), SELENIUM HEXABROMIDE(2-),
TELLURIUM HEXABROMIDE(2-), IODINE
HEXAFLUORIDE(1-), XENON HEXAFLUORIDE, ETC.
URCH DS
J CHEM SOC 1964: 5775-81

1310 MAGNETIC SUSCEPTIBILITY OF XENON HEXAFLUORIDE.
VOLAVSEK B
MONATSH CHEM 97(5): 1531-2 (1966) (IN GERMAN)

1311 CRYSTAL FIELD MODEL STUDY OF XENON HEXAFLUORIDE.
PART-1: ENERGY LEVELS AND MOLECULAR GEOMETRY.
PART-2: COMPARISONS WITH OTHER HEXAVALENT XENON
MOLECULES. PART-3: ELECTRONIC TRANSITIONS AND
BAND SHAPES.
WANG SY + LOHR LL
J CHEM PHYS 60: 3901-15 (1974); 3916-19 (1974);
61: 4110-18 (1974); 62(5): 2013-14 (1975)

1312 XENON HEXAFLUORIDE. (SPECTRUM)
WEAVER EE + WEINSTOCK B + KNOP CP
J AMER CHEM SOC 85: 111-12 (1963)

THE XENON- FLUORINE SYSTEM.
WEINSTOCK B + WEAVER EE + KNOP CP
INORG CHEM 5: 2189-203 (1966)

1313 BONDING IN XENON HEXAFLUORIDE.
WILLETT RD
THEOR CHIM ACTA 6: 186-8 (1966)

YTTRIUM MONOFLUORIDE   YF

ELECTRONIC STATES OF GASEOUS SCANDIUM
MONOFLUORIDE, YTTRIUM MONOFLUORIDE, AND
LANTHANUM MONOFLUORIDE.
BARROW RF + BASTIN MW + MOORE DLG + POTT CJ
NATURE 215: 1072-3 (1967)

GROUND STATES OF SCANDIUM MONOFLUORIDE AND
YTTRIUM MONOFLUORIDE.
BARROW RF + GISSANE WJM
PROC PHYS SOC 84: 615-6 (1964)

EVALUATION OF MOLECULAR VIBRATION FREQUENCIES OF
SCANDIUM MONOFLUORIDE, YTTRIUM MONOFLUORIDE, AND
LANTHANUM MONOFLUORIDE.
KRASNOV KS
IZV VYSSH UCHEB ZAVED, KHIM 1: 594-6 (1967) (IN
RUSSIAN)

DISSOCIATION ENERGIES AND STABILITIES OF
SCANDIUM MONOFLUORIDE, YTTRIUM MONOFLUORIDE, AND
LANTHANUM MONOFLUORIDE.
KRASNOV KS
IZV VYSSH UCHEB ZAVED, KHIM 12: 578-82 (1969)
(IN RUSSIAN)

MOLECULAR CONSTANTS OF SCANDIUM, YTTRIUM, AND
LANTHANUM HALIDES. (SCANDIUM DIFLUORIDE,
SCANDIUM MONOFLUORIDE, YTTRIUM DIFLUORIDE,
YTTRIUM MONOFLUORIDE, LANTHANUM DIFLUORIDE, AND
LANTHANUM MONOFLUORIDE)
KRASNOV KS + TIMOSHININ VS
HIGH TEMP 7(2): 333-4 (1969)

THERMODYNAMIC FUNCTIONS OF GASEOUS SCANDIUM
MONOFLUORIDE, SCANDIUM DIFLUORIDE, SCANDIUM
TRIFLUORIDE, YTTRIUM MONOFLUORIDE, YTTRIUM
DIFLUORIDE, YTTRIUM TRIFLUORIDE, LANTHANUM
MONOFLUORIDE, LANTHANUM DIFLUORIDE, AND
LANTHANUM TRIFLUORIDE.
KRASNOV KS + DANILOVA TG
HIGH TEMP 7(6): 1131-3 (1969)

1314   ELECTRONIC BAND SPECTRUM OF YTTRIUM
       MONOFLUORIDE.
       MANN DE + ACQUISTA N + LINEVSKY MJ
       P363-72 OF CONF ON RARE EARTH RES, 4TH PROC,
       1964, PHOENIX, ARIZ.

       VIBRONIC SPECTRUM OF SCANDIUM MONOFLUORIDE,
       YTTRIUM MONOFLUORIDE, AND LANTHANUM
       MONOFLUORIDE.
       SHENYAVSKAYA EA + MAL'TSEV AA
       VEST MOSK UNIV KHIM 22: 104-5 (1967) (IN
       RUSSIAN)

1315   VIBRONIC SPECTRA OF YTTRIUM MONOFLUORIDE.
       SHENYAVSKAYA EA + MAL'TSEV AA + GURVICH LV
       OPT SPECTROSC 21(6): 374-6 (1966)

       MASS SPECTROMETRIC STUDIES AT HIGH TEMPERATURES.
       PART-18: THE STABILITIES OF THE MONO- AND
       DIFLUORIDES OF SCANDIUM AND YTTRIUM.
       ZMBOV KF + MARGRAVE JL
       J CHEM PHYS 47(9): 3122-5 (1967)

YTTRIUM DIFLUORIDE YF_2

THERMODYNAMIC FUNCTIONS OF GASEOUS SCANDIUM
MONOFLUORIDE, SCANDIUM DIFLUORIDE, SCANDIUM
TRIFLUORIDE, YTTRIUM MONOFLUORIDE, YTTRIUM
DIFLUORIDE, YTTRIUM TRIFLUORIDE, LANTHANUM
MONOFLUORIDE, LANTHANUM DIFLUORIDE, AND
LANTHANUM TRIFLUORIDE.
KRASNOV KS + DANILOVA TG
HIGH TEMP 7(6): 1131-3 (1969)

MOLECULAR CONSTANTS OF SCANDIUM, YTTRIUM, AND
LANTHANUM HALIDES. (SCANDIUM DIFLUORIDE,
SCANDIUM MONOFLUORIDE, YTTRIUM DIFLUORIDE,
YTTRIUM MONOFLUORIDE, LANTHANUM DIFLUORIDE, AND
LANTHANUM MONOFLUORIDE)
KRASNOV KS + TIMOSHININ VS
HIGH TEMP 7(2): 333-4 (1969)

1316   INFRARED SPECTRA AND GEOMETRIES OF MATRIX
       ISOLATED YTTRIUM DIFLUORIDE AND YTTRIUM
       TRIFLUORIDE.
       WESLEY RD + DEKOCK CW
       J PHYS CHEM 77: 466-8 (1973)

       MASS SPECTROMETRIC STUDIES AT HIGH TEMPERATURES.
       PART-18: THE STABILITIES OF THE MONO- AND
       DIFLUORIDES OF SCANDIUM AND YTTRIUM.
       ZMBOV KF + MARGRAVE JL
       J CHEM PHYS 47(9): 3122-5 (1967)

YTTRIUM TRIFLUORIDE YF_3

1317   ELECTRON DIFFRACTION STUDY OF THE MOLECULAR
       STRUCTURE OF THE VAPOR PHASE HALIDES OF GALLIUM,
       YTTRIUM, LANTHANUM, AND NEODYMIUM.
       AKISHIN PA + NAUMOV VA + TATAEVSKII VM
       VEST MOSK UNIV 1959(1): 229-36 (IN RUSSIAN)

1318   STRUCTURES OF YTTRIUM AND BISMUTH TRIFLUORIDES
       BY NEUTRON DIFFRACTION.
       CHEETHAM AK + NORMAN N
       ACTA CHEM SCAND 28(1): 55-60 (1974)

STANDARD ENTHALPIES OF FORMATION OF SCANDIUM
TRIFLUORIDE, YTTRIUM TRIFLUORIDE, AND LANTHANUM
TRIFLUORIDE.
FINOGENOV AD
RUSS J PHYS CHEM 45: 900-1 (1971)

GEOMETRIES AND ENTROPIES OF METAL TRIFLUORIDES
FROM INFRARED SPECTRA. (SCANDIUM, YTTRIUM,
LANTHANUM, CERIUM, NEODYMIUM, EUROPIUM, AND
GADOLINIUM TRIFLUORIDES)
HASTIE JW + HAUGE RH + MARGRAVE JL
J LESS COMMON METALS 39(2): 309-34 (1975)

FORCE CONSTANTS AND GEOMETRIES OF MATRIX
ISOLATED RARE EARTH TRIFLUORIDES. (SCANDIUM
TRIFLUORIDE, YTTRIUM TRIFLUORIDE, AND LANTHANUM
TRIFLUORIDE)
HAUGE RH + HASTIE JW + MARGRAVE JL
J LESS COMMON METALS 23: 359-65 (1971)

ELECTRIC DEFLECTION OF MOLECULAR BEAMS OF THE
RARE EARTH DIFLUORIDES AND TRIFLUORIDES.
(LANTHANIDE TRIFLUORIDES, LANTHANIDE
DIFLUORIDES, SCANDIUM TRIFLUORIDE, AND YTTRIUM
TRIFLUORIDE)
KAISER EW + FALCONER WE + KLEMPERER W
J CHEM PHYS 56: 5392-8 (1972)

SUBLIMATION PRESSURES OF SCANDIUM TRIFLUORIDE,
YTTRIUM TRIFLUORIDE, AND LANTHANUM TRIFLUORIDE.
KENT RA + ZMBOV KF + KANA'AN AS + BESENBRUCH G +
MCDONALD JD + MARGRAVE JL
J INORG NUCL CHEM 28: 1419-27 (1966)

CALCULATION OF DISSOCIATION ENERGIES FOR THE
HALIDES OF THE SCANDIUM SUBGROUP. (SCANDIUM
TRIFLUORIDE, YTTRIUM TRIFLUORIDE, AND LANTHANUM
TRIFLUORIDE)
KRASNOV KS
HIGH TEMP 4(1): 128-30 (1965)

CALCULATION OF THE VIBRATIONAL FREQUENCIES OF
SCANDIUM SUBGROUP HALIDES.
KRASNOV KS
HIGH TEMP 5(4): 639-40 (1967)

FORCE CONSTANT OF THE NONPLANAR VIBRATION OF AN
XY_3 MOLECULE AND A MODEL WITH POLARIZABLE IONS.
KRASNOV KS
IZV VYSSH UCHEB ZAVED KHIM KHIM TEKHNOL 10(9):
997-1000 (1967) (IN RUSSIAN)

THERMODYNAMIC FUNCTIONS OF GASEOUS SCANDIUM
MONOFLUORIDE, SCANDIUM DIFLUORIDE, SCANDIUM
TRIFLUORIDE, YTTRIUM MONOFLUORIDE, YTTRIUM
DIFLUORIDE, YTTRIUM TRIFLUORIDE, LANTHANUM
MONOFLUORIDE, LANTHANUM DIFLUORIDE, AND
LANTHANUM TRIFLUORIDE.
KRASNOV KS + DANILOVA TG
HIGH TEMP 7(6): 1131-3 (1969)

1319   ENTHALPY OF FORMATION OF YTTRIUM TRIFLUORIDE.
       RUDZITIS E + FEDER HM + HUBBARD WN
       J PHYS CHEM 69: 2305-7 (1965)

1320   HIGH TEMPERATURE HEAT CONTENTS AND RELATED
       THERMODYNAMIC FUNCTIONS OF SEVEN RARE EARTH
       TRIFLUORIDES. (YTTRIUM TRIFLUORIDE)
       SPEDDING FH + HENDERSON DC
       J CHEM PHYS 54: 2476-83 (1971)

SATURATED VAPOR PRESSURE OF SCANDIUM
TRIFLUORIDE, YTTRIUM TRIFLUORIDE, AND LANTHANUM
TRIFLUORIDE.
SUVOROV AL + NOVIKOV GI
VEST LENINGRAD UNIV, FIZ KHIM 23: 83-8 (1968)
(IN RUSSIAN)

INFRARED SPECTRA AND GEOMETRIES OF MATRIX
ISOLATED YTTRIUM DIFLUORIDE AND YTTRIUM
TRIFLUORIDE.
WESLEY RD + DEKOCK CW
J PHYS CHEM 77: 466-8 (1973)

SUBLIMATION PRESSURES AND HEATS OF SUBLIMATION
OF THULIUM TRIFLUORIDE, YTTERBIUM TRIFLUORIDE,
AND LUTETIUM TRIFLUORIDE.
ZMBOV KF + MARGRAVE JL
J LESS COMMON METALS 12: 494-6 (1967)

MASS SPECTROMETRIC STUDIES OF SCANDIUM
TRIFLUORIDE, YTTRIUM TRIFLUORIDE, LANTHANUM
TRIFLUORIDE AND THE RARE EARTH TRIFLUORIDES.
ZMBOV KF + MARGRAVE JL
P267-90 OF MASS SPECTROMETRY IN ORGANIC
CHEMISTRY, MARGRAVE JL (ED), AMER CHEM SOC,
1968, 329P. (ADVANCES IN CHEMISTRY SERIES NO 72)

ZINC DIFLUORIDE ZnF_2

ELECTRON DIFFRACTION ANALYSIS OF THE MOLECULAR
STRUCTURES OF GROUP-2 ELEMENTS. (BERYLLIUM
DIFLUORIDE, MAGNESIUM DIFLUORIDE, CALCIUM
DIFLUORIDE, STRONTIUM DIFLUORIDE, BARIUM
DIFLUORIDE, ZINC DIFLUORIDE, CADMIUM DIFLUORIDE)
AKISHIN PA + SPIRIDONOV VP
SOV PHYS CRYSTALLOGR 2(4): 472-9 (1957)

1321 SUBLIMATION THERMODYNAMICS OF ZINC(II) FLUORIDE.
BIEFELD RM + EICK HA
J CHEM THERMODYN 5(3): 353-60 (1973)

1322 POLARIZED ION MODEL AND THE BENDING FORCE
CONSTANTS OF THE GROUP-2B HALIDES. (ZINC
DIFLUORIDE, CADMIUM DIFLUORIDE, AND MERCURY
DIFLUORIDE)
ELIEZER I
THEOR CHIM ACTA 18: 77-85 (1970)

INFRARED SPECTRA AND GEOMETRY OF MATRIX ISOLATED
COBALT DIFLUORIDE, NICKEL DIFLUORIDE, CCPPER
DIFLUORIDE, AND ZINC DIFLUORIDE.
HASTIE JW + HAUGE RH + MARGRAVE JL
HIGH TEMP SCI 1: 76-85 (1969)

FORCE FIELDS OF THE RUTILE COUNTERPARTS:
TITANIUM DIOXIDE, MAGNESIUM FLUORIDE, ZINC
FLUORIDE, AND FERROUS FLUORIDE.
IIISHI K + TOMISAKA T + UMEGAKI Y
MINERALOG J 6(1-2): 77-84 (1969)

1323 VIBRATIONAL SPECTRA AND THERMODYNAMICS OF THE
ZINC HALIDES. (ZINC DIFLUORIDE)
LOEWENSCHUSS A + RON A + SCHNEPP O
J CHEM PHYS 49: 272-9 (1968)

VAPOR PRESSURE OF ZINC DIFLUORIDE, CADMIUM
DIFLUORIDE, MAGNESIUM DIFLUORIDE, CALCIUM
DIFLUORIDE, STRONTIUM DIFLUORIDE, BARIUM
DIFLUORIDE, AND ALUMINUM TRIFLUORIDE.
RUFF O + LE BOUCHER L
Z ANORG ALLGEM CHEM 219: 376-81 (1934) (IN
GERMAN)

CRYSTAL STRUCTURE OF MANGANESE DIFLUORIDE, IRON
DIFLUORIDE, COBALT DIFLUORIDE, NICKEL
DIFLUORIDE, AND ZINC DIFLUORIDE.
STOUT JW + REED SA
J AMER CHEM SOC 76: 5279-81 (1954)

1324 INTERACTION OF MATRIX ISOLATED NICKEL FLUORIDE
AND NICKEL CHLORIDE WITH CARBON MONOXIDE,
MOLECULAR NITROGEN, NITRIC OXIDE, AND MOLECULAR
OXYGEN AND OF CALCIUM FLUORIDE, CHROMIUM(II)
FLUORIDE, MANGANESE(II) FLUORIDE, COPPER(II)
FLUORIDE, AND ZINC(II) FLUORIDE WITH CARBON
MONOXIDE IN ARGON MATRICES.
VAN LEIRSBURG DA + DEKOCK CW
J PHYS CHEM 78(2): 134-42 (1974)

ZIRCONIUM DIFLUORIDE ZrF_2

1325 INFRARED SPECTRA OF MATRIX ISOLATED ZIRCONIUM
DIFLUORIDE, TRIFLUORIDE, AND TETRAFLUORIDE.
HAUGE RH + MARGRAVE JL + HASTIE JW
HIGH TEMP SCI 5(2): 89-96 (1973)

ZIRCONIUM TRIFLUORIDE ZrF_3

INFRARED SPECTRA OF MATRIX ISOLATED ZIRCONIUM
DIFLUORIDE, TRIFLUORIDE, AND TETRAFLUORIDE.
HAUGE RH + MARGRAVE JL + HASTIE JW
HIGH TEMP SCI 5(2): 89-96 (1973)

ZIRCONIUM TETRAFLUORIDE ZrF_4

1326 VAPOR PRESSURE OF ZIRCONIUM TETRAFLUORIDE.
AKISHIN PA + BELOUSOV VI + SIDOROV LN
RUSS J INORG CHEM 8(6): 789-90 (1963)

1327 INFRARED SPECTRA OF GASEOUS GROUP-4 HALIDES:
ZIRCONIUM FLUORIDE, ZIRCONIUM CHLORIDE, AND
HAFNIUM CHLORIDE.
BUCHLER A
ARTHUR D LITTLE INC, 1960, 10P. PB148429

VAPOR PRESSURE AND HEAT OF SUBLIMATION OF
ZIRCONIUM AND HAFNIUM TETRACHLORIDES.
DENISOVA LN + SAFRONOV EK + BYSTROVA ON
RUSS J INORG CHEM 11(9): 1171-3 (OCT 1966)

1328 VAPOR PRESSURE OF ZIRCONIUM TETRAFLUORIDE.
GALKIN NP + TUMANOV YN + TARASOV VI
+ SHISHKOV YD
RUSS J INORG CHEM 8(9): 1054-5 (1963)

1329 FREQUENCIES OF ZIRCONIUM FLUORIDE (GAS)
VIBRATIONS ACCORDING TO THERMODYNAMIC DATA.
GODNEV IN + ALEKSANDROVSKAYA AM + SVERDLIN AS
OPT SPECTROSC 29(4): 362-3 (1970)

1330 NORMAL VIBRATION FREQUENCIES OF ZIRCONIUM
HALIDES.
GODNEV IN + ALEKSANDROVSKAYA AM + RIGINA IV
OPT SPECTROSC 7: 172-3 (1959)

INFRARED SPECTRA OF MATRIX ISOLATED ZIRCONIUM
DIFLUORIDE, TRIFLUORIDE, AND TETRAFLUORIDE.
HAUGE RH + MARGRAVE JL + HASTIE JW
HIGH TEMP SCI 5(2): 89-96 (1973)

1331 VAPOR PRESSURE OF ZIRCONIUM FLUORIDE.
SENSE KA + SNYDER MJ + FILBERT RB
J PHYS CHEM 58: 995-6 (1954)

1332 ELECTRON DIFFRACTION INVESTIGATION OF THE
STRUCTURE OF ZIRCONIUM TETRAFLUORIDE IN THE
VAPOR PHASE.
SPIRIDONOV VP
VEST MOSK UNIV KHIM 23(1): 113-14 (1968) (IN
RUSSIAN)

ACTINIDE FLUORIDES   ☐ ☐

HALIDES OF THE LANTHANIDES AND ACTINIDES.
BROW D
WILEY-INTERSCIENCE, NEW YORK, 1968, 280P.

1333 PAIRING ENERGIES IN THE ACTINIDE SERIES.
GRIFFITH JS + ORGEL LE
J CHEM PHYS 26: 988-92 (1957)

1334 LATTICE CONSTANTS OF ACTINIDE TETRAFLUORIDES
INCLUDING BERKELIUM.
KEENAN TK + ASPREY LB
INORG CHEM 8: 235-8 (1969)

1335 HEATS OF FORMATION OF SOME HALIDES OF THORIUM,
PROTACTINIUM, URANIUM, NEPTUNIUM, AND AMERICIUM.
(ACTINIDE)
MASLOV PG + MASLOV UP
J GEN CHEM 35(12): 2100-3 (DEC 1965)

1336 VAPOR PRESSURES AND THERMODYNAMIC CALCULATIONS
FOR SOME ACTINIDE FLUORIDES
WEINSTOCK B + WEAVER EE + MALM JG
J INORG NUCL CHEM 11: 104-14 (1959)

NOBLE GAS FLUORIDES (GENERAL)   ☐ ☐

1337 THEORY OF INERT GAS MOLECULES.
ALLEN LC
P317-28 OF HYMAN HH (ED), NOBLE GAS COMPOUNDS,
CHICAGO UNIV PRESS, 1963, 404P.

1338 D ORBITALS IN THE NOBLE GAS DIHALIDES. (ARGON
DIFLUORIDE, KRYPTON DIFLUORIDE, AND XENON
DIFLUORIDE)
CATTON RC + MITCHELL KAR
CAN J CHEM 48: 2695-2701 (1970)

1339 ELECTRON STRUCTURE OF COMPOUNDS OF INERT GASES.
DYATKINA ME
J STRUCT CHEM 10(1): 159-60 (1969)

1340 CHEMICAL PREDICTIONS BY MO THEORY. THE NOBLE GAS
HALIDES.
JORTNER J + RICE S
P15-47 OF MODERN QUANTUM CHEM PART-1, ACADEMIC
PRESS, NY, 1965.

1341 AN LCAO MO STUDY OF RARE GAS FLUORIDES.
LOHR LL + LIPSCOMB WN
P347-53 OF HYMAN HH (ED), NOBLE GAS COMPOUNDS,
CHICAGO UNIV PRESS, 1963, 404P.

1342 NOBLE GASES AND THE BINARY FLUORIDES OF XENON.
RAMALHO DE AZEVEDO MM
CIEN CULT 26(1): 3-18 (1974) (IN SPANISH)

CALCULATIONS OF ELECTRON STRUCTURES OF COMPOUNDS
OF HALOGENS AND INERT GASES BY THE NONEMPIRICAL
NDDO-2 (ALPHA, BETA) METHOD. PART-3: ENERGY
LEVELS, WAVE FUNCTIONS AND AO POPULATIONS OF
POLYATOMIC FLUORIDES.
SMOLYAR AE + CHARKIN OP + KLIMENKO NM
J STRUCT CHEM 15(6): 885-93 (1974)

1343 BOND ENERGIES AND IONIC CHARACTER OF INERT GAS
HALIDES. HELIUM-2 PHOTOELECTRON SPECTRA OF XENON
DIFLUORIDE AND KRYPTON DIFLUORIDE.
WATERS JH + GRAY HB + DEKOCK RL
J CHEM PHYS 58: 1267-8 (1973)

1344 MOLECULAR STRUCTURES OF RARE GAS COMPOUNDS.
YAMADA S
REV PHYS CHEM JAPAN 33: 38-40 (1963)

RARE EARTH FLUORIDES (GENERAL)   ☐ ☐

THE SYMMETRY AND CRYSTAL STRUCTURE OF RARE EARTH
TRIFLUORIDES. (LANTHANUM TRIFLUORIDE)
AFANASHEV ML + HABUDA SP + LUNDIN AG
ACTA CRYSTALLOGR 28: 2903-5 (1972)

1345 STRUCTURE AND MAGNETIC PROPERTIES OF RARE EARTH
FLUORIDES.
BAKER WA
SYRACUSE UNIV, NY, 1965, 8P. AD-638523

1346 INFRARED SPECTRA OF RARE EARTH METAL FLUORIDES.
BATSANOVA LR + GRIGORYEVA GN + BATSANOV SS
ZH STRUKT KHIM 4(1): 37-42 (1963) (IN RUSSIAN)

LATTICE VIBRATIONS AND STRUCTURE OF RARE EARTH
HALIDES. (LANTHANUM TRIFLUORIDE)
BAUMAN RP + PORTO SPS
PHYS REV 161: 842-8 (1967)

1347 VAPORIZATION AND SUBLIMATION THERMODYNAMICS OF
SELECTED LANTHANIDE FLUORIDES.
BIEFELD RM
MICH STATE UNIV, PHD THESIS, 1974, 190P.

1348 HALIDES OF THE LANTHANIDES AND ACTINIDES.
BROW D
WILEY-INTERSCIENCE, NEW YORK, 1968, 280P.

1349 MAGNETIC RESONANCE AND SUSCEPTIBILITY OF RARE
EARTH TRIFLUORIDES.
CARR SL + MOULTON WG
J MAG RESON 4: 400-6 (1971)

1350 HIGH TEMPERATURE ENTHALPY INCREMENTS FOR SOME
RARE EARTH TRIFLUORIDES.
CHARLU TV + CHAUDHURI AK + MARGRAVE JL
HIGH TEMP SCI 2: 1-8 (1970)

1351 INFRARED SPECTRA AND GEOMETRIES OF RARE EARTH
DIHALIDES.
DEKOCK CW + WESLEY RD + RADTKE DD
HIGH TEMP SCI 4: 41-7 (1972)

FORCE CONSTANTS AND GEOMETRIES OF MATRIX
ISOLATED RARE EARTH TRIFLUORIDES. (SCANDIUM
TRIFLUORIDE, YTTRIUM TRIFLUORIDE, AND LANTHANUM
TRIFLUORIDE)
HAUGE RH + HASTIE JW + BARGRAVE JL
J LESS COMMON METALS 23: 359-65 (1971)

ELECTRIC DEFLECTION OF MOLECULAR BEAMS OF THE
RARE EARTH DIFLUORIDES AND TRIFLUORIDES.
(LANTHANUM TRIFLUORIDE, LANTHANUM DIFLUORIDE,
SCANDIUM TRIFLUORIDE, AND YTTRIUM TRIFLUORIDE)
KAISER EW + FALCONER WE + KLEMPERER W
J CHEM PHYS 56: 5392-8 (1972)

1352 INFRARED SPECTRA OF RARE EARTH METAL FLUORIDES
IN THE REGION OF A CESIUM IODIDE PRISM.
KUSTOVA GN + BATSANOVA LR
J APPL SPECTROSC 4: 62-3 (1966)

1353 ANALYSES OF THERMOCHEMICAL ERRORS AND
SYSTEMATICS IN SUBLIMATION OF LANTHANIDE
TRIFLUORIDES.
MCCREARY JR + THORN RJ
CAN MET QUART 13(2): 369-71 (1974)

EXPERIMENTAL DETERMINATION OF THE ENTHALPIES OF
FORMATION OF RARE EARTH FLUORIDES. (LANTHANUM
TRIFLUORIDE)
POLYACHENOK OG
RUSS J INORG CHEM 12: 449-52 (1967)

1354 MEAN AMPLITUDES OF VIBRATION OF SOME LANTHANIDE
TRIFLUORIDES.
SANYAL NK + DIXIT L
INDIAN J PURE APPL PHYS 12(7): 495-7 (1974)

1355 NMR IN RARE EARTH TRIFLUORIDES.
SARASWATI V + VIJAYARAGHAVAN R
J PHYS CHEM SOLIDS 28: 2111-6 (1967)

1356 HIGH TEMPERATURE ENTHALPIES AND RELATED
THERMODYNAMIC FUNCTIONS OF THE TRIFLUORIDES OF
SCANDIUM, CERIUM, SAMARIUM, EUROPIUM,
GADOLINIUM, TERBIUM, DYSPROSIUM, ERBIUM,
THULIUM, AND YTTERBIUM.
SPEDDING FH + BEAUDRY BJ + HENDERSON DC
+ MOORMAN J
J CHEM PHYS 60(4): 1578-88 (1974)

HIGH TEMPERATURE HEAT CONTENTS AND RELATED
THERMODYNAMIC FUNCTIONS OF SEVEN RARE EARTH
TRIFLUORIDES. (YTTRIUM TRIFLUORIDE)
SPEDDING FH + HENDERSON DC
J CHEM PHYS 54: 2476-83 (1971)

1357 SATURATED VAPOR PRESSURE OF RARE EARTH
FLUORIDES.
SUVOROV AL + KRZHIZHANOVSKAYA EV + NOVIKOV GI
RUSS J INORG CHEM 11: 1441-3 (1966)

GEOMETRY AND INFRARED SPECTRA OF MATRIX ISOLATED
RARE EARTH HALIDES. (ALSO LANTHANUM TRIFLUORIDE)
WESLEY RD + DEKOCK CW
J CHEM PHYS 55: 3866-77 (1971)

1358 BIBLIOGRAPHY ON SPECTROSCOPY OF FLUORIDES AND
OXIDES BELONGING TO THE LANTHANIDE SERIES.
WESTLEY F
P148-88 OF THERMODYNAMICS OF CHEMICAL SPECIES
IMPORTANT TO ROCKET TECHNOLOGY, BECKETT CW (ED),
AIR FORCE OFF SCI RES, 1974. AFOSR-TR-75-0596

1359 MASS SPECTROMETRIC STUDIES OF SCANDIUM, YTTRIUM,
LANTHANUM, AND RARE EARTH FLUORIDES.
ZMBOV KF + MARGRAVE JL
USAEC, 1966, 54P. ORO-2907-16

MASS SPECTROMETRIC STUDIES OF SCANDIUM
TRIFLUORIDE, YTTRIUM TRIFLUORIDE, LANTHANUM
TRIFLUORIDE AND THE RARE EARTH TRIFLUORIDES.
ZMBOV KF + MARGRAVE JL
P267-90 OF MASS SPECTROMETRY IN INORGANIC
CHEMISTRY, MARGRAVE JL (ED), AMER CHEM SOC,
1968, 329P. (ADVANCES IN CHEMISTRY SERIES NO 72)

GROUP-2 FLUORIDES ☐☐

1360 NONEMPIRICAL LCAO MO SCF STUDIES OF THE GROUP-2A
DIHALIDES BERYLLIUM DIFLUORIDE, MAGNESIUM
DIFLUORIDE, AND CALCIUM DIFLUORIDE.
GOLE JL + SIU AKQ + HAYES EF
J CHEM PHYS 58: 857-68 (1973)

GROUP-3 FLUORIDES ☐☐

GROUND STATE PROPERTIES OF THE GROUP-3
TRIHALIDES. (BORON TRIFLUORIDE)
ARMSTRONG DR + PERKINS PG
J CHEM SOC A 1967: 1218-22

DISSOCIATION ENERGIES OF GROUP-3 MONOFLUORIDES.
THE POSSIBILITY OF POTENTIAL MAXIMA IN THEIR
EXCITED STATES.
MURAD E + HILDENBRAND DL + MAIN RP
J CHEM PHYS 45: 263-9 (1966)

1361 PHOTOELECTRON SPECTRA OF THE HALIDES IN GROUP-3,
GROUP-4, GROUP-5, AND GROUP-6.
POTTS AW + LEMPKA HJ + STREETS DG + PRICE WC
PHIL TRANS ROY SOC A 268: 59-76 (1970)

GROUP-4 FLUORIDES ☐☐

ENTHALPY OF SUBLIMATION OF GERMANIUM DIFLUORIDE
AND THE THERMODYNAMICS OF SUBLIMATION OF THE
GROUP-4A DIFLUORIDES.
ADAMS GP + MARGRAVE JL + STEIGER RP
J CHEM THERMODYN 3: 297-305 (1971)

COMPARISONS OF FLUORIDES, OXIDES, AND SULFIDES
CONTAINING DIVALENT TRANSITION ELEMENTS.
GOODENOUGH JB
P215-364 OF SOLID STATE CHEMISTRY, RAO CNR(ED),
DEKKER, NY, 1974.

1362 STUDY OF THE BONDING IN THE GROUP-4 TETRAHALIDES
BY PHOTOELECTRON SPECTROSCOPY.
GREEN JC + GREEN MLH + JOACHIM PJ + ORCHARD AF
+ TURNER DW
PHIL TRANS ROY SOC A 268: 111-30 (1970)

1363 ROOT MEAN SQUARE AMPLITUDES OF VIBRATION IN SOME
GROUP-4 TETRAHALIDES.
LONG DA + SEIBOLD EA
TRANS FARADAY SOC 56: 1105-9 (1960)

PHOTOELECTRON SPECTRA OF THE HALIDES IN GROUP-3
GROUP-4 GROUP-5 AND GROUP-6.
POTTS AW + LEMPKA HJ + STREETS DG + PRICE WC
PHIL TRANS ROY SOC A 268: 59-76 (1970)

GROUP-5 FLUORIDES ☐☐

PHOTOELECTRON SPECTRA OF HALIDES. PART-1:
TETRAFLUORIDES AND TETRACHLORIDES OF GROUP-5B.
(CARBON TETRAFLUORIDE, SILICON TETRAFLUORIDE,
GERMANIUM TETRAFLUORIDE, TIN TETRAFLUORIDE)
BASSETT PJ + LLOYD DR
J CHEM SOC A 1971: 641-54

1364 MOLECULAR ORBITAL STUDY OF ATOMIZATION ENERGIES
OF PNICOGEN FLUORIDE MOLECULES AND IONS.
COMPANION AL + HSIA YP
J MOL STRUCT 14(1): 117-26 (1972)

1365 GAS PHASE STRUCTURES AND MASS SPECTRA OF BINARY
PENTAFLUORIDES.
FALCONER WE + JONES GR + SUNDER WA + VASILE MJ
J FLUORINE CHEM 4(2): 213-34 (1974)

PHOTOELECTRON SPECTRA OF THE HALIDES IN GROUP-3
GROUP-4 GROUP-5 AND GROUP-6.
POTTS AW + LEMPKA HJ + STREETS DG + PRICE WC
PHIL TRANS ROY SOC A 268: 59-76 (1970)

INTRAMOLECULAR FORCE FIELDS IN GROUP-5
TRIHALIDES.
RAI SN + THAKER SN
INDIAN J PURE APPL PHYS 9: 61-2 (1971)

ASSOCIATION OF GROUP-5 PENTAFLUORIDES IN THE GAS
PHASE. (PHOSPHORUS PENTAFLUORIDE, ARSENIC
PENTAFLUORIDE, BISMUTH PENTAFLUORIDE AND
ANTIMONY PENTAFLUORIDE)
VASILE MJ + FALCONER WE
INORG CHEM 11: 2282-3 (1972)

GROUP-6 FLUORIDES　☐☐

FORCE FIELDS FOR SOME GROUP-6 HEXAFLUORIDES.
(SULFUR HEXAFLUORIDE AND TELLURIUM HEXAFLUORIDE)
ABRAMOWITZ S + LEVIN IW
J CHEM PHYS 44: 3353-6 (1966)

PHOTOELECTRON SPECTRA OF THE HALIDES IN GROUP-3
GROUP-4 GROUP-5 AND GROUP-6.
POTTS AW + LEMPKA HJ + STREETS DG + PRICE WC
PHIL TRANS ROY SOC A 268: 59-76 (1970)

1366 SCF- X(ALPHA) SCATTERED WAVE STUDIES ON BONDING
AND IONIZATION POTENTIALS. PART-1: HEXAFLUORIDES
OF GROUP-6 ELEMENTS.
ROESCH N + SMITH VH + WHANGBO MH
J AMER CHEM SOC 96: 5984-9 (1974)

FIRST ROW TRANSITION METAL FLUORIDES　☐☐
(GENERAL)

1367 PENTAHALIDES OF TRANSITION METALS.
BEVERIDGE AD + CLARK HC
P179-225 OF HALOGEN CHEMISTRY, VOL-3, GUTMANN
V (ED), ACADEMIC PRESS, 1967.

1368 GEOMETRY OF THE TRANSITION METAL DIHALIDES. THE
DIFLUORIDES OF SOME FIRST ROW TRANSITION METALS.
BUCHLER A + STAUFFER JL + KLEMPERER W
J CHEM PHYS 40: 3471-3 (1964)

DETERMINATION OF THE GEOMETRY OF HIGH
TEMPERATURE SPECIES BY ELECTRIC DEFLECTION AND
MASS SPECTROMETRIC DETECTION. (FIRST ROW
FLUORIDES, BERYLLIUM, MAGNESIUM, AND LEAD
DIFLUORIDES)
BUECHLER A + STAUFFER JL + KLEMPERER W
J AMER CHEM SOC 86: 4544-50 (1964)

1369 MEAN AMPLITUDES AND SHRINKAGE EFFECTS OF
VIBRATION FOR SOME LINEAR SYMMETRICAL
DIFLUORIDES. (OF THE FIRST ROW TRANSITION
METALS, ALSO BERYLLIUM DIFLUORIDE AND MAGNESIUM
DIFLUORIDE)
CYVIN SJ + VIZI B
VESZPREMI VEGYIPARI EGYETEM KOZLEMENYEI 11: 83-9
(1968)

1370 COMPARISONS OF FLUORIDES, OXIDES, AND SULFIDES
CONTAINING DIVALENT TRANSITION ELEMENTS.
GOODENOUGH JB
P215-364 OF SOLID STATE CHEMISTRY, RAC CNR(ED),
DEKKER, NY, 1974.

1371 VIBRATIONAL FREQUENCIES AND VALENCE FORCE
CONSTANTS OF FIRST ROW TRANSITION METAL
DIFLUORIDES.
HASTIE JW + HAUGE RH + MARGRAVE JL
CHEM COMMUN 1969: 1452-3

1372 MEAN AMPLITUDES OF VIBRATION AND
BASTIANSEN-MORINO SHRINKAGE EFFECT IN SOME
LINEAR SYMMETRICAL DIHALIDES. (BERYLLIUM
DIFLUORIDE, MAGNESIUM DIFLUORIDE, AND FIRST ROW
TRANSITION METAL FLUORIDES)
NAGARAJAN G
J MOL SPECTROSC 13: 361-92 (1964)

SECOND ROW TRANSITION METAL FLUORIDES　☐☐
(GENERAL)

1373 VESCF-MO STUDIES OF MOLECULES CONTAINING ATOMS
FROM THE SECOND ROW OF THE PERIODIC TABLE.
(SILICON DIFLUORIDE, SILICON TETRAFLUORIDE,
PHOSPHORUS TRIFLUORIDE, PHOSPHORUS
PENTAFLUORIDE, SULFUR DIFLUORIDE, SULFUR
TETRAFLUORIDE, SULFUR HEXAFLUORIDE, CHLORINE
MONOFLUORIDE, CHLORINE TRIFLUORIDE, AND CHLORINE
PENTAFLUORIDE)
BROWN RD + PEEL JB
AUST J CHEM 21: 2605-15 (1968)

1374 HALIDES OF THE SECOND AND THIRD ROW TRANSITION
METALS.
CANTERFORD JH + COLTON R
WILEY-INTERSCIENCE, NEW YORK, 1969, 409P.

1375 X-RAY DIFFRACTION STUDIES OF SOME TRANSITION
METAL HEXAFLUORIDES. (SECOND ROW, THIRD ROW, AND
RHENIUM HEPTAFLUORIDE)
SIEGEL S + NORTHROP DA
INORG CHEM 5: 2187-8 (1966)

1376 MOLECULAR BEAM MASS SPECTROMETRIC STUDY OF
BINARY PENTAFLUORIDES.　(ALSO ANTIMONY
PENTAFLUORIDE)
VASILE MJ + JONES GR + FALCONER WE
CHEM COMMUN 1971: 1355-6

1377 APPLICATION OF A MOLECULAR BEAM SOURCE MASS
SPECTROMETER TO THE STUDY OF REACTIVE FLUORIDES.
(SECOND ROW, THIRD ROW, ANTIMONY PENTAFLUORIDE,
AND BISMUTH PENTAFLUORIDE)
VASILE MJ + JONES GR + FALCONER WE
INT J MASS SPECTROM ION PHYS 10: 457-9 (1972)

THIRD ROW TRANSITION METAL FLUORIDES　☐☐
(GENERAL)

HALIDES OF THE SECOND AND THIRD ROW TRANSITION
METALS.
CANTERFORD JH + COLTON R
WILEY-INTERSCIENCE, NEW YORK, 1969, 409P.

1378 RAMAN SPECTRA OF 5D TRANSITION METAL
HEXAFLUORIDES IN THE VAPOR STATE. (THIRD ROW)
CLAASSEN HH + SELIG H
ISRAEL J CHEM 7: 499-504 (1969)

1379 THERMODYNAMIC PROPERTIES OF TRANSITION METAL
PENTAFLUORIDES AT HIGH TEMPERATURE. (THIRD ROW)
GALKIN NP + TUMANOV YN + BUTYLKIN YP
RUSS J PHYS CHEM 44: 1724-5 (1970)

X-RAY DIFFRACTION STUDIES OF SOME TRANSITION
METAL HEXAFLUORIDES. (SECOND ROW, THIRD ROW, AND
RHENIUM HEPTAFLUORIDE)
SIEGEL S + NORTHROP DA
INORG CHEM 5: 2187-8 (1966)

APPLICATION OF A MOLECULAR BEAM SOURCE MASS
SPECTROMETER TO THE STUDY OF REACTIVE FLUORIDES.
(SECOND ROW, THIRD ROW, ANTIMONY PENTAFLUORIDE,
AND BISMUTH PENTAFLUORIDE)
VASILE MJ + JONES GR + FALCONER WE
INT J MASS SPECTROM ION PHYS 10: 457-9 (1972)

MX2 MOLECULES (GENERAL)

MX_2

1380 BENDING MOTIONS IN THE DIHALIDES OF GROUP-2
METALS.
BERRY RS
J CHEM PHYS 30: 286-90, 1190 (1959)

1381 IONIC MODEL CALCULATIONS. PART-2: BENDING FORCE
CONSTANTS OF THE GROUP-2 HALIDES.
BUCHLER A
ARTHUR D LITTLE INC, 1960, 19P. PB148431

1382 SUBLIMATION PRESSURES OF REFRACTORY FLUORIDES.
KENT RA + ZMBOV KF + MCDONALD JD + BESENBRUCH G
+ EHLERT TC + BAUTISTA RG + KANA'AN AS
+ MARGRAVE JL
P249-55 OF CONF ON NUCLEAR APPLICATIONS OF
NONFISSIONABLE CERAMICS, 1966, WASHINGTON DC.

1383 ESTIMATION OF ANHARMONIC POTENTIAL CONSTANTS.
PART-1. LINEAR XY2 MOLECULES.
KUCHITSU K + MORINO Y
BULL CHEM SOC JAP 38(5): 805-13 (1965)

1384 MEAN AMPLITUDES OF VIBRATION FOR THE BORON
DIFLUORIDE, CARBON DIFLUORIDE, NITROGEN
DIFLUORIDE, ALUMINUM DIFLUORIDE, PHOSPHORUS
DIFLUORIDE, AND ZIRCONIUM DIFLUORIDE.
NAGARAJAN G
INDIAN J PURE APPL PHYS 2: 341-3 (1964)

1385 POTENTIAL CONSTANTS AND CORIOLIS COUPLING
CONSTANTS FOR SOME NONLINEAR SYMMETRICAL XY2
MOLECULES.
NAGARAJAN G
AUST J CHEM 16: 717-21 (1963)

1386 SIMPLE METHOD FOR EXACT SOLUTION OF 2 X 2
SECULAR EQUATION IN MOLECULAR VIBRATIONS.
THYAGARAJAN G + SUBHEDAR MK
INDIAN J PURE APPL PHYS 12(4): 309-11 (1974)

1387 UREY-BRADLEY FORCE FIELD AND THERMODYNAMIC
PROPERTIES FOR BENT SYMMETRICAL XY2 MOLECULES.
VENKATESWARLU K + THANALAKSHMI R
INDIAN J PURE APPL PHYS 1: 377-9 (1963)

MX3 MOLECULES (GENERAL)

MX_3

1388 INFLUENCE OF ATOMIC MASSES ON THE CORIOLIS
COUPLING COEFFICIENTS.
CYVIN BN + CYVIN SJ + KRISTIANSEN LA
Z NATURFORSCH 19A: 1148-50 (1964)

1389 VIBRATIONAL MEAN SQUARE AMPLITUDE MATRICES.
CYVIN SJ
ACTA CHEM SCAND 13: 1809-13 (1959)

1390 MEAN AMPLITUDES OF VIBRATION FOR SOME PYRAMIDAL
XY3 MOLECULES.
CYVIN SJ + CYVIN BN + MULLER A
J MOL STRUCT 4: 341-9 (1969)

1391 VIBRATIONAL FREQUENCIES OF SOME PLANAR XY3
MOLECULES.
HEATH DF + LINNETT JW
TRANS FARADAY SOC 44: 873-78 (1948)

SUBLIMATION PRESSURES OF REFRACTORY FLUORIDES.
KENT RA + ZMBOV VF + MCDONALD JD + BESENBRUCH G
+ EHLERT TC + BAUTISTA RG + KANA'AN AS +
MARGRAVE JL
P249-55 OF CONF ON NUCLEAR APPLICATIONS OF
NONFISSIONABLE CERAMICS, 1966, WASHINGTON DC.

1392 VIBRATIONAL MEAN SQUARE AMPLITUDE MATRICES.
KRISTIANSEN LA + CYVIN SJ
J MOL SPECTROSC 11: 185-94 (1963)

1393 CALCULATION OF UREY-BRADLEY POTENTIAL CONSTANTS.
PLANAR XY3 MOLECULES.
MEISINGSETH E
ACTA CHEM SCAND 6: 1601-6 (1962)

1394 VIBRATIONAL MEAN SQUARE AMPLITUDE MATRICES.
(BASTIANSEN-MORINO SHRINKAGE EFFECT OF
SYMMETRICAL XY3 AND XY4 MOLECULES)
MEISINGSETH E + CYVIN SJ
J MOL SPECTROSC 8: 464-9 (1962)

1395 CALCULATION OF FORCE CONSTANTS WITH THE MAXIMUM
OVERLAP METHOD. MOLECULES WITH LONE PAIR
ELECTRONS OR EMPTY ORBITALS.
MEZEI M + PULAY P
ACTA CHIM (BUDAPEST) 56: 331-5 (1968)

1396 MEAN AMPLITUDES OF VIBRATION IN SOME XY2Z
MOLECULES OF C2V SYMMETRY.
MUELLER A + NAGARAJAN G
Z PHYS CHEM 235(1-2): 113-26 (1967)

1397 MEAN AMPLITUDES OF VIBRATION AND
BASTIANSEN-MORINO SHRINKAGE EFFECT IN SOME XY3
MOLECULES OF PLANAR TRIGONAL SYMMETRY.
NAGARAJAN G
INDIAN J PURE APPL PHYS 4: 423-8 (1966)

1398 INERTIA DEFECT AND PLANARITY OF FOUR ATOM
MOLECULES. (XY3)
OKA T + MORINO Y
J MOL SPECTROSC 11: 349-67 (1963)

1399 CALCULATIONS OF MEAN AMPLITUDES OF VIBRATION AND
UREY-BRADLEY FORCE FIELDS FOR PLANAR XY3
MOLECULES.
PEACOCK CJ + MULLER A + KEBABCIOGLU R
J MOL STRUCT 2: 163-7 (1968)

1400 MOLECULAR FORCE FIELD FOR SOME XY3 MOLECULES.
(PHOSPHORUS TRIFLUORIDE AND ARSENIC TRIFLUORIDE)
PILLAI MGK + PILLAI PP
INDIAN J PURE APPL PHYS 6: 404-7 (1968)

1401 EVALUATION OF FORCE CONSTANTS OF PLANE XY3
MOLECULES FROM VIBRATIONAL DATA.
PISTORIUS CWFT
J CHEM PHYS 29: 1174-6 (1958)

1402 INTRAMOLECULAR FORCE FIELDS IN GROUP-5
TRIHALIDES.
RAI SN + THAKUR SN
INDIAN J PURE APPL PHYS 9: 61-2 (1971)

1403 UNIQUE FORCE FIELDS OF SOME XY3 TYPE MOLECULES.
RAMASWAMY K + SRIDHARAN T
INDIAN J PURE APPL PHYS 13(2): 98-100 (1975)

SIMPLE METHOD FOR EXACT SOLUTION OF 2 X 2
SECULAR EQUATION IN MOLECULAR VIBRATIONS.
THYAGARAJAN G + SUBHEDAR MK
INDIAN J PURE APPL PHYS 12(4): 309-11 (1974)

1404 CORRELATION OF INTERMOLECULAR EFFECTS IN THE
VIBRATION SPECTRA OF BORON FLUORIDE WITH
ALTERATIONS IN ITS QUADRATIC VALENCE FORCE
FIELD.
TEIXEIRA-DIAS JJC + JARDIM MDA
J MOL STRUCT 22(1): 133-40 (1974)

1405 GENERALIZED MEAN SQUARE AMPLITUDES AND CORIOLIS
CONSTANTS OF SOME PYRAMIDAL XY3 MOLECULES.
VENKATESWARLU K + PURUSHOTHAMAN C + JOSEPH KB
ACTA PHYS POLON 30: 807-12 (1966)

1406 UREY-BRADLEY FORCE FIELD AND THERMODYNAMIC
PROPERTIES OF PYRAMIDAL XY3 MOLECULES.
VENKATESWARLU K + RAJALAKSHMI KV
INDIAN J PURE APPL PHYS 1: 380-2 (1963)

1407 MEAN AMPLITUDES OF VIBRATION FOR SOME PYRAMIDAL
XY3 AND TETRAHEDRAL XY4 MOLECULES.
VENKATESWARLU K + RAJALAKSHMI KV
+ THANALAKSHMI R
PROC INDIAN ACAD SCI A 58(5): 290-5 (1963)

MX4 MOLECULES (GENERAL) MX$_4$

1408 VIBRATIONAL MEAN SQUARE AMPLITUDE MATRICES.
PART-7: TREATMENT OF TETRAHEDRAL XY4 MOLECULES.
CYVIN SJ
J MOL SPECTROSC 5: 38-43 (1960)

1409 MATHEMATICAL TECHNIQUES IN THE STUDY OF CORIOLIS
COUPLING IN TETRAHEDRAL XY4 MOLECULES.
CYVIN SJ + BRUNVOLL J + CYVIN BN
BULL SOC CHIM BELG 73(1-2): 120-6 (1964)

1410 INFLUENCE OF ATOMIC MASSES ON THE CORIOLIS
COUPLING COEFFICIENTS IN XY4 TETRAHEDRAL
MOLECULES.
CYVIN SJ + BRUNVOLL J + KRISTIANSEN LA
+ MEISINGSETH E
J CHEM PHYS 40: 96-104 (1964)

1411 FREQUENCIES, FORCE CONSTANTS, AND MASS
DEPENDENCE OF THE CORIOLIS CONSTANT FOR SOME
MOLECULES AND IONS OF D4H SYMMETRY.
CYVIN SJ + CYVIN BN
Z NATURFORSCH 23A: 479-81 (1968)

1412 MOLECULES WITH D4H SYMMETRY. NEW METHOD FOR THE
CALCULATION OF GVFF CONSTANTS.
GADINI A + MULLER A
MOL PHYS 12: 145-8 (1967)

1413 ELECTRON CORRELATION AND MOLECULAR SHAPE.
GILLESPIE RJ
CAN J CHEM 38: 818-26 (1960)

1414 ESTIMATION OF THE FORCE CONSTANTS AND
INTERATOMIC DISTANCES OF HALIDE COMPLEX
MOLECULES OF THE XY4 TYPE OF TETRAHEDRAL
SYMMETRY.
KOVRIKOV AB + XUEN VD
J APPL SPECTROSC 14: 1088-92 (1971)

1415 NORMAL COORDINATE TREATMENT OF XY4 TETRAHEDRAL
MOLECULES AND IONS. (GVFF)
KREBS B + MULLER A + FADINI A
J MOL SPECTROSC 24: 198-203 (1967)

1416 CALCULATION OF FORCE CONSTANTS WITH THE MAXIMUM
OVERLAP METHOD. TETRAHEDRAL MOLECULES.
MEZEI M + PULAY P
ACTA CHIM ACAD SCI HUNG 56: 67-73 (1968)

1417 STARK ENERGIES FOR MOLECULES WITH VIBRATIONALLY
INDUCED DIPOLE MOMENTS.
MUENTER AA + DYKE TR
J CHEM PHYS 63: 1224-30 (1975)

1418 NORMAL COORDINATE TREATMENT OF XY4 TETRAHEDRAL
MOLECULES AND IONS. (UREY-BRADLEY AND MVFF)
MULLER A + KREBS B
J MOL SPECTROSC 24: 180-97 (1967)

1419 MEAN AMPLITUDES OF VIBRATION OF SOME XY4
TETRAHEDRAL MOLECULES.
NAGARAJAN G
BULL SOC CHIM BELG 71: 347-60, 361-9 (1962);
INDIAN J PURE APPL PHYS 2: 205-8 (1964)

1420 MEAN AMPLITUDES OF VIBRATION OF SOME TETRAHEDRAL
XY4 TYPE MOLECULES. PART 3 AND 4.
NAGARAJAN G
BULL SOC CHIM BELG 72(11-12): 647-56, 657-65
(1963)

1421 CLASSICAL VIBRATION PROBLEM FOR THE PYRAMIDAL
XY4 MOLECULE.
PISTORIUS CWFT
Z PHYS CHEM 17: 292-9 (1958)

1422 FORCE CONSTANTS OF TETRAHEDRAL MOLECULES.
PISTORIUS CWFT
J CHEM PHYS 28: 514-15 (1958)

1423 USE OF CORIOLIS INTERACTION CONSTANTS IN THE
SOLUTION OF THE VIBRATIONAL PROBLEM FOR XY4
TETRAHEDRAL MOLECULES.
PONOMAREV YI
OPT SPECTROSC 27(4): 327-9 (1969)

1424 MOLECULAR CONSTANTS OF SOME TETRAHEDRAL XY4 TYPE
MOLECULES AND THEIR SUBSTITUTED DERIVATIVES BY
GREEN'S FUNCTION ANALYSIS.
RAMASWAMY K + RANGANATHAN V
INDIAN J PURE APPL PHYS 6(12): 651-5 (1968)

1425 APPLICATION OF UREY-BRADLEY FORCE FIELD TO SOME
TETRAHEDRAL XY4 TYPE MOLECULES AND IONS.
RAO VRA + RAI DK
INDIAN J PURE APPL PHYS 7: 277-9 (1969)

1426 JAHN-TELLER EFFECT. EXPERIMENTAL AND THEORETICAL
STUDIES ON VANADIUM TETRACHLORIDE.
SEIP HM
TIDSSKR KJEMI BERGV MET 28(8-9): 177-83 (1968)

1427 ESTIMATION OF CONFIGURATIONS AND INTERATOMIC
DISTANCES IN GROUP-5 AND GROUP-6 TRANSITION
ELEMENT TETRAHALIDES AND PENTAHALIDES.
SPIRIDONOV VP + ROMANOV GV
VEST MOSK UNIV KHIM 24: 65-8 (1969) (IN RUSSIAN)

SIMPLE METHOD FOR EXACT SOLUTION OF 2 X 2
SECULAR EQUATION IN MOLECULAR VIBRATIONS.
THYAGARAJAN G + SUBHEDAR MK
INDIAN J PURE APPL PHYS 12(4): 309-11 (1974)

1428 POTENTIAL CONSTANTS OF XY4 AND XY3Z TYPES OF
MOLECULES AND RADICALS.
VENKATESWARLU K + SOMASUNDARAM V + PILLAI MGK
Z PHYS CHEM 212: 145-8 (1959)

1429 UREY-BRADLEY FORCE FIELD FOR XY4 TETRAHEDRAL
MOLECULES.
VENKATESWARLU K + THANALAKSHMI R
J SCI IND RES (INDIA) 21B: 461-3 (1962)

1430 PROGRESSIVE STIFFNESS METHOD APPLIED TO XY4
MOLECULES IN THE LIGHT OF THE THEORY OF
CHARACTERISTIC FREQUENCIES.
VINOGRADOVA VN + GODNEV IN
SOV PHYS J 8(1): 43-6 (1965)

1431 NEW METHOD FOR RESOLVING WILSON'S SECULAR
EQUATIONS. APPLIED TO TETRAHEDRAL MOLECULES.
WENDLING EJL + MAHMOUDI S
BULL SOC CHIM FR 1970(12): 4248-54 (IN FRENCH)

MX5 MOLECULES (GENERAL) MX$_5$

1432 GENERALIZED MEAN SQUARE AMPLITUDES OF VIBRATION,
AND SHRINKAGE EFFECTS OF SOME TRIGONAL
BIPYRAMIDAL XY5 AND XY3Z2 MOLECULES.
BRUNVOLL J
ACTA CHEM SCAND 21: 473-80 (1967)

1433 NORMAL COORDINATE ANALYSIS OF TRIGONAL
BIPYRAMIDAL XY5 MOLECULES.
CONDRATE RA + NAKAMOTO K
BULL CHEM SOC JAP 39: 1108-13 (1966)

1434 THERMODYNAMIC PROPERTIES OF PENTAFLUORIDES AT
HIGH TEMPERATURES. PART-2: PENTAFLUORIDES OF
SOME 4P- AND 4D ELEMENTS.
GALKIN NP + TUMANOV YN + BUTYLKIN YP
RUSS J PHYS CHEM 44(11): 1516-8 (1970)

ELECTRON CORRELATION AND MOLECULAR SHAPE.
GILLESPIE RJ
CAN J CHEM 38: 818-26 (1960)

1435 BOND LENGTHS AND BOND ANGLES IN OCTAHEDRAL,
TRIGONAL BIPYRAMID, AND RELATED MOLECULES OF THE
NONTRANSITION ELEMENTS.
GILLESPIE RJ
CAN J CHEM 39: 318-23 (1961)

1436 POTENTIAL FIELD AND MOLECULAR VIBRATIONS OF THE
TRIGONAL BIPYRAMID AX(3)YZ MOLECULE.
HOLMES RR
J CHEM PHYS 46: 3274-29 (1967)

1437 POTENTIAL CONSTANTS, MEAN AMPLITUDES OF
VIBRATION, THERMODYNAMIC FUNCTIONS AND MOLECULAR
POLARIZABILITES OF SOME PENTAHALIDES OF TRIGONAL
BIPYRAMIDAL SYMMTERY.
NAGARAJAN G
INDIAN J PURE APPL PHYS 4: 151-7 (1966)

1438 THERMODYNAMIC FUNCTIONS OF SOME TRIGONAL
BIPYRAMIDAL PENTAHALIDES.
NAGARAJAN G
BULL SOC CHIM BELG 71: 324-8 (1962)

1439 POTENTIAL CONSTANTS OF SOME XY5 MOLECULES.
RADHAKRISHNAN M
INDIAN J PURE APPL PHYS 1: 437-8 (1963)

ESTIMATION OF CONFIGURATIONS AND INTERATOMIC
DISTANCES IN GROUP-5 AND GROUP-6 TRANSITION
ELEMENT TETRAHALIDES AND PENTAHALIDES.
SPIRIDONOV VP + ROMANOV GV
VEST MOSK UNIV KHIM 24: 65-8 (1969) (IN RUSSIAN)

1440 FURTHER EVIDENCE OF STEREOCHEMICAL NONRIGIDITY
IN FIVE AND SEVEN COORDINATE STRUCTURES.
TEBBE FN + MUETTERTIES EL
INORG CHEM 7: 172-4 (1968)

MX6 MOLECULES (GENERAL)

1441 APPLICATION OF THE METHOD OF PROGRESSIVE
RIGIDITY TO OCTAHEDRAL MX6 MOLECULES.
ALEKSANDROVSKAYA AM + VINOGRADOVA VN + GODNEV IN
OPT SPECTROSC SUPPL 3: 64 (1967)

1442 ROTATIONAL PROFILES OF VIBRATIONAL BANDS IN THE
GAS PHASE RAMAN SPECTRA OF MOLECULES OF THE
SPHERICAL TOP TYPE.
ALIEV MR
OPT SPECTROSC 31(3): 202-3 (SEP 1971)

1443 MOLECULAR FORCE FIELDS OF XY6 MOLECULES.
AWASTHI MN + MEHTA ML
SPECTROSC LETT 2: 327-31 (1969)

1444 VIBRATIONAL ANALYSIS OF SUBSTITUTED AND
PERTURBED MOLECULES. DIRECT DETERMINATION FORCE
CONSTANTS FROM THE CORIOLIS INTERACTION.
BASS CD + MARGOLIS JS
BULL CHEM SOC JAP 39(12): 2770-1 (1966)

1445 FORCE CONSTANTS FOR XY6 MOLECULES.
CALIFANO NS + GIORDANI SF
ATTI ACCAD NAC LINCEI REND SCI FIS MAT NAT 25:
284-91 (1958)

1446 FORCE CONSTANTS OF METAL HEXAFLUORIDES
CLAASSEN HH
J CHEM PHYS 30: 968-72 (1959)

STEREOCHEMISTRY AND SEVEN COORDINATION. (XENON
HEXAFLUORIDE)
CLAXTON TA + BENSON GC
CAN J CHEM 44: 157-63, 1730-1 (1966)

1447 REVISED MEAN AMPLITUDES OF VIBRATION FOR SOME
OCTAHEDRAL HEXAFLUORIDES.
CYVIN SJ + BRUNVOLL J + MULLER A
ACTA CHEM SCAND 22: 2739-41 (1968)

ELECTRON CORRELATION AND MOLECULAR SHAPE.
GILLESPIE RJ
CAN J CHEM 38: 818-26 (1960)

BOND LENGTHS AND BOND ANGLES IN OCTAHEDRAL,
TRIGONAL BIPYRAMID, AND RELATED MOLECULES OF THE
NONTRANSITON ELEMENTS.
GILLESPIE RJ
CAN J CHEM 39: 318-23 (1961)

1448 CALCULATION OF THE FORCE CONSTANTS OF
HEXAFLUORIDES.
GODNEV IN + VINOGRADOVA VN + ALEKSANDROVSKAYA AM
OPT SPECTROSC 26(6): 576-8 (JUN 1969)

1449 STATISTICAL THERMODYNAMIC PROPERTIES OF
HEXAFLUORIDE MOLECULES.
JACKSON D
LOS ALAMOS SCI LAB, 1975, 38P. LA-6025-MS

1450 MEAN SQUARE AMPLITUDES OF INTERATOMIC DISTANCES
IN HEXAFLUORIDE MOLECULES.
KIMURA M + KIMURA K
J MOL SPECTROSC 11: 368-77 (1963)

1451 JAHN-TELLER EFFECT IN HEXAFLUORIDES.
KISELJOV AA
J PHYS B 2: 270-3 (1969)

1452 MOLECULAR FORCE FIELDS FOR SOME HEXAFLUORIDES.
LINNETT JW + SIMPSON CJSM
TRANS FARADAY SOC 55: 857-66 (1959)

1453 RAMAN INTENSITIES FOR SPHERICALLY SYMMETRIC
MODES OF SOME XY4 TETRAHEDRAL AND XY6 OCTAHEDRAL
MOLECULES.
LONG DA + THOMAS EL
TRANS FARADAY SOC 59: 1026-32 (1963)

1454 NORMAL COORDINATE ANALYSIS OF ROTATION VIBRATION
OF OCTAHEDRAL XY6 AND Z6 MOLECULAR MODELS WITH
APPLICATION TO 15 XY6 MOLECULES.
MEISINGSETH E + BRUNVOLL J + CYVIN SJ
KGL NORSKE VIDENSKAB SELSKABS SKRIFTER NO 7: 49
(1964)

1455 MEAN AMPLITUDES OF VIBRATION AND
BASTIANSEN-MORINO SHRINKAGE EFFECTS IN SOME
OCTAHEDRAL XY6 MOLECULES.
MEISINGSETH E + CYVIN SJ
ACTA CHEM SCAND 16: 2452-3 (1962)

1456 FORCE FIELD OF WEINSTOCK AND GOODMAN FOR
OCTAHEDRAL HEXAFLUORIDES AND THE L MATRIX
METHOD.
MUELLER A + CYVIN SJ + BRUNVOLL J
J MOL SPECTROSC 30(1): 157-9 (1969)

1457 FORCE CONSTANTS FOR HEXAFLUORIDE MOLECULES.
MUELLER A + FADINI A + PEACOCK CJ
Z PHYS CHEM 238(1-2): 17-21 (1968)

1458 INFLUENCE OF ATOMIC MASS ON THE CORIOLIS
CONSTANTS OF XY6 MOLECULES.
MULLER A + KREBS B + CYVIN SJ
NATURWISSENSCHAFTEN 55: 34 (1968)

1459 MEAN AMPLITUDES OF VIBRATION AND THERMODYNAMIC
FUNCTIONS FOR SOME METAL HEXAFLUORIDES.
NAGARAJAN G
INDIAN J PURE APPL PHYS 2: 87-90 (1964)

1460 MEAN AMPLITUDES OF THERMAL MOTION AND SHRINKAGES
OF CHEMICAL BONDS: OCTAHEDRAL HEXAFLUORIDES.
NAGARAJAN G
INDIAN J PURE APPL PHYS 4(6): 237-43 (1966)

1461 MEAN AMPLITUDES OF VIBRATION OF SOME XY6
MOLECULES.
NAGARAJAN G
BULL SOC CHIM BELG 72: 537-59 (1963)

1462 POTENTIAL FIELD AND FORCE CONSTANTS FOR SOME XY6
MOLECULES.
NAGARAJAN G
BULL SOC CHIM BELG 71: 276-85 (1963)

1463 STATISTICAL THERMODYNAMICS OF SOME
HEXAFLUORIDES.
NAGARAJAN G + BRINKLEY DC
Z NATURFORSCH 26A: 1658-66 (1971)

1464 POTENTIAL FIELD AND FORCE CONSTANTS OF
OCTAHEDRAL MOLECULES.
PISTORIUS CWFT
J CHEM PHYS 29: 1328-32 (1958)

1465 MOLECULAR FORCE FIELD FOR SOME XY6 TYPE
MOLECULES. (SELENIUM AND TELLURIUM
HEXAFLUORIDES)
RAMASWAMY K + JAYARAMAN L
ACTA CHIM ACAD SCI HUNG 87(1): 7-13 (1975)

1466 MEAN SQUARE AMPLITUDES OF INTERATOMIC DISTANCES
IN SOME OCTAHEDRAL HEXAFLUORIDE MOLECULES.
SINGH ON + RAI DK
CAN J PHYS 43: 378-82 (1965)

1467 MEAN AMPLITUDES OF THERMAL VIBRATIONS AND
THERMODYNAMIC PROPERTIES OF METAL HEXAFLUORIDES.
SUNDARAM S
Z PHYS CHEM NEUE FOLGE 34: 225-32 (1962)

1468 EXTREMAL PROPERTIES OF FORCE CONSTANTS IN
OCTAHEDRAL HEXAFLUORIDES.
THAKUR SN + RAO VRA + RAI DK
INDIAN J PURE APPL PHYS 8: 196-8 (1970)

1469 UREY-BRADLEY FORCE FIELD OF SOME XY6 SYSTEMS.
VENKATESWARLU K + DEVI VM
CURR SCI 37: 371-2 (1968)

1470 EVALUATION OF FORCE CONSTANTS: MOLECULES OF THE
TYPE XY6.
VENKATESWARLU K + SUNDARAM S
Z PHYS CHEM NEUE FOLGE 9: 174-9 (1956)

1471 JAHN-TELLER EFFECT IN VIBRATIONAL SPECTRA OF
HEXAFLUORIDES.
WEINSTOCK B + CLAASSEN HH
J CHEM PHYS 31: 262-3 (1959)

1472 FORCE CONSTANTS OF OCTAHEDRAL MOLECULES
CALCULATED BY THE LOGARITHMIC STEP METHOD.
WENDLING EJL + MAHMOUDI S
COMPT REND C 275(20): 1141-3 (1972) (IN FRENCH)

MX7 MOLECULES (GENERAL) MX_7

ELECTRON CORRELATION AND MOLECULAR SHAPE.
GILLESPIE RJ
CAN J CHEM 38: 818-26 (1960)

FURTHER EVIDENCE OF STEREOCHEMICAL NONRIGIDITY
IN FIVE AND SEVEN COORDINATE STRUCTURES.
TEBBE FN + MUETTERTIES EL
INORG CHEM 7: 172-4 (1968)

MX8 MOLECULES (GENERAL) MX_8

ELECTRON CORRELATION AND MOLECULAR SHAPE.
GILLESPIE RJ
CAN J CHEM 38: 818-26 (1960)

MX9 MOLECULES (GENERAL) MX_9

ELECTRON CORRELATION AND MOLECULAR SHAPE.
GILLESPIE RJ
CAN J CHEM 38: 818-26 (1960)

VIBRATIONAL ANALYSIS (GENERAL)

1473 FORCE CONSTANTS AND NORMAL COORDINATE ANALYSIS.
BECHER HJ + BALLEIN K
Z PHYS CHEM NEUE FOLGE 54: 302-18 (1967)

1474 UREY-BRADLEY POTENTIAL CONSTANTS FOR SECOND ROW
FLUORIDES.
BROWN RD + PEEL JB
AUST J CHEM 21: 2361-5 (1968)

1475 BICENTRIC RESCALING OF CNDO/2 THEORY.
APPLICATIONS TO INORGANIC FLUORIDES.
COMPANION AL
J PHYS CHEM 77(26): 3085 (1973)

1476 BASTIANSEN-MORINO SHRINKAGE EFFECT IN THE STUDY
OF MOLECULAR STRUCTURE.
CYVIN SJ
TIDDSKR KJEM BERGVESEN OG METALLURIG 22: 73-9
(1962) (IN DUTCH)

1477 THEORY AND CALCULATION OF CENTRIFUGAL DISTORTION
CONSTANTS FOR POLYATOMIC MOLECULES.
CYVIN SJ + CYVIN BN + HAGEN G
Z NATURFORSCH 23A: 1649-55 (1968)

1478 LIMITATIONS OF THE UREY-BRADLEY MODEL IN NORMAL
COORDINATE ANALYSES.
DUNCAN JL
J MOL SPECTROSC 18: 62-72 (1965)

1479 FAR INFRARED SPECTRA OF PROLATE AND OBLATE
SYMMETRIC TOPS.
GREFFITHS PR + THOMPSON HW
SPECTROCHIM ACTA 24A: 1325-36 (1968)

1480 VIBRONIC AND SPIN ORBIT SPLITTING IN SPECTRA OF
SYSTEMS EXHIBITING JAHN-TELLER EFFECT.
HABITZ P + SCHWARZ WHE
THEOR CHIM ACTA 28(3): 267-82 (1973)

1481 POTENTIAL CONSTANTS OF PHOSPHORUS TRICHLORIDE
FROM AMPLITUDES OF VIBRATION AND NORMAL
VIBRATIONAL FREQUENCIES.
IWASAKI M + HEDBERG K
J CHEM PHYS 36: 594-8 (1962)

1482 FORMULAS FOR CORIOLIS CONSTANTS.
KEBABCIOGLU R + MULLER A
Z NATURFORSCH 23A: 1310-12 (1968)

1483 QUADRATIC POTENTIAL FUNCTION FOR AMMONIA.
(UREY-BRADLEY)
KING WT
J CHEM PHYS 36: 165-70 (1962)

1484 USE OF A SIMPLE ELECTROSTATIC MODEL FOR
CALCULATION OF ANGULAR FORCE CONSTANTS OF HIGHLY
SYMMETRICAL MOLECULES.
KOVRIKOV AB + PAHN DK
J APPL SPECTROSC 14(5): 662-4 (1971)

1485 TORSIONAL COORDINATES IN VIBRATIONAL
ANHARMONICITY. APPLICATION TO ETHYLENE.
MACHIDA K
J CHEM PHYS 44: 4186-94 (1966)

1486 CALCULATION OF MEAN SQUARE VIBRATIONAL
AMPLITUDES.
MAIOROV AN + GODNEV IN
J APPL SPECTROSC 6(4): 345-7 (1967)

1487 CORIOLIS ZETA SUMS FOR SOME SPHERICAL TOP
MOLECULES.
MCDOWELL RS
J CHEM PHYS 41(8): 2557-8 (1964)

1488 THEORETICAL INTERPRETATION OF THE MASS INFLUENCE
ON CORIOLIS COUPLING CONSTANTS IN FOUR XYN TYPE
MOLECULE MODELS.
MULLER A + KREBS B + CYVIN SJ
MOL PHYS 14: 491-4 (1968)

1489   SHRINKAGES OF THE INTERNUCLEAR DISTANCES BY
       MOLECULAR VIBRATIONS.
       NAGARAJAN G + LIPPINCOTT ER
       J CHEM PHYS 42: 1809-18 (1965)

1490   VIBRATION ROTATION ENERGIES OF MOLECULES.
       NIELSIN HH
       REV MOD PHYS 23: 90-136 (1951)

1491   LEAST SQUARES ADJUSTMENT OF ANHARMONIC POTENTIAL
       CONSTANTS. APPLICATION TO CARBON DIOXIDE.
       PARISEAU MA + SUZUKI I + OVEREND J
       J CHEM PHYS 42: 2335-44 (1965)

1492   ADAPTATION OF THE UREY-BRADLEY FORCE FIELD TO
       AMMONIA, PHOSPHINE, STIBINE, AND ARSINE.
       PARISEAU MA + WU E + OVEREND J
       J CHEM PHYS 39: 217-23 (1963)

1493   VIBRATIONAL ANHARMONICITY IN THE METHYL HALIDES.
       REICHMAN S + OVEREND J
       J CHEM PHYS 48: 3095-102 (1968)

1494   CLASSIFICATION OF NORMAL MODES OF VIBRATION.
       PART-2: INTERACTION TERMS IN ENERGY
       DISTRIBUTIONS. (SELENIUM AND TELLURIUM
       FLUORIDES)
       RYTTER E
       ACTA CHIM ACAD SCI HUNG 85(2): 147-51 (1975)

1495   FORCE CONSTANTS CF SMALL MCLECULES.
       (UREY-BRADLEY)
       SHIMANOUCHI T
       PURE APPL CHEM 7(1): 131-45 (1963)

1496   FORCE CONSTANTS CF MOLECULES WITH STRONGLY
       COUPLED VIBRATIONS.
       THAKUR SN + RAI SN
       J MOL STRUCT 5: 320-22 (1970)

1497   VIBRATIONS OF PENTATOMIC TETRAHEDRAL MCLECULES.
       UREY HC + BRADLEY CA
       PHYS REV 38: 1969-78 (1931)

1498   POTENTIAL CONSTANTS OF POLYATOMIC MOLECULES.
       PART-1.
       VENKATESWARLU K + THANALAKSHMI R
       J ANNAMALAI UNIV 24(PT B): 13-37 (1962)

PERMUTED TITLE INDEX

A

```
G-KLEIN-REES FRANCK-CONDON FACTORS AND R CENTROIDS OF THE A-X BAND SYSTEM OF BERYLLIUM FLUORIDE.          RYDBER 0127
                 VIBRATIONAL TRANSITION PROBABILITIES FOR THE A-X SYSTEM OF BORON MONOFLUORIDE.                         0168
                             ROTATIONAL ANALYSIS OF A-X1 BANDS OF LEAD MONOFLUORIDE.                                    0591
RMINATION OF THE EFFECTIVE VIBRATIONAL TEMPERATURE IN THE A-2 SYSTEM OF ANTIMONY MONOFLUORIDE.              DETE 0061
UORIDE.                                                   AB INITIO AND MZDO WAVE FUNCTIONS FOR SULFUR HEXAFL 1066
            ELECTRONIC PROPERTIES OF SULFUR HEXAFLUORIDE. AB INITIO CALCULATION FOR THE GROUND STATE.                   1077
   QUAD/ SELF CONSISTENT MOLECULAR ORBITAL METHODS. PART-5: AB INITIO CALCULATION OF EQUILIBRIUM GEOMETRIES AND 0720
N TRICHLORIDE.                                            AB INITIO CALCULATION ON BORON TRIFLUORIDE AND BORO 0213
FLUORIDE.                                                 AB INITIO CALCULATIONS OF THE BONDING IN KRYPTON DI 0553
   PHOSPHORUS TRIFLUORIDE AND TRIMETHYL PHOSPHINE.        AB INITIO CALCULATIONS OF THE BONDING IN PHOSPHINE, 0802
                                            COMPARATIVE   AB INITIO CALCULATIONS ON BORON FLUORIDES.                    0203
E UNRESTRICTED HARTREE-FOCK METHOD. (BORON DIFLUORIDE, N/ AB INITIO CALCULATIONS ON SOME SMALL RADICALS BY TH 0176
ND COUPLING CONSTANTS OF BORON DIFLUORIDE.               AB INITIO INVESTIGATION OF THE GEOMETRY, BONDING, A 0180
F INTERHALOGENS. (CHLORINE MONOFLUORIDE, CHLORINE TRIFLU/ AB INITIO MOLECULAR ORBITAL STUDY OF THE GEOMETRY O 0379
IFLUORIDE AND DINITROGEN TETRAFLUORIDE MOLECULES, AND ES/ AB INITIO MOLECULAR ORBITAL STUDY OF THE NITROGEN D 0688
DIFLUORIDE RADICAL.                                       AB INITIO MOLECULAR ORBITAL STUDY OF THE PHOSPHORUS 0790
AL ELEMENTS- MO FORMALISM. PART-2: DIRECT COMPARISON WITH AB INITIO RESULTS. /ITAL THEORY. ESSENTIAL STRUCTUR 0765
   IN PHOSPHORUS TRIFLUORIDE AND PHOSPHORUS OXYFLUORIDE. AN AB INITIO SCF STUDY.                         BONDING 0820
/F THE ELECTRONIC STRUCTURE OF ORGANIC COMPOUNDS. PART-3: AB INITIO STUDIES OF CHARGE DISTRIBUTION USING A M/ 0707
ORON TRIHYDRIDE, BORON DIHYDRIDE FLUORIDE, BORON HYDRIDE/ AB INITIO STUDIES OF THE ELECTRONIC STRUCTURES OF B 0241
   MOLECULES NITROGEN OXYFLUORIDE AND NITROGEN TRIFLUORIDE. AB INITIO STUDY OF THE FORCE CONSTANTS OF INORGANIC 0734
                                                          ABSENCE OF FERMI RESONANCE IN KRYPTON DIFLUORIDE.   0564
DE.                                                       ABSOLUTE INFRARED INTENSITIES OF CARBON TETRAFLUORI 0340
RIDE.                   MEASUREMENT AND CALCULATION OF THE ABSOLUTE INFRARED INTENSITIES OF PHOSPHORUS TRIFLUO 0808
BRATIONS OF NITROGEN TRIFLUORIDE.                         ABSOLUTE INFRARED INTENSITIES OF THE FUNDAMENTAL VI 0736
LATED XENON MONOFLUORIDE AND KRYPTON MONOFLUORIDE.        ABSORPTION AND EMISSION SPECTRA OF ARGON MATRIX ISO 0544
FLUORIDE.                                                 ABSORPTION AND FLUORESCENCE SPECTRA OF EUROPIUM TRI 0453
/RONIC PROPERTIES OF SULFUR HEXAFLUORIDE. PART-1: OPTICAL ABSORPTION AND X-RAY EMISSION FROM SCF MO LCAO COM/ 1078
ICON DIFLUORIDE.                ROTATIONAL ANALYSIS OF    ABSORPTION BANDS IN THE 2266 ANGSTROM SYSTEM OF SIL 0979
LUORIDE.                           LOW TEMPERATURE        ABSORPTION CONTOUR OF THE NU-3 BAND OF SULFUR HEXAF 1082
                                    FAR INFRARED          ABSORPTION IN GASEOUS CARBON TETRAFLUORIDE.          0356
   TETRAFLUORIDE, AND SULFUR HE/ FLUORINE RELAXATION BY NMR ABSORPTION IN GASEOUS CARBON TETRAFLUORIDE, SILICON 0346
                                       X-RAY              ABSORPTION IN SULFUR HEXAFLUORIDE.                   1067
E. RAMAN SCATTERING AND INFRARED, VISIBLE AND ULTRAVIOLET ABSORPTION IN THE VAPOR AND IN MATRIX ISOLATION. /D 1285
ULTRAVIOLET.                                              ABSORPTION OF BORON TRIFLUORIDE GAS IN THE EXTREME 0215
M MONOFLUORIDE, INDIUM MONOFLUORIDE, AND THALL/ MICROWAVE ABSORPTION SPECTRA OF ALUMINUM MONOFLUORIDE, GALLIU 0025
                            ULTRASOFT X-RAY              ABSORPTION SPECTRA OF BORON TRIFLUORIDE.             0204
                     VACUUM ULTRAVIOLET                   ABSORPTION SPECTRA OF BORON TRIHALIDES.              0237
N DIFLUORIDE.                          PREDISSOCIATION IN THE ABSORPTION SPECTRA OF CARBON MONOFLUORIDE AND CARBO 0293
   ULTRAVIOLET REGION. (KRYPTON DIFLUORIDE, NITROGEN TRIFL/ ABSORPTION SPECTRA OF CERTAIN FLUORIDES IN THE NEAR 0560
RIFLUORIDE, AND CHLORINE PENTAFLUORIDE MOLECULES.         ABSORPTION SPECTRA OF CHLORINE FLUORIDE, CHLORINE T 0388
              SULFUR K AND L AND FLUORINE K X-RAY EMISSION AND ABSORPTION SPECTRA OF GASEOUS SULFUR HEXAFLUORIDE.    1088
AD DIFLUORIDE.                           ULTRAVIOLET       ABSORPTION SPECTRA OF GASEOUS TIN DIFLUORIDE AND LE 0595
LUORIDE FROM 18 TO 50 GC.                                 ABSORPTION SPECTRA OF MATRIX ISOLATED NITROGEN TRIF 0698
DE) EVIDENCE OF EFFECTIVE POTENTIAL BARRIERS IN THE X-RAY ABSORPTION SPECTRA OF MOLECULES. (SULFUR HEXAFLUORI 1073
LUORIDES.                    RAMAN EFFECT AND ULTRAVIOLET ABSORPTION SPECTRA OF MOLYBDENUM AND TUNGSTEN HEXAF 0642
AMERICIUM TETRAFLUORIDE, CURIUM TRIFLUORIDE AND CURIUM T/ ABSORPTION SPECTRA OF SOLID AMERICIUM TRIFLUORIDE,   0445
ETHANE. (CARBON TETRAFLUORIDE) CORREL/ VACUUM ULTRAVIOLET ABSORPTION SPECTRA OF SOME HALOGEN DERIVATIVES OF M 0365
ILICON TETRAFLUORIDE, CARBON TETRAFLUORIDE, BCR/ INFRARED ABSORPTION SPECTRA OF SOME POLYATOMIC FLUORIDES. (S 0186
   HALIDES OF GROUP-5 ELEMENTS. (PHOSPHORUS TRIFLUORIDE)   ABSORPTION SPECTRA OF THE HYDRIDES, DEUTERIDES, AND 0806
OPERTIES OF PLUTONIUM HEXAFLUORIDE. (MOLECULAR STRUCTURE, ABSORPTION SPECTRUM)                             PR 0894
RIFLUORIDE.                        VACUUM ULTRAVIOLET     ABSORPTION SPECTRUM AND DIPOLE MOMENT OF NITROGEN T 0725
UM HEXAFLUORIDE.                                          ABSORPTION SPECTRUM AND MAGNETIC PROPERTIES OF OSMI 0751
OF ARSENIC MONOFLUORIDE.           ANALYSIS OF THE        ABSORPTION SPECTRUM AND OF TWO NEW SINGLET SYSTEMS 0090
                                   INFRARED               ABSORPTION SPECTRUM OF BORON TRIFLUORIDE.            0206
                          2500 ANGSTROM                   ABSORPTION SPECTRUM OF CARBON DIFLUORIDE.            0302
                                                          ABSORPTION SPECTRUM OF CARBON DIFLUORIDE.            0306
BRATIONAL ANALYSIS.                                       ABSORPTION SPECTRUM OF CARBON DIFLUORIDE AND ITS VI 0303
   AN ARGON MATRIX.                                       ABSORPTION SPECTRUM OF CARBON DIFLUORIDE TRAPPED IN 0297
                                                          ABSORPTION SPECTRUM OF GASEOUS HOLMIUM MONOFLUORIDE 0488
                                                          ABSORPTION SPECTRUM OF GASEOUS SILVER MONOFLUORIDE. 1012
AND URANIUM HEXAFLUORIDE.                        ELECTRONIC ABSORPTION SPECTRUM OF GASEOUS URANIUM HEXACHLORIDE 1194
                     ULTRAVIOLET                          ABSORPTION SPECTRUM OF GERMANIUM DIFLUORIDE.         0474
                     INFRARED                             ABSORPTION SPECTRUM OF LIQUID BORON TRIFLUORIDE.     0245
                                                          ABSORPTION SPECTRUM OF NITROGEN DIFLUORIDE.          0685
                                                          ABSORPTION SPECTRUM OF PLUTONIUM HEXAFLUORIDE.       0892
                          ULTRAVIOLET                     ABSORPTION SPECTRUM OF SILICON DIFLUORIDE.           0984
                          NEAR ULTRAVIOLET                ABSORPTION SPECTRUM OF SILVER MONOFLUORIDE.          1013
                          ULTRASOFT X-RAY                 ABSORPTION SPECTRUM OF SULFUR HEXAFLUORIDE.          1102
ECULE IN THE ULTRAVIOLET REGION.                          ABSORPTION SPECTRUM OF THE BISMUTH MONOFLUORIDE MOL 0143
CAL.                                   ELECTRONIC         ABSORPTION SPECTRUM OF THE NITROGEN DIFLUORIDE RADI 0694
FLUORIDE AT 10.6 MICROMETERS.      TEMPERATURE DEPENDENT  ABSORPTION SPECTRUM OF THE NU-3 BAND OF SULFUR HEXA 1091
00 TO 4200 ANGSTROMS.                                     ABSORPTION SPECTRUM OF URANIUM HEXAFLUORIDE FROM 20 1191
ERGY LEVELS OF URANIUM(IV).                               ABSORPTION SPECTRUM OF URANIUM TETRAFLUORIDE AND EN 1168
IDE.                                                      ABSORPTION SPECTRUM OF VAPORIZED TITANIUM MONOFLUOR 1143
SEARCH SPECTROMETER FOR FREE RADICAL MICROWAVE ROTATIONAL ABSORPTION STUDIES. (NITROGEN DIFLUORIDE) A CAVITY 0689
UORO PERIODATES(VII).                                     ACCEPTOR PROPERTIES OF IODINE HEPTAFLUORIDE. OCTAFL 0514
VED QUANTITIES. (TIN FLUORIDE)                            ACCURACY OF THE VIBRATIONAL WAVE FUNCTIONS AND DERI 1134
TUTION. PART-2: BORON TRIFLUORIDE, SILICON TETRAFLUORIDE/ ACCURATE FORCE CONSTANTS FROM HEAVY ISOTOPIC SUBSTI 0232
TUTION. PART-1. CARBON TETRAFLUORIDE.                     ACCURATE FORCE CONSTANTS FROM HEAVY ISOTOPIC SUBSTI 0320
HORIUM, PROTACTINIUM, URANIUM, NEPTUNIUM, AND AMERICIUM. (ACTINIDE) HEATS OF FORMATION OF SOME HALIDES OF T 1335
                          FLUORIDES OF THE               ACTINIDE ELEMENTS.                                  0006
VAPOR PRESSURES AND THERMODYNAMIC CALCULATIONS FOR SOME ACTINIDE FLUORIDES                                  1336
                  PAIRING ENERGIES IN THE               ACTINIDE SERIES.                                     1333
              LATTICE CONSTANTS OF                      ACTINIDE TETRAFLUORIDES INCLUDING BERKELIUM.          1334
              HALIDES OF THE LANTHANIDES AND            ACTINIDES.                                            1348
ND THIRD ROW TRANSITION METALS. VOLUME-3: LANTHANIDES AND ACTINIDES. /W TRANSITION METALS. VOLUME-2: SECOND A 0001
```

IUM HEXAFLUORIDE, AND PLUTONI/ LONG WAVE LENGTH, INFRARED ACTIVE FUNDAMENTAL FOR URANIUM HEXAFLUORIDE, NEPTUN 1192
/D FREE ENERGIES OF FORMATION OF PROTACTINIUM HALIDES AND ACTIVITY COEFFICIENTS FOR PROTACTINIUM IN LIQUID B/ 0900
IA, PHOSPHINE, STIBINE, AND ARSINE. ADAPTATION OF THE UREY-BRADLEY FORCE FIELD TO AMMON 1492
RAMAN SPECTRA OF SOLID XENON TETRAFLUORIDE AND ITS ADDUCT WITH XENON DIFLUORIDE. 1258
CATION TO CARBON DIOXIDE. LEAST SQUARES ADJUSTMENT OF ANHARMONIC POTENTIAL CONSTANTS. APPLI 1491
DISSOCIATION ENERGIES, IONIZATION POTENTIALS AND ELECTRON AFFINITIES OF SOME GERMANIUM FLUORIDE SPECIES. /ND 0464
/FUNCTION, BINDING ENERGY, IONIZATION POTENTIAL, ELECTRON AFFINITY, DIPOLE AND QUADRUPOLE MOMENTS, AND SPECT/ 0756
/FUNCTION, BINDING ENERGY, IONIZATION POTENTIAL, ELECTRON AFFINITY, DIPOLE, QUADRUPOLE MOMENT, QUADRUPOLE MOMENT, AND SP/ 0757
/A. DISSOCIATION ENTHALPY, IONIZATION POTENTIAL, ELECTRON AFFINITY, DIPOLE MOMENT, SPECTROSCOPIC CONSTANTS, / 0094
WITH LOW TEMPERATURE ANOMALIES. ALUMINUM AND ALKALI HALIDE MONOMERS MEAN AMPLITUDES OF VIBRATION 0024
TIUM DIFLUORIDE, CALCIUM DIFLUORIDE) GEOMETRY OF THE ALKALINE EARTH DIHALIDES. (BARIUM DIFLUORIDE, STRON 0126
IDE. GEOMETRY AND VIBRATIONAL SPECTRA OF THE ALKALINE EARTH DIHALIDES. PART-1: MAGNESIUM DIFLUOR 0609
NIQUE. INFRARED SPECTRA OF SOME ALKALINE EARTH HALIDES BY THE MATRIX ISOLATION TECH 0125
E, BERYLLIUM DICHLORIDE, MAGNESI/ INFRARED SPECTRA OF THE ALKALINE EARTH HALIDES. PART-1: BERYLLIUM DIFLUORID 0133
DISSOCIATION ENERGIES OF THE ALKALINE EARTH MONOFLUORIDES. 0119
A SMALL GAUSSIAN BASIS SET FOR NONEMPIRICAL ALL-ELECTRON SCF CALCULATIONS IN IODINE COMPOUNDS. 0495
R EFFECTS IN THE VIBRATION SPECTRA OF BORON FLUORIDE WITH ALTERATIONS IN ITS QUADRATIC VALENCE FORCE FIELD. / 1404
OF VIBRATION WITH LOW TEMPERATURE ANOMALIES. ALUMINUM AND ALKALI HALIDE MONOMERS MEAN AMPLITUDES 0024
HEAT OF ATOMIZATION OF ALUMINUM DIFLUORIDE. 0034
/ORON DIFLUORIDE, CARBON DIFLUORIDE, NITROGEN DIFLUORIDE, ALUMINUM DIFLUORIDE, PHOSPHORUS DIFLUORIDE, AND ZI/ 1384
THERMODYNAMICS OF THE VAPORIZATION OF ALUMINUM FLUORIDE AND CESIUM FLUORIDE. 0052
INFRARED SPECTRA OF ALUMINUM FLUORIDE IN SOLID ARGON. 0054
ZEEMAN EFFECT IN THE MICROWAVE ROTATIONAL SPECTRUM OF THE ALUMINUM FLUORIDE MOLECULE. 0026
INFRARED SPECTRA OF MAGNESIUM AND ALUMINUM FLUORIDES BY MATRIX ISOLATION. 0608
ISSOCIATION ENERGIES OF THE GASEOUS MONOHALIDES OF BORON, ALUMINUM, GALLIUM, INDIUM, AND THALLIUM. D 0152
FRACTION INVESTIGATION OF THE MOLECULAR STRUCTURES OF THE ALUMINUM HALIDES. ELECTRON DIF 0036
GY CURVES AND DISSOCIATION ENERGIES OF DIATOMIC BORON AND ALUMINUM HALIDES. POTENTIAL ENER 0173
MATRIX ISO/ INFRARED SPECTRUM OF ALUMINUM TRIFLUORIDE, DIALUMINUM HEXAFLUORIDE, AND ALUMINUM MONOFLUORIDE BY 0051
ELECTRONIC STATES OF GASEOUS ALUMINUM MONOFLUORIDE. 0027
MICROWAVE SPECTRUM OF ALUMINUM MONOFLUORIDE. 0029
AND SUBMILLETER WAVE SPECTRUM AND MOLECULAR CONSTANTS OF ALUMINUM MONOFLUORIDE. MILLIMETER 0033
SINGLET ELECTRONIC STATES OF ALUMINUM MONOFLUORIDE. 0023
SPECTROSCOPIC AND THERMODYNAMIC PROPERTIES OF ALUMINUM MONOFLUORIDE. 0022
HIGH TEMPERATURE MICROWAVE SPECTROSCOPY OF ALUMINUM MONOFLUORIDE AND ALUMINUM CHLORIDE. 0028
MONOFLUORIDE, AND THALL/ MICROWAVE ABSORPTION SPECTRA OF ALUMINUM MONOFLUORIDE, GALLIUM MONOFLUORIDE, INDIUM 0025
EMISSION SPECTRUM OF ALUMINUM MONOFLUORIDE IN THE VACUUM ULTRAVIOLET. 0030
ALUMINUM MONOFLUORIDE. (VAPOR PRESSURE) 0031
NATION OF THE HEATS OF FORMATION OF THE GASEOUS MOLECULES ALUMINUM OXYDIFLUORIDE AND ALUMINUM DIFLUORIDE. /MI 0035
E BOMB CALORIMETRY. PART-22: THE ENTHALPY OF FORMATION OF ALUMINUM TRIFLUORIDE. FLUORIN 0048
ROMETRIC INVESTIGATION OF THE THERMODYNAMIC PROPERTIES OF ALUMINUM TRIFLUORIDE. MASS SPECT 0040
NSTANTS, SHRINKAGE EFFECT, AND THERMODYNAMIC FUNCTIONS OF ALUMINUM TRIFLUORIDE. /RATION, CORIOLIS COUPLING CO 0050
STANDARD ENTHALPY OF FORMATION OF ALUMINUM TRIFLUORIDE. 0043
VAPOR PRESSURE OF ALUMINUM TRIFLUORIDE. 0041
DIFLUORIDE, STRONTIUM DIFLUORIDE, BARIUM DIFLUORIDE, AND ALUMINUM TRIFLUORIDE. /AGNESIUM DIFLUORIDE, CALCIUM 0049
NUM MONOFLUORIDE. HEAT OF SUBLIMATION OF ALUMINUM TRIFLUORIDE AND HEAT OF FORMATION OF ALUMI 0053
LUMINUM TRIFLUORIDE. SUBLIMATION OF ALUMINUM TRIFLUORIDE AND THE INFRARED SPECTRUM OF A 0037
LEMENTS. HEAT OF FORMATION OF ALUMINUM TRIFLUORIDE BY DIRECT COMBINATION OF THE E 0039
ALUMINUM MONOFLUORIDE BY MATRIX ISO/ INFRARED SPECTRUM OF ALUMINUM TRIFLUORIDE, DIALUMINUM HEXAFLUORIDE, AND 0051
E, AND SODIUM ALUMINUM TETRA/ INFRARED SPECTRA OF GASEOUS ALUMINUM TRIFLUORIDE, LITHIUM ALUMINUM TETRAFLUORID 0045
BOND ENERGY IN THE ALUMINUM TRIFLUORIDE MOLECULE. 0044
HARMONIC FORCE FIELDS AND MEAN AMPLITUDES FOR ALUMINUM TRIFLUORIDE MONOMER AND DIMER. 0038
FEROMETRY IN THE ULTRAVIOLET REGION 1700-2500 ANGSTROMS. (ALUMINUM- MAGNESIUM FLUORIDE) / MULTIPLE BEAM INTER 0042
ECULES. (NITROGEN TRIFLUORIDE, BORON TRIFLUORIDE, PHOSPH/ AMBIGUITIES IN THE HARMONIC FORCE FIELDS OF XY3 MOL 0709
HALIDES OF THORIUM, PROTACTINIUM, URANIUM, NEPTUNIUM, AND AMERICIUM. (ACTINIDE) HEATS OF FORMATION OF SOME 1335
INVESTIGATION OF THE CHEMISTRY OF AMERICIUM. (AMERICIUM TETRAFLUORIDE) 0056
MELTING POINTS OF CURIUM TRIFLUORIDE AND AMERICIUM TRIFLUORIDE. 0446
IUM TRIFLUORIDE AND CURIUM T/ ABSORPTION SPECTRA OF SOLID AMERICIUM TRIFLUORIDE, AMERICIUM TETRAFLUORIDE, CUR 0445
ATS AND FREE ENERGIES OF SUBLIMATION. VAPOR PRESSURES OF AMERICIUM TRIFLUORIDE AND PLUTONIUM TRIFLUORIDE, HE 0055
HOSPHORUS TRIFLUORIDE, PHOSPHORUS PENTAFLUORIDE, DIFLUORO AMINE, AND TETRAFLUORO HYDRAZINE. / STRUCTURES OF P 0801
OTOELECTRON SPECTRA OF COMPLEXES OF BORON TRIFLUORIDE AND AMINES. PH 0224
ERIES CHLORINE MONOFLUORIDE, HYDROGEN OXYCHLORIDE, CHLORO AMMONIA, AND CHLORO METHANE. /F THE ISOELECTRONIC S 0371
SILICON TETRAFLUORIDE, BORON TRIFLUORIDE, ETHANE, SILANE, AMMONIA, AND PHOSPHINE. / OF CARBON TETRAFLUORIDE, 0243
ADAPTATION OF THE UREY-BRADLEY FORCE FIELD TO AMMONIA, PHOSPHINE, STIBINE, AND ARSINE. 1492
QUADRATIC POTENTIAL FUNCTION FOR AMMONIA. (UREY-BRADLEY) 1483
OR / INTERMOLECULAR FLUORINE EXCHANGE BETWEEN TETRABUTYL AMMONIUM HEXAFLUORO PHOSPHATE AND BORON TRIFLUORIDE 0194
VIBRATIONAL MEAN SQUARE AMPLITUDE MATRICES. 1389
VIBRATIONAL MEAN SQUARE AMPLITUDE MATRICES. 1392
FECT OF SYMMETRICAL XY3 AND XY4 / VIBRATIONAL MEAN SQUARE AMPLITUDE MATRICES. (BASTIANSEN-MORINO SHRINKAGE EF 1394
L XY4 MOLECULES. VIBRATIONAL MEAN SQUARE AMPLITUDE MATRICES. PART-7: TREATMENT OF TETRAHEDRA 1408
FLUORI/ ELECTRIC FIELD DEFLECTION OF MOLECULES WITH LARGE AMPLITUDE MOTION. (XENON HEXAFLUORIDE, IODINE HEPTA 1281
CALCULATION OF MEAN SQUARE VIBRATIONAL AMPLITUDES. 1486
E, DIBROMO BORANE AND CHLORINE TRIFLUOR/ GENERALIZED MEAN AMPLITUDES AND CORIOLIS CONSTANTS IN DICHLORO BORAN 0411
XY3 MOLECULES. GENERALIZED MEAN SQUARE AMPLITUDES AND CORIOLIS CONSTANTS OF SOME PYRAMIDAL 1405
OME LINEAR SYMMETRICAL DIFLUORIDES. (OF THE FIRST R/ MEAN AMPLITUDES AND SHRINKAGE EFFECTS OF VIBRATION FOR S 1369
ALIDES. (ARSENIC TRIFLUORIDE) MEAN AMPLITUDES AND THERMODYNAMIC FUNCTIONS OF SOME TRIH 0109
ER. HARMONIC FORCE FIELDS AND MEAN AMPLITUDES FOR ALUMINUM TRIFLUORIDE MONOMER AND DIM 0038
ETRY. (SELENIUM AND TELLURIUM FLUORIDES) ROOT MEAN SQUARE AMPLITUDES IN SOME HEXAFLUORIDES OF OCTAHEDRAL SYMM 0962
MOLECULES. MEAN SQUARE AMPLITUDES OF INTERATOMIC DISTANCES IN HEXAFLUORIDE 0008
MOLECULES. MEAN SQUARE AMPLITUDES OF INTERATOMIC DISTANCES IN HEXAFLUORIDE 1450
RAL HEXAFLUORIDE MOLECULES. MEAN SQUARE AMPLITUDES OF INTERATOMIC DISTANCES IN SOME OCTAHED 1466
E LOGARITHMIC STEP ME/ FORCE CONSTANTS AND MEAN VIBRATION AMPLITUDES OF IODINE HEPTAFLUORIDE CALCULATED BY TH 0535
/LATION OF FORCE CONSTANTS AND ROOT MEAN SQUARE VIBRATION AMPLITUDES OF PHOSPHORUS PENTAFLUORIDE, PHOSPHORUS/ 0826
/TANTS, CENTRIFUGAL STRETCHING CONSTANTS, AND MEAN SQUARE AMPLITUDES OF THE VIBRATIONS OF THE NITROGEN TRIFL/ 0726
ICAL BONDS: OCTAHEDRAL HEXAFLUORIDES. MEAN AMPLITUDES OF THERMAL MOTION AND SHRINKAGES OF CHEM 1460
PROPERTIES OF METAL HEXAFLUORIDES. MEAN AMPLITUDES OF THERMAL VIBRATIONS AND THERMODYNAMIC 1467
KAGE EFFECT IN SOME LINEAR SYMMETRICAL DIHALIDES. (/ MEAN AMPLITUDES OF VIBRATION AND BASTIANSEN-MORINO SHRIN 1372
KAGE EFFECT IN SOME XY3 MOLECULES OF PLANAR TRIGONA/ MEAN AMPLITUDES OF VIBRATION AND BASTIANSEN-MORINO SHRIN 1397
KAGE EFFECTS IN SOME OCTAHEDRAL XY6 MOLECULES. MEAN AMPLITUDES OF VIBRATION AND BASTIANSEN-MORINO SHRIN 1455
PHOSPHORUS PENTAFLUORIDE AND ARSENIC PENTAFLUORIDE. MEAN AMPLITUDES OF VIBRATION AND CORIOLIS CONSTANTS FOR 0840
UENCI/ POTENTIAL CONSTANTS OF PHOSPHORUS TRICHLORIDE FROM AMPLITUDES OF VIBRATION AND NORMAL VIBRATIONAL FREQ 1481

/ CORIOLIS COUPLING COEFFICIENTS, GENERALIZED MEAN SQUARE AMPLITUDES OF VIBRATION AND SHRINKAGE CONSTANTS OF/ 1278
ON TRIHALIDES, CALCULATED FROM UREY-BRADLEY FORCE C/ MEAN AMPLITUDES OF VIBRATION AND SHRINKAGE EFFECT OF BOR 0233
OME TRIGONAL BIPYRAMIDAL XY5 AND/ GENERALIZED MEAN SQUARE AMPLITUDES OF VIBRATION, AND SHRINKAGE EFFECTS OF S 1432
OF SOME GROUP-5 TRIHALIDES. (NITROGEN TRIFLUCRIDE,/ MEAN AMPLITUDES OF VIBRATION AND THERMODYNAMIC FUNCTIONS 0719
FOR SILICON DIFLUORIDE AND OXYGEN DIFLUORIDE.   MEAN AMPLITUDES OF VIBRATION AND THERMODYNAMIC FUNCTIONS 0773
OF TECHNETIUM HEXAFLUORIDE.   MEAN AMPLITUDES OF VIBRATION AND THERMODYNAMIC FUNCTIONS 1104
FOR VANADIUM PENTAFLUORIDE.   MEAN AMPLITUDES OF VIBRATION AND THERMODYNAMIC FUNCTIONS 1216
OF XENON TETRAFLUORIDE.   MEAN AMPLITUDES OF VIBRATION AND THERMODYNAMIC FUNCTIONS 1274
FOR SOME METAL HEXAFLUORIDES.   MEAN AMPLITUDES OF VIBRATION AND THERMODYNAMIC FUNCTIONS 1459
DS FOR PLANAR XY3 MOLECULES.   CALCULATIONS OF MEAN AMPLITUDES OF VIBRATION AND UREY-BRADLEY FORCE FIEL 1399
E EFFECT, AND MOLECULAR POLARIZABILITIES FOR DIHALI/ MEAN AMPLITUDES OF VIBRATION, BASTIANSEN-MORINO SHRINKAG 0124
E EFFECT AND THERMODYNAMIC FUNCTIONS OF KRYPTON DIF/ MEAN AMPLITUDES OF VIBRATION, BASTIANSEN-MORINO SHRINKAG 0562
S, SHRINKAGE EFFECT, AND THERMODYNAMIC FUNCTICNS OF/ MEAN AMPLITUDES OF VIBRATION, CORIOLIS COUPLING CONSTANT 0050
MOLECULE WITH APPLICATION TO PHOSPHORUS PENTAFLUORI/ MEAN AMPLITUDES OF VIBRATION FOR A TRIGONAL BIPYRAMIDAL 0864
ORIDES.   REVISED MEAN AMPLITUDES OF VIBRATION FOR SOME OCTAHEDRAL HEXAFLU 1447
TETRAHEDRAL XY4 MOLECULES.   MEAN AMPLITUDES OF VIBRATION FOR SOME PYRAMIDAL XY3 AND 1407
CULES.   (NITROGEN TRIFLUORIDE, PHOSPHORUS TRIFLUORI/ MEAN AMPLITUDES OF VIBRATION FOR SOME PYRAMIDAL XY3 MOLE 0706
CULES.   (IODINE PENTAFLUORIDE, BROMINE PENTAFLUORID/ MEAN AMPLITUDES OF VIBRATION FOR SOME PYRAMIDAL XY4Z MOL 0421
                                                     MEAN AMPLITUDES OF VIBRATION FOR SULFUR TETRAFLUORIDE. 1038
ARBON DIFLUORIDE, NITROGEN DIFLUORIDE, ALUMINUM DIF/ MEAN AMPLITUDES OF VIBRATION FOR THE BORON DIFLUORIDE, C 1384
Y, BERYLLIUM, AND MAGNESIUM.   MEAN AMPLITUDES OF VIBRATION FOR THE DIHALIDES OF MERCUR 0619
                                      CALCULATED MEAN AMPLITUDES OF VIBRATION IN BORON TRIHALIDES.   0197
S.   (CARBON TETRAFLUORIDE)   ROOT MEAN SQUARE AMPLITUDES OF VIBRATION IN SOME GROUP-4 TETRAHALIDE 0342
S.   ROOT MEAN SQUARE AMPLITUDES OF VIBRATION IN SOME GROUP-4 TETRAHALIDE 1363
2V SYMMETRY.   MEAN AMPLITUDES OF VIBRATION IN SOME XY2Z MOLECULES OF C 1396
ON TETR/ A THEORETICAL EVALUATION OF THE ROOT MEAN SQUARE AMPLITUDES OF VIBRATION IN XENON DIFLUORIDE AND XEN 1257
ORIDE FROM EL/ INTERATOMIC DISTANCES AND ROOT MEAN SQUARE AMPLITUDES OF VIBRATION OF GASEOUS SILICON TETRAFLU 1002
C, PHOSPHORUS, AND VANADIU/ SPECTROSCOPIC STUDIES IN MEAN AMPLITUDES OF VIBRATION OF PENTAFLUORIDES OF ARSENI 0871
LORINE TRIFLUORIDE.   MEAN AMPLITUDES OF VIBRATION OF SELENIUM TRIOXIDE AND CH 0406
IDES.   MEAN AMPLITUDES OF VIBRATION OF SOME LANTHANIDE TRIFLUOR 1354
E MOLECULES. PART-5: TETRAHALIDES OF HAFNIUM AND LE/ MEAN AMPLITUDES OF VIBRATION OF SOME TETRAHEDRAL XY4 TYP 0487
E MOLECULES. PART 3 AND 4.   MEAN AMPLITUDES OF VIBRATION OF SOME TETRAHEDRAL XY4 TYP 1420
/IZED VALENCE FORCE FIELD FORCE CONSTANTS AND MEAN SQUARE AMPLITUDES OF VIBRATION OF SOME TRIGONAL BIPYRAMID/ 0878
ECULES. (SILICON TETRAFLUORIDE)   MEAN AMPLITUDES OF VIBRATION OF SOME XY4 TETRAHEDRAL MOL 1008
ECULES.   MEAN AMPLITUDES OF VIBRATION OF SOME XY4 TETRAHEDRAL MOL 1419
                                                     MEAN AMPLITUDES OF VIBRATION OF SOME XY6 MOLECULES.   1461
LIS CONSTANTS IN IODI/ UREY-BRADLEY FORCE CONSTANTS, MEAN AMPLITUDES OF VIBRATION, SHRINKAGE EFFECT AND CORIO 0513
S CONSTANTS, AND THERMODYNAMIC FUN/ FORCE CONSTANTS, MEAN AMPLITUDES OF VIBRATION, SHRINKAGE EFFECTS, CORIOLI 1219
ULFUR PENTAFLUORO CHLORIDE MOLECULES.   MEAN AMPLITUDES OF VIBRATION: SULFUR TETRAFLUORIDE AND S 1057
D MOLECULAR POLARIZABILITES OF/ POTENTIAL CONSTANTS, MEAN AMPLITUDES OF VIBRATION, THERMODYNAMIC FUNCTIONS AN 1437
IES.   ALUMINUM AND ALKALI HALIDE MONOMERS MEAN AMPLITUDES OF VIBRATION WITH LOW TEMPERATURE ANOMAL L 0024
IMITATIONS OF THE UREY-BRADLEY MODEL IN NORMAL COORDINATE ANALYSES.   1478
OP MOLECUL/ VAPOR PHASE RAMAN SPECTRA, RAMAN BAND CONTOUR ANALYSES, AND CORIOLIS CONSTANTS OF THE SPHERICAL T 1069
VALUES FOR/ VAPOR PHASE RAMAN SPECTRA, RAMAN BAND CONTOUR ANALYSES, CORIOLIS CONSTANTS, FORCE CONSTANTS, AND 0795
/AMAN SPECTRA, MOLECULAR STRUCTURES AND NORMAL COORDINATE ANALYSES OF GERMANIUM DIFLUORIDE, MONOMERIC GERMAN/ 0475
RO METHANE, PHOSPHORUS PENTAFLUORIDE, AND CYCLOPROPANE B/ ANALYSES OF THE VIBRATION ROTATION BANDS OF TRIFLUO 0880
N SUBLIMATION OF LANTHANIDE TRIFLUORIDES.   ANALYSES OF THERMOCHEMICAL ERRORS AND SYSTEMATICS I 1353
/POINTS-ON-A-SPHERE AND EXTENDED HUCKEL MOLECULAR ORBITAL ANALYSES OF TRIGONAL BIPYRAMIDS. (PHOSPHORUS PENTA/ 0828
ORPTION SPECTRUM OF CARBON DIFLUORIDE AND ITS VIBRATIONAL ANALYSIS.   ABS 0303
FORCE CONSTANTS AND NORMAL COORDINATE ANALYSIS.   1473
LES AND THEIR SUBSTITUTED DERIVATIVES BY GREEN'S FUNCTION ANALYSIS. /ANTS OF SOME TETRAHEDRAL XY4 TYPE MOLECU 1424
VIBRATIONAL ASSIGNMENTS AND NORMAL COORDINATE ANALYSIS FOR ANTIMONY PENTAFLUORIDE.   0087
ROTATIONAL ANALYSIS OF A-X1 BANDS OF LEAD MONOFLUORIDE.   0591
YSTEM OF SILICON DIFLUCRIDE.   ROTATIONAL ANALYSIS OF ABSORPTION BANDS IN THE 2266 ANGSTROM S 0979
ROTATIONAL ANALYSIS OF B-X2 SYSTEM OF LEAD-208 MONOFLUORIDE. 0590
PLET PI(1) TO X SINGLET SIGMA(+) SYSTEMS OF G/ ROTATIONAL ANALYSIS OF BANDS OF THE A TRIPLET PI ZERO(+) B TRI 0457
IPLET PI(1) TO X SINGLET SIGMA(+) SYSTEMS OF / ROTATIONAL ANALYSIS OF BANDS OF THE A TRIPLET PI ZERO(+), B TR 1111
SIGMA: A TRIPLET PI SYSTEM OF BORON MONOFLUC/ ROTATIONAL ANALYSIS OF BANDS OF THE C TRIPLET SIGMA, B TRIPLET 0153
F DISSOCIATION IN GERMANIUM MONOFLUORIDE.   VIBRATIONAL ANALYSIS OF C-X AND C'-X' BAND SYSTEMS AND ENERGY O 0470
MOLECULE.   ROTATIONAL ANALYSIS OF C3 SYSTEM OF THE ANTIMONY MONOFLUORIDE 0057
/RT GASES BY THE SCF LCAO MO METHOD.   PART-2: COMPARATIVE ANALYSIS OF DIFFERENT SCHEMES OF NEGLECT OF DIFFER/ 1231
2: IN/ MOLECULAR STRUCTURE OF XENON HEXAFLUORIDE. PART-1: ANALYSIS OF ELECTRON DIFFRACTION INTENSITIES. PART- 1291
ONOFLUORIDE.   ROTATIONAL ANALYSIS OF ELECTRONIC BANDS OF GASEOUS MAGNESIUM M 0600
ORIDE.   ROTATIONAL ANALYSIS OF FOUR SINGLET SYSTEMS OF ARSENIC MONOFLU 0092
ELECTRON DIFFRACTION ANALYSIS OF GALLIUM HALIDES. (GALLIUM TRIFLUORIDE) 0460
ELECTRONIC POPULATION ANALYSIS OF MOLECULAR WAVE FUNCTIONS.   0157
LORIDE.   GREEN'S FUNCTION ANALYSIS OF NITROGEN TRIFLUORIDE AND NITROGEN TRICH 0729
ULAR GEOMETRY.   BORON TRIFLUORIDE. THE ANALYSIS OF NU-2 AND THE DETERMINATION OF THE MOLEC 0209
FLUORIDE DICHLORIDE, AND PHOSPHORUS PE/ NORMAL COORDINATE ANALYSIS OF PHOSPHORUS PENTAFLUORIDE, PHOSPHORUS DI 0877
D Z6 MOLECULAR MODELS WITH APPLICATION/ NORMAL COORDINATE ANALYSIS OF ROTATION VIBRATION OF OCTAHEDRAL XY6 AN 1454
TRUM OF THE LUTETIUM MONOFLUORIDE MOLECULE. ROTATIONAL ANALYSIS OF SELECTED BANDS FROM THE ELECTRONIC SPEC 0599
FLUORIDE.   VIBRATIONAL ANALYSIS OF SELENIUM HEXAFLUORIDE AND TUNGSTEN HEXA 0960
MICROWAVE SPECTRUM, DIPOLE MOMENT, AND STRUCTURE ANALYSIS OF SELENIUM TETRAFLUORIDE.   0957
VIBRATIONAL ASSIGNMENT AND NORMAL COORDINATE ANALYSIS OF SELENIUM TETRAFLUORIDE.   0958
DE.   ROTATIONAL ANALYSIS OF SOME BANDS OF GASEOUS BARIUM MONOFLUORI 0118
ORIDE)   NORMAL COORDINATE ANALYSIS OF SOME NITROGEN HALIDES. (NITROGEN TRIFLU 0730
M OF GASEOUS SCANDIUM MONOFLUORIDE.   ANALYSIS OF SOME SINGLET TRANSITIONS IN THE SPECTRU 0930
RECT DETERMINATION FORCE CONSTANTS FROM THE / VIBRATIONAL ANALYSIS OF SUBSTITUTED AND PERTURBED MOLECULES. DI 1444
ES.   (SELENIUM HEXAFLUORIDE, TELLURIUM / GREEN'S FUNCTION ANALYSIS OF SULFUR HEXAFLUORIDE AND RELATED MOLECUL 1093
INFRARED SPECTRUM AND NORMAL COORDINATE ANALYSIS OF SULFUR TETRAFLUORIDE.   1045
ATIONAL SPECTRUM OF BORON TRIFLUORIDE.   ANALYSIS OF SYMMETRIC ROTOR RAMAN SPECTRA. PURE ROT 0205
B DOUBLET SIGMA(+) TO X DOUBLET PI SYSTEMS O/ ROTATIONAL ANALYSIS OF THE A DOUBLET SIGMA(+) TO X DOUBLET PI, 1133
SINGLET SYSTEMS OF ARSENIC MONOFLUORIDE.   ANALYSIS OF THE ABSORPTION SPECTRUM AND OF TWO NEW 0090
SYSTEMS OF GASEOUS SILVER MONOFLUORIDE.   ROTATIONAL ANALYSIS OF THE AO(+), BO(+), TO X SINGLET SIGMA(+ 1011
ULTRAVIOLET.   ROTATIONAL ANALYSIS OF THE B-X SYSTEM OF LEAD MONOFLUORIDE IN 0586
3 SIGMA SYSTEM OF THE BORON MONOFLUORIDE MOLE/ ROTATIONAL ANALYSIS OF THE BAND AT 14900 CM-1 OF A 3 SIGMA TO 0156
SIGMA. SYSTEM OF CALCIUM FLUORIDE MOLECULE IN EMISSION / ROTATIONAL ANALYSIS OF THE BANDS OF B DOUBLET SIGMA- X DOUBLET 0280
ORIDE MOLECULE.   ROTATIONAL ANALYSIS OF THE BETA BAND SYSTEM OF SILICON MONOFLU 0976
/2 AND G DOUBLET DELTA 5/2- X DOUBLET PI 3/2 / ROTATIONAL ANALYSIS OF THE C DOUBLET DELTA 5/2- X DOUBLET PI 3 1136

91

```
STEM OF INDIUM MONOFLUORIDE.                                 ROTATIONAL ANALYSIS OF THE C SINGLET PI- X SINGLET SIGMA(+) SY 0491
                                    FINE STRUCTURE ANALYSIS OF THE C1 SYSTEM OF ANTIMONY MONOFLUORIDE.       0059
                                              ROTATIONAL ANALYSIS OF THE C1 SYSTEM OF ANTIMONY MONOFLUORIDE.  0069
                                              ROTATIONAL ANALYSIS OF THE C2 SYSTEM OF ANTIMONY MONOFLUORIDE.  0068
                                              ROTATIONAL ANALYSIS OF THE GAMMA BANDS OF SILICON MONOFLUORIDE  0999
LUORIDE.                                                     ANALYSIS OF THE INFRARED SPECTRUM OF CHLORINE MONOF 0386
MENTS. (BERYLLIUM DIFLUORIDE, MAGNE/ ELECTRON DIFFRACTION ANALYSIS OF THE MOLECULAR STRUCTURES OF GROUP-2 ELE 0131
-0 AND 0-1 IN GERMANIUM MONOFLUORIDE.                        ANALYSIS OF THE ROTATIONAL STRUCTURE OF THE BANDS 0 0465
RIDE.                                          VIBRATIONAL ANALYSIS OF THE SINGLET SYSTEMS OF ARSENIC MONOFLUO 0095
A(+) OF GASEOUS SCANDIUM MONOFLUORIDE.          ROTATIONAL ANALYSIS OF THE SYSTEM B SINGLET PI TO SINGLET SIGM 0931
ONOFLUORIDE MOLECULE.                           ROTATIONAL ANALYSIS OF THE ULTRAVIOLET BANDS OF THE ANTIMONY M 0063
MONOFLUORIDE MOLECULE.                          ROTATIONAL ANALYSIS OF THE VISIBLE BAND SYSTEM OF THE BISMUTH  0146
FLUORIDE.                                       ROTATIONAL ANALYSIS OF THE VISIBLE EMISSION BANDS OF LEAD MONO 0587
NTIMONY FLUORIDE MOLECULE.                      ROTATIONAL ANALYSIS OF THE 2550-2750 ANGSTROM BAND SYSTEM AT A 0067
                                    NORMAL COORDINATE ANALYSIS OF TRIGONAL BIPYRAMIDAL XY5 MOLECULES.         1433
IDE.                                           VIBRATIONAL ANALYSIS OF ULTRAVIOLET BANDS OF ANTIMONY MONOFLUOR 0062
  TIN MONOFLUORIDE IN THE 2500 ANGSTROM REGION.             ANALYSIS OF VIBRATIONAL AND ROTATIONAL STRUCTURE OF 1138
ORIOLIS CONSTANT.                  BORON TRIFLUORIDE. THE ANALYSIS OF 2NU-3 AND THE DETERMINATION OF THE 33 C 0192
ED MOLECULES OF THE NONTRANSITION / BOND LENGTHS AND BOND ANGLES IN OCTAHEDRAL, TRIGONAL BIPYRAMID, AND RELAT 1435
U/ USE OF A SIMPLE ELECTROSTATIC MODEL FOR CALCULATION OF ANGULAR FORCE CONSTANTS OF HIGHLY SYMMETRICAL MOLEC 1484
OF RHENIUM HEPTAFLUORIDE. STRUCTURE, PSEUDOROTATION, AND ANHARMONIC COUPLING OF MODES. /ON DIFFRACTION STUDY 0920
ON DIOXIDE.                       LEAST SQUARES ADJUSTMENT OF ANHARMONIC POTENTIAL CONSTANTS. APPLICATION TO CARB 1491
RS OF HOT BANDS FOR A SYMMETRIC TOP. (NITROGEN TRIFLUORI/ ANHARMONIC POTENTIAL CONSTANTS FROM THE BAND CONTOU 0733
MOLECULES.                          ESTIMATION OF ANHARMONIC POTENTIAL CONSTANTS. PART-1. LINEAR XY2 1383
/N THE EXCITED VIBRATIONAL STATES, EQUILIBRIUM STRUCTURE, ANHARMONIC POTENTIAL FUNCTION, AND NU-1- NU-3 CORI/ 0993
                          X-RAY STUDY OF ANHARMONIC VIBRATIONS IN CALCIUM DIFLUORIDE.          0283
             TORSIONAL COORDINATES IN VIBRATIONAL ANHARMONICITY. APPLICATION TO ETHYLENE.                1485
/SUBSTITUION DATA FOR THE CALCULATION OF FORCE CONSTANTS. ANHARMONICITY EFFECTS ON THE FORCE CONSTANTS OF BO/ 0234
                                    VIBRATIONAL ANHARMONICITY IN BORON TRIFLUORIDE.                  0210
                                    VIBRATIONAL ANHARMONICITY IN THE METHYL HALIDES.                 1493
                                    VIBRATIONAL ANHARMONICITY IN THE PYRAMIDAL TRIFLUORIDES.         0745
UDY OF IODINE HEPTAFLUORIDE AND IODINE HEXAFLUORIDE(X) IN ANHYDROUS HYDROGEN FLUORIDE.          SPECTROSCOPIC ST 0518
                STRUCTURE OF THE IODINE HEXAFLUORIDE ANION. (XENON HEXAFLUORIDE)                      1300
ONIC STRUCTURES OF BORON TRIHALIDES AND BORON TETRAHALIDE ANIONS.          A THEORETICAL COMPARISON OF THE ELECTR 0225
              PARAMAGNETIC HEXAFLUORIDE ANIONS OF GROUP-6. (SELENIUM PENTAFLUORIDE)          0959
MOLECULES FROM SPECTROSCOPIC DATA. PENTACOORDINATED MOLE/ ANISOTROPIC THERMAL MOTION OF TRIGONAL BIPYRAMIDAL 0853
/M BROMIDE, AND CESIUM IODIDE AND MAGNETIC SUSCEPTIBILITY ANISOTROPY OF THALLIUM FLUORIDE, CESIUM FLUORIDE, / 1120
ONOMERS MEAN AMPLITUDES OF VIBRATION WITH LOW TEMPERATURE ANOMALIES.          ALUMINUM AND ALKALI HALIDE M 0024
                                              ANOMALOUS NMR SHIFT IN SAMARIUM TRIFLUORIDE.          0927
VANADIUM DIOXIDE.                  ELECTRONIC STRUCTURE AND ANOMALOUS THERMAL EXPANSION ON IRON DIFLUORIDE AND 0543
ATIONAL ANALYSIS OF THE 2550-2750 ANGSTROM BAND SYSTEM AT ANTIMONY FLUORIDE MOLECULE.          ROT 0067
                     ROTATIONAL STRUCTURE OF THE C3 SYSTEM OF ANTIMONY FLUORIDE MOLECULE.          0066
DE, TANTALUM BROMIDE, AND MOL/ THERMODYNAMIC FUNCTIONS OF ANTIMONY FLUORIDE, NIOBIUM BROMIDE, TANTALUM CHLORI 0075
/MICALLY INERT LONE PAIR. A SPECULATION ON THE BONDING IN ANTIMONY HEXACHLORIDE(3-), SELENIUM HEXABROMIDE(2-/ 1309
         BAND SPECTRA OF BISMUTH MONOFLUORIDE AND ANTIMONY MONOFLUORIDE.          0065
BO(+)-X(21) AND BO(+)-X(10+) BANDS OF ANTIMONY MONOFLUORIDE.          0071
HE EFFECTIVE VIBRATIONAL TEMPERATURE IN THE A-2 SYSTEM OF ANTIMONY MONOFLUORIDE.          DETERMINATION OF T 0061
              EMISSION CF MOLECULAR ANTIMONY MONOFLUORIDE.          0060
             FINE STRUCTURE ANALYSIS OF THE C1 SYSTEM OF ANTIMONY MONOFLUORIDE.          0059
                  NEAR ULTRAVIOLET AND VISIBLE BANDS OF ANTIMONY MONOFLUORIDE.          0070
                  ROTATIONAL ANALYSIS OF THE C2 SYSTEM OF ANTIMONY MONOFLUORIDE.          0068
                  ROTATIONAL ANALYSIS OF THE C1 SYSTEM OF ANTIMONY MONOFLUORIDE.          0069
         VIBRATIONAL ANALYSIS OF ULTRAVIOLET BANDS OF ANTIMONY MONOFLUORIDE.          0062
                  EMISSION SPECTRA OF ANTIMONY MONOFLUORIDE AND BISMUTH MONOFLUORIDE.          0064
         ROTATIONAL ANALYSIS OF C3 SYSTEM OF THE ANTIMONY MONOFLUORIDE MOLECULE.          0057
      ROTATIONAL ANALYSIS OF THE ULTRAVIOLET BANDS OF THE ANTIMONY MONOFLUORIDE MOLECULE.          0063
M REGION.                          A NEW BAND SYSTEM OF THE ANTIMONY MONOFLUORIDE MOLECULE IN 4050-5450 ANGSTRO 0058
LUORIDE, ARSENIC PENTAFLUORIDE, BISMUTH PENTAFLUORIDE AND ANTIMONY PENTAFLUORIDE) / PHASE. (PHOSPHORUS PENTAF 0876
         DENSITY, MELTING POINT AND VAPOR PRESSURE OF ANTIMONY PENTAFLUORIDE.          0082
                          FLUORINE NMR OF ANTIMONY PENTAFLUORIDE.          0078
TROMETRIC EVIDENCE OF DIMERS IN BISMUTH PENTAFLUORIDE AND ANTIMONY PENTAFLUORIDE.          MASS SPEC 0084
                          MASS SPECTRUM OF ANTIMONY PENTAFLUORIDE.          0085
ASS SPECTROMETRIC STUDY OF BINARY PENTAFLUORIDES.   (ALSO ANTIMONY PENTAFLUORIDE)          MOLECULAR BEAM M 1376
                  MOLECULAR STRUCTURE OF ANTIMONY PENTAFLUORIDE.          0080
L BIPYRAMIDAL PENTAHALIDES. (PHOSPHORUS PENTAFLUORIDE AND ANTIMONY PENTAFLUORIDE) / CONSTANTS OF SOME TRIGONA 0850
                  STRUCTURE OF LIQUID ANTIMONY PENTAFLUORIDE.          0081
UORIDE, PHOSPHORUS PENTAFLUORIDE, VANADIUM PENTAFLUORIDE, ANTIMONY PENTAFLUORIDE) /TAFLUORIDE, IODINE PENTAFL 0422
IBRATIONAL ASSIGNMENTS AND NORMAL COORDINATE ANALYSIS FOR ANTIMONY PENTAFLUORIDE.          V 0087
CIES IODINE PENTAFLUORIDE, TELLURIUM PENTAFLUORIDE(-) AND ANTIMONY PENTAFLUORIDE(-2). / THE ISOELECTRONIC SPE 0501
         INFRARED SPECTRA OF MATRIX ISOLATED ANTIMONY PENTAFLUORIDE AND ARSENIC PENTAFLUORIDE.          0113
/THE STUDY OF REACTIVE FLUORIDES. (SECOND ROW, THIRD ROW, ANTIMONY PENTAFLUORIDE, AND BISMUTH PENTAFLUORIDE) 1377
         POTENTIAL FIELD AND FORCE CONSTANTS OF ANTIMONY PENTAFLUORIDE AND NIOBIUM PENTAFLUORIDE.          0086
/LUORINE NMR AND RAMAN STUDY OF COMPLEX FORMATION BETWEEN ANTIMONY PENTAFLUORIDE AND NIOBIUM PENTAFLUORIDE A/ 0079
/ ITS USE IN THE STUDY OF THE LOW FREQUENCY VIBRATIONS OF ANTIMONY PENTAFLUORIDE AND THE PURE ROTATIONAL SPE/ 0083
/ATION STUDIES. (BORON TRIFLUORIDE, SULFUR TETRAFLUORIDE, ANTIMONY PENTAFLUORIDE, ARSENIC PENTAFLUORIDE, CHL/ 0239
ON OF TEMPERATURE. EVIDENCE FOR A TRIG/ RAMAN SPECTRUM OF ANTIMONY PENTAFLUORIDE IN THE GAS PHASE AS A FUNCTI 0077
                          ASSOCIATION OF ANTIMONY PENTAFLUORIDE IN VAPOR.          0088
/IA IN NIOBIUM PENTAFLUORIDE, TANTALUM PENTAFLUORIDE, AND ANTIMONY PENTAFLUORIDE. THE SHAPE OF ANTIMONY PENT/ 0076
/ETHOD. (PHOSPHORUS PENTAFLUORIDE, ARSENIC PENTAFLUORIDE, ANTIMONY PENTAFLUORIDE, VANADIUM PENTAFLUORIDE, AN/ 0878
ALCULATION OF THE FREQUENCIES OF THE NORMAL VIBRATIONS OF ANTIMONY TRIFLUORIDE.          C 0074
STANTS. (PHOSPHORUS TRIFLUORIDE, ARSENIC TRIFLUORIDE, AND ANTIMONY TRIFLUORIDE) /F UREY-BRADLEY POTENTIAL CON 0809
                          RAMAN SPECTRUM OF ANTIMONY TRIFLUORIDE.          0073
             THERMODYNAMIC PROPERTIES OF ANTIMONY TRIFLUORIDE.          0072
OUS SILVER MONOFLUORIDE.          ROTATIONAL ANALYSIS OF THE AO(+), BO(+), TO X SINGLET SIGMA(+) SYSTEMS OF GASE 1011
  BETA) METHOD. PART-3: ENERGY LEVELS, WAVE FUNCTIONS AND AO POPULATIONS OF POLYATOMIC FLUORIDES. /O-2 (ALPHA 0016
/ORO CHLORO ETHYLENE, DIFLUORO DICHLORO ETHYLENE, AND THE APPEARANCE POTENTIAL OF CARBON DIFLUORIDE(X) FROM / 0305
LE BY THE MO LCAO SELF CONSISTENT FIELD METHOD IN AN NDDO APPROXIMATION. / OF A CHLORINE PENTAFLUORIDE MOLECU 0415
UORIDES. MODIFIED NDDO-2 (ALPHA, BETA) METHOD AS A USEFUL APPROXIMATION. /METHODS ILLUSTRATED BY THE XENON FL 1231
THE BASIS OF A CALCULATION BY THE MO LCAO SCF IN THE NDDO APPROXIMATION. /HE MOLECULE CHLORINE DIFLUORIDE ON 0382
HEXAFLUORIDE AN/ STUDIES ON THE FAILURE OF THE FIRST BORN APPROXIMATION IN ELECTRON DIFFRACTION. (MOLYBDENUM 0640
```

EXAFLUORIDE)   FAILURE OF THE FIRST BORN APPROXIMATION IN ELECTRON DIFFRACTION. (TELLURIUM H 1109
XAFLUORIDE, URANIUM HEXAFLUORI/ FAILURE OF THE FIRST BORN APPROXIMATION IN ELECTRON DIFFRACTION. (TUNGSTEN HE 1158
AFLUORIDE)   FAILURE OF THE FIRST BORN APPROXIMATION IN ELECTRON DIFFRACTION. (URANIUM HEX 1199
           INFRARED SPECTRA OF ALUMINUM FLUORIDE IN SOLID ARGON.   0054
LUORIDE)   D ORBITALS IN THE NOBLE GAS DIHALIDES. (ARGON DIFLUORIDE, KRYPTON DIFLUORIDE, AND XENON DIF 1338
) FLUORIDE, AND ZINC(II) FLUORIDE WITH CARBON MONOXIDE IN ARGON MATRICES. / MANGANESE(II) FLUORIDE, COPPER(II 1324
   ABSORPTION SPECTRUM OF CARBON DIFLUORIDE TRAPPED IN AN ARGON MATRIX.   0297
N MONOFLUORIDE.   ABSORPTION AND EMISSION SPECTRA OF ARGON MATRIX ISOLATED XENON MONOFLUORIDE AND KRYPTO 0544
                                   ARGON MATRIX RAMAN SPECTRA OF OXYGEN DIFLUORIDE AND 0759
      INFRARED SPECTRA OF SIMPLE MOLECULES IN LIQUID ARGON SOLUTION. CARBON TETRAFLUORIDE.   0338
IFLUORIDE, ARSENIC TRIF/ THERMODYNAMIC PROPERTIES OF SOME ARSENIC FLUORIDES. (ARSENIC MONOFLUORIDE, ARSENIC D 0093
RA AND THERMODYNAMIC FUNCTIONS FOR VARIOUS PHOSPHORUS AND ARSENIC HALIDES. (PHOSPHORUS TRIFLUORIDE) /ED SPECT 0811
                        LIQUID GAS ARSENIC INFRARED FREQUENCY SHIFTS.   0106
THE ABSORPTION SPECTRUM AND OF TWO NEW SINGLET SYSTEMS OF ARSENIC MONOFLUORIDE.   ANALYSIS OF   0090
         ROTATIONAL ANALYSIS OF FOUR SINGLET SYSTEMS OF ARSENIC MONOFLUORIDE.   0092
           ULTRAVIOLET SPECTRUM OF THE SINGLET SYSTEMS OF ARSENIC MONOFLUORIDE.   0089
         VIBRATIONAL ANALYSIS OF THE SINGLET SYSTEMS OF ARSENIC MONOFLUORIDE.   0095
AND NEAR INFRARED REGIONS.   SPECTRUM OF ARSENIC MONOFLUORIDE IN THE NEAR VACUUM ULTRAVIOLET 0091
Y, IONIZATION POTENTIAL, ELECTRON AFFINITY, DIPOLE MOMEN/ ARSENIC MONOFLUORIDE, 3 SIGMA. DISSOCIATION ENTHALP 0094
               ENTHALPY OF FORMATION OF ARSENIC PENTAFLUORIDE.   0117
RED SPECTRA OF MATRIX ISOLATED ANTIMONY PENTAFLUORIDE AND ARSENIC PENTAFLUORIDE.   INFRA   0113
                  INFRARED SPECTRUM OF ARSENIC PENTAFLUORIDE.   0114
N AND CORIOLIS CONSTANTS FOR PHOSPHORUS PENTAFLUORIDE AND ARSENIC PENTAFLUORIDE. MEAN AMPLITUDES OF VIBRATIO 0840
MOLECULE WITH APPLICATION TO PHOSPHORUS PENTAFLUORIDE AND ARSENIC PENTAFLUORIDE. /FOR A TRIGONAL BIPYRAMIDAL 0864
ND VIBRATIONAL POTENTIAL FUNCTION OF THE E' VIBRATIONS OF ARSENIC PENTAFLUORIDE. /RPENDICULAR BAND CONTOURS A 0116
STRUCTURES OF PENTAFLUORIDES. (PHOSPHORUS PENTAFLUORIDE, ARSENIC PENTAFLUORIDE, AND BROMINE PENTAFLUORIDE) / 0831
R TO/ VIBRATIONAL SPECTRA OF PHOSPHORUS PENTAFLUORIDE AND ARSENIC PENTAFLUORIDE AND THE HEIGHT OF THE BARRIE 0856
D THEIR FORCE CONSTANTS INCLUDING THOS/ RAMAN SPECTRA OF ARSENIC PENTAFLUORIDE AND VANADIUM PENTAFLUORIDE AN 0873
/HE LOGARITHIMIC STEPS METHOD. (PHOSPHORUS PENTAFLUORIDE, ARSENIC PENTAFLUORIDE, ANTIMONY PENTAFLUORIDE, VAN/ 0878
/TAFLUORIDES IN THE GAS PHASE. (PHOSPHORUS PENTAFLUORIDE, ARSENIC PENTAFLUORIDE, BISMUTH PENTAFLUORIDE AND A/ 0876
/IFLUORIDE, SULFUR TETRAFLUORIDE, ANTIMONY PENTAFLUORIDE, ARSENIC PENTAFLUORIDE, CHLORINE TRIFLUORIDE, BROMI/ 0239
DIES IN MEAN AMPLITUDES OF VIBRATION OF PENTAFLUORIDES OF ARSENIC, PHOSPHORUS, AND VANADIUM. /ECTROSCOPIC STU 0871
                       ESR SPECTRUM OF ARSENIC TETRAFLUORIDE.   0112
IFLUORIDE, BORON TRIFLUORIDE, PHOSPHORUS TRIFLUORIDE, AND ARSENIC TRIFLUORIDE) /F XY3 MOLECULES. (NITROGEN TR 0709
 TRIFLUORO CHLORO METHANE AND SOME OTHER SYMMETRIC TOPS. (ARSENIC TRIFLUORIDE)   CENTRIFUGAL STRETCHING OF 0101
CONSTANTS. (NITROGEN TRIFLUORIDE, PHOSPHORUS TRIFLUORIDE, ARSENIC TRIFLUORIDE) /SE OF CENTRIFUGAL STRETCHING 0727
   INFRARED SPECTRUM AND VIBRATIONAL POTENTIAL FUNCTION OF ARSENIC TRIFLUORIDE.   0100
PLITUDES AND THERMODYNAMIC FUNCTIONS OF SOME TRIHALIDES. (ARSENIC TRIFLUORIDE)   MEAN AM 0109
ULES. (NITROGEN TRIFLUORIDE, PHOSPHORUS TRIFLUORIDE, AND ARSENIC TRIFLUORIDE) / FOR SOME PYRAMIDAL XY3 MOLEC 0706
                       MICROWAVE SPECTRUM OF ARSENIC TRIFLUORIDE.   0102
FIELD FOR SOME XY3 MOLECULES. (PHOSPHORUS TRIFLUORIDE AND ARSENIC TRIFLUORIDE)   MOLECULAR FORCE 1400
NTS FOR NITROGEN TRIFLUORIDE, PHOSPHORUS TRIFLUORIDE, AND ARSENIC TRIFLUORIDE. /OP MOLECULES AND FORCE CONSTA 0708
                STANDARD HEAT OF FORMATION OF ARSENIC TRIFLUORIDE.   0110
       THE INFRARED SPECTRUM AND FORCE CONSTANTS FOR ARSENIC TRIFLUORIDE.   0105
REY-BRADLEY POTENTIAL CONSTANTS. (PHOSPHORUS TRIFLUORIDE, ARSENIC TRIFLUORIDE, AND ANTIMONY TRIFLUORIDE) /F U 0809
TERMINED BY ELECTRON DIFFRACTION. MOLECULAR STRUCTURES OF ARSENIC TRIFLUORIDE AND ARSENIC PENTAFLUORIDE AS DE 0099
IONIC CHARACTER AND HYBRIDIZATION FROM DIPOLE MOMENTS FOR ARSENIC TRIFLUORIDE AND ARSENIC TRICHLORIDE. /S ON 0107
ESR STUDY OF RADICALS IN GAMMA IRRADIATED POLYCRYSTALLINE ARSENIC TRIFLUORIDE AND ARSENIC TRICHLORIDE.   0108
ELECTRON DI/ DETERMINATION OF THE MOLECULAR STRUCTURES OF ARSENIC TRIFLUORIDE AND ARSENIC TRICHLORIDE BY GAS 0104
NED BY ELECTRO/ FORCE CONSTANTS AND AVERAGE STRUCTURES OF ARSENIC TRIFLUORIDE AND ARSENIC TRICHLORIDE DETERMI 0103
              HIGH RESOLUTION SPECTRA OF ARSENIC TRIFLUORIDE AND PHOSPHORUS TRIFLUORIDE.   0816
CTIONS OF SOME GROUP-5 TRIHALIDES. (NITROGEN TRIFLUORIDE, ARSENIC TRIFLUORIDE, AND PHOSPHORUS TRIFLUORIDE) /N 0719
ARSENIC TRIFLUORIDE, ARSENIC TRICHLORI/ RAMAN SPECTRUM OF ARSENIC TRIFLUORIDE AND THE MOLECULAR CONSTANTS OF 0111
ER MOLECULES.   AN ELECTRON DIFFRACTION INVESTIGATION OF ARSENIC TRIFLUORIDE, ARSENIC PENTAFLUORIDE, AND OTH 0098
/ELD OF NITROGEN TRIFLUORIDE, PHOSPHORUS TRIFLUORIDE, AND ARSENIC TRIFLUORIDE BY THE COMBINED USE OF VIBRATI/ 0717
E.   MICROWAVE SPECTRUM OF ARSENIC TRIFLUORIDE IN THE EXCITED VIBRATIONAL STAT 0097
/F SOME VOLATILE INORGANIC FLUORIDES IN THE SOLID STATE. (ARSENIC TRIFLUORIDE, SULFUR TETRAFLUORIDE, AND SEL/ 0096
MONOFLUORIDE, GALLIUM TRIHYDRIDE, GERMANIUM TETRAHYDRIDE, ARSENIC TRIHYDRIDE, AND SELENIUM DIHYDRIDE. /LLIUM 0459
Y-BRADLEY FORCE FIELD TO AMMONIA, PHOSPHINE, STIBINE, AND ARSINE.   ADAPTATION OF THE URE 1492
UM TETRAFLUORIDE.   VIBRATIONAL ASSIGNMENT AND NORMAL COORDINATE ANALYSIS OF SELENI 0958
MOLECULE.   VIBRATIONAL ASSIGNMENT FOR IODINE PENTAFLUORIDE, A NONRIGID D5H 0525
BETWEEN A FUNDAMENTAL AND A TER/ VIBRATIONAL SPECTRA AND ASSIGNMENT OF IODINE PENTAFLUORIDE. FERMI RESONANCE 0504
                       VIBRATIONAL ASSIGNMENT OF SULFUR TETRAFLUORIDE.   1036
/UCTURE OF MOLECULES BY A MANY BODY APPROACH. PART-6: THE ASSIGNMENT OF THE HELIUM-2 PHOTOELECTRON SPECTRUM / 1090
MONY PENTAFLUORIDE.   VIBRATIONAL ASSIGNMENTS AND NORMAL COORDINATE ANALYSIS FOR ANTI 0087
M HEXAFLUORIDE AND TUNGSTEN HEXAFLUORIDE.   ASSIGNMENTS IN THE ULTRAVIOLET SPECTRA OF MOLYBDENU 0633
RAFLUORI/ A TRANSFERABLE UREY-BRADLEY FORCE FIELD AND THE ASSIGNMENTS OF SOME MIXED HALOMETHANES. (CARBON TET 0351
                       ASSOCIATION OF ANTIMONY PENTAFLUORIDE IN VAPOR.   0088
ASE. (PHOSPHORUS PENTAFLUORIDE, ARSENIC PENTAFLUORIDE, B/ ASSOCIATION OF GROUP-5 PENTAFLUORIDES IN THE GAS PH 0876
           FLUORINE-19 NMR INVESTIGATION OF THE ASSOCIATION OF SULFUR TETRAFLUORIDE.   1044
TIAL CONSTANTS AND AVE/ CENTRIFUGAL DISTORTION EFFECTS IN ASYMMETRIC ROTOR MOLECULES. PART-1. QUADRATIC POTEN 0778
           ENVIRONMENT OF THE TIN ATOM IN ORTHORHOMBIC TIN DIFLUORIDE.   1139
OF FORCE CONSTANTS. ANHARMONICITY EFFEC/ UTILITY OF HEAVY ATOM ISOTOPIC SUBSTITUION DATA FOR THE CALCULATION 0234
          INERTIA DEFECT AND PLANARITY OF FOUR ATOM MOLECULES. (XY3)   1398
/TION OF IODINE HEPTAFLUORIDE AND AN APPLICATION OF THREE ATOM SCATTERING THEORY TO IODINE HEPTAFLUORIDE AND/ 0527
OF THE INTERACTION OF MAGNESIUM DIFLUORIDE WITH RARE GAS ATOMS.   THEORETICAL STUDIES 0605
ILICON DIFLUORI/ VESCF-MO STUDIES OF MOLECULES CONTAINING ATOMS FROM THE SECOND ROW OF THE PERIODIC TABLE. (S 1373
     BARRIER TO ELECTRON PASSAGE THROUGH ELECTRONEGATIVE ATOMS IN BORON TRIFLUORIDE.   0196
NG V/ ORBITAL MODIFICATION BY THE COULOMB FIELD OF LIGAND ATOMS OF LATER SECOND ROW ELEMENTS IN PERFECT PAIRI 0862
R DIATOMIC AND POLYATOMIC MOLECU/ DENSITY LOCALIZATION OF ATOMIC AND MOLECULAR ORBITALS. PART-3: HETERONUCLEA 0175
RBON / MOLECULAR CONSTANTS FROM INFRARED BAND SHAPES. THE ATOMIC DISPLACMENTS AND INTRAMOLECULAR FORCES IN CA 0331
LES.   INFLUENCE OF ATOMIC MASS ON THE CORIOLIS CONSTANTS OF XY6 MOLECU 1458
           INFLUENCE OF ATOMIC MASSES ON THE CORIOLIS COUPLING COEFFICIENTS 1388
IN XY4 TETRAHEDRAL MOLECULES.   INFLUENCE OF ATOMIC MASSES ON THE CORIOLIS COUPLING COEFFICIENTS 1410
RMATION OF INORGANIC FLUORIDES ESPECIALLY THE ELEMENTS OF ATOMIC NUMBER BELOW 20.   HEATS OF FO 0214
FLUORIDE AND XENON DICHLORIDE.   5D ORBITALS OF XENON IN ATOMIC VALENCE STATES AND IN THE MOLECULES XENON DI 1242
AND IONS.   MOLECULAR ORBITAL STUDY OF ATOMIZATION ENERGIES OF PNICOGEN FLUORIDE MOLECULES 1364
                       HEAT OF ATOMIZATION OF ALUMINUM DIFLUORIDE.   0034
/TOR MOLECULES. PART-1. QUADRATIC POTENTIAL CONSTANTS AND AVERAGE STRUCTURE OF OXYGEN DIFLUORIDE FROM THE GR/ 0778
IC TRICHLORIDE DETERMINED BY ELECTRO/ FORCE CONSTANTS AND AVERAGE STRUCTURES OF ARSENIC TRIFLUORIDE AND ARSEN 0103

FIELD AND MOLECULAR VIBRATIONS OF THE TRIGONAL BIPYRAMID AX(3)YZ MOLECULE.   POTENTIAL 1436

B

BLET PI / ROTATIONAL STRUCTURE OF THE A DOUBLET SIGMA(+), B DOUBLET SIGMA(+), AND A QUARTET SIGMA(-) TO X DOU 0466
/ONAL ANALYSIS OF THE A DOUBLET SIGMA(+) TO X DOUBLET PI, B DOUBLET SIGMA(+) TO X DOUBLET PI SYSTEMS OF TIN / 1133
FLUORIDE MOLECULE IN EMISSION / ANALYSIS OF THE BANDS OF B DOUBLET SIGMA- X DOUBLET SIGMA. SYSTEM OF CALCIUM 0280
M MONOFLUORIDE.   ROTATIONAL ANALYSIS OF THE SYSTEM B SINGLET PI TO SINGLET SIGMA(+) OF GASEOUS SCANDIU 0931
M IN LEAD MONOFLUORIDE.   B SINGLET SIGMA(+) TO X TRIPLET SIGMA(-) BAND SYSTE 0585
M OF NITROGEN MONOFLUORIDE.   B SINGLET SIGMA(+) TO X TRIPLET SIGMA(-) BAND SYSTE 0675
                        ROTATIONAL CONSTANTS OF THE B STATE OF THE CARBON MONOFLUORIDE MOLECULE .   0294
ROTATIONAL ANALYSIS OF BANDS OF THE A TRIPLET PI ZERO(+), B TRIPLET PI(1) TO X SINGLET SIGMA(+) SYSTEMS OF G/ 0457
/OTATIONAL ANALYSIS OF BANDS OF THE A TRIPLET PI ZERO(+), B TRIPLET PI(1) TO X SINGLET SIGMA(+) SYSTEMS OF T/ 1111
MONOFLUORI/ FRANCK-CONDON FACTORS AND R CENTROIDS FOR THE B TRIPLET PI SYSTEM OF BORON 0171
LUO/ ROTATIONAL ANALYSIS OF BANDS OF THE C TRIPLET SIGMA, B TRIPLET SIGMA: A TRIPLET PI SYSTEM OF BORON MONOF 0153
/ENT AS A FUNCTION OF THE INTERNUCLEAR SEPARATION FOR THE B-X BAND SYSTEM OF MOLECULAR SILICON MONOFLUORIDE. 0967
                              THE B-X SYSTEM OF CARBON MONOFLUORIDE.   0285
               ROTATIONAL ANALYSIS OF THE B-X SYSTEM OF LEAD MONOFLUORIDE IN ULTRAVIOLET. 0586
                 ROTATIONAL ANALYSIS OF B-X2 SYSTEM OF LEAD-208 MONOFLUORIDE.   0590
ECIFIC HEAT RATIO OF URANIUM HEXAFLUORIDE MEASURED WITH A BALLISTIC PISTON COMPRESSOR.   SP 1201
LTRAVIOLET ELECTRONIC SPECTRA OF STRONTIUM DIFLUORIDE AND BARIUM DIFLUORIDE.   FAR U 1017
FLUORIDE, MAGNESIUM DIFLUORIDE, STRONTIUM DIFLUORIDE, AND BARIUM DIFLUORIDE) /NORGANIC FLUORIDES. (CADMIUM DI 0273
IUM DIFLUORIDE, CALCIUM DIFLUORIDE, STRONTIUM DIFLUORIDE, BARIUM DIFLUORIDE, AND ALUMINUM TRIFLUORIDE. /AGNES 0049
ITATIONS IN CALCIUM DIFLUORIDE, STRONTIUM DIFLUORIDE, AND BARIUM DIFLUORIDE IN THE 8 TO 150 EV RANGE. /IC EXC 0279
FLUORIDE)   GEOMETRY OF THE ALKALINE EARTH DIHALIDES. (BARIUM DIFLUORIDE, STRONTIUM DIFLUORIDE, CALCIUM DI 0126
/UM DIFLUORIDE, CALCIUM DIFLUORIDE, STRONTIUM DIFLUORIDE, BARIUM DIFLUORIDE, ZINC DIFLUORIDE, CADMIUM DIFLUO/ 0131
                          GREEN BAND SYSTEM OF BARIUM MONOFLUORIDE.   0121
ROTATIONAL ANALYSIS OF SOME BANDS OF GASEOUS BARIUM MONOFLUORIDE.   0118
THE C DOUBLET PI AND THE X DOUBLET SIGMA STATES OF BARIUM MONOFLUORIDE.   0120
THERMALLY EXCITED EMISSION SPECTRUM OF BARIUM MONOFLUORIDE.   0123
VIOLET AND ULTRAVIOLET SPECTRUM OF BARIUM MONOFLUORIDE MOLECULE.   0122
ATOMS IN BORON TRIFLUORIDE.   BARRIER TO ELECTRON PASSAGE THROUGH ELECTRONEGATIVE 0196
FLUORIDE AND ARSENIC PENTAFLUORIDE AND THE HEIGHT OF THE BARRIER TO INTERNAL EXCHANGE OF FLUORINE NUCLEI. /A 0856
ES. (SULFUR HEXAFLUORIDE) EVIDENCE OF EFFECTIVE POTENTIAL BARRIERS IN THE X-RAY ABSORPTION SPECTRA OF MOLECUL 1073
ORIDE AND ITS COMPLEXES WITH XENON DIFLUORIDE AND ORGANIC BASES. /UPOLE RESONANCE STUDIES OF NIOBIUM PENTAFLU 0664
OLARIZABILITIES FOR DIHALI/ MEAN AMPLITUDES OF VIBRATION, BASTIANSEN-MORINO SHRINKAGE EFFECT, AND MOLECULAR P 0124
C FUNCTIONS OF KRYPTON DIF/ MEAN AMPLITUDES OF VIBRATION, BASTIANSEN-MORINO SHRINKAGE EFFECT AND THERMODYNAMI 0562
YMMETRICAL DIHALIDES. (/ MEAN AMPLITUDES OF VIBRATION AND BASTIANSEN-MORINO SHRINKAGE EFFECT IN SOME LINEAR S 1372
CULES OF PLANAR TRIGONA/ MEAN AMPLITUDES OF VIBRATION AND BASTIANSEN-MORINO SHRINKAGE EFFECT IN SOME XY3 MOLE 1397
MOLECULAR STRUCTURE.   BASTIANSEN-MORINO SHRINKAGE EFFECT IN THE STUDY OF 1476
Y3 AND XY4 / VIBRATIONAL MEAN SQUARE AMPLITUDE MATRICES. (BASTIANSEN-MORINO SHRINKAGE EFFECT OF SYMMETRICAL X 1394
RAL XY6 MOLECULES.   MEAN AMPLITUDES OF VIBRATION AND BASTIANSEN-MORINO SHRINKAGE EFFECTS IN SOME OCTAHED 1455
TETRAFLUORIDE, CARBON TETRACHLORIDE, SILICON / MOLECULAR BEAM ELECTRIC DEFLECTION OF THE TETRAHALIDES CARBON 0349
2500 ANGSTROMS. (ALUMINU/ RECENT DEVELOPMENTS IN MULTIPLE BEAM INTERFEROMETRY IN THE ULTRAVIOLET REGION 1700- 0042
DES.   (ALSO ANTIMONY PENTAFLUORIDE)   MOLECULAR BEAM MASS SPECTROMETRIC STUDY OF BINARY PENTAFLUORI 1376
IVE FLUORIDES.   APPLICATION OF A MOLECULAR BEAM SOURCE MASS SPECTROMETER TO THE STUDY OF REACT 0019
IVE FLUORIDES. (SECOND ROW, T/ APPLICATION OF A MOLECULAR BEAM SOURCE MASS SPECTROMETER TO THE STUDY OF REACT 1377
S. (LANTHANIDE TRIFLUOR/ ELECTRIC DEFLECTION OF MOLECULAR BEAMS OF THE RARE EARTH DIFLUORIDES AND TRIFLUORIDE 0946
                      IONIC MODEL CALCULATIONS. PART-2: BENDING FORCE CONSTANTS OF THE GROUP-2 HALIDES.   1381
INC DIFLUORIDE, CADMIUM DIFL/ POLARIZED ION MODEL AND THE BENDING FORCE CONSTANTS OF THE GROUP-2B HALIDES. (Z 1322
                              BENDING MOTIONS IN THE DIHALIDES OF GROUP-2 METALS. 1380
OF NONPLANAR XY3 MOLECULES. (NITROGEN TRIFLUORIDE)   A BENT BOND MODEL FOR THE VIBRATIONAL FORCE CONSTANTS 0704
UREY-BRADLEY FORCE FIELD AND THERMODYNAMIC PROPERTIES FOR BENT SYMMETRICAL XY2 MOLECULES.   1387
THIRD ORDER POTENTIAL CONSTANTS OF BENT XY2 MOLECULES. (OXYGEN DIFLUORIDE)   0770
LATTICE CONSTANTS OF ACTINIDE TETRAFLUORIDES INCLUDING BERKELIUM.   1334
RUS PE/ PSEUDOROTATION OF TRIGONAL BIPYRAMIDAL MOLECULES. BERRY ROTATION CONTRA TURNSTILE ROTATION IN PHOSPHO 0870
EAN AMPLITUDES OF VIBRATION FOR THE DIHALIDES OF MERCURY, BERYLLIUM, AND MAGNESIUM.   M 0619
/CONFIGURATION RESULTS ON THE DIATOMIC MOLECULES LITHIUM, BERYLLIUM, BORON, CARBON, NITROGEN, FLUORINE, BORO/ 0174
F VAPORIZATION / THERMODYNAMIC AND PHYSICAL PROPERTIES OF BERYLLIUM COMPOUNDS. PART-1: ENTHALPY AND ENTROPY O 0138
                      ENTHALPY OF FORMATION OF BERYLLIUM DIFLUORIDE.   0139
                          HEAT OF FORMATION OF BERYLLIUM DIFLUORIDE.   0132
NEMPIRICAL LCAO MO SCF STUDIES OF THE LOW LYING STATES OF BERYLLIUM DIFLUORIDE.   NO 0137
                        THERMODYNAMIC DATA FOR BERYLLIUM DIFLUORIDE.   0136
AL DIFLUORIDES. (OF THE FIRST ROW TRANSITION METALS, ALSO BERYLLIUM DIFLUORIDE AND MAGNESIUM DIFLUORIDE) /RIC 1369
                      VAPOR PRESSURES OF BERYLLIUM DIFLUORIDE AND NICKEL DIFLUORIDE.   0135
/ INFRARED SPECTRA OF THE ALKALINE EARTH HALIDES. PART-1: BERYLLIUM DIFLUORIDE, BERYLLIUM DICHLORIDE, MAGNESI 0133
UORIDE/ MOLECULAR PROPERTIES OF THE TRIATOMIC DIFLUORIDES BERYLLIUM DIFLUORIDE, BORON DIFLUORIDE, CARBON DIFL 0141
/NEMPIRICAL LCAO MO SCF STUDIES OF THE GROUP-2A DIHALIDES BERYLLIUM DIFLUORIDE, MAGNESIUM DIFLUORIDE, AND CA/ 1360
/ SHRINKAGE EFFECT IN SOME LINEAR SYMMETRICAL DIHALIDES. (BERYLLIUM DIFLUORIDE, MAGNESIUM DIFLUORIDE, AND FI/ 1372
/ALYSIS OF THE MOLECULAR STRUCTURES OF GROUP-2 ELEMENTS. (BERYLLIUM DIFLUORIDE, MAGNESIUM DIFLUORIDE, CALCIU/ 0131
-CONDON FACTORS AND R CENTROIDS OF THE A-X BAND SYSTEM OF BERYLLIUM FLUORIDE.   RYDBERG-KLEIN-REES FRANCK 0127
OF FORMATION OF BERYLIUM CHLOR/ DISSOCIATION ENERGIES OF BERYLLIUM FLUORIDE AND BERYLLIUM CHLORIDE AND HEAT 0128
ANSPOSED TEMPERATURE DROP CALORIMETRY.   ENTHALPY OF BERYLLIUM FLUORIDE FROM 456 TO 1083 DEGREES K BY TR 0142
EFFUSION STUDIES, MASS SPECTRA, AND THERMODYNAMICS OF BERYLLIUM FLUORIDE VAPOR.   0140
N AND MASS SPECTROMETRIC DETECTION. (FIRST ROW FLUORIDES, BERYLLIUM, MAGNESIUM, AND LEAD DIFLUORIDES) /LECTIO 0134
SPECTROMETRIC DETERMINATION OF THE DISSOCIATION ENERGY OF BERYLLIUM MONOFLUORIDE.   MASS 0129
IN MONOFL/ POTENTIAL CURVES FOR SOME DIATOMIC MOLECULES. (BERYLLIUM MONOFLUORIDE, SILICON MONOFLUORIDE, AND O 0130
IC NITROGEN, CARBON MONOXIDE, BORON FLUORIDE, AND CARBON, BERYLLIUM OXIDE, LITHIUM FLUORIDE. /THE ISOELECTRON 0150
S BELONGING TO THE LANTHANIDE SERIES.   BIBLIOGRAPHY ON SPECTROSCOPY OF FLUORIDES AND OXIDE 1358
TO INORGANIC FLUORIDES.   BICENTRIC RESCALING OF CNDO/2 THEORY. APPLICATIONS 1475
                          NOBLE GASES AND THE BINARY FLUORIDES OF XENON.   1342
UM HEXAFLUORIDE, TELLURIUM HEXAFL/ ELECTRIC DEFLECTION OF BINARY HEXAFLUORIDES. (SULFUR HEXAFLUORIDE, SELENI 1084
GAS PHASE STRUCTURES AND MASS SPECTRA OF BINARY PENTAFLUORIDES.   0003
GAS PHASE STRUCTURES AND MASS SPECTRA OF BINARY PENTAFLUORIDES.   1365
DE)   MOLECULAR BEAM MASS SPECTROMETRIC STUDY OF BINARY PENTAFLUORIDES. (ALSO ANTIMONY PENTAFLUORI 1376
DISTRIBUTION IN CHLORINE MONOFLUORIDE FROM CORE ELECTRON BINDING ENERGIES.   CHARGE 0374
NITY, D/ OXYGEN MONOFLUORIDE. HARTREE-FOCK WAVE FUNCTION, BINDING ENERGY, IONIZATION POTENTIAL, ELECTRON AFFI 0756
/GEN MONOFLUORIDE, DOUBLET PI HARTREE-FOCK WAVE FUNCTION, BINDING ENERGY, IONIZATION POTENTIAL, ELECTRON AFF/ 0757
POTENTIAL ENERGY CURVES AND NATURE OF BINDING IN GROUP-3A MONOHALIDES.   0032
MONOXIDE, B/ MOLECULAR CHARGE DISTRIBUTIONS AND CHEMICAL BINDING. PART-3: THE ISOELECTRONIC NITROGEN, CARBON 0150

ON / BOND LENGTHS AND BOND ANGLES IN OCTAHEDRAL, TRIGONAL BIPYRAMID, AND RELATED MOLECULES OF THE NONTRANSITI 1435
POTENTIAL FIELD AND MOLECULAR VIBRATIONS OF THE TRIGONAL BIPYRAMID AX(3)YZ MOLECULE. 1436
ND EXTENDED HUCKEL MOLECULAR ORBITAL ANALYSES OF TRIGONAL BIPYRAMIDS. (PHOSPHORUS PENTAFLUORIDE) /-A-SPHERE A 0828
VIBRATIONS AND STEREOCHEMICAL NONRIGIDITY OF THE TRIGONAL BIPYRAMIDAL MODEL MX3Y2. /ULES. PART-14: MOLECULAR 0854
PENTAFLUORI/ MEAN AMPLITUDES OF VIBRATION FOR A TRIGONAL BIPYRAMIDAL MOLECULE WITH APPLICATION TO PHOSPHORUS 0864
ILE ROTATION IN PHOSPHCRUS PE/ PSEUDOROTATION OF TRIGONAL BIPYRAMIDAL MOLECULES. BERRY ROTATION CONTRA TURNST 0870
/AND MEAN SQUARE AMPLITUDES OF VIBRATION OF SOME TRIGONAL BIPYRAMIDAL MOLECULES BY THE LOGARITHIMIC STEPS ME/ 0878
ACOORDINATED MOLE/ ANISOTROPIC THERMAL MOTION OF TRIGONAL BIPYRAMIDAL MOLECULES FROM SPECTROSCOPIC DATA. PENT 0853
PSEUDOROTATION IN TRIGONAL BIPYRAMIDAL MOLECULES. (PHOSPHORUS PENTAFLUORIDE) 0833
RELATIVE BOND STRENGTHS IN TRIGONAL BIPYRAMIDAL MOLECULES. (PHOSPHORUS PENTAFLUORIDE) 0839
THERMODYNAMIC FUNCTIONS OF SOME TRIGONAL BIPYRAMIDAL PENTAHALIDES. 1438
AN/ POTENTIAL FIELD AND FORCE CONSTANTS OF SOME TRIGONAL BIPYRAMIDAL PENTAHALIDES. (PHOSPHORUS PENTAFLUORIDE) 0850
ASE AS A FUNCTION OF TEMPERATURE. EVIDENCE FOR A TRIGONAL BIPYRAMIDAL SHAPE FOR THE MONOMERIC SPECIES. /AS PH 0077
OLECULAR POLARIZABILITES OF SOME PENTAHALIDES OF TRIGONAL BIPYRAMIDAL SYMMTERY. /HERMODYNAMIC FUNCTIONS AND M 1437
POTENTIAL FIELD AND MOLECULAR VIBRATIONS OF THE TRIGONAL BIPYRAMIDAL THE RAMAN SPECTRUM OF PHOSPHORUS PENTA/ 0863
UDES OF VIBRATION, AND SHRINKAGE EFFECTS OF SOME TRIGONAL BIPYRAMIDAL XY5 AND XY3Z2 MOLECULES. /SQUARE AMPLIT 1432
NORMAL COORDINATE ANALYSIS OF TRIGONAL BIPYRAMIDAL XY5 MOLECULES. 1433
E) UREY-BRADLEY FORCE CONSTANTS OF TRIGONAL BIPYRAMIDAL XY5 MOLECULES. (PHOSPHORUS PENTAFLUORID 0838
NGSTROMS. NEW BAND SYSTEM OF BISMUTH FLUORIDE MOLECULE IN THE REGION 6200-7000 A 0144
EMISSION SPECTRA OF ANTIMONY MONOFLUORIDE AND BISMUTH MONOFLUORIDE. 0064
EMISSION SPECTRUM OF BISMUTH MONOFLUORIDE. 0147
YSTEMS OF SILICON MONOFLUORIDE, CALCIUM MONOFLUORIDE, AND BISMUTH MONOFLUORIDE. /D R CENTROIDS OF SOME BAND S 0972
ON THE DISSOCIATION OF LEAD MONOFLUORIDE AND BISMUTH MONOFLUORIDE. 0589
VISIBLE EMISSION SPECTRUM OF BISMUTH MONOFLUORIDE. 0145
BAND SPECTRA OF BISMUTH MONOFLUORIDE AND ANTIMONY MONOFLUORIDE. 0065
ROTATIONAL ANALYSIS OF THE VISIBLE BAND SYSTEM OF THE BISMUTH MONOFLUORIDE MOLECULE. 0146
GION. ABSORPTION SPECTRUM OF THE BISMUTH MONOFLUORIDE MOLECULE IN THE ULTRAVIOLET RE 0143
IDES. (SECOND ROW, THIRD ROW, ANTIMONY PENTAFLUORIDE, AND BISMUTH PENTAFLUORIDE /THE STUDY OF REACTIVE FLUOR 1377
MASS SPECTROMETRIC EVIDENCE OF DIMERS IN BISMUTH PENTAFLUORIDE AND ANTIMONY PENTAFLUORIDE. 0084
PHASE. (PHOSPHORUS PENTAFLUORIDE, ARSENIC PENTAFLUORIDE, BISMUTH PENTAFLUORIDE AND ANTIMONY PENTAFLUORIDE) / 0876
IDES AND ACTIVITY COEFFICIENTS FOR PROTACTINIUM IN LIQUID BISMUTH SOLUTIONS. (PROTACTINIUM TETRAFLUORIDE) /AL 0900
THERMODYNAMIC PROPERTIES OF BISMUTH TRIFLUORIDE. 0148
STRUCTURES OF YTTRIUM AND BISMUTH TRIFLUORIDES BY NEUTRON DIFFRACTION. 1318
THERMODYNAMIC FUNCTIONS OF BISMUTH TRIHALIDE VAPORS. 0149
BO(+)-X(21) AND BO(+)-X(10+) BANDS OF ANTIMONY MONOFLUORIDE. 0071
FLUORIDE.
VER MONOFLUORIDE. ROTATIONAL ANALYSIS OF THE AO(+), BO(+)-X(21) AND BO(+)-X(10+) BANDS OF ANTIMONY MONO 0071
-2 PHOTOELEC/ ELECTRONIC STRUCTURE OF MOLECULES BY A MANY BO(+), TO X SINGLET SIGMA(+) SYSTEMS OF GASEOUS SIL 1011
/PERTIES OF CARBON TETRAFLUORIDE FROM 12 DEGREES K TO ITS BODY APPROACH. PART-6: THE ASSIGNMENT OF THE HELIUM 1090
NTHALPY OF FORMATION OF THORIUM TETRAFLUORIDE BY FLUORINE BOILING POINT. THE SIGNIFICANCE OF THE PARAMETER N/ 0361
THALPY OF FORMATION OF VANADIUM PENTAFLUORIDE BY FLUORINE BOMB CALORIMETRY. E 1130
N OF BORON TRIFLUORIDE. FLUORINE BOMB CALORIMETRY. EN 1215
MATION OF PHOSPHORUS PENTAFLUORIDE AND ENTHALPI/ FLUORINE BOMB CALORIMETRY. PART-15: THE ENTHALPY OF FORMATIO 0220
N OF ALUMINUM TRIFLUORIDE. FLUORINE BOMB CALORIMETRY. PART-18: STANDARD ENTHALPY OF FOR 0865
BORON TRIFLUORIDE. FLUORINE BOMB CALORIMETRY. PART-22: THE ENTHALPY OF FORMATIO 0048
XENON DIFLUORIDE AND THE NATURE OF THE XENON- FLUORINE BOMB CALORIMETRY. PART-3: THE HEAT OF FORMATION OF 0250
ND ELECTRON AFFINITIES OF SOME GERMANIUM FLUCRIDE SPECIE/ BOND. 1223
S ELECTRON DIFFRACTION. BORON FLUORINE BOND DISSOCIATION ENERGIES, IONIZATION POTENTIALS A 0464
DES. HELIUM-2 PHOTOELECTRON SPECTRA OF XENON DIFLUORIDE / BOND DISTANCE OF BORON TRIFLUORIDE DETERMINED BY GA 0223
FLUORIDE(X) FROM TETRAFLUORO ETHYLENE. (CARBON DIFLUORIDE BOND ENERGIES AND IONIC CHARACTER OF INERT GAS HALI 1343
BOND ENERGY) /THE APPEARANCE POTENTIAL OF CARBON DI 0305
UORIDE. STRENGTH OF THE NITRCGEN FLUORINE BOND ENERGY IN THE ALUMINUM TRIFLUORIDE MOLECULE. 0044
TETRAFLUORIDE AND HAFNIUM TETRAFLUORI/ IONICITY OF THE MX BOND IN NITROGEN TRIFLUORIDE AND DINITROGEN TETRAFL 0712
CULES. (OXYGEN DIFLUORIDE, NI/ A THEORETICAL STUDY OF THE BOND IN THE TETRAHALIDES OF GROUP-4 ELEMENTS. (TIN 1141
L BIPYRAMID, AND RELATED MOLECULES OF THE NONTRANSITION / BOND INTERACTION FORCE CONSTANTS IN DIFLUORIDE MOLE 0299
BOND LENGTHS AND BOND ANGLES IN OCTAHEDRAL, TRIGONA 1435
BOND LENGTHS IN XENON DIFLUORIDE. 1236
ULFUR, SELENIUM, TELLURIUM, MOLYBDEN/ FORCE CONSTANTS AND BOND LENGTHS OF SOME INORGANIC HEXAFLUORIDES. (OF S 0631
ONPLANAR XY3 MOLECULES. (NITROGEN TRIFLUORIDE) A BENT BOND MODEL FOR THE VIBRATIONAL FORCE CONSTANTS OF N 0704
/ORDINATE TRANSFORMATIONS. APPLICATION TO THE CALCULATION BOND MOMENT PARAMETERS AND FORCE CONSTANTS. (CARBO/ 0352
ORUS TRIFLUORIDE AND PHOSPHORUS OX/ INFRARED INTENSITIES, BOND MOMENTS AND BOND MOMENT DERIVATIVES FOR PHOSPH 0797
PHOSPHORUS PENTAFLUORIDE) RELATIVE BOND STRENGTHS IN TRIGONAL BIPYRAMIDAL MOLECULES. (0839
RIDES. HIGHLY POLARIZED BORON- FLUORINE DOUBLE AND TRIPLE BONDS. LOCALIZED ORBITALS IN BORON FLUO 0158
NATURE OF BONDS IN THALLIUM MONOHALIDE MOLECULES. 1122
N AMPLITUDES OF THERMAL MOTION AND SHRINKAGES OF CHEMICAL BONDS: OCTAHEDRAL HEXAFLUORIDES. MEA 1460
HALIDES AND THE INTERHALOGEN COMPO/ ELECTRONEGATIVITY, NONBONDED INTERACTIONS AND POLARIZABILITY IN HYDROGEN 0373
EVIDENCE FOR ORBITAL INDEPENDENCE IN THE XENON- FLUORINE BONDING. /AND XENON OXIDE TETRAFLUORIDE BY ESCA AND 1229
AB INITIO INVESTIGATION OF THE GEOMETRY, BONDING, AND COUPLING CONSTANTS OF BORON DIFLUORIDE 0180
RIDES OF GROUP-6/ SCF- X(ALPHA) SCATTERED WAVE STUDIES ON BONDING AND IONIZATION POTENTIALS. PART-1: HEXAFLUO 1366
/E STEREOCHEMICALLY INERT IONE PAIR. A SPECULATION ON THE BONDING IN ANTIMONY HEXACHLORIDE(3-), SELENIUM HEX/ 1309
BONDING IN CARBON(1) AND CARBON(2) FLUORIDES. 0300
INTERATOMIC BONDING IN CRYSTALLINE MANGANESE TRIFLUORIDE. 0617
BONDING IN KRYPTON DIFLUORIDE. 0554
BONDING IN KRYPTON DIFLUORIDE. 0555
AB INITIO CALCULATIONS OF THE BONDING IN KRYPTON DIFLUORIDE. 0553
MOLECULAR ORBITAL ENERGY LEVELS AND BONDING IN KRYPTON DIFLUORIDE. 0549
IMETHYL PHOSPHINE. AB INITIO CALCULATIONS OF THE BONDING IN PHOSPHINE, PHOSPHORUS TRIFLUORIDE AND TR 0802
YFLUORIDE. AN AB INITIO SCF STUDY. BONDING IN PHOSPHORUS TRIFLUORIDE AND PHOSPHORUS OX 0820
N SPECTROSCOPY. STUDY OF THE BONDING IN THE GROUP-4 TETRAHALIDES BY PHOTOELECTRO 1362
ON DIFLUORIDE AND XENON TETRAFLUORIDE AND A CONJECTURE ON BONDING IN THE SOLIDS. HEATS OF SUBLIMATION OF XEN 1237
BONDING IN XENON DIFLUORIDE. 1230
BONDING IN XENON HEXAFLUORIDE. 1298
BONDING IN XENON HEXAFLUORIDE. 1313
WAVE FUNCTIONS FOR THE FOUR ELECTRON THREE CENTER BONDING OF FOUR AND EIGHT PI ELECTRON SYSTEMS. 0301
ALIZED MEAN AMPLITUDES AND CORIOLIS CONSTANTS IN DICHLORO BORANE, DIBROMO BORANE AND CHLORINE TRIFLUORIDE. /R 0411
ENUM HEXAFLUORIDE AN/ STUDIES ON THE FAILURE OF THE FIRST BORN APPROXIMATION IN ELECTRON DIFFRACTION. (MOLYBD 0640
IUM HEXAFLUORIDE) FAILURE OF THE FIRST BORN APPROXIMATION IN ELECTRON DIFFRACTION. (TELLUR 1109
EN HEXAFLUORIDE, URANIUM HEXAFLUORI/ FAILURE OF THE FIRST BORN APPROXIMATION IN ELECTRON DIFFRACTION. (TUNGST 1158
M HEXAFLUORIDE) FAILURE OF THE FIRST BORN APPROXIMATION IN ELECTRON DIFFRACTION. (URANIU 1199
DISSOCIATION ENERGIES OF THE GASEOUS MONOHALIDES OF BORON, ALUMINUM, GALLIUM, INDIUM, AND THALLIUM. 0152
NTIAL ENERGY CURVES AND DISSOCIATION ENERGIES OF DIATOMIC BORON AND ALUMINUM HALIDES. POTE 0173

95

```
/ON RESULTS ON THE DIATOMIC MOLECULES LITHIUM, BERYLLIUM, BORON, CARBON, NITROGEN, FLUORINE, BORON NITRIDE, /  0174
ER INVESTIGATIONS. (BORO/ THERMODYNAMIC PROPERTIES OF THE  BORON  CHLORIDE FLUORIDE SYSTEM FROM MASS SPECTROMET  0179
            ELECTRON STRUCTURE AND NMR SPECTRA OF  BORON  COMPOUNDS.                                           0229
ATION OF THE GEOMETRY, BONDING, AND COUPLING CONSTANTS OF  BORON  DIFLUORIDE.                  AB INITIO INVESTIG  0180
                        HEAT OF FORMATION OF  BORON  DIFLUORIDE.                                               0177
/RTIES OF THE TRIATOMIC DIFLUORIDES BERYLLIUM DIFLUORIDE,  BORON  DIFLUORIDE, CARBON DIFLUORIDE, NITROGEN DIFL/  0141
ORIDE, ALUMINUM DIF/ MEAN AMPLITUDES OF VIBRATION FOR THE  BORON  DIFLUORIDE, CARBON DIFLUORIDE, NITROGEN DIFLU  1384
/SMALL RADICALS BY THE UNRESTRICTED HARTREE-FOCK METHOD. (BORON  DIFLUORIDE, NITROGEN DIFLUORIDE, OXYGEN DIFL/  0176
        ELECTRON SPIN RESONANCE AND STRUCTURE OF THE  BORON  DIFLUORIDE RADICAL.                                0178
                    ESR PARAMETERS OF THE  BORON  DIFLUORIDE RADICAL.                                           0181
        CALCULATION OF RYDBERG LEVELS IN NITROGEN OXIDE AND  BORON  FLUORIDE.                                   0164
            FLUORESCENCE SPECTRA OF PHOSPHORUS NITRIDE AND  BORON  FLUORIDE.                                    0167
                    HEAT AND ENTROPY OF FORMATION OF  BORON  FLUORIDE.                                          0154
                        MICROWAVE SPECTRUM OF  BORON  FLUORIDE.                                                 0165
                    THE ELECTRONIC SPECTRUM OF  BORON  FLUORIDE.                                                0172
    VALENCE EXCITED STATES OF NITROGEN, CARBON MONOXIDE, AND  BORON  FLUORIDE.                                  0169
                MAGNETIC PROPERTIES OF THE  BORON  FLUORIDE AND BORON HYDRIDE MOLECULES.                        0159
/NG. PART-3: THE ISOELECTRONIC NITROGEN, CARBON MONOXIDE,  BORON  FLUORIDE, AND CARBON, BERYLLIUM OXIDE, LITHI/  0150
    ISOTOPE EFFECT IN THE SINGLET SPECTRAL BANDS OF THE  BORON  FLUORIDE MOLECULE.                              0166
        FIRST IONIZATION POTENTIALS OF THE MOLECULES  BORON  FLUORIDE, SILICON OXIDE AND GERMANIUM OXIDE.       0160
/ON OF INTERMOLECULAR EFFECTS IN THE VIBRATION SPECTRA OF  BORON  FLUORIDE WITH ALTERATIONS IN ITS QUADRATIC V/  1404
                COMPARATIVE AB INITIO CALCULATIONS ON  BORON  FLUORIDES.                                        0203
OUBLE AND TRIPLE BONDS.            LOCALIZED ORBITALS IN  BORON  FLUORIDES. HIGHLY POLARIZED BORON- FLUORINE D  0158
ETERMINED BY GAS ELECTRON DIFFRACTION.            BORON  FLUORINE BOND DISTANCE OF BORON TRIFLUORIDE D  0223
        FORCE CONSTANTS AND THERMODYNAMIC CONSTANTS FOR THE  BORON  HALIDES.                                    0219
            THERMODYNAMIC PROPERTIES OF  BORON  HALIDES.                                                        0240
FLUORIDE)        UREY-BRADLEY FORCE CONSTANTS OF  BORON  HALIDES AND THEIR TRANSFERABILITY. (BORON TRI  0246
    SEMIEMPIRICAL STUDY OF THE ELECTRONIC STRUCTURE OF  BORON  HYDRIDE FLUORIDES. (BORON TRIFLUORIDE)           0226
                CAMERON SYSTEM OF  BORON  MONOFLUORIDE.                                                         0163
    ELECTRONIC STRUCTURE OF CARBON MONOXIDE AND  BORON  MONOFLUORIDE.                                           0162
OIDS FOR THE B TRIPLET SIGMA(+) TO A TRIPLET PI SYSTEM OF  BORON  MONOFLUORIDE. /NCK-CONDON FACTORS AND R CENTR  0171
C TRIPLET SIGMA, B TRIPLET SIGMA: A TRIPLET PI SYSTEM OF  BORON  MONOFLUORIDE. /IONAL ANALYSIS OF BANDS OF THE  0153
IBRATIONAL TRANSITION PROBABILITIES FOR THE A-X SYSTEM OF  BORON  MONOFLUORIDE.                              V  0168
                DISSOCIATION ENERGY OF  BORON  MONOFLUORIDE FROM MASS SPECTROMETRIC STUDIES.  0161
            ELECTRONIC SPECTRUM OF THE  BORON  MONOFLUORIDE MOLECULE.                                          0155
BAND AT 14900 CM-1 OF A 3 SIGMA TO 3 SIGMA SYSTEM OF THE  BORON  MONOFLUORIDE MOLECULE. /IONAL ANALYSIS OF THE  0156
ALIDES.                CHEMICAL SHIFTS OF  BORON, PHOSPHORUS, PHOSPHORYL, AND THIOPHOSPHORYL H  0235
ON TET/ COMPARISON OF CALCULATIONS OF FORCE CONSTANTS FOR  BORON  TETRAFLUORIDE(-), CARBON TETRAFLUORIDE, SILIC  0348
YMMETRIC ROTOR RAMAN SPECTRA. PURE ROTATIONAL SPECTRUM OF  BORON  TRIFLUORIDE.                   ANALYSIS OF S  0205
SSIVE RIGIDITY METHOD TO SOME SIMPLE XYN TYPE MOLECULES. (BORON  TRIFLUORIDE)        APPLICATION OF THE PROGRE  0211
RIER TO ELECTRON PASSAGE THROUGH ELECTRONEGATIVE ATOMS IN  BORON  TRIFLUORIDE.                           BAR  0196
OF DIPOLE MOMENT DERIVATIVES AND INFRARED INTENSITIES OF  BORON  TRIFLUORIDE.              CNDO CALCULATION  0195
    CUBIC POTENTIAL CONSTANTS AND 1-TYPE RESONANCE IN  BORON  TRIFLUORIDE.                                      0231
                ELECTRONIC STRUCTURE OF  BORON  TRIFLUORIDE.                                                    0184
                    FERMI RESONANCE IN  BORON  TRIFLUORIDE.                                                     0242
E BOMB CALORIMETRY. PART-15: THE ENTHALPY OF FORMATION OF  BORON  TRIFLUORIDE.                     FLUORIN  0220
UORINE BOMB CALORIMETRY. PART-3: THE HEAT OF FORMATION OF  BORON  TRIFLUORIDE.                          FL  0250
    GROUND STATE PROPERTIES OF THE GROUP-2 TRIHALIDES. (BORON TRIFLUORIDE)                                    0185
            INFRARED ABSORPTION SPECTRUM OF  BORON  TRIFLUORIDE.                                                0206
        INFRARED ABSORPTION SPECTRUM OF LIQUID  BORON  TRIFLUORIDE.                                             0245
                INFRARED SPECTRUM OF  BORON  TRIFLUORIDE.                                                       0249
        ISOTOPIC SPLITTING IN MATRIX ISOLATED  BORON  TRIFLUORIDE.                                              0228
                MOLECULAR STRUCTURE OF  BORON  TRIFLUORIDE.                                                     0187
                    NU-3 BANDS OF  BORON  TRIFLUORIDE.                                                          0207
                        NU-3 OF  BORON  TRIFLUORIDE.                                                            0200
            PERPENDICULAR VIBRATIONS OF  BORON  TRIFLUORIDE.                                                    0202
            PHOTOELECTRON SPECTRUM OF  BORON  TRIFLUORIDE.                                                      0189
ATION VIBRATION BAND CONTOURS IN THE INFRARED SPECTRUM OF  BORON  TRIFLUORIDE.                            ROT  0193
ORE PRECISE DETERMINATION OF THE FAR INFRARED SPECTRUM OF  BORON  TRIFLUORIDE. /Y OF THE ISOTOPE EFFECT AND A M  0230
        ULTRASOFT X-RAY ABSORPTION SPECTRA OF  BORON  TRIFLUORIDE.                                              0204
ONSTANTS. ANHARMONICITY EFFECTS ON THE FORCE CONSTANTS OF  BORON  TRIFLUORIDE. / FOR THE CALCULATION OF FORCE C  0234
            VIBRATION ROTATION BANDS OF  BORON  TRIFLUORIDE.                                                    0199
            VIBRATION- ROTATION BANDS OF  BORON  TRIFLUORIDE.                                                   0191
        VIBRATIONAL ANHARMONICITY IN  BORON  TRIFLUORIDE.                                                       0210
        PHOTOELECTRON SPECTRA OF COMPLEXES OF  BORON  TRIFLUORIDE AND AMINES.                                   0224
            AB INITIO CALCULATION ON  BORON  TRIFLUORIDE AND BORON TRICHLORIDE.                                 0213
    MOLECULAR STRUCTURE AND FORCE CONSTANTS OF  BORON  TRIFLUORIDE AND BORON TRICHLORIDE.                       0221
TS OF THE EI/ FORCE CONSTANTS OF CARBON TETRAFLUORIDE AND  BORON  TRIFLUORIDE AND COMPARISON WITH FORCE CONSTAN  0212
NU-4 BANDS OF THE TEN AND ELEVEN ISOTOPES OF  BORON  TRIFLUORIDE AT HIGH RESOLUTION.                            0208
C FREE RADICALS OBTAINED IN ELECTRICAL DISCHARGES THROUGH  BORON  TRIFLUORIDE. (BORON DIFLUORIDE) /IN POLYATOMI  0222
ENTS.                HEAT FORMATION  BORON  TRIFLUORIDE BY DIRECT COMBINATION OF THE ELEM  0198
/IVE LIFETIMES OF ULTRAVIOLET EMISSION SYSTEMS EXCITED IN  BORON  TRIFLUORIDE, CARBON TETRAFLUORIDE, AND SILIC/  0216
/ONSTANTS OF CARBON TETRAFLUORIDE, SILICON TETRAFLUORIDE,  BORON  TRIFLUORIDE, ETHANE, SILANE, AMMONIA, AND PH/  0243
                NUCLEAR MAGNETIC RELAXATION IN  BORON  TRIFLUORIDE GAS.                                         0217
                    RAMAN SPECTRUM OF  BORON  TRIFLUORIDE GAS.                                                  0251
                    ABSORPTION OF  BORON  TRIFLUORIDE GAS IN THE EXTREME ULTRAVIOLET.                           0215
FLUORIDES. (SILICON TETRAFLUORIDE, CARBON TETRAFLUORIDE,  BORON  TRIFLUORIDE, NITROGEN TRIFLUORIDE) /OLYATOMIC  0186
ANGE BETWEEN TETRABUTYL AMMONIUM HEXAFLUORO PHOSPHATE AND  BORON  TRIFLUORIDE OR PHOSPHORUS PENTAFLUORIDE. /XCH  0194
/IC FORCE FIELDS OF XY3 MOLECULES. (NITROGEN TRIFLUORIDE,  BORON  TRIFLUORIDE, PHOSPHORUS TRIFLUORIDE, AND ARS/  0709
/ORCE CONSTANTS FROM HEAVY ISOTOPIC SUBSTITUTION. PART-2:  BORON  TRIFLUORIDE, SILICON TETRAFLUORIDE, AND NITR/  0232
ENTAFLUORIDE, ARSENI/ INFRARED MATRIX ISOLATION STUDIES. (BORON  TRIFLUORIDE, SULFUR TETRAFLUORIDE, ANTIMONY P  0239
ERMINATION OF THE MOLECULAR GEOMETRY.            BORON  TRIFLUORIDE. THE ANALYSIS OF NU-2 AND THE DET  0209
TERMINATION OF THE 33 CORIOLIS CONSTANT.            BORON  TRIFLUORIDE. THE ANALYSIS OF 2NU-3 AND THE DE  0192
        CALCULATED MEAN AMPLITUDES OF VIBRATION IN  BORON  TRIHALIDES.                                          0197
D-JONES POTENTIALS ON THE UREY-BRADLEY FORCE CONSTANTS OF  BORON  TRIHALIDES.               EFFECTS OF THE LENNAR  0244
            FORCE FIELDS FOR THE  BORON  TRIHALIDES.                                                            0227
KINEMATICAL EVALUATION OF FORCE CONSTANTS. APPLICATION TO  BORON  TRIHALIDES.                                   0248
PECTRA OF HALIDES. PART-2: HIGH RESOLUTION SPECTRA OF THE  BORON  TRIHALIDES.                 PHOTOELECTRON S  0188
        THERMODYNAMIC PROPERTIES OF SOME  BORON  TRIHALIDES.                                                    0236
            UNIQUE FORCE FIELD FOR  BORON  TRIHALIDES.                                                          0238
```

VACUUM ULTRAVIOLET ABSORPTION SPECTRA OF BORON TRIHALIDES.   0237
FORCE CONSTANTS OF THE BORON TRIHALIDES - A SURVEY.   0201
A THEORETICAL COMPARISON OF THE ELECTRONIC STRUCTURES OF BORON TRIHALIDES AND BORON TETRAHALIDE ANIONS.   0225
NORMAL VIBRATIONS OF BORON TRIHALIDES. (BORON TRIFLUORIDE)   0247
E C/ MEAN AMPLITUDES OF VIBRATION AND SHRINKAGE EFFECT OF BORON TRIHALIDES, CALCULATED FROM UREY-BRADLEY FORC 0233
YDRIDE/ AB INITIO STUDIES OF THE ELECTRONIC STRUCTURES OF BORON TRIHYDRIDE, BORON DIHYDRIDE FLUORIDE, BORON H 0241
LOCALIZED ORBITALS IN BORON FLUORIDES. HIGHLY POLARIZED BORON- FLUORINE DOUBLE AND TRIPLE BONDS.   0158
G/ STUDY OF THE A SINGLET PI- X SINGLET SIGMA(+) BANDS OF BORON-11 MONOFLUORIDE WITH A VACUUM ECHELLE SPECTRO 0170
OBABLE NONEXISTENCE OF XENON MONOFLUORIDE AS A CHEMICALLY BOUND SPECIES IN THE GAS PHASE.   PR 1221
A STUDY OF THE RAMAN AND BROAD LINE NMR SPECTRA OF GERMANIUM TETRAFLUORIDE.   0484
IUM MONOFLUORIDE, THALLIUM MONOCHLORIDE, AND THALLIUM MONOBROMIDE. /OF HIGH TEMPERATURE VAPORS. PART-4: THALL 1112
/LLIUM FLUORIDE, CESIUM FLUORIDE, CESIUM CHLORIDE, CESIUM BROMIDE, AND CESIUM IODIDE AND MAGNETIC SUSCEPTIBI 1120
: MAGNESIUM DIFLUORIDE, MAGNESIUM DICHLORIDE, MAGNESIUM DIBROMIDE, AND MAGNESIUM DIIODIDE. /GNESIUM DIHALIDES 0607
INTENSITIES OF FORBIDDEN TRANSITIONS IN IODINE, OR IODINE BROMIDE (CHLORIDE, FLUORIDE) MOLECULES. / RELATIVE 0498
OL/ THERMODYNAMIC FUNCTIONS OF ANTIMONY FLUORIDE, NIOBIUM BROMIDE, TANTALUM CHLORIDE, TANTALUM BROMIDE, AND M 0075
D KRYPTON MONOFLUORIDE.   EMISSION SPECTRA OF XENON MONOBROMIDE, XENON MONOCHLORIDE, XENON MONOFLUORIDE, AN 0545
IATOMIC INTERHALOGENS FROM SPECTROSCOPIC DATA. (CHLORINE, BROMINE, AND IODINE MONOFLUORIDES) /ERTIES OF THE D 0253
ENERGIES OF DIATOMIC HALOGEN FLUORIDES. (IODINE FLUORIDE, BROMINE FLUORIDE)   DISSOCIATION 0494
RIFLUORIDE, AND BROMINE PENTAFLU/ INFRARED STUDIES OF THE BROMINE FLUORIDES. (BROMINE MONOFLUORIDE, BROMINE T 0256
XAFLUORIDES. (CHLORINE HEXAFLUORIDE, IODINE HEXAFLUORIDE, BROMINE HEXAFLUORIDE) /ONANCE SPECTRA OF HALOGEN HE 0271
ESR SPECTRUM AND STRUCTURE OF BROMINE HEXAFLUORIDE.   0270
MATRIX ISOLATION OF BROMINE MONOFLUORIDE.   0255
RK EFFECT OF LINEAR MOLECULES. (CHLORINE MONOFLUORIDE AND BROMINE MONOFLUORIDE) /IONS TO THE SECOND ORDER STA 0385
OF HALIDES. (CHLORINE MONOFLUORIDE, CHLORINE TRIFLUORIDE, BROMINE MONOFLUORIDE, AND BROMINE TRIFLUORIDE) /RA 0376
S AND THE INTERHALOGEN COMPOUNDS. (CHLORINE MONOFLUORIDE, BROMINE MONOFLUORIDE, AND IODINE MONOFLUORIDE) /IDE 0373
SPECTRUM OF BROMINE MONOFLUORIDE AND ITS DISSOCIATION ENERGY.   0252
/ND ELECTRIC QUADRUPOLE MOMENTS IN CHLORINE MONOFLUORIDE, BROMINE MONOFLUORIDE, BROMINE CYANIDE, AND IODINE 0254
DE IN J-5/2 ROTATIONAL LEVELS.   ESR SPECTRUM OF BROMINE OXIDE, IODINE OXIDE AND SELENIUM MONOFLUORI 0954
INFRARED SPECTRUM OF BROMINE PENTAFLUORIDE.   0262
K-TYPE DOUBLING IN THE MILLIMETER WAVE SPECTRUM OF BROMINE PENTAFLUORIDE.   0260
MICROWAVE MEASUREMENTS OF THE J-8 FAR-9 TRANSITION OF BROMINE PENTAFLUORIDE.   0261
MICROWAVE SPECTRUM OF BROMINE PENTAFLUORIDE.   0269
MOLECULAR FORCE FIELD FOR BROMINE PENTAFLUORIDE.   0264
S FOR INTERHALOGEN COMPOUNDS. (CHLORINE PENTAFLUORIDE AND BROMINE PENTAFLUORIDE)   MOLECULAR FORCE FIELD 0425
, AND THERMODYNAMIC FUNCTIONS OF IODINE PENTAFLUORIDE AND BROMINE PENTAFLUORIDE. / STRUCTURE, FORCE CONSTANTS 0505
ES. (PHOSPHORUS PENTAFLUORIDE, ARSENIC PENTAFLUORIDE, AND BROMINE PENTAFLUORIDE) / STRUCTURES OF PENTAFLUORID 0831
RUCTURE, FORCE CONSTANTS, AND THERMODYNAMIC PROPERTIES OF BROMINE PENTAFLUORIDE.   RAMAN SPECTRUM, ST 0267
SELF IONIZATION OF HALOGEN FLUORIDES. (BROMINE PENTAFLUORIDE)   0263
/ECULES. XENON OXIDE TETRAFLUORIDE, IODINE PENTAFLUORIDE, BROMINE PENTAFLUORIDE, AND CHLORINE PENTAFLUORIDE. 0418
/ SOME SQUARE PYRAMIDAL MOLECULES. (IODINE PENTAFLUORIDE, BROMINE PENTAFLUORIDE, AND CHLORINE PENTAFLUORIDE) 0420
/OR SOME PYRAMIDAL XY4Z MOLECULES. (IODINE PENTAFLUORIDE, BROMINE PENTAFLUORIDE, AND CHLORINE PENTAFLUORIDE) 0421
CORIOLIS COUPLING COEFFICIENTS OF BROMINE PENTAFLUORIDE AND CHLORINE PENTAFLUORIDE.   0427
HE CONDENSED PHASE. EVIDEN/ INFRARED AND RAMAN SPECTRA OF BROMINE PENTAFLUORIDE AND CHLORINE TRIFLUORIDE IN T 0266
L GAS THERMODYNAMIC PROPERTIES OF CHLORINE PENTAFLUORIDE, BROMINE PENTAFLUORIDE, AND IODINE PENTAFLUORIDE. /A 0424
ELECTRON DIFFRACTION / GAS PHASE MOLECULAR STRUCTURES OF BROMINE PENTAFLUORIDE AND IODINE PENTAFLUORIDE FROM 0265
REGION.   ROTATIONAL SPECTRA OF BROMINE PENTAFLUORIDE IN THE MILLIMETER WAVE LENGTH 0268
/S. PART-1: CALCULATION METHODS. (CHLORINE PENTAFLUORIDE, BROMINE PENTAFLUORIDE, IODINE PENTAFLUORIDE, PHOSP/ 0422
ELECTRICAL CONDUCTIVITY OF SOLID CHLORINE TRIFLUORIDE AND BROMINE TRIFLUORIDE.   0259
INFRARED AND RAMAN SPECTRA OF CHLORINE TRIFLUORIDE AND BROMINE TRIFLUORIDE.   0408
MICROWAVE SPECTRUM AND MOLECULAR STRUCTURE OF BROMINE TRIFLUORIDE.   0258
ELD FOR INTERHALOGEN COMPOUNDS. (CHLORINE TRIFLUORIDE AND BROMINE TRIFLUORIDE)   MOLECULAR FORCE FI 0407
QUADRUPOLE COUPLING CONSTANTS IN CHLORINE TRIFLUORIDE AND BROMINE TRIFLUORIDE.   0400
MAN SPECTRA AND THE STRUCTURE OF CHLORINE TRIFLUORIDE AND BROMINE TRIFLUORIDE.   RA 0398
AND THERMODYNAMIC PROPERTIES OF CHLORINE TRIFLUORIDE AND BROMINE TRIFLUORIDE.   VIBRATIONAL SPECTRA 0396
VIBRATIONAL SPECTRUM OF BROMINE TRIFLUORIDE.   0257
NTAFLUORIDE, ARSENIC PENTAFLUORIDE, CHLORINE TRIFLUORIDE, BROMINE TRIFLUORIDE, AND BROMINE PENTAFLUORIDE) /PE 0239
/NFRARED SPECTRA OF MATRIX ISOLATED CHLORINE TRIFLUORIDE, BROMINE TRIFLUORIDE, AND BROMINE PENTAFLUORIDE. FL/ 0399
/PROPERTIES OF SOME INTERHALOGENS. (CHLORINE TRIFLUORIDE, BROMINE TRIFLUORIDE, IODINE TRIFLUORIDE, CHLORINE / 0397
N AMPLITUDES AND CORIOLIS CONSTANTS IN DICHLORO BORANE, DIBROMO BORANE AND CHLORINE TRIFLUORIDE. /RALIZED MEA 0411
NTS OF THE SILICON TETRAHALIDES AND FREQUENCIES OF CHLORO BROMO SILICON.   FORCE CONSTA 1004
UORIDE OR / INTERMOLECULAR FLUORINE EXCHANGE BETWEEN TETRABUTYL AMMONIUM HEXAFLUORO PHOSPHATE AND BORON TRIFL 0194

C

RMANIUM MONOFLUORIDE.   VIBRATIONAL ANALYSIS OF C-X AND C'-X' BAND SYSTEMS AND ENERGY OF DISSOCIATION IN GE 0470
DELTA 5/2- X DOUBLET PI 3/2 / ROTATIONAL ANALYSIS OF THE C DOUBLET DELTA 5/2- X DOUBLET PI 3/2 AND G DOUBLET 1136
UM MONOFLUORIDE.   THE C DOUBLET PI AND THE X DOUBLET SIGMA STATES OF BARI 0120
ONOFLUORIDE.   ROTATIONAL ANALYSIS OF THE C SINGLET PI- X SINGLET SIGMA(+) SYSTEM OF INDIUM M 0491
EM OF BORON MONOFLUO/ ROTATIONAL ANALYSIS OF BANDS OF THE C TRIPLET SIGMA, B TRIPLET SIGMA: A TRIPLET PI SYST 0153
ON IN GERMANIUM MONOFLUORIDE.   VIBRATIONAL ANALYSIS OF C-X AND C'-X' BAND SYSTEMS AND ENERGY OF DISSOCIATI 0470
FRANCK-CONDON FACTORS AND R CENTROIDS FOR THE C-X SYSTEM OF SILICON MONOFLUORIDE MOLECULE. 0975
)   VIBRATIONAL SPECTRA OF THE DIHALIDES OF MERCURY AND CADMIUM. (CADMIUM DIFLUORIDE AND MERCURY DIFLUORIDE 0275
STRONTIUM DIFLUORIDE, BARIUM DIFLUORIDE, ZINC DIFLUORIDE, CADMIUM DIFLUORIDE) /FLUORIDE, CALCIUM DIFLUORIDE, 0131
AND LANGMUIR MEASUREMENTS OF THE SUBLIMATION PRESSURE OF CADMIUM DIFLUORIDE.   KNUDSEN 0272
RAMAN SPECTRA OF CADMIUM DIFLUORIDE AND LEAD DIFLUORIDE.   0274
ORCE CONSTANTS OF THE GROUP-2B HALIDES. (ZINC DIFLUORIDE, CADMIUM DIFLUORIDE, AND MERCURY DIFLUORIDE) /DING F 1322
IFLUORIDE, STRONTIUM / VAPOR PRESSURE OF ZINC DIFLUORIDE, CADMIUM DIFLUORIDE, MAGNESIUM DIFLUORIDE, CALCIUM D 0049
DIFLUORIDE, AND/ MELTING POINTS OF INORGANIC FLUORIDES. (CADMIUM DIFLUORIDE, MAGNESIUM DIFLUORIDE, STRONTIUM 0273
Y AND ENTHALP/ ENTROPIES AND ENTHALPIES OF SUBLIMATION OF CALCIUM AND CERIUM FLUORIDES. CORRELATION OF ENTROP 0281
CIATION ENERGY OF MONOFLUORIDE COMPOUNDS OF MAGNESIUM AND CALCIUM AND THE IONIC MODEL OF A MOLECULE.   DISSO 0278
ARTH DIHALIDES. (BARIUM DIFLUORIDE, STRONTIUM DIFLUORIDE, CALCIUM DIFLUORIDE)   GEOMETRY OF THE ALKALINE E 0126
DIHALIDES BERYLLIUM DIFLUORIDE, MAGNESIUM DIFLUORIDE, AND CALCIUM DIFLUORIDE. /O SCF STUDIES OF THE GROUP-2A 1360
X-RAY STUDY OF ANHARMONIC VIBRATIONS IN CALCIUM DIFLUORIDE.   0283
ALCIUM MONOFLUORIDE.   SUBLIMATION PRESSURE OF CALCIUM DIFLUORIDE AND THE DISSOCIATION ENERGY OF C 0276
M DIFLUORIDE IN THE/ SPECTRA OF ELECTRONIC EXCITATIONS IN CALCIUM DIFLUORIDE, STRONTIUM DIFLUORIDE, AND BARIU 0279
/NC DIFLUORIDE, CADMIUM DIFLUORIDE, MAGNESIUM DIFLUORIDE, CALCIUM DIFLUORIDE, STRONTIUM DIFLUORIDE, BARIUM D/ 0049
/2 ELEMENTS. (BERYLLIUM DIFLUORIDE, MAGNESIUM DIFLUORIDE, CALCIUM DIFLUORIDE, STRONTIUM DIFLUORIDE, BARIUM D/ 0131
VAPOR PRESSURE AND HEAT OF SUBLIMATION OF CALCIUM FLUORIDE.   0282
IDES)   REEVALUATION OF THE DISSOCIATION ENERGY OF CALCIUM FLUORIDE. (ALSO SILICON AND GERMANIUM FLUOR 0277
/ULAR NITROGEN, NITRIC OXIDE, AND MOLECULAR OXYGEN AND OF CALCIUM FLUORIDE, CHROMIUM(II) FLUORIDE, MANGANESE/ 1324

THE BANDS OF B DOUBLET SIGMA- X DOUBLET SIGMA. SYSTEM OF CALCIUM FLUORIDE MOLECULE IN EMISSION SPECTRUM. /OF 0280
R CENTROIDS OF SOME BAND SYSTEMS OF SILICON MONOFLUORIDE, CALCIUM MONOFLUORIDE, AND BISMUTH MONOFLUORIDE. /D 0972
CALCULATION OF THE DISSOCIATION ENERGY OF CALCIUM MONOFLUORIDE AND URANIUM MONOFLUORIDE. 1161
FORCE CONSTANTS OF OCTAHEDRAL MOLECULES CALCULATED BY THE LOGARITHMIC STEP METHOD. 1472
/TS AND MEAN VIBRATION AMPLITUDES OF IODINE HEPTAFLUORIDE CALCULATED BY THE LOGARITHMIC STEP METHOD IN THE R/ 0535
/LS OF PLATINUM HEXAFLUORIDE AND ELEMENT 110 HEXAFLUORIDE CALCULATED BY THE SELF CONSISTENT MULTIPLE SCATTER/ 0881
RIDE. CALCULATED ELECTRONIC STRUCTURE OF URANIUM HEXAFLUO 1188
ES OF VIBRATION AND SHRINKAGE EFFECT OF BORON TRIHALIDES, CALCULATED FROM UREY-BRADLEY FORCE CONSTANTS. /ITUD 0233
ROMETHANES, TETRAFLUORC METHANE XENON TETRAFLUORIDE, AND/ CALCULATED IONIZATION POTENTIAL OF CHLORO- AND FLUO 0344
IHALIDES. CALCULATED MEAN AMPLITUDES OF VIBRATION IN BORON TR 0197
L CHARACTERISTICS OF MOLECULES OF SOME TETRAFLUORIDES FOR CALCULATING THEIR THERMODYNAMIC FUNCTIONS. /TATIONA 0622
NIC STRUCTURE OF CHLORINE TRIFLUORIDE. AN APPROXIMATE SCF CALCULATION. ELECTRO 0404
/OF NORMAL COORDINATE TRANSFORMATIONS. APPLICATION TO THE CALCULATION BOND MOMENT PARAMETERS AND FORCE CONST/ 0352
/RE OF THE MOLECULE CHLORINE DIFLUORIDE ON THE BASIS OF A CALCULATION BY THE MO LCAO SCF IN THE NDDO APPROXI/ 0382
FORCE CONSTANT CALCULATION FOR CHLORINE TRIFLUORIDE. 0403
IFLUORIDE. PART-3: A GAUSSIAN BASIS SELF CONSISTENT FIELD CALCULATION FOR NITROGEN TRIFLUORIDE. / NITROGEN TR 0746
ELECTRONIC PROPERTIES OF SULFUR HEXAFLUORIDE. AB INITIO CALCULATION FOR THE GROUND STATE. 1077
DE, NITRITE ION AND CYANIDE IO/ CONFIGURATION INTERACTION CALCULATION FOR THE GROUND STATE OF OXYGEN DIFLUORI 0763
/OPERTIES OF PENTAFLUORIDES AT HIGH TEMPERATURES. PART-1: CALCULATION METHODS. (CHLORINE PENTAFLUORIDE, BROM/ 0422
MMETRICAL MOLECU/ USE OF A SIMPLE ELECTROSTATIC MODEL FOR CALCULATION OF ANGULAR FORCE CONSTANTS OF HIGHLY SY 1484
POLYATOMIC MOLECULES. THEORY AND CALCULATION OF CENTRIFUGAL DISTORTION CONSTANTS FOR 1477
ED INTENSITIES OF BORON TRIFLUORIDE. CNDO CALCULATION OF DIPOLE MOMENT DERIVATIVES AND INFRAR 0195
S OF THE SCANDIUM SUBGROUP. (SCANDIUM TRIFLUORIDE, YTTRI/ CALCULATION OF DISSOCIATION ENERGIES FOR THE HALIDE 0948
/ CONSISTENT MOLECULAR ORBITAL METHODS. PART-5: AB INITIO CALCULATION OF EQUILIBRIUM GEOMETRIES AND QUADRATI/ 0720
VIBRATION AMPLITUDES OF PHOSPHORUS PENTAFLUORIDE, PHOSP/ CALCULATION OF FORCE CONSTANTS AND ROOT MEAN SQUARE 0826
/ UTILITY OF HEAVY ATOM ISOTOPIC SUBSTITUION DATA FOR THE CALCULATION OF FORCE CONSTANTS. ANHARMONICITY EFFEC 0234
PECIES FORCE FIELD OF PHOSPHORUS/ PARAMETER METHOD IN THE CALCULATION OF FORCE CONSTANTS. APPLICATION TO E' S 0827
IDE AND NITROGEN TRIFLUORIDE. CALCULATION OF FORCE CONSTANTS FOR NITROGEN DIFLUOR 0693
RLAP METHOD. MOLECULES WITH LONE PAIR ELECTRONS OR EMPTY/ CALCULATION OF FORCE CONSTANTS WITH THE MAXIMUM OVE 1395
RLAP METHOD. TETRAHEDRAL MOLECULES. CALCULATION OF FORCE CONSTANTS WITH THE MAXIMUM OVE 1416
MOLECULES WITH D4H SYMMETRY. NEW METHOD FOR THE CALCULATION OF GVFF CONSTANTS. 1412
CALCULATION OF MEAN SQUARE VIBRATIONAL AMPLITUDES. 1486
: APPLICATION TO THE NU-4 BAND OF NITROGEN TRIFLUORIDE. CALCULATION OF PERPENDICULAR INFRARED BAND CONTOURS 0747
BORON FLUORIDE. CALCULATION OF RYDBERG LEVELS IN NITROGEN OXIDE AND 0164
PHOSPHORUS TRIFLUORIDE. MEASUREMENT AND CALCULATION OF THE ABSOLUTE INFRARED INTENSITIES OF 0808
ONOFLUORIDE AND URANIUM MONOFLUORIDE. CALCULATION OF THE DISSOCIATION ENERGY OF CALCIUM M 1161
NE PENTAFLUORIDE MOLECULE BY THE MO LCAO SELF CONSISTENT/ CALCULATION OF THE ELECTRONIC STRUCTURE OF A CHLORI 0415
HEXAFLUORIDE. MULTIPLE SCATTERING X(ALPHA) CALCULATION OF THE ENERGY LEVEL STRUCTURE OF XENON 1304
. CALCULATION OF THE FORCE CONSTANTS OF HEXAFLUORIDES 1448
M. CENTRIFUGAL DISTORTION EFFECTS IN SULFUR DIFLUORIDE. CALCULATION OF THE FORCE FIELD AND INFRARED SPECTRU 1031
ES OF SILICON FLUORIDE, GERMANIUM FLUORIDE, AND GERMANIU/ CALCULATION OF THE FRANCK-CONDON FACTOR FOR MOLECUL 0977
IONS OF ANTIMONY TRIFLUORIDE. CALCULATION OF THE FREQUENCIES OF THE NORMAL VIBRAT 0074
FLUORIDE USING CNDO/2 TECHNIQUES. CALCULATION OF THE INFRARED SPECTRA OF SULFUR TETRA 1047
RAL STATES OF NITROGEN MONOFLUORIDE. CALCULATION OF THE SPECTROSCOPIC CONSTANTS FOR SEVE 0676
IUM SUBGROUP HALIDES. CALCULATION OF THE VIBRATIONAL FREQUENCIES OF SCAND 0949
HOSPHORUS TRIFLUORIDE, ARSENIC TRIFLUORIDE, AND ANTIMONY/ CALCULATION OF UREY-BRADLEY POTENTIAL CONSTANTS. (P 0809
ANAR XY3 MOLECULES. CALCULATION OF UREY-BRADLEY POTENTIAL CONSTANTS. PL 1393
IDE. AB INITIO CALCULATION ON BORON TRIFLUORIDE AND BORON TRICHLOR 0213
SITION ELEMENTS FROM THE DATA OF LCAO MO SCF NONEMPIRICAL CALCULATIONS. /IC STRUCTURE OF FLUORIDES OF NONTRAN 0009
KRYPTON MONOFLUORIDE AND ITS POSITIVE ION. (SCF CALCULATIONS) 0547
AND XENON OXIDE TETRAFLUORIDE VAPORS, AND FORCE CONSTANT CALCULATIONS. /NON DIFLUORIDE, XENON TETRAFLUORIDE, 1254
YDRIDE, GERMANIUM TETRAHYDRIDE, ARSENIC TR/ MOLECULAR SCF CALCULATIONS FOR GALLIUM MONOFLUORIDE, GALLIUM TRIH 0459
DOUBLE ZETA SCF CALCULATIONS FOR NITRITE ION AND OXYGEN DIFLUORIDE. 0762
VAPOR PRESSURES AND THERMODYNAMIC CALCULATIONS FOR SOME ACTINIDE FLUORIDES 1336
ION, NITROGEN OXYFLUORIDE, AND OX/ MINIMAL BASIS SET SCF CALCULATIONS FOR THE GROUND STATE OF OZONE, NITRITE 0776
MALL GAUSSIAN BASIS SET FOR NONEMPIRICAL ALL-ELECTRON SCF CALCULATIONS IN IODINE COMPOUNDS. A S 0495
HALOGENS AND INERT GASES BY THE NONEMPIRICAL NDDO-2 (AI/ CALCULATIONS OF ELECTRON STRUCTURES OF COMPOUNDS OF 0016
OF HALOGENS AND INERT GASES BY THE SCF LCAO MC METHOD. / CALCULATIONS OF ELECTRONIC STRUCTURES OF COMPOUNDS 1231
RIDE (-), CARBON TETRAFLUORIDE, SILICON TET/ COMPARISON OF CALCULATIONS OF FORCE CONSTANTS FOR BORON TETRAFLUO 0348
EY-BRADLEY FORCE FIELDS FOR PLANAR XY3 MOLECULES. CALCULATIONS OF MEAN AMPLITUDES OF VIBRATION AND UR 1399
OELECTRONIC SERIES CHLORINE MONOFLUORIDE, HYDROG/ SCF- MO CALCULATIONS OF SOME MOLECULAR PROPERTIES OF THE IS 0371
AB INITIO CALCULATIONS OF THE BONDING IN KRYPTON DIFLUORIDE. 0553
S TRIFLUORIDE AND TRIMETHYL PHOSPHINE. AB INITIO CALCULATIONS OF THE BONDING IN PHOSPHINE, PHOSPHORU 0802
L TRANSITIONS OF URANIUM HEXAFLUORIDE AND UR/ THEORETICAL CALCULATIONS OF THE ELECTRONIC STRUCTURE AND OPTICA 1179
DIFLUORIDE. NONEMPIRICAL LCAO MO SCF CALCULATIONS OF THE ELECTRONIC STRUCTURE OF SILICON 0995
COMPARATIVE AB INITIO CALCULATIONS ON BORON FLUORIDES. 0203
CHLORINE MONOFLUORIDE. SCF CALCULATIONS ON CHLORATE ION, HYDROGEN CHLORIDE AND 0143
ROM DIPOLE MOMENTS FOR ARSENIC TRIFLUORIDE AND ARSENIC T/ CALCULATIONS ON IONIC CHARACTER AND HYBRIDIZATION F 0107
ELF CONSISTENT CHARGE AND CONFIGURATION MOLECULAR ORBITAL CALCULATIONS ON PHOSPHORUS TRIFLUORIDE. S 0821
US OR SULFUR. NONEMPIRICAL VALENCE ELECTRON CALCULATIONS ON SMALL MOLECULES CONTAINING PHOSPHOR 1029
CTED HARTREE-FOCK METHOD. (BORON DIFLUORIDE, N/ AB INITIO CALCULATIONS ON SOME SMALL RADICALS BY THE UNRESTRI 0176
SPECTROSCOPIC CALCULATIONS ON SULFUR HEXAFLUORIDE. 1071
INTERHALOGENS. (CH/ CNDO/2 AND INDO ALL VALENCE ELECTRON CALCULATIONS ON THE GEOMETRY AND PROPERTIES OF SOME 0397
SIMPLE THEORETICAL CALCULATIONS ON TITANIUM MONOFLUORIDE. 1144
E GROUP-2 HALIDES. IONIC MODEL CALCULATIONS. PART-2: BENDING FORCE CONSTANTS OF TH 1381
ORIDE. SEMIEMPIRICAL MOLECULAR ORBITAL CALCULATIONS. PSEUDOROTATION IN PHOSPHORUS PENTAFLU 0843
PACITY BETWEEN 5 AND 350 DEGREES K, ENTHALPY OF FUSION A/ CALORIMETRIC STUDY OF IODINE PENTAFLUORIDE. HEAT CA 0510
FROM 456 TO 1083 DEGREES K BY TRANSPOSED TEMPERATURE DROP CALORIMETRY. ENTHALPY OF BERYLLIUM FLUORIDE 0142
PY OF FORMATION OF THORIUM TETRAFLUORIDE BY FLUORINE BOMB CALORIMETRY. ENTHAL 1130
Y OF FORMATION OF VANADIUM PENTAFLUORIDE BY FLUORINE BOMB CALORIMETRY. ENTHALP 1215
BORON TRIFLUORIDE. FLUORINE BOMB CALORIMETRY. PART-15: THE ENTHALPY OF FORMATION OF 0220
N OF PHOSPHORUS PENTAFLUORIDE AND ENTHALPI/ FLUORINE BOMB CALORIMETRY. PART-18: STANDARD ENTHALPY OF FORMATIO 0865
ALUMINUM TRIFLUORIDE. FLUORINE BOMB CALORIMETRY. PART-22: THE ENTHALPY OF FORMATION OF 0048
TRIFLUORIDE. FLUORINE BOMB CALORIMETRY. PART-3: THE HEAT OF FORMATION OF BORON 0250
CAMERON SYSTEM OF BORON MONOFLUORIDE. 0163
ROUP METAL HEXAFLUORIDE MOLECULES DOWN TO HELIUM TE/ HEAT CAPACITIES AND ELECTRONIC SPECTRA OF THE PLATINUM G 1159
TRAFLUORIDE AND URANIUM HEXAFLUORID/ LOW TEMPERATURE HEAT CAPACITIES, ENTHALPIES, AND ENTROPIES OF URANIUM TE 1166
R MOLECULES. CONSTANT VOLUME HEAT CAPACITIES OF GASEOUS CARBON TETRAFLUORIDE AND OTHE 0335
UM-242 TETRAFLUORIDE FROM 10 TO 350 DE/ EXPERIMENTAL HEAT CAPACITIES OF PLUTONIUM-242 TRIFLUORIDE AND PLUTONI 0885
HEXAFLUORIDE FROM 5 TO 350 DEGREES K. HEAT CAPACITY AND OTHER THERMODYNAMIC FUNCTIONS OF XENON 1305

BDENUM HEXAFLUORIDE BETWEEN 4 AND 350 DEGREES K. HEAT CAPACITY AND OTHER THERMODYNAMIC PROPERTIES OF MOLY 0636
ON TETRAFLUORIDE. HEAT CAPACITY AND RELATED THERMODYNAMIC FUNCTIONS OF XEN 1271
DEGREES K. RECONCILIATION OF T/ XENON TETRAFLUORIDE. HEAT CAPACITY AND THERMODYNAMIC FUNCTIONS FROM 5 TO 350 1275
LUORIDE FROM 5 TO 445 DEGREES K. HEAT CAPACITY AND THERMODYNAMIC PROPERTIES OF THALLIUM F 1124
USION A/ CALORIMETRIC STUDY OF IODINE PENTAFLUORIDE. HEAT CAPACITY BETWEEN 5 AND 350 DEGREES K, ENTHALPY OF F 0510
PERTIES OF NEPTUNIUM HEXAFLUORIDE FROM 7 TO 350 DEG/ HEAT CAPACITY, ENTHALPY OF FUSION, AND THERMODYNAMIC PRO 0650
IFLUORIDE FROM 10 TC 350 DEGREES K. HEAT CAPACITY, ENTROPY, AND ENTHALPY OF PLUTONIUM-242 TR 0886
DENUM HEXAFLUORIDE AND NIOBIUM PENTAFLUORIDE. HEAT CAPACITY, ENTROPY, AND HEATS OF TRANSITION OF MOLYB 0662
UTONIUM-242 TETRAFLUORIDE FROM 10 TO 350 DEGREES K. HEAT CAPACITY, ENTROPY, ENTHALPY, AND GIBBS ENERGY OF PI 0889
HERMODYNAMIC PROPERTIES. XENON DIFLUORIDE. HEAT CAPACITY FROM 5 TO 350 DEGREES K AND SOME DERIVED T 1246
CONSTANT VOLUME HEAT CAPACITY OF GASEOUS CARBON TETRAFLUORIDE. 0336
RAMAN SPECTRA OF CARBON AND SILICON TETRAFLUORIDES. 0364
ORIDE MOLECULES. (OXYGEN DIFLUORIDE, NITROGEN DIFLUORIDE, CARBON DIFLUORIDE) /ACTION FORCE CONSTANTS IN DIFLU 0299
BORON DIFLUORIDE, NITROGEN DIFLUORIDE, OXYGEN DIFLUORIDE, CARBON DIFLUORIDE) /STRICTED HARTREE-FOCK METHOD. (0176
ABSORPTION SPECTRUM OF CARBON DIFLUORIDE. 0302
ELECTRONIC SPECTRA OF POLYATOMIC FREE RADICALS. (CARBON DIFLUORIDE) 0308
EMISSION BANDS OF CARBON DIFLUORIDE. 0309
HEAT OF FORMATION OF CARBON DIFLUORIDE. 0298
MICROWAVE SPECTRUM OF CARBON DIFLUORIDE. 0307
2500 ANGSTROM ABSORPTION SPECTRUM OF CARBON DIFLUORIDE. 0306
ABSORPTION SPECTRUM OF CARBON DIFLUORIDE AND ITS VIBRATIONAL ANALYSIS. 0303
F/ MEAN AMPLITUDES OF VIBRATION FOR THE BORON DIFLUORIDE, CARBON DIFLUORIDE, NITROGEN DIFLUORIDE, ALUMINUM DI 1384
/OMIC DIFLUORIDES BERYLLIUM DIFLUORIDE, BORON DIFLUORIDE, CARBON DIFLUORIDE, NITROGEN DIFLUORIDE, AND OXYGEN/ 0141
HEAT OF FORMATION OF THE CARBON DIFLUORIDE RADICAL. 0304
ABSORPTION SPECTRUM OF CARBON DIFLUORIDE TRAPPED IN AN ARGON MATRIX. 0297
/LUORO DICHLORO ETHYLENE, AND THE APPEARANCE POTENTIAL OF CARBON DIFLUORIDE(X) FROM TETRAFLUORO ETHYLENE. (C/ 0305
USTMENT OF ANHARMONIC POTENTIAL CONSTANTS. APPLICATION TO CARBON DIOXIDE. LEAST SQUARES ADJ 1491
PHITE FLUORINE COMPOUNDS OF CARBON MONOFLUORIDE AND TETRACARBON FLUORIDE. / AND SYNTHETIC STUDIES OF THE GRA 0296
RARED SPECTRA OF THE MOLECULES CARBON TETRAFLUORIDE AND DICARBON HEXAFLUORIDE. /ORY OF INTENSITIES IN THE INF 0362
EMISSION SPECTRUM OF CARBON MONOFLUORIDE. 0291
AS PHASE ELECTRON RESCNANCE SPECTRUM AND DIPOLE MOMENT OF CARBON MONOFLUORIDE. G 0284
THE B-X SYSTEM OF CARBON MONOFLUORIDE. 0285
/TERMINATION OF THE C2 (A TRIPLET PI TO X TRIPLET PI) AND CARBON MONOFLUORIDE A DOUBLET SIGMA(+) TO X DOUBLE/ 0288
PREDISSOCIATION IN THE ABSORPTION SPECTRA OF CARBON MONOFLUORIDE AND CARBON DIFLUORIDE. 0293
CULAR FLOW EFFUSION AND MASS/ THERMODYNAMIC PROPERTIES OF CARBON MONOFLUORIDE AND CARBON DIFLUORIDE FROM MOLE 0286
ION OF CARBON MONOFLUORIDE AND C/ OSCILLATOR STRENGTHS OF CARBON MONOFLUORIDE AND COMMENTS ON HEATS OF FORMAT 0295
EIR POSITIVE AND NEGA/ MOLECULAR ORBITAL INVESTIGATION OF CARBON MONOFLUORIDE AND SILICON MONOFLUORIDE AND TH 0290
SYNTHETIC STUDIES OF THE GRAPHITE FLUORINE COMPOUNDS OF CARBON MONOFLUORIDE AND TETRACARBON FLUORIDE. / AND 0296
NEW BANDS OF THE CARBON MONOFLUORIDE MOLECULE. 0292
ROTATIONAL CONSTANTS OF THE B STATE OF THE CARBON MONOFLUORIDE MOLECULE . 0294
THEORETICAL STUDY OF THE SPECTROSCOPIC STATES OF THE CARBON MONOFLUORIDE MOLECULE. 0287
DETERMINATION OF THE DISSOCIATION ENERGIES OF CARBON MONOHALIDES. 0289
VALENCE EXCITED STATES OF NITROGEN, CARBON MONOXIDE, AND BORON FLUORIDE. 0169
ELECTRONIC STRUCTURE OF CARBON MONOXIDE AND BORON MONOFLUORIDE. 0162
/ND CHEMICAL BINDING. PART-3: THE ISOELECTRONIC NITROGEN, CARBON MONOXIDE, BORON FLUORIDE, AND CARBON, BERYL/ 0150
/MATRIX ISOLATED NICKEL FLUORIDE AND NICKEL CHLORIDE WITH CARBON MONOXIDE, MOLECULAR NITROGEN, NITRIC OXIDE,/ 1324
/LTS ON THE DIATOMIC MOLECULES LITHIUM, BERYLLIUM, BORON, CARBON, NITROGEN, FLUORINE, BORON NITRIDE, BERYLLI/ 0314
RBON, SILICON, AND / VAPOR PHASE RAMAN SPECTRA OF MH4 (M= CARBON, SILICON, GERMANIUM, AND TIN) AND MF4 (M= CA 0314
CALCULATION BOND MOMENT PARAMETERS AND FORCE CONSTANTS. (CARBON TETRAFLUORIDE) /RMATIONS. APPLICATION TO THE 0352
CE FIELD AND THE ASSIGNMENTS OF SOME MIXED HALOMETHANES. (CARBON TETRAFLUORIDE) /RANSFERABLE UREY-BRADLEY FOR 0351
ABSOLUTE INFRARED INTENSITIES OF CARBON TETRAFLUORIDE. 0340
FORCE CONSTANTS FROM HEAVY ISOTOPIC SUBSTITUTION. PART-1. CARBON TETRAFLUORIDE. ACCURATE 0320
CONSTANT VOLUME HEAT CAPACITY OF GASEOUS CARBON TETRAFLUORIDE. 0336
ELECTRON IMPACT SPECTRA OF METHANE AND CARBON TETRAFLUORIDE. 0333
ENTHALPY OF FORMATION OF CARBON TETRAFLUORIDE. 0329
ON OF THE FORCE CONSTANTS OF SOME TETRAHEDRAL MOLECULES. (CARBON TETRAFLUORIDE) EVALUATI 0359
FAR INFRARED ABSORPTION IN GASEOUS CARBON TETRAFLUORIDE. 0356
NTIAL CONSTANTS OF OSMIUM TETROXIDE AND SOME TETRAHALIDE (CARBON TETRAFLUORIDE) HARMONIC POTE 0353
MEASUREMENTS FOR COMPOUNDS WHICH PRODUCE NO PARENT ION. (CARBON TETRAFLUORIDE) / PHOTOELECTRON SPECTROSCOPIC 0339
INFRARED BAND CONTOURS. PART-1: SPHERICAL TOP MOLECULES. (CARBON TETRAFLUORIDE) 0325
INFRARED SPECTRA OF A CRYOSYSTEM. CARBON TETRAFLUORIDE. 0313
INFRARED SPECTRA OF HALOGEN-SUBSTITUTED METHANES. (CARBON TETRAFLUORIDE) 0355
RED SPECTRA OF SIMPLE MOLECULES IN LIQUID ARGON SOLUTION. CARBON TETRAFLUORIDE. INFRA 0338
INFRARED SPECTRUM OF CARBON TETRAFLUORIDE. 0343
PES. THE ATOMIC DISPLACMENTS AND INTRAMOLECULAR FORCES IN CARBON TETRAFLUORIDE. /TANTS FROM INFRARED BAND SHA 0331
MOLECULAR SPECTROSCOPY BY MEANS OF ESCA. (CARBON TETRAFLUORIDE) 0328
MOLECULAR STRUCTURE OF CARBON TETRAFLUORIDE. 0334
POTENTIAL CONSTANTS AND THERMODYNAMIC FUNCTIONS OF CARBON TETRAFLUORIDE. 0350
N SPECTRA OF CHLOROFLUORO METHANES IN THE GASEOUS STATE. (CARBON TETRAFLUORIDE) RAMA 0322
RAMAN SPECTRUM OF GASEOUS CARBON TETRAFLUORIDE. 0347
E AMPLITUDES OF VIBRATION IN SOME GROUP-4 TETRAHALIDES. (CARBON TETRAFLUORIDE) ROOT MEAN SQUAR 0342
STRUCTURES OF TRIFLUORO METHYL HALIDES. (CARBON TETRAFLUORIDE) 0318
THERMODYNAMIC PROPERTIES OF CARBON TETRAFLUORIDE. 0321
UREY-BRADLEY FORCE FIELD FOR CARBON TETRAFLUORIDE. 0360
VIBRATIONAL SPECTRA OF LIQUID AND CRYSTALLINE CARBON TETRAFLUORIDE. 0327
ARISON WITH FORCE CONSTANTS OF THE EI/ FORCE CONSTANTS OF CARBON TETRAFLUORIDE AND BORON TRIFLUORIDE AND COMP 0212
Y OF INTENSITIES IN THE INFRARED SPECTRA OF THE MOLECULES CARBON TETRAFLUORIDE AND DICARBON HEXAFLUORIDE. /OR 0362
INFRARED SPECTRA OF CARBON TETRAFLUORIDE AND GERMANIUM TETRAFLUORIDE. 0363
HE TRIPLE POINT OF CARBON TETRAFLUORID/ VAPOR PRESSURE OF CARBON TETRAFLUORIDE AND NITROGEN TRIFLUORIDE AND T 0345
EXPLOSION METHOD AND THE HEATS OF FORMATION OF CARBON TETRAFLUORIDE AND OTHER CARBON HALIDES. 0315
CONSTANT VOLUME HEAT CAPACITIES OF GASEOUS CARBON TETRAFLUORIDE AND OTHER MOLECULES. 0335
ENSITY DISTRIBUTIONS IN SOME FLUORIDES AND OXYCHLORIDES. (CARBON TETRAFLUORIDE AND SILICON TETRAFLUORIDE) / D 0337
FORCE FIELDS OF CARBON TETRAFLUORIDE AND SILICON TETRAFLUORIDE. 0324
HIGH RESOLUTION PHOTOELECTRON SPECTROSCOPY OF CARBON TETRAFLUORIDE AND SILICON TETRAFLUORIDE. 0319
LTRAVIOLET EMISSION SYSTEMS EXCITED IN BORON TRIFLUORIDE, CARBON TETRAFLUORIDE, AND SILICON TETRAFLUORIDE. /U 0216
/RA OF SOME POLYATOMIC FLUORIDES. (SILICON TETRAFLUORIDE, CARBON TETRAFLUORIDE, BORON TRIFLUORIDE, NITROGEN / 0186
/ MOLECULAR BEAM ELECTRIC DEFLECTION OF THE TETRAHALIDES CARBON TETRAFLUORIDE, CARBON TETRACHLORIDE, SILICON 0349
SORPTION SPECTRA OF SOME HALOGEN DERIVATIVES OF METHANE. (CARBON TETRAFLUORIDE) CORRELATION OF THE SPECTRA. / 0365
NG POINT. THE SIGNIFICANCE O/ THERMODYNAMIC PROPERTIES OF CARBON TETRAFLUORIDE FROM 12 DEGREES K TO ITS BOILI 0361
G POINT. THERMODYNAMIC PROPERTIES OF CARBON TETRAFLUORIDE FROM 4 DEGREES K TO ITS MELTIN 0326

```
                        ELECTRONIC STRUCTURE OF THE CARBON TETRAFLUORIDE MOLECULE.                             0357
STATE.                  ELECTRONIC STRUCTURE OF THE CARBON TETRAFLUORIDE MOLECULE. PART-1.   THE GROUND        0330
/7: VARIABLE TEMPERATURE HELIUM-1 AND HELIUM-2 STUDIES OF CARBON TETRAFLUORIDE, SILICON TETRAFLUORIDE, AND G/  0341
/CULATIONS OF FORCE CONSTANTS FOR BORON TETRAFLUORIDE(-), CARBON TETRAFLUORIDE, SILICON TETRAFLUORIDE, AND S/  0348
LFUR HE/ FLUORINE RELAXATION BY NMR ABSORPTION IN GASEOUS CARBON TETRAFLUORIDE, SILICON TETRAFLUORIDE, AND SU  0346
TRIFLUORIDE, ETHANE, SILANE, AMMONIA,/ FORCE CONSTANTS OF CARBON TETRAFLUORIDE, SILICON TETRAFLUORIDE, BORON   0243
 TETRACHLORI/ INTENSITIES OF THE INFRARED FUNDAMENTALS OF CARBON TETRAFLUORIDE, SILICON TETRAFLUORIDE, CARBON 1007
/ PART-1. TETRAFLUORIDES AND TETRACHLORIDES OF GROUP-5B. (CARBON TETRAFLUORIDE, SILICON TETRAFLUORIDE, GERMA/  0316
IUM TETRAFLUO/ THE FORCE CONSTANTS OF VARIOUS ISOTOPES OF CARBON TETRAFLUORIDE, SILICON TETRAFLUORIDE, GERMAN  0358
HE REAL GAS.                                              CARBON TETRAFLUORIDE. THERMODYNAMIC PROPERTIES OF T  0332
                   FORCE CONSTANTS OF THE CARBON TETRAHALIDES. (CARBON TETRAFLUORIDE)                          0354
          ELECTRON IMPACT STUDIES OF SOME TRIHALO METHANES. (CARBON TRIFLUORIDE)                               0312
TOCHLORINATION OF / INFRARED SPECTRUM OF THE FREE RADICAL CARBON TRIFLUORIDE ISOLATED IN INERT MATRICES.  PHO 0311
                      INFRARED DETECTION OF GASEOUS CARBON TRIFLUORIDE RADICAL.                                0310
                          BONDING IN CARBON(1) AND CARBON(2) FLUORIDES.                                        0300
                        BONDING IN CARBON(1) AND CARBON(2) FLUORIDES.                                          0300
VE ROTATIONAL ABSORPTION STUDIES. (NITROGEN DIFLUORIDE) A CAVITY SEARCH SPECTROMETER FOR FREE RADICAL MICROWA 0689
S.                     WAVE FUNCTIONS FOR THE FOUR ELECTRON THREE CENTER BONDING OF FOUR AND EIGHT PI ELECTRON SYSTEM 0301
OWAVE SPECTRUM AND FORCE CONSTANTS OF SILICON DIFLUORIDE.  CENTRIFUGAL DISTORTION.                    MICR     0989
SPECTRUM OF GERMANIUM DIFLUORIDE. QUADRUPOLE COUPLING AND  CENTRIFUGAL DISTORTION.                    MICROWAVE 0477
ECULES.                          THEORY AND CALCULATION OF CENTRIFUGAL DISTORTION CONSTANTS FOR POLYATOMIC MOL 1477
ORIDE.                     MILLIMETER WAVE SPECTRUM AND CENTRIFUGAL DISTORTION CONSTANTS OF NITROGEN TRIFLU    0718
LUORIDE.                                                  CENTRIFUGAL DISTORTION CONSTANTS OF PHOSPHORUS TRIF  0814
MOLECULES. PART-1. QUADRATIC POTENTIAL CONSTANTS AND AVE/ CENTRIFUGAL DISTORTION EFFECTS IN ASYMMETRIC ROTOR   0778
. CALCULATION OF THE FORCE FIELD AND INFRARED SPECTRUM.   CENTRIFUGAL DISTORTION EFFECTS IN SULFUR DIFLUORIDE  1031
/FLUORIDE BY THE COMBINED USE OF VIBRATIONAL FREQUENCIES, CENTRIFUGAL STRETCHING, AND CORIOLIS COUPLING CONS/  0717
/ENCIES, POTENTIAL FUNCTION, CORIOLIS COUPLING CONSTANTS, CENTRIFUGAL STRETCHING CONSTANTS, AND MEAN SQUARE /  0726
/E BY SOLUTION OF THE VIBRATIONAL PROBLEM WITH THE USE OF CENTRIFUGAL STRETCHING CONSTANTS. (NITROGEN TRIFLU/  0727
AND SOME OTHER SYMMETRIC TOPS. (ARSENIC TRIFLUORIDE)      CENTRIFUGAL STRETCHING OF TRIFLUORO CHLORO METHANE   0101
MANIUM.            VIBRATIONAL TRANSITION PROBABILITIES AND R CENTROIDS FOR DIATOMIC FLUORIDES OF SILICON AND GER 0994
I SYSTEM OF BORON MONOFLUORI/ FRANCK-CONDON FACTORS AND R CENTROIDS FOR THE B TRIPLET SIGMA(+) TO A TRIPLET P  0171
E MOLECULE.                    FRANCK-CONDON FACTORS AND R CENTROIDS FOR THE C-X SYSTEM OF SILICON MONOFLUORID 0975
BAND SYSTEM OF NITROGEN MONO/ FRANCK-CONDON FACTORS AND R CENTROIDS FOR THE 6 SINGLET SIGMA- X TRIPLET SIGMA   0677
NUM OXIDE AND SILICON MONOFL/ FRANCK-CONDON FACTORS AND R CENTROIDS OF A SINGLET SIGMA- X SINGLET SIGMA PLATI  0966
IDE, CALCIUM MONOFLUORIDE, A/ FRANCK-CONDON FACTORS AND R CENTROIDS OF SOME BAND SYSTEMS OF SILICON MONOFLUOR  0972
IDE.         RYDBERG-KLEIN-REES FRANCK-CONDON FACTORS AND R CENTROIDS OF THE A-X BAND SYSTEM OF BERYLLIUM FLUOR 0127
P/ ENTROPIES AND ENTHALPIES OF SUBLIMATION OF CALCIUM AND CERIUM FLUORIDES. CORRELATION OF ENTROPY AND ENTHAL 0281
/ES FROM INFRARED SPECTRA. (SCANDIUM, YTTRIUM, LANTHANUM, CERIUM, NEODYMIUM, EUROPIUM, AND GADOLINIUM TRIFLU/  0943
/THERMODYNAMIC FUNCTIONS OF THE TRIFLUORIDES OF SCANDIUM, CERIUM, SAMARIUM, EUROPIUM, GADOLINIUM, TERBIUM, D/  1356
ELECTRIC DEFLECTION AND THERMAL DECOMPOSITION STUDIES ON  CERIUM TETRAFLUORIDE, TERBIUM TETRAFLUORIDE, AND P/  0367
                    VAPOR PRESSURE AND HEAT OF SUBLIMATION OF CERIUM TRIFLUORIDE.                              0366
ERMODYNAMICS OF THE VAPORIZATION OF ALUMINUM FLUORIDE AND CESIUM FLUORIDE.                            TH       0052
/N DIAMAGNETIC MOLECULES. GJ FACTOR OF THALLIUM FLUORIDE, CESIUM FLUORIDE, CESIUM CHLORIDE, CESIUM BROMIDE, /  1120
SPECTRA OF RARE EARTH METAL FLUORIDES IN THE REGION OF A  CESIUM IODIDE PRISM.                        INFRARED 1352
                    THERMODYNAMIC PROPERTIES OF SOME CHALCOGEN FLUORIDES.                                      0012
IONS ON PHOSPHORUS TRIFLUORIDE.        SELF CONSISTENT CHARGE AND CONFIGURATION MOLECULAR ORBITAL CALCULAT     0821
ANT.                         RELAXATION OF THE MOLECULAR CHARGE DISTRIBUTION AND THE VIBRATIONAL FORCE CONST   0151
ORE ELECTRON BINDING ENERGIES.                           CHARGE DISTRIBUTION IN CHLORINE MONOFLUORIDE FROM C  0374
/ QUADRUPOLE MOMENTS AND SECOND MOMENTS OF THE ELECTRONIC CHARGE DISTRIBUTION IN OXYGEN DIFLUORIDE, OZONE, A/  0780
/CTURE OF ORGANIC COMPOUNDS. PART-3: AB INITIO STUDIES OF CHARGE DISTRIBUTION USING A MINIMAL SLATER-TYPE BA/  0707
THE ISOELECTRONIC NITROGEN, CARBON MONOXIDE, B/ MOLECULAR CHARGE DISTRIBUTIONS AND CHEMICAL BINDING. PART-3:  0150
PECTROSCOPY. (XENON HEXAFLUORIDE)                         CHEMICAL APPLICATIONS OF CORE ELECTRON EXCITATION S 1306
N, CARBON MONOXIDE, B/ MOLECULAR CHARGE DISTRIBUTIONS AND CHEMICAL BINDING. PART-3: THE ISOELECTRONIC NITROGE 0150
         MEAN AMPLITUDES OF THERMAL MOTION AND SHRINKAGES OF CHEMICAL BONDS: OCTAHEDRAL HEXAFLUORIDES.        1460
                                                          CHEMICAL COMPOUNDS OF XENON AND OTHER NOBLE GASES.  0010
LIDES.                                                    CHEMICAL PREDICTIONS BY MO THEORY. THE NOBLE GAS HA 1340
              KRYPTON DIFLUORIDE. (PHYSICAL AND CHEMICAL PROPERTIES)                                          0563
                      PHYSICAL AND CHEMICAL PROPERTIES OF HALOGEN FLUORIDES.                                  0017
. (NITROGEN TRIFLUORIDE)               PHYSICAL AND CHEMICAL PROPERTIES OF INORGANIC NITROGEN FLUORIDES        0713
                   SOME PHYSICAL AND CHEMICAL PROPERTIES OF VANADIUM PENTAFLUORIDE.                           1212
ND THIOPHOSPHORYL HALIDES.                               CHEMICAL SHIFTS OF BORON, PHOSPHORUS, PHOSPHORYL, A  0235
ROGEN MONOFLUORIDE AND PHOSPHORUS MONOFLUORIDE)  QUANTUM CHEMICAL STUDY OF SOME PNICOGEN MONOFLUORIDES. (NIT  0678
    PROBABLE NONEXISTENCE OF XENON MONOFLUORIDE AS A CHEMICALLY BOUND SPECIES IN THE GAS PHASE.              1221
                    INVESTIGATION OF THE CHEMISTRY OF AMERICIUM. (AMERICIUM TETRAFLUORIDE)                   0056
         ELECTRONIC STRUCTURE AND CHEMISTRY OF IODINE COMPOUNDS. (FLUORIDES)                                 0499
TETRAFLUORIDE, XENON HE/ STRUCTURAL CONSIDERATIONS IN THE CHEMISTRY OF NOBLE GASES. (XENON DIFLUORIDE, XENON  1235
                       HALOGEN CHEMISTRY. (VOLUMES 1, 2, 3)                                                   0005
UORIDE.                       SCF CALCULATIONS ON CHLORATE ION, HYDROGEN CHLORIDE AND CHLORINE MONOFL          0380
  AB INITIO CALCULATION ON BORON TRIFLUORIDE AND BORON TRICHLORIDE.                                           0213
, TETRAFLUORO METHANE XENON TETRAFLUORIDE, AND XENON TETRACHLORIDE. / POTENTIAL OF CHLORO- AND FLUOROMETHANES 0344
ES OF SILICON FLUORIDE, GERMANIUM FLUORIDE, AND GERMANIUM CHLORIDE. / OF THE FRANCK-CONDON FACTOR FOR MOLECUL 0977
ROM DIPOLE MOMENTS FOR ARSENIC TRIFLUORIDE AND ARSENIC TRICHLORIDE. /S ON IONIC CHARACTER AND HYBRIDIZATION F 0107
ON TETRAFLUORIDE, SILICON TETRAFLUORIDE, AND SILICON TETRACHLORIDE. /NSTANTS FOR BORON TETRAFLUORIDE(-), CARB 0348
TIGATIONS OF PHOSPHORUS PENTAFLUORIDE AND PHOSPHORUS PENTACHLORIDE.          ELECTRON DIFFRACTION INVES       0869
DIATED POLYCRYSTALLINE ARSENIC TRIFLUORIDE AND ARSENIC TRICHLORIDE.          ESR STUDY OF RADICALS IN GAMMA IRRA 0108
FUNCTION ANALYSIS OF NITROGEN TRIFLUORIDE AND NITROGEN TRICHLORIDE.                               GREEN'S     0729
ROWAVE SPECTROSCOPY OF ALUMINUM MONOFLUORIDE AND ALUMINUM CHLORIDE.                       HIGH TEMPERATURE MIC 0028
D THALLOUS HALIDE DIMERS, THALLOUS FLUORIDE, AND THALLOUS CHLORIDE. / SPECTRA AND STRUCTURE OF MATRIX ISOLATE 1123
CT. EXPERIMENTAL AND THEORETICAL STUDIES ON VANADIUM TETRACHLORIDE.                       JAHN-TELLER EFFE     1426
URE AND FORCE CONSTANTS OF BORON TRIFLUORIDE AND BORON TRICHLORIDE.                       MOLECULAR STRUCT     0221
DE, PHOSPHORUS DIFLUORIDE DICHLORIDE, AND PHOSPHORUS PENTACHLORIDE. /INATE ANALYSIS OF PHOSPHORUS PENTAFLUORI 0877
RIDES OF NITROGEN AND PHOSPHORUS, AND PHOSPHORUS OXIDE TRICHLORIDE. /. PART-3: TRIFLUORIDES AND OXIDE TRIFLUO 0701
   PHOTOELECTRON SPECTRA OF OXYGEN DIFLUORIDE AND OXYGEN DICHLORIDE.                                          0766
SPECTRA OF MATRIX ISOLATED NICKEL DIFLUORIDE AND NICKEL DICHLORIDE.                                           0653
STATES AND IN THE MOLECULES XENON DIFLUORIDE AND XENON DICHLORIDE.        5D ORBITALS OF XENON IN ATOMIC VALENCE 1242
          SCF CALCULATIONS ON CHLORATE ION, HYDROGEN CHLORIDE AND CHLORINE MONOFLUORIDE.          INFRARED SPECTRA 0380
OF GASEOUS GROUP-4 HALIDES: ZIRCONIUM FLUORIDE, ZIRCONIUM CHLORIDE, AND HAFNIUM CHLORIDE.     INFRARED SPECTRA 1327
/ISSOCIATION ENERGIES OF BERYLLIUM FLUORIDE AND BERYLLIUM CHLORIDE AND HEAT OF FORMATION OF BERYLLIUM CHLORO/ 0128
/RESONANCE STUDIES OF PHOSPHORUS DIFLUORIDE, PHOSPHORUS DICHLORIDE, AND NITROGEN DICHLORIDE IN LOW TEMPERATU/ 0793
/ AMPLITUDES OF PHOSPHORUS PENTAFLUORIDE, PHOSPHORUS PENTACHLORIDE, AND PHOSPHORUS FLUORO CHLORIDE MOLECULES. 0826
```

YSIS OF PHOSPHORUS PENTAFLUORIDE, PHOSPHORUS DIFLUORIDE DICHLORIDE, AND PHOSPHORUS PENTACHLORIDE. /INATE ANAL 0877
HE MOLECULAR CONSTANTS OF ARSENIC TRIFLUORIDE, ARSENIC TRICHLORIDE, AND PHOSPHORUS TRICHLORIDE. /UORIDE AND T 0111
ATURE VAPORS. PART-4: THALLIUM MONOFLUORIDE, THALLIUM MONOCHLORIDE, AND THALLIUM MONOBROMIDE. /OF HIGH TEMPER 1112
SITION MOMENT WITH THE INTERNUCLEAR SEPARATION IN SILICON CHLORIDE AND TIN MONOFLUORIDE. / OF ELECTRONIC TRAN 1137
OLECULAR STRUCTURES OF ARSENIC TRIFLUORIDE AND ARSENIC TRICHLORIDE BY GAS ELECTRON DIFFRACTION. /ION OF THE M 0104
/ GJ FACTOR OF THALLIUM FLUORIDE, CESIUM FLUORIDE, CESIUM CHLORIDE, CESIUM BROMIDE, AND CESIUM IODIDE AND MA/ 1120
/AVERAGE STRUCTURES OF ARSENIC TRIFLUORIDE AND ARSENIC TRICHLORIDE DETERMINED BY ELECTRON DIFFRACTION AND SP/ 0103
ES OF FORBIDDEN TRANSITIONS IN IODINE, OR IODINE BROMIDE (CHLORIDE, FLUORIDE) MOLECULES. / RELATIVE INTENSITI 0498
ESTIGATIONS. (BORO/ THERMODYNAMIC PROPERTIES OF THE BORON CHLORIDE FLUORIDE SYSTEM FROM MASS SPECTROMETER INV 0179
BRATIONAL FREQUENCI/ POTENTIAL CONSTANTS OF PHOSPHORUS TRICHLORIDE FROM AMPLITUDES OF VIBRATION AND NORMAL VI 1481
TANTALUM CHLORIDE, TANTALUM BROMIDE, AND MOLYBDENUM PENTACHLORIDE IN THE GASEOUS STATE. /E, NIOBIUM BROMIDE, 0075
/D MAGNESIUM DIHALIDES: MAGNESIUM DIFLUORIDE, MAGNESIUM DICHLORIDE, MAGNESIUM DIBROMIDE, AND MAGNESIUM DIIOD/ 0607
EARTH HALIDES. PART-1: BERYLLIUM DIFLUORIDE, BERYLLIUM DICHLORIDE, MAGNESIUM DICHLORIDE. /RA OF THE ALKALINE 0133
FLUORIDE, PHOSPHORUS PENTAFLUORIDE, AND PHOSPHORUS FLUORO CHLORIDE MOLECULES. /AMPLITUDES OF PHOSPHORUS PENTA 0826
OF VIBRATION: SULFUR TETRAFLUORIDE AND SULFUR PENTAFLUORO CHLORIDE MOLECULES.   MEAN AMPLITUDES 1057
/CARBON TETRAFLUORIDE, SILICON TETRAFLUORIDE, CARBON TETRACHLORIDE, SILICON TETRACHLORIDE, AND GERMANIUM TET/ 1007
/ON OF THE TETRAHALIDES CARBON TETRAFLUORIDE, CARBON TETRACHLORIDE, SILICON TETRACHLORIDE, SILICON TETRACHLO/ 0349
/UNCTIONS OF ANTIMONY FLUORIDE, NIOBIUM BROMIDE, TANTALUM CHLORIDE, TANTALUM BROMIDE, AND MOLYBDENUM PENTACH/ 0075
/NTERACTION OF MATRIX ISOLATED NICKEL FLUORIDE AND NICKEL CHLORIDE WITH CARBON MONOXIDE, MOLECULAR NITROGEN,/ 1324
IDE.   EMISSION SPECTRA OF XENON MONOBROMIDE, XENON MONOCHLORIDE, XENON MONOFLUORIDE, AND KRYPTON MONOFLUOR 0545
                              ELECTRON DENSITY DISTRIBUTION IN SOME CHLORIDES.   0218
TROSCOPY OF HIGH TEMPERATURE SPECIES. (FLUORIDES, OXIDES, CHLORIDES)   MATRIX ISOLATION INFRARED SPEC 0011
URE AND HEAT OF SUBLIMATION OF ZIRCONIUM AND HAFNIUM TETRACHLORIDES.   VAPOR PRESS 0486
/RVES AND DISSOCIATION ENERGIES OF DIATOMIC FLUORIDES AND CHLORIDES OF GALLIUM, INDIUM, AND THALLIUM. (MONOF/ 0458
/TRON SPECTRA OF HALIDES. PART-1. TETRAFLUORIDES AND TETRACHLORIDES OF GROUP-5B. (CARBON TETRAFLUORIDE, SILI/ 0316
S OF THE DIATOMIC INTERHALOGENS FROM SPECTROSCOPIC DATA. (CHLORINE, BROMINE, AND IODINE MONOFLUORIDES) /ERTIE 0253
                              PURE QUADRUPOLE SPECTRA OF SOLID CHLORINE COMPOUNDS.   0383
                                       THE CHLORINE DIFLUORIDE FREE RADICAL.   0393
/-2: GEOMETRICAL AND ELECTRICAL STRUCTURE OF THE MOLECULE CHLORINE DIFLUORIDE ON THE BASIS OF A CALCULATION / 0382
THE THERMODYNAMIC PROPERTIES OF SOME HALOGEN MOLECULES. (CHLORINE FLUORIDE)   PHOTOIONIZATION STUDIES AND 0377
NE PENTAFLUORIDE MOLECULES.   ABSORPTION SPECTRA OF CHLORINE FLUORIDE, CHLORINE TRIFLUORIDE, AND CHLORI 0388
FLUORIDE AND CHLORINE OXYFLUORIDE BY NUC/ STUDY OF LIQUID CHLORINE FLUORIDES AND OXYFLUORIDES, CHLORINE PENTA 0414
E DIFLUORIDE, AND CHLORINE TRIFLU/ HEATS OF FORMATION OF CHLORINE FLUORIDES. (CHLORINE MONOFLUORIDE, CHLORIN 0370
TERIZATION OF THE CHLORINE DIFLUORIDE FREE RADICAL.   CHLORINE FLUORINE SYSTEM AT LOW TEMPERATURE. CHARAC 0392
/ECTRON SPIN RESONANCE SPECTRA OF HALOGEN HEXAFLUORIDES. (CHLORINE HEXAFLUORIDE, IODINE HEXAFLUORIDE, BROMIN/ 0271
N SPIN RESONANCE SPECTRUM, AND STRUCTURE.   CHLORINE HEXAFLUORIDE RADICAL. PREPARATION, ELECTRO 0428
             ANALYSIS OF THE INFRARED SPECTRUM OF CHLORINE MONOFLUORIDE.   0386
                        DISSOCIATION ENERGY OF CHLORINE MONOFLUORIDE.   0387
OLUTION RAMAN SPECTROSCOPY IN THE RED AND NEAR INFRARED. (CHLORINE MONOFLUORIDE)   HIGH RES 0389
        HYPERFINE STRUCTURE CONSTANTS OF CHLORINE MONOFLUORIDE.   0375
         INFRARED AND RAMAN SPECTRA OF CHLORINE MONOFLUORIDE.   0381
        MOLECULAR MAGNETIC PROPERTIES OF CHLORINE MONOFLUORIDE.   0384
O(+) TO SINGLET SIGMA(+) SYSTEMS OF BOTH ISOTOPES OF CHLORINE MONOFLUORIDE.   0390
            PHOTOELECTRON SPECTRUM OF CHLORINE MONOFLUORIDE.   0369
SCF CALCULATIONS ON CHLORATE ION, HYDROGEN CHLORIDE AND CHLORINE MONOFLUORIDE.   0380
NS TO THE SECOND ORDER STARK EFFECT OF LINEAR MOLECULES. (CHLORINE MONOFLUORIDE AND BROMINE MONOFLUORIDE) /IO 0385
             ELECTRONIC STRUCTURES OF CHLORINE MONOFLUORIDE AND CHLORINE TRIFLUORIDE.   0372
F FLUORINE.   PHOTOIONIZATION STUDY OF CHLORINE MONOFLUORIDE AND THE DISSOCIATION ENERGY O 0378
/ITY IN HYDROGEN HALIDES AND THE INTERHALOGEN COMPOUNDS. (CHLORINE MONOFLUORIDE, BROMINE MONOFLUORIDE, AND I/ 0373
/ MAGNETIC PROPERTIES, AND ELECTRIC QUADRUPOLE MOMENTS IN CHLORINE MONOFLUORIDE, BROMINE MONOFLUORIDE, BROMI/ 0254
/LECULAR ORBITAL STUDY OF THE GEOMETRY OF INTERHALOGENS. (CHLORINE MONOFLUORIDE, CHLORINE TRIFLUORIDE, AND C/ 0379
/R DIFLUORIDE, SULFUR HEXAFLUORIDE, SULFUR HEXAFLUORIDE, CHLORINE MONOFLUORIDE, CHLORINE TRIFLUORIDE, AND C/ 1373
E MONOFLUORIDE, AND B/ PHOTOELECTRON SPECTRA OF HALIDES. (CHLORINE MONOFLUORIDE, CHLORINE TRIFLUORIDE, BROMIN 0376
ERGIES.   CHARGE DISTRIBUTION IN CHLORINE MONOFLUORIDE FROM CORE ELECTRON BINDING EN 0374
/OF SOME MOLECULAR PROPERTIES OF THE ISOELECTRONIC SERIES CHLORINE MONOFLUORIDE, HYDROGEN OXYCHLORIDE, CHLOR/ 0371
/TERMINING SATURATED LIQUID AND SATURATED VAPOR ENTROPY. (CHLORINE MONOFLUORIDE, OXYGEN MONOFLUORIDE, PHOSPH/ 0391
RIOLIS COUPLING COEFFICIENTS OF BROMINE PENTAFLUORIDE AND CHLORINE PENTAFLUORIDE.   CO 0427
ERTIES, DIELECTRIC CONSTANT, AND SPECIFIC CONDUCTIVITY OF CHLORINE PENTAFLUORIDE. /OR PRESSURE, CRITICAL PROP 0426
            EQUILIBRIUM STUDIES OF CHLORINE PENTAFLUORIDE.   0417
            FLUORINE NMR OF CHLORINE PENTAFLUORIDE.   0416
       MATRIX ISOLATION STUDY OF CHLORINE PENTAFLUORIDE.   0419
ECULES. (IODINE PENTAFLUORIDE, BROMINE PENTAFLUORIDE, AND CHLORINE PENTAFLUORIDE) /OR SOME PYRAMIDAL XY4Z MOL 0421
ECULES. (IODINE PENTAFLUORIDE, BROMINE PENTAFLUORIDE, AND CHLORINE PENTAFLUORIDE) / SOME SQUARE PYRAMIDAL MOL 0420
LUORIDE, IODINE PENTAFLUORIDE, BROMINE PENTAFLUORIDE, AND CHLORINE PENTAFLUORIDE. /ECULES. XENON OXIDE TETRAF 0418
MOLECULAR FORCE FIELDS FOR INTERHALOGEN COMPOUNDS. (CHLORINE PENTAFLUORIDE AND BROMINE PENTAFLUORIDE) 0425
            MICROWAVE SPECTROSCOPY OF CHLORINE PENTAFLUORIDE AT 70 AND 140 GHZ 0423
IODINE PENTAFLUORI/ IDEAL GAS THERMODYNAMIC PROPERTIES OF CHLORINE PENTAFLUORIDE, BROMINE PENTAFLUORIDE, AND 0424
/IDES AT HIGH TEMPERATURES. PART-1: CALCULATION METHODS. (CHLORINE PENTAFLUORIDE, BROMINE PENTAFLUORIDE, IOD/ 0422
CONSISTENT/ CALCULATION OF THE ELECTRONIC STRUCTURE OF A CHLORINE PENTAFLUORIDE MOLECULE BY THE MO LCAO SELF 0415
            EPR SPECTRUM OF THE RADICAL CHLORINE TETRAFLUORIDE.   0413
               IS THE RADICAL CHLORINE TETRAFLUORIDE PLANAR OR NOT.   0412
CONSTRUCTION OF HYBRID ORBITALS. (CHLORINE TRIFLUORIDE)   0405
       FORCE CONSTANT CALCULATION FOR CHLORINE TRIFLUORIDE.   0403
CORIOLIS CONSTANTS IN DICHLORO BORANE, DIBROMO BORANE AND CHLORINE TRIFLUORIDE. /RALIZED MEAN AMPLITUDES AND 0411
MEAN AMPLITUDES OF VIBRATION OF SELENIUM TRIOXIDE AND CHLORINE TRIFLUORIDE.   0406
       MICROWAVE SPECTRUM AND STRUCTURE OF CHLORINE TRIFLUORIDE.   0410
       MOLECULAR FORCE FIELD FOR CHLORINE TRIFLUORIDE.   0402
ATES. (PHOSPHORUS PENTAFLUORIDE, SULFUR HEXAFLUORIDE, AND CHLORINE TRIFLUORIDE) /N PERFECT PAIRING VALENCE ST 0862
ES OF PHOSPHORUS PENTAFLUORIDE, SULFUR TETRAFLUORIDE, AND CHLORINE TRIFLUORIDE.   VESCF-MO STUDI 0835
N.   ELECTRONIC STRUCTURE OF CHLORINE TRIFLUORIDE. AN APPROXIMATE SCF CALCULATIO 0404
       ELECTRICAL CONDUCTIVITY OF SOLID CHLORINE TRIFLUORIDE AND BROMINE TRIFLUORIDE.   0259
       INFRARED AND RAMAN SPECTRA OF CHLORINE TRIFLUORIDE AND BROMINE TRIFLUORIDE.   0408
MOLECULAR FORCE FIELD FOR INTERHALOGEN COMPOUNDS. (CHLORINE TRIFLUORIDE AND BROMINE TRIFLUORIDE) 0407
       QUADRUPOLE COUPLING CONSTANTS IN CHLORINE TRIFLUORIDE AND BROMINE TRIFLUORIDE.   0400
       RAMAN SPECTRA AND THE STRUCTURE OF CHLORINE TRIFLUORIDE AND BROMINE TRIFLUORIDE.   0398
VIBRATIONAL SPECTRA AND THERMODYNAMIC PROPERTIES OF CHLORINE TRIFLUORIDE AND BROMINE TRIFLUORIDE.   0396
EOUS SPECTRA AND GAS TO LIQUID SHIFTS.   NMR SPECTRA OF CHLORINE TRIFLUORIDE AND CHLORINE MONOFLUORIDE. GAS 0368
STRUCTURES OF THE INTERHALOGEN COMPOUNDS. PART-1: CHLORINE TRIFLUORIDE AT -120 DEGREES C.   0395
       HEAT OF FORMATION OF CHLORINE TRIFLUORIDE AT 298.14 DEGREES K.   0401
       THERMAL DISSOCIATION OF CHLORINE TRIFLUORIDE BEHIND INCIDENT SHOCK WAVES.   0394

/FLUORIDE, ANTIMONY PENTAFLUORIDE, ARSENIC PENTAFLUORIDE, CHLORINE TRIFLUORIDE, BROMINE TRIFLUORIDE, AND BRO/ 0239
INE PENTAFLUORIDE. F/ INFRARED SPECTRA OF MATRIX ISOLATED CHLORINE TRIFLUORIDE, BROMINE TRIFLUORIDE, AND BROM 0399
/S ON THE GEOMETRY AND PROPERTIES OF SOME INTERHALOGENS. (CHLORINE TRIFLUORIDE, BROMINE TRIFLUORIDE, IODINE / 0397
/ INFRARED AND RAMAN SPECTRA OF BROMINE PENTAFLUORIDE AND CHLORINE TRIFLUORIDE IN THE CONDENSED PHASE. EVIDEN 0266
SPHORUS PENTAFLUORID/ X-RAY PHOTOELECTRON SPECTROSCOPY OF CHLORINE TRIFLUORIDE, SULFUR TETRAFLUORIDE, AND PHO 0409
SPHORUS/ OBSERVATION BY ESCA OF INEQUIVALENT FLUORINES IN CHLORINE TRIFLUORIDE, SULFUR TETRAFLUORIDE, AND PHO 0874
RONIC SERIES CHLORINE MONOFLUORIDE, HYDROGEN OXYCHLORIDE, CHLORO AMMONIA, AND CHLORO METHANE. /F THE ISOELECT 0371
/ GENERALIZED MEAN AMPLITUDES AND CORIOLIS CONSTANTS IN DICHLORO BORANE, DIBROMO BORANE AND CHLORINE TRIFLUOR 0411
CONSTANTS OF THE SILICON TETRAHALIDES AND FREQUENCIES OF CHLORO BROMO SILICON.   FORCE 1004
/RAFLUORO ETHYLENE, TRIFIUORO CHLORO ETHYLENE, DIFLUORO DICHLORO ETHYLENE, AND THE APPEARANCE POTENTIAL OF C/ 0305
/ONIZATION POTENTIALS FOR TETRAFLUORO ETHYLENE, TRIFLUORO CHLORO ETHYLENE, DIFLUORO DICHLORO ETHYLENE, AND T/ 0305
NIC TRIFLUORIDE)   CENTRIFUGAL STRETCHING CF TRIFLUORO CHLORO METHANE AND SOME OTHER SYMMETRIC TOPS. (ARSE 0101
ON TETRAFLUORIDE, AND/ CALCULATED IONIZATION POTENTIAL OF CHLORO- AND FLUOROMETHANES, TETRAFLUORO METHANE XEN 0344
AND BERYLLIUM CHLORIDE AND HEAT OF FORMATION CF BERYLLIUM CHLOROFLUORIDE. /ON ENERGIES OF BERYLLIUM FLUORIDE 0128
TETRAFLUORIDE)   RAMAN SPECTRA OF CHLOROFLUORO METHANES IN THE GASEOUS STATE. (CARBON 0322
THERMODYNAMIC PROPERTIES OF CHROMIUM DIFLUORIDE.   0432
CHROMIUM MONOFLUORIDE.   SUBLIMATION PRESSURE OF CHROMIUM DIFLUORIDE AND THE DISSOCIATION ENERGY OF 0430
/DIES OF THE VAPORIZATION OF HIGH TEMPERATURE MATERIALS. (CHROMIUM DIFLUORIDE, CHROMIUM TRIFLUORIDE, AND IRO/ 0431
D OSMIUM HEXAFLUORIDE.   INFRARED SPECTRUM OF CHROMIUM HEXAFLUORIDE, MOLYBDENUM HEXAFLUORIDE, AN 0436
/RIC OXIDE, AND MOLECULAR OXYGEN AND OF CALCIUM FLUORIDE, CHROMIUM(II) FLUORIDE, MANGANESE(II) FLUORIDE, COP/ 1324
BAND SPECTRUM OF CHROMIUM MONOFLUORIDE.   0429
VIBRATIONAL SPECTRUM OF LIQUID CHROMIUM PENTAFLUORIDE.   0435
STRUCTURE OF CRYSTALLINE CHROMIUM TRIFLUORIDE.   0433
ON TRIFLUORIDE.   SUBLIMATION PRESSURES OF CHROMIUM TRIFLUORIDE, MANGANESE TRIFLUORIDE, AND IR 0434
OLECULE.   CLASSICAL VIBRATION PROBLEM FOR THE PYRAMIDAL XY4 M 1421
: INTERACTION TERMS IN ENERGY DISTRIBUTIONS. (SELENIUM A/ CLASSIFICATION OF NORMAL MODES OF VIBRATION. PART-2 1494
(SULFUR TETRAFLUORIDE)   CLASSIFICATION OF THE STATES OF NONRIGID MOLECULES. 1039
NFRARED INTENSITIES OF BORON TRIFLUORIDE.   CNDO CALCULATION OF DIPOLE MOMENT DERIVATIVES AND I 0195
(OXYGEN DIFLUORIDE)   CNDO STUDY OF THE OZONIDE ION AND RELATED SPECIES. 0782
N THE GEOMETRY AND PROPERTIES OF SOME INTERHALOGENS. (CH/ CNDO/2 AND INDO ALL VALENCE ELECTRON CALCULATIONS O 0397
ION OF THE INFRARED SPECTRA OF SULFUR TETRAFLUORIDE USING CNDO/2 TECHNIQUES.   CALCULAT 1047
BICENTRIC RESCALING OF CNDO/2 THEORY. APPLICATIONS TO INORGANIC FLUORIDES. 1475
AND LANGMUIR MEASUREMENTS OF THE SUBLIMATION PRESSURE OF COBALT DIFLUORIDE.   KNUDSEN 0439
RIDE, A/ INFRARED SPECTRA AND GEOMETRY OF MATRIX ISOLATED COBALT DIFLUORIDE, NICKEL DIFLUORIDE, COPPER DIFLUO 0437
VAPOR PRESSURE AND HEAT OF SUBLIMATION OF COBALT DIHALIDES. (COBALT DIFLUORIDE)   0438
VAPOR PRESSURE, HEAT OF SUBLIMATION, AND EVAPORATION COEFFICIENT OF LANTHANUM TRIFLUORIDE.   0576
VAPOR PRESSURE, HEAT OF SUBLIMATION, AND THE EVAPORATION COEFFICIENT OF PRASEODYMIUM TRIFLUORIDE. 0896
INFLUENCE OF ATOMIC MASSES ON THE CORIOLIS COUPLING COEFFICIENTS.   1388
/ERGIES OF FORMATION OF PROTACTINIUM HALIDES AND ACTIVITY COEFFICIENTS FOR PROTACTINIUM IN LIQUID BISMUTH SO/ 0900
VIBRATION AND SHRINKAGE / FORCE FIELD, CORIOLIS COUPLING COEFFICIENTS, GENERALIZED MEAN SQUARE AMPLITUDES OF 1278
INFLUENCE OF ATOMIC MASSES ON THE CORIOLIS COUPLING COEFFICIENTS IN XY4 TETRAHEDRAL MOLECULES.   1410
PENTAFLUORIDE.   CORIOLIS COUPLING COEFFICIENTS OF BROMINE PENTAFLUORIDE AND CHLORINE 0427
FORBIDDEN RAMAN BANDS OF SULFUR HEXAFLUORIDE. COLLISION-INDUCED RAMAN SCATTERING.   1081
HEXAFLUORIDE, OSMIUM HEXAFLUORIDE, IRIDIUM HEXAFLUORIDE,/ COLORS OF TRANSITION METAL HEXAFLUORIDES. (RHENIUM 0916
E NEAR ULTRAVIOLET.   COMPLEX BAND SYSTEM OF MANGANESE MONOFLUORIDE IN TH 0610
D NIOBIUM PENTAFLUORID/ A FLUORINE NMR AND RAMAN STUDY OF COMPLEX FORMATION BETWEEN ANTIMONY PENTAFLUORIDE AN 0079
AFLUORIDE)   DIRECTED VALENCY IN CERTAIN MOLECULES AND COMPLEX IONS. (PHOSPHORUS PENTAFLUORIDE, SULFUR HEX 0860
/ THE FORCE CONSTANTS AND INTERATOMIC DISTANCES OF HALIDE COMPLEX MOLECULES OF THE XY4 TYPE OF TETRAHEDRAL S/ 1414
TRANSITION METAL FLUORIDES AND THEIR COMPLEXES.   0015
PHOTOELECTRON SPECTRA OF COMPLEXES OF BORON TRIFLUORIDE AND AMINES.   0224
/UPOLE RESONANCE STUDIES OF NIOBIUM PENTAFLUORIDE AND ITS COMPLEXES WITH XENON DIFLUORIDE AND ORGANIC BASES. 0664
IDE.   MASS SPECTROMETRIC DETERMINATION OF THE VAPOR COMPOSITION AND VAPOR PRESSURE OF VANADIUM TRIFLUOR 1208
MASS SPECTRUM AND COMPOSITION OF NIOBIUM PENTAFLUORIDE VAPOR.   0666
OF URANIUM HEXAFLUORIDE MEASURED WITH A BALLISTIC PISTON COMPRESSOR.   SPECIFIC HEAT RATIO 1201
/: OPTICAL ABSORPTION AND X-RAY EMISSION FROM SCF MO LCAO COMPUTATIONS. PART-2: MOLECULAR ORBITAL INTERPRETA/ 1078
S OF SULFUR MONOFLUORIDE/ HARTREE-FOCK WAVE FUNCTIONS AND COMPUTED PROPERTIES FOR THE DOUBLET PI GROUND STATE 1025
RO METHANE, PHOSPHORUS PENTAFLUORIDE, AND CYCLOPROPANE BY COMPUTER TECHNIQUES. /ION ROTATION BANDS OF TRIFLUO 0880
/OF BROMINE PENTAFLUORIDE AND CHLORINE TRIFLUCRIDE IN THE CONDENSED PHASE. EVIDENCE OF AN ASSOCIATED STATE I/ 0266
N HEXAFLUORIDE.   ELECTRICAL CONDUCTIVITY AND DIELECTRIC CONSTANT OF LIQUID XENO 1307
E, CRITICAL PROPERTIES, DIELECTRIC CONSTANT, AND SPECIFIC CONDUCTIVITY OF CHLORINE PENTAFLUORIDE. /OR PRESSUR 0426
INE TRIFLUORIDE.   ELECTRICAL CONDUCTIVITY OF SOLID CHLORINE TRIFLUORIDE AND BROM 0259
D STATE OF OXYGEN DIFLUORIDE, NITRITE ION AND CYANIDE IO/ CONFIGURATION INTERACTION CALCULATION FOR THE GROUN 0763
LUORIDE AND ITS POSITIVE ION.   CONFIGURATION INTERACTION STUDIES OF NITROGEN MONOF 0672
SPHORUS TRIFLUORIDE.   SELF CONSISTENT CHARGE AND CONFIGURATION MOLECULAR ORBITAL CALCULATIONS ON PHO 0821
IRIDIUM BY MAGNETIC RESONANCE.   CONFIGURATION OF THE PENTAFLUORIDES OF RHODIUM AND 0923
/ODS FOR CORRELATING MOLECULES AND SOME OPTIMIZED VALENCE CONFIGURATION RESULTS ON THE DIATOMIC MOLECULES LI/ 0174
AND GROUP-6 TRANSITION ELEMENT TETRAHALID/ ESTIMATION OF CONFIGURATIONS AND INTERATOMIC DISTANCES IN GROUP-5 1427
MONOFLUORIDE.   REASSIGNMENT OF MOLECULAR ORBITAL CONFIGURATIONS OF THE ELECTRONIC STATES OF SCANDIUM 0937
IMATION OF XENON DIFLUORIDE AND XENON TETRAFLUORIDE AND A CONJECTURE ON BONDING IN THE SOLIDS.   HEATS OF SUBL 1237
OF SOME SIMPLE MOLECULES. (TUNG/ SPECTROSCOPIC STUDIES IN CONNECTION WITH ELECTRON DIFFRACTION INVESTIGATION 1153
NON DIFLUORIDE, XENON TETRAFLUORIDE, XENON HE/ STRUCTURAL CONSIDERATIONS IN THE CHEMISTRY OF NOBLE GASES. (XE 1235
AL CALCULATIONS ON PHOSPHORUS TRIFLUORIDE.   SELF CONSISTENT CHARGE AND CONFIGURATION MOLECULAR ORBIT 0821
/T OF NITROGEN TRIFLUORIDE. PART-3: A GAUSSIAN BASIS SELF CONSISTENT FIELD CALCULATION FOR NITROGEN TRIFLUOR/ 0746
OF A CHLORINE PENTAFLUORIDE MOLECULE BY THE MO LCAO SELF CONSISTENT FIELD METHOD IN AN NDDO APPROXIMATION. / 0415
DE, XENON TETRAFLUORIDE, XENON HEXAFLUORIDE.   SELF CONSISTENT FIELD STUDY OF THE SERIES XENON DIFLUORI 1224
ITIO CALCULATION OF EQUILIBRIUM GEOMETRIES AND QUAD/ SELF CONSISTENT MOLECULAR ORBITAL METHODS. PART-5: AB IN 0720
ORIDE AND ELEMENT 110 HEXAFLUORIDE CALCULATED BY THE SELF CONSISTENT MULTIPLE SCATTERING X-ALPHA METHOD. /FLU 0881
EN RARE EARTH TRIFLUORIDES. (YTTRI/ HIGH TEMPERATURE HEAT CONTENTS AND RELATED THERMODYNAMIC FUNCTIONS OF SEV 1320
ERICAL TOP MOLECUL/ VAPOR PHASE RAMAN SPECTRA, RAMAN BAND CONTOUR ANALYSES, AND CORIOLIS CONSTANTS OF THE SPH 1069
TS, AND VALUES FOR/ VAPOR PHASE RAMAN SPECTRA, RAMAN BAND CONTOUR ANALYSES, CORIOLIS CONSTANTS, FORCE CONSTAN 0795
LOW TEMPERATURE ABSORPTION CONTOUR OF THE NU-3 BAND OF SULFUR HEXAFLUORIDE. 1082
PENTAFLUORID/ RAMAN SPECTRA OF GASES. PART-10: RAMAN BAND CONTOUR OF THE NU-7 (E') FUNDAMENTAL OF PHOSPHORUS 0879
E' VIBRATIONS OF ARSENIC PENTAFLUORID/ PERPENDICULAR BAND CONTOURS AND VIBRATIONAL POTENTIAL FUNCTION OF THE 0116
/ULES FROM LOW TEMPERATURE INFRARED STUDIES OF VAPOR BAND CONTOURS. APPLICATION TO THE FORCE FIELD OF TUNGST/ 1156
TRIFLUORIDE.   CALCULATION OF PERPENDICULAR INFRARED BAND CONTOURS: APPLICATION TO THE NU-4 BAND OF NITROGEN 0747
IDE.   ROTATION VIBRATION BAND CONTOURS IN THE INFRARED SPECTRUM OF BORON TRIFLUOR 0193
N TRIFLUORI/ ANHARMONIC POTENTIAL CONSTANTS FROM THE BAND CONTOURS OF HOT BANDS FOR A SYMMETRIC TOP. (NITROGE 0733
INFRARED BAND CONTOURS OF RHENIUM HEXAFLUORIDE.   0915
TETRAFLUORIDE)   INFRARED BAND CONTOURS. PART-1: SPHERICAL TOP MOLECULES. (CARBON 0325
/TATION OF TRIGONAL BIPYRAMIDAL MOLECULES. BERRY ROTATION CONTRA TURNSTILE ROTATION IN PHOSPHORUS PENTAFLUOR/ 0870

LIMITATIONS OF THE UREY-BRADLEY MODEL IN NORMAL COORDINATE ANALYSES. 1478
/AND LASER RAMAN SPECTRA, MOLECULAR STRUCTURES AND NORMAL COORDINATE ANALYSES OF GERMANIUM DIFLUORIDE, MONOM/ 0475
FORCE CONSTANTS AND NORMAL COORDINATE ANALYSIS. 1473
VIBRATIONAL ASSIGNMENTS AND NORMAL COORDINATE ANALYSIS FOR ANTIMONY PENTAFLUORIDE. 0087
OSPHORUS DIFLUORIDE DICHLORIDE, AND PHOSPHORUS PE/ NORMAL COORDINATE ANALYSIS OF PHOSPHORUS PENTAFLUORIDE, PH 0877
DRAL XY6 AND Z6 MOLECULAR MODELS WITH APPLICATION/ NORMAL COORDINATE ANALYSIS OF ROTATION VIBRATION OF OCTAHE 1454
VIBRATIONAL ASSIGNMENT AND NORMAL COORDINATE ANALYSIS OF SELENIUM TETRAFLUORIDE. 0958
OGEN TRIFLUORIDE) NORMAL COORDINATE ANALYSIS OF SOME NITROGEN HALIDES. (NITR 0730
INFRARED SPECTRUM AND NORMAL COORDINATE ANALYSIS OF SULFUR TETRAFLUORIDE. 1045
ECULES. NORMAL COORDINATE ANALYSIS OF TRIGONAL BIPYRAMIDAL XY5 MOL 1433
EVIDENCE OF STEREOCHEMICAL NONRIGIDITY IN FIVE AND SEVEN COORDINATE STRUCTURES. FURTHER 1440
ODINE HEPTAFLUOR/ INTRAMOLECULAR LIGAND EXCHANGE IN SEVEN COORDINATE STRUCTURES. (RHENIUM HEPTAFLUORIDE AND I 0922
ULATION BOND MOMENT / A GEOMETRIC VISUALIZATION OF NORMAL COORDINATE TRANSFORMATIONS. APPLICATION TO THE CALC 0352
ND IONS. (GVFF) NORMAL COORDINATE TREATMENT OF XY4 TETRAHEDRAL MOLECULES A 1415
ND IONS. (UREY-BRADLEY AND MVFF) NORMAL COORDINATE TREATMENT OF XY4 TETRAHEDRAL MOLECULES A 1418
ON TO ETHYLENE. TORSIONAL COORDINATES IN VIBRATIONAL ANHARMONICITY. APPLICATI 1485
STEREOCHEMISTRY AND SEVEN COORDINATION. (XENON HEXAFLUORIDE) 1286
NEW FLUORIDES OF TRIVALENT COPPER. 0444
IES AT HIGH TEMPERATURES. PART-9: SUBLIMATION PRESSURE OF COPPER DIFLUORIDE. MASS SPECTROMETRIC STUD 0442
IES AT HIGH TEMPERATURES. PART-9: SUBLIMATION PRESSURE OF COPPER DIFLUORIDE. (AND COPPER MONOFLUORIDE) / STUD 0441
OF MATRIX ISOLATED COBALT DIFLUORIDE, NICKEL DIFLUORIDE, COPPER DIFLUORIDE, AND ZINC DIFLUORIDE. /D GEOMETRY 0437
ES OF FLUORIDES. PART-6: PRECISE STRUCTURAL PARAMETERS IN COPPER DIFLUORIDE BY NEUTRON DIFFRACTION. STRUCTUR 0443
/FLUORIDE, CHROMIUM(II) FLUORIDE, MANGANESE(II) FLUORIDE, COPPER(II) FLUORIDE, AND ZINC(II) FLUORIDE WITH CA/ 1324
DISSOCIATION ENERGY OF COPPER MONOFLUORIDE. 0440
CHARGE DISTRIBUTION IN CHLORINE MONOFLUORIDE FROM CORE ELECTRON BINDING ENERGIES. 0374
LUORIDE) CHEMICAL APPLICATIONS OF CORE ELECTRON EXCITATION SPECTROSCOPY. (XENON HEXAF 1306
DE. THE ANALYSIS OF 2NU-3 AND THE DETERMINATION OF THE 33 CORIOLIS CONSTANT. BORON TRIFLUORI 0192
FREQUENCIES, FORCE CONSTANTS, AND MASS DEPENDENCE OF THE CORIOLIS CONSTANT FOR SOME MOLECULES AND IONS OF D/ 1411
FORMULAS FOR CORIOLIS CONSTANTS. 1482
/STANTS, MEAN AMPLITUDES OF VIBRATION, SHRINKAGE EFFECTS, CORIOLIS CONSTANTS, AND THERMODYNAMIC FUNCTIONS OF/ 1219
ARSENIC PENTAFLUORIDE. MEAN AMPLITUDES OF VIBRATION AND CORIOLIS CONSTANTS FOR PHOSPHORUS PENTAFLUORIDE AND 0840
FORCE CONSTANTS FOR NITROGEN TRIFLUO/ P-R SEPARATIONS AND CORIOLIS CONSTANTS FOR SYMMETRIC TOP MOLECULES AND 0708
/ VAPOR PHASE RAMAN SPECTRA, RAMAN BAND CONTOUR ANALYSES, CORIOLIS CONSTANTS, FORCE CONSTANTS, AND VALUES FOR 0795
NE AND CHLORINE TRIFLUOR/ GENERALIZED MEAN AMPLITUDES AND CORIOLIS CONSTANTS IN DICHLORO BORANE, DIBROMO BORA 0411
/ANTS, MEAN AMPLITUDES OF VIBRATION, SHRINKAGE EFFECT AND CORIOLIS CONSTANTS IN IODINE PENTAFLUORIDE AND IOD/ 0513
GENERALIZED MEAN SQUARE AMPLITUDES AND CORIOLIS CONSTANTS OF SOME PYRAMIDAL XY3 MOLECULES. 1405
LOW TEMPERATURE INFRARED STUDIES OF VAPOR BAND CONTOURS./ CORIOLIS CONSTANTS OF SPHERICAL TOP MOLECULES FROM 1156
/OR PHASE RAMAN SPECTRA, RAMAN BAND CONTOUR ANALYSES, AND CORIOLIS CONSTANTS OF THE SPHERICAL TOP MOLECULES,/ 1069
INFLUENCE OF ATOMIC MASS ON THE CORIOLIS CONSTANTS OF XY6 MOLECULES. 1458
INFLUENCE OF ATOMIC MASSES ON THE CORIOLIS COUPLING COEFFICIENTS. 1388
UARE AMPLITUDES OF VIBRATION AND SHRINKAGE / FORCE FIELD, CORIOLIS COUPLING COEFFICIENTS, GENERALIZED MEAN SQ 1278
OLECULES. INFLUENCE OF ATOMIC MASSES ON THE CORIOLIS COUPLING COEFFICIENTS IN XY4 TETRAHEDRAL M 1410
RIDE AND CHLORINE PENTAFLUORIDE. CORIOLIS COUPLING COEFFICIENTS OF BROMINE PENTAFLUO 0427
E OF VIBRATIONAL FREQUENCIES, CENTRIFUGAL STRETCHING, AND CORIOLIS COUPLING CONSTANTS. /DE BY THE COMBINED US 0717
CONSTANTS, AN/ HARMONIC FREQUENCIES, POTENTIAL FUNCTION, CORIOLIS COUPLING CONSTANTS, CENTRIFUGAL STRETCHING 0726
ETRICAL XY2 MOLECULES. POTENTIAL CONSTANTS AND CORIOLIS COUPLING CONSTANTS FOR SOME NONLINEAR SYMM 1385
LE M/ THEORETICAL INTERPRETATION OF THE MASS INFLUENCE ON CORIOLIS COUPLING CONSTANTS IN FOUR XYN TYPE MOLECU 1488
THERMODYNAMIC FUNCTIONS OF/ MEAN AMPLITUDES OF VIBRATION, CORIOLIS COUPLING CONSTANTS, SHRINKAGE EFFECT, AND 0050
MATHEMATICAL TECHNIQUES IN THE STUDY OF CORIOLIS COUPLING IN TETRAHEDRAL XY4 MOLECULES. 1409
MOLECULES. DIRECT DETERMINATION FORCE CONSTANTS FROM THE CORIOLIS INTERACTION. /OF SUBSTITUTED AND PERTURBED 1444
HE VIBRATIONAL PROBLEM FOR XY4 TETRAHEDRAL MOLECU/ USE OF CORIOLIS INTERACTION CONSTANTS IN THE SOLUTION OF T 1423
STRUCTURE, ANHARMONIC POTENTIAL FUNCTION, AND NU-1- NU-3 CORIOLIS RESONANCE. /IBRATIONAL STATES, EQUILIBRIUM 0993
CORIOLIS ZETA SUMS FOR SOME SPHERICAL TOP MOLECULES 1487
ION OF THE ORTHORHOMBIC IODINE HEPTAFLUORIDE STRUCTURE. A CORRECTION. A REDETERMINAT 0519
EAR MOLECULES. (CHLORINE MONOFLUORIDE AND BROMINE M/ SOME CORRECTIONS TO THE SECOND ORDER STARK EFFECT OF LIN 0385
NFIGURATION RESULTS ON THE DIATOMIC MOLECULE/ METHODS FOR CORRELATING MOLECULES AND SOME OPTIMIZED VALENCE CO 0174
ELECTRON CORRELATION AND MOLECULAR SHAPE. 1413
FFRACTION STUDY OF XENON HEXAFLUORID/ EFFECTS OF ELECTRON CORRELATION IN X-RAY DIFFRACTION AND AN ELECTRON DI 1290
NS OF XENON TETRAFLUORIDE, XENON OXIDE TETRAFLUORIDE, AN/ CORRELATION OF ELECTRONIC STATES OF THE POSITIVE IO 1264
NTHALPIES OF SUBLIMATION OF CALCIUM AND CERIUM FLUORIDES. CORRELATION OF ENTROPY AND ENTHALPY IN ERRORS. /D E 0281
NTHALPY OF SUBLIMATION OF GADOLINIUM TRIFLUORIDE. ROLE OF CORRELATION OF ENTROPY AND ENTHALPY IN ERRORS. /D E 0456
AHALIDE MOLECULES. PENTACOORDINATED MOLECULES. PART-12: CORRELATION OF EXCHANGE RATES FOR SOME GROUP-5 PENT 0115
ION SPECTRA OF BORON FLUORIDE WITH ALTERATIONS IN ITS QU/ CORRELATION OF INTERMOLECULAR EFFECTS IN THE VIBRAT 1404
H INTERNUCLEAR DISTANCE FOR BANDS OF THE X SYSTEM. CORRELATION OF THE ELECTRONIC TRANSITION MOMENT WIT 0969
ME HALOGEN DERIVATIVES OF METHANE. (CARBON TETRAFLUORIDE) CORRELATION OF THE SPECTRA. /SORPTION SPECTRA OF SO 0365
LEMENTS IN PERFECT PAIRING V/ ORBITAL MODIFICATION BY THE COULOMB FIELD OF LIGAND ATOMS OF LATER SECOND ROW E 0862
ZINC FLUORIDE, AND FERROUS F/ FORCE FIELDS OF THE RUTILE COUNTERPARTS: TITANIUM DIOXIDE, MAGNESIUM FLUORIDE, 0606
FORCE CONSTANTS OF MOLECULES WITH STRONGLY COUPLED VIBRATIONS. 1496
MICROWAVE SPECTRUM OF GERMANIUM DIFLUORIDE. QUADRUPOLE COUPLING AND CENTRIFUGAL DISTORTION. 0477
INFLUENCE OF ATOMIC MASSES ON THE CORIOLIS COUPLING COEFFICIENTS. 1388
ITUDES OF VIBRATION AND SHRINKAGE / FORCE FIELD, CORIOLIS COUPLING COEFFICIENTS, GENERALIZED MEAN SQUARE AMPL 1278
INFLUENCE OF ATOMIC MASSES ON THE CORIOLIS COUPLING COEFFICIENTS IN XY4 TETRAHEDRAL MOLECULES. 1410
CHLORINE PENTAFLUORIDE. CORIOLIS COUPLING COEFFICIENTS OF BROMINE PENTAFLUORIDE AND 0427
ATIONAL FREQUENCIES, CENTRIFUGAL STRETCHING, AND CORIOLIS COUPLING CONSTANTS. /DE BY THE COMBINED USE OF VIBR 0717
S, AN/ HARMONIC FREQUENCIES, POTENTIAL FUNCTION, CORIOLIS COUPLING CONSTANTS, CENTRIFUGAL STRETCHING CONSTANT 0726
Y2 MOLECULES. POTENTIAL CONSTANTS AND CORIOLIS COUPLING CONSTANTS FOR SOME NONLINEAR SYMMETRICAL X 1385
INE TRIFLUORIDE. QUADRUPOLE COUPLING CONSTANTS IN CHLORINE TRIFLUORIDE AND BROM 0400
/RETICAL INTERPRETATION OF THE MASS INFLUENCE ON CORIOLIS COUPLING CONSTANTS IN FOUR XYN TYPE MOLECULE MODEL/ 1488
AND DINITROGEN TETRAFLUORIDE MOLECULES, AND ESR HYPERFINE COUPLING CONSTANTS IN NITROGEN DIFLUORIDE. /UORIDE 0688
AB INITIO INVESTIGATION OF THE GEOMETRY, BONDING, AND COUPLING CONSTANTS OF BORON DIFLUORIDE. 0180
AMIC FUNCTIONS OF/ MEAN AMPLITUDES OF VIBRATION, CORIOLIS COUPLING CONSTANTS, SHRINKAGE EFFECT, AND THERMODYN 0050
ACTION S/ STRUCTURE, PSEUDOROTATION, AND VIBRATIONAL MODE COUPLING IN IODINE HEPTAFLUORIDE. AN ELECTRON DIFFR 0515
DE(2-), AND IRIDIUM HEXACHLORIDE(/ VIBRATIONAL ELECTRONIC COUPLING IN IRIDIUM HEXAFLUORIDE, OSMIUM HEXACHLORI 0538
MATHEMATICAL TECHNIQUES IN THE STUDY OF CORIOLIS COUPLING IN TETRAHEDRAL XY4 MOLECULES. 1409
HEPTAFLUORIDE. STRUCTURE, PSEUDOROTATION, AND ANHARMONIC COUPLING OF MOLES. /ON DIFFRACTION STUDY OF RHENIUM 0920
VAPOR PRESSURE AND CRITICAL CONSTANTS OF URANIUM HEXAFLUORIDE. 1197
FIC CONDUCTIVITY OF CHLORINE PE/ DENSITY, VAPOR PRESSURE, CRITICAL PROPERTIES, DIELECTRIC CONSTANT, AND SPECI 0426
UORIDE. LIQUID DENSITY, VAPOR PRESSURE AND CRITICAL TEMPERATURE AND PRESSURE OF NITROGEN TRIFL 0710
INFRARED SPECTRA OF A CRYOSYSTEM. CARBON TETRAFLUORIDE. 0313
INFRARED SPECTRA OF CRYOSYSTEMS. PART-2: SULFUR HEXAFLUORIDE. 1068

BY NEUTRON DIFFRACTION.   CRYSTAL AND MOLECULAR STRUCTURE OF XENON DIFLUORIDE 1239
IDE BY NEUTRON DIFFRACTION.   CRYSTAL AND MOLECULAR STRUCTURE OF XENON TETRAFLUOR 1260
IDE.   CRYSTAL AND MOLECULAR STRUCTURE OF XENON TETRAFLUOR 1277
RT-1: ENERGY LEVELS AND MOLECULAR GEOMETRY. PART-2: COMP/   CRYSTAL FIELD MODEL STUDY OF XENON HEXAFLUORIDE. PA 1311
OF PHOSPHORUS TETRAFLUORIDE RADICALS PRODUCED IN A SINGLE   CRYSTAL OF PHOSPHORUS TRIFLUORIDE.   ESR SPECTRA 0823
N HEXAFLUORIDE, XENON TETRAFLUORIDE, XENON TRIFLUORIDE.   CRYSTAL STRUCTURE.   ENTHALPIES OF FORMATION OF XENO 1269
     PREPARATION AND PROPERTIES OF URANIUM TRIFLUORIDE. (CRYSTAL STRUCTURE)   1164
                           PREPARATION AND CRYSTAL STRUCTURE OF GALLIUM TRIFLUORIDE.   0461
                                           CRYSTAL STRUCTURE OF KRYPTON DIFLUORIDE AT -80C.   0551
                                           CRYSTAL STRUCTURE OF MANGANESE TRIFLUORIDE.   0616
                           PREPARATION AND CRYSTAL STRUCTURE OF MOLYBDENUM TRIFLUORIDE.   0620
                           PREPARATION AND CRYSTAL STRUCTURE OF OSMIUM PENTAFLUORIDE.   0750
RIDES OF OSMIUM. (OSMIUM PENTAFLUORIDE, / PREPARATION AND CRYSTAL STRUCTURE OF OSMIUM PENTAFLUORIDE. TWO FLUO 0749
HANUM TRIFLUORIDE)   SYMMETRY AND CRYSTAL STRUCTURE OF RARE EARTH TRIFLUORIDES. (LANT 0571
                                           CRYSTAL STRUCTURE OF TELLURIUM TETRAFLUORIDE.   1106
                                           CRYSTAL STRUCTURE OF THALLIUM TRIFLUORIDE.   1127
                                           CRYSTAL STRUCTURE OF TUNGSTEN PENTAFLUORIDE.   1154
           NEUTRON DIFFRACTION STUDY OF THE CRYSTAL STRUCTURE OF URANIUM TRIFLUORIDE.   1163
         PREPARATION AND PROPERTIES (CRYSTAL STRUCTURE) OF VANADIUM TETRAFLUORIDE.   1209
       VIBRATIONAL SPECTRA OF LIQUID AND CRYSTALLINE CARBON TETRAFLUORIDE.   0327
                    STRUCTURE OF CRYSTALLINE CHROMIUM TRIFLUORIDE.   0433
            INTERATOMIC BONDING IN CRYSTALLINE MANGANESE TRIFLUORIDE.   0617
                              CRYSTALLINE MODIFICATIONS OF XENON HEXAFLUORIDE.   1297
           RAMAN SPECTRUM OF CRYSTALLINE MOLYBDENUM PENTAFLUORIDE.   0623
INFRARED AND RAMAN SPECTRA OF LIQUID AND CRYSTALLINE SILICON TETRAFLUORIDE.   1001
            INFRARED INTENSITIES IN CRYSTALLINE SULFUR HEXAFLUORIDE.   1075
INFRARED AND RAMAN SPECTRA AND THE STRUCTURE OF CRYSTALLINE THALLIUM MONOFLUORIDE.   1126
         XENON HEXAFLUORIDE. STRUCTURAL CRYSTALLOGRAPHY OF TETRAMERIC PHASES.   1284
         XENON HEXAFLUORIDE. THE STRUCTURE OF A CUBIC PHASE AT -30C.   1283
ORON TRIFLUORIDE.   CUBIC POTENTIAL CONSTANTS AND 1-TYPE RESONANCE IN B 0231
RAFLUORIDE.   SOME PHYSICAL PROPERTIES OF CURIUM AND PROTACTINIUM METALS AND PROTACTINIUM TET 0899
                    EVIDENCE FOR CURIUM TETRAFLUORIDE.   0447
         MELTING POINTS OF CURIUM TRIFLUORIDE AND AMERICIUM TRIFLUORIDE.   0446
OF SOLID AMERICIUM TRIFLUORIDE, AMERICIUM TETRAFLUORIDE, CURIUM TRIFLUORIDE AND CURIUM TETRAFLUORIDE. /ECTRA 0445
ENTAFLUORIDE AND THE PURE ROTATIONAL SPECTRUM OF HYDROGEN CYANIDE. /HE LOW FREQUENCY VIBRATIONS OF ANTIMONY P 0083
S IN CHLORINE MONOFLUORIDE, BROMINE MONOFLUORIDE, BROMINE CYANIDE, AND IODINE CYANIDE. /RIC QUADRUPOLE MOMENT 0254
OR THE GROUND STATE OF OXYGEN DIFLUORIDE, NITRITE ION AND CYANIDE ION. /NFIGURATION INTERACTION CALCULATION F 0763
BY ELECTRON IMPACT OF PHOSPHORUS TRIFLUORIDE AND DIFLUORO CYANO PHOSPHINE.   IONIZATION 0800
BANDS OF TRIFLUORO METHANE, PHOSPHORUS PENTAFLUORIDE, AND CYCLOPROPANE BY COMPUTER TECHNIQUES. /ION ROTATION 0880
PROPERTIES AND REACTIONS OF URANIUM HEXAFLUORIDE BY ION CYCLOTRON RESONANCE SPECTROSCOPY.   1185
           FINE STRUCTURE ANALYSIS OF THE C1 SYSTEM OF ANTIMONY MONOFLUORIDE.   0059
           ROTATIONAL ANALYSIS OF THE C1 SYSTEM OF ANTIMONY MONOFLUORIDE.   0069
UORIDE A DOUBLET SIGMA(+/ SHOCK TUBE DETERMINATION OF THE C2 (A TRIPLET PI TO X TRIPLET PI) AND CARBON MONOFL 0288
           ROTATIONAL ANALYSIS OF THE C2 SYSTEM OF ANTIMONY MONOFLUORIDE.   0068
MEAN AMPLITUDES OF VIBRATION IN SOME XY2Z MOLECULES OF C2V SYMMETRY.   1396
           ROTATIONAL STRUCTURE OF THE C3 SYSTEM OF ANTIMONY FLUORIDE MOLECULE.   0066
           ROTATIONAL ANALYSIS OF C3 SYSTEM OF THE ANTIMONY MONOFLUORIDE MOLECULE.   0057

D

X SINGLET SIGMA PLATINUM OXIDE AND SILICON MONOFLUORIDE (D DOUBLET SIGMA(+)- X DOUBLET PI) MOLECULES. /IGMA- 0966
ORIDE, KRYPTON DIFLUORIDE, AND XENON DIFLUORIDE) D ORBITALS IN THE NOBLE GAS DIHALIDES. (ARGON DIFLU 1338
         THERMAL STABILITY AND REACTIVITY OF D- AND F ELEMENT HEXAFLUORIDES.   0630
NCE OF THE TRAPPED RADICAL XENON MONOFLUORIDE. RADIATION DAMAGE IN XENON TETRAFLUORIDE. ELECTRON SPIN RESONA 1220
F FORCE CONSTANTS OF PLANE XY3 MOLECULES FROM VIBRATIONAL DATA.   EVALUATION O 1401
NIUM FLUORIDE (GAS) VIBRATIONS ACCORDING TO THERMODYNAMIC DATA.   FREQUENCIES OF ZIRCO 1329
LUORIDE FROM ELECTRON DIFFRACTION AND ROTATIONAL CONSTANT DATA. /S OF BROMINE PENTAFLUORIDE AND IODINE PENTAF 0265
ECONCILIATION OF THE ENTROPIES FROM MOLECULAR AND THERMAL DATA. /DYNAMIC FUNCTIONS FROM 5 TO 350 DEGREES K. R 1275
/PERTIES OF THE DIATOMIC INTERHALOGENS FROM SPECTROSCOPIC DATA. (CHLORINE, BROMINE, AND IODINE MONOFLUORIDES) 0253
           THERMODYNAMIC DATA FOR BERYLLIUM DIFLUORIDE.   0136
ONICITY EFFEC/ UTILITY OF HEAVY ATOM ISOTOPIC SUBSTITUION DATA FOR THE CALCULATION OF FORCE CONSTANTS. ANHARM 0234
STRUCTURE OF FLUORIDES OF NONTRANSITION ELEMENTS FROM THE DATA OF LCAO MO SCF NONEMPIRICAL CALCULATIONS. /IC 0009
/ION OF TRIGONAL BIPYRAMIDAL MOLECULES FROM SPECTROSCOPIC DATA. PENTACOORDINATED MOLECULES. PART-13: (PHOSPH/ 0853
           VIRIAL THEOREM DECOMPOSITION OF MOLECULAR FORCE FIELDS.   0783
IUM TETRAFLUORIDE, AND P/ ELECTRIC DEFLECTION AND THERMAL DECOMPOSITION STUDIES ON CERIUM TETRAFLUORIDE, TERB 0367
ON INTENSITIES. PART-2: INTERNAL MOTION AND MEAN GEOMETRY DEDUCED BY ELECTRON DIFFRACTION. /LECTRON DIFFRACTI 1291
           INERTIA DEFECT AND PLANARITY OF FOUR ATOM MOLECULES. (XY3) 1398
/ OF THE GEOMETRY OF HIGH TEMPERATURE SPECIES BY ELECTRIC DEFLECTION AND MASS SPECTROMETRIC DETECTION. (FIRS/ 0134
IUM TETRAFLUORIDE, TERBIUM TETRAFLUORIDE, AND P/ ELECTRIC DEFLECTION AND THERMAL DECOMPOSITION STUDIES ON CER 0367
UORIDE, SELENIUM HEXAFLUORIDE, TELLURIUM HEXAFL/ ELECTRIC DEFLECTION OF BINARY HEXAFLUORIDES. (SULFUR HEXAFL 1084
LUORIDES AND TRIFLUORIDES. (LANTHANIDE TRIFLUOR/ ELECTRIC DEFLECTION OF MOLECULAR BEAMS OF THE RARE EARTH DIF 0946
. (XENON HEXAFLUORIDE, IODINE HEPTAFLUORI/ ELECTRIC FIELD DEFLECTION OF MOLECULES WITH LARGE AMPLITUDE MOTION 1281
. CARBON TETRACHLORIDE, SILICON / MOLECULAR BEAM ELECTRIC DEFLECTION OF THE TETRAHALIDES CARBON TETRAFLUORIDE 0349
           MAGNETIC FIELD DEFLECTION OF XENON HEXAFLUORIDE.   1287
M, TIN, AND LEAD. (GERMANIUM DIFLUORI/ FREQUENCIES OF THE DEFORMATION VIBRATIONS OF THE DIHALIDES OF GERMANIU 0479
AND DIPOLE MOMENT OF NITROGEN MONOFLUORIDE IN THE SINGLET DELTA STATE. /TRON PARAMAGNETIC RESONANCE SPECTRUM 0673
- X DOUBLET PI 3/2 / ROTATIONAL ANALYSIS OF THE C DOUBLET DELTA 5/2- X DOUBLET PI 3/2 AND G DOUBLET DELTA 5/2 1136
           ELECTRON DENSITY DISTRIBUTION IN SOME CHLORIDES.   0218
RIDES. (CARBON TETRAFLUORIDE AND SILICON TETRAF/ ELECTRON DENSITY DISTRIBUTIONS IN SOME FLUORIDES AND OXYCHLO 0337
LS. PART-3: HETERONUCLEAR DIATOMIC AND POLYATOMIC MOLECU/ DENSITY LOCALIZATION OF ATOMIC AND MOLECULAR ORBITA 0175
NY PENTAFLUORIDE.   DENSITY, MELTING POINT AND VAPOR PRESSURE OF ANTIMO 0082
OBIUM AND TANTALUM PENTAFLUORIDES IN THE GAS PHASE. VAPOR DENSITY OF NIOBIUM PENTAFLUORIDE. /AL SPECTRA OF NI 0660
TERING.   MEASUREMENT OF THE GAS DENSITY OF URANIUM HEXAFLUORIDE BY LASER RAMAN SCAT 1181
D PRESSURE OF NITROGEN TRIFLUORIDE.   LIQUID DENSITY, VAPOR PRESSURE AND CRITICAL TEMPERATURE AN 0710
CTRIC CONSTANT, AND SPECIFIC CONDUCTIVITY OF CHLORINE PE/ DENSITY, VAPOR PRESSURE, CRITICAL PROPERTIES, DIELE 0426
ULFUR HEXAFLUORIDE AT 10.6 MICROMETERS.   TEMPERATURE DEPENDENT ABSORPTION SPECTRUM OF THE NU-3 BAND OF S 1091
XAFLUORIDES. / POSSIBLE ROTATIONAL RAMAN SPECTRUM AND THE DEPOLARIZATION OF RAYLEIGH SCATTERING IN CERTAIN HE 0901
NITROGEN FLUORIDES AND THEIR INORGANIC DERIVATIVES.   0703
UORIDE.   CNDO CALCULATION OF DIPOLE MOMENT DERIVATIVES AND INFRARED INTENSITIES OF BORON TRIFL 0195
SOME TETRAHEDRAL XY4 TYPE MOLECULES AND THEIR SUBSTITUTED DERIVATIVES BY GREEN'S FUNCTION ANALYSIS. /ANTS OF 1424

US OX/ INFRARED INTENSITIES, BOND MOMENTS AND BOND MOMENT DERIVATIVES FOR PHOSPHORUS TRIFLUORIDE AND PHOSPHOR 0797
EL/ VACUUM ULTRAVIOLET ABSORPTION SPECTRA OF SOME HALOGEN DERIVATIVES OF METHANE. (CARBON TETRAFLUORIDE) CORR 0365
/ K, AND OF PLUTONIUM-244 DIOXIDE FROM 4 TO 25 DEGREES K. DERIVED ENTROPIES AND OTHER THERMODYNAMIC PROPERTI/ 0885
ACCURACY OF THE VIBRATIONAL WAVE FUNCTIONS AND DERIVED QUANTITIES. (TIN FLUORIDE) 1134
IFLUORIDE. HEAT CAPACITY FROM 5 TO 350 DEGREES K AND SOME DERIVED THERMODYNAMIC PROPERTIES. XENON D 1246
/RE SPECIES BY ELECTRIC DEFLECTION AND MASS SPECTROMETRIC DETECTION. (FIRST ROW FLUORIDES, BERYLLIUM, MAGNES/ 0134
INFRARED DETECTION OF GASEOUS CARBON TRIFLUORIDE RADICAL. 0310
MATRIX INFRARED SPECTRUM OF OXYGEN MONOFLUORIDE AND DETECTION OF LITHIUM OXYFLUORIDE. 0755
HORUS TRIFLUORIDE) ABSORPTION SPECTRA OF THE HYDRIDES, DEUTERIDES, AND HALIDES OF GROUP-5 ELEMENTS. (PHOSP 0806
FERMI DIAD OF NU-1 AND 2NU-2 IN OXYGEN DIFLUORIDE. 0772
/ROWAVE SPECTROMETER FOR MEASUREMENTS OF ZEEMAN EFFECT IN DIAMAGNETIC MOLECULES. GJ FACTOR OF THALLIUM FLUOR/ 1120
N OF ATOMIC AND MOLECULAR ORBITALS. PART-3: HETERONUCLEAR DIATOMIC AND POLYATOMIC MOLECULES. /ITY LOCALIZATIO 0175
POTENTIAL ENERGY CURVES AND DISSOCIATION ENERGIES OF DIATOMIC BORON AND ALUMINUM HALIDES. 0173
, A/ POTENTIAL ENERGY CURVES AND DISSOCIATION ENERGIES OF DIATOMIC FLUORIDES AND CHLORIDES OF GALLIUM, INDIUM 0458
VIBRATIONAL TRANSITION PROBABILITIES AND R CENTROIDS FOR DIATOMIC FLUORIDES OF SILICON AND GERMANIUM. 0994
ES. (PHOSPHORUS MONOFLUORIDE) THERMODYNAMIC FUNCTIONS OF DIATOMIC GASES WITH MOLECULES IN TRIPLET SIGMA STAT 0789
NE FLUORIDE) DISSOCIATION ENERGIES OF DIATOMIC HALOGEN FLUORIDES. (IODINE FLUORIDE, BROMI 0494
LORINE, BROMINE, AND IOD/ THERMODYNAMIC PROPERTIES OF THE DIATOMIC INTERHALOGENS FROM SPECTROSCOPIC DATA. (CH 0253
ODINE MONOFLUORIDE. THEORY OF THE DISSOCIATION OF DIATOMIC MOLECULES AND A STUDY OF THE EMISSION OF I 0493
N MONOFLUORIDE, AND TIN MONOFL/ POTENTIAL CURVES FOR SOME DIATOMIC MOLECULES. (BERYLLIUM MONOFLUORIDE, SILICO 0130
/ AND SOME OPTIMIZED VALENCE CONFIGURATION RESULTS ON THE DIATOMIC MOLECULES LITHIUM, BERYLLIUM, BORON, CARB/ 0174
UM MONOFLUORIDE) ELECTRIC DIPOLE MOMENTS OF OPEN SHELL DIATOMIC MOLECULES. (SULFUR MONOFLUORIDE AND SELENI 1019
HLORINE PE/ DENSITY, VAPOR PRESSURE, CRITICAL PROPERTIES, DIELECTRIC CONSTANT, AND SPECIFIC CONDUCTIVITY OF C 0426
ELECTRICAL CONDUCTIVITY AND DIELECTRIC CONSTANT OF LIQUID XENON HEXAFLUORIDE. 1307
/ COMPARATIVE ANALYSIS OF DIFFERENT SCHEMES OF NEGLECT OF DIFFERENTIAL OVERLAP METHODS ILLUSTRATED BY THE XE/ 1231
DISTANCE OF BORON TRIFLUORIDE DETERMINED BY GAS ELECTRON DIFFRACTION. BORON FLUORINE BOND 0223
AL AND MOLECULAR STRUCTURE OF XENON DIFLUORIDE BY NEUTRON DIFFRACTION. CRYST 1239
AND MOLECULAR STRUCTURE OF XENON TETRAFLUORIDE BY NEUTRON DIFFRACTION. CRYSTAL 1260
SENIC TRIFLUORIDE AND ARSENIC TRICHLORIDE BY GAS ELECTRON DIFFRACTION. /ION OF THE MOLECULAR STRUCTURES OF AR 0104
VIBRATION OF GASEOUS SILICON TETRAFLUORIDE FROM ELECTRON DIFFRACTION. /ES AND ROOT MEAN SQUARE AMPLITUDES OF 1002
RUCTURE OF PHOSPHORUS TRIFLUORIDE STUDIED BY GAS ELECTRON DIFFRACTION. MOLECULAR ST 0810
ORIDE AND ARSENIC PENTAFLUORIDE AS DETERMINED BY ELECTRON DIFFRACTION. MOLECULAR STRUCTURES OF ARSENIC TRIFLU 0099
RUCTURE OF KRYPTON DIFLUORIDE AS INVESTIGATED BY ELECTRON DIFFRACTION. ST 0558
ISE STRUCTURAL PARAMETERS IN COPPER DIFLUORIDE BY NEUTRON DIFFRACTION. STRUCTURES OF FLUORIDES. PART-6: PREC 0443
STRUCTURES OF YTTRIUM AND BISMUTH TRIFLUORIDES BY NEUTRON DIFFRACTION. 1318
RIFLUORIDE) ELECTRON DIFFRACTION ANALYSIS OF GALLIUM HALIDES. (GALLIUM T 0460
GROUP-2 ELEMENTS. (BERYLLIUM DIFLUORIDE, MAGNE/ ELECTRON DIFFRACTION ANALYSIS OF THE MOLECULAR STRUCTURES OF 0131
NON HEXAFLUORID/ EFFECTS OF ELECTRON CORRELATION IN X-RAY DIFFRACTION AND AN ELECTRON DIFFRACTION STUDY OF XE 1290
MINE PENTAFLUORIDE AND IODINE PENTAFLUORIDE FROM ELECTRON DIFFRACTION AND ROTATIONAL CONSTANT DATA. /S OF BRO 0265
RIFLUORIDE AND ARSENIC TRICHLORIDE DETERMINED BY ELECTRON DIFFRACTION AND SPECTROSCOPY. /UCTURES OF ARSENIC T 0103
US TRIFLUORIDE, PHOSPHORUS PENTAFLUORIDE, DIFLU/ ELECTRON DIFFRACTION BY GASES AND THE STRUCTURES OF PHOSPHOR 0801
/TURE OF XENON HEXAFLUORIDE. PART-1: ANALYSIS OF ELECTRON DIFFRACTION INTENSITIES. PART-2: INTERNAL MOTION A/ 1291
RSENIC PENTAFLUORIDE, AND OTHER MOLECULES. AN ELECTRON DIFFRACTION INVESTIGATION OF ARSENIC TRIFLUORIDE, A 0098
AND TANTALUM PENTAFLUORIDE. AN ELECTRON DIFFRACTION INVESTIGATION OF NIOBIUM PENTAFLUORIDE 0671
ELECTRON DIFFRACTION INVESTIGATION OF NITROGEN TRIFLUORIDE. 0738
UORIDE. NEUTRON DIFFRACTION INVESTIGATION OF ORTHORHOMBIC LEAD DIFL 0593
(TUNG/ SPECTROSCOPIC STUDIES IN CONNECTION WITH ELECTRON DIFFRACTION INVESTIGATION OF SOME SIMPLE MOLECULES. 1153
ES OF THE ALUMINUM HALIDES. ELECTRON DIFFRACTION INVESTIGATION OF THE MOLECULAR STRUCTURE 0036
NIUM TETRAFLUORIDE IN THE VAPOR PHASE. ELECTRON DIFFRACTION INVESTIGATION OF THE STRUCTURE OF ZIRCO 1332
OSMIUM HEXAFLUORIDE, IRIDIUM HEXAFLUORIDE, URA/ ELECTRON DIFFRACTION INVESTIGATION OF TUNGSTEN HEXAFLUORIDE, 1155
IDE AND PHOSPHORUS PENTACHLORIDE. ELECTRON DIFFRACTION INVESTIGATIONS OF PHOSPHORUS PENTAFLUOR 0869
/ THE FAILURE OF THE FIRST BORN APPROXIMATION IN ELECTRON DIFFRACTION. (MOLYBDENUM HEXAFLUORIDE AND TUNGSTEN/ 0640
STRUCTURE OF URANIUM HEXAFLUORIDE AS DETERMINED BY DIFFRACTION OF ELECTRONS ON THE VAPOR. 1183
STRUCTURE OF URANIUM HEXAFLUORIDE AS DETERMINED BY THE DIFFRACTION OF ELECTRONS ON THE VAPOR. 1184
EXAFLUORIDE, SULFUR TETRAFLUOR/ INVESTIGATION BY ELECTRON DIFFRACTION OF THE MOLECULAR STRUCTURES OF SULFUR H 1041
O/ MULTIPLE INTRAMOLECULAR SCATTERING EFFECTS ON ELECTRON DIFFRACTION PATTERNS FOR THE RHENIUM HEXAFLUORIDE M 0911
AT 193/ STRUCTURES OF FLUORIDES. PART-10: NEUTRON POWDER DIFFRACTION PROFILE STUDIES OF URANIUM HEXAFLUORIDE 1202
OGEN DIFLUORIDE) ELECTRON DIFFRACTION STUDIES AT ELEVATED TEMPERATURES. (NITR 0679
UORIDES. (SECOND ROW, THIRD ROW, AND RHENIUM HEPTA/ X-RAY DIFFRACTION STUDIES OF SOME TRANSITION METAL HEXAFL 1375
N HEXAFLUORIDE AND OTHER COMPOUNDS. ELECTRON DIFFRACTION STUDIES OF XENON TETRAFLUORIDE AND XENO 1259
N- FLUORINE LENGTH IN SILICON TETRAFLUORIDE. NEW ELECTRON DIFFRACTION STUDY. SILICO 1000
TIONAL MODE COUPLING IN IODINE HEPTAFLUORIDE. AN ELECTRON DIFFRACTION STUDY. /TURE, PSEUDOROTATION, AND VIBRA 0515
RE, PSEUDOROTATION, AND ANHARMONIC COUPLING OF / ELECTRON DIFFRACTION STUDY OF RHENIUM HEPTAFLUORIDE. STRUCTU 0920
ELECTRON DIFFRACTION STUDY OF RHENIUM HEXAFLUORIDE. 0909
UM TRIFLUORIDE. NEUTRON DIFFRACTION STUDY OF THE CRYSTAL STRUCTURE OF URANI 1163
VAPOR PHASE HALIDES OF GALLIUM, YTTRIUM, LANTH/ ELECTRON DIFFRACTION STUDY OF THE MOLECULAR STRUCTURE OF THE 1317
NADIUM PENTAFLUORIDE, NIOBIUM PENTAFLUORIDE, AN/ ELECTRON DIFFRACTION STUDY OF THE MOLECULAR STRUCTURES OF VA 1217
TETRAFLUORIDE. ELECTRON DIFFRACTION STUDY OF THE STRUCTURE OF GASEOUS XENON 1282
NTAFLUORIDE. ELECTRON DIFFRACTION STUDY OF THE STRUCTURE OF PHOSPHORUS PE 0851
LUORIDE AND DINITROGEN TETRAFLUORIDE. DIFFRACTION STUDY OF THE STRUCTURES OF NITROGEN DIF 0680
ECULE IN THE VAPOR FORM. ELECTRON DIFFRACTION STUDY OF THE VANADIUM PENTAFLUORIDE MOL 1218
ES IN THE GAS PHASE. (URANIUM TETRAFLUORIDE AND/ ELECTRON DIFFRACTION STUDY OF URANIUM AND THORIUM TETRAHALID 1170
FAILURE OF THE FIRST BORN APPROXIMATION IN ELECTRON DIFFRACTION. (TELLURIUM HEXAFLUORIDE) 1109
UORI/ FAILURE OF THE FIRST BORN APPROXIMATION IN ELECTRON DIFFRACTION. (TUNGSTEN HEXAFLUORIDE, URANIUM HEXAFL 1158
FAILURE OF THE FIRST BORN APPROXIMATION IN ELECTRON DIFFRACTION. (URANIUM HEXAFLUORIDE) 1199
ADICAL MICROWAVE ROTATIONAL ABSORPTION STUDIES. (NITROGEN DIFLUORIDE) A CAVITY SEARCH SPECTROMETER FOR FREE R 0689
MUIR MEASUREMENT OF THE SUBLIMATION PRESSURE OF MANGANESE DIFLUORIDE. A LANG 0611
A REINVESTIGATION OF THE INFRARED SPECTRUM OF OXYGEN DIFLUORIDE. 0774
AB INITIO CALCULATIONS OF THE BONDING IN KRYPTON DIFLUORIDE. 0553
OF THE GEOMETRY, BONDING, AND COUPLING CONSTANTS OF BORON DIFLUORIDE. AB INITIO INVESTIGATION 0180
ABSENCE OF FERMI RESONANCE IN KRYPTON DIFLUORIDE. 0564
ABSORPTION SPECTRUM OF CARBON DIFLUORIDE. 0302
ABSORPTION SPECTRUM OF NITROGEN DIFLUORIDE. 0685
BOND LENGTHS IN XENON DIFLUORIDE. 1236
BONDING IN KRYPTON DIFLUORIDE. 0554
BONDING IN KRYPTON DIFLUORIDE. 0555
BONDING IN XENON DIFLUORIDE. 1230
NDO STUDY OF THE OZONIDE ION AND RELATED SPECIES. (OXYGEN DIFLUORIDE) C 0782
DOUBLE ZETA SCF CALCULATIONS FOR NITRITE ION AND OXYGEN DIFLUORIDE. 0762
N DIFFRACTION STUDIES AT ELEVATED TEMPERATURES. (NITROGEN DIFLUORIDE) ELECTRO 0679

ELECTRONIC SPECTRA OF POLYATOMIC FREE RADICALS. (CARBON DIFLUORIDE)   0308
ELECTRONIC STRUCTURE AND PROPERTIES OF KRYPTON DIFLUORIDE.   0548
ELECTRONIC STRUCTURE OF XENON DIFLUORIDE.   1226
EMISSION BANDS OF CARBON DIFLUORIDE.   0309
N ELECTRICAL DISCHARGES THROUGH BORON TRIFLUORIDE. (BORON DIFLUORIDE) /IN POLYATOMIC FREE RADICALS OBTAINED I 0222
ENTHALPY OF FORMATION OF BERYLLIUM DIFLUORIDE.   0139
ENTHALPY CF FORMATION OF GERMANIUM DIFLUORIDE.   0472
ENVIRONMENT OF THE TIN ATOM IN ORTHORHOMBIC TIN DIFLUORIDE.   1139
ESTIMATION OF INFRARED FREQUENCIES OF SULFUR DIFLUORIDE.   1026
XTENDED HUCKEL THEORY AND THE SHAPE OF MOLECULES. (OXYGEN DIFLUORIDE)   E   0758
FAR ULTRAVIOLET SPECTROSCOPIC STUDY OF XENON DIFLUORIDE.   1256
FERMI DIAD OF NU-1 AND 2NU-2 IN OXYGEN DIFLUORIDE.   0772
FLUORINE NMR SPECTRA OF OXYGEN FLUORIDES. (OXYGEN DIFLUORIDE)   0775
GAS PHASE STRUCTURE OF XENON DIFLUORIDE.   1248
PECTRA OF THE ALKALINE EARTH DIHALIDES. PART-1: MAGNESIUM DIFLUORIDE.   GEOMETRY AND VIBRATIONAL S 0609
GEOMETRY OF MOLECULES. (OXYGEN DIFLUORIDE)   0764
HEAT FORMATION OF KRYPTON DIFLUORIDE.   0557
HEAT OF ATOMIZATION OF ALUMINUM DIFLUORIDE.   0034
HEAT OF FORMATION CF BERYLLIUM DIFLUORIDE.   0132
HEAT OF FORMATION OF BORON DIFLUORIDE.   0177
HEAT OF FORMATION OF CARBON DIFLUORIDE.   0298
HEAT OF FORMATION OF OXYGEN DIFLUORIDE.   0769
INFRARED AND RAMAN SPECTRA OF KRYPTON DIFLUORIDE.   0552
INFRARED SPECTRA OF MATRIX ISOLATED MAGNESIUM DIFLUORIDE.   0604
FORCE CONSTANTS, AND THERMODYNAMIC FUNCTIONS OF SILICON DIFLUORIDE.   INFRARED SPECTRUM 0985
INFRARED SPECTRUM CF HEAVY METAL DIHALIDES. (EUROPIUM DIFLUORIDE)   0451
INFRARED SPECTRUM OF NITROGEN DIFLUORIDE.   0692
INFRARED VIBRATIONAL PROPERTIES OF GERMANIUM DIFLUORIDE.   0473
GMUIR MEASUREMENTS OF THE SUBLIMATION PRESSURE OF CADMIUM DIFLUORIDE.   KNUDSEN AND LAN 0272
NGMUIR MEASUREMENTS OF THE SUBLIMATION PRESSURE OF COBALT DIFLUORIDE.   KNUDSEN AND LA 0439
KNUDSEN MEASUREMENT OF THE VAPOR PRESSURE OF MANGANESE DIFLUORIDE.   0613
MAGNETIC SUSCEPTIBILITY OF SILVER DIFLUORIDE.   1015
THE GASEOUS MOLECULES ALUMINUM OXYDIFLUORIDE AND ALUMINUM DIFLUORIDE. /MINATION OF THE HEATS OF FORMATION OF 0035
HIGH TEMPERATURES. PART-9: SUBLIMATION PRESSURE OF COPPER DIFLUORIDE.   MASS SPECTROMETRIC STUDIES AT 0442
MASS SPECTROMETRIC STUDY OF MAGNESIUM HALIDES. (MAGNESIUM DIFLUORIDE)   0603
MASS SPECTRUM AND MOLECULAR ENERGETICS OF KRYPTON DIFLUORIDE.   0565
O SHRINKAGE EFFECT AND THERMODYNAMIC FUNCTIONS OF KRYPTON DIFLUORIDE. /LITUDES OF VIBRATION, BASTIANSEN-MORIN 0562
NG POINT, VAPOR PRESSURE, AND HEAT OF VAPORIZATION OF TIN DIFLUORIDE.   MELTI 1140
MICROWAVE SPECTRUM AND STRUCTURE OF SULFUR DIFLUORIDE.   1030
MICROWAVE SPECTRUM OF CARBON DIFLUORIDE.   0307
MICROWAVE SPECTRUM OF GERMANIUM DIFLUORIDE.   0478
MICROWAVE SPECTRUM OF SILICON DIFLUORIDE.   0990
ICROWAVE SPECTRUM, STRUCTURE, AND DIPOLE MOMENT OF OXYGEN DIFLUORIDE.   M 0779
MILLIMETER WAVE SPECTRUM AND STRUCTURE OF OXYGEN DIFLUORIDE.   0768
E OF OZONE, NITRITE ION, NITROGEN OXYFLUORIDE, AND OXYGEN DIFLUORIDE. /T SCF CALCULATIONS FOR THE GROUND STAT 0776
IC FUNCTIONS OF SOME LINEAR SYMMETRICAL MCLECULES. (XENON DIFLUORIDE)   MOLECULAR CONSTANTS AND THERMODYNAM 1244
MOLECULAR ORBITAL ENERGY LEVELS AND BONDING IN KRYPTON DIFLUORIDE.   0549
UDY OF NITROGEN DIHYDRIDE, NITROGEN DIOXIDE, AND NITROGEN DIFLUORIDE.   MOLECULAR ORBITAL ST 0684
ITROGEN TRIFLUORIDE, PHOSPHORUS TRIFLUORIDE, AND NITROGEN DIFLUORIDE.   MOLECULAR ORBITAL STUDY OF N 0721
NEUTRON DIFFRACTION INVESTIGATION OF ORTHORHOMBIC LEAD DIFLUORIDE.   0593
NEW ELECTRONIC EMISSION FROM SILICON DIFLUORIDE.   0987
NEW ELECTRONIC EMISSION SYSTEM OF GERMANIUM DIFLUORIDE.   0476
O SCF CALCULATIONS OF THE ELECTRONIC STRUCTURE OF SILICON DIFLUORIDE.   NONEMPIRICAL LCAO M 0995
LCAO MO SCF STUDIES OF THE LOW LYING STATES OF BERYLLIUM DIFLUORIDE.   NONEMPIRICAL 0137
S ON HEATS OF FORMATION OF CARBON MONOFLUORIDE AND CARBON DIFLUORIDE. /THS OF CARBON MONOFLUORIDE AND COMMENT 0295
OXYGEN FLUORIDES. (OXYGEN DIFLUORIDE)   0785
NVESTIGATIONS AND VAPOR PRESSURE MEASUREMENTS ON EUROPIUM DIFLUORIDE.   PHASE I 0452
PHOTOELECTRON SPECTRUM OF SILICON DIFLUORIDE.   0980
PHOTOELECTRON SPECTRUM OF SILICON DIFLUORIDE.   0996
PHOTOELECTRON SPECTRUM OF XENON DIFLUORIDE.   1227
PHOTOIONIZATION MASS SPECTROMETRY OF OXYGEN DIFLUORIDE.   0760
THE ABSORPTION SPECTRA OF CARBON MONOFLUORIDE AND CARBON DIFLUORIDE.   PREDISSOCIATION IN 0293
ATION OF INERT GAS COMPOUNDS BY MATRIX ISOLATION: KRYPTON DIFLUORIDE.   PREPAR 0566
PROBABLE STRUCTURE OF XENON TETRAFLUORIDE AND XENON DIFLUORIDE.   1250
RA OF SOLID XENON TETRAFLUORIDE AND ITS ADDUCT WITH XENON DIFLUORIDE.   RAMAN SPECT 1258
RAMAN SPECTRUM OF LIQUID OXYGEN DIFLUORIDE.   0767
RELATIVISTIC MOLECULAR WAVE FUNCTIONS: XENON DIFLUORIDE.   1249
F ABSORPTION BANDS IN THE 2266 ANGSTROM SYSTEM OF SILICON DIFLUORIDE.   ROTATIONAL ANALYSIS O 0979
MOLECULAR CONSTANTS IN SOME POLYATOMIC MOLECULES. (XENON DIFLUORIDE)   SPECTROSCOPIC STUDIES OF 1245
L HYPERFINE STRUCTURE IN THE MICROWAVE SPECTRUM OF OXYGEN DIFLUORIDE.   SPIN ROTATIONA 0777
STRUCTURE OF KRYPTON DIFLUORIDE.   0561
STRUCTURE OF THE SULFUR TRIFLUORIDE POSITIVE ION. (SULFUR DIFLUORIDE)   1032
SUBLIMATION PRESSURE OF IRON DIFLUORIDE.   0541
SYNTHESIS AND CHARACTERIZATION OF VANADIUM DIFLUORIDE.   1206
THERMOCHEMISTRY OF SILICON DIFLUORIDE.   0992
PART-1: ENTHALPY AND ENTROPY OF VAPORIZATION OF BERYLLIUM DIFLUORIDE. /AL PROPERTIES OF BERYLLIUM COMPOUNDS. 0138
SYNTHESIS BY THE REDUCTION OF THE TRIHALIDES. (SCANDIUM DIFLUORIDE) /LIDES AND THE THE POSSIBILITY OF THEIR 0940
THERMODYNAMIC DATA FOR BERYLLIUM DIFLUORIDE.   0136
THERMODYNAMIC PROPERTIES OF CHROMIUM DIFLUORIDE.   0432
RIDE SYSTEM FROM MASS SPECTROMETER INVESTIGATIONS. (BORON DIFLUORIDE) / PROPERTIES OF THE BORON CHLORIDE FLUO 0179
ORDER POTENTIAL CONSTANTS OF BENT XY2 MOLECULES. (OXYGEN DIFLUORIDE)   THIRD 0770
ULTRAVIOLET ABSORPTION SPECTRUM OF GERMANIUM DIFLUORIDE.   0474
ULTRAVIOLET ABSORPTION SPECTRUM OF SILICON DIFLUORIDE.   0984
ULTRAVIOLET BAND SPECTRUM OF SILICON DIFLUORIDE.   0991
ULTRAVIOLET SPECTRUM OF SILICON DIFLUORIDE.   0983
SURE AND HEAT OF SUBLIMATION OF COBALT DIHALIDES. (COBALT DIFLUORIDE)   VAPOR PRES 0438
VAPOR PRESSURE OF NICKEL DIFLUORIDE.   0652
NAL SPECTRA AND THERMODYNAMICS OF THE ZINC HALIDES. (ZINC DIFLUORIDE)   VIBRATIO 1323
X-RAY STUDY OF ANHARMONIC VIBRATIONS IN CALCIUM DIFLUORIDE.   0283
2500 ANGSTROM ABSORPTION SPECTRUM OF CARBON DIFLUORIDE.   0306

FAR ULTRAVIOLET ELECTRONIC SPECTRA OF STRONTIUM DIFLUORIDE AND BARIUM DIFLUORIDE. 1017
BORON TRIHYDRIDE, BORON DIHYDRIDE FLUORIDE, BORON HYDRIDE DIFLUORIDE, AND BORON TRIFLUORIDE. / STRUCTURES OF 0241
N OF CHLORINE FLUORIDES. (CHLORINE MONOFLUORIDE, CHLORINE DIFLUORIDE, AND CHLORINE TRIFLUORIDE) /OF FORMATIO 0370
HIGH TEMPERATURES. PART-9: SUBLIMATION PRESSURE OF COPPER DIFLUORIDE. (AND COPPER MONOFLUORIDE) / STUDIES AT 0441
PREPARATION AND PURIFICATION OF OXYGEN DIFLUORIDE AND DETERMINATION OF ITS VAPOR PRESSURE. 0781
DIFFRACTION STUDY OF THE STRUCTURES OF NITROGEN DIFLUORIDE AND DINITROGEN TETRAFLUORIDE. 0680
AND ES/ AB INITIO MOLECULAR ORBITAL STUDY OF THE NITROGEN DIFLUORIDE AND DINITROGEN TETRAFLUORIDE MOLECULES, 0688
VACUUM ULTRAVIOLET SPECTRA OF SILICON DIFLUORIDE AND GERMANIUM DIFLUORIDE. 0981
OF GASEOUS GERMANIUM, GERMANIUM TETRAFLUORIDE, GERMANIUM DIFLUORIDE, AND GERMANIUM MONOFLUORIDE. /PROPERTIES 0468
ABSORPTION SPECTRUM OF CARBON DIFLUORIDE AND ITS VIBRATIONAL ANALYSIS. 0303
NERT GAS HALIDES. HELIUM-2 PHOTOELECTRON SPECTRA OF XENON DIFLUORIDE AND KRYPTON DIFLUORIDE. / CHARACTER OF I 1343
HELIUM-2 PHOTOELECTRON SPECTRA OF XENON DIFLUORIDE AND KRYPTON DIFLUORIDE. 1234
ULTRAVIOLET EMISSION SPECTRA. (XENON DIFLUORIDE AND KRYPTON DIFLUORIDE) 0559
INFRARED SPECTRA OF MATRIX ISOLATED TIN DIFLUORIDE AND LEAD DIFLUORIDE. 0594
RAMAN SPECTRA OF CADMIUM DIFLUORIDE AND LEAD DIFLUORIDE. 0274
ULTRAVIOLET ABSORPTION SPECTRA OF GASEOUS TIN DIFLUORIDE AND LEAD DIFLUORIDE. 0595
IDES. (OF THE FIRST ROW TRANSITION METALS, ALSO BERYLLIUM DIFLUORIDE AND MAGNESIUM DIFLUORIDE) /RICAL DIFLUOR 1369
SUBLIMATION PRESSURE OF MANGANESE DIFLUORIDE AND MANGANESE MONOFLUORIDE. 0614
SPECTRA OF THE DIHALIDES OF MERCURY AND CADMIUM. (CADMIUM DIFLUORIDE AND MERCURY DIFLUORIDE) VIBRATIONAL 0275
MIUM TRIFLUORIDE AND THE STABILITIES OF GASEOUS NEODYMIUM DIFLUORIDE AND NEODYMIUM MONOFLUORIDE. /RE OF NEODY 0643
SPECTRA OF MATRIX ISOLATED NICKEL DIFLUORIDE AND NICKEL DICHLORIDE. 0653
VAPOR PRESSURES OF BERYLLIUM DIFLUORIDE AND NICKEL DIFLUORIDE. 0135
CALCULATION OF FORCE CONSTANTS FOR NITROGEN DIFLUORIDE AND NITROGEN TRIFLUORIDE. 0693
IES OF NIOBIUM PENTAFLUORIDE AND ITS COMPLEXES WITH XENON DIFLUORIDE AND ORGANIC BASES. /UPOLE RESONANCE STUD 0664
PHOTOELECTRON SPECTRA OF OXYGEN DIFLUORIDE AND OXYGEN DICHLORIDE. 0766
UDES OF VIBRATION AND THERMODYNAMIC FUNCTIONS FOR SILICON DIFLUORIDE AND OXYGEN DIFLUORIDE. MEAN AMPLIT 0773
ELECTRON SPIN RESONANCE OF TRAPPED PHOSPHORUS DIFLUORIDE AND PHOSPHORUS TETRAFLUORIDE RADICALS. 0792
RMATION OF XENON HEXAFLUORIDE, XENON TETRAFLUORIDE, XENON DIFLUORIDE, AND PHOSPHORUS TRIFLUORIDE. /PIES OF FO 0807
ESR SPECTRA OF PHOSPHORUS DIFLUORIDE AND SULFUR TRIFLUORIDE RADICALS. 0791
ONOFLUORIDE. SUBLIMATION PRESSURE OF CALCIUM DIFLUORIDE AND THE DISSOCIATION ENERGY OF CALCIUM M 0276
MONOFLUORIDE. SUBLIMATION PRESSURE OF CHROMIUM DIFLUORIDE AND THE DISSOCIATION ENERGY OF CHROMIUM 0430
ND. XENON DIFLUORIDE AND THE NATURE OF THE XENON- FLUORINE BO 1223
THE GROUP-4A DIFLU/ ENTHALPY OF SUBLIMATION OF GERMANIUM DIFLUORIDE AND THE OXYGEN FLUORIDE FREE RADICAL. 0759
ARGON MATRIX RAMAN SPECTRA OF OXYGEN DIFLUORIDE AND THE THERMODYNAMICS OF SUBLIMATION OF 0471
FOR TITANIUM TRIFLUORIDE AND THE STABILITIES OF TITANIUM DIFLUORIDE AND TITANIUM MONOFLUORIDE. /ON PRESSURES 1145
RICES. INFRARED SPECTRA AND GEOMETRY OF TITANIUM DIFLUORIDE AND TITANIUM TRIFLUORIDE IN RARE GAS MAT 1146
STRUCTURE OF MOLECULAR URANIUM DIFLUORIDE AND URANIUM TRIFLUORIDE. 1162
CTRONIC STRUCTURE AND ANOMALOUS THERMAL EXPANSION ON IRON DIFLUORIDE AND VANADIUM DIOXIDE. ELE 0543
XENON IN ATOMIC VALENCE STATES AND IN THE MOLECULES XENON DIFLUORIDE AND XENON DICHLORIDE. 5D ORBITALS OF 1242
FORCE FIELDS IN KRYPTON DIFLUORIDE AND XENON DIFLUORIDE. 0556
OF THE ROOT MEAN SQUARE AMPLITUDES OF VIBRATION IN XENON DIFLUORIDE AND XENON TETRAFLUORIDE. /CAL EVALUATION 1257
FORBIDDEN ELECTRONIC TRANSITIONS IN XENON DIFLUORIDE AND XENON TETRAFLUORIDE. 1247
MOLECULAR SYMMETRY OF XENON DIFLUORIDE AND XENON TETRAFLUORIDE. 1240
VAPOR PRESSURE AND MELTING POINTS OF XENON DIFLUORIDE AND XENON TETRAFLUORIDE. 1251
ON BONDING IN THE SOLIDS. HEATS OF SUBLIMATION OF XENON DIFLUORIDE AND XENON TETRAFLUORIDE AND A CONJECTURE 1237
A. THERMOCHEMICAL PROPERTIES OF XENON DIFLUORIDE AND XENON TETRAFLUORIDE FROM MASS SPECTR 1253
VIOLET REGION. SPECTRA OF XENON DIFLUORIDE AND XENON TETRAFLUORIDE IN THE FAR ULTRA 1233
NFRARED SPECTRA AND GEOMETRIES OF MATRIX ISOLATED YTTRIUM DIFLUORIDE AND YTTRIUM TRIFLUORIDE. I 1316
/F SOME ARSENIC FLUORIDES. (ARSENIC MONOFLUORIDE, ARSENIC DIFLUORIDE, ARSENIC TRIFLUORIDE, ARSENIC TETRAFLUO/ 0093
STRUCTURE OF KRYPTON DIFLUORIDE AS INVESTIGATED BY ELECTRON DIFFRACTION. 0558
CRYSTAL STRUCTURE OF KRYPTON DIFLUORIDE AT -80C. 0551
/SPECTRA OF THE ALKALINE EARTH HALIDES. PART-1: BERYLLIUM DIFLUORIDE, BERYLLIUM DICHLORIDE, MAGNESIUM DICHLO/ 0133
F CARBON DIFLUORIDE(X) FROM TETRAFLUORO ETHYLENE. (CARBON DIFLUORIDE BOND ENERGY) /THE APPEARANCE POTENTIAL O 0305
/ECULAR PROPERTIES OF THE TRIATOMIC DIFLUORIDES BERYLLIUM DIFLUORIDE, BORON DIFLUORIDE, CARBON DIFLUORIDE, N/ 0141
CRYSTAL AND MOLECULAR STRUCTURE OF XENON DIFLUORIDE BY NEUTRON DIFFRACTION. 1239
LUORIDES. PART-6: PRECISE STRUCTURAL PARAMETERS IN COPPER DIFLUORIDE BY NEUTRON DIFFRACTION. STRUCTURES OF F 0443
/E BENDING FORCE CONSTANTS OF THE GROUP-2B HALIDES. (ZINC DIFLUORIDE, CADMIUM DIFLUORIDE, AND MERCURY DIFLUO/ 1322
E, CALCIUM DIFLUORIDE, STRONTIUM / VAPOR PRESSURE OF ZINC DIFLUORIDE, CADMIUM DIFLUORIDE, MAGNESIUM DIFLUORID 0049
ARED SPECTRUM. CENTRIFUGAL DISTORTION EFFECTS IN SULFUR DIFLUORIDE. CALCULATION OF THE FORCE FIELD AND INFR 1031
ALUMINUM DIF/ MEAN AMPLITUDES OF VIBRATION FOR THE BORON DIFLUORIDE, CARBON DIFLUORIDE, NITROGEN DIFLUORIDE, 1384
MICROWAVE SPECTRUM AND FORCE CONSTANTS OF SILICON DIFLUORIDE. CENTRIFUGAL DISTORTION. 0989
/HE VAPORIZATION OF HIGH TEMPERATURE MATERIALS. (CHROMIUM DIFLUORIDE, CHROMIUM TRIFLUORIDE, AND IRON DIFLUOR/ 0431
/RDINATE ANALYSIS OF PHOSPHORUS PENTAFLUORIDE. PHOSPHORUS DIFLUORIDE DICHLORIDE, AND PHOSPHORUS PENTACHLORID/ 0877
/N STUDY OF THE VACUUM ULTRAVIOLET PHOTOLYSIS OF SILICON DIFLUORIDE DIHYDRIDE. THE INFRARED AND ULTRAVIOLET/ 0986
XENON DIFLUORIDE. (FORCE CONSTANTS, SPECTRA) 1252
STEM AT LOW TEMPERATURE. CHARACTERIZATION OF THE CHLORINE DIFLUORIDE FREE RADICAL. CHLORINE FLUORINE SY 0392
NITROGEN TRIFLUORIDE. ELECTRONIC SPECTRUM OF THE NITROGEN DIFLUORIDE FREE RADICAL. /LTRAVIOLET PHOTOLYSIS OF 0691
MICROWAVE ROTATIONAL SPECTRUM OF THE NITROGEN DIFLUORIDE FREE RADICAL. 0690
THE CHLORINE DIFLUORIDE FREE RADICAL. 0393
EATS OF FORMATION GASEOUS THORIUM TRIFLUORIDE AND THORIUM DIFLUORIDE FROM MASS SPECTROMETRIC STUDIES. H 1128
/ERMODYNAMIC PROPERTIES OF CARBON MONOFLUORIDE AND CARBON DIFLUORIDE FROM MOLECULAR FLOW EFFUSION AND MASS S/ 0286
/ATIC POTENTIAL CONSTANTS AND AVERAGE STRUCTURE OF OXYGEN DIFLUORIDE FROM THE GROUND STATE ROTATIONAL SPECTR/ 0778
INORGANIC MOLECULES. PART-1: INFRARED SPECTRUM OF OXYGEN DIFLUORIDE FROM 2.5 TO 25 MICRONS. /ND STRUCTURE OF 0761
ND SOME DERIVED THERMODYNAMIC PROPERTIES. XENON DIFLUORIDE. HEAT CAPACITY FROM 5 TO 350 DEGREES K A 1246
INFRARED SPECTRA OF XENON DIFLUORIDE IN DIFFERENT SOLVENTS. 1241
INFRARED SPECTRA OF SILICON DIFLUORIDE IN INERT GAS MATRICES. 0982
IBRIUM STRUCTURE, ANHARMON/ MICROWAVE SPECTRUM OF SILICON DIFLUORIDE IN THE EXCITED VIBRATIONAL STATES, EQUIL 0993
NU-2 FERMI RESONANCE AND E/ MICROWAVE SPECTRUM OF OXYGEN DIFLUORIDE IN VIBRATIONALLY EXCITED STATES, NU-1- 2 0771
IONIZATION AND DISSOCIATION OF XENON DIFLUORIDE INDUCED BY PHOTON IMPACT. 1243
T LOW TEMPERATURE. SULFUR TETRAFLUORIDE AND SULFUR OXIDE DIFLUORIDE. INFRARED EVIDENCE FOR DIMER FORMATION A 1054
E) D ORBITALS IN THE NOBLE GAS DIHALIDES. (ARGON DIFLUORIDE, KRYPTON DIFLUORIDE, AND XENON DIFLUORID 1338
LE/ MASS SPECTROMETRIC STUDIES AT HIGH TEMPERATURES. (TIN DIFLUORIDE, LEAD DIFLUORIDE, TIN MONOFLUORIDE, AND 0592
/CTURES OF MATRIX ISOLATED MAGNESIUM DIHALIDES: MAGNESIUM DIFLUORIDE, MAGNESIUM DICHLORIDE, MAGNESIUM DIBROM/ 0607
/ LCAO MO SCF STUDIES OF THE GROUP-2A DIHALIDES BERYLLIUM DIFLUORIDE, MAGNESIUM DIFLUORIDE, AND CALCIUM DIFL/ 1360
/ EFFECT IN SOME LINEAR SYMMETRICAL DIHALIDES. (BERYLLIUM DIFLUORIDE, MAGNESIUM DIFLUORIDE, AND FIRST ROW TR/ 1372
/THE MOLECULAR STRUCTURES OF GROUP-2 ELEMENTS. (BERYLLIUM DIFLUORIDE, MAGNESIUM DIFLUORIDE, CALCIUM DIFLUORI/ 0131
IDE, AND/ MELTING POINTS OF INORGANIC FLUORIDES. (CADMIUM DIFLUORIDE, MAGNESIUM DIFLUORIDE, STRONTIUM DIFLUOR 0273
STRUCTURE OF THE MAGNESIUM DIFLUORIDE MOLECULE. 0602
/RETICAL STUDY OF THE BOND INTERACTION FORCE CONSTANTS IN DIFLUORIDE MOLECULES. (OXYGEN DIFLUORIDE, NITROGEN/ 0299
/R STRUCTURES AND NORMAL COORDINATE ANALYSES OF GERMANIUM DIFLUORIDE, MONOMERIC GERMANIUM DIFLUORIDE, AND DI/ 0475

/ INFRARED SPECTRA AND GEOMETRY OF MATRIX ISOLATED COBALT DIFLUORIDE, NICKEL DIFLUORIDE, COPPER DIFLUORIDE, A 0437
ON INTERACTION CALCULATION FOR THE GROUND STATE OF OXYGEN DIFLUORIDE, NITRITE ION AND CYANIDE ION. /NFIGURATI 0763
/RADICALS BY THE UNRESTRICTED HARTREE-FOCK METHOD. (BORON DIFLUORIDE, NITROGEN DIFLUORIDE, OXYGEN DIFLUORIDE/ 0176
ERTAIN FLUORIDES IN THE NEAR ULTRAVIOLET REGION. (KRYPTON DIFLUORIDE, NITROGEN TRIFLUORIDE) /ION SPECTRA OF C 0560
/TRICAL AND ELECTRICAL STRUCTURE OF THE MOLECULE CHLORINE DIFLUORIDE ON THE BASIS OF A CALCULATION BY THE MO/ 0382
D MOMENTS OF THE ELECTRONIC CHARGE DISTRIBUTION IN OXYGEN DIFLUORIDE, OZONE, AND SULFUR DIOXIDE. /S AND SECON 0780
755-2179 ANGSTROMS.   EMISSION SPECTRUM OF SILICON DIFLUORIDE. PART-1: THE BAND SYSTEM IN THE REGION 2 0988
HLORIDE IN/ ELECTRON SPIN RESONANCE STUDIES OF PHOSPHORUS DIFLUORIDE, PHOSPHORUS DICHLORIDE, AND NITROGEN DIC 0793
/OSPHORUS FLUORIDES. (PHOSPHORUS MONOFLUORIDE, PHOSPHORUS DIFLUORIDE, PHOSPHORUS TRIFLUORIDE, PHOSPHORUS TET/ 0788
                                                 KRYPTON DIFLUORIDE. (PHYSICAL AND CHEMICAL PROPERTIES)   0563
HNIQUE.   KRYPTON DIFLUORIDE. PREPARATION BY THE MATRIX ISOLATION TEC 0567
TORTION.   MICROWAVE SPECTRUM OF GERMANIUM DIFLUORIDE. QUADRUPOLE COUPLING AND CENTRIFUGAL DIS 0477
        AB INITIO MOLECULAR ORBITAL STUDY OF THE PHOSPHORUS DIFLUORIDE RADICAL.   0790
        ELECTRON SPIN RESONANCE AND STRUCTURE OF THE BORON DIFLUORIDE RADICAL.   0178
        ELECTRONIC ABSORPTION SPECTRUM OF THE NITROGEN DIFLUORIDE RADICAL.   0694
        ELECTRONIC STRUCTURE OF THE NITROGEN DIFLUORIDE RADICAL.   0682
              ESR PARAMETERS OF THE BORON DIFLUORIDE RADICAL.   0181
                HEAT OF FORMATION OF THE CARBON DIFLUORIDE RADICAL.   0304
        INFRARED SPECTRUM AND STRUCTURE OF THE NITROGEN DIFLUORIDE RADICAL.   0687
ARED SPECTRUM AND THERMODYNAMIC FUNCTIONS OF THE NITROGEN DIFLUORIDE RADICAL.   INFR 0686
              MICROWAVE SPECTRUM OF THE NITROGEN DIFLUORIDE RADICAL.   0681
                 RAMAN SPECTRUM OF THE NITROGEN DIFLUORIDE RADICAL.   0695
/S OF SCANDIUM, YTTRIUM, AND LANTHANUM HALIDES. (SCANDIUM DIFLUORIDE, SCANDIUM MONOFLUORIDE, YTTRIUM DIFLUOR/ 0941
/TOMS FROM THE SECOND ROW OF THE PERIODIC TABLE. (SILICON DIFLUORIDE, SILICON TETRAFLUORIDE, PHOSPHORUS TRIF/ 1373
RIDE IN THE/ SPECTRA OF ELECTRONIC EXCITATIONS IN CALCIUM DIFLUORIDE, STRONTIUM DIFLUORIDE, AND BARIUM DIFLUO 0279
E)   GEOMETRY OF THE ALKALINE EARTH DIHALIDES. (BARIUM DIFLUORIDE, STRONTIUM DIFLUORIDE, CALCIUM DIFLUORID 0126
              ELECTRONIC STRUCTURE OF KRYPTON DIFLUORIDE STUDIED BY PHOTOELECTRON SPECTROSCOPY.   0550
ORIDE.   MOLECULAR ORBITAL DESCRIPTION OF SULFUR DIFLUORIDE, SULFUR TETRAFLUORIDE AND SULFUR HEXAFLU 1027
/HERMODYNAMIC PROPERTIES OF NIOBIUM MONOFLUORIDE, NIOBIUM DIFLUORIDE, TANTALUM MONOFLUORIDE, AND TANTALUM DI/ 0655
              ELECTRONIC SPECTRUM OF SILICON DIFLUORIDE. THEORETICAL STUDY.   0997
OF THE DIHALIDES OF GERMANIUM, TIN, AND LEAD. (GERMANIUM DIFLUORIDE, TIN DIFLUORIDE, LEAD DIFLUORIDE) /TIONS 0479
        ABSORPTION SPECTRUM OF CARBON DIFLUORIDE TRAPPED IN AN ARGON MATRIX.   0297
     INFRARED SPECTRA OF MATRIX ISOLATED ZIRCONIUM DIFLUORIDE, TRIFLUORIDE, AND TETRAFLUORIDE.   1325
/ASEOUS URANIUM FLUORIDES. (URANIUM MONOFLUORIDE, URANIUM DIFLUORIDE, URANIUM TRIFLUORIDE, URANIUM TETRAFLUO/ 1160
     THEORETICAL STUDIES OF THE INTERACTION OF MAGNESIUM DIFLUORIDE WITH RARE GAS ATOMS.   0605
/ICHLORO ETHYLENE, AND THE APPEARANCE POTENTIAL OF CARBON DIFLUORIDE(X) FROM TETRAFLUORO ETHYLENE. (CARBON D/ 0305
RIDE.   PHOTOIONIZATION MASS SPECTROMETRIC STUDY OF XENON DIFLUORIDE, AND XENON TETRAFLUORIDE.   1225
/OELECTRON SPECTRA AND THE ELECTRONIC STRUCTURES OF XENON DIFLUORIDE, XENON TETRAFLUORIDE, AND XENON HEXAFLU/ 1228
TRAFLUORIDE VAPORS, AND FORCE CO/ RAMAN SPECTRA FOR XENON DIFLUORIDE, XENON TETRAFLUORIDE, AND XENON OXIDE TE 1254
.   SELF CONSISTENT FIELD STUDY OF THE SERIES XENON DIFLUORIDE, XENON TETRAFLUORIDE, XENON HEXAFLUORIDE 1224
/L CONSIDERATIONS IN THE CHEMISTRY OF NOBLE GASES. (XENON DIFLUORIDE, XENON TETRAFLUORIDE, XENON HEXAFLUORID/ 1235
ESE DIFLUOR/ RAMAN SPECTRA OF TITANIUM DIOXIDE, MAGNESIUM DIFLUORIDE, ZINC DIFLUORIDE, DIFLUORIDE, AND MANGAN 0615
ION. (FIRST ROW FLUORIDES, BERYLLIUM, MAGNESIUM, AND LEAD DIFLUORIDES) /LECTION AND MASS SPECTROMETRIC DETECT 0134
IDE AND THE THERMODYNAMICS OF SUBLIMATION OF THE GROUP-4A DIFLUORIDES. /Y OF SUBLIMATION OF GERMANIUM DIFLUOR 0471
TABILITIES OF SAMARIUM, EUROPIUM, AND GADOLINIUM MONO AND DIFLUORIDES. /LIES AT HIGH TEMPERATURES. PART-13: S 0450
AND VALENCE FORCE CONSTANTS OF FIRST ROW TRANSITION METAL DIFLUORIDES.   VIBRATIONAL FREQUENCIES   1371
BIUM AND TANTALUM FLUORIDES AT HIGH TEMPERATURES. PART-3: DIFLUORIDES AND MONOFLUORIDES. /C PROPERTIES OF NIO 0654
  ELECTRIC DEFLECTION OF MOLECULAR BEAMS OF THE RARE EARTH DIFLUORIDES AND TRIFLUORIDES. (LANTHANIDE TRIFLUOR/ 0946
   CARBON DIFLUORIDE/ MOLECULAR PROPERTIES OF THE TRIATOMIC DIFLUORIDES BERYLLIUM DIFLUORIDE, BORON DIFLUORIDE, 0141
D QUANTUM MECHANICAL STUDIES OF POLARIZABILITIES FOR THE DIFLUORIDES OF OXYGEN AND SULFUR. /HERMODYNAMICS AN 0784
H TEMPERATURES. PART-18: THE STABILITIES OF THE MONO- AND DIFLUORIDES OF SCANDIUM AND YTTRIUM. /TUDIES AT HIG 0939
ART-2: THE DISSOCIATION ENERGIES OF THE MONOFLUORIDES AND DIFLUORIDES OF SILICON AND GERMANIUM. /PERATURES. P 0463
        GEOMETRY OF THE TRANSITION METAL DIHALIDES. THE DIFLUORIDES OF SOME FIRST ROW TRANSITION METALS.   1368
/RINKAGE EFFECTS OF VIBRATION FOR SOME LINEAR SYMMETRICAL DIFLUORIDES. (OF THE FIRST ROW TRANSITION METALS, / 1369
URES OF PHOSPHORUS TRIFLUORIDE, PHOSPHORUS PENTAFLUORIDE, DIFLUORO AMINE, AND TETRAFLUORO HYDRAZINE. / STRUCT 0801
NIZATION BY ELECTRON IMPACT OF PHOSPHORUS TRIFLUORIDE AND DIFLUORO CYANO PHOSPHINE.   IO 0800
/ALS FOR TETRAFLUORO ETHYLENE, TRIFLUORO CHLORO ETHYLENE, DIFLUORO DICHLORO ETHYLENE, AND THE APPEARANCE POT/ 0305
  MAGNESIUM DICHLORIDE, MAGNESIUM DIBROMIDE, AND MAGNESIUM DIIODIDE. /GNESIUM DIHALIDES: MAGNESIUM DIFLUORIDE, 0607
AND MEAN AMPLITUDES FOR ALUMINUM TRIFLUORIDE MONOMER AND DIMER.   HARMONIC FORCE FIELDS   0038
UORIDE AND SULFUR OXIDE DIFLUORIDE. INFRARED EVIDENCE FOR DIMER FORMATION AT LOW TEMPERATURE. SULFUR TETRAFL 1054
CTROSCOPY OF HIGH TEMPERATURE VAPORS. PART-3: MONOMER AND DIMER SPECTRA OF THALLOUS FLUORIDE. /TOELECTRON SPE 1117
LUORIDE.   MASS SPECTROMETRIC EVIDENCE OF DIMERS IN BISMUTH PENTAFLUORIDE AND ANTIMONY PENTAF 0084
  SPECTRA AND STRUCTURE OF MATRIX ISOLATED THALLOUS HALIDE DIMERS, THALLOUS FLUORIDE, AND THALLOUS CHLORIDE. / 1123
GERMANIUM DIFLUORIDE, MONOMERIC GERMANIUM DIFLUORIDE, AND DIMERIC GERMANIUM DIFLUORIDE. /RDINATE ANALYSES OF 0475
/BINDING ENERGY, IONIZATION POTENTIAL, ELECTRON AFFINITY, DIPOLE AND QUADRUPOLE MOMENTS, AND SPECTROSCOPIC / 0756
/AVE FUNCTIONS BY MEASURED ELECTRONIC PROPERTIES. PART-2: DIPOLE MOMENT AND FIELD GRADIENT OF NITROGEN TRIFL/ 0746
XAFLUORIDE.   DIPOLE MOMENT AND MOLECULAR STRUCTURE OF URANIUM HE 1200
ETRAFLUORIDE.   MICROWAVE SPECTRUM, DIPOLE MOMENT, AND STRUCTURE ANALYSIS OF SELENIUM T 0957
OF BORON TRIFLUORIDE.   CNDO CALCULATION OF DIPOLE MOMENT DERIVATIVES AND INFRARED INTENSITIES 0195
  ELECTRON TRA/ DETERMINATION OF THE MATRIX ELEMENT OF THE DIPOLE MOMENT OF AN A DOUBLET SIGMA TO X DOUBLET PI 0970
        GAS PHASE ELECTRON RESONANCE SPECTRUM AND DIPOLE MOMENT OF CARBON MONOFLUORIDE.   0284
E/ GAS PHASE ELECTRON PARAMAGNETIC RESONANCE SPECTRUM AND DIPOLE MOMENT OF NITROGEN MONOFLUORIDE IN THE SINGL 0673
        VACUUM ULTRAVIOLET ABSORPTION SPECTRUM AND DIPOLE MOMENT OF NITROGEN TRIFLUORIDE.   0725
                MICROWAVE SPECTRUM, STRUCTURE, AND DIPOLE MOMENT OF OXYGEN DIFLUORIDE.   0779
                                STRUCTURE AND DIPOLE MOMENT OF SULFUR TETRAFLUORIDE.   1056
/BINDING ENERGY, IONIZATION POTENTIAL, ELECTRON AFFINITY, DIPOLE MOMENT, QUADRUPOLE MOMENT, AND SPECTROSCOPI/ 0757
/ATION ENTHALPY, IONIZATION POTENTIAL, ELECTRON AFFINITY, DIPOLE MOMENT, SPECTROSCOPIC CONSTANTS, AND IDEAL / 0094
  STARK ENERGIES FOR MOLECULES WITH VIBRATIONALLY INDUCED DIPOLE MOMENTS.   1417
T/ CALCULATIONS ON IONIC CHARACTER AND HYBRIDIZATION FROM DIPOLE MOMENTS FOR ARSENIC TRIFLUORIDE AND ARSENIC 0107
ULFUR MONOFLUORIDE AND SELENIUM MONOFLUORIDE)   ELECTRIC DIPOLE MOMENTS OF OPEN SHELL DIATOMIC MOLECULES. (S 1019
          HEAT FORMATION BORON TRIFLUORIDE BY DIRECT COMBINATION OF THE ELEMENTS.   0198
        HEAT OF FORMATION OF ALUMINUM TRIFLUORIDE BY DIRECT COMBINATION OF THE ELEMENTS.   0039
ORY. ESSENTIAL STRUCTURAL ELEMENTS- MO FORMALISM. PART-2: DIRECT COMPARISON WITH AB INITIO RESULTS. /ITAL THE 0765
/ATIONAL ANALYSIS OF SUBSTITUTED AND PERTURBED MOLECULES. DIRECT DETERMINATION FORCE CONSTANTS FROM THE CORI/ 1444
ONS. (PHOSPHORUS PENTAFLUORIDE, SULFUR HEXAFLUORIDE) DIRECTED VALENCY IN CERTAIN MOLECULES AND COMPLEX I 0860
        BAND SYSTEM IN THE DISCHARGE SPECTRUM OF LANTHANUM TRIFLUORIDE VAPORS. 0580
/ CERTAIN POLYATOMIC FREE RADICALS OBTAINED IN ELECTRICAL DISCHARGES THROUGH BORON TRIFLUORIDE. (BORON DIFLU/ 0222
/OLECULAR CONSTANTS FROM INFRARED BAND SHAPES. THE ATOMIC DISPLACMENTS AND INTRAMOLECULAR FORCES IN CARBON T/ 0331
/ INTERMEDIATE URANIUM FLUORIDES FROM MEASUREMENTS OF THE DISPROPORTIONATION PRESSURES. (URANIUM PENTAFLUORI/ 1176
ONOFLUORIDE, YTTRIUM MONOFLUORIDE, AND LANTHANUM MONOFLU/ DISSOCIATION ENERGIES AND STABILITIES OF SCANDIUM M 0935

UM SUBGROUP. (SCANDIUM TRIFLUORIDE, YTTRI/ CALCULATION OF DISSOCIATION ENERGIES FOR THE HALIDES OF THE SCANDI 0948
ECTRON AFFINITIES OF SOME GERMANIUM FLUORIDE SPECIE/ BOND DISSOCIATION ENERGIES, IONIZATION POTENTIALS AND EL 0464
YLLIUM CHLORIDE AND HEAT OF FORMATION OF BERYLLIUM CHLOR/ DISSOCIATION ENERGIES OF BERYLLIUM FLUORIDE AND BER 0128
DETERMINATION OF THE DISSOCIATION ENERGIES OF CARBON MONOHALIDES. 0289
M HALIDES.   POTENTIAL ENERGY CURVES AND DISSOCIATION ENERGIES OF DIATOMIC BORON AND ALUMINU 0173
ORIDES OF GALLIUM, INDIUM, A/ POTENTIAL ENERGY CURVES AND DISSOCIATION ENERGIES OF DIATOMIC FLUORIDES AND CHL 0458
. (IODINE FLUORIDE, BROMINE FLUORIDE)   DISSOCIATION ENERGIES OF DIATOMIC HALOGEN FLUORIDES 0494
ORIDES.   DISSOCIATION ENERGIES OF THE ALKALINE EARTH MONOFLU 0119
BORON, ALUMINUM, GALLIUM, INDIUM, AND THALLIUM.   DISSOCIATION ENERGIES OF THE GASEOUS MONOHALIDES OF 0152
/ SPECTROMETRIC STUDIES AT HIGH TEMPERATURES. PART-2: THE DISSOCIATION ENERGIES OF THE MONOFLUORIDES AND DIF/ 0463
SPECTRUM OF BROMINE MONOFLUORIDE AND ITS DISSOCIATION ENERGY.   0252
/ MATRICES. PHOTOCHLORINATION OF METHANE AND FLUOROFORM. DISSOCIATION ENERGY AND ENTROPY OF CARBON TRIFLUOR/ 0311
MASS SPECTROMETRIC DETERMINATION OF THE DISSOCIATION ENERGY OF BERYLLIUM MONOFLUORIDE.   0129
SPECTROMETRIC STUDIES.   DISSOCIATION ENERGY OF BORON MONOFLUORIDE FROM MASS 0161
CON AND GERMANIUM FLUORIDES)   REEVALUATION OF THE DISSOCIATION ENERGY OF CALCIUM FLUORIDE. (ALSO SILI 0277
SUBLIMATION PRESSURE OF CALCIUM DIFLUORIDE AND THE DISSOCIATION ENERGY OF CALCIUM MONOFLUORIDE.   0276
NIUM MONOFLUORIDE.   CALCULATION OF THE DISSOCIATION ENERGY OF CALCIUM MONOFLUORIDE AND URA 1161
DISSOCIATION ENERGY OF CHLORINE MONOFLUORIDE.   0387
SUBLIMATION PRESSURE OF CHROMIUM DIFLUORIDE AND THE DISSOCIATION ENERGY OF CHROMIUM MONOFLUORIDE.   0430
DISSOCIATION ENERGY OF COPPER MONOFLUORIDE.   0440
PHOTOIONIZATION STUDY OF CHLORINE MONOFLUORIDE AND THE DISSOCIATION ENERGY OF FLUORINE.   0378
DISSOCIATION ENERGY OF MAGNESIUM MONOFLUORIDE.   0601
GNESIUM AND CALCIUM AND THE IONIC MODEL OF A MOLECULE. DISSOCIATION ENERGY OF MONOFLUORIDE COMPOUNDS OF MA 0278
ENTHALPY OF SUBLIMATION OF PLUTONIUM TRIFLUORIDE AND THE DISSOCIATION ENERGY OF PLUTONIUM MONOFLUORIDE. /-2: 0884
STUDIES AT HIGH TEMPERATURES. PART-14: VAPOR PRESSURE AND DISSOCIATION ENERGY OF SILVER MONOFLUORIDE. /ETRIC 1014
ON AFFINITY, DIPOLE MOMEN/ ARSENIC MONOFLUORIDE, 3 SIGMA. DISSOCIATION ENERGY OF THALLIUM MONOFLUORIDE 1115
OTHER PROPERTIES OF SELENIUM MONOFLUORIDE FROM A MOLECU/ DISSOCIATION ENTHALPY, IONIZATION POTENTIAL, ELECTR 0094
ONAL ANALYSIS OF C-X AND C'-X' BAND SYSTEMS AND ENERGY OF DISSOCIATION ENTHALPY, SPECTROSCOPIC CONSTANTS, AND 0955
NOFLUORIDE AND CARBON DIFLUORIDE.   VIBRATI DISSOCIATION IN GERMANIUM MONOFLUORIDE.   0470
T SHOCK WAVES.   PREDISSOCIATION IN THE ABSORPTION SPECTRA OF CARBON MO 0293
HE EMISSION OF IODINE MONOFLUORIDE.   THEORY OF THE DISSOCIATION OF CHLORINE TRIFLUORIDE BEHIND INCIDEN 0394
LUORIDE.   ON THE DISSOCIATION OF DIATOMIC MOLECULES AND A STUDY OF T 0493
ENTHALPY OF DISSOCIATION OF LEAD MONOFLUORIDE AND BISMUTH MONOF 0589
IMPACT.   IONIZATION AND DISSOCIATION OF NITROGEN TRIFLUORIDE.   0742
ION OF THE ELECTRONIC TRANSITION MOMENT WITH INTERNUCLEAR DISSOCIATION OF XENON DIFLUORIDE INDUCED BY PHOTON 1243
INTERNUCLEAR DISTANCE FOR BANDS OF THE X SYSTEM.   CORRELAT 0969
CTRON DIFFRACTION.   BORON FLUORINE BOND DISTANCE IN GASEOUS STRONTIUM MONOFLUORIDE.   1016
ION OF GASEOUS SILICON TETRAFLUORIDE FROM EL/ INTERATOMIC DISTANCE OF BORON TRIFLUORIDE DETERMINED BY GAS ELE 0223
SHRINKAGES OF THE INTERNUCLEAR DISTANCES AND ROOT MEAN SQUARE AMPLITUDES OF VIBRAT 1002
TETRAHALID/ ESTIMATION OF CONFIGURATIONS AND INTERATOMIC DISTANCES BY MOLECULAR VIBRATIONS.   1489
MEAN SQUARE AMPLITUDES OF INTERATOMIC DISTANCES IN GROUP-5 AND GROUP-6 TRANSITION ELEMENT 1427
MEAN SQUARE AMPLITUDES OF INTERATOMIC DISTANCES IN HEXAFLUORIDE MOLECULES.   0008
MEAN SQUARE AMPLITUDES OF INTERATOMIC DISTANCES IN HEXAFLUORIDE MOLECULES.   1450
MEAN SQUARE AMPLITUDES OF INTERATOMIC DISTANCES IN SOME OCTAHEDRAL HEXAFLUORIDE MOLECULES 1466
PE OF / ESTIMATION OF THE FORCE CONSTANTS AND INTERATOMIC DISTANCES OF HALIDE COMPLEX MOLECULES OF THE XY4 TY 1414
UM AND FORCE CONSTANTS OF SILICON DIFLUORIDE. CENTRIFUGAL DISTORTION.   MICROWAVE SPECTR 0989
GERMANIUM DIFLUORIDE. QUADRUPOLE COUPLING AND CENTRIFUGAL DISTORTION.   MICROWAVE SPECTRUM OF 0477
THEORY AND CALCULATION OF CENTRIFUGAL DISTORTION CONSTANTS FOR POLYATOMIC MOLECULES.   1477
MILLIMETER WAVE SPECTRUM AND CENTRIFUGAL DISTORTION CONSTANTS OF NITROGEN TRIFLUORIDE.   0718
CENTRIFUGAL DISTORTION CONSTANTS OF PHOSPHORUS TRIFLUORIDE. 0814
ART-1. QUADRATIC POTENTIAL CONSTANTS AND AVE/ CENTRIFUGAL DISTORTION EFFECTS IN ASYMMETRIC ROTOR MOLECULES. P 0778
N OF THE FORCE FIELD AND INFRARED SPECTRUM.   CENTRIFUGAL DISTORTION EFFECTS IN SULFUR TETRAFLUORIDE. CALCULATIO 1031
TAFLUORIDE.   POLAR DISTORTIONS IN RHENIUM HEPTAFLUORIDE AND IODINE HEP 0529
RELAXATION OF THE MOLECULAR CHARGE DISTRIBUTION AND THE VIBRATIONAL FORCE CONSTANT.   0151
CTRON BINDING ENERGIES.   CHARGE DISTRIBUTION IN CHLORINE MONOFLUORIDE FROM CORE ELE 0374
/POLE MOMENTS AND SECOND MOMENTS OF THE ELECTRONIC CHARGE DISTRIBUTION IN OXYGEN DIFLUORIDE, OZONE, AND SULF/ 0780
ELECTRON DENSITY DISTRIBUTION IN SOME CHLORIDES.   0218
TETRAFLUORIDE BY ESCA AND EVIDENCE FOR ORBITAL/ ELECTRON DISTRIBUTION IN THE XENON FLUORIDES AND XENON OXIDE 1229
SPECTRUM OF SILICON MONOFLUORIDE.   DISTRIBUTION OF INTENSITY IN THE ELECTRON EMISSION 0968
/F ORGANIC COMPOUNDS. PART-3: AB INITIO STUDIES OF CHARGE DISTRIBUTION USING A MINIMAL SLATER-TYPE BASIS. (N/ 0707
ELECTRONIC NITROGEN, CARBON MONOXIDE, B/ MOLECULAR CHARGE DISTRIBUTIONS AND CHEMICAL BINDING. PART-3: THE ISO 0337
CARBON TETRAFLUORIDE AND SILICON TETRAF/ ELECTRON DENSITY DISTRIBUTIONS IN SOME FLUORIDES AND OXYCHLORIDES. (1494
L MODES OF VIBRATION. PART-2: INTERACTION TERMS IN ENERGY DISTRIBUTIONS. (SELENIUM AND TELLURIUM FLUORIDES) / 1370
COMPARISONS OF FLUORIDES, OXIDES, AND SULFIDES CONTAINING DIVALENT TRANSITION ELEMENTS.   0158
TALS IN BORON FLUORIDES. HIGHLY POLARIZED BORON- FLUORINE DOUBLE AND TRIPLE BONDS.   LOCALIZED ORBI 1099
INFRARED DOUBLE RESONANCE IN SULFUR HEXAFLUORIDE.   0762
YGEN DIFLUORIDE.   DOUBLE ZETA SCF CALCULATIONS FOR NITRITE ION AND OX 1136
ELTA 5/2- X DOUBLET PI 3/2 / ROTATIONAL ANALYSIS OF THE C DOUBLET DELTA 5/2- X DOUBLET PI 3/2 AND G DOUBLET D 0120
MONOFLUORIDE.   THE C DOUBLET PI AND THE X DOUBLET SIGMA STATES OF BARIUM 0971
ORIDE.   A QUARTET SIGMA TO X DOUBLET PI ELECTRONIC TRANSITION OF SILICON MONOFLU 1025
/TREE-FOCK WAVE FUNCTIONS AND COMPUTED PROPERTIES FOR THE DOUBLET PI GROUND STATES OF SULFUR MONOFLUORIDE AN/ 0757
GY, IONIZATION POTENTIAL, ELECTRON / OXYGEN MONOFLUORIDE, DOUBLET PI HARTREE-FOCK WAVE FUNCTION, BINDING ENER 0978
OLECULE.   A QUARTET SIGMA TO DOUBLET PI TRANSITION OF THE SILICON MONOFLUORIDE M 0966
SINGLET SIGMA PLATINUM OXIDE AND SILICON MONOFLUORIDE (D DOUBLET SIGMA(+)- X DOUBLET PI) MOLECULES. /IGMA-X 0466
SIGMA(-) TO X DOUBLET PI / ROTATIONAL STRUCTURE OF THE A DOUBLET SIGMA(+), B DOUBLET SIGMA(+), AND A QUARTET 1133
) TO X DOUBLET PI SYSTEMS O/ ROTATIONAL ANALYSIS OF THE A DOUBLET SIGMA(+) TO X DOUBLET PI, B DOUBLET SIGMA(+ 0288
/(A TRIPLET PI TO X TRIPLET PI) AND CARBON MONOFLUORIDE A DOUBLET SIGMA(+) TO X DOUBLET PI BAND SYSTEM OSCIL/ 0970
/ATION OF THE MATRIX ELEMENT OF THE DIPOLE MOMENT OF AN A DOUBLET SIGMA TO X DOUBLET PI ELECTRON TRANSITON I/ 1135
/ STRUCTURE AND ISOTOPIC SHIFT IN THE (1,0) BAND OF THE A DOUBLET SIGMA- X DOUBLET PI SYSTEM OF THE TIN MONO/ 0280
LUORIDE MOLECULE IN EMISSION / ANALYSIS OF THE BANDS OF B DOUBLET SIGMA- X DOUBLET SIGMA. SYSTEM OF CALCIUM F 0723
TRIFLUORIDE IN THE EXCITED VIBRATIONAL STATES. (ZETA TYPE DOUBLING AND RESONANCE IN THE V3 AND V4 STATES) /N 0743
SIGNS OF L DOUBLING CONSTANTS IN NITROGEN TRIFLUORIDE.   0260
PENTAFLUORIDE.   K-TYPE DOUBLING IN THE MILLIMETER WAVE SPECTRUM OF BROMINE 0803
HE NU-4 EQUALS 1 STATE.   L-TYPE DOUBLING TRANSITIONS OF PHOSPHORUS TRIFLUORIDE IN T 0804
HE NU/ EFFECTS OF THE THIRD ORDER CONSTANTS ON THE L-TYPE DOUBLING TRANSITIONS OF PHOSPHORUS TRIFLUORIDE IN T 0142
RIDE FROM 456 TO 1083 DEGREES K BY TRANSPOSED TEMPERATURE DROP CALORIMETRY.   ENTHALPY OF BERYLLIUM FLUO 1356
CANDIUM, CERIUM, SAMARIUM, EUROPIUM, GADOLINIUM, TERBIUM, DYSPROSIUM, ERBIUM, THULIUM, AND YTTERBIUM. /S OF S 0448
TEMPERATURES. PART-17: SUBLIMATION AND VAPOR PRESSURES OF DYSPROSIUM, HOLMIUM, AND ERBIUM TRIFLUORIDES. /IGH 0449
BIUM TRIFLUORIDE.   STABILITIES OF DYSPROSIUM TRIFLUORIDE, HOLMIUM TRIFLUORIDE, AND ER 0190
FORCE CONSTANTS FOR MOLECULES WITH D3H SYMMETRY.

E OF THE CORIOLIS CONSTANT FOR SOME MOLECULES AND IONS OF D4H SYMMETRY. / FORCE CONSTANTS, AND MASS DEPENDENC 1411
F CONSTANTS.   MOLECULES WITH D4H SYMMETRY. NEW METHOD FOR THE CALCULATION OF GVF 1412
BRATIONAL ASSIGNMENT FOR IODINE PENTAFLUORIDE, A NONRIGID D5H MOLECULE.   VI 0525

E

E" FORCE FIELD OF PHOSPHORUS PENTAFLUORIDE.   0861
PECTRA OF GASES. PART-10: RAMAN BAND CONTOUR CF THE NU-7 (E') FUNDAMENTAL OF PHOSPHORUS PENTAFLUORIDE. /MAN S 0879
/OD IN THE CALCULATION OF FORCE CONSTANTS. APPLICATION TO E' SPECIES FORCE FIELD OF PHOSPHORUS PENTAFLUORIDE. 0827
R BAND CONTOURS AND VIBRATIONAL POTENTIAL FUNCTION OF THE E' VIBRATIONS OF ARSENIC PENTAFLUORIDE. /RPENDICULA 0116
E INFRARED SPECTRA OF RUTHENIU/ JAHN-TELLER EFFECT IN THE E(8) VIBRATIONAL MODE OF HEXAFLUORIDE MOLECULES. TH 0926
FLUOR/ ELECTRIC DEFLECTION OF MOLECULAR BEAMS OF THE RARE EARTH DIFLUORIDES AND TRIFLUORIDES. (LANTHANIDE TRI 0946
INFRARED SPECTRA AND GEOMETRIES OF RARE EARTH DIHALIDES.   1351
UORIDE, CALCIUM DIFLUORIDE)   GEOMETRY OF THE ALKALINE EARTH DIHALIDES. (BARIUM DIFLUORIDE, STRONTIUM DIFL 0126
GEOMETRY AND VIBRATIONAL SPECTRA OF THE ALKALINE EARTH DIHALIDES. PART-1: MAGNESIUM DIFLUORIDE.   0609
OMETRIC STUDIES OF SCANDIUM, YTTRIUM, LANTHANUM, AND RARE EARTH FLUORIDES.   MASS SPECTR 1359
SATURATED VAPOR PRESSURE OF RARE EARTH FLUORIDES.   1357
STRUCTURE AND MAGNETIC PROPERTIES OF RARE EARTH FLUORIDES.   1345
NTAL DETERMINATION OF THE ENTHALPIES OF FORMATION OF RARE EARTH FLUORIDES. (LANTHANUM TRIFLUORIDE)   EXPERIME 0578
GEOMETRY AND INFRARED SPECTRA OF MATRIX ISOLATED RARE EARTH HALIDES. (ALSO LANTHANUM TRIFLUORIDE)   0583
INFRARED SPECTRA OF SOME ALKALINE EARTH HALIDES BY THE MATRIX ISOLATION TECHNIQUE.   0125
LATTICE VIBRATIONS AND STRUCTURE OF RARE EARTH HALIDES. (LANTHANUM TRIFLUORIDE)   0572
IUM DICHLORIDE, MAGNESI/ INFRARED SPECTRA OF THE ALKALINE EARTH HALIDES. PART-1: BERYLLIUM DIFLUORIDE, BERYLL 0133
INFRARED SPECTRA OF RARE EARTH METAL FLUORIDES.   1346
IDE PRISM.   INFRARED SPECTRA OF RARE EARTH METAL FLUORIDES IN THE REGION OF A CESIUM IOD 1352
DISSOCIATION ENERGIES OF THE ALKALINE EARTH MONOFLUORIDES.   0119
HIGH TEMPERATURE ENTHALPY INCREMENTS FOR SCME RARE EARTH TRIFLUORIDES.   1350
MAGNETIC RESONANCE AND SUSCEPTIBILITY OF RARE EARTH TRIFLUORIDES.   1349
, YTTRIUM TRIFLUORIDE, LANTHANUM TRIFLUORIDE AND THE RARE EARTH TRIFLUORIDES. /TUDIES OF SCANDIUM TRIFLUORIDE 0953
NMR IN RARE EARTH TRIFLUORIDES.   1355
SYMMETRY AND CRYSTAL STRUCTURE OF RARE EARTH TRIFLUORIDES. (LANTHANUM TRIFLUORIDE)   0571
T/ FORCE CONSTANTS AND GEOMETRIES OF MATRIX ISOLATED RARE EARTH TRIFLUORIDES. (SCANDIUM TRIFLUORIDE, YTTRIUM 0944
ONTENTS AND RELATED THERMODYNAMIC FUNCTIONS OF SEVEN RARE EARTH TRIFLUORIDES. (YTTRIUM TRIFLUORIDE) /E HEAT C 1320
LET SIGMA(+) BANDS OF BORON-11 MONOFLUORIDE WITH A VACUUM ECHELLE SPECTROGRAPH. / OF THE A SINGLET PI- X SING 0170
N SPECTRA OF MOLECULES. (SULFUR HEXAFLUORIDE) EVIDENCE OF EFFECTIVE POTENTIAL BARRIERS IN THE X-RAY ABSORPTIO 1073
S OF MOLECULES OF SOME / EMPIRICAL METHOD FOR DETERMINING EFFECTIVE VIBRATIONAL AND ROTATIONAL CHARACTERISTIC 0622
OF ANTIMONY MONOFLUORIDE.   DETERMINATION OF THE EFFECTIVE VIBRATIONAL TEMPERATURE IN THE A-2 SYSTEM 0061
ON MONOFLUORIDE AND CARBON DIFLUORIDE FROM MOLECULAR FLOW EFFUSION AND MASS SPECTROMETER INVESTIGATIONS. /ARB 0286
OF BERYLLIUM FLUORIDE VAPOR.   EFFUSION STUDIES, MASS SPECTRA, AND THERMODYNAMICS 0140
ON TRIFLUORIDE AND COMPARISON WITH FORCE CONSTANTS OF THE EIGHTH PERIOD. /NTS OF CARBON TETRAFLUORIDE AND BOR 0212
/RMINATION OF THE GEOMETRY OF HIGH TEMPERATURE SPECIES BY ELECTRIC DEFLECTION AND MASS SPECTROMETRIC DETECTI/ 0134
ES ON CERIUM TETRAFLUORIDE, TERBIUM TRIFLUORIDE, AND P/ ELECTRIC DEFLECTION AND THERMAL DECOMPOSITION STUDI 0367
UR HEXAFLUORIDE, SELENIUM HEXAFLUORIDE, TELLURIUM HEXAFL/ ELECTRIC DEFLECTION OF BINARY HEXAFLUORIDES. (SULF 1084
EARTH DIFLUORIDES AND TRIFLUORIDES. (LANTHANIDE TRIFLUOR/ ELECTRIC DEFLECTION OF MOLECULAR BEAMS OF THE RARE 0946
AFLUORIDE, CARBON TETRACHLORIDE, SILICON / MCLECULAR BEAM ELECTRIC DEFLECTION OF THE TETRAHALIDES CARBON TETR 0349
CULES. (SULFUR MONOFLUCRIDE AND SELENIUM MONOFLUORIDE) ELECTRIC DIFOLE MOMENTS OF OPEN SHELL DIATOMIC MOLE 1019
MPLITUDE MOTION. (XENON HEXAFLUORIDE, IODINE HEPTAFLUORI/ ELECTRIC FIELD DEFLECTION OF MOLECULES WITH LARGE A 1281
E, BRO/ MOLECULAR ZEEMAN EFFECT, MAGNETIC PROPERTIES, AND ELECTRIC QUADRUPOLE MOMENTS IN CHLORINE MONOFLUORID 0254
LIQUID XENON HEXAFLUORIDE.   ELECTRICAL CONDUCTIVITY AND DIELECTRIC CONSTANT OF 1307
DE AND BROMINE TRIFLUORIDE.   ELECTRICAL CONDUCTIVITY OF SOLID CHLORINE TRIFLUORI 0259
/ SPECTRA OF CERTAIN POLYATOMIC FREE RADICALS OBTAINED IN ELECTRICAL DISCHARGES THROUGH BORON TRIFLUORIDE. (/ 0222
ORIDE ON / GEOMETRY OF MOLECULES. PART-2: GEOMETRICAL AND ELECTRICAL STRUCTURE OF THE MOLECULE CHLORINE DIFLU 0382
TE FLUORINE COMPOUNDS OF CARBON MONOFLUO/ THERMODYNAMIC, ELECTROCHEMICAL AND SYNTHETIC STUDIES OF THE GRAPHI 0296
IE/ BOND DISSOCIATION ENERGIES, IONIZATION POTENTIALS AND ELECTRON AFFINITIES OF SOME GERMANIUM FLUORIDE SPEC 0464
/OCK WAVE FUNCTION, BINDING ENERGY, IONIZATION POTENTIAL, ELECTRON AFFINITY, DIPOLE AND QUADRUPOLE MOMENTS, / 0756
/OCK WAVE FUNCTION, BINDING ENERGY, IONIZATION POTENTIAL, ELECTRON AFFINITY, DIPOLE MOMENT, QUADRUPOLE MOMEN/ 0757
/E, 3 SIGMA. DISSOCIATION ENTHALPY, IONIZATION POTENTIAL, ELECTRON AFFINITY, DIPOLE MOMENT, SPECTROSCOPIC CO/ 0094
CHARGE DISTRIBUTION IN CHLORINE MONOFLUORIDE FROM CORE ELECTRON BINDING ENERGIES.   0374
PHOSPHORUS OR SULFUR.   NONEMPIRICAL VALENCE ELECTRON CALCULATIONS ON SMALL MOLECULES CONTAINING 1029
S OF SOME INTERHALOGENS. (CH/ CNDO/2 AND INDO ALL VALENCE ELECTRON CALCULATIONS ON THE GEOMETRY AND PROPERTIE 0397
ELECTRON CORRELATION AND MOLECULAR SHAPE.   1413
ECTRON DIFFRACTION STUDY OF XENON HEXAFLUORID/ EFFECTS OF ELECTRON CORRELATION IN X-RAY DIFFRACTION AND AN EL 1290
ELECTRON DENSITY DISTRIBUTION IN SOME CHLORIDES.   0218
D OXYCHLORIDES. (CARBON TETRAFLUORIDE AND SILICON TETRAF/ ELECTRON DENSITY DISTRIBUTIONS IN SOME FLUORIDES AN 0337
RINE BOND DISTANCE OF BORON TRIFLUORIDE DETERMINED BY GAS ELECTRON DIFFRACTION.   BORON FLUO 0223
RES OF ARSENIC TRIFLUORIDE AND ARSENIC TRICHLORIDE BY GAS ELECTRON DIFFRACTION. /ION OF THE MOLECULAR STRUCTU 0104
ITUDES OF VIBRATION OF GASEOUS SILICON TETRAFLUORIDE FROM ELECTRON DIFFRACTION. /ES AND ROOT MEAN SQUARE AMPL 1002
ECULAR STRUCTURE OF PHOSPHORUS TRIFLUORIDE STUDIED BY GAS ELECTRON DIFFRACTION.   MOL 0810
IC TRIFLUORIDE AND ARSENIC PENTAFLUORIDE AS DETERMINED BY ELECTRON DIFFRACTION. MOLECULAR STRUCTURES OF ARSEN 0099
STRUCTURE OF KRYPTON DIFLUORIDE AS INVESTIGATED BY ELECTRON DIFFRACTION.   0558
GALLIUM TRIFLUORIDE)   ELECTRON DIFFRACTION ANALYSIS OF GALLIUM HALIDES. (0460
CTURES OF GROUP-2 ELEMENTS. (BERYLLIUM DIFLUORIDE, MAGNE/ ELECTRON DIFFRACTION ANALYSIS OF THE MOLECULAR STRU 0131
/S OF BROMINE PENTAFLUORIDE AND IODINE PENTAFLUORIDE FROM ELECTRON DIFFRACTION AND ROTATIONAL CONSTANT DATA. 0265
ARSENIC TRIFLUORIDE AND ARSENIC TRICHLORIDE DETERMINED BY ELECTRON DIFFRACTION AND SPECTROSCOPY. /UCTURES OF 0103
PHOSPHORUS TRIFLUORIDE, PHOSPHORUS PENTAFLUORIDE, DIFLU/ ELECTRON DIFFRACTION BY GASES AND THE STRUCTURES OF 0801
/LAR STRUCTURE OF XENON HEXAFLUORIDE. PART-1: ANALYSIS OF ELECTRON DIFFRACTION INTENSITIES. PART-2: INTERNAL/ 1291
UORIDE, ARSENIC PENTAFLUORIDE, AND OTHER MOLECULES.   AN ELECTRON DIFFRACTION INVESTIGATION OF ARSENIC TRIFL 0098
FLUORIDE AND TANTALUM PENTAFLUORIDE.   ELECTRON DIFFRACTION INVESTIGATION OF NIOBIUM PENTA 0671
LUORIDE.   AN ELECTRON DIFFRACTION INVESTIGATION OF NITROGEN TRIF 0738
OLECULES. (TUNG/ SPECTROSCOPIC STUDIES IN CONNECTION WITH ELECTRON DIFFRACTION INVESTIGATION OF SOME SIMPLE M 1153
STRUCTURES OF THE ALUMINUM HALIDES.   ELECTRON DIFFRACTION INVESTIGATION OF THE MOLECULAR 0036
OF ZIRCONIUM TETRAFLUORIDE IN THE VAPOR PHASE.   ELECTRON DIFFRACTION INVESTIGATION OF THE STRUCTURE 1332
FLUORIDE, OSMIUM HEXAFLUORIDE, IRIDIUM HEXAFLUORIDE, URA/ ELECTRON DIFFRACTION INVESTIGATION OF TUNGSTEN HEXA 1155
ENTAFLUORIDE AND PHOSPHORUS PENTACHLORIDE.   ELECTRON DIFFRACTION INVESTIGATIONS OF PHOSPHORUS P 0869
/TUDIES ON THE FAILURE OF THE FIRST BORN APPROXIMATION IN ELECTRON DIFFRACTION (MOLYBDENUM HEXAFLUORIDE AND/ 0640
SULFUR HEXAFLUORIDE, SULFUR TETRAFLUOR/ INVESTIGATION BY ELECTRON DIFFRACTION OF THE MOLECULAR STRUCTURES OF 1041
LUORIDE MO/ MULTIPLE INTRAMOLECULAR SCATTERING EFFECTS ON ELECTRON DIFFRACTION PATTERNS FOR THE RHENIUM HEXAF 0911
ES. (NITROGEN DIFLUORIDE)   ELECTRON DIFFRACTION STUDIES AT ELEVATED TEMPERATUR 0679
AND XENON HEXAFLUORIDE AND OTHER COMPOUNDS.   ELECTRON DIFFRACTION STUDIES OF XENON TETRAFLUORIDE 1259
SILICON- FLUORINE LENGTH IN SILICON TETRAFLUORIDE. NEW ELECTRON DIFFRACTION STUDY.   1000

AND VIBRATIONAL MODE COUPLING IN IODINE HEPTAFLUORIDE. AN ELECTRON DIFFRACTION STUDY. /TURE, PSEUDOROTATION, 0515
STRUCTURE, PSEUDOROTATION, AND ANHARMONIC COUPLING OF / ELECTRON DIFFRACTION STUDY OF RHENIUM HEPTAFLUORIDE 0920
ELECTRON DIFFRACTION STUDY OF RHENIUM HEXAFLUORIDE. 0909
RES OF VANADIUM PENTAFLUORIDE, NIOBIUM PENTAFLUORIDE, AN/ ELECTRON DIFFRACTION STUDY OF THE MOLECULAR STRUCTU 1217
RE OF THE VAPOR PHASE HALIDES OF GALLIUM, YTTRIUM, LANTH/ ELECTRON DIFFRACTION STUDY OF THE MOLECULAR STRUCTU 1317
OUS XENON TETRAFLUORIDE. ELECTRON DIFFRACTION STUDY OF THE STRUCTURE OF GASE 1282
PHORUS PENTAFLUORIDE. ELECTRON DIFFRACTION STUDY OF THE STRUCTURE OF PHOS 0851
ORIDE MOLECULE IN THE VAPOR FORM. ELECTRON DIFFRACTION STUDY OF THE VANADIUM PENTAFLU 1218
ETRAHALIDES IN THE GAS PHASE. (URANIUM TETRAFLUORIDE AND/ ELECTRON DIFFRACTION STUDY OF URANIUM AND THORIUM T 1170
FAILURE OF THE FIRST BORN APPROXIMATION IN ELECTRON DIFFRACTION. (TELLURIUM HEXAFLUORIDE) 1109
UM HEXAFLUORI/ FAILURE OF THE FIRST BORN APPROXIMATION IN ELECTRON DIFFRACTION. (TUNGSTEN HEXAFLUORIDE, URANI 1158
FAILURE OF THE FIRST BORN APPROXIMATION IN ELECTRON DIFFRACTION. (URANIUM HEXAFLUORIDE) 1199
NON OXIDE TETRAFLUORIDE BY ESCA AND EVIDENCE FOR ORBITAL/ ELECTRON DISTRIBUTION IN THE XENON FLUORIDES AND XE 1229
DISTRIBUTION OF INTENSITY IN THE ELECTRON EMISSION SPECTRUM OF SILICON MONOFLUORIDE. 0968
DE) CHEMICAL APPLICATIONS OF CORE ELECTRON EXCITATION SPECTROSCOPY. (XENON HEXAFLUORI 1306
ORO CYANO PHOSPHINE. IONIZATION BY ELECTRON IMPACT OF PHOSPHORUS TRIFLUORIDE AND DIFLU 0800
FLUORIDE. ELECTRON IMPACT SPECTRA OF METHANE AND CARBON TETRA 0333
CARBON TRIFLUORIDE) ELECTRON IMPACT STUDIES OF SOME TRIHALO METHANES. (0312
MOMENT OF NITROGEN MONOFLUORIDE IN THE SINGLE/ GAS PHASE ELECTRON PARAMAGNETIC RESONANCE SPECTRUM AND DIPOLE 0673
ICAL CHLORINE TETRAFLUORIDE. ELECTRON PARAMAGNETIC RESONANCE SPECTRUM OF THE RAD 0413
ORON TRIFLUORIDE. BARRIER TO ELECTRON PASSAGE THROUGH ELECTRONEGATIVE ATOMS IN B 0196
ND SELENIUM MONOFLUORIDE. GAS PHASE ELECTRON RESONANCE SPECTRA OF SULFUR MONOFLUORIDE A 1021
SELENIUM MONOFLUORIDE, SELENIUM OXIDE AND IODI/ GAS PHASE ELECTRON RESONANCE SPECTRA OF SULFUR MONOFLUORIDE, 1020
RBON MONOFLUORIDE. GAS PHASE ELECTRON RESONANCE SPECTRUM AND DIPOLE MOMENT OF CA 0284
DIFLUORIDE RADICAL. ELECTRON SPIN RESONANCE AND STRUCTURE OF THE BORON 0178
/N DIFLUORIDE AND DINITROGEN TETRAFLUORIDE MOLECULES, AND ELECTRON SPIN RESONANCE HYPERFINE COUPLING CONSTAN/ 0688
N MONOFLUORIDE. RADIATION DAMAGE IN XENON TETRAFLUORIDE. ELECTRON SPIN RESONANCE OF THE TRAPPED RADICAL XENO 1220
AL. ELECTRON SPIN RESONANCE OF THE XENON FLUORIDE RADIC 1288
ORIDE AND PHOSPHORUS TETRAFLUORIDE RADICALS. ELECTRON SPIN RESONANCE OF TRAPPED PHOSPHORUS DIFLU 0792
LUORIDE RADICAL. ELECTRON SPIN RESONANCE PARAMETERS OF THE BORON DIF 0181
RIDES. (CHLORINE HEXAFLUORIDE, IODINE HEXAFLUORIDE, BROM/ ELECTRON SPIN RESONANCE SPECTRA OF HALOGEN HEXAFLUO 0271
ORIDE AND SULFUR TRIFLUORIDE RADICALS. ELECTRON SPIN RESONANCE SPECTRA OF PHOSPHORUS DIFLU 0791
FLUORIDE RADICALS PRODUCED IN A SINGLE CRYSTAL OF PHOSPH/ ELECTRON SPIN RESONANCE SPECTRA OF PHOSPHORUS TETRA 0823
CHLORINE HEXAFLUORIDE RADICAL. PREPARATION, ELECTRON SPIN RESONANCE SPECTRUM, AND STRUCTURE. 0428
ROMINE HEXAFLUORIDE. ELECTRON SPIN RESONANCE SPECTRUM AND STRUCTURE OF B 0270
UORIDE. ELECTRON SPIN RESONANCE SPECTRUM OF ARSENIC TETRAFL 0112
IODINE OXIDE AND SELENIUM MONOFLUORIDE IN J-5/2 ROTATION/ ELECTRON SPIN RESONANCE SPECTRUM OF BROMINE OXIDE, 0954
ORIDE. ELECTRON SPIN RESONANCE SPECTRUM OF KRYPTON MONOFLU 0546
IDE IN GAMMA IRRADIATED XENON TETRAFLUORIDE. ELECTRON SPIN RESONANCE SPECTRUM OF XENON MONOFLUOR 1222
ORIDE, PHOSPHORUS DICHLORIDE, AND NITROGEN DICHLORIDE IN/ ELECTRON SPIN RESONANCE STUDIES OF PHOSPHORUS DIFLU 0793
IRRADIATED POLYCRYSTALLINE ARSENIC TRIFLUORIDE AND ARSEN/ ELECTRON SPIN RESONANCE STUDY OF RADICALS IN GAMMA 0108
DS. ELECTRON STRUCTURE AND NMR SPECTRA OF BORON COMPOUN 0229

ERT GASES BY THE NONEMPIRICAL NDDO-2 (AL/ CALCULATIONS OF ELECTRON STRUCTURE OF COMPOUNDS OF INERT GASES. 1339
ELECTRON SYSTEMS. WAVE FUNCTIONS FOR THE FOUR ELECTRON STRUCTURES OF COMPOUNDS OF HALOGENS AND IN 0016
/ THE DIPOLE MOMENT CF AN A DOUBLET SIGMA TO X DOUBLET PI ELECTRON THREE CENTER BONDING OF FOUR AND EIGHT PI 0301
E OF URANIUM HEXAFLUORIDE AS DETERMINED BY DIFFRACTION OF ELECTRON TRANSITON IN A SILICON MONOFLUORIDE MOLEC/ 0970
URANIUM HEXAFLUORIDE AS DETERMINED BY THE DIFFRACTION OF ELECTRONS ON THE VAPOR. STRUCTUR 1183
WITH THE MAXIMUM OVERLAP METHOD. MOLECULES WITH LONE PAIR ELECTRONS ON THE VAPOR. STRUCTURE OF 1184
BARRIER TO ELECTRON PASSAGE THROUGH ELECTRONS OR EMPTY ORBITALS. /N OF FORCE CONSTANTS 1395
IZABILITY IN HYDROGEN HALIDES AND THE INTERHALOGEN COMPO/ ELECTRONEGATIVE ATOMS IN BORON TRIFLUORIDE. 0196
EXACHLORIDE AND URANIUM HEXAFLUORIDE. ELECTRONEGATIVITY, NONBONDED INTERACTIONS AND POLAR 0373
ELECTRONIC ABSORPTION SPECTRUM OF GASEOUS URANIUM H 1194
UORIDE RADICAL. ELECTRONIC ABSORPTION SPECTRUM OF THE NITROGEN DIFL 0694
N HEXAFLUORIDE MOLECULE. ELECTRONIC AND GEOMETRIC STRUCTURE OF THE FREE XENO 1302
ELECTRONIC BAND SPECTRUM OF YTTRIUM MONOFLUORIDE. 1314

ROTATIONAL ANALYSIS OF ELECTRONIC BANDS OF GASEOUS MAGNESIUM MONOFLUORIDE. 0600
/, MOLECULAR QUADRUPOLE MOMENTS AND SECOND MOMENTS OF THE ELECTRONIC CHARGE DISTRIBUTION IN OXYGEN DIFLUORID/ 0780
HEXACHLORIDE(2-), AND IRIDIUM HEXACHLORIDE(/ VIBRATIONAL ELECTRONIC COUPLING IN IRIDIUM HEXAFLUORIDE, OSMIUM 0538
NEW ELECTRONIC EMISSION FROM SILICON DIFLUORIDE. 0987
TS OF IODINE MONOFLUORIDE. ELECTRONIC EMISSION SPECTRUM AND MOLECULAR CONSTAN 0496
NEW ELECTRONIC EMISSION SYSTEM OF GERMANIUM DIFLUORIDE. 0476
TIUM DIFLUORIDE, AND BARIUM DIFLUORIDE IN THE/ SPECTRA OF ELECTRONIC EXCITATIONS IN CALCIUM DIFLUORIDE, STRON 0279
NCTIONS. ELECTRONIC POPULATION ANALYSIS OF MOLECULAR WAVE FU 0157
ITIO CALCULATION FOR THE GROUND STATE. ELECTRONIC PROPERTIES OF SULFUR HEXAFLUORIDE. AB IN 1077
1: OPTICAL ABSORPTION AND X-RAY EMISSION FROM SCF MO LCA/ ELECTRONIC PROPERTIES OF SULFUR HEXAFLUORIDE. PART- 1078
/RACTERIZATION OF GROUND STATE WAVE FUNCTIONS BY MEASURED ELECTRONIC PROPERTIES. PART-2: DIPOLE MOMENT AND F/ 0746
REME ULTRAVIOLET. ELECTRONIC SPECTRA OF LANTHANUM FLUORIDE IN THE EXT 0570
UM PENTAFLUORIDES. ELECTRONIC SPECTRA OF LIQUID RUTHENIUM AND MOLYBDEN 0925
TS POSITIVE ION. ELECTRONIC SPECTRA OF PHOSPHORUS MONOFLUORIDE AND I 0786
RBON DIFLUORIDE) ELECTRONIC SPECTRA OF POLYATOMIC FREE RADICALS. (CA 0308
UM DIFLUORIDE. FAR ULTRAVIOLET ELECTRONIC SPECTRA OF STRONTIUM DIFLUORIDE AND BARI 1017
FLUORIDE MOLECULES DOWN TO HELIUM TE/ HEAT CAPACITIES AND ELECTRONIC SPECTRA OF THE PLATINUM GROUP METAL HEXA 1159
THE ELECTRONIC SPECTRUM OF BORON FLUORIDE. 0172
VIBRONICALLY INDUCED JAHN-TELLER PROGRESSIONS IN THE ELECTRONIC SPECTRUM OF IRIDIUM HEXAFLUORIDE. 0537
ELECTRONIC SPECTRUM OF LANTHANUM MONOFLUORIDE. 0569
ELECTRONIC SPECTRUM OF LANTHANUM TRIFLUORIDE. 0582
ELECTRONIC SPECTRUM OF LUTETIUM MONOFLUORIDE. 0598
ELECTRONIC SPECTRUM OF SCANDIUM MONOFLUORIDE. 0934
CAL STUDY. ELECTRONIC SPECTRUM OF SILICON DIFLUORIDE. THEORETI 0997
ULE. ELECTRONIC SPECTRUM OF THE BORON MONOFLUORIDE MOLEC 0155
LECULE. ROTATIONAL ANALYSIS OF SELECTED BANDS FROM THE ELECTRONIC SPECTRUM OF THE LUTETIUM MONOFLUORIDE MO 0599
/E VACUUM ULTRAVIOLET PHOTOLYSIS OF NITROGEN TRIFLUORIDE. ELECTRONIC SPECTRUM OF THE NITROGEN DIFLUORIDE FRE/ 0691
EXAFLUORIDE: A CASE OF ELECTRONIC ISOMERISM. ELECTRONIC STATES AND MOLECULAR GEOMETRY OF XENON H 1294
SINGLET ELECTRONIC STATES OF ALUMINUM MONOFLUORIDE. 0023
ELECTRONIC STATES OF GASEOUS ALUMINUM MONOFLUORIDE. 0027
YTTRIUM MONOFLUORIDE, AND LANTHANUM MONOFLUORIDE. ELECTRONIC STATES OF GASEOUS SCANDIUM MONOFLUORIDE, 0928
HIGHER ELECTRONIC STATES OF RHENIUM HEXAFLUORIDE. 0914
REASSIGNMENT OF MOLECULAR ORBITAL CONFIGURATIONS OF THE ELECTRONIC STATES OF SCANDIUM MONOFLUORIDE. 0937
RAFLUORIDE, XENON OXIDE TETRAFLUORIDE, AN/ CORRELATION OF ELECTRONIC STATES OF THE POSITIVE IONS OF XENON TET 1264
N ON IRON DIFLUORIDE AND VANADIUM DIOXIDE. ELECTRONIC STRUCTURE AND ANOMALOUS THERMAL EXPANSIO 0543

NDS. (FLUORIDES)   ELECTRONIC STRUCTURE AND CHEMISTRY OF IODINE COMPOU 0499
ONOFLUORIDE.   ELECTRONIC STRUCTURE AND GROUND STATE OF SCANDIUM M 0932
NIUM HEXAFLUORIDE AND UR/ THEORETICAL CALCULATIONS OF THE   ELECTRONIC STRUCTURE AND OPTICAL TRANSITIONS OF URA 1179
UORIDE.   ELECTRONIC STRUCTURE AND PROPERTIES OF KRYPTON DIFL 0548
LECULE BY THE MO LCAO SELF CONSISTENT/ CALCULATION OF THE   ELECTRONIC STRUCTURE OF A CHLORINE PENTAFLUORIDE MO 0415
ORON TRIFLUORIDE)   SEMIEMPIRICAL STUDY OF THE   ELECTRONIC STRUCTURE OF BORON HYDRIDE FLUORIDES. (B 0226
                                                                ELECTRONIC STRUCTURE OF BORON TRIFLUORIDE.   0184
ONOFLUORIDE.   ELECTRONIC STRUCTURE OF CARBON MONOXIDE AND BORON M 0162
PROXIMATE SCF CALCULATION.   ELECTRONIC STRUCTURE OF CHLORINE TRIFLUORIDE. AN AP 0404
ELEMENTS FROM THE DATA OF LCAO MO/ CHARACTERISTICS OF THE   ELECTRONIC STRUCTURE OF FLUORIDES OF NONTRANSITION 0009
                                                                ELECTRONIC STRUCTURE OF IRON TRIFLUORIDE.   0542
BY PHOTOELECTRON SPECTROSCOPY.   ELECTRONIC STRUCTURE OF KRYPTON DIFLUORIDE STUDIED 0550
PROACH. PART-6: THE ASSIGNMENT OF THE HELIUM-2 PHOTOELEC/   ELECTRONIC STRUCTURE OF MOLECULES BY A MANY BODY AP 1090
                                                                ELECTRONIC STRUCTURE OF NEPTUNIUM HEXAFLUORIDE.   0648
AB INITIO STUDIES OF CHA/ MOLECULAR ORBITAL THEORY OF THE   ELECTRONIC STRUCTURE OF ORGANIC COMPOUNDS. PART-3: 0707
D POLYTOPAL REARRANGEMENT IN PHOSPHORANES.   ELECTRONIC STRUCTURE OF PHOSPHORUS PENTAFLUORIDE AN 0875
        NONEMPIRICAL LCAO MO SCF CALCULATIONS OF THE   ELECTRONIC STRUCTURE OF SILICON DIFLUORIDE.   0995
BY ULTRALONG WAVE X-RAY SPECTROSCOPY.   ELECTRONIC STRUCTURE OF SULFUR HEXAFLUORIDE STUDIED 1065
LECULE. PART-1. THE GROUND STATE.   ELECTRONIC STRUCTURE OF THE CARBON TETRAFLUORIDE MO 0330
LECULE.   ELECTRONIC STRUCTURE OF THE CARBON TETRAFLUORIDE MO 0357
ICAL.   ELECTRONIC STRUCTURE OF THE NITROGEN DIFLUORIDE RAD 0682
        THEORETICAL AND EXPERIMENTAL STUDIES OF THE   ELECTRONIC STRUCTURE OF THE XENON FLUORIDES.   1238
                                CALCULATED   ELECTRONIC STRUCTURE OF URANIUM HEXAFLUORIDE.   1188
                                                                ELECTRONIC STRUCTURE OF XENON DIFLUORIDE.   1226
TETRAHALIDE ANIONS.   A THEORETICAL COMPARISON OF THE   ELECTRONIC STRUCTURES OF BORON TRIHALIDES AND BORON 0225
HYDRIDE FLUORIDE, BORON HYDRIDE/ AB INITIO STUDIES OF THE   ELECTRONIC STRUCTURES OF BORON TRIHYDRIDE, BORON DI 0241
CHLORINE TRIFLUORIDE.   ELECTRONIC STRUCTURES OF CHLORINE MONOFLUORIDE AND 0372
INERT GASES BY THE SCF LCAO MO METHOD. / CALCULATIONS OF   ELECTRONIC STRUCTURES OF COMPOUNDS OF HALOGENS AND 1231
ND TETRAFLUORO PHOSPHORANE.   ELECTRONIC STRUCTURES OF PHOSPHORUS PENTAFLUORIDE A 0857
TRAF/ HELIUM-1 AND HELIUM-2 PHOTOELECTRON SPECTRA AND THE   ELECTRONIC STRUCTURES OF XENON FLUORIDES, XENON TE 1228
NTERNUCLEAR SEPARATION FOR THE B-X BAND SYSTEM OF MOLECU/   ELECTRONIC TRANSITION MOMENT AS A FUNCTION OF THE I 0967
ANCE FOR BANDS OF THE X SYSTEM.   CORRELATION OF THE   ELECTRONIC TRANSITION MOMENT WITH INTERNUCLEAR DIST 0969
SEPARATION IN SILICON CHLORIDE AND TIN MONO/ VARIATION OF   ELECTRONIC TRANSITION MOMENT WITH THE INTERNUCLEAR 1137
        A QUARTET SIGMA TO X DOUBLET PI   ELECTRONIC TRANSITION OF SILICON MONOFLUORIDE.   0971
OLECULE.   A NEW   ELECTRONIC TRANSITION OF THE MERCURY MONOFLUORIDE M 0618
OMPARISONS WITH OTHER HEXAVALENT XENON MOLECULES. PART-3:   ELECTRONIC TRANSITIONS AND BAND SHAPES. / PART-2: C 1311
N TETRAFLUORIDE.   FORBIDDEN   ELECTRONIC TRANSITIONS IN XENON DIFLUORIDE AND XENO 1247
THE MOLECULES OF NEODYMIUM HALIDES.   ELECTRONOGRAPHIC INVESTIGATION OF THE STRUCTURE OF 0644
E CONSTANTS OF HIGHLY SYMMETRICAL MOLECU/ USE OF A SIMPLE   ELECTROSTATIC MODEL FOR CALCULATION OF ANGULAR FORC 1484
                ELECTRON DIFFRACTION STUDIES AT   ELEVATED TEMPERATURES. (NITROGEN DIFLUORIDE)   0679
TION.   NU-4 BANDS OF THE TEN AND   ELEVEN ISOTOPES OF BORON TRIFLUORIDE AT HIGH RESOLU 0208
EXAFLUORIDE.   SULFUR K AND L AND FLUORINE K X-RAY   EMISSION AND ABSORPTION SPECTRA OF GASEOUS SULFUR H 1088
                                                                EMISSION BANDS OF CARBON DIFLUORIDE.   0309
        ROTATIONAL ANALYSIS OF THE VISIBLE   EMISSION BANDS OF LEAD MONOFLUORIDE.   0587
/ULFUR HEXAFLUORIDE. PART-1: OPTICAL ABSORPTION AND X-RAY   EMISSION FROM SCF MO LCAO COMPUTATIONS. PART-2: MO/ 1078
        NEW ELECTRONIC   EMISSION FROM SILICON DIFLUORIDE.   0987
THE DISSOCIATION OF DIATOMIC MOLECULES AND A STUDY OF THE   EMISSION OF IODINE MONOFLUORIDE.   THEORY OF 0493
                                                                EMISSION OF MOLECULAR ANTIMONY MONOFLUORIDE.   0060
TH MONOFLUORIDE.   EMISSION SPECTRA OF ANTIMONY MONOFLUORIDE AND BISMU 0064
OFLUORIDE AND KRYPTON MONOFLUORIDE.   ABSORPTION AND   EMISSION SPECTRA OF ARGON MATRIX ISOLATED XENON MON 0544
S OBTAINED IN ELECTRICAL DISCHARGES THROUGH BORON TRIFLU/   EMISSION SPECTRA OF CERTAIN POLYATOMIC FREE RADICAL 0222
LORIDE, XENON MONOFLUORIDE, AND KRYPTON MONOFLUORIDE.   EMISSION SPECTRA OF XENON MONOBROMIDE, XENON MONOCH 0545
LUORIDE)   ULTRAVIOLET   EMISSION SPECTRA. (XENON DIFLUORIDE AND KRYPTON DIF 0559
- X DOUBLET SIGMA. SYSTEM OF CALCIUM FLUORIDE MOLECULE IN   EMISSION SPECTRUM. /OF THE BANDS OF B DOUBLET SIGMA 0280
E MONOFLUORIDE.   ELECTRONIC   EMISSION SPECTRUM AND MOLECULAR CONSTANTS OF IODIN 0496
ACUUM ULTRAVIOLET.   EMISSION SPECTRUM OF ALUMINUM MONOFLUORIDE IN THE V 0030
        THERMALLY EXCITED   EMISSION SPECTRUM OF BARIUM MONOFLUORIDE.   0123
                                                                EMISSION SPECTRUM OF BISMUTH MONOFLUORIDE.   0147
        VISIBLE   EMISSION SPECTRUM OF BISMUTH MONOFLUORIDE.   0145
                                                                EMISSION SPECTRUM OF CARBON MONOFLUORIDE.   0291
E BAND SYSTEM IN THE REGION 2755-2179 ANGSTROMS.   EMISSION SPECTRUM OF LEAD MONOFLUORIDE.   0588
        DISTRIBUTION OF INTENSITY IN THE ELECTRON   EMISSION SPECTRUM OF SILICON DIFLUORIDE. PART-1: TH 0988
                                                                EMISSION SPECTRUM OF SILICON MONOFLUORIDE.   0968
NOFLUORIDE.   EMISSION SPECTRUM OF SILICON TRIFLUORIDE.   0998
                                                                EMISSION SPECTRUM OF THE GAMMA SYSTEM OF SILICON MO 0974
                                                                EMISSION SPECTRUM OF TITANIUM MONOFLUORIDE.   1142
        NEW ELECTRONIC   EMISSION SYSTEM OF GERMANIUM DIFLUORIDE.   0476
ON TETRAFLUORIDE, AND/ RADIATIVE LIFETIMES OF ULTRAVIOLET   EMISSION SYSTEMS EXCITED IN BORON TRIFLUORIDE, CARB 0216
OMPOUNDS.   A SMALL GAUSSIAN BASIS SET FOR NONEMPIRICAL   ALL-ELECTRON SCF CALCULATIONS IN IODINE C 0495
OF NONTRANSITION ELEMENTS FROM THE DATA OF LCAO MO SCF   NONEMPIRICAL CALCULATIONS. /IC STRUCTURE OF FLUORIDES 0009
C STRUCTURE OF SILICON DIFLUORIDE.   NONEMPIRICAL LCAO MO SCF CALCULATIONS OF THE ELECTRONI 0995
IDES BERYLLIUM DIFLUORIDE, MAGNESIUM DIFLUORIDE, AND /   NONEMPIRICAL LCAO MO SCF STUDIES OF THE GROUP-2A DIHAL 1360
ES OF BERYLLIUM DIFLUORIDE.   NONEMPIRICAL LCAO MO SCF STUDIES OF THE LOW LYING STAT 0137
NAL AND ROTATIONAL CHARACTERISTICS OF MOLECULES OF SOME /   EMPIRICAL METHOD FOR DETERMINING EFFECTIVE VIBRATIC 0622
/TURES OF COMPOUNDS OF HALOGENS AND INERT GASES BY THE   NONEMPIRICAL NDDO-2 (ALPHA, BETA) METHOD. PART-3: ENE/ 0016
LECULES CONTAINING PHOSPHORUS OR SULFUR.   NONEMPIRICAL VALENCE ELECTRON CALCULATIONS ON SMALL MO 1029
MUM OVERLAP METHOD. MOLECULES WITH LONE PAIR ELECTRONS OR   EMPTY ORBITALS. /N OF FORCE CONSTANTS WITH THE MAXI 1395
RIDES.   ENERGETICS AND STABILITY OF GAS PHASE SCANDIUM FLUO 0951
        MASS SPECTRUM AND MOLECULAR   ENERGETICS OF KRYPTON DIFLUORIDE.   0565
UTION IN CHLORINE MONOFLUORIDE FROM CORE ELECTRON BINDING   ENERGIES.   CHARGE DISTRIB 0374
HELIUM-2 PHOTOELECTRON SPECTRA OF XENON DIFLUORIDE / BOND   ENERGIES AND IONIC CHARACTER OF INERT GAS HALIDES. 1343
YTTRIUM MONOFLUORIDE, AND LANTHANUM MONOFLU/ DISSOCIATION   ENERGIES AND STABILITIES OF SCANDIUM MONOFLUORIDE, 0935
IPOLE MOMENTS.   STARK   ENERGIES FOR MOLECULES WITH VIBRATIONALLY INDUCED D 1417
(SCANDIUM TRIFLUORIDE, YTTRI/ CALCULATION OF DISSOCIATION   ENERGIES FOR THE HALIDES OF THE SCANDIUM SUBGROUP. 0948
        PAIRING   ENERGIES IN THE ACTINIDE SERIES.   1333
TIES OF SOME GERMANIUM FLUORIDE SPECIE/ BOND DISSOCIATION   ENERGIES, IONIZATION POTENTIALS AND ELECTRON AFFINI 0464
DE AND HEAT OF FORMATION OF BERYLLIUM CHLOR/ DISSOCIATION   ENERGIES OF BERYLLIUM FLUORIDE AND BERYLLIUM CHLORI 0128
        DETERMINATION OF THE DISSOCIATION   ENERGIES OF CARBON MONOHALIDES.   0289
        POTENTIAL ENERGY CURVES AND DISSOCIATION   ENERGIES OF DIATOMIC BORON AND ALUMINUM HALIDES.   0173
LIUM, INDIUM, A/ POTENTIAL ENERGY CURVES AND DISSOCIATION   ENERGIES OF DIATOMIC FLUORIDES AND CHLORIDES OF GAL 0458

ORIDE, BROMINE FLUORIDE)   DISSOCIATION ENERGIES OF DIATOMIC HALOGEN FLUORIDES. (IODINE FLU 0494
CTIVITY COEFFICIENTS FCR PROTACTINIUM IN / ESTIMATED FREE ENERGIES OF FORMATION OF PROTACTINIUM HALIDES AND A 0900
VIBRATION ROTATION ENERGIES OF MOLECULES.   1490
MOLECULAR ORBITAL STUDY OF ATOMIZATION ENERGIES OF PNICOGEN FLUORIDE MOLECULES AND IONS.   1364
IUM TRIFLUORIDE AND PLUTONIUM TRIFLUORIDE, HEATS AND FREE ENERGIES OF SUBLIMATION.   VAPOR PRESSURES OF AMERIC 0055
DISSOCIATION ENERGIES OF THE ALKALINE EARTH MONOFLUORIDES.   0119
NUM, GALLIUM, INDIUM, AND THALLIUM.   DISSOCIATION ENERGIES OF THE GASEOUS MONOHALIDES OF BORON, ALUMI 0152
/C STUDIES AT HIGH TEMPERATURES. PART-2: THE DISSOCIATION ENERGIES OF THE MONOFLUORIDES AND DIFLUORIDES OF S/ 0463
IDE(X) FROM TETRAFLUORC ETHYLENE. (CARBON DIFLUORIDE BOND ENERGY) /THE APPEARANCE POTENTIAL OF CARBON DIFLUOR 0305
SPECTRUM OF BROMINE MONOFLUORIDE AND ITS DISSOCIATION ENERGY.   0252
PHOTOCHLORINATION OF METHANE AND FLUOROFORM. DISSOCIATION ENERGY AND ENTROPY OF CARBON TRIFLUORIDE. /RICES.   0311
BORON AND ALUMINUM HALIDES.   POTENTIAL ENERGY CURVES AND DISSOCIATION ENERGIES OF DIATOMIC 0173
FLUORIDES AND CHLORIDES OF GALLIUM, INDIUM, A/ POTENTIAL ENERGY CURVES AND DISSOCIATION ENERGIES OF DIATOMIC 0458
OHALIDES.   POTENTIAL ENERGY CURVES AND NATURE OF BINDING IN GROUP-3A MON 0032
/ NORMAL MODES OF VIBRATION. PART-2: INTERACTION TERMS IN ENERGY DISTRIBUTIONS. (SELENIUM AND TELLURIUM FLUO/ 1494
TEMPERATURE. (THALLIUM / ENTHALPIES, ENTROPIES, AND FREE ENERGY FUNCTIONS OF THALLIUM MONOHALIDES ABOVE ROOM 1116
BOND ENERGY IN THE ALUMINUM TRIFLUORIDE MOLECULE.   0044
OXYGEN MONOFLUORIDE. HARTREE-FOCK WAVE FUNCTION, BINDING ENERGY, IONIZATION POTENTIAL, ELECTRON AFFINITY, D/ 0756
/FLUORIDE, DOUBLET PI HARTREE-FOCK WAVE FUNCTION, BINDING ENERGY, ICNIZATION POTENTIAL, ELECTRON AFFINITY, D/ 0757
MULTIPLE SCATTERING X(ALPHA) CALCULATION OF THE ENERGY LEVEL STRUCTURE OF XENON HEXAFLUORIDE.   1304
MOLECULAR ORBITAL ENERGY LEVELS AND BONDING IN KRYPTON DIFLUORIDE.   0549
CRYSTAL FIELD MODEL STUDY OF XENON HEXAFLUORIDE. PART-1: ENERGY LEVELS AND MOLECULAR GEOMETRY. PART-2: COMP/ 1311
DE.   NONRIGID MOLECULE EFFECTS ON THE ROVIBRONIC ENERGY LEVELS AND SPECTRA OF PHOSPHORUS PENTAFLUORI 0841
MIUM TRIFLUORIDES.   ENERGY LEVELS AND SPECTROSCOPIC PARAMETERS OF NEODY 0646
ABSORPTION SPECTRUM OF URANIUM TETRAFLUORIDE AND ENERGY LEVELS CF URANIUM(IV).   1168
/BY THE NONEMPIRICAL NDDO-2 (ALPHA, BETA) METHOD. PART-3: ENERGY LEVELS, WAVE FUNCTIONS AND AO POPULATIONS O/ 0016
MASS SPECTROMETRIC DETERMINATION OF THE DISSOCIATION ENERGY OF BERYLLIUM MONOFLUORIDE.   0129
C STUDIES.   DISSOCIATION ENERGY OF BORON MONOFLUORIDE FROM MASS SPECTROMETRI 0161
NIUM FLUORIDES)   REEVALUATION OF THE DISSOCIATION ENERGY OF CALCIUM FLUORIDE. (ALSO SILICON AND GERMA 0277
ATION PRESSURE OF CALCIUM DIFLUORIDE AND THE DISSOCIATION ENERGY OF CALCIUM MONOFLUORIDE.   SUBLIM 0276
RIDE.   CALCULATION OF THE DISSOCIATION ENERGY OF CALCIUM MONOFLUORIDE AND URANIUM MONOFLUO 1161
DISSOCIATION ENERGY OF CHLORINE MONOFLUORIDE.   0387
TION PRESSURE OF CHROMIUM DIFLUORIDE AND THE DISSOCIATION ENERGY OF CHROMIUM MONOFLUORIDE.   SUBLIMA 0430
DISSOCIATION ENERGY OF COPPER MONOFLUORIDE.   0440
VIBRATIONAL ANALYSIS OF C-X AND C'-X' BAND SYSTEMS AND ENERGY OF DISSOCIATION IN GERMANIUM MONOFLUORIDE.   0470
ATION STUDY OF CHLORINE MONOFLUORIDE AND THE DISSOCIATION ENERGY OF FLUORINE.   PHOTOIONIZ 0378
DISSOCIATION ENERGY OF MAGNESIUM MONOFLUORIDE.   0601
ALCIUM AND THE IONIC MCDEL OF A MOLECULE.   DISSOCIATION ENERGY OF MONOFLUORIDE COMPOUNDS OF MAGNESIUM AND C 0278
SUBLIMATION OF PLUTCNIUM TRIFLUORIDE AND THE DISSOCIATION ENERGY OF PLUTONIUM MONOFLUORIDE. /-2: ENTHALPY OF 0884
0 DEGREES K.   HEAT CAPACITY, ENTROPY, ENTHALPY, AND GIBBS ENERGY OF PLUTONIUM-242 TETRAFLUORIDE FROM 10 TO 35 0889
GH TEMPERATURES. PART-14: VAPOR PRESSURE AND DISSOCIATION ENERGY OF SILVER MONOFLUORIDE. /ETRIC STUDIES AT HI 1014
ENTHALPY OF FORMATION AND THE DISSOCIATION ENERGY OF THALLIUM MONOFLUORIDE 1115
AND URANIUM HEXAFLUORID/ LCW TEMPERATURE HEAT CAPACITIES, ENTHALPIES, AND ENTROPIES OF URANIUM TETRAFLUORIDE 1166
HE TRIFLUORIDES OF SCANDIUM, CERIUM, SA/ HIGH TEMPERATURE ENTHALPIES AND RELATED THERMODYNAMIC FUNCTIONS OF T 1356
THALLIUM MONOHALIDES ABOVE ROOM TEMPERATURE. (THALLIUM / ENTHALPIES, ENTROPIES, AND FREE ENERGY FUNCTIONS OF 1116
D TANTALUM PENTAFLUORIDE.   ENTHALPIES OF FORMATION OF NIOBIUM PENTAFLUORIDE AN 0667
ANTHANUM TRIFLUORIDE)   EXPERIMENTAL DETERMINATION OF THE ENTHALPIES OF FORMATION OF RARE EARTH FLUORIDES. (L 0578
TRIUM TRIFLUORIDE, AND LANTHANUM TRIFLUORIDE.   STANDARD ENTHALPIES OF FORMATION OF SCANDIUM TRIFLUORIDE, YT 0942
ENIUM HEXAFLUORIDE, TELLURIUM HEXAFLUORIDE AND THEIR THE/ ENTHALPIES OF FORMATION OF SULFUR HEXAFLUORIDE, SEI 1092
N TETRAFLUORIDE, XENON DIFLUORIDE, AND PHOSPHORUS TRIFLU/ ENTHALPIES OF FORMATION OF XENON HEXAFLUORIDE, XENO 0807
N TETRAFLUORIDE, XENON TETRAFLUORIDE. CRYSTAL STRUCTURE. ENTHALPIES OF FORMATION OF XENON HEXAFLUORIDE, XENO 1269
ORIDES. CORRELATION OF ENTROPY AND ENTHALP/ ENTROPIES AND ENTHALPIES OF SUBLIMATION OF CALCIUM AND CERIUM FLU 0281
TRIFLUORIDES.   ENTROPIES AND ENTHALPIES OF SUBLIMATION OF NEODYMIUM AND TERBIUM 0645
/RD ENTHALPY OF FORMATION OF PHOSPHORUS PENTAFLUORIDE AND ENTHALPIES OF TRANSITION BETWEEN VARIOUS FORMS OF / 0865
/ AND PHYSICAL PROPERTIES OF BERYLLIUM COMPOUNDS. PART-1: ENTHALPY AND ENTROPY OF VAPORIZATION OF BERYLLIUM / 0138
UORIDE FROM 10 TO 350 DEGREES K.   HEAT CAPACITY, ENTROPY, ENTHALPY, AND GIBBS ENERGY OF PLUTONIUM-242 TETRAFL 0889
CALCIUM AND CERIUM FLUORIDES. CORRELATION OF ENTROPY AND ENTHALPY IN ERRORS. /D ENTHALPIES OF SUBLIMATION OF 0281
S.   HIGH TEMPERATURE ENTHALPY INCREMENTS FOR SOME RARE EARTH TRIFLUORIDE 1350
DIPOLE MOMEN/ ARSENIC MONOFLUORIDE, 3 SIGMA. DISSOCIATION ENTHALPY, IONIZATION POTENTIAL, ELECTRON AFFINITY, 0094
REES K BY TRANSPOSED TEMPERATURE DROP CALORIMETRY.   ENTHALPY OF BERYLLIUM FLUORIDE FROM 456 TO 1083 DEG 0142
ENTHALPY OF DISSOCIATION OF NITROGEN TRIFLUORIDE. 0742
F THALLIUM MONOFLUORIDE   ENTHALPY OF FORMATION AND THE DISSOCIATION ENERGY O 1115
FLUORINE BOMB CALORIMETRY. PART-22: THE   ENTHALPY OF FORMATION OF ALUMINUM TRIFLUORIDE.   0048
STANDARD ENTHALPY OF FORMATION OF ALUMINUM TRIFLUORIDE.   0043
ENTHALPY OF FORMATION OF ARSENIC PENTAFLUORIDE.   0117
ENTHALPY OF FORMATION OF BERYLLIUM DIFLUORIDE.   0139
FLUORINE BOMB CALORIMETRY. PART-15: THE ENTHALPY OF FORMATION OF BORON TRIFLUORIDE.   0220
ENTHALPY OF FORMATION OF CARBON TETRAFLUORIDE.   0329
ENTHALPY OF FORMATION OF GERMANIUM DIFLUORIDE.   0472
ENTHALPY OF FORMATION OF GERMANIUM TETRAFLUORIDE.   0469
ENTHALPY OF FORMATION OF GERMANIUM TRIFLUORIDE.   0480
PRASEODYMIUM TRIFLUORIDE.   ENTHALPY OF FORMATION OF LANTHANUM TRIFLUORIDE AND 0577
ND ENTHALPI/ FLUORINE BOMB CALORIMETRY. PART-18: STANDARD ENTHALPY OF FORMATION OF PHOSPHORUS PENTAFLUORIDE A 0865
LUORINE BOMB CALORIMETRY.   ENTHALPY OF FORMATION OF THORIUM TETRAFLUORIDE BY F 1130
ENTHALPY OF FORMATION OF TUNGSTEN HEXAFLUORIDE.   1157
TUNGSTEN HEXAFLUORIDE.   ENTHALPY OF FORMATION OF TUNGSTEN PENTAFLUORIDE AND 1151
ENTHALPY OF FORMATION OF URANIUM TETRAFLUORIDE.   1171
FLUORINE BOMB CALORIMETRY.   ENTHALPY OF FORMATION OF VANADIUM PENTAFLUORIDE BY 1215
ENTHALPY CF FORMATION OF YTTRIUM TRIFLUORIDE.   1319
ENTHALPY OF FORMATION PHOSPHORUS TRIFLUORIDE.   0818
NEPTUNIUM HEXAFLUORIDE FROM 7 TO 350 DEG/ HEAT CAPACITY, ENTHALPY OF FUSION, AND THERMODYNAMIC PROPERTIES OF 0650
/ENTAFLUORIDE. HEAT CAPACITY BETWEEN 5 AND 350 DEGREES K. ENTHALPY OF FUSION AND VAPORIZATION, STANDARD ENTR/ 0510
0 DEGREES K.   HEAT CAPACITY, ENTROPY, AND ENTHALPY OF PLUTONIUM-242 TRIFLUORIDE FROM 10 TO 35 0886
ROLE OF CORRELATION OF ENTROPY AND ENTHALPY / ENTROPY AND ENTHALPY OF SUBLIMATION OF GADOLINIUM TRIFLUORIDE. 0456
THE THERMODYNAMICS OF SUBLIMATION OF THE GROUP-4A DIFLU/ ENTHALPY OF SUBLIMATION OF GERMANIUM DIFLUORIDE AND 0471
/IES OF PLUTONIUM COMPOUNDS AT HIGH TEMPERATURES. PART-2: ENTHALPY OF SUBLIMATION OF PLUTONIUM TRIFLUORIDE A/ 0884
REES K. ENTHALPY AND ENTROPY OF FUSION.   ENTHALPY OF URANIUM TETRAFLUORIDE FROM 298-1400 DEG 1169
TIES OF SELENIUM MONOFLUORIDE FROM A MOLECU/ DISSOCIATION ENTHALPY, SPECTROSCOPIC CONSTANTS, AND OTHER PROPER 0955
AND CERIUM FLUORIDES. CORRELATION OF ENTROPY AND ENTHALP/ ENTROPIES AND ENTHALPIES OF SUBLIMATION OF CALCIUM 0281
M AND TERBIUM TRIFLUORIDES.   ENTROPIES AND ENTHALPIES OF SUBLIMATION OF NEODYMIU 0645

NOHALIDES ABOVE ROOM TEMPERATURE. (THALLIUM / ENTHALPIES, ENTROPIES, AND FREE ENERGY FUNCTIONS OF THALLIUM MO 1116
/OF PLUTONIUM-244 DIOXIDE FROM 4 TO 25 DEGREES K. DERIVED ENTROPIES AND OTHER THERMODYNAMIC PROPERTIES AT 29/ 0885
 FUNCTIONS FROM 5 TO 350 DEGREES K. RECONCILIATION OF THE ENTROPIES FROM MOLECULAR AND THERMAL DATA. /DYNAMIC 1275
RA. (SCANDIUM, YTTRIUM, LANTHANUM, CERIUM/ GEOMETRIES AND ENTROPIES OF METAL TRIFLUORIDES FROM INFRARED SPECT 0943
FLUORID/ LOW TEMPERATURE HEAT CAPACITIES, ENTHALPIES, AND ENTROPIES OF URANIUM TETRAFLUORIDE AND URANIUM HEXA 1166
BLIMATION OF CALCIUM AND CERIUM FLUORIDES. CORRELATION OF ENTROPY AND ENTHALPY IN ERRORS. /D ENTHALPIES OF SU 0281
FROM 10 TO 350 DEGREES K.   HEAT CAPACITY, ENTROPY, AND ENTHALPY OF PLUTONIUM-242 TRIFLUORIDE 0886
RIFLUORIDE. ROLE OF CORRELATION OF ENTROPY AND ENTHALPY / ENTROPY AND ENTHALPY OF SUBLIMATION OF GADOLINIUM T 0456
FLUORIDE AND NIOBIUM PENTAFLUORIDE.   HEAT CAPACITY, ENTROPY, AND HEATS OF TRANSITION OF MOLYBDENUM HEXA 0662
/THOD OF DETERMINING SATURATED LIQUID AND SATURATED VAPOR ENTROPY. (CHLORINE MONOFLUORIDE, OXYGEN MONOFLUORI/ 0391
2 TETRAFLUORIDE FROM 10 TO 350 DEGREES K.   HEAT CAPACITY, ENTROPY, ENTHALPY, AND GIBBS ENERGY OF PLUTONIUM-24 0889
NATION OF METHANE AND FLUOROFORM. DISSOCIATION ENERGY AND ENTROPY OF CARBON TRIFLUORIDE. /RICES. PHOTOCHLORI 0311
                                               HEAT AND ENTROPY OF FORMATION OF BORON FLUORIDE. 0154
ANIUM TETRAFLUORIDE FROM 298-1400 DEGREES K. ENTHALPY AND ENTROPY OF FUSION.   ENTHALPY OF UR 1169
/DEGREES K, ENTHALPY OF FUSION AND VAPORIZATION, STANDARD ENTROPY OF THE VAPOR, AND OTHER THERMODYNAMIC PROP/ 0510
L PROPERTIES OF BERYLLIUM COMPOUNDS. PART-1: ENTHALPY AND ENTROPY OF VAPORIZATION OF BERYLLIUM DIFLUORIDE. /A 0138
LUORIDE.   ENVIRONMENT OF THE TIN ATOM IN ORTHORHOMBIC TIN DIF 1139
                                                EPR SEE ELECTRON PARAMAGNETIC RESONANCE
          SIMPLE METHOD FOR EXACT SOLUTION OF 2 X 2 SECULAR EQUATION IN MOLECULAR VIBRATIONS. 1386
             NEW METHOD FOR RESOLVING WILSON'S SECULAR EQUATIONS. APPLIED TO TETRAHEDRAL MOLECULES. 1431
LYBDENUM TUNGSTEN/ THERMODYNAMIC SIMILARITY AND UNIVERSAL EQUATIONS OF STATE OF HEXAFLUORIDES. (OF SULFUR, MO 0632
FLUORIDE, AND ANTIMONY PENTAFLUORIDE. TH/ POLYMER MONOMER EQUILIBRIA IN NIOBIUM PENTAFLUORIDE, TANTALUM PENTA 0076
/ECULAR ORBITAL METHODS. PART-5: AB INITIO CALCULATION OF EQUILIBRIUM GEOMETRIES AND QUADRATIC FORCE CONSTAN/ 0720
NITROGEN TRIFLUORIDE IN THE EXCITED VIBRATIONAL STATES. (EQUILIBRIUM STRUCTURE)   MICROWAVE SPECTRA OF 0722
TIONALLY EXCITED STATES, NU-1- 2 NU-2 FERMI RESONANCE AND EQUILIBRIUM STRUCTURE. / OXYGEN DIFLUORIDE IN VIBRA 0771
/OF SILICON DIFLUORIDE IN THE EXCITED VIBRATIONAL STATES, EQUILIBRIUM STRUCTURE, ANHARMONIC POTENTIAL FUNCTI/ 0993
                                                EQUILIBRIUM STUDIES OF CHLORINE PENTAFLUORIDE. 0417
      VAPOR PRESSURE OF URANIUM HEXAFLUORIDE IN EQUILIBRIUM WITH URANIUM PENTAFLUORIDE. 1180
R TETRAFLUORIDE, AND PHOSPHORUS/ OBSERVATION BY ESCA OF INEQUIVALENT FLUORINES IN CHLORINE TRIFLUORIDE, SULFU 0874
IUM, SAMARIUM, EUROPIUM, GADOLINIUM, TERBIUM, DYSPROSIUM, ERBIUM, THULIUM, AND YTTERBIUM. /S OF SCANDIUM, CER 1356
ITIES OF DYSPROSIUM TRIFLUORIDE, HOLMIUM TRIFLUORIDE, AND ERBIUM TRIFLUORIDE.   STABIL 0449
BLIMATION AND VAPOR PRESSURES OF DYSPROSIUM, HOLMIUM, AND ERBIUM TRIFLUORIDES. /IGH TEMPERATURES. PART-17: SU 0448
CERIUM FLUORIDES. CORRELATION OF ENTROPY AND ENTHALPY IN ERRORS. /D ENTHALPIES OF SUBLIMATION OF CALCIUM AND 0281
IFLUORIDE. ROLE OF CORRELATION OF ENTROPY AND ENTHALPY IN ERRORS. /D ENTHALPY OF SUBLIMATION OF GADOLINIUM TR 0456
TRIFLUORIDES.   ANALYSES OF THERMOCHEMICAL ERRORS AND SYSTEMATICS IN SUBLIMATION OF LANTHANIDE 1353
/ IN THE XENON FLUORIDES AND XENON OXIDE TETRAFLUORIDE BY ESCA AND EVIDENCE FOR ORBITAL INDEPENDENCE IN THE / 1229
          MOLECULAR SPECTROSCOPY BY MEANS OF ESCA. (CARBON TETRAFLUORIDE) 0328
IDE, SULFUR TETRAFLUORIDE, AND PHOSPHORUS/ OBSERVATION BY ESCA OF INEQUIVALENT FLUORINES IN CHLORINE TRIFLUOR 0874
                                                ESR SEE ELECTRON SPIN RESONANCE
TETRAFLUORIDE, SILICON TETRAFLUORIDE, BORON TRIFLUORIDE, ETHANE, SILANE, AMMONIA, AND PHOSPHINE. / OF CARBON 0243
COORDINATES IN VIBRATIONAL ANHARMONICITY. APPLICATION TO ETHYLENE.   TORSIONAL 1485
ORO ETHYLENE, AND / IONIZATION POTENTIALS FOR TETRAFLUORO ETHYLENE, TRIFLUORO CHLORO ETHYLENE, DIFLUORO DICHL 0305
S AT HIGH TEMPERATURES. PART-13: STABILITIES OF SAMARIUM, EUROPIUM, AND GADOLINIUM MONO AND DIFLUORIDES. /DIE 0450
PECTRA. (SCANDIUM, YTTRIUM, LANTHANUM, CERIUM, NEODYMIUM, EUROPIUM, AND GADOLINIUM TRIFLUORIDES) / INFRARED S 0943
          INFRARED SPECTRUM OF HEAVY METAL DIHALIDES. (EUROPIUM DIFLUORIDE) 0451
     PHASE INVESTIGATIONS AND VAPOR PRESSURE MEASUREMENTS ON EUROPIUM DIFLUORIDE. 0452
/TIONS OF THE TRIFLUORIDES OF SCANDIUM, CERIUM, SAMARIUM, EUROPIUM, GADOLINIUM, TERBIUM, DYSPROSIUM, ERBIUM,/ 1356
          MAGNETIC SUSCEPTIBILITY OF EUROPIUM TRIFLUORIDE. 0454
      OPTICAL ABSORPTION AND FLUORESCENCE SPECTRA OF EUROPIUM TRIFLUORIDE. 0453
          VAPOR PRESSURE, HEAT OF SUBLIMATION, AND EVAPORATION COEFFICIENT OF LANTHANUM TRIFLUORIDE. 0576
      VAPOR PRESSURE, HEAT OF SUBLIMATION, AND THE EVAPORATION COEFFICIENT OF PRASEODYMIUM TRIFLUORIDE 0896
AR VIBRATIONS.   SIMPLE METHOD FOR EXACT SOLUTION OF 2 X 2 SECULAR EQUATION IN MOLECUL 1386
YDROLYSIS AND INTRA- INTERMOLECULAR MECHANISM OF FLUORINE EXCHANGE. /UR, PHOSPHORUS, AND SILICON FLUORIDES. H 1042
SPHATE AND BORON TRIFLUORIDE OR / INTERMOLECULAR FLUORINE EXCHANGE BETWEEN TETRABUTYL AMMONIUM HEXAFLUORO PHO 0194
               INTRAMOLECULAR EXCHANGE IN PHOSPHORUS PENTAHALIDE MOLECULES. 0852
EPTAFLUORIDE AND IODINE HEPTAFLUOR/ INTRAMOLECULAR LIGAND EXCHANGE IN SEVEN COORDINATE STRUCTURES. (RHENIUM H 0922
               FLUORINE EXCHANGE IN SULFUR TETRAFLUORIDE. 1049
               FLUORINE EXCHANGE IN SULFUR TETRAFLUORIDE. 1053
          INTRAMOLECULAR EXCHANGE IN SULFUR TETRAFLUORIDE. 1046
          PROBLEM OF FLUORINE EXCHANGE IN SULFUR TETRAFLUORIDE. 1055
          INTRAMOLECULAR EXCHANGE IN VANADIUM PENTAFLUORIDE. 1214
/BROMINE TRIFLUORIDE, AND BROMINE PENTAFLUORIDE. FLUORINE EXCHANGE MECHANISM OF LIQUID CHLORINE TRIFLUORIDE,/ 0399
                                                EXCHANGE OF FLUORINE IN PHOSPHORUS PENTAFLUORIDE. 0836
PENTAFLUORIDE AND THE HEIGHT OF THE BARRIER TO INTERNAL EXCHANGE OF FLUORINE NUCLEI. /AFLUORIDE AND ARSENIC 0856
MR. (SULFUR TETRAFLUORIDE AND SELENIUM TETR/ STRUCTURE OF EXCHANGE PROCESSES IN SOME INORGANIC FLUORIDES BY N 1048
ES.   PENTACOORDINATED MOLECULES. PART-12: CORRELATION OF EXCHANGE RATES FOR SOME GROUP-5 PENTAHALIDE MOLECUL 0115
          CHEMICAL APPLICATIONS OF CORE ELECTRON EXCITATION SPECTROSCOPY. (XENON HEXAFLUORIDE) 1306
RIDE, AND BARIUM DIFLUORIDE IN THE/ SPECTRA OF ELECTRONIC EXCITATIONS IN CALCIUM DIFLUORIDE, STRONTIUM DIFLUO 0279
                             THERMALLY EXCITED EMISSION SPECTRUM OF BARIUM MONOFLUORIDE. 0123
AND/ RADIATIVE LIFETIMES OF ULTRAVIOLET EMISSION SYSTEMS EXCITED IN BORON TRIFLUORIDE, CARBON TETRAFLUORIDE, 0216
UM TRIFLUORIDE.   LASER EXCITED RAMAN SPECTRA OF MATRIX ISOLATED PRASEODYMI 0895
MICROWAVE SPECTRUM OF OXYGEN DIFLUORIDE IN VIBRATIONALLY EXCITED STATES, NU-1- 2 NU-2 FERMI RESONANCE AND E/ 0771
RON FLUORIDE.   VALENCE EXCITED STATES OF NITROGEN, CARBON MONOXIDE, AND BO 0169
E.   ROTATIONAL STRUCTURE IN SOME HIGHER EXCITED STATES OF THE GERMANIUM(I) FLUORIDE MOLECUL 0467
     MICROWAVE SPECTRUM OF ARSENIC TRIFLUORIDE IN THE EXCITED VIBRATIONAL STATE. 0097
     MICROWAVE SPECTRA OF NITROGEN TRIFLUORIDE IN THE EXCITED VIBRATIONAL STATES. (EQUILIBRIUM STRUCTURE) 0722
ANHARMON/ MICROWAVE SPECTRUM OF SILICON DIFLUORIDE IN THE EXCITED VIBRATIONAL STATES, EQUILIBRIUM STRUCTURE, 0993
RESONA/ MICROWAVE SPECTRA OF NITROGEN TRIFLUORIDE IN THE EXCITED VIBRATIONAL STATES. (ZETA TYPE DOUBLING AND 0723
VIBRONIC AND SPIN ORBIT SPLITTING IN SPECTRA OF SYSTEMS EXHIBITING JAHN-TELLER EFFECT. 1480
MINATION OF ITS STABILITY. (LANTHAN/ DEMONSTRATION OF THE EXISTENCE OF DILANTHANUM HEXAFLUORIDE GAS AND DETER 0581
ND SPECIES IN THE GAS PHASE.   PROBABLE NONEXISTENCE OF XENON MONOFLUORIDE AS A CHEMICALLY BOU 1221
          ELECTRONIC STRUCTURE AND ANOMALOUS THERMAL EXPANSION ON IRON DIFLUORIDE AND VANADIUM DIOXIDE. 0543
ON TETRAFLUORIDE AND OTHER CARBON HALIDES.   EXPLOSION METHOD AND THE HEATS OF FORMATION OF CARB 0315
/ ASPECTS OF FORCE FIELDS. PART-1. POINTS-ON-A-SPHERE AND EXTENDED HUCKEL MOLECULAR ORBITAL ANALYSES OF TRIG/ 0828
(OXYGEN DIFLUORIDE)   EXTENDED HUCKEL THEORY AND THE SHAPE OF MOLECULES. 0758
L HEXAFLUORIDES.   EXTREMAL PROPERTIES OF FORCE CONSTANTS IN OCTAHEDRA 1468
          ABSORPTION OF BORON TRIFLUORIDE GAS IN THE EXTREME ULTRAVIOLET. 0215
     ELECTRONIC SPECTRA OF LANTHANUM FLUORIDE IN THE EXTREME ULTRAVIOLET. 0570

F

THERMAL STABILITY AND REACTIVITY OF D- AND F ELEMENT HEXAFLUORIDES. 0630
FLUORIDE, AND GERMANIU/ CALCULATION OF THE FRANCK-CONDON FACTOR FOR MOLECULES OF SILICON FLUORIDE, GERMANIUM 0977
/ASUREMENTS OF ZEEMAN EFFECT IN DIAMAGNETIC MOLECULES. GJ FACTOR OF THALLIUM FLUORIDE, CESIUM FLUORIDE, CESI/ 1120
DIFFRACTION. (MOLYBDENUM HEXAFLUORIDE AN/ STUDIES ON THE FAILURE OF THE FIRST BORN APPROXIMATION IN ELECTRON 0640
DIFFRACTION. (TELLURIUM HEXAFLUORIDE) FAILURE OF THE FIRST BORN APPROXIMATION IN ELECTRON 1109
DIFFRACTION. (TUNGSTEN HEXAFLUORIDE, URANIUM HEXAFLUORI/ FAILURE OF THE FIRST BORN APPROXIMATION IN ELECTRON 1158
DIFFRACTION. (URANIUM HEXAFLUORIDE) FAILURE OF THE FIRST BORN APPROXIMATION IN ELECTRON 1199
MICROWAVE MEASUREMENTS OF THE J-8 FAR-9 TRANSITION OF BROMINE PENTAFLUORIDE. 0261
FERMI DIAD OF NU-1 AND 2NU-2 IN OXYGEN DIFLUORIDE. 0772
DIFLUORIDE IN VIBRATIONALLY EXCITED STATES, NU-1- 2 NU-2 FERMI RESONANCE AND EQUILIBRIUM STRUCTURE. / OXYGEN 0771
/RATIONAL SPECTRA AND ASSIGNMENT OF IODINE PENTAFLUORIDE. FERMI RESONANCE BETWEEN A FUNDAMENTAL AND A TERNAR/ 0504
FERMI RESONANCE IN BORON TRIFLUORIDE. 0242
ABSENCE OF FERMI RESONANCE IN KRYPTON DIFLUORIDE. 0564
TITANIUM DIOXIDE, MAGNESIUM FLUORIDE, ZINC FLUORIDE, AND FERROUS FLUORIDE. /ELDS OF THE RUTILE COUNTERPARTS: 0606
FLUORIDE WITH ALTERATIONS IN ITS QUADRATIC VALENCE FORCE FIELD. /R EFFECTS IN THE VIBRATION SPECTRA OF BORON 1404
STEP METHOD IN THE RANGE OF THE GENERALIZED VALENCE FORCE FIELD. /EPTAFLUORIDE CALCULATED BY THE LOGARITHMIC 0535
INFRARED SPECTRUM OF NITROGEN TRIFLUORIDE AND FORCE FIELD. 0697
POTENTIAL FIELD AND FORCE CONSTANTS FOR SOME XY6 MOLECULES. 1462
AND NIOBIUM PENTAFLUORIDE. POTENTIAL FIELD AND FORCE CONSTANTS OF ANTIMONY PENTAFLUORIDE 0086
POTENTIAL FIELD AND FORCE CONSTANTS OF OCTAHEDRAL MOLECULES. 1464
DAL PENTAHALIDES. (PHOSPHORUS PENTAFLUORIDE AN/ POTENTIAL FIELD AND FORCE CONSTANTS OF SOME TRIGONAL BIPYRAMI 0850
ON EFFECTS IN SULFUR DIFLUORIDE. CALCULATION CF THE FORCE FIELD AND INFRARED SPECTRUM.   CENTRIFUGAL DISTORTI 1031
DE. FORCE FIELD AND MOLECULAR CONSTANTS OF NITROGEN TRIFLUORI 0735
RAMIDAL THE RAMAN SPECTRUM OF PHOSPHORUS PENTA/ POTENTIAL FIELD AND MOLECULAR VIBRATIONS OF THE TRIGONAL BIPY 0863
RAMID AX(3)YZ MOLECULE. POTENTIAL FIELD AND MOLECULAR VIBRATIONS OF THE TRIGONAL BIPY 1436
S. (CARBON TETRAFLUORI/ A TRANSFERABLE UREY-BRADLEY FORCE FIELD AND THE ASSIGNMENTS OF SOME MIXED HALOMETHANE 0351
ICAL XY2 MOLECULES. UREY-BRADLEY FORCE FIELD AND THERMODYNAMIC PROPERTIES FOR BENT SYMMETR 1387
3 MOLECULES. UREY-BRADLEY FORCE FIELD AND THERMODYNAMIC PROPERTIES OF PYRAMIDAL XY 1406
MEAN SQUARE AMPLITUDES OF VIBRATION AND SHRINKAGE / FORCE FIELD, CORIOLIS COUPLING COEFFICIENTS, GENERALIZED 1278
MOTION. (XENON HEXAFLUORIDE, IODINE HEPTAFLUORI/ ELECTRIC FIELD DEFLECTION OF MOLECULES WITH LARGE AMPLITUDE 1281
MAGNETIC FIELD DEFLECTION OF XENON HEXAFLUORIDE. 1287
UNIQUE FORCE FIELD FOR BORON TRIHALIDES. 0238
MOLECULAR FORCE FIELD FOR BROMINE PENTAFLUORIDE. 0264
UREY-BRADLEY FORCE FIELD FOR CARBON TETRAFLUORIDE. 0360
MOLECULAR FORCE FIELD FOR CHLORINE TRIFLUORIDE. 0402
RIDE AND BROMINE TRIFLUORIDE) MOLECULAR FORCE FIELD FOR INTERHALOGEN COMPOUNDS. (CHLORINE TRIFLUO 0407
IDE, IODINE OXYGEN PENTAFLUORIDE, AND IO/ MOLECULAR FORCE FIELD FOR INTERHALOGEN COMPOUNDS. IODINE PENTAFLUOR 0511
INFRARED INTENSITIES AND FORCE FIELD FOR NITROGEN TRIFLUORIDE. 0714
FORCE FIELD FOR SILICON TETRAFLUORIDE. 1006
DE AND ARSENIC TRIFLUORIDE) MOLECULAR FORCE FIELD FOR SOME XY3 MOLECULES. (PHOSPHORUS TRIFLUORI 1400
LLURIUM HEXAFLUORIDES) MOLECULAR FORCE FIELD FOR SOME XY6 TYPE MOLECULES. (SELENIUM AND TE 1465
MOLECULAR FORCE FIELD FOR SULFUR TETRAFLUORIDE. 1050
UREY-BRADLEY FORCE FIELD FOR SULFUR TETRAFLUORIDE. 1059
GENERAL VALENCE FORCE FIELD FOR XENON TETRAFLUORIDE. 1265
UREY-BRADLEY FORCE FIELD FOR XY4 TETRAHEDRAL MOLECULES. 1429
VIBRATION OF SOME TRIGONAL BI/ GENERALIZED VALENCE FORCE FIELD FORCE CONSTANTS AND MEAN SQUARE AMPLITUDES OF 0878
/EASURED ELECTRONIC PROPERTIES. PART-2: DIPOLE MOMENT AND FIELD GRADIENT OF NITROGEN TRIFLUORIDE. PART-3: A / 0746
UREY-BRADLEY FORCE FIELD. (LEAD TETRAFLUORIDE) 0596
INE PENTAFLUORIDE MOLECULE BY THE MO LCAO SELF CONSISTENT FIELD METHOD IN AN NDDO APPROXIMATION. / OF A CHLOR 0415
ERGY LEVELS AND MOLECULAR GEOMETRY. PART-2: COMP/ CRYSTAL FIELD MODEL STUDY OF XENON HEXAFLUORIDE. PART-1: EN 1311
IN PERFECT PAIRING V/ ORBITAL MODIFICATION BY THE COULOMB FIELD OF LIGAND ATOMS OF LATER SECOND ROW ELEMENTS 0862
VIBRATIONAL SPECTRUM AND FORCE FIELD OF MOLYBDENUM HEXAFLUORIDE. 0634
DE, AND ARSENIC TRIFLUORIDE BY THE COMBINE/ GENERAL FORCE FIELD OF NITROGEN TRIFLUORIDE, PHOSPHORUS TRIFLUORI 0717
E˜ FORCE FIELD OF PHOSPHORUS PENTAFLUORIDE. 0861
ATION OF FORCE CONSTANTS. APPLICATION TO E˜ SPECIES FORCE FIELD OF PHOSPHORUS PENTAFLUORIDE. /D IN THE CALCUL 0827
RAMAN INTENSITIES AND THE FORCE FIELD OF PHOSPHORUS PENTAFLUORIDE. 0844
UREY-BRADLEY FORCE FIELD OF SOME XY6 SYSTEMS. 1469
STUDIES OF VAPOR BAND CONTOURS. APPLICATION TO THE FORCE FIELD OF TUNGSTEN HEXAFLUORIDE. /MPERATURE INFRARED 1156
VIBRATIONAL SPECTRUM AND FORCE FIELD OF URANIUM HEXAFLUORIDE. 1196
LUORIDES AND THE L MATRIX METHOD. FORCE FIELD OF WEINSTOCK AND GOODMAN FOR OCTAHEDRAL HEXAF 1456
FORCE FIELD STUDY FOR SULFUR TETRAFLUORIDE. 1052
ETRAFLUORIDE, XENON HEXAFLUORIDE. SELF CONSISTENT FIELD STUDY OF THE SERIES XENON DIFLUORIDE, XENON T 1224
ADAPTATION OF THE UREY-BRADLEY FORCE FIELD TO AMMONIA, PHOSPHINE, STIBINE, AND ARSINE. 1492
NS. APPLICATION OF UREY-BRADLEY FORCE FIELD TO SOME TETRAHEDRAL XY4 TYPE MOLECULES AND IO 1425
VIRIAL THEOREM DECOMPOSITION OF MOLECULAR FORCE FIELDS. 0783
MONOMER AND DIMER. HARMONIC FORCE FIELDS AND MEAN AMPLITUDES FOR ALUMINUM TRIFLUORIDE 0038
UORIDES. FORCE FIELDS FOR GROUP-4 TETRAFLUORIDES AND GROUP-5 TRIFL 0715
LUORIDE AND BROMINE PENTAFLUORIDE) MOLECULAR FORCE FIELDS FOR INTERHALOGEN COMPOUNDS. (CHLORINE PENTAF 0425
NS OF MEAN AMPLITUDES OF VIBRATION AND UREY-BRADLEY FORCE FIELDS FOR PLANAR XY3 MOLECULES.   CALCULATIO 1399
FLUORIDE AND TELLURIUM HEXAFLUORIDE) FORCE FIELDS FOR SOME GROUP-6 HEXAFLUORIDES. (SULFUR HEXA 1061
MOLECULAR FORCE FIELDS FOR SOME HEXAFLUORIDES. 1452
FORCE FIELDS FOR THE BORON TRIHALIDES. 0227
INTRAMOLECULAR FORCE FIELDS IN GROUP-5 TRIHALIDES. 1402
FORCE FIELDS IN KRYPTON DIFLUORIDE AND XENON DIFLUORIDE. 0556
ORIDE. FORCE FIELDS OF CARBON TETRAFLUORIDE AND SILICON TETRAFLU 0324
UNIQUE FORCE FIELDS OF SOME XY3 TYPE MOLECULES. 1403
, MAGNESIUM FLUORIDE, ZINC FLUORIDE, AND FERROUS F/ FORCE FIELDS OF THE RUTILE COUNTERPARTS: TITANIUM DIOXIDE 0606
ON TRIFLUORIDE, PHOSPH/ AMBIGUITIES IN THE HARMONIC FORCE FIELDS OF XY3 MOLECULES. (NITROGEN TRIFLUORIDE, BOR 0709
MOLECULAR FORCE FIELDS OF XY6 MOLECULES. 1443
KEL MOLECULAR ORBITAL / GILLESPIE-NYHOLM ASPECTS OF FORCE FIELDS. PART-1. POINTS-ON-A-SPHERE AND EXTENDED HUC 0828
EDRAL XY6 MOLECULES. (SULFUR HEXAFLUORID/ MOLECULAR FORCE FIELDS. PART-8: VIBRATION FREQUENCIES OF SOME OCTAH 1080
Y MONOFLUORIDE. FINE STRUCTURE ANALYSIS OF THE C1 SYSTEM OF ANTIMON 0059
S MONOFLUORIDE. ROTATIONAL FINE STRUCTURE OF THE TRIPLET PI STATE OF PHOSPHORU 0787
ORIDE. OCTAHEDRAL FINE STRUCTURE SPLITTINGS IN NU-3 OF SULFUR HEXAFLU 1063
MOLYBDENUM HEXAFLUORIDE AN/ STUDIES ON THE FAILURE OF THE FIRST BORN APPROXIMATION IN ELECTRON DIFFRACTION. (0640
TELLURIUM HEXAFLUORIDE) FAILURE OF THE FIRST BORN APPROXIMATION IN ELECTRON DIFFRACTION. (1109
TUNGSTEN HEXAFLUORIDE, URANIUM HEXAFLUORI/ FAILURE OF THE FIRST BORN APPROXIMATION IN ELECTRON DIFFRACTION. (1158

URANIUM HEXAFLUORIDE)   FAILURE OF THE FIRST BORN APPROXIMATION IN ELECTRON DIFFRACTION. (1199
FLUORIDE, SILICON OXIDE AND GERMANIUM OXIDE.   FIRST IONIZATION POTENTIALS OF THE MOLECULES BORON 0160
/Y ELECTRIC DEFLECTION AND MASS SPECTROMETRIC DETECTION. (FIRST ROW FLUORIDES, BERYLLIUM, MAGNESIUM, AND LEA/ 0134
    VIBRATIONAL FREQUENCIES AND VALENCE FORCE CONSTANTS OF FIRST ROW TRANSITION METAL DIFLUORIDES.   1371
HALIDES. (BERYLLIUM DIFLUORIDE, MAGNESIUM DIFLUORIDE, AND FIRST ROW TRANSITION METAL FLUORIDES) /MMETRICAL DI 1372
F THE TRANSITION METAL DIHALIDES. THE DIFLUORIDES OF SOME FIRST ROW TRANSITION METALS.   GEOMETRY O 1368
/BRATION FOR SOME LINEAR SYMMETRICAL DIFLUORIDES. (OF THE FIRST ROW TRANSITION METALS, ALSO BERYLLIUM DIFLUO/ 1369
HIRD ROW T/ HALIDES OF THE TRANSITION ELEMENTS. VOLUME-1: FIRST ROW TRANSITION METALS. VOLUME-2: SECOND AND T 0001
/CARBON MONOFLUORIDE AND CARBON DIFLUORIDE FROM MOLECULAR FLOW EFFUSION AND MASS SPECTROMETER INVESTIGATIONS. 0286
                          OPTICAL ABSORPTION AND FLUORESCENCE SPECTRA OF EUROPIUM TRIFLUORIDE.   0453
N FLUORIDE.   FLUORESCENCE SPECTRA OF PHOSPHORUS NITRIDE AND BORO 0167
E VIBRATIONAL WAVE FUNCTIONS AND DERIVED QUANTITIES. (TIN FLUORIDE)   ACCURACY OF TH 1134
CALCULATION OF RYDBERG LEVELS IN NITROGEN OXIDE AND BORON FLUORIDE.   0164
    FLUORESCENCE SPECTRA OF PHOSPHORUS NITRIDE AND BORON FLUORIDE.   0167
            HEAT AND ENTROPY OF FORMATION OF BORON FLUORIDE.   0154
                    MICROWAVE SPECTRUM OF BORON FLUORIDE.   0165
URE VAPORS. PART-3: MONOMER AND DIMER SPECTRA OF THALLOUS FLUORIDE. /TOELECTRON SPECTROSCOPY OF HIGH TEMPERAT 1117
MODYNAMIC PROPERTIES OF SOME HALOGEN MOLECULES. (CHLORINE FLUORIDE)   PHOTOIONIZATION STUDIES AND THE THER 0377
RAVIOLET REGION 1700-2500 ANGSTROMS. (ALUMINUM- MAGNESIUM FLUORIDE) / MULTIPLE BEAM INTERFEROMETRY IN THE ULT 0042
                    ROTATIONAL SPECTRUM OF IODINE FLUORIDE.   0497
CTORS AND R CENTROIDS OF THE A-X BAND SYSTEM OF BERYLLIUM FLUORIDE.   RYDBERG-KLEIN-REES FRANCK-CONDON FA 0127
FLUORIDE AND IODINE HEXAFLUORIDE(X) IN ANHYDROUS HYDROGEN FLUORIDE.   SPECTROSCOPIC STUDY OF IODINE HEPTA 0518
    STARK EFFECT FOR THALLIUM MONOFLUORIDE AND POTASSIUM FLUORIDE.   1118
            SUBLIMATION THERMODYNAMICS OF ZINC(II) FLUORIDE.   1321
                THE ELECTRONIC SPECTRUM OF BORON FLUORIDE.   0172
FLUORINE COMPOUNDS OF CARBON MONOFLUORIDE AND TETRACARBON FLUORIDE. / AND SYNTHETIC STUDIES OF THE GRAPHITE 0296
CE EXCITED STATES OF NITROGEN, CARBON MONOXIDE, AND BORON FLUORIDE.   VALEN 0169
    VAPOR PRESSURE AND HEAT OF SUBLIMATION OF CALCIUM FLUORIDE.   0282
                    VAPOR PRESSURE OF ZIRCONIUM FLUORIDE.   1331
    REEVALUATION OF THE DISSOCIATION ENERGY OF CALCIUM FLUORIDE. (ALSO SILICON AND GERMANIUM FLUORIDES) 0277
ON OF BERYLLIUM CHLOR/ DISSOCIATION ENERGIES OF BERYLLIUM FLUORIDE AND BERYLLIUM CHLORIDE AND HEAT OF FORMATI 0128
    MAGNETIC PROPERTIES OF THE BORON FLUORIDE AND BORON HYDRIDE MOLECULES.   0159
/RT-3: THE ISOELECTRONIC NITROGEN, CARBON MONOXIDE, BORON FLUORIDE, AND CARBON, BERYLLIUM OXIDE, LITHIUM FLU/ 0150
    THERMODYNAMICS OF THE VAPORIZATION OF ALUMINUM FLUORIDE AND CESIUM FLUORIDE.   0052
MOLECULAR NITROGEN/ INTERACTION OF MATRIX ISOLATED NICKEL FLUORIDE AND NICKEL CHLORIDE WITH CARBON MONOXIDE, 1324
CTURE OF MATRIX ISOLATED THALLOUS HALIDE DIMERS, THALLOUS FLUORIDE, AND THALLOUS CHLORIDE. / SPECTRA AND STRU 1123
/ECTRONIC STRUCTURES OF BORON TRIHYDRIDE, BORON DIHYDRIDE FLUORIDE, BORON HYDRIDE DIFLUORIDE, AND BORON TRIF/ 0241
SOCIATION ENERGIES OF DIATOMIC HALOGEN FLUORIDES. (IODINE FLUORIDE, BROMINE FLUORIDE)   DIS 0494
/N EFFECT IN DIAMAGNETIC MOLECULES. GJ FACTOR OF THALLIUM FLUORIDE, CESIUM FLUORIDE, CESIUM CHLORIDE, CESIUM/ 1120
LUORIDE MOLECULES.   ABSORPTION SPECTRA OF CHLORINE FLUORIDE, CHLORINE TRIFLUORIDE, AND CHLORINE PENTAF 0388
    MICROWAVE INVESTIGATIONS OF METHYL FLUORIDE, FLUOROFORM, AND PHOSPHORUS TRIFLUORIDE.   0798
    MATRIX RAMAN SPECTRA OF OXYGEN DIFLUORIDE AND THE OXYGEN FLUORIDE FREE RADICAL.   ARGON 0759
EMPERATURE DROP CALORIMETRY.   ENTHALPY OF BERYLLIUM FLUORIDE FROM 456 TO 1083 DEGREES K BY TRANSPOSED T 0142
    HEAT CAPACITY AND THERMODYNAMIC PROPERTIES OF THALLIUM FLUORIDE FROM 5 TO 445 DEGREES K.   1124
C DATA.   FREQUENCIES OF ZIRCONIUM FLUORIDE (GAS) VIBRATIONS ACCORDING TO THERMODYNAMI 1329
/ION OF THE FRANCK-CONDON FACTOR FOR MOLECULES OF SILICON FLUORIDE, GERMANIUM FLUORIDE, AND GERMANIUM CHLORI/ 0977
/ROGEN, FLUORINE, BORON NITRIDE, BERYLLIUM OXIDE, LITHIUM FLUORIDE, HELIUM- NEON, CARBON MONOXIDE, AND BORON/ 0174
    INFRARED SPECTRA OF ALUMINUM FLUORIDE IN SOLID ARGON.   0054
    ELECTRONIC SPECTRA OF LANTHANUM FLUORIDE IN THE EXTREME ULTRAVIOLET.   0570
ISOTOPE EFFECT IN THE SINGLET SPECTRAL BANDS OF THE BORON FLUORIDE MOLECULE.   0166
NALYSIS OF THE 2550-2750 ANGSTROM BAND SYSTEM AT ANTIMONY FLUORIDE MOLECULE.   ROTATIONAL A 0067
RUCTURE IN SOME HIGHER EXCITED STATES OF THE GERMANIUM(I) FLUORIDE MOLECULE.   ROTATIONAL ST 0467
    ROTATIONAL STRUCTURE OF THE C3 SYSTEM OF ANTIMONY FLUORIDE MOLECULE.   0066
FECT IN THE MICROWAVE ROTATIONAL SPECTRUM OF THE ALUMINUM FLUORIDE MOLECULE.   ZEEMAN EF 0026
DS OF B DOUBLET SIGMA- X DOUBLET SIGMA. SYSTEM OF CALCIUM FLUORIDE MOLECULE IN EMISSION SPECTRUM. /OF THE BAN 0280
.   NEW BAND SYSTEM OF BISMUTH FLUORIDE MOLECULE IN THE REGION 6200-7000 ANGSTROMS 0144
IDDEN TRANSITIONS IN IODINE, OR IODINE BROMIDE (CHLORIDE, FLUORIDE) MOLECULES. / RELATIVE INTENSITIES OF FORB 0498
LECULAR ORBITAL STUDY OF ATOMIZATION ENERGIES OF PNICOGEN FLUORIDE MOLECULES AND IONS.   MO 1364
LUM BROMIDE, AND MOL/ THERMODYNAMIC FUNCTIONS OF ANTIMONY FLUORIDE, NIOBIUM BROMIDE, TANTALUM CHLORIDE, TANTA 0075
    URANIUM PENTAFLUORIDE, A NEW FLUORIDE OF URANIUM.   1178
    ELECTRON SPIN RESONANCE OF THE XENON FLUORIDE RADICAL.   1288
FIRST IONIZATION POTENTIALS OF THE MOLECULES BORON FLUORIDE, SILICON OXIDE AND GERMANIUM OXIDE.   0160
TION POTENTIALS AND ELECTRON AFFINITIES OF SOME GERMANIUM FLUORIDE SPECIES. /ND DISSOCIATION ENERGIES, IONIZA 0464
NS. (BORO/ THERMODYNAMIC PROPERTIES OF THE BORON CHLORIDE FLUORIDE SYSTEM FROM MASS SPECTROMETER INVESTIGATIO 0179
ON STUDIES, MASS SPECTRA, AND THERMODYNAMICS OF BERYLLIUM FLUORIDE VAPOR.   EFFUSI 1404
/INTERMOLECULAR EFFECTS IN THE VIBRATION SPECTRA OF BORON FLUORIDE WITH ALTERATIONS IN ITS QUADRATIC VALENCE/ 1404
S OF THE RUTILE COUNTERPARTS: TITANIUM DIOXIDE, MAGNESIUM FLUORIDE, ZINC FLUORIDE, AND FERROUS FLUORIDE. /ELD 0606
    INFRARED SPECTRA OF GASEOUS GROUP-4 HALIDES: ZIRCONIUM FLUORIDE, ZIRCONIUM CHLORIDE, AND HAFNIUM CHLORIDE. 1327
    AN LCAO MO STUDY OF RARE GAS FLUORIDES.   1341
AR BEAM SOURCE MASS SPECTROMETER TO THE STUDY OF REACTIVE FLUORIDES.   APPLICATION OF A MOLECUL 0019
RIC RESCALING OF CNDO/2 THEORY. APPLICATIONS TO INORGANIC FLUORIDES.   BICENT 1475
    BONDING IN CARBON(1) AND CARBON(2) FLUORIDES.   0300
Y LEVELS, WAVE FUNCTIONS AND AO POPULATIONS OF POLYATOMIC FLUORIDES. /O-2 (ALPHA, BETA) METHOD. PART-3: ENERG 0016
ON TERMS IN ENERGY DISTRIBUTIONS. (SELENIUM AND TELLURIUM FLUORIDES) /L MODES OF VIBRATION. PART-2: INTERACTI 1494
    COMPARATIVE AB INITIO CALCULATIONS ON BORON FLUORIDES.   0203
ELECTRONIC STRUCTURE AND CHEMISTRY OF IODINE COMPOUNDS. (FLUORIDES)   0499
    ENERGETICS AND STABILITY OF GAS PHASE SCANDIUM FLUORIDES.   0951
    HIGH TEMPERATURE NEGATIVE IONS. GASEOUS GROUP-3 FLUORIDES.   0047
    INFRARED SPECTRA OF RARE EARTH METAL FLUORIDES.   1346
SPECTROMETRIC AND THERMOCHEMICAL STUDIES OF THE MANGANESE FLUORIDES.   MASS 0612
C STUDIES OF SCANDIUM, YTTRIUM, LANTHANUM, AND RARE EARTH FLUORIDES.   MASS SPECTROMETRI 1359
    MASS SPECTROMETRIC STUDIES OF SOME GASEOUS SULFUR FLUORIDES.   1023
    MASS SPECTROMETRIC STUDIES OF SOME GASEOUS SULFUR FLUORIDES.   1028
IDE, MAGNESIUM DIFLUORIDE, AND FIRST ROW TRANSITION METAL FLUORIDES) /MMETRICAL DIHALIDES. (BERYLLIUM DIFLUOR 1372
    PHYSICAL AND CHEMICAL PROPERTIES OF HALOGEN FLUORIDES.   0017
N ENERGY OF CALCIUM FLUORIDE. (ALSO SILICON AND GERMANIUM FLUORIDES)   REEVALUATION OF THE DISSOCIATIO 0277
FLUORIDES OF OCTAHEDRAL SYMMETRY. (SELENIUM AND TELLURIUM FLUORIDES) ROOT MEAN SQUARE AMPLITUDES IN SOME HEXA 0962
    SATURATED VAPOR PRESSURE OF RARE EARTH FLUORIDES.   1357
    STRUCTURE AND MAGNETIC PROPERTIES OF RARE EARTH FLUORIDES.   1345
    SUBLIMATION PRESSURES OF REFRACTORY FLUORIDES.   1382

RIMENTAL STUDIES OF THE ELECTRONIC STRUCTURE OF THE XENON FLUORIDES.   THEORETICAL AND EXPE 1238
PROPERTIES OF NIOBIUM(V), TANTALUM(V), AND MOLYBDENUM(V) FLUORIDES.   THERMODYNAMIC 0668
THERMODYNAMIC PROPERTIES OF SOME CHALCOGEN FLUORIDES.   0012
UREY-BRADLEY POTENTIAL CONSTANTS FOR SECOND ROW FLUORIDES.   1474
RESSURES AND THERMODYNAMIC CALCULATIONS FOR SOME ACTINIDE FLUORIDES   VAPOR P 1336
ION AND SUBLIMATION THERMODYNAMICS OF SELECTED LANTHANIDE FLUORIDES.   VAPORIZAT 1347
VIBRATIONAL SPECTRA OF INORGANIC FLUORIDES.   0014
ITROGEN TRIFLUORIDE, AND PHOSPHORYL, THIONYL AND SULFURYL FLUORIDES. /LE MOMENTS IN PHOSPHORUS TRIFLUORIDE, N 0744
/TIAL ENERGY CURVES AND DISSOCIATION ENERGIES OF DIATOMIC FLUORIDES AND CHLORIDES OF GALLIUM, INDIUM, AND TH/ 0458
RIES.   BIBLIOGRAPHY ON SPECTROSCOPY OF FLUORIDES AND OXIDES BELONGING TO THE LANTHANIDE SE 1358
ND SILICON TETRAF/ ELECTRON DENSITY DISTRIBUTIONS IN SOME FLUORIDES AND OXYCHLORIDES. (CARBON TETRAFLUORIDE A 0337
AND CHLORINE OXYFLUORIDE BY NUC/ STUDY OF LIQUID CHLORINE FLUORIDES AND OXYFLUORIDES, CHLORINE PENTAFLUORIDE 0414
TRANSITION METAL FLUORIDES AND THEIR COMPLEXES.   0015
NITROGEN FLUORIDES AND THEIR INORGANIC DERIVATIVES.   0703
EVIDENCE FOR ORBITAL/ ELECTRON DISTRIBUTION IN THE XENON FLUORIDES AND XENON OXIDE TETRAFLUORIDE BY ESCA AND 1229
E, ARSENIC TRIF/ THERMODYNAMIC PROPERTIES OF SOME ARSENIC FLUORIDES. (ARSENIC MONOFLUORIDE, ARSENIC DIFLUORID 0093
DES.   THERMODYNAMIC PROPERTIES OF NIOBIUM AND TANTALUM FLUORIDES AT HIGH TEMPERATURES. PART-1: PENTAFLUORI 0665
DES AND/ THERMODYNAMIC PROPERTIES OF NIOBIUM AND TANTALUM FLUORIDES AT HIGH TEMPERATURES. PART-2: TETRAFLUORI 0657
AND MO/ THERMODYNAMIC PROPERTIES OF NIOBIUM AND TANTALUM FLUORIDES AT HIGH TEMPERATURES. PART-3: DIFLUORIDES 0654
/ DEFLECTION AND MASS SPECTROMETRIC DETECTION. (FIRST ROW FLUORIDES, BERYLLIUM, MAGNESIUM, AND LEAD DIFLUORI/ 0134
IRICAL STUDY OF THE ELECTRONIC STRUCTURE OF BORON HYDRIDE FLUORIDES. (BORON TRIFLUORIDE)   SEMIEMP 0226
DE, AND BROMINE PENTAFLU/ INFRARED STUDIES OF THE BROMINE FLUORIDES. (BROMINE MONOFLUORIDE, BROMINE TRIFLUORI 0256
SELF IONIZATION OF HALOGEN FLUORIDES. (BROMINE PENTAFLUORIDE)   0263
INFRARED SPECTRA OF MAGNESIUM AND ALUMINUM FLUORIDES BY MATRIX ISOLATION.   0608
M TETR/ STRUCTURE OF EXCHANGE PROCESSES IN SOME INORGANIC FLUORIDES BY NMR. (SULFUR TETRAFLUORIDE AND SELENIU 1048
E, STRONTIUM DIFLUORIDE, AND/ MELTING POINTS OF INORGANIC FLUORIDES. (CADMIUM DIFLUORIDE, MAGNESIUM DIFLUORID 0273
IDE, AND CHLORINE TRIFLU/ HEATS OF FORMATION OF CHLORINE FLUORIDES. (CHLORINE MONOFLUORIDE, CHLORINE DIFLUOR 0370
/PIES AND ENTHALPIES OF SUBLIMATION OF CALCIUM AND CERIUM FLUORIDES. CORRELATION OF ENTROPY AND ENTHALPY IN / 0281
BELOW 20.   HEATS OF FORMATION OF INORGANIC FLUORIDES ESPECIALLY THE ELEMENTS OF ATOMIC NUMBER 0214
ON PRESSURES. (UR/ THERMODYNAMICS OF INTERMEDIATE URANIUM FLUORIDES FROM MEASUREMENTS OF THE DISPROPORTIONATI 1176
AND TRIPLE BONDS.   LOCALIZED ORBITALS IN BORON FLUORIDES. HIGHLY POLARIZED BORON- FLUORINE DOUBLE 0158
/ FLUORINE NMR SPECTRA OF SULFUR, PHOSPHORUS, AND SILICON FLUORIDES. HYDROLYSIS AND INTRA- INTERMOLECULAR MEC 1042
DIFLUORIDE, NITROGEN TRIFL/ ABSORPTION SPECTRA OF CERTAIN FLUORIDES IN THE NEAR ULTRAVIOLET REGION. (KRYPTON 0560
INFRARED SPECTRA OF RARE EARTH METAL FLUORIDES IN THE REGION OF A CESIUM IODIDE PRISM.   1352
SULFUR / THE INFRARED SPECTRA OF SOME VOLATILE INORGANIC FLUORIDES IN THE SOLID STATE. (ARSENIC TRIFLUORIDE, 0096
DISSOCIATION ENERGIES OF DIATOMIC HALOGEN FLUORIDES. (IODINE FLUORIDE, BROMINE FLUORIDE) 0494
ETERMINATION OF THE ENTHALPIES OF FORMATION OF RARE EARTH FLUORIDES. (LANTHANUM TRIFLUORIDE)   EXPERIMENTAL D 0578
/OF DIFFERENTIAL OVERLAP METHODS ILLUSTRATED BY THE XENON FLUORIDES. MODIFIED NDDO-2 (ALPHA, BETA) METHOD AS/ 1231
PHYSICAL CHEMICAL PROPERTIES OF INORGANIC NITROGEN FLUORIDES. (NITROGEN TRIFLUORIDE)   0713
EMICAL AND THERMODYNAMIC PROPERTIES OF INORGANIC NITROGEN FLUORIDES. (NITROGEN TRIFLUORIDE)   THERMOCH 0696
F LCAO MO/ CHARACTERISTICS OF THE ELECTRONIC STRUCTURE OF FLUORIDES OF NONTRANSITION ELEMENTS FROM THE DATA O 0009
/ATION AND CRYSTAL STRUCTURE OF OSMIUM PENTAFLUORIDE. TWO FLUORIDES OF OSMIUM. (OSMIUM PENTAFLUORIDE, AND OS/ 0749
FLUORIDES OF PHOSPHORUS.   0819
NAL TRANSITION PROBABILITIES AND R CENTROIDS FOR DIATOMIC FLUORIDES OF SILICON AND GERMANIUM.   VIBRATIO 0994
FLUORIDES OF THE ACTINIDE ELEMENTS.   0006
FLUORIDES OF THE HEAVY TRANSITION METALS. (REVIEW)   0002
FLUORIDES OF THE MAIN GROUP ELEMENTS.   0007
NEW FLUORIDES OF TRIVALENT COPPER.   0444
NOBLE GASES AND THE BINARY FLUORIDES OF XENON.   1342
TRANSITION ELEMENTS.   COMPARISONS OF FLUORIDES, OXIDES, AND SULFIDES CONTAINING DIVALENT 1370
ATION INFRARED SPECTROSCOPY OF HIGH TEMPERATURE SPECIES. (FLUORIDES, OXIDES, CHLORIDES)   MATRIX ISOL 0011
FLUORINE NMR SPECTRA OF OXYGEN FLUORIDES. (OXYGEN DIFLUORIDE)   0775
OXYGEN FLUORIDES. (OXYGEN DIFLUORIDE)   0785
ILE STUDIES OF URANIUM HEXAFLUORIDE AT 193/ STRUCTURES OF FLUORIDES. PART-10: NEUTRON POWDER DIFFRACTION PROF 1202
COPPER DIFLUORIDE BY NEUTRON DIFFRACTION.   STRUCTURES OF FLUORIDES. PART-6: PRECISE STRUCTURAL PARAMETERS IN 0443
/IES OF PHOSPHORUS(2), PHOSPHORUS(4), AND SOME PHOSPHORUS FLUORIDES. (PHOSPHORUS MONOFLUORIDE, PHOSPHORUS DI/ 0788
/R BEAM SOURCE MASS SPECTROMETER TO THE STUDY OF REACTIVE FLUORIDES. (SECOND ROW, THIRD ROW, ANTIMONY PENTAF/ 1377
RIDE, BOR/ INFRARED ABSORPTION SPECTRA OF SOME POLYATOMIC FLUORIDES. (SILICON TETRAFLUORIDE, CARBON TETRAFLUO 0186
E, URANIUM TRIFLUO/ HEATS OF FORMATION OF GASEOUS URANIUM FLUORIDES. (URANIUM MONOFLUORIDE, URANIUM DIFLUORID 1160
Y OF CHLORINE MONOFLUORIDE AND THE DISSOCIATION ENERGY OF FLUORINE.   PHOTOIONIZATION STUD 0378
ENTHALPY OF FORMATION OF THORIUM TETRAFLUORIDE BY FLUORINE BOMB CALORIMETRY.   1130
ENTHALPY OF FORMATION OF VANADIUM PENTAFLUORIDE BY FLUORINE BOMB CALORIMETRY.   1215
FORMATION OF BORON TRIFLUORIDE.   FLUORINE BOMB CALORIMETRY. PART-15: THE ENTHALPY OF 0220
PY OF FORMATION OF PHOSPHORUS PENTAFLUORIDE AND ENTHALPI/ FLUORINE BOMB CALORIMETRY. PART-18: STANDARD ENTHAL 0865
FORMATION OF ALUMINUM TRIFLUORIDE.   FLUORINE BOMB CALORIMETRY. PART-22: THE ENTHALPY OF 0048
ATION OF BORON TRIFLUORIDE.   FLUORINE BOMB CALORIMETRY. PART-3: THE HEAT OF FORM 0250
XENON DIFLUORIDE AND THE NATURE OF THE XENON- FLUORINE BOND.   1223
NED BY GAS ELECTRON DIFFRACTION.   BORON FLUORINE BOND DISTANCE OF BORON TRIFLUORIDE DETERMI 0223
N TETRAFLUORIDE.   STRENGTH OF THE NITROGEN FLUORINE BOND IN NITROGEN TRIFLUORIDE AND DINITROGE 0712
ESCA AND EVIDENCE FOR ORBITAL INDEPENDENCE IN THE XENON- FLUORINE BONDING. /AND XENON OXIDE TETRAFLUORIDE BY 1229
/C MOLECULES LITHIUM, BERYLLIUM, BORON, CARBON, NITROGEN, FLUORINE, BORON NITRIDE, BERYLLIUM OXIDE, LITHIUM / 0174
/, ELECTROCHEMICAL AND SYNTHETIC STUDIES OF THE GRAPHITE FLUORINE COMPOUNDS OF CARBON MONOFLUORIDE AND TETR/ 0296
IZED ORBITALS IN BORON FLUORIDES. HIGHLY POLARIZED BORON- FLUORINE DOUBLE AND TRIPLE BONDS.   LOCAL 0158
LUORO PHOSPHATE AND BORON TRIFLUORIDE OR / INTERMOLECULAR FLUORINE EXCHANGE BETWEEN TETRABUTYL AMMONIUM HEXAF 0194
FLUORINE EXCHANGE IN SULFUR TETRAFLUORIDE.   1049
FLUORINE EXCHANGE IN SULFUR TETRAFLUORIDE.   1053
PROBLEM OF FLUORINE EXCHANGE IN SULFUR TETRAFLUORIDE.   1055
/LUORIDE, BROMINE TRIFLUORIDE, AND BROMINE PENTAFLUORIDE. FLUORINE EXCHANGE MECHANISM OF LIQUID CHLORINE TRI/ 0399
EXCHANGE OF FLUORINE IN PHOSPHORUS PENTAFLUORIDE.   0836
X-RAY SPECTRA OF SULFUR AND FLUORINE IN SULFUR HEXAFLUORIDE.   1101
GASEOUS SULFUR HEXAFLUORIDE.   SULFUR K AND L AND FLUORINE K X-RAY EMISSION AND ABSORPTION SPECTRA OF 1088
RON DIFFRACTION STUDY.   SILICON- FLUORINE LENGTH IN SILICON TETRAFLUORIDE. NEW ELECT 1000
ETWEEN ANTIMONY PENTAFLUORIDE AND NIOBIUM PENTAFLUORID/ A FLUORINE NMR AND RAMAN STUDY OF COMPLEX FORMATION B 0079
FLUORINE NMR OF ANTIMONY PENTAFLUORIDE.   0078
FLUORINE NMR OF CHLORINE PENTAFLUORIDE.   0416
FLUORINE NMR OF IODINE HEPTAFLUORIDE, RHENIUM HEPTA 0517
FLUORIDE, AND OXIDE PENTAFLUORIDES.   FLUORINE NMR OF PLATINUM HEXAFLUORIDE.   0883
SUSCEPTIBILITY MEASUREMENTS AND FLUORINE NMR OF PLATINUM HEXAFLUORIDE. 0883
IFLUORIDE)   FLUORINE NMR SPECTRA OF OXYGEN FLUORIDES. (OXYGEN D 0775
ICON FLUORIDES. HYDROLYSIS AND INTRA- INTERMOLECULAR MEC/ FLUORINE NMR SPECTRA OF SULFUR, PHOSPHORUS, AND SIL 1042
DE AND THE HEIGHT OF THE BARRIER TO INTERNAL EXCHANGE OF FLUORINE NUCLEI. /AFLUORIDE AND ARSENIC PENTAFLUORI 0856

RBON TETRAFLUORIDE, SILICON TETRAFLUORIDE, AND SULFUR HE/ FLUORINE RELAXATION BY NMR ABSORPTION IN GASEOUS CA 0346
XAFLUORIDE, TUNGSTEN HEXAFLUORIDE, / MOLECULAR MOTION AND FLUORINE RELAXATION IN THE LIQUIDS OF MOLYBDENUM HE 0638
FRARED SPECTRA OF MATRIX ISOLATED SPECIES IN THE GALLIUM- FLUORINE SYSTEM. IN 0462
THE XENON- FLUORINE SYSTEM. 1255
VAPORIZATION REACTIONS IN THE THULIUM- FLUORINE SYSTEM. 1131
N OF THE CHLORINE DIFLUORIDE FREE RADICAL. CHLORINE FLUORINE SYSTEM AT LOW TEMPERATURE. CHARACTERIZATIO 0392
RIDE, AND PHOSPHORUS/ CBSERVATION BY ESCA OF INEQUIVALENT FLUORINES IN CHLORINE TRIFLUORIDE, SULFUR TETRAFLUO 0874
SULFUR TETRAFLUORIDE. FLUORINE-19 NMR INVESTIGATION OF THE ASSOCIATION OF 1044
S PENTAFLUORIDE, PHOSPHORUS PENTACHLORIDE, AND PHOSPHORUS FLUORO CHLORIDE MOLECULES. /AMPLITUDES OF PHOSPHORU 0826
MICROWAVE INVESTIGATIONS OF METHYL FLUORIDE, FLUOROFORM, AND PHOSPHORUS TRIFLUORIDE. 0798
/TED IN INERT MATRICES. PHOTOCHLORINATION OF METHANE AND FLUOROFORM. DISSOCIATION ENERGY AND ENTROPY OF CAR/ 0311
RIDE, AND/ CALCULATED IONIZATION POTENTIAL OF CHLORO- AND FLUOROMETHANES, TETRAFLUORO METHANE XENON TETRAFLUO 0344
E AND XENON TETRAFLUCRIDE. FORBIDDEN ELECTRONIC TRANSITIONS IN XENON DIFLUORIC 1247
SION-INDUCED RAMAN SCATTERING. FORBIDDEN RAMAN BANDS OF SULFUR HEXAFLUORIDE. COLLI 1081
(CH/ SPIN ORBITAL INTERACTION AND RELATIVE INTENSITIES OF FORBIDDEN TRANSITIONS IN IODINE, OR IODINE BROMIDE 0498
OF THE MOLECULAR CHARGE DISTRIBUTION AND THE VIBRATIONAL FORCE CONSTANT. RELAXATION 0151
FORCE CONSTANT CALCULATION FOR CHLORINE TRIFLUORIDE 0403
TETRAFLUORIDE, AND XENON OXIDE TETRAFLUORIDE VAPORS, AND FORCE CONSTANT CALCULATIONS. /NON DIFLUORIDE, XENON 1254
MOLECULE AND A MODEL WITH POLARIZABLE IONS. FORCE CONSTANT OF THE NONPLANAR VIBRATION OF AN XY3 0573
EFFECT OF BORON TRIHALIDES, CALCULATED FROM UREY-BRADLEY FORCE CONSTANTS. /ITUDES OF VIBRATION AND SHRINKAGE 0233
RIFLUORIDE AND ARSENIC TRICHLORIDE DETERMINED BY ELECTRO/ FORCE CONSTANTS AND AVERAGE STRUCTURES OF ARSENIC T 0103
HEXAFLUORIDES. (OF SULFUR, SELENIUM, TELLURIUM, MOLYBDEN/ FORCE CONSTANTS AND BOND LENGTHS OF SOME INORGANIC 0631
ARE EARTH TRIFLUORIDES. (SCANDIUM TRIFLUORIDE, YTTRIUM T/ FORCE CONSTANTS AND GEOMETRIES OF MATRIX ISOLATED R 0944
COMPLEX MOLECULES OF THE XY4 TYPE OF / ESTIMATION OF FORCE CONSTANTS AND INTERATOMIC DISTANCES OF HALIDE 1414
S CONSTANT FOR SOME MOLECULES AND IONS OF D/ FREQUENCIES, FORCE CONSTANTS, AND MASS DEPENDENCE OF THE CORIOLI 1411
DINE HEPTAFLUORIDE CALCULATED BY THE LOGARITHMIC STEP ME/ FORCE CONSTANTS AND MEAN VIBRATION AMPLITUDES OF IO 0535
FORCE CONSTANTS AND NORMAL COORDINATE ANALYSIS. 1473
ITUDES OF PHOSPHORUS PENTAFLUORIDE, PHOSP/ CALCULATION OF FORCE CONSTANTS AND ROOT MEAN SQUARE VIBRATION AMPL 0826
BORON HALIDES. FORCE CONSTANTS AND THERMODYNAMIC CONSTANTS FOR THE 0219
INE PENTAFLUORIDE AND BROMINE PENTA/ MOLECULAR STRUCTURE, FORCE CONSTANTS, AND THERMODYNAMIC FUNCTIONS OF IOD 0505
NE HEPTAFLUORIDE. FORCE CONSTANTS, AND THERMODYNAMIC FUNCTIONS OF IODI 0528
ICON DIFLUORIDE. INFRARED SPECTRUM, FORCE CONSTANTS, AND THERMODYNAMIC FUNCTIONS OF SIL 0985
OMINE PENTAFLUORIDE. RAMAN SPECTRUM, STRUCTURE, FORCE CONSTANTS, AND THERMODYNAMIC PROPERTIES OF BR 0267
/PECTRA, RAMAN BAND CONTOUR ANALYSES, CORIOLIS CONSTANTS, FORCE CONSTANTS, AND VALUES FOR THERMODYNAMIC FUNC/ 0795
/VY ATOM ISOTOPIC SUBSTITUION DATA FOR THE CALCULATION OF FORCE CONSTANTS. ANHARMONICITY EFFECTS ON THE FORC/ 0234
KINEMATICAL EVALUATION OF FORCE CONSTANTS. APPLICATION TO BORON TRIHALIDES. 0248
ELD OF PHOSPHORUS/ PARAMETER METHOD IN THE CALCULATION OF FORCE CONSTANTS. APPLICATION TO E' SPECIES FORCE FI 0827
APPLICATION TO THE CALCULATION BCND MOMENT PARAMETERS AND FORCE CONSTANTS. (CARBON TETRAFLUORIDE) /RMATIONS. 0352
THE INFRARED SPECTRUM AND FORCE CONSTANTS FOR ARSENIC TRIFLUORIDE. 0105
TETRAFLUORIDE, SILICON TET/ COMPARISON OF CALCULATIONS OF FORCE CONSTANTS FOR BORON TETRAFLUORIDE(-), CARBON 0348
FORCE CONSTANTS FOR HEXAFLUORIDE MOLECULES. 1457
FORCE CONSTANTS FOR MOLECULES WITH D3H SYMMETRY. 0190
N TRIFLUORIDE. CALCULATION OF FORCE CONSTANTS FOR NITROGEN DIFLUORIDE AND NITROGE 0693
/S AND CORIOLIS CONSTANTS FOR SYMMETRIC TOP MOLECULES AND FORCE CONSTANTS FOR NITROGEN TRIFLUORIDE, PHOSPHOR/ 0708
POTENTIAL FIELD AND FORCE CONSTANTS FOR SOME XY6 MOLECULES. 1462
RAMAN SPECTRUM AND FORCE CONSTANTS FOR SULFUR HEXAFLUORIDE. 1096
FORCE CONSTANTS FOR SULFUR TETRAFLUORIDE. 1035
FORCE CONSTANTS FOR XY6 MOLECULES. 1445
ART-2: BORON TRIFLUORIDE, SILICON TETRAFLUORIDE/ ACCURATE FORCE CONSTANTS FROM HEAVY ISOTOPIC SUBSTITUTION. P 0232
ART-1. CARBON TETRAFLUCRIDE. ACCURATE FORCE CONSTANTS FROM HEAVY ISOTOPIC SUBSTITUTION. P 0320
SUBSTITUTED AND PERTURBED MOLECULES. DIRECT DETERMINATION FORCE CONSTANTS FROM THE CORIOLIS INTERACTION. /OF 1444
FLUORIDE, NI/ A THEORETICAL STUDY OF THE BOND INTERACTION FORCE CONSTANTS IN DIFLUORIDE MOLECULES. (OXYGEN DI 0299
EXTREMAL PROPERTIES OF FORCE CONSTANTS IN OCTAHEDRAL HEXAFLUORIDES. 1468
/ENIC PENTAFLUORIDE AND VANADIUM PENTAFLUORIDE AND THEIR FORCE CONSTANTS INCLUDING THOSE OF PHOSPHORUS PENT/ 0873
NKAGE EFFECT AND CORIOLIS CONSTANTS IN IODI/ UREY-BRADLEY FORCE CONSTANTS, MEAN AMPLITUDES OF VIBRATION, SHRI 0513
NKAGE EFFECTS, CORIOLIS CONSTANTS, AND THERMODYNAMIC FUN/ FORCE CONSTANTS, MEAN AMPLITUDES OF VIBRATION, SHRI 1219
EVALUATION OF FORCE CONSTANTS: MOLECULES OF THE TYPE XY6. 1470
NITIO CALCULATION OF EQUILIBRIUM GEOMETRIES AND QUADRATIC FORCE CONSTANTS. (NITROGEN TRIFLUORIDE) /RT-5: AB I 0720
UM PENTAFLUORIDE. POTENTIAL FIELD AND FORCE CONSTANTS OF ANTIMONY PENTAFLUORIDE AND NIOBI 0086
ABILITY. (BORON TRIFLUORIDE) UREY-BRADLEY FORCE CONSTANTS OF BORON HALIDES AND THEIR TRANSFER 0246
HLORIDE. MOLECULAR STRUCTURE AND FORCE CONSTANTS OF BORON TRIFLUORIDE AND BORON TRIC 0221
FECTS OF THE LENNARD-JONES POTENTIALS ON THE UREY-BRADLEY FORCE CONSTANTS OF BORON TRIHALIDES. EF 0244
RIFLUORIDE AND COMPARISON WITH FORCE CONSTANTS OF THE EI/ FORCE CONSTANTS OF CARBON TETRAFLUORIDE AND BORON T 0212
TRAFLUORIDE, BORON TRIFLUORIDE, ETHANE, SILANE, AMMONIA,/ FORCE CONSTANTS OF CARBON TETRAFLUORIDE, SILICON TE 0243
ORIDES. VIBRATIONAL FREQUENCIES AND VALENCE FORCE CONSTANTS OF FIRST ROW TRANSITION METAL DIFLU 1371
CALCULATION OF THE FORCE CONSTANTS OF HEXAFLUORIDES. 1448
F A SIMPLE ELECTROSTATIC MODEL FOR CALCULATION OF ANGULAR FORCE CONSTANTS OF HIGHLY SYMMETRICAL MOLECULES. /O 1484
FLUORIDE AND NITROGEN TRIFLUORIDE. AB INITIO STUDY OF THE FORCE CONSTANTS OF INORGANIC MOLECULES NITROGEN OXY 0734
FORCE CONSTANTS OF METAL HEXAFLUORIDES. 0018
FORCE CONSTANTS OF METAL HEXAFLUORIDES 1446
VIBRATIONS. FORCE CONSTANTS OF MOLECULES WITH STRONGLY COUPLED 1496
LONE PAIR MODEL AND THE VIBRATIONAL FORCE CONSTANTS OF NITROGEN TRIFLUORIDE. 0705
EN TRIFLUORIDE) A BENT BOND MODEL FOR THE VIBRATIONAL FORCE CONSTANTS OF NONPLANAR XY3 MOLECULES. (NITROG 0704
POTENTIAL FIELD AND FORCE CONSTANTS OF OCTAHEDRAL MOLECULES. 1464
BY THE LOGARITHMIC STEP METHOD. FORCE CONSTANTS OF OCTAHEDRAL MOLECULES CALCULATED 1472
ONAL DATA. EVALUATION OF FORCE CONSTANTS OF PLANE XY3 MOLECULES FROM VIBRATI 1401
DISTORTION. MICROWAVE SPECTRUM AND FORCE CONSTANTS OF SILICON DIFLUORIDE. CENTRIFUGAL 0989
FORCE CONSTANTS OF SMALL MOLECULES. (UREY-BRADLEY) 1495
BON TETRAFLUORIDE) EVALUATION OF THE FORCE CONSTANTS OF SOME TETRAHEDRAL MOLECULES. (CAR 0359
ALIDES. (PHOSPHORUS PENTAFLUORIDE AN/ POTENTIAL FIELD AND FORCE CONSTANTS OF SOME TRIGONAL BIPYRAMIDAL PENTAH 0850
N OXIDE TETRAFLUORIDE, I/ VIBRATIONAL SPECTRA AND VALENCE FORCE CONSTANTS OF SQUARE PYRAMIDAL MOLECULES. XENO 0418
C SUBSTITUTION. FORCE CONSTANTS OF SULFUR HEXAFLUORIDE FROM ISOTOPI 1100
FORCE CONSTANTS OF TETRAHEDRAL MOLECULES. 1422
ANIUM TETRAFLUORIDES) ORBITAL VALENCE FORCE CONSTANTS OF TETRAHEDRAL XY4 MOLECULES. (GERM 0481
FORCE CONSTANTS OF THE BORON TRIHALIDES - A SURVEY. 0201
TETRAFLUORIDE) FORCE CONSTANTS OF THE CARBON TETRAHALIDES. (CARBON 0354
IONIC MODEL CALCULATIONS. PART-2: BENDING FORCE CONSTANTS OF THE GROUP-2 HALIDES. 1381
UORIDE, CADMIUM DIFL/ POLARIZED ION MODEL AND THE BENDING FORCE CONSTANTS OF THE GROUP-2B HALIDES. (ZINC DIFL 1322
TECHNETIUM, AND RHENIUM. VALENCE FORCE CONSTANTS OF THE HEXAFLUORIDES OF MOLYBDENUM, 0641
QUENCIES OF CHLORO BROMO SILICON. FORCE CONSTANTS OF THE SILICON TETRAHALIDES AND FRE 1004

ES. (PHOSPHORUS PENTAFLUORIDE)   UREY-BRADLEY FORCE CONSTANTS OF TRIGONAL BIPYRAMIDAL XY5 MOLECUL 0838
                              VIBRATIONAL SPECTRA AND FORCE CONSTANTS OF URANIUM TETRAFLUORIDE.   1172
FLUORIDE, SILICON TETRAFLUORIDE, GERMANIUM TETRAFLUO/ THE FORCE CONSTANTS OF VARIOUS ISOTOPES OF CARBON TETRA 0358
SOME SQUARE PYRAMIDAL MOLECU/ REPARAMETERIZED VIBRATIONAL FORCE CONSTANTS. PART-1: METHOD AND APPLICATION TO 0420
                                  XENON DIFLUORIDE. (FORCE CONSTANTS, SPECTRA)   1252
LECULES WITH LONE PAIR ELECTRONS OR EMPTY/ CALCULATION OF FORCE CONSTANTS WITH THE MAXIMUM OVERLAP METHOD. MO 1395
TRAHEDRAL MOLECULES.   CALCULATION OF FORCE CONSTANTS WITH THE MAXIMUM OVERLAP METHOD. TE 1416
  BORON FLUORIDE WITH ALTERATIONS IN ITS QUADRATIC VALENCE FORCE FIELD. /R EFFECTS IN THE VIBRATION SPECTRA OF 1404
            INFRARED SPECTRUM OF NITROGEN TRIFLUORIDE AND FORCE FIELD.   0697
STORTION EFFECTS IN SULFUR DIFLUORIDE. CALCULATION OF THE FORCE FIELD AND INFRARED SPECTRUM.   CENTRIFUGAL DI 1031
FLUORIDE.   FORCE FIELD AND MOLECULAR CONSTANTS OF NITROGEN TRI 0735
ETHANES. (CARBON TETRAFLUORI/ A TRANSFERABLE UREY-BRADLEY FORCE FIELD AND THE ASSIGNMENTS OF SOME MIXED HALOM 0351
YMMETRICAL XY2 MOLECULES.   UREY-BRADLEY FORCE FIELD AND THERMODYNAMIC PROPERTIES FOR BENT S 1387
DAL XY3 MOLECULES.   UREY-BRADLEY FORCE FIELD AND THERMODYNAMIC PROPERTIES OF PRYRAMI 1406
LIZED MEAN SQUARE AMPLITUDES OF VIBRATION AND SHRINKAGE / FORCE FIELD, CORIOLIS COUPLING COEFFICIENTS, GENERA 1278
                                 UNIQUE FORCE FIELD FOR BORON TRIHALIDES.   0238
                               MOLECULAR FORCE FIELD FOR BROMINE PENTAFLUORIDE.   0264
                            UREY-BRADLEY FORCE FIELD FOR CARBON TETRAFLUORIDE.   0360
                               MOLECULAR FORCE FIELD FOR CHLORINE TRIFLUORIDE.   0402
RIFLUORIDE AND BROMINE TRIFLUORIDE)   MOLECULAR FORCE FIELD FOR INTERHALOGEN COMPOUNDS. (CHLORINE T 0407
AFLUORIDE, IODINE OXYGEN PENTAFLUORIDE, AND IO/ MOLECULAR FORCE FIELD FOR INTERHALOGEN COMPOUNDS. IODINE PENT 0511
                 INFRARED INTENSITIES AND FORCE FIELD FOR NITROGEN TRIFLUORIDE.   0714
                                          FORCE FIELD FOR SILICON TETRAFLUORIDE.   1006
FLUORIDE AND ARSENIC TRIFLUORIDE)   MOLECULAR FORCE FIELD FOR SOME XY3 MOLECULES. (PHOSPHORUS TRI 1400
AND TELLURIUM HEXAFLUORIDES)   MOLECULAR FORCE FIELD FOR SOME XY6 TYPE MOLECULES. (SELENIUM 1465
                               MOLECULAR FORCE FIELD FOR SULFUR TETRAFLUORIDE.   1050
                            UREY-BRADLEY FORCE FIELD FOR SULFUR TETRAFLUORIDE.   1059
                         GENERAL VALENCE FORCE FIELD FOR XENON TETRAFLUORIDE.   1265
                            UREY-BRADLEY FORCE FIELD FOR XY4 TETRAHEDRAL MOLECULES.   1429
DES OF VIBRATION OF SOME TRIGONAL BI/ GENERALIZED VALENCE FORCE FIELD FORCE CONSTANTS AND MEAN SQUARE AMPLITU 0878
                            UREY-BRADLEY FORCE FIELD. (LEAD TETRAFLUORIDE)   0596
                   VIBRATIONAL SPECTRUM AND FORCE FIELD OF MOLYBDENUM HEXAFLUORIDE.   0634
FLUORIDE, AND ARSENIC TRIFLUORIDE BY THE COMBINE/ GENERAL FORCE FIELD OF NITROGEN TRIFLUORIDE, PHOSPHORUS TRI 0717
                                      E' FORCE FIELD OF PHOSPHORUS PENTAFLUORIDE.   0861
        RAMAN INTENSITIES AND THE FORCE FIELD OF PHOSPHORUS PENTAFLUORIDE.   0844
                            UREY-BRADLEY FORCE FIELD OF SOME XY6 SYSTEMS.   1469
FRARED STUDIES OF VAPOR BAND CONTOURS. APPLICATION TO THE FORCE FIELD OF TUNGSTEN HEXAFLUORIDE. /MPERATURE IN 1156
                   VIBRATIONAL SPECTRUM AND FORCE FIELD OF URANIUM HEXAFLUORIDE.   1196
HEXAFLUORIDES AND THE L MATRIX METHOD.   FORCE FIELD OF WEINSTOCK AND GOODMAN FOR OCTAHEDRAL 1456
                                          FORCE FIELD STUDY FOR SULFUR TETRAFLUORIDE.   1052
INE.   ADAPTATION OF THE UREY-BRADLEY FORCE FIELD TO AMMONIA, PHOSPHINE, STIBINE, AND ARS 1492
AND IONS.   APPLICATION OF UREY-BRADLEY FORCE FIELD TO SOME TETRAHEDRAL XY4 TYPE MOLECULES 1425
            VIRIAL THEOREM DECOMPOSITION OF MOLECULAR FORCE FIELDS.   0783
UORIDE MONOMER AND DIMER.   HARMONIC FORCE FIELDS AND MEAN AMPLITUDES FOR ALUMINUM TRIFL 0038
 TRIFLUORIDES.   FORCE FIELDS FOR GROUP-4 TETRAFLUORIDES AND GROUP-5 0715
PENTAFLUORIDE AND BROMINE PENTAFLUORIDE)   MOLECULAR FORCE FIELDS FOR INTERHALOGEN COMPOUNDS. (CHLORINE 0425
ULATIONS OF MEAN AMPLITUDES OF VIBRATION AND UREY-BRADLEY FORCE FIELDS FOR PLANAR XY3 MOLECULES.   CALC 1399
R HEXAFLUORIDE AND TELLURIUM HEXAFLUORIDE)   FORCE FIELDS FOR SOME GROUP-6 HEXAFLUORIDES. (SULFU 1061
                               MOLECULAR FORCE FIELDS FOR SOME HEXAFLUORIDES.   1452
                                          FORCE FIELDS FOR THE BORON TRIHALIDES.   0227
                          INTRAMOLECULAR FORCE FIELDS IN GROUP-5 TRIHALIDES.   1402
RIDE.   FORCE FIELDS IN KRYPTON DIFLUORIDE AND XENON DIFLUO 0556
TRAFLUORIDE.   FORCE FIELDS OF CARBON TETRAFLUORIDE AND SILICON TE 0324
                                  UNIQUE FORCE FIELDS OF SOME XY3 TYPE MOLECULES.   1403
IOXIDE, MAGNESIUM FLUORIDE, ZINC FLUORIDE, AND FERROUS F/ FORCE FIELDS OF THE RUTILE COUNTERPARTS: TITANIUM D 0606
E, BORON TRIFLUORIDE, PHOSPH/ AMBIGUITIES IN THE HARMONIC FORCE FIELDS OF XY3 MOLECULES. (NITROGEN TRIFLUORID 0709
                               MOLECULAR FORCE FIELDS OF XY6 MOLECULES.   1443
ED HUCKEL MOLECULAR ORBITAL / GILLESPIE-NYHOLM ASPECTS OF FORCE FIELDS. PART-1. POINTS-ON-A-SPHERE AND EXTEND 0828
OCTAHEDRAL XY6 MOLECULES. (SULFUR HEXAFLUORID/ MOLECULAR FORCE FIELDS. PART-8: VIBRATION FREQUENCIES OF SOME 1080
D BAND SHAPES. THE ATOMIC DISPLACMENTS AND INTRAMOLECULAR FORCES IN CARBON TETRAFLUORIDE. /TANTS FROM INFRARE 0331
STUDY OF THE VANADIUM PENTAFLUORIDE MOLECULE IN THE VAPOR FORM.   ELECTRON DIFFRACTION 1218
/NTAFLUORIDE AND ENTHALPIES OF TRANSITION BETWEEN VARIOUS FORMS OF PHOSPHORUS. THERMODYNAMIC FUNCTIONS OF PH/ 0865
/ECULAR ORBITAL THEORY. ESSENTIAL STRUCTURAL ELEMENTS- MO FORMALISM. PART-2: DIRECT COMPARISON WITH AB INITI/ 0765
            XENON TETRAFLUORIDE: HEAT OF FORMATION.   1267
ONOFLUORIDE   ENTHALPY OF FORMATION AND THE DISSOCIATION ENERGY OF THALLIUM M 1115
AND SULFUR OXIDE DIFLUORIDE. INFRARED EVIDENCE FOR DIMER FORMATION AT LOW TEMPERATURE.   SULFUR TETRAFLUORIDE 1054
M PENTAFLUORID/ A FLUORINE NMR AND RAMAN STUDY OF COMPLEX FORMATION BETWEEN ANTIMONY PENTAFLUORIDE AND NIOBIU 0079
F THE ELEMENTS.   HEAT FORMATION BORON TRIFLUORIDE BY DIRECT COMBINATION O 0198
IFLUORIDE FROM MASS SPECTROMETRIC STUDIES.   HEATS OF FORMATION GASEOUS THORIUM TRIFLUORIDE AND THORIUM D 1128
  HEAT OF SUBLIMATION OF ALUMINUM TRIFLUORIDE AND HEAT OF FORMATION OF ALUMINUM MONOFLUORIDE.   0053
     FLUORINE BOMB CALORIMETRY. PART-22: THE ENTHALPY OF FORMATION OF ALUMINUM TRIFLUORIDE.   0048
            STANDARD ENTHALPY OF FORMATION OF ALUMINUM TRIFLUORIDE.   0043
TION OF THE ELEMENTS.   HEAT OF FORMATION OF ALUMINUM TRIFLUORIDE BY DIRECT COMBINA 0039
                          ENTHALPY OF FORMATION OF ARSENIC PENTAFLUORIDE.   0117
                     STANDARD HEAT OF FORMATION OF ARSENIC TRIFLUORIDE.   0110
OF BERYLLIUM FLUORIDE AND BERYLLIUM CHLORIDE AND HEAT OF FORMATION OF BERYLLIUM CHLOROFLUORIDE. /ON ENERGIES 0128
                          ENTHALPY OF FORMATION OF BERYLLIUM DIFLUORIDE.   0139
                             HEAT OF FORMATION OF BERYLLIUM DIFLUORIDE.   0132
                             HEAT OF FORMATION OF BORON DIFLUORIDE.   0177
            HEAT AND ENTROPY OF FORMATION OF BORON FLUORIDE.   0154
  FLUORINE BOMB CALORIMETRY. PART-15: THE ENTHALPY OF FORMATION OF BORON TRIFLUORIDE.   0220
     FLUORINE BOMB CALORIMETRY. PART-3: THE HEAT OF FORMATION OF BORON TRIFLUORIDE.   0250
                             HEAT OF FORMATION OF CARBON DIFLUORIDE.   0298
/TRENGTHS OF CARBON MONOFLUORIDE AND COMMENTS ON HEATS OF FORMATION OF CARBON MONOFLUORIDE AND CARBON DIFLUO/ 0295
                          ENTHALPY OF FORMATION OF CARBON TETRAFLUORIDE.   0329
HALIDES.   EXPLOSION METHOD AND THE HEATS OF FORMATION OF CARBON TETRAFLUORIDE AND OTHER CARBON 0315
RIDE, CHLORINE DIFLUORIDE, AND CHLORINE TRIFLU/ HEATS OF FORMATION OF CHLORINE FLUORIDES. (CHLORINE MONOFLUO 0370
K.   HEAT OF FORMATION OF CHLORINE TRIFLUORIDE AT 298.14 DEGREES 0401
NOFLUORIDE, URANIUM DIFLUORIDE, URANIUM TRIFLUO/ HEATS OF FORMATION OF GASEOUS URANIUM FLUORIDES. (URANIUM MO 1160
                          ENTHALPY OF FORMATION OF GERMANIUM DIFLUORIDE.   0472

ENTHALPY OF FORMATION OF GERMANIUM TETRAFLUORIDE. 0469
MANIUM DIOXIDES.   HEATS OF FORMATION OF GERMANIUM TETRAFLUORIDE AND OF THE GER 0483
ENTHALPY OF FORMATION OF GERMANIUM TRIFLUORIDE. 0480
MENTS OF ATOMIC NUMBER BELOW 20.   HEATS OF FORMATION OF INORGANIC FLUORIDES ESPECIALLY THE ELE 0214
HEAT FORMATION OF KRYPTON DIFLUORIDE. 0557
TRIFLUORIDE.   ENTHALPY OF FORMATION OF LANTHANUM TRIFLUORIDE AND PRASEODYMIUM 0577
TAFLUORIDE.   ENTHALPIES OF FORMATION OF NIOBIUM PENTAFLUORIDE AND TANTALUM PEN 0667
HEAT OF FORMATION OF OXYGEN DIFLUORIDE. 0769
FLUORINE BOMB CALORIMETRY. PART-18: STANDARD ENTHALPY OF FORMATION OF PHOSPHORUS PENTAFLUORIDE AND ENTHALPI/ 0865
FICIENTS FOR PROTACTINIUM IN / ESTIMATED FREE ENERGIES OF FORMATION OF PROTACTINIUM HALIDES AND ACTIVITY COEF 0900
UORIDE)   EXPERIMENTAL DETERMINATION OF THE ENTHALPIES OF FORMATION OF RARE EARTH FLUORIDES. (LANTHANUM TRIFL 0578
IDE, AND LANTHANUM TRIFLUORIDE.   STANDARD ENTHALPIES OF FORMATION OF SCANDIUM TRIFLUORIDE, YTTRIUM TRIFLUOR 0942
URANIUM, NEPTUNIUM, AND AMERICIUM. (ACTINIDE)   HEATS OF FORMATION OF SOME HALIDES OF THORIUM, PROTACTINIUM, 1335
RIDE, TELLURIUM HEXAFLUORIDE AND THEIR THE/ ENTHALPIES OF FORMATION OF SULFUR HEXAFLUORIDE, SELENIUM HEXAFLUO 1092
HEAT OF FORMATION OF THE CARBON DIFLUORIDE RADICAL. 0304
UORIDE / MASS SPECTROMETRIC DETERMINATION OF THE HEATS OF FORMATION OF THE GASEOUS MOLECULES ALUMINUM OXYDIFL 0035
CALORIMETRY.   ENTHALPY OF FORMATION OF THORIUM TETRAFLUORIDE BY FLUORINE BOMB 1130
RAFLUORIDE.   HEATS OF FORMATION OF TITANIUM TETRAFLUORIDE AND HAFNIUM TET 1147
XAFLUORIDE.   ENTHALPY OF FORMATION OF TUNGSTEN HEXAFLUORIDE. 1157
ENTHALPY OF FORMATION OF TUNGSTEN PENTAFLUORIDE AND TUNGSTEN HE 1151
ENTHALPY OF FORMATION OF URANIUM TETRAFLUORIDE. 1171
B CALORIMETRY.   ENTHALPY OF FORMATION OF VANADIUM PENTAFLUORIDE BY FLUORINE BOM 1215
E, XENON DIFLUORIDE, AND PHOSPHORUS TRIFLU/ ENTHALPIES OF FORMATION OF XENON HEXAFLUORIDE, XENON TETRAFLUORID 0807
E, XENON TETRAFLUORIDE. CRYSTAL STRUCTURE.   ENTHALPIES OF FORMATION OF XENON HEXAFLUORIDE, XENON TETRAFLUORID 1269
HEAT OF FORMATION OF XENON TETRAFLUORIDE. 1266
ENTHALPY CF FORMATION OF YTTRIUM TRIFLUORIDE. 1319
ENTHALPY OF FORMATION PHOSPHORUS TRIFLUORIDE. 0818
FORMULAS FOR CORIOLIS CONSTANTS. 1482
IDE, GERMANIUM FLUORIDE, AND GERMANIU/ CALCULATION OF THE FRANCK-CONDON FACTOR FOR MOLECULES OF SILICON FLUOR 0977
PLET SIGMA(+) TO A TRIPLET PI SYSTEM OF BORON MONOFLUORI/ FRANCK-CONDON FACTORS AND R CENTROIDS FOR THE B TRI 0171
YSTEM OF SILICON MONOFLUORIDE MOLECULE.   FRANCK-CONDON FACTORS AND R CENTROIDS FOR THE C-X S 0975
GLET SIGMA- X TRIPLET SIGMA BAND SYSTEM OF NITROGEN MONO/ FRANCK-CONDON FACTORS AND R CENTROIDS FOR THE 6 SIN 0677
SIGMA- X SINGLET SIGMA PLATINUM OXIDE AND SILICON MONOFL/ FRANCK-CONDON FACTORS AND R CENTROIDS OF A SINGLET 0966
SYSTEMS OF SILICON MONCFLUORIDE, CALCIUM MONOFLUORIDE, A/ FRANCK-CONDON FACTORS AND R CENTROIDS OF SOME BAND 0972
ND SYSTEM OF BERYLLIUM FLUORIDE.   RYDBERG-KLEIN-REES FRANCK-CONDON FACTORS AND R CENTROIDS OF THE A-X BA 0127
AND ACTIVITY COEFFICIENTS FCR PROTACTINIUM IN / ESTIMATED FREE ENERGIES OF FORMATION OF PROTACTINIUM HALIDES 0900
MERICIUM TRIFLUORIDE AND PLUTONIUM TRIFLUORIDE, HEATS AND FREE ENERGIES OF SUBLIMATION.   VAPOR PRESSURES OF A 0055
ROCM TEMPERATURE. (THALLIUM / ENTHALPIES, ENTROPIES, AND FREE ENERGY FUNCTIONS OF THALLIUM MONOHALIDES ABOVE 1116
AMAN SPECTRA OF OXYGEN DIFLUORIDE AND THE OXYGEN FLUORIDE FREE RADICAL.   ARGON MATRIX R 0759
TEMPERATURE. CHARACTERIZATION OF THE CHLORINE DIFLUORIDE FREE RADICAL.   CHLORINE FLUORINE SYSTEM AT LOW 0392
IFLUORIDE. ELECTRONIC SPECTRUM OF THE NITROGEN DIFLUORIDE FREE RADICAL. /LTRAVIOLET PHOTOLYSIS OF NITROGEN TR 0691
MICROWAVE ROTATIONAL SPECTRUM OF THE NITROGEN DIFLUORIDE FREE RADICAL. 0690
THE CHLORINE DIFLUORIDE RADICAL. 0393
ATRICES. PHOTOCHLORINATION OF / INFRARED SPECTRUM OF THE FREE RADICAL CARBON TRIFLUORIDE ISOLATED IN INERT M 0311
S. (NITROGEN DIFLUORIDE) A CAVITY SEARCH SPECTROMETER FOR FREE RADICAL MICROWAVE ROTATIONAL ABSORPTION STUDIE 0689
DE DIHYDRIDE. THE INFRARED AND ULTRAVIOLET SPECTRA OF THE FREE RADICAL SILICON DIFLUORIDE. /SILICON DIFLUORI 0986
ELECTRONIC SPECTRA OF POLYATOMIC FREE RADICALS. (CARBON DIFLUORIDE) 0308
OUGH BORON TRIFLU/ EMISSION SPECTRA OF CERTAIN POLYATOMIC FREE RADICALS OBTAINED IN ELECTRICAL DISCHARGES THR 0222
ELECTRONIC AND GEOMETRIC STRUCTURE OF THE FREE XENON HEXAFLUORIDE MOLECULE. 1302
ORIDE FROM AMPLITUDES OF VIBRATION AND NORMAL VIBRATIONAL FREQUENCIES. /ENTIAL CONSTANTS OF PHOSPHORUS TRICHL 1481
Y4 MOLECULES IN THE LIGHT OF THE THEORY OF CHARACTERISTIC FREQUENCIES. /KESSIVE STIFFNESS METHOD APPLIED TO X 1430
W TRANSITION METAL DIFLUORIDES.   VIBRATIONAL FREQUENCIES AND VALENCE FORCE CONSTANTS OF FIRST RO 1371
/D ARSENIC TRIFLUORIDE BY THE COMBINED USE OF VIBRATIONAL FREQUENCIES, CENTRIFUGAL STRETCHING, AND CORIOLIS / 0717
F THE CORIOLIS CONSTANT FOR SOME MOLECULES AND IONS OF D/ FREQUENCIES, FORCE CONSTANTS, AND MASS DEPENDENCE C 1411
FORCE CONSTANTS OF THE SILICON TETRAHALIDES AND FREQUENCIES OF CHLORO BROMO SILICON. 1004
LUORIDE, AND LANTHANUM/ EVALUATION OF MOLECULAR VIBRATION FREQUENCIES OF SCANDIUM MONOFLUORIDE, YTTRIUM MONOF 0936
CALCULATION OF THE VIBRATIONAL FREQUENCIES OF SCANDIUM SUBGROUP HALIDES. 0949
UR HEXAFLUORID/ MOLECULAR FORCE FIELDS. PART-8: VIBRATION FREQUENCIES OF SOME OCTAHEDRAL XY6 MOLECULES. (SULF 1080
VIBRATIONAL FREQUENCIES OF SOME PLANAR XY3 MOLECULES. 1391
ESTIMATION OF INFRARED FREQUENCIES OF SULFUR DIFLUORIDE. 1026
HALIDES OF GERMANIUM, TIN, AND LEAD. (GERMANIUM DIFLUORI/ FREQUENCIES OF THE DEFORMATION VIBRATIONS OF THE DI 0479
IFLUORIDE.   CALCULATION OF THE FREQUENCIES OF THE NORMAL VIBRATIONS OF ANTIMONY TR 0074
IDES.   VALENCE VIBRATION FREQUENCIES OF TRIHALIDE MOLECULES OF LANTHANUM HAL 0574
ACCORDING TO THERMODYNAMIC DATA.   FREQUENCIES OF ZIRCONIUM FLUORIDE (GAS) VIBRATIONS 1329
NORMAL VIBRATION FREQUENCIES OF ZIRCONIUM HALIDES. 1330
CONSTANTS, CENTRIFUGAL STRETCHING CONSTANTS, AN/ HARMONIC FREQUENCIES, POTENTIAL FUNCTION, CORIOLIS COUPLING 0726
LIQUID GAS ARSENIC INFRARED FREQUENCY SHIFTS. 0106
/NFRARED SPECTROMETER AND ITS USE IN THE STUDY OF THE LOW FREQUENCY VIBRATIONS OF ANTIMONY PENTAFLUORIDE AND/ 0083
GNMENT OF IODINE PENTAFLUORIDE. FERMI RESONANCE BETWEEN A FUNDAMENTAL AND A TERNARY COMBINATION. /RA AND ASSI 0504
AFLUORIDE, AND PLUTONI/ LONG WAVE LENGTH, INFRARED ACTIVE FUNDAMENTAL FOR URANIUM HEXAFLUORIDE, NEPTUNIUM HEX 1192
NU-7 FUNDAMENTAL OF PHOSPHORUS PENTAFLUORIDE. 0848
RA OF GASES. PART-10: RAMAN BAND CONTOUR OF THE NU-7 (E') FUNDAMENTAL OF PHOSPHORUS PENTAFLUORIDE. /MAN SPECT 0879
ON OF A SINGLE VIBRATION- ROTATION TRANSITION IN THE NU-3 FUNDAMENTAL OF SULFUR HEXAFLUORIDE. /TICAL SATURATI 1074
ABSOLUTE INFRARED INTENSITIES OF THE FUNDAMENTAL VIBRATIONS OF NITROGEN TRIFLUORIDE. 0736
FLUORIDE, CARBON TETRACHLORI/ INTENSITIES OF THE INFRARED FUNDAMENTALS OF CARBON TETRAFLUORIDE, SILICON TETRA 1007
FLUORIDE FROM 298-1400 DEGREES K. ENTHALPY AND ENTROPY OF FUSION.   ENTHALPY OF URANIUM TETRA 1169
EXAFLUORIDE FROM 7 TO 350 DEG/ HEAT CAPACITY, ENTHALPY OF FUSION, AND THERMODYNAMIC PROPERTIES OF NEPTUNIUM H 0650
/. HEAT CAPACITY BETWEEN 5 AND 350 DEGREES K, ENTHALPY OF FUSION AND VAPORIZATION, STANDARD ENTROPY OF THE V/ 0510

G

/NALYSIS OF THE C DOUBLET DELTA 5/2- X DOUBLET PI 3/2 AND G DOUBLET DELTA 5/2- X DOUBLET PI 3/2 SYSTEMS OF T/ 1136
RUPOLE MOMENTS AND SECOND MOMENTS OF THE ELECT/ MOLECULAR G VALUES, MAGNETIC SUSCEPTIBILITIES, MOLECULAR QUAD 0780
ADRUPOLE MOMENTS / ZEEMAN STUDIES INCLUDING THE MOLECULAR G VALUES, MAGNETIC SUSCEPTIBILITIES, AND MOLECULAR QU 0744
ERATURES. PART-13: STABILITIES OF SAMARIUM, EUROPIUM, AND GADOLINIUM MONO AND DIFLUORIDES. /DIES AT HIGH TEMP 0450
REFLECTION SPECTRA OF GADOLINIUM OXIDE AND GADOLINIUM TRIFLUORIDE POWDERS 0455
/HE TRIFLUORIDES OF SCANDIUM, CERIUM, SAMARIUM, EUROPIUM, GADOLINIUM, TERBIUM, DYSPROSIUM, ERBIUM, THULIUM, / 1356
OPY AND ENTHALPY / ENTROPY AND ENTHALPY OF SUBLIMATION OF GADOLINIUM TRIFLUORIDE. ROLE OF CORRELATION OF ENTR 0456
IUM, YTTRIUM, LANTHANUM, CERIUM, NEODYMIUM, EUROPIUM, AND GADOLINIUM TRIFLUORIDES) / INFRARED SPECTRA. (SCAND 0943
ELECTRON DIFFRACTION ANALYSIS OF GALLIUM HALIDES. (GALLIUM TRIFLUORIDE) 0460

N ENERGIES OF THE GASEOUS MONOHALIDES OF BORON, ALUMINUM, GALLIUM, INDIUM, AND THALLIUM.   DISSOCIATIO 0152
SOCIATION ENERGIES OF DIATOMIC FLUORIDES AND CHLORIDES OF GALLIUM, INDIUM, AND THALLIUM. (MONOFLUORIDES) /DIS 0458
ZERO(+) B TRIPLET PI(1) TO X SINGLET SIGMA(+) SYSTEMS OF GALLIUM MONOFLUORIDE. /OF BANDS OF THE A TRIPLET PI 0457
TETRAHYDRIDE, ARSENIC TR/ MOLECULAR SCF CALCULATIONS FOR GALLIUM MONOFLUORIDE, GALLIUM TRIHYDRIDE, GERMANIUM 0459
L/ MICROWAVE ABSORPTION SPECTRA OF ALUMINUM MONOFLUORIDE. GALLIUM MONOFLUORIDE, INDIUM MONOFLUORIDE, AND THAL 0025
MICROWAVE SPECTRA OF THE THALLIUM, INDIUM AND GALLIUM MONOHALIDES. 1110
PREPARATION AND CRYSTAL STRUCTURE OF GALLIUM TRIFLUORIDE. 0461
OF THE MOLECULAR STRUCTURE OF THE VAPOR PHASE HALIDES OF GALLIUM, YTTRIUM, LANTHANUM, AND NEODYMIUM. / STUDY 1317
INFRARED SPECTRA OF MATRIX ISOLATED SPECIES IN THE GALLIUM- FLUORINE SYSTEM. 0462
ROTATIONAL ANALYSIS OF THE GAMMA BANDS OF SILICON MONOFLUORIDE. 0999
E AND ARSENIC TRICHLORIDE.   ESR STUDY OF RADICALS IN GAMMA IRRADIATED POLYCRYSTALLINE ARSENIC TRIFLUORID 0108
ELECTRON SPIN RESONANCE SPECTRUM OF XENON MONOFLUORIDE IN GAMMA IRRADIATED XENON TETRAFLUORIDE. 1222
EMISSION SPECTRUM OF THE GAMMA SYSTEM OF SILICON MONOFLUORIDE. 0974
ARBON TETRAFLUORIDE. THERMODYNAMIC PROPERTIES OF THE REAL GAS.   C 0332
NUCLEAR MAGNETIC RELAXATION IN BORON TRIFLUORIDE GAS. 0217
IPYRAMIDAL THE RAMAN SPECTRUM OF PHOSPHORUS PENTAFLUORIDE GAS. /LD AND MOLECULAR VIBRATIONS OF THE TRIGONAL B 0863
RAMAN SPECTRUM OF BORON TRIFLUORIDE GAS. 0251
TRUM AND THERMODYNAMIC PROPERTIES OF URANIUM HEXAFLUORIDE GAS.   VIBRATIONAL SPEC 1186
TRUM AND THERMODYNAMIC PROPERTIES OF URANIUM HEXAFLUORIDE GAS.   VIBRATIONAL SPEC 1187
/MONSTRATION OF THE EXISTENCE OF DILANTHANUM HEXAFLUORIDE GAS AND DETERMINATION OF ITS STABILITY. (LANTHANUM/ 0581
LIQUID GAS ARSENIC INFRARED FREQUENCY SHIFTS. 0106
DIES OF THE INTERACTION OF MAGNESIUM DIFLUORIDE WITH RARE GAS ATOMS.   THEORETICAL STU 0605
MOLECULAR STRUCTURES OF RARE GAS COMPOUNDS. 1344
DE.   PREPARATION OF INERT GAS COMPOUNDS BY MATRIX ISOLATION: KRYPTON DIFLUORI 0566
STRUCTURES OF NOBLE GAS COMPOUNDS. (XENON HEXAFLUORIDE) 1292
SCATTERING.   MEASUREMENT OF THE GAS DENSITY OF URANIUM HEXAFLUORIDE BY LASER RAMAN 1181
E, AND XENON DIFLUORIDE)   D ORBITALS IN THE NOBLE GAS DIHALIDES. (ARGON DIFLUORIDE, KRYPTON DIFLUORID 1338
FLUORINE BOND DISTANCE OF BORON TRIFLUORIDE DETERMINED BY GAS ELECTRON DIFFRACTION.   BORON 0223
UCTURES OF ARSENIC TRIFLUORIDE AND ARSENIC TRICHLORIDE BY GAS ELECTRON DIFFRACTION. /ION OF THE MOLECULAR STR 0104
MOLECULAR STRUCTURE OF PHOSPHORUS TRIFLUORIDE STUDIED BY GAS ELECTRON DIFFRACTION. 0810
AN LCAO MO STUDY OF RARE GAS FLUORIDES. 1341
CHEMICAL PREDICTIONS BY MO THEORY. THE NOBLE GAS HALIDES. 1340
N DIFLUORIDE / BOND ENERGIES AND IONIC CHARACTER OF INERT GAS HALIDES. HELIUM-2 PHOTOELECTRON SPECTRA OF XENO 1343
ABSORPTION OF BORON TRIFLUORIDE GAS IN THE EXTREME ULTRAVIOLET. 0215
Y OF TITANIUM DIFLUORIDE AND TITANIUM TRIFLUORIDE IN RARE GAS MATRICES.   INFRARED SPECTRA AND GEOMETR 1146
INFRARED SPECTRA OF SILICON DIFLUORIDE IN INERT GAS MATRICES. 0982
THEORY OF INERT GAS MOLECULES. 1337
F XENON MONOFLUORIDE AS A CHEMICALLY BOUND SPECIES IN THE GAS PHASE.   PROBABLE NONEXISTENCE O 1221
R A TRIG/ RAMAN SPECTRUM OF ANTIMONY PENTAFLUORIDE IN THE GAS PHASE AS A FUNCTION OF TEMPERATURE. EVIDENCE FO 0077
AND DIPOLE MOMENT OF NITROGEN MONOFLUORIDE IN THE SINGLE/ GAS PHASE ELECTRON PARAMAGNETIC RESONANCE SPECTRUM 0673
FLUORIDE, SELENIUM MONOFLUORIDE, SELENIUM OXIDE AND IODI/ GAS PHASE ELECTRON RESONANCE SPECTRA OF SULFUR MONO 1020
FLUORIDE AND SELENIUM MONOFLUORIDE. GAS PHASE ELECTRON RESONANCE SPECTRA OF SULFUR MONO 1021
MENT OF CARBON MONOFLUORIDE. GAS PHASE ELECTRON RESONANCE SPECTRUM AND DIPOLE MO 0284
RIDE AND IODINE PENTAFLUORIDE FROM ELECTRON DIFFRACTION / GAS PHASE MOLECULAR STRUCTURES OF BROMINE PENTAFLUO 0265
FLUORIDE, B/ ASSOCIATION OF GROUP-5 PENTAFLUORIDES IN THE GAS PHASE. (PHOSPHORUS PENTAFLUORIDE, ARSENIC PENTA 0876
AL TOP T/ ROTATIONAL PROFILES OF VIBRATIONAL BANDS IN THE GAS PHASE RAMAN SPECTRA OF MOLECULES OF THE SPHERIC 1442
ENERGETICS AND STABILITY OF GAS PHASE SCANDIUM FLUORIDES. 0951
GAS PHASE STRUCTURE OF XENON DIFLUORIDE. 1248
GAS PHASE STRUCTURE OF XENON HEXAFLUORIDE. 1293
TAFLUORIDES. GAS PHASE STRUCTURES AND MASS SPECTRA OF BINARY PEN 0003
TAFLUORIDES. GAS PHASE STRUCTURES AND MASS SPECTRA OF BINARY PEN 1365
/RACTION STUDY OF URANIUM AND THORIUM TETRAHALIDES IN THE GAS PHASE. (URANIUM TETRAFLUORIDE AND THORIUM TETR/ 1170
/AL SPECTRA OF NIOBIUM AND TANTALUM PENTAFLUORIDES IN THE GAS PHASE. VAPOR DENSITY OF NIOBIUM PENTAFLUORIDE. 0660
/INITY, DIPOLE MOMENT, SPECTROSCOPIC CONSTANTS, AND IDEAL GAS THERMODYNAMIC FUNCTIONS FROM A HARTREE-FOCK MO/ 0094
IDE, BROMINE PENTAFLUORIDE, AND IODINE PENTAFLUORI/ IDEAL GAS THERMODYNAMIC PROPERTIES OF CHLORINE PENTAFLUOR 0424
ENON TETRAFLUORIDE AND XENON HEXAFLUORIDE)   IDEAL GAS THERMODYNAMIC PROPERTIES OF XENON COMPOUNDS. (X 1273
RIFLUORIDE AND CHLORINE MONOFLUORIDE. GASEOUS SPECTRA AND GAS TO LIQUID SHIFTS.   NMR SPECTRA OF CHLORINE T 0368
FREQUENCIES OF ZIRCONIUM FLUORIDE (GAS) VIBRATIONS ACCORDING TO THERMODYNAMIC DATA. 1329
CHEMICAL COMPOUNDS OF XENON AND OTHER NOBLE GASES. 0010
ELECTRON STRUCTURE OF COMPOUNDS OF INERT GASES. 1339
NOBLE GASES AND THE BINARY FLUORIDES OF XENON. 1342
PHOSPHORUS PENTAFLUORIDE, DIFLU/ ELECTRON DIFFRACTION BY GASES AND THE STRUCTURES OF PHOSPHORUS TRIFLUORIDE, 0801
/F ELECTRON STRUCTURES OF COMPOUNDS OF HALOGENS AND INERT GASES BY THE NONEMPIRICAL NDDO-2 (ALPHA, BETA) MET/ 0016
/ELECTRONIC STRUCTURES OF COMPOUNDS OF HALOGENS AND INERT GASES BY THE SCF LCAO MO METHOD.   PART-2: COMPARAT/ 1231
FUNDAMENTAL OF PHOSPHORUS PENTAFLUORID/ RAMAN SPECTRA OF GASES. PART-10: RAMAN BAND CONTOUR OF THE NU-7 (E') 0879
SPHORUS MONOFLUORIDE) THERMODYNAMIC FUNCTIONS OF DIATOMIC GASES WITH MOLECULES IN TRIPLET SIGMA STATES. (PHO 0789
N HE/ STRUCTURAL CONSIDERATIONS IN THE CHEMISTRY OF NOBLE GASES. (XENON DIFLUORIDE, XENON TETRAFLUORIDE, XENO 1235
ELECTRONIC STATES OF GASEOUS ALUMINUM MONOFLUORIDE. 0027
AFLUORIDE, AND SODIUM ALUMINUM TETRA/ INFRARED SPECTRA OF GASEOUS ALUMINUM TRIFLUORIDE, LITHIUM ALUMINUM TETR 0045
FAR INFRARED SPECTRA OF GASEOUS AND LIQUID SULFUR HEXAFLUORIDE. 1095
RAMAN SPECTRUM OF SULFUR HEXAFLUORIDE IN THE GASEOUS AND SOLID STATES. 1079
ROTATIONAL ANALYSIS OF SOME BANDS OF GASEOUS BARIUM MONOFLUORIDE. 0118
CONSTANT VOLUME HEAT CAPACITY OF GASEOUS CARBON TETRAFLUORIDE. 0336
FAR INFRARED ABSORPTION IN GASEOUS CARBON TETRAFLUORIDE. 0356
RAMAN SPECTRUM OF GASEOUS CARBON TETRAFLUORIDE. 0347
CONSTANT VOLUME HEAT CAPACITIES OF GASEOUS CARBON TETRAFLUORIDE AND OTHER MOLECULES. 0335
, AND SULFUR HE/ FLUORINE RELAXATION BY NMR ABSORPTION IN GASEOUS CARBON TETRAFLUORIDE, SILICON TETRAFLUORIDE 0346
INFRARED DETECTION OF GASEOUS CARBON TRIFLUORIDE RADICAL. 0310
UM DIFLUORIDE, AND GERMANIUM/ THERMODYNAMIC PROPERTIES OF GASEOUS GERMANIUM, GERMANIUM TETRAFLUORIDE, GERMANI 0468
HIGH TEMPERATURE NEGATIVE IONS. GASEOUS GROUP-3 FLUORIDES. 0047
IUM CHLORIDE, AND HAFNIUM CHLORIDE.   INFRARED SPECTRA OF GASEOUS GROUP-4 HALIDES: ZIRCONIUM FLUORIDE, ZIRCON 1327
ABSORPTION SPECTRUM OF GASEOUS HOLMIUM MONOFLUORIDE. 0488
ROTATIONAL ANALYSIS OF ELECTRONIC BANDS OF GASEOUS MAGNESIUM MONOFLUORIDE. 0600
/TROMETRIC DETERMINATION OF THE HEATS OF FORMATION OF THE GASEOUS MOLECULES ALUMINUM OXYDIFLUORIDE AND ALUMI/ 0035
DIUM, AND THALLIUM.   DISSOCIATION ENERGIES OF THE GASEOUS MONOHALIDES OF BORON, ALUMINUM, GALLIUM, IN 0152
/PRESSURE OF NEODYMIUM TRIFLUORIDE AND THE STABILITIES OF GASEOUS NEODYMIUM DIFLUORIDE AND NEODYMIUM MONOFLU/ 0643
RAMAN SPECTRUM OF GASEOUS NITROGEN TRIFLUORIDE. 0740
L ANALYSIS OF SOME SINGLET TRANSITIONS IN THE SPECTRUM OF GASEOUS SCANDIUM MONOFLUORIDE.   ROTATIONA 0930
NALYSIS OF THE SYSTEM B SINGLET PI TO SINGLET SIGMA(+) OF GASEOUS SCANDIUM MONOFLUORIDE.   ROTATIONAL A 0931
, AND LANTHANUM MONOFLUORIDE.   ELECTRONIC STATES OF GASEOUS SCANDIUM MONOFLUORIDE, YTTRIUM MONOFLUORIDE 0928

THE 293-3000 DEGREES K RANGE. THERMODYNAMIC FUNCTIONS OF GASEOUS SCANDIUM, YTTRIUM, AND LANTHANUM HALIDES IN 0568
/ISTANCES AND ROOT MEAN SQUARE AMPLITUDES OF VIBRATION OF GASEOUS SILICON TETRAFLUORIDE FROM ELECTRON DIFFRA/ 1002
ABSORPTION SPECTRUM OF GASEOUS SILVER MONOFLUORIDE. 1012
SIS OF THE AO(+), BO(+), TO X SINGLET SIGMA(+) SYSTEMS OF GASEOUS SILVER MONOFLUORIDE. ROTATIONAL ANALY 1011
PECTRA OF CHLORINE TRIFLUORIDE AND CHLORINE MONOFLUORIDE. GASEOUS SPECTRA AND GAS TO LIQUID SHIFTS. NMR S 0368
DE, TANTALUM BROMIDE, AND MOLYBDENUM PENTACHLORIDE IN THE GASEOUS STATE. /E, NIOBIUM BROMIDE, TANTALUM CHLORI 0075
RAMAN SPECTRA OF CHLOROFLUORO METHANES IN THE GASEOUS STATE. (CARBON TETRAFLUORIDE) 0322
INTERNUCLEAR DISTANCE IN GASEOUS STRONTIUM MONOFLUORIDE. 1016
MASS SPECTROMETRIC STUDIES OF SOME GASEOUS SULFUR FLUORIDES. 1023
MASS SPECTROMETRIC STUDIES OF SOME GASEOUS SULFUR FLUORIDES. 1028
L AND FLUORINE K X-RAY EMISSION AND ABSORPTION SPECTRA OF GASEOUS SULFUR HEXAFLUORIDE. SULFUR K AND 1088
FROM MASS SPECTROMETRIC STUDIES. HEATS OF FORMATION GASEOUS THORIUM TRIFLUORIDE AND THORIUM DIFLUORIDE 1128
ULTRAVIOLET ABSORPTION SPECTRA OF GASEOUS TIN DIFLUORIDE AND LEAD DIFLUORIDE. 0595
E. GASEOUS TRIMER AND TETRAMER OF THALLIUM MONOFLUORID 1114
RANIUM DIFLUORIDE, URANIUM TRIFLUO/ HEATS OF FORMATION OF GASEOUS URANIUM FLUORIDES. (URANIUM MONOFLUORIDE, U 1160
DE. ELECTRONIC ABSORPTION SPECTRUM OF GASEOUS URANIUM HEXACHLORIDE AND URANIUM HEXAFLUORI 1194
THERMODYNAMIC PROPERTIES OF GASEOUS URANIUM HEXAFLUORIDE. 1198
STRUCTURE OF GASEOUS XENON HEXAFLUORIDE. 1295
ELECTRON DIFFRACTION STUDY OF THE STRUCTURE OF GASEOUS XENON TETRAFLUORIDE. 1282
/NT AND FIELD GRADIENT OF NITROGEN TRIFLUORIDE. PART-3: A GAUSSIAN BASIS SELF CONSISTENT FIELD CALCULATION F/ 0746
F CALCULATIONS IN IODINE COMPOUNDS. A SMALL GAUSSIAN BASIS SET FOR NONEMPIRICAL ALL-ELECTRON SC 0495
IN DICHLORO BORANE, DIBROMO BORANE AND CHLORINE TRIFLUOR/ GENERALIZED MEAN AMPLITUDES AND CORIOLIS CONSTANTS 0411
STANTS OF SOME PYRAMIDAL XY3 MOLECULES. GENERALIZED MEAN SQUARE AMPLITUDES AND CORIOLIS CON 1405
SHRINKAGE / FORCE FIELD, CORIOLIS COUPLING COEFFICIENTS, GENERALIZED MEAN SQUARE AMPLITUDES OF VIBRATION AND 1278
D SHRINKAGE EFFECTS OF SOME TRIGONAL BIPYRAMIDAL XY5 AND/ GENERALIZED MEAN SQUARE AMPLITUDES OF VIBRATION, AN 1432
ULATED BY THE LOGARITHMIC STEP METHOD IN THE RANGE OF THE GENERALIZED VALENCE FORCE FIELD. /EPTAFLUORIDE CALC 0535
MEAN SQUARE AMPLITUDES OF VIBRATION OF SOME TRIGONAL BI/ GENERALIZED VALENCE FORCE FIELD FORCE CONSTANTS AND 0878
MOLECULE. ELECTRONIC AND GEOMETRIC STRUCTURE OF THE FREE XENON HEXAFLUORIDE 1302
ORMATIONS. APPLICATION TO THE CALCULATION BOND MOMENT / A GEOMETRIC VISUALIZATION OF NORMAL COORDINATE TRANSF 0352
E CHLORINE DIFLUORIDE ON / GEOMETRY OF MOLECULES. PART-2: GEOMETRICAL AND ELECTRICAL STRUCTURE OF THE MOLECUL 0382
INFRARED SPECTRA. (SCANDIUM, YTTRIUM, LANTHANUM, CERIUM/ GEOMETRIES AND ENTROPIES OF METAL TRIFLUORIDES FROM 0943
/AL METHODS. PART-5: AB INITIO CALCULATION OF EQUILIBRIUM GEOMETRIES AND QUADRATIC FORCE CONSTANTS, (NITROGE/ 0720
ES. (SCANDIUM TRIFLUORIDE, YTTRIUM T/ FORCE CONSTANTS AND GEOMETRIES OF MATRIX ISOLATED RARE EARTH TRIFLUORID 0944
D YTTRIUM TRIFLUORIDE. INFRARED SPECTRA AND GEOMETRIES OF MATRIX ISOLATED YTTRIUM DIFLUORIDE AN 1316
INFRARED SPECTRA AND GEOMETRIES OF RARE EARTH DIHALIDES. 1351
E ANALYSIS OF NU-2 AND THE DETERMINATION OF THE MOLECULAR GEOMETRY. BORON TRIFLUORIDE. TH 0209
RE EARTH HALIDES. (ALSO LANTHANUM TRIFLUORIDE) GEOMETRY AND INFRARED SPECTRA OF MATRIX ISOLATED RA 0583
CNDO/2 AND INDO ALL VALENCE ELECTRON CALCULATIONS ON THE GEOMETRY AND PROPERTIES OF SOME INTERHALOGENS. (CH/ 0397
RTH DIHALIDES. PART-1: MAGNESIUM DIFLUORIDE. GEOMETRY AND VIBRATIONAL SPECTRA OF THE ALKALINE EA 0609
DIFLUORIDE. AB INITIO INVESTIGATION OF THE GEOMETRY, BONDING, AND COUPLING CONSTANTS OF BORON 0180
DIFFRACTION INTENSITIES. PART-2: INTERNAL MOTION AND MEAN GEOMETRY DEDUCED BY ELECTRON DIFFRACTION. /LECTRON 1291
FLECTION AND MASS SPECTROMETRIC DET/ DETERMINATION OF THE GEOMETRY OF HIGH TEMPERATURE SPECIES BY ELECTRIC DE 0134
CHLORINE TRIFLU/ AB INITIO MOLECULAR ORBITAL STUDY OF THE GEOMETRY OF INTERHALOGENS. (CHLORINE MONOFLUORIDE, 0379
EL DIFLUORIDE, COPPER DIFLUORIDE, A/ INFRARED SPECTRA AND GEOMETRY OF MATRIX ISOLATED COBALT DIFLUORIDE, NICK 0437
GEOMETRY OF MOLECULES. (OXYGEN DIFLUORIDE) 0764
TRICAL STRUCTURE OF THE MOLECULE CHLORINE DIFLUORIDE ON / GEOMETRY OF MOLECULES. PART-2: GEOMETRICAL AND ELEC 0382
ND THEIR GROUND IONIC STATES. THEORETICAL STUDY OF THE GEOMETRY OF PHOSPHINE AND PHOSPHORUS TRIFLUORIDE A 0794
IFLUORIDE, STRONTIUM DIFLUORIDE, CALCIUM DIFLUORIDE) GEOMETRY OF THE ALKALINE EARTH DIHALIDES. (BARIUM D 1210
LUORIDES OF SOME FIRST ROW TRANSITION METALS. GEOMETRY OF THE TRANSITION METAL DIHALIDES. THE DIF 1368
ORIDE IN RARE GAS MATRICES. INFRARED SPECTRA AND GEOMETRY OF TITANIUM DIFLUORIDE AND TITANIUM TRIFLU 1146
C ISOMERISM. ELECTRONIC STATES AND MOLECULAR GEOMETRY OF XENON HEXAFLUORIDE: A CASE OF ELECTRONI 1294
/ XENON HEXAFLUORIDE. PART-1: ENERGY LEVELS AND MOLECULAR GEOMETRY. PART-2: COMPARISONS WITH OTHER HEXAVALEN/ 1311
RGIES OF THE MONOFLUORIDES AND DIFLUORIDES OF SILICON AND GERMANIUM. /PERATURES. PART-2: THE DISSOCIATION ENE 0463
IES AND R CENTROIDS FOR DIATOMIC FLUORIDES OF SILICON AND GERMANIUM. VIBRATIONAL TRANSITION PROBABILIT 0994
D / VAPOR PHASE RAMAN SPECTRA OF MH4 (M= CARBON, SILICON, GERMANIUM, AND TIN) AND MF4 (M= CARBON, SILICON, AN 0314
ENTHALPY OF FORMATION OF GERMANIUM DIFLUORIDE. 0472
INFRARED VIBRATIONAL PROPERTIES OF GERMANIUM DIFLUORIDE. 0473
MICROWAVE SPECTRUM OF GERMANIUM DIFLUORIDE. 0478
NEW ELECTRONIC EMISSION SYSTEM OF GERMANIUM DIFLUORIDE. 0476
ULTRAVIOLET ABSORPTION SPECTRUM OF GERMANIUM DIFLUORIDE. 0474
VACUUM ULTRAVIOLET SPECTRA OF SILICON DIFLUORIDE AND GERMANIUM DIFLUORIDE. 0981
IMATION OF THE GROUP-4A DIFLU/ ENTHALPY OF SUBLIMATION OF GERMANIUM DIFLUORIDE AND THE THERMODYNAMICS OF SUBL 0471
/, MOLECULAR STRUCTURES AND NORMAL COORDINATE ANALYSES OF GERMANIUM DIFLUORIDE, MONOMERIC GERMANIUM DIFLUORI/ 0475
IFUGAL DISTORTION. MICROWAVE SPECTRUM OF GERMANIUM DIFLUORIDE. QUADRUPOLE COUPLING AND CENTR 0477
E FRANCK-CONDON FACTOR FOR MOLECULES OF SILICON FLUORIDE, GERMANIUM FLUORIDE, AND GERMANIUM CHLORIDE. / OF TH 0977
ES, IONIZATION POTENTIALS AND ELECTRON AFFINITIES OF SOME GERMANIUM FLUORIDE SPECIES. /ND DISSOCIATION ENERGI 0464
ISSOCIATION ENERGY OF CALCIUM FLUORIDE. (ALSO SILICON AND GERMANIUM FLUORIDES) REEVALUATION OF THE D 0277
ORIDE, AND GERMANIUM/ THERMODYNAMIC PROPERTIES OF GASEOUS GERMANIUM, GERMANIUM TETRAFLUORIDE, GERMANIUM DIFLU 0468
ROTATIONAL STRUCTURE IN SOME HIGHER EXCITED STATES OF THE GERMANIUM(I) FLUORIDE MOLECULE. 0467
S OF THE ROTATIONAL STRUCTURE OF THE BANDS 0-0 AND 0-1 IN GERMANIUM MONOFLUORIDE. ANALYSI 0465
+), AND A QUARTET SIGMA(-) TO X DOUBLET PI TRANSITIONS OF GERMANIUM MONOFLUORIDE. /SIGMA(+), B DOUBLET SIGMA(0466
C-X AND C´-X´ BAND SYSTEMS AND ENERGY OF DISSOCIATION IN GERMANIUM MONOFLUORIDE. VIBRATIONAL ANALYSIS OF 0470
MONOFLUORIDE. SPECTRA OF SILICON MONOFLUORIDE, GERMANIUM MONOFLUORIDE, TIN MONOFLUORIDE, AND LEAD 0964
NTIALS OF THE MOLECULES BORON FLUORIDE, SILICON OXIDE AND GERMANIUM OXIDE. FIRST IONIZATION POTE 0160
LUORIDE, CARBON TETRACHLORIDE, SILICON TETRACHLORIDE, AND GERMANIUM TETRACHLORIDE. /AFLUORIDE, SILICON TETRAF 1007
/ACHLORIDE, SILICON TETRAFLUORIDE, SILICON TETRACHLORIDE, GERMANIUM TETRACHLORIDE, TITANIUM TETRAFLUORIDE, T/ 0349
A STUDY OF THE RAMAN AND BROAD LINE NMR SPECTRA OF GERMANIUM TETRAFLUORIDE. 0484
ENTHALPY OF FORMATION OF GERMANIUM TETRAFLUORIDE. 0469
INFRARED SPECTRA OF CARBON TETRAFLUORIDE AND GERMANIUM TETRAFLUORIDE. 0363
UDIES OF CARBON TETRAFLUORIDE, SILICON TETRAFLUORIDE, AND GERMANIUM TETRAFLUORIDE. / HELIUM-1 AND HELIUM-2 ST 0341
RAMAN AND INFRARED SPECTRA OF GERMANIUM TETRAFLUORIDE. 0482
ES. HEATS OF FORMATION OF GERMANIUM TETRAFLUORIDE AND OF THE GERMANIUM DIOXID 0483
ISOTOPES OF CARBON TETRAFLUORIDE, SILICON TETRAFLUORIDE, GERMANIUM TETRAFLUORIDE AND SULFUR TRIOXIDE. /RIOUS 0358
F GROUP-5B. (CARBON TETRAFLUORIDE, SILICON TETRAFLUORIDE, GERMANIUM TETRAFLUORIDE, TIN TETRAFLUORIDE) /IDES O 0316
AL VALENCE FORCE CONSTANTS OF TETRAHEDRAL XY4 MOLECULES. (GERMANIUM TETRAFLUORIDES) ORBIT 0481
/LCULATIONS FOR GALLIUM MONOFLUORIDE, GALLIUM TRIHYDRIDE, GERMANIUM TETRAHYDRIDE, ARSENIC TRIHYDRIDE, AND SE/ 0459
/ENCIES OF THE DEFORMATION VIBRATIONS OF THE DIHALIDES OF GERMANIUM, TIN, AND LEAD. (GERMANIUM DIFLUORIDE, T/ 0480
ENTHALPY OF FORMATION OF GERMANIUM TRIFLUORIDE. 0480
TO 350 DEGREES K. HEAT CAPACITY, ENTROPY, ENTHALPY, AND GIBBS ENERGY OF PLUTONIUM-242 TETRAFLUORIDE FROM 10 0889
OINTS-ON-A-SPHERE AND EXTENDED HUCKEL MOLECULAR ORBITAL / GILLESPIE-NYHOLM ASPECTS OF FORCE FIELDS. PART-1. P 0828

/ MEASUREMENTS OF ZEEMAN EFFECT IN DIAMAGNETIC MOLECULES. GJ FACTOR OF THALLIUM FLUORIDE, CESIUM FLUORIDE, C/ 1120
PREPARATION AND CHARACTERIZATION OF GOLD PENTAFLUORIDE. 0485
IX METHOD. FORCE FIELD OF WEINSTOCK AND GOODMAN FOR OCTAHEDRAL HEXAFLUORIDES AND THE L MATR 1456
/D ELECTRONIC PROPERTIES. PART-2: DIPOLE MOMENT AND FIELD GRADIENT OF NITROGEN TRIFLUORIDE. PART-3: A GAUSSI/ 0746
/RMODYNAMIC, ELECTROCHEMICAL AND SYNTHETIC STUDIES OF THE GRAPHITE FLUORINE COMPOUNDS OF CARBON MONOFLUORID/ 0296
L XY4 TYPE MOLECULES AND THEIR SUBSTITUTED DERIVATIVES BY GREEN'S FUNCTION ANALYSIS. /ANTS OF SOME TETRAHEDRA 1424
ND NITROGEN TRICHLORIDE. GREEN'S FUNCTION ANALYSIS OF NITROGEN TRIFLUORIDE A 0729
D RELATED MOLECULES. (SELENIUM HEXAFLUORIDE, TELLURIUM / GREEN'S FUNCTION ANALYSIS OF SULFUR HEXAFLUORIDE AN 1093
GREEN BAND SYSTEM OF BARIUM MONOFLUORIDE. 0121
OMETRY OF PHOSPHINE AND PHOSPHORUS TRIFLUORIDE AND THEIR GROUND IONIC STATES. THEORETICAL STUDY OF THE GE 0794
IES OF SULFUR HEXAFLUORIDE. AB INITIO CALCULATION FOR THE GROUND STATE. ELECTRONIC PROPERT 1077
UCTURE OF THE CARBON TETRAFLUORIDE MOLECULE. PART-1. THE GROUND STATE. ELECTRONIC STR 0330
CYANIDE IO/ CONFIGURATION INTERACTION CALCULATION FOR THE GROUND STATE OF OXYGEN DIFLUORIDE, NITRITE ION AND 0763
ORIDE, AND OX/ MINIMAL BASIS SET SCF CALCULATIONS FOR THE GROUND STATE OF OZONE, NITRITE ION, NITROGEN OXYFLU 0776
ELECTRONIC STRUCTURE AND GROUND STATE OF SCANDIUM MONOFLUORIDE. 0932
(BORON TRIFLUORIDE) GROUND STATE PROPERTIES OF THE GROUP-2 TRIHALIDES. 0185
TANTS AND AVERAGE STRUCTURE OF OXYGEN DIFLUORIDE FROM THE GROUND STATE ROTATIONAL SPECTRUM. /C POTENTIAL CONS 0778
PROPERTIES. PART-2: DIPOLE MOMENT AN/ CHARACTERIZATION OF GROUND STATE WAVE FUNCTIONS BY MEASURED ELECTRONIC 0746
MONOFLUORIDE. GROUND STATES OF SCANDIUM MONOFLUORIDE AND YTTRIUM 0929
/AVE FUNCTIONS AND COMPUTED PROPERTIES FOR THE DOUBLET PI GROUND STATES OF SULFUR MONOFLUORIDE AND SELENIUM / 1025
FLUORIDES OF THE MAIN GROUP ELEMENTS. 0007
E/ HEAT CAPACITIES AND ELECTRONIC SPECTRA OF THE PLATINUM GROUP METAL HEXAFLUORIDE MOLECULES DOWN TO HELIUM T 1159
SYMMETRY GROUP OF PHOSPHORUS PENTAFLUORIDE. 0837
/TRON DIFFRACTION ANALYSIS OF THE MOLECULAR STRUCTURES OF GROUP-2 ELEMENTS. (BERYLLIUM DIFLUORIDE, MAGNESIUM/ 0131
ODEL CALCULATIONS. PART-2: BENDING FORCE CONSTANTS OF THE GROUP-2 HALIDES. IONIC M 1381
BENDING MOTIONS IN THE DIHALIDES OF GROUP-2 METALS. 1380
GROUND STATE PROPERTIES OF THE GROUP-2 TRIHALIDES. (BORON TRIFLUORIDE) 0185
DIFLUORIDE, AND / NONEMPIRICAL LCAO MO SCF STUDIES OF THE GROUP-2A DIHALIDES BERYLLIUM DIFLUORIDE, MAGNESIUM 1360
ECT, AND MOLECULAR POLARIZABILITIES FOR DIHALIDES OF SOME GROUP-2A ELEMENTS. /BASTIANSEN-MORINO SHRINKAGE EFF 0124
/LARIZED ION MODEL AND THE BENDING FORCE CONSTANTS OF THE GROUP-2B HALIDES. (ZINC DIFLUORIDE, CADMIUM DIFLUO/ 1322
HIGH TEMPERATURE NEGATIVE IONS. GASEOUS GROUP-3 FLUORIDES. 0047
PHOTOELECTRON SPECTRA OF THE HALIDES IN GROUP-3, GROUP-4, GROUP-5, AND GROUP-6. 1361
POTENTIAL ENERGY CURVES AND NATURE OF BINDING IN GROUP-3A MONOHALIDES. 0032
E, AND A REASSIGNMENT OF THE SPECTRA OF TETRAFLUORIDES OF GROUP-4. /TRIFLUORIDE AND NITROGEN OXIDE TRIFLUORID 0700
TRAFLUORI/ IONICITY OF THE MX BOND IN THE TETRAHALIDES OF GROUP-4 ELEMENTS. (TIN TETRAFLUORIDE AND HAFNIUM TE 1141
PHOTOELECTRON SPECTRA OF THE HALIDES IN GROUP-3, GROUP-4, GROUP-5, AND GROUP-6. 1361
RIDE, AND HAFNIUM CHLORIDE. INFRARED SPECTRA OF GASEOUS GROUP-4 HALIDES: ZIRCONIUM FLUORIDE, ZIRCONIUM CHLO 1327
HELIUM-1 RESONANCE PHOTOELECTRON SPECTRA OF GROUP-4 TETRAFLUORIDES. 0317
FORCE FIELDS FOR GROUP-4 TETRAFLUORIDES AND GROUP-5 TRIFLUORIDES. 0715
ROOT MEAN SQUARE AMPLITUDES OF VIBRATION IN SOME GROUP-4 TETRAHALIDES. 1363
STUDY OF THE BONDING IN THE GROUP-4 TETRAHALIDES BY PHOTOELECTRON SPECTROSCOPY. 1362
ROOT MEAN SQUARE AMPLITUDES OF VIBRATION IN SOME GROUP-4 TETRAHALIDES. (CARBON TETRAFLUORIDE) 0342
M DIFLUORIDE AND THE THERMODYNAMICS OF SUBLIMATION OF THE GROUP-4A DIFLUORIDES. /Y OF SUBLIMATION OF GERMANIU 0471
D VALUES FOR THERMODYNAMIC FUNCTIONS OF THE TRIHALIDES OF GROUP-5, /, CORIOLIS CONSTANTS, FORCE CONSTANTS, AN 0795
PHOTOELECTRON SPECTRA OF THE HALIDES IN GROUP-3, GROUP-4, GROUP-5, AND GROUP-6. 1361
/STIMATION OF CONFIGURATIONS AND INTERATOMIC DISTANCES IN GROUP-5 AND GROUP-6 TRANSITION ELEMENT TETRAHALIDE/ 1427
PTION SPECTRA OF THE HYDRIDES, DEUTERIDES, AND HALIDES OF GROUP-5 ELEMENTS. (PHOSPHORUS TRIFLUORIDE) ABSOR 0806
S PENTAFLUORIDE, ARSENIC PENTAFLUORIDE, B/ ASSOCIATION OF GROUP-5 PENTAFLUORIDES IN THE GAS PHASE. (PHOSPHORU 0876
OLECULES. PART-12: CORRELATION OF EXCHANGE RATES FOR SOME GROUP-5 PENTAHALIDE MOLECULES. PENTACOORDINATED M 0115
FORCE FIELDS FOR GROUP-4 TETRAFLUORIDES AND GROUP-5 TRIFLUORIDES. 0715
INTRAMOLECULAR FORCE FIELDS IN GROUP-5 TRIHALIDES. 1402
/LITUDES OF VIBRATION AND THERMODYNAMIC FUNCTIONS OF SOME GROUP-5 TRIHALIDES. (NITROGEN TRIFLUORIDE, ARSENIC/ 0719
/OF HALIDES. PART-1. TETRAFLUORIDES AND TETRACHLORIDES OF GROUP-5B. (CARBON TETRAFLUORIDE, SILICON TETRAFLUO/ 0316
SPECTRA OF THE HALIDES IN GROUP-3, GROUP-4, GROUP-5, AND GROUP-6. PHOTOELECTRON 1361
NDING AND IONIZATION POTENTIALS. PART-1: HEXAFLUORIDES OF GROUP-6 ELEMENTS. /HA) SCATTERED WAVE STUDIES ON BO 1366
IUM, MO/ INFRARED SPECTRA AND MOLECULAR STRUCTURE OF SOME GROUP-6 HEXAFLUORIDES. (OF SULFUR, SELENIUM, TELLUR 0961
LURIUM HEXAFLUORIDE) FORCE FIELDS FOR SOME GROUP-6 HEXAFLUORIDES. (SULFUR HEXAFLUORIDE AND TEL 1061
PARAMAGNETIC HEXAFLUORIDE ANIONS OF GROUP-6. (SELENIUM PENTAFLUORIDE) 0959
/ CONFIGURATIONS AND INTERATOMIC DISTANCES IN GROUP-5 AND GROUP-6 TRANSITION ELEMENT TETRAHALIDES AND PENTAH/ 1427
RDINATE TREATMENT OF XY4 TETRAHEDRAL MOLECULES AND IONS. (GVFF) NORMAL COO 1415
ULES WITH D4H SYMMETRY. NEW METHOD FOR THE CALCULATION OF GVFF CONSTANTS. MOLEC 1412

H

E TETRAHEDRAL XY4 TYPE MOLECULES. PART-5: TETRAHALIDES OF HAFNIUM AND LEAD. /N AMPLITUDES OF VIBRATION OF SOM 0487
UP-4 HALIDES: ZIRCONIUM FLUORIDE, ZIRCONIUM CHLORIDE, AND HAFNIUM CHLORIDE. INFRARED SPECTRA OF GASEOUS GRO 1327
VAPOR PRESSURE AND HEAT OF SUBLIMATION OF ZIRCONIUM AND HAFNIUM TETRACHLORIDES. 0486
HEATS OF FORMATION OF TITANIUM TETRAFLUORIDE AND HAFNIUM TETRAFLUORIDE. 1147
TETRAHALIDES OF GROUP-4 ELEMENTS. (TIN TETRAFLUORIDE AND HAFNIUM TETRAFLUORIDE) /ICITY OF THE MX BOND IN THE 1141
ELECTRONIC STRUCTURES OF BORON TRIHALIDES AND BORON TETRAHALIDE ANIONS. A THEORETICAL COMPARISON OF THE 0225
NIC POTENTIAL CONSTANTS OF OSMIUM TETROXIDE AND SOME TETRAHALIDE (CARBON TETRAFLUORIDE) HARMO 0353
/TION OF THE FORCE CONSTANTS AND INTERATOMIC DISTANCES OF HALIDE COMPLEX MOLECULES OF THE XY4 TYPE OF TETRAH/ 1414
/ RAMAN SPECTRA AND STRUCTURE OF MATRIX ISOLATED THALLOUS HALIDE DIMERS, THALLOUS FLUORIDE, AND THALLOUS CHL/ 1123
INTRAMOLECULAR EXCHANGE IN PHOSPHORUS PENTAHALIDE MOLECULES. 0852
NATURE OF BONDS IN THALLIUM MONOHALIDE MOLECULES. 1122
T-12: CORRELATION OF EXCHANGE RATES FOR SOME GROUP-5 PENTAHALIDE MOLECULES. PENTACOORDINATED MOLECULES. PAR 0115
OW TEMPERATURE ANOMALIES. ALUMINUM AND ALKALI HALIDE MONOMERS MEAN AMPLITUDES OF VIBRATION WITH L 0024
CALCULATED MEAN AMPLITUDES OF VIBRATION IN BORON TRIHALIDES. 0197
ATION OF THE VIBRATIONAL FREQUENCIES OF SCANDIUM SUBGROUP HALIDES. CALCUL 0949
CHEMICAL PREDICTIONS BY MO THEORY. THE NOBLE GAS HALIDES. 1340
IFTS OF BORON, PHOSPHORUS, PHOSPHORYL, AND THIOPHOSPHORYL HALIDES. CHEMICAL SH 0235
DETERMINATION OF THE DISSOCIATION ENERGIES OF CARBON MONOHALIDES. 0289
OTENTIALS ON THE UREY-BRADLEY FORCE CONSTANTS OF BORON TRIHALIDES. EFFECTS OF THE LENNARD-JONES P 0244
INVESTIGATION OF THE MOLECULAR STRUCTURES OF THE ALUMINUM HALIDES. ELECTRON DIFFRACTION 0036
ESTIGATION OF THE STRUCTURE OF THE MOLECULES OF NEODYMIUM HALIDES. ELECTRONOGRAPHIC INV 0644
UP-5 AND GROUP-6 TRANSITION ELEMENT TETRAHALIDES AND PENTAHALIDES. /URATIONS AND INTERATOMIC DISTANCES IN GRO 1427
ATS OF FORMATION OF CARBON TETRAFLUORIDE AND OTHER CARBON HALIDES. EXPLOSION METHOD AND THE HE 0315
FORCE CONSTANTS AND THERMODYNAMIC CONSTANTS FOR THE BORON HALIDES. 0219
FORCE FIELDS FOR THE BORON TRIHALIDES. 0227
INFRARED SPECTRA AND GEOMETRIES OF RARE EARTH DIHALIDES. 1351

NFRARED SPECTRA AND STRUCTURE OF MATRIX ISOLATED THALLIUM HALIDES.   I 1113
                          INTRAMOLECULAR FORCE FIELDS IN GROUP-5 TRIHALIDES.   1402
CULATIONS. PART-2: BENDING FORCE CONSTANTS OF THE GROUP-2 HALIDES.   IONIC MODEL CAL 1381
AL EVALUATION OF FORCE CONSTANTS. APPLICATION TO BORON TRIHALIDES.   KINEMATIC 0248
MICROWAVE SPECTRA OF THE THALLIUM, INDIUM AND GALLIUM MONOHALIDES.   1110
                         NORMAL VIBRATION FREQUENCIES OF ZIRCONIUM HALIDES.   1330
        HALIDES. PART-2: HIGH RESOLUTION SPECTRA OF THE BORON TRIHALIDES.   PHOTOELECTRON SPECTRA OF 0188
        AND DISSOCIATION ENERGIES OF DIATOMIC BORON AND ALUMINUM HALIDES.   POTENTIAL ENERGY CURVES 0173
NTIAL ENERGY CURVES AND NATURE OF BINDING IN GROUP-3A MONOHALIDES.   POTE 0032
        MEAN SQUARE AMPLITUDES OF VIBRATION IN SOME GROUP-4 TETRAHALIDES.   ROOT 1363
                        THERMODYNAMIC FUNCTIONS OF LEAD TETRAHALIDES.   0597
THERMODYNAMIC FUNCTIONS OF SOME TRIGONAL BIPYRAMIDAL PENTAHALIDES.   1438
                      THERMODYNAMIC PROPERTIES OF BORON HALIDES.   0240
        THERMODYNAMIC PROPERTIES OF SOME BORON TRIHALIDES.   0236
                          UNIQUE FORCE FIELD FOR BORON TRIHALIDES.   0238
        VACUUM ULTRAVIOLET ABSORPTION SPECTRA OF BORON TRIHALIDES.   0237
VIBRATION FREQUENCIES OF TRIHALIDE MOLECULES OF LANTHANUM HALIDES.   VALENCE 0574
                        VIBRATIONAL ANHARMONICITY IN THE METHYL HALIDES.   1493
                          FORCE CONSTANTS OF THE BORON TRIHALIDES - A SURVEY.   0201
/ES, ENTROPIES, AND FREE ENERGY FUNCTIONS OF THALLIUM MONOHALIDES ABOVE ROOM TEMPERATURE. (THALLIUM MONOFLUO/ 1116
OMETRY AND INFRARED SPECTRA OF MATRIX ISOLATED RARE EARTH HALIDES. (ALSO LANTHANUM TRIFLUORIDE)   GE 0583
IN / ESTIMATED FREE ENERGIES OF FORMATION OF PROTACTINIUM HALIDES AND ACTIVITY COEFFICIENTS FOR PROTACTINIUM 0900
TICAL COMPARISON OF THE ELECTRONIC STRUCTURES OF BORON TRIHALIDES AND BORON TETRAHALIDE ANIONS.   A THEORE 0225
          FORCE CONSTANTS OF THE SILICON HALIDES AND FREQUENCIES OF CHLORO BROMO SILICON.   1004
DISTANCES IN GROUP-5 AND GROUP-6 TRANSITION ELEMENT TETRAHALIDES AND PENTAHALIDES. /URATIONS AND INTERATOMIC 1427
LOW TEMPERATU/ MOLECULAR STRUCTURES OF VARIOUS PHOSPHORUS HALIDES AND PHOSPHORUS PENTAFLUORIDE. INFRARED AND 0849
/Y, NONBONDED INTERACTIONS AND POLARIZABILITY IN HYDROGEN HALIDES AND THE INTERHALOGEN COMPOUNDS. (CHLORINE / 0373
BY THE REDUC/ THERMODYNAMIC CHARACTERISTICS OF SCANDIUM DIHALIDES AND THE THE POSSIBILITY OF THEIR SYNTHESIS 0940
DE)   UREY-BRADLEY FORCE CONSTANTS OF BORON HALIDES AND THEIR TRANSFERABILITY. (BORON TRIFLUORI 0246
XENON DIFLUORIDE)   D ORBITALS IN THE NOBLE GAS DIHALIDES. (ARGON DIFLUORIDE, KRYPTON DIFLUORIDE, AND 1338
        MEAN AMPLITUDES AND THERMODYNAMIC FUNCTIONS OF SOME HALIDES. (ARSENIC TRIFLUORIDE)   0109
CALCIUM DIFLUORIDE)   GEOMETRY OF THE ALKALINE EARTH DIHALIDES. (BARIUM DIFLUORIDE, STRONTIUM DIFLUORIDE, 0126
        AND / NONEMPIRICAL LCAO MO SCF STUDIES OF THE GROUP-2A DIHALIDES BERYLLIUM DIFLUORIDE, MAGNESIUM DIFLUORIDE, 1360
/SEN-MORINO SHRINKAGE EFFECT IN SOME LINEAR SYMMETRICAL DIHALIDES. (BERYLLIUM DIFLUORIDE, MAGNESIUM DIFLUORI/ 1372
          GROUND STATE PROPERTIES OF THE GROUP-2 TRIHALIDES. (BORON TRIFLUORIDE)   0185
                NORMAL VIBRATIONS OF BORON TRIHALIDES. (BORON TRIFLUORIDE)   0247
        STUDY OF THE BONDING IN THE GROUP-4 TETRAHALIDES BY PHOTOELECTRON SPECTROSCOPY.   1362
        INFRARED SPECTRA OF SOME ALKALINE EARTH HALIDES BY THE MATRIX ISOLATION TECHNIQUE.   0125
/AMPLITUDES OF VIBRATION AND SHRINKAGE EFFECT OF BORON TRIHALIDES, CALCULATED FROM UREY-BRADLEY FORCE CONSTA/ 0233
                FORCE CONSTANTS OF THE CARBON HALIDES. (CARBON TETRAFLUORIDE)   0354
MEAN SQUARE AMPLITUDES OF VIBRATION IN SOME GROUP-4 TETRAHALIDES.   (CARBON TETRAFLUORIDE)   ROOT 0342
                STRUCTURES OF TRIFLUORO METHYL HALIDES. (CARBON TETRAFLUORIDE)   0318
SILICON / MOLECULAR BEAM ELECTRIC DEFLECTION OF THE TETRAHALIDES CARBON TETRAFLUORIDE, CARBON TETRACHLORIDE, 0349
DE, BROMINE MONOFLUORIDE, AND B/ PHOTOELECTRON SPECTRA OF HALIDES. (CHLORINE MONOFLUORIDE, CHLORINE TRIFLUORI 0376
        VAPOR PRESSURE AND HEAT OF SUBLIMATION OF COBALT DIHALIDES. (COBALT DIFLUORIDE)   0438
                INFRARED SPECTRUM OF HEAVY METAL DIHALIDES. (EUROPIUM DIFLUORIDE)   0451
        ELECTRON DIFFRACTION ANALYSIS OF GALLIUM HALIDES. (GALLIUM TRIFLUORIDE)   0460
FLUORIDE / BOND ENERGIES AND IONIC CHARACTER OF INERT GAS HALIDES. HELIUM-2 PHOTOELECTRON SPECTRA OF XENON DI 1343
                PHOTOELECTRON SPECTRA OF THE HALIDES IN GROUP-3, GROUP-4, GROUP-5, AND GROUP-6.   1361
D/ ELECTRON DIFFRACTION STUDY OF URANIUM AND THORIUM TETRAHALIDES IN THE GAS PHASE. (URANIUM TETRAFLUORIDE AN 1170
MIC FUNCTIONS OF GASEOUS SCANDIUM, YTTRIUM, AND LANTHANUM HALIDES IN THE 293-3000 DEGREES K RANGE. THERMODYNA 0568
        LATTICE VIBRATIONS AND STRUCTURE OF RARE EARTH HALIDES. (LANTHANUM TRIFLUORIDE)   0572
                MASS SPECTROMETRIC STUDY OF MAGNESIUM HALIDES. (MAGNESIUM DIFLUORIDE)   0603
/AN SPECTRA AND STRUCTURES OF MATRIX ISOLATED MAGNESIUM DIHALIDES: MAGNESIUM DIFLUORIDE, MAGNESIUM DICHLORID/ 0607
        NORMAL COORDINATE ANALYSIS OF SOME NITROGEN HALIDES. (NITROGEN TRIFLUORIDE)   0730
/VIBRATION AND THERMODYNAMIC FUNCTIONS OF SOME GROUP-5 TRIHALIDES. (NITROGEN TRIFLUORIDE, ARSENIC TRIFLUORID/ 0719
ALLIUM.   DISSOCIATION ENERGIES OF THE GASEOUS MONOHALIDES OF BORON, ALUMINUM, GALLIUM, INDIUM, AND TH 0152
/TION STUDY OF THE MOLECULAR STRUCTURE OF THE VAPOR PHASE HALIDES OF GALLIUM, YTTRIUM, LANTHANUM, AND NEODYM/ 1317
LUORI/ FREQUENCIES OF THE DEFORMATION VIBRATIONS OF THE DIHALIDES OF GERMANIUM, TIN, AND LEAD. (GERMANIUM DIF 0479
                BENDING MOTIONS IN THE DIHALIDES OF GROUP-2 METALS.   1380
HAFNIUM TETRAFLUORI/ IONICITY OF THE MX BOND IN THE TETRAHALIDES OF GROUP-4 ELEMENTS. (TIN TETRAFLUORIDE AND 1141
NSTANTS, AND VALUES FOR THERMODYNAMIC FUNCTIONS OF THE TETRAHALIDES OF GROUP-5. /, CORIOLIS CONSTANTS, FORCE CO 0795
E)   ABSORPTION SPECTRA OF THE HYDRIDES, DEUTERIDES, AND HALIDES OF GROUP-5 ELEMENTS. (PHOSPHORUS TRIFLUORID 0806
TION OF SOME TETRAHEDRAL XY4 TYPE MOLECULES. PART-5: TETRAHALIDES OF HAFNIUM AND LEAD. /N AMPLITUDES OF VIBRA 0487
AND MERCURY DIFLUORIDE)   VIBRATIONAL SPECTRA OF THE DIHALIDES OF MERCURY AND CADMIUM. (CADMIUM DIFLUORIDE 0275
                MEAN AMPLITUDES OF VIBRATION FOR THE DIHALIDES OF MERCURY, BERYLLIUM, AND MAGNESIUM.   0619
SURES OF NIOBIUM AND TANTALUM PENTAFLUORIDES.   HALIDES OF NIOBIUM AND TANTALUM. PART-3: VAPOR PRES 0663
NO SHRINKAGE EFFECT, AND MOLECULAR POLARIZABILITIES FOR DIHALIDES OF SOME GROUP-2A ELEMENTS. /BASTIANSEN-MORI 0124
                HALIDES OF THE LANTHANIDES AND ACTINIDES.   1348
RIDE, YTTRI/ CALCULATION OF DISSOCIATION ENERGIES FOR THE HALIDES OF THE SCANDIUM SUBGROUP. (SCANDIUM TRIFLUO 0948
LS.   HALIDES OF THE SECOND AND THIRD ROW TRANSITION META 1374
ROW TRANSITION METALS. VOLUME-2: SECOND AND THIRD ROW T/ HALIDES OF THE TRANSITION ELEMENTS. VOLUME-1: FIRST 0001
M, AND AMERICIUM. (ACTINIDE)   HEATS OF FORMATION OF SOME HALIDES OF THORIUM, PROTACTINIUM, URANIUM, NEPTUNIU 1335
                PENTAHALIDES OF TRANSITION METALS.   1367
AMIC FUNCTIONS AND MOLECULAR POLARIZABILITES OF SOME PENTAHALIDES OF TRIGONAL BIPYRAMIDAL SYMMTERY. /HERMODYN 1437
CHLORIDE, MAGNESI/ INFRARED SPECTRA OF THE ALKALINE EARTH HALIDES. PART-1: BERYLLIUM DIFLUORIDE, BERYLLIUM DI 0133
GEOMETRY AND VIBRATIONAL SPECTRA OF THE ALKALINE EARTH DIHALIDES. PART-1: MAGNESIUM DIFLUORIDE.   0609
OF GROUP-5B. (CARBON TETRAFLUOR/ PHOTOELECTRON SPECTRA OF HALIDES. PART-1: TETRAFLUORIDES AND TETRACHLORIDES 0316
ON TRIHALIDES.   PHOTOELECTRON SPECTRA OF HALIDES. PART-2: HIGH RESOLUTION SPECTRA OF THE BOR 0188
S OF NITROGEN AND PHOSPHORUS, A/ PHOTOELECTRON SPECTRA OF HALIDES. PART-3: TRIFLUORIDES AND OXIDE TRIFLUORIDE 0701
                MOLECULAR CONSTANTS OF PHOSPHORUS HALIDES. PART-4. (PHOSPHORUS PENTAFLUORIDE)   0867
HELIUM-2 STUDIES OF CARBON TETR/ PHOTOELECTRON SPECTRA OF HALIDES. PART-7: VARIABLE TEMPERATURE HELIUM-1 AND 0341
                VIBRATIONAL SPECTRA OF MIXED PHOSPHORUS HALIDES. (PHOSPHORUS PENTAFLUORIDE)   0829
/LD AND FORCE CONSTANTS OF SOME TRIGONAL BIPYRAMIDAL PENTAHALIDES. (PHOSPHORUS PENTAFLUORIDE AND ANTIMONY PE/ 0850
HERMODYNAMIC FUNCTIONS FOR VARIOUS PHOSPHORUS AND ARSENIC HALIDES. (PHOSPHORUS TRIFLUORIDE) /ED SPECTRA AND T 0811
                VAPOR PRESSURE OF PLUTONIUM HALIDES. (PLUTONIUM TRIFLUORIDE)   0887
POSSIBILITY OF THEIR SYNTHESIS BY THE REDUCTION OF THE TRIHALIDES.   (SCANDIUM DIFLUORIDE) /LIDES AND THE THE 0940
/ MOLECULAR CONSTANTS OF SCANDIUM, YTTRIUM, AND LANTHANUM HALIDES. (SCANDIUM DIFLUORIDE, SCANDIUM MONOFLUORID 0941
ON METALS.   GEOMETRY OF THE TRANSITION METAL DIHALIDES. THE DIFLUORIDES OF SOME FIRST ROW TRANSITI 1368
        VIBRATIONAL SPECTRA AND THERMODYNAMICS OF THE ZINC HALIDES. (ZINC DIFLUORIDE)   1323

/ON MODEL AND THE BENDING FORCE CONSTANTS OF THE GROUP-2B HALIDES. (ZINC DIFLUORIDE, CADMIUM DIFLUORIDE, AND/ 1322
D HAFNIUM CHLORIDE.   INFRARED SPECTRA OF GASEOUS GROUP-4 HALIDES: ZIRCONIUM FLUORIDE, ZIRCONIUM CHLORIDE, AN 1327
ELECTRON IMPACT STUDIES OF SOME TRIHALO METHANES. (CARBON TRIFLUORIDE) 0312
HALOGEN CHEMISTRY. (VOLUMES 1, 2, 3) 0005
MOLECULAR CONSTANTS OF SOME PHOSPHORUS HALOGEN COMPOUNDS. (PHOSPHORUS PENTAFLUORIDE) 0868
MOLECULAR CONSTANTS OF PHOSPHORUS HALOGEN COMPOUNDS. (PHOSPHORUS TRIFLUORIDE) 0813
DE) CORREL/ VACUUM ULTRAVIOLET ABSORPTION SPECTRA OF SOME HALOGEN DERIVATIVES OF METHANE. (CARBON TETRAFLUORI 0365
PHYSICAL AND CHEMICAL PROPERTIES OF HALOGEN FLUORIDES. 0017
SELF IONIZATION OF HALOGEN FLUORIDES. (BROMINE PENTAFLUORIDE) 0263
DE) DISSOCIATION ENERGIES OF DIATOMIC HALOGEN FLUORIDES. (IODINE FLUORIDE, BROMINE FLUORI 0494
NE HEXAFLUORIDE, BROM/ ELECTRON SPIN RESONANCE SPECTRA OF HALOGEN HEXAFLUORIDES. (CHLORINE HEXAFLUORIDE, IODI 0271
NIZATION STUDIES AND THE THERMODYNAMIC PROPERTIES OF SOME HALOGEN MOLECULES. (CHLORINE FLUORIDE)   PHOTOIO 0377
(AL/ CALCULATIONS OF ELECTRON STRUCTURES OF COMPOUNDS OF HALOGENS AND INERT GASES BY THE NONEMPIRICAL NDDO-2 0016
/ CALCULATIONS OF ELECTRONIC STRUCTURES OF COMPOUNDS OF HALOGENS AND INERT GASES BY THE SCF LCAO MO METHOD. 1231
) INFRARED SPECTRA OF HALOGEN-SUBSTITUTED METHANES. (CARBON TETRAFLUORIDE 0355
REY-BRADLEY FORCE FIELD AND THE ASSIGNMENTS OF SOME MIXED HALOMETHANES. (CARBON TETRAFLUORIDE) /RANSFERABLE U 0351
NUM TRIFLUORIDE MONOMER AND DIMER. HARMONIC FORCE FIELDS AND MEAN AMPLITUDES FOR ALUMI 0038
RIFLUORIDE, BORON TRIFLUORIDE, PHOSPH/ AMBIGUITIES IN THE HARMONIC FORCE FIELDS OF XY3 MOLECULES. (NITROGEN T 0709
COUPLING CONSTANTS, CENTRIFUGAL STRETCHING CONSTANTS, AN/ HARMONIC FREQUENCIES, POTENTIAL FUNCTION, CORIOLIS 0726
D SOME TETRAHALIDE (CARBON TETRAFLUORIDE) HARMONIC POTENTIAL CONSTANTS OF OSMIUM TETROXIDE AN 0353
/ CALCULATIONS ON SOME SMALL RADICALS BY THE UNRESTRICTED HARTREE-FOCK METHOD. (BORON DIFLUORIDE, NITROGEN D/ 0176
C CONSTANTS, AND IDEAL GAS THERMODYNAMIC FUNCTIONS FROM A HARTREE-FOCK MOLECULAR ORBITAL INVESTIGATION. /COPI 0094
ION POTENTIAL, ELECTRON AFFINITY, D/ OXYGEN MONOFLUORIDE. HARTREE-FOCK WAVE FUNCTION, BINDING ENERGY, IONIZAT 0756
ION POTENTIAL, ELECTRON / OXYGEN MONOFLUORIDE, DOUBLET PI HARTREE-FOCK WAVE FUNCTION, BINDING ENERGY, IONIZAT 0757
FOR THE DOUBLET PI GROUND STATES OF SULFUR MONOFLUORIDE/ HARTREE-FOCK WAVE FUNCTIONS AND COMPUTED PROPERTIES 1025
HEAT AND ENTROPY OF FORMATION OF BORON FLUORIDE. 0154
NUM GROUP METAL HEXAFLUORIDE MOLECULES DOWN TO HELIUM TE/ HEAT CAPACITIES AND ELECTRONIC SPECTRA OF THE PLATI 1159
UM TETRAFLUORIDE AND URANIUM HEXAFLUORID/ LOW TEMPERATURE HEAT CAPACITIES, ENTHALPIES, AND ENTROPIES OF URANI 1166
OTHER MOLECULES.   CONSTANT VOLUME HEAT CAPACITIES OF GASEOUS CARBON TETRAFLUORIDE AND 0335
UTONIUM-242 TETRAFLUORIDE FROM 10 TO 350 DE/ EXPERIMENTAL HEAT CAPACITIES OF PLUTONIUM-242 TRIFLUORIDE AND PL 0885
XENON HEXAFLUORIDE FROM 5 TO 350 DEGREES K. HEAT CAPACITY AND OTHER THERMODYNAMIC FUNCTIONS OF 1305
MOLYBDENUM HEXAFLUORIDE BETWEEN 4 AND 350 DEGREES K. HEAT CAPACITY AND OTHER THERMODYNAMIC PROPERTIES OF 0636
F XENON TETRAFLUORIDE. HEAT CAPACITY AND RELATED THERMODYNAMIC FUNCTIONS O 1271
350 DEGREES K. RECONCILIATION OF T/ XENON TETRAFLUORIDE. HEAT CAPACITY AND THERMODYNAMIC FUNCTIONS FROM 5 TO 1275
IUM FLUORIDE FROM 5 TO 445 DEGREES K. HEAT CAPACITY AND THERMODYNAMIC PROPERTIES OF THALL 1124
OF FUSION A/ CALORIMETRIC STUDY OF IODINE PENTAFLUORIDE. HEAT CAPACITY BETWEEN 5 AND 350 DEGREES K, ENTHALPY 0510
C PROPERTIES OF NEPTUNIUM HEXAFLUORIDE FROM 7 TO 350 DEG/ HEAT CAPACITY, ENTHALPY OF FUSION, AND THERMODYNAMI 0650
42 TRIFLUORIDE FROM 10 TO 350 DEGREES K. HEAT CAPACITY, ENTROPY, AND ENTHALPY OF PLUTONIUM-2 0886
MOLYBDENUM HEXAFLUORIDE AND NIOBIUM PENTAFLUORIDE. HEAT CAPACITY, ENTROPY, AND HEATS OF TRANSITION OF 0662
OF PLUTONIUM-242 TETRAFLUORIDE FROM 10 TO 350 DEGREES K. HEAT CAPACITY, ENTROPY, ENTHALPY, AND GIBBS ENERGY 0889
VED THERMODYNAMIC PROPERTIES.   XENON DIFLUORIDE. HEAT CAPACITY FROM 5 TO 350 DEGREES K AND SOME DERI 1246
CONSTANT VOLUME HEAT CAPACITY OF GASEOUS CARBON TETRAFLUORIDE. 0336
HIGH TEMPERATURE HEAT CONTENT OF URANIUM TETRAFLUORIDE. 1173
F SEVEN RARE EARTH TRIFLUORIDES. (YTTRI/ HIGH TEMPERATURE HEAT CONTENTS AND RELATED THERMODYNAMIC FUNCTIONS O 1320
ION OF THE ELEMENTS. HEAT FORMATION BORON TRIFLUORIDE BY DIRECT COMBINAT 0198
HEAT FORMATION OF KRYPTON DIFLUORIDE. 0557
HEAT OF ATOMIZATION OF ALUMINUM DIFLUORIDE. 0034
XENON TETRAFLUORIDE: HEAT OF FORMATION. 1267
COMBINATION OF THE ELEMENTS. HEAT OF FORMATION OF ALUMINUM TRIFLUORIDE BY DIRECT 0039
STANDARD HEAT OF FORMATION OF ARSENIC TRIFLUORIDE. 0110
ENERGIES OF BERYLLIUM FLUORIDE AND BERYLLIUM CHLORIDE AND HEAT OF FORMATION OF BERYLLIUM CHLOROFLUORIDE. /ON 0128
HEAT OF FORMATION OF BERYLLIUM DIFLUORIDE. 0132
HEAT OF FORMATION OF BORON DIFLUORIDE. 0177
FLUORINE BOMB CALORIMETRY. PART-3: THE HEAT OF FORMATION OF BORON TRIFLUORIDE. 0250
HEAT OF FORMATION OF CARBON DIFLUORIDE. 0298
DEGREES K. HEAT OF FORMATION OF CHLORINE TRIFLUORIDE AT 298.14 0401
HEAT OF FORMATION OF OXYGEN DIFLUORIDE. 0769
HEAT OF FORMATION OF THE CARBON DIFLUORIDE RADICAL. 0304
HEAT OF FORMATION OF XENON TETRAFLUORIDE. 1266
LANTHANUM TRIFLUORIDE.   VAPOR PRESSURE, HEAT OF SUBLIMATION, AND EVAPORATION COEFFICIENT OF 0576
T OF PRASEODYMIUM TRIFLUORIDE.   VAPOR PRESSURE, HEAT OF SUBLIMATION, AND THE EVAPORATION COEFFICIEN 0896
T OF FORMATION OF ALUMINUM MONOFLUORIDE.   HEAT OF SUBLIMATION OF ALUMINUM TRIFLUORIDE AND HEA 0053
VAPOR PRESSURE AND HEAT OF SUBLIMATION OF CALCIUM FLUORIDE. 0282
VAPOR PRESSURE AND HEAT OF SUBLIMATION OF CERIUM TRIFLUORIDE. 0366
FLUORIDE)   VAPOR PRESSURE AND HEAT OF SUBLIMATION OF COBALT DIHALIDES. (COBALT DI 0438
MASS SPECTROMETRIC METHOD.   DETERMINATION OF HEAT OF SUBLIMATION OF URANIUM TETRAFLUORIDE BY THE 1165
HLORIDES.   VAPOR PRESSURE AND HEAT OF SUBLIMATION OF ZIRCONIUM AND HAFNIUM TETRAC 0486
MELTING POINT, VAPOR PRESSURE, AND HEAT OF VAPORIZATION OF TIN DIFLUORIDE. 1140
BALLISTIC PISTON COMPRESSOR.   SPECIFIC HEAT RATIO OF URANIUM HEXAFLUORIDE MEASURED WITH A 1201
SURES OF AMERICIUM TRIFLUORIDE AND PLUTONIUM TRIFLUORIDE, HEATS AND FREE ENERGIES OF SUBLIMATION. VAPOR PRES 0055
THORIUM DIFLUORIDE FROM MASS SPECTROMETRIC STUDIES. HEATS OF FORMATION GASEOUS THORIUM TRIFLUORIDE AND 1128
/ILLATOR STRENGTHS OF CARBON MONOFLUORIDE AND COMMENTS ON HEATS OF FORMATION OF CARBON MONOFLUORIDE AND CARB/ 0295
R CARBON HALIDES.   EXPLOSION METHOD AND THE HEATS OF FORMATION OF CARBON TETRAFLUORIDE AND OTHE 0315
MONOFLUORIDE, CHLORINE DIFLUORIDE, AND CHLORINE TRIFLU/ HEATS OF FORMATION OF CHLORINE FLUORIDES. (CHLORINE 0370
RANIUM MONOFLUORIDE, URANIUM DIFLUORIDE, URANIUM TRIFLUO/ HEATS OF FORMATION OF GASEOUS URANIUM FLUORIDES. (U 1160
F THE GERMANIUM DIOXIDES. HEATS OF FORMATION OF GERMANIUM TETRAFLUORIDE AND O 0483
Y THE ELEMENTS OF ATOMIC NUMBER BELOW 20. HEATS OF FORMATION OF INORGANIC FLUORIDES ESPECIALL 0214
ACTINIUM, URANIUM, NEPTUNIUM, AND AMERICIUM. (ACTINIDE) HEATS OF FORMATION OF SOME HALIDES OF THORIUM, PROT 1335
M OXYDIFLUORIDE / MASS SPECTROMETRIC DETERMINATION OF THE HEATS OF FORMATION OF THE GASEOUS MOLECULES ALUMINU 0035
FNIUM TETRAFLUORIDE. HEATS OF FORMATION OF TITANIUM TETRAFLUORIDE AND HA 1147
IUM TRIFLUORIDE, AND LUTETIUM / SUBLIMATION PRESSURES AND HEATS OF SUBLIMATION OF THULIUM TRIFLUORIDE, YTTERB 1132
TETRAFLUORIDE AND A CONJECTURE ON BONDING IN THE SOLIDS. HEATS OF SUBLIMATION OF XENON DIFLUORIDE AND XENON 1237
NIOBIUM PENTAFLUORIDE.   HEAT CAPACITY, ENTROPY, AND HEATS OF TRANSITION OF MOLYBDENUM HEXAFLUORIDE AND 0662
ATION OF FORCE CONSTANTS. ANHARMONICITY EFFEC/ UTILITY OF HEAVY ATOM ISOTOPIC SUBSTITUTION DATA FOR THE CALCUL 0234
UORIDE.   ACCURATE FORCE CONSTANTS FROM HEAVY ISOTOPIC SUBSTITUTION. PART-1. CARBON TETRAFL 0320
IDE, SILICON TETRAFLUORIDE/ ACCURATE FORCE CONSTANTS FROM HEAVY ISOTOPIC SUBSTITUTION. PART-2: BORON TRIFLUOR 0232
INFRARED SPECTRUM OF HEAVY METAL DIHALIDES. (EUROPIUM DIFLUORIDE) 0451
AFLUORIDE, MOLYBDENUM HEXAFLUORI/ VAPOR PRESSURES OF SOME HEAVY TRANSITION METAL HEXAFLUORIDES. (TUNGSTEN HEX 0903
FLUORIDES OF THE HEAVY TRANSITION METALS. (REVIEW) 0002
/SPHORUS PENTAFLUORIDE AND ARSENIC PENTAFLUORIDE AND THE HEIGHT OF THE BARRIER TO INTERNAL EXCHANGE OF FLUO/ 0856

F THE PLATINUM GROUP METAL HEXAFLUORIDE MOLECULES DOWN TO HELIUM TEMPERATURES. /TIES AND ELECTRONIC SPECTRA O 1159
/ORINE, BORON NITRIDE, BERYLLIUM OXIDE, LITHIUM FLUORIDE, HELIUM- NEON, CARBON MONOXIDE, AND BORON MONOFLUOR/ 0174
ELECTRONIC STRUCTURES OF XENON DIFLUORIDE, XENON TETRAF/ HELIUM-1 AND HELIUM-2 PHOTOELECTRON SPECTRA AND THE 1228
/LECTRON SPECTRA OF HALIDES. PART-7: VARIABLE TEMPERATURE HELIUM-1 AND HELIUM-2 STUDIES OF CARBON TETRAFLUOR/ 0341
TETRAFLUORIDES.   HELIUM-1 RESONANCE PHOTOELECTRON SPECTRA OF GROUP-4 0317
TRUCTURES OF XENON DIFLUORIDE, XENON TETRAF/ HELIUM-1 AND HELIUM-2 PHOTOELECTRON SPECTRA AND THE ELECTRONIC S 1228
AND KRYPTON DIFLUORIDE.   HELIUM-2 PHOTOELECTRON SPECTRA OF XENON DIFLUORIDE 1234
/ BOND ENERGIES AND IONIC CHARACTER OF INERT GAS HALIDES. HELIUM-2 PHOTOELECTRON SPECTRA OF XENON DIFLUORIDE 1343
/S BY A MANY BODY APPROACH. PART-6: THE ASSIGNMENT OF THE HELIUM-2 PHOTOELECTRON SPECTRUM OF SULFUR HEXAFLUO/ 1090
/RA OF HALIDES. PART-7: VARIABLE TEMPERATURE HELIUM-1 AND HELIUM-2 STUDIES OF CARBON TETRAFLUORIDE, SILICON / 0341
EVIDENCE FOR THE MOLECULAR SYMMETRY OF IODINE HEPTAFLUORIDE.   0523
FORCE CONSTANTS AND THERMODYNAMIC FUNCTIONS OF IODINE HEPTAFLUORIDE.   0528
NE PENTAFLUORIDE, IODINE OXYGEN PENTAFLUORIDE, AND IODINE HEPTAFLUORIDE. /LD FOR INTERHALOGEN COMPOUNDS. IODI 0511
MOLECULAR ORBITAL TREATMENT OF IODINE HEPTAFLUORIDE.   0533
MOLECULAR STRUCTURE OF IODINE HEPTAFLUORIDE.   0530
MOLECULAR STRUCTURE OF XENON HEXAFLUORIDE AND IODINE HEPTAFLUORIDE.   0526
MOLECULAR SYMMETRY OF IODINE HEPTAFLUORIDE.   0524
MOLECULAR SYMMETRY OF IODINE HEPTAFLUORIDE.   0531
MOSSBAUER STUDY OF IODINE PENTAFLUORIDE AND IODINE HEPTAFLUORIDE.   0503
ERMODYNAMIC FUNCTIONS FOR IODINE PENTAFLUORIDE AND IODINE HEPTAFLUORIDE.   POTENTIAL CONSTANTS AND TH 0509
POTENTIAL CONSTANTS OF IODINE HEPTAFLUORIDE.   0532
PREPARATION AND PROPERTIES (INFRARED SPECTRUM) OF OSMIUM HEPTAFLUORIDE.   0754
PREPARATION AND PROPERTIES OF RHENIUM HEPTAFLUORIDE.   0921
AND OTHER PROPERTIES OF RHENIUM HEXAFLUORIDE AND RHENIUM HEPTAFLUORIDE.   VAPOR PRESSURES 0912
TIONAL SPECTRA AND STRUCTURES OF IODINE PENTAFLUORIDE AND HEPTAFLUORIDE.   VIBRA 0506
VIBRATIONAL SPECTRA OF RHENIUM HEPTAFLUORIDE.   0919
METAL HEXAFLUORIDES. (SECOND ROW, THIRD ROW, AND RHENIUM HEPTAFLUORIDE) /FRACTION STUDIES OF SOME TRANSITION 1375
, PSEUDOROTATION, AND VIBRATIONAL MODE COUPLING IN IODINE HEPTAFLUORIDE. AN ELECTRON DIFFRACTION STUDY. /TURE 0515
/RIMETHYL SILYL) SILANE, INFRARED INVESTIGATION OF IODINE HEPTAFLUORIDE AND AN APPLICATION OF THREE ATOM SCA/ 0527
LIGAND EXCHANGE IN SEVEN COORDINATE STRUCTURES. (RHENIUM HEPTAFLUORIDE AND IODINE HEPTAFLUORIDE) /AMOLECULAR 0922
POLAR DISTORTIONS IN RHENIUM HEPTAFLUORIDE AND IODINE HEPTAFLUORIDE.   0529
US HYDROGEN FLUORIDE.   SPECTROSCOPIC STUDY OF IODINE HEPTAFLUORIDE AND IODINE HEXAFLUORIDE(X) IN ANHYDRO 0518
MASS SPECTRA AND SUBLIMATION PRESSURES OF IODINE HEPTAFLUORIDE AND IODINE OXYGEN PENTAFLUORIDE.   0534
NMR SPECTRA OF IODINE HEPTAFLUORIDE AND IODINE OXYGEN PENTAFLUORIDE.   0516
VIBRATIONAL SPECTRA OF IODINE HEPTAFLUORIDE AND RHENIUM HEPTAFLUORIDE.   0522
INTRAMOLECULAR REARRANGEMENT IN IODINE HEPTAFLUORIDE AND XENON HEXAFLUORIDE.   0520
C.   STRUCTURE OF INTERHALOGEN COMPOUNDS. PART-2: IODINE HEPTAFLUORIDE AT -110 DEGREES C AND AT -145 DEGREES 0521
/E, BROMINE PENTAFLUORIDE, IODINE PENTAFLUORIDE, CHLORINE HEPTAFLUORIDE, BROMINE HEPTAFLUORIDE, AND IODINE H/ 0397
/ FORCE CONSTANTS AND MEAN VIBRATION AMPLITUDES OF IODINE HEPTAFLUORIDE CALCULATED BY THE LOGARITHMIC STEP ME 0535
MOLECULAR CONSTANTS OF RHENIUM HEPTAFLUORIDE MOLECULE.   0918
ACCEPTOR PROPERTIES OF IODINE HEPTAFLUORIDE. OCTAFLUORO PERIODATES(VII).   0514
WITH LARGE AMPLITUDE MOTION. (XENON HEXAFLUORIDE, IODINE HEPTAFLUORIDE, RHENIUM HEPTAFLUORIDE) /OF MOLECULES 1281
TAFLUORIDES.   FLUORINE NMR OF IODINE HEPTAFLUORIDE, RHENIUM HEPTAFLUORIDE, AND OXIDE PEN 0517
A REDETERMINATION OF THE ORTHORHOMBIC IODINE HEPTAFLUORIDE STRUCTURE. A CORRECTION.   0519
MONIC COUPLING OF / ELECTRON DIFFRACTION STUDY OF RHENIUM HEPTAFLUORIDE. STRUCTURE, PSEUDOROTATION, AND ANHAR 0920
TY LOCALIZATION OF ATOMIC AND MOLECULAR ORBITALS. PART-3: HETERONUCLEAR DIATOMIC AND POLYATOMIC MOLECULES. /I 0175
AB INITIO AND MZDO WAVE FUNCTIONS FOR SULFUR HEXAFLUORIDE.   1066
ABSORPTION SPECTRUM AND MAGNETIC PROPERTIES OF OSMIUM HEXAFLUORIDE.   0751
ABSORPTION SPECTRUM OF PLUTONIUM HEXAFLUORIDE.   0892
TO MOLECULAR SPECTROSCOPY. MOLECULAR NITROGEN AND SULFUR HEXAFLUORIDE. /TION OF SYNCHROTON ORBITAL RADIATION 1083
BONDING IN XENON HEXAFLUORIDE.   1298
BONDING IN XENON HEXAFLUORIDE.   1313
CALCULATED ELECTRONIC STRUCTURE OF URANIUM HEXAFLUORIDE.   1188
ICATIONS OF CORE ELECTRON EXCITATION SPECTROSCOPY. (XENON HEXAFLUORIDE)   CHEMICAL APPL 1306
BAND CONTOURS. APPLICATION TO THE FORCE FIELD OF TUNGSTEN HEXAFLUORIDE. /MPERATURE INFRARED STUDIES OF VAPOR 1156
CRYSTALLINE MODIFICATIONS OF XENON HEXAFLUORIDE.   1297
DIPOLE MOMENT AND MOLECULAR STRUCTURE OF URANIUM HEXAFLUORIDE.   1200
CULES AND COMPLEX IONS. (PHOSPHORUS PENTAFLUORIDE, SULFUR HEXAFLUORIDE)   DIRECTED VALENCY IN CERTAIN MOLE 0860
AY DIFFRACTION AND AN ELECTRON DIFFRACTION STUDY OF XENON HEXAFLUORIDE. /FECTS OF ELECTRON CORRELATION IN X-R 1290
ICAL CONDUCTIVITY AND DIELECTRIC CONSTANT OF LIQUID XENON HEXAFLUORIDE.   ELECTR 1307
ELECTRON DIFFRACTION STUDY OF RHENIUM HEXAFLUORIDE.   0909
TION SPECTRUM OF GASEOUS URANIUM HEXACHLORIDE AND URANIUM HEXAFLUORIDE.   ELECTRONIC ABSORP 1194
SIGNMENT OF THE HELIUM-2 PHOTOELECTRON SPECTRUM OF SULFUR HEXAFLUORIDE. /A MANY BODY APPROACH. PART-6: THE AS 1090
ELECTRONIC STRUCTURE OF NEPTUNIUM HEXAFLUORIDE.   0648
ENTHALPY OF FORMATION OF TUNGSTEN HEXAFLUORIDE.   1157
HALPY OF FORMATION OF TUNGSTEN PENTAFLUORIDE AND TUNGSTEN HEXAFLUORIDE.   ENT 1151
ESR SPECTRUM AND STRUCTURE OF BROMINE HEXAFLUORIDE.   0270
EVIDENCE FOR PSEUDO JAHN-TELLER EFFECT IN XENON HEXAFLUORIDE.   1279
ERS IN THE X-RAY ABSORPTION SPECTRA OF MOLECULES. (SULFUR HEXAFLUORIDE) EVIDENCE OF EFFECTIVE POTENTIAL BARRI 1073
ST BORN APPROXIMATION IN ELECTRON DIFFRACTION. (TELLURIUM HEXAFLUORIDE)   FAILURE OF THE FIR 1109
IRST BORN APPROXIMATION IN ELECTRON DIFFRACTION. (URANIUM HEXAFLUORIDE)   FAILURE OF THE F 1199
FAR INFRARED AND MILLIMETER WAVE STUDIES OF XENON HEXAFLUORIDE.   1299
FAR INFRARED SPECTRA OF GASEOUS AND LIQUID SULFUR HEXAFLUORIDE.   1095
S CARBON TETRAFLUORIDE, SILICON TETRAFLUORIDE, AND SULFUR HEXAFLUORIDE. /LAXATION BY NMR ABSORPTION IN GASEOU 0346
FORCE CONSTANTS FOR SULFUR HEXAFLUORIDE.   1096
GAS PHASE STRUCTURE OF XENON HEXAFLUORIDE.   1293
URANIUM TETRAFLUORIDE, URANIUM PENTAFLUORIDE, AND URANIUM HEXAFLUORIDE) /UM DIFLUORIDE, URANIUM TRIFLUORIDE, 1160
TURES OF XENON DIFLUORIDE, XENON TETRAFLUORIDE, AND XENON HEXAFLUORIDE. /RON SPECTRA AND THE ELECTRONIC STRUC 1228
HIGHER ELECTRONIC STATES OF RHENIUM HEXAFLUORIDE.   0914
ERTIES OF XENON COMPOUNDS. (XENON TETRAFLUORIDE AND XENON HEXAFLUORIDE)   IDEAL GAS THERMODYNAMIC PROP 1273
INFRARED BAND CONTOURS OF RHENIUM HEXAFLUORIDE.   0915
INFRARED DOUBLE RESONANCE IN SULFUR HEXAFLUORIDE.   1099
INFRARED INTENSITIES IN CRYSTALLINE SULFUR HEXAFLUORIDE.   1075
INFRARED SPECTRA OF CRYOSYSTEMS. PART-2: SULFUR HEXAFLUORIDE.   1068
INFRARED SPECTRUM OF IRIDIUM HEXAFLUORIDE.   0540
INFRARED SPECTRUM OF PLUTONIUM HEXAFLUORIDE.   0891
INFRARED SPECTRUM OF SULFUR HEXAFLUORIDE.   1086
INFRARED SPECTRUM OF URANIUM HEXAFLUORIDE.   1182
MOLECULAR REARRANGEMENT IN IODINE HEPTAFLUORIDE AND XENON HEXAFLUORIDE.   INTRA 0520
JAHN-TELLER EFFECT IN RHENIUM HEXAFLUORIDE.   0904

LER EFFECTS IN THE 2E5/2G TO 2G3/2G TRANSITION OF RHENIUM HEXAFLUORIDE.   JAHN-TEL 0913
                    JAHN-TELLER VIBRATIONS OF RHENIUM HEXAFLUORIDE.   0910
TEMPERATURE ABSORPTION CONTOUR OF THE NU-3 BAND OF SULFUR HEXAFLUORIDE.   LOW 1082
LPIES, AND ENTROPIES OF URANIUM TETRAFLUORIDE AND URANIUM HEXAFLUORIDE. /W TEMPERATURE HEAT CAPACITIES, ENTHA 1166
                    MAGNETIC FIELD DEFLECTION OF XENON HEXAFLUORIDE.   1287
                    MAGNETIC SUSCEPTIBILITY OF RHENIUM HEXAFLUORIDE.   0917
                    MAGNETIC SUSCEPTIBILITY OF XENON HEXAFLUORIDE.   1308
                    MAGNETIC SUSCEPTIBILITY OF XENON HEXAFLUORIDE.   1310
ES OF VIBRATION AND THERMODYNAMIC FUNCTIONS OF TECHNETIUM HEXAFLUORIDE.   MEAN AMPLITUD 1104
ION OF SULFUR DIFLUORIDE, SULFUR TETRAFLUORIDE AND SULFUR HEXAFLUORIDE.   MOLECULAR ORBITAL DESCRIPT 1027
                    MOLECULAR STRUCTURE OF XENON HEXAFLUORIDE.   1280
                    MOLECULAR STRUCTURE OF XENON HEXAFLUORIDE.   1303
ALPHA) CALCULATION OF THE ENERGY LEVEL STRUCTURE OF XENON HEXAFLUORIDE.   MULTIPLE SCATTERING X (1304
                    NEAR INFRARED SYSTEM OF RHENIUM HEXAFLUORIDE.   0902
                    NU-3 AND NU-4 BANDS OF SULFUR HEXAFLUORIDE.   1070
     OCTAHEDRAL FINE STRUCTURE SPLITTINGS IN NU-3 OF SULFUR HEXAFLUORIDE.   1063
ON- ROTATION TRANSITION IN THE NU-3 FUNDAMENTAL OF SULFUR HEXAFLUORIDE. /TICAL SATURATION OF A SINGLE VIBRATI 1074
STUDY OF XENON DIFLUORIDE, XENON TETRAFLUORIDE, AND XENON HEXAFLUORIDE. PHOTOIONIZATION MASS SPECTROMETRIC 1225
                    POLYMORPHISM IN XENON HEXAFLUORIDE.   1296
OF RAYLEIGH SCATTERING IN CERTAIN HEXAFLUORIDES. (RHENIUM HEXAFLUORIDE) /MAN SPECTRUM AND THE DEPOLARIZATION 0901
                    PREPARATION AND SOME PROPERTIES OF PLATINUM HEXAFLUORIDE.   0882
     PREPARATION AND SOME THERMAL PROPERTIES OF PURE XENON HEXAFLUORIDE.   1301
        RAMAN SPECTRA OF MATRIX ISOLATED MOLECULES. (SULFUR HEXAFLUORIDE)   1097
                    RAMAN SPECTRUM OF SOLID SULFUR HEXAFLUORIDE.   1098
                    RAMAN SPECTRUM OF URANIUM HEXAFLUORIDE.   1190
                    RAMAN SPECTRUM OF XENON HEXAFLUORIDE.   1289
          SEARCH FOR A JAHN-TELLER EFFECT IN IRIDIUM HEXAFLUORIDE.   0539
F THE SERIES XENON DIFLUORIDE, XENON TETRAFLUORIDE, XENON HEXAFLUORIDE.   SELF CONSISTENT FIELD STUDY O 1224
OFT X-RAY SPECTROSCOPY ON SULFUR TETRAFLUORIDE AND SULFUR HEXAFLUORIDE.   S 1033
              SPECTROSCOPIC CALCULATIONS ON SULFUR HEXAFLUORIDE.   1071
                    SPECTRUM OF RHENIUM HEXAFLUORIDE.   0906
          STEREOCHEMISTRY AND SEVEN COORDINATION. (XENON HEXAFLUORIDE)   1286
OBLE GASES. (XENON DIFLUORIDE, XENON TETRAFLUORIDE, XENON HEXAFLUORIDE) /CONSIDERATIONS IN THE CHEMISTRY OF N 1235
                    STRUCTURE OF GASEOUS XENON HEXAFLUORIDE.   1295
           STRUCTURES OF NOBLE GAS COMPOUNDS. (XENON HEXAFLUORIDE)   1292
K X-RAY EMISSION AND ABSORPTION SPECTRA OF GASEOUS SULFUR HEXAFLUORIDE.   SULFUR K AND L AND FLUORINE 1088
SUSCEPTIBILITY MEASUREMENTS AND FLUORINE NMR OF PLATINUM HEXAFLUORIDE.   0883
PECTRA OF THE MOLECULES CARBON TETRAFLUORIDE AND DICARBON HEXAFLUORIDE. /ORY OF INTENSITIES IN THE INFRARED S 0362
THERMOCHEMICAL STUDIES OF XENON TETRAFLUORIDE AND XENON HEXAFLUORIDE.   1276
                    THERMODYNAMIC FUNCTIONS FOR SULFUR HEXAFLUORIDE.   1089
           THERMODYNAMIC PROPERTIES OF GASEOUS URANIUM HEXAFLUORIDE.   1198
                    THERMODYNAMIC PROPERTIES OF SULFUR HEXAFLUORIDE.   1087
                    THERMODYNAMIC PROPERTIES OF URANIUM HEXAFLUORIDE.   1204
                    THERMODYNAMIC STABILITY OF URANIUM HEXAFLUORIDE.   1203
ROPORTIONATION PRESSURES. (URANIUM PENTAFLUORIDE, URANIUM HEXAFLUORIDE) /ORIDES FROM MEASUREMENTS OF THE DISP 1176
          ULTRASOFT X-RAY ABSORPTION SPECTRUM OF SULFUR HEXAFLUORIDE.   1102
       VAPOR PRESSURE AND CRITICAL CONSTANTS OF URANIUM HEXAFLUORIDE.   1197
     VAPOR PRESSURE AND TRANSITION POINTS OF TECHNETIUM HEXAFLUORIDE.   1105
                    VIBRATIONAL SPECTRA OF RHENIUM HEXAFLUORIDE.   0905
        VIBRATIONAL SPECTRUM AND FORCE FIELD OF MOLYBDENUM HEXAFLUORIDE.   0634
          VIBRATIONAL SPECTRUM AND FORCE FIELD OF URANIUM HEXAFLUORIDE.   1196
     VIBRATIONAL SPECTRUM AND STRUCTURE OF SOLID URANIUM HEXAFLUORIDE.   1189
VIBRATIONAL STRUCTURE AND MOLECULAR STRUCTURE OF RHENIUM HEXAFLUORIDE.   0907
TELLER PROGRESSIONS IN THE ELECTRONIC SPECTRUM OF IRIDIUM HEXAFLUORIDE.   VIBRONICALLY INDUCED JAHN- 0537
                    X-RAY ABSORPTION IN SULFUR HEXAFLUORIDE.   1067
     X-RAY SPECTRA OF SULFUR AND FLUORINE IN SULFUR HEXAFLUORIDE.   1101
ELECTRONIC STATES AND MOLECULAR GEOMETRY OF XENON HEXAFLUORIDE: A CASE OF ELECTRONIC ISOMERISM. 1294
STATE.   ELECTRONIC PROPERTIES OF SULFUR HEXAFLUORIDE. AB INITIO CALCULATION FOR THE GROUND 1077
SO/ INFRARED SPECTRUM OF ALUMINUM TRIFLUORIDE, DIALUMINUM HEXAFLUORIDE, AND ALUMINUM MONOFLUORIDE BY MATRIX I 0051
PAIRING VALENCE STATES. (PHOSPHORUS PENTAFLUORIDE, SULFUR HEXAFLUORIDE, AND CHLORINE TRIFLUORIDE) /N PERFECT 0862
D BY THE SELF CONSISTENT / MOLECULAR ORBITALS OF PLATINUM HEXAFLUORIDE AND ELEMENT 110 HEXAFLUORIDE CALCULATE 0881
/PARATION AND PROPERTIES OF (VAPOR PRESSURE) OF PLUTONIUM HEXAFLUORIDE AND IDENTIFICATION OF PLUTONIUM(VI) O/ 0890
              MOLECULAR STRUCTURE OF XENON HEXAFLUORIDE AND IODINE HEPTAFLUORIDE. 0526
EPORTED OCTAFLUORIDE.   OSMIUM HEXAFLUORIDE AND ITS IDENTITY WITH THE PREVIOUSLY R 0753
TANTS (VAPOR PRESSURE) OF SILICON TETRAFLUORIDE, TUNGSTEN HEXAFLUORIDE, AND MOLYBDENUM HEXAFLUORIDE. /AL CONS 0639
CAPACITY, ENTROPY, AND HEATS OF TRANSITION OF MOLYBDENUM HEXAFLUORIDE AND NIOBIUM PENTAFLUORIDE. HEAT 0662
TRON DIFFRACTION STUDIES OF XENON TETRAFLUORIDE AND XENON HEXAFLUORIDE AND OTHER COMPOUNDS. ELEC 1259
          VIBRATIONAL SPECTRA OF OSMIUM HEXAFLUORIDE AND PLATINUM HEXAFLUORIDE. 0752
              INFRARED SPECTRA OF NEPTUNIUM HEXAFLUORIDE AND PLUTONIUM HEXAFLUORIDE. 0649
FLUORIDE, TELLURIUM / GREEN'S FUNCTION ANALYSIS OF SULFUR HEXAFLUORIDE AND RELATED MOLECULES. (SELENIUM HEXA 1093
     VAPOR PRESSURES AND OTHER PROPERTIES OF RHENIUM HEXAFLUORIDE AND RHENIUM HEPTAFLUORIDE. 0912
          POTENTIAL CONSTANTS FOR MOLYBDENUM HEXAFLUORIDE AND RHENIUM HEXAFLUORIDE. 0635
          MOLECULAR VIBRATIONS OF SULFUR HEXAFLUORIDE AND SULFUR TETRAFLUORIDE. 1034
     VIBRATIONAL SPECTRA OF MOLYBDENUM HEXAFLUORIDE AND TECHNETIUM HEXAFLUORIDE. 0628
FORCE FIELDS FOR SOME GROUP-6 HEXAFLUORIDES. (SULFUR HEXAFLUORIDE AND TELLURIUM HEXAFLUORIDE) 1061
              INFRARED SPECTRA OF SELENIUM HEXAFLUORIDE AND TELLURIUM HEXAFLUORIDE. 1107
     ASSIGNMENTS IN THE ULTRAVIOLET SPECTRA OF MOLYBDENUM HEXAFLUORIDE AND TUNGSTEN HEXAFLUORIDE. 0633
T BORN APPROXIMATION IN ELECTRON DIFFRACTION. (MOLYBDENUM HEXAFLUORIDE AND TUNGSTEN HEXAFLUORIDE) /F THE FIRS 0640
          VIBRATIONAL ANALYSIS OF SELENIUM HEXAFLUORIDE AND TUNGSTEN HEXAFLUORIDE. 0960
E ELECTRONIC STRUCTURE AND OPTICAL TRANSITIONS OF URANIUM HEXAFLUORIDE AND URANIUM PENTAFLUORIDE. /IONS OF TH 1179
              STRUCTURE OF THE IODINE HEXAFLUORIDE ANION. (XENON HEXAFLUORIDE) 1300
                   PARAMAGNETIC HEXAFLUORIDE ANIONS OF GROUP-6. (SELENIUM PENTAFLUO 0959
RIDE)
ONS ON THE VAPOR.   STRUCTURE OF URANIUM HEXAFLUORIDE AS DETERMINED BY DIFFRACTION OF ELECTR 1183
ECTRONS ON THE VAPOR.   STRUCTURE OF URANIUM HEXAFLUORIDE AS DETERMINED BY THE DIFFRACTION OF EL 1184
          THERMODYNAMIC PROPERTIES OF URANIUM HEXAFLUORIDE AT HIGH TEMPERATURES. 1193
DEPENDENT ABSORPTION SPECTRUM OF THE NU-3 BAND OF SULFUR HEXAFLUORIDE AT 10.6 MICROMETERS. TEMPERATURE 1091
10: NEUTRON POWDER DIFFRACTION PROFILE STUDIES OF URANIUM HEXAFLUORIDE AT 193 DEGREES AND 293 DEGREES K. /RT- 1202
              PHOTOELECTRON SPECTRUM OF SULFUR HEXAFLUORIDE AT 584 ANGSTROMS. 1076
CAPACITY AND OTHER THERMODYNAMIC PROPERTIES OF MOLYBDENUM HEXAFLUORIDE BETWEEN 4 AND 350 DEGREES K. HEAT 0636
LECTIVE PHOTODISSOCIATION OF POLYATOMIC MOLECULES (SULFUR HEXAFLUORIDE) BY INFRARED RADIATION. SE 1064

Y.	PROPERTIES AND REACTIONS OF URANIUM	HEXAFLUORIDE BY ION CYCLOTRON RESONANCE SPECTROSCOP	1185
	MEASUREMENT OF THE GAS DENSITY OF URANIUM	HEXAFLUORIDE BY LASER RAMAN SCATTERING.	1181
/LUORIDE, SULFUR DIFLUORIDE, SULFUR TETRAFLUORIDE, SULFUR	HEXAFLUORIDE, CHLORINE MONOFLUORIDE, CHLORINE TRIF/	1373
FORBIDDEN RAMAN BANDS OF SULFUR	HEXAFLUORIDE. COLLISION-INDUCED RAMAN SCATTERING.	1081
TELLURIUM HEXABROMIDE(2-), IODINE HEXAFLUORIDE(1-), XENON	HEXAFLUORIDE, ETC. /3-), SELENIUM HEXABROMIDE(2-),	1309
FORCE CONSTANTS OF SULFUR	HEXAFLUORIDE FROM ISOTOPIC SUBSTITUTION.	1100
ABSORPTION SPECTRUM OF URANIUM	HEXAFLUORIDE FROM 2000 TO 4200 ANGSTROMS.	1191
INFRARED SPECTRUM OF TELLURIUM	HEXAFLUORIDE FROM 25 TO 40 MICRONS.	1108
HEAT CAPACITY AND OTHER THERMODYNAMIC FUNCTIONS OF XENON	HEXAFLUORIDE FROM 5 TO 350 DEGREES K.	1305
ALPY OF FUSION, AND THERMODYNAMIC PROPERTIES OF NEPTUNIUM	HEXAFLUORIDE FROM 7 TO 350 DEGREES K. /PACITY, ENTH	0650
RATIONAL SPECTRUM AND THERMODYNAMIC PROPERTIES OF URANIUM	HEXAFLUORIDE GAS. VIB	1186
RATIONAL SPECTRUM AND THERMODYNAMIC PROPERTIES OF URANIUM	HEXAFLUORIDE GAS. VIB	1187
. (LANTHAN/ DEMONSTRATION OF THE EXISTENCE OF DILANTHANUM	HEXAFLUORIDE GAS AND DETERMINATION OF ITS STABILITY	0581
IDE. VAPOR PRESSURE OF URANIUM	HEXAFLUORIDE IN EQUILIBRIUM WITH URANIUM PENTAFLUOR	1180
RAMAN SPECTRUM OF SULFUR	HEXAFLUORIDE IN THE GASEOUS AND SOLID STATES.	1079
/LECTION OF MOLECULES WITH LARGE AMPLITUDE MOTION. (XENON	HEXAFLUORIDE, IODINE HEPTAFLUORIDE, RHENIUM HEPTAF/	1281
/IN RESONANCE SPECTRA OF HALOGEN HEXAFLUORIDES. (CHLORINE	HEXAFLUORIDE, IODINE HEXAFLUORIDE, BROMINE HEXAFLU/	0271
PLUTONIUM	HEXAFLUORIDE: ITS PREPARATION AND PROPERTIES.	0893
ESSOR. SPECIFIC HEAT RATIO OF URANIUM	HEXAFLUORIDE MEASURED WITH A BALLISTIC PISTON COMPR	1201
TRUM) PROPERTIES OF PLUTONIUM	HEXAFLUORIDE. (MOLECULAR STRUCTURE, ABSORPTION SPEC	0894
ELECTRONIC AND GEOMETRIC STRUCTURE OF THE FREE XENON	HEXAFLUORIDE MOLECULE.	1302
EFFECTS ON ELECTRON DIFFRACTION PATTERNS FOR THE RHENIUM	HEXAFLUORIDE MOLECULE. /E INTRAMOLECULAR SCATTERING	0911
FORCE CONSTANTS FOR	HEXAFLUORIDE MOLECULES.	1457
MEAN SQUARE AMPLITUDES OF INTERATOMIC DISTANCES IN	HEXAFLUORIDE MOLECULES.	0008
MEAN SQUARE AMPLITUDES OF INTERATOMIC DISTANCES IN	HEXAFLUORIDE MOLECULES.	1450
RE AMPLITUDES OF INTERATOMIC DISTANCES IN SOME OCTAHEDRAL	HEXAFLUORIDE MOLECULES. MEAN SQUA	1466
STATISTICAL THERMODYNAMIC PROPERTIES OF	HEXAFLUORIDE MOLECULES.	1449
VIBRATIONAL PROPERTIES OF	HEXAFLUORIDE MOLECULES.	0021
VIBRONIC INTERACTIONS IN METAL	HEXAFLUORIDE MOLECULES.	0908
/ITIES AND ELECTRONIC SPECTRA OF THE PLATINUM GROUP METAL	HEXAFLUORIDE MOLECULES DOWN TO HELIUM TEMPERATURES.	1159
SOME PROPERTIES OF THE	HEXAFLUORIDE MOLECULES. (REVIEW)	0020
HENIU/ JAHN-TELLER EFFECT IN THE E(8) VIBRATIONAL MODE OF	HEXAFLUORIDE MOLECULES. THE INFRARED SPECTRA OF RUT	0926
HEXAFLUORIDE. INFRARED SPECTRUM OF CHROMIUM	HEXAFLUORIDE, MOLYBDENUM HEXAFLUORIDE, AND OSMIUM	0436
/ OF SOME HEAVY TRANSITION METAL HEXAFLUORIDES. (TUNGSTEN	HEXAFLUORIDE, MOLYBDENUM HEXAFLUORIDE, RHENIUM HEX/	0903
/ONG WAVE LENGTH, INFRARED ACTIVE FUNDAMENTAL FOR URANIUM	HEXAFLUORIDE, NEPTUNIUM HEXAFLUORIDE, AND PLUTONIU/	1192
HEXACHLORIDE(/ VIBRATIONAL ELECTRONIC COUPLING IN IRIDIUM	HEXAFLUORIDE, OSMIUM HEXACHLORIDE(2-), AND IRIDIUM	0538
RIDE,/ COLORS OF TRANSITION METAL HEXAFLUORIDES. (RHENIUM	HEXAFLUORIDE, OSMIUM HEXAFLUORIDE, IRIDIUM HEXAFLUO	0916
RIDE, URA/ ELECTRON DIFFRACTION INVESTIGATION OF TUNGSTEN	HEXAFLUORIDE, OSMIUM HEXAFLUORIDE, IRIDIUM HEXAFLUO	1155
ALPHA SPECTROMETRIC INVESTIGATION OF URANIUM	HEXAFLUORIDE. PART-1.	1195
ION INTENSITIES. PART-2: IN/ MOLECULAR STRUCTURE OF XENON	HEXAFLUORIDE. PART-1: ANALYSIS OF ELECTRON DIFFRACT	1291
EOMETRY. PART-2: COMP/ CRYSTAL FIELD MODEL STUDY OF XENON	HEXAFLUORIDE. PART-1: ENERGY LEVELS AND MOLECULAR G	1311
EMISSION FROM SCF MO LCA/ ELECTRONIC PROPERTIES OF SULFUR	HEXAFLUORIDE. PART-1: OPTICAL ABSORPTION AND X-RAY	1078
/OF HEXAFLUORIDES AT HIGH TEMPERATURES. PART-2: PLUTONIUM	HEXAFLUORIDE. PART-3: URANIUM HEXAFLUORIDE. PART-4/	0629
HEXAFL/ SOME RECENT STUDIES WITH HEXAFLUORIDES. (URANIUM	HEXAFLUORIDE, PLUTONIUM HEXAFLUORIDE, AND NEPTUNIUM	1205
SONANCE SPECTRUM, AND STRUCTURE. CHLORINE	HEXAFLUORIDE RADICAL. PREPARATION, ELECTRON SPIN RE	0428
E AND ULTRAVIOL/ SPECTRAL OBSERVATIONS ON MOLECULAR XENON	HEXAFLUORIDE. RAMAN SCATTERING AND INFRARED, VISIBL	1285
PHYSICAL AND THERMODYNAMIC PROPERTIES OF SULFUR	HEXAFLUORIDE. (REVIEW)	1062
/ORIOLIS CONSTANTS OF THE SPHERICAL TOP MOLECULES, SULFUR	HEXAFLUORIDE, SELENIUM HEXAFLUORIDE, TELLURIUM HEX/	1069
/ON FREQUENCIES OF SOME OCTAHEDRAL XY6 MOLECULES. (SULFUR	HEXAFLUORIDE, SELENIUM HEXAFLUORIDE, TELLURIUM HEX/	1080
FL/ ELECTRIC DEFLECTION OF BINARY HEXAFLUORIDES. (SULFUR	HEXAFLUORIDE, SELENIUM HEXAFLUORIDE, TELLURIUM HEXA	1084
FLUORIDE AND THEIR THE/ ENTHALPIES OF FORMATION OF SULFUR	HEXAFLUORIDE, SELENIUM HEXAFLUORIDE, TELLURIUM HEXA	1092
LASER ISOTOPE SEPARATION. (SULFUR	HEXAFLUORIDE SPECTRUM)	1094
XENON	HEXAFLUORIDE. (SPECTRUM)	1312
RIC PHASES. XENON	HEXAFLUORIDE. STRUCTURAL CRYSTALLOGRAPHY OF TETRAME	1284
OSCOPY. ELECTRONIC STRUCTURE OF SULFUR	HEXAFLUORIDE STUDIED BY ULTRALONG WAVE X-RAY SPECTR	1065
/ECTRON DIFFRACTION OF THE MOLECULAR STRUCTURES OF SULFUR	HEXAFLUORIDE, SULFUR TETRAFLUORIDE, SELENIUM HEXAF/	1041
FLUORIDE, URANIUM HEXAFLUORI/ RAMAN SPECTRA OF MOLYBDENUM	HEXAFLUORIDE, TECHNETIUM HEXAFLUORIDE, RHENIUM HEXA	1072
/ACTION INVESTIGATION OF SOME SIMPLE MOLECULES. (TUNGSTEN	HEXAFLUORIDE, TELLURIUM HEXAFLUORIDE, AND MOLYBDEN/	1153
C. XENON	HEXAFLUORIDE. THE STRUCTURE OF A CUBIC PHASE AT -30	1283
XAFLUO/ HIGH RESOLUTION NMR SPECTRUM OF LIQUID MOLYBDENUM	HEXAFLUORIDE, TUNGSTEN HEXAFLUORIDE, AND URANIUM HE	0637
/ION AND FLUORINE RELAXATION IN THE LIQUIDS OF MOLYBDENUM	HEXAFLUORIDE, TUNGSTEN HEXAFLUORIDE, AND URANIUM H/	0638
/ST BORN APPROXIMATION IN ELECTRON DIFFRACTION. (TUNGSTEN	HEXAFLUORIDE, URANIUM HEXAFLUORIDE, TELLURIUM HEXA/	1158
SOME PHYSICAL PROPERTIES OF TUNGSTEN	HEXAFLUORIDE. (VAPOR PRESSURE)	1152
SPECTROSCOPIC STUDY OF IODINE HEPTAFLUORIDE AND IODINE	HEXAFLUORIDE(X) IN ANHYDROUS HYDROGEN FLUORIDE.	0518
, AND PHOSPHORUS TRIFLU/ ENTHALPIES OF FORMATION OF XENON	HEXAFLUORIDE, XENON TETRAFLUORIDE, XENON DIFLUORIDE	0807
IDE. CRYSTAL STRUCTURE. ENTHALPIES OF FORMATION OF XENON	HEXAFLUORIDE, XENON TETRAFLUORIDE, XENON TETRAFLUOR	1269
LENIUM HEXABROMIDE(2-), TELLURIUM HEXABROMIDE(2-), IODINE	HEXAFLUORIDE(1-), XENON HEXAFLUORIDE, ETC. /3-), SE	1309
CALCULATION OF THE FORCE CONSTANTS OF	HEXAFLUORIDES.	1448
EXTREMAL PROPERTIES OF FORCE CONSTANTS IN OCTAHEDRAL	HEXAFLUORIDES.	1468
FORCE CONSTANTS OF METAL	HEXAFLUORIDES.	0018
FORCE CONSTANTS OF METAL	HEXAFLUORIDES.	1446
JAHN-TELLER EFFECT IN	HEXAFLUORIDES.	1451
JAHN-TELLER EFFECT IN VIBRATIONAL SPECTRA OF	HEXAFLUORIDES.	1471
ERMAL MOTION AND SHRINKAGES OF CHEMICAL BONDS: OCTAHEDRAL	HEXAFLUORIDES. MEAN AMPLITUDES OF TH	1460
THERMAL VIBRATIONS AND THERMODYNAMIC PROPERTIES OF METAL	HEXAFLUORIDES. MEAN AMPLITUDES OF	1467
S OF VIBRATION AND THERMODYNAMIC FUNCTIONS FOR SOME METAL	HEXAFLUORIDES. MEAN AMPLITUDE	1459
IELD FOR SOME XY6 TYPE MOLECULES. (SELENIUM AND TELLURIUM	HEXAFLUORIDES) MOLECULAR FORCE F	1465
MOLECULAR FORCE FIELDS FOR SOME	HEXAFLUORIDES.	1452
ULTRAVIOLET ABSORPTION SPECTRA OF MOLYBDENUM AND TUNGSTEN	HEXAFLUORIDES. RAMAN EFFECT AND	0642
REACTIVITY AND THERMAL STABILITY OF	HEXAFLUORIDES.	0004
REVISED MEAN AMPLITUDES OF VIBRATION FOR SOME OCTAHEDRAL	HEXAFLUORIDES.	1447
STATISTICAL THERMODYNAMICS OF SOME	HEXAFLUORIDES.	1463
THERMAL STABILITY AND REACTIVITY OF D- AND F ELEMENT	HEXAFLUORIDES.	0630
FORCE FIELD OF WEINSTOCK AND GOODMAN FOR OCTAHEDRAL	HEXAFLUORIDES AND THE L MATRIX METHOD.	1456
UM HEXAFLUORIDE. PART-3: URAN/ THERMODYNAMIC STABILITY OF	HEXAFLUORIDES AT HIGH TEMPERATURES. PART-2: PLUTONI	0629
LUORIDE, BROM/ ELECTRON SPIN RESONANCE SPECTRA OF HALOGEN	HEXAFLUORIDES. (CHLORINE HEXAFLUORIDE, IODINE HEXAF	0271
MOLECULAR STRUCTURE OF MOLYBDENUM, TUNGSTEN, AND URANIUM	HEXAFLUORIDES FROM INFRARED AND RAMAN SPECTRA.	0627
RAMAN SPECTRA OF 5D TRANSITION METAL	HEXAFLUORIDES IN THE VAPOR STATE. (THIRD ROW)	1378
AVE STUDIES ON BONDING AND IONIZATION POTENTIALS. PART-1:	HEXAFLUORIDES OF GROUP-6 ELEMENTS. /HA) SCATTERED W	1366
M. VALENCE FORCE CONSTANTS OF THE	HEXAFLUORIDES OF MOLYBDENUM, TECHNETIUM, AND RHENIU	0641

TELLURIUM FLUORIDES) ROOT MEAN SQUARE AMPLITUDES IN SOME HEXAFLUORIDES OF OCTAHEDRAL SYMMETRY. (SELENIUM AND 0962
/MODYNAMIC SIMILARITY AND UNIVERSAL EQUATIONS OF STATE OF HEXAFLUORIDES. (OF SULFUR, MOLYBDENUM TUNGSTEN, AN/ 0632
RAMAN SPECTRA AND MOLECULAR CONSTANTS OF THE HEXAFLUORIDES OF SULFUR, SELENIUM, AND TELLURIUM. 0963
YBDEN/ FORCE CONSTANTS AND BOND LENGTHS OF SOME INORGANIC HEXAFLUORIDES. (OF SULFUR, SELENIUM, TELLURIUM, MOL 0631
INFRARED SPECTRA AND MOLECULAR STRUCTURE OF SOME GROUP-6 HEXAFLUORIDES. (OF SULFUR, SELENIUM, TELLURIUM, MO/ 0961
AND THE DEPOLARIZATION OF RAYLEIGH SCATTERING IN CERTAIN HEXAFLUORIDES. (RHENIUM HEXAFLUORIDE) /MAN SPECTRUM 0901
UORIDE, IRIDIUM HEXAFLUORIDE,/ COLORS OF TRANSITION METAL HEXAFLUORIDES. (RHENIUM HEXAFLUORIDE, OSMIUM HEXAFL 0916
HEPTA/ X-RAY DIFFRACTION STUDIES OF SOME TRANSITION METAL HEXAFLUORIDES. (SECOND ROW, THIRD ROW, AND RHENIUM 1375
EXAFLUORIDE) FORCE FIELDS FOR SOME GROUP-6 HEXAFLUORIDES. (SULFUR HEXAFLUORIDE AND TELLURIUM H 1061
FLUORIDE, TELLURIUM HEXAFL/ ELECTRIC DEFLECTION OF BINARY HEXAFLUORIDES. (SULFUR HEXAFLUORIDE, SELENIUM HEXA 1084
EXAFLUORI/ VAPOR PRESSURES OF SOME HEAVY TRANSITION METAL HEXAFLUORIDES. (TUNGSTEN HEXAFLUORIDE, MOLYBDENUM H 0903
AFLUORIDE, AND NEPTUNIUM HEXAFL/ SOME RECENT STUDIES WITH HEXAFLUORIDES. (URANIUM HEXAFLUORIDE, PLUTONIUM HEX 1205
/RMOLECULAR FLUORINE EXCHANGE BETWEEN TETRABUTYL AMMONIUM HEXAFLUORO PHOSPHATE AND BORON TRIFLUORIDE OR PHOS/ 0194
/S AND MOLECULAR GEOMETRY. PART-2: COMPARISONS WITH OTHER HEXAVALENT XENON MOLECULES. PART-3: ELECTRONIC TRA/ 1311
BONDS. LOCALIZED ORBITALS IN BORON FLUORIDES. HIGHLY POLARIZED BORON- FLUORINE DOUBLE AND TRIPLE 0158
TATIC MODEL FOR CALCULATION OF ANGULAR FORCE CONSTANTS OF HIGHLY SYMMETRICAL MOLECULES. /OF A SIMPLE ELECTROS 1484
MULTIPLET SPLITTING IN 1S HOLE STATES OF MOLECULES. 0683
. PART-17: SUBLIMATION AND VAPOR PRESSURES OF DYSPROSIUM, HOLMIUM, AND ERBIUM TRIFLUORIDES. /IGH TEMPERATURES 0448
ABSORPTION SPECTRUM OF GASEOUS HOLMIUM MONOFLUORIDE. 0488
STABILITIES OF DYSPROSIUM TRIFLUORIDE, HOLMIUM TRIFLUORIDE, AND ERBIUM TRIFLUORIDE. 0449
ANHARMONIC POTENTIAL CONSTANTS FROM THE BAND CONTOURS OF HOT BANDS FOR A SYMMETRIC TOP. (NITROGEN TRIFLUORI/ 0733
/OF FORCE FIELDS. PART-1. POINTS-ON-A-SPHERE AND EXTENDED HUCKEL MOLECULAR ORBITAL ANALYSES OF TRIGONAL BIPY/ 0828
IFLUORIDE) EXTENDED HUCKEL THEORY AND THE SHAPE OF MOLECULES. (OXYGEN D 0758
CONSTRUCTION OF HYBRID ORBITALS. (CHLORINE TRIFLUORIDE) 0405
UORIDE AND ARSENIC T/ CALCULATIONS ON IONIC CHARACTER AND HYBRIDIZATION FROM DIPOLE MOMENTS FOR ARSENIC TRIFL 0107
PHOSPHORUS PENTAFLUORIDE, DIFLUORO AMINE, AND TETRAFLUORO HYDRAZINE. / STRUCTURES OF PHOSPHORUS TRIFLUORIDE, 0801
ERMANIUM TETRAHYDRIDE, ARSENIC TRIHYDRIDE, AND SELENIUM DIHYDRIDE. /LLIUM MONOFLUORIDE, GALLIUM TRIHYDRIDE, G 0459
/GALLIUM MONOFLUORIDE, GALLIUM TRIHYDRIDE, GERMANIUM TETRAHYDRIDE, ARSENIC TRIHYDRIDE, AND SELENIUM DIHYDRID/ 0459
/ INITIO STUDIES OF THE ELECTRONIC STRUCTURES OF BORON TRIHYDRIDE, BORON DIHYDRIDE FLUORIDE, BORON HYDRIDE D/ 0241
URES OF BORON TRIHYDRIDE, BORON DIHYDRIDE FLUORIDE, BORON HYDRIDE DIFLUORIDE, AND BORON TRIFLUORIDE. / STRUCT 0241
/F THE ELECTRONIC STRUCTURES OF BORON TRIHYDRIDE, BORON DIHYDRIDE FLUORIDE, BORON HYDRIDE DIFLUORIDE, AND BO/ 0241
SEMIEMPIRICAL STUDY OF THE ELECTRONIC STRUCTURE OF BORON HYDRIDE FLUORIDES. (BORON TRIFLUORIDE) 0226
/AR SCF CALCULATIONS FOR GALLIUM MONOFLUORIDE, GALLIUM TRIHYDRIDE, GERMANIUM TETRAHYDRIDE, ARSENIC TRIHYDRID/ 0459
MAGNETIC PROPERTIES OF THE BORON FLUORIDE AND BORON HYDRIDE MOLECULES. 0159
MOLECULAR ORBITAL STUDY OF NITROGEN DIHYDRIDE, NITROGEN DIOXIDE, AND NITROGEN DIFLUORIDE. 0684
/E VACUUM ULTRAVIOLET PHOTOLYSIS OF SILICON DIFLUORIDE. THE INFRARED AND ULTRAVIOLET SPECTRA OF T/ 0986
TS. (PHOSPHORUS TRIFLUORIDE) ABSORPTION SPECTRA OF THE HYDRIDES, DEUTERIDES, AND HALIDES OF GROUP-5 ELEMEN 0806
SCF CALCULATIONS ON CHLORATE ION, HYDROGEN CHLORIDE AND CHLORINE MONOFLUORIDE. 0380
NTIMONY PENTAFLUORIDE AND THE PURE ROTATIONAL SPECTRUM OF HYDROGEN CYANIDE. /HE LOW FREQUENCY VIBRATIONS OF A 0083
INE HEPTAFLUORIDE AND IODINE HEXAFLUORIDE(X) IN ANHYDROUS HYDROGEN FLUORIDE. SPECTROSCOPIC STUDY OF IOD 0518
/NEGATIVITY, NONBONDED INTERACTIONS AND POLARIZABILITY IN HYDROGEN HALIDES AND THE INTERHALOGEN COMPOUNDS. (/ 0373
/MR SPECTRA OF SULFUR, PHOSPHORUS, AND SILICON FLUORIDES. HYDROGEN OXYCHLORIDE, CHLORO AMMONIA, AND CHLORO M/ 0371
/FLUORIDE AND DINITROGEN TETRAFLUORIDE MOLECULES, AND ESR HYDROLYSIS AND INTRA- INTERMOLECULAR MECHANISM OF / 1042
PHOSPHORUS TETRAFLUORIDE RADICAL. ISOTROPIC HYPERFINE COUPLING CONSTANTS IN NITROGEN DIFLUORID/ 0688
IDE. HYPERFINE SPLITTINGS AND MOLECULAR STRUCTURE IN THE 0824
YGEN DIFLUORIDE. HYPERFINE STRUCTURE CONSTANTS OF CHLORINE MONOFLUOR 0375
SPIN ROTATIONAL HYPERFINE STRUCTURE IN THE MICROWAVE SPECTRUM OF OX 0777
HYPERFINE STRUCTURE OF INDIUM MONOFLUORIDE. 0489

I

/ON AFFINITY, DIPOLE MOMENT, SPECTROSCOPIC CONSTANTS, AND IDEAL GAS THERMODYNAMIC FUNCTIONS FROM A HARTREE-F/ 0094
AFLUORIDE, BROMINE PENTAFLUORIDE, AND IODINE PENTAFLUORI/ IDEAL GAS THERMODYNAMIC PROPERTIES OF CHLORINE PENT 0424
DS. (XENON TETRAFLUORIDE AND XENON HEXAFLUORIDE) IDEAL GAS THERMODYNAMIC PROPERTIES OF XENON COMPOUN 1273
PERTIES OF (VAPOR PRESSURE) OF PLUTONIUM HEXAFLUORIDE AND IDENTIFICATION OF PLUTONIUM(VI) OXYFLUORIDE. /D PRO 0890
OSMIUM HEXAFLUORIDE AND ITS IDENTITY WITH THE PREVIOUSLY REPORTED OCTAFLUORIDE. 0753
/ERENT SCHEMES OF NEGLECT OF DIFFERENTIAL OVERLAP METHODS ILLUSTRATED BY THE XENON FLUORIDES. MODIFIED NDDO-/ 1231
ON AND DISSOCIATION OF XENON DIFLUORIDE INDUCED BY PHOTON IMPACT. IONIZATI 1243
PHOSPHINE. IONIZATION BY ELECTRON IMPACT OF PHOSPHORUS TRIFLUORIDE AND DIFLUORO CYANO 0800
ELECTRON IMPACT SPECTRA OF METHANE AND CARBON TETRAFLUORIDE. 0333
IFLUORIDE) ELECTRON IMPACT STUDIES OF SOME TRIHALO METHANES. (CARBON TR 0312
THERMAL DISSOCIATION OF CHLORINE TRIFLUORIDE BEHIND INCIDENT SHOCK WAVES. 0394
HIGH TEMPERATURE ENTHALPY INCREMENTS FOR SOME RARE EARTH TRIFLUORIDES. 1350
ENON OXIDE TETRAFLUORIDE BY ESCA AND EVIDENCE FOR ORBITAL INDEPENDENCE IN THE XENON- FLUORINE BONDING. /AND X 1229
MICROWAVE SPECTRA OF THE THALLIUM, INDIUM AND GALLIUM MONOHALIDES. 1110
S OF THE GASEOUS MONOHALIDES OF BORON, ALUMINUM, GALLIUM, INDIUM, AND THALLIUM. DISSOCIATION ENERGIE 0152
ENERGIES OF DIATOMIC FLUORIDES AND CHLORIDES OF GALLIUM, INDIUM, AND THALLIUM. (MONOFLUORIDES) /DISSOCIATION 0458
HYPERFINE STRUCTURE OF INDIUM MONOFLUORIDE. 0489
MICROWAVE ROTATIONAL SPECTRUM OF INDIUM MONOFLUORIDE. 0490
NALYSIS OF THE C SINGLET PI- X SINGLET SIGMA(+) SYSTEM OF INDIUM MONOFLUORIDE. ROTATIONAL A 0491
N SPECTRA OF ALUMINUM MONOFLUORIDE, GALLIUM MONOFLUORIDE, AND THALLIUM MONOFLUORIDE. /IO 0025
VAPORIZATION THERMODYNAMICS OF INDIUM TRIFLUORIDE. 0492
TRY AND PROPERTIES OF SOME INTERHALOGENS. (CH/ CNDO/2 AND INDO ALL VALENCE ELECTRON CALCULATIONS ON THE GEOME 0397
FLUORIDE. PREPARATION OF INERT GAS COMPOUNDS BY MATRIX ISOLATION: KRYPTON DI 0566
F XENON DIFLUORIDE / BOND ENERGIES AND IONIC CHARACTER OF INERT GAS HALIDES. HELIUM-2 PHOTOELECTRON SPECTRA O 1343
INFRARED SPECTRA OF SILICON DIFLUORIDE IN INERT GAS MATRICES. 0982
THEORY OF INERT GAS MOLECULES. 1337
ELECTRON STRUCTURE OF COMPOUNDS OF INERT GASES. 1339
/IONS OF ELECTRON STRUCTURES OF COMPOUNDS OF HALOGENS AND INERT GASES BY THE NONEMPIRICAL NDDO-2 (ALPHA, BET/ 0016
/NS OF ELECTRONIC STRUCTURES OF COMPOUNDS OF HALOGENS AND INERT GASES BY THE SCF LCAO MO METHOD. PART-2: CO/ 1231
TIMONY HEXACHLORIDE(3-), SELENIUM H/ THE STEREOCHEMICALLY INERT LONE PAIR. A SPECULATION ON THE BONDING IN AN 1309
/CTRUM OF THE FREE RADICAL CARBON TRIFLUORIDE ISOLATED IN INERT MATRICES. PHOTOCHLORINATION OF METHANE AND / 0311
. (XY3) INERTIA DEFECT AND PLANARITY OF FOUR ATOM MOLECULES 1398
FAR INFRARED ABSORPTION IN GASEOUS CARBON TETRAFLUORIDE 0356
RIDES. (SILICON TETRAFLUORIDE, CARBON TETRAFLUORIDE, BOR/ INFRARED ABSORPTION SPECTRA OF SOME POLYATOMIC FLUO 0186
INFRARED ABSORPTION SPECTRUM OF BORON TRIFLUORIDE. 0206
ORIDE. INFRARED ABSORPTION SPECTRUM OF LIQUID BORON TRIFLU 0245
E, NEPTUNIUM HEXAFLUORIDE, AND PLUTONI/ LONG WAVE LENGTH, INFRARED ACTIVE FUNDAMENTAL FOR URANIUM HEXAFLUORID 1192
RES AND NORMAL COORDINATE ANALYSES OF GERMANIUM D/ MATRIX INFRARED AND LASER RAMAN SPECTRA, MOLECULAR STRUCTU 0475
/VARIOUS PHOSPHORUS HALIDES AND PHOSPHORUS PENTAFLUORIDE. INFRARED AND LOW TEMPERATURE RAMAN VIBRATIONAL SPE/ 0849

LUORIDE. FAR	INFRARED AND MILLIMETER WAVE STUDIES OF XENON HEXAF 1299
E OF MOLYBDENUM, TUNGSTEN, AND URANIUM HEXAFLUORIDES FROM	INFRARED AND RAMAN SPECTRA. MOLECULAR STRUCTUR 0627
ISOLATED THALLOUS HALIDE DIMERS, THALLOUS FLUORIDE, AND /	INFRARED AND RAMAN SPECTRA AND STRUCTURE OF MATRIX 1123
RUS TETRAFLUORIDE.	INFRARED AND RAMAN SPECTRA AND STRUCTURE OF PHOSPHO 0825
ISOLATED MAGNESIUM DIHALIDES: MAGNESIUM DIFLUORIDE, MAG/	INFRARED AND RAMAN SPECTRA AND STRUCTURES OF MATRIX 0607
STALLINE THALLIUM MONOFLUORIDE.	INFRARED AND RAMAN SPECTRA AND THE STRUCTURE OF CRY 1126
AND CHLORINE TRIFLUORIDE IN THE CONDENSED PHASE. EVIDEN/	INFRARED AND RAMAN SPECTRA OF BROMINE PENTAFLUORIDE 0266
.	INFRARED AND RAMAN SPECTRA OF CHLORINE MONOFLUORIDE 0381
AND BROMINE TRIFLUORIDE.	INFRARED AND RAMAN SPECTRA OF CHLORINE TRIFLUORIDE 0408
	INFRARED AND RAMAN SPECTRA OF KRYPTON DIFLUORIDE. 0552
E SILICON TETRAFLUORIDE.	INFRARED AND RAMAN SPECTRA OF LIQUID AND CRYSTALLIN 1001
.	INFRARED AND RAMAN SPECTRA OF SILICON TETRAFLUORIDE 1005
	INFRARED AND RAMAN SPECTRUM OF NITROGEN TRIFLUORIDE 0724
/AVIOLET PHOTOLYSIS OF SILICON DIFLUORIDE DIHYDRIDE. THE	INFRARED AND ULTRAVIOLET SPECTRA OF THE FREE RADIC/ 0986
D OF NITROGEN TRIFLUORIDE. CALCULATION OF PERPENDICULAR	INFRARED BAND CONTOURS: APPLICATION TO THE NU-4 BAN 0747
ULES. (CARBON TETRAFLUORIDE)	INFRARED BAND CONTOURS OF RHENIUM HEXAFLUORIDE. 0915
FLUORIDE.	INFRARED BAND CONTOURS. PART-1: SPHERICAL TOP MOLEC 0325
NTRAMOLECULAR FORCES IN CARBON / MOLECULAR CONSTANTS FROM	INFRARED BAND SHAPES AND VIBRATIONS OF NITROGEN TRI 0699
IFLUORIDE NU-4 AND NU-4 + NU-2 AND TRIFL/ PSEUDO-PARALLEL	INFRARED BAND SHAPES. THE ATOMIC DISPLACMENTS AND I 0331
HIGH RESOLUTION RAMAN SPECTROSCOPY IN THE RED AND NEAR	INFRARED BANDS OF OBLATE SYMMETRIC TOPS NITROGEN TR 0716
DICAL.	INFRARED. (CHLORINE MONOFLUORIDE) 0389
	INFRARED DETECTION OF GASEOUS CARBON TRIFLUORIDE RA 0310
ATURE. SULFUR TETRAFLUORIDE AND SULFUR OXIDE DIFLUORIDE.	INFRARED DOUBLE RESONANCE IN SULFUR HEXAFLUORIDE. 1099
ESTIMATION OF	INFRARED EVIDENCE FOR DIMER FORMATION AT LOW TEMPER 1054
LIQUID GAS ARSENIC	INFRARED FREQUENCIES OF SULFUR DIFLUORIDE. 1026
CON TETRAFLUORIDE, CARBON TETRACHLORI/ INTENSITIES OF THE	INFRARED FREQUENCY SHIFTS. 0106
RIFLUORIDE.	INFRARED FUNDAMENTALS OF CARBON TETRAFLUORIDE, SILI 1007
DERIVATIVES FOR PHOSPHORUS TRIFLUORIDE AND PHOSPHORUS OX/	INFRARED INTENSITIES AND FORCE FIELD FOR NITROGEN T 0714
RIDE.	INFRARED INTENSITIES, BOND MOMENTS AND BOND MOMENT 0797
CNDO CALCULATION OF DIPOLE MOMENT DERIVATIVES AND	INFRARED INTENSITIES IN CRYSTALLINE SULFUR HEXAFLUO 1075
ABSOLUTE	INFRARED INTENSITIES OF BORON TRIFLUORIDE. 0195
MEASUREMENT AND CALCULATION OF THE ABSOLUTE	INFRARED INTENSITIES OF CARBON TETRAFLUORIDE. 0340
OF NITROGEN TRIFLUORIDE. ABSOLUTE	INFRARED INTENSITIES OF PHOSPHORUS TRIFLUORIDE. 0808
/TRUCTURE DETERMINATION OF TRIS (TRIMETHYL SILYL) SILANE,	INFRARED INTENSITIES OF THE FUNDAMENTAL VIBRATIONS 0736
DE, SULFUR TETRAFLUORIDE, ANTIMONY PENTAFLUORIDE, ARSENI/	INFRARED INVESTIGATION OF IODINE HEPTAFLUORIDE AND/ 0527
OCIATION OF POLYATOMIC MOLECULES (SULFUR HEXAFLUORIDE) BY	INFRARED MATRIX ISOLATION STUDIES. (BORON TRIFLUORI 0839
ENIC MONOFLUORIDE IN THE NEAR VACUUM ULTRAVIOLET AND NEAR	INFRARED RADIATION. SELECTIVE PHOTODISS 1064
YTTRIUM DIFLUORIDE AND YTTRIUM TRIFLUORIDE.	INFRARED REGIONS. SPECTRUM OF ARS 0091
IDES.	INFRARED SPECTRA AND GEOMETRIES OF MATRIX ISOLATED 1316
BALT DIFLUORIDE, NICKEL DIFLUORIDE, COPPER DIFLUORIDE, A/	INFRARED SPECTRA AND GEOMETRIES OF RARE EARTH DIHAL 1351
E AND TITANIUM TRIFLUORIDE IN RARE GAS MATRICES.	INFRARED SPECTRA AND GEOMETRY OF MATRIX ISOLATED CO 0437
OUP-6 HEXAFLUORIDES. (OF SULFUR, SELENIUM, TELLURIUM, MO/	INFRARED SPECTRA AND GEOMETRY OF TITANIUM DIFLUORID 1146
HALLIUM HALIDES.	INFRARED SPECTRA AND MOLECULAR STRUCTURE OF SOME GR 0961
RIOUS PHOSPHORUS AND ARSENIC HALIDES. (PHOSPHO/ RAMAN AND	INFRARED SPECTRA AND STRUCTURE OF MATRIX ISOLATED T 1113
IDE.	INFRARED SPECTRA AND THERMODYNAMIC FUNCTIONS FOR VA 0811
N.	INFRARED SPECTRA OF A CRYOSYSTEM. CARBON TETRAFLUOR 0313
IUM TETRAFLUORIDE.	INFRARED SPECTRA OF ALUMINUM FLUORIDE IN SOLID ARGO 0054
AFLUORIDE.	INFRARED SPECTRA OF CARBON TETRAFLUORIDE AND GERMAN 0363
ITHIUM ALUMINUM TETRAFLUORIDE, AND SODIUM ALUMINUM TETRA/	INFRARED SPECTRA OF CRYOSYSTEMS. PART-2: SULFUR HEX 1068
LUORIDE. FAR	INFRARED SPECTRA OF GASEOUS ALUMINUM TRIFLUORIDE, L 0045
IUM FLUORIDE, ZIRCONIUM CHLORIDE, AND HAFNIUM CHLORIDE.	INFRARED SPECTRA OF GASEOUS AND LIQUID SULFUR HEXAF 1095
RAMAN AND	INFRARED SPECTRA OF GASEOUS GROUP-4 HALIDES: ZIRCON 1327
CARBON TETRAFLUORIDE)	INFRARED SPECTRA OF GERMANIUM TETRAFLUORIDE. 0482
S BY MATRIX ISOLATION.	INFRARED SPECTRA OF HALOGEN-SUBSTITUTED METHANES. (0355
LUORIDE AND ARSENIC PENTAFLUORIDE.	INFRARED SPECTRA OF MAGNESIUM AND ALUMINUM FLUORIDE 0608
ORIDE, BROMINE TRIFLUORIDE, AND BROMINE PENTAFLUORIDE. F/	INFRARED SPECTRA OF MATRIX ISOLATED ANTIMONY PENTAF 0113
ORIDE.	INFRARED SPECTRA OF MATRIX ISOLATED CHLORINE TRIFLU 0399
DES. (ALSO LANTHANUM TRIFLUORIDE) GEOMETRY AND	INFRARED SPECTRA OF MATRIX ISOLATED MAGNESIUM DIFLU 0604
GALLIUM- FLUORINE SYSTEM.	INFRARED SPECTRA OF MATRIX ISOLATED RARE EARTH HALI 0583
AND LEAD DIFLUORIDE.	INFRARED SPECTRA OF MATRIX ISOLATED SPECIES IN THE 0462
ORIDE, TRIFLUORIDE, AND TETRAFLUORIDE.	INFRARED SPECTRA OF MATRIX ISOLATED TIN DIFLUORIDE 0594
ONIUM HEXAFLUORIDE.	INFRARED SPECTRA OF MATRIX ISOLATED ZIRCONIUM DIFLU 1325
	INFRARED SPECTRA OF NEPTUNIUM HEXAFLUORIDE AND PLUT 0649
ORUS TRIFLUORIDE.	INFRARED SPECTRA OF NIOBIUM TETRAFLUORIDE. 0658
RUS OXIDE TRIFLUORIDE, AND PHOSPHORUS PENTAFLUORIDE.	INFRARED SPECTRA OF NITROGEN TRIFLUORIDE AND PHOSPH 0748
PS. FAR	INFRARED SPECTRA OF PHOSPHORUS TRIFLUORIDE, PHOSPHO 0799
	INFRARED SPECTRA OF PROLATE AND OBLATE SYMMETRIC TO 1479
HE REGION OF A CESIUM IODIDE PRISM.	INFRARED SPECTRA OF RARE EARTH METAL FLUORIDES. 1346
/THE E(8) VIBRATIONAL MODE OF HEXAFLUORIDE MOLECULES. THE	INFRARED SPECTRA OF RARE EARTH METAL FLUORIDES IN T 1352
RIUM HEXAFLUORIDE.	INFRARED SPECTRA OF RUTHENIUM HEXAFLUORIDE AND RHO/ 0926
MATRICES.	INFRARED SPECTRA OF SELENIUM HEXAFLUORIDE AND TELLU 1107
N SOLUTION. CARBON TETRAFLUORIDE.	INFRARED SPECTRA OF SILICON DIFLUORIDE IN INERT GAS 0982
THE MATRIX ISOLATION TECHNIQUE.	INFRARED SPECTRA OF SIMPLE MOLECULES IN LIQUID ARGO 0338
ES IN THE SOLID STATE. (ARSENIC TRIFLUORIDE, SULFUR / THE	INFRARED SPECTRA OF SOME ALKALINE EARTH HALIDES BY 0125
/2 TECHNIQUES. CALCULATION OF THE	INFRARED SPECTRA OF SOME VOLATILE INORGANIC FLUORID 0096
T-1: BERYLLIUM DIFLUORIDE, BERYLLIUM DICHLORIDE, MAGNESI/	INFRARED SPECTRA OF SULFUR TETRAFLUORIDE USING CNDO 1047
IDE AND DICARBON HEXAFLUORI/ THEORY OF INTENSITIES IN THE	INFRARED SPECTRA OF THE ALKALINE EARTH HALIDES. PAR 0133
S PENTAFLUORIDE. HIGH RESOLUTION	INFRARED SPECTRA OF THE MOLECULES CARBON TETRAFLUOR 0362
M TETRAFLUORIDE, AND VANADIUM TRIFLUORIDE.	INFRARED SPECTRA OF THE PARALLEL BANDS OF PHOSPHORU 0872
OLVENTS.	INFRARED SPECTRA OF VANADIUM PENTAFLUORIDE, VANADIU 1210
RIUM/ GEOMETRIES AND ENTROPIES OF METAL TRIFLUORIDES FROM	INFRARED SPECTRA OF XENON DIFLUORIDE IN DIFFERENT S 1241
HE LOW FREQUENCY VIBRATIONS OF ANT/ CONSTRUCTION OF A FAR	INFRARED SPECTRA. (SCANDIUM, YTTRIUM, LANTHANUM, CE 0943
(FLUORIDES, OXIDES, CHLORIDES) MATRIX ISOLATION	INFRARED SPECTROMETER AND ITS USE IN THE STUDY OF T 0083
IN SULFUR DIFLUORIDE. CALCULATION OF THE FORCE FIELD AND	INFRARED SPECTROSCOPY OF HIGH TEMPERATURE SPECIES. 0011
XENON TETRAFLUORIDE. (INFRARED	INFRARED SPECTRUM. CENTRIFUGAL DISTORTION EFFECTS 1031
	INFRARED SPECTRUM) 1263
RIFLUORIDE. THE	INFRARED SPECTRUM AND FORCE CONSTANTS FOR ARSENIC T 0105
EN TRIFLUORIDE.	INFRARED SPECTRUM AND MOLECULAR CONSTANTS OF NITROG 0728
SULFUR TETRAFLUORIDE.	INFRARED SPECTRUM AND NORMAL COORDINATE ANALYSIS OF 1045
LUORIDE RADICAL.	INFRARED SPECTRUM AND STRUCTURE OF THE NITROGEN DIF 0687

E NITROGEN DIFLUORIDE RADICAL. | INFRARED SPECTRUM AND THERMODYNAMIC FUNCTIONS OF TH 0686
N OF ARSENIC TRIFLUORIDE. | INFRARED SPECTRUM AND VIBRATIONAL POTENTIAL FUNCTIO 0100
IC FUNCTIONS OF SILICON DIFLUORIDE. | INFRARED SPECTRUM, FORCE CONSTANTS, AND THERMODYNAM 0985
SUBLIMATION CF ALUMINUM TRIFLUORIDE AND THE | INFRARED SPECTRUM OF ALUMINUM TRIFLUORIDE. 0037
UM HEXAFLUORIDE, AND ALUMINUM MONOFLUORIDE BY MATRIX ISO/ | INFRARED SPECTRUM OF ALUMINUM TRIFLUORIDE, DIALUMIN 0051
| INFRARED SPECTRUM OF ARSENIC PENTAFLUORIDE. 0114
| INFRARED SPECTRUM OF BORON TRIFLUORIDE. 0249
ROTATION VIBRATION BAND CONTOURS IN THE | INFRARED SPECTRUM OF BORON TRIFLUORIDE. 0193
SOTOPE EFFECT AND A MORE PRECISE DETERMINATION OF THE FAR | INFRARED SPECTRUM OF BORON TRIFLUORIDE. /Y OF THE I 0230
| INFRARED SPECTRUM OF BROMINE PENTAFLUORIDE. 0262
| INFRARED SPECTRUM OF CARBON TETRAFLUORIDE. 0343
ANALYSIS OF THE | INFRARED SPECTRUM OF CHLORINE MONOFLUORIDE. 0386
NUM HEXAFLUORIDE, AND OSMIUM HEXAFLUORIDE. | INFRARED SPECTRUM OF CHROMIUM HEXAFLUORIDE, MOLYBDE 0436
UM DIFLUORIDE) | INFRARED SPECTRUM OF HEAVY METAL DIHALIDES. (EUROPI 0451
| INFRARED SPECTRUM OF IRIDIUM HEXAFLUORIDE. 0540
LUORIDE. | INFRARED SPECTRUM OF MATRIX ISOLATED NIOBIUM PENTAF 0659
| INFRARED SPECTRUM OF NIOBIUM PENTAFLUORIDE. 0661
| INFRARED SPECTRUM OF NITROGEN DIFLUORIDE. 0692
FAR | INFRARED SPECTRUM OF NITROGEN TRIFLUORIDE. 0702
FIELD. | INFRARED SPECTRUM OF NITROGEN TRIFLUORIDE AND FORCE 0697
PREPARATION AND PROPERTIES (INFRARED SPECTRUM) OF OSMIUM HEPTAFLUORIDE. 0754
A REINVESTIGATION OF THE | INFRARED SPECTRUM OF OXYGEN DIFLUORIDE. 0774
/AL SPECTRA AND STRUCTURE OF INORGANIC MOLECULES. PART-1: | INFRARED SPECTRUM OF OXYGEN DIFLUORIDE FROM 2.5 TO/ 0761
ION OF LITHIUM OXYFLUORIDE. MATRIX | INFRARED SPECTRUM OF OXYGEN MONOFLUORIDE AND DETECT 0755
| INFRARED SPECTRUM OF PHOSPHORUS PENTAFLUORIDE. 0866
FAR | INFRARED SPECTRUM OF PHOSPHORUS PENTAFLUORIDE. 0842
FAR | INFRARED SPECTRUM OF PHOSPHORUS PENTAFLUORIDE. 0855
| INFRARED SPECTRUM OF PLUTONIUM HEXAFLUORIDE. 0891
| INFRARED SPECTRUM OF SILICON TETRAFLUORIDE. 1003
| INFRARED SPECTRUM OF SULFUR HEXAFLUORIDE. 1086
RAMAN AND | INFRARED SPECTRUM OF SULFUR TETRAFLUORIDE. 1040
TO 40 MICRONS. | INFRARED SPECTRUM OF TELLURIUM HEXAFLUORIDE FROM 25 1108
ORIDE ISOLATED IN INERT MATRICES. PHOTOCHLORINATION OF / | INFRARED SPECTRUM OF THE FREE RADICAL CARBON TRIFLU 0311
ORUS PENTAFLUORIDE) RAMAN AND | INFRARED SPECTRUM OF TRIFLUORO PHOSPHORANE. (PHOSPH 0847
| INFRARED SPECTRUM OF URANIUM HEXAFLUORIDE. 1182
MONOFLUORIDE, BROMINE TRIFLUORIDE, AND BROMINE PENTAFLU/ | INFRARED STUDIES OF THE BROMINE FLUORIDES. (BROMINE 0256
/ONSTANTS OF SPHERICAL TOP MOLECULES FROM LOW TEMPERATURE | INFRARED STUDIES OF VAPOR BAND CONTOURS. APPLICATI/ 1156
NEAR | INFRARED SYSTEM OF RHENIUM HEXAFLUORIDE. 0902
RIDE. | INFRARED VIBRATIONAL PROPERTIES OF GERMANIUM DIFLUO 0473
/NS ON MOLECULAR XENON HEXAFLUORIDE. RAMAN SCATTERING AND | INFRARED, VISIBLE AND ULTRAVIOLET ABSORPTION IN TH/ 1285
NITROGEN FLUORIDES AND THEIR | INORGANIC DERIVATIVES. 0703
BICENTRIC RESCALING OF CNDO/2 THEORY. APPLICATIONS TO | INORGANIC FLUORIDES. 1475
VIBRATIONAL SPECTRA OF | INORGANIC FLUORIDES. 0014
ND SELENIUM TETR/ STRUCTURE OF EXCHANGE PROCESSES IN SOME | INORGANIC FLUORIDES BY NMR. (SULFUR TETRAFLUORIDE A 1048
DIFLUORIDE, STRONTIUM DIFLUORIDE, AND/ MELTING POINTS OF | INORGANIC FLUORIDES. (CADMIUM DIFLUORIDE, MAGNESIUM 0273
IC NUMBER BELOW 20. HEATS OF FORMATION OF | INORGANIC FLUORIDES ESPECIALLY THE ELEMENTS OF ATOM 0214
IFLUORIDE, SULFUR / THE INFRARED SPECTRA OF SOME VOLATILE | INORGANIC FLUORIDES IN THE SOLID STATE. (ARSENIC TR 0096
URIUM, MOLYBDEN/ FORCE CONSTANTS AND BOND LENGTHS OF SOME | INORGANIC HEXAFLUORIDES. (OF SULFUR, SELENIUM, TELL 0631
EN TRIFLUORIDE. AB INITIO STUDY CF THE FORCE CONSTANTS OF | INORGANIC MOLECULES NITROGEN OXYFLUORIDE AND NITROG 0734
XYGEN DIFLUORIDE FR/ VIBRATIONAL SPECTRA AND STRUCTURE OF | INORGANIC MOLECULES. PART-1: INFRARED SPECTRUM OF O 0761
) THERMOCHEMICAL AND THERMODYNAMIC PROPERTIES OF | INORGANIC NITROGEN FLUORIDES. (NITROGEN TRIFLUORIDE 0696
) PHYSICAL CHEMICAL PROPERTIES OF | INORGANIC NITROGEN FLUORIDES. (NITROGEN TRIFLUORIDE 0713
E. INFRARED | INTENSITIES AND FORCE FIELD FOR NITROGEN TRIFLUORID 0714
FLUORIDE. RAMAN | INTENSITIES AND THE FORCE FIELD OF PHOSPHORUS PENTA 0844
ES FOR PHOSPHORUS TRIFLUORIDE AND PHOSPHORUS OX/ INFRARED | INTENSITIES, BOND MOMENTS AND BOND MOMENT DERIVATIV 0797
XY4 TETRAHEDRAL AND XY6 OCTAHEDRAL MOLECULES. RAMAN | INTENSITIES FOR SPHERICALLY SYMMETRIC MODES OF SOME 1453
INFRARED | INTENSITIES IN CRYSTALLINE SULFUR HEXAFLUORIDE. 1075
S CARBON TETRAFLUORIDE AND DICARBON HEXAFLUORI/ THEORY OF | INTENSITIES IN THE INFRARED SPECTRA OF THE MOLECULE 0362
NDO CALCULATION OF DIPCLE MOMENT DERIVATIVES AND INFRARED | INTENSITIES OF BORON TRIFLUORIDE. C 0195
ABSOLUTE INFRARED | INTENSITIES OF CARBON TETRAFLUORIDE. 0340
IODINE BROMIDE (CH/ SPIN ORBITAL INTERACTION AND RELATIVE | INTENSITIES OF FORBIDDEN TRANSITIONS IN IODINE, OR 0498
MEASUREMENT AND CALCULATION OF THE ABSOLUTE INFRARED | INTENSITIES OF PHOSPHORUS TRIFLUORIDE. 0808
EN TRIFLUORIDE. ABSOLUTE INFRARED | INTENSITIES OF THE FUNDAMENTAL VIBRATIONS OF NITROG 0736
TETRAFLUORIDE, SILICON TETRAFLUORIDE, CARBON TETRACHLORI/ | INTENSITIES OF THE INFRARED FUNDAMENTALS OF CARBON 1007
/N HEXAFLUORIDE. PART-1: ANALYSIS OF ELECTRON DIFFRACTION | INTENSITIES. PART-2: INTERNAL MOTION AND MEAN GEOM/ 1291
CON MONOFLUORIDE. DISTRIBUTION OF | INTENSITY IN THE ELECTRON EMISSION SPECTRUM OF SILI 0968
S. DIRECT DETERMINATION FORCE CONSTANTS FROM THE CORIOLIS | INTERACTION. /OF SUBSTITUTED AND PERTURBED MOLECULE 1444
RANSITIONS IN IODINE, OR IODINE BROMIDE (CH/ SPIN ORBITAL | INTERACTION AND RELATIVE INTENSITIES OF FORBIDDEN T 0498
GEN DIFLUORIDE, NITRITE ION AND CYANIDE IO/ CONFIGURATION | INTERACTION CALCULATION FOR THE GROUND STATE OF OXY 0763
IONAL PROBLEM FOR XY4 TETRAHEDRAL MOLECU/ USE OF CORIOLIS | INTERACTION CONSTANTS IN THE SOLUTION OF THE VIBRAT 1423
. (OXYGEN DIFLUORIDE, NI/ A THEORETICAL STUDY OF THE BOND | INTERACTION FORCE CONSTANTS IN DIFLUORIDE MOLECULES 0299
NUCLEAR SPIN RELAXATION STUDY OF THE SPIN ROTATION | INTERACTION IN SPHERICAL TOP MOLECULES. 0323
NUCLEAR SPIN RELAXATION STUDY OF THE SPIN ROTATION | INTERACTION IN SYMMETRIC TOP MOLECULES. 0183
TOMS. THEORETICAL STUDIES OF THE | INTERACTION OF MAGNESIUM DIFLUORIDE WITH RARE GAS A 0605
NICKEL CHLORIDE WITH CARBON MONOXIDE, MOLECULAR NITROGEN/ | INTERACTION OF MATRIX ISOLATED NICKEL FLUORIDE AND 1324
S POSITIVE ION. CONFIGURATION | INTERACTION STUDIES OF NITROGEN MONOFLUORIDE AND IT 0672
M A/ CLASSIFICATION OF NORMAL MODES OF VIBRATION. PART-2: | INTERACTION TERMS IN ENERGY DISTRIBUTIONS. (SELENIU 1494
AND THE INTERHALOGEN COMPO/ ELECTRONEGATIVITY, NONBONDED | INTERACTIONS AND POLARIZABILITY IN HYDROGEN HALIDES 0373
VIBRONIC | INTERACTIONS IN METAL HEXAFLUORIDE MOLECULES. 0908
ORIDE. | INTERATOMIC BONDING IN CRYSTALLINE MANGANESE TRIFLU 0617
ES OF VIBRATION OF GASEOUS SILICON TETRAFLUORIDE FROM EL/ | INTERATOMIC DISTANCES AND ROOT MEAN SQUARE AMPLITUD 1002
TION ELEMENT TETRAHALID/ ESTIMATION OF CONFIGURATIONS AND | INTERATOMIC DISTANCES IN GROUP-5 AND GROUP-6 TRANSI 1427
MEAN SQUARE AMPLITUDES OF | INTERATOMIC DISTANCES IN HEXAFLUORIDE MOLECULES. 0008
MEAN SQUARE AMPLITUDES OF | INTERATOMIC DISTANCES IN HEXAFLUORIDE MOLECULES. 1450
DE MOLECULES. MEAN SQUARE AMPLITUDES OF | INTERATOMIC DISTANCES IN SOME OCTAHEDRAL HEXAFLUORI 1466
F THE XY4 TYPE OF / ESTIMATION OF THE FORCE CONSTANTS AND | INTERATOMIC DISTANCES OF HALIDE COMPLEX MOLECULES O 1414
ANGSTROMS. (ALUMINU/ RECENT DEVELOPMENTS IN MULTIPLE BEAM | INTERFEROMETRY IN THE ULTRAVIOLET REGION 1700-2500 0042
/ERACTIONS AND POLARIZABILITY IN HYDROGEN HALIDES AND THE | INTERHALOGEN COMPOUNDS. (CHLORINE MONOFLUORIDE, BR/ 0373
BROMINE PENTAFLUORIDE) MOLECULAR FORCE FIELDS FOR | INTERHALOGEN COMPOUNDS. (CHLORINE PENTAFLUORIDE AND 0425
ROMINE TRIFLUORIDE) MOLECULAR FORCE FIELD FOR | INTERHALOGEN COMPOUNDS. (CHLORINE TRIFLUORIDE AND B 0407

E OXYGEN PENTAFLUORIDE, AND IO/ MOLECULAR FORCE FIELD FOR INTERHALOGEN COMPOUNDS. IODINE PENTAFLUORIDE, IODIN 0511
E AT -120 DEGREES C.   STRUCTURES OF THE INTERHALOGEN COMPOUNDS. PART-1: CHLORINE TRIFLUORID 0395
E AT -110 DEGREES C AND AT -145 DEGREES C.   STRUCTURE OF INTERHALOGEN COMPOUNDS. PART-2: IODINE HEPTAFLUORID 0521
FLU/ AB INITIO MOLECULAR ORBITAL STUDY OF THE GECMETRY OF INTERHALOGENS. (CHLORINE MONOFLUORIDE, CHLORINE TRI 0379
/TRON CALCULATIONS ON THE GEOMETRY AND PROPERTIES OF SOME INTERHALOGENS. (CHLORINE TRIFLUORIDE, BROMINE TRIF/ 0397
ROMINE, AND IOD/ THERMCDYNAMIC PROPERTIES OF THE DIATOMIC INTERHALOGENS FROM SPECTROSCOPIC DATA. (CHLORINE, B 0253
THE DISPROPORTIONATION PRESSURES. (UR/ THERMCDYNAMICS OF INTERMEDIATE URANIUM FLUORIDES FROM MEASUREMENTS OF 1176
BORON FLUORIDE WITH ALTERATIONS IN ITS QU/ CORRELATION OF INTERMOLECULAR EFFECTS IN THE VIBRATION SPECTRA OF 1404
AMMONIUM HEXAFLUORC PHOSPHATE AND BORON TRIFLUORIDE OR / INTERMOLECULAR FLUORINE EXCHANGE BETWEEN TETRABUTYL 0194
PHOSPHORUS, AND SILICON FLUORIDES. HYDROLYSIS AND INTRA- INTERMOLECULAR MECHANISM OF FLUORINE EXCHANGE. /UR, 1042
                                                         INTERMOLECULAR POTENTIAL FOR SILICON TETRAFLUORIDE. 1010
D ARSENIC PENTAFLUORIDE AND THE HEIGHT OF THE BARRIER TO INTERNAL EXCHANGE OF FLUORINE NUCLEI. /AFLUORIDE AN 0856
/1: ANALYSIS OF ELECTRON DIFFRACTION INTENSITIES. PART-2: INTERNAL MOTION AND MEAN GEOMETRY DEDUCED BY ELECT/ 1291
     CORRELATION OF THE ELECTRONIC TRANSITION MOMENT WITH INTERNUCLEAR DISTANCE FOR BANDS OF THE X SYSTEM. 0969
RIDE.   INTERNUCLEAR DISTANCE IN GASEOUS STRONTIUM MONOFLUO 1016
                                  SHRINKAGES OF THE INTERNUCLEAR DISTANCES BY MOLECULAR VIBRATIONS. 1489
MOLECU/ ELECTRONIC TRANSITICN MOMENT AS A FUNCTION OF THE INTERNUCLEAR SEPARATION FOR THE B-X BAND SYSTEM OF 0967
MONO/ VARIATION OF ELECTRONIC TRANSITION MOMENT WITH THE INTERNUCLEAR SEPARATION IN SILICON CHLORIDE AND TIN 1137
FROM SCF MO LCAO COMPUTATIONS. PART-2: MOLECULAR ORBITAL INTERPRETATION OF ITS X-RAY ABSORPTION SPECTRA. /ON 1078
UPLING CONSTANTS IN FOUR XYN TYPE MOLECULE M/ THEORETICAL INTERPRETATION OF THE MASS INFLUENCE ON CORIOLIS CO 1488
/ULFUR, PHOSPHORUS, ANC SILICON FLUORIDES. HYDROLYSIS AND INTRA- INTERMOLECULAR MECHANISM OF FLUORINE EXCHAN/ 1042
OLECULES.   INTRAMOLECULAR EXCHANGE IN PHOSPHORUS PENTAHALIDE M 0852
                                                          INTRAMOLECULAR EXCHANGE IN SULFUR TETRAFLUORIDE. 1046
                                                          INTRAMOLECULAR EXCHANGE IN VANADIUM PENTAFLUORIDE. 1214
                                                          INTRAMOLECULAR FORCE FIELDS IN GROUP-5 TRIHALIDES. 1402
TS FROM INFRARED BAND SHAPES. THE ATOMIC DISPLACMENTS AND INTRAMOLECULAR FORCES IN CARBON TETRAFLUORIDE. /TAN 0331
STRUCTURES. (RHENIUM HEPTAFLUORIDE AND IODINE HEPTAFLUOR/ INTRAMOLECULAR LIGAND EXCHANGE IN SEVEN COORDINATE 0922
E AND XENON HEXAFLUORIDE.   INTRAMOLECULAR REARRANGEMENT IN IODINE HEPTAFLUORID 0911
ACTION PATTERNS FOR THE RHENIUM HEXAFLUORIDE MO/ MULTIPLE INTRAMOLECULAR SCATTERING EFFECTS ON ELECTRON DIFFR 0558
               STRUCTURE OF KRYPTON DIFLUORIDE AS INVESTIGATED BY ELECTRON DIFFRACTION.   0558
/UM FLUORIDE, CESIUM CHLORIDE, CESIUM BROMIDE, CESIUM IODIDE AND MAGNETIC SUSCEPTIBILITY ANISOTROPY OF T/ 1120
           BAND SPECTRA OF THALLIUM MONOIODIDE AND MONOFLUORIDE.   1125
A OF RARE EARTH METAL FLUORIDES IN THE REGION OF A CESIUM IODIDE PRISM.   INFRARED SPECTR 1352
SIS SET FOR NONEMPIRICAL ALL-ELECTRON SCF CALCULATIONS IN IODINE COMPOUNDS.   A SMALL GAUSSIAN BA 0495
           ELECTRONIC STRUCTURE AND CHEMISTRY OF IODINE COMPOUNDS. (FLUORIDES)   0499
MONOFLUORIDE, BROMINE MONOFLUORIDE, BROMINE CYANIDE, AND IODINE CYANIDE. /RIC QUADRUPOLE MOMENTS IN CHLORINE 0254
               ROTATIONAL SPECTRUM OF IODINE FLUORIDE.   0497
DISSOCIATION ENERGIES OF DIATOMIC HALOGEN FLUORIDES. (IODINE FLUORIDE, BROMINE FLUORIDE)   0494
          EVIDENCE FOR THE MOLECULAR SYMMETRY OF IODINE HEPTAFLUORIDE.   0523
     FORCE CONSTANTS AND THERMODYNAMIC FUNCTIONS OF IODINE HEPTAFLUORIDE.   0528
N SEVEN COORDINATE STRUCTURES. (RHENIUM HEPTAFLUORIDE AND IODINE HEPTAFLUORIDE) /AMOLECULAR LIGAND EXCHANGE I 0922
          MOLECULAR ORBITAL TREATMENT OF IODINE HEPTAFLUORIDE.   0533
                MOLECULAR STRUCTURE OF IODINE HEPTAFLUORIDE.   0530
   MOLECULAR STRUCTURE OF XENON HEXAFLUORIDE AND IODINE HEPTAFLUORIDE.   0526
                MOLECULAR SYMMETRY OF IODINE HEPTAFLUORIDE.   0524
                MOLECULAR SYMMETRY OF IODINE HEPTAFLUORIDE.   0531
   POLAR DISTORTIONS IN RHENIUM HEPTAFLUORIDE AND IODINE HEPTAFLUORIDE.   0529
                POTENTIAL CONSTANTS OF IODINE HEPTAFLUORIDE.   0532
/UCTURE, PSEUDOROTATION, AND VIBRATIONAL MODE COUPLING IN IODINE HEPTAFLUORIDE. AN ELECTRON DIFFRACTION STUD/ 0515
/TRIS (TRIMETHYL SILYL) SILANE, INFRARED INVESTIGATION OF IODINE HEPTAFLUORIDE AND AN APPLICATION OF THREE A/ 0527
ANHYDROUS HYDROGEN FLUORIDE.   SPECTROSCOPIC STUDY OF IODINE HEPTAFLUORIDE AND IODINE HEXAFLUORIDE(X) IN 0518
E.   NMR SPECTRA OF IODINE HEPTAFLUORIDE AND IODINE OXYGEN PENTAFLUORID 0516
E.   MASS SPECTRA AND SUBLIMATION PRESSURES OF IODINE HEPTAFLUORIDE AND IODINE OXYGEN PENTAFLUORID 0534
               VIBRATIONAL SPECTRA OF IODINE HEPTAFLUORIDE AND RHENIUM HEPTAFLUORIDE.   0522
     INTRAMOLECULAR REARRANGEMENT IN IODINE HEPTAFLUORIDE AND XENON HEXAFLUORIDE.   0520
DEGREES C.   STRUCTURE OF INTERHALOGEN COMPCUNDS. PART-2: IODINE HEPTAFLUORIDE AT -110 DEGREES C AND AT -145 0521
STEP ME/ FORCE CONSTANTS AND MEAN VIBRATION AMPLITUDES OF IODINE HEPTAFLUORIDE CALCULATED BY THE LOGARITHMIC 0535
          ACCEPTOR PROPERTIES OF IODINE HEPTAFLUORIDE. OCTAFLUORO PERIODATES(VII).   0514
LECULES WITH LARGE AMPLITUDE MOTION. (XENON HEXAFLUORIDE, IODINE HEPTAFLUORIDE, RHENIUM HEPTAFLUORIDE) /OF MO 1281
IDE PENTAFLUORIDES.   FLUORINE NMR OF IODINE HEPTAFLUORIDE, RHENIUM HEPTAFLUORIDE, AND OX 0517
A REDETERMINATION OF THE CRTHORHOMBIC IODINE HEPTAFLUORIDE STRUCTURE. A CORRECTION.   0519
               STRUCTURE OF THE IODINE HEXAFLUORIDE ANION. (XENON HEXAFLUORIDE)   1300
SPECTRA OF HALOGEN HEXAFLUORIDES. (CHLORINE HEXAFLUORIDE, IODINE HEXAFLUCRIDE, BROMINE HEXAFLUORIDE) /ONANCE 0271
3-), SELENIUM HEXABROMIDE(2-), TELLURIUM HEXABROMIDE(2-), IODINE HEXAFLUORIDE(1-), XENON HEXAFLUORIDE, ETC. / 1309
POUNDS. (CHLORINE MONOFLUORIDE, BROMINE MONOFLUORIDE, AND IODINE MONOFLUORIDE) /IDES AND THE INTERHALOGEN COM 0373
ELECTRONIC EMISSION SPECTRUM AND MOLECULAR CONSTANTS OF IODINE MONOFLUORIDE.   0496
TION OF DIATOMIC MOLECULES AND A STUDY OF THE EMISSION OF IODINE MONOFLUORIDE.   THEORY OF THE DISSOCIA 0493
HALOGENS FROM SPECTROSCOPIC DATA. (CHLORINE, EROMINE, AND IODINE MONOFLUORIDES) /ERTIES OF THE DIATOMIC INTER 0253
/ION AND RELATIVE INTENSITIES OF FORBIDDEN TRANSITIONS IN IODINE, OR IODINE BROMIDE (CHLORIDE, FLUORIDE) MOL/ 0498
R MONOFLUORIDE, SELENIUM MONOFLUCRIDE, SELENIUM OXIDE AND IODINE OXIDE. / ELECTRON RESONANCE SPECTRA OF SULFU 1020
ATIONAL LEVELS.   ESR SPECTRUM OF BROMINE OXIDE, IODINE OXIDE AND SELENIUM MONOFLUORIDE IN J-5/2 ROT 0954
NS OF XENON TETRAFLUORIDE, XENON OXIDE TETRAFLUORIDE, AND IODINE PENTAFLUORIDE. /IC STATES OF THE POSITIVE IO 1264
IES OF CHLORINE PENTAFLUORIDE, BROMINE PENTAFLUORIDE, AND IODINE PENTAFLUORIDE. /AL GAS THERMODYNAMIC PROPERT 0424
          MICROWAVE SPECTRUM OF IODINE PENTAFLUORIDE.   0502
     POTENTIAL CONSTANTS AND THERMODYNAMIC FUNCTIONS OF IODINE PENTAFLUORIDE.   0507
     POTENTIAL CONSTANTS AND THERMODYNAMIC PROPERTIES OF IODINE PENTAFLUORIDE.   0508
          VIBRATIONAL ASSIGNMENT FOR IODINE PENTAFLUORIDE, A NONRIGID D5H MOLECULE.   0525
TRUCTURE, FORCE CONSTANTS, AND THERMODYNAMIC FUNCTIONS OF IODINE PENTAFLUORIDE AND BROMINE PENTAFLUORIDE. / S 0505
     VIBRATIONAL SPECTRA AND STRUCTURES OF IODINE PENTAFLUORIDE AND HEPTAFLUORIDE.   0506
          MOSSBAUER STUDY OF IODINE PENTAFLUORIDE AND IODINE HEPTAFLUORIDE.   0503
     POTENTIAL CONSTANTS AND THERMODYNAMIC FUNCTIONS FOR IODINE PENTAFLUORIDE AND IODINE HEPTAFLUORIDE.   0509
/OF VIBRATION, SHRINKAGE EFFECT AND CORIOLIS CONSTANTS IN IODINE PENTAFLUORIDE AND IODINE OXYGEN PENTAFLUORI/ 0513
/F SQUARE PYRAMIDAL MOLECULES. XENON OXIDE TETRAFLUORIDE, IODINE PENTAFLUORIDE, BROMINE PENTAFLUORIDE, AND C/ 0418
/HOD AND APPLICATION TO SOME SQUARE PYRAMIDAL MOLECULES. (IODINE PENTAFLUORIDE, BROMINE PENTAFLUORIDE, AND C/ 0420
/LITUDES OF VIBRATION FOR SOME PYRAMIDAL XY4Z MOLECULES. (IODINE PENTAFLUORIDE, BROMINE PENTAFLUORIDE, AND C/ 0421
N THE LIQUID STATE.   RAMAN SPECTRUM OF IODINE PENTAFLUORIDE. EVIDENCE FOR POLYMERIZATION I 0512
DAMENTAL AND A TER/ VIBRATIONAL SPECTRA AND ASSIGNMENT OF IODINE PENTAFLUORIDE. FERMI RESONANCE BETWEEN A FUN 0504
/ PHASE MOLECULAR STRUCTURES OF BROMINE PENTAFLUORIDE AND IODINE PENTAFLUORIDE FROM ELECTRON DIFFRACTION AND/ 0265
50 DEGREES K, ENTHALPY OF FUSION A/ CALORIMETRIC STUDY OF IODINE PENTAFLUORIDE. HEAT CAPACITY BETWEEN 5 AND 3 0510
AND IO/ MOLECULAR FORCE FIELD FOR INTERHALOGEN COMPOUNDS. IODINE PENTAFLUORIDE, IODINE OXYGEN PENTAFLUORIDE, 0511

/METHODS. (CHLORINE PENTAFLUORIDE, BROMINE PENTAFLUORIDE, IODINE PENTAFLUORIDE, PHOSPHORUS PENTAFLUORIDE, VA/ 0422
D ANTIM/ VIBRATIONAL SPECTRA OF THE ISOELECTRONIC SPECIES IODINE PENTAFLUORIDE, TELLURIUM PENTAFLUORIDE(-) AN 0501
VIBRATIONAL SPECTRUM OF IODINE TRIFLUORIDE. 0500
/TERHALOGENS. (CHLORINE TRIFLUORIDE, BROMINE TRIFLUORIDE, IODINE TRIFLUORIDE, CHLORINE PENTAFLUORIDE, BROMIN/ 0397
RACTION STUDIES OF NITROGEN MONOFLUORIDE AND ITS POSITIVE ION. CONFIGURATION INTE 0672
RONIC SPECTRA OF PHOSPHORUS MONOFLUORIDE AND ITS POSITIVE ION. ELECT 0786
LATION FOR THE GROUND STATE OF OXYGEN DIFLUORIDE, NITRITE ION AND CYANIDE ION. /NFIGURATION INTERACTION CALCU 0763
DOUBLE ZETA SCF CALCULATIONS FOR NITRITE ION AND OXYGEN DIFLUORIDE. 0762
CNDO STUDY OF THE OZONIDE ION AND RELATED SPECIES. (OXYGEN DIFLUORIDE) 0782
SCOPIC MEASUREMENTS FOR COMPOUNDS WHICH PRODUCE NO PARENT ION. (CARBON TETRAFLUORIDE) / PHOTOELECTRON SPECTRO 0339
PROPERTIES AND REACTIONS OF URANIUM HEXAFLUORIDE BY ION CYCLOTRON RESONANCE SPECTROSCOPY. 1185
OUP-2B HALIDES. (ZINC DIFLUORIDE, CADMIUM DIFL/ POLARIZED ION MODEL AND THE BENDING FORCE CONSTANTS OF THE GR 1322
T SCF CALCULATIONS FOR THE GROUND STATE OF OZONE, NITRITE ION, NITROGEN OXYFLUORIDE, AND OXYGEN DIFLUORIDE. / 0776
KRYPTON MONOFLUORIDE AND ITS POSITIVE ION. (SCF CALCULATIONS) 0547
STRUCTURE OF THE SULFUR TRIFLUORIDE POSITIVE ION. (SULFUR DIFLUORIDE) 1032
EY FORCE FIELD TO SOME TETRAHEDRAL XY4 TYPE MOLECULES AND IONS. APPLICATION OF UREY-BRADL 1425
VIBRATION OF AN XY3 MOLECULE AND A MODEL WITH POLARIZABLE IONS. FORCE CONSTANT OF THE NONPLANAR 0573
AND SILICON MONOFLUORIDE AND THEIR POSITIVE AND NEGATIVE IONS. /ORBITAL INVESTIGATION OF CARBON MONOFLUORIDE 0290
F ATOMIZATION ENERGIES OF PNICOGEN FLUORIDE MOLECULES AND IONS. MOLECULAR ORBITAL STUDY O 1364
/ND SELENIUM MONOFLUORIDE AND THEIR POSITIVE AND NEGATIVE IONS. A COMPARISON OF THE THEORETICAL AND EXPERIME/ 1025
HIGH TEMPERATURE NEGATIVE IONS. GASEOUS GROUP-3 FLUORIDES. 0047
MAL COORDINATE TREATMENT OF XY4 TETRAHEDRAL MOLECULES AND IONS. (GVFF) NOR 1415
EPENDENCE OF THE CORIOLIS CONSTANT FOR SOME MOLECULES AND IONS OF D4H SYMMETRY. / FORCE CONSTANTS, AND MASS D 1411
IDE, AN/ CORRELATION OF ELECTRONIC STATES OF THE POSITIVE IONS OF XENON TETRAFLUORIDE, XENON OXIDE TETRAFLUOR 1264
E) DIRECTED VALENCY IN CERTAIN MOLECULES AND COMPLEX IONS. (PHOSPHORUS PENTAFLUORIDE, SULFUR HEXAFLUORID 0860
MAL COORDINATE TREATMENT OF XY4 TETRAHEDRAL MOLECULES AND IONS. (UREY-BRADLEY AND MVFF) NOR 1418
TS FOR ARSENIC TRIFLUORIDE AND ARSENIC T/ CALCULATIONS ON IONIC CHARACTER AND HYBRIDIZATION FROM DIPOLE MOMEN 0107
OELECTRON SPECTRA OF XENON DIFLUORIDE / BOND ENERGIES AND IONIC CHARACTER OF INERT GAS HALIDES. HELIUM-2 PHOT 1343
STANTS OF THE GROUP-2 HALIDES. IONIC MODEL CALCULATIONS. PART-2: BENDING FORCE CON 1381
F MONOFLUORIDE COMPOUNDS OF MAGNESIUM AND CALCIUM AND THE IONIC MODEL OF A MOLECULE. DISSOCIATION ENERGY O 0278
OF PHOSPHINE AND PHOSPHORUS TRIFLUORIDE AND THEIR GROUND IONIC STATES. THEORETICAL STUDY OF THE GEOMETRY 0794
P-4 ELEMENTS. (TIN TETRAFLUORIDE AND HAFNIUM TETRAFLUORI/ IONICITY OF THE MX BOND IN THE TETRAHALIDES OF GROU 1141
UCED BY PHOTON IMPACT. IONIZATION AND DISSOCIATION OF XENON DIFLUORIDE IND 1243
RIDE AND DIFLUORO CYANO PHOSPHINE. IONIZATION BY ELECTRON IMPACT OF PHOSPHORUS TRIFLUO 0800
RIDE) SELF IONIZATION OF HALOGEN FLUORIDES. (BROMINE PENTAFLUO 0263
/ONOFLUORIDE. HARTREE-FOCK WAVE FUNCTION, BINDING ENERGY, IONIZATION POTENTIAL, ELECTRON AFFINITY, DIPOLE AN/ 0756
EN/ ARSENIC MONOFLUORIDE, 3 SIGMA. DISSOCIATION ENTHALPY, IONIZATION POTENTIAL, ELECTRON AFFINITY, DIPOLE MOM/ 0094
/, DOUBLET PI HARTREE-FOCK WAVE FUNCTION, BINDING ENERGY, IONIZATION POTENTIAL, ELECTRON AFFINITY, DIPOLE MO/ 0757
TETRAFLUORO METHANE XENON TETRAFLUORIDE, AND/ CALCULATED IONIZATION POTENTIAL OF CHLORO- AND FLUOROMETHANES, 0305
ME GERMANIUM FLUORIDE SPECIE/ BOND DISSOCIATION ENERGIES, IONIZATION POTENTIALS AND ELECTRON AFFINITIES OF SO 0464
FLUORO CHLORO ETHYLENE, DIFLUORO DICHLORO ETHYLENE, AND / IONIZATION POTENTIALS FOR TETRAFLUORO ETHYLENE, TRI 0305
DE, SILICON OXIDE AND GERMANIUM OXIDE. FIRST IONIZATION POTENTIALS OF THE MOLECULES BORON FLUORI 0160
UP-6/ SCF- X(ALPHA) SCATTERED WAVE STUDIES ON BONDING AND IONIZATION POTENTIALS. PART-1: HEXAFLUORIDES OF GRO 1366
CONFIGURATION OF THE PENTAFLUORIDES OF RHODIUM AND IRIDIUM BY MAGNETIC RESONANCE. 0923
INFRARED SPECTRUM OF IRIDIUM HEXAFLUORIDE. 0540
SEARCH FOR A JAHN-TELLER EFFECT IN IRIDIUM HEXAFLUORIDE. 0539
AFLUORIDE, RHENIUM HEXAFLUORIDE, OSMIUM HEXAFLUORIDE, AND IRIDIUM HEXAFLUORIDE) /HEXAFLUORIDE, MOLYBDENUM HEX 0903
ED JAHN-TELLER PROGRESSIONS IN THE ELECTRONIC SPECTRUM OF IRIDIUM HEXAFLUORIDE. VIBRONICALLY INDUC 0537
EXAFLUORIDE. (RHENIUM HEXAFLUORIDE, OSMIUM HEXAFLUORIDE, IRIDIUM HEXAFLUORIDE, AND PLATINUM HEXAFLUORIDE) /H 0916
HEXAFLUORIDE, RHODIUM HEXAFLUORIDE, OSMIUM HEXAFLUORIDE, IRIDIUM HEXAFLUORIDE, AND PLATINUM HEXAFLUORIDE) /M 1084
IRIDIUM HEXACHLORIDE(/ VIBRATIONAL ELECTRONIC COUPLING IN IRIDIUM HEXAFLUORIDE, OSMIUM HEXACHLORIDE(2-), AND 0538
/STIGATION OF TUNGSTEN HEXAFLUORIDE, OSMIUM HEXAFLUORIDE, IRIDIUM HEXAFLUORIDE, URANIUM HEXAFLUORIDE, NEPTUN/ 1155
ATERIALS. (CHROMIUM DIFLUORIDE, CHROMIUM TRIFLUCRIDE, AND IRON DIFLUORIDE) /APORIZATION OF HIGH TEMPERATURE M 0431
SUBLIMATION PRESSURE OF IRON DIFLUORIDE. 0541
ELECTRONIC STRUCTURE AND ANOMALOUS THERMAL EXPANSION ON IRON DIFLUORIDE AND VANADIUM DIOXIDE. 0543
ELECTRONIC STRUCTURE OF IRON TRIFLUORIDE. 0542
SURES OF CHROMIUM TRIFLUORIDE, MANGANESE TRIFLUORIDE, AND IRON TRIFLUORIDE. SUBLIMATION PRES 0434
ARSENIC TRICHLORIDE. ESR STUDY OF RADICALS IN GAMMA IRRADIATED POLYCRYSTALLINE ARSENIC TRIFLUORIDE AND 0108
ON SPIN RESONANCE SPECTRUM OF XENON MONOFLUORIDE IN GAMMA IRRADIATED XENON TETRAFLUORIDE. ELECTR 1222
/R CHARGE DISTRIBUTIONS AND CHEMICAL BINDING. PART-3: THE ISOELECTRONIC NITROGEN, CARBON MONOXIDE, BORON FLU/ 0150
SCF- MO CALCULATIONS OF SOME MOLECULAR PROPERTIES OF THE ISOELECTRONIC SERIES CHLORINE MONOFLUORIDE, HYDROG/ 0371
UM PENTAFLUORIDE(-) AND ANTIM/ VIBRATIONAL SPECTRA OF THE ISOELECTRONIC SPECIES IODINE PENTAFLUORIDE, TELLURI 0501
UORIDE. INFRARED SPECTRA OF MATRIX ISOLATED ANTIMONY PENTAFLUORIDE AND ARSENIC PENTAFL 0113
ISOTOPIC SPLITTING IN MATRIX ISOLATED BORON TRIFLUORIDE. 0228
AND BROMINE PENTAFLUORIDE. F/ INFRARED SPECTRA OF MATRIX ISOLATED CHLORINE TRIFLUORIDE, BROMINE TRIFLUORIDE, 0399
ER DIFLUORIDE, A/ INFRARED SPECTRA AND GEOMETRY OF MATRIX ISOLATED COBALT DIFLUORIDE, NICKEL DIFLUORIDE, COPP 0437
INFRARED SPECTRUM OF THE FREE RADICAL CARBON TRIFLUORIDE ISOLATED IN INERT MATRICES. PHOTOCHLORINATION OF / 0311
INFRARED SPECTRA OF MATRIX ISOLATED MAGNESIUM DIFLUORIDE. 0604
MAG/ INFRARED AND RAMAN SPECTRA AND STRUCTURES OF MATRIX ISOLATED MAGNESIUM DIHALIDES: MAGNESIUM DIFLUORIDE, 0607
RAMAN SPECTRA OF MATRIX ISOLATED MOLECULES. (SULFUR HEXAFLUORIDE) 1097
SPECTRA OF MATRIX ISOLATED NICKEL DIFLUORIDE AND NICKEL DICHLORIDE. 0653
ARBON MONOXIDE, MOLECULAR NITROGEN/ INTERACTION OF MATRIX ISOLATED NICKEL FLUORIDE AND NICKEL CHLORIDE WITH C 1324
INFRARED SPECTRUM OF MATRIX ISOLATED NIOBIUM PENTAFLUORIDE. 0659
ABSORPTION SPECTRA OF MATRIX ISOLATED NITROGEN TRIFLUORIDE FROM 18 TO 50 GC. 0698
LASER EXCITED RAMAN SPECTRA OF MATRIX ISOLATED PRASEODYMIUM TRIFLUORIDE. 0895
ORIDE) GEOMETRY AND INFRARED SPECTRA OF MATRIX ISOLATED RARE EARTH HALIDES. (ALSO LANTHANUM TRIFLU 0583
RIDE, YTTRIUM T/ FORCE CONSTANTS AND GEOMETRIES OF MATRIX ISOLATED RARE EARTH TRIFLUORIDES. (SCANDIUM TRIFLUO 0944
INFRARED SPECTRA OF MATRIX ISOLATED SPECIES IN THE GALLIUM- FLUORINE SYSTEM. 0462
INFRARED SPECTRA AND STRUCTURE OF MATRIX ISOLATED THALLIUM HALIDES. 1113
AND / INFRARED AND RAMAN SPECTRA AND STRUCTURE OF MATRIX ISOLATED THALLOUS HALIDE DIMERS, THALLOUS FLUORIDE, 1123
INFRARED SPECTRA OF MATRIX ISOLATED TIN DIFLUORIDE AND LEAD DIFLUORIDE. 0594
E. ABSORPTION AND EMISSION SPECTRA OF ARGON MATRIX ISOLATED XENON MONOFLUORIDE AND KRYPTON MONOFLUORID 0544
. INFRARED SPECTRA AND GEOMETRIES OF MATRIX ISOLATED YTTRIUM DIFLUORIDE AND YTTRIUM TRIFLUORIDE 1316
RAFLUORIDE. INFRARED SPECTRA OF MATRIX ISOLATED ZIRCONIUM DIFLUORIDE, TRIFLUORIDE, AND TET 1325
RED SPECTRA OF MAGNESIUM AND ALUMINUM FLUORIDES BY MATRIX ISOLATION. INFRA 0608
LUMINUM HEXAFLUORIDE, AND ALUMINUM MONOFLUORIDE BY MATRIX ISOLATION. /D SPECTRUM OF ALUMINUM TRIFLUORIDE, DIA 0051
BLE AND ULTRAVIOLET ABSORPTION IN THE VAPOR AND IN MATRIX ISOLATION. /DE. RAMAN SCATTERING AND INFRARED, VISI 1285
SPECIES. (FLUORIDES, OXIDES, CHLORIDES) MATRIX ISOLATION INFRARED SPECTROSCOPY OF HIGH TEMPERATURE 0011
PREPARATION OF INERT GAS COMPOUNDS BY MATRIX ISOLATION: KRYPTON DIFLUORIDE. 0566
MATRIX ISOLATION OF BROMINE MONOFLUORIDE. 0255

FLUORIDE, ANTIMONY PENTAFLUORIDE, ARSENI/ INFRARED MATRIX ISOLATION STUDIES. (BORON TRIFLUORIDE, SULFUR TETRA 0239
AFLUORIDE AND TELLURIUM TETRAFLUORIDE, INCLUDING A MATRIX ISOLATION STUDY. /A AND STRUCTURES OF SELENIUM TETR 0956
MATRIX ISOLATION STUDY OF CHLORINE PENTAFLUORIDE. 0419
S OF NITROGEN TRIFLUORIDE. ELECTRONIC SPECTRUM OF/ MATRIX ISOLATION STUDY OF THE VACUUM ULTRAVIOLET PHOTOLYSI 0691
S OF SILICON DIFLUORIDE DIHYDRIDE. THE INFRARED / MATRIX ISOLATION STUDY OF THE VACUUM ULTRAVIOLET PHOTOLYSI 0986
ARED SPECTRA OF SOME ALKALINE EARTH HALIDES BY THE MATRIX ISOLATION TECHNIQUE. INFR 0125
KRYPTON DIFLUORIDE. PREPARATION BY THE MATRIX ISOLATION TECHNIQUE. 0567
ULAR GEOMETRY OF XENON HEXAFLUORIDE: A CASE OF ELECTRONIC ISOMERISM. ELECTRONIC STATES AND MOLEC 1294
THE FAR INFRARED SPECTRUM OF BORON TRIFLUOR/ STUDY OF THE ISOTOPE EFFECT AND A MORE PRECISE DETERMINATION OF 0230
BORON FLUORIDE MOLECULE. ISOTOPE EFFECT IN THE SINGLET SPECTRAL BANDS OF THE 0166
LASER ISOTOPE SEPARATION. (SULFUR HEXAFLUORIDE SPECTRUM) 1094
NU-4 BANDS OF THE TEN AND ELEVEN ISOTOPES OF BORON TRIFLUORIDE AT HIGH RESOLUTION. 0208
RIDE, GERMANIUM TETRAFLUO/ THE FORCE CONSTANTS OF VARIOUS ISOTOPES OF CARBON TETRAFLUORIDE, SILICON TETRAFLUO 0358
O(+) TO SINGLET SIGMA(+) SYSTEMS OF BOTH ISOTOPES OF CHLORINE MONOFLUORIDE. 0390
IGMA- X DOUBLET PI SYSTEM OF TH/ ROTATIONAL STRUCTURE AND ISOTOPIC SHIFT IN THE (1,0) BAND OF THE A DOUBLET S 1135
RIDE. ISOTOPIC SPLITTING IN MATRIX ISOLATED BORON TRIFLUO 0228
RCE CONSTANTS. ANHARMONICITY EFFEC/ UTILITY OF HEAVY ATOM ISOTOPIC SUBSTITUION DATA FOR THE CALCULATION OF FO 0234
FORCE CONSTANTS OF SULFUR HEXAFLUORIDE FROM ISOTOPIC SUBSTITUTION. 1100
. ACCURATE FORCE CONSTANTS FROM HEAVY ISOTOPIC SUBSTITUTION. PART-1. CARBON TETRAFLUORIDE 0320
ILICON TETRAFLUORIDE/ ACCURATE FORCE CONSTANTS FROM HEAVY ISOTOPIC SUBSTITUTION. PART-2: BORON TRIFLUORIDE, S 0232
URE IN THE PHOSPHORUS TETRAFLUORIDE RADICAL. ISOTROPIC HYPERFINE SPLITTINGS AND MOLECULAR STRUCT 0824

J

BROMINE OXIDE, IODINE OXIDE AND SELENIUM MONOFLUORIDE IN J-5/2 ROTATIONAL LEVELS. ESR SPECTRUM OF 0954
MICROWAVE MEASUREMENTS OF THE J-8 FAR-9 TRANSITION OF BROMINE PENTAFLUORIDE. 0261
AND SPIN ORBIT SPLITTING IN SPECTRA OF SYSTEMS EXHIBITING JAHN-TELLER EFFECT. VIBRONIC 1480
UDIES ON VANADIUM TETRACHLORIDE. JAHN-TELLER EFFECT. EXPERIMENTAL AND THEORETICAL ST 1426
JAHN-TELLER EFFECT IN HEXAFLUORIDES. 1451
SEARCH FOR A JAHN-TELLER EFFECT IN IRIDIUM HEXAFLUORIDE. 0539
HEXAFLUORIDE MOLECULES. THE INFRARED SPECTRA OF RUTHENIU/ JAHN-TELLER EFFECT IN RHENIUM HEXAFLUORIDE. 0904
UORIDES. JAHN-TELLER EFFECT IN THE E(8) VIBRATIONAL MODE OF 0926
EVIDENCE FOR PSEUDO JAHN-TELLER EFFECT IN VIBRATIONAL SPECTRA OF HEXAFL 1471
ION OF RHENIUM HEXAFLUORIDE. JAHN-TELLER EFFECT IN XENON HEXAFLUORIDE. 1279
OF IRIDIUM HEXAFLUORIDE. VIBRONICALLY INDUCED JAHN-TELLER EFFECTS IN THE 2E5/2G TO 2G3/2G TRANSIT 0913
JAHN-TELLER PROGRESSIONS IN THE ELECTRONIC SPECTRUM 0537
JAHN-TELLER VIBRATIONS OF RHENIUM HEXAFLUORIDE. 0910

K

N SPECTRA OF GASEOUS SULFUR HEXAFLUORIDE. SULFUR K AND L AND FLUORINE K X-RAY EMISSION AND ABSORPTIO 1088
BROMINE PENTAFLUORIDE. K-TYPE DOUBLING IN THE MILLIMETER WAVE SPECTRUM OF 0260
ION TO BORON TRIHALIDES. KINEMATICAL EVALUATION OF FORCE CONSTANTS. APPLICAT 0248
N PRESSURE OF CADMIUM DIFLUORIDE. KNUDSEN AND LANGMUIR MEASUREMENTS OF THE SUBLIMATIC 0272
N PRESSURE OF COBALT DIFLUORIDE. KNUDSEN AND LANGMUIR MEASUREMENTS OF THE SUBLIMATIO 0439
ESE DIFLUORIDE. KNUDSEN MEASUREMENT OF THE VAPOR PRESSURE OF MANGAN 0613
AB INITIO CALCULATIONS OF THE BONDING IN KRYPTON DIFLUORIDE. 0553
ABSENCE OF FERMI RESONANCE IN KRYPTON DIFLUORIDE. 0564
S. HELIUM-2 PHOTOELECTRON SPECTRA OF XENON DIFLUORIDE AND KRYPTON DIFLUORIDE. / CHARACTER OF INERT GAS HALIDE 1343
BONDING IN KRYPTON DIFLUORIDE. 0554
BONDING IN KRYPTON DIFLUORIDE. 0555
ELECTRONIC STRUCTURE AND PROPERTIES OF KRYPTON DIFLUORIDE. 0548
HEAT FORMATION OF KRYPTON DIFLUORIDE. 0557
HELIUM-2 PHOTOELECTRON SPECTRA OF XENON DIFLUORIDE AND KRYPTON DIFLUORIDE. 1234
INFRARED AND RAMAN SPECTRA OF KRYPTON DIFLUORIDE. 0552
MASS SPECTRUM AND MOLECULAR ENERGETICS OF KRYPTON DIFLUORIDE. 0565
EN-MORINO SHRINKAGE EFFECT AND THERMODYNAMIC FUNCTIONS OF KRYPTON DIFLUORIDE. /LITUDES OF VIBRATION, BASTIANS 0562
MOLECULAR ORBITAL ENERGY LEVELS AND BONDING IN KRYPTON DIFLUORIDE. 0549
PREPARATION OF INERT GAS COMPOUNDS BY MATRIX ISOLATION: KRYPTON DIFLUORIDE. 0566
STRUCTURE OF KRYPTON DIFLUORIDE. 0561
ULTRAVIOLET EMISSION SPECTRA. (XENON DIFLUORIDE AND KRYPTON DIFLUORIDE) 0559
D ORBITALS IN THE NOBLE GAS DIHALIDES. (ARGON DIFLUORIDE, KRYPTON DIFLUORIDE, AND XENON DIFLUORIDE) 1338
FORCE FIELDS IN KRYPTON DIFLUORIDE AND XENON DIFLUORIDE. 0556
RACTION. STRUCTURE OF KRYPTON DIFLUORIDE AS INVESTIGATED BY ELECTRON DIFF 0558
CRYSTAL STRUCTURE OF KRYPTON DIFLUORIDE AT -80C. 0551
TRA OF CERTAIN FLUORIDES IN THE NEAR ULTRAVIOLET REGION. (KRYPTON DIFLUORIDE, NITROGEN TRIFLUORIDE) /ION SPEC 0560
ES) KRYPTON DIFLUORIDE. (PHYSICAL AND CHEMICAL PROPERTI 0563
TION TECHNIQUE. KRYPTON DIFLUORIDE. PREPARATION BY THE MATRIX ISOLA 0567
SCOPY. ELECTRONIC STRUCTURE OF KRYPTON DIFLUORIDE STUDIED BY PHOTOELECTRON SPECTRO 0550
N SPECTRA OF ARGON MATRIX ISOLATED XENON MONOFLUORIDE AND KRYPTON MONOFLUORIDE. ABSORPTION AND EMISSIO 0544
ELECTRON SPIN RESONANCE SPECTRUM OF KRYPTON MONOFLUORIDE. 0546
MONOBROMIDE, XENON MONOCHLORIDE, XENON MONOFLUORIDE, AND KRYPTON MONOFLUORIDE. EMISSION SPECTRA OF XENON 0545
CULATIONS) KRYPTON MONOFLUORIDE AND ITS POSITIVE ION. (SCF CAL 0547

L

TRA OF GASEOUS SULFUR HEXAFLUORIDE. SULFUR K AND L AND FLUORINE K X-RAY EMISSION AND ABSORPTION SPEC 1088
SIGNS OF L DOUBLING CONSTANTS IN NITROGEN TRIFLUORIDE. 0743
EINSTOCK AND GOODMAN FOR OCTAHEDRAL HEXAFLUORIDES AND THE L MATRIX METHOD. FORCE FIELD OF W 1456
DE IN THE NU-4 EQUALS 1 STATE. L-TYPE DOUBLING TRANSITIONS OF PHOSPHORUS TRIFLUORI 0803
DE IN THE NU/ EFFECTS OF THE THIRD ORDER CONSTANTS ON THE L-TYPE DOUBLING TRANSITIONS OF PHOSPHORUS TRIFLUORI 0804
UORIDE. HIGH RESOLUTION STUDY OF L-TYPE RESONANCE IN THE NU-4 BAND OF NITROGEN TRIFL 0731
MANGANESE DIFLUORIDE. A LANGMUIR MEASUREMENT OF THE SUBLIMATION PRESSURE OF 0611
F CADMIUM DIFLUORIDE. KNUDSEN AND LANGMUIR MEASUREMENTS OF THE SUBLIMATION PRESSURE O 0272
F COBALT DIFLUORIDE. KNUDSEN AND LANGMUIR MEASUREMENTS OF THE SUBLIMATION PRESSURE O 0439
VAPORIZATION AND SUBLIMATION THERMODYNAMICS OF SELECTED LANTHANIDE FLUORIDES. 1347
ON SPECTROSCOPY OF FLUORIDES AND OXIDES BELONGING TO THE LANTHANIDE SERIES. BIBLIOGRAPHY 1358
F THERMOCHEMICAL ERRORS AND SYSTEMATICS IN SUBLIMATION OF LANTHANIDE TRIFLUORIDES. ANALYSES O 1353
MEAN AMPLITUDES OF VIBRATION OF SOME LANTHANIDE TRIFLUORIDES. 1354
/R BEAMS OF THE RARE EARTH DIFLUORIDES AND TRIFLUORIDES. (LANTHANIDE TRIFLUORIDES, LANTHANIDE DIFLUORIDES, S/ 0946

HALIDES OF THE LANTHANIDES AND ACTINIDES. 1348
LUME-2: SECOND AND THIRD ROW TRANSITION METALS. VOLUME-3: LANTHANIDES AND ACTINIDES. /W TRANSITION METALS. VO 0001
STRUCTURE OF THE VAPOR PHASE HALIDES OF GALLIUM, YTTRIUM, LANTHANUM, AND NEODYMIUM. / STUDY OF THE MOLECULAR 1317
MASS SPECTROMETRIC STUDIES OF SCANDIUM, YTTRIUM, LANTHANUM, AND RARE EARTH FLUORIDES. 1359
/ TRIFLUORIDES FROM INFRARED SPECTRA. (SCANDIUM, YTTRIUM, LANTHANUM, CERIUM, NEODYMIUM, EUROPIUM, AND GADOLI/ 0943
ELECTRONIC SPECTRA OF LANTHANUM FLUORIDE IN THE EXTREME ULTRAVIOLET. 0570
VALENCE VIBRATION FREQUENCIES OF TRIHALIDE MOLECULES OF LANTHANUM HALIDES. 0574
THERMODYNAMIC FUNCTIONS OF GASEOUS SCANDIUM, YTTRIUM, AND LANTHANUM HALIDES IN THE 293-3000 DEGREES K RANGE. 0568
ONOFLUORID/ MOLECULAR CONSTANTS OF SCANDIUM, YTTRIUM, AND LANTHANUM HALIDES. (SCANDIUM DIFLUORIDE, SCANDIUM M 0941
STABILITY. (LANTHAN/ DEMONSTRATION OF THE EXISTENCE OF DILANTHANUM HEXAFLUORIDE GAS AND DETERMINATION OF ITS 0581
ITIES OF SCANDIUM MONOFLUORIDE, YTTRIUM MONOFLUORIDE, AND LANTHANUM MONOFLUORIDE. /IATION ENERGIES AND STABIL 0935
ELECTRONIC SPECTRUM OF LANTHANUM MONOFLUORIDE. 0569
GASEOUS SCANDIUM MONOFLUORIDE, YTTRIUM MONOFLUORIDE, AND LANTHANUM MONOFLUORIDE. ELECTRONIC STATES OF 0928
NCIES OF SCANDIUM MONOFLUORIDE, YTTRIUM MONOFLUORIDE, AND LANTHANUM MONOFLUORIDE. /MOLECULAR VIBRATION FREQUE 0936
CTRUM OF SCANDIUM MONOFLUORIDE, YTTRIUM MONOFLUORIDE, AND LANTHANUM MONOFLUORIDE. VIBRONIC SPE 0938
SUBGROUP. (SCANDIUM TRIFLUORIDE, YTTRIUM TRIFLUORIDE, AND LANTHANUM TRIFLUORIDE) /HE HALIDES OF THE SCANDIUM 0948
NUM HEXAFLUORIDE GAS AND DETERMINATION OF ITS STABILITY. (LANTHANUM TRIFLUORIDE) /F THE EXISTENCE OF DILANTHA 0581
ELECTRONIC SPECTRUM OF LANTHANUM TRIFLUORIDE. 0582
OF THE ENTHALPIES OF FORMATION OF RARE EARTH FLUORIDES. (LANTHANUM TRIFLUORIDE) EXPERIMENTAL DETERMINATION 0578
LUORIDES. (SCANDIUM TRIFLUORIDE, YTTRIUM TRIFLUORIDE, AND LANTHANUM TRIFLUORIDE) /IX ISOLATED RARE EARTH TRIF 0944
ARED SPECTRA OF MATRIX ISOLATED RARE EARTH HALIDES. (ALSO LANTHANUM TRIFLUORIDE) GEOMETRY AND INFR 0583
LATTICE VIBRATIONS AND STRUCTURE OF RARE EARTH HALIDES. (LANTHANUM TRIFLUORIDE) 0572
NEAR ULTRAVIOLET OPTICAL CONSTANTS OF LANTHANUM TRIFLUORIDE. 0584
NUCLEAR RELAXATION TIMES OF LANTHANUM TRIFLUORIDE. 0579
RESSURE OF SCANDIUM TRIFLUORIDE, YTTRIUM TRIFLUORIDE, AND LANTHANUM TRIFLUORIDE. SATURATED VAPOR P 0952
RMATION OF SCANDIUM TRIFLUORIDE, YTTRIUM TRIFLUORIDE, AND LANTHANUM TRIFLUORIDE. STANDARD ENTHALPIES OF FO 0942
ESSURES OF SCANDIUM TRIFLUORIDE, YTTRIUM TRIFLUORIDE, AND LANTHANUM TRIFLUORIDE. SUBLIMATION PR 0947
MMETRY AND CRYSTAL STRUCTURE OF RARE EARTH TRIFLUORIDES. (LANTHANUM TRIFLUORIDE) SY 0571
SURE, HEAT OF SUBLIMATION, AND EVAPORATION COEFFICIENT OF LANTHANUM TRIFLUORIDE. VAPOR PRES 0576
VAPORIZATION STUDIES OF LANTHANUM TRIFLUORIDE. 0575
ENTHALPY OF FORMATION OF LANTHANUM TRIFLUORIDE AND PRASEODYMIUM TRIFLUORIDE. 0577
/IC STUDIES OF SCANDIUM TRIFLUORIDE, YTTRIUM TRIFLUORIDE, LANTHANUM TRIFLUORIDE AND THE RARE EARTH TRIFLUORI/ 0953
BAND SYSTEM IN THE DISCHARGE SPECTRUM OF LANTHANUM TRIFLUORIDE VAPORS. 0580
EODYMIUM TRIFLUORIDE. LASER EXCITED RAMAN SPECTRA OF MATRIX ISOLATED PRAS 0895
TRUM) LASER ISOTOPE SEPARATION. (SULFUR HEXAFLUORIDE SPEC 1094
MEASUREMENT OF THE GAS DENSITY OF URANIUM HEXAFLUORIDE BY LASER RAMAN SCATTERING. 1181
L COORDINATE ANALYSES OF GERMANIUM D/ MATRIX INFRARED AND LASER RAMAN SPECTRA, MOLECULAR STRUCTURES AND NORMA 0475
LASER SPECTROSCOPY. (PHOSPHORUS PENTAFLUORIDE) 0832
/TAL MODIFICATION BY THE COULOMB FIELD OF LIGAND ATOMS OF LATER SECOND ROW ELEMENTS IN PERFECT PAIRING VALEN/ 0862
ING BERKELIUM. LATTICE CONSTANTS OF ACTINIDE TETRAFLUORIDES INCLUD 1334
DES. (LANTHANUM TRIFLUORIDE) LATTICE VIBRATIONS AND STRUCTURE OF RARE EARTH HALI 0572
/ART-1: OPTICAL ABSORPTION AND X-RAY EMISSION FROM SCF MO LCAO COMPUTATIONS. PART-2: MOLECULAR ORBITAL INTER/ 1078
/URES OF COMPOUNDS OF HALOGENS AND INERT GASES BY THE SCF LCAO MO METHOD. PART-2: COMPARATIVE ANALYSIS OF D/ 1231
E OF SILICON DIFLUORIDE. NONEMPIRICAL LCAO MO SCF CALCULATIONS OF THE ELECTRONIC STRUCTUR 0995
E OF FLUORIDES OF NONTRANSITION ELEMENTS FROM THE DATA OF LCAO MO SCF NONEMPIRICAL CALCULATIONS. /IC STRUCTUR 0009
LIUM DIFLUORIDE, MAGNESIUM DIFLUORIDE, AND / NONEMPIRICAL LCAO MO SCF STUDIES OF THE GROUP-2A DIHALIDES BERYI 1360
LLIUM DIFLUORIDE. NONEMPIRICAL LCAO MO SCF STUDIES OF THE LOW LYING STATES OF BERY 0137
AN LCAO MO STUDY OF RARE GAS FLUORIDES. 1341
LORINE DIFLUORIDE ON THE BASIS OF A CALCULATION BY THE MO LCAO SCF IN THE NDDO APPROXIMATION. /HE MOLECULE CH 0382
/STRUCTURE OF A CHLORINE PENTAFLUORIDE MOLECULE BY THE MO LCAO SELF CONSISTENT FIELD METHOD IN AN NDDO APPRO/ 0415
L XY4 TYPE MOLECULES. PART-5: TETRAHALIDES OF HAFNIUM AND LEAD. /N AMPLITUDES OF VIBRATION OF SOME TETRAHEDRA 0487
INFRARED SPECTRA OF MATRIX ISOLATED TIN DIFLUORIDE AND LEAD DIFLUORIDE. 0594
NEUTRON DIFFRACTION INVESTIGATION OF ORTHORHOMBIC LEAD DIFLUORIDE. 0593
RAMAN SPECTRA OF CADMIUM DIFLUORIDE AND LEAD DIFLUORIDE. 0274
RAVIOLET ABSORPTION SPECTRA OF GASEOUS TIN DIFLUORIDE AND LEAD DIFLUORIDE. ULT 0595
/TROMETRIC STUDIES AT HIGH TEMPERATURES. (TIN DIFLUORIDE, LEAD DIFLUORIDE, TIN MONOFLUORIDE, AND LEAD MONOFL/ 0592
ETECTION. (FIRST ROW FLUORIDES, BERYLLIUM, MAGNESIUM, AND LEAD DIFLUORIDES) /LECTION AND MASS SPECTROMETRIC D 0134
/ATION VIBRATIONS OF THE DIHALIDES OF GERMANIUM, TIN, AND LEAD. (GERMANIUM DIFLUORIDE, TIN DIFLUORIDE, LEAD / 0479
B SINGLET SIGMA(+) TO X TRIPLET SIGMA(-) BAND SYSTEM IN LEAD MONOFLUORIDE. 0585
EMISSION SPECTRUM OF LEAD MONOFLUORIDE. 0588
ROTATIONAL ANALYSIS OF A-X1 BANDS OF LEAD MONOFLUORIDE. 0591
ROTATIONAL ANALYSIS OF THE VISIBLE EMISSION BANDS OF LEAD MONOFLUORIDE. 0587
NOFLUORIDE, GERMANIUM MONOFLUORIDE, TIN MONOFLUORIDE, AND LEAD MONOFLUORIDE. SPECTRA OF SILICON MO 0964
ON THE DISSOCIATION OF LEAD MONOFLUORIDE AND BISMUTH MONOFLUORIDE. 0589
ROTATIONAL ANALYSIS OF THE B-X SYSTEM OF LEAD MONOFLUORIDE IN ULTRAVIOLET. 0586
UREY-BRADLEY FORCE FIELD. (LEAD TETRAFLUORIDE) 0596
THERMODYNAMIC FUNCTIONS OF LEAD TETRAHALIDES. 0597
ROTATIONAL ANALYSIS OF B-X2 SYSTEM OF LEAD-208 MONOFLUORIDE. 0590
NSTANTS. APPLICATION TO CARBON DIOXIDE. LEAST SQUARES ADJUSTMENT OF ANHARMONIC POTENTIAL CO 1491
ACTION STUDY. SILICON- FLUORINE LENGTH IN SILICON TETRAFLUORIDE. NEW ELECTRON DIFFR 1000
AFLUORIDE, NEPTUNIUM HEXAFLUORIDE, AND PLUTONI/ LONG WAVE LENGTH, INFRARED ACTIVE FUNDAMENTAL FOR URANIUM HEX 1192
L SPECTRA OF BROMINE PENTAFLUORIDE IN THE MILLIMETER WAVE LENGTH REGION. ROTATIONA 0268
YRAMID, AND RELATED MOLECULES OF THE NONTRANSITION / BOND LENGTHS AND BOND ANGLES IN OCTAHEDRAL, TRIGONAL BIP 1435
BOND LENGTHS IN XENON DIFLUORIDE. 1236
SELENIUM, TELLURIUM, MOLYBDEN/ FORCE CONSTANTS AND BOND LENGTHS OF SOME INORGANIC HEXAFLUORIDES. (OF SULFUR 0631
CONSTANTS OF BORON TRIHALIDES. EFFECTS OF THE LENNARD-JONES POTENTIALS ON THE UREY-BRADLEY FORCE 0244
MULTIPLE SCATTERING X(ALPHA) CALCULATION OF THE ENERGY LEVEL STRUCTURE OF XENON HEXAFLUORIDE. 1304
ODINE OXIDE AND SELENIUM MONOFLUORIDE IN J-5/2 ROTATIONAL LEVELS. ESR SPECTRUM OF BROMINE OXIDE, I 0954
MOLECULAR ORBITAL ENERGY LEVELS AND BONDING IN KRYPTON DIFLUORIDE. 0549
/ FIELD MODEL STUDY OF XENON HEXAFLUORIDE. PART-1: ENERGY LEVELS AND MOLECULAR GEOMETRY. PART-2: COMPARISONS/ 1311
NONRIGID MOLECULE EFFECTS ON THE ROVIBRONIC ENERGY LEVELS AND SPECTRA OF PHOSPHORUS PENTAFLUORIDE. 0841
IFLUORIDES. ENERGY LEVELS AND SPECTROSCOPIC PARAMETERS OF NEODYMIUM TR 0646
CALCULATION OF RYDBERG LEVELS IN NITROGEN OXIDE AND BORON FLUORIDE. 0164
ABSORPTION SPECTRUM OF URANIUM TETRAFLUORIDE AND ENERGY LEVELS OF URANIUM(IV). 1168
/NONEMPIRICAL NDDO-2 (ALPHA BETA) METHOD. PART-3: ENERGY LEVELS, WAVE FUNCTIONS AND AO POPULATIONS OF POLYA/ 0016
N BORON TRIFLUORIDE, CARBON TETRAFLUORIDE, AND/ RADIATIVE LIFETIMES OF ULTRAVIOLET EMISSION SYSTEMS EXCITED I 0216
T PAIRING V/ ORBITAL MODIFICATION BY THE COULOMB FIELD OF LIGAND ATOMS OF LATER SECOND ROW ELEMENTS IN PERFEC 0862
ENIUM HEPTAFLUORIDE AND IODINE HEPTAFLUOR/ INTRAMOLECULAR LIGAND EXCHANGE IN SEVEN COORDINATE STRUCTURES. (RH 0922
/RESSIVE STIFFNESS METHOD APPLIED TO XY4 MOLECULES IN THE LIGHT OF THE THEORY OF CHARACTERISTIC FREQUENCIES. 1430
RDINATE ANALYSES. LIMITATIONS OF THE UREY-BRADLEY MODEL IN NORMAL COO 1478

A STUDY OF THE RAMAN AND BROAD LINE NMR SPECTRA OF GERMANIUM TETRAFLUORIDE. 0484
FLUORIDE) ZEEMAN EFFECT OF SOME LINEAR AND SYMMETRIC TOP MOLECULES. (PHOSPHORUS TRI 0796
E M/ SOME CORRECTIONS TO THE SECOND ORDER STARK EFFECT OF LINEAR MOLECULES. (CHLORINE MONOFLUORIDE AND BROMIN 0385
/N AMPLITUDES AND SHRINKAGE EFFECTS OF VIBRATION FOR SOME LINEAR SYMMETRICAL DIFLUORIDES. (OF THE FIRST ROW / 1369
/VIBRATION AND BASTIANSEN-MORINO SHRINKAGE EFFECT IN SOME LINEAR SYMMETRICAL DIHALIDES. (BERYLLIUM DIFLUORID/ 1372
MOLECULAR CONSTANTS AND THERMODYNAMIC FUNCTIONS OF SOME LINEAR SYMMETRICAL MOLECULES. (XENON DIFLUORIDE) 1244
IAL CONSTANTS AND CORIOLIS COUPLING CONSTANTS FOR SOME NONLINEAR SYMMETRICAL XY2 MOLECULES. POTENT 1385
ESTIMATION OF ANHARMONIC POTENTIAL CONSTANTS. PART-1. LINEAR XY2 MOLECULES. 1383
VIBRATIONAL SPECTRA OF LIQUID AND CRYSTALLINE CARBON TETRAFLUORIDE. 0327
INFRARED AND RAMAN SPECTRA OF LIQUID AND CRYSTALLINE SILICON TETRAFLUORIDE. 1001
LUORIDE, OXYGEN MONOFLUO/ METHOD OF DETERMINING SATURATED LIQUID AND SATURATED VAPOR ENTROPY. (CHLORINE MONOF 0391
STRUCTURE OF LIQUID ANTIMONY PENTAFLUORIDE. 0081
INFRARED SPECTRA OF SIMPLE MOLECULES IN LIQUID ARGON SOLUTION. CARBON TETRAFLUORIDE. 0338
/UM HALIDES AND ACTIVITY COEFFICIENTS FOR PROTACTINIUM IN LIQUID BISMUTH SOLUTIONS. (PROTACTINIUM TETRAFLUOR/ 0900
INFRARED ABSORPTICN SPECTRUM OF LIQUID BORON TRIFLUORIDE. 0245
E PENTAFLUORIDE AND CHLORINE OXYFLUORIDE BY NUC/ STUDY OF LIQUID CHLORINE FLUORIDES AND OXYFLUORIDES, CHLORIN 0414
N THE CONDENSED PHASE. EVIDENCE OF AN ASSOCIATED STATE IN LIQUID CHLORINE TRIFLUORIDE. /HLORINE TRIFLUORIDE I 0266
/ND BROMINE PENTAFLUORIDE. FLUORINE EXCHANGE MECHANISM OF LIQUID CHLORINE TRIFLUORIDE, BROMINE TRIFLUORIDE, / 0399
VIBRATIONAL SPECTRUM OF LIQUID CHROMIUM PENTAFLUORIDE. 0435
TURE AND PRESSURE OF NITROGEN TRIFLUORIDE. LIQUID DENSITY, VAPOR PRESSURE AND CRITICAL TEMPERA 0710
LIQUID GAS ARSENIC INFRARED FREQUENCY SHIFTS. 0106
DE, AND URANIUM HEXAFLUO/ HIGH RESOLUTION NMR SPECTRUM OF LIQUID MOLYBDENUM HEXAFLUORIDE, TUNGSTEN HEXAFLUORI 0637
RIDE. RAMAN SPECTRA OF LIQUID NIOBIUM PENTAFLUORIDE AND TANTALUM PENTAFLUO 0670
RAMAN SPECTRUM OF LIQUID OXYGEN DIFLUORIDE. 0767
ELECTRONIC SPECTRA OF LIQUID RUTHENIUM AND MOLYBDENUM PENTAFLUORIDES. 0925
IDE AND CHLORINE MONOFLUORIDE. GASEOUS SPECTRA AND GAS TO LIQUID SHIFTS. NMR SPECTRA OF CHLORINE TRIFLUOR 0368
IODINE PENTAFLUORIDE. EVIDENCE FOR POLYMERIZATION IN THE LIQUID STATE. RAMAN SPECTRUM OF 0512
FAR INFRARED SPECTRA OF GASEOUS AND LIQUID SULFUR HEXAFLUORIDE. 1095
ELECTRICAL CONDUCTIVITY AND DIELECTRIC CONSTANT OF LIQUID XENON HEXAFLUORIDE. 1307
UORIDE, / MOLECULAR MOTION AND FLUORINE RELAXATION IN THE LIQUIDS OF MOLYBDENUM HEXAFLUORIDE, TUNGSTEN HEXAFL 0638
TETRA/ INFRARED SPECTRA OF GASEOUS ALUMINUM TRIFLUORIDE, LITHIUM ALUMINUM TETRAFLUORIDE, AND SODIUM ALUMINUM 0045
/ VALENCE CONFIGURATION RESULTS ON THE DIATOMIC MOLECULES LITHIUM, BERYLLIUM, BORON, CARBON, NITROGEN, FLUOR/ 0174
ON MONOXIDE, BORON FLUORIDE, AND CARBON, BERYLLIUM OXIDE, LITHIUM FLUORIDE. /THE ISOELECTRONIC NITROGEN, CARB 0150
INFRARED SPECTRUM OF OXYGEN MONOFLUORIDE AND DETECTION OF LITHIUM OXYFLUORIDE. MATRIX 0755
-3: HETERONUCLEAR DIATOMIC AND POLYATOMIC MOLECU/ DENSITY LOCALIZATION OF ATOMIC AND MOLECULAR ORBITALS. PART 0175
IZED BORON- FLUORINE DOUBLE AND TRIPLE BONDS. LOCALIZED ORBITALS IN BORON FLUORIDES. HIGHLY POLAR 0158
/ VIBRATION OF SOME TRIGONAL BIPYRAMIDAL MOLECULES BY THE LOGARITHMIC STEPS METHOD. (PHOSPHORUS PENTAFLUORI/ 0878
FORCE CONSTANTS OF OCTAHEDRAL MOLECULES CALCULATED BY THE LOGARITHMIC STEP METHOD. 1472
/ION AMPLITUDES OF IODINE HEPTAFLUORIDE CALCULATED BY THE LOGARITHMIC STEP METHOD IN THE RANGE OF THE GENERA/ 0535
HEXACHLORIDE(3-), SELENIUM H/ THE STEREOCHEMICALLY INERT LONE PAIR. A SPECULATION ON THE BONDING IN XENON 1309
CONSTANTS WITH THE MAXIMUM OVERLAP METHOD. MOLECULES WITH LONE PAIR ELECTRONS OR EMPTY ORBITALS. /N OF FORCE 1395
OF NITROGEN TRIFLUORIDE. LONE PAIR MODEL AND THE VIBRATIONAL FORCE CONSTANTS 0705
RANIUM HEXAFLUORIDE, NEPTUNIUM HEXAFLUORIDE, AND PLUTONI/ LONG WAVE LENGTH, INFRARED ACTIVE FUNDAMENTAL FOR U 1192
ELECTRONIC SPECTRUM OF LUTETIUM MONOFLUORIDE. 0598
SIS OF SELECTED BANDS FROM THE ELECTRONIC SPECTRUM OF THE LUTETIUM MONOFLUORIDE MOLECULE. ROTATIONAL ANALY 0599
MATION OF THULIUM TRIFLUORIDE, YTTERBIUM TRIFLUORIDE, AND LUTETIUM TRIFLUORIDE. /PRESSURES AND HEATS OF SUBLI 1132
NONEMPIRICAL LCAO MO SCF STUDIES OF THE LOW LYING STATES OF BERYLLIUM DIFLUORIDE. 0137

M

OF VIBRATION FOR THE DIHALIDES OF MERCURY, BERYLLIUM, AND MAGNESIUM. MEAN AMPLITUDES 0619
N. INFRARED SPECTRA OF MAGNESIUM AND ALUMINUM FLUORIDES BY MATRIX ISOLATIO 0608
CULE. DISSOCIATION ENERGY OF MONOFLUORIDE COMPOUNDS OF MAGNESIUM AND CALCIUM AND THE IONIC MODEL OF A MOLE 0278
SPECTROMETRIC DETECTION. (FIRST ROW FLUORIDES, BERYLLIUM, MAGNESIUM, AND LEAD DIFLUORIDES. /LECTION AND MASS 0134
IDES. PART-1: BERYLLIUM DIFLUORIDE, BERYLLIUM DICHLORIDE, MAGNESIUM DICHLORIDE. /RA OF THE ALKALINE EARTH HAL 0133
RATIONAL SPECTRA OF THE ALKALINE EARTH DIHALIDES. PART-1: MAGNESIUM DIFLUORIDE. GEOMETRY AND VIB 0609
INFRARED SPECTRA OF MATRIX ISOLATED MAGNESIUM DIFLUORIDE. 0604
IRST ROW TRANSITION METALS, ALSO BERYLLIUM DIFLUORIDE AND MAGNESIUM DIFLUORIDE) /RICAL DIFLUORIDES. (OF THE F 1369
F STUDIES OF THE GROUP-2A DIHALIDES BERYLLIUM DIFLUORIDE, MAGNESIUM DIFLUORIDE, AND CALCIUM DIFLUORIDE. /O SC 1360
/OME LINEAR SYMMETRICAL DIHALIDES. (BERYLLIUM DIFLUORIDE, MAGNESIUM DIFLUORIDE, AND FIRST ROW TRANSITION MET/ 1372
/ VAPOR PRESSURE OF ZINC DIFLUORIDE, CADMIUM DIFLUORIDE, MAGNESIUM DIFLUORIDE, CALCIUM DIFLUORIDE, STRONTIUM 0049
/R STRUCTURES OF GROUP-2 ELEMENTS. (BERYLLIUM DIFLUORIDE, MAGNESIUM DIFLUORIDE, CALCIUM DIFLUORIDE, STRONTIU/ 0131
STRUCTURE OF THE MAGNESIUM DIFLUORIDE MOLECULE. 0602
/TING POINTS OF INORGANIC FLUORIDES. (CADMIUM DIFLUORIDE, MAGNESIUM DIFLUORIDE, STRONTIUM DIFLUORIDE, AND BA/ 0273
THEORETICAL STUDIES OF THE INTERACTION OF MAGNESIUM DIFLUORIDE WITH RARE GAS ATOMS. 0605
AND MANGANESE DIFLUOR/ RAMAN SPECTRA OF TITANIUM DIOXIDE, MAGNESIUM DIFLUORIDE, ZINC DIFLUORIDE, DIFLUORIDE, 0615
/ARED AND RAMAN SPECTRA AND STRUCTURES OF MATRIX ISOLATED MAGNESIUM DIHALIDES: MAGNESIUM DIFLUORIDE, MAGNESI/ 0607
IN THE ULTRAVIOLET REGION 1700-2500 ANGSTROMS. (ALUMINUM- MAGNESIUM FLUORIDE) / MULTIPLE BEAM INTERFEROMETRY 0042
/RCE FIELDS OF THE RUTILE COUNTERPARTS: TITANIUM DIOXIDE, MAGNESIUM FLUORIDE, ZINC FLUORIDE, AND FERROUS FLU/ 0606
MASS SPECTROMETRIC STUDY OF MAGNESIUM HALIDES. (MAGNESIUM DIFLUORIDE) 0603
DISSOCIATION ENERGY OF MAGNESIUM MONOFLUORIDE. 0601
ROTATIONAL ANALYSIS OF ELECTRONIC BANDS OF GASEOUS MAGNESIUM MONOFLUORIDE. 0600
SPECTROSCOPIC STUDIES OF SCANDIUM MONOFLUORIDE AND MAGNESIUM OXIDE. 0933
MAGNETIC FIELD DEFLECTION OF XENON HEXAFLUORIDE. 1287
S IN CHLORINE MONOFLUORIDE, BRO/ MOLECULAR ZEEMAN EFFECT, MAGNETIC PROPERTIES, AND ELECTRIC QUADRUPOLE MOMENT 0254
MOLECULAR MAGNETIC PROPERTIES OF CHLORINE MONOFLUORIDE. 0384
ABSORPTION SPECTRUM AND MAGNETIC PROPERTIES OF OSMIUM HEXAFLUORIDE. 0751
STRUCTURE AND MAGNETIC PROPERTIES OF RARE EARTH FLUORIDES. 1345
HYDRIDE MOLECULES. MAGNETIC PROPERTIES OF THE BORON FLUORIDE AND BORON 0159
NUCLEAR MAGNETIC RELAXATION IN BORON TRIFLUORIDE GAS. 0217
IGURATION OF THE PENTAFLUORIDES OF RHODIUM AND IRIDIUM BY MAGNETIC RESONANCE. CONF 0923
HLORINE PENTAFLUORIDE AND CHLORINE OXYFLUORIDE BY NUCLEAR MAGNETIC RESONANCE. / FLUORIDES AND OXYFLUORIDES, C 0414
RAFLUORIDE, SILICON TETRA/ FLUORINE RELAXATION BY NUCLEAR MAGNETIC RESONANCE ABSORPTION IN GASEOUS CARBON TET 0346
TION BETWEEN ANTIMONY PENTAFLUORIDE A/ A FLUORINE NUCLEAR MAGNETIC RESONANCE AND RAMAN STUDY OF COMPLEX FORMA 0079
TRIFLUORIDES. MAGNETIC RESONANCE AND SUSCEPTIBILITY OF RARE EARTH 1349
NUCLEAR MAGNETIC RESONANCE IN RARE EARTH TRIFLUORIDES. 1355
OF SULFUR TETRAFLUORIDE. FLUORINE-19 NUCLEAR MAGNETIC RESONANCE INVESTIGATION OF THE ASSOCIATION 1044
FLUORINE NUCLEAR MAGNETIC RESONANCE OF ANTIMONY PENTAFLUORIDE. 0078
FLUORINE NUCLEAR MAGNETIC RESONANCE OF CHLORINE PENTAFLUORIDE. 0416
HEPTAFLUORIDE, AND OXIDE PENTAFLUORIDES FLUORINE NUCLEAR MAGNETIC RESONANCE OF IODINE HEPTAFLUORIDE, RHENIUM 0517

SUSCEPTIBILITY MEASUREMENTS AND FLUORINE NUCLEAR MAGNETIC RESONANCE OF PLATINUM HEXAFLUORIDE. 0883
ANOMALOUS NUCLEAR MAGNETIC RESONANCE SHIFT IN SAMARIUM TRIFLUORIDE. 0927
ELECTRON STRUCTURE AND NUCLEAR MAGNETIC RESONANCE SPECTRA OF BORON COMPOUNDS. 0229
AND CHLORINE MONOFLUORIDE. GASEOUS SPECTRA AND G/ NUCLEAR MAGNETIC RESONANCE SPECTRA OF CHLORINE TRIFLUORIDE 0368
DE. A STUDY OF THE RAMAN AND BROAD LINE NUCLEAR MAGNETIC RESONANCE SPECTRA OF GERMANIUM TETRAFLUORI 0484
AND IODINE OXYGEN PENTAFLUORIDE. NUCLEAR MAGNETIC RESONANCE SPECTRA OF IODINE HEPTAFLUORIDE 0516
YGEN DIFLUORIDE) FLUORINE NUCLEAR MAGNETIC RESONANCE SPECTRA OF OXYGEN FLUORIDES. (OX 0775
ND SILICON FLUORIDES. HYDROLYSIS AND IN/ FLUORINE NUCLEAR MAGNETIC RESONANCE SPECTRA OF SULFUR, PHOSPHORUS, A 1042
R TETRAFLUORIDE. NUCLEAR MAGNETIC RESONANCE SPECTRUM AND STRUCTURE OF SULFU 1037
XAFLUORIDE, TUNGSTEN HEXAFLUORID/ HIGH RESOLUTION NUCLEAR MAGNETIC RESONANCE SPECTRUM OF LIQUID MOLYBDENUM HE 0637
/XCHANGE PROCESSES IN SOME INORGANIC FLUORIDES BY NUCLEAR MAGNETIC RESONANCE (SULFUR TETRAFLUORIDE AND SELE/ 1048
ENTS AND SECOND MOMENTS OF THE ELECT/ MOLECULAR G VALUES, MAGNETIC SUSCEPTIBILITIES, MOLECULAR QUADRUPOLE MOM 0780
OMENTS / ZEEMAN STUDIES INCLUDING THE MOLECULAR G VALUES, MAGNETIC SUSCEPTIBILITY, AND MOLECULAR QUADRUPOLE M 0744
/, CESIUM CHLORIDE, CESIUM BROMIDE, AND CESIUM IODIDE AND MAGNETIC SUSCEPTIBILITY ANISOTROPY OF THALLIUM FLU/ 1120
MAGNETIC SUSCEPTIBILITY OF EUROPIUM TRIFLUORIDE. 0454
MAGNETIC SUSCEPTIBILITY OF RHENIUM HEXAFLUORIDE. 0917
MAGNETIC SUSCEPTIBILITY OF SILVER DIFLUORIDE. 1015
MAGNETIC SUSCEPTIBILITY OF XENON HEXAFLUORIDE. 1308
MAGNETIC SUSCEPTIBILITY OF XENON HEXAFLUORIDE. 1310
FLUORIDES OF THE MAIN GROUP ELEMENTS. 0007
A LANGMUIR MEASUREMENT OF THE SUBLIMATION PRESSURE OF MANGANESE DIFLUORIDE. 0611
KNUDSEN MEASUREMENT OF THE VAPOR PRESSURE OF MANGANESE DIFLUORIDE. 0613
E, MAGNESIUM DIFLUORIDE, ZINC DIFLUORIDE, DIFLUORIDE, AND MANGANESE DIFLUORIDE. /N SPECTRA OF TITANIUM DIOXID 0615
SUBLIMATION PRESSURE OF MANGANESE DIFLUORIDE AND MANGANESE MONOFLUORIDE. 0614
MASS SPECTROMETRIC AND THERMOCHEMICAL STUDIES OF THE MANGANESE FLUORIDES. 0612
/R OXYGEN AND OF CALCIUM FLUORIDE, CHROMIUM(II) FLUORIDE, MANGANESE(II) FLUORIDE, COPPER(II) FLUORIDE, AND Z/ 1324
COMPLEX BAND SYSTEM OF MANGANESE MONOFLUORIDE IN THE NEAR ULTRAVIOLET. 0610
CRYSTAL STRUCTURE OF MANGANESE TRIFLUORIDE. 0616
INTERATOMIC BONDING IN CRYSTALLINE MANGANESE TRIFLUORIDE. 0617
SUBLIMATION PRESSURES OF CHROMIUM TRIFLUORIDE, MANGANESE TRIFLUORIDE, AND IRON TRIFLUORIDE. 0434
OLECULES AND IONS OF D/ FREQUENCIES, FORCE CONSTANTS, AND MASS DEPENDENCE OF THE CORIOLIS CONSTANT FOR SOME M 1411
UR XYN TYPE MOLECULE M/ THEORETICAL INTERPRETATION OF THE MASS INFLUENCE ON CORIOLIS COUPLING CONSTANTS IN FO 1488
INFLUENCE OF ATOMIC MASS ON THE CORIOLIS CONSTANTS OF XY6 MOLECULES. 1458
OPERTIES OF XENON DIFLUORIDE AND XENON TETRAFLUORIDE FROM MASS SPECTRA. THERMOCHEMICAL PR 1253
PTAFLUORIDE AND IODINE OXYGEN PENTAFLUORIDE. MASS SPECTRA AND SUBLIMATION PRESSURES OF IODINE HE 0534
IDE VAPOR. MASS SPECTRA, AND THERMODYNAMICS OF BERYLLIUM FLUOR 0140
GAS PHASE STRUCTURES AND MASS SPECTRA OF BINARY PENTAFLUORIDES. 0003
GAS PHASE STRUCTURES AND MASS SPECTRA OF BINARY PENTAFLUORIDES. 1365
MICROWAVE AND MASS SPECTRA OF SULFUR MONOFLUORIDE. 1024
DE AND CARBON DIFLUORIDE FROM MOLECULAR FLOW EFFUSION AND MASS SPECTROMETER INVESTIGATIONS. /ARBON MONOFLUORI 0286
/IC PROPERTIES OF THE BORON CHLORIDE FLUORIDE SYSTEM FROM MASS SPECTROMETER INVESTIGATIONS. (BORON DIFLUORID/ 0179
S. APPLICATION OF A MOLECULAR BEAM SOURCE MASS SPECTROMETER TO THE STUDY OF REACTIVE FLUORIDE 0019
S. (SECOND ROW, T/ APPLICATION OF A MOLECULAR BEAM SOURCE MASS SPECTROMETER TO THE STUDY OF REACTIVE FLUORIDE 1377
E MANGANESE FLUORIDES. MASS SPECTROMETRIC AND THERMOCHEMICAL STUDIES OF TH 0612
/Y OF HIGH TEMPERATURE SPECIES BY ELECTRIC DEFLECTION AND MASS SPECTROMETRIC DETECTION. (FIRST ROW FLUORIDES/ 0134
N ENERGY OF BERYLLIUM MONOFLUORIDE. MASS SPECTROMETRIC DETERMINATION OF THE DISSOCIATIO 0129
RMATION OF THE GASEOUS MOLECULES ALUMINUM OXYDIFLUORIDE / MASS SPECTROMETRIC DETERMINATION OF THE HEATS OF FO 0035
SITION AND VAPOR PRESSURE OF VANADIUM TRIFLUORIDE. MASS SPECTROMETRIC DETERMINATION OF THE VAPOR COMPO 1208
NTAFLUORIDE AND ANTIMONY PENTAFLUORIDE. MASS SPECTROMETRIC EVIDENCE OF DIMERS IN BISMUTH PE 0084
IC PROPERTIES OF ALUMINUM TRIFLUORIDE. MASS SPECTROMETRIC INVESTIGATION OF THE THERMODYNAM 0040
ON OF HEAT OF SUBLIMATION OF URANIUM TETRAFLUORIDE BY THE MASS SPECTROMETRIC METHOD. DETERMINATI 1165
DISSOCIATION ENERGY OF BORON MONOFLUORIDE FROM MASS SPECTROMETRIC STUDIES. 0161
N GASEOUS THORIUM TRIFLUORIDE AND THORIUM DIFLUORIDE FROM MASS SPECTROMETRIC STUDIES. HEATS OF FORMATIO 1128
RT-9: SUBLIMATION PRESSURE OF COPPER DIFLUORIDE. (AND CO/ MASS SPECTROMETRIC STUDIES AT HIGH TEMPERATURES. PA 0441
RT-9: SUBLIMATION PRESSURE OF COPPER DIFLUORIDE. MASS SPECTROMETRIC STUDIES AT HIGH TEMPERATURES. PA 0442
RT-17: SUBLIMATION AND VAPOR PRESSURES OF DYSPROSIUM, HO/ MASS SPECTROMETRIC STUDIES AT HIGH TEMPERATURES. PA 0448
RT-13: STABILITIES OF SAMARIUM, EUROPIUM, AND GADOLINIUM/ MASS SPECTROMETRIC STUDIES AT HIGH TEMPERATURES. PA 0450
RT-2: THE DISSOCIATION ENERGIES OF THE MONOFLUORIDES AND/ MASS SPECTROMETRIC STUDIES AT HIGH TEMPERATURES. PA 0463
RT-18: THE STABILITIES OF THE MONO- AND DIFLUORIDES OF S/ MASS SPECTROMETRIC STUDIES AT HIGH TEMPERATURES. PA 0939
RT-14: VAPOR PRESSURE AND DISSOCIATION ENERGY OF SILVER / MASS SPECTROMETRIC STUDIES AT HIGH TEMPERATURES. PA 1014
RT-16: SUBLIMATION PRESSURES FOR TITANIUM TRIFLUORIDE AN/ MASS SPECTROMETRIC STUDIES AT HIGH TEMPERATURES. PA 1145
IN DIFLUORIDE, LEAD DIFLUORIDE, TIN MONOFLUORIDE, AND LE/ MASS SPECTROMETRIC STUDIES AT HIGH TEMPERATURES. (T 0592
T HIGH TEMPERATURES. PART-2: ENTHALPY OF SUBLIMATION OF / MASS SPECTROMETRIC STUDIES OF PLUTONIUM COMPOUNDS A 0884
YTTRIUM TRIFLUORIDE, LANTHANUM TRIFLUORIDE AND THE RARE/ MASS SPECTROMETRIC STUDIES OF SCANDIUM TRIFLUORIDE, 0953
NTHANUM, AND RARE EARTH FLUORIDES. MASS SPECTROMETRIC STUDIES OF SCANDIUM, YTTRIUM, LA 1359
LUORIDES. MASS SPECTROMETRIC STUDIES OF SOME GASEOUS SULFUR F 1023
LUORIDES. MASS SPECTROMETRIC STUDIES OF SOME GASEOUS SULFUR F 1028
(ALSO ANTIMONY PENTAFLUORIDE) MOLECULAR BEAM MASS SPECTROMETRIC STUDY OF BINARY PENTAFLUORIDES. 1376
NESIUM DIFLUORIDE) MASS SPECTROMETRIC STUDY OF MAGNESIUM HALIDES. (MAG 0603
TETRAFLUORIDE, AND XENON HEXAFLUORIDE. PHOTOIONIZATION MASS SPECTROMETRIC STUDY OF XENON DIFLUORIDE, XENON 1225
PHOTOIONIZATION MASS SPECTROMETRY OF OXYGEN DIFLUORIDE. 0760
IDE VAPOR. MASS SPECTRUM AND COMPOSITION OF NIOBIUM PENTAFLUOR 0666
IFLUORIDE. MASS SPECTRUM AND MOLECULAR ENERGETICS OF KRYPTON D 0565
MASS SPECTRUM OF ANTIMONY PENTAFLUORIDE. 0085
INFLUENCE OF ATOMIC MASSES ON THE CORIOLIS COUPLING COEFFICIENTS. 1388
TETRAHEDRAL MOLECULES. INFLUENCE OF ATOMIC MASSES ON THE CORIOLIS COUPLING COEFFICIENTS IN XY4 1410
UPLING IN TETRAHEDRAL XY4 MOLECULES. MATHEMATICAL TECHNIQUES IN THE STUDY OF CORIOLIS CO 1409
US DICHLORIDE, AND NITROGEN DICHLORIDE IN LOW TEMPERATURE MATRICES. /UDIES OF PHOSPHORUS DIFLUORIDE, PHOSPHOR 0793
TITANIUM DIFLUORIDE AND TITANIUM TRIFLUORIDE IN RARE GAS MATRICES. INFRARED SPECTRA AND GEOMETRY OF 1146
INFRARED SPECTRA OF SILICON DIFLUORIDE IN INERT GAS MATRICES. 0982
RIDE, AND ZINC(II) FLUORIDE WITH CARBON MONOXIDE IN ARGON MATRICES. / MANGANESE(II) FLUORIDE, COPPER(II) FLUO 1324
VIBRATIONAL MEAN SQUARE AMPLITUDE MATRICES. 1389
VIBRATIONAL MEAN SQUARE AMPLITUDE MATRICES. 1392
MMETRICAL XY3 AND XY4 / VIBRATIONAL MEAN SQUARE AMPLITUDE MATRICES. (BASTIANSEN-MORINO SHRINKAGE EFFECT OF SY 1394
CULES. VIBRATIONAL MEAN SQUARE AMPLITUDE MATRICES. PART-7: TREATMENT OF TETRAHEDRAL XY4 MOLE 1408
/OF THE FREE RADICAL CARBON TRIFLUORIDE ISOLATED IN INERT MATRICES. PHOTOCHLORINATION OF METHANE AND FLUORO/ 0311
ORPTION SPECTRUM OF CARBON DIFLUORIDE TRAPPED IN AN ARGON MATRIX. ABS 0297
SIGMA TO X DOUBLET PI ELECTRON TRA/ DETERMINATION OF THE MATRIX ELEMENT OF THE DIPOLE MOMENT OF AN A DOUBLET 0970
STRUCTURES AND NORMAL COORDINATE ANALYSES OF GERMANIUM D/ MATRIX INFRARED AND LASER RAMAN SPECTRA, MOLECULAR 0475
DETECTION OF LITHIUM OXYFLUORIDE. MATRIX INFRARED SPECTRUM OF OXYGEN MONOFLUORIDE AND 0755

PENTAFLUORIDE.   INFRARED SPECTRA OF MATRIX ISOLATED ANTIMONY PENTAFLUORIDE AND ARSENIC 0113
                       ISOTOPIC SPLITTING IN MATRIX ISOLATED BORON TRIFLUORIDE.   0228
UORIDE, AND BROMINE PENTAFLUORIDE. F/ INFRARED SPECTRA OF MATRIX ISOLATED CHLORINE TRIFLUORIDE, BROMINE TRIFLI 0399
E, COPPER DIFLUORIDE, A/ INFRARED SPECTRA AND GEOMETRY OF MATRIX ISOLATED COBALT DIFLUORIDE, NICKEL DIFLUORID 0437
                        INFRARED SPECTRA OF MATRIX ISOLATED MAGNESIUM DIFLUORIDE.   0604
UORIDE, MAG/ INFRARED AND RAMAN SPECTRA AND STRUCTURES OF MATRIX ISOLATED MAGNESIUM DIHALIDES: MAGNESIUM DIFL 0607
               RAMAN SPECTRA OF MATRIX ISOLATED MOLECULES. (SULFUR HEXAFLUORIDE) 1097
RIDE.   SPECTRA OF MATRIX ISOLATED NICKEL DIFLUORIDE AND NICKEL DICHLO 0653
 WITH CARBON MONOXIDE, MOLECULAR NITROGEN/ INTERACTION OF MATRIX ISOLATED NICKEL FLUORIDE AND NICKEL CHLORIDE 1324
               INFRARED SPECTRUM OF MATRIX ISOLATED NIOBIUM PENTAFLUORIDE.   0659
GC.   ABSORPTION SPECTRA OF MATRIX ISOLATED NITROGEN TRIFLUORIDE FROM 18 TO 50 0698
               LASER EXCITED RAMAN SPECTRA OF MATRIX ISOLATED PRASEODYMIUM TRIFLUORIDE.   0895
 TRIFLUORIDE)   GEOMETRY AND INFRARED SPECTRA OF MATRIX ISOLATED RARE EARTH HALIDES. (ALSO LANTHANUM 0583
TRIFLUORIDE, YTTRIUM T/ FORCE CONSTANTS AND GEOMETRIES OF MATRIX ISOLATED RARE EARTH TRIFLUORIDES. (SCANDIUM 0944
STEM.   INFRARED SPECTRA OF MATRIX ISOLATED SPECIES IN THE GALLIUM- FLUORINE SY 0462
                INFRARED SPECTRA AND STRUCTURE OF MATRIX ISOLATED THALLIUM HALIDES.   1113
UORIDE, AND / INFRARED AND RAMAN SPECTRA AND STRUCTURE OF MATRIX ISOLATED THALLOUS HALIDE DIMERS, THALLOUS FL 1123
                         INFRARED SPECTRA OF MATRIX ISOLATED TIN DIFLUORIDE AND LEAD DIFLUORIDE. 0594
FLUORIDE.   ABSORPTION AND EMISSION SPECTRA OF ARGON MATRIX ISOLATED XENON MONOFLUORIDE AND KRYPTON MONO 0544
LUORIDE.   INFRARED SPECTRA AND GEOMETRIES OF MATRIX ISOLATED YTTRIUM DIFLUORIDE AND YTTRIUM TRIF 1316
AND TETRAFLUORIDE.   INFRARED SPECTRA OF MATRIX ISOLATED ZIRCONIUM DIFLUORIDE, TRIFLUORIDE, 1325
   INFRARED SPECTRA OF MAGNESIUM AND ALUMINUM FLUORIDES BY MATRIX ISOLATION.   0608
DE, DIALUMINUM HEXAFLUORIDE, AND ALUMINUM MONOFLUORIDE IN MATRIX ISOLATION. /D SPECTRUM OF ALUMINUM TRIFLUORI 0051
D, VISIBLE AND ULTRAVIOLET ABSORPTION IN THE VAPOR AND IN MATRIX ISOLATION. /DE. RAMAN SCATTERING AND INFRARE 1285
ERATURE SPECIES. (FLUORIDES, OXIDES, CHLORIDES)   MATRIX ISOLATION INFRARED SPECTROSCOPY OF HIGH TEMP 0011
          PREPARATION OF INERT GAS COMPOUNDS BY MATRIX ISOLATION: KRYPTON DIFLUORIDE.   0566
                                                          MATRIX ISOLATION OF BROMINE MONOFLUORIDE.   0255
R TETRAFLUORIDE, ANTIMONY PENTAFLUORIDE, ARSENI/ INFRARED MATRIX ISOLATION STUDIES. (BORON TRIFLUORIDE, SULFU 0239
UM TETRAFLUORIDE AND TELLURIUM TETRAFLUORIDE, INCLUDING A MATRIX ISOLATION STUDY. /A AND STRUCTURES OF SELENI 0956
                                                          MATRIX ISOLATION STUDY OF CHLORINE PENTAFLUORIDE. 0419
OTOLYSIS OF NITROGEN TRIFLUORIDE. ELECTRONIC SPECTRUM OF/ MATRIX ISOLATION STUDY OF THE VACUUM ULTRAVIOLET PH 0691
OTOLYSIS OF SILICON DIFLUORIDE DIHYDRIDE. THE INFRARED / MATRIX ISOLATION STUDY OF THE VACUUM ULTRAVIOLET PH 0986
   INFRARED SPECTRA OF SOME ALKALINE EARTH HALIDES BY THE MATRIX ISOLATION TECHNIQUE.   0125
          KRYPTON DIFLUORIDE. PREPARATION BY THE MATRIX ISOLATION TECHNIQUE.   0567
NSTOCK AND GOODMAN FOR OCTAHEDRAL HEXAFLUORIDES AND THE L MATRIX METHOD.   FORCE FIELD OF WEI 1456
XYGEN FLUORIDE FREE RADICAL.   ARGON MATRIX RAMAN SPECTRA OF OXYGEN DIFLUORIDE AND THE O 0759
BORANE, DIBROMO BORANE AND CHLORINE TRIFLUOR/ GENERALIZED MEAN AMPLITUDES AND CORIOLIS CONSTANTS IN DICHLORO 0411
FOR SOME LINEAR SYMMETRICAL DIFLUORIDES. (OF THE FIRST R/ MEAN AMPLITUDES AND SHRINKAGE EFFECTS OF VIBRATION 1369
 TRIHALIDES. (ARSENIC TRIFLUORIDE)   MEAN AMPLITUDES AND THERMODYNAMIC FUNCTIONS OF SOME 0109
D DIMER.   HARMONIC FORCE FIELDS AND MEAN AMPLITUDES FOR ALUMINUM TRIFLUORIDE MONOMER AN 0038
 CHEMICAL BONDS: OCTAHEDRAL HEXAFLUORIDES.   MEAN AMPLITUDES OF THERMAL MOTION AND SHRINKAGES OF 1460
AMIC PROPERTIES OF METAL HEXAFLUORIDES.   MEAN AMPLITUDES OF THERMAL VIBRATIONS AND THERMODYN 1467
SHRINKAGE EFFECT IN SOME LINEAR SYMMETRICAL DIHALIDES. (/ MEAN AMPLITUDES OF VIBRATION AND BASTIANSEN-MORINO 1372
SHRINKAGE EFFECT IN SOME XY3 MOLECULES OF PLANAR TRIGONA/ MEAN AMPLITUDES OF VIBRATION AND BASTIANSEN-MORINO 1397
SHRINKAGE EFFECTS IN SOME OCTAHEDRAL XY6 MOLECULES.   MEAN AMPLITUDES OF VIBRATION AND BASTIANSEN-MORINO 1455
 FOR PHOSPHORUS PENTAFLUORIDE AND ARSENIC PENTAFLUORIDE. MEAN AMPLITUDES OF VIBRATION AND CORIOLIS CONSTANTS 0840
F BORON TRIHALIDES, CALCULATED FROM UREY-BRADLEY FORCE C/ MEAN AMPLITUDES OF VIBRATION AND SHRINKAGE EFFECT O 0233
TIONS OF SOME GROUP-5 TRIHALIDES. (NITROGEN TRIFLUORIDE,/ MEAN AMPLITUDES OF VIBRATION AND THERMODYNAMIC FUNC 0719
TIONS FOR SILICON DIFLUORIDE AND OXYGEN DIFLUORIDE.   MEAN AMPLITUDES OF VIBRATION AND THERMODYNAMIC FUNC 0773
TIONS OF TECHNETIUM HEXAFLUORIDE.   MEAN AMPLITUDES OF VIBRATION AND THERMODYNAMIC FUNC 1104
TIONS FOR VANADIUM PENTAFLUORIDE.   MEAN AMPLITUDES OF VIBRATION AND THERMODYNAMIC FUNC 1216
TIONS OF XENON TETRAFLUORIDE.   MEAN AMPLITUDES OF VIBRATION AND THERMODYNAMIC FUNC 1274
TIONS FOR SOME METAL HEXAFLUORIDES.   MEAN AMPLITUDES OF VIBRATION AND THERMODYNAMIC FUNC 1459
 FIELDS FOR PLANAR XY3 MOLECULES.   CALCULATIONS OF MEAN AMPLITUDES OF VIBRATION AND UREY-BRADLEY FORCE 1399
INKAGE EFFECT, AND MOLECULAR POLARIZABILITIES FOR DIHALI/ MEAN AMPLITUDES OF VIBRATION, BASTIANSEN-MORINO SHR 0124
INKAGE EFFECT AND THERMODYNAMIC FUNCTIONS OF KRYPTON DIF/ MEAN AMPLITUDES OF VIBRATION, BASTIANSEN-MORINO SHR 0562
STANTS, SHRINKAGE EFFECT, AND THERMODYNAMIC FUNCTIONS OF/ MEAN AMPLITUDES OF VIBRATION, CORIOLIS COUPLING CON 0050
IDAL MOLECULE WITH APPLICATION TO PHOSPHORUS PENTAFLUORI/ MEAN AMPLITUDES OF VIBRATION FOR A TRIGONAL BIPYRAM 0864
XAFLUORIDES.   REVISED MEAN AMPLITUDES OF VIBRATION FOR SOME OCTAHEDRAL HE 1447
 MOLECULES. (NITROGEN TRIFLUORIDE, PHOSPHORUS TRIFLUORI/ MEAN AMPLITUDES OF VIBRATION FOR SOME PYRAMIDAL XY3 0706
 MOLECULES.   MEAN AMPLITUDES OF VIBRATION FOR SOME PYRAMIDAL XY3 1390
 AND TETRAHEDRAL XY4 MOLECULES.   MEAN AMPLITUDES OF VIBRATION FOR SOME PYRAMIDAL XY3 1407
Z MOLECULES. (IODINE PENTAFLUORIDE, BROMINE PENTAFLUORID/ MEAN AMPLITUDES OF VIBRATION FOR SOME PYRAMIDAL XY4 0421
DE.   MEAN AMPLITUDES OF VIBRATION FOR SULFUR TETRAFLUORI 1038
DE, CARBON DIFLUORIDE, NITROGEN DIFLUORIDE, ALUMINUM DIF/ MEAN AMPLITUDES OF VIBRATION FOR THE BORON DIFLUORI 1384
ERCURY, BERYLLIUM, AND MAGNESIUM.   MEAN AMPLITUDES OF VIBRATION FOR THE DIHALIDES OF M 0619
                                   CALCULATED MEAN AMPLITUDES OF VIBRATION IN BORON TRIHALIDES.   0197
 OF C2V SYMMETRY.   MEAN AMPLITUDES OF VIBRATION IN SOME XY2Z MOLECULES 1396
RSENIC, PHOSPHORUS, AND VANADIU/ SPECTROSCOPIC STUDIES IN MEAN AMPLITUDES OF VIBRATION OF PENTAFLUORIDES OF A 0871
ND CHLORINE TRIFLUORIDE.   MEAN AMPLITUDES OF VIBRATION OF SELENIUM TRIOXIDE A 0406
FLUORIDES.   MEAN AMPLITUDES OF VIBRATION OF SOME LANTHANIDE TRI 1354
4 TYPE MOLECULES. PART-5: TETRAHALIDES OF HAFNIUM AND LE/ MEAN AMPLITUDES OF VIBRATION OF SOME TETRAHEDRAL XY 0487
4 TYPE MOLECULES. PART 3 AND 4.   MEAN AMPLITUDES OF VIBRATION OF SOME TETRAHEDRAL XY 1420
L MOLECULES. (SILICON TETRAFLUORIDE)   MEAN AMPLITUDES OF VIBRATION OF SOME XY4 TETRAHEDRA 1008
L MOLECULES.   MEAN AMPLITUDES OF VIBRATION OF SOME XY4 TETRAHEDRA 1419
                                                MEAN AMPLITUDES OF VIBRATION OF SOME XY6 MOLECULES. 1461
CORIOLIS CONSTANTS IN IODI/ UREY-BRADLEY FORCE CONSTANTS, MEAN AMPLITUDES OF VIBRATION, SHRINKAGE EFFECT AND 0513
RIOLIS CONSTANTS, AND THERMODYNAMIC FUN/ FORCE CONSTANTS, MEAN AMPLITUDES OF VIBRATION, SHRINKAGE EFFECTS, CO 1219
AND SULFUR PENTAFLUORO CHLORIDE MOLECULES.   MEAN AMPLITUDES OF VIBRATION: SULFUR TETRAFLUORIDE 1057
NS AND MOLECULAR POLARIZABILITES OF/ POTENTIAL CONSTANTS, MEAN AMPLITUDES OF VIBRATION, THERMODYNAMIC FUNCTIO 1437
NOMALIES.   ALUMINUM AND ALKALI HALIDE MONOMERS MEAN AMPLITUDES OF VIBRATION WITH LOW TEMPERATURE A 0024
TRON DIFFRACTION INTENSITIES. PART-2: INTERNAL MOTION AND MEAN GEOMETRY DEDUCED BY ELECTRON DIFFRACTION. /LEC 1291
                                    VIBRATIONAL MEAN SQUARE AMPLITUDE MATRICES.   1389
                                    VIBRATIONAL MEAN SQUARE AMPLITUDE MATRICES.   1392
SHRINKAGE EFFECT OF SYMMETRICAL XY3 AND XY4 / VIBRATIONAL MEAN SQUARE AMPLITUDE MATRICES. (BASTIANSEN-MORINO 1394
F TETRAHEDRAL XY4 MOLECULES.   VIBRATIONAL MEAN SQUARE AMPLITUDE MATRICES. PART-7: TREATMENT O 1408
ME PYRAMIDAL XY3 MOLECULES.   GENERALIZED MEAN SQUARE AMPLITUDES AND CORIOLIS CONSTANTS OF PY 0405
AHEDRAL SYMMETRY. (SELENIUM AND TELLURIUM FLUORIDES) ROOT MEAN SQUARE AMPLITUDES IN SOME HEXAFLUORIDES OF OCT 0962
HEXAFLUORIDE MOLECULES.   MEAN SQUARE AMPLITUDES OF INTERATOMIC DISTANCES IN 0008
HEXAFLUORIDE MOLECULES.   MEAN SQUARE AMPLITUDES OF INTERATOMIC DISTANCES IN 1450

SOME OCTAHEDRAL HEXAFLUORIDE MOLECULES. | MEAN SQUARE AMPLITUDES OF INTERATOMIC DISTANCES IN | 1466
/OUPLING CONSTANTS, CENTRIFUGAL STRETCHING CONSTANTS, AND | MEAN SQUARE AMPLITUDES OF THE VIBRATIONS OF THE NI/ | 0726
FORCE FIELD, CORIOLIS COUPLING COEFFICIENTS, GENERALIZED | MEAN SQUARE AMPLITUDES OF VIBRATION AND SHRINKAGE / | 1278
EFFECTS OF SOME TRIGONAL BIPYRAMIDAL XY5 AND/ GENERALIZED | MEAN SQUARE AMPLITUDES OF VIBRATION, AND SHRINKAGE | 1432
TETRAHALIDES. (CARBON TETRAFLUORIDE) ROOT | MEAN SQUARE AMPLITUDES OF VIBRATION IN SOME GROUP-4 | 0342
TETRAHALIDES. ROOT | MEAN SQUARE AMPLITUDES OF VIBRATION IN SOME GROUP-4 | 1363
RIDE AND XENON TETR/ A THEORETICAL EVALUATION OF THE ROOT | MEAN SQUARE AMPLITUDES OF VIBRATION IN XENON DIFLUO | 1257
CON TETRAFLUORIDE FROM EL/ INTERATOMIC DISTANCES AND ROOT | MEAN SQUARE AMPLITUDES OF VIBRATION OF GASEOUS SILI | 1002
L BI/ GENERALIZED VALENCE FORCE FIELD FORCE CONSTANTS AND | MEAN SQUARE AMPLITUDES OF VIBRATION OF SOME TRIGONA | 0878
AFLUORIDE, PHOSP/ CALCULATION OF FORCE CONSTANTS AND ROOT | MEAN SQUARE VIBRATION AMPLITUDES OF PHOSPHORUS PENT | 0826
CALCULATION OF | MEAN SQUARE VIBRATIONAL AMPLITUDES. | 1486
ALCULATED BY THE LOGARITHMIC STEP ME/ FORCE CONSTANTS AND | MEAN VIBRATION AMPLITUDES OF IODINE HEPTAFLUORIDE C | 0535
/IONAL CONSTANTS, STATISTCIAL THERMODYNAMICS AND QUANTUM | MECHANICAL STUDIES OF POLARIZABILITIES FOR THE DI/ | 0784
D SILICON FLUORIDES. HYDROLYSIS AND INTRA- INTERMOLECULAR | MECHANISM OF FLUORINE EXCHANGE. /UR, PHOSPHORUS, AN | 1042
/RIFLUORIDE, AND BROMINE PENTAFLUORIDE. FLUORINE EXCHANGE | MECHANISM OF LIQUID CHLORINE TRIFLUORIDE, BROMINE / | 0399
SPECTROSCOPIC CONSEQUENCES OF DIFFERENT TUNNELING | MECHANISMS IN NONRIGID PHOSPHORUS PENTAFLUORIDE. | 0834
ROPERTIES OF CARBON TETRAFLUORIDE FROM 4 DEGREES K TO ITS | MELTING POINT. THERMODYNAMIC P | 0326
LUORIDE. DENSITY, | MELTING POINT AND VAPOR PRESSURE OF ANTIMONY PENTAF | 0082
TION OF TIN DIFLUORIDE. | MELTING POINT, VAPOR PRESSURE, AND HEAT OF VAPORIZA | 1140
TRIFLUORIDE. | MELTING POINTS OF CURIUM TRIFLUORIDE AND AMERICIUM | 0446
LUORIDE, MAGNESIUM DIFLUORIDE, STRONTIUM DIFLUORIDE, AND/ | MELTING POINTS OF INORGANIC FLUORIDES. (CADMIUM DIF | 0273
LUORIDE. VAPOR PRESSURE AND | MELTING POINTS OF XENON DIFLUORIDE AND XENON TETRAF | 1251
Y DIFLUORIDE) VIBRATIONAL SPECTRA OF THE DIHALIDES OF | MERCURY AND CADMIUM. (CADMIUM DIFLUORIDE AND MERCUR | 0275
MEAN AMPLITUDES OF VIBRATION FOR THE DIHALIDES OF | MERCURY, BERYLLIUM, AND MAGNESIUM. | 0619
OUP-2B HALIDES. (ZINC DIFLUORIDE, CADMIUM DIFLUORIDE, AND | MERCURY DIFLUORIDE) /DING FORCE CONSTANTS OF THE GR | 1322
A NEW ELECTRONIC TRANSITION OF THE | MERCURY MONOFLUORIDE MOLECULE. | 0618
NCIES AND VALENCE FORCE CONSTANTS OF FIRST ROW TRANSITION | METAL DIFLUORIDES. VIBRATIONAL FREQUE | 1371
INFRARED SPECTRUM OF HEAVY | METAL DIHALIDES. (EUROPIUM DIFLUORIDE) | 0451
TRANSITION METALS. GEOMETRY OF THE TRANSITION | METAL DIHALIDES. THE DIFLUORIDES OF SOME FIRST ROW | 1368
IFLUORIDE, MAGNESIUM DIFLUORIDE, AND FIRST ROW TRANSITION | METAL FLUORIDES. | 1346
TRANSITION | METAL FLUORIDES) /MMETRICAL DIHALIDES. (BERYLLIUM D | 1372
ISM. INFRARED SPECTRA OF RARE EARTH | METAL FLUORIDES AND THEIR COMPLEXES. | 0015
VIBRONIC INTERACTIONS IN | METAL FLUORIDES IN THE REGION OF A CESIUM IODIDE PR | 1352
/ CAPACITIES AND ELECTRONIC SPECTRA OF THE PLATINUM GROUP | METAL HEXAFLUORIDE MOLECULES. | 0908
FORCE CONSTANTS OF | METAL HEXAFLUORIDE MOLECULES DOWN TO HELIUM TEMPER/ | 1159
FORCE CONSTANTS OF | METAL HEXAFLUORIDES. | 0018
DES OF THERMAL VIBRATIONS AND THERMODYNAMIC PROPERTIES OF | METAL HEXAFLUORIDES | 1446
LITUDES OF VIBRATION AND THERMODYNAMIC FUNCTIONS FOR SOME | METAL HEXAFLUORIDES. MEAN AMPLITU | 1467
RAMAN SPECTRA OF 5D TRANSITION | METAL HEXAFLUORIDES. MEAN AMP | 1459
HEXAFLUORIDE, IRIDIUM HEXAFLUORIDE,/ COLORS OF TRANSITION | METAL HEXAFLUORIDES IN THE VAPOR STATE. (THIRD ROW) | 1378
ENIUM HEPTA/ X-RAY DIFFRACTION STUDIES OF SOME TRANSITION | METAL HEXAFLUORIDES. (RHENIUM HEXAFLUORIDE, OSMIUM | 0916
ENUM HEXAFLUORI/ VAPOR PRESSURES OF SOME HEAVY TRANSITION | METAL HEXAFLUORIDES. (SECOND ROW, THIRD ROW, AND RH | 1375
TRANSITION | METAL HEXAFLUORIDES. (TUNGSTEN HEXAFLUORIDE, MOLYBD | 0903
W) THERMODYNAMIC PROPERTIES OF TRANSITION | METAL PENTAFLUORIDES AND RELATED COMPOUNDS. | 0013
, YTTRIUM, LANTHANUM, CERIUM/ GEOMETRIES AND ENTROPIES OF | METAL PENTAFLUORIDES AT HIGH TEMPERATURE. (THIRD RO | 1379
BENDING MOTIONS IN THE DIHALIDES OF GROUP-2 | METAL TRIFLUORIDES FROM INFRARED SPECTRA. (SCANDIUM | 0943
L DIHALIDES. THE DIFLUORIDES OF SOME FIRST ROW TRANSITION | METALS. | 1380
HALIDES OF THE SECOND AND THIRD ROW TRANSITION | METALS. GEOMETRY OF THE TRANSITION META | 1368
PENTAHALIDES OF TRANSITION | METALS. | 1374
/AR SYMMETRICAL DIFLUORIDES. (OF THE FIRST ROW TRANSITION | METALS, ALSO BERYLLIUM DIFLUORIDE AND MAGNESIUM DI/ | 1369
SOME PHYSICAL PROPERTIES OF CURIUM AND PROTACTINIUM | METALS AND PROTACTINIUM TETRAFLUORIDE. | 0899
FLUORIDES OF THE HEAVY TRANSITION | METALS. (REVIEW) | 0002
/ THE TRANSITION ELEMENTS. VOLUME-1: FIRST ROW TRANSITION | METALS. VOLUME-2: SECOND AND THIRD ROW TRANSITION / | 0001
LUORIDE, HYDROGEN OXYCHLORIDE, CHLORO AMMONIA, AND CHLORO | METHANE. /F THE ISOELECTRONIC SERIES CHLORINE MONOF | 0371
ELECTRON IMPACT SPECTRA OF | METHANE AND CARBON TETRAFLUORIDE. | 0333
/UORIDE ISOLATED IN INERT MATRICES. PHOTOCHLORINATION OF | METHANE AND FLUOROFORM. DISSOCIATION ENERGY AND EN/ | 0311
FLUORIDE) CENTRIFUGAL STRETCHING OF TRIFLUORO CHLORO | METHANE AND SOME OTHER SYMMETRIC TOPS. (ARSENIC TRI | 0101
/VIOLET ABSORPTION SPECTRA OF SOME HALOGEN DERIVATIVES OF | METHANE. (CARBON TETRAFLUORIDE) CORRELATION OF THE/ | 0365
S NITROGEN TRIFLUORIDE NU-4 AND NU-4 + NU-2 AND TRIFLUORO | METHANE NU-6. /FRARED BANDS OF OBLATE SYMMETRIC TOP | 0716
B/ ANALYSES OF THE VIBRATION ROTATION BANDS OF TRIFLUORO | METHANE, PHOSPHORUS PENTAFLUORIDE, AND CYCLOPROPANE | 0880
/ION POTENTIAL OF CHLORO- AND FLUOROMETHANES, TETRAFLUORO | METHANE XENON TETRAFLUORIDE, AND XENON TETRACHLORI/ | 0344
INFRARED SPECTRA OF HALOGEN-SUBSTITUTED | METHANES. (CARBON TETRAFLUORIDE) | 0355
ELECTRON IMPACT STUDIES OF SOME TRIHALO | METHANES. (CARBON TRIFLUORIDE) | 0312
E) RAMAN SPECTRA OF CHLOROFLUORO | METHANES IN THE GASEOUS STATE. (CARBON TETRAFLUORID | 0322
RIDE. MICROWAVE INVESTIGATIONS OF | METHYL FLUORIDE, FLUOROFORM, AND PHOSPHORUS TRIFLUO | 0798
VIBRATIONAL ANHARMONICITY IN THE | METHYL HALIDES. | 1493
STRUCTURES OF TRIFLUORO | METHYL HALIDES. (CARBON TETRAFLUORIDE) | 0318
F THE BONDING IN PHOSPHINE, PHOSPHORUS TRIFLUORIDE AND TRIMETHYL PHOSPHINE. | AB INITIO CALCULATIONS O | 0802
ORUS TRIFLUORIDE. STRUCTURES OF PHOSPHORUS PENTAFLUORIDE, | METHYL PHOSPHORUS TETRAFLUORIDE AND DIMETHYL PHOSPH | 0845
ORUS PENTAFLUORIDE, METHYL PHOSPHORUS TETRAFLUORIDE AND DIMETHYL PHOSPHORUS TRIFLUORIDE. STRUCTURES OF PHOSPH | 0845
INE HEPTAFLUORIDE AN/ STRUCTURE DETERMINATION OF TRIS (TRIMETHYL SILYL) SILANE, INFRARED INVESTIGATION OF IOD | 0527
ECTRA OF MH4 (M= CARBON, SILICON, GERMANIUM, AND TIN) AND | MF4 (M= CARBON, SILICON, AND GERMANIUM). / RAMAN SP | 0314
4 (M= CARBON, SILICON, AND / VAPOR PHASE RAMAN SPECTRA OF | MH4 (M= CARBON, SILICON, GERMANIUM, AND TIN) AND MF | 0314
-1: INFRARED SPECTRUM OF OXYGEN DIFLUORIDE FROM 2.5 TO 25 | MICRONS. /ND STRUCTURE OF INORGANIC MOLECULES. | 0761
DE, GALLIUM MONOFLUORIDE, INDIUM MONOFLUORIDE, AND THALL/ | MICROWAVE ABSORPTION SPECTRA OF ALUMINUM MONOFLUORI | 0025
MICROWAVE AND MASS SPECTRA OF SULFUR MONOFLUORIDE. | 1024
FORM, AND PHOSPHORUS TRIFLUORIDE. | MICROWAVE INVESTIGATIONS OF METHYL FLUORIDE, FLUORO | 0798
OF BROMINE PENTAFLUORIDE. | MICROWAVE MEASUREMENTS OF THE J-8 FAR-9 TRANSITION | 0261
DIFLUORIDE) A CAVITY SEARCH SPECTROMETER FOR FREE RADICAL | MICROWAVE ROTATIONAL ABSORPTION STUDIES. (NITROGEN | 0689
E. | MICROWAVE ROTATIONAL SPECTRUM OF INDIUM MONOFLUORID | 0490
IDES. | MICROWAVE ROTATIONAL SPECTRUM OF THALLIUM MONOFLUOR | 1119
IDE MOLECULE. ZEEMAN EFFECT IN THE | MICROWAVE ROTATIONAL SPECTRUM OF THE ALUMINUM FLUOR | 0026
ORIDE FREE RADICAL. | MICROWAVE ROTATIONAL SPECTRUM OF THE NITROGEN DIFLU | 0690
ITROGEN-14 AND THE STRUCTURE OF NITROGEN TRIFLUORIDE FROM | MICROWAVE SPECTRA. NUCLEAR QUADRUPOLE MOMENT OF N | 0741
CITED VIBRATIONAL STATES. (EQUILIBRIUM STRUCTURE) | MICROWAVE SPECTRA OF NITROGEN TRIFLUORIDE IN THE EX | 0722
CITED VIBRATIONAL STATES. (ZETA TYPE DOUBLING AND RESONA/ | MICROWAVE SPECTRA OF NITROGEN TRIFLUORIDE IN THE EX | 0723
UM MONOHALIDES. | MICROWAVE SPECTRA OF THE THALLIUM, INDIUM AND GALLI | 1110
FFECT IN DIAMAGNETIC MOLECULES. GJ FACT/ HIGH TEMPERATURE | MICROWAVE SPECTROMETER FOR MEASUREMENTS OF ZEEMAN E | 1120
REE MILLIMETERS, PART-2: (NITROGEN TRIFLUORIDE AND PHOSP/ | MICROWAVE SPECTROSCOPY IN THE REGION FROM TWO TO TH | 0711

ALUMINUM CHLORIDE.   HIGH TEMPERATURE MICROWAVE SPECTROSCOPY OF ALUMINUM MONOFLUORIDE AND 0028
70 AND 140 GHZ   MICROWAVE SPECTROSCOPY OF CHLORINE PENTAFLUORIDE AT 0423
IFLUORIDE. CENTRIFUGAL DISTORTION.   MICROWAVE SPECTRUM AND FORCE CONSTANTS OF SILICON D 0989
NE TRIFLUORIDE.   MICROWAVE SPECTRUM AND MOLECULAR STRUCTURE OF BROMI 0258
ORIDE.   MICROWAVE SPECTRUM AND STRUCTURE OF CHLORINE TRIFLU 0410
DE.   MICROWAVE SPECTRUM AND STRUCTURE OF SULFUR DIFLUORI 1030
ALYSIS OF SELENIUM TETRAFLUORIDE.   MICROWAVE SPECTRUM, DIPOLE MOMENT, AND STRUCTURE AN 0957
                                                            MICROWAVE SPECTRUM OF ALUMINUM MONOFLUORIDE.   0029
                                                            MICROWAVE SPECTRUM OF ARSENIC TRIFLUORIDE.   0102
CITED VIBRATIONAL STATE.   MICROWAVE SPECTRUM OF ARSENIC TRIFLUORIDE IN THE EX 0097
                                                            MICROWAVE SPECTRUM OF BORON FLUORIDE.   0165
                                                            MICROWAVE SPECTRUM OF BROMINE PENTAFLUORIDE.   0269
                                                            MICROWAVE SPECTRUM OF CARBON DIFLUORIDE.   0307
                                                            MICROWAVE SPECTRUM OF GERMANIUM DIFLUORIDE.   0478
OLE COUPLING AND CENTRIFUGAL DISTORTION.   MICROWAVE SPECTRUM OF GERMANIUM DIFLUORIDE. QUADRUP 0477
                                                            MICROWAVE SPECTRUM OF IODINE PENTAFLUORIDE.   0502
                                                            MICROWAVE SPECTRUM OF NITROGEN DIFLUORIDE RADICAL. 0681
                 SPIN ROTATIONAL HYPERFINE STRUCTURE IN THE MICROWAVE SPECTRUM OF OXYGEN DIFLUORIDE.   0777
NALLY EXCITED STATES, NU-1- 2 NU-2 FERMI RESONANCE AND E/   MICROWAVE SPECTRUM OF OXYGEN DIFLUORIDE IN VIBRATIO 0771
                                                            MICROWAVE SPECTRUM OF PHOSPHORUS TRIFLUORIDE.   0805
ITED VIBRATIONAL STATES, EQUILIBRIUM STRUCTURE, ANHARMON/   MICROWAVE SPECTRUM OF SILICON DIFLUORIDE.   0990
AL.   MICROWAVE SPECTRUM OF SILICON DIFLUORIDE IN THE EXC 0993
OXYGEN DIFLUORIDE.   MICROWAVE SPECTRUM OF THE SULFUR MONOFLUORIDE RADIC 1018
LAR CONSTANTS OF ALUMINUM MONOFLUORIDE.   MICROWAVE SPECTRUM, STRUCTURE, AND DIPOLE MOMENT OF 0779
            ROTATIONAL SPECTRA OF BROMINE PENTAFLUORIDE IN THE MILLIMETER AND SUBMILLETER WAVE SPECTRUM AND MOLECU 0033
CONSTANTS OF NITROGEN TRIFLUORIDE.   MILLIMETER WAVE LENGTH REGION.   0268
FLUORIDE.   MILLIMETER WAVE SPECTRUM AND CENTRIFUGAL DISTORTION 0718
                            K-TYPE DOUBLING IN THE   MILLIMETER WAVE SPECTRUM AND STRUCTURE OF OXYGEN DI 0768
                            FAR INFRARED AND   MILLIMETER WAVE SPECTRUM OF BROMINE PENTAFLUORIDE. 0260
P/ MICROWAVE SPECTROSCOPY IN THE REGION FROM TWO TO THREE   MILLIMETER WAVE STUDIES OF XENON HEXAFLUORIDE.   1299
TATE OF OZONE, NITRITE ION, NITROGEN OXYFLUORIDE, AND OX/   MILLIMETERS, PART-2: (NITROGEN TRIFLUORIDE AND PHOS 0711
PART-3: AB INITIO STUDIES OF CHARGE DISTRIBUTION USING A   MINIMAL BASIS SET SCF CALCULATIONS FOR THE GROUND S 0776
ABLE UREY-BRADLEY FORCE FIELD AND THE ASSIGNMENTS OF SOME   MINIMAL SLATER-TYPE BASIS. (NITROGEN TRIFLUORIDE) / 0707
)   VIBRATIONAL SPECTRA OF   MIXED HALOMETHANES. (CARBON TETRAFLUORIDE) /RANSFER 0351
                                 SEE BOTH   MIXED PHOSPHORUS HALIDES. (PHOSPHORUS PENTAFLUORIDE 0829
ISOELECTRONIC SERIES CHLORINE MONOFLUORIDE, HYDROG/ SCF-   MO AND MOLECULAR ORBITAL.
/MOLECULAR ORBITAL THEORY. ESSENTIAL STRUCTURAL ELEMENTS-   MO CALCULATIONS OF SOME MOLECULAR PROPERTIES OF THE 0371
/. PART-1: OPTICAL ABSORPTION AND X-RAY EMISSION FROM SCF   MO FORMALISM. PART-2: DIRECT COMPARISON WITH AB IN/ 0765
CHLORINE DIFLUORIDE ON THE BASIS OF A CALCULATION BY THE   MO LCAO COMPUTATIONS. PART-2: MOLECULAR ORBITAL IN/ 1078
/IC STRUCTURE OF A CHLORINE PENTAFLUORIDE MOLECULE BY THE   MO LCAO SCF IN THE NDDO APPROXIMATION. /HE MOLECULE 0382
/OF COMPOUNDS OF HALOGENS AND INERT GASES BY THE SCF LCAO   MO LCAO SELF CONSISTENT FIELD METHOD IN AN NDDO AP/ 0415
SILICON DIFLUORIDE.   NONEMPIRICAL LCAO MO METHOD. PART-2: COMPARATIVE ANALYSIS OF DIFFER/ 1231
FLUORIDES OF NONTRANSITION ELEMENTS FROM THE DATA OF LCAO   MO SCF CALCULATIONS OF THE ELECTRONIC STRUCTURE OF 0995
DIFLUORIDE, MAGNESIUM DIFLUORIDE, AND / NONEMPIRICAL LCAO   MO SCF NONEMPIRICAL CALCULATIONS. /IC STRUCTURE OF 0009
DIFLUORIDE.   NONEMPIRICAL LCAO MO SCF STUDIES OF THE GROUP-2A DIHALIDES BERYLLIUM 1360
                                                            MO SCF STUDIES OF THE LOW LYING STATES OF BERYLLIUM 0137
                                            AN LCAO   MO STUDY OF RARE GAS FLUORIDES.   1341
                                CHEMICAL PREDICTIONS BY   MO THEORY. THE NOBLE GAS HALIDES.   1340
DIFFRACTION S/ STRUCTURE, PSEUDOROTATION, AND VIBRATIONAL   MODE COUPLING IN IODINE HEPTAFLUORIDE. AN ELECTRON 0515
A OF RUTHENIU/ JAHN-TELLER EFFECT IN THE E(8) VIBRATIONAL   MODE OF HEXAFLUORIDE MOLECULES. THE INFRARED SPECTR 0926
DE. STRUCTURE, PSEUDOROTATION, AND ANHARMONIC COUPLING OF   MODES. /ON DIFFRACTION STUDY OF RHENIUM HEPTAFLUORI 0920
LECULES.   RAMAN INTENSITIES FOR SPHERICALLY SYMMETRIC   MODES OF SOME XY4 TETRAHEDRAL AND XY6 OCTAHEDRAL MO 1453
ERGY DISTRIBUTIONS. (SELENIUM A/ CLASSIFICATION OF NORMAL   MODES OF VIBRATION. PART-2: INTERACTION TERMS IN 1494
2B HALIDES. (ZINC DIFLUORIDE, CADMIUM DIFL/ POLARIZED ION   MODEL AND THE BENDING FORCE CONSTANTS OF THE GROUP- 1322
EN TRIFLUORIDE.   LONE PAIR   MODEL AND THE VIBRATIONAL FORCE CONSTANTS OF NITROG 0705
OF THE GROUP-2 HALIDES.   IONIC   MODEL CALCULATIONS. PART-2: BENDING FORCE CONSTANTS 1381
HIGHLY SYMMETRICAL MOLECU/ USE OF A SIMPLE ELECTROSTATIC   MODEL FOR CALCULATION OF ANGULAR FORCE CONSTANTS OF 1484
NAR XY3 MOLECULES. (NITROGEN TRIFLUORIDE)   A BENT BOND   MODEL FOR THE VIBRATIONAL FORCE CONSTANTS OF NONPLA 0704
                  LIMITATIONS OF THE UREY-BRADLEY   MODEL IN NORMAL COORDINATE ANALYSES.   1478
ND STEREOCHEMICAL NONRIGIDITY OF THE TRIGONAL BIPYRAMIDAL   MODEL MX3Y2. /ULES. PART-14: MOLECULAR VIBRATIONS A 0854
FLUORIDE COMPOUNDS OF MAGNESIUM AND CALCIUM AND THE IONIC   MODEL OF A MOLECULE.   DISSOCIATION ENERGY OF MONO 0278
EVELS AND MOLECULAR GEOMETRY. PART-2: COMP/ CRYSTAL FIELD   MODEL STUDY OF XENON HEXAFLUORIDE. PART-1: ENERGY L 1311
STANT OF THE NONPLANAR VIBRATION OF AN XY3 MOLECULE AND A   MODEL WITH POLARIZABLE IONS.   FORCE CON 0573
ON CORIOLIS COUPLING CONSTANTS IN FOUR XYN TYPE MOLECULE   MODELS. /TICAL INTERPRETATION OF THE MASS INFLUENCE 1488
OF ROTATION VIBRATION OF OCTAHEDRAL XY6 AND Z6 MOLECULAR   MODELS WITH APPLICATION TO 15 XY6 MOLECULES. /LYSIS 1454
5 TO 350 DEGREES K. RECONCILIATION OF THE ENTROPIES FROM   MOLECULAR AND THERMAL DATA. /DYNAMIC FUNCTIONS FROM 1275
                                EMISSION OF   MOLECULAR ANTIMONY MONOFLUORIDE.   0060
DES CARBON TETRAFLUORIDE, CARBON TETRACHLORIDE, SILICON /   MOLECULAR BEAM ELECTRIC DEFLECTION OF THE TETRAHALI 0349
ENTAFLUORIDES.   (ALSO ANTIMONY PENTAFLUORIDE)   MOLECULAR BEAM MASS SPECTROMETRIC STUDY OF BINARY P 1376
Y OF REACTIVE FLUORIDES.   APPLICATION OF A MOLECULAR BEAM SOURCE MASS SPECTROMETER TO THE STUD 0019
Y OF REACTIVE FLUORIDES. (SECOND ROW, T/ APPLICATION OF A   MOLECULAR BEAM SOURCE MASS SPECTROMETER TO THE STUD 1377
RIFLUORIDES. (LANTHANIDE TRIFLUOR/ ELECTRIC DEFLECTION OF   MOLECULAR BEAMS OF THE RARE EARTH DIFLUORIDES AND T 0946
ORCE CONSTANT.   RELAXATION OF THE MOLECULAR CHARGE DISTRIBUTION AND THE VIBRATIONAL F 0151
. PART-3: THE ISOELECTRONIC NITROGEN, CARBON MONOXIDE, B/   MOLECULAR CHARGE DISTRIBUTIONS AND CHEMICAL BINDING 0150
SOME LINEAR SYMMETRICAL MOLECULES. (XENON DIFLUORIDE)   MOLECULAR CONSTANTS AND THERMODYNAMIC FUNCTIONS OF 1244
ATOMIC DISPLACMENTS AND INTRAMOLECULAR FORCES IN CARBON /   MOLECULAR CONSTANTS FROM INFRARED BAND SHAPES. THE 0331
                           SPECTROSCOPIC STUDIES OF   MOLECULAR CONSTANTS IN SOME POLYATOMIC MOLECULES.   0046
XENON DIFLUORIDE.   SPECTROSCOPIC STUDIES OF   MOLECULAR CONSTANTS IN SOME POLYATOMIC MOLECULES. (1245
         MILLIMETER AND SUBMILLETER WAVE SPECTRUM AND   MOLECULAR CONSTANTS OF ALUMINUM MONOFLUORIDE.   0033
TRICHLORI/ RAMAN SPECTRUM OF ARSENIC TRIFLUORIDE AND THE   MOLECULAR CONSTANTS OF ARSENIC TRIFLUORIDE, ARSENIC 0111
             ELECTRONIC EMISSION SPECTRUM AND   MOLECULAR CONSTANTS OF IODINE MONOFLUORIDE.   0496
                           FORCE FIELD AND   MOLECULAR CONSTANTS OF NITROGEN TRIFLUORIDE.   0735
                     INFRARED SPECTRUM AND   MOLECULAR CONSTANTS OF NITROGEN TRIFLUORIDE.   0728
(PHOSPHORUS PENTAFLUORIDE)   MOLECULAR CONSTANTS OF PHOSPHORUS HALIDES. PART-4. 0867
. (PHOSPHORUS TRIFLUORIDE)   MOLECULAR CONSTANTS OF PHOSPHORUS HALOGEN COMPOUNDS 0813
HOSPHINE.   RAMAN SPECTRA AND MOLECULAR CONSTANTS OF PHOSPHORUS TRIFLUORIDE AND P 0822
LE.   MOLECULAR CONSTANTS OF RHENIUM HEPTAFLUORIDE MOLECU 0918
ANUM HALIDES. (SCANDIUM DIFLUORIDE, SCANDIUM MONOFLUORID/   MOLECULAR CONSTANTS OF SCANDIUM, YTTRIUM, AND LANTH 0941
OUNDS. (PHOSPHORUS PENTAFLUORIDE)   MOLECULAR CONSTANTS OF SOME PHOSPHORUS HALOGEN COMP 0868
LECULES AND THEIR SUBSTITUTED DERIVATIVES BY GREEN'S FUN/   MOLECULAR CONSTANTS OF SOME TETRAHEDRAL XY4 TYPE MO 1424

SELENIUM, AND TELLURIUM.   RAMAN SPECTRA AND MOLECULAR CONSTANTS OF THE HEXAFLUORIDES OF SULFUR, 0963
                                   MASS SPECTRUM AND MOLECULAR ENERGETICS OF KRYPTON DIFLUORIDE.   0565
/ERTIES OF CARBON MONOFLUORIDE AND CARBON DIFLUORIDE FROM MOLECULAR FLOW EFFUSION AND MASS SPECTROMETER INVE/ 0286
                                                  MOLECULAR FORCE FIELD FOR BROMINE PENTAFLUORIDE.   0264
                                                  MOLECULAR FORCE FIELD FOR CHLORINE TRIFLUORIDE.   0402
CHLORINE TRIFLUORIDE AND BROMINE TRIFLUORIDE)   MOLECULAR FORCE FIELD FOR INTERHALOGEN COMPOUNDS. (0407
ODINE PENTAFLUORIDE, ICDINE OXYGEN PENTAFLUORIDE, AND IO/ MOLECULAR FORCE FIELD FOR INTERHALOGEN COMPOUNDS. I 0511
PHORUS TRIFLUORIDE AND ARSENIC TRIFLUORIDE)   MOLECULAR FORCE FIELD FOR SOME XY3 MOLECULES. (PHOS   1400
(SELENIUM AND TELLURIUM HEXAFLUORIDES)   MOLECULAR FORCE FIELD FOR SOME XY6 TYPE MOLECULES.   1465
                                                  MOLECULAR FORCE FIELD FOR SULFUR TETRAFLUORIDE.   1050
                        VIRIAL THEOREM DECOMPOSITION OF MOLECULAR FORCE FIELDS.   0783
(CHLORINE PENTAFLUORIDE AND BROMINE PENTAFLUORIDE) MOLECULAR FORCE FIELDS FOR INTERHALOGEN COMPOUNDS.   0425
                                                  MOLECULAR FORCE FIELDS FOR SOME HEXAFLUORIDES.   1452
                                                  MOLECULAR FORCE FIELDS OF XY6 MOLECULES.   1443
ES CF SOME OCTAHEDRAL XY6 MOLECULES. (SULFUR HEXAFLUORID/ MOLECULAR FORCE FIELDS. PART-8: VIBRATION FREQUENCI 1080
CULAR QUADRUPOLE MCMENTS AND SECOND MOMENTS OF THE ELECT/ MOLECULAR G VALUES, MAGNETIC SUSCEPTIBILITIES, MOLE 0780
LECULAR QUADRUPOLE MOMENTS / ZEEMAN STUDIES INCLUDING THE MOLECULAR G VALUES, MAGNETIC SUSCEPTIBILITY, AND MO 0744
UORIDE. THE ANALYSIS OF NU-2 AND THE DETERMINATION OF THE MOLECULAR GEOMETRY.   BORON TRIFL 0209
 ELECTRONIC ISOMERISM.   ELECTRONIC STATES AND MOLECULAR GEOMETRY OF XENON HEXAFLUORIDE: A CASE OF 1294
/L STUDY OF XENON HEXAFLUORIDE. PART-1: ENERGY LEVELS AND MOLECULAR GEOMETRY. PART-2: COMPARISONS WITH CTHER/ 1311
IDE.   MOLECULAR MAGNETIC PROPERTIES OF CHLORINE MONOFLUOR 0384
/ ANALYSIS OF ROTATION VIBRATION OF OCTAHEDRAL XY6 AND Z6 MOLECULAR MODELS WITH APPLICATION TO 15 XY6 MOLECU/ 1454
UIDS OF MOLYBDENUM HEXAFLUORIDE, TUNGSTEN HEXAFLUORIDE, / MOLECULAR MOTION AND FLUORINE RELAXATION IN THE LIQ 0638
/ICKEL FLUORIDE AND NICKEL CHLORIDE WITH CARBCN MONOXIDE, MOLECULAR NITROGEN, NITRIC OXIDE, AND MOLECULAR OX/ 1324
                             SEE BOTH MO AND MOLECULAR ORBITAL.
/E FIELDS. PART-1. PCINTS-ON-A-SPHERE AND EXTENDED HUCKEL MOLECULAR ORBITAL ANALYSES OF TRIGONAL BIPYRAMIDS./ 0828
ORIDE.   SELF CONSISTENT CHARGE AND CCNFIGURATION MOLECULAR ORBITAL CALCULATIONS ON PHOSPHORUS TRIFLU 0821
HOSPHORUS PENTAFLUORIDE.   SEMIEMPIRICAL MOLECULAR ORBITAL CALCULATIONS. PSEUDOROTATION IN P 0843
STATES OF SCANDIUM MONCFLUORIDE.   REASSIGNMENT OF MOLECULAR ORBITAL CONFIGURATIONS OF THE ELECTRONIC 0937
F VALENCES 2, 4, AND 6.   MOLECULAR ORBITAL DESCRIPTION OF SULFUR COMPOUNDS O 1085
SULFUR TETRAFLUORIDE AND SULFUR HEXAFLUORIDE.   MOLECULAR ORBITAL DESCRIPTION OF SULFUR DIFLUORIDE, 1027
TON DIFLUORIDE.   MOLECULAR ORBITAL ENERGY LEVELS AND BONDING IN KRYP 0549
/ND X-RAY EMISSION FROM SCF MO LCAO COMPUTATIONS. PART-2: MOLECULAR ORBITAL INTERPRETATION OF ITS X-RAY ABSO/ 1078
AND IDEAL GAS THERMODYNAMIC FUNCTIONS FROM A HARTREE-FOCK MOLECULAR ORBITAL INVESTIGATION. /COPIC CONSTANTS, 0094
IDE AND SILICON MONOFLUORIDE AND THEIR POSITIVE AND NEGA/ MOLECULAR ORBITAL INVESTIGATION OF CARBON MONOFLUOR 0290
ATION OF EQUILIBRIUM GEOMETRIES AND QUAD/ SELF CONSISTENT MOLECULAR ORBITAL METHODS. PART-5: AB INITIO CALCUL 0720
NTS, AND OTHER PROPERTIES OF SELENIUM MONOFLUORIDE FROM A MOLECULAR ORBITAL STUDY. /LPY, SPECTROSCOPIC CONSTA 0955
, PHOSPHORUS MONOFLUORIDE, AND SULFUR MONOFLUORIDE FROM A MOLECULAR ORBITAL STUDY. /IDE, SILICON MONOFLUORIDE 0973
PNICOGEN FLUORIDE MOLECULES AND IONS.   MOLECULAR ORBITAL STUDY OF ATOMIZATION ENERGIES OF 1364
OGEN DIOXIDE, AND NITRCGEN DIFLUCRIDE.   MOLECULAR ORBITAL STUDY OF NITROGEN DIHYDRIDE, NITR 0684
OSPHORUS TRIFLUORIDE, AND NITROGEN DIFLUORIDE.   MOLECULAR ORBITAL STUDY OF NITROGEN TRIFLUORIDE, PH 0721
OGENS. (CHLORINE MONOFLUORIDE, CHLORINE TRIFLU/ AB INITIO MOLECULAR ORBITAL STUDY OF THE GEOMETRY OF INTERHAL 0379
AND DINITROGEN TETRAFLUORIDE MOLECULES, AND ES/ AB INITIO MOLECULAR ORBITAL STUDY OF THE NITROGEN DIFLUORIDE 0688
E RADICAL.   AB INITIO MOLECULAR ORBITAL STUDY OF THE PHOSPHORUS DIFLUORID 0790
ENTS- MO FORMALISM. PART-2: DIRECT COMPARISO/ APPROXIMATE MOLECULAR ORBITAL THEORY. ESSENTIAL STRUCTURAL ELEM 0765
E OF ORGANIC COMPOUNDS. PART-3: AB INITIO STUDIES OF CHA/ MOLECULAR ORBITAL THEORY OF THE ELECTRONIC STRUCTUR 0707
.   MOLECULAR ORBITAL TREATMENT OF IODINE HEPTAFLUORIDE 0533
MENT 110 HEXAFLUORIDE CALCULATED BY THE SELF CONSISTENT / MOLECULAR ORBITALS OF PLATINUM HEXAFLUORIDE AND ELE 0881
AND POLYATOMIC MOLECU/ DENSITY LOCALIZATION OF ATOMIC AND MOLECULAR ORBITALS. PART-3: HETERONUCLEAR DIATOMIC 0175
/EAN AMPLITUDES OF VIBRATION, THERMODYNAMIC FUNCTIONS AND MOLECULAR POLARIZABILITES OF SOME PENTAHALIDES OF / 1437
/ES OF VIBRATION, BASTIANSEN-MORINO SHRINKAGE EFFECT, AND MOLECULAR POLARIZABILITIES FOR DIHALIDES OF SOME G/ 0124
LORINE MONOFLUORIDE, HYDROG/ SCF- MO CALCULATIONS OF SOME MOLECULAR PROPERTIES OF THE ISOELECTRONIC SERIES CH 0371
ERYLLIUM DIFLUORIDE, BORON DIFLUCRIDE, CARBON DIFLUORIDE/ MOLECULAR PROPERTIES OF THE TRIATOMIC DIFLUORIDES B 0141
, GALLIUM TRIHYDRIDE, GERMANIUM TETRAHYDRIDE, ARSENIC TR/ MOLECULAR SCF CALCULATIONS FOR GALLIUM MONOFLUORIDE 0459
                     ELECTRON CORRELATION AND MOLECULAR SHAPE.   1413
OF THE INTERNUCLEAR SEPARATION FOR THE B-X BAND SYSTEM OF MOLECULAR SILICON MONOFLUORIDE. /ENT AS A FUNCTION 0967
TRAFLUORIDE)   MOLECULAR SPECTROSCOPY BY MEANS OF ESCA. (CARBON TE 0328
UR HEXAFL/ APPLICATION OF SYNCHROTON ORBITAL RADIATION TO MOLECULAR SPECTROSCOPY. MOLECULAR NITROGEN AND SULF 1083
BASTIANSEN-MORINO SHRINKAGE EFFECT IN THE STUDY OF MOLECULAR STRUCTURE.   1476
PROPERTIES OF PLUTONIUM HEXAFLUORIDE. (MOLECULAR STRUCTURE, ABSORPTION SPECTRUM)   0894
IFLUORIDE AND BORON TRICHLORIDE.   MOLECULAR STRUCTURE AND FORCE CONSTANTS OF BORON TR 0221
AMIC FUNCTIONS OF IODINE PENTAFLUORIDE AND BRCMINE PENTA/ MOLECULAR STRUCTURE, FORCE CONSTANTS, AND THERMODYN 0505
RADICAL.   ISOTROPIC HYPERFINE SPLITTINGS AND MOLECULAR STRUCTURE IN THE PHOSPHORUS TETRAFLUORIDE 0824
                                                  MOLECULAR STRUCTURE OF ANTIMONY PENTAFLUORIDE.   0080
                                                  MOLECULAR STRUCTURE OF BORON TRIFLUORIDE.   0187
                      MICROWAVE SPECTRUM AND MOLECULAR STRUCTURE OF BROMINE TRIFLUORIDE.   0258
                                                  MOLECULAR STRUCTURE OF CARBON TETRAFLUORIDE.   0334
                                                  MOLECULAR STRUCTURE OF IODINE HEPTAFLUORIDE.   0530
ANIUM HEXAFLUORIDES FRCM INFRARED AND RAMAN SPECTRA. MOLECULAR STRUCTURE OF MOLYBDENUM, TUNGSTEN, AND UR 0627
ED BY GAS ELECTRON DIFFRACTION.   MOLECULAR STRUCTURE OF PHOSPHORUS TRIFLUORIDE STUDI 0810
VIBRATIONAL STRUCTURE AND MOLECULAR STRUCTURE OF RHENIUM HEXAFLUORIDE.   0907
(OF SULFUR, SELENIUM, TELLURIUM, MO/ INFRARED SPECTRA AND MOLECULAR STRUCTURE OF SOME GROUP-6 HEXAFLUORIDES. 0961
ALLIUM, YTTRIUM, LANTH/ ELECTRON DIFFRACTION STUDY OF THE MOLECULAR STRUCTURE OF THE VAPOR PHASE HALIDES OF G 1317
                   DIPOLE MOMENT AND MOLECULAR STRUCTURE OF URANIUM HEXAFLUORIDE.   1200
DIFFRACTION.   CRYSTAL AND MOLECULAR STRUCTURE OF XENON DIFLUORIDE BY NEUTRON 1239
                                                  MOLECULAR STRUCTURE OF XENON HEXAFLUORIDE.   1280
                                                  MOLECULAR STRUCTURE OF XENON HEXAFLUORIDE.   1303
E HEPTAFLUORIDE.   MOLECULAR STRUCTURE OF XENON HEXAFLUORIDE AND IODIN 0526
ANALYSIS OF ELECTRON DIFFRACTION INTENSITIES. PART-2: IN/ MOLECULAR STRUCTURE OF XENON HEXAFLUORIDE. PART-1: 1291
                         CRYSTAL AND MOLECULAR STRUCTURE OF XENON TETRAFLUORIDE.   1277
ON DIFFRACTION.   CRYSTAL AND MOLECULAR STRUCTURE OF XENON TETRAFLUORIDE BY NEUTR 1260
OF GERMANIUM D/ MATRIX INFRARED AND LASER RAMAN SPECTRA. MOLECULAR STRUCTURES AND NORMAL COORDINATE ANALYSES 0475
ENIC PENTAFLUORIDE AS DETERMINED BY ELECTRON DIFFRACTION. MOLECULAR STRUCTURES OF ARSENIC TRIFLUORIDE AND ARS 0099
ENIC TRICHLORIDE BY GAS ELECTRON DI/ DETERMINATION OF THE MOLECULAR STRUCTURES OF ARSENIC TRIFLUORIDE AND ARS 0104
ODINE PENTAFLUORIDE FRCM ELECTRON DIFFRACTION / GAS PHASE MOLECULAR STRUCTURES OF BROMINE PENTAFLUORIDE AND I 0265
M DIFLUORIDE, MAGNE/ ELECTRON DIFFRACTION ANALYSIS OF THE MOLECULAR STRUCTURES CF GROUP-2 ELEMENTS. (BERYLLIU 0131
                                                  MOLECULAR STRUCTURES OF RARE GAS COMPOUNDS.   1344
TETRAFLUOR/ INVESTIGATION BY ELECTRON DIFFRACTION OF THE MOLECULAR STRUCTURES OF SULFUR HEXAFLUORIDE, SULFUR 1041
ELECTRON DIFFRACTION INVESTIGATION OF THE MOLECULAR STRUCTURES OF THE ALUMINUM HALIDES.   0036
BIUM PENTAFLUORIDE, AN/ ELECTRON DIFFRACTION STUDY OF THE MOLECULAR STRUCTURES OF VANADIUM PENTAFLUORIDE, NIO 1217

AND PHOSPHORUS PENTAFLUORIDE. INFRARED AND LOW TEMPERATU/ MOLECULAR STRUCTURES OF VARIOUS PHOSPHORUS HALIDES 0849
                                                        MOLECULAR SYMMETRY OF IODINE HEPTAFLUORIDE.   0524
                                                        MOLECULAR SYMMETRY OF IODINE HEPTAFLUORIDE.   0531
                                    EVIDENCE FOR THE MOLECULAR SYMMETRY OF IODINE HEPTAFLUORIDE.   0523
TRAFLUORIDE.   MOLECULAR SYMMETRY OF XENON DIFLUORIDE AND XENON TE 1240
E.   STRUCTURE OF MOLECULAR URANIUM DIFLUORIDE AND URANIUM TRIFLUORID 1162
ORIDE, YTTRIUM MONOFLUORIDE, AND LANTHANUM/ EVALUATION OF MOLECULAR VIBRATION FREQUENCIES OF SCANDIUM MONOFLU 0936
SHRINKAGES OF THE INTERNUCLEAR DISTANCES BY MOLECULAR VIBRATIONS.   1489
LE METHOD FOR EXACT SOLUTION OF 2 X 2 SECULAR EQUATION IN MOLECULAR VIBRATIONS.   SIMP 1386
OF THE TRIGONAL BI/ PENTACOORDINATED MOLECULES. PART-14: MOLECULAR VIBRATIONS AND STEREOCHEMICAL NONRIGIDITY 0854
FUR TETRAFLUORIDE.   MOLECULAR VIBRATIONS OF SULFUR HEXAFLUORIDE AND SUL 1034
                                                        MOLECULAR VIBRATIONS OF SULFUR TETRAFLUORIDE.   1058
)YZ MOLECULE.   POTENTIAL FIELD AND MOLECULAR VIBRATIONS OF THE TRIGONAL BIPYRAMID AX(3 1436
E RAMAN SPECTRUM OF PHOSPHORUS PENTA/ POTENTIAL FIELD AND MOLECULAR VIBRATIONS OF THE TRIGONAL BIPYRAMIDAL TH 0863
ELECTRONIC POPULATION ANALYSIS OF MOLECULAR WAVE FUNCTIONS.   0157
RELATIVISTIC MOLECULAR WAVE FUNCTIONS: XENON DIFLUORIDE.   1249
INFRARED, VISIBLE AND ULTRAVIOL/ SPECTRAL OBSERVATIONS ON MOLECULAR XENON HEXAFLUORIDE. RAMAN SCATTERING AND 1285
LECTRIC QUADRUPOLE MOMENTS IN CHLORINE MONOFLUORIDE, BRO/ MOLECULAR ZEEMAN EFFECT, MAGNETIC PROPERTIES, AND E 0254
RAMAN EFFECT AND ULTRAVIOLET ABSORPTION SPECTRA OF MOLYBDENUM AND TUNGSTEN HEXAFLUORIDES.   0642
ONIUM HEXAFLUORIDE. PART-3: URANIUM HEXAFLUORIDE. PART-4: MOLYBDENUM AND TUNGSTEN HEXAFLUORIDES. /ART-2: PLUT 0629
VIBRATIONAL SPECTRA OF MOLYBDENUM AND TUNGSTEN PENTAFLUORIDES.   0625
XAFLUORIDE, URANIUM HEXAFLUORIDE, TELLURIUM HEXAFLUORIDE, MOLYBDENUM HEXAFLUORIDE) /DIFFRACTION. (TUNGSTEN HE 1158
URE) OF SILICON TETRAFLUORIDE, TUNGSTEN HEXAFLUORIDE, AND MOLYBDENUM HEXAFLUORIDE. /AL CONSTANTS (VAPOR PRESS 0639
ULES. (TUNGSTEN HEXAFLUORIDE, TELLURIUM HEXAFLUORIDE, AND MOLYBDENUM HEXAFLUORIDE) /TION OF SOME SIMPLE MOLEC 1153
VIBRATIONAL SPECTRUM AND FORCE FIELD OF MOLYBDENUM HEXAFLUORIDE.   0634
HEAT CAPACITY, ENTROPY, AND HEATS OF TRANSITION OF MOLYBDENUM HEXAFLUORIDE AND NIOBIUM PENTAFLUORIDE.   0662
INFRARED SPECTRUM OF CHROMIUM HEXAFLUORIDE, MOLYBDENUM HEXAFLUORIDE, AND OSMIUM HEXAFLUORIDE.   0436
POTENTIAL CONSTANTS FOR MOLYBDENUM HEXAFLUORIDE AND RHENIUM HEXAFLUORIDE.   0635
VIBRATIONAL SPECTRA OF MOLYBDENUM HEXAFLUORIDE AND TECHNETIUM HEXAFLUORIDE   0628
ASSIGNMENTS IN THE ULTRAVIOLET SPECTRA OF MOLYBDENUM HEXAFLUORIDE AND TUNGSTEN HEXAFLUORIDE.   0633
/F THE FIRST BORN APPROXIMATION IN ELECTRON DIFFRACTION. (MOLYBDENUM HEXAFLUORIDE AND TUNGSTEN HEXAFLUORIDE) 0640
HEAT CAPACITY AND OTHER THERMODYNAMIC PROPERTIES OF MOLYBDENUM HEXAFLUORIDE BETWEEN 4 AND 350 DEGREES K   0636
/ TRANSITION METAL HEXAFLUORIDES. (TUNGSTEN HEXAFLUORIDE, MOLYBDENUM HEXAFLUORIDE, RHENIUM HEXAFLUORIDE, OSM/ 0903
HENIUM HEXAFLUORIDE, URANIUM HEXAFLUORI/ RAMAN SPECTRA OF MOLYBDENUM HEXAFLUORIDE, TECHNETIUM HEXAFLUORIDE, R 1072
URANIUM HEXAFLUO/ HIGH RESOLUTION NMR SPECTRUM OF LIQUID MOLYBDENUM HEXAFLUORIDE, TUNGSTEN HEXAFLUORIDE, AND 0637
/LECULAR MOTION AND FLUORINE RELAXATION IN THE LIQUIDS OF MOLYBDENUM HEXAFLUORIDE, TUNGSTEN HEXAFLUORIDE, AN/ 0638
/FLUORIDE, SELENIUM HEXAFLUORIDE, TELLURIUM HEXAFLUORIDE, MOLYBDENUM HEXAFLUORIDE, TUNGSTEN HEXAFLUORIDE, AN/ 1069
/FLUORIDE, SELENIUM HEXAFLUORIDE, TELLURIUM HEXAFLUORIDE, MOLYBDENUM HEXAFLUORIDE, TUNGSTEN HEXAFLUORIDE, UR/ 1084
NIOBIUM BROMIDE, TANTALUM CHLORIDE, TANTALUM BROMIDE, AND MOLYBDENUM PENTACHLORIDE IN THE GASEOUS STATE. /E, 0075
RIDE, ANTIMONY PENTAFLUORIDE, VANADIUM PENTAFLUORIDE, AND MOLYBDENUM PENTAFLUORIDE) /ORIDE, ARSENIC PENTAFLUO 0878
RAMAN SPECTRUM OF CRYSTALLINE MOLYBDENUM PENTAFLUORIDE.   0623
VIBRATIONAL SPECTRA OF MOLYBDENUM PENTAFLUORIDE AND TUNGSTEN PENTAFLUORIDE   0624
HENIUM PENTAFLUORIDE, AND OSMIUM PENT/ VAPOR PRESSURES OF MOLYBDENUM PENTAFLUORIDE, TUNGSTEN PENTAFLUORIDE, R 0624
ELECTRONIC SPECTRA OF LIQUID RUTHENIUM AND MOLYBDENUM PENTAFLUORIDES.   0925
VALENCE FORCE CONSTANTS OF THE HEXAFLUORIDES OF MOLYBDENUM, TECHNETIUM, AND RHENIUM.   0641
RAMAN SPECTRUM OF SOLID MOLYBDENUM TETRAFLUORIDE.   0621
PREPARATION AND CRYSTAL STRUCTURE OF MOLYBDENUM TRIFLUORIDE.   0620
E GROUP-6 HEXAFLUORIDES. (OF SULFUR, SELENIUM, TELLURIUM, MOLYBDENUM, TUNGSTEN, AND URANIUM) /TRUCTURE OF SOM 0961
NIVERSAL EQUATIONS OF STATE OF HEXAFLUORIDES. (OF SULFUR, MOLYBDENUM TUNGSTEN, AND URANIUM) /SIMILARITY AND U 0632
M INFRARED AND RAMAN SPECTRA.   MOLECULAR STRUCTURE OF MOLYBDENUM, TUNGSTEN, AND URANIUM HEXAFLUORIDES FRO 0627
INORGANIC HEXAFLUORIDES. (OF SULFUR, SELENIUM, TELLURIUM, MOLYBDENUM, TUNGSTEN, URANIUM, AND RHENIUM) / SOME 0631
THERMODYNAMIC PROPERTIES OF NIOBIUM(V), TANTALUM(V), AND MOLYBDENUM(V) FLUORIDES.   0668
/CTIONS BY MEASURED ELECTRONIC PROPERTIES. PART-2: DIPOLE MOMENT AND FIELD GRADIENT OF NITROGEN TRIFLUORIDE./ 0746
IDE.   DIPOLE MOMENT AND MOLECULAR STRUCTURE OF URANIUM HEXAFLUOR 1200
ORIDE.   MICROWAVE SPECTRUM, DIPOLE MOMENT, AND STRUCTURE ANALYSIS OF SELENIUM TETRAFLU 0957
FOR THE B-X BAND SYSTEM OF MOLECU/ ELECTRONIC TRANSITION MOMENT AS A FUNCTION OF THE INTERNUCLEAR SEPARATION 0967
N TRIFLUORIDE.   CNDO CALCULATION OF DIPOLE MOMENT DERIVATIVES AND INFRARED INTENSITIES OF BORO   0195
HOSPHORUS OX/ INFRARED INTENSITIES, BOND MOMENTS AND DIPOLE MOMENT DERIVATIVES FOR PHOSPHORUS TRIFLUORIDE AND P 0797
ON TRA/ DETERMINATION OF THE MATRIX ELEMENT OF THE DIPOLE MOMENT OF AN A DOUBLET SIGMA TO X DOUBLET PI ELECTR 0970
GAS PHASE ELECTRON RESONANCE SPECTRUM AND DIPOLE MOMENT OF CARBON MONOFLUORIDE.   0284
/HASE ELECTRON PARAMAGNETIC RESONANCE SPECTRUM AND DIPOLE MOMENT OF NITROGEN MONOFLUORIDE IN THE SINGLET DEL/ 0673
VACUUM ULTRAVIOLET ABSORPTION SPECTRUM AND DIPOLE MOMENT OF NITROGEN TRIFLUORIDE.   0725
TRIFLUORIDE FROM MICROWAVE SPECTRA.   NUCLEAR QUADRUPOLE MOMENT OF NITROGEN-14 AND THE STRUCTURE OF NITROGEN 0741
MICROWAVE SPECTRUM, STRUCTURE, AND DIPOLE MOMENT OF OXYGEN DIFLUORIDE.   0779
STRUCTURE AND DIPOLE MOMENT OF SULFUR TETRAFLUORIDE.   1056
/ATE TRANSFORMATIONS. APPLICATION TO THE CALCULATION BOND MOMENT PARAMETERS AND FORCE CONSTANTS. (CARBON TET/ 0352
/ ENERGY, IONIZATION POTENTIAL, ELECTRON AFFINITY, DIPOLE MOMENT, QUADRUPOLE MOMENT, AND SPECTROSCOPIC CONST/ 0757
/NTHALPY, IONIZATION POTENTIAL, ELECTRON AFFINITY, DIPOLE MOMENT, SPECTROSCOPIC CONSTANTS, AND IDEAL GAS THE/ 0094
X SYSTEM.   CORRELATION OF THE ELECTRONIC TRANSITION MOMENT WITH INTERNUCLEAR DISTANCE FOR BANDS OF THE   0969
CHLORIDE AND TIN MONO/ VARIATION OF ELECTRONIC TRANSITION MOMENT WITH THE INTERNUCLEAR SEPARATION IN SILICON 1137
ENERGIES FOR MOLECULES WITH VIBRATIONALLY INDUCED DIPOLE MOMENTS.   STARK 1417
TRIFLUORIDE AND PHOSPHORUS OX/ INFRARED INTENSITIES, BOND MOMENTS AND BOND MOMENT DERIVATIVES FOR PHOSPHORUS 0797
/ VALUES, MAGNETIC SUSCEPTIBILITIES, MOLECULAR QUADRUPOLE MOMENTS AND SECOND MOMENTS OF THE ELECTRONIC CHARG/ 0780
/TION POTENTIAL, ELECTRON AFFINITY, DIPOLE AND QUADRUPOLE MOMENTS, AND SPECTROSCOPIC CONSTANTS. COMPARISON O/ 0756
/LATIONS ON IONIC CHARACTER AND HYBRIDIZATION FROM DIPOLE MOMENTS FOR ARSENIC TRIFLUORIDE AND ARSENIC TRICHL/ 0107
/MAN EFFECT, MAGNETIC PROPERTIES, AND ELECTRIC QUADRUPOLE MOMENTS IN CHLORINE MONOFLUORIDE, BROMINE MONOFLUO/ 0254
/ALUES, MAGNETIC SUSCEPTIBILITY, AND MOLECULAR QUADRUPOLE MOMENTS IN PHOSPHORUS TRIFLUORIDE, NITROGEN TRIFLU/ 0744
ONOFLUORIDE AND SELENIUM MONOFLUORIDE)   ELECTRIC DIPOLE MOMENTS OF OPEN SHELL DIATOMIC MOLECULES. (SULFUR M 1019
ART-13: STABILITIES OF SAMARIUM, EUROPIUM, AND GADOLINIUM MONO AND DIFLUORIDES. /DIES AT HIGH TEMPERATURES. P 0450
IES AT HIGH TEMPERATURES. PART-18: THE STABILITIES OF THE MONO- AND DIFLUORIDES OF SCANDIUM AND YTTRIUM. /TUD 0939
ET SIGMA TO X DOUBLET PI ELECTRONIC TRANSITION OF SILICON MONOFLUORIDE.   A QUART 0971
ABSORPTION SPECTRUM OF GASEOUS HOLMIUM MONOFLUORIDE.   0488
ABSORPTION SPECTRUM OF GASEOUS SILVER MONOFLUORIDE.   1012
ABSORPTION SPECTRUM OF VAPORIZED TITANIUM MONOFLUORIDE.   1143
RPTION SPECTRUM AND OF TWO NEW SINGLET SYSTEMS OF ARSENIC MONOFLUORIDE.   ANALYSIS OF THE ABSO 0090
ANALYSIS OF THE INFRARED SPECTRUM OF CHLORINE MONOFLUORIDE.   0386
OTATIONAL STRUCTURE OF THE BANDS 0-0 AND 0-1 IN GERMANIUM MONOFLUORIDE.   ANALYSIS OF THE R 0465
INGLET SIGMA(+) TO X TRIPLET SIGMA(-) BAND SYSTEM IN LEAD MONOFLUORIDE.   B S 0585
ET SIGMA(+) TO X TRIPLET SIGMA(-) BAND SYSTEM OF NITROGEN MONOFLUORIDE.   B SINGL 0675
BAND SPECTRA OF THALLIUM MONOIODIDE AND MONOFLUORIDE.   1125

BAND SPECTRUM OF CHROMIUM MONOFLUORIDE. 0429
BAND SPECTRUM OF NICKEL MONOFLUORIDE. 0651
BAND SPECTRUM OF SILICON MONOFLUORIDE. 0965
BO(+)-X(21) AND BO(+)-X(10+) BANDS OF ANTIMONY MONOFLUORIDE. 0071
HE SPECTROSCOPIC CONSTANTS FOR SEVERAL STATES OF NITROGEN MONOFLUORIDE. CALCULATION OF T 0676
CAMERON SYSTEM OF BORON MONOFLUORIDE. 0163
IVE VIBRATIONAL TEMPERATURE IN THE A-2 SYSTEM OF ANTIMONY MONOFLUORIDE. DETERMINATION OF THE EFFECT 0061
DISSOCIATION ENERGY OF CHLORINE MONOFLUORIDE. 0387
DISSOCIATION ENERGY OF COPPER MONOFLUORIDE. 0440
DISSOCIATION ENERGY OF MAGNESIUM MONOFLUORIDE. 0601
OF INTENSITY IN THE ELECTRON EMISSION SPECTRUM OF SILICON MONOFLUORIDE. DISTRIBUTION 0968
ELECTRON SPIN RESONANCE SPECTRUM OF KRYPTON MONOFLUORIDE. 0546
ELECTRONIC BAND SPECTRUM OF YTTRIUM MONOFLUORIDE. 1314
ONIC EMISSION SPECTRUM AND MOLECULAR CONSTANTS OF IODINE MONOFLUORIDE. ELECTR 0496
ELECTRONIC SPECTRUM OF LANTHANUM MONOFLUORIDE. 0569
ELECTRONIC SPECTRUM OF LUTETIUM MONOFLUORIDE. 0598
ELECTRONIC SPECTRUM OF SCANDIUM MONOFLUORIDE. 0934
ELECTRONIC STATES OF GASEOUS ALUMINUM MONOFLUORIDE. 0027
ELECTRONIC STRUCTURE AND GROUND STATE OF SCANDIUM MONOFLUORIDE. 0932
ELECTRONIC STRUCTURE OF CARBON MONOXIDE AND BORON MONOFLUORIDE. 0162
R SEPARATION FOR THE B-X BAND SYSTEM OF MOLECULAR SILICON MONOFLUORIDE. /ENT AS A FUNCTION OF THE INTERNUCLEA 0967
EMISSION OF MOLECULAR ANTIMONY MONOFLUORIDE. 0060
EMISSION SPECTRUM OF BISMUTH MONOFLUORIDE. 0147
EMISSION SPECTRUM OF CARBON MONOFLUORIDE. 0291
EMISSION SPECTRUM OF LEAD MONOFLUORIDE. 0588
EMISSION SPECTRUM OF THE GAMMA SYSTEM OF SILICON MONOFLUORIDE. 0974
EMISSION SPECTRUM OF TITANIUM MONOFLUORIDE. 1142
OF THALLIUM MONOHALIDES ABOVE ROOM TEMPERATURE. (THALLIUM MONOFLUORIDE) /NTROPIES, AND FREE ENERGY FUNCTIONS 1116
ALPY OF FORMATION AND THE DISSOCIATION ENERGY OF THALLIUM MONOFLUORIDE ENTH 1115
FINE STRUCTURE ANALYSIS OF THE C1 SYSTEM OF ANTIMONY MONOFLUORIDE. /NCK-CONDON FACTORS AND R CENTROIDS F 0059
OR THE B TRIPLET SIGMA(+) TO A TRIPLET PI SYSTEM OF BORON MONOFLUORIDE. /NDON FACTORS AND R CENTROIDS FOR THE 0171
6 SINGLET SIGMA- X TRIPLET SIGMA BAND SYSTEM OF NITROGEN MONOFLUORIDE. GAS PHAS 0677
E ELECTRON RESONANCE SPECTRUM AND DIPOLE MOMENT OF CARBON MONOFLUORIDE. 0284
GASEOUS TRIMER AND TETRAMER OF THALLIUM MONOFLUORIDE. 1114
GREEN BAND SYSTEM OF BARIUM MONOFLUORIDE. 0121
OF ALUMINUM TRIFLUORIDE AND HEAT OF FORMATION OF ALUMINUM MONOFLUORIDE. HEAT OF SUBLIMATION 0053
AMAN SPECTROSCOPY IN THE RED AND NEAR INFRARED. (CHLORINE MONOFLUORIDE) HIGH RESOLUTION R 0389
HYPERFINE STRUCTURE CONSTANTS OF CHLORINE MONOFLUORIDE. 0375
HYPERFINE STRUCTURE OF INDIUM MONOFLUORIDE. 0489
D RAMAN SPECTRA AND THE STRUCTURE OF CRYSTALLINE THALLIUM MONOFLUORIDE. INFRARED AN 1126
INFRARED AND RAMAN SPECTRA OF CHLORINE MONOFLUORIDE. 0381
INTERNUCLEAR DISTANCE IN GASEOUS STRONTIUM MONOFLUORIDE. 1016
RIC DETERMINATION OF THE DISSOCIATION ENERGY OF BERYLLIUM MONOFLUORIDE. MASS SPECTROMET 0129
9: SUBLIMATION PRESSURE OF COPPER DIFLUORIDE. (AND COPPER MONOFLUORIDE) / STUDIES AT HIGH TEMPERATURES. PART- 0441
PART-14: VAPOR PRESSURE AND DISSOCIATION ENERGY OF SILVER MONOFLUORIDE. /ETRIC STUDIES AT HIGH TEMPERATURES. 1014
E AND THE STABILITIES OF TITANIUM DIFLUORIDE AND TITANIUM MONOFLUORIDE. /ON PRESSURES FOR TITANIUM TRIFLUORID 1145
NIUM TRIFLUORIDE AND THE DISSOCIATION ENERGY OF PLUTONIUM MONOFLUORIDE. /-2: ENTHALPY OF SUBLIMATION OF PLUTO 0884
MATRIX ISOLATION OF BROMINE MONOFLUORIDE. 0255
ITHIUM FLUORIDE, HELIUM- NEON, CARBON MONOXIDE, AND BORON MONOFLUORIDE. /E, BORON NITRIDE, BERYLLIUM OXIDE, L 0174
MICROWAVE AND MASS SPECTRA OF SULFUR MONOFLUORIDE. 1024
MICROWAVE ROTATIONAL SPECTRUM OF INDIUM MONOFLUORIDE. 0490
MICROWAVE SPECTRUM OF ALUMINUM MONOFLUORIDE. 0029
ILLETER WAVE SPECTRUM AND MOLECULAR CONSTANTS OF ALUMINUM MONOFLUORIDE. MILLIMETER AND SUBM 0033
MOLECULAR MAGNETIC PROPERTIES OF CHLORINE MONOFLUORIDE. 0384
NEAR ULTRAVIOLET ABSORPTION SPECTRUM OF SILVER MONOFLUORIDE. 1013
NEAR ULTRAVIOLET AND VISIBLE BANDS OF ANTIMONY MONOFLUORIDE. 0070
NEW SPECTRA OF OXYGEN LIKE MOLECULES. (NITROGEN MONOFLUORIDE) 0674
TO SINGLET SIGMA(+) SYSTEMS OF BOTH ISOTOPES OF CHLORINE MONOFLUORIDE. O(+) 0390
PHOTOELECTRON SPECTRUM OF CHLORINE MONOFLUORIDE. 0369
IDE. ELECTRON SPIN RESONANCE OF THE TRAPPED RADICAL XENON MONOFLUORIDE. RADIATION DAMAGE IN XENON TETRAFLUOR 1220
BITAL CONFIGURATIONS OF THE ELECTRONIC STATES OF SCANDIUM MONOFLUORIDE. REASSIGNMENT OF MOLECULAR OR 0937
ROTATIONAL ANALYSIS OF A-X1 BANDS OF LEAD MONOFLUORIDE. 0591
ROTATIONAL ANALYSIS OF B-X2 SYSTEM OF LEAD-208 MONOFLUORIDE. 0590
PLET SIGMA, B TRIPLET SIGMA: A TRIPLET PI SYSTEM OF BORON MONOFLUORIDE. /IONAL ANALYSIS OF BANDS OF THE C TRI 0153
B TRIPLET PI(1) TO X SINGLET SIGMA(+) SYSTEMS OF GALLIUM MONOFLUORIDE. /OF BANDS OF THE A TRIPLET PI ZERO(+) 0457
B TRIPLET PI(1) TO X SINGLET SIGMA(+) SYSTEMS OF THALLIUM MONOFLUORIDE. / BANDS OF THE A TRIPLET PI ZERO(+), 1111
ATIONAL ANALYSIS OF ELECTRONIC BANDS OF GASEOUS MAGNESIUM MONOFLUORIDE. ROT 0600
ROTATIONAL ANALYSIS OF FOUR SINGLET SYSTEMS OF ARSENIC MONOFLUORIDE. 0092
ROTATIONAL ANALYSIS OF SOME BANDS OF GASEOUS BARIUM MONOFLUORIDE. 0118
E SINGLET TRANSITIONS IN THE SPECTRUM OF GASEOUS SCANDIUM MONOFLUORIDE. ROTATIONAL ANALYSIS OF SOM 0930
LET PI, B DOUBLET SIGMA(+) TO X DOUBLET PI SYSTEMS OF TIN MONOFLUORIDE. / OF THE A DOUBLET SIGMA(+) TO X DOUB 1133
), BO(+), TO X SINGLET SIGMA(+) SYSTEMS OF GASEOUS SILVER MONOFLUORIDE. ROTATIONAL ANALYSIS OF THE AO(+ 1011
OF THE C SINGLET PI- X SINGLET SIGMA(+) SYSTEM OF INDIUM MONOFLUORIDE. ROTATIONAL ANALYSIS 0491
ROTATIONAL ANALYSIS OF THE C1 SYSTEM OF ANTIMONY MONOFLUORIDE. 0069
ROTATIONAL ANALYSIS OF THE C2 SYSTEM OF ANTIMONY MONOFLUORIDE. 0068
ROTATIONAL ANALYSIS OF THE GAMMA BANDS OF SILICON MONOFLUORIDE. 0999
STEM B SINGLET PI TO SINGLET SIGMA(+) OF GASEOUS SCANDIUM MONOFLUORIDE. ROTATIONAL ANALYSIS OF THE SY 0931
ROTATIONAL ANALYSIS OF THE VISIBLE EMISSION BANDS OF LEAD MONOFLUORIDE. 0587
ONAL FINE STRUCTURE OF THE TRIPLET PI STATE OF PHOSPHORUS MONOFLUORIDE. ROTATI 0787
QUARTET SIGMA(-) TO X DOUBLET PI TRANSITIONS OF GERMANIUM MONOFLUORIDE. /SIGMA(+), B DOUBLET SIGMA(+), AND A 0466
CULATIONS ON CHLORATE ION, HYDROGEN CHLORIDE AND CHLORINE MONOFLUORIDE. SCF CAL 0380
SIMPLE THEORETICAL CALCULATIONS ON TITANIUM MONOFLUORIDE. 1144
SINGLET ELECTRONIC STATES OF ALUMINUM MONOFLUORIDE. 0023
SPECTROSCOPIC AND THERMODYNAMIC PROPERTIES OF ALUMINUM MONOFLUORIDE. 0022
SPECTRUM OF SULFUR MONOFLUORIDE. 1022
CALCIUM DIFLUORIDE AND THE DISSOCIATION ENERGY OF CALCIUM MONOFLUORIDE. SUBLIMATION PRESSURE OF 0276
ROMIUM DIFLUORIDE AND THE DISSOCIATION ENERGY OF CHROMIUM MONOFLUORIDE. SUBLIMATION PRESSURE OF CH 0430
UBLIMATION PRESSURE OF MANGANESE DIFLUORIDE AND MANGANESE MONOFLUORIDE. S 0614
STABILITIES OF GASEOUS NEODYMIUM DIFLUORIDE AND NEODYMIUM MONOFLUORIDE. /RE OF NEODYMIUM TRIFLUORIDE AND THE 0643
THE B-X SYSTEM OF CARBON MONOFLUORIDE. 0285

THE C DOUBLET PI AND THE X DOUBLET SIGMA STATES OF BARIUM MONOFLUORIDE. 0120
DIATOMIC MOLECULES AND A STUDY OF THE EMISSION OF IODINE MONOFLUORIDE. THEORY OF THE DISSOCIATION OF 0493
THERMALLY EXCITED EMISSION SPECTRUM OF BARIUM MONOFLUORIDE. 0123
ASES WITH MOLECULES IN TRIPLET SIGMA STATES. (PHOSPHORUS MONOFLUORIDE) THERMODYNAMIC FUNCTIONS OF DIATOMIC G 0789
MANIUM TETRAFLUORIDE, GERMANIUM DIFLUORIDE, AND GERMANIUM MONOFLUORIDE. /PROPERTIES OF GASEOUS GERMANIUM, GER 0468
THERMODYNAMICS OF VAPORIZATION OF THALLIUM MONOFLUORIDE. 1121
ULTRAVIOLET SPECTRUM OF ARSENIC MONOFLUORIDE. 0089
H THE INTERNUCLEAR SEPARATION IN SILICON CHLORIDE AND TIN MONOFLUORIDE. / OF ELECTRONIC TRANSITION MOMENT WIT 1137
'-X' BAND SYSTEMS AND ENERGY OF DISSOCIATION IN GERMANIUM MONOFLUORIDE. VIBRATIONAL ANALYSIS OF C-X AND C 0470
VIBRATIONAL ANALYSIS OF THE SINGLET SYSTEMS OF ARSENIC MONOFLUORIDE. 0095
VIBRATIONAL ANALYSIS OF ULTRAVIOLET BANDS OF ANTIMONY MONOFLUORIDE. 0062
ONAL TRANSITION PROBABILITIES FOR THE A-X SYSTEM OF BORON MONOFLUORIDE. VIBRATI 0168
VIBRONIC SPECTRA OF YTTRIUM MONOFLUORIDE. 1315
VISIBLE EMISSION SPECTRUM OF BISMUTH MONOFLUORIDE. 0145
/TION OF THE C2 (A TRIPLET PI TO X TRIPLET PI) AND CARBON MONOFLUORIDE A DOUBLET SIGMA(+) TO X DOUBLET PI BA/ 0288
HIGH TEMPERATURE MICROWAVE SPECTROSCOPY OF ALUMINUM MONOFLUORIDE AND ALUMINUM CHLORIDE. 0028
BAND SPECTRA OF BISMUTH MONOFLUORIDE AND ANTIMONY MONOFLUORIDE. 0065
EMISSION SPECTRA OF ANTIMONY MONOFLUORIDE AND BISMUTH MONOFLUORIDE. 0064
ON THE DISSOCIATION OF LEAD MONOFLUORIDE AND BISMUTH MONOFLUORIDE. 0589
SECOND ORDER STARK EFFECT OF LINEAR MOLECULES. (CHLORINE MONOFLUORIDE AND BROMINE MONOFLUORIDE) /IONS TO THE 0385
PREDISSOCIATION IN THE ABSORPTION SPECTRA OF CARBON MONOFLUORIDE AND CARBON DIFLUORIDE. 0293
LOW EFFUSION AND MASS/ THERMODYNAMIC PROPERTIES OF CARBON MONOFLUORIDE AND CARBON DIFLUORIDE FROM MOLECULAR F 0286
ELECTRONIC STRUCTURES OF CHLORINE MONOFLUORIDE AND CHLORINE TRIFLUORIDE. 0372
CARBON MONOFLUORIDE AND C/ OSCILLATOR STRENGTHS OF CARBON MONOFLUORIDE AND COMMENTS ON HEATS OF FORMATION OF 0295
MATRIX INFRARED SPECTRUM OF OXYGEN MONOFLUORIDE AND DETECTION OF LITHIUM OXYFLUORIDE. 0755
SPECTRUM OF BROMINE MONOFLUORIDE AND ITS DISSOCIATION ENERGY. 0252
CONFIGURATION INTERACTION STUDIES OF NITROGEN MONOFLUORIDE AND ITS POSITIVE ION. 0672
ELECTRONIC SPECTRA OF PHOSPHORUS MONOFLUORIDE AND ITS POSITIVE ION. 0786
S) KRYPTON MONOFLUORIDE AND ITS POSITIVE ION. (SCF CALCULATION 0547
PTION AND EMISSION SPECTRA OF ARGON MATRIX ISOLATED XENON MONOFLUORIDE AND KRYPTON MONOFLUORIDE. ABSOR 0544
N SPECTRA OF XENON MONOBROMIDE, XENON MONOCHLORIDE, XENON MONOFLUORIDE, AND KRYPTON MONOFLUORIDE. EMISSIO 0545
HIGH TEMPERATURES. (TIN DIFLUORIDE, LEAD DIFLUORIDE, TIN MONOFLUORIDE, AND LEAD MONOFLUORIDE) /IC STUDIES AT 0592
SPECTROSCOPIC STUDIES OF SCANDIUM MONOFLUORIDE AND MAGNESIUM OXIDE. 0933
CHEMICAL STUDY OF SOME PNICOGEN MONOFLUORIDES. (NITROGEN MONOFLUORIDE AND PHOSPHORUS MONOFLUORIDE) QUANTUM 0678
STARK EFFECT FOR THALLIUM MONOFLUORIDE AND POTASSIUM FLUORIDE. 1118
DIPOLE MOMENTS OF OPEN SHELL DIATOMIC MOLECULES. (SULFUR MONOFLUORIDE AND SELENIUM MONOFLUORIDE) ELECTRIC 1019
GAS PHASE ELECTRON RESONANCE SPECTRA OF SULFUR MONOFLUORIDE AND SELENIUM MONOFLUORIDE. 1021
/ED PROPERTIES FOR THE DOUBLET PI GROUND STATES OF SULFUR MONOFLUORIDE AND SELENIUM MONOFLUORIDE AND THEIR P/ 1025
ITIVE AND NEGA/ MOLECULAR ORBITAL INVESTIGATION OF CARBON MONOFLUORIDE AND SILICON MONOFLUORIDE AND THEIR POS 0290
TIC STUDIES OF THE GRAPHITE FLUORINE COMPOUNDS OF CARBON MONOFLUORIDE AND TETRACARBON FLUORIDE. / AND SYNTHE 0296
E. PHOTOIONIZATION STUDY OF CHLORINE MONOFLUORIDE AND THE DISSOCIATION ENERGY OF FLUORIN 0378
CALCULATION OF THE DISSOCIATION ENERGY OF CALCIUM MONOFLUORIDE AND URANIUM MONOFLUORIDE. 1161
GROUND STATES OF SCANDIUM MONOFLUORIDE AND YTTRIUM MONOFLUORIDE. 0929
/MODYNAMIC PROPERTIES OF SOME ARSENIC FLUORIDES. (ARSENIC MONOFLUORIDE, ARSENIC DIFLUORIDE, ARSENIC TRIFLUOR/ 0093
AS PHASE. PROBABLE NONEXISTENCE OF XENON MONOFLUORIDE AS A CHEMICALLY BOUND SPECIES IN THE G 1221
/DROGEN HALIDES AND THE INTERHALOGEN COMPOUNDS. (CHLORINE MONOFLUORIDE, BROMINE MONOFLUORIDE, AND IODINE MON/ 0373
/ PROPERTIES, AND ELECTRIC QUADRUPOLE MOMENTS IN CHLORINE MONOFLUORIDE, BROMINE MONOFLUORIDE, BROMINE CYANID/ 0254
AFLU/ INFRARED STUDIES OF THE BROMINE FLUORIDES. (BROMINE MONOFLUORIDE, BROMINE TRIFLUORIDE, AND BROMINE PENT 0256
UMINUM TRIFLUORIDE, DIALUMINUM HEXAFLUORIDE, AND ALUMINUM MONOFLUORIDE BY MATRIX ISOLATION. /D SPECTRUM OF AL 0051
/ FACTORS AND R CENTROIDS OF SOME BAND SYSTEMS OF SILICON MONOFLUORIDE, CALCIUM MONOFLUORIDE, AND BISMUTH MO/ 0972
IFLU/ HEATS OF FORMATION OF CHLORINE FLUORIDES. (CHLORINE MONOFLUORIDE, CHLORINE DIFLUORIDE, AND CHLORINE TR/ 0370
/RBITAL STUDY OF THE GEOMETRY OF INTERHALOGENS. (CHLORINE MONOFLUORIDE, CHLORINE TRIFLUORIDE, AND CHLORINE P/ 0379
/IDE, SULFUR TETRAFLUORIDE, SULFUR HEXAFLUORIDE, CHLORINE MONOFLUORIDE, CHLORINE TRIFLUORIDE, AND CHLORINE P/ 1373
ORIDE, AND B/ PHOTOELECTRON SPECTRA OF HALIDES. (CHLORINE MONOFLUORIDE, CHLORINE TRIFLUORIDE, BROMINE MONOFLU 0376
THE IONIC MODEL OF A MOLECULE. DISSOCIATION ENERGY OF MONOFLUORIDE COMPOUNDS OF MAGNESIUM AND CALCIUM AND 0278
/INGLET SIGMA- X SINGLET SIGMA PLATINUM OXIDE AND SILICON MONOFLUORIDE (D DOUBLET SIGMA(+)- X DOUBLET PI) MO/ 0966
, BINDING ENERGY, IONIZATION POTENTIAL, ELECTRON / OXYGEN MONOFLUORIDE, DOUBLET PI HARTREE-FOCK WAVE FUNCTION 0757
SPECTROSCOPIC CONSTANTS, AND OTHER PROPERTIES OF SELENIUM MONOFLUORIDE FROM A MOLECULAR ORBITAL STUDY. /LPY, 0955
CHARGE DISTRIBUTION IN CHLORINE MONOFLUORIDE FROM CORE ELECTRON BINDING ENERGIES. 0374
DISSOCIATION ENERGY OF BORON MONOFLUORIDE FROM MASS SPECTROMETRIC STUDIES. 0161
RIDE, AND THALL/ MICROWAVE ABSORPTION SPECTRA OF ALUMINUM MONOFLUORIDE, GALLIUM MONOFLUORIDE, INDIUM MONOFLUO 0025
DRIDE, ARSENIC TR/ MOLECULAR SCF CALCULATIONS FOR GALLIUM MONOFLUORIDE, GALLIUM TRIHYDRIDE, GERMANIUM TETRAHY 0459
FTS. NMR SPECTRA OF CHLORINE TRIFLUORIDE AND CHLORINE MONOFLUORIDE. GASEOUS SPECTRA AND GAS TO LIQUID SHI 0368
IDE, AND LEAD MONOFLUORIDE. SPECTRA OF SILICON MONOFLUORIDE, GERMANIUM MONOFLUORIDE, TIN MONOFLUOR 0964
NERGY, IONIZATION POTENTIAL, ELECTRON AFFINITY, D/ OXYGEN MONOFLUORIDE. HARTREE-FOCK WAVE FUNCTION, BINDING E 0756
/OLECULAR PROPERTIES OF THE ISOELECTRONIC SERIES CHLORINE MONOFLUORIDE, HYDROGEN OXYCHLORIDE, CHLORO AMMONIA/ 0371
E. ELECTRON SPIN RESONANCE SPECTRUM OF XENON MONOFLUORIDE IN GAMMA IRRADIATED XENON TETRAFLUORID 1222
ESR SPECTRUM OF BROMINE OXIDE, IODINE OXIDE AND SELENIUM MONOFLUORIDE IN J-5/2 ROTATIONAL LEVELS. 0954
COMPLEX BAND SYSTEM OF MANGANESE MONOFLUORIDE IN THE NEAR ULTRAVIOLET. 0610
R INFRARED REGIONS. SPECTRUM OF ARSENIC MONOFLUORIDE IN THE NEAR VACUUM ULTRAVIOLET AND NEA 0091
MAGNETIC RESONANCE SPECTRUM AND DIPOLE MOMENT OF NITROGEN MONOFLUORIDE IN THE SINGLET DELTA STATE. /TRON PARA 0673
EMISSION SPECTRUM OF ALUMINUM MONOFLUORIDE IN THE VACUUM ULTRAVIOLET. 0030
ANALYSIS OF VIBRATIONAL AND ROTATIONAL STRUCTURE OF TIN MONOFLUORIDE IN THE 2500 ANGSTROM REGION. 1138
ROTATIONAL ANALYSIS OF THE B-X SYSTEM OF LEAD MONOFLUORIDE IN ULTRAVIOLET. 0586
A NEW ELECTRONIC TRANSITION OF THE MERCURY MONOFLUORIDE MOLECULE. 0618
A QUARTET SIGMA TO DOUBLET PI TRANSITION OF THE SILICON MONOFLUORIDE MOLECULE. 0978
LET SIGMA TO X DOUBLET PI ELECTRON TRANSITON IN A SILICON MONOFLUORIDE MOLECULE. / DIPOLE MOMENT OF AN A DOUB 0970
ELECTRONIC SPECTRUM OF THE BORON MONOFLUORIDE MOLECULE. 0155
DON FACTORS AND R CENTROIDS FOR THE C-X SYSTEM OF SILICON MONOFLUORIDE MOLECULE. FRANCK-CON 0975
NEW BANDS OF THE CARBON MONOFLUORIDE MOLECULE. 0292
ROTATIONAL ANALYSIS OF C3 SYSTEM OF THE ANTIMONY MONOFLUORIDE MOLECULE. 0057
LECTED BANDS FROM THE ELECTRONIC SPECTRUM OF THE LUTETIUM MONOFLUORIDE MOLECULE. ROTATIONAL ANALYSIS OF SE 0599
ATIONAL ANALYSIS OF THE ULTRAVIOLET BANDS OF THE ANTIMONY MONOFLUORIDE MOLECULE. ROT 0063
TIONAL ANALYSIS OF THE VISIBLE BAND SYSTEM OF THE BISMUTH MONOFLUORIDE MOLECULE. ROTA 0146
AT 14900 CM-1 OF A 3 SIGMA TO 3 SIGMA SYSTEM OF THE BORON MONOFLUORIDE MOLECULE. /IONAL ANALYSIS OF THE BAND 0156
ROTATIONAL ANALYSIS OF THE BETA BAND SYSTEM OF SILICON MONOFLUORIDE MOLECULE. 0976
G DOUBLET DELTA 5/2- X DOUBLET PI 3/2 SYSTEMS OF THE TIN MONOFLUORIDE MOLECULE. /A 5/2- X DOUBLET PI 3/2 AND 1136
ROTATIONAL CONSTANTS OF THE B STATE OF THE CARBON MONOFLUORIDE MOLECULE. 0294
ND OF THE A DOUBLET SIGMA- X DOUBLET PI SYSTEM OF THE TIN MONOFLUORIDE MOLECULE. /TOPIC SHIFT IN THE (1,0) BA 1135
EORETICAL STUDY OF THE SPECTROSCOPIC STATES OF THE CARBON MONOFLUORIDE MOLECULE. TH 0287

VIOLET AND ULTRAVIOLET SPECTRUM OF BARIUM MONOFLUORIDE MOLECULE. 0122
ABSORPTION SPECTRUM OF THE BISMUTH MONOFLUORIDE MOLECULE IN THE ULTRAVIOLET REGION. 0143
A NEW BAND SYSTEM OF THE ANTIMONY MONOFLUORIDE MOLECULE IN 4050-5450 ANGSTROM REGION. 0058
RIDE, AND TANTALUM D/ THERMODYNAMIC PROPERTIES OF NIOBIUM MONOFLUORIDE, NIOBIUM DIFLUORIDE, TANTALUM MONOFLUO 0655
/ SATURATED LIQUID AND SATURATED VAPOR ENTROPY. (CHLORINE MONOFLUORIDE, OXYGEN MONOFLUORIDE, PHOSPHORUS TRIF/ 0391
/HOSPHORUS(4), AND SOME PHOSPHORUS FLUORIDES. (PHOSPHORUS MONOFLUORIDE, PHOSPHORUS DIFLUORIDE, PHOSPHORUS TR/ 0788
MICROWAVE SPECTRUM OF THE SULFUR MONOFLUORIDE RADICAL. 1018
AND IODI/ GAS PHASE ELECTRON RESONANCE SPECTRA OF SULFUR MONOFLUORIDE, SELENIUM MONOFLUORIDE, SELENIUM OXIDE 1020
POTENTIAL CURVES FOR SOME DIATOMIC MOLECULES. (BERYLLIUM MONOFLUORIDE, SILICON MONOFLUORIDE, AND TIN MONOFL/ 0130
FLUORIDE, AND SULFUR MONOFLUORIDE/ PROPERTIES OF NITROGEN MONOFLUORIDE, SILICON MONOFLUORIDE, PHOSPHORUS MONO 0973
/OIONIZATION OF HIGH TEMPERATURE VAPORS. PART-4: THALLIUM MONOFLUORIDE, THALLIUM MONOCHLORIDE, AND THALLIUM / 1112
/EATS OF FORMATION OF GASEOUS URANIUM FLUORIDES. (URANIUM MONOFLUORIDE, URANIUM DIFLUORIDE, URANIUM TRIFLUOR/ 1160
ALUMINUM MONOFLUORIDE. (VAPOR PRESSURE) 0031
OF THE A SINGLET PI- X SINGLET SIGMA(+) BANDS OF BORON-11 MONOFLUORIDE WITH A VACUUM ECHELLE SPECTROGRAPH. / 0170
/M, AND LANTHANUM HALIDES. (SCANDIUM DIFLUORIDE, SCANDIUM MONOFLUORIDE, YTTRIUM DIFLUORIDE, YTTRIUM MONOFLUO/ 0941
ONOFLUORIDE. ELECTRONIC STATES OF GASEOUS SCANDIUM MONOFLUORIDE, YTTRIUM MONOFLUORIDE, AND LANTHANUM M 0928
ONOFLU/ DISSOCIATION ENERGIES AND STABILITIES OF SCANDIUM MONOFLUORIDE, YTTRIUM MONOFLUORIDE, AND LANTHANUM M 0935
/VALUATION OF MOLECULAR VIBRATION FREQUENCIES OF SCANDIUM MONOFLUORIDE, YTTRIUM MONOFLUORIDE, AND LANTHANUM / 0936
ONOFLUORIDE. VIBRONIC SPECTRUM OF SCANDIUM MONOFLUORIDE, YTTRIUM MONOFLUORIDE, AND LANTHANUM M 0938
ATION POTENTIAL, ELECTRON AFFINITY, DIPOLE MOMEN/ ARSENIC MONOFLUORIDE, 3 SIGMA. DISSOCIATION ENTHALPY, IONIZ 0094
DISSOCIATION ENERGIES OF THE ALKALINE EARTH MONOFLUORIDES. 0119
MICROWAVE ROTATIONAL SPECTRUM OF THALLIUM MONOFLUORIDES. 1119
LUORIDES AND CHLORIDES OF GALLIUM, INDIUM, AND THALLIUM. (MONOFLUORIDES) /DISSOCIATION ENERGIES OF DIATOMIC F 0458
M FLUORIDES AT HIGH TEMPERATURES. PART-3: DIFLUORIDES AND MONOFLUORIDES) /C PROPERTIES OF NIOBIUM AND TANTALU 0654
S FROM SPECTROSCOPIC DATA. (CHLORINE, BROMINE, AND IODINE MONOFLUORIDES) /ERTIES OF THE DIATOMIC INTERHALOGEN 0253
/H TEMPERATURES. PART-2: THE DISSOCIATION ENERGIES OF THE MONOFLUORIDES AND DIFLUORIDES OF SILICON AND GERMA/ 0463
S MONOFLUORIDE) QUANTUM CHEMICAL STUDY OF SOME PNICOGEN MONOFLUORIDES. (NITROGEN MONOFLUORIDE AND PHOSPHORU 0678
FORCE FIELDS AND MEAN AMPLITUDES FOR ALUMINUM TRIFLUORIDE MONOMER AND DIMER. HARMONIC 0038
ELECTRON SPECTROSCOPY OF HIGH TEMPERATURE VAPORS. PART-3: MONOMER AND DIMER SPECTRA OF THALLOUS FLUORIDE. /TO 1117
UM PENTAFLUORIDE, AND ANTIMONY PENTAFLUORIDE. TH/ POLYMER MONOMER EQUILIBRIA IN NIOBIUM PENTAFLUORIDE, TANTAL 0076
ERATURE ANOMALIES. ALUMINUM AND ALKALI HALIDE MONOMERS MEAN AMPLITUDES OF VIBRATION WITH LOW TEMP 0024
/ AND NORMAL COORDINATE ANALYSES OF GERMANIUM DIFLUORIDE, MONOMERIC GERMANIUM DIFLUORIDE, AND DIMERIC GERMAN/ 0475
RATURE. EVIDENCE FOR A TRIGONAL BIPYRAMIDAL SHAPE FOR THE MONOMERIC SPECIES. /AS PHASE AS A FUNCTION OF TEMPE 0077
HEPTAFLUORIDE. MOSSBAUER STUDY OF IODINE PENTAFLUORIDE AND IODINE 0503
LYBDENUM HEXAFLUORIDE, TUNGSTEN HEXAFLUORIDE, / MOLECULAR MOTION AND FLUORINE RELAXATION IN THE LIQUIDS OF MO 0638
/IS OF ELECTRON DIFFRACTION INTENSITIES. PART-2: INTERNAL MOTION AND MEAN GEOMETRY DEDUCED BY ELECTRON DIFFR/ 1291
HEXAFLUORIDES. MEAN AMPLITUDES OF THERMAL MOTION AND SHRINKAGES OF CHEMICAL BONDS: OCTAHEDRAL 1460
ROSCOPIC DATA. PENTACOORDINATED MOLE/ ANISOTROPIC THERMAL MOTION OF TRIGONAL BIPYRAMIDAL MOLECULES FROM SPECT 0853
/CTRIC FIELD DEFLECTION OF MOLECULES WITH LARGE AMPLITUDE MOTION. (XENON HEXAFLUORIDE, IODINE HEPTAFLUORIDE,/ 1281
BENDING MOTIONS IN THE DIHALIDES OF GROUP-2 METALS. 1380
ION 1700-2500 ANGSTROMS. (ALUMINU/ RECENT DEVELOPMENTS IN MULTIPLE BEAM INTERFEROMETRY IN THE ULTRAVIOLET REG 0042
RON DIFFRACTION PATTERNS FOR THE RHENIUM HEXAFLUORIDE MO/ MULTIPLE INTRAMOLECULAR SCATTERING EFFECTS ON ELECT 0911
RGY LEVEL STRUCTURE OF XENON HEXAFLUORIDE. MULTIPLE SCATTERING X(ALPHA) CALCULATION OF THE ENE 1304
LEMENT 110 HEXAFLUORIDE CALCULATED BY THE SELF CONSISTENT MULTIPLE SCATTERING X-ALPHA METHOD. /FLUORIDE AND E 0881
MULTIPLET SPLITTING IN 1S HOLE STATES OF MOLECULES. 0683
OF XY4 TETRAHEDRAL MOLECULES AND IONS. (UREY-BRADLEY AND MVFF) NORMAL COORDINATE TREATMENT 1418
IN TETRAFLUORIDE AND HAFNIUM TETRAFLUORI/ IONICITY OF THE MX BOND IN THE TETRAHALIDES OF GROUP-4 ELEMENTS. (T 1141
REOCHEMICAL NONRIGIDITY OF THE TRIGONAL BIPYRAMIDAL MODEL MX3Y2. /ULES. PART-14: MOLECULAR VIBRATIONS AND STE 0854
ATION OF THE METHOD OF PROGRESSIVE RIGIDITY TO OCTAHEDRAL MX6 MOLECULES. APPLIC 1441
AB INITIO AND MZDO WAVE FUNCTIONS FOR SULFUR HEXAFLUORIDE. 1066

N

POTENTIAL ENERGY CURVES AND NATURE OF BINDING IN GROUP-3A MONOHALIDES. 0032
NATURE OF BONDS IN THALLIUM MONOHALIDE MOLECULES. 1122
XENON DIFLUORIDE AND THE NATURE OF THE XENON- FLUORINE BOND. 1223
OLECULE BY THE MO LCAO SELF CONSISTENT FIELD METHOD IN AN NDDO APPROXIMATION. / OF A CHLORINE PENTAFLUORIDE M 0415
E ON THE BASIS OF A CALCULATION BY THE MO LCAO SCF IN THE NDDO APPROXIMATION. /HE MOLECULE CHLORINE DIFLUORID 0382
/LAP METHODS ILLUSTRATED BY THE XENON FLUORIDES. MODIFIED NDDO-2 (ALPHA, BETA) METHOD AS A USEFUL APPROXIMAT/ 1231
/OMPOUNDS OF HALOGENS AND INERT GASES BY THE NONEMPIRICAL NDDO-2 (ALPHA, BETA) METHOD. PART-3: ENERGY LEVELS/ 0016
OFLUORIDE AND SILICON MONOFLUORIDE AND THEIR POSITIVE AND NEGATIVE IONS. /ORBITAL INVESTIGATION OF CARBON MON 0290
/LUORIDE AND SELENIUM MONOFLUORIDE AND THEIR POSITIVE AND NEGATIVE IONS. A COMPARISON OF THE THEORETICAL AND/ 1025
HIGH TEMPERATURE NEGATIVE IONS. GASEOUS GROUP-3 FLUORIDES. 0047
/D. PART-2: COMPARATIVE ANALYSIS OF DIFFERENT SCHEMES OF NEGLECT OF DIFFERENTIAL OVERLAP METHODS ILLUSTRATE/ 1231
E VAPOR PHASE HALIDES OF GALLIUM, YTTRIUM, LANTHANUM, AND NEODYMIUM. / STUDY OF THE MOLECULAR STRUCTURE OF TH 1317
ENTROPIES AND ENTHALPIES OF SUBLIMATION OF NEODYMIUM AND TERBIUM TRIFLUORIDES. 0645
INFRARED SPECTRA. (SCANDIUM, YTTRIUM, LANTHANUM, CERIUM, NEODYMIUM, EUROPIUM, AND GADOLINIUM TRIFLUORIDES) / 0943
RAPHIC INVESTIGATION OF THE STRUCTURE OF THE MOLECULES OF NEODYMIUM HALIDES. ELECTRONOG 0644
S NEODYMIUM DIFLUORIDE AND NEODY/ SUBLIMATION PRESSURE OF NEODYMIUM TRIFLUORIDE AND THE STABILITIES OF GASEOU 0643
ENERGY LEVELS AND SPECTROSCOPIC PARAMETERS OF NEODYMIUM TRIFLUORIDES. 0646
BORON NITRIDE, BERYLLIUM OXIDE, LITHIUM FLUORIDE, HELIUM- NEON, CARBON MONOXIDE, AND BORON MONOFLUORIDE. /E, 0174
MATION OF SOME HALIDES OF THORIUM, PROTACTINIUM, URANIUM, NEPTUNIUM, AND AMERICIUM. (ACTINIDE) HEATS OF FOR 1335
ELECTRONIC STRUCTURE OF NEPTUNIUM HEXAFLUORIDE. 0648
RIDES. (URANIUM HEXAFLUORIDE, PLUTONIUM HEXAFLUORIDE, AND NEPTUNIUM HEXAFLUORIDE) /CENT STUDIES WITH HEXAFLUO 1205
INFRARED SPECTRA OF NEPTUNIUM HEXAFLUORIDE AND PLUTONIUM HEXAFLUORIDE. 0649
/EXAFLUORIDE, IRIDIUM HEXAFLUORIDE, URANIUM HEXAFLUORIDE, NEPTUNIUM HEXAFLUORIDE, AND PLUTONIUM HEXAFLUORIDE. 1155
/H, INFRARED ACTIVE FUNDAMENTAL FOR URANIUM HEXAFLUORIDE, NEPTUNIUM HEXAFLUORIDE, AND PLUTONIUM HEXAFLUORIDE. 1192
CITY, ENTHALPY OF FUSION, AND THERMODYNAMIC PROPERTIES OF NEPTUNIUM HEXAFLUORIDE FROM 7 TO 350 DEGREES K. /PA 0650
E. THERMODYNAMIC PROPERTIES OF URANIUM PENTAFLUORIDE, NEPTUNIUM PENTAFLUORIDE, AND PLUTONIUM PENTAFLUORID 1177
CRYSTAL AND MOLECULAR STRUCTURE OF XENON DIFLUORIDE BY NEUTRON DIFFRACTION. 1239
CRYSTAL AND MOLECULAR STRUCTURE OF XENON TETRAFLUORIDE BY NEUTRON DIFFRACTION. 1260
-6: PRECISE STRUCTURAL PARAMETERS IN COPPER DIFLUORIDE BY NEUTRON DIFFRACTION. STRUCTURES OF FLUORIDES. PART 0443
STRUCTURES OF YTTRIUM AND BISMUTH TRIFLUORIDES BY NEUTRON DIFFRACTION. 1318
NEUTRON DIFFRACTION INVESTIGATION OF ORTHORHOMBIC L 0593
EAD DIFLUORIDE. NEUTRON DIFFRACTION STUDY OF THE CRYSTAL STRUCTURE 1163
OF URANIUM TRIFLUORIDE. NEUTRON POWDER DIFFRACTION PROFILE STUDIES OF URANI 1202
UM HEXAFLUORIDE AT 193/ STRUCTURES OF FLUORIDES. PART-10: VAPOR PRESSURE OF NICKEL DIFLUORIDE. 0652
VAPOR PRESSURES OF BERYLLIUM DIFLUORIDE AND NICKEL DIFLUORIDE. 0135
SPECTRA OF MATRIX ISOLATED NICKEL DIFLUORIDE AND NICKEL DICHLORIDE. 0653
/ECTRA AND GEOMETRY OF MATRIX ISOLATED COBALT DIFLUORIDE, NICKEL DIFLUORIDE, COPPER DIFLUORIDE, AND ZINC DIF/ 0437

OXIDE, MOLECULAR NITROGEN/ INTERACTION OF MATRIX ISOLATED NICKEL FLUORIDE AND NICKEL CHLORIDE WITH CARBON MON 1324
BAND SPECTRUM OF NICKEL MONOFLUORIDE. 0651
. PART-3: DIFLUORIDES AND MO/ THERMODYNAMIC PROPERTIES OF NIOBIUM AND TANTALUM FLUORIDES AT HIGH TEMPERATURES 0654
. PART-2: TETRAFLUORIDES AND/ THERMODYNAMIC PROPERTIES OF NIOBIUM AND TANTALUM FLUORIDES AT HIGH TEMPERATURES 0657
. PART-1: PENTAFLUORIDES. THERMODYNAMIC PROPERTIES OF NIOBIUM AND TANTALUM FLUORIDES AT HIGH TEMPERATURES 0665
OBIUM AND TANTALUM PENTAFLUORIDES. HALIDES OF NIOBIUM AND TANTALUM. PART-3: VAPOR PRESSURES OF NI 0663
THERMODYNAMIC PROPERTIES OF NIOBIUM AND TANTALUM PENTAFLUORIDES. 0669
E. VAPOR DENSITY OF NIOBIUM PENTA/ VIBRATIONAL SPECTRA OF NIOBIUM AND TANTALUM PENTAFLUORIDES IN THE GAS PHAS 0660
E, AND MOL/ THERMODYNAMIC FUNCTIONS OF ANTIMONY FLUORIDE, NIOBIUM BROMIDE, TANTALUM CHLORIDE, TANTALUM BROMID 0075
MONOFLUORIDE, AND TANTALUM D/ THERMODYNAMIC PROPERTIES OF NIOBIUM MONOFLUORIDE, NIOBIUM DIFLUORIDE, TANTALUM 0655
Y, AND HEATS OF TRANSITION OF MOLYBDENUM HEXAFLUORIDE AND NIOBIUM PENTAFLUORIDE. HEAT CAPACITY, ENTROP 0662
INFRARED SPECTRUM OF MATRIX ISOLATED NIOBIUM PENTAFLUORIDE. 0659
INFRARED SPECTRUM OF NIOBIUM PENTAFLUORIDE. 0661
L FIELD AND FORCE CONSTANTS OF ANTIMONY PENTAFLUORIDE AND NIOBIUM PENTAFLUORIDE. POTENTIA 0086
DIFLU/ NIOBIUM-93 NUCLEAR QUADRUPOLE RESONANCE STUDIES OF NIOBIUM PENTAFLUORIDE AND ITS COMPLEXES WITH XENON 0664
Y OF COMPLEX FORMATION BETWEEN ANTIMONY PENTAFLUORIDE AND NIOBIUM PENTAFLUORIDE AND TANTALUM PENTAFLUORIDE. / 0079
ENTHALPIES OF FORMATION OF NIOBIUM PENTAFLUORIDE AND TANTALUM PENTAFLUORIDE. 0667
RAMAN SPECTRA OF LIQUID NIOBIUM PENTAFLUORIDE AND TANTALUM PENTAFLUORIDE. 0670
ELECTRON DIFFRACTION INVESTIGATION OF NIOBIUM PENTAFLUORIDE AND TANTALUM PENTAFLUORIDE. 0671
/Y OF THE MOLECULAR STRUCTURES OF VANADIUM PENTAFLUORIDE, NIOBIUM PENTAFLUORIDE, AND TANTALUM PENTAFLUORIDE. 1217
ANTIMONY PENTAFLUORIDE. TH/ POLYMER MONOMER EQUILIBRIA IN NIOBIUM PENTAFLUORIDE, TANTALUM PENTAFLUORIDE, AND 0076
MASS SPECTRUM AND COMPOSITION OF NIOBIUM PENTAFLUORIDE VAPOR. 0666
INFRARED SPECTRA OF NIOBIUM TETRAFLUORIDE. 0658
M TRIFLUORIDE, AND TANTALUM / THERMODYNAMIC PROPERTIES OF NIOBIUM TRIFLUORIDE, NIOBIUM TETRAFLUORIDE, TANTALU 0656
S. THERMODYNAMIC PROPERTIES OF NIOBIUM(V), TANTALUM(V), AND MOLYBDENUM(V) FLUORIDE 0668
NIOBIUM PENTAFLUORIDE AND ITS COMPLEXES WITH XENON DIFLU/ NIOBIUM-93 NUCLEAR QUADRUPOLE RESONANCE STUDIES OF 0664
/ICKEL CHLORIDE WITH CARBON MONOXIDE, MOLECULAR NITROGEN, NITRIC OXIDE, AND MOLECULAR OXYGEN AND OF CALCIUM / 1324
FLUORESCENCE SPECTRA OF PHOSPHORUS NITRIDE AND BORON FLUORIDE. 0167
/IUM, BERYLLIUM, BORON, CARBON, NITROGEN, FLUORINE, BORON NITRIDE, BERYLLIUM OXIDE, LITHIUM FLUORIDE, HELIUM/ 0174
ON CALCULATION FOR THE GROUND STATE OF OXYGEN DIFLUORIDE, NITRITE ION AND CYANIDE ION. /NFIGURATION INTERACTI 0763
DOUBLE ZETA SCF CALCULATIONS FOR NITRITE ION AND OXYGEN DIFLUORIDE. 0762
/ASIS SET SCF CALCULATIONS FOR THE GROUND STATE OF OZONE, NITRITE ION, NITROGEN OXYFLUORIDE, AND OXYGEN DIFL/ 0776
/ HALIDES. PART-3: TRIFLUORIDES AND OXIDE TRIFLUORIDES OF NITROGEN AND PHOSPHORUS, AND PHOSPHORUS OXIDE TRIC/ 0701
ON ORBITAL RADIATION TO MOLECULAR SPECTROSCOPY. MOLECULAR NITROGEN AND SULFUR HEXAFLUORIDE. /TION OF SYNCHROT 1083
VALENCE EXCITED STATES OF NITROGEN, CARBON MONOXIDE, AND BORON FLUORIDE. 0169
/IBUTIONS AND CHEMICAL BINDING. PART-3: THE ISOELECTRONIC NITROGEN, CARBON MONOXIDE, BORON FLUORIDE, AND CAR/ 0150
DIES OF PHOSPHORUS DIFLUORIDE, PHOSPHORUS DICHLORIDE, AND NITROGEN DICHLORIDE IN LOW TEMPERATURE MATRICES. /U 0793
OR FREE RADICAL MICROWAVE ROTATIONAL ABSORPTION STUDIES. (NITROGEN DIFLUORIDE) A CAVITY SEARCH SPECTROMETER F 0689
ABSORPTION SPECTRUM OF NITROGEN DIFLUORIDE. 0685
ELECTRON DIFFRACTION STUDIES AT ELEVATED TEMPERATURES. (NITROGEN DIFLUORIDE) 0679
INFRARED SPECTRUM OF NITROGEN DIFLUORIDE. 0692
/F VIBRATION FOR THE BORON DIFLUORIDE, CARBON DIFLUORIDE, NITROGEN DIFLUORIDE, ALUMINUM DIFLUORIDE, PHOSPHOR/ 1384
DIFFRACTION STUDY OF THE STRUCTURES OF NITROGEN DIFLUORIDE AND DINITROGEN TETRAFLUORIDE. 0680
LECULES, AND ES/ AB INITIO MOLECULAR ORBITAL STUDY OF THE NITROGEN DIFLUORIDE AND DINITROGEN TETRAFLUORIDE MO 0688
CALCULATION OF FORCE CONSTANTS FOR NITROGEN DIFLUORIDE AND NITROGEN TRIFLUORIDE. 0693
ERYLLIUM DIFLUORIDE, BORON DIFLUORIDE, CARBON DIFLUORIDE, NITROGEN DIFLUORIDE, AND OXYGEN DIFLUORIDE. (SCF) / 0141
CE CONSTANTS IN DIFLUORIDE MOLECULES. (OXYGEN DIFLUORIDE, NITROGEN DIFLUORIDE, CARBON DIFLUORIDE) /ACTION FOR 0299
MICROWAVE ROTATIONAL SPECTRUM OF THE NITROGEN DIFLUORIDE FREE RADICAL. 0690
/THE UNRESTRICTED HARTREE-FOCK METHOD. (BORON DIFLUORIDE, NITROGEN DIFLUORIDE, OXYGEN DIFLUORIDE, CARBON DIF/ 0176
ELECTRONIC ABSORPTION SPECTRUM OF THE NITROGEN DIFLUORIDE RADICAL. 0694
ELECTRONIC STRUCTURE OF THE NITROGEN DIFLUORIDE RADICAL. 0682
INFRARED SPECTRUM AND THERMODYNAMIC FUNCTIONS OF THE NITROGEN DIFLUORIDE RADICAL. 0686
INFRARED SPECTRUM AND STRUCTURE OF THE NITROGEN DIFLUORIDE RADICAL. 0687
MICROWAVE SPECTRUM OF NITROGEN DIFLUORIDE RADICAL. 0681
RAMAN SPECTRUM OF THE NITROGEN DIFLUORIDE RADICAL. 0695
DIFLUORIDE. MOLECULAR ORBITAL STUDY OF NITROGEN DIHYDRIDE, NITROGEN DIOXIDE, AND NITROGEN 0684
ON. PART-2: BORON TRIFLUORIDE, SILICON TETRAFLUORIDE, AND NITROGEN DIOXIDE. /S FROM HEAVY ISOTOPIC SUBSTITUTI 0232
NITROGEN FLUORIDES AND THEIR INORGANIC DERIVATIVES. 0703
PHYSICAL CHEMICAL PROPERTIES OF INORGANIC NITROGEN FLUORIDES. (NITROGEN TRIFLUORIDE) 0713
THERMOCHEMICAL AND THERMODYNAMIC PROPERTIES OF INORGANIC NITROGEN FLUORIDES. (NITROGEN TRIFLUORIDE) 0696
DINITROGEN TETRAFLUORIDE. STRENGTH OF THE NITROGEN FLUORINE BOND IN NITROGEN TRIFLUORIDE AND 0712
/HE DIATOMIC MOLECULES LITHIUM, BERYLLIUM, BORON, CARBON, NITROGEN, FLUORINE, BORON NITRIDE, BERYLLIUM OXIDE/ 0174
NORMAL COORDINATE ANALYSIS OF SOME NITROGEN HALIDES. (NITROGEN TRIFLUORIDE) 0730
B SINGLET SIGMA(+) TO X TRIPLET SIGMA(-) BAND SYSTEM OF NITROGEN MONOFLUORIDE. 0675
TION OF THE SPECTROSCOPIC CONSTANTS FOR SEVERAL STATES OF NITROGEN MONOFLUORIDE. CALCULA 0676
S FOR THE 6 SINGLET SIGMA- X TRIPLET SIGMA BAND SYSTEM OF NITROGEN MONOFLUORIDE. /NDON FACTORS AND R CENTROID 0677
NEW SPECTRA OF OXYGEN LIKE MOLECULES. (NITROGEN MONOFLUORIDE) 0674
CONFIGURATION INTERACTION STUDIES OF NITROGEN MONOFLUORIDE AND ITS POSITIVE ION. 0672
QUANTUM CHEMICAL STUDY OF SOME PNICOGEN MONOFLUORIDES. (NITROGEN MONOFLUORIDE AND PHOSPHORUS MONOFLUORIDE) 0678
TRON PARAMAGNETIC RESONANCE SPECTRUM AND DIPOLE MOMENT OF NITROGEN MONOFLUORIDE IN THE SINGLET DELTA STATE. / 0673
ORUS MONOFLUORIDE, AND SULFUR MONOFLUORIDE/ PROPERTIES OF NITROGEN MONOFLUORIDE, SILICON MONOFLUORIDE, PHOSPH 0973
/RIDE AND NICKEL CHLORIDE WITH CARBON MONOXIDE, MOLECULAR NITROGEN, NITRIC OXIDE, AND MOLECULAR OXYGEN AND O/ 1324
CALCULATION OF RYDBERG LEVELS IN NITROGEN OXIDE AND BORON FLUORIDE. 0164
NITIO STUDY OF THE FORCE CONSTANTS OF INORGANIC MOLECULES NITROGEN OXYFLUORIDE AND NITROGEN TRIFLUORIDE. AB I 0734
CALCULATIONS FOR THE GROUND STATE OF OZONE, NITRITE ION, NITROGEN OXYFLUORIDE, AND OXYGEN DIFLUORIDE. /T SCF 0776
TION STUDY OF THE STRUCTURES OF NITROGEN DIFLUORIDE AND DINITROGEN TETRAFLUORIDE. DIFFRAC 0680
THE NITROGEN FLUORINE BOND IN NITROGEN TRIFLUORIDE AND DINITROGEN TETRAFLUORIDE. STRENGTH OF 0712
/MOLECULAR ORBITAL STUDY OF THE NITROGEN DIFLUORIDE AND DINITROGEN TETRAFLUORIDE MOLECULES, AND ESR HYPERFIN/ 0688
VIBRATIONAL FORCE CONSTANTS OF NONPLANAR XY3 MOLECULES. (NITROGEN TRIFLUORIDE) A BENT BOND MODEL FOR THE 0704
UTE INFRARED INTENSITIES OF THE FUNDAMENTAL VIBRATIONS OF NITROGEN TRIFLUORIDE. ABSOL 0736
IDES IN THE NEAR ULTRAVIOLET REGION. (KRYPTON DIFLUORIDE, NITROGEN TRIFLUORIDE) /ION SPECTRA OF CERTAIN FLUOR 0560
AN ELECTRON DIFFRACTION INVESTIGATION OF NITROGEN TRIFLUORIDE. 0738
FROM THE BAND CONTOURS OF HOT BANDS FOR A SYMMETRIC TOP. (NITROGEN TRIFLUORIDE) /ARMONIC POTENTIAL CONSTANTS 0733
R INFRARED BAND CONTOURS: APPLICATION TO THE NU-4 BAND OF NITROGEN TRIFLUORIDE. CALCULATION OF PERPENDICULA 0747
ENTHALPY OF DISSOCIATION OF NITROGEN TRIFLUORIDE. 0742
FAR INFRARED SPECTRUM OF NITROGEN TRIFLUORIDE. 0702
FORCE FIELD AND MOLECULAR CONSTANTS OF NITROGEN TRIFLUORIDE. 0735
RESOLUTION STUDY OF L-TYPE RESONANCE IN THE NU-4 BAND OF NITROGEN TRIFLUORIDE. HIGH 0731
N TETRAFLUORIDE, CARBON TETRAFLUORIDE, BORON TRIFLUORIDE, NITROGEN TRIFLUORIDE) /OLYATOMIC FLUORIDES. (SILICO 0186
INFRARED AND RAMAN SPECTRUM OF NITROGEN TRIFLUORIDE. 0724

INFRARED BAND SHAPES AND VIBRATIONS OF NITROGEN TRIFLUORIDE. 0699
INFRARED INTENSITIES AND FORCE FIELD FOR NITROGEN TRIFLUORIDE. 0714
INFRARED SPECTRUM AND MOLECULAR CONSTANTS OF NITROGEN TRIFLUORIDE. 0728
, VAPOR PRESSURE AND CRITICAL TEMPERATURE AND PRESSURE OF NITROGEN TRIFLUORIDE. LIQUID DENSITY 0710
LONE PAIR MODEL AND THE VIBRATIONAL FORCE CONSTANTS OF NITROGEN TRIFLUORIDE. 0705
TER WAVE SPECTRUM AND CENTRIFUGAL DISTORTION CONSTANTS OF NITROGEN TRIFLUORIDE. MILLIME 0718
F CHARGE DISTRIBUTION USING A MINIMAL SLATER-TYPE BASIS. (NITROGEN TRIFLUORIDE) / PART-3: AB INITIO STUDIES O 0707
POTENTIAL FUNCTION OF NITROGEN TRIFLUORIDE. 0737
RAMAN SPECTRUM OF GASEOUS NITROGEN TRIFLUORIDE. 0740
OF EQUILIBRIUM GEOMETRIES AND QUADRATIC FORCE CONSTANTS. (NITROGEN TRIFLUORIDE) /RT-5: AB INITIO CALCULATION 0720
SIGNS OF L DOUBLING CONSTANTS IN NITROGEN TRIFLUORIDE. 0743
THERMODYNAMIC PROPERTIES OF NITROGEN TRIFLUORIDE. 0739
CUUM ULTRAVIOLET ABSORPTION SPECTRUM AND DIPOLE MOMENT OF NITROGEN TRIFLUORIDE. VA 0725
TYPE RESONANCE IN AN OVERTONE BAND. THE 2NU-4 SPECTRUM OF NITROGEN TRIFLUORIDE. 1- 0732
INFRARED SPECTRUM OF NITROGEN TRIFLUORIDE AND FORCE FIELD. 0697
, AND A REASSIGNMENT OF THE SPE/ PHOTOELECTRON SPECTRA OF NITROGEN TRIFLUORIDE AND NITROGEN OXIDE TRIFLUORIDE 0700
GREEN'S FUNCTION ANALYSIS OF NITROGEN TRIFLUORIDE AND NITROGEN TRICHLORIDE. 0729
OPY IN THE REGION FROM TWO TO THREE MILLIMETERS, PART-2: (NITROGEN TRIFLUORIDE AND PHOSPHORUS TRIFLUORIDE) /C 0711
INFRARED SPECTRA OF NITROGEN TRIFLUORIDE AND PHOSPHORUS TRIFLUORIDE. 0748
/ MOLECULAR QUADRUPOLE MOMENTS IN PHOSPHORUS TRIFLUORIDE, NITROGEN TRIFLUORIDE, AND PHOSPHORYL, THIONYL AND / 0744
TETRAFLUORID/ VAPOR PRESSURE OF CARBON TETRAFLUORIDE AND NITROGEN TRIFLUORIDE AND THE TRIPLE POINT OF CARBON 0345
/AND THERMODYNAMIC FUNCTIONS OF SOME GROUP-5 TRIHALIDES. (NITROGEN TRIFLUORIDE, ARSENIC TRIFLUORIDE, AND PHO/ 0719
/IGUITIES IN THE HARMONIC FORCE FIELDS OF XY3 MOLECULES. (NITROGEN TRIFLUORIDE, BORON TRIFLUORIDE, PHOSPHORU/ 0709
/ ISOLATION STUDY OF THE VACUUM ULTRAVIOLET PHOTOLYSIS OF NITROGEN TRIFLUORIDE. ELECTRONIC SPECTRUM OF THE N/ 0691
EAR QUADRUPOLE MOMENT OF NITROGEN-14 AND THE STRUCTURE OF NITROGEN TRIFLUORIDE FROM MICROWAVE SPECTRA. NUCL 0741
ABSORPTION SPECTRA OF MATRIX ISOLATED NITROGEN TRIFLUORIDE FROM 18 TO 50 GC. 0698
TES. (EQUILIBRIUM STRUCTURE) MICROWAVE SPECTRA OF NITROGEN TRIFLUORIDE IN THE EXCITED VIBRATIONAL STA 0722
TES. (ZETA TYPE DOUBLING AND RESONA/ MICROWAVE SPECTRA OF NITROGEN TRIFLUORIDE IN THE EXCITED VIBRATIONAL STA 0723
ANTS, AND MEAN SQUARE AMPLITUDES OF THE VIBRATIONS OF THE NITROGEN TRIFLUORIDE MOLECULE. /AL STRETCHING CONST 0726
/ PSEUDO-PARALLEL INFRARED BANDS OF OBLATE SYMMETRIC TOPS NITROGEN TRIFLUORIDE NU-4 AND NU-4 + NU-2 AND TRIFL 0716
/ PROPERTIES. PART-2: DIPOLE MOMENT AND FIELD GRADIENT OF NITROGEN TRIFLUORIDE. PART-3: A GAUSSIAN BASIS SEL/ 0746
/LITUDES OF VIBRATION FOR SOME PYRAMIDAL XY3 MOLECULES. (NITROGEN TRIFLUORIDE, PHOSPHORUS TRIFLUORIDE, AND / 0706
/ANTS FOR SYMMETRIC TOP MOLECULES AND FORCE CONSTANTS FOR NITROGEN TRIFLUORIDE, PHOSPHORUS TRIFLUORIDE, AND / 0708
RSENIC TRIFLUORIDE BY THE COMBINE/ GENERAL FORCE FIELD OF NITROGEN TRIFLUORIDE, PHOSPHORUS TRIFLUORIDE, AND A 0717
ITROGEN DIFLUORIDE. MOLECULAR ORBITAL STUDY OF NITROGEN TRIFLUORIDE, PHOSPHORUS TRIFLUORIDE, AND N 0721
/OBLEM WITH THE USE OF CENTRIFUGAL STRETCHING CONSTANTS. (NITROGEN TRIFLUORIDE, PHOSPHORUS TRIFLUORIDE, ARSE/ 0727
DE FROM MICROWAVE SPECTRA. NUCLEAR QUADRUPOLE MOMENT OF NITROGEN-14 AND THE STRUCTURE OF NITROGEN TRIFLUORI 0741
NMR SEE NUCLEAR MAGNETIC RESONANCE
STRUCTURES OF NOBLE GAS COMPOUNDS. (XENON HEXAFLUORIDE) 1292
LUORIDE, AND XENON DIFLUORIDE) D ORBITALS IN THE NOBLE GAS DIHALIDES. (ARGON DIFLUORIDE, KRYPTON DIF 1338
CHEMICAL PREDICTIONS BY MO THEORY. THE NOBLE GAS HALIDES. 1340
CHEMICAL COMPOUNDS OF XENON AND OTHER NOBLE GASES. 0010
NOBLE GASES AND THE BINARY FLUORIDES OF XENON. 1342
, XENON HE/ STRUCTURAL CONSIDERATIONS IN THE CHEMISTRY OF NOBLE GASES. (XENON DIFLUORIDE, XENON TETRAFLUORIDE 1235
LIMITATIONS OF THE UREY-BRADLEY MODEL IN NORMAL COORDINATE ANALYSES. 1478
/FRARED AND LASER RAMAN SPECTRA, MOLECULAR STRUCTURES AND NORMAL COORDINATE ANALYSES OF GERMANIUM DIFLUORIDE/ 0475
FORCE CONSTANTS AND NORMAL COORDINATE ANALYSIS. 1473
DE. VIBRATIONAL ASSIGNMENTS AND NORMAL COORDINATE ANALYSIS FOR ANTIMONY PENTAFLUORI 0087
IDE, PHOSPHORUS DIFLUORIDE DICHLORIDE, AND PHOSPHORUS PE/ NORMAL COORDINATE ANALYSIS OF PHOSPHORUS PENTAFLUOR 0877
OCTAHEDRAL XY6 AND Z6 MOLECULAR MODELS WITH APPLICATION/ NORMAL COORDINATE ANALYSIS OF ROTATION VIBRATION OF 1454
E. VIBRATIONAL ASSIGNMENT AND NORMAL COORDINATE ANALYSIS OF SELENIUM TETRAFLUORID 0958
. (NITROGEN TRIFLUORIDE) NORMAL COORDINATE ANALYSIS OF SOME NITROGEN HALIDES 0730
INFRARED SPECTRUM AND NORMAL COORDINATE ANALYSIS OF SULFUR TETRAFLUORIDE 1045
XY5 MOLECULES. NORMAL COORDINATE ANALYSIS OF TRIGONAL BIPYRAMIDAL 1433
HE CALCULATION BOND MOMENT / A GEOMETRIC VISUALIZATION OF NORMAL COORDINATE TRANSFORMATIONS. APPLICATION TO T 0352
CULES AND IONS. (GVFF) NORMAL COORDINATE TREATMENT OF XY4 TETRAHEDRAL MOLE 1415
CULES AND IONS. (UREY-BRADLEY AND MVFF) NORMAL COORDINATE TREATMENT OF XY4 TETRAHEDRAL MOLE 1418
S IN ENERGY DISTRIBUTIONS. (SELENIUM A/ CLASSIFICATION OF NORMAL MODES OF VIBRATION. PART-2: INTERACTION TERM 1494
NORMAL VIBRATION FREQUENCIES OF ZIRCONIUM HALIDES. 1330
CALCULATION OF THE FREQUENCIES OF THE NORMAL VIBRATIONS OF ANTIMONY TRIFLUORIDE. 0074
UORIDE) NORMAL VIBRATIONS OF BORON TRIHALIDES. (BORON TRIFL 0247
F PHOSPHORUS TRICHLORIDE FROM AMPLITUDES OF VIBRATION AND NORMAL VIBRATIONAL FREQUENCIES. /ENTIAL CONSTANTS O 1481
K TO ITS BOILING POINT. THE SIGNIFICANCE OF THE PARAMETER NU. /RTIES OF CARBON TETRAFLUORIDE FROM 12 DEGREES 0361
FERMI DIAD OF NU-1 AND 2NU-2 IN OXYGEN DIFLUORIDE. 0772
EQUILIBRIUM STRUCTURE, ANHARMONIC POTENTIAL FUNCTION, AND NU-1- NU-3 CORIOLIS RESONANCE. /IBRATIONAL STATES, 0993
/UM OF OXYGEN DIFLUORIDE IN VIBRATIONALLY EXCITED STATES, NU-1- 2 NU-2 FERMI RESONANCE AND EQUILIBRIUM STRUC/ 0771
Y. BORON TRIFLUORIDE. THE ANALYSIS OF NU-2 AND THE DETERMINATION OF THE MOLECULAR GEOMETR 0209
BLATE SYMMETRIC TOPS NITROGEN TRIFLUORIDE NU-4 AND NU-4 + NU-2 AND TRIFLUORO METHANE NU-6. /FRARED BANDS OF O 0716
XYGEN DIFLUORIDE IN VIBRATIONALLY EXCITED STATES, NU-1- 2 NU-2 FERMI RESONANCE AND EQUILIBRIUM STRUCTURE. / O 0771
FERMI DIAD OF NU-1 AND 2NU-2 IN OXYGEN DIFLUORIDE. 0772
NU-3 AND NU-4 BANDS OF SULFUR HEXAFLUORIDE. 1070
ANT. BORON TRIFLUORIDE. THE ANALYSIS OF 2NU-3 AND THE DETERMINATION OF THE 33 CORIOLIS CONST 0192
NU-3 BAND OF PHOSPHORUS TRIFLUORIDE. 0815
LOW TEMPERATURE ABSORPTION CONTOUR OF THE NU-3 BAND OF SULFUR HEXAFLUORIDE. 1082
S. TEMPERATURE DEPENDENT ABSORPTION SPECTRUM OF THE NU-3 BAND OF SULFUR HEXAFLUORIDE AT 10.6 MICROMETER 1091
NU-3 BANDS OF BORON TRIFLUORIDE. 0207
BRIUM STRUCTURE, ANHARMONIC POTENTIAL FUNCTION, AND NU-1- NU-3 CORIOLIS RESONANCE. /IBRATIONAL STATES, EQUILI 0993
URATION OF A SINGLE VIBRATION- ROTATION TRANSITION IN THE NU-3 FUNDAMENTAL OF SULFUR HEXAFLUORIDE. /TICAL SAT 1074
NU-3 OF BORON TRIFLUORIDE. 0200
OCTAHEDRAL FINE STRUCTURE SPLITTINGS IN NU-3 OF SULFUR HEXAFLUORIDE. 1063
RARED BANDS OF OBLATE SYMMETRIC TOPS NITROGEN TRIFLUORIDE NU-4 AND NU-4 + NU-2 AND TRIFLUORO METHANE NU-6. /F 0716
PERPENDICULAR INFRARED BAND CONTOURS: APPLICATION TO THE NU-4 BAND OF NITROGEN TRIFLUORIDE. CALCULATION OF 0747
HIGH RESOLUTION STUDY OF L-TYPE RESONANCE IN THE NU-4 BAND OF NITROGEN TRIFLUORIDE. 0731
NU-3 AND NU-4 BANDS OF SULFUR HEXAFLUORIDE. 1070
TRIFLUORIDE AT HIGH RESOLUTION. NU-4 BANDS OF THE TEN AND ELEVEN ISOTOPES OF BORON 0208
YPE DOUBLING TRANSITIONS OF PHOSPHORUS TRIFLUORIDE IN THE NU-4 EQUALS 1 STATE. /RD ORDER CONSTANTS ON THE L-T 0804
YPE DOUBLING TRANSITIONS OF PHOSPHORUS TRIFLUORIDE IN THE NU-4 EQUALS 1 STATE. L-T 0803
1-TYPE RESONANCE IN AN OVERTONE BAND. THE 2NU-4 SPECTRUM OF NITROGEN TRIFLUORIDE. 0732
2NU-4 SPECTRUM OF PHOSPHORUS TRIFLUORIDE. 0817
EN TRIFLUORIDE NU-4 AND NU-4 + NU-2 AND TRIFLUORO METHANE NU-6. /FRARED BANDS OF OBLATE SYMMETRIC TOPS NITROG 0716

/MAN SPECTRA OF GASES. PART-10: RAMAN BAND CONTOUR OF THE NU-7 (E') FUNDAMENTAL OF PHOSPHORUS PENTAFLUORIDE. 0879
NU-7 FUNDAMENTAL OF PHOSPHORUS PENTAFLUORIDE. 0848
POTENTIAL FUNCTION FOR THE NU-7 VIBRATION OF PHOSPHORUS PENTAFLUORIDE. 0830
S. NUCLEAR MAGNETIC RELAXATION IN BORON TRIFLUORIDE GA 0217
RIDES, CHLORINE PENTAFLUORIDE AND CHLORINE OXYFLUORIDE BY NUCLEAR MAGNETIC RESONANCE. / FLUORIDES AND OXYFLUO 0414
RBON TETRAFLUORIDE, SILICON TETRA/ FLUORINE RELAXATION BY NUCLEAR MAGNETIC RESONANCE ABSORPTION IN GASEOUS CA 0346
EX FORMATION BETWEEN ANTIMONY PENTAFLUORIDE A/ A FLUORINE NUCLEAR MAGNETIC RESONANCE AND RAMAN STUDY OF COMPL 0079
ES. NUCLEAR MAGNETIC RESONANCE IN RARE EARTH TRIFLUORID 1355
OCIATION OF SULFUR TETRAFLUORIDE. FLUORINE-19 NUCLEAR MAGNETIC RESONANCE INVESTIGATION OF THE ASS 1044
E. FLUORINE NUCLEAR MAGNETIC RESONANCE OF ANTIMONY PENTAFLUORID 0078
E. FLUORINE NUCLEAR MAGNETIC RESONANCE OF CHLORINE PENTAFLUORID 0416
RHENIUM HEPTAFLUORIDE, AND OXIDE PENTAFLUORIDES FLUORINE NUCLEAR MAGNETIC RESONANCE OF IODINE HEPTAFLUORIDE, 0517
. SUSCEPTIBILITY MEASUREMENTS AND FLUORINE NUCLEAR MAGNETIC RESONANCE OF PLATINUM HEXAFLUORIDE 0883
ORIDE. ANOMALOUS NUCLEAR MAGNETIC RESONANCE SHIFT IN SAMARIUM TRIFLU 0927
DS. ELECTRON STRUCTURE AND NUCLEAR MAGNETIC RESONANCE SPECTRA OF BORON COMPOUN 0229
LUORIDE AND CHLORINE MONOFLUORIDE. GASEOUS SPECTRA AND G/ NUCLEAR MAGNETIC RESONANCE SPECTRA OF CHLORINE TRIF 0368
RAFLUORIDE. A STUDY OF THE RAMAN AND BROAD LINE NUCLEAR MAGNETIC RESONANCE SPECTRA OF GERMANIUM TET 0484
LUORIDE AND IODINE OXYGEN PENTAFLUORIDE. NUCLEAR MAGNETIC RESONANCE SPECTRA OF IODINE HEPTAF 0516
DES. (OXYGEN DIFLUORIDE) FLUORINE NUCLEAR MAGNETIC RESONANCE SPECTRA OF OXYGEN FLUORI 0775
HORUS, AND SILICON FLUORIDES. HYDROLYSIS AND IN/ FLUORINE NUCLEAR MAGNETIC RESONANCE SPECTRA OF SULFUR, PHOSP 1042
OF SULFUR TETRAFLUORIDE. NUCLEAR MAGNETIC RESONANCE SPECTRUM AND STRUCTURE 1037
DENUM HEXAFLUORIDE, TUNGSTEN HEXAFLUORID/ HIGH RESOLUTION NUCLEAR MAGNETIC RESONANCE SPECTRUM OF LIQUID MOLYB 0637
/URE OF EXCHANGE PROCESSES IN SOME INORGANIC FLUORIDES BY NUCLEAR MAGNETIC RESONANCE (SULFUR TETRAFLUORIDE / 1048
RUCTURE OF NITROGEN TRIFLUORIDE FROM MICROWAVE SPECTRA. NUCLEAR QUADRUPOLE MOMENT OF NITROGEN-14 AND THE ST 0741
TAFLUORIDE AND ITS COMPLEXES WITH XENON DIFLU/ NIOBIUM-93 NUCLEAR QUADRUPOLE RESONANCE STUDIES OF NIOBIUM PEN 0664
NUCLEAR RELAXATION TIMES OF LANTHANUM TRIFLUORIDE. 0579
INTERACTION IN SYMMETRIC TOP MOLECULES. NUCLEAR SPIN RELAXATION STUDY OF THE SPIN ROTATION 0183
INTERACTION IN SPHERICAL TOP MOLECULES. NUCLEAR SPIN RELAXATION STUDY OF THE SPIN ROTATION 0533
HE HEIGHT OF THE BARRIER TO INTERNAL EXCHANGE OF FLUORINE NUCLEI. /AFLUORIDE AND ARSENIC PENTAFLUORIDE AND T 0856
OF INORGANIC FLUORIDES ESPECIALLY THE ELEMENTS OF ATOMIC NUMBER BELOW 20. HEATS OF FORMATION 0214

O

F CHLORINE MONOFLUORIDE. O(+) TO SINGLET SIGMA(+) SYSTEMS OF BOTH ISOTOPES O 0390
FAR INFRARED SPECTRA OF PROLATE AND OBLATE SYMMETRIC TOPS. 1479
NU-4 + NU-2 AND TRIFL/ PSEUDO-PARALLEL INFRARED BANDS OF OBLATE SYMMETRIC TOPS NITROGEN TRIFLUORIDE NU-4 AND 0716
EXAFLUORIDE AND ITS IDENTITY WITH THE PREVIOUSLY REPORTED OCTAFLUORIDE. OSMIUM H 0753
ACCEPTOR PROPERTIES OF IODINE HEPTAFLUORIDE. OCTAFLUORO PERIODATES(VII). 0514
FUR HEXAFLUORIDE. OCTAHEDRAL FINE STRUCTURE SPLITTINGS IN NU-3 OF SUL 1063
MEAN SQUARE AMPLITUDES OF INTERATOMIC DISTANCES IN SOME OCTAHEDRAL HEXAFLUORIDE MOLECULES. 1466
EXTREMAL PROPERTIES OF FORCE CONSTANTS IN OCTAHEDRAL HEXAFLUORIDES. 1468
TUDES OF THERMAL MOTION AND SHRINKAGES OF CHEMICAL BONDS: OCTAHEDRAL HEXAFLUORIDES. MEAN AMPLI 1460
REVISED MEAN AMPLITUDES OF VIBRATION FOR SOME OCTAHEDRAL HEXAFLUORIDES. 1447
FORCE FIELD OF WEINSTOCK AND GOODMAN FOR OCTAHEDRAL HEXAFLUORIDES AND THE L MATRIX METHOD. 1456
POTENTIAL FIELD AND FORCE CONSTANTS OF OCTAHEDRAL MOLECULES. 1464
HERICALLY SYMMETRIC MODES OF SOME XY4 TETRAHEDRAL AND XY6 OCTAHEDRAL MOLECULES. RAMAN INTENSITIES FOR SP 1453
STEP METHOD. FORCE CONSTANTS OF OCTAHEDRAL MOLECULES CALCULATED BY THE LOGARITHMIC 1472
APPLICATION OF THE METHOD OF PROGRESSIVE RIGIDITY TO OCTAHEDRAL MX6 MOLECULES. 1441
DES) ROOT MEAN SQUARE AMPLITUDES IN SOME HEXAFLUORIDES OF OCTAHEDRAL SYMMETRY. (SELENIUM AND TELLURIUM FLUORI 0962
ES OF THE NONTRANSITION / BOND LENGTHS AND BOND ANGLES IN OCTAHEDRAL, TRIGONAL BIPYRAMID, AND RELATED MOLECUL 1435
TION/ NORMAL COORDINATE ANALYSIS OF ROTATION VIBRATION OF OCTAHEDRAL XY6 AND Z6 MOLECULAR MODELS WITH APPLICA 1454
VIBRATION AND BASTIANSEN-MORINO SHRINKAGE EFFECTS IN SOME OCTAHEDRAL XY6 MOLECULES. MEAN AMPLITUDES OF 1455
/ULAR FORCE FIELDS. PART-8: VIBRATION FREQUENCIES OF SOME OCTAHEDRAL XY6 MOLECULES. (SULFUR HEXAFLUORIDE, SE/ 1080
AND SELENIUM MONOFLUORIDE) ELECTRIC DIPOLE MOMENTS OF OPEN SHELL DIATOMIC MOLECULES. (SULFUR MONOFLUORIDE 1019
PIUM TRIFLUORIDE. OPTICAL ABSORPTION AND FLUORESCENCE SPECTRA OF EURO 0453
CA/ ELECTRONIC PROPERTIES OF SULFUR HEXAFLUORIDE. PART-1: OPTICAL ABSORPTION AND X-RAY EMISSION FROM SCF MO L 1078
NEAR ULTRAVIOLET OPTICAL CONSTANTS OF LANTHANUM TRIFLUORIDE. 0584
TRANSITION IN THE NU-3 FUNDAMENTAL OF SULFUR HEXAFLUORID/ OPTICAL SATURATION OF A SINGLE VIBRATION- ROTATION 1074
THEORETICAL CALCULATIONS OF THE ELECTRONIC STRUCTURE AND OPTICAL TRANSITIONS OF URANIUM HEXAFLUORIDE AND UR/ 1179
OMIC MOLECULE/ METHODS FOR CORRELATING MOLECULES AND SOME OPTIMIZED VALENCE CONFIGURATION RESULTS ON THE DIAT 0174
HN-TELLER EFFECT. VIBRONIC AND SPIN ORBIT SPLITTING IN SPECTRA OF SYSTEMS EXHIBITING JA 1480
SEE BOTH MO AND MOLECULAR ORBITAL.
/PART-1. POINTS-ON-A-SPHERE AND EXTENDED HUCKEL MOLECULAR ORBITAL ANALYSES OF TRIGONAL BIPYRAMIDS. (PHOSPHOR/ 0828
SELF CONSISTENT CHARGE AND CONFIGURATION MOLECULAR ORBITAL CALCULATIONS ON PHOSPHORUS TRIFLUORIDE. 0821
PENTAFLUORIDE. SEMIEMPIRICAL MOLECULAR ORBITAL CALCULATIONS. PSEUDOROTATION IN PHOSPHORUS 0843
SCANDIUM MONOFLUORIDE. REASSIGNMENT OF MOLECULAR ORBITAL CONFIGURATIONS OF THE ELECTRONIC STATES OF 0937
2, 4, AND 6. MOLECULAR ORBITAL DESCRIPTION OF SULFUR COMPOUNDS OF VALENCES 1085
TRAFLUORIDE AND SULFUR HEXAFLUORIDE. MOLECULAR ORBITAL DESCRIPTION OF SULFUR DIFLUORIDE, SULFUR TE 1027
RIDE. MOLECULAR ORBITAL ENERGY LEVELS AND BONDING IN KRYPTON DIFLUO 0549
/S AND XENON OXIDE TETRAFLUORIDE BY ESCA AND EVIDENCE FOR ORBITAL INDEPENDENCE IN THE XENON- FLUORINE BONDIN/ 1229
BIDDEN TRANSITIONS IN IODINE, OR IODINE BROMIDE (CH/ SPIN ORBITAL INTERACTION AND RELATIVE INTENSITIES OF FOR 0498
/MISSION FROM SCF MO LCAO COMPUTATIONS. PART-2: MOLECULAR ORBITAL INTERPRETATION OF ITS X-RAY ABSORPTION SPE/ 1078
GAS THERMODYNAMIC FUNCTIONS FROM A HARTREE-FOCK MOLECULAR ORBITAL INVESTIGATION. /COPIC CONSTANTS, AND IDEAL 0094
LICON MONOFLUORIDE AND THEIR POSITIVE AND NEGA/ MOLECULAR ORBITAL INVESTIGATION OF CARBON MONOFLUORIDE AND SI 0290
QUILIBRIUM GEOMETRIES AND QUAD/ SELF CONSISTENT MOLECULAR ORBITAL METHODS. PART-5: AB INITIO CALCULATION OF E 0720
ATOMS OF LATER SECOND ROW ELEMENTS IN PERFECT PAIRING V/ ORBITAL MODIFICATION BY THE COULOMB FIELD OF LIGAND 0862
LAR NITROGEN AND SULFUR HEXAFL/ APPLICATION OF SYNCHROTON ORBITAL RADIATION TO MOLECULAR SPECTROSCOPY. MOLECU 1083
THER PROPERTIES OF SELENIUM MONOFLUORIDE FROM A MOLECULAR ORBITAL STUDY. /LPY, SPECTROSCOPIC CONSTANTS, AND O 0955
US MONOFLUORIDE, AND SULFUR MONOFLUORIDE FROM A MOLECULAR ORBITAL STUDY. /IDE, SILICON MONOFLUORIDE, PHOSPHOR 0973
LUORIDE MOLECULES AND IONS. MOLECULAR ORBITAL STUDY OF ATOMIZATION ENERGIES OF PNICOGEN F 1364
DE, AND NITROGEN DIFLUORIDE. MOLECULAR ORBITAL STUDY OF NITROGEN DIHYDRIDE, NITROGEN DIOXI 0684
RIFLUORIDE, AND NITROGEN DIFLUORIDE. MOLECULAR ORBITAL STUDY OF NITROGEN TRIFLUORIDE, PHOSPHORUS T 0721
LORINE MONOFLUORIDE, CHLORINE TRIFLU/ AB INITIO MOLECULAR ORBITAL STUDY OF THE GEOMETRY OF INTERHALOGENS. (CH 0379
OGEN TETRAFLUORIDE MOLECULES, AND ES/ AB INITIO MOLECULAR ORBITAL STUDY OF THE NITROGEN DIFLUORIDE AND DINITR 0688
AB INITIO MOLECULAR ORBITAL STUDY OF THE PHOSPHORUS DIFLUORIDE RADICAL. 0790
ORMALISM. PART-2: DIRECT COMPARISO/ APPROXIMATE MOLECULAR ORBITAL THEORY. ESSENTIAL STRUCTURAL ELEMENTS- MO F 0765
IC COMPOUNDS. PART-3: AB INITIO STUDIES OF CHA/ MOLECULAR ORBITAL THEORY OF THE ELECTRONIC STRUCTURE OF ORGAN 0707
MOLECULAR ORBITAL TREATMENT OF IODINE HEPTAFLUORIDE. 0533
MOLECULES. (GERMANIUM TETRAFLUORIDES) ORBITAL VALENCE FORCE CONSTANTS OF TETRAHEDRAL XY4 0481
ERLAP METHOD. MOLECULES WITH LONE PAIR ELECTRONS OR EMPTY ORBITALS. /N OF FORCE CONSTANTS WITH THE MAXIMUM OV 1395

RUS PENTAFLUORIDE, ARSENIC PENTAFLUORIDE, AND BROMINE PE/ ORBITALS AND STRUCTURES OF PENTAFLUORIDES. (PHOSPHO 0831
CONSTRUCTION OF HYBRID ORBITALS. (CHLORINE TRIFLUORIDE) 0405
- FLUORINE DOUBLE AND TRIPLE BONDS.   LOCALIZED ORBITALS IN BORON FLUORIDES. HIGHLY POLARIZED BORON 0158
IDE, KRYPTON DIFLUORIDE, AND XENON DIFLUORIDE)   D ORBITALS IN THE NOBLE GAS DIHALIDES. (ARGON DIFLUOR 1338
EXAFLUORIDE CALCULATED BY THE SELF CONSISTENT / MOLECULAR ORBITALS OF PLATINUM HEXAFLUORIDE AND ELEMENT 110 H 0881
HE MOLECULES XENON DIFLUORIDE AND XENON DICHLORIDE.   5D ORBITALS OF XENON IN ATOMIC VALENCE STATES AND IN T 1242
OMIC MOLECU/ DENSITY LOCALIZATION OF ATOMIC AND MOLECULAR ORBITALS. PART-3: HETERONUCLEAR DIATOMIC AND POLYAT 0175
OF PHOSPHORUS TRIFLUORIDE IN THE NU/ EFFECTS OF THE THIRD ORDER CONSTANTS ON THE L-TYPE DOUBLING TRANSITIONS 0804
XYGEN DIFLUORIDE)   THIRD ORDER POTENTIAL CONSTANTS OF BENT XY2 MOLECULES. (O 0770
ONOFLUORIDE AND BROMINE M/ SOME CORRECTIONS TO THE SECOND ORDER STARK EFFECT OF LINEAR MOLECULES. (CHLORINE M 0385
PENTAFLUORIDE AND ITS COMPLEXES WITH XENON DIFLUORIDE AND ORGANIC BASES. /UPOLE RESONANCE STUDIES OF NIOBIUM 0664
/ MOLECULAR ORBITAL THEORY OF THE ELECTRONIC STRUCTURE OF ORGANIC COMPOUNDS. PART-3: AB INITIO STUDIES OF CHA 0707
ECTION.   A REDETERMINATION OF THE ORTHORHOMBIC IODINE HEPTAFLUORIDE STRUCTURE. A CORR 0519
NEUTRON DIFFRACTION INVESTIGATION OF ORTHORHOMBIC LEAD DIFLUORIDE. 0593
ENVIRONMENT OF THE TIN ATOM IN ORTHORHOMBIC TIN DIFLUORIDE. 1139
NOFLUORIDE A DOUBLET SIGMA(+) TO X DOUBLET PI BAND SYSTEM OSCILLATOR STRENGTHS. / X TRIPLET PI) AND CARBON MO 0288
MENTS ON HEATS OF FORMATION OF CARBON MONOFLUORIDE AND C/ OSCILLATOR STRENGTHS OF CARBON MONOFLUORIDE AND COM 0295
PREPARATION AND PROPERTIES (INFRARED SPECTRUM) OF OSMIUM HEPTAFLUORIDE. 0754
VIBRATIONAL ELECTRONIC COUPLING IN IRIDIUM HEXAFLUORIDE, OSMIUM HEXACHLORIDE(2-), AND IRIDIUM HEXACHLORIDE(/ 0538
ABSORPTION SPECTRUM AND MAGNETIC PROPERTIES OF OSMIUM HEXAFLUORIDE. 0751
M OF CHROMIUM HEXAFLUORIDE, MOLYBDENUM HEXAFLUORIDE, AND OSMIUM HEXAFLUORIDE.   INFRARED SPECTRU 0436
AFLUORIDE, MOLYBDENUM HEXAFLUORIDE, RHENIUM HEXAFLUORIDE, OSMIUM HEXAFLUORIDE, AND IRIDIUM HEXAFLUORIDE) /HEX 0903
OUSLY REPORTED OCTAFLUORIDE. OSMIUM HEXAFLUORIDE AND ITS IDENTITY WITH THE PREVI 0753
VIBRATIONAL SPECTRA OF OSMIUM HEXAFLUORIDE AND PLATINUM HEXAFLUORIDE. 0752
/F TRANSITION METAL HEXAFLUORIDES. (RHENIUM HEXAFLUORIDE, OSMIUM HEXAFLUORIDE, IRIDIUM HEXAFLUORIDE, AND PLA/ 0916
/EXAFLUORIDE, RHENIUM HEXAFLUORIDE, RHODIUM HEXAFLUORIDE, OSMIUM HEXAFLUORIDE, IRIDIUM HEXAFLUORIDE, AND PLA/ 1084
/TRON DIFFRACTION INVESTIGATION OF TUNGSTEN HEXAFLUORIDE, OSMIUM HEXAFLUORIDE, IRIDIUM HEXAFLUORIDE, URANIUM/ 1155
PREPARATION AND CRYSTAL STRUCTURE OF OSMIUM PENTAFLUORIDE. 0750
ORIDE, TUNGSTEN PENTAFLUORIDE, RHENIUM PENTAFLUORIDE, AND OSMIUM PENTAFLUORIDE. /SURES OF MOLYBDENUM PENTAFLU 0624
IUM PENTAFLUORIDE, / PREPARATION AND CRYSTAL STRUCTURE OF OSMIUM PENTAFLUORIDE. TWO FLUORIDES OF OSMIUM. (OSM 0749
FLUORIDE)   HARMONIC POTENTIAL CONSTANTS OF OSMIUM TETROXIDE AND SOME TETRAHALIDE (CARBON TETRA 0353
OR EMPTY/ CALCULATION OF FORCE CONSTANTS WITH THE MAXIMUM OVERLAP METHOD. MOLECULES WITH LONE PAIR ELECTRONS 1395
CALCULATION OF FORCE CONSTANTS WITH THE MAXIMUM OVERLAP METHOD. TETRAHEDRAL MOLECULES. 1416
/ANALYSIS OF DIFFERENT SCHEMES OF NEGLECT OF DIFFERENTIAL OVERLAP METHODS ILLUSTRATED BY THE XENON FLUORIDES/ 1231
UORIDE.   1-TYPE RESONANCE IN AN OVERTONE BAND. THE 2NU-4 SPECTRUM OF NITROGEN TRIFL 0732
BORON TRIFLUORIDE, SILICON TETRAFLUORIDE, AND NITROGEN DIOXIDE. /S FROM HEAVY ISOTOPIC SUBSTITUTION. PART-2: 0232
ALOUS THERMAL EXPANSION ON IRON DIFLUORIDE AND VANADIUM DIOXIDE.   ELECTRONIC STRUCTURE AND ANOM 0543
F ANHARMONIC POTENTIAL CONSTANTS. APPLICATION TO CARBON DIOXIDE.   LEAST SQUARES ADJUSTMENT O 1491
GE DISTRIBUTION IN OXYGEN DIFLUORIDE, OZONE, AND SULFUR DIOXIDE. /S AND SECOND MOMENTS OF THE ELECTRONIC CHAR 0780
CTROSCOPIC STUDIES OF SCANDIUM MONOFLUORIDE AND MAGNESIUM OXIDE.   SPE 0933
ICON TETRAFLUORIDE, GERMANIUM TETRAFLUORIDE AND SULFUR TRIOXIDE. /RIOUS ISOTOPES OF CARBON TETRAFLUORIDE, SIL 0358
CALCULATION OF RYDBERG LEVELS IN NITROGEN OXIDE AND BORON FLUORIDE. 0164
VALENCE EXCITED STATES OF NITROGEN, CARBON MONOXIDE, AND BORON FLUORIDE. 0169
ELECTRONIC STRUCTURE OF CARBON MONOXIDE AND BORON MONOFLUORIDE. 0162
ERYLLIUM OXIDE, LITHIUM FLUORIDE, HELIUM- NEON, CARBON MONOXIDE, AND BORON MONOFLUORIDE. /E, BORON NITRIDE, B 0174
MEAN AMPLITUDES OF VIBRATION OF SELENIUM TRIOXIDE AND CHLORINE TRIFLUORIDE. 0406
REFLECTION SPECTRA OF GADOLINIUM OXIDE AND GADOLINIUM TRIFLUORIDE POWDERS. 0455
ATION POTENTIALS OF THE MOLECULES BORON FLUORIDE, SILICON OXIDE AND GERMANIUM OXIDE.   FIRST IONIZ 0160
A OF SULFUR MONOFLUORIDE, SELENIUM MONOFLUORIDE, SELENIUM OXIDE AND IODINE OXIDE. / ELECTRON RESONANCE SPECTR 1020
/HLORIDE WITH CARBON MONOXIDE, MOLECULAR NITROGEN, NITRIC OXIDE, AND MOLECULAR OXYGEN AND OF CALCIUM FLUORID/ 1324
MOLECULAR ORBITAL STUDY OF NITROGEN DIHYDRIDE, NITROGEN DIOXIDE, AND NITROGEN DIFLUORIDE. 0684
/R CENTROIDS OF A SINGLET SIGMA- X SINGLET SIGMA PLATINUM OXIDE AND SILICON MONOFLUORIDE (D DOUBLET SIGMA(+) / 0966
HARMONIC POTENTIAL CONSTANTS OF OSMIUM TETROXIDE AND SOME TETRAHALIDE (CARBON TETRAFLUORIDE) 0353
/L BINDING. PART-3: THE ISOELECTRONIC NITROGEN, CARBON MONOXIDE, BORON FLUORIDE, AND CARBON, BERYLLIUM OXIDE/ 0150
TION AT LOW TEMPERATURE.   SULFUR TETRAFLUORIDE AND SULFUR OXIDE DIFLUORIDE. INFRARED EVIDENCE FOR DIMER FORMA 1054
/LUORIDE FROM 10 TO 350 DEGREES K, AND OF PLUTONIUM-244 DIOXIDE FROM 4 TO 25 DEGREES K. DERIVED ENTROPIES AN/ 0885
5/2 ROTATIONAL LEVELS.   ESR SPECTRUM OF BROMINE OXIDE, IODINE OXIDE AND SELENIUM MONOFLUORIDE IN J- 0954
N, CARBON MONOXIDE, BORON FLUORIDE, AND CARBON, BERYLLIUM OXIDE, LITHIUM FLUORIDE. /THE ISOELECTRONIC NITROGE 0150
/ON, CARBON, NITROGEN, FLUORINE, BORON NITRIDE, BERYLLIUM OXIDE, LITHIUM FLUORIDE, HELIUM- NEON, CARBON MONO/ 0174
ORIDE, AND MANGANESE DIFLUOR/ RAMAN SPECTRA OF TITANIUM DIOXIDE, MAGNESIUM DIFLUORIDE, ZINC DIFLUORIDE, DIFLU 0615
US F/ FORCE FIELDS OF THE RUTILE COUNTERPARTS: TITANIUM DIOXIDE, MAGNESIUM FLUORIDE, ZINC FLUORIDE, AND FERRO 0606
/LATED NICKEL FLUORIDE AND NICKEL CHLORIDE WITH CARBON MONOXIDE, MOLECULAR NITROGEN, NITRIC OXIDE, AND MOLEC/ 1324
E NMR OF IODINE HEPTAFLUORIDE, RHENIUM HEPTAFLUORIDE, AND OXIDE PENTAFLUORIDES.   FLUORIN 0517
AND SHRINKAGE CONSTANTS OF XENON TETRAFLUORIDE AND XENON OXIDE TETRAFLUORIDE. /QUARE AMPLITUDES OF VIBRATION 1278
ERMODYNAMIC PROPERTIES OF SULFUR TETRAFLUORIDE AND SULFUR OXIDE TETRAFLUORIDE.   TH 1051
STATES OF THE POSITIVE IONS OF XENON TETRAFLUORIDE, XENON OXIDE TETRAFLUORIDE, AND IODINE PENTAFLUORIDE. /IC 1264
L/ ELECTRON DISTRIBUTION IN THE XENON FLUORIDES AND XENON OXIDE TETRAFLUORIDE BY ESCA AND EVIDENCE FOR ORBITA 1229
/NCE FORCE CONSTANTS OF SQUARE PYRAMIDAL MOLECULES. XENON OXIDE TETRAFLUORIDE, IODINE PENTAFLUORIDE, BROMINE/ 0418
/TRA FOR XENON DIFLUORIDE, XENON TETRAFLUORIDE, AND XENON OXIDE TETRAFLUORIDE VAPORS, AND FORCE CONSTANT CAL/ 1254
STABILITIES OF TANTALUM PENTAFLUORIDE AND TANTALUM OXIDE TRIFLUORIDE. 1103
/OTOELECTRON SPECTRA OF NITROGEN TRIFLUORIDE AND NITROGEN OXIDE TRIFLUORIDE, AND A REASSIGNMENT OF THE SPECT/ 0700
INFRARED SPECTRA OF PHOSPHORUS TRIFLUORIDE, PHOSPHORUS OXIDE TRIFLUORIDE, AND PHOSPHORUS PENTAFLUORIDE. 0799
/OTOELECTRON SPECTRA OF HALIDES. PART-3: TRIFLUORIDES AND OXIDE TRIFLUORIDES OF NITROGEN AND PHOSPHORUS, AND/ 0701
RMATION OF GERMANIUM TETRAFLUORIDE AND OF THE GERMANIUM DIOXIDES.   HEATS OF FO 0483
ELEMENTS.   COMPARISONS OF FLUORIDES, OXIDES, AND SULFIDES CONTAINING DIVALENT TRANSITION 1370
BIBLIOGRAPHY ON SPECTROSCOPY OF FLUORIDES AND OXIDES BELONGING TO THE LANTHANIDE SERIES. 1358
RED SPECTROSCOPY OF HIGH TEMPERATURE SPECIES. (FLUORIDES, OXIDES, CHLORIDES)   MATRIX ISOLATION INFRA 0011
THE ISOELECTRONIC SERIES CHLORINE MONOFLUORIDE, HYDROGEN OXYCHLORIDE, CHLORO AMMONIA, AND CHLORO METHANE. /F 0371
RAF/ ELECTRON DENSITY DISTRIBUTIONS IN SOME FLUORIDES AND OXYCHLORIDES. (CARBON TETRAFLUORIDE AND SILICON TET 0337
THE HEATS OF FORMATION OF THE GASEOUS MOLECULES ALUMINUM OXYDIFLUORIDE AND ALUMINUM DIFLUORIDE. /MINATION OF 0035
SPECTRUM OF OXYGEN MONOFLUORIDE AND DETECTION OF LITHIUM OXYFLUORIDE.   MATRIX INFRARED 0755
LUTONIUM HEXAFLUORIDE AND IDENTIFICATION OF PLUTONIUM(VI) OXYFLUORIDE. /D PROPERTIES OF (VAPOR PRESSURE) OF P 0890
BONDING IN PHOSPHORUS TRIFLUORIDE AND PHOSPHORUS OXYFLUORIDE. AN AB INITIO SCF STUDY. 0820
DY OF THE FORCE CONSTANTS OF INORGANIC MOLECULES NITROGEN OXYFLUORIDE AND NITROGEN TRIFLUORIDE. AB INITIO STU 0734
IONS FOR THE GROUND STATE OF OZONE, NITRITE ION, NITROGEN OXYFLUORIDE, AND OXYGEN DIFLUORIDE. /T SCF CALCULAT 0776
DES AND OXYFLUORIDES, CHLORINE PENTAFLUORIDE AND CHLORINE OXYFLUORIDE BY NUCLEAR MAGNETIC RESONANCE. / FLUORI 0414
XYFLUORIDE BY NUC/ STUDY OF LIQUID CHLORINE FLUORIDES AND OXYFLUORIDES, CHLORINE PENTAFLUORIDE AND CHLORINE O 0414
/ONOXIDE, MOLECULAR NITROGEN, NITRIC OXIDE, AND MOLECULAR OXYGEN AND OF CALCIUM FLUORIDE, CHROMIUM(II) FLUOR/ 1324
NICAL STUDIES OF POLARIZABILITIES FOR THE DIFLUORIDES OF OXYGEN AND SULFUR. /HERMODYNAMICS AND QUANTUM MECHA 0784
A REINVESTIGATION OF THE INFRARED SPECTRUM OF OXYGEN DIFLUORIDE. 0774

149

CNDO STUDY OF THE OZONIDE ION AND RELATED SPECIES. (OXYGEN DIFLUORIDE) 0782
DOUBLE ZETA SCF CALCULATIONS FOR NITRITE ION AND OXYGEN DIFLUORIDE. 0762
EXTENDED HUCKEL THEORY AND THE SHAPE OF MOLECULES. (OXYGEN DIFLUORIDE) 0758
FERMI DIAD OF NU-1 AND 2NU-2 IN OXYGEN DIFLUORIDE. 0772
GEOMETRY OF MOLECULES. (OXYGEN DIFLUORIDE) 0764
HEAT OF FORMATION OF OXYGEN DIFLUORIDE. 0769
ON AND THERMODYNAMIC FUNCTIONS FOR SILICON DIFLUORIDE AND OXYGEN DIFLUORIDE.   MEAN AMPLITUDES OF VIBRATI 0773
MICROWAVE SPECTRUM, STRUCTURE, AND DIPOLE MOMENT OF OXYGEN DIFLUORIDE. 0779
MILLIMETER WAVE SPECTRUM AND STRUCTURE OF OXYGEN DIFLUORIDE. 0768
ND STATE OF OZONE, NITRITE ION, NITROGEN OXYFLUORIDE, AND OXYGEN DIFLUORIDE. /T SCF CALCULATIONS FOR THE GROU 0776
PHOTOIONIZATION MASS SPECTROMETRY OF OXYGEN DIFLUORIDE. 0760
RAMAN SPECTRUM OF LIQUID OXYGEN DIFLUORIDE. 0767
TATIONAL HYPERFINE STRUCTURE IN THE MICROWAVE SPECTRUM OF OXYGEN DIFLUORIDE.   SPIN RO 0777
THIRD ORDER POTENTIAL CONSTANTS OF BENT XY2 MOLECULES. (OXYGEN DIFLUORIDE) 0770
ESSURE.   PREPARATION AND PURIFICATION OF OXYGEN DIFLUORIDE AND DETERMINATION OF ITS VAPOR PR 0781
PHOTOELECTRON SPECTRA OF OXYGEN DIFLUORIDE AND OXYGEN DICHLORIDE. 0766
CAL.   ARGON MATRIX RAMAN SPECTRA OF OXYGEN DIFLUORIDE AND THE OXYGEN FLUORIDE FREE RADI 0759
TREE-FOCK METHOD. (BORON DIFLUORIDE, NITROGEN DIFLUORIDE, OXYGEN DIFLUORIDE, CARBON DIFLUORIDE) /STRICTED HAR 0176
/. QUADRATIC POTENTIAL CONSTANTS AND AVERAGE STRUCTURE OF OXYGEN DIFLUORIDE FROM THE GROUND STATE ROTATIONAL/ 0778
TURE OF INORGANIC MOLECULES. PART-1: INFRARED SPECTRUM OF OXYGEN DIFLUORIDE FROM 2.5 TO 25 MICRONS. /ND STRUC 0761
NU-1- 2 NU-2 FERMI RESONANCE AND E/ MICROWAVE SPECTRUM OF OXYGEN DIFLUORIDE IN VIBRATIONALLY EXCITED STATES, 0771
IGURATION INTERACTION CALCULATION FOR THE GROUND STATE OF OXYGEN DIFLUORIDE, NITRITE ION AND CYANIDE ION. /NF 0763
/ND INTERACTION FORCE CONSTANTS IN DIFLUORIDE MOLECULES. (OXYGEN DIFLUORIDE, NITROGEN DIFLUORIDE, CARBON DIF/ 0299
D SECOND MOMENTS OF THE ELECTRONIC CHARGE DISTRIBUTION IN OXYGEN DIFLUORIDE, OZONE, AND SULFUR DIOXIDE. /S AN 0780
N DIFLUORIDE, CARBON DIFLUORIDE, NITROGEN DIFLUORIDE, AND OXYGEN DIFLUORIDE. (SCF) /ERYLLIUM DIFLUORIDE, BORO 0141
OXYGEN FLUORIDES. (OXYGEN DIFLUORIDE) 0785
FLUORINE NMR SPECTRA OF OXYGEN FLUORIDES. (OXYGEN DIFLUORIDE) 0775
NEW SPECTRA OF OXYGEN LIKE MOLECULES. (NITROGEN MONOFLUORIDE) 0674
COMPARISON OF THE THEORETICAL AND EXPERIMENTAL RESULTS. (OXYGEN MONOFLUORIDE) /POSITIVE AND NEGATIVE IONS. A 1025
ORIDE.   MATRIX INFRARED SPECTRUM OF OXYGEN MONOFLUORIDE AND DETECTION OF LITHIUM OXYFLU 0755
UNCTION, BINDING ENERGY, IONIZATION POTENTIAL, ELECTRON / OXYGEN MONOFLUORIDE, DOUBLET PI HARTREE-FOCK WAVE F 0757
NDING ENERGY, IONIZATION POTENTIAL, ELECTRON AFFINITY, D/ OXYGEN MONOFLUORIDE. HARTREE-FOCK WAVE FUNCTION, BI 0756
QUID AND SATURATED VAPOR ENTROPY. (CHLORINE MONOFLUORIDE, OXYGEN MONOFLUORIDE, PHOSPHORUS TRIFLUORIDE) /ED LI 0391
SUBLIMATION PRESSURES OF IODINE HEPTAFLUORIDE AND IODINE OXYGEN PENTAFLUORIDE.   MASS SPECTRA AND 0534
NMR SPECTRA OF IODINE HEPTAFLUORIDE AND IODINE OXYGEN PENTAFLUORIDE. 0516
AND CORIOLIS CONSTANTS IN IODINE PENTAFLUORIDE AND IODINE OXYGEN PENTAFLUORIDE. /VIBRATION, SHRINKAGE EFFECT 0513
FOR INTERHALOGEN COMPOUNDS. IODINE PENTAFLUORIDE, IODINE OXYGEN PENTAFLUORIDE, AND IODINE HEPTAFLUORIDE. /LD 0511
ENT DERIVATIVES FOR PHOSPHORUS TRIFLUORIDE AND PHOSPHORUS OXYGEN TRIFLUORIDE. /IES, BOND MOMENTS AND BOND MOM 0797
L SPECTRA AND STRUCTURES OF XENON TETRAFLUORIDE AND XENON OXYTETRAFLUORIDE.   VIBRATIONA 1261
RUTHENIUM PENTAFLUORIDE AND RUTHENIUM OXYTETRAFLUORIDE. (VAPOR PRESSURES) 0924
THE ELECTRONIC CHARGE DISTRIBUTION IN OXYGEN DIFLUORIDE, OZONE, AND SULFUR DIOXIDE. /S AND SECOND MOMENTS OF 0780
/NIMAL BASIS SET SCF CALCULATIONS FOR THE GROUND STATE OF OZONE, NITRITE ION, NITROGEN OXYFLUORIDE, AND OXYG/ 0776
)   CNDO STUDY OF THE OZONIDE ION AND RELATED SPECIES. (OXYGEN DIFLUORIDE 0782

P

C TOP MOLECULES AND FORCE CONSTANTS FOR NITROGEN TRIFLUO/ P-R SEPARATIONS AND CORIOLIS CONSTANTS FOR SYMMETRI 0708
CHLORIDE(3-), SELENIUM H/ THE STEREOCHEMICALLY INERT LONE PAIR. A SPECULATION ON THE BONDING IN ANTIMONY HEXA 1309
ANTS WITH THE MAXIMUM OVERLAP METHOD. MOLECULES WITH LONE PAIR ELECTRONS OR EMPTY ORBITALS. /N OF FORCE CONST 1395
ITROGEN TRIFLUORIDE.   LONE PAIR MODEL AND THE VIBRATIONAL FORCE CONSTANTS OF N 0705
PAIRING ENERGIES IN THE ACTINIDE SERIES. 1333
/ OF LIGAND ATOMS OF LATER SECOND ROW ELEMENTS IN PERFECT PAIRING VALENCE STATES. (PHOSPHORUS PENTAFLUORIDE,/ 0862
HIGH RESOLUTION INFRARED SPECTRA OF THE PARALLEL BANDS OF PHOSPHORUS PENTAFLUORIDE. 0872
IUM PENTAFLUORIDE)   PARAMAGNETIC HEXAFLUORIDE ANIONS OF GROUP-6. (SELEN 0959
F NITROGEN MONOFLUORIDE IN THE SINGLE/ GAS PHASE ELECTRON PARAMAGNETIC RESONANCE SPECTRUM AND DIPOLE MOMENT O 0673
RINE TETRAFLUORIDE.   ELECTRON PARAMAGNETIC RESONANCE SPECTRUM OF THE RADICAL CHLO 0413
SPECTROSCOPIC MEASUREMENTS FOR COMPOUNDS WHICH PRODUCE NO PARENT ION. (CARBON TETRAFLUORIDE) / PHOTOELECTRON 0339
LUORIDE.   BARRIER TO ELECTRON PASSAGE THROUGH ELECTRONEGATIVE ATOMS IN BORON TRIF 0196
INTRAMOLECULAR SCATTERING EFFECTS ON ELECTRON DIFFRACTION PATTERNS FOR THE RHENIUM HEXAFLUORIDE MOLECULE. /E 0911
EXCHANGE RATES FOR SOME GROUP-5 PENTAHALIDE MOLECULES. PENTACOORDINATED MOLECULES. PART-12: CORRELATION OF 0115
/ TRIGONAL BIPYRAMIDAL MOLECULES FROM SPECTROSCOPIC DATA. PENTACOORDINATED MOLECULES. PART-13: (PHOSPHORUS P/ 0853
ATIONS AND STEREOCHEMICAL NONRIGIDITY OF THE TRIGONAL BI/ PENTACOORDINATED MOLECULES. PART-14: MOLECULAR VIBR 0854
CHLORINE MONOFLUORIDE, CHLORINE TRIFLUORIDE, AND CHLORINE PENTAFLUORIDE) /OF THE GEOMETRY OF INTERHALOGENS. (0379
IC DATA. PENTACOORDINATED MOLECULES. PART-13: (PHOSPHORUS PENTAFLUORIDE) /YRAMIDAL MOLECULES FROM SPECTROSCOP 0853
ENON TETRAFLUORIDE, XENON OXIDE TETRAFLUORIDE, AND IODINE PENTAFLUORIDE. /IC STATES OF THE POSITIVE IONS OF X 1264
CRYSTAL STRUCTURE OF TUNGSTEN PENTAFLUORIDE. 1154
DENSITY, MELTING POINT AND VAPOR PRESSURE OF ANTIMONY PENTAFLUORIDE. 0082
IELECTRIC CONSTANT, AND SPECIFIC CONDUCTIVITY OF CHLORINE PENTAFLUORIDE. /OR PRESSURE, CRITICAL PROPERTIES, D 0426
E' FORCE FIELD OF PHOSPHORUS PENTAFLUORIDE. 0861
ELECTRON DIFFRACTION STUDY OF THE STRUCTURE OF PHOSPHORUS PENTAFLUORIDE. 0851
ENTHALPY OF FORMATION OF ARSENIC PENTAFLUORIDE. 0117
EQUILIBRIUM STUDIES OF CHLORINE PENTAFLUORIDE. 0417
EXCHANGE OF FLUORINE IN PHOSPHORUS PENTAFLUORIDE. 0836
FAR INFRARED SPECTRUM OF PHOSPHORUS PENTAFLUORIDE. 0842
FAR INFRARED SPECTRUM OF PHOSPHORUS PENTAFLUORIDE. 0855
FLUORINE NMR OF ANTIMONY PENTAFLUORIDE. 0078
FLUORINE NMR OF CHLORINE PENTAFLUORIDE. 0416
RIOLIS CONSTANTS, AND THERMODYNAMIC FUNCTIONS OF VANADIUM PENTAFLUORIDE. /OF VIBRATION, SHRINKAGE EFFECTS, CO 1219
ULAR ORBITAL ANALYSES OF TRIGONAL BIPYRAMIDS. (PHOSPHORUS PENTAFLUORIDE) /-A-SPHERE AND EXTENDED HUCKEL MOLEC 0828
EATS OF TRANSITION OF MOLYBDENUM HEXAFLUORIDE AND NIOBIUM PENTAFLUORIDE.   HEAT CAPACITY, ENTROPY, AND H 0662
TION INFRARED SPECTRA OF THE PARALLEL BANDS OF PHOSPHORUS PENTAFLUORIDE.   HIGH RESOLU 0872
TRIFLUORIDE, PHOSPHORUS OXIDE TRIFLUORIDE, AND PHOSPHORUS PENTAFLUORIDE.   INFRARED SPECTRA OF PHOSPHORUS 0799
INFRARED SPECTRUM OF ARSENIC PENTAFLUORIDE. 0114
INFRARED SPECTRUM OF BROMINE PENTAFLUORIDE. 0262
INFRARED SPECTRUM OF MATRIX ISOLATED NIOBIUM PENTAFLUORIDE. 0659
INFRARED SPECTRUM OF NIOBIUM PENTAFLUORIDE. 0661
INFRARED SPECTRUM OF PHOSPHORUS PENTAFLUORIDE. 0866
. (BROMINE MONOFLUORIDE, BROMINE TRIFLUORIDE, AND BROMINE PENTAFLUORIDE) /ED STUDIES OF THE BROMINE FLUORIDES 0256
HEXAFLUORO PHOSPHATE AND BORON TRIFLUORIDE OR PHOSPHORUS PENTAFLUORIDE. /XCHANGE BETWEEN TETRABUTYL AMMONIUM 0194
INTRAMOLECULAR EXCHANGE IN VANADIUM PENTAFLUORIDE. 1214

-TYPE DOUBLING IN THE MILLIMETER WAVE SPECTRUM OF BROMINE PENTAFLUORIDE.   K 0260
LASER SPECTROSCOPY. (PHOSPHORUS PENTAFLUORIDE)   0832
ATION PRESSURES OF IODINE HEPTAFLUORIDE AND IODINE OXYGEN PENTAFLUORIDE.   MASS SPECTRA AND SUBLIM 0534
MASS SPECTRUM OF ANTIMONY PENTAFLUORIDE.   0085
MATRIX ISOLATION STUDY OF CHLORINE PENTAFLUORIDE.   0419
DES OF VIBRATION AND THERMODYNAMIC FUNCTIONS FOR VANADIUM PENTAFLUORIDE.   MEAN AMPLITU 1216
OWAVE MEASUREMENTS OF THE J-8 FAR-9 TRANSITION OF BROMINE PENTAFLUORIDE.   MICR 0261
MICROWAVE SPECTRUM OF BROMINE PENTAFLUORIDE.   0269
MICROWAVE SPECTRUM OF IODINE PENTAFLUORIDE.   0502
ROMETRIC STUDY OF BINARY PENTAFLUORIDES. (ALSO ANTIMONY PENTAFLUORIDE)   MOLECULAR BEAM MASS SPECT 1376
ULAR CONSTANTS OF PHOSPHORUS HALIDES. PART-4. (PHOSPHORUS PENTAFLUORIDE)   MOLEC 0867
NSTANTS OF SOME PHOSPHORUS HALOGEN COMPOUNDS. (PHOSPHORUS PENTAFLUORIDE)   MOLECULAR CO 0868
MOLECULAR FORCE FIELD FOR BROMINE PENTAFLUORIDE.   0264
MOLECULAR STRUCTURE OF ANTIMONY PENTAFLUORIDE.   0080
NMR SPECTRA OF IODINE HEPTAFLUORIDE AND IODINE OXYGEN PENTAFLUORIDE.   0516
ON THE ROVIBRONIC ENERGY LEVELS AND SPECTRA OF PHOSPHORUS PENTAFLUORIDE.   NONRIGID MOLECULE EFFECTS 0841
NU-7 FUNDAMENTAL OF PHOSPHORUS PENTAFLUORIDE.   0848
HLORINE TRIFLUORIDE, SULFUR TETRAFLUORIDE, AND PHOSPHORUS PENTAFLUORIDE. /ESCA OF INEQUIVALENT FLUORINES IN C 0874
PARAMAGNETIC HEXAFLUORIDE ANIONS OF GROUP-6. (SELENIUM PENTAFLUORIDE)   0959
ANTS. APPLICATION TO E' SPECIES FORCE FIELD OF PHOSPHORUS PENTAFLUORIDE. /D IN THE CALCULATION OF FORCE CONST 0827
TIONAL POTENTIAL FUNCTION OF THE E' VIBRATIONS OF ARSENIC PENTAFLUORIDE. /RPENDICULAR BAND CONTOURS AND VIBRA 0116
PHOTOELECTRON SPECTRUM OF PHOSPHORUS PENTAFLUORIDE.   0846
POLAR STATES IN VANADIUM PENTAFLUORIDE.   1213
POTENTIAL CONSTANTS AND THERMODYNAMIC FUNCTIONS OF IODINE PENTAFLUORIDE.   0507
OTENTIAL CONSTANTS AND THERMODYNAMIC PROPERTIES OF IODINE PENTAFLUORIDE.   P 0508
POTENTIAL FUNCTION FOR PHOSPHORUS PENTAFLUORIDE.   0859
POTENTIAL FUNCTION FOR THE NU-7 VIBRATION OF PHOSPHORUS PENTAFLUORIDE.   0830
PREPARATION AND CHARACTERIZATION OF GOLD PENTAFLUORIDE.   0485
PREPARATION AND CRYSTAL STRUCTURE OF OSMIUM PENTAFLUORIDE.   0750
PREPARATION AND PROPERTIES (VAPOR PRESSURE) OF TUNGSTEN PENTAFLUORIDE.   1150
DOROTATION IN TRIGONAL BIPYRAMIDAL MOLECULES. (PHOSPHORUS PENTAFLUORIDE) /ON OF TRIGONAL BIPYRAMIDAL MOLECULE PSEU 0833
S. BERRY ROTATION CONTRA TURNSTILE ROTATION IN PHOSPHORUS PENTAFLUORIDE. /ON OF TRIGONAL BIPYRAMIDAL MOLECULE 0870
D INFRARED SPECTRUM OF TRIFLUORO PHOSPHORANE. (PHOSPHORUS PENTAFLUORIDE) RAMAN AN 0847
RAMAN INTENSITIES AND THE FORCE FIELD OF PHOSPHORUS PENTAFLUORIDE.   0844
N BAND CONTOUR OF THE NU-7 (E') FUNDAMENTAL OF PHOSPHORUS PENTAFLUORIDE. /MAN SPECTRA OF GASES. PART-10: RAMA 0879
RAMAN SPECTRUM OF CRYSTALLINE MOLYBDENUM PENTAFLUORIDE.   0623
RAMAN SPECTRUM OF SOLID PHOSPHORUS PENTAFLUORIDE   0858
FORCE CONSTANTS, AND THERMODYNAMIC PROPERTIES OF BROMINE PENTAFLUORIDE.   RAMAN SPECTRUM, STRUCTURE, 0267
STRENGTHS IN TRIGONAL BIPYRAMIDAL MOLECULES. (PHOSPHORUS PENTAFLUORIDE) RELATIVE BOND 0839
SELF IONIZATION OF HALOGEN FLUORIDES. (BROMINE PENTAFLUORIDE)   0263
ECULAR ORBITAL CALCULATIONS. PSEUDOROTATION IN PHOSPHORUS PENTAFLUORIDE.   SEMIEMPIRICAL MOL 0843
SOME PHYSICAL AND CHEMICAL PROPERTIES OF VANADIUM PENTAFLUORIDE.   1212
OF DIFFERENT TUNNELING MECHANISMS IN NONRIGID PHOSPHORUS PENTAFLUORIDE.   SPECTROSCOPIC CONSEQUENCES 0834
STRUCTURE OF LIQUID ANTIMONY PENTAFLUORIDE.   0081
STRUCTURE OF THE RADICAL SULFUR PENTAFLUORIDE.   1060
SYMMETRY GROUP OF PHOSPHORUS PENTAFLUORIDE.   0837
D OPTICAL TRANSITIONS OF URANIUM HEXAFLUORIDE AND URANIUM PENTAFLUORIDE. /IONS OF THE ELECTRONIC STRUCTURE AN 1179
RUS TRIFLUORIDE, PHOSPHORUS TETRAFLUORIDE, AND PHOSPHORUS PENTAFLUORIDE) /IDE, PHOSPHORUS DIFLUORIDE, PHOSPHO 0788
, ARSENIC TRIFLUORIDE, ARSENIC TETRAFLUORIDE, AND ARSENIC PENTAFLUORIDE) /IC MONOFLUORIDE, ARSENIC DIFLUORIDE 0093
THERMODYNAMIC PROPERTIES OF TUNGSTEN PENTAFLUORIDE.   1149
STANTS OF TRIGONAL BIPYRAMIDAL XY5 MOLECULES. (PHOSPHORUS PENTAFLUORIDE) UREY-BRADLEY FORCE CON 0838
SSURE OF URANIUM HEXAFLUORIDE IN EQUILIBRIUM WITH URANIUM PENTAFLUORIDE.   VAPOR PRE 1180
L ASSIGNMENTS AND NORMAL COORDINATE ANALYSIS FOR ANTIMONY PENTAFLUORIDE.   VIBRATIONA 0087
RATIONAL SPECTRA OF MIXED PHOSPHORUS HALIDES. (PHOSPHORUS PENTAFLUORIDE) VIB 0829
PENTAFLUORIDES IN THE GAS PHASE. VAPOR DENSITY OF NIOBIUM PENTAFLUORIDE. /AL SPECTRA OF NIOBIUM AND TANTALUM 0660
VIBRATIONAL SPECTRA OF VANADIUM PENTAFLUORIDE.   1211
VIBRATIONAL SPECTRUM OF LIQUID CHROMIUM PENTAFLUORIDE.   0435
HLORINE TRIFLUORIDE, SULFUR TETRAFLUORIDE, AND PHOSPHORUS PENTAFLUORIDE. /RAY PHOTOELECTRON SPECTROSCOPY OF C 0409
THE ISOELECTRONIC SPECIES IODINE PENTAFLUORIDE, TELLURIUM PENTAFLUORIDE(-) AND ANTIMONY PENTAFLUORIDE(-2) . / 0501
NE PENTAFLUORIDE, TELLURIUM PENTAFLUORIDE(-) AND ANTIMONY PENTAFLUORIDE(-2). / THE ISOELECTRONIC SPECIES IODI 0501
URANIUM PENTAFLUORIDE, A NEW FLUORIDE OF URANIUM.   1178
VIBRATIONAL ASSIGNMENT FOR IODINE PENTAFLUORIDE, A NONRIGID D5H MOLECULE.   0525
MASS SPECTROMETRIC EVIDENCE OF DIMERS IN BISMUTH PENTAFLUORIDE AND ANTIMONY PENTAFLUORIDE.   0084
TS OF SOME TRIGONAL BIPYRAMIDAL PENTAHALIDES. (PHOSPHORUS PENTAFLUORIDE AND ANTIMONY PENTAFLUORIDE) / CONSTAN 0850
INFRARED SPECTRA OF MATRIX ISOLATED ANTIMONY PENTAFLUORIDE AND ARSENIC PENTAFLUORIDE.   0113
ITUDES OF VIBRATION AND CORIOLIS CONSTANTS FOR PHOSPHORUS PENTAFLUORIDE AND ARSENIC PENTAFLUORIDE. MEAN AMPL 0840
GONAL BIPYRAMIDAL MOLECULE WITH APPLICATION TO PHOSPHORUS PENTAFLUORIDE AND ARSENIC PENTAFLUORIDE. /FOR A TRI 0864
IGHT OF THE BARRIER TO/ VIBRATIONAL SPECTRA OF PHOSPHORUS PENTAFLUORIDE AND ARSENIC PENTAFLUORIDE AND THE HE 0856
Y OF REACTIVE FLUORIDES. (SECOND ROW, THIRD ROW, ANTIMONY PENTAFLUORIDE, AND BISMUTH PENTAFLUORIDE) /THE STUD 1377
ECULAR FORCE FIELDS FOR INTERHALOGEN COMPOUNDS. (CHLORINE PENTAFLUORIDE AND BROMINE PENTAFLUORIDE) MOL 0425
E, FORCE CONSTANTS, AND THERMODYNAMIC FUNCTIONS OF IODINE PENTAFLUORIDE AND BROMINE PENTAFLUORIDE. / STRUCTUR 0505
/ OF LIQUID CHLORINE FLUORIDES AND OXYFLUORIDES, CHLORINE PENTAFLUORIDE AND CHLORINE OXYFLUORIDE BY NUCLEAR / 0414
CORIOLIS COUPLING COEFFICIENTS OF BROMINE PENTAFLUORIDE AND CHLORINE PENTAFLUORIDE.   0427
NSED PHASE. EVIDEN/ INFRARED AND RAMAN SPECTRA OF BROMINE PENTAFLUORIDE AND CHLORINE TRIFLUORIDE IN THE CONDE 0266
/IBRATION ROTATION BANDS OF TRIFLUORO METHANE, PHOSPHORUS PENTAFLUORIDE, AND CYCLOPROPANE BY COMPUTER TECHNI 0880
/Y. PART-18: STANDARD ENTHALPY OF FORMATION OF PHOSPHORUS PENTAFLUORIDE AND ENTHALPIES OF TRANSITION BETWEEN/ 0865
VIBRATIONAL SPECTRA AND STRUCTURES OF IODINE PENTAFLUORIDE AND HEPTAFLUORIDE.   0506
MOSSBAUER STUDY OF IODINE PENTAFLUORIDE AND IODINE HEPTAFLUORIDE.   0503
OTENTIAL CONSTANTS AND THERMODYNAMIC FUNCTIONS FOR IODINE PENTAFLUORIDE AND IODINE HEPTAFLUORIDE. P 0509
RATION, SHRINKAGE EFFECT AND CORIOLIS CONSTANTS IN IODINE PENTAFLUORIDE AND IODINE OXYGEN PENTAFLUORIDE. /VIB 0513
N DIFFRACTION / GAS PHASE MOLECULAR STRUCTURES OF BROMINE PENTAFLUORIDE AND IODINE PENTAFLUORIDE FROM ELECTRO 0265
/OBIUM-93 NUCLEAR QUADRUPOLE RESONANCE STUDIES OF NIOBIUM PENTAFLUORIDE AND ITS COMPLEXES WITH XENON DIFLUOR/ 0664
POTENTIAL FIELD AND FORCE CONSTANTS OF ANTIMONY PENTAFLUORIDE AND NIOBIUM PENTAFLUORIDE.   0086
/MR AND RAMAN STUDY OF COMPLEX FORMATION BETWEEN ANTIMONY PENTAFLUORIDE AND NIOBIUM PENTAFLUORIDE AND TANTAL/ 0079
DIFFRACTION INVESTIGATION OF ARSENIC TRIFLUORIDE, ARSENIC PENTAFLUORIDE, AND OTHER MOLECULES. AN ELECTRON 0098
ELECTRON DIFFRACTION INVESTIGATIONS OF PHOSPHORUS PENTAFLUORIDE AND PHOSPHORUS PENTACHLORIDE.   0869
ORANES. ELECTRONIC STRUCTURE OF PHOSPHORUS PENTAFLUORIDE AND POLYTOPAL REARRANGEMENT IN PHOSPH 0875
R PRESSURES) RUTHENIUM PENTAFLUORIDE AND RUTHENIUM OXYTETRAFLUORIDE. (VAPO 0924
STABILITIES OF TANTALUM PENTAFLUORIDE AND TANTALUM OXIDE TRIFLUORIDE.   1103
ELECTRON DIFFRACTION INVESTIGATION OF NIOBIUM PENTAFLUORIDE AND TANTALUM PENTAFLUORIDE.   0671

ENTHALPIES OF FORMATION OF NIOBIUM PENTAFLUORIDE AND TANTALUM PENTAFLUORIDE. 0667
RAMAN SPECTRA OF LIQUID NIOBIUM PENTAFLUORIDE AND TANTALUM PENTAFLUORIDE. 0670
ELECTRONIC STRUCTURES OF PHOSPHORUS PENTAFLUORIDE AND TETRAFLUORO PHOSPHORANE. 0857
/IN THE STUDY OF THE LOW FREQUENCY VIBRATIONS OF ANTIMONY PENTAFLUORIDE AND THE PURE ROTATIONAL SPECTRUM OF / 0083
ENTHALPY OF FORMATION OF TUNGSTEN PENTAFLUORIDE AND TUNGSTEN HEXAFLUORIDE. 1151
VIBRATIONAL SPECTRA OF MOLYBDENUM PENTAFLUORIDE AND TUNGSTEN PENTAFLUORIDE. 0626
RIDE, URANIUM TRIFLUORIDE, URANIUM TETRAFLUORIDE, URANIUM PENTAFLUORIDE, AND URANIUM HEXAFLUORIDE) /UM DIFLUO 1160
FORCE CONSTANTS INCLUDING THOS/ RAMAN SPECTRA OF ARSENIC PENTAFLUORIDE AND VANADIUM PENTAFLUORIDE AND THEIR 0873
E/ ORBITALS AND STRUCTURES OF PENTAFLUORIDES. (PHOSPHORUS PENTAFLUORIDE, ARSENIC PENTAFLUORIDE, AND BROMINE P 0831
/ MOLECULES BY THE LOGARITHIMIC STEPS METHOD. (PHOSPHORUS PENTAFLUORIDE, ARSENIC PENTAFLUORIDE, ANTIMONY PEN/ 0878
/ OF GROUP-5 PENTAFLUORIDES IN THE GAS PHASE. (PHOSPHORUS PENTAFLUORIDE, ARSENIC PENTAFLUORIDE, BISMUTH PENT/ 0876
/DIES. (BORON TRIFLUORIDE, SULFUR TETRAFLUORIDE, ANTIMONY PENTAFLUORIDE, ARSENIC PENTAFLUORIDE, CHLORINE TRI/ 0239
. MOLECULAR STRUCTURES OF ARSENIC TRIFLUORIDE AND ARSENIC PENTAFLUORIDE AS DETERMINED BY ELECTRON DIFFRACTION 0099
MICROWAVE SPECTROSCOPY OF CHLORINE PENTAFLUORIDE AT 70 AND 140 GHZ 0423
/E PYRAMIDAL MOLECULES. XENON OXIDE TETRAFLUORIDE, IODINE PENTAFLUORIDE, BROMINE PENTAFLUORIDE, AND CHLORINE/ 0418
/ APPLICATION TO SOME SQUARE PYRAMIDAL MOLECULES. (IODINE PENTAFLUORIDE, BROMINE PENTAFLUORIDE, AND CHLORINE/ 0420
/ OF VIBRATION FOR SOME PYRAMIDAL XY4Z MOLECULES. (IODINE PENTAFLUORIDE, BROMINE PENTAFLUORIDE, AND CHLORINE/ 0421
NTAFLUORI/ IDEAL GAS THERMODYNAMIC PROPERTIES OF CHLORINE PENTAFLUORIDE, BROMINE PENTAFLUORIDE, AND IODINE PE 0424
/ORIDE, BROMINE TRIFLUORIDE, IODINE TRIFLUORIDE, CHLORINE PENTAFLUORIDE, BROMINE PENTAFLUORIDE, IODINE PENTA/ 0397
/IGH TEMPERATURES. PART-1: CALCULATION METHODS. (CHLORINE PENTAFLUORIDE, BROMINE PENTAFLUORIDE, IODINE PENTA/ 0422
ENTHALPY OF FORMATION OF VANADIUM PENTAFLUORIDE BY FLUORINE BOMB CALORIMETRY. 1215
/AND THE STRUCTURES OF PHOSPHORUS TRIFLUORIDE, PHOSPHORUS PENTAFLUORIDE, DIFLUORO AMINE, AND TETRAFLUORO HYD/ 0801
IQUID STATE. RAMAN SPECTRUM OF IODINE PENTAFLUORIDE. EVIDENCE FOR POLYMERIZATION IN THE L 0512
L AND A TER/ VIBRATIONAL SPECTRA AND ASSIGNMENT OF IODINE PENTAFLUORIDE. FERMI RESONANCE BETWEEN A FUNDAMENTA 0504
/D CHLORINE TRIFLUORIDE, BROMINE TRIFLUORIDE, AND BROMINE PENTAFLUORIDE. FLUORINE EXCHANGE MECHANISM OF LIQU/ 0399
THE TRIGONAL BIPYRAMIDAL THE RAMAN SPECTRUM OF PHOSPHORUS PENTAFLUORIDE GAS. /LD AND MOLECULAR VIBRATIONS OF 0863
EES K, ENTHALPY OF FUSION A/ CALORIMETRIC STUDY OF IODINE PENTAFLUORIDE. HEAT CAPACITY BETWEEN 5 AND 350 DEGR 0510
PERATURE. EVIDENCE FOR A TRIG/ RAMAN SPECTRUM OF ANTIMONY PENTAFLUORIDE IN THE GAS PHASE AS A FUNCTION OF TEM 0077
ROTATIONAL SPECTRA OF BROMINE PENTAFLUORIDE IN THE MILLIMETER WAVE LENGTH REGION. 0268
ASSOCIATION OF ANTIMONY PENTAFLUORIDE IN VAPOR. 0088
/ STRUCTURES OF VARIOUS PHOSPHORUS HALIDES AND PHOSPHORUS PENTAFLUORIDE. INFRARED AND LOW TEMPERATURE RAMAN / 0849
MOLECULAR FORCE FIELD FOR INTERHALOGEN COMPOUNDS. IODINE PENTAFLUORIDE, IODINE OXYGEN PENTAFLUORIDE, AND IO/ 0511
DIMETHYL PHOSPHORUS TRIFLUORIDE. STRUCTURES OF PHOSPHORUS PENTAFLUORIDE, METHYL PHOSPHORUS TETRAFLUORIDE AND 0845
NT/ CALCULATION OF THE ELECTRONIC STRUCTURE OF A CHLORINE PENTAFLUORIDE MOLECULE BY THE MO LCAO SELF CONSISTE 0415
ELECTRON DIFFRACTION STUDY OF THE VANADIUM PENTAFLUORIDE MOLECULE IN THE VAPOR FORM. 1218
OF CHLORINE FLUORIDE, CHLORINE TRIFLUORIDE, AND CHLORINE PENTAFLUORIDE MOLECULES. ABSORPTION SPECTRA 0388
UM PENTAFLUORIDE. THERMODYNAMIC PROPERTIES OF URANIUM PENTAFLUORIDE, NEPTUNIUM PENTAFLUORIDE, AND PLUTONI 1217
/IFFRACTION STUDY OF THE MOLECULAR STRUCTURES OF VANADIUM PENTAFLUORIDE, NIOBIUM PENTAFLUORIDE, AND TANTALUM/ 1217
D PHOSPHORUS PE/ NORMAL COORDINATE ANALYSIS OF PHOSPHORUS PENTAFLUORIDE, PHOSPHORUS DIFLUORIDE DICHLORIDE, AN 0877
/ AND ROOT MEAN SQUARE VIBRATION AMPLITUDES OF PHOSPHORUS PENTAFLUORIDE, PHOSPHORUS PENTACHLORIDE, AND PHOSP/ 0826
/ILICON TETRAFLUORIDE, PHOSPHORUS TRIFLUORIDE, PHOSPHORUS PENTAFLUORIDE, SULFUR DIFLUORIDE, SULFUR TETRAFLUO/ 1373
ALENCY IN CERTAIN MOLECULES AND COMPLEX IONS. (PHOSPHORUS PENTAFLUORIDE, SULFUR HEXAFLUORIDE) DIRECTED V 0860
/ ELEMENTS IN PERFECT PAIRING VALENCE STATES. (PHOSPHORUS PENTAFLUORIDE, SULFUR HEXAFLUORIDE, AND CHLORINE T/ 0862
RIFLUORIDE. VESCF-MO STUDIES OF PHOSPHORUS PENTAFLUORIDE, SULFUR TETRAFLUORIDE, AND CHLORINE T 0835
PENTAFLUORIDE. TH/ POLYMER MONOMER EQUILIBRIA IN NIOBIUM PENTAFLUORIDE, TANTALUM PENTAFLUORIDE, AND ANTIMONY 0076
/ VIBRATIONAL SPECTRA OF THE ISOELECTRONIC SPECIES IODINE PENTAFLUORIDE, TELLURIUM PENTAFLUORIDE(-) AND ANTIM 0501
AFLUORIDE, AND OSMIUM PENT/ VAPOR PRESSURES OF MOLYBDENUM PENTAFLUORIDE, TUNGSTEN PENTAFLUORIDE, RHENIUM PENT 0624
TAFLUORIDE, / PREPARATION AND CRYSTAL STRUCTURE OF OSMIUM PENTAFLUORIDE. TWO FLUORIDES OF OSMIUM. (OSMIUM PEN 0749
EASUREMENTS OF THE DISPROPORTIONATION PRESSURES. (URANIUM PENTAFLUORIDE, URANIUM HEXAFLUORIDE) /ORIDES FROM M 1176
TRIFLUORIDE. INFRARED SPECTRA OF VANADIUM PENTAFLUORIDE, VANADIUM TETRAFLUORIDE, AND VANADIUM 1210
MASS SPECTRUM AND COMPOSITION OF NIOBIUM PENTAFLUORIDE VAPOR. 0666
ELECTRONIC SPECTRA OF LIQUID RUTHENIUM AND MOLYBDENUM PENTAFLUORIDES. 0925
OF IODINE HEPTAFLUORIDE, RHENIUM HEPTAFLUORIDE, AND OXIDE PENTAFLUORIDES. FLUORINE NMR 0517
GAS PHASE STRUCTURES AND MASS SPECTRA OF BINARY PENTAFLUORIDES. 0003
GAS PHASE STRUCTURES AND MASS SPECTRA OF BINARY PENTAFLUORIDES. 1365
TANTALUM. PART-3: VAPOR PRESSURES OF NIOBIUM AND TANTALUM PENTAFLUORIDES. HALIDES OF NIOBIUM AND 0663
BIUM AND TANTALUM FLUORIDES AT HIGH TEMPERATURES. PART-1: PENTAFLUORIDES. THERMODYNAMIC PROPERTIES OF NIO 0665
THERMODYNAMIC PROPERTIES OF NIOBIUM AND TANTALUM PENTAFLUORIDES. 0669
VIBRATIONAL SPECTRA OF MOLYBDENUM AND TUNGSTEN PENTAFLUORIDES. 0625
MOLECULAR BEAM MASS SPECTROMETRIC STUDY OF BINARY PENTAFLUORIDES. (ALSO ANTIMONY PENTAFLUORIDE) 1376
TRANSITION METAL PENTAFLUORIDES AND RELATED COMPOUNDS. 0013
THERMODYNAMIC PROPERTIES OF TRANSITION METAL PENTAFLUORIDES AT HIGH TEMPERATURE. (THIRD ROW) 1379
ATION METHODS. (CHLORINE PEN/ THERMODYNAMIC PROPERTIES OF PENTAFLUORIDES AT HIGH TEMPERATURES. PART-1: CALCUL 0422
LUORIDES OF SOME 4P- AND 4D / THERMODYNAMIC PROPERTIES OF PENTAFLUORIDES AT HIGH TEMPERATURES. PART-2: PENTAF 1434
LUORIDES OF SOME 5D-ELEMENTS. THERMODYNAMIC PROPERTIES OF PENTAFLUORIDES AT HIGH TEMPERATURES. PART-3: PENTAF 0536
LUORIDES OF 5F ELEMENTS. THERMODYNAMIC PROPERTIES OF PENTAFLUORIDES AT HIGH TEMPERATURES. PART-4: PENTAF 0647
LUORIDE, ARSENIC PENTAFLUORIDE, B/ ASSOCIATION OF GROUP-5 PENTAFLUORIDES IN THE GAS PHASE. (PHOSPHORUS PENTAF 0876
IOBIUM PENTA/ VIBRATIONAL SPECTRA OF NIOBIUM AND TANTALUM PENTAFLUORIDES IN THE GAS PHASE. VAPOR DENSITY OF N 0660
SPECTROSCOPIC STUDIES IN MEAN AMPLITUDES OF VIBRATION OF PENTAFLUORIDES OF ARSENIC, PHOSPHORUS, AND VANADIU/ 0871
ESONANCE. CONFIGURATION OF THE PENTAFLUORIDES OF RHODIUM AND IRIDIUM BY MAGNETIC R 0923
PENTAFLUORIDE, AND BROMINE PE/ ORBITALS AND STRUCTURES OF PENTAFLUORIDES. (PHOSPHORUS PENTAFLUORIDE, ARSENIC 0831
AMPLITUDES OF VIBRATION: SULFUR TETRAFLUORIDE AND SULFUR PENTAFLUORO CHLORIDE MOLECULES. MEAN 1057
VIBRATIONS OF PENTATOMIC TETRAHEDRAL MOLECULES. 1497
/MB FIELD OF LIGAND ATOMS OF LATER SECOND ROW ELEMENTS IN PERFECT PAIRING VALENCE STATES. (PHOSPHORUS PENTAF/ 0862
LUORIDE AND COMPARISON WITH FORCE CONSTANTS OF THE EIGHTH PERIOD. /NTS OF CARBON TETRAFLUORIDE AND BORON TRIF 0212
ACCEPTOR PROPERTIES OF IODINE HEPTAFLUORIDE. OCTAFLUORO PERIODATES(VII). 0514
/OF MOLECULES CONTAINING ATOMS FROM THE SECOND ROW OF THE PERIODIC TABLE. (SILICON DIFLUORIDE, SILICON TETRA/ 1373
AL FUNCTION OF THE E' VIBRATIONS OF ARSENIC PENTAFLUORID/ PERPENDICULAR BAND CONTOURS AND VIBRATIONAL POTENTI 0116
O THE NU-4 BAND OF NITROGEN TRIFLUORIDE. CALCULATION OF PERPENDICULAR INFRARED BAND CONTOURS: APPLICATION T 0747
PERPENDICULAR VIBRATIONS OF BORON TRIFLUORIDE. 0202
STANTS FROM THE / VIBRATIONAL ANALYSIS OF SUBSTITUTED AND PERTURBED MOLECULES. DIRECT DETERMINATION FORCE CON 1444
OF THE STRUCTURE OF ZIRCONIUM TETRAFLUORIDE IN THE VAPOR PHASE. ELECTRON DIFFRACTION INVESTIGATION 1332
NON MONOFLUORIDE AS A CHEMICALLY BOUND SPECIES IN THE GAS PHASE. PROBABLE NONEXISTENCE OF XE 1221
TRIG/ RAMAN SPECTRUM OF ANTIMONY PENTAFLUORIDE IN THE GAS PHASE AS A FUNCTION OF TEMPERATURE. EVIDENCE FOR A 0077
XENON HEXAFLUORIDE. THE STRUCTURE OF A CUBIC PHASE AT -30C. 1283
DIPOLE MOMENT OF NITROGEN MONOFLUORIDE IN THE SINGLE/ GAS PHASE ELECTRON PARAMAGNETIC RESONANCE SPECTRUM AND 0673
RIDE, SELENIUM MONOFLUORIDE, SELENIUM OXIDE AND IODI/ GAS PHASE ELECTRON RESONANCE SPECTRA OF SULFUR MONOFLUC 1020
RIDE AND SELENIUM MONOFLUORIDE. GAS PHASE ELECTRON RESONANCE SPECTRA OF SULFUR MONOFLUO 1021
OF CARBON MONOFLUORIDE. GAS PHASE ELECTRON RESONANCE SPECTRUM AND DIPOLE MOMENT 0284
/ PENTAFLUORIDE AND CHLORINE TRIFLUORIDE IN THE CONDENSED PHASE. EVIDENCE OF AN ASSOCIATED STATE IN LIQUID C/ 0266

/IFFRACTION STUDY OF THE MOLECULAR STRUCTURE CF THE VAPOR PHASE HALIDES OF GALLIUM, YTTRIUM, LANTHANUM, AND / 1317
S ON EUROPIUM DIFLUORIDE. PHASE INVESTIGATIONS AND VAPOR PRESSURE MEASUREMENT 0452
AND IODINE PENTAFLUORIDE FROM ELECTRON DIFFRACTION / GAS PHASE MOLECULAR STRUCTURES OF BROMINE PENTAFLUORIDE 0265
RIDE, B/ ASSOCIATION OF GROUP-5 PENTAFLUORIDES IN THE GAS PHASE. (PHOSPHORUS PENTAFLUORIDE, ARSENIC PENTAFLUO 0876
MANIUM, AND TIN) AND MF4 (M= CARBON, SILICON, AND / VAPOR PHASE RAMAN SPECTRA OF MH4 (M= CARBON, SILICON, GER 0314
OP T/ ROTATIONAL PROFILES OF VIBRATIONAL BANDS IN THE GAS PHASE RAMAN SPECTRA OF MOLECULES OF THE SPHERICAL T 1442
ND CORIOLIS CONSTANTS OF THE SPHERICAL TOP MOLECUL/ VAPOR PHASE RAMAN SPECTRA, RAMAN BAND CONTOUR ANALYSES, A 1069
ORIOLIS CONSTANTS, FORCE CONSTANTS, AND VALUES FOR/ VAPOR PHASE RAMAN SPECTRA, RAMAN BAND CONTOUR ANALYSES, C 0795
ENERGETICS AND STABILITY OF GAS PHASE SCANDIUM FLUORIDES. 0951
GAS PHASE STRUCTURE OF XENON DIFLUORIDE. 1248
GAS PHASE STRUCTURE OF XENON HEXAFLUORIDE. 1293
UORIDES. GAS PHASE STRUCTURES AND MASS SPECTRA OF BINARY PENTAFL 0003
UORIDES. GAS PHASE STRUCTURES AND MASS SPECTRA OF BINARY PENTAFL 1365
/ION STUDY OF URANIUM AND THORIUM TETRAHALIDES IN THE GAS PHASE. (URANIUM TETRAFLUORIDE AND THORIUM TETRAFLU/ 1170
SPECTRA OF NIOBIUM AND TANTALUM PENTAFLUORIDES IN THE GAS PHASE. VAPOR DENSITY OF NIOBIUM PENTAFLUORIDE. /AL 0660
ON HEXAFLUORIDE. STRUCTURAL CRYSTALLOGRAPHY OF TETRAMERIC PHASES. XEN 1284
/FLUORINE EXCHANGE BETWEEN TETRABUTYL AMMONIUM HEXAFLUORO PHOSPHATE AND BORON TRIFLUORIDE OR PHOSPHORUS PENT/ 0194
FLUORIDE, BORON TRIFLUORIDE, ETHANE, SILANE, AMMONIA, AND PHOSPHINE. / OF CARBON TETRAFLUORIDE, SILICON TETRA 0243
CTRON IMPACT OF PHOSPHORUS TRIFLUORIDE AND DIFLUORO CYANO PHOSPHINE. IONIZATION BY ELE 0800
TRA AND MOLECULAR CONSTANTS OF PHOSPHORUS TRIFLUORIDE AND PHOSPHINE. RAMAN SPEC 0822
UND IONIC STATES. THEORETICAL STUDY OF THE GEOMETRY OF PHOSPHINE AND PHOSPHORUS TRIFLUORIDE AND THEIR GRO 0794
SPHINE. AB INITIO CALCULATIONS OF THE BONDING IN PHOSPHINE, PHOSPHORUS TRIFLUORIDE AND TRIMETHYL PHO 0802
ADAPTATION OF THE UREY-BRADLEY FORCE FIELD TO AMMONIA, PHOSPHINE, STIBINE, AND ARSINE. 1492
IC STRUCTURES OF PHOSPHORUS PENTAFLUORIDE AND TETRAFLUORO PHOSPHORANE. ELECTRON 0857
RAMAN AND INFRARED SPECTRUM CF TRIFLUORO PHOSPHORANE. (PHOSPHORUS PENTAFLUORIDE) 0847
F PHOSPHORUS PENTAFLUORIDE AND POLYTOPAL REARRANGEMENT IN PHOSPHORANES. ELECTRONIC STRUCTURE O 0875
FLUORIDES OF PHOSPHORUS. 0819
/INFRARED SPECTRA AND THERMODYNAMIC FUNCTIONS FOR VARIOUS PHOSPHORUS AND ARSENIC HALIDES. (PHOSPHORUS TRIFLU/ 0811
RT-3: TRIFLUORIDES AND OXIDE TRIFLUORIDES OF NITROGEN AND PHOSPHORUS, AND PHOSPHORUS OXIDE TRICHLORIDE. /. PA 0701
NTRA- INTERMOLECULAR MEC/ FLUORINE NMR SPECTRA OF SULFUR, PHOSPHORUS, AND SILICON FLUORIDES. HYDROLYSIS AND I 1042
EAN AMPLITUDES OF VIBRATION OF PENTAFLUORIDES OF ARSENIC, PHOSPHORUS, AND VANADIUM. /ECTROSCOPIC STUDIES IN M 0871
THERMODYNAMIC FUNCTIONS OF SOME PHOSPHORUS COMPOUNDS. (PHOSPHORUS TRIFLUORIDE) 0812
RADICALS. ELECTRON SPIN RESONANCE OF TRAPPED PHOSPHORUS DIFLUORIDE AND PHOSPHORUS TETRAFLUORIDE 0792
LS. ESR SPECTRA OF PHOSPHORUS DIFLUORIDE AND SULFUR TRIFLUORIDE RADICA 0791
BON DIFLUORIDE, NITROGEN DIFLUORIDE, ALUMINUM DIFLUORIDE, PHOSPHORUS DIFLUORIDE, AND ZIRCONIUM DIFLUORIDE. /R 1384
ITROGEN DICHLORIDE IN/ ELECTRON SPIN RESONANCE STUDIES OF PHOSPHORUS DIFLUORIDE, PHOSPHORUS DICHLORIDE, AND N 0793
AB INITIO MOLECULAR ORBITAL STUDY OF THE PHOSPHORUS DIFLUORIDE RADICAL. 0790
/MIC PROPERTIES OF PHOSPHORUS(2), PHOSPHORUS(4), AND SOME PHOSPHORUS FLUORIDES. (PHOSPHORUS MONOFLUORIDE, PH/ 0788
FRARED AND LOW TEMPERATU/ MOLECULAR STRUCTURES OF VARIOUS PHOSPHORUS HALIDES AND PHOSPHORUS PENTAFLUORIDE. IN 0849
DE) MOLECULAR CONSTANTS OF PHOSPHORUS HALIDES. PART-4. (PHOSPHORUS PENTAFLUORI 0867
VIBRATIONAL SPECTRA OF MIXED PHOSPHORUS HALIDES. (PHOSPHORUS PENTAFLUORIDE) 0829
RIDE) MOLECULAR CONSTANTS OF SOME PHOSPHORUS HALOGEN COMPOUNDS. (PHOSPHORUS PENTAFLUO 0868
DE) MOLECULAR CONSTANTS OF PHOSPHORUS HALOGEN COMPOUNDS. (PHOSPHORUS TRIFLUORI 0813
F SOME PNICOGEN MONOFLUORIDES. (NITROGEN MONOFLUORIDE AND PHOSPHORUS MONOFLUORIDE) QUANTUM CHEMICAL STUDY O 0678
ROTATIONAL FINE STRUCTURE OF THE TRIPLET PI STATE OF PHOSPHORUS MONOFLUORIDE. 0787
DIATOMIC GASES WITH MOLECULES IN TRIPLET SIGMA STATES. (PHOSPHORUS MONOFLUORIDE) THERMODYNAMIC FUNCTIONS OF 0789
ELECTRONIC SPECTRA OF PHOSPHORUS MONOFLUORIDE AND ITS POSITIVE ION. 0786
/OPERTIES OF NITROGEN MONOFLUORIDE, SILICON MONOFLUORIDE, PHOSPHORUS MONOFLUORIDE, AND SULFUR MONOFLUORIDE F/ 0973
FLUORESCENCE SPECTRA OF PHOSPHORUS NITRIDE AND BORON FLUORIDE. 0167
LENCE ELECTRON CALCULATIONS ON SMALL MOLECULES CONTAINING PHOSPHORUS OR SULFUR. NONEMPIRICAL VA 1029
SPECTROSCOPIC DATA. PENTACOORDINATED MOLECULES. PART-13: (PHOSPHORUS PENTAFLUORIDE) /YRAMIDAL MOLECULES FROM 0853
E' FORCE FIELD OF PHOSPHORUS PENTAFLUORIDE. 0861
ELECTRON DIFFRACTION STUDY OF THE STRUCTURE OF PHOSPHORUS PENTAFLUORIDE. 0851
EXCHANGE OF FLUORINE IN PHOSPHORUS PENTAFLUORIDE. 0836
FAR INFRARED SPECTRUM OF PHOSPHORUS PENTAFLUORIDE. 0842
FAR INFRARED SPECTRUM OF PHOSPHORUS PENTAFLUORIDE. 0855
UCKEL MOLECULAR ORBITAL ANALYSES OF TRIGONAL BIPYRAMIDS. (PHOSPHORUS PENTAFLUORIDE) /-A-SPHERE AND EXTENDED H 0828
HIGH RESOLUTION INFRARED SPECTRA OF THE PARALLEL BANDS OF PHOSPHORUS PENTAFLUORIDE. 0872
INFRARED SPECTRUM OF PHOSPHORUS PENTAFLUORIDE. 0866
YL AMMONIUM HEXAFLUORO PHOSPHATE AND BORON TRIFLUORIDE OR PHOSPHORUS PENTAFLUORIDE. /XCHANGE BETWEEN TETRABUT 0194
LASER SPECTROSCOPY. (PHOSPHORUS PENTAFLUORIDE) 0832
LE EFFECTS ON THE ROVIBRONIC ENERGY LEVELS AND SPECTRA OF PHOSPHORUS PENTAFLUORIDE. NONRIGID MOLECU 0841
NU-7 FUNDAMENTAL OF PHOSPHORUS PENTAFLUORIDE. 0848
ORINES IN CHLORINE TRIFLUORIDE, SULFUR TETRAFLUORIDE, AND PHOSPHORUS PENTAFLUORIDE. /ESCA OF INEQUIVALENT FLU 0874
FORCE CONSTANTS. APPLICATION TO E' SPECIES FORCE FIELD OF PHOSPHORUS PENTAFLUORIDE. /D IN THE CALCULATION OF 0827
PHOTOELECTRON SPECTRUM OF PHOSPHORUS PENTAFLUORIDE. 0846
POTENTIAL FUNCTION FOR THE NU-7 VIBRATION OF PHOSPHORUS PENTAFLUORIDE. 0830
POTENTIAL FUNCTION FOR PHOSPHORUS PENTAFLUORIDE. 0859
PSEUDOROTATION IN TRIGONAL BIPYRAMIDAL MOLECULES. (PHOSPHORUS PENTAFLUORIDE) 0833
AL MOLECULES. BERRY ROTATION CONTRA TURNSTILE ROTATION IN PHOSPHORUS PENTAFLUORIDE. /ON OF TRIGONAL BIPYRAMID 0870
RAMAN AND INFRARED SPECTRUM OF TRIFLUORO PHOSPHORANE. (PHOSPHORUS PENTAFLUORIDE) 0847
RAMAN INTENSITIES AND THE FORCE FIELD OF PHOSPHORUS PENTAFLUORIDE. 0844
NTAFLUORIDE AND THEIR FORCE CONSTANTS INCLUDING THOSE OF PHOSPHORUS PENTAFLUORIDE. /FLUORIDE AND VANADIUM PE 0873
RT-10: RAMAN BAND CONTOUR OF THE NU-7 (E') FUNDAMENTAL OF PHOSPHORUS PENTAFLUORIDE. /MAN SPECTRA OF GASES. PA 0879
RAMAN SPECTRUM OF SOLID PHOSPHORUS PENTAFLUORIDE 0858
LATIVE BOND STRENGTHS IN TRIGONAL BIPYRAMIDAL MOLECULES. (PHOSPHORUS PENTAFLUORIDE) RE 0839
PIRICAL MOLECULAR ORBITAL CALCULATIONS. PSEUDOROTATION IN PHOSPHORUS PENTAFLUORIDE. SEMIEM 0843
ONSEQUENCES OF DIFFERENT TUNNELING MECHANISMS IN NONRIGID PHOSPHORUS PENTAFLUORIDE. SPECTROSCOPIC C 0834
SYMMETRY GROUP OF PHOSPHORUS PENTAFLUORIDE. 0837
Y FORCE CONSTANTS OF TRIGONAL BIPYRAMIDAL XY5 MOLECULES. (PHOSPHORUS PENTAFLUORIDE. UREY-BRADLE 0838
OSCOPY OF CHLORINE TRIFLUORIDE, SULFUR TETRAFLUORIDE, AND PHOSPHORUS PENTAFLUORIDE. /RAY PHOTOELECTRON SPECTR 0409
/CE CONSTANTS OF SOME TRIGONAL BIPYRAMIDAL PENTAHALIDES. (PHOSPHORUS PENTAFLUORIDE AND ANTIMONY PENTAFLUORID/ 0850
MEAN AMPLITUDES OF VIBRATION AND CORIOLIS CONSTANTS FOR PHOSPHORUS PENTAFLUORIDE AND ARSENIC PENTAFLUORIDE. 0840
AND THE HEIGHT OF THE BARRIER TO/ VIBRATIONAL SPECTRA OF PHOSPHORUS PENTAFLUORIDE AND ARSENIC PENTAFLUORIDE. 0856
/ FOR A TRIGONAL BIPYRAMIDAL MOLECULE WITH APPLICATION TO PHOSPHORUS PENTAFLUORIDE AND ARSENIC PENTAFLUORIDE. 0864
/ES OF THE VIBRATION ROTATION BANDS OF TRIFLUORO METHANE, PHOSPHORUS PENTAFLUORIDE, AND CYCLOPROPANE BY COMP/ 0880
/ CALORIMETRY. PART-18: STANDARD ENTHALPY OF FORMATION OF PHOSPHORUS PENTAFLUORIDE AND ENTHALPIES OF TRANSIT/ 0865
DE. ELECTRON DIFFRACTION INVESTIGATIONS OF PHOSPHORUS PENTAFLUORIDE AND PHOSPHORUS PENTACHLORI 0869
T IN PHOSPHORANES. ELECTRONIC STRUCTURE OF PHOSPHORUS PENTAFLUORIDE AND POLYTOPAL REARRANGEMEN 0875
E. ELECTRONIC STRUCTURES OF PHOSPHORUS PENTAFLUORIDE AND TETRAFLUORO PHOSPHORAN 0857

D BROMINE PE/ ORBITALS AND STRUCTURES OF PENTAFLUORIDES. (PHOSPHORUS PENTAFLUORIDE, ARSENIC PENTAFLUORIDE, AN 0831
/BIPYRAMIDAL MOLECULES BY THE LOGARITHIMIC STEPS METHOD. (PHOSPHORUS PENTAFLUORIDE, ARSENIC PENTAFLUORIDE, A/ 0878
ASSOCIATION OF GROUP-5 PENTAFLUORIDES IN THE GAS PHASE. (PHOSPHORUS PENTAFLUORIDE, ARSENIC PENTAFLUORIDE, B/ 0876
RATIONS OF THE TRIGONAL BIPYRAMIDAL THE RAMAN SPECTRUM OF PHOSPHORUS PENTAFLUORIDE GAS. /LD AND MOLECULAR VIB 0863
UORIDE AND DIMETHYL PHOSPHORUS TRIFLUORIDE. STRUCTURES OF PHOSPHORUS PENTAFLUORIDE, METHYL PHOSPHORUS TETRAFL 0845
HLORIDE, AND PHOSPHORUS PE/ NORMAL COORDINATE ANALYSIS OF PHOSPHORUS PENTAFLUORIDE, PHOSPHORUS DIFLUORIDE DIC 0877
/E CONSTANTS AND ROOT MEAN SQUARE VIBRATION AMPLITUDES OF PHOSPHORUS PENTAFLUORIDE, PHOSPHORUS PENTACHLORIDE/ 0826
DIRECTED VALENCY IN CERTAIN MOLECULES AND COMPLEX IONS. (PHOSPHORUS PENTAFLUORIDE, SULFUR HEXAFLUORIDE) 0860
/ SECOND ROW ELEMENTS IN PERFECT PAIRING VALENCE STATES. (PHOSPHORUS PENTAFLUORIDE, SULFUR HEXAFLUORIDE, AND/ 0862
CHLORINE TRIFLUORIDE. VESCF-MO STUDIES OF PHOSPHORUS PENTAFLUORIDE, SULFUR TETRAFLUORIDE, AND 0835
/TAFLUORIDE, BROMINE PENTAFLUORIDE, IODINE PENTAFLUORIDE, PHOSPHORUS PENTAFLUORIDE, VANADIUM PENTAFLUORIDE, / 0422
INTRAMOLECULAR EXCHANGE IN PHOSPHORUS PENTAHALIDE MOLECULES. 0852
CHEMICAL SHIFTS OF BORON, PHOSPHORUS, PHOSPHORYL, AND THIOPHOSPHORYL HALIDES. 0235
INFRARED AND RAMAN SPECTRA AND STRUCTURE OF PHOSPHORUS TETRAFLUORIDE. 0825
ROPIC HYPERFINE SPLITTINGS AND MOLECULAR STRUCTURE IN THE PHOSPHORUS TETRAFLUORIDE RADICAL. ISOT 0824
GLE CRYSTAL OF PHOSPHORUS TRIFLUORIDE. ESR SPECTRA OF PHOSPHORUS TETRAFLUORIDE RADICALS PRODUCED IN A SIN 0823
ONSTANTS OF ARSENIC TRIFLUORIDE, ARSENIC TRICHLORIDE, AND PHOSPHORUS TRICHLORIDE. /UORIDE AND THE MOLECULAR C 0111
AND NORMAL VIBRATIONAL FREQUENCI/ POTENTIAL CONSTANTS OF PHOSPHORUS TRICHLORIDE FROM AMPLITUDES OF VIBRATION 1481
E HYDRIDES, DEUTERIDES, AND HALIDES OF GROUP-5 ELEMENTS. (PHOSPHORUS TRIFLUORIDE) ABSORPTION SPECTRA OF TH 0806
CENTRIFUGAL DISTORTION CONSTANTS OF PHOSPHORUS TRIFLUORIDE. 0814
HEXAFLUORIDE, XENON TETRAFLUORIDE, XENON DIFLUORIDE, AND PHOSPHORUS TRIFLUORIDE. /PIES OF FORMATION OF XENON 0807
ENTHALPY OF FORMATION PHOSPHORUS TRIFLUORIDE. 0818
HIGH RESOLUTION SPECTRA OF ARSENIC TRIFLUORIDE AND PHOSPHORUS TRIFLUORIDE. 0816
INFRARED SPECTRA OF NITROGEN TRIFLUORIDE AND PHOSPHORUS TRIFLUORIDE. 0748
IHALIDES. (NITROGEN TRIFLUORIDE, ARSENIC TRIFLUORIDE, AND PHOSPHORUS TRIFLUORIDE) /NCTIONS OF SOME GROUP-5 TR 0719
T AND CALCULATION OF THE ABSOLUTE INFRARED INTENSITIES OF PHOSPHORUS TRIFLUORIDE. MEASUREMEN 0808
POR ENTROPY. (CHLORINE MONOFLUORIDE, OXYGEN MONOFLUORIDE, PHOSPHORUS TRIFLUORIDE) /ED LIQUID AND SATURATED VA 0391
ROWAVE INVESTIGATIONS OF METHYL FLUORIDE, FLUOROFORM, AND PHOSPHORUS TRIFLUORIDE. MIC 0798
O TO THREE MILLIMETERS, PART-2: (NITROGEN TRIFLUORIDE AND PHOSPHORUS TRIFLUORIDE) /COPY IN THE REGION FROM TW 0711
MICROWAVE SPECTRUM OF PHOSPHORUS TRIFLUORIDE. 0805
NU-3 BAND OF PHOSPHORUS TRIFLUORIDE. 0815
HARGE AND CONFIGURATION MOLECULAR ORBITAL CALCULATIONS ON PHOSPHORUS TRIFLUORIDE. SELF CONSISTENT C 0821
EEMAN EFFECT OF SOME LINEAR AND SYMMETRIC TOP MOLECULES. (PHOSPHORUS TRIFLUORIDE) Z 0796
2NU-4 SPECTRUM OF PHOSPHORUS TRIFLUORIDE. 0817
XY3 MOLECULES. (NITROGEN TRIFLUORIDE, BORON TRIFLUORIDE, PHOSPHORUS TRIFLUORIDE, AND ARSENIC TRIFLUORIDE) /F 0709
FOR SOME PYRAMIDAL XY3 MOLECULES. (NITROGEN TRIFLUORIDE, PHOSPHORUS TRIFLUORIDE, AND ARSENIC TRIFLUORIDE) / 0706
MOLECULAR FORCE FIELD FOR SOME XY3 MOLECULES. (PHOSPHORUS TRIFLUORIDE AND ARSENIC TRIFLUORIDE) 1400
P MOLECULES AND FORCE CONSTANTS FOR NITROGEN TRIFLUORIDE, PHOSPHORUS TRIFLUORIDE, AND ARSENIC TRIFLUORIDE. /O 0708
THE COMBINE/ GENERAL FORCE FIELD OF NITROGEN TRIFLUORIDE, PHOSPHORUS TRIFLUORIDE, AND ARSENIC TRIFLUORIDE BY 0717
IONIZATION BY ELECTRON IMPACT OF PHOSPHORUS TRIFLUORIDE AND DIFLUORO CYANO PHOSPHINE 0800
MOLECULAR ORBITAL STUDY OF NITROGEN TRIFLUORIDE, PHOSPHORUS TRIFLUORIDE, AND NITROGEN DIFLUORIDE. 0721
RAMAN SPECTRA AND MOLECULAR CONSTANTS OF PHOSPHORUS TRIFLUORIDE AND PHOSPHINE. 0822
AN AB INITIO SCF STUDY. BONDING IN PHOSPHORUS TRIFLUORIDE AND PHOSPHORUS OXYFLUORIDE. 0820
/NTENSITIES, BOND MOMENTS AND BOND MOMENT DERIVATIVES FOR PHOSPHORUS TRIFLUORIDE AND PHOSPHORUS OXYGEN TRIFL/ 0797
S. THEORETICAL STUDY OF THE GEOMETRY OF PHOSPHINE AND PHOSPHORUS TRIFLUORIDE AND THEIR GROUND IONIC STATE 0794
AB INITIO CALCULATIONS OF THE BONDING IN PHOSPHINE, PHOSPHORUS TRIFLUORIDE AND TRIMETHYL PHOSPHINE. 0802
CENTRIFUGAL STRETCHING CONSTANTS. (NITROGEN TRIFLUORIDE, PHOSPHORUS TRIFLUORIDE, ARSENIC TRIFLUORIDE) /SE OF 0727
TIMONY/ CALCULATION OF UREY-BRADLEY POTENTIAL CONSTANTS. (PHOSPHORUS TRIFLUORIDE, ARSENIC TRIFLUORIDE, AND AN 0809
L-TYPE DOUBLING TRANSITIONS OF PHOSPHORUS TRIFLUORIDE IN THE NU-4 EQUALS 1 STATE. 0803
/RD ORDER CONSTANTS ON THE L-TYPE DOUBLING TRANSITIONS OF PHOSPHORUS TRIFLUORIDE IN THE NU-4 EQUALS 1 STATE. 0804
/ETIC SUSCEPTIBILITY, AND MOLECULAR QUADRUPOLE MOMENTS IN PHOSPHORUS TRIFLUORIDE, NITROGEN TRIFLUORIDE, AND / 0744
E, AND PHOSPHORUS PENTAFLUORIDE. INFRARED SPECTRA OF PHOSPHORUS TRIFLUORIDE, PHOSPHORUS OXIDE TRIFLUORID 0799
IFLU/ ELECTRON DIFFRACTION BY GASES AND THE STRUCTURES OF PHOSPHORUS TRIFLUORIDE, PHOSPHORUS PENTAFLUORIDE, D 0801
/IODIC TABLE. (SILICON DIFLUORIDE, SILICON TETRAFLUORIDE, PHOSPHORUS TRIFLUORIDE, PHOSPHORUS PENTAFLUORIDE, / 1373
RACTION. MOLECULAR STRUCTURE OF PHOSPHORUS TRIFLUORIDE STUDIED BY GAS ELECTRON DIFF 0810
LUORIDES. (PHOSPHORUS MONOFL/ THERMODYNAMIC PROPERTIES OF PHOSPHORUS(2), PHOSPHORUS(4), AND SOME PHOSPHORUS F 0788
PHORUS MONOFL/ THERMODYNAMIC PROPERTIES OF PHOSPHORUS(2), PHOSPHORUS(4), AND SOME PHOSPHORUS FLUORIDES. (PHOS 0788
CHEMICAL SHIFTS OF BORON, PHOSPHORUS, PHOSPHORYL, AND THIOPHOSPHORYL HALIDES. 0235
ENTS IN PHOSPHORUS TRIFLUORIDE, NITROGEN TRIFLUORIDE, AND PHOSPHORYL, THIONYL AND SULFURYL FLUORIDES. /LE MOM 0744
/ RADICAL CARBON TRIFLUORIDE ISOLATED IN INERT MATRICES. PHOTOCHLORINATION OF METHANE AND FLUOROFORM. DISSO/ 0311
EXAFLUORIDE) BY INFRARED RADIATION. SELECTIVE PHOTODISSOCIATION OF POLYATOMIC MOLECULES (SULFUR H 1064
OF XENON DIFLUORIDE, XENON TETRAF/ HELIUM-1 AND HELIUM-2 PHOTOELECTRON SPECTRA AND THE ELECTRONIC STRUCTURES 1228
RIDE AND AMINES. PHOTOELECTRON SPECTRA OF COMPLEXES OF BORON TRIFLUO 0224
HELIUM-1 RESONANCE PHOTOELECTRON SPECTRA OF GROUP-4 TETRAFLUORIDES. 0317
ORIDE, CHLORINE TRIFLUORIDE, BROMINE MONOFLUORIDE, AND B/ PHOTOELECTRON SPECTRA OF HALIDES. (CHLORINE MONOFLU 0376
RIDES AND TETRACHLORIDES OF GROUP-5B. (CARBON TETRAFLUOR/ PHOTOELECTRON SPECTRA OF HALIDES. PART-1. TETRAFLUO 0316
LUTION SPECTRA OF THE BORON TRIHALIDES. PHOTOELECTRON SPECTRA OF HALIDES. PART-2: HIGH RESO 0188
DES AND OXIDE TRIFLUORIDES OF NITROGEN AND PHOSPHORUS, A/ PHOTOELECTRON SPECTRA OF HALIDES. PART-3: TRIFLUORI 0701
TEMPERATURE HELIUM-1 AND HELIUM-2 STUDIES OF CARBON TETR/ PHOTOELECTRON SPECTRA OF HALIDES. PART-7: VARIABLE 0341
ITROGEN OXIDE TRIFLUORIDE, AND A REASSIGNMENT OF THE SPE/ PHOTOELECTRON SPECTRA OF NITROGEN TRIFLUORIDE AND N 0700
EN DICHLORIDE. PHOTOELECTRON SPECTRA OF OXYGEN DIFLUORIDE AND OXYG 0766
OUP-4, GROUP-5, AND GROUP-6. PHOTOELECTRON SPECTRA OF THE HALIDES IN GROUP-3, GR 1361
ON DIFLUORIDE. HELIUM-2 PHOTOELECTRON SPECTRA OF XENON DIFLUORIDE AND KRYPT 1234
/RGIES AND IONIC CHARACTER OF INERT GAS HALIDES. HELIUM-2 PHOTOELECTRON SPECTRA OF XENON DIFLUORIDE AND KRYP/ 1343
NDS WHICH PRODUCE NO PARENT ION. (CARBON/ IMPLICATIONS OF PHOTOELECTRON SPECTROSCOPIC MEASUREMENTS FOR COMPOU 0339
ELECTRONIC STRUCTURE OF KRYPTON DIFLUORIDE STUDIED BY PHOTOELECTRON SPECTROSCOPY. 0550
STUDY OF THE BONDING IN THE GROUP-4 TETRAHALIDES BY PHOTOELECTRON SPECTROSCOPY. 1362
AND SILICON TETRAFLUORIDE. HIGH RESOLUTION PHOTOELECTRON SPECTROSCOPY OF CARBON TETRAFLUORIDE 0319
SULFUR TETRAFLUORIDE, AND PHOSPHORUS PENTAFLUORID/ X-RAY PHOTOELECTRON SPECTROSCOPY OF CHLORINE TRIFLUORIDE, 0409
RS. PART-3: MONOMER AND DIMER SPECTRA OF THALLOUS FLUORI/ PHOTOELECTRON SPECTROSCOPY OF HIGH TEMPERATURE VAPO 1117
PHOTOELECTRON SPECTRUM OF BORON TRIFLUORIDE. 0189
PHOTOELECTRON SPECTRUM OF CHLORINE MONOFLUORIDE. 0369
PHOTOELECTRON SPECTRUM OF PHOSPHORUS PENTAFLUORIDE. 0846
PHOTOELECTRON SPECTRUM OF SILICON DIFLUORIDE. 0980
PHOTOELECTRON SPECTRUM OF SILICON DIFLUORIDE. 0996
ANY BODY APPROACH. PART-6: THE ASSIGNMENT OF THE HELIUM-2 PHOTOELECTRON SPECTRUM OF SULFUR HEXAFLUORIDE. /A M 1090
4 ANGSTROMS. PHOTOELECTRON SPECTRUM OF SULFUR HEXAFLUORIDE AT 58 1076
PHOTOELECTRON SPECTRUM OF XENON DIFLUORIDE. 1227
IFLUORIDE, XENON TETRAFLUORIDE, AND XENON HEXAFLUORIDE. PHOTOIONIZATION MASS SPECTROMETRIC STUDY OF XENON D 1225
IDE. PHOTOIONIZATION MASS SPECTROMETRY OF OXYGEN DIFLUOR 0760

THALLIUM MONOFLUORIDE, THALLIUM MONOCHLORIDE, AND THALL/ PHOTOIONIZATION OF HIGH TEMPERATURE VAPORS. PART-4: 1112
RTIES OF SOME HALOGEN MOLECULES. (CHLORINE FLUORIDE) PHOTOIONIZATION STUDIES AND THE THERMODYNAMIC PROPE 0377
THE DISSOCIATION ENERGY OF FLUORINE. PHOTOIONIZATION STUDY OF CHLORINE MONOFLUORIDE AND 0378
TRUM OF/ MATRIX ISOLATION STUDY OF THE VACUUM ULTRAVIOLET PHOTOLYSIS OF NITROGEN TRIFLUORIDE. ELECTRONIC SPEC 0691
FRARED / MATRIX ISOLATION STUDY OF THE VACUUM ULTRAVIOLET PHOTOLYSIS OF SILICON DIFLUORIDE DIHYDRIDE. THE IN 0986
ONIZATION AND DISSOCIATION OF XENON DIFLUORIDE INDUCED BY PHOTON IMPACT. I 1243
KRYPTON DIFLUORIDE. (PHYSICAL AND CHEMICAL PROPERTIES) 0563
ES. PHYSICAL AND CHEMICAL PROPERTIES OF HALOGEN FLUORID 0017
LUORIDE. SOME PHYSICAL AND CHEMICAL PROPERTIES OF VANADIUM PENTAF 1212
AFLUORIDE. (REVIEW) PHYSICAL AND THERMODYNAMIC PROPERTIES OF SULFUR HEX 1062
FLUORIDES. (NITROGEN TRIFLUORIDE) PHYSICAL CHEMICAL PROPERTIES OF INORGANIC NITROGEN 0713
AFLUORIDE, TUNGSTEN HEXAFLUORIDE, AND MOLYBDENUM HEXAFLU/ PHYSICAL CONSTANTS (VAPOR PRESSURE) OF SILICON TETR 0639
ENTHALPY AND ENTROPY OF VAPORIZATION / THERMODYNAMIC AND PHYSICAL PROPERTIES OF BERYLLIUM COMPOUNDS. PART-1: 0138
LS AND PROTACTINIUM TETRAFLUORIDE. SOME PHYSICAL PROPERTIES OF CURIUM AND PROTACTINIUM META 0899
R PRESSURE) SOME PHYSICAL PROPERTIES OF TUNGSTEN HEXAFLUORIDE. (VAPO 1152
ORIDE. THE C DOUBLET PI AND THE X DOUBLET SIGMA STATES OF BARIUM MONOFLU 0120
/TATIONAL ANALYSIS OF THE A DOUBLET SIGMA(+) TO X DOUBLET PI, B DOUBLET SIGMA(+) TO X DOUBLET PI SYSTEMS OF / 1133
THE FOUR ELECTRON THREE CENTER BONDING OF FOUR AND EIGHT PI ELECTRON SYSTEMS. WAVE FUNCTIONS FOR 0301
/ OF THE DIPOLE MOMENT OF AN A DOUBLET SIGMA TO X DOUBLET PI ELECTRON TRANSITON IN A SILICON MONOFLUORIDE MO/ 0970
A QUARTET SIGMA TO X DOUBLET PI ELECTRONIC TRANSITION OF SILICON MONOFLUORIDE. 0971
/K WAVE FUNCTIONS AND COMPUTED PROPERTIES FOR THE DOUBLET PI GROUND STATES OF SULFUR MONOFLUORIDE AND SELENI/ 1025
ZATION POTENTIAL, ELECTRON / OXYGEN MONOFLUORIDE, DOUBLET PI HARTREE-FOCK WAVE FUNCTION, BINDING ENERGY, IONI 0757
E AND SILICON MONOFLUORIDE (D DOUBLET SIGMA(+)- X DOUBLET PI) MOLECULES. /IGMA- X SINGLET SIGMA PLATINUM OXID 0966
ROTATIONAL FINE STRUCTURE OF THE TRIPLET PI STATE OF PHOSPHORUS MONOFLUORIDE. 0787
S AND R CENTROIDS FOR THE B TRIPLET SIGMA(+) TO A TRIPLET PI SYSTEM OF BORON MONOFLUORIDE. /NCK-CONDON FACTOR 0171
BANDS OF THE C TRIPLET SIGMA, B TRIPLET SIGMA: A TRIPLET PI SYSTEM OF BORON MONOFLUORIDE. /IONAL ANALYSIS OF 0153
SHIFT IN THE (1,0) BAND OF THE A DOUBLET SIGMA- X DOUBLET PI SYSTEM OF THE TIN MONOFLUORIDE MOLECULE. /TOPIC 1135
RIDE. ROTATIONAL ANALYSIS OF THE SYSTEM B SINGLET PI TO SINGLET SIGMA(+) OF GASEOUS SCANDIUM MONOFLUO 0931
ET SIGMA(+/ SHOCK TUBE DETERMINATION OF THE C2 (A TRIPLET PI TO X TRIPLET PI) AND CARBON MONOFLUORIDE A DOUBL 0288
A QUARTET SIGMA TO DOUBLET PI TRANSITION OF THE SILICON MONOFLUORIDE MOLECULE. 0978
, B DOUBLET SIGMA(+), AND A QUARTET SIGMA(-) TO X DOUBLET PI TRANSITIONS OF GERMANIUM MONOFLUORIDE. /SIGMA(+) 0466
STEMS OF G/ ROTATIONAL ANALYSIS OF BANDS OF THE A TRIPLET PI ZERO(+) B TRIPLET PI(1) TO X SINGLET SIGMA(+) SY 0457
YSTEMS OF / ROTATIONAL ANALYSIS OF BANDS OF THE A TRIPLET PI ZERO(+), B TRIPLET PI(1) TO X SINGLET SIGMA(+) S 1111
/ ANALYSIS OF BANDS OF THE A TRIPLET PI ZERO(+), B TRIPLET PI(1) TO X SINGLET SIGMA(+) SYSTEMS OF GALLIUM MON/ 0457
/ANALYSIS OF BANDS OF THE A TRIPLET PI ZERO(+), B TRIPLET PI(1) TO X SINGLET SIGMA(+) SYSTEMS OF THALLIUM MO/ 1111
/OTATIONAL ANALYSIS OF THE C DOUBLET DELTA 5/2- X DOUBLET PI 3/2 AND G DOUBLET DELTA 5/2- X DOUBLET PI 3/2 S/ 1136
DE WITH A VACUUM ECHELLE SPECTROG/ STUDY OF THE A SINGLET PI- X SINGLET SIGMA(+) BANDS OF BORON-11 MONOFLUORI 0170
E. ROTATIONAL ANALYSIS OF THE C SINGLET PI- X SINGLET SIGMA(+) SYSTEM OF INDIUM MONOFLUORID 0491
T RATIO OF URANIUM HEXAFLUORIDE MEASURED WITH A BALLISTIC PISTON COMPRESSOR. SPECIFIC HEA 1201
IS THE RADICAL CHLORINE TETRAFLUORIDE PLANAR OR NOT. 0412
ENON TETRAFLUORIDE. VIBRATIONS OF PLANAR SYMMETRICAL XY4 MOLECULES WITH APPLICATION X 1268
STIANSEN-MORINO SHRINKAGE EFFECT IN SOME XY3 MOLECULES OF PLANAR TRIGONAL SYMMETRY. /UDES OF VIBRATION AND BA 1397
H POLARIZABLE IONS. FORCE CONSTANT OF THE NONPLANAR VIBRATION OF AN XY3 MOLECULE AND A MODEL WIT 0573
CALCULATION OF UREY-BRADLEY POTENTIAL CONSTANTS. PLANAR XY3 MOLECULES. 1393
AMPLITUDES OF VIBRATION AND UREY-BRADLEY FORCE FIELDS FOR PLANAR XY3 MOLECULES. CALCULATIONS OF MEAN 1399
VIBRATIONAL FREQUENCIES OF SOME PLANAR XY3 MOLECULES. 1391
BENT BOND MODEL FOR THE VIBRATIONAL FORCE CONSTANTS OF NONPLANAR XY3 MOLECULES. (NITROGEN TRIFLUORIDE) A 0704
INERTIA DEFECT AND PLANARITY OF FOUR ATOM MOLECULES. (XY3) 1398
EVALUATION OF FORCE CONSTANTS OF PLANE XY3 MOLECULES FROM VIBRATIONAL DATA. 1401
HELIUM TE/ HEAT CAPACITIES AND ELECTRONIC SPECTRA OF THE PLATINUM GROUP METAL HEXAFLUORIDE MOLECULES DOWN TO 1159
AFLUORIDE, OSMIUM HEXAFLUORIDE, IRIDIUM HEXAFLUORIDE, AND PLATINUM HEXAFLUORIDE) /HEXAFLUORIDES. (RHENIUM HEX 0916
AFLUORIDE, OSMIUM HEXAFLUORIDE, IRIDIUM HEXAFLUORIDE, AND PLATINUM HEXAFLUORIDE) /M HEXAFLUORIDE, RHODIUM HEX 1084
PREPARATION AND SOME PROPERTIES OF PLATINUM HEXAFLUORIDE. 0882
SUSCEPTIBILITY MEASUREMENTS AND FLUORINE NMR OF PLATINUM HEXAFLUORIDE. 0883
VIBRATIONAL SPECTRA OF OSMIUM HEXAFLUORIDE AND PLATINUM HEXAFLUORIDE. 0752
CALCULATED BY THE SELF CONSISTENT / MOLECULAR ORBITALS OF PLATINUM HEXAFLUORIDE AND ELEMENT 110 HEXAFLUORIDE 0881
/TORS AND R CENTROIDS OF A SINGLET SIGMA- X SINGLET SIGMA PLATINUM OXIDE AND SILICON MONOFLUORIDE (D DOUBLET/ 0966
NTHALPY OF SUBLIMATION OF / MASS SPECTROMETRIC STUDIES OF PLUTONIUM COMPOUNDS AT HIGH TEMPERATURES. PART-2: E 0884
VAPOR PRESSURE OF PLUTONIUM HALIDES. (PLUTONIUM TRIFLUORIDE) 0887
ABSORPTION SPECTRUM OF PLUTONIUM HEXAFLUORIDE. 0892
UORIDE, URANIUM HEXAFLUORIDE, NEPTUNIUM HEXAFLUORIDE, AND PLUTONIUM HEXAFLUORIDE. /XAFLUORIDE, IRIDIUM HEXAFL 1155
INFRARED SPECTRA OF NEPTUNIUM HEXAFLUORIDE AND PLUTONIUM HEXAFLUORIDE. 0649
INFRARED SPECTRUM OF PLUTONIUM HEXAFLUORIDE. 0891
TAL FOR URANIUM HEXAFLUORIDE, NEPTUNIUM HEXAFLUORIDE, AND PLUTONIUM HEXAFLUORIDE. /, INFRARED ACTIVE FUNDAMEN 1192
IUM(VI/ PREPARATION AND PROPERTIES OF (VAPOR PRESSURE) OF PLUTONIUM HEXAFLUORIDE AND IDENTIFICATION OF PLUTON 0890
/ECENT STUDIES WITH HEXAFLUORIDES. (URANIUM HEXAFLUORIDE, PLUTONIUM HEXAFLUORIDE, AND NEPTUNIUM HEXAFLUORIDE) 1205
IES. PLUTONIUM HEXAFLUORIDE: ITS PREPARATION AND PROPERT 0893
PTION SPECTRUM) PROPERTIES OF PLUTONIUM HEXAFLUORIDE. (MOLECULAR STRUCTURE, ABSOR 0894
/STABILITY OF HEXAFLUORIDES AT HIGH TEMPERATURES. PART-2: PLUTONIUM HEXAFLUORIDE. PART-3: URANIUM HEXAFLUORI/ 0629
ES OF URANIUM PENTAFLUORIDE, NEPTUNIUM PENTAFLUORIDE, AND PLUTONIUM PENTAFLUORIDE. THERMODYNAMIC PROPERTI 1177
SUBLIMATION OF PLUTONIUM TETRAFLUORIDE. 0888
UBLIMATION. VAPOR PRESSURES OF AMERICIUM TRIFLUORIDE AND PLUTONIUM TRIFLUORIDE, HEATS AND FREE ENERGIES OF S 0055
PRESSURE) OF PLUTONIUM HEXAFLUORIDE AND IDENTIFICATION OF PLUTONIUM(VI) OXYFLUORIDE. /D PROPERTIES OF (VAPOR 0890
K. HEAT CAPACITY, ENTROPY, ENTHALPY, AND GIBBS ENERGY OF PLUTONIUM-242 TETRAFLUORIDE FROM 10 TO 350 DEGREES 0889
UORIDE FROM 10 TO 350 DE/ EXPERIMENTAL HEAT CAPACITIES OF PLUTONIUM-242 TRIFLUORIDE AND PLUTONIUM-242 TETRAFL 0885
HEAT CAPACITY, ENTROPY, AND ENTHALPY OF PLUTONIUM-242 TRIFLUORIDE FROM 10 TO 350 DEGREES K. 0886
/ONIUM-242 TETRAFLUORIDE FROM 10 TO 350 DEGREES K, AND OF PLUTONIUM-244 DIOXIDE FROM 4 TO 25 DEGREES K. DERI/ 0885
MOLECULAR ORBITAL STUDY OF ATOMIZATION ENERGIES OF PNICOGEN FLUORIDE MOLECULES AND IONS. 1364
PHOSPHORUS MONOFLUORIDE) QUANTUM CHEMICAL STUDY OF SOME PNICOGEN MONOFLUORIDES. (NITROGEN MONOFLUORIDE AND 0678
S OF CARBON TETRAFLUORIDE FROM 4 DEGREES K TO ITS MELTING POINT. THERMODYNAMIC PROPERTIE 0326
DENSITY, MELTING POINT AND VAPOR PRESSURE OF ANTIMONY PENTAFLUORIDE. 0082
BON TETRAFLUORIDE AND NITROGEN TRIFLUORIDE AND THE TRIPLE POINT OF CARBON TETRAFLUORIDE. /POR PRESSURE OF CAR 0345
OF CARBON TETRAFLUORIDE FROM 12 DEGREES K TO ITS BOILING POINT. THE SIGNIFICANCE OF THE PARAMETER NU. /RTIES 0361
TIN DIFLUORIDE. MELTING POINT, VAPOR PRESSURE, AND HEAT OF VAPORIZATION OF 1140
IDE. MELTING POINTS OF CURIUM TRIFLUORIDE AND AMERICIUM TRIFLUOR 0446
MAGNESIUM DIFLUORIDE, STRONTIUM DIFLUORIDE, AND/ MELTING POINTS OF INORGANIC FLUORIDES. (CADMIUM DIFLUORIDE, 0273
VAPOR PRESSURE AND TRANSITION POINTS OF TECHNETIUM HEXAFLUORIDE. 1105
VAPOR PRESSURE AND MELTING POINTS OF XENON DIFLUORIDE AND XENON TETRAFLUORIDE. 1251
BITAL / GILLESPIE-NYHOLM ASPECTS OF FORCE FIELDS. PART-1. POINTS-ON-A-SPHERE AND EXTENDED HUCKEL MOLECULAR OR 0828
NE HEPTAFLUORIDE. POLAR DISTORTIONS IN RHENIUM HEPTAFLUORIDE AND IODI 0529

ROTATIONAL SPECTRA OF NONPOLAR MOLECULES.   0182
                                        POLAR STATES IN VANADIUM PENTAFLUORIDE. 1213
/UDES OF VIBRATION, THERMODYNAMIC FUNCTIONS AND MOLECULAR POLARIZABILITES OF SOME PENTAHALIDES OF TRIGONAL B/ 1437
/ATION, BASTIANSEN-MORINO SHRINKAGE EFFECT, AND MOLECULAR POLARIZABILITIES FOR DIHALIDES OF SOME GROUP-2A EL/ 0124
/STCIAL THERMODYNAMICS AND QUANTUM MECHANICAL STUDIES OF POLARIZABILITIES FOR THE DIFLUORIDES OF OXYGEN AN/ 0784
OGEN COMPO/ ELECTRONEGATIVITY, NONBONDED INTERACTIONS AND POLARIZABILITY IN HYDROGEN HALIDES AND THE INTERHAL 0373
E NONPLANAR VIBRATION CF AN XY3 MOLECULE AND A MODEL WITH POLARIZABLE IONS.   FORCE CONSTANT OF TH 0573
          LOCALIZED ORBITALS IN BORON FLUORIDES. HIGHLY POLARIZED BORON- FLUORINE DOUBLE AND TRIPLE BONDS. 0158
OF THE GROUP-2B HALIDES. (ZINC DIFLUORIDE, CADMIUM DIFL/ POLARIZED ION MODEL AND THE BENDING FORCE CONSTANTS 1322
RT-3: ENERGY LEVELS, WAVE FUNCTICNS AND AC PCPULATIONS OF POLYATOMIC FLUORIDES. /O-2 (ALPHA, BETA) METHOD. PA 0016
N TETRAFLUORIDE, BOR/ INFRARED ABSORPTION SPECTRA OF SOME POLYATOMIC FLUORIDES. (SILICON TETRAFLUORIDE, CARBO 0186
                              ELECTRONIC SPECTRA OF POLYATOMIC FREE RADICALS. (CARBON DIFLUORIDE) 0308
CHARGES THROUGH BORCN TRIFLU/ EMISSION SPECTRA OF CERTAIN POLYATOMIC FREE RADICALS OBTAINED IN ELECTRICAL DIS 0222
ND MOLECULAR ORBITALS. PART-3: HETERONUCLEAR DIATOMIC AND POLYATOMIC MOLECULES. /ITY LOCALIZATION OF ATOMIC A 0175
      SPECTROSCOPIC STUDIES OF MOLECULAR CONSTANTS IN SOME POLYATOMIC MOLECULES.   0046
Y AND CALCULATION OF CENTRIFUGAL DISTCRTION CONSTANTS FOR POLYATOMIC MOLECULES.   THEOR 1477
            POTENTIAL CONSTANTS OF POLYATOMIC MOLECULES. PART-1.   1498
REC RADIATION.   SELECTIVE PHOTODISSOCIATION OF POLYATOMIC MOLECULES (SULFUR HEXAFLUORIDE) BY INFRA 1064
      SPECTROSCOPIC STUDIES OF MOLECULAR CONSTANTS IN SOME POLYATOMIC MOLECULES. (XENON DIFLUORIDE) 1245
CHLORIDE.   ESR STUDY OF RADICALS IN GAMMA IRRADIATED POLYCRYSTALLINE ARSENIC TRIFLUORIDE AND ARSENIC TRI 0108
, TANTALUM PENTAFLUORIDE, AND ANTIMONY PENTAFLUORIDE. TH/ POLYMER MONOMER EQUILIBRIA IN NIOBIUM PENTAFLUORIDE 0076
    RAMAN SPECTRUM OF IODINE PENTAFLUORIDE. EVIDENCE FOR POLYMERIZATION IN THE LIQUID STATE. 0512
                                        POLYMORPHISM IN XENON HEXAFLUORIDE. 1296
      ELECTRONIC STRUCTURE OF PHOSPHORUS PENTAFLUORIDE AND POLYTOPAL REARRANGEMENT IN PHOSPHORANES. 0875
                                  ELECTRONIC POPULATION ANALYSIS OF MOLECULAR WAVE FUNCTIONS. 0157
ETA) METHOD. PART-3: ENERGY LEVELS, WAVE FUNCTIONS AND AO POPULATIONS OF POLYATOMIC FLUORIDES. /O-2 (ALPHA, B 0016
OF CARBON MONOFLUORIDE AND SILICON MONOFLUORIDE AND THEIR POSITIVE AND NEGATIVE IONS. /ORBITAL INVESTIGATION 0290
/ SULFUR MONOFLUORIDE AND SELENIUM MONOFLUORIDE AND THEIR POSITIVE AND NEGATIVE IONS. A COMPARISON OF THE TH/ 1025
TION INTERACTION STUDIES OF NITROGEN MONOFLUORIDE AND ITS POSITIVE ION.   CONFIGURA 0672
      ELECTRONIC SPECTRA OF PHOSPHORUS MONOFLUORIDE AND ITS POSITIVE ION.   0786
            KRYPTON MONOFLUCRIDE AND ITS POSITIVE ION. (SCF CALCULATIONS) 0547
            STRUCTURE OF THE SULFUR TRIFLUORIDE POSITIVE ION. (SULFUR DIFLUORIDE) 1032
ETRAFLUORIDE, AN/ CORRELATION OF ELECTRONIC STATES OF THE POSITIVE IONS OF XENON TETRAFLUORIDE, XENON OXIDE T 1264
      STARK EFFECT FOR THALLIUM MONOFLUORIDE AND POTASSIUM FLUORIDE. 1118
OF MOLECULES. (SULFUR HEXAFLUORIDE) EVIDENCE OF EFFECTIVE POTENTIAL BARRIERS IN THE X-RAY ABSORPTION SPECTRA 1073
/EFFECTS IN ASYMMETRIC ROTOR MOLECULES. PART-1. QUADRATIC POTENTIAL CONSTANTS AND AVERAGE STRUCTURE OF OXYGE/ 0778
FOR SOME NONLINEAR SYMMETRICAL XY2 MOLECULES. POTENTIAL CONSTANTS AND CORIOLIS COUPLING CONSTANTS 1385
IODINE PENTAFLUORIDE AND IODINE HEPTAFLUCRIDE. POTENTIAL CONSTANTS AND THERMODYNAMIC FUNCTIONS FOR 0509
CARBON TETRAFLUORIDE. POTENTIAL CONSTANTS AND THERMODYNAMIC FUNCTIONS OF 0350
IODINE PENTAFLUORIDE. POTENTIAL CONSTANTS AND THERMODYNAMIC FUNCTIONS OF 0507
IODINE PENTAFLUORIDE. POTENTIAL CONSTANTS AND THERMODYNAMIC PROPERTIES OF 0508
RIFLUORIDE.   CUBIC POTENTIAL CONSTANTS AND 1-TYPE RESONANCE IN BORON T 0231
      LEAST SQUARES ADJUSTMENT OF ANHARMONIC POTENTIAL CONSTANTS. APPLICATION TO CARBON DIOXIDE. 1491
RHENIUM HEXAFLUORIDE. POTENTIAL CONSTANTS FOR MOLYBDENUM HEXAFLUORIDE AND 0635
            UREY-BRADLEY POTENTIAL CONSTANTS FOR SECOND ROW FLUORIDES. 1474
ANDS FOR A SYMMETRIC TOP. (NITROGEN TRIFLUORI/ ANHARMONIC POTENTIAL CONSTANTS FROM THE BAND CONTOURS OF HOT B 0733
THERMODYNAMIC FUNCTIONS AND MOLECULAR POLARIZABILITIES OF/ POTENTIAL CONSTANTS, MEAN AMPLITUDES OF VIBRATION, 1437
DIFLUORIDE)   THIRD ORDER POTENTIAL CONSTANTS OF BENT XY2 MOLECULES. (OXYGEN 0770
                                        POTENTIAL CONSTANTS OF IODINE HEPTAFLUORIDE. 0532
TRAHALIDE (CARBON TETRAFLUORIDE)   HARMONIC POTENTIAL CONSTANTS OF OSMIUM TETROXIDE AND SOME TE 0353
AMPLITUDES OF VIBRATION AND NORMAL VIBRATIONAI FREQUENCI/ POTENTIAL CONSTANTS OF PHOSPHORUS TRICHLORIDE FROM 1481
.   POTENTIAL CONSTANTS OF POLYATOMIC MOLECULES. PART-1 1498
OUNDS. (SILICON TETRAFLUORIDE) POTENTIAL CONSTANTS OF SOME XY4 TYPE OF SILANE COMP 1009
                                        POTENTIAL CONSTANTS OF SOME XY5 MOLECULES. 1439
LES AND RADICALS. POTENTIAL CONSTANTS OF XY4 AND XY3Z TYPES OF MOLECU 1428
      ESTIMATION OF ANHARMONIC POTENTIAL CONSTANTS. PART-1. LINEAR XY2 MOLECULES. 1383
IC TRIFLUORIDE, AND ANTIMONY/ CALCULATION OF UREY-BRADLEY POTENTIAL CONSTANTS. (PHOSPHORUS TRIFLUORIDE, ARSEN 0809
            CALCULATION OF UREY-BRADLEY POTENTIAL CONSTANTS. PLANAR XY3 MOLECULES. 1393
LLIUM MONOFLUORIDE, SILICON MONOFLUORIDE, AND TIN MONOFL/ POTENTIAL CURVES FOR SOME DIATOMIC MOLECULES. (BERY 0130
/. HARTREE-FOCK WAVE FUNCTION, BINDING ENERGY, IONIZATION POTENTIAL, ELECTRON AFFINITY, DIPOLE AND QUADRUPOL/ 0756
/I HARTREE-FOCK WAVE FUNCTION, BINDING ENERGY, IONIZATION POTENTIAL, ELECTRON AFFINITY, DIPOLE MOMENT, QUADR/ 0757
/MONOFLUORIDE, 3 SIGMA. DISSOCIATION ENTHALPY, IONIZATION POTENTIAL, ELECTRON AFFINITY, DIPOLE MOMENT, SPECT/ 0094
F DIATOMIC BORON AND ALUMINUM HALIDES. POTENTIAL ENERGY CURVES AND DISSOCIATION ENERGIES O 0173
F DIATOMIC FLUORIDES AND CHLORIDES OF GALLIUM, INDIUM, A/ POTENTIAL ENERGY CURVES AND DISSOCIATION ENERGIES O 0458
OUP-3A MONOHALIDES. POTENTIAL ENERGY CURVES AND NATURE OF BINDING IN GR 0032
LECULES. POTENTIAL FIELD AND FORCE CONSTANTS FOR SOME XY6 MO 1462
TAFLUORIDE AND NIOBIUM PENTAFLUORIDE. POTENTIAL FIELD AND FORCE CONSTANTS OF ANTIMONY PEN 0086
OLECULES. POTENTIAL FIELD AND FORCE CONSTANTS OF OCTAHEDRAL M 1464
L BIPYRAMIDAL PENTAHALIDES. (PHOSPHORUS PENTAFLUORIDE AN/ POTENTIAL FIELD AND FORCE CONSTANTS OF SOME TRIGONA 0850
GONAL BIPYRAMIDAL THE RAMAN SPECTRUM OF PHOSPHORUS PENTA/ POTENTIAL FIELD AND MOLECULAR VIBRATIONS OF THE TRI 0863
GONAL BIPYRAMID AX(3)YZ MOLECULE. POTENTIAL FIELD AND MOLECULAR VIBRATIONS OF THE TRI 1436
            INTERMOLECULAR POTENTIAL FOR SILICON TETRAFLUORIDE. 1010
/ED VIBRATIONAL STATES, EQUILIBRIUM STRUCTURE, ANHARMONIC POTENTIAL FUNCTION, AND NU-1- NU-3 CORIOLIS RESONA/ 0993
NTRIFUGAL STRETCHING CONSTANTS, AN/ HARMONIC FREQUENCIES, POTENTIAL FUNCTION, CORIOLIS COUPLING CONSTANTS, CE 0726
            QUADRATIC POTENTIAL FUNCTION FOR AMMONIA. (UREY-BRADLEY) 1483
                                        POTENTIAL FUNCTION FOR PHOSPHORUS PENTAFLUORIDE. 0859
ORUS PENTAFLUORIDE. POTENTIAL FUNCTION FOR THE NU-7 VIBRATION OF PHOSPH 0830
            INFRARED SPECTRUM AND VIBRATIONAL POTENTIAL FUNCTION OF ARSENIC TRIFLUORIDE. 0100
                                        POTENTIAL FUNCTION OF NITROGEN TRIFLUORIDE. 0737
PENTAFLUORID/ PERPENDICULAR BAND CONTOURS AND VIBRATIONAL POTENTIAL FUNCTION OF THE E' VIBRATIONS OF ARSENIC 0116
/ETHYLENE, DIFLUORC DICHLORO ETHYLENE, AND THE APPEARANCE POTENTIAL OF CARBON DIFLUORIDE(X) FROM TETRAFLUORO/ 0305
O METHANE XENON TETRAFLUORIDE, AND/ CALCULATED IONIZATION POTENTIAL OF CHLORO- AND FLUOROMETHANES, TETRAFLUOR 0344
M FLUORIDE SPECIE/ BOND DISSOCIATION ENERGIES, IONIZATION POTENTIALS AND ELECTRON AFFINITIES OF SOME GERMANIU 0464
RO ETHYLENE, DIFLUORC DICHLORO ETHYLENE, AND / IONIZATION POTENTIALS FOR TETRAFLUORO ETHYLENE, TRIFLUORO CHLO 0305
OXIDE AND GERMANIUM OXIDE.   FIRST IONIZATION POTENTIALS OF THE MOLECULES BORON FLUORIDE, SILICON 0160
ORON TRIHALIDES.   EFFECTS OF THE LENNARD-JONES POTENTIALS ON THE UREY-BRADLEY FORCE CONSTANTS OF B 0244
/(ALPHA) SCATTERED WAVE STUDIES ON BONDING AND IONIZATION POTENTIALS. PART-1: HEXAFLUORIDES OF GROUP-6 ELEME/ 1366
LUORIDE AT 193/ STRUCTURES OF FLUORIDES. PART-10: NEUTRCN POWDER DIFFRACTION PROFILE STUDIES OF URANIUM HEXAF 1202
ON SPECTRA OF GADOLINIUM OXIDE AND GADOLINIUM TRIFLUORIDE POWDERS.   REFLECTI 0455
UDIES ON CERIUM TETRAFLUORIDE, TERBIUM TETRAFLUORIDE, AND PRASEODYMIUM TETRAFLUORIDE. /ERMAL DECOMPOSITION ST 0367

PREPARATION, STRUCTURE AND SPECTRA OF PRASEODYMIUM TETRAFLUORIDE. 0897
ENTHALPY OF FORMATION OF LANTHANUM TRIFLUORIDE AND PRASEODYMIUM TRIFLUORIDE. 0577
LASER EXCITED RAMAN SPECTRA OF MATRIX ISOLATED PRASEODYMIUM TRIFLUORIDE. 0895
, HEAT OF SUBLIMATION, AND THE EVAPORATION COEFFICIENT OF PRASEODYMIUM TRIFLUORIDE. VAPOR PRESSURE 0896
OF BORON TRIFLUOR/ STUDY OF THE ISOTOPE EFFECT AND A MORE PRECISE DETERMINATION OF THE FAR INFRARED SPECTRUM 0230
BY NEUTRON DIFFRACTION. STRUCTURES OF FLUORIDES. PART-6: PRECISE STRUCTURAL PARAMETERS IN COPPER DIFLUORIDE 0443
CHEMICAL PREDICTIONS BY MO THEORY. THE NOBLE GAS HALIDES. 1340
IDE. PREPARATION AND CHARACTERIZATION OF GOLD PENTAFLUOR 0485
ORIDE. PREPARATION AND CRYSTAL STRUCTURE OF GALLIUM TRIFLU 0461
FLUORIDE. PREPARATION AND CRYSTAL STRUCTURE OF MOLYBDENUM TRI 0620
UORIDE. TWO FLUORIDES OF OSMIUM. (OSMIUM PENTAFLUORIDE, / PREPARATION AND CRYSTAL STRUCTURE OF OSMIUM PENTAFL 0749
UORIDE. PREPARATION AND CRYSTAL STRUCTURE OF OSMIUM PENTAFL 0750
PLUTONIUM HEXAFLUORIDE: ITS PREPARATION AND PROPERTIES. 0893
ANADIUM TETRAFLUORIDE. PREPARATION AND PROPERTIES (CRYSTAL STRUCTURE) OF V 1209
SMIUM HEPTAFLUORIDE. PREPARATION AND PROPERTIES (INFRARED SPECTRUM) OF O 0754
. PREPARATION AND PROPERTIES OF RHENIUM HEPTAFLUORIDE 0921
(CRYSTAL STRUCTURE) PREPARATION AND PROPERTIES OF URANIUM TRIFLUORIDE. 1164
LUTONIUM HEXAFLUORIDE AND IDENTIFICATION OF PLUTONIUM(VI/ PREPARATION AND PROPERTIES OF (VAPOR PRESSURE) OF P 0890
STEN PENTAFLUORIDE. PREPARATION AND PROPERTIES (VAPOR PRESSURE) OF TUNG 1150
ND DETERMINATION OF ITS VAPOR PRESSURE. PREPARATION AND PURIFICATION OF OXYGEN DIFLUORIDE A 0781
ORIDE. PREPARATION AND SOME PROPERTIES OF PLATINUM HEXAFLU 0882
ON HEXAFLUORIDE. PREPARATION AND SOME THERMAL PROPERTIES OF PURE XEN 1301
KRYPTON DIFLUORIDE. PREPARATION BY THE MATRIX ISOLATION TECHNIQUE. 0567
STRUCTURE. CHLORINE HEXAFLUORIDE RADICAL. PREPARATION, ELECTRON SPIN RESONANCE SPECTRUM, AND 0428
ION: KRYPTON DIFLUORIDE. PREPARATION OF INERT GAS COMPOUNDS BY MATRIX ISOLAT 0566
TETRAFLUORIDE. PREPARATION, STRUCTURE AND SPECTRA OF PRASEODYMIUM 0897
ALUMINUM MONOFLUORIDE. (VAPOR PRESSURE) 0031
ATION OF OXYGEN DIFLUORIDE AND DETERMINATION OF ITS VAPOR PRESSURE. PREPARATION AND PURIFIC 0781
SOME PHYSICAL PROPERTIES OF TUNGSTEN HEXAFLUORIDE. (VAPOR PRESSURE) 1152
RIDE. VAPOR PRESSURE AND CRITICAL CONSTANTS OF URANIUM HEXAFLUO 1197
ITROGEN TRIFLUORIDE. LIQUID DENSITY, VAPOR PRESSURE AND CRITICAL TEMPERATURE AND PRESSURE OF N 0710
/ECTROMETRIC STUDIES AT HIGH TEMPERATURES. PART-14: VAPOR PRESSURE AND DISSOCIATION ENERGY OF SILVER MONOFLU/ 1014
E. VAPOR PRESSURE AND HEAT OF SUBLIMATION OF CALCIUM FLUORID 0282
IDE. VAPOR PRESSURE AND HEAT OF SUBLIMATION OF CERIUM TRIFLUOR 0366
S. (COBALT DIFLUORIDE) VAPOR PRESSURE AND HEAT OF SUBLIMATION OF COBALT DIHALIDE 0438
AFNIUM TETRACHLORIDES. VAPOR PRESSURE AND HEAT OF SUBLIMATION OF ZIRCONIUM AND H 0486
E. MELTING POINT, VAPOR PRESSURE, AND HEAT OF VAPORIZATION OF TIN DIFLUORID 1140
XENON TETRAFLUORIDE. VAPOR PRESSURE AND MELTING POINTS OF XENON DIFLUORIDE AND 1251
UORIDE. VAPOR PRESSURE AND TRANSITION POINTS OF TECHNETIUM HEXAFL 1105
AND SPECIFIC CONDUCTIVITY OF CHLORINE PE/ DENSITY, VAPOR PRESSURE, CRITICAL PROPERTIES, DIELECTRIC CONSTANT, 0426
FICIENT OF LANTHANUM TRIFLUORIDE. VAPOR PRESSURE, HEAT OF SUBLIMATION, AND EVAPORATION COEF 0576
COEFFICIENT OF PRASEODYMIUM TRIFLUORIDE. VAPOR PRESSURE, HEAT OF SUBLIMATION, AND THE EVAPORATION 0896
PHASE INVESTIGATIONS AND VAPOR PRESSURE MEASUREMENTS ON EUROPIUM DIFLUORIDE. 0452
VAPOR PRESSURE OF ALUMINUM TRIFLUORIDE. 0041
DENSITY, MELTING POINT AND VAPOR PRESSURE OF ANTIMONY PENTAFLUORIDE. 0082
KNUDSEN AND LANGMUIR MEASUREMENTS OF THE SUBLIMATION PRESSURE OF CADMIUM DIFLUORIDE. 0272
ENERGY OF CALCIUM MONOFLUORIDE. SUBLIMATION PRESSURE OF CALCIUM DIFLUORIDE AND THE DISSOCIATION 0276
UORIDE AND THE TRIPLE POINT OF CARBON TETRAFLUORID/ VAPOR PRESSURE OF CARBON TETRAFLUORIDE AND NITROGEN TRIFL 0345
N ENERGY OF CHROMIUM MONOFLUORIDE. SUBLIMATION PRESSURE OF CHROMIUM DIFLUORIDE AND THE DISSOCIATIO 0430
KNUDSEN AND LANGMUIR MEASUREMENTS OF THE SUBLIMATION PRESSURE OF COBALT DIFLUORIDE. 0439
OMETRIC STUDIES AT HIGH TEMPERATURES. PART-9: SUBLIMATION PRESSURE OF COPPER DIFLUORIDE. MASS SPECTR 0442
/METRIC STUDIES AT HIGH TEMPERATURES. PART-9: SUBLIMATION PRESSURE OF COPPER DIFLUORIDE. (AND COPPER MONOFLU/ 0441
SUBLIMATION PRESSURE OF IRON DIFLUORIDE. 0541
A LANGMUIR MEASUREMENT OF THE SUBLIMATION PRESSURE OF MANGANESE DIFLUORIDE. 0611
KNUDSEN MEASUREMENT OF THE VAPOR PRESSURE OF MANGANESE DIFLUORIDE. 0613
FLUORIDE. SUBLIMATION PRESSURE OF MANGANESE DIFLUORIDE AND MANGANESE MONO 0614
ES OF GASEOUS NEODYMIUM DIFLUORIDE AND NEODY/ SUBLIMATION PRESSURE OF NEODYMIUM TRIFLUORIDE AND THE STABILITI 0643
VAPOR PRESSURE OF NICKEL DIFLUORIDE. 0652
DE) VAPOR PRESSURE OF PLUTONIUM HALIDES. (PLUTONIUM TRIFLUORI 0887
ION OF PLUTONIUM(VI/ PREPARATION AND PROPERTIES OF (VAPOR PRESSURE) OF PLUTONIUM HEXAFLUORIDE AND IDENTIFICAT 0890
SATURATED VAPOR PRESSURE OF RARE EARTH FLUORIDES. 1357
DE, AND LANTHANUM TRIFLUORIDE. SATURATED VAPOR PRESSURE OF SCANDIUM TRIFLUORIDE, YTTRIUM TRIFLUORI 0952
UORIDE, AND MOLYBDENUM HEXAFLU/ PHYSICAL CONSTANTS (VAPOR PRESSURE) OF SILICON TETRAFLUORIDE, TUNGSTEN HEXAFL 0639
SATURATED VAPOR PRESSURE OF SOLID URANIUM TETRAFLUORIDE. 1167
VAPOR PRESSURE OF THORIUM TETRAFLUORIDE. 1129
VAPOR PRESSURE OF TITANIUM TETRAFLUORIDE. 1148
PREPARATION AND PROPERTIES (VAPOR PRESSURE) OF TUNGSTEN PENTAFLUORIDE. 1150
H URANIUM PENTAFLUORIDE. VAPOR PRESSURE OF URANIUM HEXAFLUORIDE IN EQUILIBRIUM WIT 1180
VAPOR PRESSURE OF URANIUM TETRAFLUORIDE. 1174
VAPOR PRESSURE OF URANIUM TETRAFLUORIDE. 1175
ROMETRIC DETERMINATION OF THE VAPOR COMPOSITION AND VAPOR PRESSURE OF VANADIUM TRIFLUORIDE. MASS SPECT 1208
GNESIUM DIFLUORIDE, CALCIUM DIFLUORIDE, STRONTIUM / VAPOR PRESSURE OF ZINC DIFLUORIDE, CADMIUM DIFLUORIDE, MA 0049
VAPOR PRESSURE OF ZIRCONIUM FLUORIDE. 1331
VAPOR PRESSURE OF ZIRCONIUM TETRAFLUORIDE. 1326
VAPOR PRESSURE OF ZIRCONIUM TETRAFLUORIDE. 1328
NIUM PENTAFLUORIDE AND RUTHENIUM OXYTETRAFLUORIDE. (VAPOR PRESSURES) RUTHE 0924
UORIDE, YTTERBIUM TRIFLUORIDE, AND LUTETIUM / SUBLIMATION PRESSURES AND HEATS OF SUBLIMATION OF THULIUM TRIFL 1132
IDE AND RHENIUM HEPTAFLUORIDE. VAPOR PRESSURES AND OTHER PROPERTIES OF RHENIUM HEXAFLUOR 0912
CTINIDE FLUORIDES VAPOR PRESSURES AND THERMODYNAMIC CALCULATIONS FOR SOME A 1336
/ETRIC STUDIES AT HIGH TEMPERATURES. PART-16: SUBLIMATION PRESSURES FOR TITANIUM TRIFLUORIDE AND THE STABILI/ 1145
IFLUORIDE, HEATS AND FREE ENERGIES OF SUBLIMATION. VAPOR PRESSURES OF AMERICIUM TRIFLUORIDE AND PLUTONIUM TR 0055
RIDE. VAPOR PRESSURES OF BERYLLIUM DIFLUORIDE AND NICKEL DIFLUO 0135
ORIDE, AND IRON TRIFLUORIDE. SUBLIMATION PRESSURES OF CHROMIUM TRIFLUORIDE, MANGANESE TRIFLU 0434
/IES AT HIGH TEMPERATURES. PART-17: SUBLIMATION AND VAPOR PRESSURES OF DYSPROSIUM, HOLMIUM, AND ERBIUM TRIFL/ 0448
PENTAFLUORIDE. MASS SPECTRA AND SUBLIMATION PRESSURES OF IODINE HEPTAFLUORIDE AND IODINE OXYGEN 0534
TAFLUORIDE, RHENIUM PENTAFLUORIDE, AND OSMIUM PENT/ VAPOR PRESSURES OF MOLYBDENUM PENTAFLUORIDE, TUNGSTEN PEN 0624
HALIDES OF NIOBIUM AND TANTALUM. PART-3: VAPOR PRESSURES OF NIOBIUM AND TANTALUM PENTAFLUORIDES. 0663
SUBLIMATION PRESSURES OF REFRACTORY FLUORIDES. 1382
IDE, AND LANTHANUM TRIFLUORIDE. SUBLIMATION PRESSURES OF SCANDIUM TRIFLUORIDE, YTTRIUM TRIFLUOR 0947
DES. (TUNGSTEN HEXAFLUORIDE, MOLYBDENUM HEXAFLUORI/ VAPOR PRESSURES OF SOME HEAVY TRANSITION METAL HEXAFLUORI 0903

/UM FLUORIDES FROM MEASUREMENTS OF THE DISPROPORTIONATION PRESSURES. (URANIUM PENTAFLUORIDE, URANIUM HEXAFLU/ 1176
OSMIUM HEXAFLUORIDE AND ITS IDENTITY WITH THE PREVIOUSLY REPORTED OCTAFLUORIDE. 0753
RE EARTH METAL FLUORIDES IN THE REGION OF A CESIUM IODIDE PRISM. INFRARED SPECTRA OF RA 1352
S OF SILICON AND GERMANIUM. VIBRATIONAL TRANSITION PROBABILITIES AND R CENTROIDS FOR DIATOMIC FLUORIDE 0994
IDE. VIBRATIONAL TRANSITION PROBABILITIES FOR THE A-X SYSTEM OF BORON MONOFLUOR 0168
TOELECTRON SPECTROSCOPIC MEASUREMENTS FOR COMPOUNDS WHICH PRODUCE NO PARENT ION. (CARBON TETRAFLUORIDE) / PHO 0339
/CTURES OF FLUORIDES. PART-10: NEUTRON POWDER DIFFRACTION PROFILE STUDIES OF URANIUM HEXAFLUORIDE AT 193 DEG/ 1202
N SPECTRA OF MOLECULES OF THE SPHERICAL TOP T/ ROTATIONAL PROFILES OF VIBRATIONAL BANDS IN THE GAS PHASE RAMA 1442
HEXAFLUORIDE. VIBRONICALLY INDUCED JAHN-TELLER PROGRESSIONS IN THE ELECTRONIC SPECTRUM OF IRIDIUM 0537
MOLECULES. (BORON TRIFLUORIDE) APPLICATION OF THE PROGRESSIVE RIGIDITY METHOD TO SOME SIMPLE XYN TYPE 0211
APPLICATION OF THE METHOD OF PROGRESSIVE RIGIDITY TO OCTAHEDRAL MX6 MOLECULES. 1441
ES IN THE LIGHT OF THE THEORY OF CHARACTERISTIC FREQUENC/ PROGRESSIVE STIFFNESS METHOD APPLIED TO XY4 MOLECUL 1430
FAR INFRARED SPECTRA OF PROLATE AND OBLATE SYMMETRIC TOPS. 1479
PROPERTIES OF PROMETHIUM COMPOUNDS. (PROMETHIUM TRIFLUORIDE) 0898
PROTACTINIUM IN / ESTIMATED FREE ENERGIES OF FORMATION OF PROTACTINIUM HALIDES AND ACTIVITY COEFFICIENTS FOR 0900
SOME PHYSICAL PROPERTIES OF CURIUM AND PROTACTINIUM METALS AND PROTACTINIUM TETRAFLUORIDE. 0899
CTINIDE) HEATS OF FORMATION OF SOME HALIDES OF THORIUM, PROTACTINIUM, URANIUM, NEPTUNIUM, AND AMERICIUM. (A 1335
UREY-BRADLEY FORCE FIELD AND THERMODYNAMIC PROPERTIES OF PYRAMIDAL XY3 MOLECULES. 1406
EVIDENCE FOR PSEUDO JAHN-TELLER EFFECT IN XENON HEXAFLUORIDE. 1279
TOPS NITROGEN TRIFLUORIDE NU-4 AND NU-4 + NU-2 AND TRIFL/ PSEUDO-PARALLEL INFRARED BANDS OF OBLATE SYMMETRIC 0716
ON DIFFRACTION STUDY OF RHENIUM HEPTAFLUORIDE. STRUCTURE, PSEUDOROTATION, AND ANHARMONIC COUPLING OF MODES. / 0920
DINE HEPTAFLUORIDE. AN ELECTRON DIFFRACTION S/ STRUCTURE, PSEUDOROTATION, AND VIBRATIONAL MODE COUPLING IN IO 0515
SEMIEMPIRICAL MOLECULAR ORBITAL CALCULATIONS. PSEUDOROTATION IN PHOSPHORUS PENTAFLUORIDE. 0843
PHOSPHORUS PENTAFLUORIDE) PSEUDOROTATION IN TRIGONAL BIPYRAMIDAL MOLECULES. (0833
ERRY ROTATION CONTRA TURNSTILE ROTATION IN PHOSPHORUS PE/ PSEUDOROTATION OF TRIGONAL BIPYRAMIDAL MOLECULES. B 0870
PURE QUADRUPOLE SPECTRA OF SOLID CHLORINE COMPOUNDS 0383
ANALYSIS OF SYMMETRIC ROTOR RAMAN SPECTRA. PURE ROTATIONAL SPECTRUM OF BORON TRIFLUORIDE. 0205
OW FREQUENCY VIBRATIONS OF ANTIMONY PENTAFLUORIDE AND THE PURE ROTATIONAL SPECTRUM OF HYDROGEN CYANIDE. /HE L 0083
PREPARATION AND SOME THERMAL PROPERTIES OF PURE XENON HEXAFLUORIDE. 1301
OF ITS VAPOR PRESSURE. PREPARATION AND PURIFICATION OF OXYGEN DIFLUORIDE AND DETERMINATION 0781
/CONSTANTS. PART-1: METHOD AND APPLICATION TO SOME SQUARE PYRAMIDAL MOLECULES. (IODINE PENTAFLUORIDE, BROMIN/ 0420
THE VIBRATIONAL PROBLE/ DETERMINATION OF THE STRUCTURE OF PYRAMIDAL MOLECULES OF THE XY3 TYPE BY SOLUTION OF 0727
/IBRATIONAL SPECTRA AND VALENCE FORCE CONSTANTS OF SQUARE PYRAMIDAL MOLECULES. XENON OXIDE TETRAFLUORIDE, IO/ 0418
VIBRATIONAL ANHARMONICITY IN THE PYRAMIDAL TRIFLUORIDES. 0745
MEAN AMPLITUDES OF VIBRATION FOR SOME PYRAMIDAL XY3 AND TETRAHEDRAL XY4 MOLECULES. 1407
ZED MEAN SQUARE AMPLITUDES AND CORIOLIS CONSTANTS OF SOME PYRAMIDAL XY3 MOLECULES. GENERALI 1405
MEAN AMPLITUDES OF VIBRATION FOR SOME PYRAMIDAL XY3 MOLECULES. 1390
OSPHORUS TRIFLUORI/ MEAN AMPLITUDES OF VIBRATION FOR SOME PYRAMIDAL XY3 MOLECULES. (NITROGEN TRIFLUORIDE, PH 0706
CLASSICAL VIBRATION PROBLEM FOR THE PYRAMIDAL XY4 MOLECULE. 1421
OMINE PENTAFLUORID/ MEAN AMPLITUDES OF VIBRATION FOR SOME PYRAMIDAL XY4Z MOLECULES. (IODINE PENTAFLUORIDE, BR 0421

Q

RT-5: AB INITIO CALCULATION OF EQUILIBRIUM GEOMETRIES AND QUADRATIC FORCE CONSTANTS. (NITROGEN TRIFLUORIDE) / 0720
/ISTORTION EFFECTS IN ASYMMETRIC ROTOR MOLECULES. PART-1. QUADRATIC POTENTIAL CONSTANTS AND AVERAGE STRUCTUR/ 0778
DLEY) QUADRATIC POTENTIAL FUNCTION FOR AMMONIA. (UREY-BRA 1483
BRATION SPECTRA OF BORON FLUORIDE WITH ALTERATIONS IN ITS QUADRATIC VALENCE FORCE FIELD. /R EFFECTS IN THE VI 1404
MICROWAVE SPECTRUM OF GERMANIUM DIFLUORIDE. QUADRUPOLE COUPLING AND CENTRIFUGAL DISTORTION. 0477
DE AND BROMINE TRIFLUORIDE. QUADRUPOLE COUPLING CONSTANTS IN CHLORINE TRIFLUORI 0400
/ IONIZATION POTENTIAL, ELECTRON AFFINITY, DIPOLE MOMENT, QUADRUPOLE MOMENT, AND SPECTROSCOPIC CONSTANTS. A / 0757
OF NITROGEN TRIFLUORIDE FROM MICROWAVE SPECTRA. NUCLEAR QUADRUPOLE MOMENT OF NITROGEN-14 AND THE STRUCTURE 0741
MOLECULAR G VALUES, MAGNETIC SUSCEPTIBILITIES, MOLECULAR QUADRUPOLE MOMENTS AND SECOND MOMENTS OF THE ELECT/ 0780
/RGY, IONIZATION POTENTIAL, ELECTRON AFFINITY, DIPOLE QUADRUPOLE MOMENTS, AND SPECTROSCOPIC CONSTANTS. C/ 0756
/LECULAR ZEEMAN EFFECT, MAGNETIC PROPERTIES, AND ELECTRIC QUADRUPOLE MOMENTS IN CHLORINE MONOFLUORIDE, BROMI/ 0254
/LECULAR G VALUES, MAGNETIC SUSCEPTIBILITY, AND MOLECULAR QUADRUPOLE MOMENTS IN PHOSPHORUS TRIFLUORIDE, NITR/ 0744
DE AND ITS COMPLEXES WITH XENON DIFLU/ NIOBIUM-93 NUCLEAR QUADRUPOLE RESONANCE STUDIES OF NIOBIUM PENTAFLUORI 0664
PURE QUADRUPOLE SPECTRA OF SOLID CHLORINE COMPOUNDS. 0383
ES. (NITROGEN MONOFLUORIDE AND PHOSPHORUS MONOFLUORIDE) QUANTUM CHEMICAL STUDY OF SOME PNICOGEN MONOFLUORID 0678
/F VIBRATIONAL CONSTANTS, STATISTCIAL THERMODYNAMICS AND QUANTUM MECHANICAL STUDIES OF POLARIZABILITIES FO/ 0784
/URE OF THE A DOUBLET SIGMA(+), B DOUBLET SIGMA(+), AND A QUARTET SIGMA(-) TO X DOUBLET PI TRANSITIONS OF GE/ 0466
ON MONOFLUORIDE MOLECULE. A QUARTET SIGMA TO DOUBLET PI TRANSITION OF THE SILIC 0978
OF SILICON MONOFLUORIDE. A QUARTET SIGMA TO X DOUBLET PI ELECTRONIC TRANSITION 0971

R

ERMANIUM. VIBRATIONAL TRANSITION PROBABILITIES AND R CENTROIDS FOR DIATOMIC FLUORIDES OF SILICON AND G 0994
PI SYSTEM OF BORON MONOFLUORI/ FRANCK-CONDON FACTORS AND R CENTROIDS FOR THE B TRIPLET SIGMA(+) TO A TRIPLET 0171
IDE MOLECULE. FRANCK-CONDON FACTORS AND R CENTROIDS FOR THE C-X SYSTEM OF SILICON MONOFLUOR 0975
A BAND SYSTEM OF NITROGEN MONO/ FRANCK-CONDON FACTORS AND R CENTROIDS FOR THE 6 SINGLET SIGMA- X TRIPLET SIGM 0677
TINUM OXIDE AND SILICON MONOFL/ FRANCK-CONDON FACTORS AND R CENTROIDS OF A SINGLET SIGMA- X SINGLET SIGMA PLA 0966
ORIDE, CALCIUM MONOFLUORIDE, A/ FRANCK-CONDON FACTORS AND R CENTROIDS OF SOME BAND SYSTEMS OF SILICON MONOFLU 0972
ORIDE. RYDBERG-KLEIN-REES FRANCK-CONDON FACTORS AND R CENTROIDS OF THE A-X BAND SYSTEM OF BERYLLIUM FLU 0127
OF POLYATOMIC MOLECULES (SULFUR HEXAFLUORIDE) BY INFRARED RADIATION. SELECTIVE PHOTODISSOCIATION 1064
PIN RESONANCE OF THE TRAPPED RADICAL XENON MONOFLUORIDE. RADIATION DAMAGE IN XENON TETRAFLUORIDE. ELECTRON S 1220
OGEN AND SULFUR HEXAFL/ APPLICATION OF SYNCHROTON ORBITAL RADIATION TO MOLECULAR SPECTROSCOPY. MOLECULAR NITR 1083
EXCITED IN BORON TRIFLUORIDE, CARBON TETRAFLUORIDE, AND/ RADIATIVE LIFETIMES OF ULTRAVIOLET EMISSION SYSTEMS 0216
ITIO MOLECULAR ORBITAL STUDY OF THE PHOSPHORUS DIFLUORIDE RADICAL. AB IN 0790
SPECTRA OF OXYGEN DIFLUORIDE AND THE OXYGEN FLUORIDE FREE RADICAL. ARGON MATRIX RAMAN 0759
ERATURE. CHARACTERIZATION OF THE CHLORINE DIFLUORIDE FREE RADICAL. CHLORINE FLUORINE SYSTEM AT LOW TEMP 0392
TRON SPIN RESONANCE AND STRUCTURE OF THE BORON DIFLUORIDE RADICAL. ELEC 0178
ELECTRON SPIN RESONANCE OF THE XENON FLUORIDE RADICAL. 1288
ELECTRONIC ABSORPTION SPECTRUM OF THE NITROGEN DIFLUORIDE RADICAL. 0694
ELECTRONIC STRUCTURE OF THE NITROGEN DIFLUORIDE RADICAL. 0682
ESR PARAMETERS OF THE BORON DIFLUORIDE RADICAL. 0181
HEAT OF FORMATION OF THE CARBON DIFLUORIDE RADICAL. 0304
INFRARED DETECTION OF GASEOUS CARBON TRIFLUORIDE RADICAL. 0310
NFRARED SPECTRUM AND STRUCTURE OF THE NITROGEN DIFLUORIDE RADICAL. I 0687
UM AND THERMODYNAMIC FUNCTIONS OF THE NITROGEN DIFLUORIDE RADICAL. INFRARED SPECTR 0686
S AND MOLECULAR STRUCTURE IN THE PHOSPHORUS TETRAFLUORIDE RADICAL. ISOTROPIC HYPERFINE SPLITTING 0824
RIDE. ELECTRONIC SPECTRUM OF THE NITROGEN DIFLUORIDE FREE RADICAL. /LTRAVIOLET PHOTOLYSIS OF NITROGEN TRIFLUO 0691

OWAVE ROTATIONAL SPECTRUM OF THE NITROGEN DIFLUORIDE FREE RADICAL.   MICR 0690
                 MICROWAVE SPECTRUM OF NITROGEN DIFLUORIDE RADICAL.   0681
                 MICROWAVE SPECTRUM OF THE SULFUR MONOFLUORIDE RADICAL.   1018
                 RAMAN SPECTRUM OF THE NITROGEN DIFLUORIDE RADICAL.   0695
                 THE CHLORINE DIFLUORIDE FREE RADICAL.   0393
ES.   PHOTOCHLORINATION OF / INFRARED SPECTRUM OF THE FREE RADICAL CARBON TRIFLUORIDE ISOLATED IN INERT MATRIC   0311
                          EPR SPECTRUM OF THE RADICAL CHLORINE TETRAFLUORIDE.   0413
                                 IS THE RADICAL CHLORINE TETRAFLUORIDE PLANAR OR NOT.   0412
ITROGEN DIFLUORIDE) A CAVITY SEARCH SPECTROMETER FOR FREE RADICAL MICROWAVE ROTATIONAL ABSORPTION STUDIES. (N   0689
RUM, AND STRUCTURE.   CHLORINE HEXAFLUORIDE RADICAL. PREPARATION, ELECTRON SPIN RESONANCE SPECT   0428
HYDRIDE. THE INFRARED AND ULTRAVIOLET SPECTRA OF THE FREE RADICAL SILICON DIFLUORIDE. /SILICON DIFLUORIDE DI   0986
                              STRUCTURE OF THE RADICAL SULFUR PENTAFLUORIDE.   1060
NON TETRAFLUORIDE. ELECTRON SPIN RESONANCE OF THE TRAPPED RADICAL XENON MONOFLUORIDE. RADIATION DAMAGE IN XE   1220
RAPPED PHOSPHORUS DIFLUORIDE AND PHOSPHORUS TETRAFLUORIDE RADICALS.   ELECTRON SPIN RESONANCE OF T   0792
R SPECTRA OF PHOSPHORUS DIFLUORIDE AND SULFUR TRIFLUORIDE RADICALS.   ES   0791
OTENTIAL CONSTANTS OF XY4 AND XY3Z TYPES OF MOLECULES AND RADICALS.   P   1428
BORON DIFLUORIDE, N/ AB INITIO CALCULATIONS ON SOME SMALL RADICALS BY THE UNRESTRICTED HARTREE-FOCK METHOD. (0176
           ELECTRONIC SPECTRA OF POLYATOMIC FREE RADICALS. (CARBON DIFLUORIDE)   0308
C TRIFLUORIDE AND ARSENIC TRICHLORIDE.   ESR STUDY OF RADICALS IN GAMMA IRRADIATED POLYCRYSTALLINE ARSENI 0108
BORON TRIFLU/ EMISSION SPECTRA OF CERTAIN POLYATOMIC FREE RADICALS OBTAINED IN ELECTRICAL DISCHARGES THROUGH   0222
TRIFLUORIDE.   ESR SPECTRA OF PHOSPHORUS TETRAFLUORIDE RADICALS PRODUCED IN A SINGLE CRYSTAL OF PHOSPHORUS 0823
FLUORIDE.   A STUDY OF THE RAMAN AND BROAD LINE NMR SPECTRA OF GERMANIUM TETRA   0484
ONS FOR VARIOUS PHOSPHORUS AND ARSENIC HALIDES. (PHOSPHO/ RAMAN AND INFRARED SPECTRA AND THERMODYNAMIC FUNCTI   0811
DE.   RAMAN AND INFRARED SPECTRA OF GERMANIUM TETRAFLUORI   0482
.   RAMAN AND INFRARED SPECTRUM OF SULFUR TETRAFLUORIDE   1040
                                                RAMAN AND INFRARED SPECTRUM OF TRIFLUORO PHOSPHORAN   0847
E. (PHOSPHORUS PENTAFLUORIDE)   RAMAN AND INFRARED SPECTRUM OF SULFUR TETRAFLUORIDE
ED RAMAN SCATTERING.   FORBIDDEN RAMAN BANDS OF SULFUR HEXAFLUORIDE. COLLISION-INDUC   1081
MOLYBDENUM AND TUNGSTEN HEXAFLUORIDES.   RAMAN EFFECT AND ULTRAVIOLET ABSORPTION SPECTRA OF   0642
 PENTAFLUORIDE.   RAMAN INTENSITIES AND THE FORCE FIELD OF PHOSPHORUS   0844
F SOME XY4 TETRAHEDRAL AND XY6 OCTAHEDRAL MOLECULES. RAMAN INTENSITIES FOR SPHERICALLY SYMMETRIC MODES O   1453
EMENT OF THE GAS DENSITY OF URANIUM HEXAFLUORIDE BY LASER RAMAN SCATTERING.   MEASUR 1181
L/ SPECTRAL OBSERVATIONS ON MOLECULAR XENON HEXAFLUORIDE. RAMAN SCATTERING AND INFRARED, VISIBLE AND ULTRAVIO   1285
UM, TUNGSTEN, AND URANIUM HEXAFLUORIDES FROM INFRARED AND RAMAN SPECTRA.   MOLECULAR STRUCTURE OF MOLYBDEN   0627
TRIFLUORIDE AND PHOSPHINE.   RAMAN SPECTRA AND MOLECULAR CONSTANTS OF PHOSPHORUS   0822
UORIDES OF SULFUR, SELENIUM, AND TELLURIUM.   RAMAN SPECTRA AND MOLECULAR CONSTANTS OF THE HEXAFL   0963
LOUS HALIDE DIMERS, THALLOUS FLUORIDE, AND / INFRARED AND RAMAN SPECTRA AND STRUCTURE OF MATRIX ISOLATED THAL   1123
RIDE.   INFRARED AND RAMAN SPECTRA AND STRUCTURE OF PHOSPHORUS TETRAFLUO   0825
NESIUM DIHALIDES: MAGNESIUM DIFLUORIDE, MAG/ INFRARED AND RAMAN SPECTRA AND STRUCTURES OF MATRIX ISOLATED MAG   0607
RIDE AND BROMINE TRIFLUORIDE.   RAMAN SPECTRA AND THE STRUCTURE OF CHLORINE TRIFLUO   0398
LIUM MONOFLUORIDE.   INFRARED AND RAMAN SPECTRA AND THE STRUCTURE OF CRYSTALLINE THAL   1126
RIDE, AND XENON OXIDE TETRAFLUORIDE VAPORS, AND FORCE CO/ RAMAN SPECTRA FOR XENON DIFLUORIDE, XENON TETRAFLUO   1254
DINATE ANALYSES OF GERMANIUM D/ MATRIX INFRARED AND LASER RAMAN SPECTRA MOLECULAR STRUCTURES AND NORMAL COOR   0475
 PENTAFLUORIDE AND   THEIR FORCE CONSTANTS INCLUDING THOS/ RAMAN SPECTRA OF ARSENIC PENTAFLUORIDE AND VANADIUM   0873
TRIFLUORIDE IN THE CONDENSED PHASE. EVIDEN/ INFRARED AND RAMAN SPECTRA OF BROMINE PENTAFLUORIDE AND CHLORINE   0266
RIDE.   RAMAN SPECTRA OF CADMIUM DIFLUORIDE AND LEAD DIFLUO   0274
                                                RAMAN SPECTRA OF CARBON AND SILICON TETRAFLUORIDES.   0364
                              INFRARED AND RAMAN SPECTRA OF CHLORINE MONOFLUORIDE.   0381
RIFLUORIDE.   INFRARED AND RAMAN SPECTRA OF CHLORINE TRIFLUORIDE AND BROMINE T   0408
US STATE. (CARBON TETRAFLUORIDE)   RAMAN SPECTRA OF CHLOROFLUORO METHANES IN THE GASEO   0322
 OF THE NU-7 (E') FUNDAMENTAL OF PHOSPHORUS PENTAFLUORID/ RAMAN SPECTRA OF GASES. PART-10: RAMAN BAND CONTOUR   0879
                              INFRARED AND RAMAN SPECTRA OF KRYPTON DIFLUORIDE.   0552
RAFLUORIDE.   INFRARED AND RAMAN SPECTRA OF LIQUID AND CRYSTALLINE SILICON TET   1001
ANTALUM PENTAFLUORIDE.   RAMAN SPECTRA OF LIQUID NIOBIUM PENTAFLUORIDE AND T   0670
 HEXAFLUORIDE)   RAMAN SPECTRA OF MATRIX ISOLATED MOLECULES. (SULFUR   1097
UORIDE.   LASER EXCITED RAMAN SPECTRA OF MATRIX ISOLATED PRASEODYMIUM TRIFI   0895
, AND TIN) AND MF4 (M= CARBON, SILICON, AND / VAPOR PHASE RAMAN SPECTRA OF MH4 (M= CARBON, SILICON, GERMANIUM   0314
/OTATIONAL PROFILES OF VIBRATIONAL BANDS IN THE GAS PHASE RAMAN SPECTRA OF MOLECULES OF THE SPHERICAL TOP TY/   1442
M HEXAFLUORIDE, RHENIUM HEXAFLUORIDE, URANIUM HEXAFLUORI/ RAMAN SPECTRA OF MOLYBDENUM HEXAFLUORIDE, TECHNETIU   1072
LUORIDE FREE RADICAL.   ARGON MATRIX RAMAN SPECTRA OF OXYGEN DIFLUORIDE AND THE OXYGEN F   0759
                              INFRARED AND RAMAN SPECTRA OF SILICON TETRAFLUORIDE.   1005
ADDUCT WITH XENON DIFLUORIDE.   RAMAN SPECTRA OF SOLID XENON TETRAFLUORIDE AND ITS   1258
RIDE, ZINC DIFLUORIDE, DIFLUORIDE, AND MANGANESE DIFLUOR/ RAMAN SPECTRA OF TITANIUM DIOXIDE, MAGNESIUM DIFLUO   0615
IN THE VAPOR STATE. (THIRD ROW)   RAMAN SPECTRA OF 5D TRANSITION METAL HEXAFLUORIDES   1378
IFLUORIDE.   ANALYSIS OF SYMMETRIC ROTOR RAMAN SPECTRA. PURE ROTATIONAL SPECTRUM OF BORON TR   0205
IOLIS CONSTANTS OF THE SPHERICAL TOP MOLECUL/ VAPOR PHASE RAMAN SPECTRA, RAMAN BAND CONTOUR ANALYSES, AND COR   1069
S CONSTANTS, FORCE CONSTANTS, AND VALUES FOR/ VAPOR PHASE RAMAN SPECTRA, RAMAN BAND CONTOUR ANALYSES, CORIOLI   0795
HLORINE MONOFLUORIDE)   HIGH RESOLUTION RAMAN SPECTROSCOPY IN THE RED AND NEAR INFRARED. (C   0389
FLUORIDE.   RAMAN SPECTRUM AND FORCE CONSTANTS FOR SULFUR TETRA   1035
CATTERING IN CERTAIN HEXAFLUORIDES. / POSSIBLE ROTATIONAL RAMAN SPECTRUM AND THE DEPOLARIZATION OF RAYLEIGH S   0901
 PHASE AS A FUNCTION OF TEMPERATURE. EVIDENCE FOR A TRIG/ RAMAN SPECTRUM OF ANTIMONY PENTAFLUORIDE IN THE GAS   0077
                                                RAMAN SPECTRUM OF ANTIMONY TRIFLUORIDE.   0073
ULAR CONSTANTS OF ARSENIC TRIFLUORIDE, ARSENIC TRICHLORI/ RAMAN SPECTRUM OF ARSENIC TRIFLUORIDE AND THE MOLEC   0111
IDE.   RAMAN SPECTRUM OF BORON TRIFLUORIDE GAS.   0251
                                                RAMAN SPECTRUM OF CRYSTALLINE MOLYBDENUM PENTAFLUOR   0623
                                                RAMAN SPECTRUM OF GASEOUS CARBON TETRAFLUORIDE.   0347
                                                RAMAN SPECTRUM OF GASEOUS NITROGEN TRIFLUORIDE.   0740
R POLYMERIZATION IN THE LIQUID STATE.   RAMAN SPECTRUM OF IODINE PENTAFLUORIDE. EVIDENCE FO   0512
                                                RAMAN SPECTRUM OF LIQUID OXYGEN DIFLUORIDE.   0767
                              INFRARED AND RAMAN SPECTRUM OF NITROGEN TRIFLUORIDE.   0724
 AND MOLECULAR VIBRATIONS OF THE TRIGONAL BIPYRAMIDAL THE RAMAN SPECTRUM OF PHOSPHORUS PENTAFLUORIDE GAS. /LD   0863
                                                RAMAN SPECTRUM OF SOLID MOLYBDENUM TETRAFLUORIDE.   0621
                                                RAMAN SPECTRUM OF SOLID PHOSPHORUS PENTAFLUORIDE   0858
                                                RAMAN SPECTRUM OF SOLID SULFUR HEXAFLUORIDE.   1098
S AND SOLID STATES.   RAMAN SPECTRUM OF SULFUR HEXAFLUORIDE IN THE GASEOU   1079
                                                RAMAN SPECTRUM OF THE NITROGEN DIFLUORIDE RADICAL.   0695
                                                RAMAN SPECTRUM OF URANIUM HEXAFLUORIDE.   1190
                                                RAMAN SPECTRUM OF XENON HEXAFLUORIDE.   1289
RMODYNAMIC PROPERTIES OF BROMINE PENTAFLUORIDE.   RAMAN SPECTRUM, STRUCTURE, FORCE CONSTANTS, AND THE   0267
ENTAFLUORIDE AND NIOBIUM PENTAFLUORID/ A FLUORINE NMR AND RAMAN STUDY OF COMPLEX FORMATION BETWEEN ANTIMONY P   0079
ND PHOSPHORUS PENTAFLUORIDE. INFRARED AND LOW TEMPERATURE RAMAN VIBRATIONAL SPECTRA. /US PHOSPHORUS HALIDES A   0849

E TRIFLUOR/ ELECTRIC DEFLECTION OF MOLECULAR BEAMS OF THE RARE EARTH DIFLUORIDES AND TRIFLUORIDES. (LANTHANID 0946
INFRARED SPECTRA AND GEOMETRIES OF RARE EARTH DIHALIDES. 1351
PECTROMETRIC STUDIES OF SCANDIUM, YTTRIUM, LANTHANUM, AND RARE EARTH FLUORIDES. MASS S 1359
SATURATED VAPOR PRESSURE OF RARE EARTH FLUORIDES. 1357
STRUCTURE AND MAGNETIC PROPERTIES OF RARE EARTH FLUORIDES. 1345
ERIMENTAL DETERMINATION OF THE ENTHALPIES OF FORMATION OF RARE EARTH FLUORIDES. (LANTHANUM TRIFLUORIDE) EXP 0578
GEOMETRY AND INFRARED SPECTRA OF MATRIX ISOLATED RARE EARTH HALIDES. (ALSO LANTHANUM TRIFLUORIDE) 0583
LATTICE VIBRATIONS AND STRUCTURE OF RARE EARTH HALIDES. (LANTHANUM TRIFLUORIDE) 0572
INFRARED SPECTRA OF RARE EARTH METAL FLUORIDES. 1346
M IODIDE PRISM. INFRARED SPECTRA OF RARE EARTH METAL FLUORIDES IN THE REGION OF A CESIU 1352
HIGH TEMPERATURE ENTHALPY INCREMENTS FOR SOME RARE EARTH TRIFLUORIDES. 1350
MAGNETIC RESONANCE AND SUSCEPTIBILITY OF RARE EARTH TRIFLUORIDES. 1349
ORIDE, YTTRIUM TRIFLUORIDE, LANTHANUM TRIFLUORIDE AND THE RARE EARTH TRIFLUORIDES. /TUDIES OF SCANDIUM TRIFLU 0953
NMR IN RARE EARTH TRIFLUORIDES. 1355
SYMMETRY AND CRYSTAL STRUCTURE OF RARE EARTH TRIFLUORIDES. (LANTHANUM TRIFLUORIDE) 0571
RIUM T/ FORCE CONSTANTS AND GEOMETRIES OF MATRIX ISOLATED RARE EARTH TRIFLUORIDES. (SCANDIUM TRIFLUORIDE, YTT 0944
EAT CONTENTS AND RELATED THERMODYNAMIC FUNCTIONS OF SEVEN RARE EARTH TRIFLUORIDES. (YTTRIUM TRIFLUORIDE) /E H 1320
L STUDIES OF THE INTERACTION OF MAGNESIUM DIFLUORIDE WITH RARE GAS ATOMS. THEORETICA 0605
MOLECULAR STRUCTURES OF RARE GAS COMPOUNDS. 1344
AN LCAO MO STUDY OF RARE GAS FLUORIDES. 1341
OMETRY OF TITANIUM DIFLUORIDE AND TITANIUM TRIFLUORIDE IN RARE GAS MATRICES. INFRARED SPECTRA AND GE 1146
TACOORDINATED MOLECULES. PART-12: CORRELATION OF EXCHANGE RATES FOR SOME GROUP-5 PENTAHALIDE MOLECULES. PEN 0115
/IBLE ROTATIONAL RAMAN SPECTRUM AND THE DEPOLARIZATION OF RAYLEIGH SCATTERING IN CERTAIN HEXAFLUORIDES. (RHE/ 0901
VAPORIZATION REACTIONS IN THE THULIUM- FLUORINE SYSTEM. 1131
RESONANCE SPECTROSCOPY. PROPERTIES AND REACTIONS OF URANIUM HEXAFLUORIDE BY ION CYCLOTRON 1185
A MOLECULAR BEAM SOURCE MASS SPECTROMETER TO THE STUDY OF REACTIVE FLUORIDES. APPLICATION OF 0019
/ MOLECULAR BEAM SOURCE MASS SPECTROMETER TO THE STUDY OF REACTIVE FLUORIDES. (SECOND ROW, THIRD ROW, ANTIMO/ 1377
REACTIVITY AND THERMAL STABILITY OF HEXAFLUORIDES. 0004
THERMAL STABILITY AND REACTIVITY OF D- AND F ELEMENT HEXAFLUORIDES. 0630
CARBON TETRAFLUORIDE. THERMODYNAMIC PROPERTIES OF THE REAL GAS. 0332
AFLUORIDE. INTRAMOLECULAR REARRANGEMENT IN IODINE HEPTAFLUORIDE AND XENON HEX 0520
RONIC STRUCTURE OF PHOSPHORUS PENTAFLUORIDE AND POLYTOPAL REARRANGEMENT IN PHOSPHORANES. ELECT 0875
THE ELECTRONIC STATES OF SCANDIUM MONOFLUORIDE. REASSIGNMENT OF MOLECULAR ORBITAL CONFIGURATIONS OF 0937
/TROGEN TRIFLUORIDE AND NITROGEN OXIDE TRIFLUORIDE, AND A REASSIGNMENT OF THE SPECTRA OF TETRAFLUORIDES OF G/ 0700
/ITY AND THERMODYNAMIC FUNCTIONS FROM 5 TO 350 DEGREES K. RECONCILIATION OF THE ENTROPIES FROM MOLECULAR AND/ 1275
HIGH RESOLUTION RAMAN SPECTROSCOPY IN THE RED AND NEAR INFRARED. (CHLORINE MONOFLUORIDE) 0389
ORIDE STRUCTURE. A CORRECTION. A REDETERMINATION OF THE ORTHORHOMBIC IODINE HEPTAFLU 0519
/ALIDES AND THE THE POSSIBILITY OF THEIR SYNTHESIS BY THE REDUCTION OF THE TRIHALIDES. (SCANDIUM DIFLUORIDE) 0940
FLUORIDE. (ALSO SILICON AND GERMANIUM FLUORIDES) REEVALUATION OF THE DISSOCIATION ENERGY OF CALCIUM 0277
UM TRIFLUORIDE POWDERS. REFLECTION SPECTRA OF GADOLINIUM OXIDE AND GADOLINI 0455
SUBLIMATION PRESSURES OF REFRACTORY FLUORIDES. 1382
DIFLUORIDE. A REINVESTIGATION OF THE INFRARED SPECTRUM OF OXYGEN 0774
RIDE. RELATIVISTIC MOLECULAR WAVE FUNCTIONS: XENON DIFLUO 1249
AFLUORIDE, SILICON TETRAFLUORIDE, AND SULFUR HE/ FLUORINE RELAXATION BY NMR ABSORPTION IN GASEOUS CARBON TETR 0346
NUCLEAR MAGNETIC RELAXATION IN BORON TRIFLUORIDE GAS. 0217
E, TUNGSTEN HEXAFLUORIDE, / MOLECULAR MOTION AND FLUORINE RELAXATION IN THE LIQUIDS OF MOLYBDENUM HEXAFLUORID 0638
THE VIBRATIONAL FORCE CONSTANT. RELAXATION OF THE MOLECULAR CHARGE DISTRIBUTION AND 0151
N SYMMETRIC TOP MOLECULES. NUCLEAR SPIN RELAXATION STUDY OF THE SPIN ROTATION INTERACTION I 0183
N SPHERICAL TOP MOLECULES. NUCLEAR SPIN RELAXATION STUDY OF THE SPIN ROTATION INTERACTION I 0323
NUCLEAR RELAXATION TIMES OF LANTHANUM TRIFLUORIDE. 0579
: METHOD AND APPLICATION TO SOME SQUARE PYRAMIDAL MOLECU/ REPARAMETERIZED VIBRATIONAL FORCE CONSTANTS. PART-1 0420
IC FLUORIDES. BICENTRIC RESCALING OF CNDO/2 THEORY. APPLICATIONS TO INORGAN 1475
THE TEN AND ELEVEN ISOTOPES OF BORON TRIFLUORIDE AT HIGH RESOLUTION. NU-4 BANDS OF 0208
F PHOSPHORUS PENTAFLUORIDE. HIGH RESOLUTION INFRARED SPECTRA OF THE PARALLEL BANDS O 0872
UORIDE, TUNGSTEN HEXAFLUORIDE, AND URANIUM HEXAFLUO/ HIGH RESOLUTION NMR SPECTRUM OF LIQUID MOLYBDENUM HEXAFL 0637
RAFLUORIDE AND SILICON TETRAFLUORIDE. HIGH RESOLUTION PHOTOELECTRON SPECTROSCOPY OF CARBON TET 0319
NFRARED. (CHLORINE MONOFLUORIDE) HIGH RESOLUTION RAMAN SPECTROSCOPY IN THE RED AND NEAR I 0389
HORUS TRIFLUORIDE. HIGH RESOLUTION SPECTRA OF ARSENIC TRIFLUORIDE AND PHOSP 0816
PHOTOELECTRON SPECTRA OF HALIDES. PART-2: HIGH RESOLUTION SPECTRA OF THE BORON TRIHALIDES. 0188
ND OF NITROGEN TRIFLUORIDE. HIGH RESOLUTION STUDY OF L-TYPE RESONANCE IN THE NU-4 BA 0731
TRAHEDRAL MOLECULES. NEW METHOD FOR RESOLVING WILSON'S SECULAR EQUATIONS. APPLIED TO TE 1431
OF THE PENTAFLUORIDES OF RHODIUM AND IRIDIUM BY MAGNETIC RESONANCE. CONFIGURATION 0923
E, ANHARMONIC POTENTIAL FUNCTION, AND NU-1- NU-3 CORIOLIS RESONANCE. /IBRATIONAL STATES, EQUILIBRIUM STRUCTUR 0993
ENTAFLUORIDE AND CHLORINE OXYFLUORIDE BY NUCLEAR MAGNETIC RESONANCE. / FLUORIDES AND OXYFLUORIDES, CHLORINE P 0414
E, SILICON TETRA/ FLUORINE RELAXATION BY NUCLEAR MAGNETIC RESONANCE ABSORPTION IN GASEOUS CARBON TETRAFLUORID 0346
ORIDE IN VIBRATIONALLY EXCITED STATES, NU-1- 2 NU-2 FERMI RESONANCE AND EQUILIBRIUM STRUCTURE. / OXYGEN DIFLU 0771
EEN ANTIMONY PENTAFLUORIDE A/ A FLUORINE NUCLEAR MAGNETIC RESONANCE AND RAMAN STUDY OF COMPLEX FORMATION BETW 0079
ICAL. ELECTRON SPIN RESONANCE AND STRUCTURE OF THE BORON DIFLUORIDE RAD 0178
IDES. MAGNETIC RESONANCE AND SUSCEPTIBILITY OF RARE EARTH TRIFLUOR 1349
/AL SPECTRA AND ASSIGNMENT OF IODINE PENTAFLUORIDE. FERMI RESONANCE BETWEEN A FUNDAMENTAL AND A TERNARY COMB/ 0504
/ND DINITROGEN TETRAFLUORIDE MOLECULES, AND ELECTRON SPIN RESONANCE HYPERFINE COUPLING CONSTANTS IN NITROGEN/ 0688
F NITROGEN TRIFLUORIDE. 1-TYPE RESONANCE IN AN OVERTONE BAND. THE 2NU-4 SPECTRUM O 0732
CUBIC POTENTIAL CONSTANTS AND 1-TYPE RESONANCE IN BORON TRIFLUORIDE. 0231
FERMI RESONANCE IN BORON TRIFLUORIDE. 0242
ABSENCE OF FERMI RESONANCE IN KRYPTON DIFLUORIDE. 0564
NUCLEAR MAGNETIC RESONANCE IN RARE EARTH TRIFLUORIDES. 1355
INFRARED DOUBLE RESONANCE IN SULFUR HEXAFLUORIDE. 1099
HIGH RESOLUTION STUDY OF L-TYPE RESONANCE IN THE NU-4 BAND OF NITROGEN TRIFLUORIDE. 0731
N THE EXCITED VIBRATIONAL STATES. (ZETA TYPE DOUBLING AND RESONANCE IN THE V3 AND V4 STATES) /N TRIFLUORIDE I 0723
R TETRAFLUORIDE. FLUORINE-19 NUCLEAR MAGNETIC RESONANCE INVESTIGATION OF THE ASSOCIATION OF SULFU 1044
FLUORINE NUCLEAR MAGNETIC RESONANCE OF ANTIMONY PENTAFLUORIDE. 0078
FLUORINE NUCLEAR MAGNETIC RESONANCE OF CHLORINE PENTAFLUORIDE. 0416
ORIDE, AND OXIDE PENTAFLUORIDES FLUORINE NUCLEAR MAGNETIC RESONANCE OF IODINE HEPTAFLUORIDE, RHENIUM HEPTAFLU 0517
SUSCEPTIBILITY MEASUREMENTS AND FLUORINE NUCLEAR MAGNETIC RESONANCE OF PLATINUM HEXAFLUORIDE. 0883
. RADIATION DAMAGE IN XENON TETRAFLUORIDE. ELECTRON SPIN RESONANCE OF THE TRAPPED RADICAL XENON MONOFLUORIDE 1220
ELECTRON SPIN RESONANCE OF THE XENON FLUORIDE RADICAL. 1288
PHORUS TETRAFLUORIDE RADICALS. ELECTRON SPIN RESONANCE OF TRAPPED PHOSPHORUS DIFLUORIDE AND PHOS 0792
L. ELECTRON SPIN RESONANCE PARAMETERS OF THE BORON DIFLUORIDE RADICA 0181
ORIDES. HELIUM-1 RESONANCE PHOTOELECTRON SPECTRA OF GROUP-4 TETRAFLU 0317
ANOMALOUS NUCLEAR MAGNETIC RESONANCE SHIFT IN SAMARIUM TRIFLUORIDE. 0927

ELECTRON STRUCTURE AND NUCLEAR MAGNETIC RESONANCE SPECTRA OF BORON COMPOUNDS.   0229
INE MONOFLUORIDE. GASEOUS SPECTRA AND G/ NUCLEAR MAGNETIC RESONANCE SPECTRA OF CHLORINE TRIFLUORIDE AND CHLOR 0368
    A STUDY OF THE RAMAN AND BROAD LINE NUCLEAR MAGNETIC RESONANCE SPECTRA OF GERMANIUM TETRAFLUORIDE.   0484
NE HEXAFLUORIDE, IODINE HEXAFLUORIDE, BROM/ ELECTRON SPIN RESONANCE SPECTRA OF HALOGEN HEXAFLUORIDES. (CHLORI 0271
E OXYGEN PENTAFLUORIDE.   NUCLEAR MAGNETIC RESONANCE SPECTRA OF IODINE HEPTAFLUORIDE AND IODIN 0516
UORIDE)   FLUORINE NUCLEAR MAGNETIC RESONANCE SPECTRA OF OXYGEN FLUORIDES. (OXYGEN DIFL 0775
UR TRIFLUORIDE RADICALS.   ELECTRON SPIN RESONANCE SPECTRA OF PHOSPHORUS DIFLUORIDE AND SULF 0791
ALS PRODUCED IN A SINGLE CRYSTAL OF PHOSPH/ ELECTRON SPIN RESONANCE SPECTRA OF PHOSPHORUS TETRAFLUORIDE RADIC 0823
UM MONOFLUORIDE.   GAS PHASE ELECTRON RESONANCE SPECTRA OF SULFUR MONOFLUORIDE AND SELENI 1021
MONOFLUORIDE, SELENIUM OXIDE AND IODI/ GAS PHASE ELECTRON RESONANCE SPECTRA OF SULFUR MONOFLUORIDE, SELENIUM 1020
N FLUORIDES. HYDROLYSIS AND IN/ FLUORINE NUCLEAR MAGNETIC RESONANCE SPECTRA OF SULFUR, PHOSPHORUS, AND SILICO 1042
ES AND REACTIONS OF URANIUM HEXAFLUORIDE BY ION CYCLOTRON RESONANCE SPECTROSCOPY.   PROPERTI 1185
FLUORIDE.   GAS PHASE ELECTRON RESONANCE SPECTRUM AND DIPOLE MOMENT OF CARBON MONO 0284
NOFLUORIDE IN THE SINGLE/ GAS PHASE ELECTRON PARAMAGNETIC RESONANCE SPECTRUM AND DIPOLE MOMENT OF NITROGEN MO 0673
CHLORINE HEXAFLUORIDE RADICAL. PREPARATION, ELECTRON SPIN RESONANCE SPECTRUM, AND STRUCTURE.   0428
ORIDE.   ELECTRON SPIN RESONANCE SPECTRUM AND STRUCTURE OF BROMINE HEXAFLU 0270
UORIDE.   NUCLEAR MAGNETIC RESONANCE SPECTRUM AND STRUCTURE OF SULFUR TETRAFL 1037
           ELECTRON SPIN RESONANCE SPECTRUM OF ARSENIC TETRAFLUORIDE.   0112
ND SELENIUM MONOFLUORIDE IN J-5/2 ROTATION/ ELECTRON SPIN RESONANCE SPECTRUM OF BROMINE OXIDE, IODINE OXIDE A 0954
           ELECTRON SPIN RESONANCE SPECTRUM OF KRYPTON MONOFLUORIDE.   0546
E, TUNGSTEN HEXAFLUORID/ HIGH RESOLUTION NUCLEAR MAGNETIC RESONANCE SPECTRUM OF LIQUID MOLYBDENUM HEXAFLUORID 0637
ORIDE.   ELECTRON PARAMAGNETIC RESONANCE SPECTRUM OF THE RADICAL CHLORINE TETRAFLU 0413
RRADIATED XENON TETRAFLUORIDE.   ELECTRON SPIN RESONANCE SPECTRUM OF XENON MONOFLUORIDE IN GAMMA I 1222
COMPLEXES WITH XENON DIFLU/ NIOBIUM-93 NUCLEAR QUADRUPOLE RESONANCE STUDIES OF NIOBIUM PENTAFLUORIDE AND ITS 0664
RUS DICHLORIDE, AND NITROGEN DICHLORIDE IN/ ELECTRON SPIN RESONANCE STUDIES OF PHOSPHORUS DIFLUORIDE, PHOSPHO 0793
YCRYSTALLINE ARSENIC TRIFLUORIDE AND ARSEN/ ELECTRON SPIN RESONANCE STUDY OF RADICALS IN GAMMA IRRADIATED POL 0108
/ROCESSES IN SOME INORGANIC FLUORIDES BY NUCLEAR MAGNETIC RESONANCE (SULFUR TETRAFLUORIDE AND SELENIUM TETR/ 1048
N/ AB INITIO CALCULATIONS ON SOME SMALL RADICALS BY THE UNRESTRICTED HARTREE-FOCK METHOD. (BORON DIFLUORIDE, 0176
, SELENIUM, TELLURIUM, MOLYBDENUM, TUNGSTEN, URANIUM, AND RHENIUM) / SOME INORGANIC HEXAFLUORIDES. (OF SULFUR 0631
TANTS OF THE HEXAFLUORIDES OF MOLYBDENUM, TECHNETIUM, AND RHENIUM.   VALENCE FORCE CONS 0641
LITUDE MOTION. (XENON HEXAFLUORIDE, IODINE HEPTAFLUORIDE, RHENIUM HEPTAFLUORIDE) /OF MOLECULES WITH LARGE AMP 1281
           PREPARATION AND PROPERTIES OF RHENIUM HEPTAFLUORIDE.   0921
    THREE ATOM SCATTERING THEORY TO IODINE HEPTAFLUORIDE AND RHENIUM HEPTAFLUORIDE. /ORIDE AND AN APPLICATION OF 0527
       VIBRATIONAL SPECTRA OF IODINE HEPTAFLUORIDE AND RHENIUM HEPTAFLUORIDE.   0522
           VIBRATIONAL SPECTRA OF RHENIUM HEPTAFLUORIDE.   0919
ANSITION METAL HEXAFLUORIDES. (SECOND ROW, THIRD ROW, AND RHENIUM HEPTAFLUORIDE) /FRACTION STUDIES OF SOME TR 1375
OLECULAR LIGAND EXCHANGE IN SEVEN COORDINATE STRUCTURES. (RHENIUM HEPTAFLUORIDE AND IODINE HEPTAFLUORIDE) /AM 0922
           POLAR DISTORTIONS IN RHENIUM HEPTAFLUORIDE AND IODINE HEPTAFLUORIDE.   0529
        FLUORINE NMR OF IODINE HEPTAFLUORIDE, RHENIUM HEPTAFLUORIDE, AND OXIDE PENTAFLUORIDES.   0517
           MOLECULAR CONSTANTS OF RHENIUM HEPTAFLUORIDE MOLECULE.   0918
ND ANHARMONIC COUPLING OF / ELECTRON DIFFRACTION STUDY OF RHENIUM HEPTAFLUORIDE. STRUCTURE, PSEUDOROTATION, A 0920
           ELECTRON DIFFRACTION STUDY OF RHENIUM HEXAFLUORIDE.   0909
           HIGHER ELECTRONIC STATES OF RHENIUM HEXAFLUORIDE.   0914
           INFRARED BAND CONTOURS OF RHENIUM HEXAFLUORIDE.   0915
           JAHN-TELLER EFFECT IN RHENIUM HEXAFLUORIDE.   0904
JAHN-TELLER EFFECTS IN THE 2E5/2G TO 2G3/2G TRANSITION OF RHENIUM HEXAFLUORIDE.   0913
           JAHN-TELLER VIBRATIONS OF RHENIUM HEXAFLUORIDE.   0910
           MAGNETIC SUSCEPTIBILITY OF RHENIUM HEXAFLUORIDE.   0917
           NEAR INFRARED SYSTEM OF RHENIUM HEXAFLUORIDE.   0902
IZATION OF RAYLEIGH SCATTERING IN CERTAIN HEXAFLUORIDES. (RHENIUM HEXAFLUORIDE) /MAN SPECTRUM AND THE DEPOLAR 0901
       POTENTIAL CONSTANTS FOR MOLYBDENUM HEXAFLUORIDE AND RHENIUM HEXAFLUORIDE.   0635
           SPECTRUM OF RHENIUM HEXAFLUORIDE.   0906
           VIBRATIONAL SPECTRA OF RHENIUM HEXAFLUORIDE.   0905
    VIBRATIONAL STRUCTURE AND MOLECULAR STRUCTURE OF RHENIUM HEXAFLUORIDE.   0907
       VAPOR PRESSURES AND OTHER PROPERTIES OF RHENIUM HEXAFLUORIDE AND RHENIUM HEPTAFLUORIDE.   0912
ATTERING EFFECTS ON ELECTRON DIFFRACTION PATTERNS FOR THE RHENIUM HEXAFLUORIDE MOLECULE. /E INTRAMOLECULAR SC 0911
/ORIDES. (TUNGSTEN HEXAFLUORIDE, MOLYBDENUM HEXAFLUORIDE, RHENIUM HEXAFLUORIDE, OSMIUM HEXAFLUORIDE, AND IRI/ 0903
HEXAFLUORIDE,/ COLORS OF TRANSITION METAL HEXAFLUORIDES. (RHENIUM HEXAFLUORIDE, OSMIUM HEXAFLUORIDE, IRIDIUM 0916
/AFLUORIDE, URANIUM HEXAFLUORIDE, RUTHENIUM HEXAFLUORIDE, RHENIUM HEXAFLUORIDE, RHODIUM HEXAFLUORIDE, OSMIUM/ 1084
/TRA OF MOLYBDENUM HEXAFLUORIDE, TECHNETIUM HEXAFLUORIDE, RHENIUM HEXAFLUORIDE, URANIUM HEXAFLUORIDE, SULFUR/ 1072
URES OF MOLYBDENUM PENTAFLUORIDE, TUNGSTEN PENTAFLUORIDE, RHENIUM PENTAFLUORIDE, AND OSMIUM PENTAFLUORIDE. /S 0624
       CONFIGURATION OF THE PENTAFLUORIDES OF RHODIUM AND IRIDIUM BY MAGNETIC RESONANCE.   0923
CULES. THE INFRARED SPECTRA OF RUTHENIUM HEXAFLUORIDE AND RHODIUM HEXAFLUORIDE. /AL MODE OF HEXAFLUORIDE MOLE 0926
/AFLUORIDE, RUTHENIUM HEXAFLUORIDE, RHENIUM HEXAFLUORIDE, RHODIUM HEXAFLUORIDE, OSMIUM HEXAFLUORIDE, IRIDIUM/ 1084
    VIBRATIONAL ASSIGNMENT FOR IODINE PENTAFLUORIDE, A NONRIGID D5H MOLECULE.   0525
ELS AND SPECTRA OF PHOSPHORUS PENTAFLUORIDE.   NONRIGID MOLECULE EFFECTS ON THE ROVIBRONIC ENERGY LEV 0841
    CLASSIFICATION OF THE STATES OF NONRIGID MOLECULES. (SULFUR TETRAFLUORIDE)   1039
OPIC CONSEQUENCES OF DIFFERENT TUNNELING MECHANISMS IN NONRIGID PHOSPHORUS PENTAFLUORIDE.   SPECTROSC 0834
       FURTHER EVIDENCE OF STEREOCHEMICAL NONRIGIDITY IN FIVE AND SEVEN COORDINATE STRUCTURES.   1440
(BORON TRIFLUORIDE)   APPLICATION OF THE PROGRESSIVE RIGIDITY METHOD TO SOME SIMPLE XYN TYPE MOLECULES.   0211
ULES. PART-14: MOLECULAR VIBRATIONS AND STEREOCHEMICAL NONRIGIDITY OF THE TRIGONAL BIPYRAMIDAL MODEL MX3Y2. / 0854
    APPLICATION OF THE METHOD OF PROGRESSIVE RIGIDITY TO OCTAHEDRAL MX6 MOLECULES.   1441
, AND FREE ENERGY FUNCTIONS OF THALLIUM MONOHALIDES ABOVE ROOM TEMPERATURE. (THALLIUM MONOFLUORIDE) /NTROPIES 1116
F OCTAHEDRAL SYMMETRY. (SELENIUM AND TELLURIUM FLUORIDES) ROOT MEAN SQUARE AMPLITUDES IN SOME HEXAFLUORIDES O 0962
OUP-4 TETRAHALIDES. (CARBON TETRAFLUORIDE)   ROOT MEAN SQUARE AMPLITUDES OF VIBRATION IN SOME GR 0342
OUP-4 TETRAHALIDES.   ROOT MEAN SQUARE AMPLITUDES OF VIBRATION IN SOME GR 1363
IFLUORIDE AND XENON TETR/ A THEORETICAL EVALUATION OF THE ROOT MEAN SQUARE AMPLITUDES OF VIBRATION IN XENON D 1257
SILICON TETRAFLUORIDE FROM EL/ INTERATOMIC DISTANCES AND ROOT MEAN SQUARE AMPLITUDES OF VIBRATION OF GASEOUS 1002
PENTAFLUORIDE, PHOSP/ CALCULATION OF FORCE CONSTANTS AND ROOT MEAN SQUARE VIBRATION AMPLITUDES OF PHOSPHORUS 0826
           VIBRATION ROTATION BANDS OF BORON TRIFLUORIDE.   0199
           VIBRATION- ROTATION BANDS OF BORON TRIFLUORIDE.   0191
TAFLUORIDE, AND CYCLOPROPANE B/ ANALYSES OF THE VIBRATION ROTATION BANDS OF TRIFLUORO METHANE, PHOSPHORUS PEN 0880
/ PSEUDOROTATION OF TRIGONAL BIPYRAMIDAL MOLECULES. BERRY ROTATION CONTRA TURNSTILE ROTATION IN PHOSPHORUS PE 0870
           VIBRATION ROTATION ENERGIES OF MOLECULES.   1490
       NUCLEAR SPIN RELAXATION STUDY OF THE SPIN ROTATION INTERACTION IN SPHERICAL TOP MOLECULES.   0323
       NUCLEAR SPIN RELAXATION STUDY OF THE SPIN ROTATION INTERACTION IN SYMMETRIC TOP MOLECULES.   0183
UR HEXAFLUORID/ OPTICAL SATURATION OF A SINGLE VIBRATION- ROTATION TRANSITION IN THE NU-3 FUNDAMENTAL OF SULF 1074
ECTRUM OF BORON TRIFLUORIDE.   ROTATION VIBRATION BAND CONTOURS IN THE INFRARED SP 0193
AR MODELS WITH APPLICATION/ NORMAL COORDINATE ANALYSIS OF ROTATION VIBRATION OF OCTAHEDRAL XY6 AND Z6 MOLECUL 1454
) A CAVITY SEARCH SPECTROMETER FOR FREE RADICAL MICROWAVE ROTATIONAL ABSORPTION STUDIES. (NITROGEN DIFLUORIDE 0689

IDE.   ROTATIONAL ANALYSIS OF A-X1 BANDS OF LEAD MONOFLUOR 0591
  ANGSTROM SYSTEM OF SILICON DIFLUORIDE.   ROTATIONAL ANALYSIS OF ABSORPTION BANDS IN THE 2266 0979
FLUORIDE.   ROTATIONAL ANALYSIS OF B-X2 SYSTEM OF LEAD-208 MONO 0590
RO(+) B TRIPLET PI(1) TO X SINGLET SIGMA(+) SYSTEMS OF G/   ROTATIONAL ANALYSIS OF BANDS OF THE A TRIPLET PI ZE 0457
RO(+), B TRIPLET PI(1) TO X SINGLET SIGMA(+) SYSTEMS OF /   ROTATIONAL ANALYSIS OF BANDS OF THE A TRIPLET PI 1111
, B TRIPLET SIGMA: A TRIPLET PI SYSTEM OF BORON MONOFLUO/   ROTATIONAL ANALYSIS OF BANDS OF THE C TRIPLET SIGMA 0153
NOFLUORIDE MOLECULE.   ROTATIONAL ANALYSIS OF C3 SYSTEM OF THE ANTIMONY MO 0057
MAGNESIUM MONOFLUORIDE.   ROTATIONAL ANALYSIS OF ELECTRONIC BANDS OF GASEOUS 0600
NIC MONOFLUORIDE.   ROTATIONAL ANALYSIS OF FOUR SINGLET SYSTEMS OF ARSE 0092
TRONIC SPECTRUM OF THE LUTETIUM MONOFLUORIDE MOLECULE.   ROTATIONAL ANALYSIS OF SELECTED BANDS FROM THE ELEC 0599
  MONOFLUORIDE.   ROTATIONAL ANALYSIS OF SOME BANDS OF GASEOUS BARIUM 0118
THE SPECTRUM OF GASEOUS SCANDIUM MONOFLUORIDE.   ROTATIONAL ANALYSIS OF SOME SINGLET TRANSITIONS IN 0930
DOUBLET PI, B DOUBLET SIGMA(+) TO X DOUBLET PI SYSTEMS O/   ROTATIONAL ANALYSIS OF THE A DOUBLET SIGMA(+) TO X 1133
ET SIGMA(+) SYSTEMS OF GASEOUS SILVER MONOFLUORIDE.   ROTATIONAL ANALYSIS OF THE AO(+), BO(+), TO X SINGL 1011
LUORIDE IN ULTRAVIOLET.   ROTATIONAL ANALYSIS OF THE B-X SYSTEM OF LEAD MONOF 0586
3 SIGMA TO 3 SIGMA SYSTEM OF THE BORON MONOFLUORIDE MOLE/   ROTATIONAL ANALYSIS OF THE BAND AT 14900 CM-1 OF A 0156
CON MONOFLUORIDE MOLECULE.   ROTATIONAL ANALYSIS OF THE BETA BAND SYSTEM OF SILI 0976
OUBLET PI 3/2 AND G DOUBLET DELTA 5/2- X DOUBLET PI 3/2 /   ROTATIONAL ANALYSIS OF THE C DOUBLET DELTA 5/2- X D 1136
SIGMA(+) SYSTEM OF INDIUM MONOFLUORIDE.   ROTATIONAL ANALYSIS OF THE C SINGLET PI- X SINGLET 0491
NOFLUORIDE.   ROTATIONAL ANALYSIS OF THE C1 SYSTEM OF ANTIMONY MO 0069
NOFLUORIDE.   ROTATIONAL ANALYSIS OF THE C2 SYSTEM OF ANTIMONY MO 0068
ONOFLUORIDE.   ROTATIONAL ANALYSIS OF THE GAMMA BANDS OF SILICON M 0999
INGLET SIGMA(+) OF GASEOUS SCANDIUM MONOFLUORIDE.   ROTATIONAL ANALYSIS OF THE SYSTEM B SINGLET PI TO S 0931
  ANTIMONY MONOFLUORIDE MOLECULE.   ROTATIONAL ANALYSIS OF THE ULTRAVIOLET BANDS OF THE 0063
HE BISMUTH MONOFLUORIDE MOLECULE.   ROTATIONAL ANALYSIS OF THE VISIBLE BAND SYSTEM OF T 0146
F LEAD MONOFLUORIDE.   ROTATIONAL ANALYSIS OF THE VISIBLE EMISSION BANDS O 0587
SYSTEM AT ANTIMONY FLUORIDE MOLECULE.   ROTATIONAL ANALYSIS OF THE 2550-2750 ANGSTROM BAND 0067
/PIRICAL METHOD FOR DETERMINING EFFECTIVE VIBRATIONAL AND   ROTATIONAL CHARACTERISTICS OF MOLECULES OF SOME TE/ 0622
DE AND IODINE PENTAFLUORIDE FROM ELECTRON DIFFRACTION AND   ROTATIONAL CONSTANT DATA. /S OF BROMINE PENTAFLUORI 0265
ONOFLUORIDE MOLECULE .   ROTATIONAL CONSTANTS OF THE B STATE OF THE CARBON M 0294
F PHOSPHORUS MONOFLUORIDE.   ROTATIONAL FINE STRUCTURE OF THE TRIPLET PI STATE O 0787
CTRUM OF OXYGEN DIFLUORIDE.   SPIN   ROTATIONAL HYPERFINE STRUCTURE IN THE MICROWAVE SPE 0777
NE OXIDE, IODINE OXIDE AND SELENIUM MONOFLUORIDE IN J-5/2   ROTATIONAL LEVELS.   ESR SPECTRUM OF BROMI 0954
  PHASE RAMAN SPECTRA OF MOLECULES OF THE SPHERICAL TOP T/   ROTATIONAL PROFILES OF VIBRATIONAL BANDS IN THE GAS 1442
  RAYLEIGH SCATTERING IN CERTAIN HEXAFLUORIDES. / POSSIBLE   ROTATIONAL RAMAN SPECTRUM AND THE DEPOLARIZATION OF 0901
MILLIMETER WAVE LENGTH REGION.   ROTATIONAL SPECTRA OF BROMINE PENTAFLUORIDE IN THE 0268
                                                              ROTATIONAL SPECTRA OF NONPOLAR MOLECULES. 0182
RAGE STRUCTURE OF OXYGEN DIFLUORIDE FROM THE GROUND STATE   ROTATIONAL SPECTRUM. /C POTENTIAL CONSTANTS AND AVE 0778
              ANALYSIS OF SYMMETRIC ROTOR RAMAN SPECTRA. PURE ROTATIONAL SPECTRUM OF BORON TRIFLUORIDE. 0205
EQUENCY VIBRATIONS OF ANTIMONY PENTAFLUORIDE AND THE PURE   ROTATIONAL SPECTRUM OF HYDROGEN CYANIDE. /HE LOW FR 0083
                                                  MICROWAVE   ROTATIONAL SPECTRUM OF INDIUM MONOFLUORIDE. 0490
                                                              ROTATIONAL SPECTRUM OF IODINE FLUORIDE. 0497
                                                  MICROWAVE   ROTATIONAL SPECTRUM OF THALLIUM MONOFLUORIDES. 1119
LE.   ZEEMAN EFFECT IN THE MICROWAVE   ROTATIONAL SPECTRUM OF THE ALUMINUM FLUORIDE MOLECU 0026
RADICAL.   MICROWAVE   ROTATIONAL SPECTRUM OF THE NITROGEN DIFLUORIDE FREE 0690
) BAND OF THE A DOUBLET SIGMA- X DOUBLET PI SYSTEM OF TH/   ROTATIONAL STRUCTURE AND ISOTOPIC SHIFT IN THE (1,0 1135
OF THE GERMANIUM(I) FLUORIDE MOLECULE.   ROTATIONAL STRUCTURE IN SOME HIGHER EXCITED STATES 0467
OUBLET SIGMA(+), AND A QUARTET SIGMA(-) TO X DOUBLET PI /   ROTATIONAL STRUCTURE OF THE A DOUBLET SIGMA(+), B D 0466
RMANIUM MONOFLUORIDE.   ANALYSIS OF THE   ROTATIONAL STRUCTURE OF THE BANDS 0-0 AND 0-1 IN GE 0465
LUORIDE MOLECULE.   ROTATIONAL STRUCTURE OF THE C3 SYSTEM OF ANTIMONY F 0066
0 ANGSTROM REGION.   ANALYSIS OF VIBRATIONAL AND   ROTATIONAL STRUCTURE OF TIN MONOFLUORIDE IN THE 250 1138
NTS AND AVE/ CENTRIFUGAL DISTORTION EFFECTS IN ASYMMETRIC   ROTOR MOLECULES. PART-1. QUADRATIC POTENTIAL CONSTA 0778
RON TRIFLUORIDE.   ANALYSIS OF SYMMETRIC   ROTOR RAMAN SPECTRA. PURE ROTATIONAL SPECTRUM OF BO 0205
PENTAFLUORIDE.   NONRIGID MOLECULE EFFECTS ON THE   ROVIBRONIC ENERGY LEVELS AND SPECTRA OF PHOSPHORUS 0841
TRANSITION METAL HEXAFLUORIDES IN THE VAPOR STATE. (THIRD   ROW)   RAMAN SPECTRA OF 5D 1378
ANSITION METAL PENTAFLUORIDES AT HIGH TEMPERATURE. (THIRD   ROW)   THERMODYNAMIC PROPERTIES OF TR 1379
/ION BY THE COULOMB FIELD OF LIGAND ATOMS OF LATER SECOND   ROW ELEMENTS IN PERFECT PAIRING VALENCE STATES. (P/ 0862
              UREY-BRADLEY POTENTIAL CONSTANTS FOR SECOND   ROW FLUORIDES. 1474
/TRIC DEFLECTION AND MASS SPECTROMETRIC DETECTION. (FIRST   ROW FLUORIDES, BERYLLIUM, MAGNESIUM, AND LEAD DIFL/ 0134
/MO STUDIES OF MOLECULES CONTAINING ATOMS FROM THE SECOND   ROW OF THE PERIODIC TABLE. (SILICON DIFLUORIDE, SI/ 1373
N STUDIES OF SOME TRANSITION METAL HEXAFLUORIDES. (SECOND   ROW, THIRD ROW, AND RHENIUM HEPTAFLUORIDE) /FRACTIO 1375
/SPECTROMETER TO THE STUDY OF REACTIVE FLUORIDES. (SECOND   ROW, THIRD ROW, ANTIMONY PENTAFLUORIDE, AND BISMUT/ 1377
RATIONAL FREQUENCIES AND VALENCE FORCE CONSTANTS OF FIRST   ROW TRANSITION METAL DIFLUORIDES.   VIB 1371
S. (BERYLLIUM DIFLUORIDE, MAGNESIUM DIFLUORIDE, AND FIRST   ROW TRANSITION METAL FLUORIDES) /MMETRICAL DIHALIDE 1372
TRANSITION METAL DIHALIDES. THE DIFLUORIDES OF SOME FIRST   ROW TRANSITION METALS.   GEOMETRY OF THE 1368
                     HALIDES OF THE SECOND AND THIRD   ROW TRANSITION METALS. 1374
/N FOR SOME LINEAR SYMMETRICAL DIFLUORIDES. (CF THE FIRST   ROW TRANSITION METALS, ALSO BERYLLIUM DIFLUORIDE A/ 1369
OW T/ HALIDES OF THE TRANSITION ELEMENTS. VOLUME-1: FIRST   ROW TRANSITION METALS. VOLUME-2: SECOND AND THIRD R 0001
                     ELECTRONIC SPECTRA OF LIQUID   RUTHENIUM AND MOLYBDENUM PENTAFLUORIDES. 0925
L MODE OF HEXAFLUORIDE MOLECULES. THE INFRARED SPECTRA OF   RUTHENIUM HEXAFLUORIDE AND RHODIUM HEXAFLUORIDE. /A 0926
/XAFLUORIDE, TUNGSTEN HEXAFLUORIDE, URANIUM HEXAFLUORIDE,   RUTHENIUM HEXAFLUORIDE, RHENIUM HEXAFLUORIDE, RHOD/ 1084
IDE. (VAPOR PRESSURES)   RUTHENIUM PENTAFLUORIDE AND RUTHENIUM OXYTETRAFLUOR 0924
UORIDE, ZINC FLUORIDE, AND FERROUS F/ FORCE FIELDS OF THE   RUTILE COUNTERPARTS: TITANIUM DIOXIDE, MAGNESIUM FL 0606
.   CALCULATION OF   RYDBERG LEVELS IN NITROGEN OXIDE AND BORON FLUORIDE 0164
ROIDS OF THE A-X BAND SYSTEM OF BERYLLIUM FLUORIDE.   RYDBERG-KLEIN-REES FRANCK-CONDON FACTORS AND R CENT 0127

                                        S

/IC STUDIES AT HIGH TEMPERATURES. PART-13: STABILITIES OF   SAMARIUM, EUROPIUM, AND GADOLINIUM MONO AND DIFLUO/ 0450
/NAMIC FUNCTIONS OF THE TRIFLUORIDES OF SCANDIUM, CERIUM,   SAMARIUM, EUROPIUM, GADOLINIUM, TERBIUM, DYSPROSIU/ 1356
                      ANOMALOUS NMR SHIFT IN   SAMARIUM TRIFLUORIDE. 0927
RINE MONOFLUORIDE, OXYGEN MONOFLUO/ METHOD OF DETERMINING   SATURATED LIQUID AND SATURATED VAPOR ENTROPY. (CHLO 0391
                                                              SATURATED VAPOR PRESSURE OF RARE EARTH FLUORIDES. 1357
TTRIUM TRIFLUORIDE, AND LANTHANUM TRIFLUORIDE.   SATURATED VAPOR PRESSURE OF SCANDIUM TRIFLUORIDE, Y 0952
RIDE.   SATURATED VAPOR PRESSURE OF SOLID URANIUM TETRAFLUO 1167
ON IN THE NU-3 FUNDAMENTAL OF SULFUR HEXAFLUORID/ OPTICAL   SATURATION OF A SINGLE VIBRATION- ROTATION TRANSITI 1074
PART-18: THE STABILITIES OF THE MONO- AND DIFLUORIDES OF   SCANDIUM AND YTTRIUM. /TUDIES AT HIGH TEMPERATURES. 0939
/D RELATED THERMODYNAMIC FUNCTIONS OF THE TRIFLUORIDES OF   SCANDIUM, CERIUM, SAMARIUM, EUROPIUM, GADOLINIUM, / 1356
SYNTHESIS BY THE REDUC/ THERMODYNAMIC CHARACTERISTICS OF   SCANDIUM DIHALIDES AND THE THE POSSIBILITY OF THEIR 0940
              ENERGETICS AND STABILITY OF GAS PHASE   SCANDIUM FLUORIDES. 0951

ELECTRONIC SPECTRUM OF	SCANDIUM MONOFLUORIDE.	0934
ELECTRONIC STRUCTURE AND GROUND STATE OF	SCANDIUM MONOFLUORIDE.	0932
ECULAR ORBITAL CONFIGURATIONS OF THE ELECTRONIC STATES OF	SCANDIUM MONOFLUORIDE. REASSIGNMENT OF MOL	0937
IS OF SOME SINGLET TRANSITIONS IN THE SPECTRUM OF GASEOUS	SCANDIUM MONOFLUORIDE. ROTATIONAL ANALYS	0930
OF THE SYSTEM B SINGLET PI TO SINGLET SIGMA (+) OF GASEOUS	SCANDIUM MONOFLUORIDE. ROTATIONAL ANALYSIS	0931
SPECTROSCOPIC STUDIES OF	SCANDIUM MONOFLUORIDE AND MAGNESIUM OXIDE.	0933
GROUND STATES OF	SCANDIUM MONOFLUORIDE AND YTTRIUM MONOFLUORIDE.	0929
NTHANUM MONOFLUORIDE. ELECTRONIC STATES OF GASEOUS	SCANDIUM MONOFLUORIDE, YTTRIUM MONOFLUORIDE, AND LA	0928
NTHANUM MONOFLU/ DISSOCIATION ENERGIES AND STABILITIES OF	SCANDIUM MONOFLUORIDE, YTTRIUM MONOFLUORIDE, AND LA	0935
NTHANUM/ EVALUATION OF MOLECULAR VIBRATION FREQUENCIES OF	SCANDIUM MONOFLUORIDE, YTTRIUM MONOFLUORIDE, AND LA	0936
NTHANUM MONOFLUORIDE. VIBRONIC SPECTRUM OF	SCANDIUM MONOFLUORIDE, YTTRIUM MONOFLUORIDE, AND LA	0938
CALCULATION OF THE VIBRATIONAL FREQUENCIES OF	SCANDIUM SUBGROUP HALIDES.	0949
/CULATION OF DISSOCIATION ENERGIES FOR THE HALIDES OF THE	SCANDIUM SUBGROUP. (SCANDIUM TRIFLUORIDE, YTTRIUM /	0948
STRUCTURE AND SOME PROPERTIES OF	SCANDIUM TRIFLUORIDE.	0945
THERMODYNAMIC PROPERTIES OF	SCANDIUM TRIFLUORIDE.	0950
ORIDES. (LANTHANIDE TRIFLUORIDES, LANTHANIDE DIFLUORIDES,	SCANDIUM TRIFLUORIDE, AND YTTRIUM TRIFLUORIDE)/FLU	0946
HANUM TRIFLUORIDE. STANDARD ENTHALPIES OF FORMATION OF	SCANDIUM TRIFLUORIDE, YTTRIUM TRIFLUORIDE, AND LANT	0942
/ GEOMETRIES OF MATRIX ISOLATED RARE EARTH TRIFLUORIDES.	(SCANDIUM TRIFLUORIDE, YTTRIUM TRIFLUORIDE, AND LAN/	0944
HANUM TRIFLUORIDE. SUBLIMATION PRESSURES OF	SCANDIUM TRIFLUORIDE, YTTRIUM TRIFLUORIDE, AND LANT	0947
HANUM TRIFLUORIDE. SATURATED VAPOR PRESSURE OF	SCANDIUM TRIFLUORIDE, YTTRIUM TRIFLUORIDE, AND LANT	0952
M TRIFLUORIDE AND THE RARE/ MASS SPECTROMETRIC STUDIES OF	SCANDIUM TRIFLUORIDE, YTTRIUM TRIFLUORIDE, LANTHANU	0953
-3000 DEGREES K RANGE. THERMODYNAMIC FUNCTIONS OF GASEOUS	SCANDIUM, YTTRIUM, AND LANTHANUM HALIDES IN THE 293	0568
DIFLUORIDE, SCANDIUM MONOFLUORID/ MOLECULAR CONSTANTS OF	SCANDIUM, YTTRIUM, AND LANTHANUM HALIDES. (SCANDIUM	0941
DES. MASS SPECTROMETRIC STUDIES OF	SCANDIUM, YTTRIUM, LANTHANUM, AND RARE EARTH FLUORI	1359
/ ENTROPIES OF METAL TRIFLUORIDES FROM INFRARED SPECTRA.	(SCANDIUM, YTTRIUM, LANTHANUM, CERIUM, NEODYMIUM, /	0943
TENTIALS. PART-1: HEXAFLUORIDES OF GROUP-6/ SCF- X(ALPHA)	SCATTERED WAVE STUDIES ON BONDING AND IONIZATION PO	1366
MAN BANDS OF SULFUR HEXAFLUORIDE. COLLISION-INDUCED RAMAN	SCATTERING. FORBIDDEN RA	1081
OF THE GAS DENSITY OF URANIUM HEXAFLUORIDE BY LASER RAMAN	SCATTERING. MEASUREMENT	1181
/TRAL OBSERVATIONS ON MOLECULAR XENON HEXAFLUORIDE. RAMAN	SCATTERING AND INFRARED, VISIBLE AND ULTRAVIOLET A/	1285
FOR THE RHENIUM HEXAFLUORIDE MO/ MULTIPLE INTRAMOLECULAR	SCATTERING EFFECTS ON ELECTRON DIFFRACTION PATTERNS	0911
/TIONAL RAMAN SPECTRUM AND THE DEPOLARIZATION OF RAYLEIGH	SCATTERING IN CERTAIN HEXAFLUORIDES. (RHENIUM HEXA/	0901
/OF IODINE HEPTAFLUORIDE AND AN APPLICATION OF THREE ATOM	SCATTERING THEORY TO IODINE HEPTAFLUORIDE AND RHEN/	0527
STRUCTURE OF XENON HEXAFLUORIDE. MULTIPLE	SCATTERING X(ALPHA) CALCULATION OF THE ENERGY LEVEL	1304
0 HEXAFLUORIDE CALCULATED BY THE SELF CONSISTENT MULTIPLE	SCATTERING X-ALPHA METHOD. /FLUORIDE AND ELEMENT 11	0881
DIFLUORIDE, NITROGEN DIFLUORIDE, AND OXYGEN DIFLUORIDE.	(SCF) /ERYLLIUM DIFLUORIDE, BORON DIFLUORIDE, CARBON	0141
CTRONIC STRUCTURE OF CHLORINE TRIFLUORIDE. AN APPROXIMATE	SCF CALCULATION. ELE	0404
KRYPTON MONOFLUORIDE AND ITS POSITIVE ION.	(SCF CALCULATIONS)	0547
TRIHYDRIDE, GERMANIUM TETRAHYDRIDE, ARSENIC TR/ MOLECULAR	SCF CALCULATIONS FOR GALLIUM MONOFLUORIDE, GALLIUM	0459
IDE. DOUBLE ZETA	SCF CALCULATIONS FOR NITRITE ION AND OXYGEN DIFLUOR	0762
RITE ION, NITROGEN OXYFLUORIDE, AND OX/ MINIMAL BASIS SET	SCF CALCULATIONS FOR THE GROUND STATE OF OZONE, NIT	0776
A SMALL GAUSSIAN BASIS SET FOR NONEMPIRICAL ALL-ELECTRON	SCF CALCULATIONS IN IODINE COMPOUNDS.	0495
ICON DIFLUORIDE. NONEMPIRICAL LCAO MO	SCF CALCULATIONS OF THE ELECTRONIC STRUCTURE OF SIL	0995
AND CHLORINE MONOFLUORIDE.	SCF CALCULATIONS ON CHLORATE ION, HYDROGEN CHLORIDE	0380
E DIFLUORIDE ON THE BASIS OF A CALCULATION BY THE MO LCAO	SCF IN THE NDDO APPROXIMATION. /HE MOLECULE CHLORIN	0382
/RUCTURES OF COMPOUNDS OF HALOGENS AND INERT GASES BY THE	SCF LCAO MO METHOD. PART-2: COMPARATIVE ANALYSIS /	1231
/RIDE. PART-1: OPTICAL ABSORPTION AND X-RAY EMISSION FROM	SCF MO LCAO COMPUTATIONS. PART-2: MOLECULAR ORBITA/	1078
ORIDES OF NONTRANSITION ELEMENTS FROM THE DATA OF LCAO MO	SCF NONEMPIRICAL CALCULATIONS. /IC STRUCTURE OF FLU	0009
LUORIDE, MAGNESIUM DIFLUORIDE, AND / NONEMPIRICAL LCAO MO	SCF STUDIES OF THE GROUP-2A DIHALIDES BERYLLIUM DIF	1360
FLUORIDE. NONEMPIRICAL LCAO MO	SCF STUDIES OF THE LOW LYING STATES OF BERYLLIUM DI	0137
ORUS TRIFLUORIDE AND PHOSPHORUS OXYFLUORIDE. AN AB INITIO	SCF STUDY. BONDING IN PHOSPH	0820
F THE ISOELECTRONIC SERIES CHLORINE MONOFLUORIDE, HYDROG/	SCF- MO CALCULATIONS OF SOME MOLECULAR PROPERTIES O	0371
IONIZATION POTENTIALS. PART-1: HEXAFLUORIDES OF GROUP-6/	SCF- X(ALPHA) SCATTERED WAVE STUDIES ON BONDING AND	1366
/AO MO METHOD. PART-2: COMPARATIVE ANALYSIS CF DIFFERENT	SCHEMES OF NEGLECT OF DIFFERENTIAL OVERLAP METHODS/	1231
RIDE.	SEARCH FOR A JAHN-TELLER EFFECT IN IRIDIUM HEXAFLUO	0539
TIONAL ABSORPTION STUDIES. (NITROGEN DIFLUORIDE) A CAVITY	SEARCH SPECTROMETER FOR FREE RADICAL MICROWAVE ROTA	0689
HALIDES OF THE	SECOND AND THIRD ROW TRANSITION METALS.	1374
/EMENTS. VOLUME-1: FIRST ROW TRANSITION METALS. VOLUME-2:	SECOND AND THIRD ROW TRANSITION METALS. VOLUME-3: /	0001
/NETIC SUSCEPTIBILITIES, MOLECULAR QUADRUPOLE MOMENTS AND	SECOND MOMENTS OF THE ELECTRONIC CHARGE DISTRIBUTI/	0780
ORINE MONOFLUORIDE AND BROMINE M/ SOME CORRECTIONS TO THE	SECOND ORDER STARK EFFECT OF LINEAR MOLECULES. (CHL	0385
/DIFICATION BY THE COULOMB FIELD OF LIGAND ATOMS OF LATER	SECOND ROW ELEMENTS IN PERFECT PAIRING VALENCE STA/	0862
UREY-BRADLEY POTENTIAL CONSTANTS FOR	SECOND ROW FLUORIDES.	1474
/ VESCF-MO STUDIES OF MOLECULES CONTAINING ATOMS FROM THE	SECOND ROW OF THE PERIODIC TABLE. (SILICON DIFLUORI	1373
FRACTION STUDIES OF SOME TRANSITION METAL HEXAFLUORIDES.	(SECOND ROW, THIRD ROW, AND RHENIUM HEPTAFLUORIDE) /	1375
/E MASS SPECTROMETER TO THE STUDY OF REACTIVE FLUORIDES.	(SECOND ROW, THIRD ROW, ANTIMONY PENTAFLUORIDE, AND/	1377
SIMPLE METHOD FOR EXACT SOLUTION OF 2 X 2	SECULAR EQUATION IN MOLECULAR VIBRATIONS.	1386
NEW METHOD FOR RESOLVING WILSON'S	SECULAR EQUATIONS. APPLIED TO TETRAHEDRAL MOLECULES	1431
LUTETIUM MONOFLUORIDE MOLECULE. ROTATIONAL ANALYSIS OF	SELECTED BANDS FROM THE ELECTRONIC SPECTRUM OF THE	0599
VAPORIZATION AND SUBLIMATION THERMODYNAMICS OF	SELECTED LANTHANIDE FLUORIDES.	1347
(SULFUR HEXAFLUORIDE) BY INFRARED RADIATION.	SELECTIVE PHOTODISSOCIATION OF POLYATOMIC MOLECULES	1064
A AND MOLECULAR CONSTANTS OF THE HEXAFLUORIDES OF SULFUR,	SELENIUM, AND TELLURIUM. RAMAN SPECTR	0963
TION. PART-2: INTERACTION TERMS IN ENERGY DISTRIBUTIONS.	(SELENIUM AND TELLURIUM FLUORIDES) /L MODES OF VIBRA	1494
AMPLITUDES IN SOME HEXAFLUORIDES OF OCTAHEDRAL SYMMETRY.	(SELENIUM AND TELLURIUM FLUORIDES) ROOT MEAN SQUARE	0962
MOLECULAR FORCE FIELD FOR SOME XY6 TYPE MOLECULES.	(SELENIUM AND TELLURIUM HEXAFLUORIDES)	1465
IHYDRIDE, GERMANIUM TETRAHYDRIDE, ARSENIC TRIHYDRIDE, AND	SELENIUM DIHYDRIDE. /LLIUM MONOFLUORIDE, GALLIUM TR	0459
/SPECULATION ON THE BONDING IN ANTIMONY HEXACHLORIDE (3-),	SELENIUM HEXABROMIDE (2-), TELLURIUM HEXABROMIDE (2-/	1309
/STRUCTURES OF SULFUR HEXAFLUORIDE, SULFUR TETRAFLUORIDE,	SELENIUM HEXABROMIDE, AND SELENIUM TETRAFLUORIDE.	1041
INFRARED SPECTRA OF	SELENIUM HEXAFLUORIDE AND TELLURIUM HEXAFLUORIDE.	1107
/HEXAFLUORIDE, URANIUM HEXAFLUORIDE, SULFUR HEXAFLUORIDE,	SELENIUM HEXAFLUORIDE, AND TELLURIUM HEXAFLUORIDE /	1072
VIBRATIONAL ANALYSIS OF	SELENIUM HEXAFLUORIDE AND TUNGSTEN HEXAFLUORIDE.	0960
ANALYSIS OF SULFUR HEXAFLUORIDE AND RELATED MOLECULES.	(SELENIUM HEXAFLUORIDE, TELLURIUM HEXAFLUORIDE) /ION	1093
S OF SOME OCTAHEDRAL XY6 MOLECULES. (SULFUR HEXAFLUORIDE,	SELENIUM HEXAFLUORIDE, TELLURIUM HEXAFLUORIDE) /CIE	1080
HEIR THE/ ENTHALPIES OF FORMATION OF SULFUR HEXAFLUORIDE,	SELENIUM HEXAFLUORIDE, TELLURIUM HEXAFLUORIDE AND T	1092
/NTS OF THE SPHERICAL TOP MOLECULES, SULFUR HEXAFLUORIDE,	SELENIUM HEXAFLUORIDE, TELLURIUM HEXAFLUORIDE, MOL/	1069
/FLECTION CF BINARY HEXAFLUORIDES. (SULFUR HEXAFLUORIDE,	SELENIUM HEXAFLUORIDE, TELLURIUM HEXAFLUORIDE, MOL/	1084
F OPEN SHELL DIATOMIC MOLECULES. (SULFUR MONOFLUORIDE AND	SELENIUM MONOFLUORIDE) ELECTRIC DIPOLE MOMENTS O	1019
ASE ELECTRON RESONANCE SPECTRA OF SULFUR MONOFLUORIDE AND	SELENIUM MONOFLUORIDE. GAS PH	1021
/ THE DOUBLET PI GROUND STATES OF SULFUR MONOFLUORIDE AND	SELENIUM MONOFLUORIDE AND THEIR POSITIVE AND NEGAT/	1025
/THALPY, SPECTROSCOPIC CONSTANTS, AND OTHER PROPERTIES OF	SELENIUM MONOFLUORIDE FROM A MOLECULAR ORBITAL STU/	0955
ESR SPECTRUM OF BROMINE OXIDE, IODINE OXIDE AND	SELENIUM MONOFLUORIDE IN J-5/2 ROTATIONAL LEVELS.	0954
/PHASE ELECTRON RESONANCE SPECTRA OF SULFUR MONOFLUORIDE,	SELENIUM MONOFLUORIDE, SELENIUM OXIDE AND IODINE O/	1020

PARAMAGNETIC HEXAFLUORIDE ANIONS OF GROUP-6. (SELENIUM PENTAFLUORIDE) 0959
/LAR STRUCTURE OF SOME GROUP-6 HEXAFLUORIDES. (OF SULFUR, SELENIUM, TELLURIUM, MOLYBDENUM, TUNGSTEN, AND URA/ 0961
/OND LENGTHS OF SOME INORGANIC HEXAFLUORIDES. (OF SULFUR, SELENIUM, TELLURIUM, MOLYBDENUM, TUNGSTEN, URANIUM/ 0631
ROWAVE SPECTRUM, DIPOLE MOMENT, AND STRUCTURE ANALYSIS OF SELENIUM TETRAFLUORIDE. MIC 0957
OME INORGANIC FLUORIDES BY NMR. (SULFUR TETRAFLUORIDE AND SELENIUM TETRAFLUORIDE) /OF EXCHANGE PROCESSES IN S 1048
ID STATE. (ARSENIC TRIFLUORIDE, SULFUR TETRAFLUORIDE, AND SELENIUM TETRAFLUORIDE) /GANIC FLUORIDES IN THE SOL 0096
VIBRATIONAL ASSIGNMENT AND NORMAL COORDINATE ANALYSIS OF SELENIUM TETRAFLUORIDE. 0958
INCLUDING A MATRI/ VIBRATIONAL SPECTRA AND STRUCTURES OF SELENIUM TETRAFLUORIDE AND TELLURIUM TETRAFLUORIDE, 0956
MEAN AMPLITUDES OF VIBRATION OF SELENIUM TRIOXIDE AND CHLORINE TRIFLUORIDE. 0406
ORBITAL CALCULATIONS ON PHOSPHORUS TRIFLUORIDE. SELF CONSISTENT CHARGE AND CONFIGURATION MOLECULAR 0821
/ADIENT OF NITROGEN TRIFLUORIDE. PART-3: A GAUSSIAN BASIS SELF CONSISTENT FIELD CALCULATION FOR NITROGEN TRI/ 0746
/TURE OF A CHLORINE PENTAFLUORIDE MOLECULE BY THE MO LCAO SELF CONSISTENT FIELD METHOD IN AN NDDO APPROXIMAT/ 0415
LUORIDE, XENON TETRAFLUORIDE, XENON HEXAFLUORIDE. SELF CONSISTENT FIELD STUDY OF THE SERIES XENON DIF 1224
AB INITIO CALCULATION OF EQUILIBRIUM GEOMETRIES AND QUAD/ SELF CONSISTENT MOLECULAR ORBITAL METHODS. PART-5: 0720
/AFLUORIDE AND ELEMENT 110 HEXAFLUORIDE CALCULATED BY THE SELF CONSISTENT MULTIPLE SCATTERING X-ALPHA METHOD. 0881
AFLUORIDE) SELF IONIZATION OF HALOGEN FLUORIDES. (BROMINE PENT 0263
OROTATION IN PHOSPHORUS PENTAFLUORIDE. SEMIEMPIRICAL MOLECULAR ORBITAL CALCULATIONS. PSEUD 0843
BORON HYDRIDE FLUORIDES. (BORON TRIFLUORIDE) SEMIEMPIRICAL STUDY OF THE ELECTRONIC STRUCTURE OF 0226
/ONIC TRANSITION MOMENT AS A FUNCTION OF THE INTERNUCLEAR SEPARATION FOR THE B-X BAND SYSTEM OF MOLECULAR SI/ 0967
/ON OF ELECTRONIC TRANSITION MOMENT WITH THE INTERNUCLEAR SEPARATION IN SILICON CHLORIDE AND TIN MONOFLUORID/ 1137
LASER ISOTOPE SEPARATION. (SULFUR HEXAFLUORIDE SPECTRUM) 1094
P MOLECULES AND FORCE CONSTANTS FOR NITROGEN TRIFLUO/ P-R SEPARATIONS AND CORIOLIS CONSTANTS FOR SYMMETRIC TO 0708
PAIRING ENERGIES IN THE ACTINIDE SERIES. 1333
/ATIONS OF SOME MOLECULAR PROPERTIES OF THE ISOELECTRONIC SERIES CHLORINE MONOFLUORIDE, HYDROGEN OXYCHLORIDE/ 0371
HEXAFLUORIDE. SELF CONSISTENT FIELD STUDY OF THE SERIES XENON DIFLUORIDE, XENON TETRAFLUORIDE, XENON 1224
IN IODINE COMPOUNDS. A SMALL GAUSSIAN BASIS SET FOR NONEMPIRICAL ALL-ELECTRON SCF CALCULATIONS 0495
NITRITE ION, NITROGEN OXYFLUORIDE, AND OX/ MINIMAL BASIS SET SCF CALCULATIONS FOR THE GROUND STATE OF OZONE, 0776
ELECTRON CORRELATION AND MOLECULAR SHAPE. 1413
CTION OF TEMPERATURE. EVIDENCE FOR A TRIGONAL BIPYRAMIDAL SHAPE FOR THE MONOMERIC SPECIES. /AS PHASE AS A FUN 0077
, TANTALUM PENTAFLUORIDE, AND ANTIMONY PENTAFLUORIDE. THE SHAPE OF ANTIMONY PENTAFLUORIDE. /IUM PENTAFLUORIDE 0076
EXTENDED HUCKEL THEORY AND THE SHAPE OF MOLECULES. (OXYGEN DIFLUORIDE) 0758
XENON MOLECULES. PART-3: ELECTRONIC TRANSITIONS AND BAND SHAPES. / PART-2: COMPARISONS WITH OTHER HEXAVALENT 1311
INFRARED BAND SHAPES AND VIBRATIONS OF NITROGEN TRIFLUORIDE. 0699
FORCES IN CARBON / MOLECULAR CONSTANTS FROM INFRARED BAND SHAPES. THE ATOMIC DISPLACMENTS AND INTRAMOLECULAR 0331
SELENIUM MONOFLUORIDE) ELECTRIC DIPOLE MOMENTS OF OPEN SHELL DIATOMIC MOLECULES. (SULFUR MONOFLUORIDE AND 1019
ANOMALOUS NMR SHIFT IN SAMARIUM TRIFLUORIDE. 0927
OUBLET PI SYSTEM OF TH/ ROTATIONAL STRUCTURE AND ISOTOPIC SHIFT IN THE (1,0) BAND OF THE A DOUBLET SIGMA- X D 1135
LIQUID GAS ARSENIC INFRARED FREQUENCY SHIFTS. 0106
CHLORINE MONOFLUORIDE. GASEOUS SPECTRA AND GAS TO LIQUID SHIFTS. NMR SPECTRA OF CHLORINE TRIFLUORIDE AND 0368
OSPHORYL HALIDES. CHEMICAL SHIFTS OF BORON, PHOSPHORUS, PHOSPHORYL, AND THIOPH 0235
X TRIPLET PI) AND CARBON MONOFLUORIDE A DOUBLET SIGMA(+/ SHOCK TUBE DETERMINATION OF THE C2 (A TRIPLET PI TO 0288
RMAL DISSOCIATION OF CHLORINE TRIFLUORIDE BEHIND INCIDENT SHOCK WAVES. THE 0394
/NTS, GENERALIZED MEAN SQUARE AMPLITUDES OF VIBRATION AND SHRINKAGE CONSTANTS OF XENON TETRAFLUORIDE AND XEN/ 1278
/Y-BRADLEY FORCE CONSTANTS, MEAN AMPLITUDES OF VIBRATION, SHRINKAGE EFFECT AND CORIOLIS CONSTANTS IN IODINE / 0513
R DIHALI/ MEAN AMPLITUDES OF VIBRATION, BASTIANSEN-MORINO SHRINKAGE EFFECT, AND MOLECULAR POLARIZABILITIES FO 0124
/AN AMPLITUDES OF VIBRATION, CORIOLIS COUPLING CONSTANTS, SHRINKAGE EFFECT, AND THERMODYNAMIC FUNCTIONS OF A/ 0050
PTON DIF/ MEAN AMPLITUDES OF VIBRATION, BASTIANSEN-MORINO SHRINKAGE EFFECT AND THERMODYNAMIC FUNCTIONS OF KRY 0562
ES. (/ MEAN AMPLITUDES OF VIBRATION AND BASTIANSEN-MORINO SHRINKAGE EFFECT IN SOME LINEAR SYMMETRICAL DIHALID 1372
IGONA/ MEAN AMPLITUDES OF VIBRATION AND BASTIANSEN-MORINO SHRINKAGE EFFECT IN SOME XY3 MOLECULES OF PLANAR TR 1397
E. BASTIANSEN-MORINO SHRINKAGE EFFECT IN THE STUDY OF MOLECULAR STRUCTUR 1476
OM UREY-BRADLEY FORCE C/ MEAN AMPLITUDES OF VIBRATION AND SHRINKAGE EFFECT OF BORON TRIHALIDES, CALCULATED FR 0233
/IONAL MEAN SQUARE AMPLITUDE MATRICES. (BASTIANSEN-MORINO SHRINKAGE EFFECT OF SYMMETRICAL XY3 AND XY4 MOLECU/ 1394
NAMIC FUN/ FORCE CONSTANTS, MEAN AMPLITUDES OF VIBRATION, SHRINKAGE EFFECTS, CORIOLIS CONSTANTS, AND THERMODY 1219
MEAN AMPLITUDES OF VIBRATION AND BASTIANSEN-MORINO SHRINKAGE EFFECTS IN SOME OCTAHEDRAL XY6 MOLECULES. 1455
AND/ GENERALIZED MEAN SQUARE AMPLITUDES OF VIBRATION, AND SHRINKAGE EFFECTS OF SOME TRIGONAL BIPYRAMIDAL XY5 1432
ETRICAL DIFLUORIDES. (OF THE FIRST R/ MEAN AMPLITUDES AND SHRINKAGE EFFECTS OF VIBRATION FOR SOME LINEAR SYMM 1369
DES. MEAN AMPLITUDES OF THERMAL MOTION AND SHRINKAGES OF CHEMICAL BONDS: OCTAHEDRAL HEXAFLUORI 1460
AR VIBRATIONS. SHRINKAGES OF THE INTERNUCLEAR DISTANCES BY MOLECUL 1489
SIGMA PLATINUM OXIDE AND SILICON MONOFLUORIDE (D DOUBLET SIGMA(+)- X DOUBLET PI) MOLECULES. /IGMA- X SINGLET 0966
) TO X DOUBLET PI / ROTATIONAL STRUCTURE OF THE A DOUBLET SIGMA(+), B DOUBLET SIGMA(+), AND A QUARTET SIGMA(- 0466
UM ECHELLE SPECTROG/ STUDY OF THE A SINGLET PI- X SINGLET SIGMA(+) BANDS OF BORON-11 MONOFLUORIDE WITH A VACU 0170
ROTATIONAL ANALYSIS OF THE SYSTEM B SINGLET PI TO SINGLET SIGMA(+) OF GASEOUS SCANDIUM MONOFLUORIDE. 0931
ROTATIONAL ANALYSIS OF THE C SINGLET PI- X SINGLET SIGMA(+) SYSTEM OF INDIUM MONOFLUORIDE. 0491
LUORIDE. O(+) TO SINGLET SIGMA(+) SYSTEMS OF BOTH ISOTOPES OF CHLORINE MONOF 0390
OF THE A TRIPLET PI ZERO(+) B TRIPLET PI(1) TO X SINGLET SIGMA(+) SYSTEMS OF GALLIUM MONOFLUORIDE. /OF BANDS 0457
ROTATIONAL ANALYSIS OF THE AO(+), BO(+), TO X SINGLET SIGMA(+) SYSTEMS OF GASEOUS SILVER MONOFLUORIDE. 1011
OF THE A TRIPLET PI ZERO(+), B TRIPLET PI(1) TO X SINGLET SIGMA(+) SYSTEMS OF THALLIUM MONOFLUORIDE. / BANDS 1111
/ FRANCK-CONDON FACTORS AND R CENTROIDS FOR THE B TRIPLET SIGMA(+) TO A TRIPLET PI SYSTEM OF BORON MONOFLUORI 0171
OUBLET PI SYSTEMS O/ ROTATIONAL ANALYSIS OF THE A DOUBLET SIGMA(+) TO X DOUBLET PI, B DOUBLET SIGMA(+) TO X D 1133
/ET PI TO X TRIPLET PI) AND CARBON MONOFLUORIDE A DOUBLET SIGMA(+) TO X DOUBLET PI BAND SYSTEM OSCILLATOR ST/ 0288
MONOFLUORIDE. B SINGLET SIGMA(+) TO X TRIPLET SIGMA(-) BAND SYSTEM IN LEAD 0585
GEN MONOFLUORIDE. B SINGLET SIGMA(+) TO X TRIPLET SIGMA(-) BAND SYSTEM OF NITRO 0675
B SINGLET SIGMA(+) TO X TRIPLET SIGMA(-) BAND SYSTEM IN LEAD MONOFLUORIDE. 0585
B SINGLET SIGMA(+) TO X TRIPLET SIGMA(-) BAND SYSTEM OF NITROGEN MONOFLUORIDE. 0675
/HE A DOUBLET SIGMA(+), B DOUBLET SIGMA(+), AND A QUARTET SIGMA(-) TO X DOUBLET PI TRANSITIONS OF GERMANIUM / 0466
N MONOFLUO/ ROTATIONAL ANALYSIS OF BANDS OF THE C TRIPLET SIGMA, B TRIPLET SIGMA: A TRIPLET PI SYSTEM OF BORO 0153
ACTORS AND R CENTROIDS FOR THE 6 SINGLET SIGMA- X TRIPLET SIGMA BAND SYSTEM OF NITROGEN MONOFLUORIDE. /NDON F 0677
ELECTRON AFFINITY, DIPOLE MOMEN/ ARSENIC MONOFLUORIDE, 3 SIGMA. DISSOCIATION ENTHALPY, IONIZATION POTENTIAL, 0120
/ON FACTORS AND R CENTROIDS OF A SINGLET SIGMA- X SINGLET SIGMA PLATINUM OXIDE AND SILICON MONOFLUORIDE (D D/ 0966
THE C DOUBLET PI AND THE X DOUBLET SIGMA STATES OF BARIUM MONOFLUORIDE. 0120
MIC FUNCTIONS OF DIATOMIC GASES WITH MOLECULES IN TRIPLET SIGMA STATES. (PHOSPHORUS MONOFLUORIDE) THERMODYNA 0789
ION / ANALYSIS OF THE BANDS OF B DOUBLET SIGMA- X DOUBLET SIGMA. SYSTEM OF CALCIUM FLUORIDE MOLECULE IN EMISS 0280
LUORIDE MOLECULE. A QUARTET SIGMA TO DOUBLET PI TRANSITION OF THE SILICON MONOF 0978
/ THE MATRIX ELEMENT OF THE DIPOLE MOMENT OF AN A DOUBLET SIGMA TO X DOUBLET PI ELECTRON TRANSITON IN A SILI/ 0970
CON MONOFLUORIDE. A QUARTET SIGMA TO X DOUBLET PI ELECTRONIC TRANSITION OF SILI 0971
OLE/ ROTATIONAL ANALYSIS OF THE BAND AT 14900 CM-1 OF A 3 SIGMA TO 3 SIGMA SYSTEM OF THE BORON MONOFLUORIDE M 0156
/RE AND ISOTOPIC SHIFT IN THE (1,0) BAND OF THE A DOUBLET SIGMA- X DOUBLET PI SYSTEM OF THE TIN MONOFLUORIDE/ 1135
MOLECULE IN EMISSION / ANALYSIS OF THE BANDS OF B DOUBLET SIGMA- X DOUBLET SIGMA. SYSTEM OF CALCIUM FLUORIDE 0280
ONOFL/ FRANCK-CONDON FACTORS AND R CENTROIDS OF A SINGLET SIGMA- X SINGLET SIGMA PLATINUM OXIDE AND SILICON M 0966
/ FRANCK-CONDON FACTORS AND R CENTROIDS FOR THE 6 SINGLET SIGMA- X TRIPLET SIGMA BAND SYSTEM OF NITROGEN MONO 0677

DE.	SIGNS OF L DOUBLING CONSTANTS IN NITROGEN TRIFLUORI	0743
UORIDE, SILICON TETRAFLUORIDE, BORON TRIFLUORIDE, ETHANE,	SILANE, AMMONIA, AND PHOSPHINE. / OF CARBON TETRAFL	0243
POTENTIAL CONSTANTS OF SOME XY4 TYPE OF	SILANE COMPOUNDS. (SILICON TETRAFLUORIDE)	1009
IDE AN/ STRUCTURE DETERMINATION OF TRIS (TRIMETHYL SILYL)	SILANE, INFRARED INVESTIGATION OF IODINE HEPTAFLUOR	0527
OCIATION ENERGIES OF THE MONOFLUORIDES AND DIFLUORIDES OF	SILICON AND GERMANIUM. /PERATURES. PART-2: THE DISS	0463
N PROBABILITIES AND R CENTROIDS FOR DIATOMIC FLUORIDES OF	SILICON AND GERMANIUM. VIBRATIONAL TRANSITIC	0994
ION OF THE DISSOCIATION ENERGY OF CALCIUM FLUORIDE. (ALSO	SILICON AND GERMANIUM FLUORIDES) REEVALUAT	0277
NIC TRANSITION MOMENT WITH THE INTERNUCLEAR SEPARATION IN	SILICON CHLORIDE AND TIN MONOFLUORIDE. / OF ELECTRO	1137
SPECTRUM, FORCE CONSTANTS, AND THERMODYNAMIC FUNCTIONS OF	SILICON DIFLUORIDE. INFRARED	0985
MICROWAVE SPECTRUM OF	SILICON DIFLUORIDE.	0990
NEW ELECTRONIC EMISSION FROM	SILICON DIFLUORIDE.	0987
L LCAO MO SCF CALCULATIONS OF THE ELECTRONIC STRUCTURE OF	SILICON DIFLUORIDE. NONEMPIRICA	0995
PHOTOELECTRON SPECTRUM OF	SILICON DIFLUORIDE.	0980
PHOTOELECTRON SPECTRUM OF	SILICON DIFLUORIDE.	0996
ALYSIS OF ABSORPTION BANDS IN THE 2266 ANGSTROM SYSTEM OF	SILICON DIFLUORIDE. ROTATIONAL AN	0979
THERMOCHEMISTRY OF	SILICON DIFLUORIDE.	0992
ULTRAVIOLET ABSORPTION SPECTRUM OF	SILICON DIFLUORIDE.	0984
ULTRAVIOLET BAND SPECTRUM OF	SILICON DIFLUORIDE.	0991
ULTRAVIOLET SPECTRUM OF	SILICON DIFLUORIDE.	0983
VACUUM ULTRAVIOLET SPECTRA OF	SILICON DIFLUORIDE AND GERMANIUM DIFLUORIDE.	0981
N AMPLITUDES OF VIBRATION AND THERMODYNAMIC FUNCTIONS FOR	SILICON DIFLUORIDE AND OXYGEN DIFLUORIDE. MEA	0773
MICROWAVE SPECTRUM AND FORCE CONSTANTS OF	SILICON DIFLUORIDE. CENTRIFUGAL DISTORTION.	0989
/ ISOLATION STUDY OF THE VACUUM ULTRAVIOLET PHOTOLYSIS OF	SILICON DIFLUORIDE DIHYDRIDE. THE INFRARED AND UL/	0986
INFRARED SPECTRA OF	SILICON DIFLUORIDE IN INERT GAS MATRICES.	0982
S, EQUILIBRIUM STRUCTURE, ANHARMON/ MICROWAVE SPECTRUM OF	SILICON DIFLUORIDE IN THE EXCITED VIBRATIONAL STATE	0993
REGION 2755-2179 ANGSTROMS. EMISSION SPECTRUM OF	SILICON DIFLUORIDE. PART-1: THE BAND SYSTEM IN THE	0988
/AINING ATOMS FROM THE SECOND ROW OF THE PERIODIC TABLE.	(SILICON DIFLUORIDE, SILICON TETRAFLUORIDE, PHOSPHO/	1373
ELECTRONIC SPECTRUM OF	SILICON DIFLUORIDE. THEORETICAL STUDY.	0997
CALCULATION OF THE FRANCK-CONDON FACTOR FOR MOLECULES OF	SILICON FLUORIDE, GERMANIUM FLUORIDE, AND GERMANIU/	0977
ULAR MEC/ FLUORINE NMR SPECTRA OF SULFUR, PHOSPHORUS, AND	SILICON FLUORIDES. HYDROLYSIS AND INTRA- INTERMOLEC	1042
LICON, AND / VAPOR PHASE RAMAN SPECTRA OF MH4 (M= CARBON,	SILICON, GERMANIUM, AND TIN) AND MF4 (M= CARBON, SI	0314
A QUARTET SIGMA TO X DOUBLET PI ELECTRONIC TRANSITION OF	SILICON MONOFLUORIDE.	0971
BAND SPECTRUM OF	SILICON MONOFLUORIDE.	0965
IBUTION OF INTENSITY IN THE ELECTRON EMISSION SPECTRUM OF	SILICON MONOFLUORIDE. DISTR	0968
ERNUCLEAR SEPARATION FOR THE B-X BAND SYSTEM OF MOLECULAR	SILICON MONOFLUORIDE. /ENT AS A FUNCTION OF THE INT	0967
EMISSION SPECTRUM OF THE GAMMA SYSTEM OF	SILICON MONOFLUORIDE.	0974
ROTATIONAL ANALYSIS OF THE GAMMA BANDS OF	SILICON MONOFLUORIDE.	0999
/LECULAR ORBITAL INVESTIGATION OF CARBON MONOFLUORIDE AND	SILICON MONOFLUORIDE AND THEIR POSITIVE AND NEGATI/	0290
VES FOR SOME DIATOMIC MOLECULES. (BERYLLIUM MONOFLUORIDE,	SILICON MONOFLUORIDE, AND TIN MONOFLUORIDE) /AL CUR	0130
/K-CONDON FACTORS AND R CENTROIDS OF SOME BAND SYSTEMS OF	SILICON MONOFLUORIDE, CALCIUM MONOFLUORIDE, AND BI/	0972
/S OF A SINGLET SIGMA- X SINGLET SIGMA PLATINUM OXIDE AND	SILICON MONOFLUORIDE (D DOUBLET SIGMA(+)- X DOUBLE/	0966
ONOFLUORIDE, AND LEAD MONOFLUORIDE. SPECTRA OF	SILICON MONOFLUORIDE, GERMANIUM MONOFLUORIDE, TIN M	0964
A QUARTET SIGMA TO DOUBLET PI TRANSITION OF THE	SILICON MONOFLUORIDE MOLECULE.	0978
N A DOUBLET SIGMA TO X DOUBLET PI ELECTRON TRANSITON IN A	SILICON MONOFLUORIDE MOLECULE. / DIPOLE MOMENT OF A	0970
ANCK-CONDON FACTORS AND R CENTROIDS FOR THE C-X SYSTEM OF	SILICON MONOFLUORIDE MOLECULE. FR	0975
ROTATIONAL ANALYSIS OF THE BETA BAND SYSTEM OF	SILICON MONOFLUORIDE MOLECULE.	0976
SULFUR MONOFLUORIDE/ PROPERTIES OF NITROGEN MONOFLUORIDE,	SILICON MONOFLUORIDE, PHOSPHORUS MONOFLUORIDE, AND	0973
ST IONIZATION POTENTIALS OF THE MOLECULES BORON FLUORIDE,	SILICON OXIDE AND GERMANIUM OXIDE. FIR	0160
OME FLUORIDES AND OXYCHLORIDES. (CARBON TETRAFLUORIDE AND	SILICON TETRAFLUORIDE) / DENSITY DISTRIBUTIONS IN S	0337
FORCE FIELD FOR	SILICON TETRAFLUORIDE.	1006
FORCE FIELDS OF CARBON TETRAFLUORIDE AND	SILICON TETRAFLUORIDE.	0324
N PHOTOELECTRON SPECTROSCOPY OF CARBON TETRAFLUORIDE AND	SILICON TETRAFLUORIDE. HIGH RESOLUTIO	0319
INFRARED AND RAMAN SPECTRA OF LIQUID AND CRYSTALLINE	SILICON TETRAFLUORIDE.	1001
INFRARED AND RAMAN SPECTRA OF	SILICON TETRAFLUORIDE.	1005
INFRARED SPECTRUM OF	SILICON TETRAFLUORIDE.	1003
INTERMOLECULAR POTENTIAL FOR	SILICON TETRAFLUORIDE.	1010
PLITUDES OF VIBRATION OF SOME XY4 TETRAHEDRAL MOLECULES.	(SILICON TETRAFLUORIDE) MEAN AM	1008
OTENTIAL CONSTANTS OF SOME XY4 TYPE OF SILANE COMPOUNDS.	(SILICON TETRAFLUORIDE) P	1009
S EXCITED IN BORON TRIFLUORIDE, CARBON TETRAFLUORIDE, AND	SILICON TETRAFLUORIDE. /ULTRAVIOLET EMISSION SYSTEM	0216
/E HELIUM-1 AND HELIUM-2 STUDIES OF CARBON TETRAFLUORIDE,	SILICON TETRAFLUORIDE, AND GERMANIUM TETRAFLUORIDE.	0341
M HEAVY ISOTOPIC SUBSTITUTION. PART-2: BORON TRIFLUORIDE,	SILICON TETRAFLUORIDE, AND NITROGEN DIOXIDE. /S FRO	0232
NSTANTS FOR BORON TETRAFLUORIDE(-), CARBON TETRAFLUORIDE,	SILICON TETRAFLUORIDE, AND SILICON TETRACHLORIDE. /	0348
XATION BY NMR ABSORPTION IN GASEOUS CARBON TETRAFLUORIDE,	SILICON TETRAFLUORIDE, AND SULFUR HEXAFLUORIDE. /LA	0346
ILANE, AMMONIA, / FORCE CONSTANTS OF CARBON TETRAFLUORIDE,	SILICON TETRAFLUORIDE, BORON TRIFLUORIDE, ETHANE, S	0243
/ES OF THE INFRARED FUNDAMENTALS OF CARBON TETRAFLUORIDE,	SILICON TETRAFLUORIDE, CARBON TETRACHLORIDE, SILIC/	1007
/FRARED ABSORPTION SPECTRA OF SOME POLYATOMIC FLUORIDES.	(SILICON TETRAFLUORIDE, CARBON TETRAFLUORIDE, BORON/	0186
S AND ROOT MEAN SQUARE AMPLITUDES OF VIBRATION OF GASEOUS	SILICON TETRAFLUORIDE FROM ELECTRON DIFFRACTION. /E	1002
/E CONSTANTS OF VARIOUS ISOTOPES OF CARBON TETRAFLUORIDE,	SILICON TETRAFLUORIDE, GERMANIUM TETRAFLUORIDE AND/	0358
/S AND TETRACHLORIDES OF GROUP-5B. (CARBON TETRAFLUORIDE,	SILICON TETRAFLUORIDE, GERMANIUM TETRAFLUORIDE, TI/	0316
DY. SILICON- FLUORINE LENGTH IN	SILICON TETRAFLUORIDE. NEW ELECTRON DIFFRACTION STU	1000
/TETRAHALIDES CARBON TETRAFLUORIDE, CARBON TETRACHLORIDE,	SILICON TETRAFLUORIDE, SILICON TETRACHLORIDE, GERM/	0349
OLYBDENUM HEXAFLU/ PHYSICAL CONSTANTS (VAPOR PRESSURE) OF	SILICON TETRAFLUORIDE, TUNGSTEN HEXAFLUORIDE, AND M	0639
RAMAN SPECTRA OF CARBON AND	SILICON TETRAFLUORIDES.	0364
O SILICON. FORCE CONSTANTS OF THE	SILICON TETRAHALIDES AND FREQUENCIES OF CHLORO BROM	1004
EMISSION SPECTRUM OF	SILICON TRIFLUORIDE.	0998
NEW ELECTRON DIFFRACTION STUDY.	SILICON- FLUORINE LENGTH IN SILICON TETRAFLUORIDE.	1000
MAGNETIC SUSCEPTIBILITY OF	SILVER DIFLUORIDE.	1015
ABSORPTION SPECTRUM OF GASEOUS	SILVER MONOFLUORIDE.	1012
TURES. PART-14: VAPOR PRESSURE AND DISSOCIATION ENERGY OF	SILVER MONOFLUORIDE. /ETRIC STUDIES AT HIGH TEMPERA	1014
NEAR ULTRAVIOLET ABSORPTION SPECTRUM OF	SILVER MONOFLUORIDE.	1013
HE AO(+), BO(+), TO X SINGLET SIGMA(+) SYSTEMS OF GASEOUS	SILVER MONOFLUORIDE. ROTATIONAL ANALYSIS OF T	1011
TAFLUORIDE AN/ STRUCTURE DETERMINATION OF TRIS (TRIMETHYL	SILYL) SILANE, INFRARED INVESTIGATION OF IODINE HEP	0527
FLUORIDES. (OF SULFUR, MOLYBDENUM TUNGSTEN/ THERMODYNAMIC	SIMILARITY AND UNIVERSAL EQUATIONS OF STATE OF HEXA	0632
AR FORCE CONSTANTS OF HIGHLY SYMMETRICAL MOLECU/ USE OF A	SIMPLE ELECTROSTATIC MODEL FOR CALCULATION OF ANGUL	1484
QUATION IN MOLECULAR VIBRATIONS.	SIMPLE METHOD FOR EXACT SOLUTION OF 2 X 2 SECULAR E	1386
ETRAFLUORIDE. INFRARED SPECTRA OF	SIMPLE MOLECULES IN LIQUID ARGON SOLUTION. CARBON T	0338
/NNECTION WITH ELECTRON DIFFRACTION INVESTIGATION OF SOME	SIMPLE MOLECULES. (TUNGSTEN HEXAFLUORIDE, TELLURIU/	1153
ORIDE.	SIMPLE THEORETICAL CALCULATIONS ON TITANIUM MONOFLU	1144
APPLICATION OF THE PROGRESSIVE RIGIDITY METHOD TO SOME	SIMPLE XYN TYPE MOLECULES. (BORON TRIFLUORIDE)	0211
PECTRA OF PHOSPHORUS TETRAFLUORIDE RADICALS PRODUCED IN A	SINGLE CRYSTAL OF PHOSPHORUS TRIFLUORIDE. ESR S	0823

```
UNDAMENTAL OF SULFUR HEXAFLUORID/ OPTICAL SATURATION OF A     SINGLE VIBRATION- ROTATION TRANSITION IN THE NU-3 F  1074
PECTRUM AND DIPOLE MOMENT OF NITROGEN MONOFLUORIDE IN THE      SINGLET DELTA STATE. /TRON PARAMAGNETIC RESONANCE S  0673
                                                              SINGLET ELECTRONIC STATES OF ALUMINUM MONOFLUORIDE.  0023
MONOFLUORIDE.          ROTATIONAL ANALYSIS OF THE SYSTEM B     SINGLET PI TO SINGLET SIGMA(+) OF GASEOUS SCANDIUM   0931
NOFLUORIDE WITH A VACUUM ECHELLE SPECTROG/ STUDY OF THE A      SINGLET PI- X SINGLET SIGMA(+) BANDS OF BORON-11 MO  0170
OFLUORIDE.             ROTATIONAL ANALYSIS OF THE C            SINGLET PI- X SINGLET SIGMA(+) SYSTEM OF INDIUM MON  0491
NE MONOFLUORIDE.                         O(+) TO              SINGLET SIGMA(+) SYSTEMS OF BOTH ISOTOPES OF CHLORI  0390
OF BANDS OF THE A TRIPLET PI ZERO(+) B TRIPLET PI(1) TO X      SINGLET SIGMA(+) SYSTEMS OF GALLIUM MONOFLUORIDE. /  0457
RIDE.       ROTATIONAL ANALYSIS OF THE AO(+), BO(+), TO X      SINGLET SIGMA(+) SYSTEMS OF GASEOUS SILVER MONOFLUO  1011
/ BANDS OF THE A TRIPLET PI ZERO(+), B TRIPLET PI(1) TO X      SINGLET SIGMA(+) SYSTEMS OF THALLIUM MONOFLUORIDE.   1111
IN LEAD MONOFLUORIDE.                                  B      SINGLET SIGMA(+) TO X TRIPLET SIGMA(-) BAND SYSTEM   0585
OF NITROGEN MONOFLUORIDE.                              B      SINGLET SIGMA(+) TO X TRIPLET SIGMA(-) BAND SYSTEM   0675
ILICON MONOFL/ FRANCK-CONDON FACTORS AND R CENTROIDS OF A      SINGLET SIGMA- X SINGLET SIGMA PLATINUM OXIDE AND S  0966
GEN MONO/ FRANCK-CONDON FACTORS AND R CENTROIDS FOR THE 6      SINGLET SIGMA- X TRIPLET SIGMA BAND SYSTEM OF NITRO  0677
LE.                    ISOTOPE EFFECT IN THE                  SINGLET SPECTRAL BANDS OF THE BORON FLUORIDE MOLECU  0166
     ANALYSIS OF THE ABSORPTION SPECTRUM AND OF TWO NEW       SINGLET SYSTEMS OF ARSENIC MONOFLUORIDE.             0090
               ROTATIONAL ANALYSIS OF FOUR                   SINGLET SYSTEMS OF ARSENIC MONOFLUORIDE.             0092
               VIBRATIONAL ANALYSIS OF THE                   SINGLET SYSTEMS OF ARSENIC MONOFLUORIDE.             0095
DIUM MONOFLUORIDE.        ROTATIONAL ANALYSIS OF SOME         SINGLET TRANSITIONS IN THE SPECTRUM OF GASEOUS SCAN  0930
     AB INITIO STUDIES OF CHARGE DISTRIBUTION USING A MINIMAL SLATER-TYPE BASIS. (NITROGEN TRIFLUORIDE) / PART-3:  0707
ALUMINUM TRIFLUORIDE, LITHIUM ALUMINUM TETRAFLUORIDE, AND     SODIUM ALUMINUM TETRAFLUORIDE. /SPECTRA OF GASEOUS   0045
SULFUR HEXAFLUORIDE.                                          SOFT X-RAY SPECTROSCOPY ON SULFUR TETRAFLUORIDE AND  1033
E, CURIUM TRIFLUORIDE AND CURIUM T/ ABSORPTION SPECTRA OF     SOLID AMERICIUM TRIFLUORIDE, AMERICIUM TETRAFLUORID  0445
     INFRARED SPECTRA OF ALUMINUM FLUORIDE IN               SOLID ARGON.                                         0054
          PURE QUADRUPOLE SPECTRA OF                         SOLID CHLORINE COMPOUNDS.                            0383
          ELECTRICAL CONDUCTIVITY OF                         SOLID CHLORINE TRIFLUORIDE AND BROMINE TRIFLUORIDE.  0259
               RAMAN SPECTRUM OF                             SOLID MOLYBDENUM TETRAFLUORIDE.                      0621
               RAMAN SPECTRUM OF                             SOLID PHOSPHORUS PENTAFLUORIDE                       0858
/ARED SPECTRA OF SOME VOLATILE INORGANIC FLUORIDES IN THE     SOLID STATE. (ARSENIC TRIFLUORIDE, SULFUR TETRAFLU/  0096
RAMAN SPECTRUM OF SULFUR HEXAFLUORIDE IN THE GASEOUS AND      SOLID STATES.                                        1079
               RAMAN SPECTRUM OF                             SOLID SULFUR HEXAFLUORIDE.                           1098
     VIBRATIONAL SPECTRUM AND STRUCTURE OF                   SOLID URANIUM HEXAFLUORIDE.                          1189
          SATURATED VAPOR PRESSURE OF                        SOLID URANIUM TETRAFLUORIDE.                         1167
DIFLUORIDE.            RAMAN SPECTRA OF                       SOLID XENON TETRAFLUORIDE AND ITS ADDUCT WITH XENON  1258
ND XENON TETRAFLUORIDE AND A CONJECTURE ON BONDING IN THE     SOLIDS.  HEATS OF SUBLIMATION OF XENON DIFLUORIDE A  1237
     INFRARED SPECTRA OF SIMPLE MOLECULES IN LIQUID ARGON     SOLUTION. CARBON TETRAFLUORIDE.                      0338
DRAL MOLECU/ USE OF CORIOLIS INTERACTION CONSTANTS IN THE     SOLUTION OF THE VIBRATIONAL PROBLEM FOR XY4 TETRAHE  1423
/ THE STRUCTURE OF PYRAMIDAL MOLECULES OF THE XY3 TYPE BY     SOLUTION OF THE VIBRATIONAL PROBLEM WITH THE USE O/  0727
RATIONS.          SIMPLE METHOD FOR EXACT                    SOLUTION OF 2 X 2 SECULAR EQUATION IN MOLECULAR VIB  1386
     ACTIVITY COEFFICIENTS FOR PROTACTINIUM IN LIQUID BISMUTH SOLUTIONS. (PROTACTINIUM TETRAFLUORIDE) /ALIDES AND   0900
          INFRARED SPECTRA OF XENON DIFLUORIDE IN DIFFERENT   SOLVENTS.                                            1241
LUORIDES.          APPLICATION OF A MOLECULAR BEAM           SOURCE MASS SPECTROMETER TO THE STUDY OF REACTIVE F  0019
LUORIDES. (SECOND ROW, T/ APPLICATION OF A MOLECULAR BEAM     SOURCE MASS SPECTROMETER TO THE STUDY OF REACTIVE F  1377
NTIALS AND ELECTRON AFFINITIES OF SOME GERMANIUM FLUORIDE     SPECIES. /ND DISSOCIATION ENERGIES, IONIZATION POTE  0464
IDENCE FOR A TRIGONAL BIPYRAMIDAL SHAPE FOR THE MONOMERIC     SPECIES. /AS PHASE AS A FUNCTION OF TEMPERATURE. EV  0077
IC DET/ DETERMINATION OF THE GEOMETRY OF HIGH TEMPERATURE     SPECIES BY ELECTRIC DEFLECTION AND MASS SPECTROMETR  0134
ATRIX ISOLATION INFRARED SPECTROSCOPY OF HIGH TEMPERATURE     SPECIES. (FLUORIDES, OXIDES, CHLORIDES)          M  0011
     IN THE CALCULATION OF FORCE CONSTANTS. APPLICATION TO E' SPECIES FORCE FIELD OF PHOSPHORUS PENTAFLUORIDE. /D  0827
          INFRARED SPECTRA OF MATRIX ISOLATED               SPECIES IN THE GALLIUM- FLUORINE SYSTEM.             0462
NONEXISTENCE OF XENON MONOFLUORIDE AS A CHEMICALLY BOUND      SPECIES IN THE GAS PHASE.               PROBABLE 1221
DE(-) AND ANTIM/ VIBRATIONAL SPECTRA OF THE ISOELECTRONIC     SPECIES IODINE PENTAFLUORIDE, TELLURIUM PENTAFLUORI  0501
     CNDO STUDY OF THE OZONIDE ION AND RELATED               SPECIES. (OXYGEN DIFLUORIDE)                         0782
R PRESSURE, CRITICAL PROPERTIES, DIELECTRIC CONSTANT, AND     SPECIFIC CONDUCTIVITY OF CHLORINE PENTAFLUORIDE. /O  0426
D WITH A BALLISTIC PISTON COMPRESSOR.                         SPECIFIC HEAT RATIO OF URANIUM HEXAFLUORIDE MEASURE  1201
     MOLECULAR ORBITAL INTERPRETATION OF ITS X-RAY ABSORPTION SPECTRA. /ON FROM SCF MO LCAO COMPUTATIONS. PART-2:  1078
NGSTEN, AND URANIUM HEXAFLUORIDES FROM INFRARED AND RAMAN     SPECTRA.     MOLECULAR STRUCTURE OF MOLYBDENUM, TU  0627
AFLUORIDE. INFRARED AND LOW TEMPERATURE RAMAN VIBRATIONAL     SPECTRA. /US PHOSPHORUS HALIDES AND PHOSPHORUS PENT  0849
     AND THE STRUCTURE OF NITROGEN TRIFLUORIDE FROM MICROWAVE SPECTRA.   NUCLEAR QUADRUPOLE MOMENT OF NITROGEN-14  0741
IES OF XENON DIFLUORIDE AND XENON TETRAFLUORIDE FROM MASS     SPECTRA.            THERMOCHEMICAL PROPERT 1253
          XENON DIFLUORIDE. (FORCE CONSTANTS,                SPECTRA)                                             1252
MI RESONANCE BETWEEN A FUNDAMENTAL AND A TER/ VIBRATIONAL     SPECTRA AND ASSIGNMENT OF IODINE PENTAFLUORIDE. FER  0504
E.                                          VIBRATIONAL      SPECTRA AND FORCE CONSTANTS OF URANIUM TETRAFLUORID  1172
IFLUORIDE AND YTTRIUM TRIFLUORIDE.          INFRARED         SPECTRA AND GEOMETRIES OF MATRIX ISOLATED YTTRIUM D  1316
UORIDE, NICKEL DIFLUORIDE, COPPER DIFLUORIDE, A/ INFRARED     SPECTRA AND GEOMETRIES OF RARE EARTH DIHALIDES.      1351
ANIUM TRIFLUORIDE IN RARE GAS MATRICES.     INFRARED         SPECTRA AND GEOMETRY OF MATRIX ISOLATED COBALT DIFI  0437
UORIDE AND PHOSPHINE.                       RAMAN           SPECTRA AND GEOMETRY OF TITANIUM DIFLUORIDE AND TIT  1146
S OF SULFUR, SELENIUM, AND TELLURIUM.       RAMAN           SPECTRA AND MOLECULAR CONSTANTS OF PHOSPHORUS TRIFL  0822
AFLUORIDES. (OF SULFUR, SELENIUM, TELLURIUM, MO/ INFRARED     SPECTRA AND MOLECULAR CONSTANTS OF THE HEXAFLUORIDE  0963
1: INFRARED SPECTRUM OF OXYGEN DIFLUORIDE FR/ VIBRATIONAL     SPECTRA AND MOLECULAR STRUCTURE OF SOME GROUP-6 HEX  0961
ALIDES.                                     INFRARED        SPECTRA AND STRUCTURE OF INORGANIC MOLECULES. PART-  0761
ALIDE DIMERS, THALLOUS FLUORIDE, AND / INFRARED AND RAMAN     SPECTRA AND STRUCTURE OF MATRIX ISOLATED THALLIUM H  1113
               INFRARED AND RAMAN                           SPECTRA AND STRUCTURE OF MATRIX ISOLATED THALLOUS H  1123
                          VIBRATIONAL                       SPECTRA AND STRUCTURE OF PHOSPHORUS TETRAFLUORIDE.   0825
HEPTAFLUORIDE.                              VIBRATIONAL      SPECTRA AND STRUCTURE OF XENON TETRAFLUORIDE.        1262
     DIHALIDES: MAGNESIUM DIFLUORIDE, MAG/ INFRARED AND RAMAN SPECTRA AND STRUCTURES OF IODINE PENTAFLUORIDE AND    0506
D TELLURIUM TETRAFLUORIDE, INCLUDING A MATRI/ INFRARED AND RAMAN SPECTRA AND STRUCTURES OF MATRIX ISOLATED MAGNESIUM 0607
ENON OXYTETRAFLUORIDE.                      VIBRATIONAL      SPECTRA AND STRUCTURES OF SELENIUM TETRAFLUORIDE AN  0956
UORIDE AND IODINE OXYGEN PENTAFLUORIDE.          MASS        SPECTRA AND STRUCTURES OF XENON TETRAFLUORIDE AND X  1261
UORIDE, XENON TETRAF/ HELIUM-1 AND HELIUM-2 PHOTOELECTRON     SPECTRA AND SUBLIMATION PRESSURES OF IODINE HEPTAFL  0534
ND BROMINE TRIFLUORIDE.                     RAMAN           SPECTRA AND THE ELECTRONIC STRUCTURES OF XENON DIFL  1228
ONOFLUORIDE.                  INFRARED AND RAMAN            SPECTRA AND THE STRUCTURE OF CHLORINE TRIFLUORIDE A  0398
SPHORUS AND ARSENIC HALIDES. (PHOSPHO/ RAMAN AND INFRARED     SPECTRA AND THE STRUCTURE OF CRYSTALLINE THALLIUM M  1126
IFLUORIDE AND BROMINE TRIFLUORIDE.          VIBRATIONAL      SPECTRA AND THERMODYNAMIC FUNCTIONS FOR VARIOUS PHO  0811
APOR.              EFFUSION STUDIES, MASS                    SPECTRA AND THERMODYNAMIC PROPERTIES OF CHLORINE TR  0396
                                                            SPECTRA, AND THERMODYNAMICS OF BERYLLIUM FLUORIDE V  0140
NC DIFLUORIDE)                              VIBRATIONAL      SPECTRA AND THERMODYNAMICS OF THE ZINC HALIDES. (ZI  1323
IDAL MOLECULES. XENON OXIDE TETRAFLUORIDE, I/ VIBRATIONAL     SPECTRA FOR XENON DIFLUORIDE, XENON TETRAFLUORIDE,   1254
AND XENON OXIDE TETRAFLUORIDE VAPORS, AND FORCE CO/ RAMAN     SPECTRA AND VALENCE FORCE CONSTANTS OF SQUARE PYRAM  0418
     ANALYSES OF GERMANIUM D/ MATRIX INFRARED AND LASER RAMAN SPECTRA, MOLECULAR STRUCTURES AND NORMAL COORDINATE  0475
               INFRARED                                      SPECTRA OF A CRYOSYSTEM. CARBON TETRAFLUORIDE.       0313
```

INFRARED SPECTRA OF ALUMINUM FLUORIDE IN SOLID ARGON.		0054
IDE, INDIUM MONOFLUORIDE, AND THALL/ MICROWAVE ABSORPTION SPECTRA OF ALUMINUM MONOFLUORIDE, GALLIUM MONOFLUOR		0025
UORIDE. EMISSION SPECTRA OF ANTIMONY MONOFLUORIDE AND BISMUTH MONOFL		0064
AND KRYPTON MONOFLUORIDE. ABSORPTION AND EMISSION SPECTRA OF ARGON MATRIX ISOLATED XENON MONOFLUORIDE		0544
FLUORIDE AND THEIR FORCE CONSTANTS INCLUDING THOS/ RAMAN SPECTRA OF ARSENIC PENTAFLUORIDE AND VANADIUM PENTA		0873
UORIDE. HIGH RESOLUTION SPECTRA OF ARSENIC TRIFLUORIDE AND PHOSPHORUS TRIFL		0816
GAS PHASE STRUCTURES AND MASS SPECTRA OF BINARY PENTAFLUORIDES.		0003
GAS PHASE STRUCTURES AND MASS SPECTRA OF BINARY PENTAFLUORIDES.		1365
UORIDE. BAND SPECTRA OF BISMUTH MONOFLUORIDE AND ANTIMONY MONOFL		0065
ELECTRON STRUCTURE AND NMR SPECTRA OF BORON COMPOUNDS.		0229
U/ CORRELATION OF INTERMOLECULAR EFFECTS IN THE VIBRATION SPECTRA OF BORON FLUORIDE WITH ALTERATIONS IN ITS Q		1404
ULTRASOFT X-RAY ABSORPTION SPECTRA OF BORON TRIFLUORIDE.		0204
VACUUM ULTRAVIOLET ABSORPTION SPECTRA OF BORON TRIHALIDES.		0237
UORIDE IN THE CONDENSED PHASE. EVIDEN/ INFRARED AND RAMAN SPECTRA OF BROMINE PENTAFLUORIDE AND CHLORINE TRIFL		0266
WAVE LENGTH REGION. ROTATIONAL SPECTRA OF BROMINE PENTAFLUORIDE IN THE MILLIMETER		0268
RAMAN SPECTRA OF CADMIUM DIFLUORIDE AND LEAD DIFLUORIDE.		0274
RAMAN SPECTRA OF CARBON AND SILICON TETRAFLUORIDES.		0364
E. PREDISSOCIATION IN THE ABSORPTION SPECTRA OF CARBON MONOFLUORIDE AND CARBON DIFLUORID		0293
FLUORIDE. INFRARED SPECTRA OF CARBON TETRAFLUORIDE AND GERMANIUM TETRA		0363
T REGION. (KRYPTON DIFLUORIDE, NITROGEN TRIFL/ ABSORPTION SPECTRA OF CERTAIN FLUORIDES IN THE NEAR ULTRAVIOLE		0560
D IN ELECTRICAL DISCHARGES THROUGH BORON TRIFLU/ EMISSION SPECTRA OF CERTAIN POLYATOMIC FREE RADICALS OBTAINE		0222
AND CHLORINE PENTAFLUORIDE MOLECULES. ABSORPTION SPECTRA OF CHLORINE FLUORIDE, CHLORINE TRIFLUORIDE,		0388
INFRARED AND RAMAN SPECTRA OF CHLORINE MONOFLUORIDE.		0381
RIDE. INFRARED AND RAMAN SPECTRA OF CHLORINE TRIFLUORIDE AND BROMINE TRIFLUO		0408
UORIDE. GASEOUS SPECTRA AND GAS TO LIQUID SHIFTS. NMR SPECTRA OF CHLORINE TRIFLUORIDE AND CHLORINE MONOFL		0368
TE. (CARBON TETRAFLUORIDE) RAMAN SPECTRA OF CHLOROFLUORO METHANES IN THE GASEOUS STA		0322
S. PHOTOELECTRON SPECTRA OF COMPLEXES OF BORON TRIFLUORIDE AND AMINE		0224
. INFRARED SPECTRA OF CRYOSYSTEMS. PART-2: SULFUR HEXAFLUORIDE		1068
RIDE, STRONTIUM DIFLUORIDE, AND BARIUM DIFLUORIDE IN THE/ SPECTRA OF ELECTRONIC EXCITATIONS IN CALCIUM DIFLUO		0279
OPTICAL ABSORPTION AND FLUORESCENCE SPECTRA OF EUROPIUM TRIFLUORIDE.		0453
IDE POWDERS. REFLECTION SPECTRA OF GADOLINIUM OXIDE AND GADOLINIUM TRIFLUOR		0455
E NU-7 (E´) FUNDAMENTAL OF PHOSPHORUS PENTAFLUORID/ RAMAN SPECTRA OF GASES. PART-10: RAMAN BAND CONTOUR OF TH		0879
UMINUM TETRAFLUORIDE, AND SODIUM ALUMINUM TETRA/ INFRARED SPECTRA OF GASEOUS ALUMINUM TRIFLUORIDE, LITHIUM AL		0045
FAR INFRARED SPECTRA OF GASEOUS AND LIQUID SULFUR HEXAFLUORIDE.		1095
IDE, ZIRCONIUM CHLORIDE, AND HAFNIUM CHLORIDE. INFRARED SPECTRA OF GASEOUS GROUP-4 HALIDES: ZIRCONIUM FLUOR		1327
LFUR K AND L AND FLUORINE K X-RAY EMISSION AND ABSORPTION SPECTRA OF GASEOUS SULFUR HEXAFLUORIDE. SU		1088
DE. ULTRAVIOLET ABSORPTION SPECTRA OF GASEOUS TIN DIFLUORIDE AND LEAD DIFLUORI		0595
A STUDY OF THE RAMAN AND BROAD LINE NMR SPECTRA OF GERMANIUM TETRAFLUORIDE.		0484
RAMAN AND INFRARED SPECTRA OF GERMANIUM TETRAFLUORIDE.		0482
HELIUM-1 RESONANCE PHOTOELECTRON SPECTRA OF GROUP-4 TETRAFLUORIDES.		0317
E TRIFLUORIDE, BROMINE MONOFLUORIDE, AND B/ PHOTOELECTRON SPECTRA OF HALIDES. (CHLORINE MONOFLUORIDE, CHLORIN		0376
ACHLORIDES OF GROUP-5B. (CARBON TETRAFLUOR/ PHOTOELECTRON SPECTRA OF HALIDES. PART-1. TETRAFLUORIDES AND TETR		0316
OF THE BORON TRIHALIDES. PHOTOELECTRON SPECTRA OF HALIDES. PART-2: HIGH RESOLUTION SPECTRA		0188
TRIFLUORIDES OF NITROGEN AND PHOSPHORUS, A/ PHOTOELECTRON SPECTRA OF HALIDES. PART-3: TRIFLUORIDES AND OXIDE		0701
LIUM-1 AND HELIUM-2 STUDIES OF CARBON TETR/ PHOTOELECTRON SPECTRA OF HALIDES. PART-7: VARIABLE TEMPERATURE HE		0341
ORIDE, IODINE HEXAFLUORIDE, BROM/ ELECTRON SPIN RESONANCE SPECTRA OF HALOGEN HEXAFLUORIDES. (CHLORINE HEXAFLU		0271
TRAFLUORIDE) INFRARED SPECTRA OF HALOGEN-SUBSTITUTED METHANES. (CARBON TE		0355
JAHN-TELLER EFFECT IN VIBRATIONAL SPECTRA OF HEXAFLUORIDES.		1471
VIBRATIONAL SPECTRA OF INORGANIC FLUORIDES.		0014
ENTAFLUORIDE. NMR SPECTRA OF IODINE HEPTAFLUORIDE AND IODINE OXYGEN P		0516
UORIDE. VIBRATIONAL SPECTRA OF IODINE HEPTAFLUORIDE AND RHENIUM HEPTAFL		0522
INFRARED AND RAMAN SPECTRA OF KRYPTON DIFLUORIDE.		0552
IOLET. ELECTRONIC SPECTRA OF LANTHANUM FLUORIDE IN THE EXTREME ULTRAV		0570
IDE. VIBRATIONAL SPECTRA OF LIQUID AND CRYSTALLINE CARBON TETRAFLUOR		0327
RIDE. INFRARED AND RAMAN SPECTRA OF LIQUID AND CRYSTALLINE SILICON TETRAFLUO		1001
M PENTAFLUORIDE. RAMAN SPECTRA OF LIQUID NIOBIUM PENTAFLUORIDE AND TANTALU		0670
ORIDES. ELECTRONIC SPECTRA OF LIQUID RUTHENIUM AND MOLYBDENUM PENTAFLU		0925
IX ISOLATION. INFRARED SPECTRA OF MAGNESIUM AND ALUMINUM FLUORIDES BY MATR		0608
ND ARSENIC PENTAFLUORIDE. INFRARED SPECTRA OF MATRIX ISOLATED ANTIMONY PENTAFLUORIDE A		0113
OMINE TRIFLUORIDE, AND BROMINE PENTAFLUORIDE. F/ INFRARED SPECTRA OF MATRIX ISOLATED CHLORINE TRIFLUORIDE, BR		0399
LUORIDE) INFRARED SPECTRA OF MATRIX ISOLATED MAGNESIUM DIFLUORIDE.		0604
CKEL DICHLORIDE. RAMAN SPECTRA OF MATRIX ISOLATED MOLECULES. (SULFUR HEXAF		1097
M 18 TO 50 GC. SPECTRA OF MATRIX ISOLATED NICKEL DIFLUORIDE AND NI		0653
. LASER EXCITED RAMAN ABSORPTION SPECTRA OF MATRIX ISOLATED NITROGEN TRIFLUORIDE FRO		0698
O LANTHANUM TRIFLUORIDE) GEOMETRY AND INFRARED SPECTRA OF MATRIX ISOLATED PRASEODYMIUM TRIFLUORIDE		0895
FLUORINE SYSTEM. INFRARED SPECTRA OF MATRIX ISOLATED RARE EARTH HALIDES. (ALS		0583
DIFLUORIDE. INFRARED SPECTRA OF MATRIX ISOLATED SPECIES IN THE GALLIUM-		0462
IFLUORIDE, AND TETRAFLUORIDE. INFRARED SPECTRA OF MATRIX ISOLATED TIN DIFLUORIDE AND LEAD		0594
ELECTRON IMPACT SPECTRA OF MATRIX ISOLATED ZIRCONIUM DIFLUORIDE, TR		1325
TIN) AND MF4 (M= CARBON, SILICON, AND / VAPOR PHASE RAMAN SPECTRA OF METHANE AND CARBON TETRAFLUORIDE.		0333
NTAFLUORIDE) VIBRATIONAL SPECTRA OF MH4 (M= CARBON, SILICON, GERMANIUM, AND		0314
ONAI PROFILES OF VIBRATIONAL BANDS IN THE GAS PHASE RAMAN SPECTRA OF MIXED PHOSPHORUS HALIDES. (PHOSPHORUS PE		0829
E OF EFFECTIVE POTENTIAL BARRIERS IN THE X-RAY ABSORPTION SPECTRA OF MOLECULES OF THE SPHERICAL TOP TYPE. /TI		1442
RAMAN EFFECT AND ULTRAVIOLET ABSORPTION SPECTRA OF MOLECULES. (SULFUR HEXAFLUORIDE) EVIDENC		1073
VIBRATIONAL SPECTRA OF MOLYBDENUM AND TUNGSTEN HEXAFLUORIDES.		0642
EXAFLUORIDE. VIBRATIONAL SPECTRA OF MOLYBDENUM AND TUNGSTEN PENTAFLUORIDES.		0625
AFLUORIDE. ASSIGNMENTS IN THE ULTRAVIOLET SPECTRA OF MOLYBDENUM HEXAFLUORIDE AND TECHNETIUM H		0628
FLUORIDE, RHENIUM HEXAFLUORIDE, URANIUM HEXAFLUORI/ RAMAN SPECTRA OF MOLYBDENUM HEXAFLUORIDE AND TUNGSTEN HEX		0633
NTAFLUORIDE. VIBRATIONAL SPECTRA OF MOLYBDENUM HEXAFLUORIDE, TECHNETIUM HEXA		1072
AFLUORIDE. INFRARED SPECTRA OF MOLYBDENUM PENTAFLUORIDE AND TUNGSTEN PE		0626
HE GAS PHASE. VAPOR DENSITY OF NIOBIUM PENTA/ VIBRATIONAL SPECTRA OF NEPTUNIUM HEXAFLUORIDE AND PLUTONIUM HEX		0649
INFRARED SPECTRA OF NIOBIUM AND TANTALUM PENTAFLUORIDES IN T		0660
TRIFLUORIDE, AND A REASSIGNMENT OF THE SPE/ PHOTOELECTRON SPECTRA OF NIOBIUM TETRAFLUORIDE.		0658
LUORIDE. INFRARED SPECTRA OF NITROGEN TRIFLUORIDE AND NITROGEN OXIDE		0700
ATIONAL STATES. (EQUILIBRIUM STRUCTURE) MICROWAVE SPECTRA OF NITROGEN TRIFLUORIDE AND PHOSPHORUS TRIF		0748
ATIONAL STATES. (ZETA TYPE DOUBLING AND RESONA/ MICROWAVE SPECTRA OF NITROGEN TRIFLUORIDE IN THE EXCITED VIBR		0722
ROTATIONAL SPECTRA OF NITROGEN TRIFLUORIDE IN THE EXCITED VIBR		0723
ORIDE. VIBRATIONAL SPECTRA OF NONPOLAR MOLECULES.		0182
PHOTOELECTRON SPECTRA OF OSMIUM HEXAFLUORIDE AND PLATINUM HEXAFLU		0752
SPECTRA OF OXYGEN DIFLUORIDE AND OXYGEN DICHLORIDE.		0766

E FREE RADICAL.   ARGON MATRIX RAMAN SPECTRA OF OXYGEN DIFLUORIDE AND THE OXYGEN FLUORID 0759
                                                         FLUORINE NMR SPECTRA OF OXYGEN FLUORIDES. (OXYGEN DIFLUORIDE) 0775
ORIDE)   NEW SPECTRA OF OXYGEN LIKE MOLECULES. (NITROGEN MONOFLU 0674
RIDE RADICALS.   ESR SPECTRA OF PHOSPHORUS DIFLUORIDE AND SULFUR TRIFLUO 0791
ION.   ELECTRONIC SPECTRA OF PHOSPHORUS MONOFLUORIDE AND ITS POSITIVE 0786
                                                        FLUORESCENCE SPECTRA OF PHOSPHORUS NITRIDE AND BORON FLUORIDE. 0167
IGID MOLECULE EFFECTS ON THE ROVIBRONIC ENERGY LEVELS AND SPECTRA OF PHOSPHORUS PENTAFLUORIDE.   NONR 0841
TAFLUORIDE AND THE HEIGHT OF THE BARRIER TO/ VIBRATIONAL SPECTRA OF PHOSPHORUS PENTAFLUORIDE AND ARSENIC PEN 0856
ED IN A SINGLE CRYSTAL OF PHOSPHORUS TRIFLUORIDE.   ESR SPECTRA OF PHOSPHORUS TETRAFLUORIDE RADICALS PRODUC 0823
TRIFLUORIDE, AND PHOSPHORUS PENTAFLUORIDE.   INFRARED SPECTRA OF PHOSPHORUS TRIFLUORIDE, PHOSPHORUS OXIDE 0799
RIDE)   ELECTRONIC SPECTRA OF POLYATOMIC FREE RADICALS. (CARBON DIFLUO 0308
                                    PREPARATION, STRUCTURE AND SPECTRA OF PRASEODYMIUM TETRAFLUORIDE. 0897
                                                  FAR INFRARED SPECTRA OF PROLATE AND OBLATE SYMMETRIC TOPS. 1479
                                                      INFRARED SPECTRA OF RARE EARTH METAL FLUORIDES. 1346
OF A CESIUM IODIDE PRISM.   INFRARED SPECTRA OF RARE EARTH METAL FLUORIDES IN THE REGION 1352
                                                   VIBRATIONAL SPECTRA OF RHENIUM HEPTAFLUORIDE. 0919
                                                   VIBRATIONAL SPECTRA OF RHENIUM HEXAFLUORIDE. 0905
/VIBRATIONAL MODE OF HEXAFLUORIDE MOLECULES. THE INFRARED SPECTRA OF RUTHENIUM HEXAFLUORIDE AND RHODIUM HEXA/ 0926
FLUORIDE.   INFRARED SPECTRA OF SELENIUM HEXAFLUORIDE AND TELLURIUM HEXA 1107
IDE.   VACUUM ULTRAVIOLET SPECTRA OF SILICON DIFLUORIDE AND GERMANIUM DIFLUOR 0981
.   INFRARED SPECTRA OF SILICON DIFLUORIDE IN INERT GAS MATRICES 0982
RIDE, TIN MONOFLUORIDE, AND LEAD MONOFLUORIDE.   SPECTRA OF SILICON MONOFLUORIDE, GERMANIUM MONOFLUO 0964
                                        INFRARED AND RAMAN SPECTRA OF SILICON TETRAFLUORIDE. 1005
N. CARBON TETRAFLUORIDE.   INFRARED SPECTRA OF SIMPLE MOLECULES IN LIQUID ARGON SOLUTIO 0338
ETRAFLUORIDE, CURIUM TRIFLUORIDE AND CURIUM T/ ABSORPTION SPECTRA OF SOLID AMERICIUM TRIFLUORIDE, AMERICIUM T 0445
                                       PURE QUADRUPOLE SPECTRA OF SOLID CHLORINE COMPOUNDS. 0383
WITH XENON DIFLUORIDE.   RAMAN SPECTRA OF SOLID XENON TETRAFLUORIDE AND ITS ADDUCT 1258
X ISOLATION TECHNIQUE.   INFRARED SPECTRA OF SOME ALKALINE EARTH HALIDES BY THE MATRI 0125
RBON TETRAFLUORIDE) CORREL/ VACUUM ULTRAVIOLET ABSORPTION SPECTRA OF SOME HALOGEN DERIVATIVES OF METHANE. (CA 0365
AFLUORIDE, CARBON TETRAFLUORIDE, BOR/ INFRARED ABSORPTION SPECTRA OF SOME POLYATOMIC FLUORIDES. (SILICON TETR 0186
SOLID STATE. (ARSENIC TRIFLUORIDE, SULFUR / THE INFRARED SPECTRA OF SOME VOLATILE INORGANIC FLUORIDES IN THE 0096
DE.   FAR ULTRAVIOLET ELECTRONIC SPECTRA OF STRONTIUM DIFLUORIDE AND BARIUM DIFLUORI 1017
DE.   X-RAY SPECTRA OF SULFUR AND FLUORINE IN SULFUR HEXAFLUORI 1101
                                    MICROWAVE AND MASS SPECTRA OF SULFUR MONOFLUORIDE. 1024
ORIDE.   GAS PHASE ELECTRON RESONANCE SPECTRA OF SULFUR MONOFLUORIDE AND SELENIUM MONOFLU 1021
DE, SELENIUM OXIDE AND IODI/ GAS PHASE ELECTRON RESONANCE SPECTRA OF SULFUR MONOFLUORIDE, SELENIUM MONOFLUORI 1020
S. HYDROLYSIS AND INTRA- INTERMOLECULAR MEC/ FLUORINE NMR SPECTRA OF SULFUR, PHOSPHORUS, AND SILICON FLUORIDE 1042
QUES.   CALCULATION OF THE INFRARED SPECTRA OF SULFUR TETRAFLUORIDE USING CNDO/2 TECHNI 1047
             VIBRONIC AND SPIN ORBIT SPLITTING IN SPECTRA OF SYSTEMS EXHIBITING JAHN-TELLER EFFECT. 1480
                                                       BAND SPECTRA OF THALLIUM MONOIODIDE AND MONOFLUORIDE. 1125
OPY OF HIGH TEMPERATURE VAPORS. PART-3: MONOMER AND DIMER SPECTRA OF THALLOUS FLUORIDE. /TOELECTRON SPECTROSC
GNESIUM DIFLUORIDE.   GEOMETRY AND VIBRATIONAL SPECTRA OF THE ALKALINE EARTH DIHALIDES. PART-1: MA 0609
LLIUM DIFLUORIDE, BERYLLIUM DICHLORIDE, MAGNESI/ INFRARED SPECTRA OF THE ALKALINE EARTH HALIDES. PART-1: BERY 0133
ADMIUM DIFLUORIDE AND MERCURY DIFLUORIDE)   VIBRATIONAL SPECTRA OF THE DIHALIDES OF MERCURY AND CADMIUM. (C 0275
LICON DIFLUORIDE DIHYDRATE. THE INFRARED AND ULTRAVIOLET SPECTRA OF THE FREE RADICAL SILICON DIFLUORIDE. /SI 0986
, AND GROUP-6.   PHOTOELECTRON SPECTRA OF THE HALIDES IN GROUP-3, GROUP-4, GROUP-5 1361
GROUP-5 ELEMENTS. (PHOSPHORUS TRIFLUORIDE)   ABSORPTION SPECTRA OF THE HYDRIDES, DEUTERIDES, AND HALIDES OF 0806
UORIDE, TELLURIUM PENTAFLUORIDE(-) AND ANTIM/ VIBRATIONAL SPECTRA OF THE ISOELECTRONIC SPECIES IODINE PENTAFL 0501
ICARBON HEXAFLUORI/ THEORY OF INTENSITIES IN THE INFRARED SPECTRA OF THE MOLECULES CARBON TETRAFLUORIDE AND D 0362
UORIDE.   HIGH RESOLUTION INFRARED SPECTRA OF THE PARALLEL BANDS OF PHOSPHORUS PENTAFL 0872
LECULES DOWN TO HELIUM TE/ HEAT CAPACITIES AND ELECTRONIC SPECTRA OF THE PLATINUM GROUP METAL HEXAFLUORIDE MO 1159
IDES.   MICROWAVE SPECTRA OF THE THALLIUM, INDIUM AND GALLIUM MONOHAL 1110
ZINC DIFLUORIDE, DIFLUORIDE, AND MANGANESE DIFLUOR/ RAMAN SPECTRA OF TITANIUM DIOXIDE, MAGNESIUM DIFLUORIDE, 0615
                                                   VIBRATIONAL SPECTRA OF VANADIUM PENTAFLUORIDE. 1211
UORIDE, AND VANADIUM TRIFLUORIDE.   INFRARED SPECTRA OF VANADIUM PENTAFLUORIDE, VANADIUM TETRAFL 1210
                                                   VIBRATIONAL SPECTRA OF XENON COMPOUNDS. 1232
                                  HELIUM-2 PHOTOELECTRON SPECTRA OF XENON DIFLUORIDE AND KRYPTON DIFLUORIDE. 1234
/C CHARACTER OF INERT GAS HALIDES. HELIUM-2 PHOTOELECTRON SPECTRA OF XENON DIFLUORIDE AND KRYPTON DIFLUORIDE. 1343
IN THE FAR ULTRAVIOLET REGION.   SPECTRA OF XENON DIFLUORIDE AND XENON TETRAFLUORIDE 1233
                                                      INFRARED SPECTRA OF XENON DIFLUORIDE IN DIFFERENT SOLVENTS. 1241
ENON MONOFLUORIDE, AND KRYPTON MONOFLUORIDE.   EMISSION SPECTRA OF XENON MONOBROMIDE, XENON MONOCHLORIDE, X 0545
                                                      VIBRONIC SPECTRA OF YTTRIUM MONOFLUORIDE. 1315
VAPOR STATE. (THIRD ROW)   RAMAN SPECTRA OF 5D TRANSITION METAL HEXAFLUORIDES IN THE 1378
IDE.   ANALYSIS OF SYMMETRIC ROTOR RAMAN SPECTRA. PURE ROTATIONAL SPECTRUM OF BORON TRIFLUOR 0205
CONSTANTS OF THE SPHERICAL TOP MOLECUL/ VAPOR PHASE RAMAN SPECTRA, RAMAN BAND CONTOUR ANALYSES, AND CORIOLIS 1069
TANTS, FORCE CONSTANTS, AND VALUES FOR/ VAPOR PHASE RAMAN SPECTRA, RAMAN BAND CONTOUR ANALYSES, CORIOLIS CONS 0795
/ETRIES AND ENTROPIES OF METAL TRIFLUORIDES FROM INFRARED SPECTRA. (SCANDIUM, YTTRIUM, LANTHANUM, CERIUM, NE/ 0943
                               ULTRAVIOLET EMISSION SPECTRA. (XENON DIFLUORIDE AND KRYPTON DIFLUORIDE) 0559
                        ISOTOPE EFFECT IN THE SINGLET SPECTRAL BANDS OF THE BORON FLUORIDE MOLECULE. 0166
DE. RAMAN SCATTERING AND INFRARED, VISIBLE AND ULTRAVIOL/ SPECTRAL OBSERVATIONS ON MOLECULAR XENON HEXAFLUORI 1285
A(+) BANDS OF BORON-11 MONOFLUORIDE WITH A VACUUM ECHELLE SPECTROGRAPH. / OF THE A SINGLET PI- X SINGLET SIGM 0170
EQUENCY VIBRATIONS OF ANT/ CONSTRUCTION OF A FAR INFRARED SPECTROMETER AND ITS USE IN THE STUDY OF THE LOW FR 0083
ABSORPTION STUDIES. (NITROGEN DIFLUORIDE) A CAVITY SEARCH SPECTROMETER FOR FREE RADICAL MICROWAVE ROTATIONAL 0689
IAMAGNETIC MOLECULES. GJ FACT/ HIGH TEMPERATURE MICROWAVE SPECTROMETER FOR MEASUREMENTS OF ZEEMAN EFFECT IN D 1120
D CARBON DIFLUORIDE FROM MOLECULAR FLOW EFFUSION AND MASS SPECTROMETER INVESTIGATIONS. /ARBON MONOFLUORIDE AN 0286
ROPERTIES OF THE BORON CHLORIDE FLUORIDE SYSTEM FROM MASS SPECTROMETER INVESTIGATIONS. (BORON DIFLUORIDE) / P 0179
              APPLICATION OF A MOLECULAR BEAM SOURCE MASS SPECTROMETER TO THE STUDY OF REACTIVE FLUORIDES. 0019
ECOND ROW, T/ APPLICATION OF A MOLECULAR BEAM SOURCE MASS SPECTROMETER TO THE STUDY OF REACTIVE FLUORIDES. (S 1377
GANESE FLUORIDES.   MASS SPECTROMETRIC AND THERMOCHEMICAL STUDIES OF THE MAN 0612
/HIGH TEMPERATURE SPECIES BY ELECTRIC DEFLECTION AND MASS SPECTROMETRIC DETECTION. (FIRST ROW FLUORIDES, BER/ 0134
RGY OF BERYLLIUM MONOFLUORIDE.   MASS SPECTROMETRIC DETERMINATION OF THE DISSOCIATION ENE 0129
ON OF THE GASEOUS MOLECULES ALUMINUM OXYDIFLUORIDE / MASS SPECTROMETRIC DETERMINATION OF THE HEATS OF FORMATI 0035
N AND VAPOR PRESSURE OF VANADIUM TRIFLUORIDE.   MASS SPECTROMETRIC DETERMINATION OF THE VAPOR COMPOSITIO 1208
UORIDE AND ANTIMONY PENTAFLUORIDE.   MASS SPECTROMETRIC EVIDENCE OF DIMERS IN BISMUTH PENTAFL 0084
OPERTIES OF ALUMINUM TRIFLUORIDE.   MASS SPECTROMETRIC INVESTIGATION OF THE THERMODYNAMIC PR 0040
. PART-1.   ALPHA SPECTROMETRIC INVESTIGATION OF URANIUM HEXAFLUORIDE 1195
HEAT OF SUBLIMATION OF URANIUM TETRAFLUORIDE BY THE MASS SPECTROMETRIC METHOD.   DETERMINATION OF 1165
DISSOCIATION ENERGY OF BORON MONOFLUORIDE FROM MASS SPECTROMETRIC STUDIES. 0161
EOUS THORIUM TRIFLUORIDE AND THORIUM DIFLUORIDE FROM MASS SPECTROMETRIC STUDIES.   HEATS OF FORMATION GAS 1128
: STABILITIES OF SAMARIUM, EUROPIUM, AND GADOLINIUM/ MASS SPECTROMETRIC STUDIES AT HIGH TEMPERATURES. PART-13 0450

: VAPOR PRESSURE AND DISSOCIATION ENERGY OF SILVER / MASS SPECTROMETRIC STUDIES AT HIGH TEMPERATURES. PART-14 1014
: SUBLIMATION PRESSURES FOR TITANIUM TRIFLUORIDE AN/ MASS SPECTROMETRIC STUDIES AT HIGH TEMPERATURES. PART-16 1145
: SUBLIMATION AND VAPOR PRESSURES OF DYSPROSIUM, HO/ MASS SPECTROMETRIC STUDIES AT HIGH TEMPERATURES. PART-17 0448
: THE STABILITIES OF THE MONO- AND DIFLUORIDES OF S/ MASS SPECTROMETRIC STUDIES AT HIGH TEMPERATURES. PART-18 0939
THE DISSOCIATION ENERGIES OF THE MONOFLUORIDES AND/ MASS SPECTROMETRIC STUDIES AT HIGH TEMPERATURES. PART-2: 0463
SUBLIMATION PRESSURE OF COPPER DIFLUORIDE. (AND CO/ MASS SPECTROMETRIC STUDIES AT HIGH TEMPERATURES. PART-9: 0441
SUBLIMATION PRESSURE OF COPPER DIFLUORIDE. MASS SPECTROMETRIC STUDIES AT HIGH TEMPERATURES. PART-9: 0442
FLUORIDE, LEAD DIFLUORIDE, TIN MONOFLUORIDE, AND LE/ MASS SPECTROMETRIC STUDIES AT HIGH TEMPERATURES. (TIN DI 0592
H TEMPERATURES. PART-2: ENTHALPY OF SUBLIMATION OF / MASS SPECTROMETRIC STUDIES OF PLUTONIUM COMPOUNDS AT HIG 0884
IUM TRIFLUORIDE, LANTHANUM TRIFLUORIDE AND THE RARE/ MASS SPECTROMETRIC STUDIES OF SCANDIUM TRIFLUORIDE, YTTR 0953
UM, AND RARE EARTH FLUORIDES. MASS SPECTROMETRIC STUDIES OF SCANDIUM, YTTRIUM, LANTHAN 1359
DES. MASS SPECTROMETRIC STUDIES OF SOME GASEOUS SULFUR FLUORI 1023
DES. MASS SPECTROMETRIC STUDIES OF SOME GASEOUS SULFUR FLUORI 1028
SO ANTIMONY PENTAFLUORIDE) MOLECULAR BEAM MASS SPECTROMETRIC STUDY OF BINARY PENTAFLUORIDES. (AL 1376
M DIFLUORIDE) MASS SPECTROMETRIC STUDY OF MAGNESIUM HALIDES. (MAGNESIU 0603
AFLUORIDE, AND XENON HEXAFLUORIDE. PHOTOIONIZATION MASS SPECTROMETRIC STUDY OF XENON DIFLUORIDE, XENON TETR 1225
PHOTOIONIZATION MASS SPECTROMETRY OF OXYGEN DIFLUORIDE. 0760
NUM MONOFLUORIDE. SPECTROSCOPIC AND THERMODYNAMIC PROPERTIES OF ALUMI 0022
SPECTROSCOPIC CALCULATIONS ON SULFUR HEXAFLUORIDE. 1071
ECHANISMS IN NONRIGID PHOSPHORUS PENTAFLUORIDE. SPECTROSCOPIC CONSEQUENCES OF DIFFERENT TUNNELING M 0834
/ELECTRON AFFINITY, DIPOLE MOMENT, QUADRUPOLE MOMENT, AND SPECTROSCOPIC CONSTANTS. A COMPARISON OF THEORETIC/ 0757
/ IONIZATION POTENTIAL, ELECTRON AFFINITY, DIPOLE MOMENT, SPECTROSCOPIC CONSTANTS, AND IDEAL GAS THERMODYNAM/ 0094
LENIUM MONOFLUORIDE FROM A MOLECU/ DISSOCIATION ENTHALPY, SPECTROSCOPIC CONSTANTS, AND OTHER PROPERTIES OF SE 0955
/L, ELECTRON AFFINITY, DIPOLE AND QUADRUPOLE MOMENTS, AND SPECTROSCOPIC CONSTANTS. COMPARISON OF THEORETICAL/ 0756
GEN MONOFLUORIDE. CALCULATION OF THE SPECTROSCOPIC CONSTANTS FOR SEVERAL STATES OF NITRO 0676
/RMODYNAMIC PROPERTIES OF THE DIATOMIC INTERHALOGENS FROM SPECTROSCOPIC DATA. (CHLORINE, BROMINE, AND IODINE/ 0253
/IC THERMAL MOTION OF TRIGONAL BIPYRAMIDAL MOLECULES FROM SPECTROSCOPIC DATA. PENTACOORDINATED MOLECULES. PA/ 0853
UCE NO PARENT ION. (CARBON/ IMPLICATIONS OF PHOTOELECTRON SPECTROSCOPIC MEASUREMENTS FOR COMPOUNDS WHICH PROD 0339
ENERGY LEVELS AND SPECTROSCOPIC PARAMETERS OF NEODYMIUM TRIFLUORIDES. 0646
ECULE. THEORETICAL STUDY OF THE SPECTROSCOPIC STATES OF THE CARBON MONOFLUORIDE MOL 0287
IFFRACTION INVESTIGATION OF SOME SIMPLE MOLECULES. (TUNG/ SPECTROSCOPIC STUDIES IN CONNECTION WITH ELECTRON D 1153
ON OF PENTAFLUORIDES OF ARSENIC, PHOSPHORUS, AND VANADIU/ SPECTROSCOPIC STUDIES IN MEAN AMPLITUDES OF VIBRATI 0871
E POLYATOMIC MOLECULES. SPECTROSCOPIC STUDIES OF MOLECULAR CONSTANTS IN SOM 0046
E POLYATOMIC MOLECULES. (XENON DIFLUORIDE) SPECTROSCOPIC STUDIES OF MOLECULAR CONSTANTS IN SOM 1245
MAGNESIUM OXIDE. SPECTROSCOPIC STUDIES OF SCANDIUM MONOFLUORIDE AND 0933
EMPERATURE MATERIALS. (CHROMIUM DIFLUORIDE, CHROMIUM TRI/ SPECTROSCOPIC STUDIES OF THE VAPORIZATION OF HIGH T 0431
TISTCIAL THERMODYNAMICS AND QUANTUM MECHANICAL STUDIES / SPECTROSCOPIC STUDIES OF VIBRATIONAL CONSTANTS, STA 0784
INE HEXAFLUORIDE(X) IN ANHYDROUS HYDROGEN FLUORIDE. SPECTROSCOPIC STUDY OF IODINE HEPTAFLUORIDE AND IOD 0518
FAR ULTRAVIOLET SPECTROSCOPIC STUDY OF XENON DIFLUORIDE. 1256
FAR ULTRAVIOLET SPECTROSCOPIC STUDY OF XENON TETRAFLUORIDE. 1272
STRUCTURE OF KRYPTON DIFLUORIDE STUDIED BY PHOTOELECTRON SPECTROSCOPY. ELECTRONIC 0550
RE OF SULFUR HEXAFLUORIDE STUDIED BY ULTRALONG WAVE X-RAY SPECTROSCOPY. ELECTRONIC STRUCTU 1065
RSENIC TRICHLORIDE DETERMINED BY ELECTRON DIFFRACTION AND SPECTROSCOPY. /UCTURES OF ARSENIC TRIFLUORIDE AND A 0103
CTIONS OF URANIUM HEXAFLUORIDE BY ION CYCLOTRON RESONANCE SPECTROSCOPY. PROPERTIES AND REA 1185
THE BONDING IN THE GROUP-4 TETRAHALIDES BY PHOTOELECTRON SPECTROSCOPY. STUDY OF 1362
E) MOLECULAR SPECTROSCOPY BY MEANS OF ESCA. (CARBON TETRAFLUORID 0328
E MONOFLUORIDE) HIGH RESOLUTION RAMAN SPECTROSCOPY IN THE RED AND NEAR INFRARED. (CHLORIN 0389
ETERS, PART-2: (NITROGEN TRIFLUORIDE AND PHOSP/ MICROWAVE SPECTROSCOPY IN THE REGION FROM TWO TO THREE MILLIM 0711
APPLICATION OF SYNCHROTON ORBITAL RADIATION TO MOLECULAR SPETROSCOPY. MOLECULAR NITROGEN AND SULFUR HEXAFL/ 1083
CHLORIDE. HIGH TEMPERATURE MICROWAVE SPECTROSCOPY OF ALUMINUM MONOFLUORIDE AND ALUMINUM 0028
ETRAFLUORIDE. HIGH RESOLUTION PHOTOELECTRON SPECTROSCOPY OF CARBON TETRAFLUORIDE AND SILICON T 0319
0 GHZ MICROWAVE SPECTROSCOPY OF CHLORINE PENTAFLUORIDE AT 70 AND 14 0423
LUORIDE, AND PHOSPHORUS PENTAFLUORID/ X-RAY PHOTOELECTRON SPECTROSCOPY OF CHLORINE TRIFLUORIDE, SULFUR TETRAF 0409
HE LANTHANIDE SERIES. BIBLIOGRAPHY ON SPECTROSCOPY OF FLUORIDES AND OXIDES BELONGING TO T 1358
S, OXIDES, CHLORIDES) MATRIX ISOLATION INFRARED SPECTROSCOPY OF HIGH TEMPERATURE SPECIES. (FLUORIDE 0011
NOMER AND DIMER SPECTRA OF THALLOUS FLUORI/ PHOTOELECTRON SPECTROSCOPY OF HIGH TEMPERATURE VAPORS. PART-3: MO 1117
AFLUORIDE. SOFT X-RAY SPECTROSCOPY ON SULFUR TETRAFLUORIDE AND SULFUR HEX 1033
LASER SPECTROSCOPY. (PHOSPHORUS PENTAFLUORIDE) 0832
CHEMICAL APPLICATIONS OF CORE ELECTRON EXCITATION SPECTROSCOPY. (XENON HEXAFLUORIDE) 1306
ET SIGMA. SYSTEM OF CALCIUM FLUORIDE MOLECULE IN EMISSION SPECTRUM. /OF THE BANDS OF B DOUBLET SIGMA- X DOUBL 0280
URE OF OXYGEN DIFLUORIDE FROM THE GROUND STATE ROTATIONAL SPECTRUM. /C POTENTIAL CONSTANTS AND AVERAGE STRUCT 0778
R DIFLUORIDE. CALCULATION OF THE FORCE FIELD AND INFRARED SPECTRUM. CENTRIFUGAL DISTORTION EFFECTS IN SULFU 1031
LASER ISOTOPE SEPARATION. (SULFUR HEXAFLUORIDE SPECTRUM) 1094
PLUTONIUM HEXAFLUORIDE. (MOLECULAR STRUCTURE, ABSORPTION SPECTRUM) PROPERTIES OF 0894
XENON HEXAFLUORIDE. (SPECTRUM) 1312
XENON TETRAFLUORIDE. (INFRARED SPECTRUM) 1263
TROGEN TRIFLUORIDE. MILLIMETER WAVE SPECTRUM AND CENTRIFUGAL DISTORTION CONSTANTS OF NI 0718
APOR. MASS SPECTRUM AND COMPOSITION OF NIOBIUM PENTAFLUORIDE V 0666
GAS PHASE ELECTRON RESONANCE SPECTRUM AND DIPOLE MOMENT OF CARBON MONOFLUORIDE. 0284
IN THE SINGLE/ GAS PHASE ELECTRON PARAMAGNETIC RESONANCE SPECTRUM AND DIPOLE MOMENT OF NITROGEN MONOFLUORIDE 0673
VACUUM ULTRAVIOLET ABSORPTION SPECTRUM AND DIPOLE MOMENT OF NITROGEN TRIFLUORIDE. 0725
E. THE INFRARED SPECTRUM AND FORCE CONSTANTS FOR ARSENIC TRIFLUORID 0105
DE. RAMAN SPECTRUM AND FORCE CONSTANTS FOR SULFUR TETRAFLUORI 1035
CENTRIFUGAL DISTORTION. MICROWAVE SPECTRUM AND FORCE CONSTANTS OF SILICON DIFLUORIDE. 0989
VIBRATIONAL SPECTRUM AND FORCE FIELD OF MOLYBDENUM HEXAFLUORIDE 0634
VIBRATIONAL SPECTRUM AND FORCE FIELD OF URANIUM HEXAFLUORIDE. 1196
RIDE. ABSORPTION SPECTRUM AND MAGNETIC PROPERTIES OF OSMIUM HEXAFLUO 0751
UORIDE. MILLIMETER AND SUBMILLETER WAVE SPECTRUM AND MOLECULAR CONSTANTS OF ALUMINUM MONOFL 0033
ORIDE. ELECTRONIC EMISSION SPECTRUM AND MOLECULAR CONSTANTS OF IODINE MONOFLU 0496
ORIDE. INFRARED SPECTRUM AND MOLECULAR CONSTANTS OF NITROGEN TRIFLU 0728
RIDE. MASS SPECTRUM AND MOLECULAR ENERGETICS OF KRYPTON DIFLUO 0565
RIDE. MICROWAVE SPECTRUM AND MOLECULAR STRUCTURE OF BROMINE TRIFLUO 0258
ETRAFLUORIDE. INFRARED SPECTRUM AND NORMAL COORDINATE ANALYSIS OF SULFUR T 1045
MONOFLUORIDE. ANALYSIS OF THE ABSORPTION SPECTRUM, AND OF TWO NEW SINGLET SYSTEMS OF ARSENIC 0090
EXAFLUORIDE RADICAL. PREPARATION, ELECTRON SPIN RESONANCE SPECTRUM, AND STRUCTURE. CHLORINE H 0428
ESR SPECTRUM AND STRUCTURE OF BROMINE HEXAFLUORIDE. 0270
MICROWAVE SPECTRUM AND STRUCTURE OF CHLORINE TRIFLUORIDE. 0410
MILLIMETER WAVE SPECTRUM AND STRUCTURE OF OXYGEN DIFLUORIDE. 0768
E. VIBRATIONAL SPECTRUM AND STRUCTURE OF SOLID URANIUM HEXAFLUORID 1189
MICROWAVE SPECTRUM AND STRUCTURE OF SULFUR DIFLUORIDE. 1030

	NMR SPECTRUM AND STRUCTURE OF SULFUR TETRAFLUORIDE.	1037
ADICAL.	INFRARED SPECTRUM AND STRUCTURE OF THE NITROGEN DIFLUORIDE R	0687
ING IN CERTAIN HEXAFLUCRIDES. / POSSIBLE ROTATIONAL RAMAN	SPECTRUM AND THE DEPOLARIZATION OF RAYLEIGH SCATTER	0901
N DIFLUORIDE RADICAL.	INFRARED SPECTRUM AND THERMODYNAMIC FUNCTIONS OF THE NITROGE	0686
XAFLUORIDE GAS.	VIBRATIONAL SPECTRUM AND THERMODYNAMIC PROPERTIES OF URANIUM HE	1186
XAFLUORIDE GAS.	VIBRATIONAL SPECTRUM AND THERMODYNAMIC PROPERTIES OF URANIUM HE	1187
NIC TRIFLUORIDE.	INFRARED SPECTRUM AND VIBRATIONAL POTENTIAL FUNCTION OF ARSE	0100
SELENIUM TETRAFLUORIDE.	MICROWAVE SPECTRUM, DIPOLE MOMENT, AND STRUCTURE ANALYSIS OF	0957
ONS OF SILICON DIFLUORIDE.	INFRARED SPECTRUM, FORCE CONSTANTS, AND THERMODYNAMIC FUNCTI	0985
	MICROWAVE SPECTRUM OF ALUMINUM MONOFLUORIDE.	0029
RAVIOLET.	EMISSION SPECTRUM OF ALUMINUM MONOFLUORIDE IN THE VACUUM ULT	0030
SUBLIMATION OF ALUMINUM TRIFLUORIDE AND THE INFRARED	SPECTRUM OF ALUMINUM TRIFLUORIDE.	0037
UORIDE, AND ALUMINUM MONOFLUORIDE BY MATRIX ISO/ INFRARED	SPECTRUM OF ALUMINUM TRIFLUORIDE, DIALUMINUM HEXAFL	0051
	MASS SPECTRUM OF ANTIMONY PENTAFLUORIDE.	0085
AS A FUNCTION OF TEMPERATURE. EVIDENCE FOR A TRIG/ RAMAN	SPECTRUM OF ANTIMONY PENTAFLUORIDE IN THE GAS PHASE	0077
	RAMAN SPECTRUM OF ANTIMONY TRIFLUORIDE.	0073
	ULTRAVIOLET SPECTRUM OF ARSENIC MONOFLUORIDE.	0089
ULTRAVIOLET AND NEAR INFRARED REGIONS.	SPECTRUM OF ARSENIC MONOFLUORIDE IN THE NEAR VACUUM	0091
	INFRARED SPECTRUM OF ARSENIC PENTAFLUORIDE.	0114
	ESR SPECTRUM OF ARSENIC TETRAFLUORIDE.	0112
	MICROWAVE SPECTRUM OF ARSENIC TRIFLUORIDE.	0102
ONSTANTS OF ARSENIC TRIFLUORIDE, ARSENIC TRICHLORI/ RAMAN	SPECTRUM OF ARSENIC TRIFLUORIDE AND THE MOLECULAR C	0111
ATIONAL STATE.	MICROWAVE SPECTRUM OF ARSENIC TRIFLUORIDE IN THE EXCITED VIBR	0097
THERMALLY EXCITED EMISSION	SPECTRUM OF BARIUM MONOFLUORIDE.	0123
VIOLET AND ULTRAVIOLET	SPECTRUM OF BARIUM MONOFLUORIDE MOLECULE.	0122
EMISSION	SPECTRUM OF BISMUTH MONOFLUORIDE.	0147
VISIBLE EMISSION	SPECTRUM OF BISMUTH MONOFLUORIDE.	0145
	MICROWAVE SPECTRUM OF BORON FLUORIDE.	0165
THE ELECTRONIC	SPECTRUM OF BORON FLUORIDE.	0172
NALYSIS OF SYMMETRIC ROTOR RAMAN SPECTRA. PURE ROTATIONAL	SPECTRUM OF BORON TRIFLUORIDE. A	0205
INFRARED ABSORPTION	SPECTRUM OF BORON TRIFLUORIDE.	0206
INFRARED	SPECTRUM OF BORON TRIFLUORIDE.	0249
PHOTOELECTRON	SPECTRUM OF BORON TRIFLUORIDE.	0189
ROTATION VIBRATION BAND CONTOURS IN THE INFRARED	SPECTRUM OF BORON TRIFLUORIDE.	0193
FECT AND A MORE PRECISE DETERMINATION OF THE FAR INFRARED	SPECTRUM OF BORON TRIFLUORIDE. /Y OF THE ISOTOPE EF	0230
	RAMAN SPECTRUM OF BORON TRIFLUORIDE GAS.	0251
ON ENERGY.	SPECTRUM OF BROMINE MONOFLUORIDE AND ITS DISSOCIATI	0252
M MONOFLUORIDE IN J=5/2 ROTATIONAL LEVELS. ESR	SPECTRUM OF BROMINE OXIDE, IODINE OXIDE AND SELENIU	0954
	INFRARED SPECTRUM OF BROMINE PENTAFLUORIDE.	0262
K-TYPE DOUBLING IN THE MILLIMETER WAVE	SPECTRUM OF BROMINE PENTAFLUORIDE.	0260
MICROWAVE	SPECTRUM OF BROMINE PENTAFLUORIDE.	0269
VIBRATIONAL	SPECTRUM OF BROMINE TRIFLUORIDE.	0257
ABSORPTION	SPECTRUM OF CARBON DIFLUORIDE.	0302
MICROWAVE	SPECTRUM OF CARBON DIFLUORIDE.	0307
2500 ANGSTROM ABSORPTION	SPECTRUM OF CARBON DIFLUORIDE.	0306
NALYSIS. ABSORPTION	SPECTRUM OF CARBON DIFLUORIDE AND ITS VIBRATIONAL A	0303
ATRIX. ABSORPTION	SPECTRUM OF CARBON DIFLUORIDE TRAPPED IN AN ARGON M	0297
EMISSION	SPECTRUM OF CARBON MONOFLUORIDE.	0291
INFRARED	SPECTRUM OF CARBON TETRAFLUORIDE.	0343
ANALYSIS OF THE INFRARED	SPECTRUM OF CHLORINE MONOFLUORIDE.	0386
PHOTOELECTRON	SPECTRUM OF CHLORINE MONOFLUORIDE.	0369
LUORIDE, AND OSMIUM HEXAFLUORIDE. INFRARED	SPECTRUM OF CHROMIUM HEXAFLUORIDE, MOLYBDENUM HEXAF	0436
BAND	SPECTRUM OF CHROMIUM MONOFLUORIDE.	0429
RAMAN	SPECTRUM OF CRYSTALLINE MOLYBDENUM PENTAFLUORIDE.	0623
RAMAN	SPECTRUM OF GASEOUS CARBON TETRAFLUORIDE.	0347
ABSORPTION	SPECTRUM OF GASEOUS HOLMIUM MONOFLUORIDE.	0488
RAMAN	SPECTRUM OF GASEOUS NITROGEN TRIFLUORIDE.	0740
ROTATIONAL ANALYSIS OF SOME SINGLET TRANSITIONS IN THE	SPECTRUM OF GASEOUS SCANDIUM MONOFLUORIDE.	0930
ABSORPTION	SPECTRUM OF GASEOUS SILVER MONOFLUORIDE.	1012
M HEXAFLUORIDE. ELECTRONIC ABSORPTION	SPECTRUM OF GASEOUS URANIUM HEXACHLORIDE AND URANIU	1194
MICROWAVE	SPECTRUM OF GERMANIUM DIFLUORIDE.	0478
ULTRAVIOLET ABSORPTION	SPECTRUM OF GERMANIUM DIFLUORIDE.	0474
NG AND CENTRIFUGAL DISTORTION. MICROWAVE	SPECTRUM OF GERMANIUM DIFLUORIDE. QUADRUPOLE COUPLI	0477
RIDE) INFRARED	SPECTRUM OF HEAVY METAL DIHALIDES. (EUROPIUM DIFLUO	0451
RATIONS OF ANTIMONY PENTAFLUORIDE AND THE PURE ROTATIONAL	SPECTRUM OF HYDROGEN CYANIDE. /HE LOW FREQUENCY VIB	0083
MICROWAVE ROTATIONAL	SPECTRUM OF INDIUM MONOFLUORIDE.	0490
ROTATIONAL	SPECTRUM OF IODINE FLUORIDE.	0497
MICROWAVE	SPECTRUM OF IODINE PENTAFLUORIDE.	0502
MERIZATION IN THE LIQUID STATE. RAMAN	SPECTRUM OF IODINE PENTAFLUORIDE. EVIDENCE FOR POLY	0512
VIBRATIONAL	SPECTRUM OF IODINE TRIFLUORIDE.	0500
INFRARED	SPECTRUM OF IRIDIUM HEXAFLUORIDE.	0540
ICALLY INDUCED JAHN-TELLER PROGRESSIONS IN THE ELECTRONIC	SPECTRUM OF IRIDIUM HEXAFLUORIDE. VIBRON	0537
ELECTRON SPIN RESONANCE	SPECTRUM OF KRYPTON MONOFLUORIDE.	0546
ELECTRONIC	SPECTRUM OF LANTHANUM MONOFLUORIDE.	0569
ELECTRONIC	SPECTRUM OF LANTHANUM TRIFLUORIDE.	0582
BAND SYSTEM IN THE DISCHARGE	SPECTRUM OF LANTHANUM TRIFLUORIDE VAPORS.	0580
EMISSION	SPECTRUM OF LEAD MONOFLUORIDE.	0588
INFRARED ABSORPTION	SPECTRUM OF LIQUID BORON TRIFLUORIDE.	0245
VIBRATIONAL	SPECTRUM OF LIQUID CHROMIUM PENTAFLUORIDE.	0435
N HEXAFLUORIDE, AND URANIUM HEXAFLUO/ HIGH RESOLUTION NMR	SPECTRUM OF LIQUID MOLYBDENUM HEXAFLUORIDE, TUNGSTE	0637
RAMAN	SPECTRUM OF LIQUID OXYGEN DIFLUORIDE.	0767
ELECTRONIC	SPECTRUM OF LUTETIUM MONOFLUORIDE.	0598
INFRARED	SPECTRUM OF MATRIX ISOLATED NIOBIUM PENTAFLUORIDE.	0659
BAND	SPECTRUM OF NICKEL MONOFLUORIDE.	0651
INFRARED	SPECTRUM OF NIOBIUM PENTAFLUORIDE.	0661
ABSORPTION	SPECTRUM OF NITROGEN DIFLUORIDE.	0685
INFRARED	SPECTRUM OF NITROGEN DIFLUORIDE.	0692
MICROWAVE	SPECTRUM OF NITROGEN DIFLUORIDE RADICAL.	0681
FAR INFRARED	SPECTRUM OF NITROGEN TRIFLUORIDE.	0702
INFRARED AND RAMAN	SPECTRUM OF NITROGEN TRIFLUORIDE.	0724

1-TYPE RESONANCE IN AN OVERTONE BAND. THE 2NU-4 SPECTRUM OF NITROGEN TRIFLUORIDE. 0732
INFRARED SPECTRUM OF NITROGEN TRIFLUORIDE AND FORCE FIELD. 0697
PREPARATION AND PROPERTIES (INFRARED SPECTRUM) OF OSMIUM HEPTAFLUORIDE. 0754
A REINVESTIGATION OF THE INFRARED SPECTRUM OF OXYGEN DIFLUORIDE. 0774
SPIN ROTATIONAL HYPERFINE STRUCTURE IN THE MICROWAVE SPECTRUM OF OXYGEN DIFLUORIDE. 0777
/A AND STRUCTURE OF INORGANIC MOLECULES. PART-1: INFRARED SPECTRUM OF OXYGEN DIFLUORIDE FROM 2.5 TO 25 MICRO/ 0761
TED STATES, NU-1- 2 NU-2 FERMI RESONANCE AND E/ MICROWAVE SPECTRUM OF OXYGEN DIFLUORIDE IN VIBRATIONALLY EXCI 0771
THIUM OXYFLUORIDE. MATRIX INFRARED SPECTRUM OF OXYGEN MONOFLUORIDE AND DETECTION OF LI 0755
FAR INFRARED SPECTRUM OF PHOSPHORUS PENTAFLUORIDE. 0842
FAR INFRARED SPECTRUM OF PHOSPHORUS PENTAFLUORIDE. 0855
INFRARED SPECTRUM OF PHOSPHORUS PENTAFLUORIDE. 0866
PHOTOELECTRON SPECTRUM OF PHOSPHORUS PENTAFLUORIDE. 0846
OLECULAR VIBRATIONS OF THE TRIGONAL BIPYRAMIDAL THE RAMAN SPECTRUM OF PHOSPHORUS PENTAFLUORIDE GAS. /LD AND M 0863
MICROWAVE SPECTRUM OF PHOSPHORUS TRIFLUORIDE. 0805
2NU-4 SPECTRUM OF PHOSPHORUS TRIFLUORIDE. 0817
ABSORPTION SPECTRUM OF PLUTONIUM HEXAFLUORIDE. 0892
INFRARED SPECTRUM OF PLUTONIUM HEXAFLUORIDE. 0891
SPECTRUM OF RHENIUM HEXAFLUORIDE. 0906
ELECTRONIC SPECTRUM OF SCANDIUM MONOFLUORIDE. 0934
RIDE, AND LANTHANUM MONOFLUORIDE. VIBRONIC SPECTRUM OF SCANDIUM MONOFLUORIDE, YTTRIUM MONOFLUO 0938
MICROWAVE SPECTRUM OF SILICON DIFLUORIDE. 0990
PHOTOELECTRON SPECTRUM OF SILICON DIFLUORIDE. 0980
PHOTOELECTRON SPECTRUM OF SILICON DIFLUORIDE. 0996
ULTRAVIOLET ABSORPTION SPECTRUM OF SILICON DIFLUORIDE. 0984
ULTRAVIOLET BAND SPECTRUM OF SILICON DIFLUORIDE. 0991
ULTRAVIOLET SPECTRUM OF SILICON DIFLUORIDE. 0983
TIONAL STATES, EQUILIBRIUM STRUCTURE, ANHARMON/ MICROWAVE SPECTRUM OF SILICON DIFLUORIDE IN THE EXCITED VIBRA 0993
STEM IN THE REGION 2755-2179 ANGSTROMS. EMISSION SPECTRUM OF SILICON DIFLUORIDE. PART-1: THE BAND SY 0988
ELECTRONIC SPECTRUM OF SILICON DIFLUORIDE. THEORETICAL STUDY. 0997
BAND SPECTRUM OF SILICON MONOFLUORIDE. 0965
DISTRIBUTION OF INTENSITY IN THE ELECTRON EMISSION SPECTRUM OF SILICON MONOFLUORIDE. 0968
INFRARED SPECTRUM OF SILICON TETRAFLUORIDE. 1003
EMISSION SPECTRUM OF SILICON TRIFLUORIDE. 0998
NEAR ULTRAVIOLET ABSORPTION SPECTRUM OF SILVER MONOFLUORIDE. 1013
RAMAN SPECTRUM OF SOLID MOLYBDENUM TETRAFLUORIDE. 0621
RAMAN SPECTRUM OF SOLID PHOSPHORUS PENTAFLUORIDE 0858
RAMAN SPECTRUM OF SOLID SULFUR HEXAFLUORIDE. 1098
ACH. PART-6: THE ASSIGNMENT OF THE HELIUM-2 PHOTOELECTRON SPECTRUM OF SULFUR HEXAFLUORIDE. /A MANY BODY APPRO 1090
INFRARED SPECTRUM OF SULFUR HEXAFLUORIDE. 1086
ULTRASOFT X-RAY ABSORPTION SPECTRUM OF SULFUR HEXAFLUORIDE. 1102
PHOTOELECTRON SPECTRUM OF SULFUR HEXAFLUORIDE AT 584 ANGSTROMS. 1076
SOLID STATES. RAMAN SPECTRUM OF SULFUR HEXAFLUORIDE IN THE GASEOUS AND 1079
SPECTRUM OF SULFUR MONOFLUORIDE. 1022
RAMAN AND INFRARED SPECTRUM OF SULFUR TETRAFLUORIDE. 1040
CRONS. INFRARED SPECTRUM OF TELLURIUM HEXAFLUORIDE FROM 25 TO 40 MI 1108
MICROWAVE ROTATIONAL SPECTRUM OF THALLIUM MONOFLUORIDES. 1119
ZEEMAN EFFECT IN THE MICROWAVE ROTATIONAL SPECTRUM OF THE ALUMINUM FLUORIDE MOLECULE. 0026
E ULTRAVIOLET REGION. ABSORPTION SPECTRUM OF THE BISMUTH MONOFLUORIDE MOLECULE IN TH 0143
ELECTRONIC SPECTRUM OF THE BORON MONOFLUORIDE MOLECULE. 0155
LATED IN INERT MATRICES. PHOTOCHLORINATION OF / INFRARED SPECTRUM OF THE FREE RADICAL CARBON TRIFLUORIDE ISO 0311
E. EMISSION SPECTRUM OF THE GAMMA SYSTEM OF SILICON MONOFLUORID 0974
ROTATIONAL ANALYSIS OF SELECTED BANDS FROM THE ELECTRONIC SPECTRUM OF THE LUTETIUM MONOFLUORIDE MOLECULE. 0599
LTRAVIOLET PHOTOLYSIS OF NITROGEN TRIFLUORIDE. MICROWAVE ROTATIONAL SPECTRUM OF THE NITROGEN DIFLUORIDE FREE RADICAL. 0691
ELECTRONIC SPECTRUM OF THE NITROGEN DIFLUORIDE FREE RADICAL. / 0691
ELECTRONIC ABSORPTION SPECTRUM OF THE NITROGEN DIFLUORIDE RADICAL. 0694
RAMAN SPECTRUM OF THE NITROGEN DIFLUORIDE RADICAL. 0695
10.6 MICROMETERS. TEMPERATURE DEPENDENT ABSORPTION SPECTRUM OF THE NU-3 BAND OF SULFUR HEXAFLUORIDE AT 1091
EPR SPECTRUM OF THE RADICAL CHLORINE TETRAFLUORIDE. 0413
MICROWAVE SPECTRUM OF THE SULFUR MONOFLUORIDE RADICAL. 1018
EMISSION SPECTRUM OF TITANIUM MONOFLUORIDE. 1142
AFLUORIDE) RAMAN AND INFRARED SPECTRUM OF TRIFLUORO PHOSPHORANE. (PHOSPHORUS PENT 0847
INFRARED SPECTRUM OF URANIUM HEXAFLUORIDE. 1182
RAMAN SPECTRUM OF URANIUM HEXAFLUORIDE. 1190
ANGSTROMS. ABSORPTION SPECTRUM OF URANIUM HEXAFLUORIDE FROM 2000 TO 4200 1191
OF URANIUM(IV). ABSORPTION SPECTRUM OF URANIUM TETRAFLUORIDE AND ENERGY LEVELS 1168
ABSORPTION SPECTRUM OF VAPORIZED TITANIUM MONOFLUORIDE. 1143
PHOTOELECTRON SPECTRUM OF XENON DIFLUORIDE. 1227
RAMAN SPECTRUM OF XENON HEXAFLUORIDE. 1289
XENON TETRAFLUORIDE. ELECTRON SPIN RESONANCE SPECTRUM OF XENON MONOFLUORIDE IN GAMMA IRRADIATED 1222
ULTRAVIOLET SPECTRUM OF XENON TETRAFLUORIDE. 1270
ELECTRONIC BAND SPECTRUM OF YTTRIUM MONOFLUORIDE. 1314
FLUORIDE. MICROWAVE SPECTRUM, STRUCTURE, AND DIPOLE MOMENT OF OXYGEN DI 0779
AMIC PROPERTIES OF BROMINE PENTAFLUORIDE. RAMAN SPECTRUM, STRUCTURE, FORCE CONSTANTS, AND THERMODYN 0267
(3-), SELENIUM H/ THE STEREOCHEMICALLY INERT LONE PAIR. A SPECULATION ON THE BONDING IN ANTIMONY HEXACHLORIDE 1309
CORIOLIS ZETA SUMS FOR SOME SPHERICAL TOP MOLECULES. 1487
SPIN RELAXATION STUDY OF THE SPIN ROTATION INTERACTION IN SPHERICAL TOP MOLECULES. NUCLEAR 0323
INFRARED BAND CONTOURS. PART-1: SPHERICAL TOP MOLECULES. (CARBON TETRAFLUORIDE) 0325
ED STUDIES OF VAPOR BAND CONTOURS./ CORIOLIS CONSTANTS OF SPHERICAL TOP MOLECULES FROM LOW TEMPERATURE INFRAR 1156
/MAN BAND CONTOUR ANALYSES, AND CORIOLIS CONSTANTS OF THE SPHERICAL TOP MOLECULES, SULFUR HEXAFLUORIDE, SELE/ 1069
BANDS IN THE GAS PHASE RAMAN SPECTRA OF MOLECULES OF THE SPHERICAL TOP TYPE. /TIONAL PROFILES OF VIBRATIONAL 1442
AND XY6 OCTAHEDRAL MOLECULES. RAMAN INTENSITIES FOR SPHERICALLY SYMMETRIC MODES OF SOME XY4 TETRAHEDRA 1453
NG JAHN-TELLER EFFECT. VIBRONIC AND SPIN ORBIT SPLITTING IN SPECTRA OF SYSTEMS EXHIBITI 1480
F FORBIDDEN TRANSITIONS IN IODINE, OR IODINE BROMIDE (CH/ SPIN ORBITAL INTERACTION AND RELATIVE INTENSITIES C 0498
ION IN SYMMETRIC TOP MOLECULES. NUCLEAR SPIN RELAXATION STUDY OF THE SPIN ROTATION INTERACT 0183
ION IN SPHERICAL TOP MOLECULES. NUCLEAR SPIN RELAXATION STUDY OF THE SPIN ROTATION INTERACT 0323
E RADICAL. ELECTRON SPIN RESONANCE AND STRUCTURE OF THE BORON DIFLUORID 0178
/IDE AND DINITROGEN TETRAFLUORIDE MOLECULES, AND ELECTRON SPIN RESONANCE HYPERFINE COUPLING CONSTANTS IN NIT/ 0688
ORIDE. RADIATION DAMAGE IN XENON TETRAFLUORIDE. ELECTRON SPIN RESONANCE OF THE TRAPPED RADICAL XENON MONOFLU 1220
ELECTRON SPIN RESONANCE OF THE XENON FLUORIDE RADICAL. 1288
PHOSPHORUS TETRAFLUORIDE RADICALS. ELECTRON SPIN RESONANCE OF TRAPPED PHOSPHORUS DIFLUORIDE AND 0792

ADICAL. ELECTRON SPIN RESONANCE PARAMETERS OF THE BORON DIFLUORIDE R 0181
HLORINE HEXAFLUORIDE, IODINE HEXAFLUORIDE, BROM/ ELECTRON SPIN RESONANCE SPECTRA OF HALOGEN HEXAFLUORIDES. (C 0271
SULFUR TRIFLUORIDE RADICALS. ELECTRON SPIN RESONANCE SPECTRA OF PHOSPHORUS DIFLUORIDE AND 0791
RADICALS PRODUCED IN A SINGLE CRYSTAL OF PHOSPH/ ELECTRON SPIN RESONANCE SPECTRA OF PHOSPHORUS TETRAFLUORIDE 0823
CHLORINE HEXAFLUORIDE RADICAL. PREPARATION, ELECTRON SPIN RESONANCE SPECTRUM, AND STRUCTURE. 0428
XAFLUORIDE. ELECTRON SPIN RESONANCE SPECTRUM AND STRUCTURE OF BROMINE HE 0270
ELECTRON SPIN RESONANCE SPECTRUM OF ARSENIC TETRAFLUORIDE. 0112
IDE AND SELENIUM MONCFLUORIDE IN J-5/2 ROTATION/ ELECTRON SPIN RESONANCE SPECTRUM OF BROMINE OXIDE, IODINE OX 0954
ELECTRON SPIN RESONANCE SPECTRUM OF KRYPTON MONOFLUORIDE. 0546
MMA IRRADIATED XENON TETRAFLUORIDE. ELECTRON SPIN RESONANCE SPECTRUM OF XENON MONOFLUORIDE IN GA 1222
OSPHORUS DICHLORIDE, AND NITROGEN DICHLORIDE IN/ ELECTRON SPIN RESONANCE STUDIES OF PHOSPHORUS DIFLUORIDE, PH 0793
D POLYCRYSTALLINE ARSENIC TRIFLUCRIDE AND ARSEN/ ELECTRON SPIN RESONANCE STUDY OF RADICALS IN GAMMA IRRADIATE 0108
E SPECTRUM OF OXYGEN DIFLUORIDE. SPIN ROTATIONAL HYPERFINE STRUCTURE IN THE MICROWAV 0777
ISOTOPIC SPLITTING IN MATRIX ISOLATED BORON TRIFLUORIDE. 0228
LER EFFECT. VIBRONIC AND SPIN ORBIT SPLITTING IN SPECTRA OF SYSTEMS EXHIBITING JAHN-TEL 1480
MULTIPLET SPLITTING IN 1S HOLE STATES OF MOLECULES. 0683
S TETRAFLUORIDE RADICAL. ISOTROPIC HYPERFINE SPLITTINGS AND MOLECULAR STRUCTURE IN THE PHOSPHORU 0824
OCTAHEDRAL FINE STRUCTURE SPLITTINGS IN NU-3 OF SULFUR HEXAFLUORIDE. 1063
VIBRATIONAL MEAN SQUARE AMPLITUDE MATRICES. 1389
VIBRATIONAL MEAN SQUARE AMPLITUDE MATRICES. 1392
KAGE EFFECT OF SYMMETRICAL XY3 AND XY4 / VIBRATIONAL MEAN SQUARE AMPLITUDE MATRICES. (BASTIANSEN-MORINO SHRIN 1394
RAHEDRAL XY4 MOLECULES. VIBRATIONAL MEAN SQUARE AMPLITUDE MATRICES. PART-7: TREATMENT OF TET 1408
RAMIDAL XY3 MOLECULES. GENERALIZED MEAN SQUARE AMPLITUDES AND CORIOLIS CONSTANTS OF SOME PY 1405
AL SYMMETRY. (SELENIUM AND TELLURIUM FLUORIDES) ROOT MEAN SQUARE AMPLITUDES IN SOME HEXAFLUORIDES OF OCTAHEDR 0962
LUORIDE MOLECULES. MEAN SQUARE AMPLITUDES OF INTERATOMIC DISTANCES IN HEXAF 0008
LUORIDE MOLECULES. MEAN SQUARE AMPLITUDES OF INTERATOMIC DISTANCES IN HEXAF 1450
OCTAHEDRAL HEXAFLUORIDE MOLECULES. MEAN SQUARE AMPLITUDES OF INTERATOMIC DISTANCES IN SOME 1466
/NG CONSTANTS, CENTRIFUGAL STRETCHING CONSTANTS, AND MEAN SQUARE AMPLITUDES OF THE VIBRATIONS OF THE NITROGE/ 0726
/ FIELD, CORIOLIS COUPLING COEFFICIENTS, GENERALIZED MEAN SQUARE AMPLITUDES OF VIBRATION AND SHRINKAGE CONST/ 1278
TS OF SOME TRIGONAL BIPYRAMIDAL XY5 AND/ GENERALIZED MEAN SQUARE AMPLITUDES OF VIBRATION, AND SHRINKAGE EFFEC 1432
AHALIDES. (CARBON TETRAFLUORIDE) ROOT MEAN SQUARE AMPLITUDES OF VIBRATION IN SOME GROUP-4 TETR 0342
AHALIDES. ROOT MEAN SQUARE AMPLITUDES OF VIBRATION IN SOME GROUP-4 TETR 1363
AND XENON TETR/ A THEORETICAL EVALUATION OF THE ROOT MEAN SQUARE AMPLITUDES OF VIBRATION IN XENON DIFLUORIDE 1257
ETRAFLUORIDE FROM EL/ INTERATOMIC DISTANCES AND ROOT MEAN SQUARE AMPLITUDES OF VIBRATION OF GASEOUS SILICON T 1002
GENERALIZED VALENCE FORCE FIELD FORCE CONSTANTS AND MEAN SQUARE AMPLITUDES OF VIBRATION OF SOME TRIGONAL BI/ 0878
/ FORCE CONSTANTS. PART-1: METHOD AND APPLICATION TO SOME SQUARE PYRAMIDAL MOLECULES. (IODINE PENTAFLUORIDE,/ 0420
DE, I/ VIBRATIONAL SPECTRA AND VALENCE FORCE CONSTANTS OF SQUARE PYRAMIDAL MOLECULES. XENON OXIDE TETRAFLUORI 0418
RIDE, PHOSP/ CALCULATION OF FORCE CONSTANT AND ROOT MEAN SQUARE VIBRATION AMPLITUDES OF PHOSPHORUS PENTAFLUO 0826
CALCULATION OF MEAN SQUARE VIBRATIONAL AMPLITUDES. 1486
S. APPLICATION TO CARBON DIOXIDE. LEAST SQUARES ADJUSTMENT OF ANHARMONIC POTENTIAL CONSTANT 1491
LUORIDE, AND ERBIUM TRIFLUORIDE. STABILITIES OF DYSPROSIUM TRIFLUORIDE, HOLMIUM TRIF 0449
DY/ SUBLIMATION PRESSURE OF NEODYMIUM TRIFLUORIDE AND THE STABILITIES OF GASEOUS NEODYMIUM DIFLUORIDE AND NEO 0643
/ASS SPECTROMETRIC STUDIES AT HIGH TEMPERATURES. PART-13: STABILITIES OF SAMARIUM, EUROPIUM, AND GADOLINIUM / 0450
LUORIDE, AND LANTHANUM MONOFLU/ DISSOCIATION ENERGIES AND STABILITIES OF SCANDIUM MONOFLUORIDE, YTTRIUM MONOF 0935
OXIDE TRIFLUORIDE. STABILITIES OF TANTALUM PENTAFLUORIDE AND TANTALUM 1103
/SPECTROMETRIC STUDIES AT HIGH TEMPERATURES. PART-18: THE STABILITIES OF THE MONO- AND DIFLUORIDES OF SCANDI/ 0939
/: SUBLIMATION PRESSURES FOR TITANIUM TRIFLUORIDE AND THE STABILITIES OF TITANIUM DIFLUORIDE AND TITANIUM MO/ 1145
UORIDES. THERMAL STABILITY AND REACTIVITY OF D- AND F ELEMENT HEXAFL 0630
OF DILANTHANUM HEXAFLUORIDE GAS AND DETERMINATION OF ITS STABILITY. (LANTHANUM TRIFLUORIDE) /F THE EXISTENCE 0581
ENERGETICS AND STABILITY OF GAS PHASE SCANDIUM FLUORIDES. 0951
REACTIVITY AND THERMAL STABILITY OF HEXAFLUORIDES. 0004
RT-2: PLUTONIUM HEXAFLUORIDE. PART-3: URAN/ THERMODYNAMIC STABILITY OF HEXAFLUORIDES AT HIGH TEMPERATURES. PA 0629
THERMODYNAMIC STABILITY OF URANIUM HEXAFLUORIDE. 1203
ORIDE, YTTRIUM TRIFLUORIDE, AND LANTHANUM TRIFLUORIDE. STANDARD ENTHALPIES OF FORMATION OF SCANDIUM TRIFLU 0942
IDE. STANDARD ENTHALPY OF FORMATION OF ALUMINUM TRIFLUOR 0043
LUORIDE AND ENTHALPI/ FLUORINE BOMB CALORIMETRY. PART-18: STANDARD ENTHALPY OF FORMATION OF PHOSPHORUS PENTAF 0865
/ AND 350 DEGREES K, ENTHALPY OF FUSION AND VAPORIZATION, STANDARD ENTROPY OF THE VAPOR, AND OTHER THERMODYN/ 0510
STANDARD HEAT OF FORMATION OF ARSENIC TRIFLUORIDE. 0110
M FLUORIDE. STARK EFFECT FOR THALLIUM MONOFLUORIDE AND POTASSIU 1118
ORIDE AND BROMINE M/ SOME CORRECTIONS TO THE SECOND ORDER STARK EFFECT OF LINEAR MOLECULES. (CHLORINE MONOFLU 0385
UCED DIPOLE MOMENTS. STARK ENERGIES FOR MOLECULES WITH VIBRATIONALLY IND 1417
STUDIES / SPECTROSCOPIC STUDIES OF VIBRATIONAL CONSTANTS, STATISTCIAL THERMODYNAMICS AND QUANTUM MECHANICAL 0784
E MOLECULES. STATISTICAL THERMODYNAMIC PROPERTIES OF HEXAFLUORID 1449
STATISTICAL THERMODYNAMICS OF SOME HEXAFLUORIDES. 1463
NTS OF OCTAHEDRAL MOLECULES CALCULATED BY THE LOGARITHMIC STEP METHOD. FORCE CONSTA 1472
/ES OF IODINE HEPTAFLUORIDE CALCULATED BY THE LOGARITHMIC STEP METHOD IN THE RANGE OF THE GENERALIZED VALENC/ 0535
/ SOME TRIGONAL BIPYRAMIDAL MOLECULES BY THE LOGARITHIMIC STEPS METHOD. (PHOSPHORUS PENTAFLUORIDE, ARSENIC P/ 0878
NATE STRUCTURES. FURTHER EVIDENCE OF STEREOCHEMICAL NONRIGIDITY IN FIVE AND SEVEN COORDI 1440
/COORDINATED MOLECULES. PART-14: MOLECULAR VIBRATIONS AND STEREOCHEMICAL NONRIGIDITY OF THE TRIGONAL BIPYRAM/ 0854
THE BONDING IN ANTIMONY HEXACHLORIDE(3-), SELENIUM H/ THE STEREOCHEMICALLY INERT LONE PAIR. A SPECULATION ON 1309
FLUORIDE) STEREOCHEMISTRY AND SEVEN COORDINATION. (XENON HEXA 1286
ON OF THE UREY-BRADLEY FORCE FIELD TO AMMONIA, PHOSPHINE, STIBINE, AND ARSINE. ADAPTATI 1492
GHT OF THE THEORY OF CHARACTERISTIC FREQUENC/ PROGRESSIVE STIFFNESS METHOD APPLIED TO XY4 MOLECULES IN THE LI 1430
TRIFLUORIDE AND DINITROGEN TETRAFLUORIDE. STRENGTH OF THE NITROGEN FLUORINE BOND IN NITROGEN 0712
A DOUBLET SIGMA(+) TO X DOUBLET PI BAND SYSTEM OSCILLATOR STRENGTHS. / X TRIPLET PI) AND CARBON MONOFLUORIDE 0288
HORUS PENTAFLUORIDE. RELATIVE BOND STRENGTHS IN TRIGONAL BIPYRAMIDAL MOLECULES. (PHOSP 0839
ATS OF FORMATION OF CARBON MONOFLUORIDE AND C/ OSCILLATOR STRENGTHS OF CARBON MONOFLUORIDE AND COMMENTS ON HE 0295
THE COMBINED USE OF VIBRATIONAL FREQUENCIES, CENTRIFUGAL STRETCHING, AND CORIOLIS COUPLING CONSTANTS. /DE BY 0717
/NTIAL FUNCTION, CORIOLIS COUPLING CONSTANTS, CENTRIFUGAL STRETCHING CONSTANTS, AND MEAN SQUARE AMPLITUDES O/ 0726
/N OF THE VIBRATIONAL PROBLEM WITH THE USE OF CENTRIFUGAL STRETCHING CONSTANTS. (NITROGEN TRIFLUORIDE, PHOSP/ 0727
ER SYMMETRIC TOPS. (ARSENIC TRIFLUORIDE) CENTRIFUGAL STRETCHING OF TRIFLUORO CHLORO METHANE AND SOME OTH 0101
FORCE CONSTANTS OF MOLECULES WITH STRONGLY COUPLED VIBRATIONS. 1496
FAR ULTRAVIOLET ELECTRONIC SPECTRA OF STRONTIUM DIFLUORIDE AND BARIUM DIFLUORIDE. 1017
NIC FLUORIDES. (CADMIUM DIFLUORIDE, MAGNESIUM DIFLUORIDE, STRONTIUM DIFLUORIDE, AND BARIUM DIFLUORIDE) /NORGA 0273
SPECTRA OF ELECTRONIC EXCITATIONS IN CALCIUM DIFLUORIDE, STRONTIUM DIFLUORIDE, AND BARIUM DIFLUORIDE IN THE/ 0279
/UM DIFLUORIDE, MAGNESIUM DIFLUORIDE, CALCIUM DIFLUORIDE, STRONTIUM DIFLUORIDE, BARIUM DIFLUORIDE, AND ALUMI/ 0049
/UM DIFLUORIDE, MAGNESIUM DIFLUORIDE, CALCIUM DIFLUORIDE, STRONTIUM DIFLUORIDE, BARIUM DIFLUORIDE, ZINC DIFL/ 0131
ETRY OF THE ALKALINE EARTH DIHALIDES. (BARIUM DIFLUORIDE, STRONTIUM DIFLUORIDE, CALCIUM DIFLUORIDE) GEOM 0126
INTERNUCLEAR DISTANCE IN GASEOUS STRONTIUM MONOFLUORIDE. 1016
GASES. (XENON DIFLUORIDE, XENON TETRAFLUORIDE, XENON HE/ STRUCTURAL CONSIDERATIONS IN THE CHEMISTRY OF NOBLE 1235

```
                            XENON HEXAFLUORIDE. STRUCTURAL CRYSTALLOGRAPHY OF TETRAMERIC PHASES.        1284
OMPARISO/ APPROXIMATE MOLECULAR ORBITAL THEORY. ESSENTIAL STRUCTURAL ELEMENTS- MO FORMALISM. PART-2: DIRECT C  0765
ON DIFFRACTION.   STRUCTURES OF FLUORIDES. PART-6: PRECISE STRUCTURAL PARAMETERS IN COPPER DIFLUORIDE BY NEUTR  0443
TIANSEN-MORINO SHRINKAGE EFFECT IN THE STUDY CF MOLECULAR STRUCTURE.                                       BAS 1476
DICAL. PREPARATION, ELECTRON SPIN RESONANCE SPECTRUM, AND STRUCTURE.                                           0428
UORIDE, XENON TETRAFLUCRIDE, XENON TETRAFLUORIDE. CRYSTAL STRUCTURE.  CHLORINE HEXAFLUORIDE RA            1269
IFLUORIDE IN THE EXCITED VIBRATIONAL STATES. (EQUILIBRIUM STRUCTURE)  ENTHALPIES OF FORMATION OF XENON HEXAFL 0722
ITED STATES, NU-1- 2 NU-2 FERMI RESONANCE AND EQUILIBRIUM STRUCTURE.  MICROWAVE SPECTRA OF NITROGEN TR       0771
EPARATION AND PROPERTIES OF URANIUM TRIFLUORIDE. (CRYSTAL STRUCTURE) / OXYGEN DIFLUORIDE IN VIBRATIONALLY EXC PR 1164
      REDETERMINATION OF THE ORTHORHOMBIC IODINE HEPTAFLUORIDE STRUCTURE. A CORRECTION.                     A 0519
          PROPERTIES OF PLUTONIUM HEXAFLUORIDE. (MOLECULAR STRUCTURE, ABSORPTION SPECTRUM)                     0894
                    MICROWAVE SPECTRUM, DIPOLE MOMENT, AND STRUCTURE ANALYSIS OF SELENIUM TETRAFLUORIDE.       0957
OFLUORIDE.                                            FINE STRUCTURE ANALYSIS OF THE C1 SYSTEM OF ANTIMONY MON  0059
IFLUORIDE AND VANADIUM DIOXIDE.                 ELECTRONIC STRUCTURE AND ANOMALOUS THERMAL EXPANSION ON IRON D  0543
IDES)                                           ELECTRONIC STRUCTURE AND CHEMISTRY OF IODINE COMPOUNDS. (FLUOR  0499
                           MICROWAVE SPECTRUM, STRUCTURE, AND DIPOLE MOMENT OF OXYGEN DIFLUORIDE.              0779
                                                       STRUCTURE AND DIPOLE MOMENT OF SULFUR TETRAFLUORIDE.    1056
AND BORON TRICHLORIDE.                            MOLECULAR STRUCTURE AND FORCE CONSTANTS OF BORON TRIFLUORIDE  0221
.                                                ELECTRONIC STRUCTURE AND GROUND STATE OF SCANDIUM MONOFLUORIDE 0932
HE A DOUBLET SIGMA- X DOUBLET PI SYSTEM OF TH/ ROTATIONAL STRUCTURE AND ISOTOPIC SHIFT IN THE (1,0) BAND OF T 1135
ORIDES.                                                    STRUCTURE AND MAGNETIC PROPERTIES OF RARE EARTH FLU 1345
UORIDE.                                         VIBRATIONAL STRUCTURE AND MOLECULAR STRUCTURE OF RHENIUM HEXAFL 0907
                                                  ELECTRON STRUCTURE AND NMR SPECTRA OF BORON COMPOUNDS.       0229
UORIDE AND UR/ THEORETICAL CALCULATIONS OF THE ELECTRONIC STRUCTURE AND OPTICAL TRANSITIONS OF URANIUM HEXAFL 1179
                                                ELECTRONIC STRUCTURE AND PROPERTIES OF KRYPTON DIFLUORIDE.     0548
DE.                                                        STRUCTURE AND SOME PROPERTIES OF SCANDIUM TRIFLUORI 0945
.                                               PREPARATION, STRUCTURE AND SPECTRA OF PRASEODYMIUM TETRAFLUORIDE 0897
/IFLUORIDE IN THE EXCITED VIBRATIONAL STATES, EQUILIBRIUM STRUCTURE, ANHARMONIC POTENTIAL FUNCTION, AND NU-1/ 0993
                                                HYPERFINE STRUCTURE CONSTANTS OF CHLORINE MONOFLUORIDE.       0375
ILANE, INFRARED INVESTIGATION OF IODINE HEPTAFLUORIDE AN/ STRUCTURE DETERMINATION OF TRIS (TRIMETHYL SILYL) S 0527
IONS OF IODINE PENTAFLUORIDE AND BROMINE PENTA/ MOLECULAR STRUCTURE, FORCE CONSTANTS, AND THERMODYNAMIC FUNCT 0505
RTIES OF BROMINE PENTAFLUORIDE.           RAMAN SPECTRUM, STRUCTURE, FORCE CONSTANTS, AND THERMODYNAMIC PROPE 0267
ANIUM(I) FLUORIDE MOLECULE.                    ROTATIONAL STRUCTURE IN SOME HIGHER EXCITED STATES OF THE GERM 0467
ORIDE.                         SPIN ROTATIONAL HYPERFINE STRUCTURE IN THE MICROWAVE SPECTRUM OF OXYGEN DIFLU 0777
        ISOTROPIC HYPERFINE SPLITTINGS AND MOLECULAR STRUCTURE IN THE PHOSPHORUS TETRAFLUORIDE RADICAL.       0824
HE MO LCAO SELF CONSISTENT/ CALCULATION OF THE ELECTRONIC STRUCTURE OF A CHLORINE PENTAFLUORIDE MOLECULE BY T 0415
                 XENON HEXAFLUORIDE. THE STRUCTURE OF A CUBIC PHASE AT -30C.                            1283
                                                MOLECULAR STRUCTURE OF ANTIMONY PENTAFLUORIDE.               0080
ORIDE)       SEMIEMPIRICAL STUDY OF THE ELECTRONIC STRUCTURE OF BORON HYDRIDE FLUORIDES. (BORON TRIFLU 0226
                                                ELECTRONIC STRUCTURE OF BORON TRIFLUORIDE.                   0184
                                                MOLECULAR STRUCTURE OF BORON TRIFLUORIDE.                   0187
                        ESR SPECTRUM AND STRUCTURE OF BROMINE HEXAFLUORIDE.                             0270
            MICROWAVE SPECTRUM AND MOLECULAR STRUCTURE OF BROMINE TRIFLUORIDE.                          0258
                                                ELECTRONIC STRUCTURE OF CARBON MONOXIDE AND BORON MONOFLUORIDE 0162
                                                MOLECULAR STRUCTURE OF CARBON TETRAFLUORIDE.               0334
                    MICROWAVE SPECTRUM AND STRUCTURE OF CHLORINE TRIFLUORIDE.                           0410
CF CALCULATION.                                 ELECTRONIC STRUCTURE OF CHLORINE TRIFLUORIDE. AN APPROXIMATE S 0404
UORIDE.                          RAMAN SPECTRA AND THE STRUCTURE OF CHLORINE TRIFLUORIDE AND BROMINE TRIFLI 0398
                                                  ELECTRON STRUCTURE OF COMPOUNDS OF INERT GASES.            1339
                                                       STRUCTURE OF CRYSTALLINE CHROMIUM TRIFLUORIDE.         0433
                INFRARED AND RAMAN SPECTRA AND THE STRUCTURE OF CRYSTALLINE THALLIUM MONOFLUORIDE.      1126
LUORIDES BY NMR. (SULFUR TETRAFLUORIDE AND SELENIUM TETR/ STRUCTURE OF EXCHANGE PROCESSES IN SOME INORGANIC F 1048
OM THE DATA OF LCAO MO/ CHARACTERISTICS OF THE ELECTRONIC STRUCTURE OF FLUORIDES OF NONTRANSITION ELEMENTS FR 0009
                        PREPARATION AND CRYSTAL STRUCTURE OF GALLIUM TRIFLUORIDE.                       0461
                                                       STRUCTURE OF GASEOUS XENON HEXAFLUORIDE.              1295
            ELECTRON DIFFRACTION STUDY OF THE STRUCTURE OF GASEOUS XENON TETRAFLUORIDE.                1282
                                HYPERFINE STRUCTURE OF INDIUM MONOFLUORIDE.                           0489
SPECTRUM OF OXYGEN DIFLUORIDE FR/ VIBRATIONAL SPECTRA AND STRUCTURE OF INORGANIC MOLECULES. PART-1: INFRARED 0761
HEPTAFLUORIDE AT -110 DEGREES C AND AT -145 DEGREES C.   STRUCTURE OF INTERHALOGEN COMPOUNDS. PART-2: IODINE 0521
                                MOLECULAR STRUCTURE OF IODINE HEPTAFLUORIDE.                          0530
                                                ELECTRON STRUCTURE OF IRON TRIFLUORIDE.                    0542
                                                       STRUCTURE OF KRYPTON DIFLUORIDE.                      0561
ELECTRON DIFFRACTION.                                      STRUCTURE OF KRYPTON DIFLUORIDE AS INVESTIGATED BY 0558
                                                   CRYSTAL STRUCTURE OF KRYPTON DIFLUORIDE AT -80C.          0551
CTRON SPECTROSCOPY.                             ELECTRONIC STRUCTURE OF KRYPTON DIFLUORIDE STUDIED BY PHOTOELE 0550
                                                       STRUCTURE OF LIQUID ANTIMONY PENTAFLUORIDE.           0081
                                                   CRYSTAL STRUCTURE OF MANGANESE TRIFLUORIDE.               0616
                        INFRARED SPECTRA AND STRUCTURE OF MATRIX ISOLATED THALLIUM HALIDES.            1113
, THALLOUS FLUORIDE, AND / INFRARED AND RAMAN SPECTRA AND STRUCTURE OF MATRIX ISOLATED THALLOUS HALIDE DIMERS 1123
UM TRIFLUORIDE.                                            STRUCTURE OF MOLECULAR URANIUM DIFLUORIDE AND URANI 1162
T-6: THE ASSIGNMENT OF THE HELIUM-2 PHOTOELEC/ ELECTRONIC STRUCTURE OF MOLECULES BY A MANY BODY APPROACH. PAR 1090
                    PREPARATION AND CRYSTAL STRUCTURE OF MOLYBDENUM TRIFLUORIDE.                        0620
FLUORIDES FROM INFRARED AND RAMAN SPECTRA.     MOLECULAR STRUCTURE OF MOLYBDENUM, TUNGSTEN, AND URANIUM HEXA 0627
                                                       STRUCTURE OF NEPTUNIUM HEXAFLUORIDE.                  1068
ECTRA.   NUCLEAR QUADRUPOLE MOMENT OF NITROGEN-14 AND THE STRUCTURE OF NITROGEN TRIFLUORIDE FROM MICROWAVE SP 0741
TUDIES OF CHA/ MOLECULAR ORBITAL THEORY OF THE ELECTRONIC STRUCTURE OF ORGANIC COMPOUNDS. PART-3: AB INITIO S 0707
                    PREPARATION AND CRYSTAL STRUCTURE OF OSMIUM PENTAFLUORIDE.                          0750
OSMIUM. (OSMIUM PENTAFLUORIDE, / PREPARATION AND CRYSTAL STRUCTURE OF OSMIUM PENTAFLUORIDE. TWO FLUORIDES OF 0749
                  MILLIMETER WAVE SPECTRUM AND STRUCTURE OF OXYGEN DIFLUORIDE.                           0768
/CULES. PART-1. QUADRATIC POTENTIAL CONSTANTS AND AVERAGE STRUCTURE OF OXYGEN DIFLUORIDE FROM THE GROUND STA/ 0778
            ELECTRON DIFFRACTION STUDY OF THE STRUCTURE OF PHOSPHORUS PENTAFLUORIDE.                    0851
REARRANGEMENT IN PHOSPHORANES.                  ELECTRONIC STRUCTURE OF PHOSPHORUS PENTAFLUORIDE AND POLYTOPAL 0875
                INFRARED AND RAMAN SPECTRA AND STRUCTURE OF PHOSPHORUS TETRAFLUORIDE.                   0825
ELECTRON DIFFRACTION.                            MOLECULAR STRUCTURE OF PHOSPHORUS TRIFLUORIDE STUDIED BY GAS 0810
SOLUTION OF THE VIBRATIONAL PROBLE/ DETERMINATION OF THE STRUCTURE OF PYRAMIDAL MOLECULES OF THE XY3 TYPE BY 0727
ORIDE)                LATTICE VIBRATIONS AND STRUCTURE OF RARE EARTH HALIDES. (LANTHANUM  TRIFLU 0572
IFLUORIDE)             SYMMETRY AND CRYSTAL STRUCTURE OF RARE EARTH TRIFLUORIDES. (LANTHANUM TR 0571
        NONEMPIRICAL LCAO MO SCF CALCULATIONS OF THE ELECTRONIC STRUCTURE OF SILICON DIFLUORIDE.                0995
                VIBRATIONAL SPECTRUM AND STRUCTURE OF SOLID URANIUM HEXAFLUORIDE.                   1189
, SELENIUM, TELLURIUM, MO/ INFRARED SPECTRA AND MOLECULAR STRUCTURE OF SOME GROUP-6 HEXAFLUORIDES. (OF SULFUR 0961
```

```
                              MICROWAVE SPECTRUM AND  STRUCTURE OF SULFUR DIFLUORIDE.                              1030
NG WAVE X-RAY SPECTROSCOPY.              ELECTRONIC  STRUCTURE OF SULFUR HEXAFLUORIDE STUDIED BY ULTRALO 1065
                              NMR SPECTRUM  AND  STRUCTURE OF SULFUR TETRAFLUORIDE.                          1037
ULES.                                              STRUCTURE OF SULFUR TETRAFLUORIDE AND RELATED MOLEC 1043
                                          CRYSTAL  STRUCTURE OF TELLURIUM TETRAFLUORIDE.                      1106
                                          CRYSTAL  STRUCTURE OF THALLIUM TRIFLUORIDE.                          1127
A(+), AND A QUARTET SIGMA(-) TO X DOUBLET PI / ROTATIONAL  STRUCTURE OF THE A DOUBLET SIGMA(+), B DOUBLET SIGM 0466
OFLUORIDE.                    ANALYSIS OF THE ROTATIONAL  STRUCTURE OF THE BANDS 0-0 AND 0-1 IN GERMANIUM MON 0465
                ELECTRON SPIN RESONANCE AND  STRUCTURE OF THE BORON DIFLUORIDE RADICAL.            0178
                                       ELECTRONIC  STRUCTURE OF THE CARBON TETRAFLUORIDE MOLECULE.         0357
T-1.  THE GROUND STATE.                 ELECTRONIC  STRUCTURE OF THE CARBON TETRAFLUORIDE MOLECULE. PAR 0330
ECULE.                                   ROTATIONAL  STRUCTURE OF THE C3 SYSTEM OF ANTIMONY FLUORIDE MOL 0066
            ELECTRONIC AND GEOMETRIC  STRUCTURE OF THE FREE XENON HEXAFLUORIDE MOLECULE.   1302
HEXAFLUORIDE)                                        STRUCTURE OF THE IODINE HEXAFLUORIDE ANION. (XENON 1300
                                                     STRUCTURE OF THE MAGNESIUM DIFLUORIDE MOLECULE.     0602
/EOMETRY OF MOLECULES. PART-2: GEOMETRICAL AND ELECTRICAL  STRUCTURE OF THE MOLECULE CHLORINE DIFLUORIDE ON T/ 0382
        ELECTRONOGRAPHIC INVESTIGATION OF THE  STRUCTURE OF THE MOLECULES OF NEODYMIUM HALIDES.   0644
                                       ELECTRONIC  STRUCTURE OF THE NITROGEN DIFLUORIDE RADICAL.         0682
                      INFRARED SPECTRUM AND  STRUCTURE OF THE NITROGEN DIFLUORIDE RADICAL.         0687
                                                     STRUCTURE OF THE RADICAL SULFUR PENTAFLUORIDE.      1060
SULFUR DIFLUORIDE)                                   STRUCTURE OF THE SULFUR TRIFLUORIDE POSITIVE ION. ( 1032
OFLUORIDE.                           ROTATIONAL FINE  STRUCTURE OF THE TRIPLET PI STATE OF PHOSPHORUS MON 0787
TRIUM, LANTH/ ELECTRON DIFFRACTION STUDY OF THE MOLECULAR  STRUCTURE OF THE VAPOR PHASE HALIDES OF GALLIUM, YT 1317
    THEORETICAL AND EXPERIMENTAL STUDIES OF THE ELECTRONIC  STRUCTURE OF THE XENON FLUORIDES.                    1238
REGION.           ANALYSIS OF VIBRATIONAL AND ROTATIONAL  STRUCTURE OF TIN MONOFLUORIDE IN THE 2500 ANGSTROM 1138
                                          CRYSTAL  STRUCTURE OF TUNGSTEN PENTAFLUORIDE.                   1154
                             CALCULATED ELECTRONIC  STRUCTURE OF URANIUM HEXAFLUORIDE.                       1188
                    DIPOLE MOMENT AND MOLECULAR  STRUCTURE OF URANIUM HEXAFLUORIDE.                       1200
DIFFRACTION OF ELECTRONS ON THE VAPOR.               STRUCTURE OF URANIUM HEXAFLUORIDE AS DETERMINED BY 1183
THE DIFFRACTION OF ELECTRONS ON THE VAPOR.           STRUCTURE OF URANIUM HEXAFLUORIDE AS DETERMINED BY 1184
          NEUTRON DIFFRACTION STUDY OF THE CRYSTAL  STRUCTURE OF URANIUM TRIFLUORIDE.                      1163
            PREPARATION AND PROPERTIES (CRYSTAL  STRUCTURE) OF VANADIUM TETRAFLUORIDE.                 1209
                                       ELECTRONIC  STRUCTURE OF XENON DIFLUORIDE.                         1226
                                       GAS PHASE  STRUCTURE OF XENON DIFLUORIDE.                         1248
N.                      CRYSTAL AND MOLECULAR  STRUCTURE OF XENON DIFLUORIDE BY NEUTRON DIFFRACTIO 1239
                                       GAS PHASE  STRUCTURE OF XENON HEXAFLUORIDE.                     1293
                                       MOLECULAR  STRUCTURE OF XENON HEXAFLUORIDE.                     1280
                                       MOLECULAR  STRUCTURE OF XENON HEXAFLUORIDE.                     1303
TIPLE SCATTERING X(ALPHA) CALCULATION OF THE ENERGY LEVEL  STRUCTURE OF XENON HEXAFLUORIDE.              MUL 1304
ORIDE.                                 MOLECULAR  STRUCTURE OF XENON HEXAFLUORIDE AND IODINE HEPTAFLU 0526
F ELECTRON DIFFRACTION INTENSITIES. PART-2: IN/ MOLECULAR  STRUCTURE OF XENON HEXAFLUORIDE. PART-1: ANALYSIS O 1291
                                 CRYSTAL AND MOLECULAR  STRUCTURE OF XENON TETRAFLUORIDE.                   1277
                      VIBRATIONAL SPECTRA AND  STRUCTURE OF XENON TETRAFLUORIDE.                   1262
DE.                                      PROBABLE  STRUCTURE OF XENON TETRAFLUORIDE AND XENON DIFLUORI 1250
TION.                        CRYSTAL AND MOLECULAR  STRUCTURE OF XENON TETRAFLUORIDE BY NEUTRON DIFFRAC 1260
HASE.        ELECTRON DIFFRACTION INVESTIGATION OF THE  STRUCTURE OF ZIRCONIUM TETRAFLUORIDE IN THE VAPOR P 1332
OF / ELECTRON DIFFRACTION STUDY OF RHENIUM HEPTAFLUORIDE.  STRUCTURE, PSEUDOROTATION, AND ANHARMONIC COUPLING 0920
PLING IN IODINE HEPTAFLUORIDE. AN ELECTRON DIFFRACTION S/  STRUCTURE, PSEUDOROTATION, AND VIBRATIONAL MODE COU 0515
.                            OCTAHEDRAL FINE  STRUCTURE SPLITTINGS IN NU-3 OF SULFUR HEXAFLUORIDE 1063
F STEREOCHEMICAL NONRIGIDITY IN FIVE AND SEVEN COORDINATE  STRUCTURES.         FURTHER EVIDENCE O 1440
S.                                       GAS PHASE  STRUCTURES AND MASS SPECTRA OF BINARY PENTAFLUORIDE 0003
S.                                       GAS PHASE  STRUCTURES AND MASS SPECTRA OF BINARY PENTAFLUORIDE 1365
IUM D/ MATRIX INFRARED AND LASER RAMAN SPECTRA, MOLECULAR  STRUCTURES AND NORMAL COORDINATE ANALYSES OF GERMAN 0475
FLUORIDE AS DETERMINED BY ELECTRON DIFFRACTION. MOLECULAR  STRUCTURES OF ARSENIC TRIFLUORIDE AND ARSENIC PENTA 0099
LORIDE DETERMINED BY ELECTRO/ FORCE CONSTANTS AND AVERAGE  STRUCTURES OF ARSENIC TRIFLUORIDE AND ARSENIC TRICH 0103
LORIDE BY GAS ELECTRON DI/ DETERMINATION OF THE MOLECULAR  STRUCTURES OF ARSENIC TRIFLUORIDE AND ARSENIC TRICH 0104
E ANIONS.     A THEORETICAL COMPARISON OF THE ELECTRONIC  STRUCTURES OF BORON TRIHALIDES AND BORON TETRAHALID 0225
ORIDE, BORON HYDRIDE/ AB INITIO STUDIES OF THE ELECTRONIC  STRUCTURES OF BORON TRIHYDRIDE, BORON DIHYDRIDE FLU 0241
AFLUORIDE FROM ELECTRON DIFFRACTION / GAS PHASE MOLECULAR  STRUCTURES OF BROMINE PENTAFLUORIDE AND IODINE PENT 0265
IFLUORIDE.                           ELECTRONIC  STRUCTURES OF CHLORINE MONOFLUORIDE AND CHLORINE TR 0372
BY THE NONEMPIRICAL NDDO-2 (AL/ CALCULATIONS OF ELECTRON  STRUCTURES OF COMPOUNDS OF HALOGENS AND INERT GASES 0016
BY THE SCF LCAO MO METHOD.  / CALCULATIONS OF ELECTRONIC  STRUCTURES OF COMPOUNDS OF HALOGENS AND INERT GASES 1231
FFRACTION PROFILE STUDIES OF URANIUM HEXAFLUORIDE AT 193/  STRUCTURES OF FLUORIDES. PART-10: NEUTRON POWDER DI 1202
PARAMETERS IN COPPER DIFLUORIDE BY NEUTRON DIFFRACTION.  STRUCTURES OF FLUORIDES. PART-6: PRECISE STRUCTURAL 0443
DE, MAGNE/ ELECTRON DIFFRACTION ANALYSIS OF THE MOLECULAR  STRUCTURES OF GROUP-2 ELEMENTS. (BERYLLIUM DIFLUORI 0131
E.                        VIBRATIONAL SPECTRA AND  STRUCTURES OF IODINE PENTAFLUORIDE AND HEPTAFLUORID 0506
MAGNESIUM DIFLUORIDE, MAG/ INFRARED AND RAMAN SPECTRA AND  STRUCTURES OF MATRIX ISOLATED MAGNESIUM DIHALIDES:  0607
TRAFLUORIDE.                DIFFRACTION STUDY OF THE  STRUCTURES OF NITROGEN DIFLUORIDE AND DINITROGEN TE 0680
IDE)                                             STRUCTURES OF NOBLE GAS COMPOUNDS. (XENON HEXAFLUOR 1292
RIDE, ARSENIC PENTAFLUORIDE, AND BROMINE PE/ ORBITALS AND  STRUCTURES OF PENTAFLUORIDES. (PHOSPHORUS PENTAFLUO 0831
ORO PHOSPHORANE.                         ELECTRONIC  STRUCTURES OF PHOSPHORUS PENTAFLUORIDE AND TETRAFLU 0857
PHORUS TETRAFLUORIDE AND DIMETHYL PHOSPHORUS TRIFLUORIDE.  STRUCTURES OF PHOSPHORUS PENTAFLUORIDE, METHYL PHOS 0845
NTAFLUORIDE, DIFLU/ ELECTRON DIFFRACTION BY GASES AND THE  STRUCTURES OF PHOSPHORUS TRIFLUORIDE, PHOSPHORUS PE 0801
                                        MOLECULAR  STRUCTURES OF RARE GAS COMPOUNDS.                   1344
TETRAFLUORIDE, INCLUDING A MATRI/ VIBRATIONAL SPECTRA AND  STRUCTURES OF SELENIUM TETRAFLUORIDE AND TELLURIUM 0956
R/ INVESTIGATION BY ELECTRON DIFFRACTION OF THE MOLECULAR  STRUCTURES OF SULFUR HEXAFLUORIDE, SULFUR TETRAFLUO 1041
    ELECTRON DIFFRACTION INVESTIGATION OF THE MOLECULAR  STRUCTURES OF THE ALUMINUM HALIDES.               0036
HLORINE TRIFLUORIDE AT -120 DEGREES C.               STRUCTURES OF THE INTERHALOGEN COMPOUNDS. PART-1: C 0395
RAFLUORIDE)                                          STRUCTURES OF TRIFLUORO METHYL HALIDES. (CARBON TET 0318
FLUORIDE, AN/ ELECTRON DIFFRACTION STUDY OF THE MOLECULAR  STRUCTURES OF VANADIUM PENTAFLUORIDE, NIOBIUM PENTA 1217
ORUS PENTAFLUORIDE. INFRARED AND LOW TEMPERATU/ MOLECULAR  STRUCTURES OF VARIOUS PHOSPHORUS HALIDES AND PHOSPH 0849
/-1 AND HELIUM-2 PHOTOELECTRON SPECTRA AND THE ELECTRONIC  STRUCTURES OF XENON DIFLUORIDE, XENON TETRAFLUORID/ 1228
AFLUORIDE.                      VIBRATIONAL SPECTRA AND  STRUCTURES OF XENON TETRAFLUORIDE AND XENON OXYTETR 1261
EUTRON DIFFRACTION.                                  STRUCTURES OF YTTRIUM AND BISMUTH TRIFLUORIDES BY N 1318
FLUOR/ INTRAMOLECULAR LIGAND EXCHANGE IN SEVEN COORDINATE  STRUCTURES. (RHENIUM HEPTAFLUORIDE AND IODINE HEPTA 0922
    CALCULATION OF THE VIBRATIONAL FREQUENCIES OF SCANDIUM  SUBGROUP HALIDES.                              0949
/OF DISSOCIATION ENERGIES FOR THE HALIDES OF THE SCANDIUM  SUBGROUP. (SCANDIUM TRIFLUORIDE, YTTRIUM TRIFLUORI 0948
IDE AND PLUTONIUM TRIFLUORIDE, HEATS AND FREE ENERGIES OF  SUBLIMATION. VAPOR PRESSURES OF AMERICIUM TRIFLUOR 0055
UM TRIFLUORIDE.            VAPOR PRESSURE, HEAT OF  SUBLIMATION, AND EVAPORATION COEFFICIENT OF LANTHAN 0576
SEODYMIUM TRIFLUORIDE.     VAPOR PRESSURE, HEAT OF  SUBLIMATION, AND THE EVAPORATION COEFFICIENT OF PRA 0896
```

/ASS SPECTROMETRIC STUDIES AT HIGH TEMPERATURES. PART-17: SUBLIMATION AND VAPOR PRESSURES OF DYSPROSIUM, HOL/ 0448
MATION OF ALUMINUM MONOFLUORIDE.   HEAT OF SUBLIMATION OF ALUMINUM TRIFLUORIDE AND HEAT OF FOR 0053
D SPECTRUM OF ALUMINUM TRIFLUORIDE.   SUBLIMATION OF ALUMINUM TRIFLUORIDE AND THE INFRARE 0037
ATION OF ENTROPY AND ENTHALP/ ENTROPIES AND ENTHALPIES OF SUBLIMATION OF CALCIUM AND CERIUM FLUORIDES. CORREL 0281
                        VAPOR PRESSURE AND HEAT OF SUBLIMATION OF CALCIUM FLUORIDE.   0282
                        VAPOR PRESSURE AND HEAT OF SUBLIMATION OF CERIUM TRIFLUORIDE.   0366
)   VAPOR PRESSURE AND HEAT OF SUBLIMATION OF COBALT DIHALIDES. (COBALT DIFLUORIDE 0438
ELATION OF ENTROPY AND ENTHALPY / ENTROPY AND ENTHALPY OF SUBLIMATION OF GADOLINIUM TRIFLUORIDE. ROLE OF CORR 0456
YNAMICS OF SUBLIMATION OF THE GROUP-4A DIFLU/ ENTHALPY OF SUBLIMATION OF GERMANIUM DIFLUORIDE AND THE THERMOD 0471
        ANALYSES OF THERMOCHEMICAL ERRORS AND SYSTEMATICS IN SUBLIMATION OF LANTHANIDE TRIFLUORIDES.   1353
                        ENTROPIES AND ENTHALPIES OF SUBLIMATION OF NEODYMIUM AND TERBIUM TRIFLUORIDES.   0645
                                                    SUBLIMATION OF PLUTONIUM TETRAFLUORIDE.   0888
/NIUM COMPOUNDS AT HIGH TEMPERATURES. PART-2: ENTHALPY OF SUBLIMATION OF PLUTONIUM TRIFLUORIDE AND THE DISSO/ 0884
ECTROMETRIC METHOD.   DETERMINATION OF HEAT OF SUBLIMATION OF URANIUM TETRAFLUORIDE BY THE MASS SP 1165
RIDE AND A CONJECTURE ON BONDING IN THE SOLIDS. HEATS OF SUBLIMATION OF XENON DIFLUORIDE AND XENON TETRAFLUO 1237
                        VAPOR PRESSURE AND HEAT OF SUBLIMATION OF ZIRCONIUM AND HAFNIUM TETRACHLORIDES 0486
         KNUDSEN AND LANGMUIR MEASUREMENTS OF THE SUBLIMATION PRESSURE OF CADMIUM DIFLUORIDE.   0272
DISSOCIATION ENERGY OF CALCIUM MONOFLUORIDE.   SUBLIMATION PRESSURE OF CALCIUM DIFLUORIDE AND THE   0276
   DISSOCIATION ENERGY OF CHROMIUM MONOFLUORIDE.   SUBLIMATION PRESSURE OF CHROMIUM DIFLUORIDE AND THE   0430
         KNUDSEN AND LANGMUIR MEASUREMENTS OF THE SUBLIMATION PRESSURE OF COBALT DIFLUORIDE.   0439
MASS SPECTROMETRIC STUDIES AT HIGH TEMPERATURES. PART-9: SUBLIMATION PRESSURE OF COPPER DIFLUORIDE.   0442
MASS SPECTROMETRIC STUDIES AT HIGH TEMPERATURES. PART-9: SUBLIMATION PRESSURE OF COPPER DIFLUORIDE. (AND CO/ 0441
                                                    SUBLIMATION PRESSURE OF IRON DIFLUORIDE.   0541
           A LANGMUIR MEASUREMENT OF THE SUBLIMATION PRESSURE OF MANGANESE DIFLUORIDE.   0611
NGANESE MONOFLUORIDE.   SUBLIMATION PRESSURE OF MANGANESE DIFLUORIDE AND MA 0614
HE STABILITIES OF GASEOUS NEODYMIUM DIFLUORIDE AND NEODY/ SUBLIMATION PRESSURE OF NEODYMIUM TRIFLUORIDE AND T 0643
HULIUM TRIFLUORIDE, YTTERBIUM TRIFLUORIDE, AND LUTETIUM / SUBLIMATION PRESSURES AND HEATS OF SUBLIMATION OF T 1132
/ASS SPECTROMETRIC STUDIES AT HIGH TEMPERATURES. PART-16: SUBLIMATION PRESSURES FOR TITANIUM TRIFLUORIDE AND/ 1145
ANESE TRIFLUORIDE, AND IRON TRIFLUORIDE.   SUBLIMATION PRESSURES OF CHROMIUM TRIFLUORIDE, MANG 0434
ODINE OXYGEN PENTAFLUORIDE.   MASS SPECTRA AND SUBLIMATION PRESSURES OF IODINE HEPTAFLUORIDE AND I 0534
                                                    SUBLIMATION PRESSURES OF REFRACTORY FLUORIDES.   1382
IUM TRIFLUORIDE, AND LANTHANUM TRIFLUORIDE.   SUBLIMATION PRESSURES OF SCANDIUM TRIFLUORIDE, YTTR 0947
LUORIDES.   VAPORIZATION AND SUBLIMATION THERMODYNAMICS OF SELECTED LANTHANIDE F 1347
                                                    SUBLIMATION THERMODYNAMICS OF ZINC(II) FLUORIDE.   1321
F ALUMINUM MONOFLUORIDE.   MILLIMETER AND SUBMILLETER WAVE SPECTRUM AND MOLECULAR CONSTANTS O 0033
ANTS. ANHARMONICITY EFFEC/ UTILITY OF HEAVY ATOM ISOTOPIC SUBSTITUION DATA FOR THE CALCULATION OF FORCE CONST 0234
NATION FORCE CONSTANTS FROM THE / VIBRATIONAL ANALYSIS OF SUBSTITUTED AND PERTURBED MOLECULES. DIRECT DETERMI 1444
/NSTANTS OF SOME TETRAHEDRAL XY4 TYPE MOLECULES AND THEIR SUBSTITUTED DERIVATIVES BY GREEN'S FUNCTION ANALYS/ 1424
         FORCE CONSTANTS OF SULFUR HEXAFLUORIDE FROM ISOTOPIC SUBSTITUTION.   1100
         ACCURATE FORCE CONSTANTS FROM HEAVY ISOTOPIC SUBSTITUTION. PART-1. CARBON TETRAFLUORIDE.   0320
TRAFLUORIDE/ ACCURATE FORCE CONSTANTS FROM HEAVY ISOTOPIC SUBSTITUTION. PART-2: BORON TRIFLUORIDE, SILICON TE 0232
         COMPARISONS OF FLUORIDES, OXIDES, AND SULFIDES CONTAINING DIVALENT TRANSITION ELEMENTS.   1370
CALCULATIONS ON SMALL MOLECULES CONTAINING PHOSPHORUS OR SULFUR.   NONEMPIRICAL VALENCE ELECTRON 1029
ES OF POLARIZABILITIES FOR THE DIFLUORIDES OF OXYGEN AND SULFUR. /HERMODYNAMICS AND QUANTUM MECHANICAL STUDI 0784
                        X-RAY SPECTRA OF SULFUR AND FLUORINE IN SULFUR HEXAFLUORIDE.   1101
         MOLECULAR ORBITAL DESCRIPTION OF SULFUR COMPOUNDS OF VALENCES 2, 4, AND 6.   1085
         ESTIMATION OF INFRARED FREQUENCIES OF SULFUR DIFLUORIDE.   1026
         MICROWAVE SPECTRUM AND STRUCTURE OF SULFUR DIFLUORIDE.   1030
ND INFRARED SPECTRUM. CENTRIFUGAL DISTORTION EFFECTS IN SULFUR DIFLUORIDE. CALCULATION OF THE FORCE FIELD A 1031
HEXAFLUORIDE.   MOLECULAR ORBITAL DESCRIPTION OF SULFUR DIFLUORIDE, SULFUR TETRAFLUORIDE AND SULFUR 1027
/ORIDE, PHOSPHORUS TRIFLUORIDE, PHOSPHORUS PENTAFLUORIDE, SULFUR DIFLUORIDE, SULFUR TETRAFLUORIDE, SULFUR HE/ 1373
ONIC CHARGE DISTRIBUTION IN OXYGEN DIFLUORIDE, OZONE, AND SULFUR DIOXIDE. /S AND SECOND MOMENTS OF THE ELECTR 0780
         MASS SPECTROMETRIC STUDIES OF SOME GASEOUS SULFUR FLUORIDES.   1023
         MASS SPECTROMETRIC STUDIES OF SOME GASEOUS SULFUR FLUORIDES.   1028
         AB INITIO AND MZDO WAVE FUNCTIONS FOR SULFUR HEXAFLUORIDE.   1066
DIATION TO MOLECULAR SPECTROSCOPY. MOLECULAR NITROGEN AND SULFUR HEXAFLUORIDE. /TION OF SYNCHROTON ORBITAL RA 1083
IN MOLECULES AND COMPLEX IONS. (PHOSPHORUS PENTAFLUORIDE, SULFUR HEXAFLUORIDE)   DIRECTED VALENCY IN CERTA 0860
THE ASSIGNMENT OF THE HELIUM-2 PHOTOELECTRON SPECTRUM OF SULFUR HEXAFLUORIDE. /A MANY BODY APPROACH. PART-6: 1090
L BARRIERS IN THE X-RAY ABSORPTION SPECTRA OF MOLECULES. (SULFUR HEXAFLUORIDE) EVIDENCE OF EFFECTIVE POTENTIA 1073
         FAR INFRARED SPECTRA OF GASEOUS AND LIQUID SULFUR HEXAFLUORIDE.   1095
GASEOUS CARBON TETRAFLUORIDE, SILICON TETRAFLUORIDE, AND SULFUR HEXAFLUORIDE. /LAXATION BY NMR ABSORPTION IN 0346
                        FORCE CONSTANTS FOR SULFUR HEXAFLUORIDE.   1096
                        INFRARED DOUBLE RESONANCE IN SULFUR HEXAFLUORIDE.   1099
                        INFRARED INTENSITIES IN CRYSTALLINE SULFUR HEXAFLUORIDE.   1075
         INFRARED SPECTRA OF CRYOSYSTEMS. PART-2: SULFUR HEXAFLUORIDE.   1068
                        INFRARED SPECTRUM OF SULFUR HEXAFLUORIDE.   1086
LOW TEMPERATURE ABSORPTION CONTOUR OF THE NU-3 BAND OF SULFUR HEXAFLUORIDE.   1082
                        NU-3 AND NU-4 BANDS OF SULFUR HEXAFLUORIDE.   1070
         OCTAHEDRAL FINE STRUCTURE SPLITTINGS IN NU-3 OF SULFUR HEXAFLUORIDE.   1063
VIBRATION- ROTATION TRANSITION IN THE NU-3 FUNDAMENTAL OF SULFUR HEXAFLUORIDE. /TICAL SATURATION OF A SINGLE 1074
         RAMAN SPECTRA OF MATRIX ISOLATED MOLECULES. (SULFUR HEXAFLUORIDE)   1097
                        RAMAN SPECTRUM OF SOLID SULFUR HEXAFLUORIDE.   1098
                        SPECTROSCOPIC CALCULATIONS ON SULFUR HEXAFLUORIDE.   1071
                        THERMODYNAMIC FUNCTIONS FOR SULFUR HEXAFLUORIDE.   1089
                        THERMODYNAMIC PROPERTIES OF SULFUR HEXAFLUORIDE.   1087
         ULTRASOFT X-RAY ABSORPTION SPECTRUM OF SULFUR HEXAFLUORIDE.   1102
                        X-RAY ABSORPTION IN SULFUR HEXAFLUORIDE.   1067
GROUND STATE.   ELECTRONIC PROPERTIES OF SULFUR HEXAFLUORIDE. AB INITIO CALCULATION FOR THE 1077
ERFECT PAIRING VALENCE STATES. (PHOSPHORUS PENTAFLUORIDE, SULFUR HEXAFLUORIDE, AND CHLORINE TRIFLUORIDE) /N P 0862
UM HEXAFLUORIDE, TELLURIUM / GREEN'S FUNCTION ANALYSIS OF SULFUR HEXAFLUORIDE AND RELATED MOLECULES. (SELENI 1093
         MOLECULAR VIBRATIONS OF SULFUR HEXAFLUORIDE AND SULFUR TETRAFLUORIDE.   1034
         FORCE FIELDS FOR SOME GROUP-6 HEXAFLUORIDES. (SULFUR HEXAFLUORIDE AND TELLURIUM HEXAFLUORIDE) 1061
ERATURE DEPENDENT ABSORPTION SPECTRUM OF THE NU-3 BAND OF SULFUR HEXAFLUORIDE AT 10.6 MICROMETERS. TEMP 1091
         PHOTOELECTRON SPECTRUM OF SULFUR HEXAFLUORIDE AT 584 ANGSTROMS.   1076
SELECTIVE PHOTODISSOCIATION OF POLYATOMIC MOLECULES (SULFUR HEXAFLUORIDE) BY INFRARED RADIATION.   1064
RING.   FORBIDDEN RAMAN BANDS OF SULFUR HEXAFLUORIDE. COLLISION-INDUCED RAMAN SCATTE 1081
         FORCE CONSTANTS OF SULFUR HEXAFLUORIDE FROM ISOTOPIC SUBSTITUTION.   1100
         RAMAN SPECTRUM OF SULFUR HEXAFLUORIDE IN THE GASEOUS AND SOLID STATES 1079
X-RAY EMISSION FROM SCF MO LCA/ ELECTRONIC PROPERTIES OF SULFUR HEXAFLUORIDE. PART-1: OPTICAL ABSORPTION AND 1078
         PHYSICAL AND THERMODYNAMIC PROPERTIES OF SULFUR HEXAFLUORIDE. (REVIEW)   1062

/EXAFLUORIDE, RHENIUM HEXAFLUORIDE, URANIUM HEXAFLUORIDE, SULFUR HEXAFLUORIDE, SELENIUM HEXAFLUORIDE, AND TE/ 1072
/, AND CORIOLIS CONSTANTS OF THE SPHERICAL TOP MOLECULES, SULFUR HEXAFLUORIDE, SELENIUM HEXAFLUORIDE, TELLUR/ 1069
/VIBRATION FREQUENCIES OF SOME OCTAHEDRAL XY6 MOLECULES. (SULFUR HEXAFLUORIDE, SELENIUM HEXAFLUORIDE, TELLUR/ 1080
UM HEXAFL/ ELECTRIC DEFLECTION OF BINARY HEXAFLUORIDES. (SULFUR HEXAFLUORIDE, SELENIUM HEXAFLUORIDE, TELLURI 1084
UM HEXAFLUORIDE AND THEIR THE/ ENTHALPIES OF FORMATION OF SULFUR HEXAFLUORIDE, SELENIUM HEXAFLUORIDE, TELLURI 1092
LASER ISOTOPE SEPARATION. (SULFUR HEXAFLUORIDE SPECTRUM) 1094
SPECTROSCOPY.   ELECTRONIC STRUCTURE OF SULFUR HEXAFLUORIDE STUDIED BY ULTRALONG WAVE X-RAY 1065
/N BY ELECTRON DIFFRACTION OF THE MOLECULAR STRUCTURES OF SULFUR HEXAFLUORIDE, SULFUR TETRAFLUORIDE, SELENIU/ 1041
SORPTION SPECTRA OF GASEOUS SULFUR HEXAFLUORIDE.   SULFUR K AND L AND FLUORINE K X-RAY EMISSION AND AB 1088
TY AND UNIVERSAL EQUATIONS OF STATE OF HEXAFLUORIDES. (OF SULFUR, MOLYBDENUM TUNGSTEN, AND URANIUM) /SIMILARI 0632
MICROWAVE AND MASS SPECTRA OF SULFUR MONOFLUORIDE. 1024
SPECTRUM OF SULFUR MONOFLUORIDE. 1022
LECTRIC DIPOLE MOMENTS OF OPEN SHELL DIATOMIC MOLECULES. (SULFUR MONOFLUORIDE AND SELENIUM MONOFLUORIDE)   E 1019
GAS PHASE ELECTRON RESONANCE SPECTRA OF SULFUR MONOFLUORIDE AND SELENIUM MONOFLUORIDE. 1021
/ COMPUTED PROPERTIES FOR THE DOUBLET PI GROUND STATES OF SULFUR MONOFLUORIDE AND SELENIUM MONOFLUORIDE AND / 1025
/RIDE, SILICON MONOFLUORIDE, PHOSPHORUS MONOFLUORIDE, AND SULFUR MONOFLUORIDE FROM A MOLECULAR ORBITAL STUDY. 0973
MICROWAVE SPECTRUM OF THE SULFUR MONOFLUORIDE RADICAL. 1018
M OXIDE AND IODI/ GAS PHASE ELECTRON RESONANCE SPECTRA OF SULFUR MONOFLUORIDE, SELENIUM MONOFLUORIDE, SELENIU 1020
STRUCTURE OF THE RADICAL SULFUR PENTAFLUORIDE. 1060
IS AND INTRA- INTERMOLECULAR MEC/ FLUORINE NMR SPECTRA OF SULFUR, PHOSPHORUS, AND SILICON FLUORIDES. HYDROLYS 1042
N SPECTRA AND MOLECULAR CONSTANTS OF THE HEXAFLUORIDES OF SULFUR, SELENIUM, AND TELLURIUM.   RAMA 0963
/D MOLECULAR STRUCTURE OF SOME GROUP-6 HEXAFLUORIDES. (OF SULFUR, SELENIUM, TELLURIUM, MOLYBDENUM, TUNGSTEN,/ 0961
/TS AND BOND LENGTHS OF SOME INORGANIC HEXAFLUORIDES. (OF SULFUR, SELENIUM, TELLURIUM, MOLYBDENUM, TUNGSTEN,/ 0631
CLASSIFICATION OF THE STATES OF NONRIGID MOLECULES. (SULFUR TETRAFLUORIDE) 1039
FLUORINE EXCHANGE IN SULFUR TETRAFLUORIDE. 1049
FLUORINE EXCHANGE IN SULFUR TETRAFLUORIDE. 1053
FLUORINE-19 NMR INVESTIGATION OF THE ASSOCIATION OF SULFUR TETRAFLUORIDE. 1044
FORCE FIELD STUDY FOR SULFUR TETRAFLUORIDE. 1052
OF LIQUID CHLORINE TRIFLUORIDE, BROMINE TRIFLUORIDE, AND SULFUR TETRAFLUORIDE. / FLUORINE EXCHANGE MECHANISM 0399
INFRARED SPECTRUM AND NORMAL COORDINATE ANALYSIS OF SULFUR TETRAFLUORIDE. 1045
INTRAMOLECULAR EXCHANGE IN SULFUR TETRAFLUORIDE. 1046
MEAN AMPLITUDES OF VIBRATION FOR SULFUR TETRAFLUORIDE. 1038
MOLECULAR FORCE FIELD FOR SULFUR TETRAFLUORIDE. 1050
MOLECULAR VIBRATIONS OF SULFUR TETRAFLUORIDE. 1058
NMR SPECTRUM AND STRUCTURE OF SULFUR TETRAFLUORIDE. 1037
PROBLEM OF FLUORINE EXCHANGE IN SULFUR TETRAFLUORIDE. 1055
RAMAN AND INFRARED SPECTRUM OF SULFUR TETRAFLUORIDE. 1040
RAMAN SPECTRUM AND FORCE CONSTANTS FOR SULFUR TETRAFLUORIDE. 1035
STRUCTURE AND DIPOLE MOMENT OF SULFUR TETRAFLUORIDE. 1056
UREY-BRADLEY FORCE FIELD FOR SULFUR TETRAFLUORIDE. 1059
VIBRATIONAL ASSIGNMENT OF SULFUR TETRAFLUORIDE. 1036
VESCF-MO STUDIES OF PHOSPHORUS PENTAFLUORIDE, SULFUR TETRAFLUORIDE, AND CHLORINE TRIFLUORIDE. 0835
/-RAY PHOTOELECTRON SPECTROSCOPY OF CHLORINE TRIFLUORIDE, SULFUR TETRAFLUORIDE, AND PHOSPHORUS PENTAFLUORIDE. 0409
/ ESCA OF INEQUIVALENT FLUORINES IN CHLORINE TRIFLUORIDE, SULFUR TETRAFLUORIDE, AND PHOSPHORUS PENTAFLUORIDE. 0874
STRUCTURE OF SULFUR TETRAFLUORIDE AND RELATED MOLECULES. 1043
GANIC FLUORIDES IN THE SOLID STATE. (ARSENIC TRIFLUORIDE, SULFUR TETRAFLUORIDE, AND SELENIUM TETRAFLUORIDE) / 0096
F EXCHANGE PROCESSES IN SOME INORGANIC FLUORIDES BY NMR. (SULFUR TETRAFLUORIDE AND SELENIUM TETRAFLUORIDE) /O 1048
SOFT X-RAY SPECTROSCOPY ON SULFUR TETRAFLUORIDE AND SULFUR HEXAFLUORIDE. 1033
NFRARED EVIDENCE FOR DIMER FORMATION AT LOW TEMPERATURE. SULFUR TETRAFLUORIDE AND SULFUR OXIDE DIFLUORIDE. I 1054
.   THERMODYNAMIC PROPERTIES OF SULFUR TETRAFLUORIDE AND SULFUR OXIDE TETRAFLUORIDE 1051
E MOLECULES.   MEAN AMPLITUDES OF VIBRATION: SULFUR TETRAFLUORIDE AND SULFUR PENTAFLUORO CHLORID 1057
I/ INFRARED MATRIX ISOLATION STUDIES. (BORON TRIFLUORIDE, SULFUR TETRAFLUORIDE, ANTIMONY PENTAFLUORIDE, ARSEN 0239
CALCULATION OF THE INFRARED SPECTRA OF SULFUR TETRAFLUORIDE USING CNDO/2 TECHNIQUES. 1047
)   STRUCTURE OF THE SULFUR TRIFLUORIDE POSITIVE ION. (SULFUR DIFLUORIDE 1032
ESR SPECTRA OF PHOSPHORUS DIFLUORIDE AND SULFUR TRIFLUORIDE RADICALS. 0791
ORIDE, SILICON TETRAFLUORIDE, GERMANIUM TETRAFLUORIDE AND SULFUR TRIOXIDE. /RIOUS ISOTOPES OF CARBON TETRAFLU 0358
UORIDE, NITROGEN TRIFLUORIDE, AND PHOSPHORYL, THIONYL AND SULFURYL FLUORIDES. /LE MOMENTS IN PHOSPHORUS TRIFL 0744
CORIOLIS ZETA SUMS FOR SOME SPHERICAL TOP MOLECULES. 1487
SECOND MOMENTS OF THE ELECT/ MOLECULAR G VALUES, MAGNETIC SUSCEPTIBILITIES, MOLECULAR QUADRUPOLE MOMENTS AND 0780
/EEMAN STUDIES INCLUDING THE MOLECULAR G VALUES, MAGNETIC SUSCEPTIBILITY, AND MOLECULAR QUADRUPOLE MOMENTS I/ 0744
/CHLORIDE, CESIUM BROMIDE, AND CESIUM IODIDE AND MAGNETIC SUSCEPTIBILITY ANISOTROPY OF THALLIUM FLUORIDE, CE/ 1120
TINUM HEXAFLUORIDE.   SUSCEPTIBILITY MEASUREMENTS AND FLUORINE NMR OF PLA 0883
MAGNETIC SUSCEPTIBILITY OF EUROPIUM TRIFLUORIDE. 0454
MAGNETIC RESONANCE AND SUSCEPTIBILITY OF RARE EARTH TRIFLUORIDES. 1349
MAGNETIC SUSCEPTIBILITY OF RHENIUM HEXAFLUORIDE. 0917
MAGNETIC SUSCEPTIBILITY OF SILVER DIFLUORIDE. 1015
MAGNETIC SUSCEPTIBILITY OF XENON HEXAFLUORIDE. 1308
MAGNETIC SUSCEPTIBILITY OF XENON HEXAFLUORIDE. 1310
AHEDRAL MOLECULES.   RAMAN INTENSITIES FOR SPHERICALLY SYMMETRIC MODES OF SOME XY4 TETRAHEDRAL AND XY6 OCT 1453
TRUM OF BORON TRIFLUORIDE.   ANALYSIS OF SYMMETRIC ROTOR RAMAN SPECTRA. PURE ROTATIONAL SPEC 0205
SPIN RELAXATION STUDY OF THE SPIN ROTATION INTERACTION IN SYMMETRIC TOP MOLECULES.   NUCLEAR 0183
ROGEN TRIFLUO/ P-R SEPARATIONS AND CORIOLIS CONSTANTS FOR SYMMETRIC TOP MOLECULES AND FORCE CONSTANTS FOR NIT 0708
ZEEMAN EFFECT OF SOME LINEAR AND SYMMETRIC TOP MOLECULES. (PHOSPHORUS TRIFLUORIDE) 0796
NTIAL CONSTANTS FROM THE BAND CONTOURS OF HOT BANDS FOR A SYMMETRIC TOP. (NITROGEN TRIFLUORIDE) /ARMONIC POTE 0733
FAR INFRARED SPECTRA OF PROLATE AND OBLATE SYMMETRIC TOPS. 1479
GAL STRETCHING OF TRIFLUORO CHLORO METHANE AND SOME OTHER SYMMETRIC TOPS. (ARSENIC TRIFLUORIDE)   CENTRIFU 0101
NU-2 AND TRIFL/ PSEUDO-PARALLEL INFRARED BANDS OF OBLATE SYMMETRIC TOPS NITROGEN TRIFLUORIDE NU-4 AND NU-4 + 0716
/TUDES AND SHRINKAGE EFFECTS OF VIBRATION FOR SOME LINEAR SYMMETRICAL DIFLUORIDES. (OF THE FIRST ROW TRANSIT/ 1369
/ON AND BASTIANSEN-MORINO SHRINKAGE EFFECT IN SOME LINEAR SYMMETRICAL DIHALIDES. (BERYLLIUM DIFLUORIDE, MAGN/ 1372
ODEL FOR CALCULATION OF ANGULAR FORCE CONSTANTS OF HIGHLY SYMMETRICAL MOLECULES. /OF A SIMPLE ELECTROSTATIC M 1484
ULAR CONSTANTS AND THERMODYNAMIC FUNCTIONS OF SOME LINEAR SYMMETRICAL MOLECULES. (XENON DIFLUORIDE)   MOLEC 1244
STANTS AND CORIOLIS COUPLING CONSTANTS FOR SOME NONLINEAR SYMMETRICAL XY2 MOLECULES.   POTENTIAL CON 1385
BRADLEY FORCE FIELD AND THERMODYNAMIC PROPERTIES FOR BENT SYMMETRICAL XY2 MOLECULES.   UREY- 1387
MPLITUDE MATRICES. (BASTIANSEN-MORINO SHRINKAGE EFFECT OF SYMMETRICAL XY3 AND XY4 MOLECULES) /L MEAN SQUARE A 1394
TRAFLUORIDE.   VIBRATIONS OF PLANAR SYMMETRICAL XY4 MOLECULES WITH APPLICATION XENON TE 1268
F HALIDE COMPLEX MOLECULES OF THE XY4 TYPE OF TETRAHEDRAL SYMMETRY. /CE CONSTANTS AND INTERATOMIC DISTANCES O 1414
FORCE CONSTANTS FOR MOLECULES WITH D3H SYMMETRY. 0190
THE CORIOLIS CONSTANT FOR SOME MOLECULES AND IONS OF D4H SYMMETRY. / FORCE CONSTANTS, AND MASS DEPENDENCE OF 1411
SHRINKAGE EFFECT IN SOME XY3 MOLECULES OF PLANAR TRIGONAL SYMMETRY. /UDES OF VIBRATION AND BASTIANSEN-MORINO 1397
EAN AMPLITUDES OF VIBRATION IN SOME XY2Z MOLECULES OF C2V SYMMETRY.   M 1396

ORIDES. (LANTHANUM TRIFLUORIDE)

NSTANTS.

EVIDENCE FOR THE MOLECULAR

E.

EAN SQUARE AMPLITUDES IN SOME HEXAFLUORIDES OF OCTAHEDRAL
ARIZABILITES OF SOME PENTAHALIDES OF TRIGONAL BIPYRAMIDAL
OPY. MOLECULAR NITROGEN AND SULFUR HEXAFL/ APPLICATION OF
DE.
/S OF SCANDIUM DIHALIDES AND THE THE POSSIBILITY OF THEIR
DS OF CARBON MONOFLUO/ THERMODYNAMIC, ELECTROCHEMICAL AND
ES.   ANALYSES OF THERMOCHEMICAL ERRORS AND

MOLECULES WITH D4H SYMMETRY. NEW METHOD FOR THE CALCULATION OF GVFF CO   1412
SYMMETRY AND CRYSTAL STRUCTURE OF RARE EARTH TRIFLU   0571
SYMMETRY GROUP OF PHOSPHORUS PENTAFLUORIDE.   0837
MOLECULAR SYMMETRY OF IODINE HEPTAFLUORIDE.   0523
MOLECULAR SYMMETRY OF IODINE HEPTAFLUORIDE.   0524
MOLECULAR SYMMETRY OF IODINE HEPTAFLUORIDE.   0531
MOLECULAR SYMMETRY OF XENON DIFLUORIDE AND XENON TETRAFLUORID   1240
SYMMETRY. (SELENIUM AND TELLURIUM FLUORIDES) ROOT M   0962
SYMMTERY. /HERMODYNAMIC FUNCTIONS AND MOLECULAR POL   1437
SYNCHROTON ORBITAL RADIATION TO MOLECULAR SPECTROSC   1083
SYNTHESIS AND CHARACTERIZATION OF VANADIUM DIFLUORI   1206
SYNTHESIS BY THE REDUCTION OF THE TRIHALIDES.   (SC/   0940
SYNTHETIC STUDIES OF THE GRAPHITE   FLUORINE COMPOUN   0296
SYSTEMATICS IN SUBLIMATION OF LANTHANIDE TRIFLUORID   1353

T

/LES CONTAINING ATOMS FROM THE SECOND ROW OF THE PERIODIC TABLE. (SILICON DIFLUORIDE, SILICON TETRAFLUORIDE,/   1373
/DYNAMIC FUNCTIONS OF ANTIMONY FLUORIDE, NIOBIUM BROMIDE, TANTALUM CHLORIDE, TANTALUM BROMIDE, AND MOLYBDENU/   0075
NTAFLUORIDES.   THERMODYNAMIC PROPERTIES OF NIOBIUM AND TANTALUM FLUORIDES AT HIGH TEMPERATURES. PART-1: PE   0665
TRAFLUORIDES AND/ THERMODYNAMIC PROPERTIES OF NIOBIUM AND TANTALUM FLUORIDES AT HIGH TEMPERATURES. PART-2: TE   0657
FLUORIDES AND MO/ THERMODYNAMIC PROPERTIES OF NIOBIUM AND TANTALUM FLUORIDES AT HIGH TEMPERATURES. PART-3: DI   0654
/ PROPERTIES OF NIOBIUM MONOFLUORIDE, NIOBIUM DIFLUORIDE, TANTALUM MONOFLUORIDE, AND TANTALUM DIFLUORIDE AT /   0655
NTALUM PENTAFLUORIDES.   HALIDES OF NIOBIUM AND TANTALUM. PART-3: VAPOR PRESSURES OF NIOBIUM AND TA   0663
WEEN ANTIMONY PENTAFLUORIDE AND NIOBIUM PENTAFLUORIDE AND TANTALUM PENTAFLUORIDE. /Y OF COMPLEX FORMATION BET   0079
ON DIFFRACTION INVESTIGATION OF NIOBIUM PENTAFLUORIDE AND TANTALUM PENTAFLUORIDE.   ELECTR   0671
RES OF VANADIUM PENTAFLUORIDE, NIOBIUM PENTAFLUORIDE, AND TANTALUM PENTAFLUORIDE. /Y OF THE MOLECULAR STRUCTU   1217
ENTHALPIES OF FORMATION OF NIOBIUM PENTAFLUORIDE AND TANTALUM PENTAFLUORIDE.   0667
RAMAN SPECTRA OF LIQUID NIOBIUM PENTAFLUORIDE AND TANTALUM PENTAFLUORIDE.   0670
TH/ POLYMER MONOMER EQUILIBRIA IN NIOBIUM PENTAFLUORIDE, TANTALUM PENTAFLUORIDE, AND ANTIMONY PENTAFLUORIDE.   0076
DE.   STABILITIES OF TANTALUM PENTAFLUORIDE AND TANTALUM OXIDE TRIFLUORI   1103
THERMODYNAMIC PROPERTIES OF NIOBIUM AND TANTALUM PENTAFLUORIDES.   0669
SITY OF NIOBIUM PENTA/ VIBRATIONAL SPECTRA OF NIOBIUM AND TANTALUM PENTAFLUORIDES IN THE GAS PHASE. VAPOR DEN   0660
/ROPERTIES OF NIOBIUM TRIFLUORIDE, NIOBIUM TETRAFLUORIDE, TANTALUM TRIFLUORIDE, AND TANTALUM TETRAFLUORIDE A/   0656
THERMODYNAMIC PROPERTIES OF NIOBIUM(V), TANTALUM(V), AND MOLYBDENUM(V) FLUORIDES.   0668
LENCE FORCE CONSTANTS OF THE HEXAFLUORIDES OF MOLYBDENUM, TECHNETIUM, AND RHENIUM.   VA   0641
AN AMPLITUDES OF VIBRATION AND THERMODYNAMIC FUNCTIONS OF TECHNETIUM HEXAFLUORIDE.   ME   1104
VAPOR PRESSURE AND TRANSITION POINTS OF TECHNETIUM HEXAFLUORIDE.   1105
VIBRATIONAL SPECTRA OF MOLYBDENUM HEXAFLUORIDE AND TECHNETIUM HEXAFLUORIDE.   0628
IUM HEXAFLUORI/ RAMAN SPECTRA OF MOLYBDENUM HEXAFLUORIDE, TECHNETIUM HEXAFLUORIDE, RHENIUM HEXAFLUORIDE, URAN   1072
R CONSTANTS OF THE HEXAFLUORIDES OF SULFUR, SELENIUM, AND TELLURIUM.   RAMAN SPECTRA AND MOLECULA   0963
INTERACTION TERMS IN ENERGY DISTRIBUTIONS. (SELENIUM AND TELLURIUM FLUORIDES) /L MODES OF VIBRATION. PART-2:   1494
SOME HEXAFLUORIDES OF OCTAHEDRAL SYMMETRY. (SELENIUM AND TELLURIUM FLUORIDES) ROOT MEAN SQUARE AMPLITUDES IN   0962
/ IN ANTIMONY HEXACHLORIDE(3-), SELENIUM HEXABROMIDE(2-), TELLURIUM HEXABROMIDE(2-), IODINE HEXAFLUORIDE(1-)/   1309
OF THE FIRST BORN APPROXIMATION IN ELECTRON DIFFRACTION. (TELLURIUM HEXAFLUORIDE)   FAILURE   1109
FOR SOME GROUP-6 HEXAFLUORIDES. (SULFUR HEXAFLUORIDE AND TELLURIUM HEXAFLUORIDE)   FORCE FIELDS   1061
AFLUORIDE AND RELATED MOLECULES. (SELENIUM HEXAFLUORIDE AND TELLURIUM HEXAFLUORIDE) /ION ANALYSIS OF SULFUR HEX   1093
INFRARED SPECTRA OF SELENIUM HEXAFLUORIDE AND TELLURIUM HEXAFLUORIDE.   1107
6 MOLECULES. (SULFUR HEXAFLUORIDE, SELENIUM HEXAFLUORIDE, TELLURIUM HEXAFLUORIDE) /CIES OF SOME OCTAHEDRAL XY   1080
/GATION OF SOME SIMPLE MOLECULES. (TUNGSTEN HEXAFLUORIDE, TELLURIUM HEXAFLUORIDE, AND MOLYBDENUM HEXAFLUORID/   1153
/FORMATION OF SULFUR HEXAFLUORIDE, SELENIUM HEXAFLUORIDE, TELLURIUM HEXAFLUORIDE AND THEIR THERMODYNAMIC PRO/   1092
INFRARED SPECTRUM OF TELLURIUM HEXAFLUORIDE FROM 25 TO 40 MICRONS.   1108
FLUORIDE, SULFUR HEXAFLUORIDE, SELENIUM HEXAFLUORIDE, AND TELLURIUM HEXAFLUORIDE IN THE VAPOR STATE. /UM HEXA   1072
IFFRACTION. (TUNGSTEN HEXAFLUORIDE, URANIUM HEXAFLUORIDE, TELLURIUM HEXAFLUORIDE, MOLYBDENUM HEXAFLUORIDE) /D   1158
/P MOLECULES. (SULFUR HEXAFLUORIDE, SELENIUM HEXAFLUORIDE, TELLURIUM HEXAFLUORIDE, MOLYBDENUM HEXAFLUORIDE, T/   1069
/FLUORIDES. (SULFUR HEXAFLUORIDE, SELENIUM HEXAFLUORIDE, TELLURIUM HEXAFLUORIDE, MOLYBDENUM HEXAFLUORIDE, T/   1084
AR FORCE FIELD FOR SOME XY6 TYPE MOLECULES. (SELENIUM AND TELLURIUM HEXAFLUORIDES)   MOLECUL   1465
TURE OF SOME GROUP-6 HEXAFLUORIDES. (OF SULFUR, SELENIUM, TELLURIUM, MOLYBDENUM, TUNGSTEN, AND URANIUM) /TRUC   0961
/S OF SOME INORGANIC HEXAFLUORIDES. (OF SULFUR, SELENIUM, TELLURIUM, MOLYBDENUM, TUNGSTEN, URANIUM, AND RHEN/   0631
/ECTRA OF THE ISOELECTRONIC SPECIES IODINE PENTAFLUORIDE, TELLURIUM PENTAFLUORIDE(-) AND ANTIMONY PENTAFLUOR/   0501
CRYSTAL STRUCTURE OF TELLURIUM TETRAFLUORIDE.   1106
/NAL SPECTRA AND STRUCTURES OF SELENIUM TETRAFLUORIDE AND TELLURIUM TETRAFLUORIDE, INCLUDING A MATRIX ISOLAT/   0956
/DES OF SCANDIUM, CERIUM, SAMARIUM, EUROPIUM, GADOLINIUM, TERBIUM, DYSPROSIUM, ERBIUM, THULIUM, AND YTTERBIU/   1356
/D THERMAL DECOMPOSITION STUDIES ON CERIUM TETRAFLUORIDE, TERBIUM TETRAFLUORIDE, AND PRASEODYMIUM TETRAFLUOR/   0367
ENTROPIES AND ENTHALPIES OF SUBLIMATION OF NEODYMIUM AND TERBIUM TRIFLUORIDES.   0645
/CATION OF NORMAL MODES OF VIBRATION. PART-2: INTERACTION TERMS IN ENERGY DISTRIBUTIONS. (SELENIUM AND TELLU/   1494
ENTAFLUORIDE. FERMI RESONANCE BETWEEN A FUNDAMENTAL AND A TERNARY COMBINATION. /RA AND ASSIGNMENT OF IODINE P   0504
ATION BOND MOMENT PARAMETERS AND FORCE CONSTANTS. (CARBON TETRAFLUORIDE) /RMATIONS. APPLICATION TO THE CALCUL   0352
TUDY OF THE RAMAN AND BROAD LINE NMR SPECTRA OF GERMANIUM TETRAFLUORIDE.   A S   0484
ARE AMPLITUDES OF VIBRATION IN XENON DIFLUORIDE AND XENON TETRAFLUORIDE. /CAL EVALUATION OF THE ROOT MEAN SQU   1257
D THE ASSIGNMENTS OF SOME MIXED HALOMETHANES. (CARBON TETRAFLUORIDE) /RANSFERABLE UREY-BRADLEY FORCE FIEL   0351
ABSOLUTE INFRARED INTENSITIES OF CARBON TETRAFLUORIDE.   0340
ONSTANTS FROM HEAVY ISOTOPIC SUBSTITUTION. PART-1. CARBON TETRAFLUORIDE.   ACCURATE FORCE C   0320
ASSIFICATION OF THE STATES OF NONRIGID MOLECULES. (SULFUR TETRAFLUORIDE)   CL   1039
CONSTANT VOLUME HEAT CAPACITY OF GASEOUS CARBON TETRAFLUORIDE.   0336
CRYSTAL AND MOLECULAR STRUCTURE OF XENON TETRAFLUORIDE.   1277
CRYSTAL STRUCTURE OF TELLURIUM TETRAFLUORIDE.   1106
Y OF THE STRUCTURES OF NITROGEN DIFLUORIDE AND DINITROGEN TETRAFLUORIDE.   DIFFRACTION STUD   0680
CTRON DIFFRACTION STUDY OF THE STRUCTURE OF GASEOUS XENON TETRAFLUORIDE.   ELE   1282
ELECTRON IMPACT SPECTRA OF METHANE AND CARBON TETRAFLUORIDE.   0333
SPECTRUM OF XENON MONOFLUORIDE IN GAMMA IRRADIATED XENON TETRAFLUORIDE.   ELECTRON SPIN RESONANCE   1222
ENTHALPY OF FORMATION OF CARBON TETRAFLUORIDE.   0329
ENTHALPY OF FORMATION OF GERMANIUM TETRAFLUORIDE.   0469
ENTHALPY OF FORMATION OF URANIUM TETRAFLUORIDE.   1171
EPR SPECTRUM OF THE RADICAL CHLORINE TETRAFLUORIDE.   0413
ESR SPECTRUM OF ARSENIC TETRAFLUORIDE.   0112
R PROTACTINIUM IN LIQUID BISMUTH SOLUTIONS. (PROTACTINIUM TETRAFLUORIDE) /ALIDES AND ACTIVITY COEFFICIENTS FO   0900
HE FORCE CONSTANTS OF SOME TETRAHEDRAL MOLECULES. (CARBON TETRAFLUORIDE)   EVALUATION OF T   0359
EVIDENCE FOR CURIUM TETRAFLUORIDE.   0447
FAR INFRARED ABSORPTION IN GASEOUS CARBON TETRAFLUORIDE.   0356

```
                  FAR ULTRAVIOLET SPECTROSCOPIC STUDY OF XENON TETRAFLUORIDE.                                            1272
                             FLUORINE EXCHANGE IN SULFUR TETRAFLUORIDE.                                                  1049
                             FLUORINE EXCHANGE IN SULFUR TETRAFLUORIDE.                                                  1053
     LUORINE-19 NMR INVESTIGATION OF THE ASSOCIATION OF SULFUR TETRAFLUORIDE.                                      F     1044
    DDEN ELECTRONIC TRANSITIONS IN XENON DIFLUORIDE AND XENON TETRAFLUORIDE.                                   FORBI     1247
                FORCE CONSTANTS OF THE CARBON TETRAHALIDES. (CARBON TETRAFLUORIDE)                                       0354
                                  FORCE FIELD FOR SILICON TETRAFLUORIDE.                                                 1006
                             FORCE FIELD STUDY FOR SULFUR TETRAFLUORIDE.                                                 1052
                      GENERAL VALENCE FORCE FIELD FOR XENON TETRAFLUORIDE.                                               1265
   ONSTANTS OF OSMIUM TETROXIDE AND SOME TETRAHALIDE (CARBON TETRAFLUORIDE)           HARMONIC POTENTIAL C     0353
   EAT CAPACITY AND RELATED THERMODYNAMIC FUNCTIONS OF XENON TETRAFLUORIDE.                               H     1271
                            HEAT OF FORMATION OF XENON TETRAFLUORIDE.                                                    1266
                       HIGH TEMPERATURE HEAT CONTENT OF URANIUM TETRAFLUORIDE.                                           1173
   EMENTS FOR COMPOUNDS WHICH PRODUCE NO PARENT ION. / PHOTOELECTRON SPECTROSCOPIC MEASUR     0339
        INFRARED AND RAMAN SPECTRA AND STRUCTURE OF PHOSPHORUS TETRAFLUORIDE.                                            0825
   RARED AND RAMAN SPECTRA OF LIQUID AND CRYSTALLINE SILICON TETRAFLUORIDE.                                    INF       1001
                INFRARED AND RAMAN SPECTRA OF SILICON TETRAFLUORIDE.                                                     1005
   D BAND CONTOURS. PART-1: SPHERICAL TCP MOLECULES. (CARBON TETRAFLUORIDE)                              INFRARE 0325
                    INFRARED SPECTRA OF A CRYOSYSTEM. CARBON TETRAFLUORIDE.                                              0313
             INFRARED SPECTRA OF HALOGEN-SUBSTITUTED METHANES. (CARBON TETRAFLUORIDE)                                    0355
   UID CHLORINE TRIFLUORIDE, BROMINE TRIFLUORIDE, AND SULFUR TETRAFLUORIDE. / FLUORINE EXCHANGE MECHANISM OF LIQ 0399
       OF MATRIX ISOLATED ZIRCONIUM DIFLUORIDE, TRIFLUORIDE, AND TETRAFLUORIDE.                  INFRARED SPECTRA 1325
                            INFRARED SPECTRA OF NIOBIUM TETRAFLUORIDE.                                                   0658
   CTRA OF SIMPLE MOLECULES IN LIQUID ARGON SOLUTION. CARBON TETRAFLUORIDE.                          INFRARED SPE 0338
   NFRARED SPECTRUM AND NORMAL COORDINATE ANALYSIS OF SULFUR TETRAFLUORIDE.                                    I 1045
                       INFRARED SPECTRUM OF CARBON TETRAFLUORIDE.                                                        0343
                       INFRARED SPECTRUM OF SILICON TETRAFLUORIDE.                                                       1003
               INTERMOLECULAR POTENTIAL FOR SILICON TETRAFLUORIDE.                                                       1010
                      INTRAMOLECULAR EXCHANGE IN SULFUR TETRAFLUORIDE.                                                   1046
         INVESTIGATION OF THE CHEMISTRY OF AMERICIUM. (AMERICIUM TETRAFLUORIDE)                                          0056
   LITUDES OF VIBRATION AND THERMODYNAMIC FUNCTIONS OF XENON TETRAFLUORIDE.                            MEAN AMP 1274
                    MEAN AMPLITUDES OF VIBRATION FOR SULFUR TETRAFLUORIDE.                                               1038
   OF VIBRATION OF SOME XY4 TETRAHEDRAL MOLECULES. (SILICON TETRAFLUORIDE)                       MEAN AMPLITUDES 1008
   ECTRUM, DIPOLE MOMENT, AND STRUCTURE ANALYSIS OF SELENIUM TETRAFLUORIDE.                          MICROWAVE SP 0957
   E ATOMIC DISPLACMENTS AND INTRAMOLECULAR FORCES IN CARBON TETRAFLUORIDE. /TANTS FROM INFRARED BAND SHAPES. TH 0331
                          MOLECULAR FORCE FIELD FOR SULFUR TETRAFLUORIDE.                                                1050
          MOLECULAR SPECTROSCOPY BY MEANS OF ESCA. (CARBON TETRAFLUORIDE)                                                0328
                       MOLECULAR STRUCTURE OF CARBON TETRAFLUORIDE.                                                      0334
           MOLECULAR SYMMETRY OF XENON DIFLUORIDE AND XENON TETRAFLUORIDE.                                               1240
   MOLECULAR VIBRATIONS OF SULFUR HEXAFLUORIDE AND SULFUR TETRAFLUORIDE.                                                 1034
                   MOLECULAR VIBRATIONS OF SULFUR TETRAFLUORIDE.                                                         1058
                   NMR SPECTRUM AND STRUCTURE OF SULFUR TETRAFLUORIDE.                                                   1037
   POTENTIAL CONSTANTS AND THERMODYNAMIC FUNCTIONS OF CARBON TETRAFLUORIDE.                                              0350
        CONSTANTS OF SOME XY4 TYPE OF SILANE COMPOUNDS. (SILICON TETRAFLUORIDE)                         POTENTIAL 1009
   WO FLUORIDES OF OSMIUM. (OSMIUM PENTAFLUORIDE, AND OSMIUM TETRAFLUORIDE) /TRUCTURE OF OSMIUM PENTAFLUORIDE. T 0749
   REPARATION AND PROPERTIES (CRYSTAL STRUCTURE) OF VANADIUM TETRAFLUORIDE.                                     P 1209
          PREPARATION, STRUCTURE AND SPECTRA OF PRASEODYMIUM TETRAFLUORIDE.                                              0897
                  PROBLEM OF FLUORINE EXCHANGE IN SULFUR TETRAFLUORIDE.                                                  1055
                 RAMAN AND INFRARED SPECTRA OF GERMANIUM TETRAFLUORIDE.                                                  0482
                    RAMAN AND INFRARED SPECTRUM OF SULFUR TETRAFLUORIDE.                                                 1040
   RA OF CHLOROFLUORO METHANES IN THE GASEOUS STATE. (CARBON TETRAFLUORIDE)                          RAMAN SPECT 0322
          RAMAN SPECTRUM AND FORCE CONSTANTS FOR SULFUR TETRAFLUORIDE.                                                   1035
                   RAMAN SPECTRUM OF GASEOUS CARBON TETRAFLUORIDE.                                                       0347
                   RAMAN SPECTRUM OF SOLID MOLYBDENUM TETRAFLUORIDE.                                                     0621
   TUDES OF VIBRATION IN SOME GROUP-4 TETRAHALIDES. (CARBON TETRAFLUORIDE)            ROOT MEAN SQUARE AMPLI 0342
              SATURATED VAPOR PRESSURE OF SOLID URANIUM TETRAFLUORIDE.                                                   1167
   ERTIES OF CURIUM AND PROTACTINIUM METALS AND PROTACTINIUM TETRAFLUORIDE.                   SOME PHYSICAL PROP 0899
   OGEN FLUORINE BOND IN NITROGEN TRIFLUORIDE AND DINITROGEN TETRAFLUORIDE.               STRENGTH OF THE NITR 0712
              STRUCTURE AND DIPOLE MOMENT OF SULFUR TETRAFLUORIDE.                                                       1056
            STRUCTURES OF TRIFLUORO METHYL HALIDES. (CARBON TETRAFLUORIDE)                                               0318
                    SUBLIMATION OF PLUTONIUM TETRAFLUORIDE.                                                              0888
               THERMODYNAMIC PROPERTIES OF CARBON TETRAFLUORIDE.                                                         0321
                   ULTRAVIOLET SPECTRUM OF XENON TETRAFLUORIDE.                                                          1270
                UREY-BRADLEY FORCE FIELD FOR CARBON TETRAFLUORIDE.                                                       0360
                UREY-BRADLEY FORCE FIELD FOR SULFUR TETRAFLUORIDE.                                                       1059
                   UREY-BRADLEY FORCE FIELD. (LEAD TETRAFLUORIDE)                                                        0596
   PRESSURE AND MELTING POINTS OF XENON DIFLUORIDE AND XENON TETRAFLUORIDE.                                 VAPOR 1251
                   VAPOR PRESSURE OF THORIUM TETRAFLUORIDE.                                                              1129
                   VAPOR PRESSURE OF TITANIUM TETRAFLUORIDE.                                                             1148
                   VAPOR PRESSURE OF URANIUM TETRAFLUORIDE.                                                              1174
                   VAPOR PRESSURE OF URANIUM TETRAFLUORIDE.                                                              1175
                   VAPOR PRESSURE OF ZIRCONIUM TETRAFLUORIDE.                                                            1326
                   VAPOR PRESSURE OF ZIRCONIUM TETRAFLUORIDE.                                                            1328
   F PLANAR SYMMETRICAL XY4 MOLECULES WITH APPLICATION XENON TETRAFLUORIDE.                         VIBRATIONS O 1268
   NAL ASSIGNMENT AND NORMAL COORDINATE ANALYSIS OF SELENIUM TETRAFLUORIDE.                           VIBRATIO 0958
               VIBRATIONAL ASSIGNMENT OF SULFUR TETRAFLUORIDE.                                                          1036
        VIBRATIONAL SPECTRA AND FORCE CONSTANTS OF URANIUM TETRAFLUORIDE.                                               1172
         VIBRATIONAL SPECTRA AND STRUCTURE OF XENON TETRAFLUORIDE.                                                      1262
   VIBRATIONAL SPECTRA OF LIQUID AND CRYSTALLINE CARBON TETRAFLUORIDE.                                                  0327
   / COMPARISON OF CALCULATIONS OF FORCE CONSTANTS FOR BORON TETRAFLUORIDE(-), CARBON TETRAFLUORIDE, SILICON TET 0348
   LIDS. HEATS OF SUBLIMATION OF XENON DIFLUORIDE AND XENON TETRAFLUORIDE AND A CONJECTURE ON BONDING IN THE SO 1237
   LUORIDE, ARSENIC DIFLUORIDE, ARSENIC TRIFLUORIDE, ARSENIC TETRAFLUORIDE, AND ARSENIC PENTAFLUORIDE) /IC MONOF 0093
   WITH FORCE CONSTANTS OF THE EI/ FORCE CONSTANTS OF CARBON TETRAFLUORIDE AND BORON TRIFLUORIDE AND COMPARISON 0212
   VESCF-MO STUDIES OF PHOSPHORUS PENTAFLUORIDE, SULFUR TETRAFLUORIDE, AND CHLORINE TRIFLUORIDE.                0835
   TENSITIES IN THE INFRARED SPECTRA OF THE MOLECULES CARBON TETRAFLUORIDE AND DICARBON HEXAFLUORIDE. /ORY OF IN 0362
   STRUCTURES OF PHOSPHORUS PENTAFLUORIDE, METHYL PHOSPHORUS TETRAFLUORIDE AND DIMETHYL PHOSPHORUS TRIFLUORIDE. 0845
   ABSORPTION SPECTRUM OF URANIUM TETRAFLUORIDE AND ENERGY LEVELS OF URANIUM(IV).                                1168
   INFRARED SPECTRA OF CARBON TETRAFLUORIDE AND GERMANIUM TETRAFLUORIDE.                                          0363
   HEATS OF FORMATION OF TITANIUM TETRAFLUORIDE AND HAFNIUM TETRAFLUORIDE.                                        1147
   THE MX BOND IN THE TETRAHALIDES OF GROUP-4 ELEMENTS. (TIN TETRAFLUORIDE AND HAFNIUM TETRAFLUORIDE) /ICITY OF 1141
```

RAMAN SPECTRA OF SOLID XENON TETRAFLUORIDE AND ITS ADDUCT WITH XENON DIFLUORIDE. 1258
ISOTOPIC SUBSTITUTION. PART-2: BORON TRIFLUORIDE, SILICON TETRAFLUORIDE, AND NITROGEN DIOXIDE. /S FROM HEAVY 0232
LE POINT OF CARBON TETRAFLUORID/ VAPOR PRESSURE OF CARBON TETRAFLUORIDE AND NITROGEN TRIFLUORIDE AND THE TRIP 0345
HEATS OF FORMATION OF GERMANIUM TETRAFLUORIDE AND OF THE GERMANIUM DIOXIDES. 0483
EXPLOSION METHOD AND THE HEATS OF FORMATION OF CARBON TETRAFLUORIDE AND OTHER CARBON HALIDES. 0315
CONSTANT VOLUME HEAT CAPACITIES OF GASEOUS CARBON TETRAFLUORIDE AND OTHER MOLECULES. 0335
OF INEQUIVALENT FLUORINES IN CHLORINE TRIFLUORIDE, SULFUR TETRAFLUORIDE, AND PHOSPHORUS PENTAFLUORIDE. /ESCA 0874
PHOSPHORUS DIFLUORIDE, PHOSPHORUS TRIFLUORIDE, PHOSPHORUS TETRAFLUORIDE, AND PHOSPHORUS PENTAFLUORIDE) /IDE, 0788
HOTOELECTRON SPECTROSCCPY OF CHLORINE TRIFLUORIDE, SULFUR TETRAFLUORIDE, AND PHOSPHORUS PENTAFLUORIDE. /RAY P 0409
STRUCTURE OF SULFUR TETRAFLUORIDE AND RELATED MOLECULES. 1043
NGE PROCESSES IN SOME INORGANIC FLUORIDES BY NMR. (SULFUR TETRAFLUORIDE AND SELENIUM TETRAFLUORIDE) /OF EXCHA 1048
LUORIDES IN THE SOLID STATE. (ARSENIC TRIFLUORIDE, SULFUR TETRAFLUORIDE, AND SELENIUM TETRAFLUORIDE) /GANIC F 0096
DISTRIBUTIONS IN SOME FLUORIDES AND OXYCHLORIDES. (CARBON TETRAFLUORIDE AND SILICON TETRAFLUORIDE) / DENSITY 0337
FORCE FIELDS OF CARBON TETRAFLUORIDE AND SILICON TETRAFLUORIDE. 0324
HIGH RESOLUTION PHOTOELECTRON SPECTROSCOPY OF CARBON TETRAFLUORIDE AND SILICON TETRAFLUORIDE. 0319
LET EMISSION SYSTEMS EXCITED IN BORON TRIFLUORIDE, CARBON TETRAFLUORIDE, AND SILICON TETRAFLUORIDE. /ULTRAVIO 0216
SPECTRA OF GASEOUS ALUMINUM TRIFLUORIDE, LITHIUM ALUMINUM TETRAFLUORIDE, AND SODIUM ALUMINUM TETRAFLUORIDE. / 0045
OLECULAR ORBITAL DESCRIPTION OF SULFUR TETRAFLUORIDE, SULFUR TETRAFLUORIDE AND SULFUR HEXAFLUORIDE. M 1027
SOFT X-RAY SPECTROSCOPY ON SULFUR TETRAFLUORIDE AND SULFUR HEXAFLUORIDE. 1033
EVIDENCE FOR DIMER FORMATION AT LOW TEMPERATURE. SULFUR TETRAFLUORIDE AND SULFUR OXIDE DIFLUORIDE. INFRARED 1054
THERMODYNAMIC PROPERTIES OF SULFUR TETRAFLUORIDE AND SULFUR OXIDE TETRAFLUORIDE. 1051
ULES. MEAN AMPLITUDES OF VIBRATION: SULFUR TETRAFLUORIDE AND SULFUR PENTAFLUORO CHLORIDE MOLEC 1057
G A MATRI/ VIBRATIONAL SPECTRA AND STRUCTURES OF SELENIUM TETRAFLUORIDE AND TELLURIUM TETRAFLUORIDE, INCLUDIN 0956
ANIUM AND THORIUM TETRAHALIDES IN THE GAS PHASE. (URANIUM TETRAFLUORIDE AND THORIUM TETRAFLUORIDE) /UDY OF UR 1170
URE HEAT CAPACITIES, ENTHALPIES, AND ENTROPIES OF URANIUM TETRAFLUORIDE AND URANIUM HEXAFLUORIDE. /W TEMPERAT 1166
INFRARED SPECTRA OF VANADIUM PENTAFLUORIDE, VANADIUM TETRAFLUORIDE, AND VANADIUM TRIFLUORIDE. 1210
PROBABLE STRUCTURE OF XENON TETRAFLUORIDE AND XENON DIFLUORIDE. 1250
AND THE ELECTRONIC STRUCTURES OF XENON DIFLUORIDE, XENON TETRAFLUORIDE, AND XENON HEXAFLUORIDE. /RON SPECTRA 1228
L GAS THERMODYNAMIC PROPERTIES OF XENON COMPOUNDS. (XENON TETRAFLUORIDE AND XENON HEXAFLUORIDE) IDEA 1273
ATION MASS SPECTROMETRIC STUDY OF XENON DIFLUORIDE, XENON TETRAFLUORIDE, AND XENON HEXAFLUORIDE. PHOTOIONIZ 1225
THERMOCHEMICAL STUDIES OF XENON TETRAFLUORIDE AND XENON HEXAFLUORIDE. 1276
OUNDS. ELECTRON DIFFRACTION STUDIES OF XENON TETRAFLUORIDE AND XENON HEXAFLUORIDE AND OTHER COMP 1259
AMPLITUDES OF VIBRATION AND SHRINKAGE CONSTANTS OF XENON TETRAFLUORIDE AND XENON OXIDE TETRAFLUORIDE. /QUARE 1278
AND FORCE CO/ RAMAN SPECTRA FOR XENON DIFLUORIDE, XENON TETRAFLUORIDE, AND XENON OXIDE TETRAFLUORIDE VAPORS 1254
VIBRATIONAL SPECTRA AND STRUCTURES OF XENON TETRAFLUORIDE, AND XENON OXYTETRAFLUORIDE. 1261
OF CHLORO- AND FLUOROMETHANES, TETRAFLUORO METHANE XENON TETRAFLUORIDE, AND XENON TETRACHLORIDE. / POTENTIAL 0344
/RED MATRIX ISOLATION STUDIES. (BORON TRIFLUORIDE, SULFUR TETRAFLUORIDE, ANTIMONY PENTAFLUORIDE, ARSENIC PEN/ 0239
/TRON DISTRIBUTION IN THE XENON FLUORIDES AND XENON OXIDE TETRAFLUORIDE BY ESCA AND EVIDENCE FOR ORBITAL IND/ 1229
ENTHALPY OF FORMATION OF THORIUM TETRAFLUORIDE BY FLUORINE BOMB CALORIMETRY. 1130
CRYSTAL AND MOLECULAR STRUCTURE OF XENON TETRAFLUORIDE BY NEUTRON DIFFRACTION. 1260
DETERMINATION OF HEAT OF SUBLIMATICN OF URANIUM TETRAFLUORIDE BY THE MASS SPECTROMETRIC METHOD. 1165
/ULAR BEAM ELECTRIC DEFLECTION OF THE TETRAHALIDES CARBON TETRAFLUORIDE, CARBON TETRACHLORIDE, SILICON TETRA/ 0349
/BSORPTION SPECTRA OF SOME POLYATOMIC FLUORIDES. (SILICON TETRAFLUORIDE, CARBON TETRAFLUORIDE, BORON TRIFLUO/ 0186
N SPECTRA OF SOME HALOGEN DERIVATIVES OF METHANE. (CARBON TETRAFLUORIDE) CORRELATION OF THE SPECTRA. /SORPTIO 0365
/RPTION SPECTRA OF SOLID AMERICIUM TRIFLUORIDE, AMERICIUM TETRAFLUORIDE, CURIUM TRIFLUORIDE AND CURIUM TETRA/ 0445
ED RADICAL XENON MONOFLUORIDE. RADIATION DAMAGE IN XENON TETRAFLUORIDE. ELECTRON SPIN RESONANCE OF THE TRAPP 1220
OT MEAN SQUARE AMPLITUDES OF VIBRATION OF GASEOUS SILICON TETRAFLUORIDE FROM ELECTRON DIFFRACTION. /ES AND RO 1002
THERMOCHEMICAL PROPERTIES OF XENON DIFLUORIDE AND XENON TETRAFLUORIDE FROM MASS SPECTRA. 1253
ITY, ENTROPY, ENTHALPY, AND GIBBS ENERGY OF PLUTONIUM-242 TETRAFLUORIDE FROM 10 TO 350 DEGREES K. HEAT CAPAC 0889
/APACITIES OF PLUTONIUM-242 TRIFLUORIDE AND PLUTONIUM-242 TETRAFLUORIDE FROM 10 TO 350 DEGREES K, AND OF PLU/ 0885
T. THE SIGNIFICANCE O/ THERMODYNAMIC PROPERTIES OF CARBON TETRAFLUORIDE FROM 12 DEGREES K TO ITS BOILING POIN 0361
ENTROPY OF FUSION. ENTHALPY OF URANIUM TETRAFLUORIDE FROM 298-1400 DEGREES K. ENTHALPY AND 1169
THERMODYNAMIC PROPERTIES OF XENON TETRAFLUORIDE FROM 4 DEGREES K TO ITS MELTING POINT 0326
THERMODYNAMIC PROPERTIES OF GASEOUS GERMANIUM, GERMANIUM TETRAFLUORIDE, GERMANIUM DIFLUORIDE, AND GERMANIU/ 0468
TIONS FROM 5 TO 350 DEGREES K. RECONCILIATION OF T/ XENON TETRAFLUORIDE. HEAT CAPACITY AND THERMODYNAMIC FUNC 1275
XENON TETRAFLUORIDE: HEAT OF FORMATION. 1267
SPECTRA OF XENON DIFLUORIDE AND XENON TETRAFLUORIDE IN THE FAR ULTRAVIOLET REGION. 1233
N DIFFRACTION INVESTIGATION OF THE STRUCTURE OF ZIRCONIUM TETRAFLUORIDE IN THE VAPOR PHASE. ELECTRO 1332
XENON TETRAFLUORIDE. (INFRARED SPECTRUM) 1263
/RCE CONSTANTS OF SQUARE PYRAMIDAL MOLECULES. XENON OXIDE TETRAFLUORIDE, IODINE PENTAFLUORIDE, BROMINE PENTA/ 0418
ELECTRONIC STRUCTURE OF THE CARBON TETRAFLUORIDE MOLECULE. 0357
ELECTRONIC STRUCTURE OF THE CARBON TETRAFLUORIDE MOLECULE. PART-1. THE GROUND STATE. 0330
/ ORBITAL STUDY OF THE NITROGEN DIFLUORIDE AND DINITROGEN TETRAFLUORIDE MOLECULES, AND ESR HYPERFINE COUPLIN/ 0688
SILICON- FLUORINE LENGTH IN SILICON TETRAFLUORIDE. NEW ELECTRON DIFFRACTION STUDY. 1000
/ ROW OF THE PERIODIC TABLE. (SILICON DIFLUORIDE, SILICON TETRAFLUORIDE, PHOSPHORUS TRIFLUORIDE, PHOSPHORUS / 1373
IS THE RADICAL CHLORINE TETRAFLUORIDE PLANAR OR NOT. 0412
FINE SPLITTINGS AND MOLECULAR STRUCTURE IN THE PHOSPHORUS TETRAFLUORIDE RADICAL. ISOTROPIC HYPER 0824
RESONANCE OF TRAPPED PHOSPHORUS DIFLUORIDE AND PHOSPHORUS TETRAFLUORIDE RADICALS. ELECTRON SPIN 0792
OF PHOSPHORUS TRIFLUORIDE. ESR SPECTRA OF PHOSPHORUS TETRAFLUORIDE RADICALS PRODUCED IN A SINGLE CRYSTAL 0823
/ THE MOLECULAR STRUCTURES OF SULFUR HEXAFLUORIDE, SULFUR TETRAFLUORIDE, SELENIUM HEXAFLUORIDE, AND SELENIUM/ 1041
/ABLE TEMPERATURE HELIUM-1 AND HELIUM-2 STUDIES OF CARBON TETRAFLUORIDE, SILICON TETRAFLUORIDE, AND GERMANIU/ 0341
/NS OF FORCE CONSTANTS FOR BORON TETRAFLUORIDE(-), CARBON TETRAFLUORIDE, SILICON TETRAFLUORIDE, AND SILICON / 0348
/ FLUORINE RELAXATION BY NMR ABSORPTION IN GASEOUS CARBON TETRAFLUORIDE, SILICON TETRAFLUORIDE, AND SULFUR HE 0346
RIDE, ETHANE, SILANE, AMMONIA,/ FORCE CONSTANTS OF CARBON TETRAFLUORIDE, SILICON TETRAFLUORIDE, BORON TRIFLUO 0243
HLORI/ INTENSITIES OF THE INFRARED FUNDAMENTALS OF CARBON TETRAFLUORIDE, SILICON TETRAFLUORIDE, CARBON TETRAC 1007
/. TETRAFLUORIDES AND TETRACHLORIDES OF GROUP-5B. (CARBON TETRAFLUORIDE, SILICON TETRAFLUORIDE, GERMANIUM TE/ 0316
RAFLUO/ THE FORCE CONSTANTS OF VARIOUS ISOTOPES OF CARBON TETRAFLUORIDE, SILICON TETRAFLUORIDE, GERMANIUM TET 0358
THERMODYNAMIC PROPERTIES OF NIOBIUM TRIFLUORIDE, NIOBIUM TETRAFLUORIDE, TANTALUM TRIFLUORIDE, AND TANTALUM / 0656
/C DEFLECTION AND THERMAL DECOMPOSITION STUDIES ON CERIUM TETRAFLUORIDE, TERBIUM TETRAFLUORIDE, AND PRASEODY/ 0367
GAS. CARBON TETRAFLUORIDE. THERMODYNAMIC PROPERTIES OF THE REAL 0332
M HEXAFLU/ PHYSICAL CONSTANTS (VAPOR PRESSURE) OF SILICON TETRAFLUORIDE, TUNGSTEN HEXAFLUORIDE, AND MOLYBDENU 0639
/UORIDE, URANIUM DIFLUORIDE, URANIUM TRIFLUORIDE, URANIUM TETRAFLUORIDE, URANIUM PENTAFLUORIDE, AND URANIUM / 1160
CALCULATION OF THE INFRARED SPECTRA OF SULFUR TETRAFLUORIDE USING CNLO/2 TECHNIQUES. 1047
FLU/ ENTHALPIES OF FORMATION OF XENON HEXAFLUORIDE, XENON TETRAFLUORIDE, XENON DIFLUORIDE, AND PHOSPHORUS TRI 0807
SISTENT FIELD STUDY OF THE SERIES XENON DIFLUORIDE, XENON TETRAFLUORIDE, XENON HEXAFLUORIDE. SELF CON 1224
IN THE CHEMISTRY OF NOBLE GASES. (XENON DIFLUORIDE, XENON TETRAFLUORIDE, XENON HEXAFLUORIDE) /CONSIDERATIONS 1235
/ATION OF ELECTRONIC STATES OF THE POSITIVE ICNS OF XENON TETRAFLUORIDE, XENON OXIDE TETRAFLUORIDE, AND IODI/ 1264
RE. ENTHALPIES OF FORMATION OF XENON HEXAFLUORIDE, XENON TETRAFLUORIDE. CRYSTAL STRUCTU 1269
HELIUM-1 RESONANCE PHOTOELECTRON SPECTRA OF GROUP-4 TETRAFLUORIDES. 0317
FORCE CONSTANTS OF TETRAHEDRAL XY4 MOLECULES. (GERMANIUM TETRAFLUORIDES) ORBITAL VALENCE 0481

RAMAN SPECTRA OF CARBON AND SILICON TETRAFLUORIDES. 0364
FORCE FIELDS FOR GROUP-4 TETRAFLUORIDES AND GROUP-5 TRIFLUORIDES. 0715
BON TETRAFLUOR/ PHOTOELECTRON SPECTRA OF HALIDES. PART-1. TETRAFLUORIDES AND TETRACHLORIDES OF GROUP-5B. (CAR 0316
BIUM AND TANTALUM FLUORIDES AT HIGH TEMPERATURES. PART-2: TETRAFLUORIDES AND TRIFLUORIDES. /PROPERTIES OF NIO 0657
/ONAL AND ROTATIONAL CHARACTERISTICS OF MOLECULES OF SOME TETRAFLUORIDES FOR CALCULATING THEIR THERMODYNAMIC/ 0622
LATTICE CONSTANTS OF ACTINIDE TETRAFLUORIDES INCLUDING BERKELIUM. 1334
N OXIDE TRIFLUORIDE, AND A REASSIGNMENT OF THE SPECTRA OF TETRAFLUORIDES OF GROUP-4. /TRIFLUORIDE AND NITROGE 0700
FLUORO DICHLORO ETHYLENE, AND / IONIZATION POTENTIALS FOR TETRAFLUORO ETHYLENE, TRIFLUORO CHLORO ETHYLENE, DI 0305
RIFLUORIDE, PHOSPHORUS PENTAFLUORIDE, DIFLUORO AMINE, AND TETRAFLUORO HYDRAZINE. / STRUCTURES OF PHOSPHORUS T 0801
/ATED IONIZATION POTENTIAL OF CHLORO- AND FLUOROMETHANES, TETRAFLUORO METHANE XENON TETRAFLUORIDE, AND XENON/ 0344
ELECTRONIC STRUCTURES OF PHOSPHORUS PENTAFLUORIDE AND TETRAFLUORO PHOSPHORANE. 0857
N INTENSITIES FOR SPHERICALLY SYMMETRIC MODES OF SOME XY4 TETRAHEDRAL AND XY6 OCTAHEDRAL MOLECULES.   RAMA 1453
ATION OF FORCE CONSTANTS WITH THE MAXIMUM OVERLAP METHOD. TETRAHEDRAL MOLECULES.   CALCUL 1416
FORCE CONSTANTS OF TETRAHEDRAL MOLECULES. 1422
TOMIC MASSES ON THE CORIOLIS COUPLING COEFFICIENTS IN XY4 TETRAHEDRAL MOLECULES.   INFLUENCE OF A 1410
MEAN AMPLITUDES OF VIBRATION OF SOME XY4 TETRAHEDRAL MOLECULES. 1419
THOD FOR RESOLVING WILSON'S SECULAR EQUATIONS. APPLIED TO TETRAHEDRAL MOLECULES.   NEW ME 1431
UREY-BRADLEY FORCE FIELD FOR XY4 TETRAHEDRAL MOLECULES. 1429
STANTS IN THE SOLUTION OF THE VIBRATIONAL PROBLEM FOR XY4 TETRAHEDRAL MOLECULES. /OF CORIOLIS INTERACTION CON 1423
VIBRATIONS OF PENTATOMIC TETRAHEDRAL MOLECULES. 1497
NORMAL COORDINATE TREATMENT OF XY4 TETRAHEDRAL MOLECULES AND IONS. (GVFF) 1415
VFF) NORMAL COORDINATE TREATMENT OF XY4 TETRAHEDRAL MOLECULES AND IONS. (UREY-BRADLEY AND M 1418
EVALUATION OF THE FORCE CONSTANTS OF SOME TETRAHEDRAL MOLECULES. (CARBON TETRAFLUORIDE) 0359
MEAN AMPLITUDES OF VIBRATION OF SOME XY4 TETRAHEDRAL MOLECULES. (SILICON TETRAFLUORIDE) 1008
DISTANCES OF HALIDE COMPLEX MOLECULES OF THE XY4 TYPE OF TETRAHEDRAL SYMMETRY. /CE CONSTANTS AND INTERATOMIC 1414
HEMATICAL TECHNIQUES IN THE STUDY OF CORIOLIS COUPLING IN TETRAHEDRAL XY4 MOLECULES.   MAT 1409
MEAN AMPLITUDES OF VIBRATION FOR SOME PYRAMIDAL XY3 AND TETRAHEDRAL XY4 MOLECULES. 1407
ONAL MEAN SQUARE AMPLITUDE MATRICES. PART-7: TREATMENT OF TETRAHEDRAL XY4 MOLECULES.   VIBRATI 1408
S) ORBITAL VALENCE FORCE CONSTANTS OF TETRAHEDRAL XY4 MOLECULES. (GERMANIUM TETRAFLUORIDE) 0481
APPLICATION OF UREY-BRADLEY FORCE FIELD TO SOME TETRAHEDRAL XY4 TYPE MOLECULES AND IONS. 1425
D DERIVATIVES BY GREEN'S FUN/ MOLECULAR CONSTANTS OF SOME TETRAHEDRAL XY4 TYPE MOLECULES AND THEIR SUBSTITUTE 1424
MEAN AMPLITUDES OF VIBRATION OF SOME TETRAHEDRAL XY4 TYPE MOLECULES. PART 3 AND 4. 1420
S OF HAFNIUM AND LE/ MEAN AMPLITUDES OF VIBRATION OF SOME TETRAHEDRAL XY4 TYPE MOLECULES. PART-5: TETRAHALIDE 0487
GASEOUS TRIMER OF THALLIUM MONOFLUORIDE. 1114
XENON HEXAFLUORIDE. STRUCTURAL CRYSTALLOGRAPHY OF TETRAMERIC PHASES. 1284
EOUS MONOHALIDES OF BORON, ALUMINUM, GALLIUM, INDIUM, AND THALLIUM.   DISSOCIATION ENERGIES OF THE GAS 0152
/ OF ZEEMAN EFFECT IN DIAMAGNETIC MOLECULES. GJ FACTOR OF THALLIUM FLUORIDE, CESIUM FLUORIDE, CESIUM CHLORID/ 1120
HEAT CAPACITY AND THERMODYNAMIC PROPERTIES OF THALLIUM FLUORIDE FROM 5 TO 445 DEGREES K. 1124
INFRARED SPECTRA AND STRUCTURE OF MATRIX ISOLATED THALLIUM HALIDES. 1113
MICROWAVE SPECTRA OF THE THALLIUM, INDIUM AND GALLIUM MONOHALIDES. 1110
ENTHALPY OF FORMATION AND THE DISSOCIATION ENERGY OF THALLIUM MONOFLUORIDE 1115
GASEOUS TRIMER AND TETRAMER OF THALLIUM MONOFLUORIDE. 1114
FRARED AND RAMAN SPECTRA AND THE STRUCTURE OF CRYSTALLINE THALLIUM MONOFLUORIDE.   IN 1126
OFLUORIDE, GALLIUM MONOFLUORIDE, INDIUM MONOFLUORIDE, AND THALLIUM MONOFLUORIDE. /ION SPECTRA OF ALUMINUM MON 0025
ZERO(+), B TRIPLET PI(1) TO X SINGLET SIGMA(+) SYSTEMS OF THALLIUM MONOFLUORIDE. / BANDS OF THE A TRIPLET PI 1111
THERMODYNAMICS OF VAPORIZATION OF THALLIUM MONOFLUORIDE. 1121
STARK EFFECT FOR THALLIUM MONOFLUORIDE AND POTASSIUM FLUORIDE. 1118
HALL/ PHOTOIONIZATION OF HIGH TEMPERATURE VAPORS. PART-4: THALLIUM MONOFLUORIDE, THALLIUM MONOCHLORIDE, AND T 1112
MICROWAVE ROTATIONAL SPECTRUM OF THALLIUM MONOFLUORIDE. 1119
DIATOMIC FLUORIDES AND CHLORIDES OF GALLIUM, INDIUM, AND THALLIUM. (MONOFLUORIDES) /DISSOCIATION ENERGIES OF 0458
NATURE OF BONDS IN THALLIUM MONOHALIDE MOLECULES. 1122
IUM / ENTHALPIES, ENTROPIES, AND FREE ENERGY FUNCTIONS OF THALLIUM MONOHALIDES ABOVE ROOM TEMPERATURE. (THALL 1116
BAND SPECTRA OF THALLIUM MONOIODIDE AND MONOFLUORIDE. 1125
CRYSTAL STRUCTURE OF THALLIUM TRIFLUORIDE. 1127
TEMPERATURE VAPORS. PART-3: MONOMER AND DIMER SPECTRA OF THALLOUS FLUORIDE. /TOELECTRON SPECTROSCOPY OF HIGH 1117
/RARED AND RAMAN SPECTRA AND STRUCTURE OF MATRIX ISOLATED THALLOUS HALIDE DIMERS, THALLOUS FLUORIDE, AND THA/ 1123
VIRIAL THEOREM DECOMPOSITION OF MOLECULAR FORCE FIELDS. 0783
POLE MOMENT, AND SPECTROSCOPIC CONSTANTS. A COMPARISON OF THEORETICAL AND EXPERIMENTAL RESULTS. /MENT, QUADRU 0757
UPOLE MOMENTS, AND SPECTROSCOPIC CONSTANTS. COMPARISON OF THEORETICAL AND EXPERIMENTAL RESULTS. COMMENTS. /DR 0756
/ND THEIR POSITIVE AND NEGATIVE IONS. A COMPARISON OF THE THEORETICAL AND EXPERIMENTAL RESULTS. (OXYGEN MONO/ 1025
NIC STRUCTURE OF THE XENON FLUORIDES. THEORETICAL AND EXPERIMENTAL STUDIES OF THE ELECTRO 1238
E AND OPTICAL TRANSITIONS OF URANIUM HEXAFLUORIDE AND UR/ THEORETICAL CALCULATIONS OF THE ELECTRONIC STRUCTUR 1179
SIMPLE THEORETICAL CALCULATIONS ON TITANIUM MONOFLUORIDE. 1144
OF BORON TRIHALIDES AND BORON TETRAHALIDE ANIONS.   A THEORETICAL COMPARISON OF THE ELECTRONIC STRUCTURES 0225
ITUDES OF VIBRATION IN XENON DIFLUORIDE AND XENON TETR/ THEORETICAL EVALUATION OF THE ROOT MEAN SQUARE AMPL 1488
CORIOLIS COUPLING CONSTANTS IN FOUR XYN TYPE MOLECULE M/ THEORETICAL INTERPRETATION OF THE MASS INFLUENCE ON 0605
DIFLUORIDE WITH RARE GAS ATOMS. THEORETICAL STUDIES OF THE INTERACTION OF MAGNESIUM 1426
JAHN-TELLER EFFECT. EXPERIMENTAL AND THEORETICAL STUDIES ON VANADIUM TETRACHLORIDE. 0997
ELECTRONIC SPECTRUM OF SILICON DIFLUORIDE. THEORETICAL STUDY. 0299
STANTS IN DIFLUORIDE MOLECULES. (OXYGEN DIFLUORIDE, NI/ A THEORETICAL STUDY OF THE BOND INTERACTION FORCE CON 0794
PHOSPHORUS TRIFLUORIDE AND THEIR GROUND IONIC STATES. THEORETICAL STUDY OF THE GEOMETRY OF PHOSPHINE AND 0287
E CARBON MONOFLUORIDE MOLECULE. THEORETICAL STUDY OF THE SPECTROSCOPIC STATES OF TH 1477
NSTANTS FOR POLYATOMIC MOLECULES. THEORY AND CALCULATION OF CENTRIFUGAL DISTORTION CO 0758
DE) EXTENDED HUCKEL THEORY AND THE SHAPE OF MOLECULES. (OXYGEN DIFLUORI 1475
BICENTRIC RESCALING OF CNDO/2 THEORY. APPLICATIONS TO INORGANIC FLUORIDES. 0765
. PART-2: DIRECT COMPARISO/ APPROXIMATE MOLECULAR ORBITAL THEORY. ESSENTIAL STRUCTURAL ELEMENTS- MO FORMALISM 1430
FNESS METHOD APPLIED TO XY4 MOLECULES IN THE LIGHT OF THE THEORY OF CHARACTERISTIC FREQUENCIES. /RESSIVE STIF 1337
THEORY OF INERT GAS MOLECULES. 0362
E MOLECULES CARBON TETRAFLUORIDE AND DICARBON HEXAFLUORI/ THEORY OF INTENSITIES IN THE INFRARED SPECTRA OF TH 0493
D A STUDY OF THE EMISSION OF IODINE MONOFLUORIDE. THEORY OF THE DISSOCIATION OF DIATOMIC MOLECULES AN 0707
UNDS. PART-3: AB INITIO STUDIES OF CHA/ MOLECULAR ORBITAL THEORY OF THE ELECTRONIC STRUCTURE OF ORGANIC COMPO 1340
CHEMICAL PREDICTIONS BY MO THEORY. THE NOBLE GAS HALIDES. 0527
/EPTAFLUORIDE AND AN APPLICATION OF THREE ATOM SCATTERING THEORY TO IODINE HEPTAFLUORIDE AND RHENIUM HEPTAFL/ 1275
EES K. RECONCILIATION OF THE ENTROPIES FROM MOLECULAR AND THERMAL DATA. /DYNAMIC FUNCTIONS FROM 5 TO 350 DEGR 0367
DE, TERBIUM TETRAFLUORIDE, AND P/ ELECTRIC DEFLECTION AND THERMAL DECOMPOSITION STUDIES ON CERIUM TETRAFLUORI 0394
INCIDENT SHOCK WAVES. THERMAL DISSOCIATION OF CHLORINE TRIFLUORIDE BEHIND 0543
IOXIDE. ELECTRONIC STRUCTURE AND ANOMALOUS THERMAL EXPANSION ON IRON DIFLUORIDE AND VANADIUM D 1460
TAHEDRAL HEXAFLUORIDES. MEAN AMPLITUDES OF THERMAL MOTION AND SHRINKAGES OF CHEMICAL BONDS: OC 0853
OM SPECTROSCOPIC DATA. PENTACOORDINATED MOLE/ ANISOTROPIC THERMAL MOTION OF TRIGONAL BIPYRAMIDAL MOLECULES FR 1301
PREPARATION AND SOME THERMAL PROPERTIES OF PURE XENON HEXAFLUORIDE.

T HEXAFLUORIDES.   THERMAL STABILITY AND REACTIVITY OF D- AND F ELEMEN 0630

REACTIVITY AND THERMAL STABILITY OF HEXAFLUORIDES. 0004

METAL HEXAFLUORIDES.   MEAN AMPLITUDES OF THERMAL VIBRATIONS AND THERMODYNAMIC PROPERTIES OF 1467

LUORIDE.   THERMALLY EXCITED EMISSION SPECTRUM OF BARIUM MONOF 0123

GANIC NITROGEN FLUORIDES. (NITROGEN TRIFLUORIDE) THERMOCHEMICAL AND THERMODYNAMIC PROPERTIES OF INOR 0696

N OF LANTHANIDE TRIFLUORIDES.   ANALYSES OF THERMOCHEMICAL ERRORS AND SYSTEMATICS IN SUBLIMATIO 1353

ENON TETRAFLUORIDE FROM MASS SPECTRA. THERMOCHEMICAL PROPERTIES OF XENON DIFLUORIDE AND X 1253

MASS SPECTROMETRIC AND THERMOCHEMICAL STUDIES OF THE MANGANESE FLUORIDES. 0612

ENON HEXAFLUORIDE.   THERMOCHEMICAL STUDIES OF XENON TETRAFLUORIDE AND X 1276

THERMOCHEMISTRY OF SILICON DIFLUORIDE. 0992

COMPOUNDS. PART-1: ENTHALPY AND ENTROPY OF VAPORIZATION / THERMODYNAMIC AND PHYSICAL PROPERTIES OF BERYLLIUM 0138

DES   VAPOR PRESSURES AND THERMODYNAMIC CALCULATIONS FOR SOME ACTINIDE FLUORI 1336

AND THE THE POSSIBILITY OF THEIR SYNTHESIS BY THE REDUC/ THERMODYNAMIC CHARACTERISTICS OF SCANDIUM DIHALIDES 0940

FORCE CONSTANTS AND THERMODYNAMIC CONSTANTS FOR THE BORON HALIDES. 0219

NCIES OF ZIRCONIUM FLUORIDE (GAS) VIBRATIONS ACCORDING TO THERMODYNAMIC DATA.   FREQUE 1329

THERMODYNAMIC DATA FOR BERYLLIUM DIFLUORIDE. 0136

S OF THE GRAPHITE FLUORINE COMPOUNDS OF CARBON MONOFLUO/ THERMODYNAMIC, ELECTROCHEMICAL AND SYNTHETIC STUDIE 0296

OF MOLECULES OF SOME TETRAFLUORIDES FOR CALCULATING THEIR THERMODYNAMIC FUNCTIONS. /TATIONAL CHARACTERISTICS 0622

ES OF/ POTENTIAL CONSTANTS, MEAN AMPLITUDES OF VIBRATION, THERMODYNAMIC FUNCTIONS AND MOLECULAR POLARIZABILIT 1437

D IODINE HEPTAFLUORIDE.   POTENTIAL CONSTANTS AND THERMODYNAMIC FUNCTIONS FOR IODINE PENTAFLUORIDE AN 0509

OXYGEN DIFLUORIDE.   MEAN AMPLITUDES OF VIBRATION AND THERMODYNAMIC FUNCTIONS FOR SILICON DIFLUORIDE AND 0773

S.   MEAN AMPLITUDES OF VIBRATION AND THERMODYNAMIC FUNCTIONS FOR SOME METAL HEXAFLUORIDE 1459

THERMODYNAMIC FUNCTIONS FOR SULFUR HEXAFLUORIDE. 1089

MEAN AMPLITUDES OF VIBRATION AND THERMODYNAMIC FUNCTIONS FOR VANADIUM PENTAFLUORIDE. 1216

ARSENIC HALIDES. (PHOSPHO/ RAMAN AND INFRARED SPECTRA AND THERMODYNAMIC FUNCTIONS FOR VARIOUS PHOSPHORUS AND 0811

/Y, DIPOLE MOMENT, SPECTROSCOPIC CONSTANTS, AND IDEAL GAS THERMODYNAMIC FUNCTIONS FROM A HARTREE-FOCK MOLECU/ 0094

CONCILIATION OF T/ XENON TETRAFLUORIDE. HEAT CAPACITY AND THERMODYNAMIC FUNCTIONS FROM 5 TO 350 DEGREES K. RE 1275

ATION, CORIOLIS COUPLING CONSTANTS, SHRINKAGE EFFECT, AND THERMODYNAMIC FUNCTIONS OF ALUMINUM TRIFLUORIDE. /R 0050

UM BROMIDE, TANTALUM CHLORIDE, TANTALUM BROMIDE, AND MOL/ THERMODYNAMIC FUNCTIONS OF ANTIMONY FLUORIDE, NIOBI 0075

.   THERMODYNAMIC FUNCTIONS OF BISMUTH TRIHALIDE VAPORS 0149

POTENTIAL CONSTANTS AND THERMODYNAMIC FUNCTIONS OF CARBON TETRAFLUORIDE. 0350

CULES IN TRIPLET SIGMA STATES. (PHOSPHORUS MONOFLUORIDE) THERMODYNAMIC FUNCTIONS OF DIATOMIC GASES WITH MOLE 0789

M, AND LANTHANUM HALIDES IN THE 293-3000 DEGREES K RANGE. THERMODYNAMIC FUNCTIONS OF GASEOUS SCANDIUM, YTTRIU 0568

FORCE CONSTANTS AND THERMODYNAMIC FUNCTIONS OF IODINE HEPTAFLUORIDE. 0528

POTENTIAL CONSTANTS AND THERMODYNAMIC FUNCTIONS OF IODINE PENTAFLUORIDE. 0507

BROMINE PENTA/ MOLECULAR STRUCTURE, FORCE CONSTANTS, AND THERMODYNAMIC FUNCTIONS OF IODINE PENTAFLUORIDE AND 0505

UDES OF VIBRATION, BASTIANSEN-MORINO SHRINKAGE EFFECT AND THERMODYNAMIC FUNCTIONS OF KRYPTON DIFLUORIDE. /LIT 0562

THERMODYNAMIC FUNCTIONS OF LEAD TETRAHALIDES. 0597

/LPIES OF TRANSITION BETWEEN VARIOUS FORMS OF PHOSPHORUS. THERMODYNAMIC FUNCTIONS OF PHOSPHORUS PENTAFLUORID/ 0865

RIDES. (YTTRI/ HIGH TEMPERATURE HEAT CONTENTS AND RELATED THERMODYNAMIC FUNCTIONS OF SEVEN RARE EARTH TRIFLUO 1320

INFRARED SPECTRUM, FORCE CONSTANTS, AND THERMODYNAMIC FUNCTIONS OF SILICON DIFLUORIDE. 0985

(NITROGEN TRIFLUORIDE,/ MEAN AMPLITUDES OF VIBRATION AND THERMODYNAMIC FUNCTIONS OF SOME GROUP-5 TRIHALIDES. 0719

MOLECULES. (XENON DIFLUORIDE)   MOLECULAR CONSTANTS AND THERMODYNAMIC FUNCTIONS OF SOME LINEAR SYMMETRICAL 1244

S. (PHOSPHORUS TRIFLUORIDE) THERMODYNAMIC FUNCTIONS OF SOME PHOSPHORUS COMPOUND 0812

L PENTAHALIDES. THERMODYNAMIC FUNCTIONS OF SOME TRIGONAL BIPYRAMIDA 1438

C TRIFLUORIDE)   MEAN AMPLITUDES AND THERMODYNAMIC FUNCTIONS OF SOME TRIHALIDES. (ARSENI 0109

MEAN AMPLITUDES OF VIBRATION AND THERMODYNAMIC FUNCTIONS OF TECHNETIUM HEXAFLUORIDE. 1104

RADICAL.   INFRARED SPECTRUM AND THERMODYNAMIC FUNCTIONS OF THE NITROGEN DIFLUORIDE 0686

DIUM, CERIUM, SA/ HIGH TEMPERATURE ENTHALPIES AND RELATED THERMODYNAMIC FUNCTIONS OF THE TRIFLUORIDES OF SCAN 1356

/SES, CORIOLIS CONSTANTS, FORCE CONSTANTS, AND VALUES FOR THERMODYNAMIC FUNCTIONS OF THE TRIHALIDES OF GROUP/ 0795

/OF VIBRATION, SHRINKAGE EFFECTS, CORIOLIS CONSTANTS, AND THERMODYNAMIC FUNCTIONS OF VANADIUM PENTAFLUORIDE. 1219

5 TO 350 DEGREES K.   HEAT CAPACITY AND OTHER THERMODYNAMIC FUNCTIONS OF XENON HEXAFLUORIDE FROM 1305

HEAT CAPACITY AND RELATED THERMODYNAMIC FUNCTIONS OF XENON TETRAFLUORIDE. 1271

MEAN AMPLITUDES OF VIBRATION AND THERMODYNAMIC FUNCTIONS OF XENON TETRAFLUORIDE. 1274

ND VAPORIZATION, STANDARD ENTROPY OF THE VAPOR, AND OTHER THERMODYNAMIC PROPERTIES. / K, ENTHALPY OF FUSION A 0510

, SELENIUM HEXAFLUORIDE, TELLURIUM HEXAFLUORIDE AND THEIR THERMODYNAMIC PROPERTIES. /N OF SULFUR HEXAFLUORIDE 1092

E. HEAT CAPACITY FROM 5 TO 350 DEGREES K AND SOME DERIVED THERMODYNAMIC PROPERTIES.   XENON DIFLUORID 1246

OXIDE FROM 4 TO 25 DEGREES K. DERIVED ENTROPIES AND OTHER THERMODYNAMIC PROPERTIES AT 298.15 DEGREES K. /4 DI 0885

OLECULES.   UREY-BRADLEY FORCE FIELD AND THERMODYNAMIC PROPERTIES FOR BENT SYMMETRICAL XY2 M 1387

SPECTROSCOPIC AND THERMODYNAMIC PROPERTIES OF ALUMINUM MONOFLUORIDE. 0022

MASS SPECTROMETRIC INVESTIGATION OF THE THERMODYNAMIC PROPERTIES OF ALUMINUM TRIFLUORIDE. 0040

THERMODYNAMIC PROPERTIES OF ANTIMONY TRIFLUORIDE. 0072

THERMODYNAMIC PROPERTIES OF BISMUTH TRIFLUORIDE. 0148

THERMODYNAMIC PROPERTIES OF BORON HALIDES. 0240

RAMAN SPECTRUM, STRUCTURE, FORCE CONSTANTS, AND THERMODYNAMIC PROPERTIES OF BROMINE PENTAFLUORIDE. 0267

CARBON DIFLUORIDE FROM MOLECULAR FLOW EFFUSION AND MASS/ THERMODYNAMIC PROPERTIES OF CARBON MONOFLUORIDE AND 0286

THERMODYNAMIC PROPERTIES OF CARBON TETRAFLUORIDE. 0321

OM 4 DEGREES K TO ITS MELTING POINT. THERMODYNAMIC PROPERTIES OF CARBON TETRAFLUORIDE FR 0326

OM 12 DEGREES K TO ITS BOILING POINT. THE SIGNIFICANCE O/ THERMODYNAMIC PROPERTIES OF CARBON TETRAFLUORIDE FR 0361

BROMINE PENTAFLUORIDE, AND IODINE PENTAFLUORI/ IDEAL GAS THERMODYNAMIC PROPERTIES OF CHLORINE PENTAFLUORIDE, 0424

D BROMINE TRIFLUORIDE.   VIBRATIONAL SPECTRA AND THERMODYNAMIC PROPERTIES OF CHLORINE TRIFLUORIDE AN 0396

THERMODYNAMIC PROPERTIES OF CHROMIUM DIFLUORIDE. 0432

ANIUM TETRAFLUORIDE, GERMANIUM DIFLUORIDE, AND GERMANIUM/ THERMODYNAMIC PROPERTIES OF GASEOUS GERMANIUM, GERM 0468

ORIDE.   THERMODYNAMIC PROPERTIES OF GASEOUS URANIUM HEXAFLU 1198

STATISTICAL THERMODYNAMIC PROPERTIES OF HEXAFLUORIDE MOLECULES. 1449

RIDES. (NITROGEN TRIFLUORIDE)   THERMOCHEMICAL AND THERMODYNAMIC PROPERTIES OF INORGANIC NITROGEN FLUO 0696

POTENTIAL CONSTANTS AND THERMODYNAMIC PROPERTIES OF IODINE PENTAFLUORIDE. 0508

MEAN AMPLITUDES OF THERMAL VIBRATIONS AND THERMODYNAMIC PROPERTIES OF METAL HEXAFLUORIDES. 1467

BETWEEN 4 AND 350 DEGREES K.   HEAT CAPACITY AND OTHER THERMODYNAMIC PROPERTIES OF MOLYBDENUM HEXAFLUORIDE 0636

FROM 7 TO 350 DEG/ HEAT CAPACITY, ENTHALPY OF FUSION, AND THERMODYNAMIC PROPERTIES OF NEPTUNIUM HEXAFLUORIDE 0650

UORIDES AT HIGH TEMPERATURES. PART-3: DIFLUORIDES AND MO/ THERMODYNAMIC PROPERTIES OF NIOBIUM AND TANTALUM FL 0654

UORIDES AT HIGH TEMPERATURES. PART-2: TETRAFLUORIDES AND/ THERMODYNAMIC PROPERTIES OF NIOBIUM AND TANTALUM FL 0657

UORIDES AT HIGH TEMPERATURES. PART-1: PENTAFLUORIDES. THERMODYNAMIC PROPERTIES OF NIOBIUM AND TANTALUM FL 0665

NTAFLUORIDES.   THERMODYNAMIC PROPERTIES OF NIOBIUM AND TANTALUM PE 0669

IOBIUM DIFLUORIDE, TANTALUM MONOFLUORIDE, AND TANTALUM D/ THERMODYNAMIC PROPERTIES OF NIOBIUM MONOFLUORIDE, N 0655

OBIUM TETRAFLUORIDE, TANTALUM TRIFLUORIDE, AND TANTALUM / THERMODYNAMIC PROPERTIES OF NIOBIUM TRIFLUORIDE, NI 0656

, AND MOLYBDENUM (V) FLUORIDES.   THERMODYNAMIC PROPERTIES OF NIOBIUM (V), TANTALUM (V) 0668

THERMODYNAMIC PROPERTIES OF NITROGEN TRIFLUORIDE. 0739

TEMPERATURES. PART-1: CALCULATION METHODS. (CHLORINE PEN/ THERMODYNAMIC PROPERTIES OF PENTAFLUORIDES AT HIGH 0422

TEMPERATURES. PART-3: PENTAFLUORIDES OF SOME 5D-ELEMENTS. THERMODYNAMIC PROPERTIES OF PENTAFLUORIDES AT HIGH 0536

TEMPERATURES. PART-4: PENTAFLUORIDES OF 5F ELEMENTS. THERMODYNAMIC PROPERTIES OF PENTAFLUORIDES AT HIGH 0647
TEMPERATURES. PART-2: PENTAFLUORIDES OF SOME 4P- AND 4D / THERMODYNAMIC PROPERTIES OF PENTAFLUORIDES AT HIGH 1434
US(4), AND SOME PHOSPHORUS FLUORIDES. (PHOSPHORUS MONOFL/ THERMODYNAMIC PROPERTIES OF PHOSPHORUS(2), PHOSPHOR 0788
S.   UREY-BRADLEY FORCE FIELD AND THERMODYNAMIC PROPERTIES OF PRYRAMIDAL XY3 MOLECULE 1406
                                                               THERMODYNAMIC PROPERTIES OF SCANDIUM TRIFLUORIDE. 0950
(ARSENIC MONOFLUORIDE, ARSENIC DIFLUORIDE, ARSENIC TRIF/ THERMODYNAMIC PROPERTIES OF SOME ARSENIC FLUORIDES. 0093
                                                               THERMODYNAMIC PROPERTIES OF SOME BORON TRIHALIDES. 0236
S.   THERMODYNAMIC PROPERTIES OF SOME CHALCOGEN FLUORIDE 0012
(CHLORINE FLUORIDE)   PHOTOIONIZATION STUDIES AND THE THERMODYNAMIC PROPERTIES OF SOME HALOGEN MOLECULES. 0377
                                                               THERMODYNAMIC PROPERTIES OF SULFUR HEXAFLUORIDE. 1087
EVIEW)   PHYSICAL AND THERMODYNAMIC PROPERTIES OF SULFUR HEXAFLUORIDE. (R 1062
D SULFUR OXIDE TETRAFLUORIDE.   THERMODYNAMIC PROPERTIES OF SULFUR TETRAFLUORIDE AN 1051
5 TO 445 DEGREES K.   HEAT CAPACITY AND THERMODYNAMIC PROPERTIES OF THALLIUM FLUORIDE FROM 1124
RIDE SYSTEM FROM MASS SPECTROMETER INVESTIGATIONS. (BORO/ THERMODYNAMIC PROPERTIES OF THE BORON CHLORIDE FLUO 0179
ENS FROM SPECTROSCOPIC DATA. (CHLORINE, BROMINE, AND IOD/ THERMODYNAMIC PROPERTIES OF THE DIATOMIC INTERHALOG 0253
                          CARBON TETRAFLUORIDE. THERMODYNAMIC PROPERTIES OF THE REAL GAS. 0332
LUORIDES AT HIGH TEMPERATURE. (THIRD ROW)   THERMODYNAMIC PROPERTIES OF TRANSITION METAL PENTAF 1379
                                                               THERMODYNAMIC PROPERTIES OF TUNGSTEN PENTAFLUORIDE. 1149
                                                               THERMODYNAMIC PROPERTIES OF URANIUM HEXAFLUORIDE. 1204
HIGH TEMPERATURES.   THERMODYNAMIC PROPERTIES OF URANIUM HEXAFLUORIDE AT 1193
S.   VIBRATIONAL SPECTRUM AND THERMODYNAMIC PROPERTIES OF URANIUM HEXAFLUORIDE GA 1186
S.   VIBRATIONAL SPECTRUM AND THERMODYNAMIC PROPERTIES OF URANIUM HEXAFLUORIDE GA 1187
NEPTUNIUM PENTAFLUORIDE, AND PLUTONIUM PENTAFLUORIDE. THERMODYNAMIC PROPERTIES OF URANIUM PENTAFLUORIDE, 1177
TETRAFLUORIDE AND XENON HEXAFLUORIDE)   IDEAL GAS THERMODYNAMIC PROPERTIES OF XENON COMPOUNDS. (XENON 1273
STATE OF HEXAFLUORIDES. (OF SULFUR, MOLYBDENUM TUNGSTEN/ THERMODYNAMIC SIMILARITY AND UNIVERSAL EQUATIONS OF 0632
MPERATURES. PART-2: PLUTONIUM HEXAFLUORIDE. PART-3: URAN/ THERMODYNAMIC STABILITY OF HEXAFLUORIDES AT HIGH TE 0629
                                                               THERMODYNAMIC STABILITY OF URANIUM HEXAFLUORIDE. 1203
/TROSCOPIC STUDIES OF VIBRATIONAL CONSTANTS, STATISTCIAL THERMODYNAMICS AND QUANTUM MECHANICAL STUDIES OF P/ 0784
                          EFFUSION STUDIES, MASS SPECTRA, AND THERMODYNAMICS OF BERYLLIUM FLUORIDE VAPOR. 0140
                                                   VAPORIZATION THERMODYNAMICS OF INDIUM TRIFLUORIDE. 0492
OM MEASUREMENTS OF THE DISPROPORTIONATION PRESSURES. (UR/ THERMODYNAMICS OF INTERMEDIATE URANIUM FLUORIDES FR 1176
                          VAPORIZATION AND SUBLIMATION THERMODYNAMICS OF SELECTED LANTHANIDE FLUORIDES. 1347
                                                   STATISTICAL THERMODYNAMICS OF SOME HEXAFLUORIDES. 1463
/ ENTHALPY OF SUBLIMATION OF GERMANIUM DIFLUORIDE AND THE THERMODYNAMICS OF SUBLIMATION OF THE GROUP-4A DIFLU 0471
RIDE AND CESIUM FLUORIDE.   THERMODYNAMICS OF THE VAPORIZATION OF ALUMINUM FLUO 0052
E)   VIBRATIONAL SPECTRA AND THERMODYNAMICS OF THE ZINC HALIDES. (ZINC DIFLUORID 1323
RIDE.   THERMODYNAMICS OF VAPORIZATION OF THALLIUM MONOFLUO 1121
                          SUBLIMATION THERMODYNAMICS OF ZINC(II) FLUORIDE. 1321
PHORUS TRIFLUORIDE, NITROGEN TRIFLUORIDE, AND PHOSPHORYL, THIONYL AND SULFURYL FLUORIDES. /LE MOMENTS IN PHOS 0744
CHEMICAL SHIFTS OF BORON, PHOSPHORUS, PHOSPHORYL, AND THIOPHOSPHORYL HALIDES. 0235
TIONS OF PHOSPHORUS TRIFLUORIDE IN THE NU/ EFFECTS OF THE THIRD ORDER CONSTANTS ON THE L-TYPE DOUBLING TRANSI 0804
ES. (OXYGEN DIFLUORIDE)   THIRD ORDER POTENTIAL CONSTANTS OF BENT XY2 MOLECUL 0770
OF 5D TRANSITION METAL HEXAFLUORIDES IN THE VAPOR STATE. (THIRD ROW)   RAMAN SPECTRA 1378
OF TRANSITION METAL PENTAFLUORIDES AT HIGH TEMPERATURE. (THIRD ROW)   THERMODYNAMIC PROPERTIES 1379
DIES OF SOME TRANSITION METAL HEXAFLUORIDES. (SECOND ROW, THIRD ROW, AND RHENIUM HEPTAFLUORIDE) /FRACTION STU 1375
/ROMETER TO THE STUDY OF REACTIVE FLUORIDES. (SECOND ROW, THIRD ROW, ANTIMONY PENTAFLUORIDE, AND BISMUTH PEN/ 1377
HALIDES OF THE SECOND AND THIRD ROW TRANSITION METALS. 1374
/UME-1: FIRST ROW TRANSITION METALS. VOLUME-2: SECOND AND THIRD ROW TRANSITION METALS. VOLUME-3: LANTHANIDES/ 0001
ICIUM. (ACTINIDE)   HEATS OF FORMATION OF SOME HALIDES OF THORIUM, PROTACTINIUM, URANIUM, NEPTUNIUM, AND AMER 1335
                          VAPOR PRESSURE OF THORIUM TETRAFLUORIDE. 1129
                          ENTHALPY OF FORMATION OF THORIUM TETRAFLUORIDE BY FLUORINE BOMB CALORIMETRY. 1130
RAFLUORIDE AND/ ELECTRON DIFFRACTION STUDY OF URANIUM AND THORIUM TETRAHALIDES IN THE GAS PHASE. (URANIUM TET 1170
S SPECTROMETRIC STUDIES.   HEATS OF FORMATION GASEOUS THORIUM TRIFLUORIDE AND THORIUM DIFLUORIDE FROM MAS 1128
ARIUM, EUROPIUM, GADOLINIUM, TERBIUM, DYSPROSIUM, ERBIUM, THULIUM, AND YTTERBIUM. /S OF SCANDIUM, CERIUM, SAM 1356
ETIUM / SUBLIMATION PRESSURES AND HEATS OF SUBLIMATION OF THULIUM TRIFLUORIDE, YTTERBIUM TRIFLUORIDE, AND LUT 1132
                          VAPORIZATION REACTIONS IN THE THULIUM- FLUORINE SYSTEM. 1131
                          NUCLEAR RELAXATION TIMES OF LANTHANUM TRIFLUORIDE. 0579
/HE DEFORMATION VIBRATIONS OF THE DIHALIDES OF GERMANIUM, TIN, AND LEAD. (GERMANIUM DIFLUORIDE, TIN DIFLUORI/ 0479
RAMAN SPECTRA OF MH4 (M= CARBON, SILICON, GERMANIUM, AND TIN) AND MF4 (M= CARBON, SILICON, AND GERMANIUM). / 0314
                          ENVIRONMENT OF THE TIN ATOM IN ORTHORHOMBIC TIN DIFLUORIDE. 1139
ELTING POINT, VAPOR PRESSURE, AND HEAT OF VAPORIZATION OF TIN DIFLUORIDE.   M 1140
                          INFRARED SPECTRA OF MATRIX ISOLATED TIN DIFLUORIDE AND LEAD DIFLUORIDE. 0594
                          ULTRAVIOLET ABSORPTION SPECTRA OF GASEOUS TIN DIFLUORIDE AND LEAD DIFLUORIDE. 0595
AND LE/ MASS SPECTROMETRIC STUDIES AT HIGH TEMPERATURES. (TIN DIFLUORIDE, LEAD DIFLUORIDE, TIN MONOFLUORIDE, 0592
F THE VIBRATIONAL WAVE FUNCTIONS AND DERIVED QUANTITIES. (TIN FLUORIDE)   ACCURACY O 1134
CULES. (BERYLLIUM MONOFLUORIDE, SILICON MONOFLUORIDE, AND TIN MONOFLUORIDE) /AL CURVES FOR SOME DIATOMIC MOLE 0130
DOUBLET PI, B DOUBLET SIGMA(+) TO X DOUBLET PI SYSTEMS OF TIN MONOFLUORIDE. / OF THE A DOUBLET SIGMA(+) TO X 1133
WITH THE INTERNUCLEAR SEPARATION IN SILICON CHLORIDE AND TIN MONOFLUORIDE. / OF ELECTRONIC TRANSITION MOMENT 1137
SPECTRA OF SILICON MONOFLUORIDE, GERMANIUM MONOFLUORIDE, TIN MONOFLUORIDE, AND LEAD MONOFLUORIDE. 0964
ANALYSIS OF VIBRATIONAL AND ROTATIONAL STRUCTURE OF TIN MONOFLUORIDE IN THE 2500 ANGSTROM REGION. 1138
AND G DOUBLET DELTA 5/2- X DOUBLET PI 3/2 SYSTEMS OF THE TIN MONOFLUORIDE MOLECULE. /A 2- X DOUBLET PI 3/2 1136
) BAND OF THE A DOUBLET SIGMA- X DOUBLET PI SYSTEM OF THE TIN MONOFLUORIDE MOLECULE. /TOPIC SHIFT IN THE (1,0 1135
FLUORIDE, SILICON TETRAFLUORIDE, GERMANIUM TETRAFLUORIDE, TIN TETRAFLUORIDE) /IDES OF GROUP-5B. (CARBON TETRA 0316
OF THE MX BOND IN THE TETRAHALIDES OF GROUP-4 ELEMENTS. (TIN TETRAFLUORIDE AND HAFNIUM TETRAFLUORIDE) /ICITY 1141
E GAS MATRICES.   INFRARED SPECTRA AND GEOMETRY OF TITANIUM DIFLUORIDE AND TITANIUM TRIFLUORIDE IN RAR 1146
RIDE, DIFLUORIDE, AND MANGANESE DIFLUOR/ RAMAN SPECTRA OF TITANIUM DIOXIDE, MAGNESIUM DIFLUORIDE, ZINC DIFLUO 0615
, AND FERROUS F/ FORCE FIELDS OF THE RUTILE COUNTERPARTS: TITANIUM DIOXIDE, MAGNESIUM FLUORIDE, ZINC FLUORIDE 0606
                          ABSORPTION SPECTRUM OF VAPORIZED TITANIUM MONOFLUORIDE. 1143
                          EMISSION SPECTRUM OF TITANIUM MONOFLUORIDE. 1142
                          SIMPLE THEORETICAL CALCULATIONS ON TITANIUM MONOFLUORIDE. 1144
                          VAPOR PRESSURE OF TITANIUM TETRAFLUORIDE. 1148
                          HEATS OF FORMATION OF TITANIUM TETRAFLUORIDE AND HAFNIUM TETRAFLUORIDE. 1147
/LUORIDE, SILICON TETRACHLORIDE, GERMANIUM TETRACHLORIDE, TITANIUM TETRAFLUORIDE, TITANIUM TETRACHLORIDE, VA/ 0349
/AT HIGH TEMPERATURES. PART-16: SUBLIMATION PRESSURES FOR TITANIUM TRIFLUORIDE AND THE STABILITIES OF TITANI 1145
CORIOLIS ZETA SUMS FOR SOME SPHERICAL TOP MOLECULES. 1487
ATION STUDY OF THE SPIN ROTATION INTERACTION IN SYMMETRIC TOP MOLECULES.   NUCLEAR SPIN RELAX 0183
ATION STUDY OF THE SPIN ROTATION INTERACTION IN SPHERICAL TOP MOLECULES.   NUCLEAR SPIN RELAX 0323
LUO/ P-R SEPARATIONS AND CORIOLIS CONSTANTS FOR SYMMETRIC TOP MOLECULES AND FORCE CONSTANTS FOR NITROGEN TRIF 0708
INFRARED BAND CONTOURS. PART-1: SPHERICAL TOP MOLECULES. (CARBON TETRAFLUORIDE) 0325
OF VAPOR BAND CONTOURS./ CORIOLIS CONSTANTS OF SPHERICAL TOP MOLECULES FROM LOW TEMPERATURE INFRARED STUDIES 1156

ZEEMAN EFFECT OF SOME LINEAR AND SYMMETRIC TOP MOLECULES. (PHOSPHORUS TRIFLUORIDE) 0796
/ONTOUR ANALYSES, AND CORIOLIS CONSTANTS OF THE SPHERICAL TOP MOLECULES, SULFUR HEXAFLUORIDE, SELENIUM HEXAF/ 1069
TANTS FROM THE BAND CONTOURS OF HOT BANDS FOR A SYMMETRIC TOP. (NITROGEN TRIFLUORIDE) /ARMONIC POTENTIAL CONS 0733
THE GAS PHASE RAMAN SPECTRA OF MOLECULES OF THE SPHERICAL TOP TYPE. /TIONAL PROFILES OF VIBRATIONAL BANDS IN 1442
FAR INFRARED SPECTRA OF PROLATE AND OBLATE SYMMETRIC TOPS. 1479
HING OF TRIFLUORO CHLORO METHANE AND SOME OTHER SYMMETRIC TOPS. (ARSENIC TRIFLUORIDE) CENTRIFUGAL STRETC 0101
TRIFL/ PSEUDO-PARALLEL INFRARED BANDS OF OBLATE SYMMETRIC TOPS NITROGEN TRIFLUORIDE NU-4 AND NU-4 + NU-2 AND 0716
APPLICATION TO ETHYLENE. TORSIONAL COORDINATES IN VIBRATIONAL ANHARMONICITY. 1485
UREY-BRADLEY FORCE CONSTANTS OF BORON HALIDES AND THEIR TRANSFERABILITY. (BORON TRIFLUORIDE) 0246
NMENTS OF SOME MIXED HALOMETHANES. (CARBON TETRAFLUORI/ A TRANSFERABLE UREY-BRADLEY FORCE FIELD AND THE ASSIG 0351
D MOMENT / A GEOMETRIC VISUALIZATION OF NORMAL COORDINATE TRANSFORMATIONS. APPLICATION TO THE CALCULATION BON 0352
/ FORMATION OF PHOSPHORUS PENTAFLUORIDE AND ENTHALPIES OF TRANSITION BETWEEN VARIOUS FORMS OF PHOSPHORUS. TH/ 0865
URATIONS AND INTERATOMIC DISTANCES IN GROUP-5 AND GROUP-6 TRANSITION ELEMENT TETRAHALIDES AND PENTAHALIDES. / 1427
DRAL, TRIGONAL BIPYRAMID, AND RELATED MOLECULES OF THE NONTRANSITION ELEMENTS. /THS AND BOND ANGLES IN OCTAHE 1435
NS OF FLUORIDES, OXIDES, AND SULFIDES CONTAINING DIVALENT TRANSITION ELEMENTS. COMPARISO 1370
/TERISTICS OF THE ELECTRONIC STRUCTURE OF FLUORIDES OF NONTRANSITION ELEMENTS FROM THE DATA OF LCAO MO SCF N/ 0009
METALS. VOLUME-2: SECOND AND THIRD ROW T/ HALIDES OF THE TRANSITION ELEMENTS. VOLUME-1: FIRST ROW TRANSITION 0001
UORID/ OPTICAL SATURATION OF A SINGLE VIBRATION- ROTATION TRANSITION IN THE NU-3 FUNDAMENTAL OF SULFUR HEXAFL 1074
ONAL FREQUENCIES AND VALENCE FORCE CONSTANTS OF FIRST ROW TRANSITION METAL DIFLUORIDES. VIBRATI 1371
FIRST ROW TRANSITION METALS. GEOMETRY OF THE TRANSITION METAL DIHALIDES. THE DIFLUORIDES OF SOME 1368
BERYLLIUM DIFLUORIDE, MAGNESIUM DIFLUORIDE, AND FIRST ROW TRANSITION METAL FLUORIDES) /MMETRICAL DIHALIDES. (1372
TRANSITION METAL FLUORIDES AND THEIR COMPLEXES. 0015
(THIRD ROW) RAMAN SPECTRA OF 5D TRANSITION METAL HEXAFLUORIDES IN THE VAPOR STATE. 1378
DE, OSMIUM HEXAFLUORIDE, IRIDIUM HEXAFLUORIDE,/ COLORS OF TRANSITION METAL HEXAFLUORIDES. (RHENIUM HEXAFLUORI 0916
ROW, AND RHENIUM HEPTA/ X-RAY DIFFRACTION STUDIES OF SOME TRANSITION METAL HEXAFLUORIDES. (SECOND ROW, THIRD 1375
IDE, MOLYBDENUM HEXAFLUORI/ VAPOR PRESSURES OF SOME HEAVY TRANSITION METAL HEXAFLUORIDES. (TUNGSTEN HEXAFLUOR 0903
DS. TRANSITION METAL PENTAFLUORIDES AND RELATED COMPOUN 0013
(THIRD ROW) THERMODYNAMIC PROPERTIES OF TRANSITION METAL PENTAFLUORIDES AT HIGH TEMPERATURE 1379
HALIDES OF THE SECOND AND THIRD ROW TRANSITION METALS. 1374
PENTAHALIDES OF TRANSITION METALS. 1367
/R SOME LINEAR SYMMETRICAL DIFLUORIDES. (OF THE FIRST ROW TRANSITION METALS, ALSO BERYLLIUM DIFLUORIDE AND M/ 1369
FLUORIDES OF THE HEAVY TRANSITION METALS. (REVIEW) 0002
SEPARATION FOR THE B-X BAND SYSTEM OF MOLECU/ ELECTRONIC TRANSITION MOMENT AS A FUNCTION OF THE INTERNUCLEAR 0967
NDS OF THE X SYSTEM. CORRELATION OF THE ELECTRONIC TRANSITION MOMENT WITH INTERNUCLEAR DISTANCE FOR BA 0969
IN SILICON CHLORIDE AND TIN MONO/ VARIATION OF ELECTRONIC TRANSITION MOMENT WITH THE INTERNUCLEAR SEPARATION 1137
MICROWAVE MEASUREMENTS OF THE J-8 FAR-9 TRANSITION OF BROMINE PENTAFLUORIDE. 0261
ENTAFLUORIDE. HEAT CAPACITY, ENTROPY, AND HEATS OF TRANSITION OF MOLYBDENUM HEXAFLUORIDE AND NIOBIUM P 0662
JAHN-TELLER EFFECTS IN THE 2E5/2G TO 2G3/2G TRANSITION OF RHENIUM HEXAFLUORIDE. 0913
A QUARTET SIGMA TO X DOUBLET PI ELECTRONIC TRANSITION OF SILICON MONOFLUORIDE. 0971
A NEW EXTENDED TRANSITION OF THE MERCURY MONOFLUORIDE MOLECULE. 0618
A QUARTET SIGMA TO DOUBLET PI TRANSITION OF THE SILICON MONOFLUORIDE MOLECULE. 0978
VAPOR PRESSURE AND TRANSITION POINTS OF TECHNETIUM HEXAFLUORIDE. 1105
IC FLUORIDES OF SILICON AND GERMANIUM. VIBRATIONAL TRANSITION PROBABILITIES AND R CENTROIDS FOR DIATOM 0994
N MONOFLUORIDE. VIBRATIONAL TRANSITION PROBABILITIES FOR THE A-X SYSTEM OF BORO 0168
WITH OTHER HEXAVALENT XENON MOLECULES. PART-3: ELECTRONIC TRANSITIONS AND BAND SHAPES. / PART-2: COMPARISONS 1311
/RBITAL INTERACTION AND RELATIVE INTENSITIES OF FORBIDDEN TRANSITIONS IN IODINE, OR IODINE BROMIDE (CHLORIDE/ 0498
OFLUORIDE. ROTATIONAL ANALYSIS OF SOME SINGLET TRANSITIONS IN THE SPECTRUM OF GASEOUS SCANDIUM MON 0930
RIDE. FORBIDDEN ELECTRONIC TRANSITIONS IN XENON DIFLUORIDE AND XENON TETRAFLUORIDE 1247
DOUBLET SIGMA(+), AND A QUARTET SIGMA(-) TO X DOUBLET PI TRANSITIONS OF GERMANIUM MONOFLUORIDE. /SIGMA(+), B 0466
QUALS 1 STATE. L-TYPE DOUBLING TRANSITIONS OF PHOSPHORUS TRIFLUORIDE IN THE NU-4 E 0803
/ECTS OF THE THIRD ORDER CONSTANTS ON THE L-TYPE DOUBLING TRANSITIONS OF PHOSPHORUS TRIFLUORIDE IN THE NU-4 / 0804
/CAL CALCULATIONS OF THE ELECTRONIC STRUCTURE AND OPTICAL TRANSITIONS OF URANIUM HEXAFLUORIDE AND URANIUM PE/ 1179
OLE MOMENT OF AN A DOUBLET SIGMA TO X DOUBLET PI ELECTRON TRANSITION IN A SILICON MONOFLUORIDE MOLECULE. / DIP 0970
HALPY OF BERYLLIUM FLUORIDE FROM 456 TO 1083 DEGREES K BY TRANSPOSED TEMPERATURE DROP CALORIMETRY. ENT 0142
ABSORPTION SPECTRUM OF CARBON DIFLUORIDE TRAPPED IN AN ARGON MATRIX. 0297
LUORIDE RADICALS. ELECTRON SPIN RESONANCE OF TRAPPED PHOSPHORUS DIFLUORIDE AND PHOSPHORUS TETRAF 0792
GE IN XENON TETRAFLUORIDE. ELECTRON SPIN RESONANCE OF THE TRAPPED RADICAL XENON MONOFLUORIDE. RADIATION DAMA 1220
ON OF IODINE HEPTAFLUORIDE AN/ STRUCTURE DETERMINATION OF TRIS (TRIMETHYL SILYL) SILANE, INFRARED INVESTIGATI 0527
IFLUORIDE, CARBON DIFLUORIDE/ MOLECULAR PROPERTIES OF THE TRIATOMIC DIFLUORIDES BERYLLIUM DIFLUORIDE, BORON D 0141
NAL FORCE CONSTANTS OF NONPLANAR XY3 MOLECULES. (NITROGEN TRIFLUORIDE) A BENT BOND MODEL FOR THE VIBRATIO 0704
N DIHYDRIDE FLUORIDE, BORON HYDRIDE DIFLUORIDE, AND BORON TRIFLUORIDE. / STRUCTURES OF BORON TRIHYDRIDE, BORO 0241
OF INORGANIC MOLECULES NITROGEN OXYFLUORIDE AND NITROGEN TRIFLUORIDE. AB INITIO STUDY OF THE FORCE CONSTANTS 0734
RED INTENSITIES OF THE FUNDAMENTAL VIBRATIONS OF NITROGEN TRIFLUORIDE. ABSOLUTE INFRA 0736
HE NEAR ULTRAVIOLET REGION. (KRYPTON DIFLUORIDE, NITROGEN TRIFLUORIDE) /ION SPECTRA OF CERTAIN FLUORIDES IN T 0560
DEUTERIDES, AND HALIDES OF GROUP-5 ELEMENTS. (PHOSPHORUS TRIFLUORIDE) ABSORPTION SPECTRA OF THE HYDRIDES, 0806
AN ELECTRON DIFFRACTION INVESTIGATION OF NITROGEN TRIFLUORIDE. 0738
IC ROTOR RAMAN SPECTRA. PURE ROTATIONAL SPECTRUM OF BORON TRIFLUORIDE. ANALYSIS OF SYMMETR 0205
BAND CONTOURS OF HOT BANDS FOR A SYMMETRIC TOP. (NITROGEN TRIFLUORIDE) /ARMONIC POTENTIAL CONSTANTS FROM THE 0733
ANOMALOUS NMR SHIFT IN SAMARIUM TRIFLUORIDE. 0927
RIGIDITY METHOD TO SOME SIMPLE XYN TYPE MOLECULES. (BORON TRIFLUORIDE) APPLICATION OF THE PROGRESSIVE 0211
O ELECTRON PASSAGE THROUGH ELECTRONEGATIVE ATOMS IN BORON TRIFLUORIDE. BARRIER T 0196
N OF FORCE CONSTANTS FOR NITROGEN DIFLUORIDE AND NITROGEN TRIFLUORIDE. CALCULATIO 0693
D BAND CONTOURS: APPLICATION TO THE NU-4 BAND OF NITROGEN TRIFLUORIDE. CALCULATIO OF PERPENDICULAR INFRARE 0747
N OF THE FREQUENCIES OF THE NORMAL VIBRATIONS OF ANTIMONY TRIFLUORIDE. CALCULATIO 0074
CENTRIFUGAL DISTORTION CONSTANTS OF PHOSPHORUS TRIFLUORIDE. 0814
RO CHLORO METHANE AND SOME OTHER SYMMETRIC TOPS. (ARSENIC TRIFLUORIDE) CENTRIFUGAL STRETCHING OF TRIFLUO 0101
POLE MOMENT DERIVATIVES AND INFRARED INTENSITIES OF BORON TRIFLUORIDE. CNDO CALCULATION OF DI 0195
CONSTRUCTION OF HYBRID ORBITALS. (CHLORINE TRIFLUORIDE) 0405
CRYSTAL STRUCTURE OF MANGANESE TRIFLUORIDE. 0616
CRYSTAL STRUCTURE OF THALLIUM TRIFLUORIDE. 1127
CUBIC POTENTIAL CONSTANTS AND 1-TYPE RESONANCE IN BORON TRIFLUORIDE. 0231
UORIDE GAS AND DETERMINATION OF ITS STABILITY. (LANTHANUM TRIFLUORIDE) /F THE EXISTENCE OF DILANTHANUM HEXAFL 0501
LECTRON DIFFRACTION ANALYSIS OF GALLIUM HALIDES. (GALLIUM TRIFLUORIDE) E 0460
ELECTRON IMPACT STUDIES OF SOME TRIHALO METHANES. (CARBON TRIFLUORIDE) 0312
ELECTRONIC SPECTRUM OF LANTHANUM TRIFLUORIDE. 0582
ELECTRONIC STRUCTURE OF BORON TRIFLUORIDE. 0184
ELECTRONIC STRUCTURE OF IRON TRIFLUORIDE. 0542
ECTRONIC STRUCTURES OF CHLORINE MONOFLUORIDE AND CHLORINE TRIFLUORIDE. EL 0372
EMISSION SPECTRUM OF SILICON TRIFLUORIDE. 0998
DE, XENON TETRAFLUORIDE, XENON DIFLUORIDE, AND PHOSPHORUS TRIFLUORIDE. /PIES OF FORMATION OF XENON HEXAFLUORI 0807

ENTHALPY OF DISSOCIATION OF NITROGEN TRIFLUORIDE. 0742
ENTHALPY OF FORMATION OF GERMANIUM TRIFLUORIDE. 0480
ENTHALPY OF FORMATION OF YTTRIUM TRIFLUORIDE. 1319
ENTHALPY OF FORMATION PHOSPHORUS TRIFLUORIDE. 0818
ORIDE RADICALS PRODUCED IN A SINGLE CRYSTAL OF PHOSPHORUS TRIFLUORIDE. ESR SPECTRA OF PHOSPHORUS TETRAFLU 0823
THALPIES OF FORMATION OF RARE EARTH FLUORIDES. (ALSO LANTHANUM TRIFLUORIDE) EXPERIMENTAL DETERMINATION OF THE EN 0578
FAR INFRARED SPECTRUM OF NITROGEN TRIFLUORIDE. 0702
FERMI RESONANCE IN BORON TRIFLUORIDE. 0242
CALORIMETRY. PART-15: THE ENTHALPY OF FORMATION OF BORON TRIFLUORIDE. FLUORINE BOMB 0220
LORIMETRY. PART-22: THE ENTHALPY OF FORMATION OF ALUMINUM TRIFLUORIDE. FLUORINE BOMB CA 0048
BOMB CALORIMETRY. PART-3: THE HEAT OF FORMATION OF BORON TRIFLUORIDE. FLUORINE 0250
FORCE CONSTANT CALCULATION FOR CHLORINE TRIFLUORIDE. 0403
FORCE FIELD AND MOLECULAR CONSTANTS OF NITROGEN TRIFLUORIDE. 0735
CONSTANTS IN DICHLORO BORANE, DIBROMO BORANE AND CHLORINE TRIFLUORIDE. /KALIZED MEAN AMPLITUDES AND CORIOLIS 0411
RA OF MATRIX ISOLATED RARE EARTH HALIDES. (ALSO LANTHANUM TRIFLUORIDE) GEOMETRY AND INFRARED SPECT 0583
GROUND STATE PROPERTIES OF THE GROUP-2 TRIHALIDES. (BORON TRIFLUORIDE) 0185
CHLORINE MONOFLUORIDE, CHLORINE DIFLUORIDE, AND CHLORINE TRIFLUORIDE) /OF FORMATION OF CHLORINE FLUORIDES. (0370
ON STUDY OF L-TYPE RESONANCE IN THE NU-4 BAND OF NITROGEN TRIFLUORIDE. HIGH RESOLUTI 0731
AMIC FUNCTIONS OF SEVEN RARE EARTH TRIFLUORIDES. (YTTRIUM TRIFLUORIDE) /E HEAT CONTENTS AND RELATED THERMODYN 1320
INFRARED ABSORPTION SPECTRUM OF BORON TRIFLUORIDE. 0206
INFRARED ABSORPTION SPECTRUM OF LIQUID BORON TRIFLUORIDE. 0245
INFRARED AND RAMAN SPECTRUM OF NITROGEN TRIFLUORIDE. 0724
INFRARED BAND SHAPES AND VIBRATIONS OF NITROGEN TRIFLUORIDE. 0699
INFRARED INTENSITIES AND FORCE FIELD FOR NITROGEN TRIFLUORIDE. 0714
METRIES OF MATRIX ISOLATED YTTRIUM DIFLUORIDE AND YTTRIUM TRIFLUORIDE. INFRARED SPECTRA AND GEO 1316
ADIUM PENTAFLUORIDE, VANADIUM TETRAFLUORIDE, AND VANADIUM TRIFLUORIDE. INFRARED SPECTRA OF VAN 1210
INFRARED SPECTRUM AND MOLECULAR CONSTANTS OF NITROGEN TRIFLUORIDE. 0728
ED SPECTRUM AND VIBRATIONAL POTENTIAL FUNCTION OF ARSENIC TRIFLUORIDE. INFRAR 0100
INFRARED SPECTRUM OF BORON TRIFLUORIDE. 0249
INTERATOMIC BONDING IN CRYSTALLINE MANGANESE TRIFLUORIDE. 0617
ISOTOPIC SPLITTING IN MATRIX ISOLATED BORON TRIFLUORIDE. 0228
SER EXCITED RAMAN SPECTRA OF MATRIX ISOLATED PRASEODYMIUM TRIFLUORIDE. LA 0895
BRATIONS AND STRUCTURE OF RARE EARTH HALIDES. (LANTHANUM TRIFLUORIDE) LATTICE VI 0572
RESSURE AND CRITICAL TEMPERATURE AND PRESSURE OF NITROGEN TRIFLUORIDE. LIQUID DENSITY, VAPOR P 0710
AIR MODEL AND THE VIBRATIONAL FORCE CONSTANTS OF NITROGEN TRIFLUORIDE. LONE P 0705
MAGNETIC SUSCEPTIBILITY OF EUROPIUM TRIFLUORIDE. 0454
N OF THE VAPOR COMPOSITION AND VAPOR PRESSURE OF VANADIUM TRIFLUORIDE. MASS SPECTROMETRIC DETERMINATIO 1208
INVESTIGATION OF THE THERMODYNAMIC PROPERTIES OF ALUMINUM TRIFLUORIDE. MASS SPECTROMETRIC 0040
AND THERMODYNAMIC FUNCTIONS OF SOME TRIHALIDES. (ARSENIC TRIFLUORIDE) MEAN AMPLITUDES 0109
SHRINKAGE EFFECT, AND THERMODYNAMIC FUNCTIONS OF ALUMINUM TRIFLUORIDE. /RATION, CORIOLIS COUPLING CONSTANTS, 0050
AMPLITUDES OF VIBRATION OF SELENIUM TRIOXIDE AND CHLORINE TRIFLUORIDE. MEAN 0406
LATION OF THE ABSOLUTE INFRARED INTENSITIES OF PHOSPHORUS TRIFLUORIDE. MEASUREMENT AND CALCU 0808
. (CHLORINE MONOFLUORIDE, OXYGEN MONOFLUORIDE, PHOSPHORUS TRIFLUORIDE) /ED LIQUID AND SATURATED VAPOR ENTROPY 0391
STIGATIONS OF METHYL FLUORIDE, FLUOROFORM, AND PHOSPHORUS TRIFLUORIDE. MICROWAVE INVE 0798
MICROWAVE SPECTRUM AND MOLECULAR STRUCTURE OF BROMINE TRIFLUORIDE. 0258
MICROWAVE SPECTRUM AND STRUCTURE OF CHLORINE TRIFLUORIDE. 0410
MICROWAVE SPECTRUM OF ARSENIC TRIFLUORIDE. 0102
MICROWAVE SPECTRUM OF PHOSPHORUS TRIFLUORIDE. 0805
SPECTRUM AND CENTRIFUGAL DISTORTION CONSTANTS OF NITROGEN TRIFLUORIDE. MILLIMETER WAVE 0718
AR CONSTANTS OF PHOSPHORUS HALOGEN COMPOUNDS. (PHOSPHORUS TRIFLUORIDE) MOLECUL 0813
MOLECULAR FORCE FIELD FOR CHLORINE TRIFLUORIDE. 0402
DISTRIBUTION USING A MINIMAL SLATER-TYPE BASIS. (NITROGEN TRIFLUORIDE) / PART-3: AB INITIO STUDIES OF CHARGE 0707
MOLECULAR STRUCTURE OF BORON TRIFLUORIDE. 0187
NEAR ULTRAVIOLET OPTICAL CONSTANTS OF LANTHANUM TRIFLUORIDE. 0584
RON DIFFRACTION STUDY OF THE CRYSTAL STRUCTURE OF URANIUM TRIFLUORIDE. NEUT 1163
L COORDINATE ANALYSIS OF SOME NITROGEN HALIDES. (NITROGEN TRIFLUORIDE) NORMA 0730
NORMAL VIBRATIONS OF BORON TRIHALIDES. (BORON TRIFLUORIDE) 0247
NU-3 BAND OF PHOSPHORUS TRIFLUORIDE. 0815
NU-3 BANDS OF PHOSPHORUS TRIFLUORIDE. 0207
NU-3 OF BORON TRIFLUORIDE. 0200
NUCLEAR RELAXATION TIMES OF LANTHANUM TRIFLUORIDE. 0579
OPTICAL ABSORPTION AND FLUORESCENCE SPECTRA OF EUROPIUM TRIFLUORIDE. 0453
OSPHORUS PENTAFLUORIDE, SULFUR HEXAFLUORIDE, AND CHLORINE TRIFLUORIDE) /N PERFECT PAIRING VALENCE STATES. (PH 0862
PERPENDICULAR VIBRATIONS OF BORON TRIFLUORIDE. 0202
PHOTOELECTRON SPECTRUM OF BORON TRIFLUORIDE. 0189
CAL PROPERTIES OF INORGANIC NITROGEN FLUORIDES. (NITROGEN TRIFLUORIDE) PHYSICAL CHEMI 0713
POTENTIAL FUNCTION OF NITROGEN TRIFLUORIDE. 0737
PREPARATION AND CRYSTAL STRUCTURE OF GALLIUM TRIFLUORIDE. 0461
PREPARATION AND CRYSTAL STRUCTURE OF MOLYBDENUM TRIFLUORIDE. 0620
PROPERTIES OF PROMETHIUM COMPOUNDS. (PROMETHIUM TRIFLUORIDE) 0898
S FOR VARIOUS PHOSPHORUS AND ARSENIC HALIDES. (PHOSPHORUS TRIFLUORIDE) /ED SPECTRA AND THERMODYNAMIC FUNCTION 0811
RAMAN SPECTRUM OF ANTIMONY TRIFLUORIDE. 0073
RAMAN SPECTRUM OF GASEOUS NITROGEN TRIFLUORIDE. 0740
VIBRATION BAND CONTOURS IN THE INFRARED SPECTRUM OF BORON TRIFLUORIDE. ROTATION 0193
ONFIGURATION MOLECULAR ORBITAL CALCULATIONS ON PHOSPHORUS TRIFLUORIDE. SELF CONSISTENT CHARGE AND C 0821
BRIUM GEOMETRIES AND QUADRATIC FORCE CONSTANTS. (NITROGEN TRIFLUORIDE) /RT-5: AB INITIO CALCULATION OF EQUILI 0720
E ELECTRONIC STRUCTURE OF BORON HYDRIDE FLUORIDES. (BORON TRIFLUORIDE) SEMIEMPIRICAL STUDY OF TH 0226
SIGNS OF L DOUBLING CONSTANTS IN NITROGEN TRIFLUORIDE. 0743
STABILITIES OF TANTALUM PENTAFLUORIDE AND TANTALUM OXIDE TRIFLUORIDE. 1103
STANDARD ENTHALPY OF FORMATION OF ALUMINUM TRIFLUORIDE. 0043
STANDARD HEAT OF FORMATION OF ARSENIC TRIFLUORIDE. 0110
STRUCTURE AND SOME PROPERTIES OF SCANDIUM TRIFLUORIDE. 0945
STRUCTURE OF CRYSTALLINE CHROMIUM TRIFLUORIDE. 0433
STRUCTURE OF MOLECULAR URANIUM DIFLUORIDE AND URANIUM TRIFLUORIDE. 1162
, METHYL PHOSPHORUS TETRAFLUORIDE AND DIMETHYL PHOSPHORUS TRIFLUORIDE. STRUCTURES OF PHOSPHORUS PENTAFLUORIDE 0845
ECISE DETERMINATION OF THE FAR INFRARED SPECTRUM OF BORON TRIFLUORIDE. /Y OF THE ISOTOPE EFFECT AND A MORE PR 0230
CRYSTAL STRUCTURE OF RARE EARTH TRIFLUORIDES. (LANTHANUM TRIFLUORIDE) SYMMETRY AND 0571
THE INFRARED SPECTRUM AND FORCE CONSTANTS FOR ARSENIC TRIFLUORIDE. 0105
MIC PROPERTIES OF INORGANIC NITROGEN FLUORIDES. (NITROGEN TRIFLUORIDE) THERMOCHEMICAL AND THERMODYNA 0696
NAMIC FUNCTIONS OF SOME PHOSPHORUS COMPOUNDS. (PHOSPHORUS TRIFLUORIDE) THERMODY 0812

THERMODYNAMIC PROPERTIES OF ANTIMONY TRIFLUORIDE. 0072
THERMODYNAMIC PROPERTIES OF BISMUTH TRIFLUORIDE. 0148
THERMODYNAMIC PROPERTIES OF NITROGEN TRIFLUORIDE. 0739
THERMODYNAMIC PROPERTIES OF SCANDIUM TRIFLUORIDE. 0950
ULTRASOFT X-RAY ABSORPTION SPECTRA OF BORON TRIFLUORIDE. 0204
STANTS OF BORON HALIDES AND THEIR TRANSFERABILITY. (BORON TRIFLUORIDE) UREY-BRADLEY FORCE CON 0246
TS. ANHARMONICITY EFFECTS ON THE FORCE CONSTANTS OF BORON TRIFLUORIDE. / FOR THE CALCULATION OF FORCE CONSTAN 0234
AVIOLET ABSORPTION SPECTRUM AND DIPOLE MOMENT OF NITROGEN TRIFLUORIDE. VACUUM ULTR 0725
VAPOR PRESSURE AND HEAT OF SUBLIMATION OF CERIUM TRIFLUORIDE. 0366
OF SUBLIMATION, AND EVAPORATION COEFFICIENT OF LANTHANUM TRIFLUORIDE. VAPOR PRESSURE, HEAT 0576
LIMATION, AND THE EVAPORATION COEFFICIENT OF PRASEODYMIUM TRIFLUORIDE. VAPOR PRESSURE, HEAT OF SUB 0896
VAPOR PRESSURE OF ALUMINUM TRIFLUORIDE. 0041
VAPOR PRESSURE OF PLUTONIUM HALIDES. (PLUTONIUM TRIFLUORIDE) 0887
DE, STRONTIUM DIFLUORIDE, BARIUM DIFLUORIDE, AND ALUMINUM TRIFLUORIDE. /AGNESIUM DIFLUORIDE, CALCIUM DIFLUORI 0049
VAPORIZATION STUDIES OF LANTHANUM TRIFLUORIDE. 0575
VAPORIZATION THERMODYNAMICS OF INDIUM TRIFLUORIDE. 0492
SPHORUS PENTAFLUORIDE, SULFUR TETRAFLUORIDE, AND CHLORINE TRIFLUORIDE. VESCF-MO STUDIES OF PHO 0835
VIBRATION ROTATION BANDS OF BORON TRIFLUORIDE. 0199
VIBRATION- ROTATION BANDS OF BORON TRIFLUORIDE. 0191
VIBRATIONAL ANHARMONICITY IN BORON TRIFLUORIDE. 0210
VIBRATIONAL SPECTRUM OF BROMINE TRIFLUORIDE. 0257
VIBRATIONAL SPECTRUM OF IODINE TRIFLUORIDE. 0500
VIBRATIONAL STUDY OF VANADIUM TRIFLUORIDE. 1207
T OF SOME LINEAR AND SYMMETRIC TOP MOLECULES. (PHOSPHORUS TRIFLUORIDE) ZEEMAN EFFEC 0796
NANCE IN AN OVERTONE BAND. THE 2NU-4 SPECTRUM OF NITROGEN TRIFLUORIDE. 1-TYPE RESO 0732
2NU-4 SPECTRUM OF PHOSPHORUS TRIFLUORIDE. 0817
ORIDE AND CURIUM T/ ABSORPTION SPECTRA OF SOLID AMERICIUM TRIFLUORIDE, AMERICIUM TETRAFLUORIDE, CURIUM TRIFLU 0445
ELECTRONIC STRUCTURE OF CHLORINE TRIFLUORIDE. AN APPROXIMATE SCF CALCULATION. 0404
MELTING POINTS OF CURIUM TRIFLUORIDE AND AMERICIUM TRIFLUORIDE. 0446
PHOTOELECTRON SPECTRA OF COMPLEXES OF BORON TRIFLUORIDE AND AMINES. 0224
BY ELECTRON DIFFRACTION. MOLECULAR STRUCTURES OF ARSENIC TRIFLUORIDE AND ARSENIC PENTAFLUORIDE AS DETERMINED 0099
ARACTER AND HYBRIDIZATION FROM DIPOLE MOMENTS FOR ARSENIC TRIFLUORIDE AND ARSENIC TRICHLORIDE. /S ON IONIC CH 0107
Y OF RADICALS IN GAMMA IRRADIATED POLYCRYSTALLINE ARSENIC TRIFLUORIDE AND ARSENIC TRICHLORIDE. ESR STUD 0108
DI/ DETERMINATION OF THE MOLECULAR STRUCTURES OF ARSENIC TRIFLUORIDE AND ARSENIC TRICHLORIDE BY GAS ELECTRON 0104
LECTRO/ FORCE CONSTANTS AND AVERAGE STRUCTURES OF ARSENIC TRIFLUORIDE AND ARSENIC TRICHLORIDE DETERMINED BY E 0103
MOLECULAR FORCE FIELD FOR SOME XY3 MOLECULES. (PHOSPHORUS TRIFLUORIDE AND ARSENIC TRIFLUORIDE) 1400
AB INITIO CALCULATION ON BORON TRIFLUORIDE AND BORON TRICHLORIDE. 0213
MOLECULAR STRUCTURE AND FORCE CONSTANTS OF BORON TRIFLUORIDE AND BORON TRICHLORIDE. 0221
OF THE BROMINE FLUORIDES. (BROMINE MONOFLUORIDE, BROMINE TRIFLUORIDE, AND BROMINE PENTAFLUORIDE) /ED STUDIES 0256
ELECTRICAL CONDUCTIVITY OF SOLID CHLORINE TRIFLUORIDE AND BROMINE TRIFLUORIDE. 0259
INFRARED AND RAMAN SPECTRA OF CHLORINE TRIFLUORIDE AND BROMINE TRIFLUORIDE. 0408
LECULAR FORCE FIELD FOR INTERHALOGEN COMPOUNDS. (CHLORINE TRIFLUORIDE AND BROMINE TRIFLUORIDE) MO 0407
QUADRUPOLE COUPLING CONSTANTS IN CHLORINE TRIFLUORIDE AND BROMINE TRIFLUORIDE. 0400
RAMAN SPECTRA AND THE STRUCTURE OF CHLORINE TRIFLUORIDE AND BROMINE TRIFLUORIDE. 0398
RATIONAL SPECTRA AND THERMODYNAMIC PROPERTIES OF CHLORINE TRIFLUORIDE AND BROMINE TRIFLUORIDE. VIB 0396
TRA AND GAS TO LIQUID SHIFTS. NMR SPECTRA OF CHLORINE TRIFLUORIDE AND CHLORINE MONOFLUORIDE. GASEOUS SPEC 0368
OMETRY OF INTERHALOGENS. (CHLORINE MONOFLUORIDE, CHLORINE TRIFLUORIDE, AND CHLORINE PENTAFLUORIDE) /OF THE GE 0379
ABSORPTION SPECTRA OF CHLORINE FLUORIDE, CHLORINE TRIFLUORIDE, AND CHLORINE PENTAFLUORIDE MOLECULES. 0388
THE EI/ FORCE CONSTANTS OF CARBON TETRAFLUORIDE AND BORON TRIFLUORIDE AND COMPARISON WITH FORCE CONSTANTS OF 0212
IONIZATION BY ELECTRON IMPACT OF PHOSPHORUS TRIFLUORIDE AND DIFLUORO CYANO PHOSPHINE. 0800
STRENGTH OF THE NITROGEN FLUORINE BOND IN NITROGEN TRIFLUORIDE AND DINITROGEN TETRAFLUORIDE. 0712
INFRARED SPECTRUM OF NITROGEN TRIFLUORIDE AND FORCE FIELD. 0697
LUORIDE. HEAT OF SUBLIMATION OF ALUMINUM TRIFLUORIDE AND HEAT OF FORMATION OF ALUMINUM MONOF 0053
IGH TEMPERATURE MATERIALS. (CHROMIUM DIFLUORIDE, CHROMIUM TRIFLUORIDE, AND IRON DIFLUORIDE) /APORIZATION OF H 0431
EASSIGNMENT OF THE SPE/ PHOTOELECTRON SPECTRA OF NITROGEN TRIFLUORIDE AND NITROGEN OXIDE TRIFLUORIDE, AND A R 0700
GREEN'S FUNCTION ANALYSIS OF NITROGEN TRIFLUORIDE AND NITROGEN TRICHLORIDE. 0729
RAMAN SPECTRA AND MOLECULAR CONSTANTS OF PHOSPHORUS TRIFLUORIDE AND PHOSPHINE. 0822
O SCF STUDY. BONDING IN PHOSPHORUS TRIFLUORIDE AND PHOSPHORUS OXYFLUORIDE. AN AB INITI 0820
, BOND MOMENTS AND BOND MOMENT DERIVATIVES FOR PHOSPHORUS TRIFLUORIDE AND PHOSPHORUS OXYGEN TRIFLUORIDE. /IES 0797
HIGH RESOLUTION SPECTRA OF ARSENIC TRIFLUORIDE AND PHOSPHORUS TRIFLUORIDE. 0816
INFRARED SPECTRA OF ARSENIC TRIFLUORIDE AND PHOSPHORUS TRIFLUORIDE. 0748
E REGION FROM TWO TO THREE MILLIMETERS, PART-2: (NITROGEN TRIFLUORIDE AND PHOSPHORUS TRIFLUORIDE) /COPY IN TH 0711
EE ENERGIES OF SUBLIMATION. VAPOR PRESSURES OF AMERICIUM TRIFLUORIDE AND PLUTONIUM TRIFLUORIDE, HEATS AND FR 0055
TO 350 DE/ EXPERIMENTAL HEAT CAPACITIES OF PLUTONIUM-242 TRIFLUORIDE AND PLUTONIUM-242 TETRAFLUORIDE FROM 10 0885
ENTHALPY OF FORMATION OF LANTHANUM TRIFLUORIDE AND PRASEODYMIUM TRIFLUORIDE. 0577
INFRARED SPECTRA OF MATRIX ISOLATED ZIRCONIUM DIFLUORIDE, TRIFLUORIDE, AND TETRAFLUORIDE. 1325
/MPERATURES. PART-2: ENTHALPY OF SUBLIMATION OF PLUTONIUM TRIFLUORIDE AND THE DISSOCIATION ENERGY OF PLUTONI/ 0884
RIFLUORIDE. SUBLIMATION OF ALUMINUM TRIFLUORIDE AND THE INFRARED SPECTRUM OF ALUMINUM T 0037
TRIFLUORIDE, ARSENIC TRICHLORI/ RAMAN SPECTRUM OF ARSENIC TRIFLUORIDE AND THE MOLECULAR CONSTANTS OF ARSENIC 0111
M DIFLUORIDE AND NEODY/ SUBLIMATION PRESSURE OF NEODYMIUM TRIFLUORIDE AND THE STABILITIES OF GASEOUS NEODYMIU 0643
/EMPERATURES. PART-16: SUBLIMATION PRESSURES FOR TITANIUM TRIFLUORIDE AND THE STABILITIES OF TITANIUM DIFLUO/ 1145
ORID/ VAPOR PRESSURE OF CARBON TETRAFLUORIDE AND NITROGEN TRIFLUORIDE AND THE TRIPLE POINT OF CARBON TETRAFLU 0345
ETICAL STUDY OF THE GEOMETRY OF PHOSPHINE AND PHOSPHORUS TRIFLUORIDE AND THEIR GROUND IONIC STATES. THEOR 0794
OMETRIC STUDIES. HEATS OF FORMATION GASEOUS THORIUM TRIFLUORIDE AND THORIUM DIFLUORIDE FROM MASS SPECTR 1128
ITIO CALCULATIONS OF THE BONDING IN PHOSPHINE, PHOSPHORUS TRIFLUORIDE AND TRIMETHYL PHOSPHINE. AB IN 0802
LANTHANIDE TRIFLUORIDES, LANTHANIDE DIFLUORIDES, SCANDIUM TRIFLUORIDE, AND YTTRIUM TRIFLUORIDE) /FLUORIDES. (0946
ULES. AN ELECTRON DIFFRACTION INVESTIGATION OF ARSENIC TRIFLUORIDE, ARSENIC PENTAFLUORIDE, AND OTHER MOLEC 0098
/IDES. (ARSENIC MONOFLUORIDE, ARSENIC DIFLUORIDE, ARSENIC TRIFLUORIDE, ARSENIC TETRAFLUORIDE, AND ARSENIC PE/ 0093
/ULATION OF UREY-BRADLEY POTENTIAL CONSTANTS. (PHOSPHORUS TRIFLUORIDE, ARSENIC TRIFLUORIDE, AND ANTIMONY TRI/ 0809
/ODYNAMIC FUNCTIONS OF SOME GROUP-5 TRIHALIDES. (NITROGEN TRIFLUORIDE, ARSENIC TRIFLUORIDE, AND PHOSPHORUS T/ 0719
TRUCTURES OF THE INTERHALOGEN COMPOUNDS. PART-1: CHLORINE TRIFLUORIDE AT -120 DEGREES C. S 0395
NU-4 BANDS OF THE TEN AND ELEVEN ISOTOPES OF BORON TRIFLUORIDE AT HIGH RESOLUTION. 0208
HEAT OF FORMATION OF CHLORINE TRIFLUORIDE AT 298.14 DEGREES K. 0401
THERMAL DISSOCIATION OF CHLORINE TRIFLUORIDE BEHIND INCIDENT SHOCK WAVES. 0394
RADICALS OBTAINED IN ELECTRICAL DISCHARGES THROUGH BORON TRIFLUORIDE. (BORON DIFLUORIDE) /IN POLYATOMIC FREE 0222
/IN THE HARMONIC FORCE FIELDS OF XY3 MOLECULES. (NITROGEN TRIFLUORIDE, BORON TRIFLUORIDE, PHOSPHORUS TRIFLUO/ 0709
/RON SPECTRA OF HALIDES. (CHLORINE MONOFLUORIDE, CHLORINE TRIFLUORIDE, BROMINE MONOFLUORIDE, AND BROMINE TRI/ 0376
FLUORIDE. F/ INFRARED SPECTRA OF MATRIX ISOLATED CHLORINE TRIFLUORIDE, BROMINE TRIFLUORIDE, AND BROMINE PENTA 0399
/GEOMETRY AND PROPERTIES OF SOME INTERHALOGENS. (CHLORINE TRIFLUORIDE, BROMINE TRIFLUORIDE, IODINE TRIFLUORI/ 0397
HEAT OF FORMATION OF ALUMINUM TRIFLUORIDE BY DIRECT COMBINATION OF THE ELEMENTS. 0039

HEAT FORMATION BORON TRIFLUORIDE BY DIRECT COMBINATION OF THE ELEMENTS. 0198
/FETIMES OF ULTRAVIOLET EMISSION SYSTEMS EXCITED IN BORON TRIFLUORIDE, CARBON TETRAFLUORIDE, AND SILICON TET/ 0216
PREPARATION AND PROPERTIES OF URANIUM TRIFLUORIDE. (CRYSTAL STRUCTURE) 1164
BORON FLUORINE BOND DISTANCE OF BORON TRIFLUORIDE DETERMINED BY GAS ELECTRON DIFFRACTION. 0223
MONOFLUORIDE BY MATRIX ISO/ INFRARED SPECTRUM OF ALUMINUM TRIFLUORIDE, DIALUMINUM HEXAFLUORIDE, AND ALUMINUM 0051
/N STUDY OF THE VACUUM ULTRAVIOLET PHOTOLYSIS OF NITROGEN TRIFLUORIDE. ELECTRONIC SPECTRUM OF THE NITROGEN D/ 0691
/TS OF CARBON TETRAFLUORIDE, SILICON TETRAFLUORIDE, BORON TRIFLUORIDE, ETHANE, SILANE, AMMONIA, AND PHOSPHIN/ 0243
UPOLE MOMENT OF NITROGEN-14 AND THE STRUCTURE OF NITROGEN TRIFLUORIDE FROM MICROWAVE SPECTRA.   NUCLEAR QUADR 0741
HEAT CAPACITY, ENTROPY, AND ENTHALPY OF PLUTONIUM-242 TRIFLUORIDE FROM 10 TO 350 DEGREES K. 0886
ABSORPTION SPECTRA OF MATRIX ISOLATED NITROGEN TRIFLUORIDE FROM 18 TO 50 GC. 0698
NUCLEAR MAGNETIC RELAXATION IN BORON TRIFLUORIDE GAS. 0217
RAMAN SPECTRUM OF BORON TRIFLUORIDE GAS. 0251
ABSORPTION CF BORON TRIFLUORIDE GAS IN THE EXTREME ULTRAVIOLET. 0215
ORIDE.   STABILITIES OF DYSPROSIUM TRIFLUORIDE, HOLMIUM TRIFLUORIDE, AND ERBIUM TRIFLU 0449
SPECTRA AND GEOMETRY CF TITANIUM DIFLUORIDE AND TITANIUM TRIFLUORIDE IN RARE GAS MATRICES.   INFRARED 1146
/ AND RAMAN SPECTRA OF BROMINE PENTAFLUORIDE AND CHLORINE TRIFLUORIDE IN THE CONDENSED PHASE. EVIDENCE OF AN/ 0266
MICROWAVE SPECTRUM OF ARSENIC TRIFLUORIDE IN THE EXCITED VIBRATIONAL STATE. 0097
ILIBRIUM STRUCTURE)   MICROWAVE SPECTRA OF NITROGEN TRIFLUORIDE IN THE EXCITED VIBRATIONAL STATES. (EQU 0722
A TYPE DOUBLING AND RESONA/ MICROWAVE SPECTRA OF NITROGEN TRIFLUORIDE IN THE EXCITED VIBRATIONAL STATES. (ZET 0723
ONSTANTS ON THE L-TYPE DOUBLING TRANSITIONS CF PHOSPHORUS TRIFLUORIDE IN THE NU-4 EQUALS 1 STATE. /RD ORDER C 0804
L-TYPE DOUBLING TRANSITIONS CF PHOSPHORUS TRIFLUORIDE IN THE NU-4 EQUALS 1 STATE. 0803
INATION OF / INFRARED SPECTRUM OF THE FREE RADICAL CARBON TRIFLUORIDE ISOLATED IN INERT MATRICES.   PHOTOCHLOR 0311
DIUM ALUMINUM TETRA/ INFRARED SPECTRA OF GASEOUS ALUMINUM TRIFLUORIDE, LITHIUM ALUMINUM TETRAFLUORIDE, AND SO 0045
ORIDE.   SUBLIMATION PRESSURES OF CHROMIUM TRIFLUORIDE, MANGANESE TRIFLUORIDE, AND IRON TRIFLU 0434
BOND ENERGY IN THE ALUMINUM TRIFLUORIDE MOLECULE. 0044
MEAN SQUARE AMPLITUDES OF THE VIBRATIONS OF THE NITROGEN TRIFLUORIDE MOLECULE. /AL STRETCHING CONSTANTS, AND 0726
HARMONIC FORCE FIELDS AND MEAN AMPLITUDES FOR ALUMINUM TRIFLUORIDE MONOMER AND DIMER. 0038
ORIDE, AND TANTALUM / THERMODYNAMIC PROPERTIES OF NIOBIUM TRIFLUORIDE, NIOBIUM TETRAFLUORIDE, TANTALUM TRIFLU 0656
IDES. (SILICON TETRAFLUORIDE, CARBON TETRAFLUORIDE, BORON TRIFLUORIDE, NITROGEN TRIFLUORIDE) /OLYATOMIC FLUOR 0186
/TIBILITY, AND MOLECULAR QUADRUPOLE MOMENTS IN PHOSPHORUS TRIFLUORIDE, NITROGEN TRIFLUORIDE, AND PHOSPHORYL,/ 0744
/ARALLEL INFRARED BANDS OF OBLATE SYMMETRIC TOPS NITROGEN TRIFLUORIDE NU-4 AND NU-4 + NU-2 AND TRIFLUORO MET/ 0716
ETWEEN TETRABUTYL AMMONIUM HEXAFLUORO PHOSPHATE AND BORON TRIFLUORIDE OR PHOSPHORUS PENTAFLUORIDE. /XCHANGE B 0194
/ES. PART-2: DIPOLE MOMENT AND FIELD GRADIENT OF NITROGEN TRIFLUORIDE. PART-3: A GAUSSIAN BASIS SELF CONSIST/ 0746
PHORUS PENTAFLUORIDE.   INFRARED SPECTRA OF PHOSPHORUS TRIFLUORIDE, PHOSPHORUS OXIDE TRIFLUORIDE, AND PHOS 0799
/ON DIFFRACTION BY GASES AND THE STRUCTURES OF PHOSPHORUS TRIFLUORIDE, PHOSPHORUS PENTAFLUORIDE, DIFLUORO AM/ 0801
/. (SILICON DIFLUORIDE, SILICON TETRAFLUORIDE, PHOSPHORUS TRIFLUORIDE, PHOSPHORUS PENTAFLUORIDE, SULFUR DIFL/ 1373
/OSPHORUS MONOFLUORIDE, PHOSPHORUS DIFLUORIDE, PHOSPHORUS TRIFLUORIDE, PHOSPHORUS TETRAFLUORIDE, AND PHOSPHO/ 0788
/F VIBRATION FOR SOME PYRAMIDAL XY3 MOLECULES.   (NITROGEN TRIFLUORIDE, PHOSPHORUS TRIFLUORIDE, AND ARSENIC T/ 0706
/SYMMETRIC TOP MOLECULES AND FORCE CONSTANTS FOR NITROGEN TRIFLUORIDE, PHOSPHORUS TRIFLUORIDE, AND ARSENIC TR 0708
IFLUORIDE BY THE COMBINE/ GENERAL FORCE FIELD OF NITROGEN TRIFLUORIDE, PHOSPHORUS TRIFLUORIDE, AND ARSENIC TR 0717
IFLUORIDE.   MOLECULAR ORBITAL STUDY OF NITROGEN TRIFLUORIDE, PHOSPHORUS TRIFLUORIDE, AND NITROGEN D 0721
/H THE USE OF CENTRIFUGAL STRETCHING CONSTANTS. (NITROGEN TRIFLUORIDE, PHOSPHORUS TRIFLUORIDE, ARSENIC TRIFL/ 0727
STRUCTURE OF THE SULFUR TRIFLUORIDE POSITIVE ION. (SULFUR DIFLUORIDE) 1032
REFLECTION SPECTRA OF GADOLINIUM OXIDE AND GADOLINIUM TRIFLUORIDE POWDERS. 0455
INFRARED DETECTION OF GASEOUS CARBON TRIFLUORIDE RADICAL. 0310
ESR SPECTRA OF PHOSPHORUS DIFLUORIDE AND SULFUR TRIFLUORIDE RADICALS. 0791
HALPY / ENTROPY AND ENTHALPY OF SUBLIMATION OF GADOLINIUM TRIFLUORIDE. ROLE OF CORRELATION OF ENTROPY AND ENT 0456
/ONSTANTS FROM HEAVY ISOTOPIC SUBSTITUTION. PART-2: BORON TRIFLUORIDE, SILICON TETRAFLUORIDE, AND NITROGEN D/ 0232
MOLECULAR STRUCTURE OF PHOSPHORUS TRIFLUORIDE STUDIED BY GAS ELECTRON DIFFRACTION. 0810
ENTAFLUORID/ X-RAY PHOTOELECTRON SPECTROSCOPY OF CHLORINE TRIFLUORIDE, SULFUR TETRAFLUORIDE, AND PHOSPHORUS P 0409
/BSERVATION BY ESCA OF INEQUIVALENT FLUORINES IN CHLORINE TRIFLUORIDE, SULFUR TETRAFLUORIDE, AND PHOSPHORUS / 0874
/OLATILE INORGANIC FLUORIDES IN THE SOLID STATE. (ARSENIC TRIFLUORIDE, SULFUR TETRAFLUORIDE, AND SELENIUM TE/ 0096
UORIDE, ARSENI/ INFRARED MATRIX ISOLATION STUDIES. (BORON TRIFLUORIDE, SULFUR TETRAFLUORIDE, ANTIMONY PENTAFL 0239
TION OF THE MOLECULAR GEOMETRY.   BORON TRIFLUORIDE. THE ANALYSIS OF NU-2 AND THE DETERMINA 0209
ATION OF THE 33 CORIOLIS CONSTANT.   BORON TRIFLUORIDE. THE ANALYSIS OF 2NU-3 AND THE DETERMIN 0192
/IDES. (URANIUM MONOFLUORIDE, URANIUM DIFLUORIDE, URANIUM TRIFLUORIDE, URANIUM TETRAFLUORIDE, URANIUM PENTAF/ 1160
BAND SYSTEM IN THE DISCHARGE SPECTRUM OF LANTHANUM TRIFLUORIDE VAPORS. 0580
/UBLIMATION PRESSURES AND HEATS OF SUBLIMATION OF THULIUM TRIFLUORIDE, YTTERBIUM TRIFLUORIDE, AND LUTETIUM T/ 0942
FLUORIDE.   STANDARD ENTHALPIES OF FORMATION OF SCANDIUM TRIFLUORIDE, YTTRIUM TRIFLUORIDE, AND LANTHANUM TRI 0944
/ES OF MATRIX ISOLATED RARE EARTH TRIFLUORIDES. (SCANDIUM TRIFLUORIDE, YTTRIUM TRIFLUORIDE, AND LANTHANUM TR/ 0947
FLUORIDE.   SUBLIMATION PRESSURES OF SCANDIUM TRIFLUORIDE, YTTRIUM TRIFLUORIDE, AND LANTHANUM TR/ 0948
/GIES FOR THE HALIDES OF THE SCANDIUM SUBGROUP. (SCANDIUM TRIFLUORIDE, YTTRIUM TRIFLUORIDE, AND LANTHANUM TRI 0952
FLUORIDE.   SATURATED VAPOR PRESSURE OF SCANDIUM TRIFLUORIDE, YTTRIUM TRIFLUORIDE, AND LANTHANUM TRI 0953
RIDE AND THE RARE/ MASS SPECTROMETRIC STUDIES OF SCANDIUM TRIFLUORIDE, YTTRIUM TRIFLUORIDE, LANTHANUM TRIFLUO 0953
MICAL ERRORS AND SYSTEMATICS IN SUBLIMATION OF LANTHANIDE TRIFLUORIDES.   ANALYSES OF THERMOCHE 1353
ENERGY LEVELS AND SPECTROSCOPIC PARAMETERS OF NEODYMIUM TRIFLUORIDES. 0646
ES AND ENTHALPIES OF SUBLIMATION OF NEODYMIUM AND TERBIUM TRIFLUORIDES.   ENTROPI 0645
FORCE FIELDS FOR GROUP-4 TETRAFLUORIDES AND GROUP-5 TRIFLUORIDES. 0715
HIGH TEMPERATURE ENTHALPY INCREMENTS FOR SOME RARE EARTH TRIFLUORIDES. 1350
MAGNETIC RESONANCE AND SUSCEPTIBILITY OF RARE EARTH TRIFLUORIDES. 1349
ON AND VAPOR PRESSURES OF DYSPROSIUM, HOLMIUM, AND ERBIUM TRIFLUORIDES. /IGH TEMPERATURES. PART-17: SUBLIMATI 0448
IUM TRIFLUORIDE, LANTHANUM TRIFLUORIDE AND THE RARE EARTH TRIFLUORIDES. /TUDIES OF SCANDIUM TRIFLUORIDE, YTTR 0953
MEAN AMPLITUDES OF VIBRATION OF SOME LANTHANIDE TRIFLUORIDES. 1354
NMR IN RARE EARTH TRIFLUORIDES. 1355
LUORIDES AT HIGH TEMPERATURES. PART-2: TETRAFLUORIDES AND TRIFLUORIDES. /PROPERTIES OF NIOBIUM AND TANTALUM F 0657
VIBRATIONAL ANHARMONICITY IN THE PYRAMIDAL TRIFLUORIDES. 0745
PHOSPHORUS, A/ PHOTOELECTRON SPECTRA OF HALIDES. PART-3: TRIFLUORIDES AND OXIDE TRIFLUORIDES OF NITROGEN AND 0701
STRUCTURES OF YTTRIUM AND BISMUTH TRIFLUORIDES BY NEUTRON DIFFRACTION. 1318
IUM, LANTHANUM, CERIUM/ GEOMETRIES AND ENTROPIES OF METAL TRIFLUORIDES FROM INFRARED SPECTRA. (SCANDIUM, YTTR 0943
/ION OF MOLECULAR BEAMS OF THE RARE EARTH DIFLUORIDES AND TRIFLUORIDES. (LANTHANIDE TRIFLUORIDES, LANTHANIDE/ 0946
SYMMETRY AND CRYSTAL STRUCTURE OF RARE EARTH TRIFLUORIDES. (LANTHANUM TRIFLUORIDE) 0571
/RE ENTHALPIES AND RELATED THERMODYNAMIC FUNCTIONS OF THE TRIFLUORIDES OF SCANDIUM, CERIUM, SAMARIUM, EUROPI/ 1356
/E CONSTANTS AND GEOMETRIES OF MATRIX ISOLATED RARE EARTH TRIFLUORIDES. (SCANDIUM TRIFLUORIDE, YTTRIUM TRIFL/ 0944
S AND RELATED THERMODYNAMIC FUNCTIONS OF SEVEN RARE EARTH TRIFLUORIDES. (YTTRIUM TRIFLUORIDE) /E HEAT CONTENT 1320
NE, AND / IONIZATION POTENTIALS FOR TETRAFLUORO ETHYLENE, TRIFLUORO CHLORO ETHYLENE, DIFLUORO DICHLORO ETHYLE 0305
OPS. (ARSENIC TRIFLUORIDE)   CENTRIFUGAL STRETCHING OF TRIFLUORO CHLORO METHANE AND SOME OTHER SYMMETRIC T 0101
METRIC TOPS NITROGEN TRIFLUORIDE NU-4 AND NU-4 + NU-2 AND TRIFLUORO METHANE NU-6. /FRARED BANDS OF OBLATE SYM 0716
CLOPROPANE B/ ANALYSES OF THE VIBRATION ROTATION BANDS OF TRIFLUORO METHANE, PHOSPHORUS PENTAFLUORIDE, AND CY 0880
STRUCTURES OF TRIFLUORO METHYL HALIDES. (CARBON TETRAFLUORIDE) 0318
RAMAN AND INFRARED SPECTRUM OF TRIFLUORO PHOSPHORANE. (PHOSPHORUS PENTAFLUORIDE) 0847

NTRANSITION / BOND LENGTHS AND BOND ANGLES IN OCTAHEDRAL, TRIGONAL BIPYRAMID, AND RELATED MOLECULES OF THE NO 1435
POTENTIAL FIELD AND MOLECULAR VIBRATIONS OF THE TRIGONAL BIPYRAMID AX(3)YZ MOLECULE. 1436
-SPHERE AND EXTENDED HUCKEL MOLECULAR ORBITAL ANALYSES OF TRIGONAL BIPYRAMIDS. (PHOSPHORUS PENTAFLUORIDE) /-A 0828
OLECULAR VIBRATIONS AND STEREOCHEMICAL NONRIGIDITY OF THE TRIGONAL BIPYRAMIDAL MODEL MX3Y2. /ULES. PART-14: M 0854
HOSPHORUS PENTAFLUORI/ MEAN AMPLITUDES OF VIBRATION FOR A TRIGONAL BIPYRAMIDAL MOLECULE WITH APPLICATION TO P 0864
RA TURNSTILE ROTATION IN PHOSPHORUS PE/ PSEUDOROTATION OF TRIGONAL BIPYRAMIDAL MOLECULES. BERRY ROTATION CONT 0870
/ONSTANTS AND MEAN SQUARE AMPLITUDES OF VIBRATION OF SOME TRIGONAL BIPYRAMIDAL MOLECULES BY THE LOGARITHIMIC/ 0878
ATA. PENTACOORDINATED MOLE/ ANISOTROPIC THERMAL MOTION OF TRIGONAL BIPYRAMIDAL MOLECULES FROM SPECTROSCOPIC D 0853
UORIDE) PSEUDOCROTATION IN TRIGONAL BIPYRAMIDAL MOLECULES. (PHOSPHORUS PENTAFL 0833
UORIDE) RELATIVE BOND STRENGTHS IN TRIGONAL BIPYRAMIDAL MOLECULES. (PHOSPHORUS PENTAFL 0839
THERMODYNAMIC FUNCTIONS OF SOME TRIGONAL BIPYRAMIDAL PENTAHALIDES. 1438
AFLUORIDE AN/ POTENTIAL FIELD AND FORCE CONSTANTS OF SOME TRIGONAL BIPYRAMIDAL PENTAHALIDES. (PHOSPHORUS PENT 0850
/E GAS PHASE AS A FUNCTION OF TEMPERATURE. EVIDENCE FOR A TRIGONAL BIPYRAMIDAL SHAPE FOR THE MONOMERIC SPECI/ 0077
ONS AND MOLECULAR POLARIZABILITES OF SOME PENTAHALIDES OF TRIGONAL BIPYRAMIDAL SYMMTERY. /HERMODYNAMIC FUNCTI 1437
US PENTA/ POTENTIAL FIELD AND MOLECULAR VIBRATIONS OF THE TRIGONAL BIPYRAMIDAL THE RAMAN SPECTRUM OF PHOSPHOR 0863
RE AMPLITUDES OF VIBRATION, AND SHRINKAGE EFFECTS OF SOME TRIGONAL BIPYRAMIDAL XY5 AND XY3Z2 MOLECULES. /SQUA 1432
NORMAL COORDINATE ANALYSIS OF TRIGONAL BIPYRAMIDAL XY5 MOLECULES. 1433
TAFLUORIDE) UREY-BRADLEY FORCE CONSTANTS OF TRIGONAL BIPYRAMIDAL XY5 MOLECULES. (PHOSPHORUS PEN 0838
N-MORINO SHRINKAGE EFFECT IN SOME XY3 MOLECULES OF PLANAR TRIGONAL SYMMETRY. /UDES OF VIBRATION AND BASTIANSE 1397
VALENCE VIBRATION FREQUENCIES OF TRIHALIDE MOLECULES OF LANTHANUM HALIDES. 0574
THERMODYNAMIC FUNCTIONS OF BISMUTH TRIHALIDE VAPORS. 0149
GASEOUS TRIMER AND TETRAMER OF THALLIUM MONOFLUORIDE. 1114
ON FLUORIDES. HIGHLY PCLARIZED BORON- FLUORINE DOUBLE AND TRIPLE BONDS. LOCALIZED ORBITALS IN BOR 0158
OF CARBON TETRAFLUORIDE AND NITROGEN TRIFLUORIDE AND THE TRIPLE POINT OF CARBON TETRAFLUORIDE. /POR PRESSURE 0345
ROTATIONAL FINE STRUCTURE OF THE TRIPLET PI STATE OF PHOSPHORUS MONOFLUORIDE. 0787
A DOUBLET SIGMA(+/ SHOCK TUBE DETERMINATION OF THE C2 (A TRIPLET PI TO X TRIPLET PI) AND CARBON MONOFLUORIDE 0288
MA(+) SYSTEMS OF G/ ROTATIONAL ANALYSIS OF BANDS OF THE A TRIPLET PI ZERO(+) B TRIPLET PI(1) TO X SINGLET SIG 0457
GMA(+) SYSTEMS OF / ROTATIONAL ANALYSIS OF BANDS OF THE A TRIPLET PI ZERO(+), B TRIPLET PI(1) TO X SINGLET SI 1111
NOFLUORI/ FRANCK-CONDON FACTORS AND R CENTROIDS FOR THE B TRIPLET SIGMA(+) TO A TRIPLET PI SYSTEM OF BORON MO 0171
B SINGLET SIGMA(+) TO X TRIPLET SIGMA(-) BAND SYSTEM IN LEAD MONOFLUORIDE. 0585
DE. B SINGLET SIGMA(+) TO X TRIPLET SIGMA(-) BAND SYSTEM OF NITROGEN MONOFLUORI 0675
OF BORON MONOFLUO/ ROTATIONAL ANALYSIS OF BANDS OF THE C TRIPLET SIGMA, B TRIPLET SIGMA: A TRIPLET PI SYSTEM 0153
/ONDON FACTORS AND R CENTROIDS FOR THE 6 SINGLET SIGMA- X TRIPLET SIGMA BAND SYSTEM OF NITROGEN MONOFLUORIDE. 0677
ERMODYNAMIC FUNCTIONS OF DIATOMIC GASES WITH MOLECULES IN TRIPLET SIGMA STATES. (PHOSPHORUS MONOFLUORIDE) TH 0789
NEW FLUORIDES OF TRIVALENT COPPER. 0444
PLET PI) AND CARBON MONOFLUORIDE A DOUBLET SIGMA(+/ SHOCK TUBE DETERMINATION OF THE C2 (A TRIPLET PI TO X TRI 0288
XAFLUORIDES. (OF SULFUR, SELENIUM, TELLURIUM, MOLYBDENUM, TUNGSTEN, AND URANIUM) /TRUCTURE OF SOME GROUP-6 HE 0961
UATIONS OF STATE OF HEXAFLUORIDES. (OF SULFUR, MOLYBDENUM TUNGSTEN, AND URANIUM) /SIMILARITY AND UNIVERSAL EQ 0632
ND RAMAN SPECTRA. MOLECULAR STRUCTURE OF MOLYBDENUM, TUNGSTEN, AND URANIUM HEXAFLUORIDES FROM INFRARED A 0627
IN THE ULTRAVIOLET SPECTRA OF MOLYBDENUM HEXAFLUORIDE AND TUNGSTEN HEXAFLUORIDE. ASSIGNMENTS 0633
OF VAPOR BAND CONTOURS. APPLICATION TO THE FORCE FIELD OF TUNGSTEN HEXAFLUORIDE. /MPERATURE INFRARED STUDIES 1156
ENTHALPY OF FORMATION OF TUNGSTEN HEXAFLUORIDE. 1157
ION IN ELECTRON DIFFRACTION. (MOLYBDENUM HEXAFLUORIDE AND TUNGSTEN HEXAFLUORIDE) /F THE FIRST BORN APPROXIMAT 0640
VIBRATIONAL ANALYSIS OF SELENIUM HEXAFLUORIDE AND TUNGSTEN HEXAFLUORIDE. 0960
/CAL CONSTANTS (VAPOR PRESSURE) OF SILICON TETRAFLUORIDE, TUNGSTEN HEXAFLUORIDE, AND MOLYBDENUM HEXAFLUORIDE. 0639
ESOLUTION NMR SPECTRUM OF LIQUID MOLYBDENUM HEXAFLUORIDE, TUNGSTEN HEXAFLUORIDE, AND URANIUM HEXAFLUORIDE. /R 0637
INE RELAXATION IN THE LIQUIDS OF MOLYBDENUM HEXAFLUORIDE, TUNGSTEN HEXAFLUORIDE, AND URANIUM HEXAFLUORIDE. /R 0638
LUORIDE. TELLURIUM HEXAFLUORIDE, MOLYBDENUM HEXAFLUORIDE, TUNGSTEN HEXAFLUORIDE, AND URANIUM HEXAFLUORIDE. /F 1069
/PRESSURES OF SOME HEAVY TRANSITION METAL HEXAFLUORIDES. (TUNGSTEN HEXAFLUORIDE, MOLYBDENUM HEXAFLUORIDE, RH/ 0903
HEXAFLUORIDE, URA/ ELECTRON DIFFRACTION INVESTIGATION OF TUNGSTEN HEXAFLUORIDE, OSMIUM HEXAFLUORIDE, IRIDIUM 1155
/RON DIFFRACTION INVESTIGATION OF SOME SIMPLE MOLECULES. (TUNGSTEN HEXAFLUORIDE, TELLURIUM HEXAFLUORIDE, AND/ 1153
/UORIDE, TELLURIUM HEXAFLUORIDE, MOLYBDENUM HEXAFLUORIDE, TUNGSTEN HEXAFLUORIDE, URANIUM HEXAFLUORIDE, RUTHE/ 1084
/F THE FIRST BORN APPROXIMATION IN ELECTRON DIFFRACTION. (TUNGSTEN HEXAFLUORIDE, URANIUM HEXAFLUORIDE, TELLU/ 1158
SOME PHYSICAL PROPERTIES OF TUNGSTEN HEXAFLUORIDE. (VAPOR PRESSURE) 1152
FECT AND ULTRAVIOLET ABSORPTION SPECTRA OF MOLYBDENUM AND TUNGSTEN HEXAFLUORIDES. RAMAN EF 0642
IDE. PART-3: URANIUM HEXAFLUORIDE. PART-4: MOLYBDENUM AND TUNGSTEN HEXAFLUORIDES. /ART-2: PLUTONIUM HEXAFLUOR 0629
CRYSTAL STRUCTURE OF TUNGSTEN PENTAFLUORIDE. 1154
PREPARATION AND PROPERTIES (VAPOR PRESSURE) OF TUNGSTEN PENTAFLUORIDE. 1150
THERMODYNAMIC PROPERTIES OF TUNGSTEN PENTAFLUORIDE. 1149
VIBRATIONAL SPECTRA OF MOLYBDENUM PENTAFLUORIDE AND TUNGSTEN PENTAFLUORIDE. 0626
ENTHALPY OF FORMATION OF TUNGSTEN PENTAFLUORIDE AND TUNGSTEN HEXAFLUORIDE. 1151
OSMIUM PENT/ VAPOR PRESSURES OF MOLYBDENUM PENTAFLUORIDE, TUNGSTEN PENTAFLUORIDE, RHENIUM PENTAFLUORIDE, AND 0624
VIBRATIONAL SPECTRA OF MOLYBDENUM AND TUNGSTEN PENTAFLUORIDES. 0625
XAFLUORIDES. (OF SULFUR, SELENIUM, TELLURIUM, MOLYBDENUM, TUNGSTEN, URANIUM, AND RHENIUM) / SOME INORGANIC HE 0631
UORIDE. SPECTROSCOPIC CONSEQUENCES OF DIFFERENT TUNNELING MECHANISMS IN NONRIGID PHOSPHORUS PENTAFL 0834
OF TRIGONAL BIPYRAMIDAL MOLECULES. BERRY ROTATION CONTRA TURNSTILE ROTATION IN PHOSPHORUS PENTAFLUORIDE. /ON 0870

U

ELECTRONIC STRUCTURE OF SULFUR HEXAFLUORIDE STUDIED BY ULTRALONG WAVE X-RAY SPECTROSCOPY. 1065
RIDE. ULTRASOFT X-RAY ABSORPTION SPECTRA OF BORON TRIFLUO 0204
LUORIDE. ULTRASOFT X-RAY ABSORPTION SPECTRUM OF SULFUR HEXAF 1102
ABSORPTION OF BORON TRIFLUORIDE GAS IN THE EXTREME ULTRAVIOLET. 0215
COMPLEX BAND SYSTEM OF MANGANESE MONOFLUORIDE IN THE NEAR ULTRAVIOLET. 0610
ELECTRONIC SPECTRA OF LANTHANUM FLUORIDE IN THE EXTREME ULTRAVIOLET. 0570
EMISSION SPECTRUM OF ALUMINUM MONOFLUORIDE IN THE VACUUM ULTRAVIOLET. 0030
TIONAL ANALYSIS OF THE B-X SYSTEM OF LEAD MONOFLUORIDE IN ULTRAVIOLET. ROTA 0586
/HEXAFLUORIDE. RAMAN SCATTERING AND INFRARED, VISIBLE AND ULTRAVIOLET ABSORPTION IN THE VAPOR AND IN MATRIX / 1285
VACUUM ULTRAVIOLET ABSORPTION SPECTRA OF BORON TRIHALIDES. 0237
ORIDE AND LEAD DIFLUORIDE. ULTRAVIOLET ABSORPTION SPECTRA OF GASEOUS TIN DIFLU 0595
NGSTEN HEXAFLUORIDES. RAMAN EFFECT AND ULTRAVIOLET ABSORPTION SPECTRA OF MOLYBDENUM AND TU 0642
VATIVES OF METHANE. (CARBON TETRAFLUORIDE) CORREL/ VACUUM ULTRAVIOLET ABSORPTION SPECTRA OF SOME HALOGEN DERI 0365
F NITROGEN TRIFLUORIDE. VACUUM ULTRAVIOLET ABSORPTION SPECTRUM AND DIPOLE MOMENT O 0725
RIDE. ULTRAVIOLET ABSORPTION SPECTRUM OF GERMANIUM DIFLUO 0474
DE. ULTRAVIOLET ABSORPTION SPECTRUM OF SILICON DIFLUORI 0984
IDE. NEAR ULTRAVIOLET ABSORPTION SPECTRUM OF SILVER MONOFLUOR 1013
SPECTRUM OF ARSENIC MONOFLUORIDE IN THE NEAR VACUUM ULTRAVIOLET AND NEAR INFRARED REGIONS. 0091
IDE. NEAR ULTRAVIOLET AND VISIBLE BANDS OF ANTIMONY MONOFLUOR 0070
ULTRAVIOLET BAND SPECTRUM OF SILICON DIFLUORIDE. 0991
VIBRATIONAL ANALYSIS OF ULTRAVIOLET BANDS OF ANTIMONY MONOFLUORIDE. 0062

CULE.   ROTATIONAL ANALYSIS OF THE ULTRAVIOLET BANDS OF THE ANTIMONY MONOFLUORIDE MOLE 0063
IDE AND BARIUM DIFLUORIDE.   FAR ULTRAVIOLET ELECTRONIC SPECTRA OF STRONTIUM DIFLUOR 1017
  KRYPTON DIFLUORIDE)   ULTRAVIOLET EMISSION SPECTRA. (XENON DIFLUORIDE AND 0559
UORIDE, CARBON TETRAFLUORIDE, AND/ RADIATIVE LIFETIMES OF ULTRAVIOLET EMISSION SYSTEMS EXCITED IN BORON TRIFL 0216
IDE.   NEAR ULTRAVIOLET OPTICAL CONSTANTS OF LANTHANUM TRIFLUOR 0584
CTRONIC SPECTRUM OF/ MATRIX ISOLATION STUDY OF THE VACUUM ULTRAVIOLET PHOTOLYSIS OF NITROGEN TRIFLUORIDE. ELE 0691
RIDE. THE INFRARED / MATRIX ISOLATION STUDY OF THE VACUUM ULTRAVIOLET PHOTOLYSIS OF SILICON DIFLUORIDE DIHYD 0986
TION SPECTRUM OF THE BISMUTH MONOFLUORIDE MOLECULE IN THE ULTRAVIOLET REGION.   ABSORP 0143
RA OF XENON DIFLUORIDE AND XENON TETRAFLUORIDE IN THE FAR ULTRAVIOLET REGION.   SPECT 1233
RIFL/ ABSORPTION SPECTRA OF CERTAIN FLUORIDES IN THE NEAR ULTRAVIOLET REGION. (KRYPTON DIFLUORIDE, NITROGEN T 0560
/CENT DEVELOPMENTS IN MULTIPLE BEAM INTERFEROMETRY IN THE ULTRAVIOLET REGION 1700-2500 ANGSTROMS. (ALUMINUM-/ 0042
TUNGSTEN HEXAFLUORIDE.   ASSIGNMENTS IN THE ULTRAVIOLET SPECTRA OF MOLYBDENUM HEXAFLUORIDE AND 0633
NIUM DIFLUORIDE.   VACUUM ULTRAVIOLET SPECTRA OF SILICON DIFLUORIDE AND GERMA 0981
.   FAR ULTRAVIOLET SPECTROSCOPIC STUDY OF XENON DIFLUORIDE 1256
IDE.   FAR ULTRAVIOLET SPECTROSCOPIC STUDY OF XENON TETRAFLUOR 1272
E.   VIOLET AND ULTRAVIOLET SPECTRUM OF ARSENIC MONOFLUORIDE.   0089
                                                           ULTRAVIOLET SPECTRUM OF BARIUM MONOFLUORIDE MOLECUL 0122
                                                           ULTRAVIOLET SPECTRUM OF SILICON DIFLUORIDE.   0983
                                                           ULTRAVIOLET SPECTRUM OF XENON TETRAFLUORIDE   1270
                                                           UNIQUE FORCE FIELD FOR BORON TRIHALIDES.   0238
                                                           UNIQUE FORCE FIELDS OF SOME XY3 TYPE MOLECULES.   1403
SULFUR, MOLYBDENUM TUNGSTEN/ THERMODYNAMIC SIMILARITY AND UNIVERSAL EQUATIONS OF STATE OF HEXAFLUORIDES. (OF 0632
OF SULFUR, SELENIUM, TELLURIUM, MOLYBDENUM, TUNGSTEN, AND URANIUM) /TRUCTURE OF SOME GROUP-6 HEXAFLUORIDES. (0961
TE OF HEXAFLUORIDES. (OF SULFUR, MOLYBDENUM TUNGSTEN, AND URANIUM) /SIMILARITY AND UNIVERSAL EQUATIONS OF STA 0632
S. (OF SULFUR, SELENIUM, TELLURIUM, MOLYBDENUM, TUNGSTEN, URANIUM, AND RHENIUM) / SOME INORGANIC HEXAFLUORIDE 0631
(URANIUM TETRAFLUORIDE AND/ ELECTRON DIFFRACTION STUDY OF URANIUM AND THORIUM TETRAHALIDES IN THE GAS PHASE. 1170
                        STRUCTURE OF MOLECULAR URANIUM DIFLUORIDE AND URANIUM TRIFLUORIDE.   1162
RTIONATION PRESSURES. (UR/ THERMODYNAMICS OF INTERMEDIATE URANIUM FLUORIDES FROM MEASUREMENTS OF THE DISPROPO 1176
IFLUORIDE, URANIUM TRIFLUO/ HEATS OF FORMATION OF GASEOUS URANIUM FLUORIDES. (URANIUM MONOFLUORIDE, URANIUM D 1160
      ELECTRONIC ABSORPTION SPECTRUM OF GASEOUS URANIUM HEXACHLORIDE AND URANIUM HEXAFLUORIDE.   1194
            CALCULATED ELECTRONIC STRUCTURE OF URANIUM HEXAFLUORIDE.   1188
            DIPOLE MOMENT AND MOLECULAR STRUCTURE OF URANIUM HEXAFLUORIDE.   1200
OF THE FIRST BORN APPROXIMATION IN ELECTRON DIFFRACTION. (URANIUM HEXAFLUORIDE)   FAILURE 1199
IQUID MOLYBDENUM HEXAFLUORIDE, TUNGSTEN HEXAFLUORIDE, AND URANIUM HEXAFLUORIDE. /RESOLUTION NMR SPECTRUM OF L 0637
                      INFRARED SPECTRUM OF URANIUM HEXAFLUORIDE.   1182
DS OF MOLYBDENUM HEXAFLUORIDE, TUNGSTEN HEXAFLUORIDE, AND URANIUM HEXAFLUORIDE. /RINE RELAXATION IN THE LIQUI 0638
                        RAMAN SPECTRUM OF URANIUM HEXAFLUORIDE.   1190
          THERMODYNAMIC PROPERTIES OF GASEOUS URANIUM HEXAFLUORIDE.   1198
                THERMODYNAMIC PROPERTIES OF URANIUM HEXAFLUORIDE.   1204
                THERMODYNAMIC STABILITY OF URANIUM HEXAFLUORIDE.   1203
RIDE, MOLYBDENUM HEXAFLUORIDE, TUNGSTEN HEXAFLUORIDE, AND URANIUM HEXAFLUORIDE. /FLUORIDE, TELLURIUM HEXAFLUO 1069
        VAPOR PRESSURE AND CRITICAL CONSTANTS OF URANIUM HEXAFLUORIDE.   1197
    VIBRATIONAL SPECTRUM AND STRUCTURE OF SOLID URANIUM HEXAFLUORIDE.   1196
NS OF THE ELECTRONIC STRUCTURE AND OPTICAL TRANSITIONS OF URANIUM HEXAFLUORIDE AND URANIUM PENTAFLUORIDE. /IO 1179
F ELECTRONS ON THE VAPOR.   STRUCTURE OF URANIUM HEXAFLUORIDE AS DETERMINED BY DIFFRACTION O 1183
ON OF ELECTRONS ON THE VAPOR.   STRUCTURE OF URANIUM HEXAFLUORIDE AS DETERMINED BY THE DIFFRACTI 1184
            THERMODYNAMIC PROPERTIES OF URANIUM HEXAFLUORIDE AT HIGH TEMPERATURES.   1193
/. PART-10: NEUTRON POWDER DIFFRACTION PROFILE STUDIES OF URANIUM HEXAFLUORIDE AT 193 DEGREES AND 293 DEGREE/ 1202
CTROSCOPY.   PROPERTIES AND REACTIONS OF URANIUM HEXAFLUORIDE BY ION CYCLOTRON RESONANCE SPE 1185
        MEASUREMENT OF THE GAS DENSITY OF URANIUM HEXAFLUORIDE BY LASER RAMAN SCATTERING.   1181
              ABSORPTION SPECTRUM OF URANIUM HEXAFLUORIDE FROM 2000 TO 4200 ANGSTROMS.   1191
    VIBRATIONAL SPECTRUM AND THERMODYNAMIC PROPERTIES OF URANIUM HEXAFLUORIDE GAS.   1186
    VIBRATIONAL SPECTRUM AND THERMODYNAMIC PROPERTIES OF URANIUM HEXAFLUORIDE GAS.   1187
NTAFLUORIDE.   VAPOR PRESSURE OF URANIUM HEXAFLUORIDE IN EQUILIBRIUM WITH URANIUM PE 1180
ON COMPRESSOR.   SPECIFIC HEAT RATIO OF URANIUM HEXAFLUORIDE MEASURED WITH A BALLISTIC PIST 1201
/HEXAFLUORIDE, OSMIUM HEXAFLUORIDE, IRIDIUM HEXAFLUORIDE, URANIUM HEXAFLUORIDE, NEPTUNIUM HEXAFLUORIDE, AND / 1155
LUTONI/ LONG WAVE LENGTH, INFRARED ACTIVE FUNDAMENTAL FOR URANIUM HEXAFLUORIDE, NEPTUNIUM HEXAFLUORIDE, AND P 1192
        ALPHA SPECTROMETRIC INVESTIGATION OF URANIUM HEXAFLUORIDE. PART-1.   1195
/GH TEMPERATURES. PART-2: PLUTONIUM HEXAFLUORIDE. PART-3: URANIUM HEXAFLUORIDE. PART-4: MOLYBDENUM AND TUNGS/ 0629
EPTNIUM HEXAFL/ SOME RECENT STUDIES WITH HEXAFLUORIDES. (URANIUM HEXAFLUORIDE, PLUTONIUM HEXAFLUORIDE, AND N 1205
/LUORIDE, MOLYBDENUM HEXAFLUORIDE, TUNGSTEN HEXAFLUORIDE, URANIUM HEXAFLUORIDE, RUTHENIUM HEXAFLUORIDE, RHEN/ 1084
/FLUORIDE, TECHNETIUM HEXAFLUORIDE, RHENIUM HEXAFLUORIDE, URANIUM HEXAFLUORIDE, SULFUR HEXAFLUORIDE, SELENIU/ 1072
/IMATION IN ELECTRON DIFFRACTION. (TUNGSTEN HEXAFLUORIDE, URANIUM HEXAFLUORIDE, TELLURIUM HEXAFLUORIDE, MOLY/ 1158
RA.   MOLECULAR STRUCTURE OF MOLYBDENUM, TUNGSTEN, AND URANIUM HEXAFLUORIDES FROM INFRARED AND RAMAN SPECT 0627
ON SPECTRUM OF URANIUM TETRAFLUORIDE AND ENERGY LEVELS OF URANIUM(IV).   ABSORPTI 1168
ON OF THE DISSOCIATION ENERGY OF CALCIUM MONOFLUORIDE AND URANIUM MONOFLUORIDE.   CALCULATI 1161
TS OF FORMATION OF SOME HALIDES OF THORIUM, PROTACTINIUM, URANIUM, NEPTUNIUM, AND AMERICIUM. (ACTINIDE) HEA 1335
                                                           URANIUM PENTAFLUORIDE, A NEW FLUORIDE OF URANIUM. 1178
PLUTONIUM PENTAFLUORIDE.   THERMODYNAMIC PROPERTIES OF URANIUM PENTAFLUORIDE, NEPTUNIUM PENTAFLUORIDE, AND 1177
              ENTHALPY OF FORMATION OF URANIUM TETRAFLUORIDE.   1171
          HIGH TEMPERATURE HEAT CONTENT OF URANIUM TETRAFLUORIDE.   1173
        SATURATED VAPOR PRESSURE OF SOLID URANIUM TETRAFLUORIDE.   1167
                    VAPOR PRESSURE OF URANIUM TETRAFLUORIDE.   1174
                    VAPOR PRESSURE OF URANIUM TETRAFLUORIDE.   1175
        VIBRATIONAL SPECTRA AND FORCE CONSTANTS OF URANIUM TETRAFLUORIDE.   1172
IV).   ABSORPTION SPECTRUM OF URANIUM TETRAFLUORIDE AND ENERGY LEVELS OF URANIUM(1168
TEMPERATURE HEAT CAPACITIES, ENTHALPIES, AND ENTROPIES OF URANIUM TETRAFLUORIDE AND URANIUM HEXAFLUORIDE. /W 1166
HOD.   DETERMINATION OF HEAT OF SUBLIMATION OF URANIUM TETRAFLUORIDE BY THE MASS SPECTROMETRIC MET 1165
ALPY AND ENTROPY OF FUSION.   ENTHALPY OF URANIUM TETRAFLUORIDE FROM 298-1400 DEGREES K. ENTH 1169
    NEUTRON DIFFRACTION STUDY OF THE CRYSTAL STRUCTURE OF URANIUM TRIFLUORIDE.   1163
        PREPARATION AND PROPERTIES OF URANIUM TRIFLUORIDE. (CRYSTAL STRUCTURE)   1164
          FORCE CONSTANTS OF SMALL MOLECULES. (UREY-BRADLEY)   1495
          QUADRATIC POTENTIAL FUNCTION FOR AMMONIA. (UREY-BRADLEY)   1483
RDINATE TREATMENT OF XY4 TETRAHEDRAL MOLECULES AND IONS. (UREY-BRADLEY AND MVFF)   NORMAL COO 1418
AND SHRINKAGE EFFECT OF BORON TRIHALIDES, CALCULATED FROM UREY-BRADLEY FORCE CONSTANTS. /ITUDES OF VIBRATION 0233
BRATION, SHRINKAGE EFFECT AND CORIOLIS CONSTANTS IN IODI/ UREY-BRADLEY FORCE CONSTANTS, MEAN AMPLITUDES OF VI 0513
HEIR TRANSFERABILITY. (BORON TRIFLUORIDE)   UREY-BRADLEY FORCE CONSTANTS OF BORON HALIDES AND T 0246
        EFFECTS OF THE LENNARD-JONES POTENTIALS ON THE UREY-BRADLEY FORCE CONSTANTS OF BORON TRIHALIDES. 0244
L XY5 MOLECULES. (PHOSPHORUS PENTAFLUORIDE)   UREY-BRADLEY FORCE CONSTANTS OF TRIGONAL BIPYRAMIDA 0838

E MIXED HALOMETHANES. (CARBON TETRAFLUORI/ A TRANSFERABLE UREY-BRADLEY FORCE FIELD AND THE ASSIGNMENTS OF SOM 0351
ES FOR BENT SYMMETRICAL XY2 MOLECULES. UREY-BRADLEY FORCE FIELD AND THERMODYNAMIC PROPERTI 1387
ES OF PRYRAMIDAL XY3 MOLECULES. UREY-BRADLEY FORCE FIELD AND THERMODYNAMIC PROPERTI 1406
UREY-BRADLEY FORCE FIELD FOR CARBON TETRAFLUORIDE. 0360
UREY-BRADLEY FORCE FIELD FOR SULFUR TETRAFLUORIDE. 1059
LES. UREY-BRADLEY FORCE FIELD FOR XY4 TETRAHEDRAL MOLECU 1429
UREY-BRADLEY FORCE FIELD. (LEAD TETRAFLUORIDE) 0596
UREY-BRADLEY FORCE FIELD OF SOME XY6 SYSTEMS. 1469
BINE, AND ARSINE. ADAPTATION OF THE UREY-BRADLEY FORCE FIELD TO AMMONIA, PHOSPHINE, STI 1492
PE MOLECULES AND IONS. APPLICATION OF UREY-BRADLEY FORCE FIELD TO SOME TETRAHEDRAL XY4 TY 1425
CALCULATIONS OF MEAN AMPLITUDES OF VIBRATION AND UREY-BRADLEY FORCE FIELDS FOR PLANAR XY3 MOLECULES. 1399
LIMITATIONS OF THE UREY-BRADLEY MODEL IN NORMAL COORDINATE ANALYSES. 1478
ORIDES. UREY-BRADLEY POTENTIAL CONSTANTS FOR SECOND ROW FLU 1474
UORIDE, ARSENIC TRIFLUORIDE, AND ANTIMONY/ CALCULATION OF UREY-BRADLEY POTENTIAL CONSTANTS. (PHOSPHORUS TRIFI 0809
LES. CALCULATION OF UREY-BRADLEY POTENTIAL CONSTANTS. PLANAR XY3 MOLECU 1393
ENON FLUORIDES. MODIFIED NDDO-2 (ALPHA, BETA) METHOD AS A USEFUL APPROXIMATION. /METHODS ILLUSTRATED BY THE X 1231
THE CALCULATION OF FORCE CONSTANTS. ANHARMONICITY EFFEC/ UTILITY OF HEAVY ATOM ISOTOPIC SUBSTITUION DATA FOR 0234

V

X SINGLET SIGMA(+) BANDS OF BORON-11 MONOFLUORIDE WITH A VACUUM ECHELLE SPECTROGRAPH. / OF THE A SINGLET PI- 0170
EMISSION SPECTRUM OF ALUMINUM MONOFLUCRIDE IN THE VACUUM ULTRAVIOLET. 0030
ALIDES. VACUUM ULTRAVIOLET ABSORPTION SPECTRA OF BORON TRIH 0237
EN DERIVATIVES OF METHANE. (CARBON TETRAFLUORIDE) CORREL/ VACUUM ULTRAVIOLET ABSORPTION SPECTRA OF SOME HALOG 0365
OMENT OF NITROGEN TRIFLUORIDE. VACUUM ULTRAVIOLET ABSORPTION SPECTRUM AND DIPOLE M 0725
SPECTRUM OF ARSENIC MONOFLUORIDE IN THE NEAR VACUUM ULTRAVIOLET AND NEAR INFRARED REGIONS. 0091
DE. ELECTRONIC SPECTRUM OF/ MATRIX ISOLATION STUDY OF THE VACUUM ULTRAVIOLET PHOTOLYSIS OF NITROGEN TRIFLUORI 0691
E DIHYDRIDE. THE INFRARED / MATRIX ISOLATION STUDY OF THE VACUUM ULTRAVIOLET PHOTOLYSIS OF SILICON DIFLUORID 0986
D GERMANIUM DIFLUORIDE. VACUUM ULTRAVIOLET SPECTRA OF SILICON DIFLUORIDE AN 0981
ULE/ METHODS FOR CORRELATING MOLECULES AND SOME OPTIMIZED VALENCE CONFIGURATION RESULTS ON THE DIATOMIC MOLEC 0174
NTAINING PHOSPHORUS CR SULFUR. NONEMPIRICAL VALENCE ELECTRON CALCULATIONS ON SMALL MOLECULES CO 1029
ROPERTIES OF SOME INTERHALOGENS. (CH/ CNDO/2 AND INDO ALL VALENCE ELECTRON CALCULATIONS ON THE GEOMETRY AND P 0397
, AND BORON FLUORIDE. VALENCE EXCITED STATES OF NITROGEN, CARBON MONOXIDE 0169
AL DIFLUORIDES. VIBRATIONAL FREQUENCIES AND VALENCE FORCE CONSTANTS OF FIRST ROW TRANSITION MET 1371
ES. XENON OXIDE TETRAFLUORIDE, I/ VIBRATIONAL SPECTRA AND VALENCE FORCE CONSTANTS OF SQUARE PYRAMIDAL MOLECUL 0418
S. (GERMANIUM TETRAFLUORIDES) ORBITAL VALENCE FORCE CONSTANTS OF TETRAHEDRAL XY4 MOLECULE 0481
YBDENUM, TECHNETIUM, AND RHENIUM. VALENCE FORCE CONSTANTS OF THE HEXAFLUORIDES OF MOL 0641
ECTRA OF BORON FLUORIDE WITH ALTERATIONS IN ITS QUADRATIC VALENCE FORCE FIELD. /R EFFECTS IN THE VIBRATION SP 1404
E LOGARITHMIC STEP METHOD IN THE RANGE OF THE GENERALIZED VALENCE FORCE FIELD. /EPTAFLUORIDE CALCULATED BY TH 0535
GENERAL VALENCE FORCE FIELD FOR XENON TETRAFLUORIDE. 1265
AMPLITUDES OF VIBRATION OF SOME TRIGONAL BI/ GENERALIZED VALENCE FORCE FIELD FORCE CONSTANTS AND MEAN SQUARE 0878
E AND XENON DICHLORIDE. 5D ORBITALS OF XENON IN ATOMIC VALENCE STATES AND IN THE MOLECULES XENON DIFLUORID 1242
/ND ATOMS OF LATER SECCND ROW ELEMENTS IN PERFECT PAIRING VALENCE STATES. (PHOSPHORUS PENTAFLUORIDE, SULFUR / 0862
S OF LANTHANUM HALIDES. VALENCE VIBRATION FREQUENCIES OF TRIHALIDE MOLECULE 0574
MOLECULAR ORBITAL DESCRIPTION OF SULFUR COMPOUNDS OF VALENCES 2, 4, AND 6. 1085
SPHORUS PENTAFLUORIDE, SULFUR HEXAFLUORIDE) DIRECTED VALENCY IN CERTAIN MOLECULES AND COMPLEX IONS. (PHO 0860
F VIBRATION OF PENTAFLUORIDES OF ARSENIC, PHCSPHORUS, AND VANADIUM. /ECTROSCOPIC STUDIES IN MEAN AMPLITUDES O 0871
SYNTHESIS AND CHARACTERIZATION OF VANADIUM DIFLUORIDE. 1206
RE AND ANOMALOUS THERMAL EXPANSION ON IRON DIFLUORIDE AND VANADIUM DIOXIDE. ELECTRONIC STRUCTU 0543
FECTS, CORIOLIS CONSTANTS, AND THERMODYNAMIC FUNCTIONS OF VANADIUM PENTAFLUORIDE. /OF VIBRATION, SHRINKAGE EF 1219
INTRAMOLECULAR EXCHANGE IN VANADIUM PENTAFLUORIDE. 1214
N AMPLITUDES OF VIBRATION AND THERMODYNAMIC FUNCTIONS FOR VANADIUM PENTAFLUORIDE. MEA 1216
POLAR STATES IN VANADIUM PENTAFLUORIDE. 1213
SOME PHYSICAL AND CHEMICAL PROPERTIES OF VANADIUM PENTAFLUORIDE. 1212
VIBRATIONAL SPECTRA OF VANADIUM PENTAFLUORIDE. 1211
/FLUORIDE, ARSENIC PENTAFLUORIDE, ANTIMONY PENTAFLUORIDE, VANADIUM PENTAFLUORIDE, AND MOLYBDENUM PENTAFLUORI/ 0878
NCLUDING THOS/ RAMAN SPECTRA OF ARSENIC PENTAFLUORIDE AND VANADIUM PENTAFLUORIDE AND THEIR FORCE CONSTANTS I 0873
FLUORIDE, IODINE PENTAFLUORIDE, PHOSPHURUS PENTAFLUORIDE, VANADIUM PENTAFLUORIDE, ANTIMONY PENTAFLUORIDE) /TA 0422
ENTHALPY OF FORMATION OF VANADIUM PENTAFLUORIDE BY FLUORINE BOMB CALORIMETRY 1215
ELECTRON DIFFRACTION STUDY OF THE VANADIUM PENTAFLUORIDE MOLECULE IN THE VAPOR FORM. 1218
/LECTRON DIFFRACTION STUDY OF THE MOLECULAR STRUCTURES OF VANADIUM PENTAFLUORIDE, NIOBIUM PENTAFLUORIDE, AND/ 1217
VANADIUM TRIFLUORIDE. INFRARED SPECTRA OF VANADIUM PENTAFLUORIDE, VANADIUM TETRAFLUORIDE, AND 1210
HN-TELLER EFFECT. EXPERIMENTAL AND THEORETICAL STUDIES ON VANADIUM TETRACHLORIDE. JA 1426
PREPARATION AND PROPERTIES (CRYSTAL STRUCTURE) OF VANADIUM TETRAFLUORIDE. 1209
/HLORIDE, TITANIUM TETRAFLUORIDE, TITANIUM TETRACHLORIDE, VANADIUM TETRAFLUORIDE, AND VANADIUM TETRACHLORIDE. 0349
ERMINATION OF THE VAPOR COMPOSITION AND VAPOR PRESSURE OF VANADIUM TRIFLUORIDE. MASS SPECTROMETRIC DET 1208
VIBRATIONAL STUDY OF VANADIUM TRIFLUORIDE. 1207
ASSOCIATION OF ANTIMONY PENTAFLUORIDE IN VAPOR. 0088
S, MASS SPECTRA, AND THERMODYNAMICS OF BERYLLIUM FLUORIDE VAPOR. EFFUSION STUDIE 0140
MASS SPECTRUM AND COMPOSITION OF NIOBIUM PENTAFLUORIDE VAPOR. 0666
FLUORIDE AS DETERMINED BY DIFFRACTION OF ELECTRONS ON THE VAPOR. STRUCTURE OF URANIUM HEXA 1183
RIDE AS DETERMINED BY THE DIFFRACTION OF ELECTRONS ON THE VAPOR. STRUCTURE OF URANIUM HEXAFLUO 1184
G AND INFRARED, VISIBLE AND ULTRAVIOLET ABSORPTION IN THE VAPOR AND IN MATRIX ISOLATION. /DE. RAMAN SCATTERIN 1285
HALPY OF FUSION AND VAPORIZATION, STANDARD ENTROPY OF THE VAPOR, AND OTHER THERMODYNAMIC PROPERTIES. / K, ENT 0510
/L TOP MOLECULES FROM LOW TEMPERATURE INFRARED STUDIES OF VAPOR BAND CONTOURS. APPLICATION TO THE FORCE FIEL/ 1156
IFLUORIDE. MASS SPECTROMETRIC DETERMINATION OF THE VAPOR COMPOSITION AND VAPOR PRESSURE OF VANADIUM TR 1208
OF NIOBIUM AND TANTALUM PENTAFLUORIDES IN THE GAS PHASE. VAPOR DENSITY OF NIOBIUM PENTAFLUORIDE. /AL SPECTRA 0660
LUO/ METHOD OF DETERMINING SATURATED LIQUID AND SATURATED VAPOR ENTROPY. (CHLORINE MONOFLUORIDE, OXYGEN MONOF 0391
CTION STUDY OF THE VANADIUM PENTAFLUORIDE MOLECULE IN THE VAPOR FORM. ELECTRON DIFFRA 1218
GATION OF THE STRUCTURE OF ZIRCONIUM TETRAFLUORIDE IN THE VAPOR PHASE. ELECTRON DIFFRACTION INVESTI 1332
/TRON DIFFRACTION STUDY OF THE MOLECULAR STRUCTURE OF THE VAPOR PHASE HALIDES OF GALLIUM, YTTRIUM, LANTHANUM/ 1317
N, GERMANIUM, AND TIN) AND MF4 (M= CARBON, SILICON, AND / VAPOR PHASE RAMAN SPECTRA OF MH4 (M= CARBON, SILICO 0314
SES, CORIOLIS CONSTANTS, FORCE CONSTANTS, AND VALUES FOR/ VAPOR PHASE RAMAN SPECTRA, RAMAN BAND CONTOUR ANALY 0795
SES, AND CORIOLIS CONSTANTS OF THE SPHERICAL TOP MOLECUL/ VAPOR PHASE RAMAN SPECTRA, RAMAN BAND CONTOUR ANALY 1069
ALUMINUM MONOFLUORIDE. (VAPOR PRESSURE) 0031
URIFICATION OF OXYGEN DIFLUORIDE AND DETERMINATION OF ITS VAPOR PRESSURE. PREPARATION AND P 0781
SOME PHYSICAL PROPERTIES OF TUNGSTEN HEXAFLUORIDE. (VAPOR PRESSURE) 1152
XAFLUORIDE. VAPOR PRESSURE AND CRITICAL CONSTANTS OF URANIUM HE 1197
E OF NITROGEN TRIFLUORIDE. LIQUID DENSITY, VAPOR PRESSURE AND CRITICAL TEMPERATURE AND PRESSUR 0710
/ASS SPECTROMETRIC STUDIES AT HIGH TEMPERATURES. PART-14: VAPOR PRESSURE AND DISSOCIATION ENERGY OF SILVER M/ 1014

LUORIDE.	VAPOR PRESSURE AND HEAT OF SUBLIMATION OF CALCIUM F 0282
IFLUORIDE.	VAPOR PRESSURE AND HEAT OF SUBLIMATION OF CERIUM TR 0366
HALIDES. (COBALT DIFLUORIDE)	VAPOR PRESSURE AND HEAT OF SUBLIMATION OF COBALT DI 0438
AND HAFNIUM TETRACHLORIDES.	VAPOR PRESSURE AND HEAT OF SUBLIMATION OF ZIRCONIUM 0486
LUORIDE.   MELTING POINT, VAPOR PRESSURE, AND HEAT OF VAPORIZATION OF TIN DIF 1140
DE AND XENON TETRAFLUORIDE.	VAPOR PRESSURE AND MELTING POINTS OF XENON DIFLUORI 1251
HEXAFLUORIDE.	VAPOR PRESSURE AND TRANSITION POINTS OF TECHNETIUM 1105
STANT, AND SPECIFIC CONDUCTIVITY OF CHLORINE PE/ DENSITY, VAPOR PRESSURE, CRITICAL PROPERTIES, DIELECTRIC CON 0426
N COEFFICIENT OF LANTHANUM TRIFLUORIDE.	VAPOR PRESSURE, HEAT OF SUBLIMATION, AND EVAPORATIO 0576
ATION COEFFICIENT OF PRASEODYMIUM TRIFLUORIDE.	VAPOR PRESSURE, HEAT OF SUBLIMATION, AND THE EVAPOR 0896
PHASE INVESTIGATIONS AND VAPOR PRESSURE MEASUREMENTS ON EUROPIUM DIFLUORIDE. 0452
	VAPOR PRESSURE OF ALUMINUM TRIFLUORIDE. 0041
DENSITY, MELTING POINT AND VAPOR PRESSURE OF ANTIMONY PENTAFLUORIDE. 0082
TRIFLUORIDE AND THE TRIPLE POINT OF CARBON TETRAFLUORID/ VAPOR PRESSURE OF CARBON TETRAFLUORIDE AND NITROGEN 0345
KNUDSEN MEASUREMENT OF THE VAPOR PRESSURE OF MANGANESE DIFLUORIDE. 0613
FLUORIDE)	VAPOR PRESSURE OF NICKEL DIFLUORIDE. 0652
IFICATION OF PLUTONIUM(VI/ PREPARATION AND PROPERTIES OF (VAPOR PRESSURE) OF PLUTONIUM HALIDES. (PLUTONIUM TRI 0887
	VAPOR PRESSURE) OF PLUTONIUM HEXAFLUORIDE AND IDENT 0890
FLUORIDE, AND LANTHANUM TRIFLUORIDE.   SATURATED VAPOR PRESSURE OF RARE EARTH FLUORIDES. 1357
HEXAFLUORIDE, AND MOLYBDENUM HEXAFLU/ PHYSICAL CONSTANTS (VAPOR PRESSURE) OF SCANDIUM TRIFLUORIDE, YTTRIUM TRI 0952
	SATURATED VAPOR PRESSURE) OF SILICON TETRAFLUORIDE, TUNGSTEN 0639
	SATURATED VAPOR PRESSURE OF SOLID URANIUM TETRAFLUORIDE. 1167
	VAPOR PRESSURE OF THORIUM TETRAFLUORIDE. 1129
	VAPOR PRESSURE OF TITANIUM TETRAFLUORIDE. 1148
PREPARATION AND PROPERTIES (VAPOR PRESSURE) OF TUNGSTEN PENTAFLUORIDE. 1150
UM WITH URANIUM PENTAFLUORIDE.	VAPOR PRESSURE OF URANIUM HEXAFLUORIDE IN EQUILIBRI 1180
	VAPOR PRESSURE OF URANIUM TETRAFLUORIDE. 1174
	VAPOR PRESSURE OF URANIUM TETRAFLUORIDE. 1175
DE, MAGNESIUM DIFLUORIDE, CALCIUM DIFLUORIDE, STRONTIUM / VAPOR PRESSURE OF ZINC DIFLUORIDE, CADMIUM DIFLUORI 0049
	VAPOR PRESSURE OF ZIRCONIUM FLUORIDE. 1331
	VAPOR PRESSURE OF ZIRCONIUM TETRAFLUORIDE. 1326
	VAPOR PRESSURE OF ZIRCONIUM TETRAFLUORIDE. 1328
	(VAPOR PRESSURES) 0924
RUTHENIUM PENTAFLUORIDE AND RUTHENIUM OXYTETRAFLUORIDE. VAPOR PRESSURES AND OTHER PROPERTIES OF RHENIUM HEX 0912
AFLUORIDE AND RHENIUM HEPTAFLUORIDE.	VAPOR PRESSURES AND THERMODYNAMIC CALCULATIONS FOR 1336
SOME ACTINIDE FLUORIDES	VAPOR PRESSURES OF AMERICIUM TRIFLUORIDE AND PLUTON 0055
IUM TRIFLUORIDE, HEATS AND FREE ENERGIES OF SUBLIMATION. VAPOR PRESSURES OF BERYLLIUM DIFLUORIDE AND NICKEL 0135
DIFLUORIDE.	VAPOR PRESSURES OF DYSPROSIUM, HOLMIUM, AND ERBIUM/ 0448
/C STUDIES AT HIGH TEMPERATURES. PART-17: SUBLIMATION AND VAPOR PRESSURES OF MOLYBDENUM PENTAFLUORIDE, TUNGST 0624
EN PENTAFLUORIDE, RHENIUM PENTAFLUORIDE, AND OSMIUM PENT/ VAPOR PRESSURES OF NIOBIUM AND TANTALUM PENTAFLUORI 0663
DES.   HALIDES OF NIOBIUM AND TANTALUM. PART-3: VAPOR PRESSURES OF SOME HEAVY TRANSITION METAL HEXA 0903
FLUORIDES. (TUNGSTEN HEXAFLUORIDE, MOLYBDENUM HEXAFLUORI/ VAPOR STATE. /UM HEXAFLUORIDE, SULFUR HEXAFLUORIDE, 1072
SELENIUM HEXAFLUORIDE, AND TELLURIUM HEXAFLUORIDE IN THE VAPOR STATE. (THIRD ROW) 1378
RAMAN SPECTRA OF 5D TRANSITION METAL HEXAFLUCRIDES IN THE VAPORS.   BAND 0580
SYSTEM IN THE DISCHARGE SPECTRUM OF LANTHANUM TRIFLUORIDE VAPORS. 0149
THERMODYNAMIC FUNCTIONS OF BISMUTH TRIHALIDE VAPORS, AND FORCE CONSTANT CALCULATIONS. /NON DIFLU 1254
ORIDE, XENON TETRAFLUORIDE, AND XENON OXIDE TETRAFLUORIDE VAPORS. PART-3: MONOMER AND DIMER SPECTRA OF THALLO 1117
US FLUORI/ PHOTOELECTRON SPECTROSCOPY OF HIGH TEMPERATURE VAPORS. PART-4: THALLIUM MONOFLUORIDE, THALLIUM MON 1112
OCHLORIDE, AND THALL/ PHOTOIONIZATION OF HIGH TEMPERATURE VAPORIZATION AND SUBLIMATION THERMODYNAMICS OF SELE 1347
CTED LANTHANIDE FLUORIDES.	VAPORIZATION OF ALUMINUM FLUORIDE AND CESIUM FLUORI 0052
DE.   THERMODYNAMICS OF THE VAPORIZATION OF BERYLLIUM DIFLUORIDE. /AL PROPERTIE 0138
S OF BERYLLIUM COMPOUNDS. PART-1: ENTHALPY AND ENTROPY OF VAPORIZATION OF HIGH TEMPERATURE MATERIALS. (CHROMI 0431
UM DIFLUORIDE, CHROMIUM TRI/ SPECTROSCOPIC STUDIES OF THE VAPORIZATION OF THALLIUM MONOFLUORIDE. 1121
THERMODYNAMICS OF VAPORIZATION OF TIN DIFLUORIDE. 1140
MELTING POINT, VAPOR PRESSURE, AND HEAT OF VAPORIZATION REACTIONS IN THE THULIUM- FLUORINE SYS 1131
TEM.	VAPORIZATION, STANDARD ENTROPY OF THE VAPOR, AND O/ 0510
/CITY BETWEEN 5 AND 350 DEGREES K, ENTHALPY OF FUSION AND VAPORIZATION STUDIES OF LANTHANUM TRIFLUORIDE. 0575
	VAPORIZATION THERMODYNAMICS OF INDIUM TRIFLUORIDE. 0492
ABSORPTION SPECTRUM OF VAPORIZED TITANIUM MONOFLUORIDE. 1143
OF CARBON TETR/ PHOTOELECTRON SPECTRA OF HALIDES. PART-7: VARIABLE TEMPERATURE HELIUM-1 AND HELIUM-2 STUDIES 0341
THE SECOND ROW OF THE PERIODIC TABLE. (SILICON DIFLUORI/ VESCF-MO STUDIES OF MOLECULES CONTAINING ATOMS FROM 1373
R TETRAFLUORIDE, AND CHLORINE TRIFLUORIDE.	VESCF-MO STUDIES OF PHOSPHORUS PENTAFLUORIDE, SULFU 0835
ATED BY THE LOGARITHMIC STEP ME/ FORCE CONSTANTS AND MEAN VIBRATION AMPLITUDES OF IODINE HEPTAFLUORIDE CALCUL 0535
HOSP/ CALCULATION OF FORCE CONSTANTS AND ROOT MEAN SQUARE VIBRATION AMPLITUDES OF PHOSPHORUS PENTAFLUORIDE, P 0826
SOME LINEAR SYMMETRICAL DIHALIDES. (/ MEAN AMPLITUDES OF VIBRATION AND BASTIANSEN-MORINO SHRINKAGE EFFECT IN 1372
SOME XY3 MOLECULES OF PLANAR TRIGONA/ MEAN AMPLITUDES OF VIBRATION AND BASTIANSEN-MORINO SHRINKAGE EFFECT IN 1397
N SOME OCTAHEDRAL XY6 MOLECULES.   MEAN AMPLITUDES OF VIBRATION AND BASTIANSEN-MORINO SHRINKAGE EFFECTS I 1455
TAFLUORIDE AND ARSENIC PENTAFLUORIDE. MEAN AMPLITUDES OF VIBRATION AND CORIOLIS CONSTANTS FOR PHOSPHORUS PEN 0840
AL CONSTANTS OF PHOSPHORUS TRICHLORIDE FROM AMPLITUDES OF VIBRATION AND NORMAL VIBRATIONAL FREQUENCIES. /ENTI 1481
/LING COEFFICIENTS, GENERALIZED MEAN SQUARE AMPLITUDES OF VIBRATION AND SHRINKAGE CONSTANTS OF XENON TETRAFL/ 1278
CALCULATED FROM UREY-BRADLEY FORCE C/ MEAN AMPLITUDES OF VIBRATION AND SHRINKAGE EFFECT OF BORON TRIHALIDES, 0233
IPYRAMIDAL XY5 AND/ GENERALIZED MEAN SQUARE AMPLITUDES OF VIBRATION, AND SHRINKAGE EFFECTS OF SOME TRIGONAL B 1432
IFLUORIDE AND OXYGEN DIFLUORIDE.   MEAN AMPLITUDES OF VIBRATION AND THERMODYNAMIC FUNCTIONS FOR SILICON D 0773
L HEXAFLUORIDES.   MEAN AMPLITUDES OF VIBRATION AND THERMODYNAMIC FUNCTIONS FOR SOME META 1459
PENTAFLUORIDE.   MEAN AMPLITUDES OF VIBRATION AND THERMODYNAMIC FUNCTIONS FOR VANADIUM 1216
-5 TRIHALIDES. (NITROGEN TRIFLUORIDE,/ MEAN AMPLITUDES OF VIBRATION AND THERMODYNAMIC FUNCTIONS OF SOME GROUP 0719
HEXAFLUORIDE.   MEAN AMPLITUDES OF VIBRATION AND THERMODYNAMIC FUNCTIONS OF TECHNETIUM 1104
AFLUORIDE.   MEAN AMPLITUDES OF VIBRATION AND THERMODYNAMIC FUNCTIONS OF XENON TETR 1274
XY3 MOLECULES.   CALCULATIONS OF MEAN AMPLITUDES OF VIBRATION AND UREY-BRADLEY FORCE FIELDS FOR PLANAR 1399
BORON TRIFLUORIDE.   ROTATION VIBRATION BAND CONTOURS IN THE INFRARED SPECTRUM OF 0193
MOLECULAR POLARIZABILITIES FOR DIHALI/ MEAN AMPLITUDES OF VIBRATION, BASTIANSEN-MORINO SHRINKAGE EFFECT, AND 0124
HERMODYNAMIC FUNCTIONS OF KRYPTON DIF/ MEAN AMPLITUDES OF VIBRATION, BASTIANSEN-MORINO SHRINKAGE EFFECT AND T 0562
FFECT, AND THERMODYNAMIC FUNCTIONS OF/ MEAN AMPLITUDES OF VIBRATION, CORIOLIS COUPLING CONSTANTS, SHRINKAGE E 0050
APPLICATION TO PHOSPHORUS PENTAFLUORI/ MEAN AMPLITUDES OF VIBRATION FOR A TRIGONAL BIPYRAMIDAL MOLECULE WITH 0864
(OF THE FIRST R/ MEAN AMPLITUDES AND SHRINKAGE EFFECTS OF VIBRATION FOR SOME LINEAR SYMMETRICAL DIFLUORIDES. 1369
REVISED MEAN AMPLITUDES OF VIBRATION FOR SOME OCTAHEDRAL HEXAFLUORIDES. 1447
4 MOLECULES.   MEAN AMPLITUDES OF VIBRATION FOR SOME PYRAMIDAL XY3 AND TETRAHEDRAL XY 1407
MEAN AMPLITUDES OF VIBRATION FOR SOME PYRAMIDAL XY3 MOLECULES. 1390
GEN TRIFLUORIDE, PHOSPHORUS TRIFLUORI/ MEAN AMPLITUDES OF VIBRATION FOR SOME PYRAMIDAL XY3 MOLECULES. (NITRO 0706
E PENTAFLUORIDE, BROMINE PENTAFLUORID/ MEAN AMPLITUDES OF VIBRATION FOR SOME PYRAMIDAL XY4Z MOLECULES. (IODIN 0421

	MEAN AMPLITUDES OF VIBRATION FOR SULFUR TETRAFLUORIDE.	1038
DE, NITROGEN DIFLUORIDE, ALUMINUM DIF/ MEAN AMPLITUDES OF VIBRATION FOR THE BORON DIFLUORIDE, CARBON DIFLUORI	1384
AND MAGNESIUM. MEAN AMPLITUDES OF VIBRATION FOR THE DIHALIDES OF MERCURY, BERYLLIUM,	0619
RIUM MONOFLUORIDE, AND LANTHANUM/ EVALUATION OF MOLECULAR VIBRATION FREQUENCIES OF SCANDIUM MONOFLUORIDE, YTT	0936
LES. (SULFUR HEXAFLUORID/ MOLECULAR FORCE FIELDS. PART-8: VIBRATION FREQUENCIES OF SOME OCTAHEDRAL XY6 MOLECU	1080
THANUM HALIDES. VALENCE VIBRATION FREQUENCIES OF TRIHALIDE MOLECULES OF LAN	0574
NORMAL VIBRATION FREQUENCIES OF ZIRCONIUM HALIDES.	1330
CALCULATED MEAN AMPLITUDES OF VIBRATION IN BORON TRIHALIDES.	0197
ROOT MEAN SQUARE AMPLITUDES OF VIBRATION IN SOME GROUP-4 TETRAHALIDES.	1363
TRAFLUORIDE) ROOT MEAN SQUARE AMPLITUDES OF VIBRATION IN SOME GROUP-4 TETRAHALIDES. (CARBON TE	0342
MEAN AMPLITUDES OF VIBRATION IN SOME XY2Z MOLECULES OF C2V SYMMETRY.	1396
/RETICAL EVALUATION OF THE ROOT MEAN SQUARE AMPLITUDES OF VIBRATION IN XENON DIFLUORIDE AND XENON TETRAFLUOR/	1257
IZABLE IONS. FORCE CONSTANT OF THE NONPLANAR VIBRATION OF AN XY3 MOLECULE AND A MODEL WITH POLAR	0573
INTERATOMIC DISTANCES AND ROOT MEAN SQUARE AMPLITUDES OF VIBRATION OF GASEOUS SILICON TETRAFLUORIDE FROM EL/	1002
WITH APPLICATION/ NORMAL COORDINATE ANALYSIS OF ROTATION VIBRATION OF OCTAHEDRAL XY6 AND Z6 MOLECULAR MODELS	1454
AND VANADIU/ SPECTROSCOPIC STUDIES IN MEAN AMPLITUDES OF VIBRATION OF PENTAFLUORIDES OF ARSENIC, PHOSPHORUS,	0871
POTENTIAL FUNCTION FOR THE NU-7 VIBRATION OF PHOSPHORUS PENTAFLUORIDE.	0830
RIDE. MEAN AMPLITUDES OF VIBRATION OF SELENIUM TRIOXIDE AND CHLORINE TRIFLUO	0406
ART-5: TETRAHALIDES OF HAFNIUM AND LE/ MEAN AMPLITUDES OF VIBRATION OF SOME LANTHANIDE TRIFLUORIDES.	1354
ART 3 AND 4. MEAN AMPLITUDES OF VIBRATION OF SOME TETRAHEDRAL XY4 TYPE MOLECULES. P	0487
/ORCE FIELD FORCE CONSTANTS AND MEAN SQUARE AMPLITUDES OF VIBRATION OF SOME TETRAHEDRAL XY4 TYPE MOLECULES. P	1420
MEAN AMPLITUDES OF VIBRATION OF SOME TRIGONAL BIPYRAMIDAL MOLECULES B/	0878
ON TETRAFLUORIDE) MEAN AMPLITUDES OF VIBRATION OF SOME XY4 TETRAHEDRAL MOLECULES.	1419
MEAN AMPLITUDES OF VIBRATION OF SOME XY4 TETRAHEDRAL MOLECULES. (SILIC	1008
MEAN AMPLITUDES OF VIBRATION OF SOME XY6 MOLECULES.	1461
RIBUTIONS. (SELENIUM A/ CLASSIFICATION OF NORMAL MODES OF VIBRATION. PART-2: INTERACTION TERMS IN ENERGY DIST	1494
CLASSICAL VIBRATION PROBLEM FOR THE PYRAMIDAL XY4 MOLECULE.	1421
VIBRATION ROTATION BANDS OF BORON TRIFLUORIDE.	0199
PHORUS PENTAFLUORIDE, AND CYCLOPROPANE B/ ANALYSES OF THE VIBRATION ROTATION BANDS OF TRIFLUORO METHANE, PHOS	0880
VIBRATION ROTATION ENERGIES OF MOLECULES.	1490
IN IODI/ UREY-BRADLEY FORCE CONSTANTS, MEAN AMPLITUDES OF VIBRATION, SHRINKAGE EFFECT AND CORIOLIS CONSTANTS	0513
ND THERMODYNAMIC FUN/ FORCE CONSTANTS, MEAN AMPLITUDES OF VIBRATION, SHRINKAGE EFFECTS, CORIOLIS CONSTANTS, A	1219
S IN ITS QU/ CORRELATION OF INTERMOLECULAR EFFECTS IN THE VIBRATION SPECTRA OF BORON FLUORIDE WITH ALTERATION	1404
ORO CHLORIDE MOLECULES. MEAN AMPLITUDES OF VIBRATION: SULFUR TETRAFLUORIDE AND SULFUR PENTAFLU	1057
LARIZABILITES OF/ POTENTIAL CONSTANTS, MEAN AMPLITUDES OF VIBRATION, THERMODYNAMIC FUNCTIONS AND MOLECULAR PO	1437
ALUMINUM AND ALKALI HALIDE MONOMERS MEAN AMPLITUDES OF VIBRATION WITH LOW TEMPERATURE ANOMALIES.	0024
FORCE CONSTANTS OF MOLECULES WITH STRONGLY COUPLED VIBRATIONS.	1496
SHRINKAGES OF THE INTERNUCLEAR DISTANCES BY MOLECULAR VIBRATIONS.	1489
FOR EXACT SOLUTION OF 2 X 2 SECULAR EQUATION IN MOLECULAR VIBRATIONS. SIMPLE METHOD	1386
FREQUENCIES OF ZIRCONIUM FLUORIDE (GAS) VIBRATIONS ACCORDING TO THERMODYNAMIC DATA.	1329
IGONAL BI/ PENTACOORDINATED MOLECULES. PART-14: MOLECULAR VIBRATIONS AND STEREOCHEMICAL NONRIGIDITY OF THE TR	0854
NTHANUM TRIFLUORIDE) LATTICE VIBRATIONS AND STRUCTURE OF RARE EARTH HALIDES. (LA	0572
XAFLUORIDES. MEAN AMPLITUDES OF THERMAL VIBRATIONS AND THERMODYNAMIC PROPERTIES OF METAL HE	1467
X-RAY STUDY OF ANHARMONIC VIBRATIONS IN CALCIUM DIFLUORIDE.	0283
/ECTROMETER AND ITS USE IN THE STUDY OF THE LOW FREQUENCY VIBRATIONS OF ANTIMONY PENTAFLUORIDE AND THE PURE /	0083
CALCULATION OF THE FREQUENCIES OF THE NORMAL VIBRATIONS OF ANTIMONY TRIFLUORIDE.	0074
AND CONTOURS AND VIBRATIONAL POTENTIAL FUNCTION OF THE E' VIBRATIONS OF ARSENIC PENTAFLUORIDE. /RPENDICULAR B	0116
PERPENDICULAR VIBRATIONS OF BORON TRIFLUORIDE.	0202
NORMAL VIBRATIONS OF BORON TRIHALIDES. (BORON TRIFLUORIDE)	0247
ABSOLUTE INFRARED INTENSITIES OF THE FUNDAMENTAL VIBRATIONS OF NITROGEN TRIFLUORIDE.	0736
INFRARED BAND SHAPES AND VIBRATIONS OF NITROGEN TRIFLUORIDE.	0699
VIBRATIONS OF PENTATOMIC TETRAHEDRAL MOLECULES.	1497
APPLICATION XENON TETRAFLUORIDE. VIBRATIONS OF PLANAR SYMMETRICAL XY4 MOLECULES WITH	1268
JAHN-TELLER VIBRATIONS OF RHENIUM HEXAFLUORIDE.	0910
LUORIDE. MOLECULAR VIBRATIONS OF SULFUR HEXAFLUORIDE AND SULFUR TETRAF	1034
MOLECULAR VIBRATIONS OF SULFUR TETRAFLUORIDE.	1058
LEAD. (GERMANIUM DIFLUORI/ FREQUENCIES OF THE DEFORMATION VIBRATIONS OF THE DIHALIDES OF GERMANIUM, TIN, AND	0479
L STRETCHING CONSTANTS, AND MEAN SQUARE AMPLITUDES OF THE VIBRATIONS OF THE NITROGEN TRIFLUORIDE MOLECULE. /A	0726
LE. POTENTIAL FIELD AND MOLECULAR VIBRATIONS OF THE TRIGONAL BIPYRAMID AX(3)YZ MOLECU	1436
ECTRUM OF PHOSPHORUS PENTA/ POTENTIAL FIELD AND MOLECULAR VIBRATIONS OF THE TRIGONAL BIPYRAMIDAL THE RAMAN SP	0863
VIBRATION- ROTATION BANDS OF BORON TRIFLUORIDE.	0191
TAL OF SULFUR HEXAFLUORID/ OPTICAL SATURATION OF A SINGLE VIBRATION- ROTATION TRANSITION IN THE NU-3 FUNDAMEN	1074
CALCULATION OF MEAN VIBRATIONAL AMPLITUDES.	1486
ABSORPTION SPECTRUM OF CARBON DIFLUORIDE AND ITS VIBRATIONAL ANALYSIS.	0303
AND ENERGY OF DISSOCIATION IN GERMANIUM MONOFLUORIDE. VIBRATIONAL ANALYSIS OF C-X AND C'-X' BAND SYSTEMS	0470
UNGSTEN HEXAFLUORIDE. VIBRATIONAL ANALYSIS OF SELENIUM HEXAFLUORIDE AND T	0960
OLECULES. DIRECT DETERMINATION FORCE CONSTANTS FROM THE / VIBRATIONAL ANALYSIS OF SUBSTITUTED AND PERTURBED M	1444
NIC MONOFLUORIDE. VIBRATIONAL ANALYSIS OF THE SINGLET SYSTEMS OF ARSE	0095
NY MONOFLUORIDE. VIBRATIONAL ANALYSIS OF ULTRAVIOLET BANDS OF ANTIMO	0062
ULES OF SOME / EMPIRICAL METHOD FOR DETERMINING EFFECTIVE VIBRATIONAL AND ROTATIONAL CHARACTERISTICS OF MOLEC	0622
ORIDE IN THE 2500 ANGSTROM REGION. ANALYSIS OF VIBRATIONAL AND ROTATIONAL STRUCTURE OF TIN MONOFLU	1138
TORSIONAL COORDINATES IN VIBRATIONAL ANHARMONICITY. APPLICATION TO ETHYLENE.	1485
VIBRATIONAL ANHARMONICITY IN BORON TRIFLUORIDE.	0210
VIBRATIONAL ANHARMONICITY IN THE METHYL HALIDES.	1493
IDES. VIBRATIONAL ANHARMONICITY IN THE PYRAMIDAL TRIFLUOR	0745
IS OF SELENIUM TETRAFLUORIDE. VIBRATIONAL ASSIGNMENT AND NORMAL COORDINATE ANALYS	0958
NONRIGID D5H MOLECULE. VIBRATIONAL ASSIGNMENT FOR IODINE PENTAFLUORIDE, A	0525
VIBRATIONAL ASSIGNMENT OF SULFUR TETRAFLUORIDE.	1036
SIS FOR ANTIMONY PENTAFLUORIDE. VIBRATIONAL ASSIGNMENTS AND NORMAL COORDINATE ANALY	0087
MOLECULES OF THE SPHERICAL TOP T/ ROTATIONAL PROFILES OF VIBRATIONAL BANDS IN THE GAS PHASE RAMAN SPECTRA OF	1442
AND QUANTUM MECHANICAL STUDIES / SPECTROSCOPIC STUDIES OF VIBRATIONAL CONSTANTS, STATISTCIAL THERMODYNAMICS	0784
EVALUATION OF FORCE CONSTANTS OF PLANE XY3 MOLECULES FROM VIBRATIONAL DATA.	1401
RIDE, OSMIUM HEXACHLORIDE(2-), AND IRIDIUM HEXACHLORIDE(/ VIBRATIONAL ELECTRONIC COUPLING IN IRIDIUM HEXAFLUO	0538
RELAXATION OF THE MOLECULAR CHARGE DISTRIBUTION AND THE VIBRATIONAL FORCE CONSTANT.	0151
. LONE PAIR MODEL AND THE VIBRATIONAL FORCE CONSTANTS OF NITROGEN TRIFLUORIDE	0705
LES. (NITROGEN TRIFLUORIDE) A BENT BOND MODEL FOR THE VIBRATIONAL FORCE CONSTANTS OF NONPLANAR XY3 MOLECU	0704
LICATION TO SOME SQUARE PYRAMIDAL MOLECU/ REPARAMETERIZED VIBRATIONAL FORCE CONSTANTS. PART-1: METHOD AND APP	0420
HORUS TRICHLORIDE FROM AMPLITUDES OF VIBRATION AND NORMAL VIBRATIONAL FREQUENCIES. /ENTIAL CONSTANTS OF PHOSF	1481
OF FIRST ROW TRANSITION METAL DIFLUORIDES. VIBRATIONAL FREQUENCIES AND VALENCE FORCE CONSTANTS	1371
/FLUORIDE, AND ARSENIC TRIFLUORIDE BY THE COMBINED USE OF VIBRATIONAL FREQUENCIES, CENTRIFUGAL STRETCHING, A/	0717

BINARY FLUORIDES--PERMUTED TITLE INDEX

```
S.                                          CALCULATION OF THE VIBRATIONAL FREQUENCIES OF SCANDIUM SUBGROUP HALIDE 0949
S.                                                             VIBRATIONAL FREQUENCIES OF SOME PLANAR XY3 MOLECULE 1391
                                                               VIBRATIONAL MEAN SQUARE AMPLITUDE MATRICES.         1389
                                                               VIBRATIONAL MEAN SQUARE AMPLITUDE MATRICES.         1392
NSEN-MORINO SHRINKAGE EFFECT OF SYMMETRICAL XY3 AND XY4 /      VIBRATIONAL MEAN SQUARE AMPLITUDE MATRICES. (BASTIA 1394
TREATMENT OF TETRAHEDRAL XY4 MOLECULES.                        VIBRATIONAL MEAN SQUARE AMPLITUDE MATRICES. PART-7: 1408
AN ELECTRON DIFFRACTION S/ STRUCTURE, PSEUDOROTATION, AND      VIBRATIONAL MODE COUPLING IN IODINE HEPTAFLUORIDE.  0515
RARED SPECTRA OF RUTHENIU/ JAHN-TELLER EFFECT IN THE E(8)      VIBRATIONAL MODE OF HEXAFLUORIDE MOLECULES. THE INF 0926
DE.                                         INFRARED SPECTRUM AND VIBRATIONAL POTENTIAL FUNCTION OF ARSENIC TRIFLUORI 0100
OF ARSENIC PENTAFLUORID/ PERPENDICULAR BAND CONTOURS AND       VIBRATIONAL POTENTIAL FUNCTION OF THE E° VIBRATIONS 0116
/OF CORIOLIS INTERACTION CONSTANTS IN THE SOLUTION OF THE      VIBRATIONAL PROBLEM FOR XY4 TETRAHEDRAL MOLECULES.  1423
/F PYRAMIDAL MOLECULES OF THE XY3 TYPE BY SOLUTION OF THE      VIBRATIONAL PROBLEM WITH THE USE OF CENTRIFUGAL ST/ 0727
                                                     INFRARED  VIBRATIONAL PROPERTIES OF GERMANIUM DIFLUORIDE.    0473
                                                               VIBRATIONAL PROPERTIES OF HEXAFLUORIDE MOLECULES.  0021
SPHORUS PENTAFLUORIDE. INFRARED AND LOW TEMPERATURE RAMAN      VIBRATIONAL SPECTRA. /US PHOSPHORUS HALIDES AND PHO 0849
LUORIDE. FERMI RESONANCE BETWEEN A FUNDAMENTAL AND A TER/      VIBRATIONAL SPECTRA AND ASSIGNMENT OF IODINE PENTAF 0504
TETRAFLUORIDE.                                                 VIBRATIONAL SPECTRA AND FORCE CONSTANTS OF URANIUM  1172
CULES. PART-1: INFRARED SPECTRUM OF OXYGEN DIFLUORIDE FR/      VIBRATIONAL SPECTRA AND STRUCTURE OF INORGANIC MOLE 0761
ORIDE.                                                         VIBRATIONAL SPECTRA AND STRUCTURE OF XENON TETRAFLU 1262
LUORIDE AND HEPTAFLUORIDE.                                     VIBRATIONAL SPECTRA AND STRUCTURES OF IODINE PENTAF 0506
AFLUORIDE AND TELLURIUM TETRAFLUORIDE, INCLUDING A MATRI/      VIBRATIONAL SPECTRA AND STRUCTURES OF SELENIUM TETR 0956
UORIDE AND XENON OXYTETRAFLUORIDE.                             VIBRATIONAL SPECTRA AND STRUCTURES OF XENON TETRAFL 1261
CHLORINE TRIFLUORIDE AND BROMINE TRIFLUORIDE.                  VIBRATIONAL SPECTRA AND THERMODYNAMIC PROPERTIES OF 0396
HALIDES. (ZINC DIFLUORIDE)                                     VIBRATIONAL SPECTRA AND THERMODYNAMICS OF THE ZINC  1323
SQUARE PYRAMIDAL MOLECULES. XENON OXIDE TETRAFLUORIDE, I/      VIBRATIONAL SPECTRA AND VALENCE FORCE CONSTANTS OF  0418
                                          JAHN-TELLER EFFECT IN VIBRATIONAL SPECTRA OF HEXAFLUORIDES.             1471
                                                               VIBRATIONAL SPECTRA OF INORGANIC FLUORIDES.        0014
NIUM HEPTAFLUORIDE.                                            VIBRATIONAL SPECTRA OF IODINE HEPTAFLUORIDE AND RHE 0522
N TETRAFLUORIDE.                                               VIBRATIONAL SPECTRA OF LIQUID AND CRYSTALLINE CARBO 0327
HOSPHORUS PENTAFLUORIDE)                                       VIBRATIONAL SPECTRA OF MIXED PHOSPHORUS HALIDES. (P 0829
AFLUORIDES.                                                    VIBRATIONAL SPECTRA OF MOLYBDENUM AND TUNGSTEN PENT 0625
TECHNETIUM HEXAFLUORIDE.                                       VIBRATIONAL SPECTRA OF MOLYBDENUM HEXAFLUORIDE AND  0628
TUNGSTEN PENTAFLUORIDE.                                        VIBRATIONAL SPECTRA OF MOLYBDENUM PENTAFLUORIDE AND 0626
UORIDES IN THE GAS PHASE. VAPOR DENSITY OF NIOBIUM PENTA/      VIBRATIONAL SPECTRA OF NIOBIUM AND TANTALUM PENTAFL 0660
INUM HEXAFLUORIDE.                                             VIBRATIONAL SPECTRA OF OSMIUM HEXAFLUORIDE AND PLAT 0752
ARSENIC PENTAFLUORIDE  AND THE HEIGHT OF THE BARRIER TO/       VIBRATIONAL SPECTRA OF PHOSPHORUS PENTAFLUORIDE AND 0856
                                                               VIBRATIONAL SPECTRA OF RHENIUM HEPTAFLUORIDE.      0919
                                                               VIBRATIONAL SPECTRA OF RHENIUM HEXAFLUORIDE.       0905
. PART-1: MAGNESIUM DIFLUORIDE.                   GEOMETRY AND VIBRATIONAL SPECTRA OF THE ALKALINE EARTH DIHALIDES 0609
CADMIUM. (CADMIUM DIFLUORIDE AND MERCURY DIFLUORIDE)           VIBRATIONAL SPECTRA OF THE DIHALIDES OF MERCURY AND 0275
DINE PENTAFLUORIDE, TELLURIUM PENTAFLUORIDE(-) AND ANTIM/      VIBRATIONAL SPECTRA OF THE ISOELECTRONIC SPECIES IO 0501
                                                               VIBRATIONAL SPECTRA OF VANADIUM PENTAFLUORIDE.     1211
HEXAFLUORIDE.                                                  VIBRATIONAL SPECTRA OF XENON COMPOUNDS.            1232
AFLUORIDE.                                                     VIBRATIONAL SPECTRUM AND FORCE FIELD OF MOLYBDENUM 0634
HEXAFLUORIDE.                                                  VIBRATIONAL SPECTRUM AND FORCE FIELD OF URANIUM HEX 1196
F URANIUM HEXAFLUORIDE GAS.                                    VIBRATIONAL SPECTRUM AND STRUCTURE OF SOLID URANIUM 1189
F URANIUM HEXAFLUORIDE GAS.                                    VIBRATIONAL SPECTRUM AND THERMODYNAMIC PROPERTIES O 1186
                                                               VIBRATIONAL SPECTRUM AND THERMODYNAMIC PROPERTIES O 1187
DE.                                                            VIBRATIONAL SPECTRUM OF BROMINE TRIFLUORIDE.       0257
MICROWAVE SPECTRUM OF ARSENIC TRIFLUORIDE IN THE EXCITED       VIBRATIONAL SPECTRUM OF IODINE TRIFLUORIDE.        0500
MICROWAVE SPECTRA OF NITROGEN TRIFLUORIDE IN THE EXCITED       VIBRATIONAL SPECTRUM OF LIQUID CHROMIUM PENTAFLUORI 0435
/ MICROWAVE SPECTRUM OF SILICON DIFLUORIDE IN THE EXCITED      VIBRATIONAL STATE.                                 0097
MICROWAVE SPECTRA OF NITROGEN TRIFLUORIDE IN THE EXCITED       VIBRATIONAL STATES. (EQUILIBRIUM STRUCTURE)        0722
ENIUM HEXAFLUORIDE.                                            VIBRATIONAL STATES, EQUILIBRIUM STRUCTURE, ANHARMON 0993
                                                               VIBRATIONAL STATES. (ZETA TYPE DOUBLING AND RESONA/ 0723
NY MONOFLUORIDE.                          DETERMINATION OF THE EFFECTIVE VIBRATIONAL STRUCTURE AND MOLECULAR STRUCTURE OF RH 0907
S FOR DIATOMIC FLUORIDES OF SILICON AND GERMANIUM.            VIBRATIONAL STUDY OF VANADIUM TRIFLUORIDE.          1207
STEM OF BORON MONOFLUORIDE.                                    VIBRATIONAL TEMPERATURE IN THE A-2 SYSTEM OF ANTIMO 0061
(TIN FLUORIDE)                                    ACCURACY OF THE VIBRATIONAL TRANSITION PROBABILITIES AND R CENTROID 0994
SONANCE AND E/ MICROWAVE SPECTRUM OF OXYGEN DIFLUORIDE IN      VIBRATIONAL TRANSITION PROBABILITIES FOR THE A-X SY 0168
                                                               VIBRATIONAL WAVE FUNCTIONS AND DERIVED QUANTITIES. 1134
TEMS EXHIBITING JAHN-TELLER EFFECT.                           VIBRATIONALLY EXCITED STATES, NU-1- 2 NU-2 FERMI RE 0771
ES.                                                            VIBRATIONALLY INDUCED DIPOLE MOMENTS.              1417
                                                               VIBRONIC AND SPIN ORBIT SPLITTING IN SPECTRA OF SYS 1480
MONOFLUORIDE, AND LANTHANUM MONOFLUORIDE.                      VIBRONIC INTERACTIONS IN METAL HEXAFLUORIDE MOLECUL 0908
E ELECTRONIC SPECTRUM OF IRIDIUM HEXAFLUORIDE.                 VIBRONIC SPECTRA OF YTTRIUM MONOFLUORIDE.           1315
IDE MOLECULE.                                                  VIBRONIC SPECTRUM OF SCANDIUM MONOFLUORIDE, YTTRIUM 0938
LDS.                                                           VIBRONICALLY INDUCED JAHN-TELLER PROGRESSIONS IN TH 0537
/CULAR XENON HEXAFLUORIDE. RAMAN SCATTERING AND INFRARED,     VIOLET AND ULTRAVIOLET SPECTRUM OF BARIUM MONOFLUOR 0122
ECULE.                                       ROTATIONAL ANALYSIS OF THE VIRIAL THEOREM DECOMPOSITION OF MOLECULAR FORCE FIE 0783
                              NEAR ULTRAVIOLET AND VISIBLE AND ULTRAVIOLET ABSORPTION IN THE VAPOR AN/ 1285
                              ROTATIONAL ANALYSIS OF THE VISIBLE BAND SYSTEM OF THE BISMUTH MONOFLUORIDE MOL 0146
                                                               VISIBLE BANDS OF ANTIMONY MONOFLUORIDE.            0070
APPLICATION TO THE CALCULATION BOND MOMENT / A GEOMETRIC      VISIBLE EMISSION BANDS OF LEAD MONOFLUORIDE.       0587
RSENIC TRIFLUORIDE, SULFUR / THE INFRARED SPECTRA OF SOME     VISIBLE EMISSION SPECTRUM OF BISMUTH MONOFLUORIDE.  0145
IDE AND OTHER MOLECULES.                                      VISUALIZATION OF NORMAL COORDINATE TRANSFORMATIONS. 0352
E.                                                CONSTANT    VOLATILE INORGANIC FLUORIDES IN THE SOLID STATE. (A 0096
RATIONAL STATES. (ZETA TYPE DOUBLING AND RESONANCE IN THE     VOLUME HEAT CAPACITIES OF GASEOUS CARBON TETRAFLUOR 0335
L STATES. (ZETA TYPE DOUBLING AND RESONANCE IN THE V3 AND     VOLUME HEAT CAPACITY OF GASEOUS CARBON TETRAFLUORID 0336
                                                               V3 AND V4 STATES) /N TRIFLUORIDE IN THE EXCITED VIB 0723
                                                               V4 STATES) /N TRIFLUORIDE IN THE EXCITED VIBRATIONA 0723
```

W

, ELECTRON AFFINITY, D/ OXYGEN MONOFLUORIDE. HARTREE-FOCK WAVE FUNCTION, BINDING ENERGY, IONIZATION POTENTIAL 0756
, ELECTRON / OXYGEN MONOFLUORIDE, DOUBLET PI HARTREE-FOCK WAVE FUNCTION, BINDING ENERGY, IONIZATION POTENTIAL 0757
ELECTRONIC POPULATION ANALYSIS CF MOLECULAR WAVE FUNCTIONS. 0157
/ICAL NDDO-2 (ALPHA, BETA) METHOD. PART-3: ENERGY LEVELS, WAVE FUNCTIONS AND AO POPULATIONS OF POLYATOMIC FL/ 0016
LET PI GROUND STATES OF SULFUR MONOFLUORIDE/ HARTREE-FOCK WAVE FUNCTIONS AND COMPUTED PROPERTIES FOR THE DOUB 1025
E) ACCURACY OF THE VIBRATIONAL WAVE FUNCTIONS AND DERIVED QUANTITIES. (TIN FLUORID 1134
ART-2: DIPOLE MOMENT AN/ CHARACTERIZATION OF GROUND STATE WAVE FUNCTIONS BY MEASURED ELECTRONIC PROPERTIES. P 0746
AB INITIO AND MZDO WAVE FUNCTIONS FOR SULFUR HEXAFLUORIDE. 1066
ONDING OF FOUR AND EIGHT PI ELECTRON SYSTEMS. WAVE FUNCTIONS FOR THE FOUR ELECTRON THREE CENTER B 0301
RELATIVISTIC MOLECULAR WAVE FUNCTIONS: XENON DIFLUORIDE. 1249
M HEXAFLUORIDE, NEPTUNIUM HEXAFLUORIDE, AND PLUTONI/ LCNG WAVE LENGTH, INFRARED ACTIVE FUNDAMENTAL FOR URANIU 1192
TIONAL SPECTRA OF BROMINE PENTAFLUORIDE IN THE MILLIMETER WAVE LENGTH REGION. ROTA 0268
OF NITROGEN TRIFLUORIDE. MILLIMETER WAVE SPECTRUM AND CENTRIFUGAL DISTORTION CONSTANTS 0718
ONOFLUORIDE. MILLIMETER AND SUBMILLIMETER WAVE SPECTRUM AND MOLECULAR CONSTANTS OF ALUMINUM M 0033
MILLIMETER WAVE SPECTRUM AND STRUCTURE OF OXYGEN DIFLUORIDE. 0768
K-TYPE DOUBLING IN THE MILLIMETER WAVE SPECTRUM OF BROMINE PENTAFLUORIDE. 0260
FAR INFRARED AND MILLIMETER WAVE STUDIES OF XENON HEXAFLUORIDE. 1299
PART-1: HEXAFLUORIDES CF GROUP-6/ SCF- X(ALPHA) SCATTERED WAVE STUDIES ON BONDING AND IONIZATION POTENTIALS. 1366
NIC STRUCTURE OF SULFUR HEXAFLUORIDE STUDIED BY ULTRALONG WAVE X-RAY SPECTROSCOPY. ELECTRO 1065
ISSOCIATION OF CHLORINE TRIFLUORIDE BEHIND INCIDENT SHOCK WAVES. THERMAL D 0394
AND THE L MATRIX METHOD. FORCE FIELD OF WEINSTOCK AND GOODMAN FOR OCTAHEDRAL HEXAFLUORIDES 1456
MOLECULES. NEW METHOD FOR RESOLVING WILSON'S SECULAR EQUATIONS. APPLIED TO TETRAHEDRAL 1431

X

OF XENON HEXAFLUORIDE. MULTIPLE SCATTERING X(ALPHA) CALCULATION OF THE ENERGY LEVEL STRUCTURE 1304
ZATION POTENTIALS. PART-1: HEXAFLUORIDES OF GROUP-6/ SCF- X(ALPHA) SCATTERED WAVE STUDIES ON BONDING AND IONI 1366
STEMS O/ ROTATIONAL ANALYSIS OF THE A DOUBLET SIGMA(+) TO X DOUBLET PI, B DOUBLET SIGMA(+) TO X DOUBLET PI SY 1133
/IX ELEMENT OF THE DIPOLE MOMENT OF AN A DOUBLET SIGMA TO X DOUBLET PI ELECTRON TRANSITON IN A SILICON MONOF/ 0970
LUORIDE. A QUARTET SIGMA TO X DOUBLET PI ELECTRONIC TRANSITION OF SILICON MONOF 0971
/ISOTOPIC SHIFT IN THE (1,0) BAND OF THE A DOUBLET SIGMA- X DOUBLET PI SYSTEM OF THE TIN MONOFLUORIDE MOLECU/ 1135
/ SIGMA(+), B DOUBLET SIGMA(+), AND A QUARTET SIGMA(-) TO X DOUBLET PI TRANSITIONS OF GERMANIUM MONOFLUORIDE. 0466
PI 3/2 / ROTATIONAL ANALYSIS OF THE C DOUBLET DELTA 5/2- X DOUBLET PI 3/2 AND G DOUBLET DELTA 5/2- X DOUBLET 1136
THE C DOUBLET PI AND THE X DOUBLET SIGMA STATES OF BARIUM MONOFLUORIDE. 0120
E IN EMISSION / ANALYSIS OF THE BANDS OF B DOUBLET SIGMA- X DOUBLET SIGMA. SYSTEM OF CALCIUM FLUORIDE MOLECUL 0280
ITH A VACUUM ECHELLE SPECTROG/ STUDY OF THE A SINGLET PI- X SINGLET SIGMA(+) BANDS OF BORON-11 MONOFLUORIDE W 0170
ROTATIONAL ANALYSIS OF THE C SINGLET PI- X SINGLET SIGMA(+) SYSTEM OF INDIUM MONOFLUORIDE. 0491
/ OF BANDS OF THE A TRIPLET PI ZERO(+) B TRIPLET PI(1) TO X SINGLET SIGMA(+) SYSTEMS OF GALLIUM MONOFLUORIDE. 0457
UORIDE. ROTATIONAL ANALYSIS OF THE AO(+), BO(+), TO X SINGLET SIGMA(+) SYSTEMS OF GASEOUS SILVER MONOFL 1011
/OF BANDS OF THE A TRIPLET PI ZERO(+), B TRIPLET PI(1) TO X SINGLET SIGMA(+) SYSTEMS OF THALLIUM MONOFLUORID/ 1111
/RANCK-CONDON FACTORS AND R CENTROIDS OF A SINGLET SIGMA- X SINGLET SIGMA PLATINUM OXIDE AND SILICON MONOFLU/ 0966
SITION MOMENT WITH INTERNUCLEAR DISTANCE FOR BANDS OF THE X SYSTEM. CORRELATION OF THE ELECTRONIC TRAN 0969
MA(+/ SHOCK TUBE DETERMINATION OF THE C2 (A TRIPLET PI TO X TRIPLET PI) AND CARBON MONOFLUORIDE A DOUBLET SIG 0588
. B SINGLET SIGMA(+) TO X TRIPLET SIGMA(-) BAND SYSTEM IN LEAD MONOFLUORIDE 0585
RIDE. B SINGLET SIGMA(+) TO X TRIPLET SIGMA(-) BAND SYSTEM OF NITROGEN MONOFLUO 0675
/-CONDON FACTORS AND R CENTROIDS FOR THE 6 SINGLET SIGMA- X TRIPLET SIGMA BAND SYSTEM OF NITROGEN MONOFLUORI/ 0677
SIMPLE METHOD FOR EXACT SOLUTION OF 2 X 2 SECULAR EQUATION IN MOLECULAR VIBRATIONS. 1386
IDE CALCULATED BY THE SELF CONSISTENT MULTIPLE SCATTERING X-ALPHA METHOD. /FLUORIDE AND ELEMENT 110 HEXAFLUOR 0881
X-RAY ABSORPTION IN SULFUR HEXAFLUORIDE. 1067
ULTRASOFT X-RAY ABSORPTION SPECTRA OF BORON TRIFLUORIDE. 0204
FLUORIDE) EVIDENCE OF EFFECTIVE POTENTIAL BARRIERS IN THE X-RAY ABSORPTION SPECTRA OF MOLECULES. (SULFUR HEXA 1073
ULTRASOFT X-RAY ABSORPTION SPECTRUM OF SULFUR HEXAFLUORIDE. 1102
OF XENON HEXAFLUORID/ EFFECTS OF ELECTRON CORRELATION IN X-RAY DIFFRACTION AND AN ELECTRON DIFFRACTION STUDY 1290
HEXAFLUORIDES. (SECOND ROW, THIRD ROW, AND RHENIUM HEPTA/ X-RAY DIFFRACTION STUDIES OF SOME TRANSITION METAL 1375
LFUR HEXAFLUORIDE. SULFUR K AND L AND FLUORINE K X-RAY EMISSION AND ABSORPTION SPECTRA OF GASEOUS SU 1088
/S OF SULFUR HEXAFLUORIDE. PART-1: OPTICAL ABSORPTION AND X-RAY EMISSION FROM SCF MO LCAO COMPUTATIONS. PART/ 1078
ORIDE, SULFUR TETRAFLUORIDE, AND PHOSPHORUS PENTAFLUORID/ X-RAY PHOTOELECTRON SPECTROSCOPY OF CHLORINE TRIFLU 0409
FLUORIDE. X-RAY SPECTRA CF SULFUR AND FLUORINE IN SULFUR HEXA 1101
TRUCTURE OF SULFUR HEXAFLUORIDE STUDIED BY ULTRALONG WAVE X-RAY SPECTROSCOPY. ELECTRONIC S 1065
UR HEXAFLUORIDE. SOFT X-RAY SPECTROSCOPY ON SULFUR TETRAFLUORIDE AND SULF 1033
LUORIDE. X-RAY STUDY OF ANHARMONIC VIBRATIONS IN CALCIUM DIF 0283
NOBLE GASES AND THE BINARY FLUORIDES OF XENON. 1342
CHEMICAL COMPOUNDS OF XENON AND OTHER NOBLE GASES. 0010
VIBRATIONAL SPECTRA OF XENON COMPOUNDS. 1232
AFLUORIDE) IDEAL GAS THERMODYNAMIC PROPERTIES OF XENON COMPOUNDS. (XENON TETRAFLUORIDE AND XENON HEX 1273
BOND LENGTHS IN XENON DIFLUORIDE. 1236
BONDING IN XENON DIFLUORIDE. 1230
GAS DIHALIDES. (ARGON DIFLUORIDE, KRYPTON DIFLUORIDE, AND XENON DIFLUORIDE) D ORBITALS IN THE NOBLE 1338
ELECTRONIC STRUCTURE OF XENON DIFLUORIDE. 1226
FAR ULTRAVIOLET SPECTROSCOPIC STUDY OF XENON DIFLUORIDE. 1256
FORCE FIELDS IN KRYPTON DIFLUORIDE AND XENON DIFLUORIDE. 0556
GAS PHASE STRUCTURE OF XENON DIFLUORIDE. 1248
ODYNAMIC FUNCTIONS CF SOME LINEAR SYMMETRICAL MOLECULES. (XENON DIFLUORIDE) MOLECULAR CONSTANTS AND THERM 1244
PHOTOELECTRON SPECTRUM OF XENON DIFLUORIDE. 1227
RELATIVISTIC MOLECULAR WAVE FUNCTIONS: XENON DIFLUORIDE. 1249
IES OF MOLECULAR CONSTANTS IN SOME POLYATOMIC MOLECULES. (XENON DIFLUORIDE) SPECTROSCOPIC STUD 1245
R OF INERT GAS HALIDES. HELIUM-2 PHOTOELECTRON SPECTRA OF XENON DIFLUORIDE AND KRYPTON DIFLUORIDE. / CHARACTE 1343
HELIUM-2 PHOTOELECTRON SPECTRA OF XENON DIFLUORIDE AND KRYPTON DIFLUORIDE. 1234
ULTRAVIOLET EMISSION SPECTRA. (XENON DIFLUORIDE AND KRYPTON DIFLUORIDE) 0559
E STUDIES OF NIOBIUM PENTAFLUORIDE AND ITS COMPLEXES WITH XENON DIFLUORIDE AND ORGANIC BASES. /UPOLE RESONANC 0664
INE BOND. XENON DIFLUORIDE AND THE NATURE OF THE XENON- FLUOR 1223
UATION OF THE ROOT MEAN SQUARE AMPLITUDES OF VIBRATION IN XENON DIFLUORIDE AND XENON TETRAFLUORIDE. /CAL EVAL 1257
FORBIDDEN ELECTRONIC TRANSITIONS IN XENON DIFLUORIDE AND XENON TETRAFLUORIDE. 1247
MOLECULAR SYMMETRY OF XENON DIFLUORIDE AND XENON TETRAFLUORIDE. 1240
VAPOR PRESSURE AND MELTING POINTS OF XENON DIFLUORIDE AND XENON TETRAFLUORIDE. 1251
ECTURE ON BONDING IN THE SOLIDS. HEATS OF SUBLIMATION OF XENON DIFLUORIDE AND XENON TETRAFLUORIDE AND A CONJ 1237
SPECTRA. THERMOCHEMICAL PROPERTIES OF XENON DIFLUORIDE AND XENON TETRAFLUORIDE FROM MASS 1253

ULTRAVIOLET REGION.   SPECTRA OF XENON DIFLUORIDE AND XENON TETRAFLUORIDE IN THE FAR 1233
                           CRYSTAL AND MOLECULAR STRUCTURE OF XENON DIFLUORIDE BY NEUTRON DIFFRACTION.   1239
                                                                     XENON DIFLUORIDE. (FORCE CONSTANTS, SPECTRA)   1252
ES K AND SOME DERIVED THERMODYNAMIC PROPERTIES.   XENON DIFLUORIDE. HEAT CAPACITY FROM 5 TO 350 DEGRE 1246
                                        INFRARED SPECTRA OF XENON DIFLUORIDE IN DIFFERENT SOLVENTS.   1241
                                IONIZATION AND DISSOCIATION OF XENON DIFLUORIDE INDUCED BY PHOTON IMPACT.   1243
XAFLUORIDE.   PHOTOIONIZATION MASS SPECTROMETRIC STUDY OF XENON DIFLUORIDE, XENON TETRAFLUORIDE, AND XENON HE 1225
/2 PHOTOELECTRON SPECTRA AND THE ELECTRONIC STRUCTURES OF XENON DIFLUORIDE, XENON TETRAFLUORIDE, AND XENON H/ 1228
IDE TETRAFLUORIDE VAPORS, AND FORCE CO/ RAMAN SPECTRA FOR XENON DIFLUORIDE, XENON TETRAFLUORIDE, AND XENON OX 1254
UORIDE.   SELF CONSISTENT FIELD STUDY OF THE SERIES XENON DIFLUORIDE, XENON TETRAFLUORIDE, XENON HEXAFL 1224
/UCTURAL CONSIDERATIONS IN THE CHEMISTRY OF NOBLE GASES. (XENON DIFLUORIDE, XENON TETRAFLUORIDE, XENON HEXAF/ 1235
                            ELECTRON SPIN RESONANCE OF THE XENON FLUORIDE RADICAL.   1288
D EXPERIMENTAL STUDIES OF THE ELECTRONIC STRUCTURE OF THE XENON FLUORIDES.   THEORETICAL AN 1238
CA AND EVIDENCE FOR ORBITAL/ ELECTRON DISTRIBUTION IN THE XENON FLUORIDES AND XENON OXIDE TETRAFLUORIDE BY ES 1229
/GLECT OF DIFFERENTIAL OVERLAP METHODS ILLUSTRATED BY THE XENON FLUORIDES. MODIFIED NDDO-2 (ALPHA, BETA) MET/ 1231
                                         BONDING IN XENON HEXAFLUORIDE.   1298
                                         BONDING IN XENON HEXAFLUORIDE.   1313
L APPLICATIONS OF CORE ELECTRON EXCITATION SPECTROSCOPY. (XENON HEXAFLUORIDE)   CHEMICA 1306
                            CRYSTALLINE MODIFICATIONS OF XENON HEXAFLUORIDE.   1297
IN X-RAY DIFFRACTION AND AN ELECTRON DIFFRACTION STUDY OF XENON HEXAFLUORIDE. /FECTS OF ELECTRON CORRELATION 1290
ELECTRICAL CONDUCTIVITY AND DIELECTRIC CONSTANT OF LIQUID XENON HEXAFLUORIDE.   1307
            EVIDENCE FOR PSEUDO JAHN-TELLER EFFECT IN XENON HEXAFLUORIDE.   1279
            FAR INFRARED AND MILLIMETER WAVE STUDIES OF XENON HEXAFLUORIDE.   1299
                              GAS PHASE STRUCTURE OF XENON HEXAFLUORIDE.   1293
INTRAMOLECULAR REARRANGEMENT IN IODINE HEPTAFLUORIDE AND XENON HEXAFLUORIDE.   0520
                         MAGNETIC FIELD DEFLECTION OF XENON HEXAFLUORIDE.   1287
                            MAGNETIC SUSCEPTIBILITY OF XENON HEXAFLUORIDE.   1308
                            MAGNETIC SUSCEPTIBILITY OF XENON HEXAFLUORIDE.   1310
                               MOLECULAR STRUCTURE OF XENON HEXAFLUORIDE.   1280
                               MOLECULAR STRUCTURE OF XENON HEXAFLUORIDE.   1303
ING X(ALPHA) CALCULATION OF THE ENERGY LEVEL STRUCTURE OF XENON HEXAFLUORIDE.   MULTIPLE SCATTER 1304
                                    POLYMORPHISM IN XENON HEXAFLUORIDE.   1296
       PREPARATION AND SOME THERMAL PROPERTIES OF PURE XENON HEXAFLUORIDE.   1301
                                   RAMAN SPECTRUM OF XENON HEXAFLUORIDE.   1289
          STEREOCHEMISTRY AND SEVEN COORDINATION. (XENON HEXAFLUORIDE)   1286
                          STRUCTURE OF GASEOUS XENON HEXAFLUORIDE.   1295
         STRUCTURE OF THE IODINE HEXAFLUORIDE ANION. (XENON HEXAFLUORIDE)   1300
              STRUCTURES OF NOBLE GAS COMPOUNDS. (XENON HEXAFLUORIDE)   1292
      ELECTRONIC STATES AND MOLECULAR GEOMETRY OF XENON HEXAFLUORIDE: A CASE OF ELECTRONIC ISOMERISM. 1294
                      MOLECULAR STRUCTURE OF XENON HEXAFLUORIDE AND IODINE HEPTAFLUORIDE.   0526
(2-), TELLURIUM HEXABROMIDE(2-), IODINE HEXAFLUORIDE(1-), XENON HEXAFLUORIDE, ETC. /3-), SELENIUM HEXABROMIDE 1309
      HEAT CAPACITY AND OTHER THERMODYNAMIC FUNCTIONS OF XENON HEXAFLUORIDE FROM 5 TO 350 DEGREES K.   1305
/LD DEFLECTION OF MOLECULES WITH LARGE AMPLITUDE MOTION. (XENON HEXAFLUORIDE, IODINE HEPTAFLUORIDE, RHENIUM / 1281
         ELECTRONIC AND GEOMETRIC STRUCTURE OF THE FREE XENON HEXAFLUORIDE MOLECULE.   1302
FFRACTION INTENSITIES. PART-2: IN/ MOLECULAR STRUCTURE OF XENON HEXAFLUORIDE. PART-1: ANALYSIS OF ELECTRON DI 1291
ULAR GEOMETRY. PART-2: COMP/ CRYSTAL FIELD MODEL STUDY OF XENON HEXAFLUORIDE. PART-1: ENERGY LEVELS AND MOLEC 1311
VISIBLE AND ULTRAVIOL/ SPECTRAL OBSERVATIONS ON MOLECULAR XENON HEXAFLUORIDE. RAMAN SCATTERING AND INFRARED, 1285
                                                         XENON HEXAFLUORIDE. (SPECTRUM)   1312
ETRAMERIC PHASES.   XENON HEXAFLUORIDE. STRUCTURAL CRYSTALLOGRAPHY OF T 1284
AT -30C.   XENON HEXAFLUORIDE. THE STRUCTURE OF A CUBIC PHASE 1283
UORIDE, AND PHOSPHORUS TRIFLU/ ENTHALPIES OF FORMATION OF XENON HEXAFLUORIDE, XENON TETRAFLUORIDE, XENON DIFL 0807
AFLUORIDE. CRYSTAL STRUCTURE.   ENTHALPIES OF FORMATION OF XENON HEXAFLUORIDE, XENON TETRAFLUORIDE, XENON TETR 1269
   XENON DIFLUORIDE AND XENON DICHLORIDE.   5D ORBITALS OF XENON IN ATOMIC VALENCE STATES AND IN THE MOLECULES 1242
UORIDE, AND KRYPTON MONOFLUORIDE.   EMISSION SPECTRA OF XENON MONOBROMIDE, XENON MONOCHLORIDE, XENON MONOFL 0545
ABSORPTION AND EMISSION SPECTRA OF ARGON MATRIX ISOLATED XENON MONOFLUORIDE AND KRYPTON MONOFLUORIDE.   0544
THE GAS PHASE.   PROBABLE NONEXISTENCE OF XENON MONOFLUORIDE AS A CHEMICALLY BOUND SPECIES IN 1221
LUORIDE.   ELECTRON SPIN RESONANCE SPECTRUM OF XENON MONOFLUORIDE IN GAMMA IRRADIATED XENON TETRAF 1222
/D VALENCE FORCE CONSTANTS OF SQUARE PYRAMIDAL MOLECULES. XENON OXIDE TETRAFLUORIDE, IODINE PENTAFLUORIDE, B/ 0418
                     CRYSTAL AND MOLECULAR STRUCTURE OF XENON TETRAFLUORIDE.   1277
   ELECTRON DIFFRACTION STUDY OF THE STRUCTURE OF GASEOUS XENON TETRAFLUORIDE.   1282
            FAR ULTRAVIOLET SPECTROSCOPIC STUDY OF XENON TETRAFLUORIDE.   1272
               GENERAL VALENCE FORCE FIELD FOR XENON TETRAFLUORIDE.   1265
      HEAT CAPACITY AND RELATED THERMODYNAMIC FUNCTIONS OF XENON TETRAFLUORIDE.   1271
                                 HEAT OF FORMATION OF XENON TETRAFLUORIDE.   1266
AN AMPLITUDES OF VIBRATION AND THERMODYNAMIC FUNCTIONS OF XENON TETRAFLUORIDE.   ME 1274
                            ULTRAVIOLET SPECTRUM OF XENON TETRAFLUORIDE.   1270
IONS OF PLANAR SYMMETRICAL XY4 MOLECULES WITH APPLICATION XENON TETRAFLUORIDE.   VIBRAT 1268
                 VIBRATIONAL SPECTRA AND STRUCTURE OF XENON TETRAFLUORIDE.   1262
ORIDE.   RAMAN SPECTRA OF SOLID XENON TETRAFLUORIDE AND ITS ADDUCT WITH XENON DIFLU 1258
                       PROBABLE STRUCTURE OF XENON TETRAFLUORIDE AND XENON DIFLUORIDE.   1250
            THERMOCHEMICAL STUDIES OF XENON TETRAFLUORIDE AND XENON HEXAFLUORIDE.   1276
R COMPOUNDS.   ELECTRON DIFFRACTION STUDIES OF XENON TETRAFLUORIDE AND XENON HEXAFLUORIDE AND OTHE 1259
/QUARE AMPLITUDES OF VIBRATION AND SHRINKAGE CONSTANTS OF XENON TETRAFLUORIDE AND XENON OXIDE TETRAFLUORIDE. 1278
        VIBRATIONAL SPECTRA AND STRUCTURES OF XENON TETRAFLUORIDE AND XENON OXYTETRAFLUORIDE.   1261
ENTIAL OF CHLORO- AND FLUOROMETHANES, TETRAFLUORO METHANE XENON TETRAFLUORIDE, AND XENON TETRACHLORIDE. / POT 0344
                     CRYSTAL AND MOLECULAR STRUCTURE OF XENON TETRAFLUORIDE BY NEUTRON DIFFRACTION.   1260
TRAPPED RADICAL XENON MONOFLUORIDE.   RADIATION DAMAGE IN XENON TETRAFLUORIDE. ELECTRON SPIN RESONANCE OF THE 1220
C FUNCTIONS FROM 5 TO 350 DEGREES K. RECONCILIATION OF T/ XENON TETRAFLUORIDE. HEAT CAPACITY AND THERMODYNAMI 1275
                                                         XENON TETRAFLUORIDE: HEAT OF FORMATION.   1267
                                                         XENON TETRAFLUORIDE. (INFRARED SPECTRUM)   1263
CORRELATION OF ELECTRONIC STATES OF THE POSITIVE IONS OF XENON TETRAFLUORIDE, XENON OXIDE TETRAFLUORIDE, AN/ 1264
            XENON DIFLUORIDE AND THE NATURE OF THE XENON- FLUORINE BOND.   1223
RIDE BY ESCA AND EVIDENCE FOR ORBITAL INDEPENDENCE IN THE XENON- FLUORINE BONDING. /AND XENON OXIDE TETRAFLUO 1229
                                        THE XENON- FLUORINE SYSTEM.   1255
THE MASS INFLUENCE ON CORIOLIS COUPLING CONSTANTS IN FOUR XYN TYPE MOLECULE MODELS. /TICAL INTERPRETATION OF 1488
ICATION OF THE PROGRESSIVE RIGIDITY METHOD TO SOME SIMPLE XYN TYPE MOLECULES. (BORON TRIFLUORIDE)   APPL 0211
IMATION OF ANHARMONIC POTENTIAL CONSTANTS. PART-1. LINEAR XY2 MOLECULES.   EST 1383
ORIOLIS COUPLING CONSTANTS FOR SOME NONLINEAR SYMMETRICAL XY2 MOLECULES.   POTENTIAL CONSTANTS AND C 1385
E FIELD AND THERMODYNAMIC PROPERTIES FOR BENT SYMMETRICAL XY2 MOLECULES.   UREY-BRADLEY FORC 1387
            THIRD ORDER POTENTIAL CONSTANTS OF BENT XY2 MOLECULES. (OXYGEN DIFLUORIDE)   0770

```
                    MEAN AMPLITUDES OF VIBRATION IN SOME XY2Z MOLECULES OF C2V SYMMETRY.                          1396
          INERTIA DEFECT AND PLANARITY OF FOUR ATCM MOLECULES. (XY3)                                             1398
                    MEAN AMPLITUDES OF VIBRATION FOR SOME PYRAMIDAL XY3 AND TETRAHEDRAL XY4 MOLECULES.           1407
     RICES. (BASTIANSEN-MORINO SHRINKAGE EFFECT OF SYMMETRICAL XY3 AND XY4 MOLECULES) /L MEAN SQUARE AMPLITUDE MAT 1394
            FORCE CONSTANT OF THE NONPLANAR VIBRATION OF AN XY3 MOLECULE AND A MODEL WITH POLARIZABLE IONS.       0573
            CALCULATION OF UREY-BRADLEY POTENTIAL CONSTANTS. PLANAR XY3 MOLECULES.                                1393
     DES OF VIBRATION AND UREY-BRADLEY FORCE FIELDS FOR PLANAR XY3 MOLECULES.           CALCULATIONS OF MEAN AMPLITU 1399
     QUARE AMPLITUDES AND CORIOLIS CONSTANTS OF SOME PYRAMIDAL XY3 MOLECULES.                   GENERALIZED MEAN S 1405
                    MEAN AMPLITUDES OF VIBRATION FOR SOME PYRAMIDAL XY3 MOLECULES.                                1390
     EY FORCE FIELD AND THERMODYNAMIC PROPERTIES CF PRYRAMIDAL XY3 MOLECULES.                       UREY-BRADL 1406
                    VIBRATIONAL FREQUENCIES OF SOME PLANAR XY3 MOLECULES.                                         1391
                    EVALUATION OF FORCE CONSTANTS OF PLANE XY3 MOLECULES FROM VIBRATIONAL DATA.                   1401
     ND MODEL FOR THE VIBRATIONAL FORCE CONSTANTS OF NONPLANAR XY3 MOLECULES. (NITROGEN TRIFLUORIDE)     A BENT BO 0704
     RIDE, PHOSPH/ AMBIGUITIES IN THE HARMONIC FORCE FIELDS OF XY3 MOLECULES. (NITROGEN TRIFLUORIDE, BORON TRIFLUO 0709
     RIFLUORI/ MEAN AMPLITUDES OF VIBRATION FOR SOME PYRAMIDAL XY3 MOLECULES. (NITROGEN TRIFLUORIDE, PHOSPHORUS T 0706
     VIBRATION AND BASTIANSEN-MCRINO SHRINKAGE EFFECT IN SOME XY3 MOLECULES OF PLANAR TRIGONAL SYMMETRY. /UDES OF 1397
     TRIFLUORIDE)            MOLECULAR FORCE FIELD FOR SOME XY3 MOLECULES. (PHOSPHORUS TRIFLUORIDE AND ARSENIC 1400
     /RMINATION OF THE STRUCTURE OF PYRAMIDAL MOLECULES OF THE XY3 TYPE BY SOLUTION OF THE VIBRATIONAL PROBLEM WI/ 0727
                    UNIQUE FORCE FIELDS OF SOME XY3 TYPE MOLECULES.                                               1403
            POTENTIAL CONSTANTS OF XY4 AND XY3Z TYPES OF MOLECULES AND RADICALS.                                  1428
     ND SHRINKAGE EFFECTS OF SOME TRIGONAL BIPYRAMIDAL XY5 AND XY3Z2 MOLECULES. /SQUARE AMPLITUDES OF VIBRATION, A 1432
                    POTENTIAL CONSTANTS OF XY4 AND XY3Z TYPES OF MOLECULES AND RADICALS.                          1428
            CLASSICAL VIBRATION PROBLEM FOR THE PYRAMIDAL XY4 MOLECULE.                                           1421
     CHNIQUES IN THE STUDY OF CORIOLIS COUPLING IN TETRAHEDRAL XY4 MOLECULES.                       MATHEMATICAL TE 1409
     TUDES OF VIBRATION FCR SOME PYRAMIDAL XY3 AND TETRAHEDRAL XY4 MOLECULES.                       MEAN AMPLI 1407
     BASTIANSEN-MORINO SHRINKAGE EFFECT OF SYMMETRICAL XY3 AND XY4 MOLECULES) /L MEAN SQUARE AMPLITUDE MATRICES. ( 1394
     UARE AMPLITUDE MATRICES. PART-7: TREATMENT OF TETRAHEDRAL XY4 MOLECULES.                       VIBRATIONAL MEAN SQ 1408
            ORBITAL VALENCE FORCE CONSTANTS OF TETRAHEDRAL XY4 MOLECULES. (GERMANIUM TETRAFLUORIDES)             0481
     ERISTIC FREQUENC/ PROGRESSIVE STIFFNESS METHOD APPLIED TO XY4 MOLECULES IN THE LIGHT OF THE THEORY OF CHARACT 1430
            VIBRATIONS OF PLANAR SYMMETRICAL XY4 MOLECULES WITH APPLICATION XENON TETRAFLUORIDE. 1268
     RAMAN INTENSITIES FOR SPHERICALLY SYMMETRIC MCDES OF SOME XY4 TETRAHEDRAL AND XY6 OCTAHEDRAL MOLECULES.      1453
     OF ATOMIC MASSES ON THE CORIOLIS COUPLING COEFFICIENTS IN XY4 TETRAHEDRAL MOLECULES.                 INFLUENCE 1410
                    MEAN AMPLITUDES OF VIBRATION OF SOME XY4 TETRAHEDRAL MOLECULES.                               1419
                    UREY-BRADLEY FORCE FIELD FOR XY4 TETRAHEDRAL MOLECULES.                                       1429
     CONSTANTS IN THE SOLUTION OF THE VIBRATIONAL PROBLEM FOR XY4 TETRAHEDRAL MOLECULES. /OF CORIOLIS INTERACTION 1423
                    NORMAL COORDINATE TREATMENT OF XY4 TETRAHEDRAL MOLECULES AND IONS. (GVFF)                     1415
     ND MVFF)             NORMAL COORDINATE TREATMENT OF XY4 TETRAHEDRAL MOLECULES AND IONS. (UREY-BRADLEY A 1418
            MEAN AMPLITUDES OF VIBRATION OF SOME XY4 TETRAHEDRAL MOLECULES. (SILICON TETRAFLUORIDE) 1008
     PLICATION OF UREY-BRADLEY FORCE FIELD TO SCME TETRAHEDRAL XY4 TYPE MOLECULES AND IONS.                    AP 1425
     S BY GREEN'S FUN/ MCLECULAR CONSTANTS OF SOME TETRAHEDRAL XY4 TYPE MOLECULES AND THEIR SUBSTITUTED DERIVATIVE 1424
            MEAN AMPLITUDES OF VIBRATION OF SOME TETRAHEDRAL XY4 TYPE MOLECULES. PART 3 AND 4.                    1420
     AND LE/ MEAN AMPLITUDES OF VIBRATION OF SOME TETRAHEDRAL XY4 TYPE MOLECULES. PART-5: TETRAHALIDES OF HAFNIUM 0487
     E)             POTENTIAL CONSTANTS OF SOME XY4 TYPE OF SILANE COMPOUNDS. (SILICON TETRAFLUORID 1009
     INTERATOMIC DISTANCES OF HALIDE COMPLEX MOLECULES OF THE XY4 TYPE OF TETRAHEDRAL SYMMETRY. /CE CONSTANTS AND 1414
     AFLUORID/ MEAN AMPLITUDES OF VIBRATION FOR SOME PYRAMIDAL XY4Z MOLECULES. (IODINE PENTAFLUORIDE, BROMINE PENT 0421
     ATION, AND SHRINKAGE EFFECTS OF SOME TRIGONAL BIPYRAMIDAL XY5 AND XY3Z2 MOLECULES. /SQUARE AMPLITUDES OF VIBR 1432
            NORMAL COORDINATE ANALYSIS OF TRIGONAL BIPYRAMIDAL XY5 MOLECULES.                                     1433
                    POTENTIAL CONSTANTS OF SOME XY5 MOLECULES.                                                    1439
     UREY-BRADLEY FORCE CONSTANTS OF TRIGONAL BIPYRAMIDAL XY5 MOLECULES. (PHOSPHORUS PENTAFLUORIDE)          0838
            EVALUATION OF FORCE CONSTANTS: MOLECULES OF THE TYPE XY6.                                             1470
     / COORDINATE ANALYSIS CF ROTATION VIBRATION OF OCTAHEDRAL XY6 AND Z6 MOLECULAR MODELS WITH APPLICATION TO 15/ 1454
                    FORCE CONSTANTS FOR XY6 MOLECULES.                                                            1445
     INFLUENCE OF ATOMIC MASS ON THE CORIOLIS CONSTANTS OF XY6 MOLECULES.                                         1458
     ND BASTIANSEN-MORINO SHRINKAGE EFFECTS IN SOME OCTAHEDRAL XY6 MOLECULES.               MEAN AMPLITUDES OF VIBRATION A 1455
            MEAN AMPLITUDES OF VIBRATION OF SOME XY6 MOLECULES.                                                   1461
            MOLECULAR FORCE FIELDS OF SOME XY6 MOLECULES.                                                         1443
            POTENTIAL FIELD AND FORCE CONSTANTS FOR SOME XY6 MOLECULES.                                           1462
     /FIELDS. PART-8: VIBRATION FREQUENCIES OF SOME OCTAHEDRAL XY6 MOLECULES. (SULFUR HEXAFLUORIDE, SELENIUM HEXA/ 1080
     R SPHERICALLY SYMMETRIC MODES OF SOME XY4 TETRAHEDRAL AND XY6 OCTAHEDRAL MOLECULES.       RAMAN INTENSITIES FO 1453
            UREY-BRADLEY FORCE FIELD OF SCME XY6 SYSTEMS.                                                         1469
     ORIDES)             MOLECULAR FORCE FIELD FOR SOME XY6 TYPE MOLECULES. (SELENIUM AND TELLURIUM HEXAFLU 1465
```

Y

```
     UM, GADOLINIUM, TERBIUM, DYSPROSIUM, ERBIUM, THULIUM, AND YTTERBIUM. /S OF SCANDIUM, CERIUM, SAMARIUM, EUROPI 1356
     RESSURES AND HEATS OF SUBLIMATION OF THULIUM TRIFLUORIDE, YTTERBIUM TRIFLUORIDE, AND LUTETIUM TRIFLUORIDE. /P 1132
     STABILITIES OF THE MONO- AND DIFLUORIDES OF SCANDIUM AND YTTRIUM. /TUDIES AT HIGH TEMPERATURES. PART-18: THE 0939
     TION.                          STRUCTURES OF YTTRIUM AND BISMUTH TRIFLUORIDES BY NEUTRON DIFFRAC 1318
     EES K RANGE. THERMODYNAMIC FUNCTIONS OF GASEOUS SCANDIUM, YTTRIUM, AND LANTHANUM HALIDES IN THE 293-3000 DEGR 0568
     E, SCANDIUM MONOFLUORID/ MOLECULAR CONSTANTS OF SCANDIUM, YTTRIUM, AND LANTHANUM HALIDES. (SCANDIUM DIFLUORID 0941
     INFRARED SPECTRA ANE GEOMETRIES OF MATRIX ISOLATED YTTRIUM DIFLUORIDE AND YTTRIUM TRIFLUORIDE.          1316
     OLECULAR STRUCTURE OF THE VAPOR PHASE HALIDES OF GALLIUM, YTTRIUM, LANTHANUM, AND NEODYMIUM. / STUDY OF THE M 1317
     MASS SPECTROMETRIC STUDIES CF SCANDIUM, YTTRIUM, LANTHANUM, AND RARE EARTH FLUORIDES.         1359
     / OF METAL TRIFLUORIDES FROM INFRARED SPECTRA. (SCANDIUM, YTTRIUM, LANTHANUM, CERIUM, NEODYMIUM, EUROPIUM, A/ 0943
            ELECTRONIC BANE SPECTRUM OF YTTRIUM MONOFLUORIDE.                                                     1314
     GROUND STATES OF SCANDIUM MONOFLUORIDE AND YTTRIUM MONOFLUORIDE.                                              0929
            VIBRONIC SPECTRA OF YTIRIUM MONOFLUORIDE.                                                             1315
     ELECTRONIC STATES OF GASEOUS SCANDIUM MONOFLUORIDE, YTTRIUM MONOFLUORIDE, AND LANTHANUM MONOFLUORIDE.    0928
     IATION ENERGIES AND STABILITIES OF SCANDIUM MONOFLUORIDE, YTTRIUM MONOFLUORIDE, AND LANTHANUM MONOFLUORIDE. / 0935
     MOLECULAR VIBRATION FREQUENCIES OF SCANDIUM MONOFLUORIDE, YTTRIUM MONOFLUORIDE, AND LANTHANUM MONOFLUORIDE.  0936
     VIBRONIC SPECTRUM OF SCANDIUM MONOFLUORIDE, YTTRIUM MONOFLUORIDE, AND LANTHANUM MONOFLUORIDE.   0938
     ORIDES, LANTHANIDE DIFLUORIDES, SCANDIUM TRIFLUORIDE, AND YTTRIUM TRIFLUORIDE. /FLUORIDES. (LANTHANIDE TRIFLU 0946
            ENTHALPY OF FORMATION OF YTTRIUM TRIFLUORIDE.                                                         1319
     HERMODYNAMIC FUNCTIONS OF SEVEN RARE EARTH TRIFLUORIDES. (YTTRIUM TRIFLUORIDE) /E HEAT CONTENTS AND RELATED T 1320
     HALIDES OF THE SCANDIUM SUBGROUP. (SCANDIUM TRIFLUORIDE, YTTRIUM TRIFLUORIDE, AND LANTHANUM TRIFLUORIDE) /HE 0948
     ISOLATED RARE EARTH TRIFLUORIDES. (SCANDIUM TRIFLUORIDE, YTTRIUM TRIFLUORIDE, AND LANTHANUM TRIFLUORIDE) /IX 0944
     SATURATED VAPOR PRESSURE OF SCANDIUM TRIFLUORIDE, YTTRIUM TRIFLUORIDE, AND LANTHANUM TRIFLUORIDE.   0952
     STANDARD ENTHALPIES OF FORMATION OF SCANDIUM TRIFLUORIDE, YTTRIUM TRIFLUORIDE, AND LANTHANUM TRIFLUORIDE.  0942
     SUBLIMATION PRESSURES OF SCANDIUM TRIFLUORIDE, YTTRIUM TRIFLUORIDE, AND LANTHANUM TRIFLUORIDE.    0947
     RARE/ MASS SPECTROMETRIC STUDIES OF SCANDIUM TRIFLUORIDE, YTTRIUM TRIFLUORIDE, LANTHANUM TRIFLUORIDE AND THE 0953
```

Z

/H TEMPERATURE MICROWAVE SPECTROMETER FOR MEASUREMENTS OF ZEEMAN EFFECT IN DIAMAGNETIC MOLECULES. GJ FACTOR / 1120
OF THE ALUMINUM FLUORIDE MOLECULE. ZEEMAN EFFECT IN THE MICROWAVE ROTATIONAL SPECTRUM 0026
ADRUPOLE MOMENTS IN CHLORINE MONOFLUORIDE, BRO/ MOLECULAR ZEEMAN EFFECT, MAGNETIC PROPERTIES, AND ELECTRIC QU 0254
CULES. (PHOSPHORUS TRIFLUORIDE) ZEEMAN EFFECT OF SOME LINEAR AND SYMMETRIC TOP MOLE 0796
GNETIC SUSCEPTIBILITY, AND MOLECULAR QUADRUPOLE MOMENTS / ZEEMAN STUDIES INCLUDING THE MOLECULAR G VALUES, MA 0744
MS OF G/ ROTATIONAL ANALYSIS OF BANDS OF THE A TRIPLET PI ZERO(+) B TRIPLET PI(1) TO X SINGLET SIGMA(+) SYSTE 0457
EMS OF / ROTATIONAL ANALYSIS OF BANDS OF THE A TRIPLET PI ZERO(+), B TRIPLET PI(1) TO X SINGLET SIGMA(+) SYST 1111
FLUORIDE. DOUBLE ZETA SCF CALCULATIONS FOR NITRITE ION AND OXYGEN DI 0762
CORIOLIS ZETA SUMS FOR SOME SPHERICAL TOP MOLECULES. 1487
/NITROGEN TRIFLUORIDE IN THE EXCITED VIBRATIONAL STATES. (ZETA TYPE DOUBLING AND RESONANCE IN THE V3 AND V4 / 0723
ALT DIFLUORIDE, NICKEL DIFLUORIDE, COPPER DIFLUORIDE, AND ZINC DIFLUORIDE. /D GEOMETRY OF MATRIX ISOLATED COB 0437
CIUM DIFLUORIDE, STRONTIUM DIFLUORIDE, BARIUM DIFLUORIDE, ZINC DIFLUORIDE, CADMIUM DIFLUORIDE) /FLUORIDE, CAL 0131
/ND THE BENDING FORCE CONSTANTS OF THE GROUP-2B HALIDES. (ZINC DIFLUORIDE, CADMIUM DIFLUORIDE, AND MERCURY D/ 1322
UORIDE, CALCIUM DIFLUORIDE, STRONTIUM / VAPOR PRESSURE OF ZINC DIFLUORIDE, CADMIUM DIFLUORIDE, MAGNESIUM DIFL 0049
RAMAN SPECTRA OF TITANIUM DIOXIDE, MAGNESIUM DIFLUORIDE, ZINC DIFLUORIDE, DIFLUORIDE, AND MANGANESE DIFLUOR/ 0615
UTILE COUNTERPARTS: TITANIUM DIOXIDE, MAGNESIUM FLUORIDE, ZINC FLUORIDE, AND FERROUS FLUORIDE. /ELDS OF THE R 0606
VIBRATIONAL SPECTRA AND THERMODYNAMICS OF THE ZINC HALIDES. (ZINC DIFLUORIDE) 1323
SUBLIMATION THERMODYNAMICS OF ZINC(II) FLUORIDE. 1321
/UORIDE, MANGANESE(II) FLUORIDE, COPPER(II) FLUORIDE, AND ZINC(II) FLUORIDE WITH CARBON MONOXIDE IN ARGON MA/ 1324
VAPOR PRESSURE AND HEAT OF SUBLIMATION OF ZIRCONIUM AND HAFNIUM TETRACHLORIDES. 0486
FLUORIDE, ALUMINUM DIFLUORIDE, PHOSPHORUS DIFLUORIDE, AND ZIRCONIUM DIFLUORIDE. /RBON DIFLUORIDE, NITROGEN DI 1384
E. INFRARED SPECTRA OF MATRIX ISOLATED ZIRCONIUM DIFLUORIDE, TRIFLUORIDE, AND TETRAFLUORID 1325
VAPOR PRESSURE OF ZIRCONIUM FLUORIDE. 1331
ERMODYNAMIC DATA. FREQUENCIES OF ZIRCONIUM FLUORIDE (GAS) VIBRATIONS ACCORDING TO TH 1329
CHLORIDE. INFRARED SPECTRA OF GASEOUS GROUP-4 HALIDES: ZIRCONIUM FLUORIDE, ZIRCONIUM CHLORIDE, AND HAFNIUM 1327
NORMAL VIBRATION FREQUENCIES OF ZIRCONIUM HALIDES. 1330
VAPOR PRESSURE OF ZIRCONIUM TETRAFLUORIDE. 1326
VAPOR PRESSURE OF ZIRCONIUM TETRAFLUORIDE. 1328
ELECTRON DIFFRACTION INVESTIGATION OF THE STRUCTURE OF ZIRCONIUM TETRAFLUORIDE IN THE VAPOR PHASE. 1332
/ATE ANALYSIS OF ROTATION VIBRATION OF OCTAHEDRAL XY6 AND Z6 MOLECULAR MODELS WITH APPLICATION TO 15 XY6 MOL/ 1454

ERMODYNAMIC FUNCTIONS OF PHOSPHORUS PENTAFLUORIDE BETWEEN 0 AND 1500 DEGREES K. /IOUS FORMS OF PHOSPHORUS. TH 0865
OF TH/ ROTATIONAL STRUCTURE AND ISOTOPIC SHIFT IN THE (1,0) BAND OF THE A DOUBLET SIGMA- X DOUBLET PI SYSTEM 1135
ANALYSIS OF THE ROTATIONAL STRUCTURE OF THE BANDS 0-0 AND 0-1 IN GERMANIUM MONOFLUORIDE. 0465
ANALYSIS OF THE ROTATIONAL STRUCTURE OF THE BANDS 0-0 AND 0-1 IN GERMANIUM MONOFLUORIDE. 0465
TRANSITIONS OF PHOSPHORUS TRIFLUORIDE IN THE NU-4 EQUALS 1 STATE. /RD ORDER CONSTANTS ON THE L-TYPE DOUBLING 0804
TRANSITIONS OF PHOSPHORUS TRIFLUORIDE IN THE NU-4 EQUALS 1 STATE. L-TYPE DOUBLING 0803
EM OF TH/ ROTATIONAL STRUCTURE AND ISOTOPIC SHIFT IN THE (1,0) BAND OF THE A DOUBLET SIGMA- X DOUBLET PI SYST 1135
CTRUM OF NITROGEN TRIFLUORIDE. 1-TYPE RESONANCE IN AN OVERTONE BAND. THE 2NU-4 SPE 0732
CUBIC POTENTIAL CONSTANTS AND 1-TYPE RESONANCE IN BORON TRIFLUORIDE. 0231
MULTIPLET SPLITTING IN 1S HOLE STATES OF MOLECULES. 0683
, ENTROPY, AND ENTHALPY OF PLUTONIUM-242 TRIFLUORIDE FROM 10 TO 350 DEGREES K. HEAT CAPACITY 0886
LPY, AND GIBBS ENERGY OF PLUTONIUM-242 TETRAFLUORIDE FROM 10 TO 350 DEGREES K. HEAT CAPACITY, ENTROPY, ENTHA 0889
/IUM-242 TRIFLUORIDE AND PLUTONIUM-242 TETRAFLUORIDE FROM 10 TO 350 DEGREES K, AND OF PLUTONIUM-244 DIOXIDE / 0885
IMETRY. ENTHALPY OF BERYLLIUM FLUORIDE FROM 456 TO 1063 DEGREES K BY TRANSPOSED TEMPERATURE DROP CALOR 0132
/ MOLECULAR ORBITALS OF PLATINUM HEXAFLUORIDE AND ELEMENT 110 HEXAFLUORIDE CALCULATED BY THE SELF CONSISTENT 0881
O/ THERMODYNAMIC PROPERTIES OF CARBON TETRAFLUORIDE FROM 12 DEGREES K TO ITS BOILING POINT. THE SIGNIFICANCE 0361
ICROWAVE SPECTROSCOPY OF CHLORINE PENTAFLUORIDE AT 70 AND 140 GHZ M 0423
RON MONOFLUORIDE MOLE/ ROTATIONAL ANALYSIS OF THE BAND AT 14900 CM-1 OF A 3 SIGMA TO 3 SIGMA SYSTEM OF THE BO 0156
, STRONTIUM DIFLUORIDE, AND BARIUM DIFLUORIDE IN THE 8 TO 150 EV RANGE. /IC EXCITATIONS IN CALCIUM DIFLUORIDE 0279
NAMIC FUNCTIONS OF PHOSPHORUS PENTAFLUORIDE BETWEEN 0 AND 1500 DEGREES K. /IOUS FORMS OF PHOSPHORUS. THERMODY 0865
/N MULTIPLE BEAM INTERFEROMETRY IN THE ULTRAVIOLET REGION 1700-2500 ANGSTROMS. (ALUMINUM- MAGNESIUM FLUORIDE) 0042
TION SPECTRA OF MATRIX ISOLATED NITROGEN TRIFLUORIDE FROM 18 TO 50 GC. ABSORP 0698
SIMPLE METHOD FOR EXACT SOLUTION OF 2 X 2 SECULAR EQUATION IN MOLECULAR VIBRATIONS. 1386
CULAR ORBITAL DESCRIPTION OF SULFUR COMPOUNDS OF VALENCES 2, 4, AND 6. MOLE 1085
ULES. PART-1: INFRARED SPECTRUM OF OXYGEN DIFLUORIDE FROM 2.5 TO 25 MICRONS. /ND STRUCTURE OF INORGANIC MOLEC 0761
. JAHN-TELLER EFFECTS IN THE 2E5/2G TO 2G3/2G TRANSITION OF RHENIUM HEXAFLUORIDE 0913
JAHN-TELLER EFFECTS IN THE 2E5/2G TO 2G3/2G TRANSITION OF RHENIUM HEXAFLUORIDE. 0913
FLUORIDES ESPECIALLY THE ELEMENTS OF ATOMIC NUMBER BELOW 20. HEATS OF FORMATION OF INORGANIC 0214
ABSORPTION SPECTRUM OF URANIUM HEXAFLUORIDE FROM 2000 TO 4200 ANGSTROMS. 1191
ROTATIONAL ANALYSIS OF ABSORPTION BANDS IN THE 2266 ANGSTROM SYSTEM OF SILICON DIFLUORIDE. 0979
/TO 350 DEGREES K, AND OF PLUTONIUM-244 DIOXIDE FROM 4 TO 25 DEGREES K. DERIVED ENTROPIES AND OTHER THERMODY/ 0885
ART-1: INFRARED SPECTRUM OF OXYGEN DIFLUORIDE FROM 2.5 TO 25 MICRONS. /ND STRUCTURE OF INORGANIC MOLECULES. P 0761
IDE. 2500 ANGSTROM ABSORPTION SPECTRUM OF CARBON DIFLUOR 0306
IONAL AND ROTATIONAL STRUCTURE OF TIN MONOFLUORIDE IN THE 2500 ANGSTROM REGION. ANALYSIS OF VIBRAT 1138
MOLECULE. ROTATIONAL ANALYSIS OF THE 2550-2750 ANGSTROM BAND SYSTEM AT ANTIMONY FLUORIDE 0067
F GASEOUS SCANDIUM, YTTRIUM, AND LANTHANUM HALIDES IN THE 293-3000 DEGREES K RANGE. THERMODYNAMIC FUNCTIONS O 0568
HEAT OF FORMATION OF CHLORINE TRIFLUORIDE AT 298.14 DEGREES K. 0401
. DERIVED ENTROPIES AND OTHER THERMODYNAMIC PROPERTIES AT 298.15 DEGREES K. /4 DIOXIDE FROM 4 TO 25 DEGREES K 0885
L, ELECTRON AFFINITY, DIPOLE MOMEN/ ARSENIC MONOFLUORIDE, 3 SIGMA. DISSOCIATION ENTHALPY, IONIZATION POTENTIA 0094
MOLE/ ROTATIONAL ANALYSIS OF THE BAND AT 14900 CM-1 OF A 3 SIGMA TO 3 SIGMA SYSTEM OF THE BORON MONOFLUORIDE 0156
ORIDE. THE ANALYSIS OF 2NU-3 AND THE DETERMINATION OF THE 33 CORIOLIS CONSTANT. BORON TRIFLU 0192
NAMIC PROPERTIES OF MOLYBDENUM HEXAFLUORIDE BETWEEN 4 AND 350 DEGREES K. HEAT CAPACITY AND OTHER THERMODY 0636
R THERMODYNAMIC FUNCTIONS OF XENON HEXAFLUORIDE FROM 5 TO 350 DEGREES K. HEAT CAPACITY AND OTHE 1305
RMODYNAMIC PROPERTIES OF NEPTUNIUM HEXAFLUORIDE FROM 7 TO 350 DEGREES K. /PACITY, ENTHALPY OF FUSION, AND THE 0650
OPY, AND ENTHALPY OF PLUTONIUM-242 TRIFLUORIDE FROM 10 TO 350 DEGREES K. HEAT CAPACITY, ENTR 0886
ND GIBBS ENERGY OF PLUTONIUM-242 TETRAFLUORIDE FROM 10 TO 350 DEGREES K. HEAT CAPACITY, ENTROPY, ENTHALPY, A 0889
/2 TRIFLUORIDE AND PLUTONIUM-242 TETRAFLUORIDE FROM 10 TO 350 DEGREES K, AND OF PLUTONIUM-244 DIOXIDE FROM 4/ 0885
TIES. XENON DIFLUORIDE. HEAT CAPACITY FROM 5 TO 350 DEGREES K AND SOME DERIVED THERMODYNAMIC PROPER 1246
/UDY OF IODINE PENTAFLUORIDE. HEAT CAPACITY BETWEEN 5 AND 350 DEGREES K, ENTHALPY OF FUSION AND VAPORIZATION/ 0510
/IDE. HEAT CAPACITY AND THERMODYNAMIC FUNCTIONS FROM 5 TO 350 DEGREES K. RECONCILIATION OF THE ENTROPIES FRO/ 1275
ERMODYNAMIC PROPERTIES OF MOLYBDENUM HEXAFLUORIDE BETWEEN 4 AND 350 DEGREES K. HEAT CAPACITY AND OTHER TH 0636
AR ORBITAL DESCRIPTION OF SULFUR COMPOUNDS OF VALENCES 2, 4, AND 6. MOLECUL 1085
THERMODYNAMIC PROPERTIES OF CARBON TETRAFLUORIDE FROM 4 DEGREES K TO ITS MELTING POINT. 0326
/M 10 TO 350 DEGREES K, AND OF PLUTONIUM-244 DIOXIDE FROM 4 TO 25 DEGREES K. DERIVED ENTROPIES AND OTHER THE/ 0885

HIGH TEMPERATURES. PART-2: PENTAFLUORIDES OF SOME 4P- AND 4D ELEMENTS. /AMIC PROPERTIES OF PENTAFLUORIDES AT 1434
IDES AT HIGH TEMPERATURES. PART-2: PENTAFLUORIDES OF SOME 4P- AND 4D ELEMENTS. /AMIC PROPERTIES OF PENTAFLUOR 1434
NEW BAND SYSTEM OF THE ANTIMONY MONOFLUORIDE MOLECULE IN 4050-5450 ANGSTROM REGION.   A 0058
ABSORPTION SPECTRUM OF URANIUM HEXAFLUORIDE FROM 2000 TO 4200 ANGSTROMS.   1191
D THERMODYNAMIC PROPERTIES OF THALLIUM FLUORIDE FROM 5 TO 445 DEGREES K.   HEAT CAPACITY AN 1124
P CALORIMETRY.   ENTHALPY OF BERYLLIUM FLUORIDE FROM 456 TO 1083 DEGREES K BY TRANSPOSED TEMPERATURE DRO 0142
/RIC STUDY OF IODINE PENTAFLUORIDE. HEAT CAPACITY BETWEEN 5 AND 350 DEGREES K, ENTHALPY OF FUSION AND VAPORI/ 0510
OTHER THERMODYNAMIC FUNCTIONS OF XENON HEXAFLUORIDE FROM 5 TO 350 DEGREES K.   HEAT CAPACITY AND 1305
ROPERTIES.   XENON DIFLUORIDE. HEAT CAPACITY FROM 5 TO 350 DEGREES K AND SOME DERIVED THERMODYNAMIC P 1246
/FLUORIDE. HEAT CAPACITY AND THERMODYNAMIC FUNCTIONS FROM 5 TO 350 DEGREES K. RECONCILIATION OF THE ENTROPIE/ 1275
TY AND THERMODYNAMIC PROPERTIES OF THALLIUM FLUORIDE FROM 5 TO 445 DEGREES K.   HEAT CAPACI 1124
N THE MOLECULES XENON DIFLUORIDE AND XENON DICHLORIDE.   5D ORBITALS OF XENON IN ATOMIC VALENCE STATES AND I 1242
E. (THIRD ROW)   RAMAN SPECTRA OF 5D TRANSITION METAL HEXAFLUORIDES IN THE VAPOR STAT 1378
IDES AT HIGH TEMPERATURES. PART-3: PENTAFLUORIDES OF SOME 5D-ELEMENTS. THERMODYNAMIC PROPERTIES OF PENTAFLUOR 0536
FLUORIDES AT HIGH TEMPERATURES. PART-4: PENTAFLUORIDES OF 5F ELEMENTS.   THERMODYNAMIC PROPERTIES OF PENTA 0647
PECTRA OF MATRIX ISOLATED NITROGEN TRIFLUORIDE FROM 18 TO 50 GC.   ABSORPTION S 0698
PHOTOELECTRON SPECTRUM OF SULFUR HEXAFLUORIDE AT 584 ANGSTROMS.   1076
TAL DESCRIPTION OF SULFUR COMPOUNDS OF VALENCES 2, 4, AND 6.   MOLECULAR ORBI 1085
ROGEN MONO/ FRANCK-CONDON FACTORS AND R CENTROIDS FOR THE 6 SINGLET SIGMA- X TRIPLET SIGMA BAND SYSTEM OF NIT 0677
EW BAND SYSTEM OF BISMUTH FLUORIDE MOLECULE IN THE REGION 6200-7000 ANGSTROMS.   N 0144
D THERMODYNAMIC PROPERTIES OF NEPTUNIUM HEXAFLUORIDE FROM 7 TO 350 DEGREES K. /PACITY, ENTHALPY OF FUSION, AN 0650
MICROWAVE SPECTROSCOPY OF CHLORINE PENTAFLUORIDE AT 70 AND 140 GHZ   0423
ORIDE, STRONTIUM DIFLUORIDE, AND BARIUM DIFLUORIDE IN THE 8 TO 150 EV RANGE. /IC EXCITATIONS IN CALCIUM DIFLU 0279

AUTHOR INDEX

AARONS LJ	0794	THEORETICAL STUDY OF THE GEOMETRY OF PHOSPHINE AND PHOSPHORUS TRIFLUORIDE AND THE
ABJEAN R	0042	RECENT DEVELOPMENTS IN MULTIPLE BEAM INTERFEROMETRY IN THE ULTRAVIOLET REGION 1700
ABRAHAM KC	0057	ROTATIONAL ANALYSIS OF C3 SYSTEM OF THE ANTIMONY MONOFLUORIDE MOLECULE.
	0058	A NEW BAND SYSTEM OF THE ANTIMONY MONOFLUORIDE MOLECULE IN 4050-5450 ANGSTROM REGI
	0059	FINE STRUCTURE ANALYSIS OF THE C1 SYSTEM OF ANTIMONY MONOFLUORIDE.
	0060	EMISSION OF MOLECULAR ANTIMONY MONOFLUORIDE.
	0062	VIBRATIONAL ANALYSIS OF ULTRAVIOLET BANDS OF ANTIMONY MONOFLUORIDE.
ABRAMOWITZ S	0227	FORCE FIELDS FOR THE BORON TRIHALIDES.
	0228	ISOTOPIC SPLITTING IN MATRIX ISOLATED BORON TRIFLUORIDE.
	0659	INFRARED SPECTRUM OF MATRIX ISOLATED NIOBIUM PENTAFLUORIDE.
	0715	FORCE FIELDS FOR GROUP-4 TETRAFLUORIDES AND GROUP-5 TRIFLUORIDES.
	0830	POTENTIAL FUNCTION FOR THE NU-7 VIBRATION OF PHOSPHORUS PENTAFLUORIDE.
	0910	JAHN-TELLER VIBRATIONS OF RHENIUM HEXAFLUORIDE.
	0960	VIBRATIONAL ANALYSIS OF SELENIUM HEXAFLUORIDE AND TUNGSTEN HEXAFLUORIDE.
	1006	FORCE FIELD FOR SILICON TETRAFLUORIDE.
	1061	FORCE FIELDS FOR SOME GROUP-6 HEXAFLUORIDES. (SULFUR HEXAFLUORIDE AND TELLURIUM HE
ACQUISTA N	0291	EMISSION SPECTRUM OF CARBON MONOFLUORIDE.
	0659	INFRARED SPECTRUM OF MATRIX ISOLATED NIOBIUM PENTAFLUORIDE.
	1314	ELECTRONIC BAND SPECTRUM OF YTTRIUM MONOFLUORIDE.
ADAMS A	0438	VAPOR PRESSURE AND HEAT OF SUBLIMATION OF COBALT DIHALIDES. (COBALT DIFLUORIDE)
ADAMS CJ	0514	ACCEPTOR PROPERTIES OF IODINE HEPTAFLUORIDE. OCTAFLUORO PERIODATES(VII).
	0956	VIBRATIONAL SPECTRA AND STRUCTURES OF SELENIUM TETRAFLUORIDE AND TELLURIUM TETRAFL
	1258	RAMAN SPECTRA OF SOLID XENON TETRAFLUORIDE AND ITS ADDUCT WITH XENON DIFLUORIDE.
ADAMS GP	0471	ENTHALPY OF SUBLIMATION OF GERMANIUM DIFLUORIDE AND THE THERMODYNAMICS OF SUBLIMAT
	0472	ENTHALPY OF FORMATION OF GERMANIUM DIFLUORIDE.
ADAMS OW	0808	MEASUREMENT AND CALCULATION OF THE ABSOLUTE INFRARED INTENSITIES OF PHOSPHORUS TRI
ADAMS TS	0962	ROOT MEAN SQUARE AMPLITUDES IN SOME HEXAFLUORIDES OF OCTAHEDRAL SYMMETRY. (SELENIU
ADAMS WJ	0515	STRUCTURE, PSEUDOROTATION, AND VIBRATIONAL MODE COUPLING IN IODINE HEPTAFLUORIDE.
ADLER LS	1062	PHYSICAL AND THERMODYNAMIC PROPERTIES OF SULFUR HEXAFLUORIDE. (REVIEW)
ADOLFSON WF	1181	MEASUREMENT OF THE GAS DENSITY OF URANIUM HEXAFLUORIDE BY LASER RAMAN SCATTERING.
AFANASHEV ML	0571	SYMMETRY AND CRYSTAL STRUCTURE OF RARE EARTH TRIFLUORIDES. (LANTHANUM TRIFLUORIDE)
AGARWAL PM	0424	IDEAL GAS THERMODYNAMIC PROPERTIES OF CHLORINE PENTAFLUORIDE, BROMINE PENTAFLUORID
AGRON PA	1176	THERMODYNAMICS OF INTERMEDIATE URANIUM FLUORIDES FROM MEASUREMENTS OF THE DISPROPO
	1223	XENON DIFLUORIDE AND THE NATURE OF THE XENON- FLUORINE BOND.
	1239	CRYSTAL AND MOLECULAR STRUCTURE OF XENON DIFLUORIDE BY NEUTRON DIFFRACTION.
	1260	CRYSTAL AND MOLECULAR STRUCTURE OF XENON TETRAFLUORIDE BY NEUTRON DIFFRACTION.
AINSCOUGH JB	0080	MOLECULAR STRUCTURE OF ANTIMONY PENTAFLUORIDE.
AKHMEDZHANOV R	0313	INFRARED SPECTRA OF A CRYOSYSTEM. CARBON TETRAFLUORIDE.
AKISHIN NA	0040	MASS SPECTROMETRIC INVESTIGATION OF THE THERMODYNAMIC PROPERTIES OF ALUMINUM TRIFL
AKISHIN PA	0036	ELECTRON DIFFRACTION INVESTIGATION OF THE MOLECULAR STRUCTURES OF THE ALUMINUM HAL
	0131	ELECTRON DIFFRACTION ANALYSIS OF THE MOLECULAR STRUCTURES OF GROUP-2 ELEMENTS. (BE
	0460	ELECTRON DIFFRACTION ANALYSIS OF GALLIUM HALIDES. (GALLIUM TRIFLUORIDE)
	0644	ELECTRONOGRAPHIC INVESTIGATION OF THE STRUCTURE OF THE MOLECULES OF NEODYMIUM HALI
	1165	DETERMINATION OF HEAT OF SUBLIMATION OF URANIUM TETRAFLUORIDE BY THE MASS SPECTROM
	1170	ELECTRON DIFFRACTION STUDY OF URANIUM AND THORIUM TETRAHALIDES IN THE GAS PHASE. (
	1208	MASS SPECTROMETRIC DETERMINATION OF THE VAPOR COMPOSITION AND VAPOR PRESSURE OF VA
	1317	ELECTRON DIFFRACTION STUDY OF THE MOLECULAR STRUCTURE OF THE VAPOR PHASE HALIDES O
	1326	VAPOR PRESSURE OF ZIRCONIUM TETRAFLUORIDE.
AL ZERCHENIOV AN	0696	THERMOCHEMICAL AND THERMODYNAMIC PROPERTIES OF INORGANIC NITROGEN FLUORIDES. (NITR
ALDRIDGE JP	1063	OCTAHEDRAL FINE STRUCTURE SPLITTINGS IN NU-3 OF SULFUR HEXAFLUORIDE.
ALEKSANDROVSKAYA AM	0211	APPLICATION OF THE PROGRESSIVE RIGIDITY METHOD TO SOME SIMPLE XYN TYPE MOLECULES.
	1329	FREQUENCIES OF ZIRCONIUM FLUORIDE (GAS) VIBRATIONS ACCORDING TO THERMODYNAMIC DATA
	1330	NORMAL VIBRATION FREQUENCIES OF ZIRCONIUM HALIDES.
	1441	APPLICATION OF THE METHOD OF PROGRESSIVE RIGIDITY TO OCTAHEDRAL MX6 MOLECULES.
	1448	CALCULATION OF THE FORCE CONSTANTS OF HEXAFLUORIDES.
ALESHONKOVA YA	0826	CALCULATION OF FORCE CONSTANTS AND ROOT MEAN SQUARE VIBRATION AMPLITUDES OF PHOSPH
ALEXAKOS LG	0368	NMR SPECTRA OF CHLORINE TRIFLUORIDE AND CHLORINE MONOFLUORIDE. GASEOUS SPECTRA AND
	0516	NMR SPECTRA OF IODINE HEPTAFLUORIDE AND IODINE OXYGEN PENTAFLUORIDE.
ALEXANDER LE	0076	POLYMER MONOMER EQUILIBRIA IN NIOBIUM PENTAFLUORIDE, TANTALUM PENTAFLUORIDE, AND A
	0077	RAMAN SPECTRUM OF ANTIMONY PENTAFLUORIDE IN THE GAS PHASE AS A FUNCTION OF TEMPERA
	0501	VIBRATIONAL SPECTRA OF THE ISOELECTRONIC SPECIES IODINE PENTAFLUORIDE, TELLURIUM P
	0660	VIBRATIONAL SPECTRA OF NIOBIUM AND TANTALUM PENTAFLUORIDES IN THE GAS PHASE. VAPOR
ALEXANDRE M	0414	STUDY OF LIQUID CHLORINE FLUORIDES AND OXYFLUORIDES, CHLORINE PENTAFLUORIDE AND CH
ALIEV MR	0182	ROTATIONAL SPECTRA OF NONPOLAR MOLECULES.
	1442	ROTATIONAL PROFILES OF VIBRATIONAL BANDS IN THE GAS PHASE RAMAN SPECTRA OF MOLECUL
ALJIBURY ALK	0113	INFRARED SPECTRA OF MATRIX ISOLATED ANTIMONY PENTAFLUORIDE AND ARSENIC PENTAFLUORI
	0399	INFRARED SPECTRA OF MATRIX ISOLATED CHLORINE TRIFLUORIDE, BROMINE TRIFLUORIDE, AND
ALLAN A	0697	INFRARED SPECTRUM OF NITROGEN TRIFLUORIDE AND FORCE FIELD.
ALLEN LC	0241	AB INITIO STUDIES OF THE ELECTRONIC STRUCTURES OF BORON TRIHYDRIDE, BORON DIHYDRID
	0758	EXTENDED HUCKEL THEORY AND THE SHAPE OF MOLECULES. (OXYGEN DIFLUORIDE)
	1337	THEORY OF INERT GAS MOLECULES.
AMANO T	1018	MICROWAVE SPECTRUM OF THE SULFUR MONOFLUORIDE RADICAL.
AMBARTSUMYAN RV	1064	SELECTIVE PHOTODISSOCIATION OF POLYATOMIC MOLECULES (SULFUR HEXAFLUORIDE) BY INFRA
ANANTHAKRISHNAN TR	0827	PARAMETER METHOD IN THE CALCULATION OF FORCE CONSTANTS. APPLICATION TO E' SPECIES
ANDERSEN A	0672	CONFIGURATION INTERACTION STUDIES OF NITROGEN MONOFLUORIDE AND ITS POSITIVE ION.
ANDERSEN B	1153	SPECTROSCOPIC STUDIES IN CONNECTION WITH ELECTRON DIFFRACTION INVESTIGATION OF SOM
ANDERSON CP	0369	PHOTOELECTRON SPECTRUM OF CHLORINE MONOFLUORIDE.
ANDERSON D	0561	STRUCTURE OF KRYPTON DIFLUORIDE.
ANDERSON RE	0689	A CAVITY SEARCH SPECTROMETER FOR FREE RADICAL MICROWAVE ROTATIONAL ABSORPTION STUD
	0690	MICROWAVE ROTATIONAL SPECTRUM OF THE NITROGEN DIFLUORIDE FREE RADICAL.
ANDERSON TF	0251	RAMAN SPECTRUM OF BORON TRIFLUORIDE GAS.
	0822	RAMAN SPECTRA AND MOLECULAR CONSTANTS OF PHOSPHORUS TRIFLUORIDE AND PHOSPHINE.
ANDREWS EB	0302	ABSORPTION SPECTRUM OF CARBON DIFLUORIDE.
ANDREWS L	0544	ABSORPTION AND EMISSION SPECTRA OF ARGON MATRIX ISOLATED XENON MONOFLUORIDE AND KR
	0755	MATRIX INFRARED SPECTRUM OF OXYGEN MONOFLUORIDE AND DETECTION OF LITHIUM OXYFLUORI
	0759	ARGON MATRIX RAMAN SPECTRA OF OXYGEN DIFLUORIDE AND THE OXYGEN FLUORIDE FREE RADIC
APPELBLAD O	0999	ROTATIONAL ANALYSIS OF THE GAMMA BANDS OF SILICON MONOFLUORIDE.
APPELMAN EH	0010	CHEMICAL COMPOUNDS OF XENON AND OTHER NOBLE GASES.
ARIN ML	0138	THERMODYNAMIC AND PHYSICAL PROPERTIES OF BERYLLIUM COMPOUNDS. PART-1: ENTHALPY AND
ARMSTRONG DR	0184	ELECTRONIC STRUCTURE OF BORON TRIFLUORIDE.

ARMSTRONG DR	0185	GROUND STATE PROPERTIES OF THE GROUP-2 TRIHALIDES. (BORON TRIFLUORIDE)
	0372	ELECTRONIC STRUCTURES OF CHLORINE MONOFLUORIDE AND CHLORINE TRIFLUORIDE.
	0376	PHOTOELECTRON SPECTRA OF HALIDES. (CHLORINE MONOFLUORIDE, CHLORINE TRIFLUORIDE, BR
ARMSTRONG GT	0039	HEAT OF FORMATION OF ALUMINUM TRIFLUORIDE BY DIRECT COMBINATION OF THE ELEMENTS.
	0132	HEAT OF FORMATION OF BERYLLIUM DIFLUORIDE.
	0198	HEAT FORMATION BORON TRIFLUORIDE BY DIRECT COMBINATION OF THE ELEMENTS.
	0401	HEAT OF FORMATION OF CHLORINE TRIFLUORIDE AT 298.14 DEGREES K.
	0769	HEAT OF FORMATION OF OXYGEN DIFLUORIDE.
ARMSTRONG RL	0183	NUCLEAR SPIN RELAXATION STUDY OF THE SPIN ROTATION INTERACTION IN SYMMETRIC TOP MO
	0323	NUCLEAR SPIN RELAXATION STUDY OF THE SPIN ROTATION INTERACTION IN SPHERICAL TOP MO
	0698	ABSORPTION SPECTRA OF MATRIX ISOLATED NITROGEN TRIFLUORIDE FROM 18 TO 50 GC.
ARMSTRONG RS	0314	VAPOR PHASE RAMAN SPECTRA OF MH4 (M= CARBON, SILICON, GERMANIUM, AND TIN) AND MF4
ARULDHAS G	0827	PARAMETER METHOD IN THE CALCULATION OF FORCE CONSTANTS. APPLICATION TO E' SPECIES
ASCHER E	0639	PHYSICAL CONSTANTS (VAPOR PRESSURE) OF SILICON TETRAFLUORIDE, TUNGSTEN HEXAFLUORID
ASPREY LB	0262	INFRARED SPECTRUM OF BROMINE PENTAFLUORIDE.
	0445	ABSORPTION SPECTRA OF SOLID AMERICIUM TRIFLUORIDE, AMERICIUM TETRAFLUORIDE, CURIUM
	0447	EVIDENCE FOR CURIUM TETRAFLUORIDE.
	0634	VIBRATIONAL SPECTRUM AND FORCE FIELD OF MOLYBDENUM HEXAFLUORIDE.
	0897	PREPARATION, STRUCTURE AND SPECTRA OF PRASEODYMIUM TETRAFLUORIDE.
	0915	INFRARED BAND CONTOURS OF RHENIUM HEXAFLUORIDE.
	1156	CORIOLIS CONSTANTS OF SPHERICAL TOP MOLECULES FROM LOW TEMPERATURE INFRARED STUDIE
	1196	VIBRATIONAL SPECTRUM AND FORCE FIELD OF URANIUM HEXAFLUORIDE.
	1334	LATTICE CONSTANTS OF ACTINIDE TETRAFLUORIDES INCLUDING BERKELIUM.
ASTHANA BP	0121	GREEN BAND SYSTEM OF BARIUM MONOFLUORIDE.
	0601	DISSOCIATION ENERGY OF MAGNESIUM MONOFLUORIDE.
ASTIER M	0602	STRUCTURE OF THE MAGNESIUM DIFLUORIDE MOLECULE.
ATHENOUR C	0599	ROTATIONAL ANALYSIS OF SELECTED BANDS FROM THE ELECTRONIC SPECTRUM OF THE LUTETIUM
AULT BS	0544	ABSORPTION AND EMISSION SPECTRA OF ARGON MATRIX ISOLATED XENON MONOFLUORIDE AND KR
AUSDAL RV	0489	HYPERFINE STRUCTURE OF INDIUM MONOFLUORIDE.
AVERILL FW	0881	MOLECULAR ORBITALS OF PLATINUM HEXAFLUORIDE AND ELEMENT 110 HEXAFLUORIDE CALCULATE
AWASTHI MN	1443	MOLECULAR FORCE FIELDS OF XY6 MOLECULES.
AYNSLEY EE	0096	THE INFRARED SPECTRA OF SOME VOLATILE INORGANIC FLUORIDES IN THE SOLID STATE. (ARS
AZMAN A	0181	ESR PARAMETERS OF THE BORON DIFLUORIDE RADICAL.
BABU YKSC	0618	A NEW ELECTRONIC TRANSITION OF THE MERCURY MONOFLUORIDE MOLECULE.
BACIS R	0598	ELECTRONIC SPECTRUM OF LUTETIUM MONOFLUORIDE.
BACON J	0078	FLUORINE NMR OF ANTIMONY PENTAFLUORIDE.
BADACHHAPE RB	0296	THERMODYNAMIC, ELECTROCHEMICAL AND SYNTHETIC STUDIES OF THE GRAPHITE FLUORINE COM
BADER RFW	0150	MOLECULAR CHARGE DISTRIBUTIONS AND CHEMICAL BINDING. PART-3: THE ISOELECTRONIC NIT
	0151	RELAXATION OF THE MOLECULAR CHARGE DISTRIBUTION AND THE VIBRATIONAL FORCE CONSTANT
BAGATURYANTS AA	0415	CALCULATION OF THE ELECTRONIC STRUCTURE OF A CHLORINE PENTAFLUORIDE MOLECULE BY TH
BAGUS PS	0548	ELECTRONIC STRUCTURE AND PROPERTIES OF KRYPTON DIFLUORIDE.
	1221	PROBABLE NONEXISTENCE OF XENON MONOFLUORIDE AS A CHEMICALLY BOUND SPECIES IN THE G
BAIBUZ VF	0315	EXPLOSION METHOD AND THE HEATS OF FORMATION OF CARBON TETRAFLUORIDE AND OTHER CARB
BAILEY CR	0186	INFRARED ABSORPTION SPECTRA OF SOME POLYATOMIC FLUORIDES. (SILICON TETRAFLUORIDE,
	0187	MOLECULAR STRUCTURE OF BORON TRIFLUORIDE.
BAKER JG	0261	MICROWAVE MEASUREMENTS OF THE J-8 FAR-9 TRANSITION OF BROMINE PENTAFLUORIDE.
BAKER WA	1345	STRUCTURE AND MAGNETIC PROPERTIES OF RARE EARTH FLUORIDES.
BAKERMAN S	0699	INFRARED BAND SHAPES AND VIBRATIONS OF NITROGEN TRIFLUORIDE.
BALLEIN K	1473	FORCE CONSTANTS AND NORMAL COORDINATE ANALYSIS.
BALLHAUSEN CH	0831	ORBITALS AND STRUCTURES OF PENTAFLUORIDES. (PHOSPHORUS PENTAFLUORIDE, ARSENIC PENT
BANDRAUK AD	0150	MOLECULAR CHARGE DISTRIBUTIONS AND CHEMICAL BINDING. PART-3: THE ISOELECTRONIC NIT
	0151	RELAXATION OF THE MOLECULAR CHARGE DISTRIBUTION AND THE VIBRATIONAL FORCE CONSTANT
BANNA MS	0683	MULTIPLET SPLITTING IN 1S HOLE STATES OF MOLECULES.
BANSAL BM	0899	SOME PHYSICAL PROPERTIES OF CURIUM AND PROTACTINIUM METALS AND PROTACTINIUM TETRAF
BANTOV DV	0416	FLUORINE NMR OF CHLORINE PENTAFLUORIDE.
BANYARD KE	0676	CALCULATION OF THE SPECTROSCOPIC CONSTANTS FOR SEVERAL STATES OF NITROGEN MONOFLUO
BARANOVSKII VI	1033	SOFT X-RAY SPECTROSCOPY ON SULFUR TETRAFLUORIDE AND SULFUR HEXAFLUORIDE.
	1065	ELECTRONIC STRUCTURE OF SULFUR HEXAFLUORIDE STUDIED BY ULTRALONG WAVE X-RAY SPECTR
BARBER EJ	1152	SOME PHYSICAL PROPERTIES OF TUNGSTEN HEXAFLUORIDE. (VAPOR PRESSURE)
BARBERI P	0370	HEATS OF FORMATION OF CHLORINE FLUORIDES. (CHLORINE MONOFLUORIDE, CHLORINE DIFLUOR
BARINSKII RL	0204	ULTRASOFT X-RAY ABSORPTION SPECTRA OF BORON TRIFLUORIDE.
BARKER EF	0206	INFRARED ABSORPTION SPECTRUM OF BORON TRIFLUORIDE.
BARRETT AH	1110	MICROWAVE SPECTRA OF THE THALLIUM, INDIUM AND GALLIUM MONOHALIDES.
BARROW RF	0022	SPECTROSCOPIC AND THERMODYNAMIC PROPERTIES OF ALUMINUM MONOFLUORIDE.
	0023	SINGLET ELECTRONIC STATES OF ALUMINUM MONOFLUORIDE.
	0027	ELECTRONIC STATES OF GASEOUS ALUMINUM MONOFLUORIDE.
	0053	HEAT OF SUBLIMATION OF ALUMINUM TRIFLUORIDE AND HEAT OF FORMATION OF ALUMINUM MONO
	0118	ROTATIONAL ANALYSIS OF SOME BANDS OF GASEOUS BARIUM MONOFLUORIDE.
	0152	DISSOCIATION ENERGIES OF THE GASEOUS MONOHALIDES OF BORON, ALUMINUM, GALLIUM, INDI
	0153	ROTATIONAL ANALYSIS OF BANDS OF THE C TRIPLET SIGMA, B TRIPLET SIGMA: A TRIPLET PI
	0302	ABSORPTION SPECTRUM OF CARBON DIFLUORIDE.
	0457	ROTATIONAL ANALYSIS OF BANDS OF THE A TRIPLET PI ZERO(+) B TRIPLET PI(1) TO X SING
	0488	ABSORPTION SPECTRUM OF GASEOUS HOLMIUM MONOFLUORIDE.
	0600	ROTATIONAL ANALYSIS OF ELECTRONIC BANDS OF GASEOUS MAGNESIUM MONOFLUORIDE.
	0928	ELECTRONIC STATES OF GASEOUS SCANDIUM MONOFLUORIDE, YTTRIUM MONOFLUORIDE, AND LANT
	0929	GROUND STATES OF SCANDIUM MONOFLUORIDE AND YTTRIUM MONOFLUORIDE.
	0930	ROTATIONAL ANALYSIS OF SOME SINGLET TRANSITIONS IN THE SPECTRUM OF GASEOUS SCANDIU
	0931	ROTATIONAL ANALYSIS OF THE SYSTEM B SINGLET PI TO SINGLET SIGMA(+) OF GASEOUS SCAN
	0964	SPECTRA OF SILICON MONOFLUORIDE, GERMANIUM MONOFLUORIDE, TIN MONOFLUORIDE, AND LEA
	0965	BAND SPECTRUM OF SILICON MONOFLUORIDE.
	0983	ULTRAVIOLET SPECTRUM OF SILICON DIFLUORIDE.
	0999	ROTATIONAL ANALYSIS OF THE GAMMA BANDS OF SILICON MONOFLUORIDE.
	1011	ROTATIONAL ANALYSIS OF THE AO(+), BO(+), TO X SINGLET SIGMA(+) SYSTEMS OF GASEOUS
	1012	ABSORPTION SPECTRUM OF GASEOUS SILVER MONOFLUORIDE.
	1016	INTERNUCLEAR DISTANCE IN GASEOUS STRONTIUM MONOFLUORIDE.
	1111	ROTATIONAL ANALYSIS OF BANDS OF THE A TRIPLET PI ZERO(+), B TRIPLET PI(1) TO X SIN
	1133	ROTATIONAL ANALYSIS OF THE A DOUBLET SIGMA(+) TO X DOUBLET PI, B DOUBLET SIGMA(+)
BAR-ZIV E	1182	INFRARED SPECTRUM OF URANIUM HEXAFLUORIDE.
BARTELL LS	0099	MOLECULAR STRUCTURES OF ARSENIC TRIFLUORIDE AND ARSENIC PENTAFLUORIDE AS DETERMINE
	0515	STRUCTURE, PSEUDOROTATION, AND VIBRATIONAL MODE COUPLING IN IODINE HEPTAFLUORIDE.

BARTELL LS	0828	GILLESPIE-NYHOLM ASPECTS OF FORCE FIELDS. PART-1. POINTS-ON-A-SPHERE AND EXTENDED
	0851	ELECTRON DIFFRACTION STUDY OF THE STRUCTURE OF PHOSPHORUS PENTAFLUORIDE.
	0909	ELECTRON DIFFRACTION STUDY OF RHENIUM HEXAFLUORIDE.
	0920	ELECTRON DIFFRACTION STUDY OF RHENIUM HEPTAFLUORIDE. STRUCTURE, PSEUDOROTATION, AN
	1279	EVIDENCE FOR PSEUDO JAHN-TELLER EFFECT IN XENON HEXAFLUORIDE.
	1280	MOLECULAR STRUCTURE OF XENON HEXAFLUORIDE.
	1291	MOLECULAR STRUCTURE OF XENON HEXAFLUORIDE. PART-1: ANALYSIS OF ELECTRON DIFFRACTIO
BARTLETT N	0517	FLUORINE NMR OF IODINE HEPTAFLUORIDE, RHENIUM HEPTAFLUORIDE, AND OXIDE PENTAFLUORI
	0520	INTRAMOLECULAR REARRANGEMENT IN IODINE HEPTAFLUORIDE AND XENON HEXAFLUORIDE.
	1229	ELECTRON DISTRIBUTION IN THE XENON FLUORIDES AND XENON OXIDE TETRAFLUORIDE BY ESCA
BASCH H	1224	SELF CONSISTENT FIELD STUDY OF THE SERIES XENON DIFLUORIDE, XENON TETRAFLUORIDE, X
	1228	HELIUM-1 AND HELIUM-2 PHOTOELECTRON SPECTRA AND THE ELECTRONIC STRUCTURES OF XENON
BASS AM	0297	ABSORPTION SPECTRUM OF CARBON DIFLUORIDE TRAPPED IN AN ARGON MATRIX.
BASS CD	1444	VIBRATIONAL ANALYSIS OF SUBSTITUTED AND PERTURBED MOLECULES. DIRECT DETERMINATION
BASSETT PJ	0188	PHOTOELECTRON SPECTRA OF HALIDES. PART-2: HIGH RESOLUTION SPECTRA OF THE BORON TRI
	0189	PHOTOELECTRON SPECTRUM OF BORON TRIFLUORIDE.
	0316	PHOTOELECTRON SPECTRA OF HALIDES. PART-1. TETRAFLUORIDES AND TETRACHLORIDES OF GRO
	0317	HELIUM-1 RESONANCE PHOTOELECTRON SPECTRA OF GROUP-4 TETRAFLUORIDES.
	0700	PHOTOELECTRON SPECTRA OF NITROGEN TRIFLUORIDE AND NITROGEN OXIDE TRIFLUORIDE, AND
	0701	PHOTOELECTRON SPECTRA OF HALIDES. PART-3: TRIFLUORIDES AND OXIDE TRIFLUORIDES OF N
BASTIN MW	0118	ROTATIONAL ANALYSIS OF SOME BANDS OF GASEOUS BARIUM MONOFLUORIDE.
	0928	ELECTRONIC STATES OF GASEOUS SCANDIUM MONOFLUORIDE, YTTRIUM MONOFLUORIDE, AND LANT
BATANA A	0094	ARSENIC MONOFLUORIDE, 3 SIGMA. DISSOCIATION ENTHALPY, IONIZATION POTENTIAL, ELECTR
BATAREV GA	0654	THERMODYNAMIC PROPERTIES OF NIOBIUM AND TANTALUM FLUORIDES AT HIGH TEMPERATURES.
	0655	THERMODYNAMIC PROPERTIES OF NIOBIUM MONOFLUORIDE, NIOBIUM DIFLUORIDE, TANTALUM MON
	0656	THERMODYNAMIC PROPERTIES OF NIOBIUM TRIFLUORIDE, NIOBIUM TETRAFLUORIDE, TANTALUM T
	0657	THERMODYNAMIC PROPERTIES OF NIOBIUM AND TANTALUM FLUORIDES AT HIGH TEMPERATURES. P
	0665	THERMODYNAMIC PROPERTIES OF NIOBIUM AND TANTALUM FLUORIDES AT HIGH TEMPERATURES. P
BATES JB	0621	RAMAN SPECTRUM OF SOLID MOLYBDENUM TETRAFLUORIDE.
	0623	RAMAN SPECTRUM OF CRYSTALLINE MOLYBDENUM PENTAFLUORIDE.
BATSANOV SS	1346	INFRARED SPECTRA OF RARE EARTH METAL FLUORIDES.
BATSANOVA LR	1346	INFRARED SPECTRA OF RARE EARTH METAL FLUORIDES.
	1352	INFRARED SPECTRA OF RARE EARTH METAL FLUORIDES IN THE REGION OF A CESIUM IODIDE PR
BATTERMAN BW	0283	X-RAY STUDY OF ANHARMONIC VIBRATIONS IN CALCIUM DIFLUORIDE.
BAUER HF	0417	EQUILIBRIUM STUDIES OF CHLORINE PENTAFLUORIDE.
BAUER SH	0530	MOLECULAR STRUCTURE OF IODINE HEPTAFLUORIDE.
	0558	STRUCTURE OF KRYPTON DIFLUORIDE AS INVESTIGATED BY ELECTRON DIFFRACTION.
	0679	ELECTRON DIFFRACTION STUDIES AT ELEVATED TEMPERATURES. (NITROGEN DIFLUORIDE)
	0680	DIFFRACTION STUDY OF THE STRUCTURES OF NITROGEN DIFLUORIDE AND DINITROGEN TETRAFLU
	1183	STRUCTURE OF URANIUM HEXAFLUORIDE AS DETERMINED BY DIFFRACTION OF ELECTRONS ON THE
	1184	STRUCTURE OF URANIUM HEXAFLUORIDE AS DETERMINED BY THE DIFFRACTION OF ELECTRONS ON
	1282	ELECTRON DIFFRACTION STUDY OF THE STRUCTURE OF GASEOUS XENON TETRAFLUORIDE.
BAUMAN RP	0572	LATTICE VIBRATIONS AND STRUCTURE OF RARE EARTH HALIDES. (LANTHANUM TRIFLUORIDE)
BAUMGARTNER R	0847	RAMAN AND INFRARED SPECTRUM OF TRIFLUORO PHOSPHORANE. (PHOSPHORUS PENTAFLUORIDE)
BAUTISTA RG	0276	SUBLIMATION PRESSURE OF CALCIUM DIFLUORIDE AND THE DISSOCIATION ENERGY OF CALCIUM
	0611	A LANGMUIR MEASUREMENT OF THE SUBLIMATION PRESSURE OF MANGANESE DIFLUORIDE.
	1382	SUBLIMATION PRESSURES OF REFRACTORY FLUORIDES.
BEAGLEY B	1000	SILICON- FLUORINE LENGTH IN SILICON TETRAFLUORIDE. NEW ELECTRON DIFFRACTION STUDY.
BEAL JB	0768	MILLIMETER WAVE SPECTRUM AND STRUCTURE OF OXYGEN DIFLUORIDE.
BEALE JR	0600	ROTATIONAL ANALYSIS OF ELECTRONIC BANDS OF GASEOUS MAGNESIUM MONOFLUORIDE.
	1016	INTERNUCLEAR DISTANCE IN GASEOUS STRONTIUM MONOFLUORIDE.
BEATON S	0517	FLUORINE NMR OF IODINE HEPTAFLUORIDE, RHENIUM HEPTAFLUORIDE, AND OXIDE PENTAFLUORI
BEATTIE IR	0077	RAMAN SPECTRUM OF ANTIMONY PENTAFLUORIDE IN THE GAS PHASE AS A FUNCTION OF TEMPERA
	0501	VIBRATIONAL SPECTRA OF THE ISOELECTRONIC SPECIES IODINE PENTAFLUORIDE, TELLURIUM P
	0660	VIBRATIONAL SPECTRA OF NIOBIUM AND TANTALUM PENTAFLUORIDES IN THE GAS PHASE. VAPOR
	0829	VIBRATIONAL SPECTRA OF MIXED PHOSPHORUS HALIDES. (PHOSPHORUS PENTAFLUORIDE)
BEAUCHAMP JL	1185	PROPERTIES AND REACTIONS OF URANIUM HEXAFLUORIDE BY ION CYCLOTRON RESONANCE SPECTR
BEAUDRY BJ	1356	HIGH TEMPERATURE ENTHALPIES AND RELATED THERMODYNAMIC FUNCTIONS OF THE TRIFLUORIDE
BECHER HJ	1473	FORCE CONSTANTS AND NORMAL COORDINATE ANALYSIS.
BECKMANN L	0190	FORCE CONSTANTS FOR MOLECULES WITH D3H SYMMETRY.
BEGUN GM	0418	VIBRATIONAL SPECTRA AND VALENCE FORCE CONSTANTS OF SQUARE PYRAMIDAL MOLECULES. XEN
	1223	XENON DIFLUORIDE AND THE NATURE OF THE XENON- FLUORINE BOND.
BELOUSOV VI	1326	VAPOR PRESSURE OF ZIRCONIUM TETRAFLUORIDE.
BENDAZZOLI GL	0371	SCF- MO CALCULATIONS OF SOME MOLECULAR PROPERTIES OF THE ISOELECTRONIC SERIES CHLO
	1066	AB INITIO AND MZDO WAVE FUNCTIONS FOR SULFUR HEXAFLUORIDE.
BENEDICT WS	0355	INFRARED SPECTRA OF HALOGEN-SUBSTITUTED METHANES. (CARBON TETRAFLUORIDE)
BENSEY FN	0395	STRUCTURES OF THE INTERHALOGEN COMPOUNDS. PART-1: CHLORINE TRIFLUORIDE AT -120 DEG
	0521	STRUCTURE OF INTERHALOGEN COMPOUNDS. PART-2: IODINE HEPTAFLUORIDE AT -110 DEGREES
BENSON GC	1286	STEREOCHEMISTRY AND SEVEN COORDINATION. (XENON HEXAFLUORIDE)
BERG RA	0126	GEOMETRY OF THE ALKALINE EARTH DIHALIDES. (BARIUM DIFLUORIDE, STRONTIUM DIFLUORIDE
BERGER R	0888	SUBLIMATION OF PLUTONIUM TETRAFLUORIDE.
BERKOWITZ J	0603	MASS SPECTROMETRIC STUDY OF MAGNESIUM HALIDES. (MAGNESIUM DIFLUORIDE)
	0760	PHOTOIONIZATION MASS SPECTROMETRY OF OXYGEN DIFLUORIDE.
	1112	PHOTOIONIZATION OF HIGH TEMPERATURE VAPORS. PART-4: THALLIUM MONOFLUORIDE, THALLIU
	1117	PHOTOELECTRON SPECTROSCOPY OF HIGH TEMPERATURE VAPORS. PART-3: MONOMER AND DIMER S
	1225	PHOTOIONIZATION MASS SPECTROMETRIC STUDY OF XENON DIFLUORIDE, XENON TETRAFLUORIDE,
BERNEY CV	1045	INFRARED SPECTRUM AND NORMAL COORDINATE ANALYSIS OF SULFUR TETRAFLUORIDE.
	1053	FLUORINE EXCHANGE IN SULFUR TETRAFLUORIDE.
	1054	SULFUR TETRAFLUORIDE AND SULFUR OXIDE DIFLUORIDE. INFRARED EVIDENCE FOR DIMER FORM
BERNSTEIN HJ	0346	FLUORINE RELAXATION BY NMR ABSORPTION IN GASEOUS CARBON TETRAFLUORIDE, SILICON TET
	0761	VIBRATIONAL SPECTRA AND STRUCTURE OF INORGANIC MOLECULES. PART-1: INFRARED SPECTRU
	1098	RAMAN SPECTRUM OF SOLID SULFUR HEXAFLUORIDE.
BERNSTEIN LS	0830	POTENTIAL FUNCTION FOR THE NU-7 VIBRATION OF PHOSPHORUS PENTAFLUORIDE.
	1281	ELECTRIC FIELD DEFLECTION OF MOLECULES WITH LARGE AMPLITUDE MOTION. (XENON HEXAFLU
	1303	MOLECULAR STRUCTURE OF XENON HEXAFLUORIDE.
BERRY RS	0831	ORBITALS AND STRUCTURES OF PENTAFLUORIDES. (PHOSPHORUS PENTAFLUORIDE, ARSENIC PENT
	1380	BENDING MOTIONS IN THE DIHALIDES OF GROUP-2 METALS.
BERSUKER IB	0901	POSSIBLE ROTATIONAL RAMAN SPECTRUM AND THE DEPOLARIZATION OF RAYLEIGH SCATTERING I
BERTHIER G	0602	STRUCTURE OF THE MAGNESIUM DIFLUORIDE MOLECULE.
BERTONICINI P	0174	METHODS FOR CORRELATING MOLECULES AND SOME OPTIMIZED VALENCE CONFIGURATION RESULTS

BERTSEV VV   0313 INFRARED SPECTRA OF A CRYOSYSTEM. CARBON TETRAFLUORIDE.
                     1068 INFRARED SPECTRA OF CRYOSYSTEMS. PART-2: SULFUR HEXAFLUORIDE.
BESENBRUCH G   0272 KNUDSEN AND LANGMUIR MEASUREMENTS OF THE SUBLIMATION PRESSURE OF CADMIUM DIFLUORID
                     0439 KNUDSEN AND LANGMUIR MEASUREMENTS OF THE SUBLIMATION PRESSURE OF COBALT DIFLUORIDE
                     0448 MASS SPECTROMETRIC STUDIES AT HIGH TEMPERATURES. PART-17: SUBLIMATION AND VAPOR PR
                     0947 SUBLIMATION PRESSURES OF SCANDIUM TRIFLUORIDE, YTTRIUM TRIFLUORIDE, AND LANTHANUM
                     0984 ULTRAVIOLET ABSORPTION SPECTRUM OF SILICON DIFLUORIDE.
                     1382 SUBLIMATION PRESSURES OF REFRACTORY FLUORIDES.
BESSETTE F   0327 VIBRATIONAL SPECTRA OF LIQUID AND CRYSTALLINE CARBON TETRAFLUORIDE.
                     1001 INFRARED AND RAMAN SPECTRA OF LIQUID AND CRYSTALLINE SILICON TETRAFLUORIDE.
BEST PE   1067 X-RAY ABSORPTION IN SULFUR HEXAFLUORIDE.
BEVERIDGE AD   1367 PENTAHALIDES OF TRANSITION METALS.
BHAVSAR GP   0918 MOLECULAR CONSTANTS OF RHENIUM HEPTAFLUORIDE MOLECULE.
BHUTRA MP   0646 ENERGY LEVELS AND SPECTROSCOPIC PARAMETERS OF NEODYMIUM TRIFLUORIDES.
BIDEAU-MEHU A   0042 RECENT DEVELOPMENTS IN MULTIPLE BEAM INTERFEROMETRY IN THE ULTRAVIOLET REGION 1700
BIEFELD RM   1131 VAPORIZATION REACTIONS IN THE THULIUM- FLUORINE SYSTEM.
                     1321 SUBLIMATION THERMODYNAMICS OF ZINC(II) FLUORIDE.
                     1347 VAPORIZATION AND SUBLIMATION THERMODYNAMICS OF SELECTED LANTHANIDE FLUORIDES.
BIGELEISEN J   1186 VIBRATIONAL SPECTRUM AND THERMODYNAMIC PROPERTIES OF URANIUM HEXAFLUORIDE GAS.
                     1187 VIBRATIONAL SPECTRUM AND THERMODYNAMIC PROPERTIES OF URANIUM HEXAFLUORIDE GAS.
BILHAM J   1226 ELECTRONIC STRUCTURE OF XENON DIFLUORIDE.
BINGHAM JT   0139 ENTHALPY OF FORMATION OF BERYLLIUM DIFLUORIDE.
                     0483 HEATS OF FORMATION OF GERMANIUM TETRAFLUORIDE AND OF THE GERMANIUM DIOXIDES.
BIRKS JW   0493 THEORY OF THE DISSOCIATION OF DIATOMIC MOLECULES AND A STUDY OF THE EMISSION OF IO
BIRNBAUM G   0356 FAR INFRARED ABSORPTION IN GASEOUS CARBON TETRAFLUORIDE.
                     1095 FAR INFRARED SPECTRA OF GASEOUS AND LIQUID SULFUR HEXAFLUORIDE.
BLANCHARD S   0114 INFRARED SPECTRUM OF ARSENIC PENTAFLUORIDE.
                     0661 INFRARED SPECTRUM OF NIOBIUM PENTAFLUORIDE.
BLANKENSHIP FF   1174 VAPOR PRESSURE OF URANIUM TETRAFLUORIDE.
BLASS WE   0716 PSEUDO-PARALLEL INFRARED BANDS OF OBLATE SYMMETRIC TOPS NITROGEN TRIFLUORIDE NU-4
BLAUER JA   0154 HEAT AND ENTROPY OF FORMATION OF BORON FLUORIDE.
                     0394 THERMAL DISSOCIATION OF CHLORINE TRIFLUORIDE BEHIND INCIDENT SHOCK WAVES.
BLIEFERT C   1026 ESTIMATION OF INFRARED FREQUENCIES OF SULFUR DIFLUORIDE.
BLOCHER JM   1148 VAPOR PRESSURE OF TITANIUM TETRAFLUORIDE.
BLOCK FE   0438 VAPOR PRESSURE AND HEAT OF SUBLIMATION OF COBALT DIHALIDES. (COBALT DIFLUORIDE)
BLUE GD   0119 DISSOCIATION ENERGIES OF THE ALKALINE EARTH MONOFLUORIDES.
                     0276 SUBLIMATION PRESSURE OF CALCIUM DIFLUORIDE AND THE DISSOCIATION ENERGY OF CALCIUM
BOATE AR   0271 ELECTRON SPIN RESONANCE SPECTRA OF HALOGEN HEXAFLUORIDES. (CHLORINE HEXAFLUORIDE,
BODERSON PH   0252 SPECTRUM OF BROMINE MONOFLUORIDE AND ITS DISSOCIATION ENERGY.
BOEHLER D   0085 MASS SPECTRUM OF ANTIMONY PENTAFLUORIDE.
BOHN RK   0558 STRUCTURE OF KRYPTON DIFLUORIDE AS INVESTIGATED BY ELECTRON DIFFRACTION.
                     0680 DIFFRACTION STUDY OF THE STRUCTURES OF NITROGEN DIFLUORIDE AND DINITROGEN TETRAFLU
                     1259 ELECTRON DIFFRACTION STUDIES OF XENON TETRAFLUORIDE AND XENON HEXAFLUORIDE AND OTH
                     1282 ELECTRON DIFFRACTION STUDY OF THE STRUCTURE OF GASEOUS XENON TETRAFLUORIDE.
BOLDRINI P   0593 NEUTRON DIFFRACTION INVESTIGATION OF ORTHORHOMBIC LEAD DIFLUORIDE.
BOLOTIN AB   0977 CALCULATION OF THE FRANCK-CONDON FACTOR FOR MOLECULES OF SILICON FLUORIDE, GERMANI
BONACCORSI R   0762 DOUBLE ZETA SCF CALCULATIONS FOR NITRITE ION AND OXYGEN DIFLUORIDE.
                     0763 CONFIGURATION INTERACTION CALCULATION FOR THE GROUND STATE OF OXYGEN DIFLUORIDE, N
BONACIC-KOUTECKY V   1027 MOLECULAR ORBITAL DESCRIPTION OF SULFUR DIFLUORIDE, SULFUR TETRAFLUORIDE AND SULFU
BONHAM RA   0911 MULTIPLE INTRAMOLECULAR SCATTERING EFFECTS ON ELECTRON DIFFRACTION PATTERNS FOR TH
BORDE C   0832 LASER SPECTROSCOPY. (PHOSPHORUS PENTAFLUORIDE)
BORDE J   0837 SYMMETRY GROUP OF PHOSPHORUS PENTAFLUORIDE.
BORING M   1188 CALCULATED ELECTRONIC STRUCTURE OF URANIUM HEXAFLUORIDE.
BORNAIS J   0194 INTERMOLECULAR FLUORINE EXCHANGE BETWEEN TETRABUTYL AMMONIUM HEXAFLUORO PHOSPHATE
BOSWORTH YM   1069 VAPOR PHASE RAMAN SPECTRA, RAMAN BAND CONTOUR ANALYSES, AND CORIOLIS CONSTANTS OF
BOUGON R   0398 RAMAN SPECTRA AND THE STRUCTURE OF CHLORINE TRIFLUORIDE AND BROMINE TRIFLUORIDE.
                     1189 VIBRATIONAL SPECTRUM AND STRUCTURE OF SOLID URANIUM HEXAFLUORIDE.
BOWATER IC   0957 MICROWAVE SPECTRUM, DIPOLE MOMENT, AND STRUCTURE ANALYSIS OF SELENIUM TETRAFLUORID
BOWEN HJM   0318 STRUCTURES OF TRIFLUORO METHYL HALIDES. (CARBON TETRAFLUORIDE)
BRADLEY CA   1497 VIBRATIONS OF PENTATOMIC TETRAHEDRAL MOLECULES.
BRADLEY RH   0260 K-TYPE DOUBLING IN THE MILLIMETER WAVE SPECTRUM OF BROMINE PENTAFLUORIDE.
                     0265 GAS PHASE MOLECULAR STRUCTURES OF BROMINE PENTAFLUORIDE AND IODINE PENTAFLUORIDE F
                     0269 MICROWAVE SPECTRUM OF BROMINE PENTAFLUORIDE.
                     0502 MICROWAVE SPECTRUM OF IODINE PENTAFLUORIDE.
BRADY AP   0662 HEAT CAPACITY, ENTROPY, AND HEATS OF TRANSITION OF MOLYBDENUM HEXAFLUORIDE AND NIO
BRAND JCD   0537 VIBRONICALLY INDUCED JAHN-TELLER PROGRESSIONS IN THE ELECTRONIC SPECTRUM OF IRIDIU
                     0902 NEAR INFRARED SYSTEM OF RHENIUM HEXAFLUORIDE.
BRAU CA   0545 EMISSION SPECTRA OF XENON MONOBROMIDE, XENON MONOCHLORIDE, XENON MONOFLUORIDE, AND
BREEZE A   0372 ELECTRONIC STRUCTURES OF CHLORINE MONOFLUORIDE AND CHLORINE TRIFLUORIDE.
                     0376 PHOTOELECTRON SPECTRA OF HALIDES. (CHLORINE MONOFLUORIDE, CHLORINE TRIFLUORIDE, BR
                     0553 AB INITIO CALCULATIONS OF THE BONDING IN KRYPTON DIFLUORIDE.
                     0554 BONDING IN KRYPTON DIFLUORIDE.
                     0555 BONDING IN KRYPTON DIFLUORIDE.
BREHM B   1227 PHOTOELECTRON SPECTRUM OF XENON DIFLUORIDE.
BREWER FM   0461 PREPARATION AND CRYSTAL STRUCTURE OF GALLIUM TRIFLUORIDE.
BREWER L   0298 HEAT OF FORMATION OF CARBON DIFLUORIDE.
BRICKMANN J   0870 PSEUDOROTATION OF TRIGONAL BIPYRAMIDAL MOLECULES. BERRY ROTATION CONTRA TURNSTILE
BRICKMANN V   0833 PSEUDOROTATION IN TRIGONAL BIPYRAMIDAL MOLECULES. (PHOSPHORUS PENTAFLUORIDE)
BRICKWEDDE FG   1166 LOW TEMPERATURE HEAT CAPACITIES, ENTHALPIES, AND ENTROPIES OF URANIUM TETRAFLUORID
BRIER PN   0260 K-TYPE DOUBLING IN THE MILLIMETER WAVE SPECTRUM OF BROMINE PENTAFLUORIDE.
                     0261 MICROWAVE MEASUREMENTS OF THE J-8 FAR-9 TRANSITION OF BROMINE PENTAFLUORIDE.
                     0265 GAS PHASE MOLECULAR STRUCTURES OF BROMINE PENTAFLUORIDE AND IODINE PENTAFLUORIDE F
                     0269 MICROWAVE SPECTRUM OF BROMINE PENTAFLUORIDE.
                     0502 MICROWAVE SPECTRUM OF IODINE PENTAFLUORIDE.
BRINKLEY DC   1463 STATISTICAL THERMODYNAMICS OF SOME HEXAFLUORIDES.
BROCAS J   0834 SPECTROSCOPIC CONSEQUENCES OF DIFFERENT TUNNELING MECHANISMS IN NONRIGID PHOSPHORU
BROM JM   1113 INFRARED SPECTRA AND STRUCTURE OF MATRIX ISOLATED THALLIUM HALIDES.
BROOKBANKS DM   0261 MICROWAVE MEASUREMENTS OF THE J-8 FAR-9 TRANSITION OF BROMINE PENTAFLUORIDE.
BROTCHIE DA   0180 AB INITIO INVESTIGATION OF THE GEOMETRY, BONDING, AND COUPLING CONSTANTS OF BORON
BROW D   1348 HALIDES OF THE LANTHANIDES AND ACTINIDES.

BROWN CW	0191	VIBRATION- ROTATION BANDS OF BORON TRIFLUORIDE.
	0192	BORON TRIFLUORIDE. THE ANALYSIS OF 2NU-3 AND THE DETERMINATION OF THE 33 CORIOLIS
	0193	ROTATION VIBRATION BAND CONTOURS IN THE INFRARED SPECTRUM OF BORON TRIFLUORIDE.
	0207	NU-3 BANDS OF BORON TRIFLUORIDE.
BROWN D	0001	HALIDES OF THE TRANSITION ELEMENTS. VOLUME-1: FIRST ROW TRANSITION METALS. VOLUME-
BROWN DP	1000	SILICON- FLUORINE LENGTH IN SILICON TETRAFLUORIDE. NEW ELECTRON DIFFRACTION STUDY.
BROWN FC	0215	ABSORPTION OF BORON TRIFLUORIDE GAS IN THE EXTREME ULTRAVIOLET.
BROWN JM	0954	ESR SPECTRUM OF BROMINE OXIDE, IODINE OXIDE AND SELENIUM MONOFLUORIDE IN J-5/2 ROT
BROWN RD	0176	AB INITIO CALCULATIONS ON SOME SMALL RADICALS BY THE UNRESTRICTED HARTREE-FOCK MET
	0681	MICROWAVE SPECTRUM OF NITROGEN DIFLUORIDE RADICAL.
	0682	ELECTRONIC STRUCTURE OF THE NITROGEN DIFLUORIDE RADICAL.
	0835	VESCF-MO STUDIES OF PHOSPHORUS PENTAFLUORIDE, SULFUR TETRAFLUORIDE, AND CHLORINE T
	0957	MICROWAVE SPECTRUM, DIPOLE MOMENT, AND STRUCTURE ANALYSIS OF SELENIUM TETRAFLUORID
	1373	VESCF-MO STUDIES OF MOLECULES CONTAINING ATOMS FROM THE SECOND ROW OF THE PERIODIC
	1474	UREY-BRADLEY POTENTIAL CONSTANTS FOR SECOND ROW FLUORIDES.
BROWN RF	0373	ELECTRONEGATIVITY, NONBONDED INTERACTIONS AND POLARIZABILITY IN HYDROGEN HALIDES A
BROWN SD	0435	VIBRATIONAL SPECTRUM OF LIQUID CHROMIUM PENTAFLUORIDE.
BROWNSTEIN M	0518	SPECTROSCOPIC STUDY OF IODINE HEPTAFLUORIDE AND IODINE HEXAFLUORIDE(X) IN ANHYDROU
BROWNSTEIN S	0194	INTERMOLECULAR FLUORINE EXCHANGE BETWEEN TETRABUTYL AMMONIUM HEXAFLUORO PHOSPHATE
	0836	EXCHANGE OF FLUORINE IN PHOSPHORUS PENTAFLUORIDE.
BRUDERRECK H	0992	THERMOCHEMISTRY OF SILICON DIFLUORIDE.
BRUNDLE CR	0549	MOLECULAR ORBITAL ENERGY LEVELS AND BONDING IN KRYPTON DIFLUORIDE.
	0550	ELECTRONIC STRUCTURE OF KRYPTON DIFLUORIDE STUDIED BY PHOTOELECTRON SPECTROSCOPY.
	1228	HELIUM-1 AND HELIUM-2 PHOTOELECTRON SPECTRA AND THE ELECTRONIC STRUCTURES OF XENON
BRUNET H	1070	NU-3 AND NU-4 BANDS OF SULFUR HEXAFLUORIDE.
BRUNS R	0195	CNDO CALCULATION OF DIPOLE MOMENT DERIVATIVES AND INFRARED INTENSITIES OF BORON TR
	0299	A THEORETICAL STUDY OF THE BOND INTERACTION FORCE CONSTANTS IN DIFLUORIDE MOLECULE
BRUNVOLL J	0421	MEAN AMPLITUDES OF VIBRATION FOR SOME PYRAMIDAL XY4Z MOLECULES. (IODINE PENTAFLUOR
	0840	MEAN AMPLITUDES OF VIBRATION AND CORIOLIS CONSTANTS FOR PHOSPHORUS PENTAFLUORIDE A
	1153	SPECTROSCOPIC STUDIES IN CONNECTION WITH ELECTRON DIFFRACTION INVESTIGATION OF SOM
	1219	FORCE CONSTANTS, MEAN AMPLITUDES OF VIBRATION, SHRINKAGE EFFECTS, CORIOLIS CONSTAN
	1409	MATHEMATICAL TECHNIQUES IN THE STUDY OF CORIOLIS COUPLING IN TETRAHEDRAL XY4 MOLEC
	1410	INFLUENCE OF ATOMIC MASSES ON THE CORIOLIS COUPLING COEFFICIENTS IN XY4 TETRAHEDRA
	1432	GENERALIZED MEAN SQUARE AMPLITUDES OF VIBRATION, AND SHRINKAGE EFFECTS OF SOME TRI
	1447	REVISED MEAN AMPLITUDES OF VIBRATION FOR SOME OCTAHEDRAL HEXAFLUORIDES.
	1454	NORMAL COORDINATE ANALYSIS OF ROTATION VIBRATION OF OCTAHEDRAL XY6 AND Z6 MOLECULA
	1456	FORCE FIELD OF WEINSTOCK AND GOODMAN FOR OCTAHEDRAL HEXAFLUORIDES AND THE L MATRIX
BUCHLER A	0037	SUBLIMATION OF ALUMINUM TRIFLUORIDE AND THE INFRARED SPECTRUM OF ALUMINUM TRIFLUOR
	0526	MOLECULAR STRUCTURE OF XENON HEXAFLUORIDE AND IODINE HEPTAFLUORIDE.
	1327	INFRARED SPECTRA OF GASEOUS GROUP-4 HALIDES: ZIRCONIUM FLUORIDE, ZIRCONIUM CHLORID
	1368	GEOMETRY OF THE TRANSITION METAL DIHALIDES. THE DIFLUORIDES OF SOME FIRST ROW TRAN
	1381	IONIC MODEL CALCULATIONS. PART-2: BENDING FORCE CONSTANTS OF THE GROUP-2 HALIDES.
BUECHLER A	0133	INFRARED SPECTRA OF THE ALKALINE EARTH HALIDES. PART-1: BERYLLIUM DIFLUORIDE, BERY
	0134	DETERMINATION OF THE GEOMETRY OF HIGH TEMPERATURE SPECIES BY ELECTRIC DEFLECTION A
BUENKER RJ	0764	GEOMETRY OF MOLECULES. (OXYGEN DIFLUORIDE)
BUES W	0212	FORCE CONSTANTS OF CARBON TETRAFLUORIDE AND BORON TRIFLUORIDE AND COMPARISON WITH
BUKSHPAN S	0503	MOSSBAUER STUDY OF IODINE PENTAFLUORIDE AND IODINE HEPTAFLUORIDE.
BULANIN MO	0313	INFRARED SPECTRA OF A CRYOSYSTEM. CARBON TETRAFLUORIDE.
BULL WE	0319	HIGH RESOLUTION PHOTOELECTRON SPECTROSCOPY OF CARBON TETRAFLUORIDE AND SILICON TE
	0369	PHOTOELECTRON SPECTRUM OF CHLORINE MONOFLUORIDE.
BURAK I	1099	INFRARED DOUBLE RESONANCE IN SULFUR HEXAFLUORIDE.
BURBANK RD	0395	STRUCTURES OF THE INTERHALOGEN COMPOUNDS. PART-1: CHLORINE TRIFLUORIDE AT -120 DEG
	0519	A REDETERMINATION OF THE ORTHORHOMBIC IODINE HEPTAFLUORIDE STRUCTURE. A CORRECTION
	0520	INTRAMOLECULAR REARRANGEMENT IN IODINE HEPTAFLUORIDE AND XENON HEXAFLUORIDE.
	0521	STRUCTURE OF INTERHALOGEN COMPOUNDS. PART-2: IODINE HEPTAFLUORIDE AT -110 DEGREES
	0551	CRYSTAL STRUCTURE OF KRYPTON DIFLUORIDE AT -80C.
	1283	XENON HEXAFLUORIDE. THE STRUCTURE OF A CUBIC PHASE AT -30C.
	1284	XENON HEXAFLUORIDE. STRUCTURAL CRYSTALLOGRAPHY OF TETRAMERIC PHASES.
	1296	POLYMORPHISM IN XENON HEXAFLUORIDE.
	1297	CRYSTALLINE MODIFICATIONS OF XENON HEXAFLUORIDE.
BURDEN FR	0681	MICROWAVE SPECTRUM OF NITROGEN DIFLUORIDE RADICAL.
	0682	ELECTRONIC STRUCTURE OF THE NITROGEN DIFLUORIDE RADICAL.
	0957	MICROWAVE SPECTRUM, DIPOLE MOMENT, AND STRUCTURE ANALYSIS OF SELENIUM TETRAFLUORID
BURGOS A	1207	VIBRATIONAL STUDY OF VANADIUM TRIFLUORIDE.
BURKE TG	0381	INFRARED AND RAMAN SPECTRA OF CHLORINE MONOFLUORIDE.
	0627	MOLECULAR STRUCTURE OF MOLYBDENUM, TUNGSTEN, AND URANIUM HEXAFLUORIDES FROM INFRAR
	1107	INFRARED SPECTRA OF SELENIUM HEXAFLUORIDE AND TELLURIUM HEXAFLUORIDE.
BURNELLE LA	0213	AB INITIO CALCULATION ON BORON TRIFLUORIDE AND BORON TRICHLORIDE.
BURNETT JL	0446	MELTING POINTS OF CURIUM TRIFLUORIDE AND AMERICIUM TRIFLUORIDE.
BURNS JH	0650	HEAT CAPACITY, ENTHALPY OF FUSION, AND THERMODYNAMIC PROPERTIES OF NEPTUNIUM HEXAF
	1260	CRYSTAL AND MOLECULAR STRUCTURE OF XENON TETRAFLUORIDE BY NEUTRON DIFFRACTION.
BURTON DW	1198	THERMODYNAMIC PROPERTIES OF GASEOUS URANIUM HEXAFLUORIDE.
BURTON PG	0765	APPROXIMATE MOLECULAR ORBITAL THEORY. ESSENTIAL STRUCTURAL ELEMENTS- MO FORMALISM.
BUTLER D	0964	SPECTRA OF SILICON MONOFLUORIDE, GERMANIUM MONOFLUORIDE, TIN MONOFLUORIDE, AND LEA
BUTSKII VD	1149	THERMODYNAMIC PROPERTIES OF TUNGSTEN PENTAFLUORIDE.
BUTYLKIN YP	0422	THERMODYNAMIC PROPERTIES OF PENTAFLUORIDES AT HIGH TEMPERATURES. PART-1: CALCULATI
	0536	THERMODYNAMIC PROPERTIES OF PENTAFLUORIDES AT HIGH TEMPERATURES. PART-3: PENTAFLUO
	0630	THERMAL STABILITY AND REACTIVITY OF D- AND F ELEMENT HEXAFLUORIDES.
	0647	THERMODYNAMIC PROPERTIES OF PENTAFLUORIDES AT HIGH TEMPERATURES. PART-4: PENTAFLUO
	1177	THERMODYNAMIC PROPERTIES OF URANIUM PENTAFLUORIDE, NEPTUNIUM PENTAFLUORIDE, AND PL
	1379	THERMODYNAMIC PROPERTIES OF TRANSITION METAL PENTAFLUORIDES AT HIGH TEMPERATURE. (
	1434	THERMODYNAMIC PROPERTIES OF PENTAFLUORIDES AT HIGH TEMPERATURES. PART-2: PENTAFLUO
BYE BH	1071	SPECTROSCOPIC CALCULATIONS ON SULFUR HEXAFLUORIDE.
BYFLEET CR	0954	ESR SPECTRUM OF BROMINE OXIDE, IODINE OXIDE AND SELENIUM MONOFLUORIDE IN J-5/2 ROT
	1019	ELECTRIC DIPOLE MOMENTS OF OPEN SHELL DIATOMIC MOLECULES. (SULFUR MONOFLUORIDE AND
BYSTROVA ON	0486	VAPOR PRESSURE AND HEAT OF SUBLIMATION OF ZIRCONIUM AND HAFNIUM TETRACHLORIDES.
CABANA A	0327	VIBRATIONAL SPECTRA OF LIQUID AND CRYSTALLINE CARBON TETRAFLUORIDE.
	1001	INFRARED AND RAMAN SPECTRA OF LIQUID AND CRYSTALLINE SILICON TETRAFLUORIDE.
CADIOLI B	0196	BARRIER TO ELECTRON PASSAGE THROUGH ELECTRONEGATIVE ATOMS IN BORON TRIFLUORIDE.
	1066	AB INITIO AND MZDO WAVE FUNCTIONS FOR SULFUR HEXAFLUORIDE.

CADY GH	0624	VAPOR PRESSURES OF MOLYBDENUM PENTAFLUORIDE, TUNGSTEN PENTAFLUORIDE, RHENIUM PENTA
	0903	VAPOR PRESSURES OF SOME HEAVY TRANSITION METAL HEXAFLUORIDES. (TUNGSTEN HEXAFLUORI
	1152	SOME PHYSICAL PROPERTIES OF TUNGSTEN HEXAFLUORIDE. (VAPOR PRESSURE)
CAFASSO FA	0917	MAGNETIC SUSCEPTIBILITY OF RHENIUM HEXAFLUORIDE.
CALDER GV	0609	GEOMETRY AND VIBRATIONAL SPECTRA OF THE ALKALINE EARTH DIHALIDES. PART-1: MAGNESIU
CALIFANO NS	1445	FORCE CONSTANTS FOR XY6 MOLECULES.
CAMPBELL IE	1148	VAPOR PRESSURE OF TITANIUM TETRAFLUORIDE.
CANNON WA	0259	ELECTRICAL CONDUCTIVITY OF SOLID CHLORINE TRIFLUORIDE AND BROMINE TRIFLUORIDE.
CANTERFORD JH	0001	HALIDES OF THE TRANSITION ELEMENTS. VOLUME-1: FIRST ROW TRANSITION METALS. VOLUME-
	0002	FLUORIDES OF THE HEAVY TRANSITION METALS. (REVIEW)
	1374	HALIDES OF THE SECOND AND THIRD ROW TRANSITION METALS.
CANTOR S	0135	VAPOR PRESSURES OF BERYLLIUM DIFLUORIDE AND NICKEL DIFLUORIDE.
CAPWELL RJ	0863	POTENTIAL FIELD AND MOLECULAR VIBRATIONS OF THE TRIGONAL BIPYRAMIDAL THE RAMAN SPE
CARLLON KD	0932	ELECTRONIC STRUCTURE AND GROUND STATE OF SCANDIUM MONOFLUORIDE.
CARLSON GA	0310	INFRARED DETECTION OF GASEOUS CARBON TRIFLUORIDE RADICAL.
CARLSON TA	0319	HIGH RESOLUTION PHOTOELECTRON SPECTROSCOPY OF CARBON TETRAFLUORIDE AND SILICON TE
CARNIGLIA SC	0055	VAPOR PRESSURES OF AMERICIUM TRIFLUORIDE AND PLUTONIUM TRIFLUORIDE, HEATS AND FREE
CARPENTER DR	0540	INFRARED SPECTRUM OF IRIDIUM HEXAFLUORIDE.
CARR SL	1349	MAGNETIC RESONANCE AND SUSCEPTIBILITY OF RARE EARTH TRIFLUORIDES.
CARREIRA LA	0879	RAMAN SPECTRA OF GASES. PART-10: RAMAN BAND CONTOUR OF THE NU-7 (E') FUNDAMENTAL O
CARRINGTON A	0284	GAS PHASE ELECTRON RESONANCE SPECTRUM AND DIPOLE MOMENT OF CARBON MONOFLUORIDE.
	1019	ELECTRIC DIPOLE MOMENTS OF OPEN SHELL DIATOMIC MOLECULES. (SULFUR MONOFLUORIDE AND
	1020	GAS PHASE ELECTRON RESONANCE SPECTRA OF SULFUR MONOFLUORIDE, SELENIUM MONOFLUORIDE
	1021	GAS PHASE ELECTRON RESONANCE SPECTRA OF SULFUR MONOFLUORIDE AND SELENIUM MONOFLUOR
CARROLL PK	0285	THE B-X SYSTEM OF CARBON MONOFLUORIDE.
CARROLL TX	0374	CHARGE DISTRIBUTION IN CHLORINE MONOFLUORIDE FROM CORE ELECTRON BINDING ENERGIES.
	0409	X-RAY PHOTOELECTRON SPECTROSCOPY OF CHLORINE TRIFLUORIDE, SULFUR TETRAFLUORIDE, AN
	0874	OBSERVATION BY ESCA OF INEQUIVALENT FLUORINES IN CHLORINE TRIFLUORIDE, SULFUR TETR
	1229	ELECTRON DISTRIBUTION IN THE XENON FLUORIDES AND XENON OXIDE TETRAFLUORIDE BY ESCA
CARTER RP	0849	MOLECULAR STRUCTURES OF VARIOUS PHOSPHORUS HALIDES AND PHOSPHORUS PENTAFLUORIDE. I
CASPERS HH	0453	OPTICAL ABSORPTION AND FLUORESCENCE SPECTRA OF EUROPIUM TRIFLUORIDE.
CATON J	0370	HEATS OF FORMATION OF CHLORINE FLUORIDES. (CHLORINE MONOFLUORIDE, CHLORINE DIFLUOR
CATON RB	0155	ELECTRONIC SPECTRUM OF THE BORON MONOFLUORIDE MOLECULE.
CATTON RC	1230	BONDING IN XENON DIFLUORIDE.
	1338	D ORBITALS IN THE NOBLE GAS DIHALIDES. (ARGON DIFLUORIDE, KRYPTON DIFLUORIDE, AND
CAUNT AD	0482	RAMAN AND INFRARED SPECTRA OF GERMANIUM TETRAFLUORIDE.
CAVELL RG	1209	PREPARATION AND PROPERTIES (CRYSTAL STRUCTURE) OF VANADIUM TETRAFLUORIDE.
	1210	INFRARED SPECTRA OF VANADIUM PENTAFLUORIDE, VANADIUM TETRAFLUORIDE, AND VANADIUM T
CAZZOLI G	0718	MILLIMETER WAVE SPECTRUM AND CENTRIFUGAL DISTORTION CONSTANTS OF NITROGEN TRIFLUOR
CEDERBAUM LS	1090	ELECTRONIC STRUCTURE OF MOLECULES BY A MANY BODY APPROACH. PART-6: THE ASSIGNMENT
CHAKRAVORTY M	0060	EMISSION OF MOLECULAR ANTIMONY MONOFLUORIDE.
CHALMERS AA	0320	ACCURATE FORCE CONSTANTS FROM HEAVY ISOTOPIC SUBSTITUTION. PART-1. CARBON TETRAFLU
CHAMPETIER G	1142	EMISSION SPECTRUM OF TITANIUM MONOFLUORIDE.
CHAMPION JP	0423	MICROWAVE SPECTROSCOPY OF CHLORINE PENTAFLUORIDE AT 70 AND 140 GHZ
CHANTRY GW	0702	FAR INFRARED SPECTRUM OF NITROGEN TRIFLUORIDE.
	0983	ULTRAVIOLET SPECTRUM OF SILICON DIFLUORIDE.
	1034	MOLECULAR VIBRATIONS OF SULFUR HEXAFLUORIDE AND SULFUR TETRAFLUORIDE.
CHANUSSOT J	0268	ROTATIONAL SPECTRA OF BROMINE PENTAFLUORIDE IN THE MILLIMETER WAVE LENGTH REGION.
	0423	MICROWAVE SPECTROSCOPY OF CHLORINE PENTAFLUORIDE AT 70 AND 140 GHZ
CHARI N	0321	THERMODYNAMIC PROPERTIES OF CARBON TETRAFLUORIDE.
CHARKIN OP	0009	CHARACTERISTICS OF THE ELECTRONIC STRUCTURE OF FLUORIDES OF NONTRANSITION ELEMENTS
	0016	CALCULATIONS OF ELECTRON STRUCTURES OF COMPOUNDS OF HALOGENS AND INERT GASES BY TH
	1231	CALCULATIONS OF ELECTRONIC STRUCTURES OF COMPOUNDS OF HALOGENS AND INERT GASES BY
CHARLU TV	0448	MASS SPECTROMETRIC STUDIES AT HIGH TEMPERATURES. PART-17: SUBLIMATION AND VAPOR PR
	1350	HIGH TEMPERATURE ENTHALPY INCREMENTS FOR SOME RARE EARTH TRIFLUORIDES.
CHARPIN P	1015	MAGNETIC SUSCEPTIBILITY OF SILVER DIFLUORIDE.
CHATALIC A	0089	ULTRAVIOLET SPECTRUM OF ARSENIC MONOFLUORIDE.
	0090	ANALYSIS OF THE ABSORPTION SPECTRUM AND OF TWO NEW SINGLET SYSTEMS OF ARSENIC MONO
	1142	EMISSION SPECTRUM OF TITANIUM MONOFLUORIDE.
CHAU JYH	0342	ROOT MEAN SQUARE AMPLITUDES OF VIBRATION IN SOME GROUP-4 TETRAHALIDES. (CARBON TE
CHAUDHURI AK	1350	HIGH TEMPERATURE ENTHALPY INCREMENTS FOR SOME RARE EARTH TRIFLUORIDES.
CHAVRET M	1124	HEAT CAPACITY AND THERMODYNAMIC PROPERTIES OF THALLIUM FLUORIDE FROM 5 TO 445 DEGR
CHEALL HFK	1111	ROTATIONAL ANALYSIS OF BANDS OF THE A TRIPLET PI ZERO(+), B TRIPLET PI(1) TO X SIN
CHEETHAM AK	1318	STRUCTURES OF YTTRIUM AND BISMUTH TRIFLUORIDES BY NEUTRON DIFFRACTION.
CHERNICK CL	0926	JAHN-TELLER EFFECT IN THE E(8) VIBRATIONAL MODE OF HEXAFLUORIDE MOLECULES. THE INF
	1251	VAPOR PRESSURE AND MELTING POINTS OF XENON DIFLUORIDE AND XENON TETRAFLUORIDE.
	1261	VIBRATIONAL SPECTRA AND STRUCTURES OF XENON TETRAFLUORIDE AND XENON OXYTETRAFLUORI
	1262	VIBRATIONAL SPECTRA AND STRUCTURE OF XENON TETRAFLUORIDE.
CHERON M	0837	SYMMETRY GROUP OF PHOSPHORUS PENTAFLUORIDE.
CHESNOKOV VI	0696	THERMOCHEMICAL AND THERMODYNAMIC PROPERTIES OF INORGANIC NITROGEN FLUORIDES. (NITR
CHICHIKALYUK EM	0240	THERMODYNAMIC PROPERTIES OF BORON HALIDES.
CHIKARAISHI T	0097	MICROWAVE SPECTRUM OF ARSENIC TRIFLUORIDE IN THE EXCITED VIBRATIONAL STATE.
CHILD MS	0538	VIBRATIONAL ELECTRONIC COUPLING IN IRIDIUM HEXAFLUORIDE, OSMIUM HEXACHLORIDE(2-),
	0904	JAHN-TELLER EFFECT IN RHENIUM HEXAFLUORIDE.
CHOPOROV DY	1167	SATURATED VAPOR PRESSURE OF SOLID URANIUM TETRAFLUORIDE.
CHRISTE KO	0257	VIBRATIONAL SPECTRUM OF BROMINE TRIFLUORIDE.
	0419	MATRIX ISOLATION STUDY OF CHLORINE PENTAFLUORIDE.
	1035	RAMAN SPECTRUM AND FORCE CONSTANTS FOR SULFUR TETRAFLUORIDE.
	1036	VIBRATIONAL ASSIGNMENT OF SULFUR TETRAFLUORIDE.
CHRISTENSEN AU	1172	VIBRATIONAL SPECTRA AND FORCE CONSTANTS OF URANIUM TETRAFLUORIDE.
	1173	HIGH TEMPERATURE HEAT CONTENT OF URANIUM TETRAFLUORIDE.
CHUDINOV EG	1167	SATURATED VAPOR PRESSURE OF SOLID URANIUM TETRAFLUORIDE.
CHUPKA WA	0760	PHOTOIONIZATION MASS SPECTROMETRY OF OXYGEN DIFLUORIDE.
	1225	PHOTOIONIZATION MASS SPECTROMETRIC STUDY OF XENON DIFLUORIDE, XENON TETRAFLUORIDE,
CLAASSEN HH	0322	RAMAN SPECTRA OF CHLOROFLUORO METHANES IN THE GASEOUS STATE. (CARBON TETRAFLUORIDE
	0396	VIBRATIONAL SPECTRA AND THERMODYNAMIC PROPERTIES OF CHLORINE TRIFLUORIDE AND BROMI
	0408	INFRARED AND RAMAN SPECTRA OF CHLORINE TRIFLUORIDE AND BROMINE TRIFLUORIDE.
	0522	VIBRATIONAL SPECTRA OF IODINE HEPTAFLUORIDE AND RHENIUM HEPTAFLUORIDE.
	0539	SEARCH FOR A JAHN-TELLER EFFECT IN IRIDIUM HEXAFLUORIDE.
	0552	INFRARED AND RAMAN SPECTRA OF KRYPTON DIFLUORIDE.

CLAASSEN HH   0628 VIBRATIONAL SPECTRA OF MOLYBDENUM HEXAFLUORIDE AND TECHNETIUM HEXAFLUORIDE.
0649 INFRARED SPECTRA OF NEPTUNIUM HEXAFLUORIDE AND PLUTONIUM HEXAFLUORIDE.
0752 VIBRATIONAL SPECTRA OF OSMIUM HEXAFLUORIDE AND PLATINUM HEXAFLUORIDE.
0873 RAMAN SPECTRA OF ARSENIC PENTAFLUORIDE AND VANADIUM PENTAFLUORIDE AND   THEIR FORCE
0905 VIBRATIONAL SPECTRA OF RHENIUM HEXAFLUORIDE.
0919 VIBRATIONAL SPECTRA OF RHENIUM HEPTAFLUORIDE.
0926 JAHN-TELLER EFFECT IN THE E(8) VIBRATIONAL MODE OF HEXAFLUORIDE MOLECULES. THE INF
1072 RAMAN SPECTRA OF MOLYBDENUM HEXAFLUORIDE, TECHNETIUM HEXAFLUORIDE, RHENIUM HEXAFLU
1097 RAMAN SPECTRA OF MATRIX ISOLATED MOLECULES. (SULFUR HEXAFLUORIDE)
1190 RAMAN SPECTRUM OF URANIUM HEXAFLUORIDE.
1192 LONG WAVE LENGTH, INFRARED ACTIVE FUNDAMENTAL FOR URANIUM HEXAFLUORIDE, NEPTUNIUM
1211 VIBRATIONAL SPECTRA OF VANADIUM PENTAFLUORIDE.
1232 VIBRATIONAL SPECTRA OF XENON COMPOUNDS.
1254 RAMAN SPECTRA FOR XENON DIFLUORIDE, XENON TETRAFLUORIDE, AND XENON OXIDE TETRAFLUO
1261 VIBRATIONAL SPECTRA AND STRUCTURES OF XENON TETRAFLUORIDE AND XENON OXYTETRAFLUORI
1262 VIBRATIONAL SPECTRA AND STRUCTURE OF XENON TETRAFLUORIDE.
1263 XENON TETRAFLUORIDE. (INFRARED SPECTRUM)
1285 SPECTRAL OBSERVATIONS ON MOLECULAR XENON HEXAFLUORIDE. RAMAN SCATTERING AND INFRAR
1289 RAMAN SPECTRUM OF XENON HEXAFLUORIDE.
1299 FAR INFRARED AND MILLIMETER WAVE STUDIES OF XENON HEXAFLUORIDE.
1378 RAMAN SPECTRA OF 5D TRANSITION METAL HEXAFLUORIDES IN THE VAPOR STATE. (THIRD ROW)
1446 FORCE CONSTANTS OF METAL HEXAFLUORIDES
1471 JAHN-TELLER EFFECT IN VIBRATIONAL SPECTRA OF HEXAFLUORIDES.

CLARK HC   1209 PREPARATION AND PROPERTIES (CRYSTAL STRUCTURE) OF VANADIUM TETRAFLUORIDE.
1210 INFRARED SPECTRA OF VANADIUM PENTAFLUORIDE, VANADIUM TETRAFLUORIDE, AND VANADIUM T
1212 SOME PHYSICAL AND CHEMICAL PROPERTIES OF VANADIUM PENTAFLUORIDE.
1367 PENTAHALIDES OF TRANSITION METALS.

CLARK RJH   0314 VAPOR PHASE RAMAN SPECTRA OF MH4 (M= CARBON, SILICON, GERMANIUM, AND TIN) AND MF4
0795 VAPOR PHASE RAMAN SPECTRA, RAMAN BAND CONTOUR ANALYSES, CORIOLIS CONSTANTS, FORCE
1069 VAPOR PHASE RAMAN SPECTRA, RAMAN BAND CONTOUR ANALYSES, AND CORIOLIS CONSTANTS OF

CLASE HJ   0504 VIBRATIONAL SPECTRA AND ASSIGNMENT OF IODINE PENTAFLUORIDE. FERMI RESONANCE BETWEE
CLAUSS JK   0662 HEAT CAPACITY, ENTROPY, AND HEATS OF TRANSITION OF MOLYBDENUM HEXAFLUORIDE AND NIO
CLAXTON TA   1286 STEREOCHEMISTRY AND SEVEN COORDINATION. (XENON HEXAFLUORIDE)
CLELAND CA   0438 VAPOR PRESSURE AND HEAT OF SUBLIMATION OF COBALT DIHALIDES. (COBALT DIFLUORIDE)
CLEMENTS RM   1011 ROTATIONAL ANALYSIS OF THE AO(+), BO(+), TO X SINGLET SIGMA(+) SYSTEMS OF GASEOUS
1012 ABSORPTION SPECTRUM OF GASEOUS SILVER MONOFLUORIDE.

CLIPPARD FB   0098 AN ELECTRON DIFFRACTION INVESTIGATION OF ARSENIC TRIFLUORIDE, ARSENIC PENTAFLUORID
0099 MOLECULAR STRUCTURES OF ARSENIC TRIFLUORIDE AND ARSENIC PENTAFLUORIDE AS DETERMINE

COBB CC   1254 RAMAN SPECTRA FOR XENON DIFLUORIDE, XENON TETRAFLUORIDE, AND XENON OXIDE TETRAFLUO
COBB JC   0688 AB INITIO MOLECULAR ORBITAL STUDY OF THE NITROGEN DIFLUORIDE AND DINITROGEN TETRAF
0790 AB INITIO MOLECULAR ORBITAL STUDY OF THE PHOSPHORUS DIFLUORIDE RADICAL.

CODE RF   1287 MAGNETIC FIELD DEFLECTION OF XENON HEXAFLUORIDE.
COHZ SN   0534 MASS SPECTRA AND SUBLIMATION PRESSURES OF IODINE HEPTAFLUORIDE AND IODINE OXYGEN P
COLBURN CB   0692 INFRARED SPECTRUM OF NITROGEN DIFLUORIDE.
0703 NITROGEN FLUORIDES AND THEIR INORGANIC DERIVATIVES.
0712 STRENGTH OF THE NITROGEN FLUORINE BOND IN NITROGEN TRIFLUORIDE AND DINITROGEN TETR

COLE LG   0253 THERMODYNAMIC PROPERTIES OF THE DIATOMIC INTERHALOGENS FROM SPECTROSCOPIC DATA. (C
COLEMAN JS   0897 PREPARATION, STRUCTURE AND SPECTRA OF PRASEODYMIUM TETRAFLUORIDE.
COLIN R   0063 ROTATIONAL ANALYSIS OF THE ULTRAVIOLET BANDS OF THE ANTIMONY MONOFLUORIDE MOLECULE
0071 BO(+) AND BO(+)-X(10+) BANDS OF ANTIMONY MONOFLUORIDE.
0585 B SINGLET SIGMA(+) TO X TRIPLET SIGMA(-) BAND SYSTEM IN LEAD MONOFLUORIDE.

COLLINS GAD   0553 AB INITIO CALCULATIONS OF THE BONDING IN KRYPTON DIFLUORIDE.
0554 BONDING IN KRYPTON DIFLUORIDE.
0555 BONDING IN KRYPTON DIFLUORIDE.

COLTON R   0001 HALIDES OF THE TRANSITION ELEMENTS. VOLUME-1: FIRST ROW TRANSITION METALS. VOLUME-
0002 FLUORIDES OF THE HEAVY TRANSITION METALS. (REVIEW)
1374 HALIDES OF THE SECOND AND THIRD ROW TRANSITION METALS.

COLUSSI AJ   0112 ESR SPECTRUM OF ARSENIC TETRAFLUORIDE.
0791 ESR SPECTRA OF PHOSPHORUS DIFLUORIDE AND SULFUR TRIFLUORIDE RADICALS.

COMES FJ   1233 SPECTRA OF XENON DIFLUORIDE AND XENON TETRAFLUORIDE IN THE FAR ULTRAVIOLET REGION.
COMPANION AL   1364 MOLECULAR ORBITAL STUDY OF ATOMIZATION ENERGIES OF PNICOGEN FLUORIDE MOLECULES AND
1475 BICENTRIC RESCALING OF CNDO/2 THEORY. APPLICATIONS TO INORGANIC FLUORIDES.

CONDRATE RA   0838 UREY-BRADLEY FORCE CONSTANTS OF TRIGONAL BIPYRAMIDAL XY5 MOLECULES. (PHOSPHORUS PE
1433 NORMAL COORDINATE ANALYSIS OF TRIGONAL BIPYRAMIDAL XY5 MOLECULES.

CONNOLLY JWD   1304 MULTIPLE SCATTERING X(ALPHA) CALCULATION OF THE ENERGY LEVEL STRUCTURE OF XENON HE
CONSTANTINE MT   0426 DENSITY, VAPOR PRESSURE, CRITICAL PROPERTIES, DIELECTRIC CONSTANT, AND SPECIFIC CO
CONWAY JG   1168 ABSORPTION SPECTRUM OF URANIUM TETRAFLUORIDE AND ENERGY LEVELS OF URANIUM(IV).
COOMBER JW   0311 INFRARED SPECTRUM OF THE FREE RADICAL CARBON TRIFLUORIDE ISOLATED IN INERT MATRICE
CORNFORD AB   0766 PHOTOELECTRON SPECTRA OF OXYGEN DIFLUORIDE AND OXYGEN DICHLORIDE.
CORNWELL CD   0368 NMR SPECTRA OF CHLORINE TRIFLUORIDE AND CHLORINE MONOFLUORIDE. GASEOUS SPECTRA AND
0516 NMR SPECTRA OF IODINE HEPTAFLUORIDE AND IODINE OXYGEN PENTAFLUORIDE.

COTTON FA   0839 RELATIVE BOND STRENGTHS IN TRIGONAL BIPYRAMIDAL MOLECULES. (PHOSPHORUS PENTAFLUORI
1037 NMR SPECTRUM AND STRUCTURE OF SULFUR TETRAFLUORIDE.

COUCH LS   1214 INTRAMOLECULAR EXCHANGE IN VANADIUM PENTAFLUORIDE.
COULSON CA   0397 CNDO/2 AND INDO ALL VALENCE ELECTRON CALCULATIONS ON THE GEOMETRY AND PROPERTIES O
0556 FORCE FIELDS IN KRYPTON DIFLUORIDE AND XENON DIFLUORIDE.

COURTNEY JA   0183 NUCLEAR SPIN RELAXATION STUDY OF THE SPIN ROTATION INTERACTION IN SYMMETRIC TOP MO
0323 NUCLEAR SPIN RELAXATION STUDY OF THE SPIN ROTATION INTERACTION IN SPHERICAL TOP MO

COWLEY AH   0846 PHOTOELECTRON SPECTRUM OF PHOSPHORUS PENTAFLUORIDE.
COX JT   0796 ZEEMAN EFFECT OF SOME LINEAR AND SYMMETRIC TOP MOLECULES. (PHOSPHORUS TRIFLUORIDE)
COXON JA   0494 DISSOCIATION ENERGIES OF DIATOMIC HALOGEN FLUORIDES. (IODINE FLUORIDE, BROMINE FLU
CRADOCK S   0464 BOND DISSOCIATION ENERGIES, IONIZATION POTENTIALS AND ELECTRON AFFINITIES OF SOME
CRAWFORD B   0352 A GEOMETRIC VISUALIZATION OF NORMAL COORDINATE TRANSFORMATIONS. APPLICATION TO THE
CRUICKSHANK DWJ   0372 ELECTRONIC STRUCTURES OF CHLORINE MONOFLUORIDE AND CHLORINE TRIFLUORIDE.
0376 PHOTOELECTRON SPECTRA OF HALIDES. (CHLORINE MONOFLUORIDE, CHLORINE TRIFLUORIDE, BR
0553 AB INITIO CALCULATIONS OF THE BONDING IN KRYPTON DIFLUORIDE.
0554 BONDING IN KRYPTON DIFLUORIDE.
0555 BONDING IN KRYPTON DIFLUORIDE.

CUBICCIOTTI D   0072 THERMODYNAMIC PROPERTIES OF ANTIMONY TRIFLUORIDE.
0148 THERMODYNAMIC PROPERTIES OF BISMUTH TRIFLUORIDE.

CUBICCIOTTI D	1114	GASEOUS TRIMER AND TETRAMER OF THALLIUM MONOFLUORIDE.
	1115	ENTHALPY OF FORMATION AND THE DISSOCIATION ENERGY OF THALLIUM MONOFLUORIDE
	1116	ENTHALPIES, ENTROPIES, AND FREE ENERGY FUNCTIONS OF THALLIUM MONOHALIDES ABOVE ROO
	1121	THERMODYNAMICS OF VAPORIZATION OF THALLIUM MONOFLUORIDE.
CUNNINGHAM BB	0055	VAPOR PRESSURES OF AMERICIUM TRIFLUORIDE AND PLUTONIUM TRIFLUORIDE, HEATS AND FREE
CURL RF	0477	MICROWAVE SPECTRUM OF GERMANIUM DIFLUORIDE. QUADRUPOLE COUPLING AND CENTRIFUGAL DI
	0478	MICROWAVE SPECTRUM OF GERMANIUM DIFLUORIDE.
	0985	INFRARED SPECTRUM, FORCE CONSTANTS, AND THERMODYNAMIC FUNCTIONS OF SILICON DIFLUOR
	0989	MICROWAVE SPECTRUM AND FORCE CONSTANTS OF SILICON DIFLUORIDE. CENTRIFUGAL DISTORTI
	0990	MICROWAVE SPECTRUM OF SILICON DIFLUORIDE.
CURRAN AH	0673	GAS PHASE ELECTRON PARAMAGNETIC RESONANCE SPECTRUM AND DIPOLE MOMENT OF NITROGEN M
CURRIE GN	1020	GAS PHASE ELECTRON RESONANCE SPECTRA OF SULFUR MONOFLUORIDE, SELENIUM MONOFLUORIDE
	1021	GAS PHASE ELECTRON RESONANCE SPECTRA OF SULFUR MONOFLUORIDE AND SELENIUM MONOFLUOR
CURTIS EC	0257	VIBRATIONAL SPECTRUM OF BROMINE TRIFLUORIDE.
	0420	REPARAMETERIZED VIBRATIONAL FORCE CONSTANTS. PART-1: METHOD AND APPLICATION TO SOM
	0704	A BENT BOND MODEL FOR THE VIBRATIONAL FORCE CONSTANTS OF NONPLANAR XY3 MOLECULES.
	0705	LONE PAIR MODEL AND THE VIBRATIONAL FORCE CONSTANTS OF NITROGEN TRIFLUORIDE.
CUSACHS LC	0843	SEMIEMPIRICAL MOLECULAR ORBITAL CALCULATIONS. PSEUDOROTATION IN PHOSPHORUS PENTAFL
	1117	PHOTOELECTRON SPECTROSCOPY OF HIGH TEMPERATURE VAPORS. PART-3: MONOMER AND DIMER S
CYR T	0923	CONFIGURATION OF THE PENTAFLUORIDES OF RHODIUM AND IRIDIUM BY MAGNETIC RESONANCE.
CYVIN BN	0706	MEAN AMPLITUDES OF VIBRATION FOR SOME PYRAMIDAL XY3 MOLECULES. (NITROGEN TRIFLUOR
	1153	SPECTROSCOPIC STUDIES IN CONNECTION WITH ELECTRON DIFFRACTION INVESTIGATION OF SOM
	1388	INFLUENCE OF ATOMIC MASSES ON THE CORIOLIS COUPLING COEFFICIENTS.
	1390	MEAN AMPLITUDES OF VIBRATION FOR SOME PYRAMIDAL XY3 MOLECULES.
	1409	MATHEMATICAL TECHNIQUES IN THE STUDY OF CORIOLIS COUPLING IN TETRAHEDRAL XY4 MOLEC
	1411	FREQUENCIES, FORCE CONSTANTS, AND MASS DEPENDENCE OF THE CORIOLIS CONSTANT FOR SOM
	1477	THEORY AND CALCULATION OF CENTRIFUGAL DISTORTION CONSTANTS FOR POLYATOMIC MOLECULE
CYVIN SJ	0024	ALUMINUM AND ALKALI HALIDE MONOMERS MEAN AMPLITUDES OF VIBRATION WITH LOW TEMPERAT
	0038	HARMONIC FORCE FIELDS AND MEAN AMPLITUDES FOR ALUMINUM TRIFLUORIDE MONOMER AND DIM
	0197	CALCULATED MEAN AMPLITUDES OF VIBRATION IN BORON TRIHALIDES.
	0421	MEAN AMPLITUDES OF VIBRATION FOR SOME PYRAMIDAL XY4 MOLECULES. (IODINE PENTAFLUOR
	0706	MEAN AMPLITUDES OF VIBRATION FOR SOME PYRAMIDAL XY3 MOLECULES. (NITROGEN TRIFLUOR
	0840	MEAN AMPLITUDES OF VIBRATION AND CORIOLIS CONSTANTS FOR PHOSPHORUS PENTAFLUORIDE A
	1038	MEAN AMPLITUDES OF VIBRATION FOR SULFUR TETRAFLUORIDE.
	1071	SPECTROSCOPIC CALCULATIONS ON SULFUR HEXAFLUORIDE.
	1153	SPECTROSCOPIC STUDIES IN CONNECTION WITH ELECTRON DIFFRACTION INVESTIGATION OF SOM
	1369	MEAN AMPLITUDES AND SHRINKAGE EFFECTS OF VIBRATION FOR SOME LINEAR SYMMETRICAL DIF
	1388	INFLUENCE OF ATOMIC MASSES ON THE CORIOLIS COUPLING COEFFICIENTS.
	1389	VIBRATIONAL MEAN SQUARE AMPLITUDE MATRICES.
	1390	MEAN AMPLITUDES OF VIBRATION FOR SOME PYRAMIDAL XY3 MOLECULES.
	1392	VIBRATIONAL MEAN SQUARE AMPLITUDE MATRICES.
	1394	VIBRATIONAL MEAN SQUARE AMPLITUDE MATRICES. (BASTIANSEN-MORINO SHRINKAGE EFFECT OF
	1408	VIBRATIONAL MEAN SQUARE AMPLITUDE MATRICES. PART-7: TREATMENT OF TETRAHEDRAL XY4 M
	1409	MATHEMATICAL TECHNIQUES IN THE STUDY OF CORIOLIS COUPLING IN TETRAHEDRAL XY4 MOLEC
	1410	INFLUENCE OF ATOMIC MASSES ON THE CORIOLIS COUPLING COEFFICIENTS IN XY4 TETRAHEDRA
	1411	FREQUENCIES, FORCE CONSTANTS, AND MASS DEPENDENCE OF THE CORIOLIS CONSTANT FOR SOM
	1447	REVISED MEAN AMPLITUDES OF VIBRATION FOR SOME OCTAHEDRAL HEXAFLUORIDES.
	1454	NORMAL COORDINATE ANALYSIS OF ROTATION VIBRATION OF OCTAHEDRAL XY6 AND Z6 MOLECULA
	1455	MEAN AMPLITUDES OF VIBRATION AND BASTIANSEN-MORINO SHRINKAGE EFFECTS IN SOME OCTAH
	1456	FORCE FIELD OF WEINSTEIN AND GOODMAN FOR OCTAHEDRAL HEXAFLUORIDES AND THE L MATRIX
	1458	INFLUENCE OF ATOMIC MASS ON THE CORIOLIS CONSTANTS OF XY6 MOLECULES.
	1476	BASTIANSEN-MORINO SHRINKAGE EFFECT IN THE STUDY OF MOLECULAR STRUCTURE.
	1477	THEORY AND CALCULATION OF CENTRIFUGAL DISTORTION CONSTANTS FOR POLYATOMIC MOLECULE
	1488	THEORETICAL INTERPRETATION OF THE MASS INFLUENCE ON CORIOLIS COUPLING CONSTANTS IN
CZARNY J	0156	ROTATIONAL ANALYSIS OF THE BAND AT 14900 CM-1 OF A 3 SIGMA TO 3 SIGMA SYSTEM OF TH
DALTON BJ	0841	NONRIGID MOLECULE EFFECTS ON THE ROVIBRONIC ENERGY LEVELS AND SPECTRA OF PHOSPHORU
	1039	CLASSIFICATION OF THE STATES OF NONRIGID MOLECULES. (SULFUR TETRAFLUORIDE)
DAMEN TC	0615	RAMAN SPECTRA OF TITANIUM DIOXIDE, MAGNESIUM DIFLUORIDE, ZINC DIFLUORIDE, DIFLUORI
DANILOVA TG	0479	FREQUENCIES OF THE DEFORMATION VIBRATIONS OF THE DIHALIDES OF GERMANIUM, TIN, AND
	0568	THERMODYNAMIC FUNCTIONS OF GASEOUS SCANDIUM, YTTRIUM, AND LANTHANUM HALIDES IN THE
DANON N	0089	ULTRAVIOLET SPECTRUM OF ARSENIC MONOFLUORIDE.
	0090	ANALYSIS OF THE ABSORPTION SPECTRUM AND OF TWO NEW SINGLET SYSTEMS OF ARSENIC MONO
DARNELL AJ	1129	VAPOR PRESSURE OF THORIUM TETRAFLUORIDE.
DAS G	0174	METHODS FOR CORRELATING MOLECULES AND SOME OPTIMIZED VALENCE CONFIGURATION RESULTS
DAVIDSON ER	0157	ELECTRONIC POPULATION ANALYSIS OF MOLECULAR WAVE FUNCTIONS.
DAVIS DW	0683	MULTIPLET SPLITTING IN 1S HOLE STATES OF MOLECULES.
DAVIS RE	0375	HYPERFINE STRUCTURE CONSTANTS OF CHLORINE MONOFLUORIDE.
DEAN PAW	0078	FLUORINE NMR OF ANTIMONY PENTAFLUORIDE.
	0079	A FLUORINE NMR AND RAMAN STUDY OF COMPLEX FORMATION BETWEEN ANTIMONY PENTAFLUORIDE
DEB BM	0397	CNDO/2 AND INDO ALL VALENCE ELECTRON CALCULATIONS ON THE GEOMETRY AND PROPERTIES O
DEESE JE	1193	THERMODYNAMIC PROPERTIES OF URANIUM HEXAFLUORIDE AT HIGH TEMPERATURES.
DEHMER JL	0196	BARRIER TO ELECTRON PASSAGE THROUGH ELECTRONEGATIVE ATOMS IN BORON TRIFLUORIDE.
	1073	EVIDENCE OF EFFECTIVE POTENTIAL BARRIERS IN THE X-RAY ABSORPTION SPECTRA OF MOLECU
	1117	PHOTOELECTRON SPECTROSCOPY OF HIGH TEMPERATURE VAPORS. PART-3: MONOMER AND DIMER S
DEHMER PM	0760	PHOTOIONIZATION MASS SPECTROMETRY OF OXYGEN DIFLUORIDE.
DEITERS RM	0842	FAR INFRARED SPECTRUM OF PHOSPHORUS PENTAFLUORIDE.
	0853	ANISOTROPIC THERMAL MOTION OF TRIGONAL BIPYRAMIDAL MOLECULES FROM SPECTROSCOPIC DA
	0854	PENTACOORDINATED MOLECULES. PART-14: MOLECULAR VIBRATIONS AND STEREOCHEMICAL NONRI
DEKOCK CW	0583	GEOMETRY AND INFRARED SPECTRA OF MATRIX ISOLATED RARE EARTH HALIDES. (ALSO LANTHAN
	0895	LASER EXCITED RAMAN SPECTRA OF MATRIX ISOLATED PRASEODYMIUM TRIFLUORIDE.
	1316	INFRARED SPECTRA AND GEOMETRIES OF MATRIX ISOLATED YTTRIUM DIFLUORIDE AND YTTRIUM
	1324	INTERACTION OF MATRIX ISOLATED NICKEL FLUORIDE AND NICKEL CHLORIDE WITH CARBON MON
	1351	INFRARED SPECTRA AND GEOMETRIES OF RARE EARTH DIHALIDES.
DEKOCK RL	0376	PHOTOELECTRON SPECTRA OF HALIDES. (CHLORINE MONOFLUORIDE, CHLORINE TRIFLUORIDE, BR
	1234	HELIUM-2 PHOTOELECTRON SPECTRA OF XENON DIFLUORIDE AND KRYPTON DIFLUORIDE.
	1264	CORRELATION OF ELECTRONIC STATES OF THE POSITIVE IONS OF XENON TETRAFLUORIDE, XENO
	1343	BOND ENERGIES AND IONIC CHARACTER OF INERT GAS HALIDES. HELIUM-2 PHOTOELECTRON SPE
DELBENE JE	0684	MOLECULAR ORBITAL STUDY OF NITROGEN DIHYDRIDE, NITROGEN DIOXIDE, AND NITROGEN DIFL
DEMORTIER A	0637	HIGH RESOLUTION NMR SPECTRUM OF LIQUID MOLYBDENUM HEXAFLUORIDE, TUNGSTEN HEXAFLUOR
DENISOV MY	1208	MASS SPECTROMETRIC DETERMINATION OF THE VAPOR COMPOSITION AND VAPOR PRESSURE OF VA

DENISOVA DN	0486	VAPOR PRESSURE AND HEAT OF SUBLIMATION OF ZIRCONIUM AND HAFNIUM TETRACHLORIDES.
DEPOORTER GL	1191	ABSORPTION SPECTRUM OF URANIUM HEXAFLUORIDE FROM 2000 TO 4200 ANGSTROMS.
DESCHAMPS P	0089	ULTRAVIOLET SPECTRUM OF ARSENIC MONOFLUORIDE.
	1142	EMISSION SPECTRUM OF TITANIUM MONOFLUORIDE.
DEVAULT D	0251	RAMAN SPECTRUM OF BORON TRIFLUORIDE GAS.
DEVENTER EHV	0818	ENTHALPY OF FORMATION PHOSPHORUS TRIFLUORIDE.
	1130	ENTHALPY OF FORMATION OF THORIUM TETRAFLUORIDE BY FLUORINE BOMB CALORIMETRY.
DEVI VM	1469	UREY-BRADLEY FORCE FIELD OF SOME XY6 SYSTEMS.
DEVILLERS J	0585	B SINGLET SIGMA(+) TO X TRIPLET SIGMA(-) BAND SYSTEM IN LEAD MONOFLUORIDE.
DEVLIN JP	0299	A THEORETICAL STUDY OF THE BOND INTERACTION FORCE CONSTANTS IN DIFLUORIDE MOLECULE
DEWAR MJS	0846	PHOTOELECTRON SPECTRUM OF PHOSPHORUS PENTAFLUORIDE.
DIANOUX AJ	1015	MAGNETIC SUSCEPTIBILITY OF SILVER DIFLUORIDE.
DIBELER VH	0377	PHOTOIONIZATION STUDIES AND THE THERMODYNAMIC PROPERTIES OF SOME HALOGEN MOLECULES
	0378	PHOTOIONIZATION STUDY OF CHLORINE MONOFLUORIDE AND THE DISSOCIATION ENERGY OF FLUO
DICIANNI N	0777	SPIN ROTATIONAL HYPERFINE STRUCTURE IN THE MICROWAVE SPECTRUM OF OXYGEN DIFLUORIDE
	0778	CENTRIFUGAL DISTORTION EFFECTS IN ASYMMETRIC ROTOR MOLECULES. PART-1. QUADRATIC PO
	0779	MICROWAVE SPECTRUM, STRUCTURE, AND DIPOLE MOMENT OF OXYGEN DIFLUORIDE.
DICKSON FE	0658	INFRARED SPECTRA OF NIOBIUM TETRAFLUORIDE.
DIEBNER RL	1143	ABSORPTION SPECTRUM OF VAPORIZED TITANIUM MONOFLUORIDE.
DIERCKSEN GHF	1090	ELECTRONIC STRUCTURE OF MOLECULES BY A MANY BODY APPROACH. PART-6: THE ASSIGNMENT
DIJKERMAN H	1118	STARK EFFECT FOR THALLIUM MONOFLUORIDE AND POTASSIUM FLUORIDE.
DILONARDO G	1022	SPECTRUM OF SULFUR MONOFLUORIDE.
DISTEFANO VN	0812	THERMODYNAMIC FUNCTIONS OF SOME PHOSPHORUS COMPOUNDS. (PHOSPHORUS TRIFLUORIDE)
DIXIT L	0871	SPECTROSCOPIC STUDIES IN MEAN AMPLITUDES OF VIBRATION OF PENTAFLUORIDES OF ARSENIC
	1354	MEAN AMPLITUDES OF VIBRATION OF SOME LANTHANIDE TRIFLUORIDES.
DIXON RN	0979	ROTATIONAL ANALYSIS OF ABSORPTION BANDS IN THE 2266 ANGSTROM SYSTEM OF SILICON DIF
DJEU N	1074	OPTICAL SATURATION OF A SINGLE VIBRATION- ROTATION TRANSITION IN THE NU-3 FUNDAMEN
DODD RE	0096	THE INFRARED SPECTRA OF SOME VOLATILE INORGANIC FLUORIDES IN THE SOLID STATE. (ARS
	1040	RAMAN AND INFRARED SPECTRUM OF SULFUR TETRAFLUORIDE.
DODSWORTH PG	0457	ROTATIONAL ANALYSIS OF BANDS OF THE A TRIPLET PI ZERO(+) B TRIPLET PI(1) TO X SING
DOMALSKI ES	0039	HEAT OF FORMATION OF ALUMINUM TRIFLUORIDE BY DIRECT COMBINATION OF THE ELEMENTS.
	0198	HEAT FORMATION BORON TRIFLUORIDE BY DIRECT COMBINATION OF THE ELEMENTS.
DONALDSON JD	1139	ENVIRONMENT OF THE TIN ATOM IN ORTHORHOMBIC TIN DIFLUORIDE.
DONOHUE J	0523	EVIDENCE FOR THE MOLECULAR SYMMETRY OF IODINE HEPTAFLUORIDE.
	0524	MOLECULAR SYMMETRY OF IODINE HEPTAFLUORIDE.
DOUGLAS AE	0155	ELECTRONIC SPECTRUM OF THE BORON MONOFLUORIDE MOLECULE.
	0674	NEW SPECTRA OF OXYGEN LIKE MOLECULES. (NITROGEN MONOFLUORIDE)
	0675	B SINGLET SIGMA(+) TO X TRIPLET SIGMA(-) BAND SYSTEM OF NITROGEN MONOFLUORIDE.
	0786	ELECTRONIC SPECTRA OF PHOSPHORUS MONOFLUORIDE AND ITS POSITIVE ION.
DOUGLAS TB	0136	THERMODYNAMIC DATA FOR BERYLLIUM DIFLUORIDE.
DOUSLIN DR	0332	CARBON TETRAFLUORIDE. THERMODYNAMIC PROPERTIES OF THE REAL GAS.
DOWNS AJ	0956	VIBRATIONAL SPECTRA AND STRUCTURES OF SELENIUM TETRAFLUORIDE AND TELLURIUM TETRAFL
DOWS DA	1075	INFRARED INTENSITIES IN CRYSTALLINE SULFUR HEXAFLUORIDE.
DRAGO RS	0877	NORMAL COORDINATE ANALYSIS OF PHOSPHORUS PENTAFLUORIDE, PHOSPHORUS DIFLUORIDE DICH
DREIZLER H	1195	ALPHA SPECTROMETRIC INVESTIGATION OF URANIUM HEXAFLUORIDE. PART-1.
DRESKA M	0199	VIBRATION ROTATION BANDS OF BORON TRIFLUORIDE.
DRESKA SN	0200	NU-3 OF BORON TRIFLUORIDE.
DRESSLER K	0216	RADIATIVE LIFETIMES OF ULTRAVIOLET EMISSION SYSTEMS EXCITED IN BORON TRIFLUORIDE,
DRIFFORD M	0266	INFRARED AND RAMAN SPECTRA OF BROMINE PENTAFLUORIDE AND CHLORINE TRIFLUORIDE IN TH
	0398	RAMAN SPECTRA AND THE STRUCTURE OF CHLORINE TRIFLUORIDE AND BROMINE TRIFLUORIDE.
D'INCAN J	0598	ELECTRONIC SPECTRUM OF LUTETIUM MONOFLUORIDE.
	0599	ROTATIONAL ANALYSIS OF SELECTED BANDS FROM THE ELECTRONIC SPECTRUM OF THE LUTETIUM
DUBB JE	0426	DENSITY, VAPOR PRESSURE, CRITICAL PROPERTIES, DIELECTRIC CONSTANT, AND SPECIFIC CO
DUBE PS	0061	DETERMINATION OF THE EFFECTIVE VIBRATIONAL TEMPERATURE IN THE A-2 SYSTEM OF ANTIMO
	1137	VARIATION OF ELECTRONIC TRANSITION MOMENT WITH THE INTERNUCLEAR SEPARATION IN SILI
DUBEY VS	0280	ANALYSIS OF THE BANDS OF B DOUBLET SIGMA- X DOUBLET SIGMA. SYSTEM OF CALCIUM FLUOR
DUDASH JJ	1140	MELTING POINT, VAPOR PRESSURE, AND HEAT OF VAPORIZATION OF TIN DIFLUORIDE.
DUFFEY GH	0533	MOLECULAR ORBITAL TREATMENT OF IODINE HEPTAFLUORIDE.
DUNCAN ABF	0365	VACUUM ULTRAVIOLET ABSORPTION SPECTRA OF SOME HALOGEN DERIVATIVES OF METHANE. (CAR
	0642	RAMAN EFFECT AND ULTRAVIOLET ABSORPTION SPECTRA OF MOLYBDENUM AND TUNGSTEN HEXAFLU
	0725	VACUUM ULTRAVIOLET ABSORPTION SPECTRUM AND DIPOLE MOMENT OF NITROGEN TRIFLUORIDE.
DUNCAN JL	0201	FORCE CONSTANTS OF THE BORON TRIHALIDES - A SURVEY.
	0202	PERPENDICULAR VIBRATIONS OF BORON TRIFLUORIDE.
	0324	FORCE FIELDS OF CARBON TETRAFLUORIDE AND SILICON TETRAFLUORIDE.
	0697	INFRARED SPECTRUM OF NITROGEN TRIFLUORIDE AND FORCE FIELD.
	1478	LIMITATIONS OF THE UREY-BRADLEY MODEL IN NORMAL COORDINATE ANALYSES.
DUNLAP JL	0797	INFRARED INTENSITIES, BOND MOMENTS AND BOND MOMENT DERIVATIVES FOR PHOSPHORUS TRIF
DUNNING TH	0746	CHARACTERIZATION OF GROUND STATE WAVE FUNCTIONS BY MEASURED ELECTRONIC PROPERTIES.
DURGAVATH BK	0429	BAND SPECTRUM OF CHROMIUM MONOFLUORIDE.
DURIE RA	0496	ELECTRONIC EMISSION SPECTRUM AND MOLECULAR CONSTANTS OF IODINE MONOFLUORIDE.
DURIG JR	0864	MEAN AMPLITUDES OF VIBRATION FOR A TRIGONAL BIPYRAMIDAL MOLECULE WITH APPLICATION
	0879	RAMAN SPECTRA OF GASES. PART-10: RAMAN BAND CONTOUR OF THE NU-7 (E') FUNDAMENTAL O
DWORKIN AS	1169	ENTHALPY OF URANIUM TETRAFLUORIDE FROM 298-1400 DEGREES K. ENTHALPY AND ENTROPY OF
DYATKINA ME	0357	ELECTRONIC STRUCTURE OF THE CARBON TETRAFLUORIDE MOLECULE.
	0382	GEOMETRY OF MOLECULES. PART-2: GEOMETRICAL AND ELECTRICAL STRUCTURE OF THE MOLECUL
	0415	CALCULATION OF THE ELECTRONIC STRUCTURE OF A CHLORINE PENTAFLUORIDE MOLECULE BY TH
	1162	STRUCTURE OF MOLECULAR URANIUM DIFLUORIDE AND URANIUM TRIFLUORIDE.
	1339	ELECTRON STRUCTURE OF COMPOUNDS OF INERT GASES.
DYER PN	1020	GAS PHASE ELECTRON RESONANCE SPECTRA OF SULFUR MONOFLUORIDE, SELENIUM MONOFLUORIDE
DYKE TR	0003	GAS PHASE STRUCTURES AND MASS SPECTRA OF BINARY PENTAFLUORIDES.
	0349	MOLECULAR BEAM ELECTRIC DEFLECTION OF THE TETRAHALIDES CARBON TETRAFLUORIDE, CARBO
	1213	POLAR STATES IN VANADIUM PENTAFLUORIDE.
	1417	STARK ENERGIES FOR MOLECULES WITH VIBRATIONALLY INDUCED DIPOLE MOMENTS.
DZEVITSKII BE	0416	FLUORINE NMR OF CHLORINE PENTAFLUORIDE.
	1033	SOFT X-RAY SPECTROSCOPY ON SULFUR TETRAFLUORIDE AND SULFUR HEXAFLUORIDE.
EDGELL WF	0325	INFRARED BAND CONTOURS. PART-1: SPHERICAL TOP MOLECULES. (CARBON TETRAFLUORIDE)
EDING H	1116	ENTHALPIES, ENTROPIES, AND FREE ENERGY FUNCTIONS OF THALLIUM MONOHALIDES ABOVE ROO
EDWARDS AJ	1106	CRYSTAL STRUCTURE OF TELLURIUM TETRAFLUORIDE.
	1154	CRYSTAL STRUCTURE OF TUNGSTEN PENTAFLUORIDE.
EDWARDS HD	0798	MICROWAVE INVESTIGATIONS OF METHYL FLUORIDE, FLUOROFORM, AND PHOSPHORUS TRIFLUORID

EFFANTIN C	0598	ELECTRONIC SPECTRUM OF LUTETIUM MONOFLUORIDE.
	0599	ROTATIONAL ANALYSIS OF SELECTED BANDS FROM THE ELECTRONIC SPECTRUM OF THE LUTETIUM
EGOROVA LF	0713	PHYSICAL CHEMICAL PROPERTIES OF INORGANIC NITROGEN FLUORIDES. (NITROGEN TRIFLUORID
EHLERT TC	0034	HEAT OF ATOMIZATION OF ALUMINUM DIFLUORIDE.
	0119	DISSOCIATION ENERGIES OF THE ALKALINE EARTH MONOFLUORIDES.
	0300	BONDING IN CARBON(1) AND CARBON(2) FLUORIDES.
	0463	MASS SPECTROMETRIC STUDIES AT HIGH TEMPERATURES. PART-2: THE DISSOCIATION ENERGIES
	0612	MASS SPECTROMETRIC AND THERMOCHEMICAL STUDIES OF THE MANGANESE FLUORIDES.
	0614	SUBLIMATION PRESSURE OF MANGANESE DIFLUORIDE AND MANGANESE MONOFLUORIDE.
	1382	SUBLIMATION PRESSURES OF REFRACTORY FLUORIDES.
EICK HA	1131	VAPORIZATION REACTIONS IN THE THULIUM- FLUORINE SYSTEM.
	1321	SUBLIMATION THERMODYNAMICS OF ZINC(II) FLUORIDE.
EISENSTEIN JC	0751	ABSORPTION SPECTRUM AND MAGNETIC PROPERTIES OF OSMIUM HEXAFLUORIDE.
	0906	SPECTRUM OF RHENIUM HEXAFLUORIDE.
ELIEZER I	1322	POLARIZED ION MODEL AND THE BENDING FORCE CONSTANTS OF THE GROUP-2B HALIDES. (ZINC
ELLINGER FH	0447	EVIDENCE FOR CURIUM TETRAFLUORIDE.
ELLIS DE	1249	RELATIVISTIC MOLECULAR WAVE FUNCTIONS: XENON DIFLUORIDE.
ELLIS DJ	0676	CALCULATION OF THE SPECTROSCOPIC CONSTANTS FOR SEVERAL STATES OF NITROGEN MONOFLUO
ELLIS HM	1015	MAGNETIC SUSCEPTIBILITY OF SILVER DIFLUORIDE.
ELVERUM GW	0253	THERMODYNAMIC PROPERTIES OF THE DIATOMIC INTERHALOGENS FROM SPECTROSCOPIC DATA. (C
EMELEUS HJ	1212	SOME PHYSICAL AND CHEMICAL PROPERTIES OF VANADIUM PENTAFLUORIDE.
ENOKIDO H	0326	THERMODYNAMIC PROPERTIES OF CARBON TETRAFLUORIDE FROM 4 DEGREES K TO ITS MELTING P
EROKHIN EV	0040	MASS SPECTROMETRIC INVESTIGATION OF THE THERMODYNAMIC PROPERTIES OF ALUMINUM TRIFL
EVSEEV EM	0041	VAPOR PRESSURE OF ALUMINUM TRIFLUORIDE.
EWING JJ	0254	MOLECULAR ZEEMAN EFFECT, MAGNETIC PROPERTIES, AND ELECTRIC QUADRUPOLE MOMENTS IN C
	0545	EMISSION SPECTRA OF XENON MONOBROMIDE, XENON MONOCHLORIDE, XENON MONOFLUORIDE, AND
EWING VC	1034	MOLECULAR VIBRATIONS OF SULFUR HEXAFLUORIDE AND SULFUR TETRAFLUORIDE.
	1041	INVESTIGATION BY ELECTRON DIFFRACTION OF THE MOLECULAR STRUCTURES OF SULFUR HEXAFL
EYSEL HH	0525	VIBRATIONAL ASSIGNMENT FOR IODINE PENTAFLUORIDE, A NONRIGID D5H MOLECULE.
EZHOV YS	1170	ELECTRON DIFFRACTION STUDY OF URANIUM AND THORIUM TETRAHALIDES IN THE GAS PHASE. (
FADINI A	0348	COMPARISON OF CALCULATIONS OF FORCE CONSTANTS FOR BORON TETRAFLUORIDE(-), CARBON T
	1265	GENERAL VALENCE FORCE FIELD FOR XENON TETRAFLUORIDE.
	1415	NORMAL COORDINATE TREATMENT OF XY4 TETRAHEDRAL MOLECULES AND IONS. (GVFF)
	1457	FORCE CONSTANTS FOR HEXAFLUORIDE MOLECULES.
FAIRBROTHER F	0663	HALIDES OF NIOBIUM AND TANTALUM. PART-3: VAPOR PRESSURES OF NIOBIUM AND TANTALUM P
FALCONER WE	0003	GAS PHASE STRUCTURES AND MASS SPECTRA OF BINARY PENTAFLUORIDES.
	0019	APPLICATION OF A MOLECULAR BEAM SOURCE MASS SPECTROMETER TO THE STUDY OF REACTIVE
	0349	MOLECULAR BEAM ELECTRIC DEFLECTION OF THE TETRAHALIDES CARBON TETRAFLUORIDE, CARBO
	0367	ELECTRIC DEFLECTION AND THERMAL DECOMPOSITION STUDIES ON CERIUM TETRAFLUORIDE, TER
	0485	PREPARATION AND CHARACTERIZATION OF GOLD PENTAFLUORIDE.
	0526	MOLECULAR STRUCTURE OF XENON HEXAFLUORIDE AND IODINE HEPTAFLUORIDE.
	0529	POLAR DISTORTIONS IN RHENIUM HEPTAFLUORIDE AND IODINE HEPTAFLUORIDE.
	0546	ELECTRON SPIN RESONANCE SPECTRUM OF KRYPTON MONOFLUORIDE.
	0551	CRYSTAL STRUCTURE OF KRYPTON DIFLUORIDE AT -80C.
	0876	ASSOCIATION OF GROUP-5 PENTAFLUORIDES IN THE GAS PHASE. (PHOSPHORUS PENTAFLUORIDE,
	0946	ELECTRIC DEFLECTION OF MOLECULAR BEAMS OF THE RARE EARTH DIFLUORIDES AND TRIFLUORI
	1084	ELECTRIC DEFLECTION OF BINARY HEXAFLUORIDES. (SULFUR HEXAFLUORIDE, SELENIUM HEXAF
	1213	POLAR STATES IN VANADIUM PENTAFLUORIDE.
	1220	RADIATION DAMAGE IN XENON TETRAFLUORIDE. ELECTRON SPIN RESONANCE OF THE TRAPPED RA
	1222	ELECTRON SPIN RESONANCE SPECTRUM OF XENON MONOFLUORIDE IN GAMMA IRRADIATED XENON T
	1235	STRUCTURAL CONSIDERATIONS IN THE CHEMISTRY OF NOBLE GASES. (XENON DIFLUORIDE, XENO
	1287	MAGNETIC FIELD DEFLECTION OF XENON HEXAFLUORIDE.
	1288	ELECTRON SPIN RESONANCE OF THE XENON FLUORIDE RADICAL.
	1296	POLYMORPHISM IN XENON HEXAFLUORIDE.
	1297	CRYSTALLINE MODIFICATIONS OF XENON HEXAFLUORIDE.
	1365	GAS PHASE STRUCTURES AND MASS SPECTRA OF BINARY PENTAFLUORIDES.
	1376	MOLECULAR BEAM MASS SPECTROMETRIC STUDY OF BINARY PENTAFLUORIDES. (ALSO ANTIMONY
	1377	APPLICATION OF A MOLECULAR BEAM SOURCE MASS SPECTROMETER TO THE STUDY OF REACTIVE
FANO U	0196	BARRIER TO ELECTRON PASSAGE THROUGH ELECTRONEGATIVE ATOMS IN BORON TRIFLUORIDE.
FARBER M	0035	MASS SPECTROMETRIC DETERMINATION OF THE HEATS OF FORMATION OF THE GASEOUS MOLECULE
	0128	DISSOCIATION ENERGIES OF BERYLLIUM FLUORIDE AND BERYLLIUM CHLORIDE AND HEAT OF FOR
	0138	THERMODYNAMIC AND PHYSICAL PROPERTIES OF BERYLLIUM COMPOUNDS. PART-1: ENTHALPY AND
	0154	HEAT AND ENTROPY OF FORMATION OF BORON FLUORIDE.
	0179	THERMODYNAMIC PROPERTIES OF THE BORON CHLORIDE FLUORIDE SYSTEM FROM MASS SPECTROME
	0286	THERMODYNAMIC PROPERTIES OF CARBON MONOFLUORIDE AND CARBON DIFLUORIDE FROM MOLECUL
	0652	VAPOR PRESSURE OF NICKEL DIFLUORIDE.
FASTENAKEL D	0834	SPECTROSCOPIC CONSEQUENCES OF DIFFERENT TUNNELING MECHANISMS IN NONRIGID PHOSPHORU
FEDER HM	0048	FLUORINE BOMB CALORIMETRY. PART-22: THE ENTHALPY OF FORMATION OF ALUMINUM TRIFLUOR
	0220	FLUORINE BOMB CALORIMETRY. PART-15: THE ENTHALPY OF FORMATION OF BORON TRIFLUORIDE
	0250	FLUORINE BOMB CALORIMETRY. PART-3: THE HEAT OF FORMATION OF BORON TRIFLUORIDE.
	1319	ENTHALPY OF FORMATION OF YTTRIUM TRIFLUORIDE.
FEDOROVA TA	0240	THERMODYNAMIC PROPERTIES OF BORON HALIDES.
FEHLNER TP	0980	PHOTOELECTRON SPECTRUM OF SILICON DIFLUORIDE.
FELENBOK P	0156	ROTATIONAL ANALYSIS OF THE BAND AT 14900 CM-1 OF A 3 SIGMA TO 3 SIGMA SYSTEM OF TH
FERRAN J	0163	CAMERON SYSTEM OF BORON MONOFLUORIDE.
FERRIS LM	0900	ESTIMATED FREE ENERGIES OF FORMATION OF PROTACTINIUM HALIDES AND ACTIVITY COEFFICI
FESSENDEN RW	0791	ESR SPECTRA OF PHOSPHORUS TRIFLUORIDE AND SULFUR TRIFLUORIDE RADICALS.
FETSCH GES	0245	INFRARED ABSORPTION SPECTRUM OF LIQUID BORON TRIFLUORIDE.
FILBERT RB	1331	VAPOR PRESSURE OF ZIRCONIUM FLUORIDE.
FILIP H	1063	OCTAHEDRAL FINE STRUCTURE SPLITTINGS IN NU-3 OF SULFUR HEXAFLUORIDE.
FINOGENOV AD	0942	STANDARD ENTHALPIES OF FORMATION OF SCANDIUM TRIFLUORIDE, YTTRIUM TRIFLUORIDE, AND
FITZKY HG	1119	MICROWAVE ROTATIONAL SPECTRUM OF THALLIUM MONOFLUORIDES.
FITZPATRICK NJ	0203	COMPARATIVE AB INITIO CALCULATIONS ON BORON FLUORIDES.
FLEGEL W	1118	STARK EFFECT FOR THALLIUM MONOFLUORIDE AND POTASSIUM FLUORIDE.
FLESCH GD	1253	THERMOCHEMICAL PROPERTIES OF XENON DIFLUORIDE AND XENON TETRAFLUORIDE FROM MASS SP
FLETCHER WH	0418	VIBRATIONAL SPECTRA AND VALENCE FORCE CONSTANTS OF SQUARE PYRAMIDAL MOLECULES. XEN
FLEURY PA	0615	RAMAN SPECTRA OF TITANIUM DIOXIDE, MAGNESIUM DIFLUORIDE, ZINC DIFLUORIDE, DIFLUORI
FLICKER H	1063	OCTAHEDRAL FINE STRUCTURE SPLITTINGS IN NU-3 OF SULFUR HEXAFLUORIDE.
FLOREY JB	0843	SEMIEMPIRICAL MOLECULAR ORBITAL CALCULATIONS. PSEUDOROTATION IN PHOSPHORUS PENTAFL
FLORIN AE	0890	PREPARATION AND PROPERTIES OF (VAPOR PRESSURE) OF PLUTONIUM HEXAFLUORIDE AND IDENT

FLOTOW HE   0885   EXPERIMENTAL HEAT CAPACITIES OF PLUTONIUM-242 TRIFLUORIDE AND PLUTONIUM-242 TETRAF
   0886   HEAT CAPACITY, ENTROPY, AND ENTHALPY OF PLUTONIUM-242 TRIFLUORIDE FROM 10 TO 350 D
   0889   HEAT CAPACITY, ENTROPY, ENTHALPY, AND GIBBS ENERGY OF PLUTONIUM-242 TETRAFLUORIDE
   1246   XENON DIFLUORIDE. HEAT CAPACITY FROM 5 TO 350 DEGREES K AND SOME DERIVED THERMODYN
   1275   XENON TETRAFLUORIDE. HEAT CAPACITY AND THERMODYNAMIC FUNCTIONS FROM 5 TO 350 DEGRE
FLYGARE WH   0254   MOLECULAR ZEEMAN EFFECT, MAGNETIC PROPERTIES, AND ELECTRIC QUADRUPOLE MOMENTS IN C
   0384   MOLECULAR MAGNETIC PROPERTIES OF CHLORINE MONOFLUORIDE.
   0744   ZEEMAN STUDIES INCLUDING THE MOLECULAR G VALUES, MAGNETIC SUSCEPTIBILITY, AND MOLE
   0780   MOLECULAR G VALUES, MAGNETIC SUSCEPTIBILITIES, MOLECULAR QUADRUPOLE MOMENTS AND SE
FOMICHEV VA   0204   ULTRASOFT X-RAY ABSORPTION SPECTRA OF BORON TRIFLUORIDE.
   1033   SOFT X-RAY SPECTROSCOPY ON SULFUR TETRAFLUORIDE AND SULFUR HEXAFLUORIDE.
   1065   ELECTRONIC STRUCTURE OF SULFUR HEXAFLUORIDE STUDIED BY ULTRALONG WAVE X-RAY SPECTR
   1102   ULTRASOFT X-RAY ABSORPTION SPECTRUM OF SULFUR HEXAFLUORIDE.
FOMICHEV VV   0073   RAMAN SPECTRUM OF ANTIMONY TRIFLUORIDE.
FONTAL B   0844   RAMAN INTENSITIES AND THE FORCE FIELD OF PHOSPHORUS PENTAFLUORIDE.
FORICHEV VA   1101   X-RAY SPECTRA OF SULFUR AND FLUORINE IN SULFUR HEXAFLUORIDE.
FORNERIS R   0389   HIGH RESOLUTION RAMAN SPECTROSCOPY IN THE RED AND NEAR INFRARED. (CHLORINE MONOFLU
FORRESTER JD   1277   CRYSTAL AND MOLECULAR STRUCTURE OF XENON TETRAFLUORIDE.
FOSTER JN   0138   THERMODYNAMIC AND PHYSICAL PROPERTIES OF BERYLLIUM COMPOUNDS. PART-1: ENTHALPY AND
FOURNIER RP   0327   VIBRATIONAL SPECTRA OF LIQUID AND CRYSTALLINE CARBON TETRAFLUORIDE.
   1001   INFRARED AND RAMAN SPECTRA OF LIQUID AND CRYSTALLINE SILICON TETRAFLUORIDE.
FOWLER W   0861   E' FORCE FIELD OF PHOSPHORUS PENTAFLUORIDE.
FOX K   1063   OCTAHEDRAL FINE STRUCTURE SPLITTINGS IN NU-3 OF SULFUR HEXAFLUORIDE.
FOX WB   0255   MATRIX ISOLATION OF BROMINE MONOFLUORIDE.
FRACKOWIAK M   0786   ELECTRONIC SPECTRA OF PHOSPHORUS MONOFLUORIDE AND ITS POSITIVE ION.
FRANDON J   0279   SPECTRA OF ELECTRONIC EXCITATIONS IN CALCIUM DIFLUORIDE, STRONTIUM DIFLUORIDE, AND
FRANKLIN JL   0047   HIGH TEMPERATURE NEGATIVE IONS. GASEOUS GROUP-3 FLUORIDES.
   0480   ENTHALPY OF FORMATION OF GERMANIUM TRIFLUORIDE.
FRANZEN HF   1113   INFRARED SPECTRA AND STRUCTURE OF MATRIX ISOLATED THALLIUM HALIDES.
FRED M   0648   ELECTRONIC STRUCTURE OF NEPTUNIUM HEXAFLUORIDE.
   0908   VIBRONIC INTERACTIONS IN METAL HEXAFLUORIDE MOLECULES.
   0916   COLORS OF TRANSITION METAL HEXAFLUORIDES. (RHENIUM HEXAFLUORIDE, OSMIUM HEXAFLUORI
FREEDMAN PA   0205   ANALYSIS OF SYMMETRIC ROTOR RAMAN SPECTRA. PURE ROTATIONAL SPECTRUM OF BORON TRIFL
FREEMAN JM   1000   SILICON- FLUORINE LENGTH IN SILICON TETRAFLUORIDE. NEW ELECTRON DIFFRACTION STUDY.
FREIBERG M   1182   INFRARED SPECTRUM OF URANIUM HEXAFLUORIDE.
FREY RA   0399   INFRARED SPECTRA OF MATRIX ISOLATED CHLORINE TRIFLUORIDE, BROMINE TRIFLUORIDE, AND
FRIEC B   1192   LONG WAVE LENGTH, INFRARED ACTIVE FUNDAMENTAL FOR URANIUM HEXAFLUORIDE, NEPTUNIUM
FRIED SM   0447   EVIDENCE FOR CURIUM TETRAFLUORIDE.
   0885   EXPERIMENTAL HEAT CAPACITIES OF PLUTONIUM-242 TRIFLUORIDE AND PLUTONIUM-242 TETRAF
   0886   HEAT CAPACITY, ENTROPY, AND ENTHALPY OF PLUTONIUM-242 TRIFLUORIDE FROM 10 TO 350 D
   0889   HEAT CAPACITY, ENTROPY, ENTHALPY, AND GIBBS ENERGY OF PLUTONIUM-242 TETRAFLUORIDE
   0921   PREPARATION AND PROPERTIES OF RHENIUM HEPTAFLUORIDE.
FRISCH MA   0286   THERMODYNAMIC PROPERTIES OF CARBON MONOFLUORIDE AND CARBON DIFLUORIDE FROM MOLECUL
FRITH WC   0663   HALIDES OF NIOBIUM AND TANTALUM. PART-3: VAPOR PRESSURES OF NIOBIUM AND TANTALUM P
FROST DC   0766   PHOTOELECTRON SPECTRA OF OXYGEN DIFLUORIDE AND OXYGEN DICHLORIDE.
   1076   PHOTOELECTRON SPECTRUM OF SULFUR HEXAFLUORIDE AT 584 ANGSTROMS.
FRY JL   0453   OPTICAL ABSORPTION AND FLUORESCENCE SPECTRA OF EUROPIUM TRIFLUORIDE.
FUGGLE JC   0664   NIOBIUM-93 NUCLEAR QUADRUPOLE RESONANCE STUDIES OF NIOBIUM PENTAFLUORIDE AND ITS C
FULEA AO   0359   EVALUATION OF THE FORCE CONSTANTS OF SOME TETRAHEDRAL MOLECULES. (CARBON TETRAFLUO
GADINI A   1412   MOLECULES WITH D4H SYMMETRY. NEW METHOD FOR THE CALCULATION OF GVFF CONSTANTS.
GAEUMANN T T   0888   SUBLIMATION OF PLUTONIUM TETRAFLUORIDE.
GAGE DM   0206   INFRARED ABSORPTION SPECTRUM OF BORON TRIFLUORIDE.
GAIDE W   1195   ALPHA SPECTROMETRIC INVESTIGATION OF URANIUM HEXAFLUORIDE. PART-1.
GALKIN NP   0004   REACTIVITY AND THERMAL STABILITY OF HEXAFLUORIDES.
   0422   THERMODYNAMIC PROPERTIES OF PENTAFLUORIDES AT HIGH TEMPERATURES. PART-1: CALCULATI
   0536   THERMODYNAMIC PROPERTIES OF PENTAFLUORIDES AT HIGH TEMPERATURES. PART-3: PENTAFLUO
   0622   EMPIRICAL METHOD FOR DETERMINING EFFECTIVE VIBRATIONAL AND ROTATIONAL CHARACTERIST
   0629   THERMODYNAMIC STABILITY OF HEXAFLUORIDES AT HIGH TEMPERATURES. PART-2: PLUTONIUM H
   0630   THERMAL STABILITY AND REACTIVITY OF D- AND F ELEMENT HEXAFLUORIDES.
   0647   THERMODYNAMIC PROPERTIES OF PENTAFLUORIDES AT HIGH TEMPERATURES. PART-4: PENTAFLUO
   0654   THERMODYNAMIC PROPERTIES OF NIOBIUM AND TANTALUM FLUORIDES AT HIGH TEMPERATURES. P
   0655   THERMODYNAMIC PROPERTIES OF NIOBIUM MONOFLUORIDE, NIOBIUM DIFLUORIDE, TANTALUM MON
   0656   THERMODYNAMIC PROPERTIES OF NIOBIUM TRIFLUORIDE, NIOBIUM TETRAFLUORIDE, TANTALUM T
   0657   THERMODYNAMIC PROPERTIES OF NIOBIUM AND TANTALUM FLUORIDES AT HIGH TEMPERATURES. P
   0665   THERMODYNAMIC PROPERTIES OF NIOBIUM AND TANTALUM FLUORIDES AT HIGH TEMPERATURES. P
   1177   THERMODYNAMIC PROPERTIES OF URANIUM PENTAFLUORIDE, NEPTUNIUM PENTAFLUORIDE, AND PL
   1328   VAPOR PRESSURE OF ZIRCONIUM TETRAFLUORIDE.
   1379   THERMODYNAMIC PROPERTIES OF TRANSITION METAL PENTAFLUORIDES AT HIGH TEMPERATURE. (
   1434   THERMODYNAMIC PROPERTIES OF PENTAFLUORIDES AT HIGH TEMPERATURES. PART-2: PENTAFLUO
GANIN VA   0498   SPIN ORBITAL INTERACTION AND RELATIVE INTENSITIES OF FORBIDDEN TRANSITIONS IN IODI
GARD GL   0435   VIBRATIONAL SPECTRUM OF LIQUID CHROMIUM PENTAFLUORIDE.
GARDINER DJ   0767   RAMAN SPECTRUM OF LIQUID OXYGEN DIFLUORIDE.
GARTON G   0461   PREPARATION AND CRYSTAL STRUCTURE OF GALLIUM TRIFLUORIDE.
GASNER EL   0522   VIBRATIONAL SPECTRA OF IODINE HEPTAFLUORIDE AND RHENIUM HEPTAFLUORIDE.
   0670   RAMAN SPECTRA OF LIQUID NIOBIUM PENTAFLUORIDE AND TANTALUM PENTAFLUORIDE.
   1289   RAMAN SPECTRUM OF XENON HEXAFLUORIDE.
GAUNT J   0080   MOLECULAR STRUCTURE OF ANTIMONY PENTAFLUORIDE.
   0631   FORCE CONSTANTS AND BOND LENGTHS OF SOME INORGANIC HEXAFLUORIDES. (OF SULFUR, SELE
   0907   VIBRATIONAL STRUCTURE AND MOLECULAR STRUCTURE OF RHENIUM HEXAFLUORIDE.
   0961   INFRARED SPECTRA AND MOLECULAR STRUCTURE OF SOME GROUP-6 HEXAFLUORIDES. (OF SULFUR
   1108   INFRARED SPECTRUM OF TELLURIUM HEXAFLUORIDE FROM 25 TO 40 MICRONS.
GAVIN RM   1280   MOLECULAR STRUCTURE OF XENON HEXAFLUORIDE.
   1290   EFFECTS OF ELECTRON CORRELATION IN X-RAY DIFFRACTION AND AN ELECTRON DIFFRACTION S
   1291   MOLECULAR STRUCTURE OF XENON HEXAFLUORIDE. PART-1: ANALYSIS OF ELECTRON DIFFRACTIO
GAY RS   0844   RAMAN INTENSITIES AND THE FORCE FIELD OF PHOSPHORUS PENTAFLUORIDE.
GEBBIE HA   0702   FAR INFRARED SPECTRUM OF NITROGEN TRIFLUORIDE.
GELIUS U   0328   MOLECULAR SPECTROSCOPY BY MEANS OF ESCA. (CARBON TETRAFLUORIDE)
GELLINGS PJ   1236   BOND LENGTHS IN XENON DIFLUORIDE.
GEORGE JW   1037   NMR SPECTRUM AND STRUCTURE OF SULFUR TETRAFLUORIDE.
GERASIMOV YI   0041   VAPOR PRESSURE OF ALUMINUM TRIFLUORIDE.

GESCHWIND S	0102	MICROWAVE SPECTRUM OF ARSENIC TRIFLUORIDE.
GIANTURCO FA	1077	ELECTRONIC PROPERTIES OF SULFUR HEXAFLUORIDE. AB INITIO CALCULATION FOR THE GROUND
	1078	ELECTRONIC PROPERTIES OF SULFUR HEXAFLUORIDE. PART-1: OPTICAL ABSORPTION AND X-RAY
GIBSON JA	1042	FLUORINE NMR SPECTRA OF SULFUR, PHOSPHORUS, AND SILICON FLUORIDES. HYDROLYSIS AND
GILBERT TL	0174	METHODS FOR CORRELATING MOLECULES AND SOME OPTIMIZED VALENCE CONFIGURATION RESULTS
GILLARD IR	0681	MICROWAVE SPECTRUM OF NITROGEN DIFLUORIDE RADICAL.
GILLESPIE RJ	0078	FLUORINE NMR OF ANTIMONY PENTAFLUORIDE.
	0079	A FLUORINE NMR AND RAMAN STUDY OF COMPLEX FORMATION BETWEEN ANTIMONY PENTAFLUORIDE
	0504	VIBRATIONAL SPECTRA AND ASSIGNMENT OF IODINE PENTAFLUORIDE. FERMI RESONANCE BETWEE
	0845	STRUCTURES OF PHOSPHORUS PENTAFLUORIDE, METHYL PHOSPHORUS TETRAFLUORIDE AND DIMETH
	1043	STRUCTURE OF SULFUR TETRAFLUORIDE AND RELATED MOLECULES.
	1292	STRUCTURES OF NOBLE GAS COMPOUNDS. (XENON HEXAFLUORIDE)
	1413	ELECTRON CORRELATION AND MOLECULAR SHAPE.
	1435	BOND LENGTHS AND BOND ANGLES IN OCTAHEDRAL, TRIGONAL BIPYRAMID, AND RELATED MOLECU
GILLIAM OR	0798	MICROWAVE INVESTIGATIONS OF METHYL FLUORIDE, FLUOROFORM, AND PHOSPHORUS TRIFLUORID
GIL'FANOV FZ	0455	REFLECTION SPECTRA OF GADOLINIUM OXIDE AND GADOLINIUM TRIFLUORIDE POWDERS.
GINN SGW	0207	NU-3 BANDS OF BORON TRIFLUORIDE.
	0208	NU-4 BANDS OF THE TEN AND ELEVEN ISOTOPES OF BORON TRIFLUORIDE AT HIGH RESOLUTION.
	0209	BORON TRIFLUORIDE. THE ANALYSIS OF NU-2 AND THE DETERMINATION OF THE MOLECULAR GEO
	0210	VIBRATIONAL ANHARMONICITY IN BORON TRIFLUORIDE.
GIORDANI SF	1445	FORCE CONSTANTS FOR XY6 MOLECULES.
GISSANE WJM	0929	GROUND STATES OF SCANDIUM MONOFLUORIDE AND YTTRIUM MONOFLUORIDE.
	0930	ROTATIONAL ANALYSIS OF SOME SINGLET TRANSITIONS IN THE SPECTRUM OF GASEOUS SCANDIU
GLASS WK	1293	GAS PHASE STRUCTURE OF XENON HEXAFLUORIDE.
GLEMSER O	0436	INFRARED SPECTRUM OF CHROMIUM HEXAFLUORIDE, MOLYBDENUM HEXAFLUORIDE, AND OSMIUM H
	0754	PREPARATION AND PROPERTIES (INFRARED SPECTRUM) OF OSMIUM HEPTAFLUORIDE.
	0811	RAMAN AND INFRARED SPECTRA AND THERMODYNAMIC FUNCTIONS FOR VARIOUS PHOSPHORUS AND
GODFREY PD	0681	MICROWAVE SPECTRUM OF NITROGEN DIFLUORIDE RADICAL.
GODNEV IN	0211	APPLICATION OF THE PROGRESSIVE RIGIDITY METHOD TO SOME SIMPLE XYN TYPE MOLECULES.
	1160	HEATS OF FORMATION OF GASEOUS URANIUM FLUORIDES. (URANIUM MONOFLUORIDE, URANIUM DI
	1329	FREQUENCIES OF ZIRCONIUM FLUORIDE (GAS) VIBRATIONS ACCORDING TO THERMODYNAMIC DATA
	1330	NORMAL VIBRATION FREQUENCIES OF ZIRCONIUM HALIDES.
	1430	PROGRESSIVE STIFFNESS METHOD APPLIED TO XY4 MOLECULES IN THE LIGHT OF THE THEORY O
	1441	APPLICATION OF THE METHOD OF PROGRESSIVE RIGIDITY TO OCTAHEDRAL MX6 MOLECULES.
	1448	CALCULATION OF THE FORCE CONSTANTS OF HEXAFLUORIDES.
	1486	CALCULATION OF MEAN SQUARE VIBRATIONAL AMPLITUDES.
GOHEL VB	0127	RYDBERG-KLEIN-REES FRANCK-CONDON FACTORS AND R CENTROIDS OF THE A-X BAND SYSTEM OF
	1134	ACCURACY OF THE VIBRATIONAL WAVE FUNCTIONS AND DERIVED QUANTITIES. (TIN FLUORIDE)
GOLDSTEIN C	0503	MOSSBAUER STUDY OF IODINE PENTAFLUORIDE AND IODINE HEPTAFLUORIDE.
GOLE JL	0137	NONEMPIRICAL LCAO MO SCF STUDIES OF THE LOW LYING STATES OF BERYLLIUM DIFLUORIDE.
	0981	VACUUM ULTRAVIOLET SPECTRA OF SILICON DIFLUORIDE AND GERMANIUM DIFLUORIDE.
	1360	NONEMPIRICAL LCAO MO SCF STUDIES OF THE GROUP-2A DIHALIDES BERYLLIUM DIFLUORIDE, M
GOLEN JA	0854	PENTACOORDINATED MOLECULES. PART-14: MOLECULAR VIBRATIONS AND STEREOCHEMICAL NONRI
GOMBLER W	1044	FLUORINE-19 NMR INVESTIGATION OF THE ASSOCIATION OF SULFUR TETRAFLUORIDE.
	1055	PROBLEM OF FLUORINE EXCHANGE IN SULFUR TETRAFLUORIDE.
GOODENOUGH JB	1370	COMPARISONS OF FLUORIDES, OXIDES, AND SULFIDES CONTAINING DIVALENT TRANSITION ELEM
GOODFRIEND PL	0685	ABSORPTION SPECTRUM OF NITROGEN DIFLUORIDE.
GOODGAME DML	0461	PREPARATION AND CRYSTAL STRUCTURE OF GALLIUM TRIFLUORIDE.
GOODMAN DW	0846	PHOTOELECTRON SPECTRUM OF PHOSPHORUS PENTAFLUORIDE.
GOODMAN GL	0021	VIBRATIONAL PROPERTIES OF HEXAFLUORIDE MOLECULES.
	0537	VIBRONICALLY INDUCED JAHN-TELLER PROGRESSIONS IN THE ELECTRONIC SPECTRUM OF IRIDIU
	0552	INFRARED AND RAMAN SPECTRA OF KRYPTON DIFLUORIDE.
	0648	ELECTRONIC STRUCTURE OF NEPTUNIUM HEXAFLUORIDE.
	0902	NEAR INFRARED SYSTEM OF RHENIUM HEXAFLUORIDE.
	0908	VIBRONIC INTERACTIONS IN METAL HEXAFLUORIDE MOLECULES.
	0916	COLORS OF TRANSITION METAL HEXAFLUORIDES. (RHENIUM HEXAFLUORIDE, OSMIUM HEXAFLUORI
	1072	RAMAN SPECTRA OF MOLYBDENUM HEXAFLUORIDE, TECHNETIUM HEXAFLUORIDE, RHENIUM HEXAFLU
	1159	HEAT CAPACITIES AND ELECTRONIC SPECTRA OF THE PLATINUM GROUP METAL HEXAFLUORIDE MO
	1285	SPECTRAL OBSERVATIONS ON MOLECULAR XENON HEXAFLUORIDE. RAMAN SCATTERING AND INFRAR
	1294	ELECTRONIC STATES AND MOLECULAR GEOMETRY OF XENON HEXAFLUORIDE: A CASE OF ELECTRON
GORDY W	0033	MILLIMETER AND SUBMILLETER WAVE SPECTRUM AND MOLECULAR CONSTANTS OF ALUMINUM MONOF
	0178	ELECTRON SPIN RESONANCE AND STRUCTURE OF THE BORON DIFLUORIDE RADICAL.
	0711	MICROWAVE SPECTROSCOPY IN THE REGION FROM TWO TO THREE MILLIMETERS, PART-2: (NITRO
	0741	NUCLEAR QUADRUPOLE MOMENT OF NITROGEN-14 AND THE STRUCTURE OF NITROGEN TRIFLUORIDE
	0792	ELECTRON SPIN RESONANCE OF TRAPPED PHOSPHORUS DIFLUORIDE AND PHOSPHORUS TETRAFLUOR
	0796	ZEEMAN EFFECT OF SOME LINEAR AND SYMMETRIC TOP MOLECULES. (PHOSPHORUS TRIFLUORIDE)
	0798	MICROWAVE INVESTIGATIONS OF METHYL FLUORIDE, FLUOROFORM, AND PHOSPHORUS TRIFLUORID
GOROKHOV LN	0666	MASS SPECTRUM AND COMPOSITION OF NIOBIUM PENTAFLUORIDE VAPOR.
GOROKHOV YA	1064	SELECTIVE PHOTODISSOCIATION OF POLYATOMIC MOLECULES (SULFUR HEXAFLUORIDE) BY INFRA
GOTKIS IS	0666	MASS SPECTRUM AND COMPOSITION OF NIOBIUM PENTAFLUORIDE VAPOR.
	1149	THERMODYNAMIC PROPERTIES OF TUNGSTEN PENTAFLUORIDE.
GOUBEAU J	0212	FORCE CONSTANTS OF CARBON TETRAFLUORIDE AND BORON TRIFLUORIDE AND COMPARISON WITH
GOUBEAU VJ	0847	RAMAN AND INFRARED SPECTRUM OF TRIFLUORO PHOSPHORANE. (PHOSPHORUS PENTAFLUORIDE)
GOUTIER D	0213	AB INITIO CALCULATION ON BORON TRIFLUORIDE AND BORON TRICHLORIDE.
GRAFF G	1118	STARK EFFECT FOR THALLIUM MONOFLUORIDE AND POTASSIUM FLUORIDE.
GRANNEC J	0444	NEW FLUORIDES OF TRIVALENT COPPER.
GRAY HB	1343	BOND ENERGIES AND IONIC CHARACTER OF INERT GAS HALIDES. HELIUM-2 PHOTOELECTRON SPE
GREEN DW	0933	SPECTROSCOPIC STUDIES OF SCANDIUM MONOFLUORIDE AND MAGNESIUM OXIDE.
GREEN JC	1362	STUDY OF THE BONDING IN THE GROUP-4 TETRAHALIDES BY PHOTOELECTRON SPECTROSCOPY.
GREEN JW	0119	DISSOCIATION ENERGIES OF THE ALKALINE EARTH MONOFLUORIDES.
	0276	SUBLIMATION PRESSURE OF CALCIUM DIFLUORIDE AND THE DISSOCIATION ENERGY OF CALCIUM
GREEN MLH	1362	STUDY OF THE BONDING IN THE GROUP-4 TETRAHALIDES BY PHOTOELECTRON SPECTROSCOPY.
GREENBAUM MA	0138	THERMODYNAMIC AND PHYSICAL PROPERTIES OF BERYLLIUM COMPOUNDS. PART-1: ENTHALPY AND
	0154	HEAT AND ENTROPY OF FORMATION OF BORON FLUORIDE.
GREENBERG E	0329	ENTHALPY OF FORMATION OF CARBON TETRAFLUORIDE.
	0667	ENTHALPIES OF FORMATION OF NIOBIUM PENTAFLUORIDE AND TANTALUM PENTAFLUORIDE.
	1147	HEATS OF FORMATION OF TITANIUM TETRAFLUORIDE AND HAFNIUM TETRAFLUORIDE.
GREFFITHS PR	1479	FAR INFRARED SPECTRA OF PROLATE AND OBLATE SYMMETRIC TOPS.
GREGORY AR	0412	IS THE RADICAL CHLORINE TETRAFLUORIDE PLANAR OR NOT.
	1060	STRUCTURE OF THE RADICAL SULFUR PENTAFLUORIDE.

GREIS O	0452	PHASE INVESTIGATIONS AND VAPOR PRESSURE MEASUREMENTS ON EUROPIUM DIFLUORIDE.
GRENNAN TP	0285	THE B-X SYSTEM OF CARBON MONOFLUORIDE.
GREWE FJ	1150	PREPARATION AND PROPERTIES (VAPOR PRESSURE) OF TUNGSTEN PENTAFLUORIDE.
GRIFFITH JS	1333	PAIRING ENERGIES IN THE ACTINIDE SERIES.
GRIFFITHS JE	0848	NU-7 FUNDAMENTAL OF PHOSPHORUS PENTAFLUORIDE.
	0849	MOLECULAR STRUCTURES OF VARIOUS PHOSPHORUS HALIDES AND PHOSPHORUS PENTAFLUORIDE. I
GRIGORYEVA GN	1346	INFRARED SPECTRA OF RARE EARTH METAL FLUORIDES.
GRIMM FA	0319	HIGH RESOLUTION PHOTOELECTRON SPECTROSCOPY OF CARBON TETRAFLUORIDE AND SILICON TE
	0369	PHOTOELECTRON SPECTRUM OF CHLORINE MONOFLUORIDE.
GRISARD JW	1197	VAPOR PRESSURE AND CRITICAL CONSTANTS OF URANIUM HEXAFLUORIDE.
GROMOV BV	0088	ASSOCIATION OF ANTIMONY PENTAFLUORIDE IN VAPOR.
GROSS P	0139	ENTHALPY OF FORMATION OF BERYLLIUM DIFLUORIDE.
	0214	HEATS OF FORMATION OF INORGANIC FLUORIDES ESPECIALLY THE ELEMENTS OF ATOMIC NUMBER
	0483	HEATS OF FORMATION OF GERMANIUM TETRAFLUORIDE AND OF THE GERMANIUM DIOXIDES.
GROSS ST	0364	RAMAN SPECTRA OF CARBON AND SILICON TETRAFLUORIDES.
	0963	RAMAN SPECTRA AND MOLECULAR CONSTANTS OF THE HEXAFLUORIDES OF SULFUR, SELENIUM, AN
GROSS U	0263	SELF IONIZATION OF HALOGEN FLUORIDES. (BROMINE PENTAFLUORIDE)
GROSSE AV	1178	URANIUM PENTAFLUORIDE, A NEW FLUORIDE OF URANIUM.
GRUEN DM	0917	MAGNETIC SUSCEPTIBILITY OF RHENIUM HEXAFLUORIDE.
GUERN Y	0042	RECENT DEVELOPMENTS IN MULTIPLE BEAM INTERFEROMETRY IN THE ULTRAVIOLET REGION 1700
GUEST MF	0379	AB INITIO MOLECULAR ORBITAL STUDY OF THE GEOMETRY OF INTERHALOGENS. (CHLORINE MONO
	0794	THEORETICAL STUDY OF THE GEOMETRY OF PHOSPHINE AND PHOSPHORUS TRIFLUORIDE AND THE
GUIDOTTI C	1077	ELECTRONIC PROPERTIES OF SULFUR HEXAFLUORIDE. AB INITIO CALCULATION FOR THE GROUND
	1078	ELECTRONIC PROPERTIES OF SULFUR HEXAFLUORIDE. PART-1: OPTICAL ABSORPTION AND X-RAY
GUILLAUME J	0089	ULTRAVIOLET SPECTRUM OF ARSENIC MONOFLUORIDE.
GUILLEMIN YGS	0330	ELECTRONIC STRUCTURE OF THE CARBON TETRAFLUORIDE MOLECULE. PART-1. THE GROUND STA
GUILLIN J	0370	HEATS OF FORMATION OF CHLORINE FLUORIDE. (CHLORINE MONOFLUORIDE, CHLORINE DIFLUOR
GUILLORY WA	0691	MATRIX ISOLATION STUDY OF THE VACUUM ULTRAVIOLET PHOTOLYSIS OF NITROGEN TRIFLUORID
GULLIKSON CW	1079	RAMAN SPECTRUM OF SULFUR HEXAFLUORIDE IN THE GASEOUS AND SOLID STATES.
GUNN SR	0557	HEAT FORMATION OF KRYPTON DIFLUORIDE.
	1266	HEAT OF FORMATION OF XENON TETRAFLUORIDE.
	1267	XENON TETRAFLUORIDE: HEAT OF FORMATION.
GUNTHER WH	0892	ABSORPTION SPECTRUM OF PLUTONIUM HEXAFLUORIDE.
GUPTA LC	0400	QUADRUPOLE COUPLING CONSTANTS IN CHLORINE TRIFLUORIDE AND BROMINE TRIFLUORIDE.
GURVICH LV	0569	ELECTRONIC SPECTRUM OF LANTHANUM MONOFLUORIDE.
	0580	BAND SYSTEM IN THE DISCHARGE SPECTRUM OF LANTHANUM TRIFLUORIDE VAPORS.
	0934	ELECTRONIC SPECTRUM OF SCANDIUM MONOFLUORIDE.
	1315	VIBRONIC SPECTRA OF YTTRIUM MONOFLUORIDE.
GUSAROV AV	0666	MASS SPECTRUM AND COMPOSITION OF NIOBIUM PENTAFLUORIDE VAPOR.
	1149	THERMODYNAMIC PROPERTIES OF TUNGSTEN PENTAFLUORIDE.
GUTJAHR L	0190	FORCE CONSTANTS FOR MOLECULES WITH D3H SYMMETRY.
GUTMANN V	0005	HALOGEN CHEMISTRY. (VOLUMES 1, 2, 3)
GUTOWSKY HS	0799	INFRARED SPECTRA OF PHOSPHORUS TRIFLUORIDE, PHOSPHORUS OXIDE TRIFLUORIDE, AND PHOS
GUYON PM	1225	PHOTOIONIZATION MASS SPECTROMETRIC STUDY OF XENON DIFLUORIDE, XENON TETRAFLUORIDE,
GWINN WD	1056	STRUCTURE AND DIPOLE MOMENT OF SULFUR TETRAFLUORIDE.
HAARHOFF PC	0353	HARMONIC POTENTIAL CONSTANTS OF OSMIUM TETROXIDE AND SOME TETRAHALIDE (CARBON TETR
	0850	POTENTIAL FIELD AND FORCE CONSTANTS OF SOME TRIGONAL BIPYRAMIDAL PENTAHALIDES. (PH
HABITZ P	1480	VIBRONIC AND SPIN ORBIT SPLITTING IN SPECTRA OF SYSTEMS EXHIBITING JAHN-TELLER EFF
HABUDA SP	0571	SYMMETRY AND CRYSTAL STRUCTURE OF RARE EARTH TRIFLUORIDES. (LANTHANUM TRIFLUORIDE)
HAENSEL R	1233	SPECTRA OF XENON DIFLUORIDE AND XENON TETRAFLUORIDE IN THE FAR ULTRAVIOLET REGION.
	1302	ELECTRONIC AND GEOMETRIC STRUCTURE OF THE FREE XENON HEXAFLUORIDE MOLECULE.
HAGEN G	1268	VIBRATIONS OF PLANAR SYMMETRICAL XY4 MOLECULES WITH APPLICATION XENON TETRAFLUORID
	1477	THEORY AND CALCULATION OF CENTRIFUGAL DISTORTION CONSTANTS FOR POLYATOMIC MOLECULE
HAGEN K	1002	INTERATOMIC DISTANCES AND ROOT MEAN SQUARE AMPLITUDES OF VIBRATION OF GASEOUS SILI
HAGENMULLER P	0444	NEW FLUORIDES OF TRIVALENT COPPER.
HALE JB	0186	INFRARED ABSORPTION SPECTRA OF SOME POLYATOMIC FLUORIDES. (SILICON TETRAFLUORIDE,
	0187	MOLECULAR STRUCTURE OF BORON TRIFLUORIDE.
HALGREN TA	0158	LOCALIZED ORBITALS IN BORON FLUORIDES. HIGHLY POLARIZED BORON- FLUORINE DOUBLE AND
HALL EH	1148	VAPOR PRESSURE OF TITANIUM TETRAFLUORIDE.
HALL JA	0287	THEORETICAL STUDY OF THE SPECTROSCOPIC STATES OF THE CARBON MONOFLUORIDE MOLECULE.
HALL JH	0158	LOCALIZED ORBITALS IN BORON FLUORIDES. HIGHLY POLARIZED BORON- FLUORINE DOUBLE AND
HALL MB	0379	AB INITIO MOLECULAR ORBITAL STUDY OF THE GEOMETRY OF INTERHALOGENS. (CHLORINE MONO
	0794	THEORETICAL STUDY OF THE GEOMETRY OF PHOSPHINE AND PHOSPHORUS TRIFLUORIDE AND THE
HALLE M	0979	ROTATIONAL ANALYSIS OF ABSORPTION BANDS IN THE 2266 ANGSTROM SYSTEM OF SILICON DIF
HAMILTON WC	1269	ENTHALPIES OF FORMATION OF XENON HEXAFLUORIDE, XENON TETRAFLUORIDE, XENON TETRAFLU
HAMMERLE RH	0489	HYPERFINE STRUCTURE OF INDIUM MONOFLUORIDE.
HAND RW	0542	ELECTRONIC STRUCTURE OF IRON TRIFLUORIDE.
HANKIN D	1224	SELF CONSISTENT FIELD STUDY OF THE SERIES XENON DIFLUORIDE, XENON TETRAFLUORIDE, X
HANNAH RW	0331	MOLECULAR CONSTANTS FROM INFRARED BAND SHAPES. THE ATOMIC DISPLACMENTS AND INTRAMO
HANNAY NB	1200	DIPOLE MOMENT AND MOLECULAR STRUCTURE OF URANIUM HEXAFLUORIDE.
HANSEN KW	0851	ELECTRON DIFFRACTION STUDY OF THE STRUCTURE OF PHOSPHORUS PENTAFLUORIDE.
HARCOURT RD	0301	WAVE FUNCTIONS FOR THE FOUR ELECTRON THREE CENTER BONDING OF FOUR AND EIGHT PI ELE
HARGREAVES GB	0624	VAPOR PRESSURES OF MOLYBDENUM PENTAFLUORIDE, TUNGSTEN PENTAFLUORIDE, RHENIUM PENTA
	0749	PREPARATION AND CRYSTAL STRUCTURE OF OSMIUM PENTAFLUORIDE. TWO FLUORIDES OF OSMIUM
	0903	VAPOR PRESSURES OF SOME HEAVY TRANSITION METAL HEXAFLUORIDES. (TUNGSTEN HEXAFLUORI
HARLAND PW	0047	HIGH TEMPERATURE NEGATIVE IONS. GASEOUS GROUP-3 FLUORIDES.
	0464	BOND DISSOCIATION ENERGIES, IONIZATION POTENTIALS AND ELECTRON AFFINITIES OF SOME
	0800	IONIZATION BY ELECTRON IMPACT OF PHOSPHORUS TRIFLUORIDE AND DIFLUORO CYANO PHOSPHI
HARMONY MD	0686	INFRARED SPECTRUM AND THERMODYNAMIC FUNCTIONS OF THE NITROGEN DIFLUORIDE RADICAL.
	0687	INFRARED SPECTRUM AND STRUCTURE OF THE NITROGEN DIFLUORIDE RADICAL.
HARRINGTON JA	0288	SHOCK TUBE DETERMINATION OF THE C2 (A TRIPLET PI TO X TRIPLET PI) AND CARBON MONOF
HARRIS WC	1046	INTRAMOLECULAR EXCHANGE IN SULFUR TETRAFLUORIDE.
HARRISON RH	0332	CARBON TETRAFLUORIDE. THERMODYNAMIC PROPERTIES OF THE REAL GAS.
HARSHBARGER WR	0333	ELECTRON IMPACT SPECTRA OF METHANE AND CARBON TETRAFLUORIDE.
	0558	STRUCTURE OF KRYPTON DIFLUORIDE AS INVESTIGATED BY ELECTRON DIFFRACTION.
HART BT	0682	ELECTRONIC STRUCTURE OF THE NITROGEN DIFLUORIDE RADICAL.
HARTMANSHENN O	0370	HEATS OF FORMATION OF CHLORINE FLUORIDE. (CHLORINE MONOFLUORIDE, CHLORINE DIFLUOR
HASEGAWA A	0823	ESR SPECTRA OF PHOSPHORUS TETRAFLUORIDE RADICALS PRODUCED IN A SINGLE CRYSTAL OF P
HASSAN HA	1193	THERMODYNAMIC PROPERTIES OF URANIUM HEXAFLUORIDE AT HIGH TEMPERATURES.
HASTIE JW	0011	MATRIX ISOLATION INFRARED SPECTROSCOPY OF HIGH TEMPERATURE SPECIES. (FLUORIDES, OX

HASTIE JW	0277	REEVALUATION OF THE DISSOCIATION ENERGY OF CALCIUM FLUORIDE. (ALSO SILICON AND GER
	0437	INFRARED SPECTRA AND GEOMETRY OF MATRIX ISOLATED COBALT DIFLUORIDE, NICKEL DIFLUOR
	0451	INFRARED SPECTRUM OF HEAVY METAL DIHALIDES. (EUROPIUM DIFLUORIDE)
	0462	INFRARED SPECTRA OF MATRIX ISOLATED SPECIES IN THE GALLIUM- FLUORINE SYSTEM.
	0473	INFRARED VIBRATIONAL PROPERTIES OF GERMANIUM DIFLUORIDE.
	0592	MASS SPECTROMETRIC STUDIES AT HIGH TEMPERATURES. (TIN DIFLUORIDE, LEAD DIFLUORIDE,
	0594	INFRARED SPECTRA OF MATRIX ISOLATED TIN DIFLUORIDE AND LEAD DIFLUORIDE.
	0595	ULTRAVIOLET ABSORPTION SPECTRA OF GASEOUS TIN DIFLUORIDE AND LEAD DIFLUORIDE.
	0943	GEOMETRIES AND ENTROPIES OF METAL TRIFLUORIDES FROM INFRARED SPECTRA. (SCANDIUM, Y
	0944	FORCE CONSTANTS AND GEOMETRIES OF MATRIX ISOLATED RARE EARTH TRIFLUORIDES. (SCANDI
	0981	VACUUM ULTRAVIOLET SPECTRA OF SILICON DIFLUORIDE AND GERMANIUM DIFLUORIDE.
	0982	INFRARED SPECTRA OF SILICON DIFLUORIDE IN INERT GAS MATRICES.
	1146	INFRARED SPECTRA AND GEOMETRY OF TITANIUM DIFLUORIDE AND TITANIUM TRIFLUORIDE IN R
	1325	INFRARED SPECTRA OF MATRIX ISOLATED ZIRCONIUM DIFLUORIDE, TRIFLUORIDE, AND TETRAFL
	1371	VIBRATIONAL FREQUENCIES AND VALENCE FORCE CONSTANTS OF FIRST ROW TRANSITION METAL
HAUGE RH	0011	MATRIX ISOLATION INFRARED SPECTROSCOPY OF HIGH TEMPERATURE SPECIES. (FLUORIDES, OX
	0437	INFRARED SPECTRA AND GEOMETRY OF MATRIX ISOLATED COBALT DIFLUORIDE, NICKEL DIFLUOR
	0451	INFRARED SPECTRUM OF HEAVY METAL DIHALIDES. (EUROPIUM DIFLUORIDE)
	0462	INFRARED SPECTRA OF MATRIX ISOLATED SPECIES IN THE GALLIUM- FLUORINE SYSTEM.
	0473	INFRARED VIBRATIONAL PROPERTIES OF GERMANIUM DIFLUORIDE.
	0474	ULTRAVIOLET ABSORPTION SPECTRUM OF GERMANIUM DIFLUORIDE.
	0594	INFRARED SPECTRA OF MATRIX ISOLATED TIN DIFLUORIDE AND LEAD DIFLUORIDE.
	0595	ULTRAVIOLET ABSORPTION SPECTRA OF GASEOUS TIN DIFLUORIDE AND LEAD DIFLUORIDE.
	0604	INFRARED SPECTRA OF MATRIX ISOLATED MAGNESIUM DIFLUORIDE.
	0943	GEOMETRIES AND ENTROPIES OF METAL TRIFLUORIDES FROM INFRARED SPECTRA. (SCANDIUM, Y
	0944	FORCE CONSTANTS AND GEOMETRIES OF MATRIX ISOLATED RARE EARTH TRIFLUORIDES. (SCANDI
	0981	VACUUM ULTRAVIOLET SPECTRA OF SILICON DIFLUORIDE AND GERMANIUM DIFLUORIDE.
	0982	INFRARED SPECTRA OF SILICON DIFLUORIDE IN INERT GAS MATRICES.
	0985	INFRARED SPECTRUM, FORCE CONSTANTS, AND THERMODYNAMIC FUNCTIONS OF SILICON DIFLUOR
	1146	INFRARED SPECTRA AND GEOMETRY OF TITANIUM DIFLUORIDE AND TITANIUM TRIFLUORIDE IN R
	1325	INFRARED SPECTRA OF MATRIX ISOLATED ZIRCONIUM DIFLUORIDE, TRIFLUORIDE, AND TETRAFL
	1371	VIBRATIONAL FREQUENCIES AND VALENCE FORCE CONSTANTS OF FIRST ROW TRANSITION METAL
HAWKINS NJ	0540	INFRARED SPECTRUM OF IRIDIUM HEXAFLUORIDE.
	0891	INFRARED SPECTRUM OF PLUTONIUM HEXAFLUORIDE.
HAYES EF	0605	THEORETICAL STUDIES OF THE INTERACTION OF MAGNESIUM DIFLUORIDE WITH RARE GAS ATOMS
	1360	NONEMPIRICAL LCAO MO SCF STUDIES OF THE GROUP-2A DIHALIDES BERYLLIUM DIFLUORIDE, M
HAYES W	0215	ABSORPTION OF BORON TRIFLUORIDE GAS IN THE EXTREME ULTRAVIOLET.
HAYMAN C	0139	ENTHALPY OF FORMATION OF BERYLLIUM DIFLUORIDE.
	0214	HEATS OF FORMATION OF INORGANIC FLUORIDES ESPECIALLY THE ELEMENTS OF ATOMIC NUMBER
	0483	HEATS OF FORMATION OF GERMANIUM TETRAFLUORIDE AND OF THE GERMANIUM DIOXIDES.
HAZONY Y	0543	ELECTRONIC STRUCTURE AND ANOMALOUS THERMAL EXPANSION ON IRON DIFLUORIDE AND VANADI
HEATH DF	1080	MOLECULAR FORCE FIELDS. PART-8: VIBRATION FREQUENCIES OF SOME OCTAHEDRAL XY6 MOLEC
	1391	VIBRATIONAL FREQUENCIES OF SOME PLANAR XY3 MOLECULES.
HEBECKER VC	1127	CRYSTAL STRUCTURE OF THALLIUM TRIFLUORIDE.
HEDBERG K	1002	INTERATOMIC DISTANCES AND ROOT MEAN SQUARE AMPLITUDES OF VIBRATION OF GASEOUS SILI
	1295	STRUCTURE OF GASEOUS XENON HEXAFLUORIDE.
	1481	POTENTIAL CONSTANTS OF PHOSPHORUS TRICHLORIDE FROM AMPLITUDES OF VIBRATION AND NOR
HEDEN PF	0328	MOLECULAR SPECTROSCOPY BY MEANS OF ESCA. (CARBON TETRAFLUORIDE)
HEDMAN J	0328	MOLECULAR SPECTROSCOPY BY MEANS OF ESCA. (CARBON TETRAFLUORIDE)
HEGSTROM RA	0159	MAGNETIC PROPERTIES OF THE BORON FLUORIDE AND BORON HYDRIDE MOLECULES.
HEHRE WJ	0707	MOLECULAR ORBITAL THEORY OF THE ELECTRONIC STRUCTURE OF ORGANIC COMPOUNDS. PART-3:
	0720	SELF CONSISTENT MOLECULAR ORBITAL METHODS. PART-5: AB INITIO CALCULATION OF EQUILI
HEICKLEN J	1003	INFRARED SPECTRUM OF SILICON TETRAFLUORIDE.
HELLBERG KH	0436	INFRARED SPECTRUM OF CHROMIUM HEXAFLUORIDE, MOLYBDENUM HEXAFLUORIDE, AND OSMIUM H
	0754	PREPARATION AND PROPERTIES (INFRARED SPECTRUM) OF OSMIUM HEPTAFLUORIDE.
HENDERSON DC	1320	HIGH TEMPERATURE HEAT CONTENTS AND RELATED THERMODYNAMIC FUNCTIONS OF SEVEN RARE E
	1356	HIGH TEMPERATURE ENTHALPIES AND RELATED THERMODYNAMIC FUNCTIONS OF THE TRIFLUORIDE
HENDERSON WD	0768	MILLIMETER WAVE SPECTRUM AND STRUCTURE OF OXYGEN DIFLUORIDE.
HEPWORTH MA	0616	CRYSTAL STRUCTURE OF MANGANESE TRIFLUORIDE.
	0617	INTERATOMIC BONDING IN CRYSTALLINE MANGANESE TRIFLUORIDE.
HERRING FG	0766	PHOTOELECTRON SPECTRA OF OXYGEN DIFLUORIDE AND OXYGEN DICHLORIDE.
HERSH OL	0801	ELECTRON DIFFRACTION BY GASES AND THE STRUCTURES OF PHOSPHORUS TRIFLUORIDE, PHOSPH
HESSER JE	0216	RADIATIVE LIFETIMES OF ULTRAVIOLET EMISSION SYSTEMS EXCITED IN BORON TRIFLUORIDE,
HEWAIDY FI	1106	CRYSTAL STRUCTURE OF TELLURIUM TETRAFLUORIDE.
HIGGINSON BR	0376	PHOTOELECTRON SPECTRA OF HALIDES. (CHLORINE MONOFLUORIDE, CHLORINE TRIFLUORIDE, BR
HILDENBRAND DL	0129	MASS SPECTROMETRIC DETERMINATION OF THE DISSOCIATION ENERGY OF BERYLLIUM MONOFLUOR
	0140	EFFUSION STUDIES, MASS SPECTRA, AND THERMODYNAMICS OF BERYLLIUM FLUORIDE VAPOR.
	0160	FIRST IONIZATION POTENTIALS OF THE MOLECULES BORON FLUORIDE, SILICON OXIDE AND GER
	0161	DISSOCIATION ENERGY OF BORON MONOFLUORIDE FROM MASS SPECTROMETRIC STUDIES.
	0440	DISSOCIATION ENERGY OF COPPER MONOFLUORIDE.
	1023	MASS SPECTROMETRIC STUDIES OF SOME GASEOUS SULFUR FLUORIDES.
	1028	MASS SPECTROMETRIC STUDIES OF SOME GASEOUS SULFUR FLUORIDES.
HILL SD	0438	VAPOR PRESSURE AND HEAT OF SUBLIMATION OF COBALT DIHALIDES. (COBALT DIFLUORIDE)
HILLIER IH	0379	AB INITIO MOLECULAR ORBITAL STUDY OF THE GEOMETRY OF INTERHALOGENS. (CHLORINE MONO
	0794	THEORETICAL STUDY OF THE GEOMETRY OF PHOSPHINE AND PHOSPHORUS TRIFLUORIDE AND THE
	0802	AB INITIO CALCULATIONS OF THE BONDING IN PHOSPHINE, PHOSPHORUS TRIFLUORIDE AND TRI
HILTON AR	0768	MILLIMETER WAVE SPECTRUM AND STRUCTURE OF OXYGEN DIFLUORIDE.
HINCHLIFFE A	0688	AB INITIO MOLECULAR ORBITAL STUDY OF THE NITROGEN DIFLUORIDE AND DINITROGEN TETRAF
	0790	AB INITIO MOLECULAR ORBITAL STUDY OF THE PHOSPHORUS DIFLUORIDE RADICAL.
HINSHAW WS	0217	NUCLEAR MAGNETIC RELAXATION IN BORON TRIFLUORIDE GAS.
HIRAISHI J	0243	FORCE CONSTANTS OF CARBON TETRAFLUORIDE, SILICON TETRAFLUORIDE, BORON TRIFLUORIDE,
HIROTA E	0097	MICROWAVE SPECTRUM OF ARSENIC TRIFLUORIDE IN THE EXCITED VIBRATIONAL STATE.
	0722	MICROWAVE SPECTRA OF NITROGEN TRIFLUORIDE IN THE EXCITED VIBRATIONAL STATES. (EQUI
	0803	L-TYPE DOUBLING TRANSITIONS OF PHOSPHORUS TRIFLUORIDE IN THE NU-4 EQUALS 1 STATE.
	0804	EFFECTS OF THE THIRD ORDER CONSTANTS ON THE L-TYPE DOUBLING TRANSITIONS OF PHOSPHO
	0805	MICROWAVE SPECTRUM OF PHOSPHORUS TRIFLUORIDE.
	0993	MICROWAVE SPECTRUM OF SILICON DIFLUORIDE IN THE EXCITED VIBRATIONAL STATES, EQUILI
	1018	MICROWAVE SPECTRUM OF THE SULFUR MONOFLUORIDE RADICAL.
HITCHINGHAM WC	0613	KNUDSEN MEASUREMENT OF THE VAPOR PRESSURE OF MANGANESE DIFLUORIDE.
HIYAMA H	0247	NORMAL VIBRATIONS OF BORON TRIHALIDES. (BORON TRIFLUORIDE)

HOBROCK DL   0312   ELECTRON IMPACT STUDIES OF SOME TRIHALO METHANES. (CARBON TRIFLUORIDE)
HODGE N   0006   FLUORIDES OF THE ACTINIDE ELEMENTS.
HOEFT J   0025   MICROWAVE ABSORPTION SPECTRA OF ALUMINUM MONOFLUORIDE, GALLIUM MONOFLUORIDE, INDIU
                  0497   ROTATIONAL SPECTRUM OF IODINE FLUORIDE.
HOFFMAN CJ   0081   STRUCTURE OF LIQUID ANTIMONY PENTAFLUORIDE.
                  0082   DENSITY, MELTING POINT AND VAPOR PRESSURE OF ANTIMONY PENTAFLUORIDE.
HOFFMAN CWW   0334   MOLECULAR STRUCTURE OF CARBON TETRAFLUORIDE.
HOFLER F   1004   FORCE CONSTANTS OF THE SILICON TETRAHALIDES AND FREQUENCIES OF CHLORO BROMO SILICO
HOGE HJ   1166   LOW TEMPERATURE HEAT CAPACITIES, ENTHALPIES, AND ENTROPIES OF URANIUM TETRAFLUORID
HOHANSEN H   0831   ORBITALS AND STRUCTURES OF PENTAFLUORIDES. (PHOSPHORUS PENTAFLUORIDE, ARSENIC PENT
HOHLNEICHER G   1090   ELECTRONIC STRUCTURE OF MOLECULES BY A MANY BODY APPROACH. PART-6: THE ASSIGNMENT
HOLDER BE   0081   STRUCTURE OF LIQUID ANTIMONY PENTAFLUORIDE.
HOLISTER C   1224   SELF CONSISTENT FIELD STUDY OF THE SERIES XENON DIFLUORIDE, XENON TETRAFLUORIDE, X
HOLLAND RF   1063   OCTAHEDRAL FINE STRUCTURE SPLITTINGS IN NU-3 OF SULFUR HEXAFLUORIDE.
HOLLOWAY JH   0408   INFRARED AND RAMAN SPECTRA OF CHLORINE TRIFLUORIDE AND BROMINE TRIFLUORIDE.
                  0664   NIOBIUM-93 NUCLEAR QUADRUPOLE RESONANCE STUDIES OF NIOBIUM PENTAFLUORIDE AND ITS C
                  0695   RAMAN SPECTRUM OF THE NITROGEN DIFLUORIDE RADICAL.
                  0697   INFRARED SPECTRUM OF NITROGEN TRIFLUORIDE AND FORCE FIELD.
                  0750   PREPARATION AND CRYSTAL STRUCTURE OF OSMIUM PENTAFLUORIDE.
                  0873   RAMAN SPECTRA OF ARSENIC PENTAFLUORIDE AND VANADIUM PENTAFLUORIDE AND THEIR FORCE
                  0924   RUTHENIUM PENTAFLUORIDE AND RUTHENIUM OXYTETRAFLUORIDE. (VAPOR PRESSURES)
                  1072   RAMAN SPECTRA OF MOLYBDENUM HEXAFLUORIDE, TECHNETIUM HEXAFLUORIDE, RHENIUM HEXAFLU
                  1225   PHOTOIONIZATION MASS SPECTROMETRIC STUDY OF XENON DIFLUORIDE, XENON TETRAFLUORIDE,
HOLMES RR   0115   PENTACOORDINATED MOLECULES. PART-12: CORRELATION OF EXCHANGE RATES FOR SOME GROUP-
                  0842   FAR INFRARED SPECTRUM OF PHOSPHORUS PENTAFLUORIDE.
                  0849   MOLECULAR STRUCTURES OF VARIOUS PHOSPHORUS HALIDES AND PHOSPHORUS PENTAFLUORIDE. I
                  0852   INTRAMOLECULAR EXCHANGE IN PHOSPHORUS PENTAHALIDE MOLECULES.
                  0853   ANISOTROPIC THERMAL MOTION OF TRIGONAL BIPYRAMIDAL MOLECULES FROM SPECTROSCOPIC DA
                  0854   PENTACOORDINATED MOLECULES. PART-14: MOLECULAR VIBRATIONS AND STEREOCHEMICAL NONRI
                  1214   INTRAMOLECULAR EXCHANGE IN VANADIUM PENTAFLUORIDE.
                  1436   POTENTIAL FIELD AND MOLECULAR VIBRATIONS OF THE TRIGONAL BIPYRAMID AX(3)YZ MOLECUL
HOLZER W   1081   FORBIDDEN RAMAN BANDS OF SULFUR HEXAFLUORIDE. COLLISION-INDUCED RAMAN SCATTERING.
HOLZMAN H   0512   RAMAN SPECTRUM OF IODINE PENTAFLUORIDE. EVIDENCE FOR POLYMERIZATION IN THE LIQUID
HONERJAEGER R   0026   ZEEMAN EFFECT IN THE MICROWAVE ROTATIONAL SPECTRUM OF THE ALUMINUM FLUORIDE MOLECU
                  1120   HIGH TEMPERATURE MICROWAVE SPECTROMETER FOR MEASUREMENTS OF ZEEMAN EFFECT IN DIAMA
HORA CJ   1214   INTRAMOLECULAR EXCHANGE IN VANADIUM PENTAFLUORIDE.
HOSKINS LC   0100   INFRARED SPECTRUM AND VIBRATIONAL POTENTIAL FUNCTION OF ARSENIC TRIFLUORIDE.
                  0116   PERPENDICULAR BAND CONTOURS AND VIBRATIONAL POTENTIAL FUNCTION OF THE E' VIBRATION
                  0708   P-R SEPARATIONS AND CORIOLIS CONSTANTS FOR SYMMETRIC TOP MOLECULES AND FORCE CONST
                  0855   FAR INFRARED SPECTRUM OF PHOSPHORUS PENTAFLUORIDE.
                  0856   VIBRATIONAL SPECTRA OF PHOSPHORUS PENTAFLUORIDE AND ARSENIC PENTAFLUORIDE AND THE
HOUSTON PL   1082   LOW TEMPERATURE ABSORPTION CONTOUR OF THE NU-3 BAND OF SULFUR HEXAFLUORIDE.
HOWARD BJ   0284   GAS PHASE ELECTRON RESONANCE SPECTRUM AND DIPOLE MOMENT OF CARBON MONOFLUORIDE.
                  0954   ESR SPECTRUM OF BROMINE OXIDE, IODINE OXIDE AND SELENIUM MONOFLUORIDE IN J-5/2 ROT
HOWELL JM   0857   ELECTRONIC STRUCTURES OF PHOSPHORUS PENTAFLUORIDE AND TETRAFLUORO PHOSPHORANE.
HOY AR   0709   AMBIGUITIES IN THE HARMONIC FORCE FIELDS OF XY3 MOLECULES. (NITROGEN TRIFLUORIDE,
HRUBESH LW   0689   A CAVITY SEARCH SPECTROMETER FOR FREE RADICAL MICROWAVE ROTATIONAL ABSORPTION STUD
                  0690   MICROWAVE ROTATIONAL SPECTRUM OF THE NITROGEN DIFLUORIDE FREE RADICAL.
HSIA M   0612   MASS SPECTROMETRIC AND THERMOCHEMICAL STUDIES OF THE MANGANESE FLUORIDES.
HSIA YP   1364   MOLECULAR ORBITAL STUDY OF ATOMIZATION ENERGIES OF PNICOGEN FLUORIDE MOLECULES AND
HUBBARD WN   0048   FLUORINE BOMB CALORIMETRY. PART-22: THE ENTHALPY OF FORMATION OF ALUMINUM TRIFLUOR
                  0117   ENTHALPY OF FORMATION OF ARSENIC PENTAFLUORIDE.
                  0220   FLUORINE BOMB CALORIMETRY. PART-15: THE ENTHALPY OF FORMATION OF BORON TRIFLUORIDE
                  0250   FLUORINE BOMB CALORIMETRY. PART-3: THE HEAT OF FORMATION OF BORON TRIFLUORIDE.
                  0329   ENTHALPY OF FORMATION OF CARBON TETRAFLUORIDE.
                  0469   ENTHALPY OF FORMATION OF GERMANIUM TETRAFLUORIDE.
                  0667   ENTHALPIES OF FORMATION OF NIOBIUM PENTAFLUORIDE AND TANTALUM PENTAFLUORIDE.
                  0807   ENTHALPIES OF FORMATION OF XENON HEXAFLUORIDE, XENON TETRAFLUORIDE, XENON DIFLUORI
                  0818   ENTHALPY OF FORMATION PHOSPHORUS TRIFLUORIDE.
                  0865   FLUORINE BOMB CALORIMETRY. PART-18: STANDARD ENTHALPY OF FORMATION OF PHOSPHORUS P
                  1092   ENTHALPIES OF FORMATION OF SULFUR HEXAFLUORIDE, SELENIUM HEXAFLUORIDE, TELLURIUM H
                  1130   ENTHALPY OF FORMATION OF THORIUM TETRAFLUORIDE BY FLUORINE BOMB CALORIMETRY.
                  1147   HEATS OF FORMATION OF TITANIUM TETRAFLUORIDE AND HAFNIUM TETRAFLUORIDE.
                  1157   ENTHALPY OF FORMATION OF TUNGSTEN HEXAFLUORIDE.
                  1215   ENTHALPY OF FORMATION OF VANADIUM PENTAFLUORIDE BY FLUORINE BOMB CALORIMETRY.
                  1319   ENTHALPY OF FORMATION OF YTTRIUM TRIFLUORIDE.
HUBER H   0475   MATRIX INFRARED AND LASER RAMAN SPECTRA, MOLECULAR STRUCTURES AND NORMAL COORDINAT
HUGO TJ   0030   EMISSION SPECTRUM OF ALUMINUM MONOFLUORIDE IN THE VACUUM ULTRAVIOLET.
HUMPHRIES CM   0806   ABSORPTION SPECTRA OF THE HYDRIDES, DEUTERIDES, AND HALIDES OF GROUP-5 ELEMENTS. (
HUNT WJ   0542   ELECTRONIC STRUCTURE OF IRON TRIFLUORIDE.
HUO WM   0162   ELECTRONIC STRUCTURE OF CARBON MONOXIDE AND BORON MONOFLUORIDE.
HURST HJ   1194   ELECTRONIC ABSORPTION SPECTRUM OF GASEOUS URANIUM HEXACHLORIDE AND URANIUM HEXAFLU
HWANG YT   0335   CONSTANT VOLUME HEAT CAPACITIES OF GASEOUS CARBON TETRAFLUORIDE AND OTHER MOLECULE
                  0336   CONSTANT VOLUME HEAT CAPACITY OF GASEOUS CARBON TETRAFLUORIDE.
HYAMA H   0246   UREY-BRADLEY FORCE CONSTANTS OF BORON HALIDES AND THEIR TRANSFERABILITY. (BORON TR
HYDE RG   1029   NONEMPIRICAL VALENCE ELECTRON CALCULATIONS ON SMALL MOLECULES CONTAINING PHOSPHORU
HYMAN HH   0740   RAMAN SPECTRUM OF GASEOUS NITROGEN TRIFLUORIDE.
IBBOTT DG   1042   FLUORINE NMR SPECTRA OF SULFUR, PHOSPHORUS, AND SILICON FLUORIDES. HYDROLYSIS AND
IBERS JA   1269   ENTHALPIES OF FORMATION OF XENON HEXAFLUORIDE, XENON TETRAFLUORIDE, XENON TETRAFLU
IIISHI K   0606   FORCE FIELDS OF THE RUTILE COUNTERPARTS: TITANIUM DIOXIDE, MAGNESIUM FLUORIDE, ZIN
IONOV SP   0218   ELECTRON DENSITY DISTRIBUTION IN SOME CHLORIDES.
                  0229   ELECTRON STRUCTURE AND NMR SPECTRA OF BORON COMPOUNDS.
                  0337   ELECTRON DENSITY DISTRIBUTIONS IN SOME FLUORIDES AND OXYCHLORIDES. (CARBON TETRAFL
IONOVA GV   0218   ELECTRON DENSITY DISTRIBUTION IN SOME CHLORIDES.
                  0337   ELECTRON DENSITY DISTRIBUTIONS IN SOME FLUORIDES AND OXYCHLORIDES. (CARBON TETRAFL
IPPOLITOV YEG   0945   STRUCTURE AND SOME PROPERTIES OF SCANDIUM TRIFLUORIDE.
                  0951   ENERGETICS AND STABILITY OF GAS PHASE SCANDIUM FLUORIDES.
ISAACSON L   0295   OSCILLATOR STRENGTHS OF CARBON MONOFLUORIDE AND COMMENTS ON HEATS OF FORMATION OF
ISHII M   0243   FORCE CONSTANTS OF CARBON TETRAFLUORIDE, SILICON TETRAFLUORIDE, BORON TRIFLUORIDE,
ISRAELI YJ   1270   ULTRAVIOLET SPECTRUM OF XENON TETRAFLUORIDE.

215

IVANOV VS	0388	ABSORPTION SPECTRA OF CHLORINE FLUORIDE, CHLORINE TRIFLUORIDE, AND CHLORINE PENTAF
IWASAKI M	1481	POTENTIAL CONSTANTS OF PHOSPHORUS TRICHLORIDE FROM AMPLITUDES OF VIBRATION AND NOR
IWATA S	1083	APPLICATION OF SYNCHROTON ORBITAL RADIATION TO MOLECULAR SPECTROSCOPY. MOLECULAR N
JACHE AW	0768	MILLIMETER WAVE SPECTRUM AND STRUCTURE OF OXYGEN DIFLUORIDE.
JACK KH	0616	CRYSTAL STRUCTURE OF MANGANESE TRIFLUORIDE.
	0617	INTERATOMIC BONDING IN CRYSTALLINE MANGANESE TRIFLUORIDE.
JACKEL G	0792	ELECTRON SPIN RESONANCE OF TRAPPED PHOSPHORUS DIFLUORIDE AND PHOSPHORUS TETRAFLUOR
JACKSON D	1449	STATISTICAL THERMODYNAMIC PROPERTIES OF HEXAFLUORIDE MOLECULES.
JACKSON R	0779	MICROWAVE SPECTRUM, STRUCTURE, AND DIPOLE MOMENT OF OXYGEN DIFLUORIDE.
JACOB EJ	0909	ELECTRON DIFFRACTION STUDY OF RHENIUM HEXAFLUORIDE.
	0920	ELECTRON DIFFRACTION STUDY OF RHENIUM HEPTAFLUORIDE. STRUCTURE, PSEUDOROTATION, AN
JACOX ME	0653	SPECTRA OF MATRIX ISOLATED NICKEL DIFLUORIDE AND NICKEL DICHLORIDE.
	0691	MATRIX ISOLATION STUDY OF THE VACUUM ULTRAVIOLET PHOTOLYSIS OF NITROGEN TRIFLUORID
	0986	MATRIX ISOLATION STUDY OF THE VACUUM ULTRAVIOLET PHOTOLYSIS OF SILICON DIFLUORIDE
JAKES J	0219	FORCE CONSTANTS AND THERMODYNAMIC CONSTANTS FOR THE BORON HALIDES.
JANZEN AF	1042	FLUORINE NMR SPECTRA OF SULFUR, PHOSPHORUS, AND SILICON FLUORIDES. HYDROLYSIS AND
JARDIM MDA	1404	CORRELATION OF INTERMOLECULAR EFFECTS IN THE VIBRATION SPECTRA OF BORON FLUORIDE W
JARRY RL	0710	LIQUID DENSITY, VAPOR PRESSURE AND CRITICAL TEMPERATURE AND PRESSURE OF NITROGEN T
JAYARAMAN L	0238	UNIQUE FORCE FIELD FOR BORON TRIHALIDES.
	1465	MOLECULAR FORCE FIELD FOR SOME XY6 TYPE MOLECULES. (SELENIUM AND TELLURIUM HEXAFLU
JAYARAMAN S	0958	VIBRATIONAL ASSIGNMENT AND NORMAL COORDINATE ANALYSIS OF SELENIUM TETRAFLUORIDE.
JAYE FC	0394	THERMAL DISSOCIATION OF CHLORINE TRIFLUORIDE BEHIND INCIDENT SHOCK WAVES.
JEANNOTTE AC	0338	INFRARED SPECTRA OF SIMPLE MOLECULES IN LIQUID ARGON SOLUTION. CARBON TETRAFLUORID
JIGUCHI J	0824	ISOTROPIC HYPERFINE SPLITTINGS AND MOLECULAR STRUCTURE IN THE PHOSPHORUS TETRAFLUO
JOACHIM PJ	1362	STUDY OF THE BONDING IN THE GROUP-4 TETRAHALIDES BY PHOTOELECTRON SPECTROSCOPY.
JOHANNIN-GILLES A	0042	RECENT DEVELOPMENTS IN MULTIPLE BEAM INTERFEROMETRY IN THE ULTRAVIOLET REGION 1700
JOHANSEN D	0208	NU-4 BANDS OF THE TEN AND ELEVEN ISOTOPES OF BORON TRIFLUORIDE AT HIGH RESOLUTION.
JOHANSEN H	0380	SCF CALCULATIONS ON CHLORATE ION, HYDROGEN CHLORIDE AND CHLORINE MONOFLUORIDE.
JOHNS JWC	0022	SPECTROSCOPIC AND THERMODYNAMIC PROPERTIES OF ALUMINUM MONOFLUORIDE.
	0964	SPECTRA OF SILICON MONOFLUORIDE, GERMANIUM MONOFLUORIDE, TIN MONOFLUORIDE, AND LEA
	0965	BAND SPECTRUM OF SILICON MONOFLUORIDE.
	0983	ULTRAVIOLET SPECTRUM OF SILICON DIFLUORIDE.
JOHNSON CM	0711	MICROWAVE SPECTROSCOPY IN THE REGION FROM TWO TO THREE MILLIMETERS, PART-2: (NITRO
JOHNSON DR	0165	MICROWAVE SPECTRUM OF BORON FLUORIDE.
	1030	MICROWAVE SPECTRUM AND STRUCTURE OF SULFUR DIFLUORIDE.
	1031	CENTRIFUGAL DISTORTION EFFECTS IN SULFUR DIFLUORIDE. CALCULATION OF THE FORCE FIEL
JOHNSON FA	0692	INFRARED SPECTRUM OF NITROGEN DIFLUORIDE.
JOHNSON GK	0220	FLUORINE BOMB CALORIMETRY. PART-15: THE ENTHALPY OF FORMATION OF BORON TRIFLUORIDE
	0807	ENTHALPIES OF FORMATION OF XENON HEXAFLUORIDE, XENON TETRAFLUORIDE, XENON DIFLUORI
	1215	ENTHALPY OF FORMATION OF VANADIUM PENTAFLUORIDE BY FLUORINE BOMB CALORIMETRY.
JOHNSON J	0469	ENTHALPY OF FORMATION OF GERMANIUM TETRAFLUORIDE.
JOHNSON RC	0101	CENTRIFUGAL STRETCHING OF TRIFLUORO CHLORO METHANE AND SOME OTHER SYMMETRIC TOPS.
JOHNSTON WV	1271	HEAT CAPACITY AND RELATED THERMODYNAMIC FUNCTIONS OF XENON TETRAFLUORIDE.
JOLLY WL	0081	STRUCTURE OF LIQUID ANTIMONY PENTAFLUORIDE.
	0082	DENSITY, MELTING POINT AND VAPOR PRESSURE OF ANTIMONY PENTAFLUORIDE.
JONES DTL	0403	FORCE CONSTANT CALCULATION FOR CHLORINE TRIFLUORIDE.
JONES EA	0267	RAMAN SPECTRUM, STRUCTURE, FORCE CONSTANTS, AND THERMODYNAMIC PROPERTIES OF BROMIN
	0381	INFRARED AND RAMAN SPECTRA OF CHLORINE MONOFLUORIDE.
	0386	ANALYSIS OF THE INFRARED SPECTRUM OF CHLORINE MONOFLUORIDE.
	1005	INFRARED AND RAMAN SPECTRA OF SILICON TETRAFLUORIDE.
	1086	INFRARED SPECTRUM OF SULFUR HEXAFLUORIDE.
JONES GR	0003	GAS PHASE STRUCTURES AND MASS SPECTRA OF BINARY PENTAFLUORIDES.
	0019	APPLICATION OF A MOLECULAR BEAM SOURCE MASS SPECTROMETER TO THE STUDY OF REACTIVE
	0549	MOLECULAR ORBITAL ENERGY LEVELS AND BONDING IN KRYPTON DIFLUORIDE.
	0550	ELECTRONIC STRUCTURE OF KRYPTON DIFLUORIDE STUDIED BY PHOTOELECTRON SPECTROSCOPY.
	1228	HELIUM-1 AND HELIUM-2 PHOTOELECTRON SPECTRA AND THE ELECTRONIC STRUCTURES OF XENON
	1283	XENON HEXAFLUORIDE. THE STRUCTURE OF A CUBIC PHASE AT -30C.
	1284	XENON HEXAFLUORIDE. STRUCTURAL CRYSTALLOGRAPHY OF TETRAMERIC PHASES.
	1296	POLYMORPHISM IN XENON HEXAFLUORIDE.
	1297	CRYSTALLINE MODIFICATIONS OF XENON HEXAFLUORIDE.
	1365	GAS PHASE STRUCTURES AND MASS SPECTRA OF BINARY PENTAFLUORIDES.
	1376	MOLECULAR BEAM MASS SPECTROMETRIC STUDY OF BINARY PENTAFLUORIDES. (ALSO ANTIMONY
	1377	APPLICATION OF A MOLECULAR BEAM SOURCE MASS SPECTROMETER TO THE STUDY OF REACTIVE
JONES PJ	0660	VIBRATIONAL SPECTRA OF NIOBIUM AND TANTALUM PENTAFLUORIDES IN THE GAS PHASE. VAPOR
JONES SR	0261	MICROWAVE MEASUREMENTS OF THE J-8 FAR-9 TRANSITION OF BROMINE PENTAFLUORIDE.
JONES WD	0083	CONSTRUCTION OF A FAR INFRARED SPECTROMETER AND ITS USE IN THE STUDY OF THE LOW FR
JONES WE	0063	ROTATIONAL ANALYSIS OF THE ULTRAVIOLET BANDS OF THE ANTIMONY MONOFLUORIDE MOLECULE
	0070	NEAR ULTRAVIOLET AND VISIBLE BANDS OF ANTIMONY MONOFLUORIDE.
	0071	BO(+)-X(21) AND BO(+)-X(10+) BANDS OF ANTIMONY MONOFLUORIDE.
	0091	SPECTRUM OF ARSENIC MONOFLUORIDE IN THE NEAR VACUUM ULTRAVIOLET AND NEAR INFRARED
	0092	ROTATIONAL ANALYSIS OF FOUR SINGLET SYSTEMS OF ARSENIC MONOFLUORIDE.
	0095	VIBRATIONAL ANALYSIS OF THE SINGLET SYSTEMS OF ARSENIC MONOFLUORIDE.
	0675	B SINGLET SIGMA(+) TO X TRIPLET SIGMA(-) BAND SYSTEM OF NITROGEN MONOFLUORIDE.
JONES WJ	0205	ANALYSIS OF SYMMETRIC ROTOR RAMAN SPECTRA. PURE ROTATIONAL SPECTRUM OF BORON TRIFL
JORDAN MW	0245	INFRARED ABSORPTION SPECTRUM OF LIQUID BORON TRIFLUORIDE.
JORTNER J	1237	HEATS OF SUBLIMATION OF XENON DIFLUORIDE AND XENON TETRAFLUORIDE AND A CONJECTURE
	1238	THEORETICAL AND EXPERIMENTAL STUDIES OF THE ELECTRONIC STRUCTURE OF THE XENON FLUO
	1247	FORBIDDEN ELECTRONIC TRANSITIONS IN XENON DIFLUORIDE AND XENON TETRAFLUORIDE.
	1256	FAR ULTRAVIOLET SPECTROSCOPIC STUDY OF XENON DIFLUORIDE.
	1272	FAR ULTRAVIOLET SPECTROSCOPIC STUDY OF XENON TETRAFLUORIDE.
	1340	CHEMICAL PREDICTIONS BY MO THEORY. THE NOBLE GAS HALIDES.
JOSEPH KB	1278	FORCE FIELD, CORIOLIS COUPLING COEFFICIENTS, GENERALIZED MEAN SQUARE AMPLITUDES OF
	1405	GENERALIZED MEAN SQUARE AMPLITUDES AND CORIOLIS CONSTANTS OF SOME PYRAMIDAL XY3 MO
JOSHI KC	0143	ABSORPTION SPECTRUM OF THE BISMUTH MONOFLUORIDE MOLECULE IN THE ULTRAVIOLET REGION
JOSHI MM	1013	NEAR ULTRAVIOLET ABSORPTION SPECTRUM OF SILVER MONOFLUORIDE.
JOY HW	0344	CALCULATED IONIZATION POTENTIAL OF CHLORO- AND FLUOROMETHANES, TETRAFLUORO METHANE
JUREK R	0268	ROTATIONAL SPECTRA OF BROMINE PENTAFLUORIDE IN THE MILLIMETER WAVE LENGTH REGION.
	0423	MICROWAVE SPECTROSCOPY OF CHLORINE PENTAFLUORIDE AT 70 AND 140 GHZ
KAISER EW	0367	ELECTRIC DEFLECTION AND THERMAL DECOMPOSITION STUDIES ON CERIUM TETRAFLUORIDE, TER
	0529	POLAR DISTORTIONS IN RHENIUM HEPTAFLUORIDE AND IODINE HEPTAFLUORIDE.

KAISER EW	0946	ELECTRIC DEFLECTION OF MOLECULAR BEAMS OF THE RARE EARTH DIFLUORIDES AND TRIFLUORI
	1084	ELECTRIC DEFLECTION OF BINARY HEXAFLUORIDES. (SULFUR HEXAFLUORIDE, SELENIUM HEXAF
KAMALASANAN MN	0491	ROTATIONAL ANALYSIS OF THE C SINGLET PI- X SINGLET SIGMA(+) SYSTEM OF INDIUM MONOF
KAMPMANN FW	0212	FORCE CONSTANTS OF CARBON TETRAFLUORIDE AND BORON TRIFLUORIDE AND COMPARISON WITH
KANA'AN AD	0604	INFRARED SPECTRA OF MATRIX ISOLATED MAGNESIUM DIFLUORIDE.
KANA'AN AS	0272	KNUDSEN AND LANGMUIR MEASUREMENTS OF THE SUBLIMATION PRESSURE OF CADMIUM DIFLUORID
	0439	KNUDSEN AND LANGMUIR MEASUREMENTS OF THE SUBLIMATION PRESSURE OF COBALT DIFLUORIDE
	0613	KNUDSEN MEASUREMENT OF THE VAPOR PRESSURE OF MANGANESE DIFLUORIDE.
	0947	SUBLIMATION PRESSURES OF SCANDIUM TRIFLUORIDE, YTTRIUM TRIFLUORIDE, AND LANTHANUM
	1382	SUBLIMATION PRESSURES OF REFRACTORY FLUORIDES.
KARRENBROCK AH	0527	STRUCTURE DETERMINATION OF TRIS (TRIMETHYL SILYL) SILANE, INFRARED INVESTIGATION O
KASTLER A	0582	ELECTRONIC SPECTRUM OF LANTHANUM TRIFLUORIDE.
	0832	LASER SPECTROSCCPY. (PHOSPHORUS PENTAFLUORIDE)
KATADA K	1282	ELECTRON DIFFRACTION STUDY OF THE STRUCTURE OF GASEOUS XENON TETRAFLUORIDE.
KATTI PH	0966	FRANCK-CONDON FACTORS AND R CENTROIDS OF A SINGLET SIGMA- X SINGLET SIGMA PLATINUM
KAUFMAN JJ	0339	IMPLICATIONS OF PHOTOELECTRON SPECTROSCOPIC MEASUREMENTS FOR COMPOUNDS WHICH PRODU
	1298	BONDING IN XENON HEXAFLUORIDE.
KAUSCHKA G	1241	INFRARED SPECTRA OF XENON DIFLUORIDE IN DIFFERENT SOLVENTS.
KAY JF	1143	ABSORPTION SPECTRUM OF VAPORIZED TITANIUM MONOFLUORIDE.
KEBABCIOGLU R	1399	CALCULATIONS OF MEAN AMPLITUDES OF VIBRATION AND UREY-BRADLEY FORCE FIELDS FOR PLA
	1482	FORMULAS FOR CORIOLIS CONSTANTS.
KEENAN TK	0445	ABSORPTION SPECTRA OF SOLID AMERICIUM TRIFLUORIDE, AMERICIUM TETRAFLUORIDE, CURIUM
	1334	LATTICE CONSTANTS OF ACTINIDE TETRAFLUORIDES INCLUDING BERKELIUM.
KEMMITT RDW	0007	FLUORIDES OF THE MAIN GROUP ELEMENTS.
KENESHEA FJ	1121	THERMODYNAMICS OF VAPORIZATION OF THALLIUM MONOFLUORIDE.
	1129	VAPOR PRESSURE OF THORIUM TETRAFLUORIDE.
KENNEDY A	0712	STRENGTH OF THE NITROGEN FLUORINE BOND IN NITROGEN TRIFLUORIDE AND DINITROGEN TETR
KENNEDY RC	0634	VIBRATIONAL SPECTRUM AND FORCE FIELD OF MOLYBDENUM HEXAFLUORIDE.
KENNEY JK	0207	NU-3 BANDS OF BORON TRIFLUORIDE.
	0209	BORON TRIFLUORIDE. THE ANALYSIS OF NU-2 AND THE DETERMINATION OF THE MOLECULAR GEO
KENT RA	0430	SUBLIMATION PRESSURE OF CHROMIUM DIFLUORIDE AND THE DISSOCIATION ENERGY OF CHROMIU
	0441	MASS SPECTROMETRIC STUDIES AT HIGH TEMPERATURES. PART-9: SUBLIMATION PRESSURE OF C
	0442	MASS SPECTROMETRIC STUDIES AT HIGH TEMPERATURES. PART-9: SUBLIMATION PRESSURE OF C
	0541	SUBLIMATION PRESSURE OF IRON DIFLUORIDE.
	0614	SUBLIMATION PRESSURE OF MANGANESE DIFLUORIDE AND MANGANESE MONOFLUORIDE.
	0884	MASS SPECTROMETRIC STUDIES OF PLUTONIUM COMPOUNDS AT HIGH TEMPERATURES. PART-2: EN
	0947	SUBLIMATION PRESSURES OF SCANDIUM TRIFLUORIDE, YTTRIUM TRIFLUORIDE, AND LANTHANUM
	1382	SUBLIMATION PRESSURES OF REFRACTORY FLUORIDES.
KERMAN E	0339	IMPLICATIONS OF PHOTOELECTRON SPECTROSCOPIC MEASUREMENTS FOR COMPOUNDS WHICH PRODU
KERN S	0454	MAGNETIC SUSCEPTIBILITY OF EUROPIUM TRIFLUORIDE.
KHANAEV EI	1171	ENTHALPY OF FORMATION OF URANIUM TETRAFLUORIDE.
KHANDOZHKO SV	0074	CALCULATION OF THE FREQUENCIES OF THE NORMAL VIBRATIONS OF ANTIMONY TRIFLUORIDE.
KHANNA BN	0168	VIBRATIONAL TRANSITION PROBABILITIES FOR THE A-X SYSTEM OF BORON MONOFLUORIDE.
	0677	FRANCK-CONDON FACTORS AND R CENTROIDS FOR THE 6 SINGLET SIGMA- X TRIPLET SIGMA BAN
KHANNA LK	0280	ANALYSIS OF THE BANDS OF B DOUBLET SIGMA- X DOUBLET SIGMA. SYSTEM OF CALCIUM FLUOR
KHANNA RK	0505	MOLECULAR STRUCTURE, FORCE CONSTANTS, AND THERMODYNAMIC FUNCTIONS OF IODINE PENTAF
	0528	FORCE CONSTANTS AND THERMODYNAMIC FUNCTIONS OF IODINE HEPTAFLUORIDE.
KHANNA VM	0474	ULTRAVIOLET ABSORPTION SPECTRUM OF GERMANIUM DIFLUORIDE.
	0984	ULTRAVIOLET ABSORPTION SPECTRUM OF SILICON DIFLUORIDE.
	0985	INFRARED SPECTRUM, FORCE CONSTANTS, AND THERMODYNAMIC FUNCTIONS OF SILICON DIFLUOR
KHODEEV YS	1165	DETERMINATION CF HEAT OF SUBLIMATION OF URANIUM TETRAFLUORIDE BY THE MASS SPECTROM
KHOKHLOV VA	0654	THERMODYNAMIC PROPERTIES OF NIOBIUM AND TANTALUM FLUORIDES AT HIGH TEMPERATURES. P
	0655	THERMODYNAMIC PROPERTIES OF NIOBIUM MONOFLUORIDE, NIOBIUM DIFLUORIDE, TANTALUM MON
	0656	THERMODYNAMIC PROPERTIES OF NIOBIUM TRIFLUORIDE, NIOBIUM TETRAFLUORIDE, TANTALUM T
	0657	THERMODYNAMIC PROPERTIES OF NIOBIUM AND TANTALUM FLUORIDES AT HIGH TEMPERATURES. P
	0665	THERMODYNAMIC PROPERTIES OF NIOBIUM AND TANTALUM FLUORIDES AT HIGH TEMPERATURES. P
KHOVRIN GV	0726	HARMONIC FREQUENCIES, POTENTIAL FUNCTION, CORIOLIS COUPLING CONSTANTS, CENTRIFUGAL
	0727	DETERMINATION OF THE STRUCTURE OF PYRAMIDAL MOLECULES OF THE XY3 TYPE BY SOLUTION
KIM H	1285	SPECTRAL OBSERVATIONS ON MOLECULAR XENON HEXAFLUORIDE. RAMAN SCATTERING AND INFRAR
	1299	FAR INFRARED AND MILLIMETER WAVE STUDIES OF XENON HEXAFLUORIDE.
KIM JJ	0830	POTENTIAL FUNCTION FOR THE NU-7 VIBRATION OF PHOSPHORUS PENTAFLUORIDE.
KIMURA K	0008	MEAN SQUARE AMPLITUDES OF INTERATOMIC DISTANCES IN HEXAFLUORIDE MOLECULES.
	1450	MEAN SQUARE AMPLITUDES OF INTERATOMIC DISTANCES IN HEXAFLUORIDE MOLECULES.
KIMURA M	0008	MEAN SQUARE AMPLITUDES OF INTERATOMIC DISTANCES IN HEXAFLUORIDE MOLECULES.
	0104	DETERMINATION OF THE MOLECULAR STRUCTURES OF ARSENIC TRIFLUORIDE AND ARSENIC TRICH
	1155	ELECTRON DIFFRACTION INVESTIGATION OF TUNGSTEN HEXAFLUORIDE, OSMIUM HEXAFLUORIDE,
	1450	MEAN SQUARE AMPLITUDES OF INTERATOMIC DISTANCES IN HEXAFLUORIDE MOLECULES.
KINAKA S	0223	BORON FLUORINE BOND DISTANCE OF BORON TRIFLUORIDE DETERMINED BY GAS ELECTRON DIFFR
KINDLE C	1229	ELECTRON DISTRIBUTION IN THE XENON FLUORIDES AND XENON OXIDE TETRAFLUORIDE BY ESCA
KING EG	1172	VIBRATIONAL SPECTRA AND FORCE CONSTANTS OF URANIUM TETRAFLUORIDE.
	1173	HIGH TEMPERATURE HEAT CONTENT OF URANIUM TETRAFLUORIDE.
KING RC	0401	HEAT OF FORMATION OF CHLORINE TRIFLUORIDE AT 298.14 DEGREES K.
	0769	HEAT OF FORMATION OF OXYGEN DIFLUORIDE.
KING WT	1483	QUADRATIC POTENTIAL FUNCTION FOR AMMONIA. (UREY-BRADLEY)
KIRBY-SMITH JS	1005	INFRARED AND RAMAN SPECTRA OF SILICON TETRAFLUORIDE.
KIRCHHOFF WH	1031	CENTRIFUGAL DISTORTION EFFECTS IN SULFUR DIFLUORIDE. CALCULATION OF THE FORCE FIEL
KISELJOV AA	1451	JAHN-TELLER EFFECT IN HEXAFLUORIDES.
KISER RW	0312	ELECTRON IMPACT STUDIES OF SOME TRIHALO METHANES. (CARBON TRIFLUORIDE)
KISKER DW	0605	THEORETICAL STUDIES OF THE INTERACTION OF MAGNESIUM DIFLUORIDE WITH RARE GAS ATOMS
KISLIUK P	0102	MICROWAVE SPECTRUM OF ARSENIC TRIFLUORIDE.
KLAMECKI B	0469	ENTHALPY CF FORMATION OF GERMANIUM TETRAFLUORIDE.
KLANM H	1300	STRUCTURE OF THE IODINE HEXAFLUORIDE ANION. (XENON HEXAFLUORIDE)
KLEIER DA	0158	LOCALIZED ORBITALS IN BORON FLUORIDES. HIGHLY POLARIZED BORON- FLUORINE DOUBLE AND
KLEMPERER W	0003	GAS PHASE STRUCTURES AND MASS SPECTRA OF BINARY PENTAFLUORIDES.
	0126	GEOMETRY OF THE ALKALINE EARTH DIHALIDES. (BARIUM DIFLUORIDE, STRONTIUM DIFLUORIDE
	0133	INFRARED SPECTRA OF THE ALKALINE EARTH HALIDES. PART-1: BERYLLIUM DIFLUORIDE, BERY
	0134	DETERMINATION OF THE GEOMETRY OF HIGH TEMPERATURE SPECIES BY ELECTRIC DEFLECTION A
	0349	MOLECULAR BEAM ELECTRIC DEFLECTION OF THE TETRAHALIDES CARBON TETRAFLUORIDE, CARBO
	0526	MOLECULAR STRUCTURE OF XENON HEXAFLUORIDE AND IODINE HEPTAFLUORIDE.
	0529	POLAR DISTORTIONS IN RHENIUM HEPTAFLUORIDE AND IODINE HEPTAFLUORIDE.

217

KLEMPERER W	0946	ELECTRIC DEFLECTION OF MOLECULAR BEAMS OF THE RARE EARTH DIFLUORIDES AND TRIFLUORI
	1084	ELECTRIC DEFLECTION OF BINARY HEXAFLUORIDES. (SULFUR HEXAFLUORIDE, SELENIUM HEXAF
	1213	POLAR STATES IN VANADIUM PENTAFLUORIDE.
	1287	MAGNETIC FIELD DEFLECTION OF XENON HEXAFLUORIDE.
	1368	GEOMETRY OF THE TRANSITION METAL DIHALIDES. THE DIFLUORIDES OF SOME FIRST ROW TRAN
KLIMENKC NM	0009	CHARACTERISTICS OF THE ELECTRONIC STRUCTURE OF FLUORIDES OF NONTRANSITION ELEMENTS
	0016	CALCULATIONS OF ELECTRON STRUCTURES OF COMPOUNDS OF HALOGENS AND INERT GASES BY TH
	0382	GEOMETRY OF MOLECULES. PART-2: GEOMETRICAL AND ELECTRICAL STRUCTURE OF THE MOLECUL
	0415	CALCULATION OF THE ELECTRONIC STRUCTURE OF A CHLORINE PENTAFLUORIDE MOLECULE BY TH
	1231	CALCULATIONS OF ELECTRONIC STRUCTURES OF COMPOUNDS OF HALOGENS AND INERT GASES BY
KLYAGINA AP	0382	GEOMETRY OF MOLECULES. PART-2: GEOMETRICAL AND ELECTRICAL STRUCTURE OF THE MOLECUL
KLYUEV LI	1149	THERMODYNAMIC PROPERTIES OF TUNGSTEN PENTAFLUORIDE.
KNIGHT V	1003	INFRARED SPECTRUM OF SILICON TETRAFLUORIDE.
KNOP CP	1255	THE XENON- FLUORINE SYSTEM.
	1312	XENON HEXAFLUORIDE. (SPECTRUM)
KNOX K	0433	STRUCTURE OF CRYSTALLINE CHROMIUM TRIFLUORIDE.
KO HC	0286	THERMODYNAMIC PROPERTIES OF CARBON MONOFLUORIDE AND CARBON DIFLUORIDE FROM MOLECUL
KOJIMA H	0273	MELTING POINTS OF INORGANIC FLUORIDES. (CADMIUM DIFLUORIDE, MAGNESIUM DIFLUORIDE,
KOLESOV VP	0043	STANDARD ENTHALPY OF FORMATION OF ALUMINUM TRIFLUORIDE.
KOLOMIITSEVA TD	1068	INFRARED SPECTRA OF CRYOSYSTEMS. PART-2: SULFUR HEXAFLUORIDE.
KOMISSAROVA LN	0940	THERMODYNAMIC CHARACTERISTICS OF SCANDIUM DIHALIDES AND THE THE POSSIBILITY OF THE
KONAKA S	0103	FORCE CONSTANTS AND AVERAGE STRUCTURES OF ARSENIC TRIFLUORIDE AND ARSENIC TRICHLOR
	0104	DETERMINATION OF THE MOLECULAR STRUCTURES OF ARSENIC TRIFLUORIDE AND ARSENIC TRICH
	0221	MOLECULAR STRUCTURE AND FORCE CONSTANTS OF BORON TRIFLUORIDE AND BORON TRICHLORIDE
KONSTANTINOV YS	0416	FLUORINE NMR OF CHLORINE PENTAFLUORIDE.
KONYAEVA GP	1204	THERMODYNAMIC PROPERTIES OF URANIUM HEXAFLUORIDE.
KOPP I	0023	SINGLET ELECTRONIC STATES OF ALUMINUM MONOFLUORIDE.
	0027	ELECTRONIC STATES OF GASEOUS ALUMINUM MONOFLUORIDE.
	1133	ROTATIONAL ANALYSIS OF THE A DOUBLET SIGMA(+) TO X DOUBLET PI, B DOUBLET SIGMA(+)
KOROBTSEV VP	0654	THERMODYNAMIC PROPERTIES OF NIOBIUM AND TANTALUM FLUORIDES AT HIGH TEMPERATURES. P
	0655	THERMODYNAMIC PROPERTIES OF NIOBIUM MONOFLUORIDE, NIOBIUM DIFLUORIDE, TANTALUM MON
	0656	THERMODYNAMIC PROPERTIES OF NIOBIUM TRIFLUORIDE, NIOBIUM TETRAFLUORIDE, TANTALUM T
	0657	THERMODYNAMIC PROPERTIES OF NIOBIUM AND TANTALUM FLUORIDES AT HIGH TEMPERATURES. P
	0665	THERMODYNAMIC PROPERTIES OF NIOBIUM AND TANTALUM FLUORIDES AT HIGH TEMPERATURES. P
KORWAR VM	0966	FRANCK-CONDON FACTORS AND R CENTROIDS OF A SINGLET SIGMA- X SINGLET SIGMA PLATINUM
KOSKE P	1195	ALPHA SPECTROMETRIC INVESTIGATION OF URANIUM HEXAFLUORIDE. PART-1.
KOSKI WS	0339	IMPLICATIONS OF PHOTOELECTRON SPECTROSCOPIC MEASUREMENTS FOR COMPOUNDS WHICH PRODU
KOSTYLEV FA	1175	VAPOR PRESSURE OF URANIUM TETRAFLUORIDE.
KOSYAKOV VN	0056	INVESTIGATION OF THE CHEMISTRY OF AMERICIUM. (AMERICIUM TETRAFLUORIDE)
KOTOV YI	0693	CALCULATION OF FORCE CONSTANTS FOR NITROGEN DIFLUORIDE AND NITROGEN TRIFLUORIDE.
KOUTECKY VB	1085	MOLECULAR ORBITAL DESCRIPTION OF SULFUR COMPOUNDS OF VALENCES 2, 4, AND 6.
KOVACS I	0787	ROTATIONAL FINE STRUCTURE OF THE TRIPLET PI STATE OF PHOSPHORUS MONOFLUORIDE.
KOVRIKOV AB	1414	ESTIMATION OF THE FORCE CONSTANTS AND INTERATOMIC DISTANCES OF HALIDE COMPLEX MOLE
	1484	USE OF A SIMPLE ELECTROSTATIC MODEL FOR CALCULATION OF ANGULAR FORCE CONSTANTS OF
KRASNOV KS	0044	BOND ENERGY IN THE ALUMINUM TRIFLUORIDE MOLECULE.
	0278	DISSOCIATION ENERGY OF MONOFLUORIDE COMPOUNDS OF MAGNESIUM AND CALCIUM AND THE ION
	0568	THERMODYNAMIC FUNCTIONS OF GASEOUS SCANDIUM, YTTRIUM, AND LANTHANUM HALIDES IN THE
	0573	FORCE CONSTANT OF THE NONPLANAR VIBRATION OF AN XY3 MOLECULE AND A MODEL WITH POLA
	0574	VALENCE VIBRATION FREQUENCIES OF TRIHALIDE MOLECULES OF LANTHANUM HALIDES.
	0935	DISSOCIATION ENERGIES AND STABILITIES OF SCANDIUM MONOFLUORIDE, YTTRIUM MONOFLUORI
	0936	EVALUATION OF MOLECULAR VIBRATION FREQUENCIES OF SCANDIUM MONOFLUORIDE, YTTRIUM MO
	0941	MOLECULAR CONSTANTS OF SCANDIUM, YTTRIUM, AND LANTHANUM HALIDES. (SCANDIUM DIFLUOR
	0948	CALCULATION OF DISSOCIATION ENERGIES FOR THE HALIDES OF THE SCANDIUM SUBGROUP. (SC
	0949	CALCULATION OF THE VIBRATIONAL FREQUENCIES OF SCANDIUM SUBGROUP HALIDES.
	1122	NATURE OF BONDS IN THALLIUM MONOHALIDE MOLECULES.
KRAUSS L	0390	O(+) TO SINGLET SIGMA(+) SYSTEMS OF BOTH ISOTOPES OF CHLORINE MONOFLUORIDE.
KREBS B	0235	CHEMICAL SHIFTS OF BORON, PHOSPHORUS, PHOSPHORYL, AND THIOPHOSPHORYL HALIDES.
	0811	RAMAN AND INFRARED SPECTRA AND THERMODYNAMIC FUNCTIONS FOR VARIOUS PHOSPHORUS AND
	1415	NORMAL COORDINATE TREATMENT OF XY4 TETRAHEDRAL MOLECULES AND IONS. (GVFF)
	1418	NORMAL COORDINATE TREATMENT OF XY4 TETRAHEDRAL MOLECULES AND IONS. (UREY-BRADLEY A
	1458	INFLUENCE OF ATOMIC MASS ON THE CORIOLIS CONSTANTS OF XY6 MOLECULES.
	1488	THEORETICAL INTERPRETATION OF THE MASS INFLUENCE ON CORIOLIS COUPLING CONSTANTS IN
KRISHNA MG	0402	MOLECULAR FORCE FIELD FOR CHLORINE TRIFLUORIDE.
KRISHNAMACHARI SLNG	0222	EMISSION SPECTRA OF CERTAIN POLYATOMIC FREE RADICALS OBTAINED IN ELECTRICAL DISCHA
	0559	ULTRAVIOLET EMISSION SPECTRA. (XENON DIFLUORIDE AND KRYPTON DIFLUORIDE)
KRISHNAMURTHY N	0274	RAMAN SPECTRA OF CADMIUM DIFLUORIDE AND LEAD DIFLUORIDE.
KRISHNAMURTY VG	0651	BAND SPECTRUM OF NICKEL MONOFLUORIDE.
KRISHNAN CN	0998	EMISSION SPECTRUM OF SILICON TRIFLUORIDE.
KRISTIANSEN LA	1388	INFLUENCE OF ATOMIC MASSES ON THE CORIOLIS COUPLING COEFFICIENTS.
	1392	VIBRATIONAL MEAN SQUARE AMPLITUDE MATRICES.
	1410	INFLUENCE OF ATOMIC MASSES ON THE CORIOLIS COUPLING COEFFICIENTS IN XY4 TETRAHEDRA
KRZHIZHANOVSKAYA EV	1357	SATURATED VAPOR PRESSURE OF RARE EARTH FLUORIDES.
KUCHITSU K	0221	MOLECULAR STRUCTURE AND FORCE CONSTANTS OF BORON TRIFLUORIDE AND BORON TRICHLORIDE
	0223	BORON FLUORINE BOND DISTANCE OF BORON TRIFLUORIDE DETERMINED BY GAS ELECTRON DIFFR
	0770	THIRD ORDER POTENTIAL CONSTANTS OF BENT XY2 MOLECULES. (OXYGEN DIFLUORIDE)
	0810	MOLECULAR STRUCTURE OF PHOSPHORUS TRIFLUORIDE STUDIED BY GAS ELECTRON DIFFRACTION.
	1383	ESTIMATION OF ANHARMONIC POTENTIAL CONSTANTS. PART-1. LINEAR XY2 MOLECULES.
KUCZKOWSKI RL	1024	MICROWAVE AND MASS SPECTRA OF SULFUR MONOFLUORIDE.
KUDCHADKER AP	0424	IDEAL GAS THERMODYNAMIC PROPERTIES OF CHLORINE PENTAFLUORIDE, BROMINE PENTAFLUORID
	1273	IDEAL GAS THERMODYNAMIC PROPERTIES OF XENON COMPOUNDS. (XENON TETRAFLUORIDE AND XE
KUDCHADKER SA	0424	IDEAL GAS THERMODYNAMIC PROPERTIES OF CHLORINE PENTAFLUORIDE, BROMINE PENTAFLUORID
	1273	IDEAL GAS THERMODYNAMIC PROPERTIES OF XENON COMPOUNDS. (XENON TETRAFLUORIDE AND XE
KUDRIN LP	1161	CALCULATION OF THE DISSOCIATION ENERGY OF CALCIUM MONOFLUORIDE AND URANIUM MONOFLU
KUENDIG EP	0475	MATRIX INFRARED AND LASER RAMAN SPECTRA, MOLECULAR STRUCTURES AND NORMAL COORDINAT
KULOOR NS	0739	THERMODYNAMIC PROPERTIES OF NITROGEN TRIFLUORIDE.
KURDYUMOVA IN	0970	DETERMINATION OF THE MATRIX ELEMENT OF THE DIPOLE MOMENT OF AN A DOUBLET SIGMA TO
KUSHAWAHA VS	0120	THE C DOUBLET PI AND THE X DOUBLET SIGMA STATES OF BARIUM MONOFLUORIDE.
	0121	GREEN BAND SYSTEM OF BARIUM MONOFLUORIDE.
KUSTOVA GN	1352	INFRARED SPECTRA OF RARE EARTH METAL FLUORIDES IN THE REGION OF A CESIUM IODIDE PR
KUZMENKO NE	0967	ELECTRONIC TRANSITION MOMENT AS A FUNCTION OF THE INTERNUCLEAR SEPARATION FOR THE

KUZNETSOVA LA   0694   ELECTRONIC ABSORPTION SPECTRUM OF THE NITROGEN DIFLUORIDE RADICAL.
KUZNETSOVA TV   0713   PHYSICAL CHEMICAL PROPERTIES OF INORGANIC NITROGEN FLUORIDES. (NITROGEN TRIFLUORID
KUZ'MENKO NE   0968   DISTRIBUTION OF INTENSITY IN THE ELECTRON EMISSION SPECTRUM OF SILICON MONOFLUORID
                       0969   CORRELATION OF THE ELECTRONIC TRANSITION MOMENT WITH INTERNUCLEAR DISTANCE FOR BAN
                       0970   DETERMINATION OF THE MATRIX ELEMENT OF THE DIPOLE MOMENT OF AN A DOUBLET SIGMA TO
                       0977   CALCULATION OF THE FRANCK-CONDON FACTOR FOR MOLECULES OF SILICON FLUORIDE, GERMANI
KUZ'YAKOV YY   0967   ELECTRONIC TRANSITION MOMENT AS A FUNCTION OF THE INTERNUCLEAR SEPARATION FOR THE
KUZYAKOV YY   0289   DETERMINATION OF THE DISSOCIATION ENERGIES OF CARBON MONOHALIDES.
                       0292   NEW BANDS OF THE CARBON MONOFLUORIDE MOLECULE.
                       0294   ROTATIONAL CONSTANTS OF THE B STATE OF THE CARBON MONOFLUORIDE MOLECULE .
                       0465   ANALYSIS OF THE ROTATIONAL STRUCTURE OF THE BANDS 0-0 AND 0-1 IN GERMANIUM MONOFLU
                       0470   VIBRATIONAL ANALYSIS OF C-X AND C'-X' BAND SYSTEMS AND ENERGY OF DISSOCIATION IN G
                       0694   ELECTRONIC ABSORPTION SPECTRUM OF THE NITROGEN DIFLUORIDE RADICAL.
                       0968   DISTRIBUTION OF INTENSITY IN THE ELECTRON EMISSION SPECTRUM OF SILICON MONOFLUORID
                       0969   CORRELATION OF THE ELECTRONIC TRANSITION MOMENT WITH INTERNUCLEAR DISTANCE FOR BAN
                       0970   DETERMINATION OF THE MATRIX ELEMENT OF THE DIPOLE MOMENT OF AN A DOUBLET SIGMA TO
                       1138   ANALYSIS OF VIBRATIONAL AND ROTATIONAL STRUCTURE OF TIN MONOFLUORIDE IN THE 2500 A
LABARRE JF   0226   SEMIEMPIRICAL STUDY OF THE ELECTRONIC STRUCTURE OF BORON HYDRIDE FLUORIDES. (BORON
                       0820   BONDING IN PHOSPHORUS TRIFLUORIDE AND PHOSPHORUS OXYFLUORIDE. AN AB INITIO SCF STU
LAGEMANN RT   1086   INFRARED SPECTRUM OF SULFUR HEXAFLUORIDE.
LAGUTKIN OD   1087   THERMODYNAMIC PROPERTIES OF SULFUR HEXAFLUORIDE.
LAHAYE B   0279   SPECTRA OF ELECTRONIC EXCITATIONS IN CALCIUM DIFLUORIDE, STRONTIUM DIFLUORIDE, AND
LAIRD RK   0302   ABSORPTION SPECTRUM OF CARBON DIFLUORIDE.
LAKE RF   0224   PHOTOELECTRON SPECTRA OF COMPLEXES OF BORON TRIFLUORIDE AND AMINES.
LAL KM   0677   FRANCK-CONDON FACTORS AND R CENTROIDS FOR THE 6 SINGLET SIGMA- X TRIPLET SIGMA BAN
LALOS GT   1201   SPECIFIC HEAT RATIO OF URANIUM HEXAFLUORIDE MEASURED WITH A BALLISTIC PISTON COMPR
LAMANNA U   1077   ELECTRONIC PROPERTIES OF SULFUR HEXAFLUORIDE. AB INITIO CALCULATION FOR THE GROUND
                       1078   ELECTRONIC PROPERTIES OF SULFUR HEXAFLUORIDE. PART-1: OPTICAL ABSORPTION AND X-RAY
LANDSBERG A   0438   VAPOR PRESSURE AND HEAT OF SUBLIMATION OF COBALT DIHALIDES. (COBALT DIFLUORIDE)
LANGER S   1174   VAPOR PRESSURE OF URANIUM TETRAFLUORIDE.
LASSETTRE EN   0251   RAMAN SPECTRUM OF BORON TRIFLUORIDE GAS.
                       0333   ELECTRON IMPACT SPECTRA OF METHANE AND CARBON TETRAFLUORIDE.
                       0364   RAMAN SPECTRA OF CARBON AND SILICON TETRAFLUORIDES.
LATHAN WA   0720   SELF CONSISTENT MOLECULAR ORBITAL METHODS. PART-5: AB INITIO CALCULATION OF EQUILI
LAVALLE DE   0620   PREPARATION AND CRYSTAL STRUCTURE OF MOLYBDENUM TRIFLUORIDE.
LAVEISSIERE J   1163   NEUTRON DIFFRACTION STUDY OF THE CRYSTAL STRUCTURE OF URANIUM TRIFLUORIDE.
LAVILLA RE   0530   MOLECULAR STRUCTURE OF IODINE HEPTAFLUORIDE.
                       1088   SULFUR K AND L AND FLUORINE K X-RAY EMISSION AND ABSORPTION SPECTRA OF GASEOUS SUL
LAWLESS EW   0084   MASS SPECTROMETRIC EVIDENCE OF DIMERS IN BISMUTH PENTAFLUORIDE AND ANTIMONY PENTAF
LE BOUCHER L   0049   VAPOR PRESSURE OF ZINC DIFLUORIDE, CADMIUM DIFLUORIDE, MAGNESIUM DIFLUORIDE, CALCI
LEBARGY RC   0930   ROTATIONAL ANALYSIS OF SOME SINGLET TRANSITIONS IN THE SPECTRUM OF GASEOUS SCANDIU
LEBRETON J   0163   CAMERON SYSTEM OF BORON MONOFLUORIDE.
LEFEBVRE-BRION H   0164   CALCULATION OF RYDBERG LEVELS IN NITROGEN OXIDE AND BORON FLUORIDE.
LEGLER D   0338   INFRARED SPECTRA OF SIMPLE MOLECULES IN LIQUID ARGON SOLUTION. CARBON TETRAFLUORID
LEHMANN E   0500   VIBRATIONAL SPECTRUM OF IODINE TRIFLUORIDE.
LEIBOVICI C   0225   A THEORETICAL COMPARISON OF THE ELECTRONIC STRUCTURES OF BORON TRIHALIDES AND BORO
                       0226   SEMIEMPIRICAL STUDY OF THE ELECTRONIC STRUCTURE OF BORON HYDRIDE FLUORIDES. (BORON
                       1032   STRUCTURE OF THE SULFUR TRIFLUORIDE POSITIVE ION. (SULFUR DIFLUORIDE)
LEMONS JF   0890   PREPARATION AND PROPERTIES OF (VAPOR PRESSURE) OF PLUTONIUM HEXAFLUORIDE AND IDENT
LEMPKA HJ   1361   PHOTOELECTRON SPECTRA OF THE HALIDES IN GROUP-3, GROUP-4, GROUP-5, AND GROUP-6.
LESIECKI ML   0607   INFRARED AND RAMAN SPECTRA AND STRUCTURES OF MATRIX ISOLATED MAGNESIUM DIHALIDES:
                       0895   LASER EXCITED RAMAN SPECTRA OF MATRIX ISOLATED PRASEODYMIUM TRIFLUORIDE.
                       1123   INFRARED AND RAMAN SPECTRA AND STRUCTURE OF MATRIX ISOLATED THALLOUS HALIDE DIMERS
LETCHER JH   0746   CHARACTERIZATION OF GROUND STATE WAVE FUNCTIONS BY MEASURED ELECTRONIC PROPERTIES.
LETOKHOV VS   1064   SELECTIVE PHOTODISSOCIATION OF POLYATOMIC MOLECULES (SULFUR HEXAFLUORIDE) BY INFRA
LEVI DL   0214   HEATS OF FORMATION OF INORGANIC FLUORIDES ESPECIALLY THE ELEMENTS OF ATOMIC NUMBER
LEVIN IW   0227   FORCE FIELDS FOR THE BORON TRIHALIDES.
                       0228   ISOTOPIC SPLITTING IN MATRIX ISOLATED BORON TRIFLUORIDE.
                       0340   ABSOLUTE INFRARED INTENSITIES OF CARBON TETRAFLUORIDE.
                       0714   INFRARED INTENSITIES AND FORCE FIELD FOR NITROGEN TRIFLUORIDE.
                       0715   FORCE FIELDS FOR GROUP-4 TETRAFLUORIDES AND GROUP-5 TRIFLUORIDES.
                       0736   ABSOLUTE INFRARED INTENSITIES OF THE FUNDAMENTAL VIBRATIONS OF NITROGEN TRIFLUORID
                       0808   MEASUREMENT AND CALCULATION OF THE ABSOLUTE INFRARED INTENSITIES OF PHOSPHORUS TRI
                       0830   POTENTIAL FUNCTION FOR THE NU-7 VIBRATION OF PHOSPHORUS PENTAFLUORIDE.
                       0858   RAMAN SPECTRUM OF SOLID PHOSPHORUS PENTAFLUORIDE
                       0859   POTENTIAL FUNCTION FOR PHOSPHORUS PENTAFLUORIDE.
                       0910   JAHN-TELLER VIBRATIONS OF RHENIUM HEXAFLUORIDE.
                       0960   VIBRATIONAL ANALYSIS OF SELENIUM HEXAFLUORIDE AND TUNGSTEN HEXAFLUORIDE.
                       1006   FORCE FIELD FOR SILICON TETRAFLUORIDE.
                       1045   INFRARED SPECTRUM AND NORMAL COORDINATE ANALYSIS OF SULFUR TETRAFLUORIDE.
                       1046   INTRAMOLECULAR EXCHANGE IN SULFUR TETRAFLUORIDE.
                       1047   CALCULATION OF THE INFRARED SPECTRA OF SULFUR TETRAFLUORIDE USING CNDO/2 TECHNIQUE
                       1061   FORCE FIELDS FOR SOME GROUP-6 HEXAFLUORIDES. (SULFUR HEXAFLUORIDE AND TELLURIUM HE
LEVY DH   1020   GAS PHASE ELECTRON RESONANCE SPECTRA OF SULFUR MONOFLUORIDE, SELENIUM MONOFLUORIDE
LEVY HA   1223   XENON DIFLUORIDE AND THE NATURE OF THE XENON- FLUORINE BOND.
                       1239   CRYSTAL AND MOLECULAR STRUCTURE OF XENON DIFLUORIDE BY NEUTRON DIFFRACTION.
                       1260   CRYSTAL AND MOLECULAR STRUCTURE OF XENON TETRAFLUORIDE BY NEUTRON DIFFRACTION.
LEWIS TP   0340   ABSOLUTE INFRARED INTENSITIES OF CARBON TETRAFLUORIDE.
LIBBY DR   0288   SHOCK TUBE DETERMINATION OF THE C2 (A TRIPLET PI TO X TRIPLET PI) AND CARBON MONOF
LIDE DR   0028   HIGH TEMPERATURE MICROWAVE SPECTROSCOPY OF ALUMINUM MONOFLUORIDE AND ALUMINUM CHLO
                       0029   MICROWAVE SPECTRUM OF ALUMINUM MONOFLUORIDE.
                       0307   MICROWAVE SPECTRUM OF CARBON DIFLUORIDE.
                       0687   INFRARED SPECTRUM AND STRUCTURE OF THE NITROGEN DIFLUORIDE RADICAL.
LIEBMAN JF   0756   OXYGEN MONOFLUORIDE. HARTREE-FOCK WAVE FUNCTION, BINDING ENERGY, IONIZATION POTENT
LIEHR AD   0799   INFRARED SPECTRA OF PHOSPHORUS TRIFLUORIDE, PHOSPHORUS OXIDE TRIFLUORIDE, AND PHOS
LIM M   0366   VAPOR PRESSURE AND HEAT OF SUBLIMATION OF CERIUM TRIFLUORIDE.
LINDBERG BJ   0328   MOLECULAR SPECTROSCOPY BY MEANS OF ESCA. (CARBON TETRAFLUORIDE)
LINEVSKY MJ   0431   SPECTROSCOPIC STUDIES OF THE VAPORIZATION OF HIGH TEMPERATURE MATERIALS. (CHROMIUM
                       0608   INFRARED SPECTRA OF MAGNESIUM AND ALUMINUM FLUORIDES BY MATRIX ISOLATION.
                       0609   GEOMETRY AND VIBRATIONAL SPECTRA OF THE ALKALINE EARTH DIHALIDES. PART-1: MAGNESIU

LINEVSKY MJ	1314	ELECTRONIC BAND SPECTRUM OF YTTRIUM MONOFLUORIDE.
LINNETT JW	0860	DIRECTED VALENCY IN CERTAIN MOLECULES AND COMPLEX IONS. (PHOSPHORUS PENTAFLUORIDE,
	1080	MOLECULAR FORCE FIELDS. PART-8: VIBRATION FREQUENCIES OF SOME OCTAHEDRAL XY6 MOLEC
	1226	ELECTRONIC STRUCTURE OF XENON DIFLUORIDE.
	1391	VIBRATIONAL FREQUENCIES OF SOME PLANAR XY3 MOLECULES.
	1452	MOLECULAR FORCE FIELDS FOR SOME HEXAFLUORIDES.
LIPPINCOTT ER	1489	SHRINKAGES OF THE INTERNUCLEAR DISTANCES BY MOLECULAR VIBRATIONS.
LIPSCOMB WN	0158	LOCALIZED ORBITALS IN BORON FLUORIDES. HIGHLY POLARIZED BORON- FLUORINE DOUBLE AND
	0159	MAGNETIC PROPERTIES OF THE BORON FLUORIDE AND BORON HYDRIDE MOLECULES.
	0459	MOLECULAR SCF CALCULATIONS FOR GALLIUM MONOFLUORIDE, GALLIUM TRIHYDRIDE, GERMANIUM
	0531	MOLECULAR SYMMETRY OF IODINE HEPTAFLUORIDE.
	1240	MOLECULAR SYMMETRY OF XENON DIFLUORIDE AND XENON TETRAFLUORIDE.
	1341	AN LCAO MO STUDY OF RARE GAS FLUORIDES.
LISKOW DH	1221	PROBABLE NONEXISTENCE OF XENON MONOFLUORIDE AS A CHEMICALLY BOUND SPECIES IN THE G
LISTE DG	0371	SCF- MO CALCULATIONS OF SOME MOLECULAR PROPERTIES OF THE ISOELECTRONIC SERIES CHLO
LITTLE R	0096	THE INFRARED SPECTRA OF SOME VOLATILE INORGANIC FLUORIDES IN THE SOLID STATE. (ARS
LIU B	0547	KRYPTON MONOFLUORIDE AND ITS POSITIVE ION. (SCF CALCULATIONS)
	0548	ELECTRONIC STRUCTURE AND PROPERTIES OF KRYPTON DIFLUORIDE.
	1221	PROBABLE NONEXISTENCE OF XENON MONOFLUORIDE AS A CHEMICALLY BOUND SPECIES IN THE G
LIU DS	0091	SPECTRUM OF ARSENIC MONOFLUORIDE IN THE NEAR VACUUM ULTRAVIOLET AND NEAR INFRARED
	0092	ROTATIONAL ANALYSIS OF FOUR SINGLET SYSTEMS OF ARSENIC MONOFLUORIDE.
	0095	VIBRATIONAL ANALYSIS OF THE SINGLET SYSTEMS OF ARSENIC MONOFLUORIDE.
LIU JW	0911	MULTIPLE INTRAMOLECULAR SCATTERING EFFECTS ON ELECTRON DIFFRACTION PATTERNS FOR TH
LIVINGSTON KMS	0829	VIBRATIONAL SPECTRA OF MIXED PHOSPHORUS HALIDES. (PHOSPHORUS PENTAFLUORIDE)
LIVINGSTON RL	0334	MOLECULAR STRUCTURE OF CARBON TETRAFLUORIDE.
	0383	PURE QUADRUPOLE SPECTRA OF SOLID CHLORINE COMPOUNDS.
LLOYD DR	0188	PHOTOELECTRON SPECTRA OF HALIDES. PART-2: HIGH RESOLUTION SPECTRA OF THE BORON TRI
	0189	PHOTOELECTRON SPECTRUM OF BORON TRIFLUORIDE.
	0316	PHOTOELECTRON SPECTRA OF HALIDES. PART-1. TETRAFLUORIDES AND TETRACHLORIDES OF GRO
	0317	HELIUM-1 RESONANCE PHOTOELECTRON SPECTRA OF GROUP-4 TETRAFLUORIDES.
	0341	PHOTOELECTRON SPECTRA OF HALIDES. PART-7: VARIABLE TEMPERATURE HELIUM-1 AND HELIUM
	0376	PHOTOELECTRON SPECTRA OF HALIDES. (CHLORINE MONOFLUORIDE, CHLORINE TRIFLUORIDE, BR
	0700	PHOTOELECTRON SPECTRA OF NITROGEN TRIFLUORIDE AND NITROGEN OXIDE TRIFLUORIDE, AND
	0701	PHOTOELECTRON SPECTRA OF HALIDES. PART-3: TRIFLUORIDES AND OXIDE TRIFLUORIDES OF N
LOCKETT P	0861	E´ FORCE FIELD OF PHOSPHORUS PENTAFLUORIDE.
LOEHR TM	0435	VIBRATIONAL SPECTRUM OF LIQUID CHROMIUM PENTAFLUORIDE.
LOEWENSCHUSS A	0275	VIBRATIONAL SPECTRA OF THE DIHALIDES OF MERCURY AND CADMIUM. (CADMIUM DIFLUORIDE A
	1323	VIBRATIONAL SPECTRA AND THERMODYNAMICS OF THE ZINC HALIDES. (ZINC DIFLUORIDE)
LOHR LL	0531	MOLECULAR SYMMETRY OF IODINE HEPTAFLUORIDE.
	1240	MOLECULAR SYMMETRY OF XENON DIFLUORIDE AND XENON TETRAFLUORIDE.
	1311	CRYSTAL FIELD MODEL STUDY OF XENON HEXAFLUORIDE. PART-1: ENERGY LEVELS AND MOLECUL
	1341	AN LCAO MO STUDY OF RARE GAS FLUORIDES.
LONG DA	0342	ROOT MEAN SQUARE AMPLITUDES OF VIBRATION IN SOME GROUP-4 TETRAHALIDES. (CARBON TE
	0403	FORCE CONSTANT CALCULATION FOR CHLORINE TRIFLUORIDE.
	1363	ROOT MEAN SQUARE AMPLITUDES OF VIBRATION IN SOME GROUP-4 TETRAHALIDES.
	1453	RAMAN INTENSITIES FOR SPHERICALLY SYMMETRIC MODES OF SOME XY4 TETRAHEDRAL AND XY6
LONGBOROUGH B	0118	ROTATIONAL ANALYSIS OF SOME BANDS OF GASEOUS BARIUM MONOFLUORIDE.
LOOPSTRA BO	0593	NEUTRON DIFFRACTION INVESTIGATION OF ORTHORHOMBIC LEAD DIFLUORIDE.
LORD RC	0100	INFRARED SPECTRUM AND VIBRATIONAL POTENTIAL FUNCTION OF ARSENIC TRIFLUORIDE.
	0506	VIBRATIONAL SPECTRA AND STRUCTURES OF IODINE PENTAFLUORIDE AND HEPTAFLUORIDE.
	0856	VIBRATIONAL SPECTRA OF PHOSPHORUS PENTAFLUORIDE AND ARSENIC PENTAFLUORIDE AND THE
LOVAS FJ	0025	MICROWAVE ABSORPTION SPECTRA OF ALUMINUM MONOFLUORIDE, GALLIUM MONOFLUORIDE, INDIU
	0165	MICROWAVE SPECTRUM OF BORON FLUORIDE.
	0490	MICROWAVE ROTATIONAL SPECTRUM OF INDIUM MONOFLUORIDE.
LU CS	0738	AN ELECTRON DIFFRACTION INVESTIGATION OF NITROGEN TRIFLUORIDE.
LUNDIN AG	0571	SYMMETRY AND CRYSTAL STRUCTURE OF RARE EARTH TRIFLUORIDES. (LANTHANUM TRIFLUORIDE)
LYMAN JL	1091	TEMPERATURE DEPENDENT ABSORPTION SPECTRUM OF THE NU-3 BAND OF SULFUR HEXAFLUORIDE
LYNCH MA	0506	VIBRATIONAL SPECTRA AND STRUCTURES OF IODINE PENTAFLUORIDE AND HEPTAFLUORIDE.
LYON WG	1124	HEAT CAPACITY AND THERMODYNAMIC PROPERTIES OF THALLIUM FLUORIDE FROM 5 TO 445 DEGR
LYUBIMOV VS	0229	ELECTRON STRUCTURE AND NMR SPECTRA OF BORON COMPOUNDS.
MACCORDICK HJ	0878	GENERALIZED VALENCE FORCE FIELD FORCE CONSTANTS AND MEAN SQUARE AMPLITUDES OF VIBR
MACDONALD RG	0673	GAS PHASE ELECTRON PARAMAGNETIC RESONANCE SPECTRUM AND DIPOLE MOMENT OF NITROGEN M
MACHIDA K	1485	TORSIONAL COORDINATES IN VIBRATIONAL ANHARMONICITY. APPLICATION TO ETHYLENE.
MACLAGAN RGAR	0862	ORBITAL MODIFICATION BY THE COULOMB FIELD OF LIGAND ATOMS OF LATER SECOND ROW ELEM
MAEKAWA T	0392	CHLORINE FLUORINE SYSTEM AT LOW TEMPERATURE. CHARACTERIZATION OF THE CHLORINE DIFL
	0393	THE CHLORINE DIFLUORIDE FREE RADICAL.
MAGNUSON DW	0258	MICROWAVE SPECTRUM AND MOLECULAR STRUCTURE OF BROMINE TRIFLUORIDE.
MAHESHWARI RC	0171	FRANCK-CONDON FACTORS AND R CENTROIDS FOR THE B TRIPLET SIGMA(+) TO A TRIPLET PI S
	0975	FRANCK-CONDON FACTORS AND R CENTROIDS FOR THE C-X SYSTEM OF SILICON MONOFLUORIDE M
MAHMOUDI S	0535	FORCE CONSTANTS AND MEAN VIBRATION AMPLITUDES OF IODINE HEPTAFLUORIDE CALCULATED B
	0878	GENERALIZED VALENCE FORCE FIELD FORCE CONSTANTS AND MEAN SQUARE AMPLITUDES OF VIBR
	1431	NEW METHOD FOR RESOLVING WILSON´S SECULAR EQUATIONS. APPLIED TO TETRAHEDRAL MOLECU
	1472	FORCE CONSTANTS OF OCTAHEDRAL MOLECULES CALCULATED BY THE LOGARITHMIC STEP METHOD.
MAIOROV AN	1486	CALCULATION OF MEAN SQUARE VIBRATIONAL AMPLITUDES.
MAKAROV AN	1064	SELECTIVE PHOTODISSOCIATION OF POLYATOMIC MOLECULES (SULFUR HEXAFLUORIDE) BY INFRA
MAKEEV GN	0560	ABSORPTION SPECTRA OF CERTAIN FLUORIDES IN THE NEAR ULTRAVIOLET REGION. (KRYPTON D
MAKI A	0343	INFRARED SPECTRUM OF CARBON TETRAFLUORIDE.
MAKLACHKOV AG	0945	STRUCTURE AND SOME PROPERTIES OF SCANDIUM TRIFLUORIDE.
MALIK SK	0927	ANOMALOUS NMR SHIFT IN SAMARIUM TRIFLUORIDE.
MALM JG	0010	CHEMICAL COMPOUNDS OF XENON AND OTHER NOBLE GASES.
	0396	VIBRATIONAL SPECTRA AND THERMODYNAMIC PROPERTIES OF CHLORINE TRIFLUORIDE AND BROMI
	0552	INFRARED AND RAMAN SPECTRA OF KRYPTON DIFLUORIDE.
	0628	VIBRATIONAL SPECTRA OF MOLYBDENUM HEXAFLUORIDE AND TECHNETIUM HEXAFLUORIDE.
	0636	HEAT CAPACITY AND OTHER THERMODYNAMIC PROPERTIES OF MOLYBDENUM HEXAFLUORIDE BETWEE
	0649	INFRARED SPECTRA OF NEPTUNIUM HEXAFLUORIDE AND PLUTONIUM HEXAFLUORIDE.
	0752	VIBRATIONAL SPECTRA OF OSMIUM HEXAFLUORIDE AND PLATINUM HEXAFLUORIDE.
	0753	OSMIUM HEXAFLUORIDE AND ITS IDENTITY WITH THE PREVIOUSLY REPORTED OCTAFLUORIDE.
	0807	ENTHALPIES OF FORMATION OF XENON HEXAFLUORIDE, XENON TETRAFLUORIDE, XENON DIFLUORI
	0882	PREPARATION AND SOME PROPERTIES OF PLATINUM HEXAFLUORIDE.
	0885	EXPERIMENTAL HEAT CAPACITIES OF PLUTONIUM-242 TRIFLUORIDE AND PLUTONIUM-242 TETRAF

MALM JG
0886 HEAT CAPACITY, ENTROPY, AND ENTHALPY OF PLUTONIUM-242 TRIFLUORIDE FROM 10 TO 350 D
0889 HEAT CAPACITY, ENTROPY, ENTHALPY, AND GIBBS ENERGY OF PLUTONIUM-242 TETRAFLUORIDE
0894 PROPERTIES OF PLUTONIUM HEXAFLUORIDE. (MOLECULAR STRUCTURE, ABSORPTION SPECTRUM)
0905 VIBRATIONAL SPECTRA OF RHENIUM HEXAFLUORIDE.
0912 VAPOR PRESSURES AND OTHER PROPERTIES OF RHENIUM HEXAFLUORIDE AND RHENIUM HEPTAFLUO
0917 MAGNETIC SUSCEPTIBILITY OF RHENIUM HEXAFLUORIDE.
0921 PREPARATION AND PROPERTIES OF RHENIUM HEPTAFLUORIDE.
1105 VAPOR PRESSURE AND TRANSITION POINTS OF TECHNETIUM HEXAFLUORIDE.
1190 RAMAN SPECTRUM OF URANIUM HEXAFLUORIDE.
1205 SOME RECENT STUDIES WITH HEXAFLUORIDES. (URANIUM HEXAFLUORIDE, PLUTONIUM HEXAFLUOR
1246 XENON DIFLUORIDE. HEAT CAPACITY FROM 5 TO 350 DEGREES K AND SOME DERIVED THERMODYN
1261 VIBRATIONAL SPECTRA AND STRUCTURES OF XENON TETRAFLUORIDE AND XENON OXYTETRAFLUORI
1262 VIBRATIONAL SPECTRA AND STRUCTURE OF XENON TETRAFLUORIDE.
1263 XENON TETRAFLUORIDE. (INFRARED SPECTRUM)
1275 XENON TETRAFLUORIDE. HEAT CAPACITY AND THERMODYNAMIC FUNCTIONS FROM 5 TO 350 DEGRE
1301 PREPARATION AND SOME THERMAL PROPERTIES OF PURE XENON HEXAFLUORIDE.
1305 HEAT CAPACITY AND OTHER THERMODYNAMIC FUNCTIONS OF XENON HEXAFLUORIDE FROM 5 TO 35
1336 VAPOR PRESSURES AND THERMODYNAMIC CALCULATIONS FOR SOME ACTINIDE FLUORIDES.

MAL'TSEV AA
0149 THERMODYNAMIC FUNCTIONS OF BISMUTH TRIHALIDE VAPORS.
0166 ISOTOPE EFFECT IN THE SINGLET SPECTRAL BANDS OF THE BORON FLUORIDE MOLECULE.
0230 STUDY OF THE ISOTOPE EFFECT AND A MORE PRECISE DETERMINATION OF THE FAR INFRARED S
0569 ELECTRONIC SPECTRUM OF LANTHANUM MONOFLUORIDE.
0580 BAND SYSTEM IN THE DISCHARGE SPECTRUM OF LANTHANUM TRIFLUORIDE VAPORS.
0938 VIBRONIC SPECTRUM OF SCANDIUM MONOFLUORIDE, YTTRIUM MONOFLUORIDE, AND LANTHANUM MO
1315 VIBRONIC SPECTRA OF YTTRIUM MONOFLUORIDE.

MALYSHEV VV
0632 THERMODYNAMIC SIMILARITY AND UNIVERSAL EQUATIONS OF STATE OF HEXAFLUORIDES. (OF SU

MAMANTOV G
0270 ESR SPECTRUM AND STRUCTURE OF BROMINE HEXAFLUORIDE.
0369 PHOTOELECTRON SPECTRUM OF CHLORINE MONOFLUORIDE.
0392 CHLORINE FLUORINE SYSTEM AT LOW TEMPERATURE. CHARACTERIZATION OF THE CHLORINE DIFL
0393 THE CHLORINE DIFLUORIDE FREE RADICAL.
0428 CHLORINE HEXAFLUORIDE RADICAL. PREPARATION, ELECTRON SPIN RESONANCE SPECTRUM, AND

MANDEL M
1110 MICROWAVE SPECTRA OF THE THALLIUM, INDIUM AND GALLIUM MONOHALIDES.

MANDIROLA OBD
0105 THE INFRARED SPECTRUM AND FORCE CONSTANTS FOR ARSENIC TRIFLUORIDE.
0106 LIQUID GAS ARSENIC INFRARED FREQUENCY SHIFTS.

MANN DE
0291 EMISSION SPECTRUM OF CARBON MONOFLUORIDE.
0297 ABSORPTION SPECTRUM OF CARBON DIFLUORIDE TRAPPED IN AN ARGON MATRIX.
0303 ABSORPTION SPECTRUM OF CARBON DIFLUORIDE AND ITS VIBRATIONAL ANALYSIS.
0609 GEOMETRY AND VIBRATIONAL SPECTRA OF THE ALKALINE EARTH DIHALIDES. PART-1: MAGNESIU
0687 INFRARED SPECTRUM AND STRUCTURE OF THE NITROGEN DIFLUORIDE RADICAL.
1314 ELECTRONIC BAND SPECTRUM OF YTTRIUM MONOFLUORIDE.

MANN RH
0351 A TRANSFERABLE UREY-BRADLEY FORCE FIELD AND THE ASSIGNMENTS OF SOME MIXED HALOMETH

MANNE R
0328 MOLECULAR SPECTROSCOPY BY MEANS OF ESCA. (CARBON TETRAFLUORIDE)
0404 ELECTRONIC STRUCTURE OF CHLORINE TRIFLUORIDE. AN APPROXIMATE SCF CALCULATION.

MAR RW
0575 VAPORIZATION STUDIES OF LANTHANUM TRIFLUORIDE.
0576 VAPOR PRESSURE, HEAT OF SUBLIMATION, AND EVAPORATION COEFFICIENT OF LANTHANUM TRIF

MARGOLIS JS
1444 VIBRATIONAL ANALYSIS OF SUBSTITUTED AND PERTURBED MOLECULES. DIRECT DETERMINATION

MARGRAVE JL
0011 MATRIX ISOLATION INFRARED SPECTROSCOPY OF HIGH TEMPERATURE SPECIES. (FLUORIDES, OX
0034 HEAT OF ATOMIZATION OF ALUMINUM DIFLUORIDE.
0045 INFRARED SPECTRA OF GASEOUS ALUMINUM TRIFLUORIDE, LITHIUM ALUMINUM TETRAFLUORIDE,
0047 HIGH TEMPERATURE NEGATIVE IONS. GASEOUS GROUP-3 FLUORIDES.
0119 DISSOCIATION ENERGIES OF THE ALKALINE EARTH MONOFLUORIDES.
0177 HEAT OF FORMATION OF BORON DIFLUORIDE.
0250 FLUORINE BOMB CALORIMETRY. PART-3: THE HEAT OF FORMATION OF BORON TRIFLUORIDE.
0272 KNUDSEN AND LANGMUIR MEASUREMENTS OF THE SUBLIMATION PRESSURE OF CADMIUM DIFLUORID
0276 SUBLIMATION PRESSURE OF CALCIUM DIFLUORIDE AND THE DISSOCIATION ENERGY OF CALCIUM
0277 REEVALUATION OF THE DISSOCIATION ENERGY OF CALCIUM FLUORIDE. (ALSO SILICON AND GER
0296 THERMODYNAMIC, ELECTROCHEMICAL AND SYNTHETIC STUDIES OF THE GRAPHITE FLUORINE COM
0298 HEAT OF FORMATION OF CARBON DIFLUORIDE.
0304 HEAT OF FORMATION OF THE CARBON DIFLUORIDE RADICAL.
0305 IONIZATION POTENTIALS FOR TETRAFLUORO ETHYLENE, TRIFLUORO CHLORO ETHYLENE, DIFLUOR
0430 SUBLIMATION PRESSURE OF CHROMIUM DIFLUORIDE AND THE DISSOCIATION ENERGY OF CHROMIU
0434 SUBLIMATION PRESSURES OF CHROMIUM TRIFLUORIDE, MANGANESE TRIFLUORIDE, AND IRON TRI
0437 INFRARED SPECTRA AND GEOMETRY OF MATRIX ISOLATED COBALT DIFLUORIDE, NICKEL DIFLUOR
0439 KNUDSEN AND LANGMUIR MEASUREMENTS OF THE SUBLIMATION PRESSURE OF COBALT DIFLUORIDE
0441 MASS SPECTROMETRIC STUDIES AT HIGH TEMPERATURES. PART-9: SUBLIMATION PRESSURE OF C
0442 MASS SPECTROMETRIC STUDIES AT HIGH TEMPERATURES. PART-9: SUBLIMATION PRESSURE OF C
0448 MASS SPECTROMETRIC STUDIES AT HIGH TEMPERATURES. PART-17: SUBLIMATION AND VAPOR PR
0449 STABILITIES OF DYSPROSIUM TRIFLUORIDE, HOLMIUM TRIFLUORIDE, AND ERBIUM TRIFLUORIDE
0450 MASS SPECTROMETRIC STUDIES AT HIGH TEMPERATURES. PART-13: STABILITIES OF SAMARIUM,
0451 INFRARED SPECTRUM OF HEAVY METAL DIHALIDES. (EUROPIUM DIFLUORIDE)
0462 INFRARED SPECTRA OF MATRIX ISOLATED SPECIES IN THE GALLIUM- FLUORINE SYSTEM.
0463 MASS SPECTROMETRIC STUDIES AT HIGH TEMPERATURES. PART-2: THE DISSOCIATION ENERGIES
0471 ENTHALPY OF SUBLIMATION OF GERMANIUM DIFLUORIDE AND THE THERMODYNAMICS OF SUBLIMAT
0472 ENTHALPY OF FORMATION OF GERMANIUM DIFLUORIDE.
0473 INFRARED VIBRATIONAL PROPERTIES OF GERMANIUM DIFLUORIDE.
0474 ULTRAVIOLET ABSORPTION SPECTRUM OF GERMANIUM DIFLUORIDE.
0480 ENTHALPY OF FORMATION OF GERMANIUM TRIFLUORIDE.
0484 A STUDY OF THE RAMAN AND BROAD LINE NMR SPECTRA OF GERMANIUM TETRAFLUORIDE.
0492 VAPORIZATION THERMODYNAMICS OF INDIUM TRIFLUORIDE.
0541 SUBLIMATION PRESSURE OF IRON DIFLUORIDE.
0592 MASS SPECTROMETRIC STUDIES AT HIGH TEMPERATURES. (TIN DIFLUORIDE, LEAD DIFLUORIDE,
0594 INFRARED SPECTRA OF MATRIX ISOLATED TIN DIFLUORIDE AND LEAD DIFLUORIDE.
0595 ULTRAVIOLET ABSORPTION SPECTRA OF GASEOUS TIN DIFLUORIDE AND LEAD DIFLUORIDE.
0604 INFRARED SPECTRA OF MATRIX ISOLATED MAGNESIUM DIFLUORIDE.
0611 A LANGMUIR MEASUREMENT OF THE SUBLIMATION PRESSURE OF MANGANESE DIFLUORIDE.
0614 SUBLIMATION PRESSURE OF MANGANESE DIFLUORIDE AND MANGANESE MONOFLUORIDE.
0643 SUBLIMATION PRESSURE OF NEODYMIUM TRIFLUORIDE AND THE STABILITIES OF GASEOUS NEODY
0652 VAPOR PRESSURE OF NICKEL DIFLUORIDE.
0939 MASS SPECTROMETRIC STUDIES AT HIGH TEMPERATURES. PART-18: THE STABILITIES OF THE M
0943 GEOMETRIES AND ENTROPIES OF METAL TRIFLUORIDES FROM INFRARED SPECTRA. (SCANDIUM, Y

MARGRAVE JL	0944	FORCE CONSTANTS AND GEOMETRIES OF MATRIX ISOLATED RARE EARTH TRIFLUORIDES. (SCANDI
	0947	SUBLIMATION PRESSURES OF SCANDIUM TRIFLUORIDE, YTTRIUM TRIFLUORIDE, AND LANTHANUM
	0953	MASS SPECTROMETRIC STUDIES OF SCANDIUM TRIFLUORIDE, YTTRIUM TRIFLUORIDE, LANTHANUM
	0981	VACUUM ULTRAVIOLET SPECTRA OF SILICON DIFLUORIDE AND GERMANIUM DIFLUORIDE.
	0982	INFRARED SPECTRA OF SILICON DIFLUORIDE IN INERT GAS MATRICES.
	0984	ULTRAVIOLET ABSORPTION SPECTRUM OF SILICON DIFLUORIDE.
	0985	INFRARED SPECTRUM, FORCE CONSTANTS, AND THERMODYNAMIC FUNCTIONS OF SILICON DIFLUOR
	0990	MICROWAVE SPECTRUM OF SILICON DIFLUORIDE.
	0998	EMISSION SPECTRUM OF SILICON TRIFLUORIDE.
	1014	MASS SPECTROMETRIC STUDIES AT HIGH TEMPERATURES. PART-14: VAPOR PRESSURE AND DISSO
	1103	STABILITIES OF TANTALUM PENTAFLUORIDE AND TANTALUM OXIDE TRIFLUORIDE.
	1132	SUBLIMATION PRESSURES AND HEATS OF SUBLIMATION OF THULIUM TRIFLUORIDE, YTTERBIUM T
	1145	MASS SPECTROMETRIC STUDIES AT HIGH TEMPERATURES. PART-16: SUBLIMATION PRESSURES FO
	1146	INFRARED SPECTRA AND GEOMETRY OF TITANIUM DIFLUORIDE AND TITANIUM TRIFLUORIDE IN R
	1325	INFRARED SPECTRA OF MATRIX ISOLATED ZIRCONIUM DIFLUORIDE, TRIFLUORIDE, AND TETRAFL
	1350	HIGH TEMPERATURE ENTHALPY INCREMENTS FOR SOME RARE EARTH TRIFLUORIDES.
	1359	MASS SPECTROMETRIC STUDIES OF SCANDIUM, YTTRIUM, LANTHANUM, AND RARE EARTH FLUORID
	1371	VIBRATIONAL FREQUENCIES AND VALENCE FORCE CONSTANTS OF FIRST ROW TRANSITION METAL
	1382	SUBLIMATION PRESSURES OF REFRACTORY FLUORIDES.
MARIAM S	1057	MEAN AMPLITUDES OF VIBRATION: SULFUR TETRAFLUORIDE AND SULFUR PENTAFLUORO CHLORIDE
MARQUART JR	0603	MASS SPECTROMETRIC STUDY OF MAGNESIUM HALIDES. (MAGNESIUM DIFLUORIDE)
MARRAM EP	0037	SUBLIMATION OF ALUMINUM TRIFLUORIDE AND THE INFRARED SPECTRUM OF ALUMINUM TRIFLUOR
MARSIGNY L	0163	CAMERON SYSTEM OF BORON MONOFLUORIDE.
MARTIN D	0398	RAMAN SPECTRA AND THE STRUCTURE OF CHLORINE TRIFLUORIDE AND BROMINE TRIFLUORIDE.
MARTIN JJ	0336	CONSTANT VOLUME HEAT CAPACITY OF GASEOUS CARBON TETRAFLUORIDE.
MARTIN RL	0683	MULTIPLET SPLITTING IN 1S HOLE STATES OF MOLECULES.
MARTIN RW	0466	ROTATIONAL STRUCTURE OF THE A DOUBLET SIGMA(+), B DOUBLET SIGMA(+), AND A QUARTET
	0467	ROTATIONAL STRUCTURE IN SOME HIGHER EXCITED STATES OF THE GERMANIUM(I) FLUORIDE MO
	0476	NEW ELECTRONIC EMISSION SYSTEM OF GERMANIUM DIFLUORIDE.
	0971	A QUARTET SIGMA TO X DOUBLET PI ELECTRONIC TRANSITION OF SILICON MONOFLUORIDE.
MARTINEZ JV	1282	ELECTRON DIFFRACTION STUDY OF THE STRUCTURE OF GASEOUS XENON TETRAFLUORIDE.
MARTYNOV AM	0043	STANDARD ENTHALPY OF FORMATION OF ALUMINUM TRIFLUORIDE.
MASHIKO Y	0326	THERMODYNAMIC PROPERTIES OF CARBON TETRAFLUORIDE FROM 4 DEGREES K TO ITS MELTING P
MASIA AP	0330	ELECTRONIC STRUCTURE OF THE CARBON TETRAFLUORIDE MOLECULE. PART-1. THE GROUND STA
MASLOV PG	0240	THERMODYNAMIC PROPERTIES OF BORON HALIDES.
	1335	HEATS OF FORMATION OF SOME HALIDES OF THORIUM, PROTACTINIUM, URANIUM, NEPTUNIUM, A
MASLOV UP	1335	HEATS OF FORMATION OF SOME HALIDES OF THORIUM, PROTACTINIUM, URANIUM, NEPTUNIUM, A
MASRI FN	0231	CUBIC POTENTIAL CONSTANTS AND 1-TYPE RESONANCE IN BORON TRIFLUORIDE.
	0716	PSEUDO-PARALLEL INFRARED BANDS OF OBLATE SYMMETRIC TOPS NITROGEN TRIFLUORIDE NU-4
	0728	INFRARED SPECTRUM AND MOLECULAR CONSTANTS OF NITROGEN TRIFLUORIDE.
MASSON CR	0273	MELTING POINTS OF INORGANIC FLUORIDES. (CADMIUM DIFLUORIDE, MAGNESIUM DIFLUORIDE,
MATHEW MP	0427	CORIOLIS COUPLING COEFFICIENTS OF BROMINE PENTAFLUORIDE AND CHLORINE PENTAFLUORIDE
MATHEWS WC	0306	2500 ANGSTROM ABSORPTION SPECTRUM OF CARBON DIFLUORIDE.
MATSUMURA C	0723	MICROWAVE SPECTRA OF NITROGEN TRIFLUORIDE IN THE EXCITED VIBRATIONAL STATES. (ZETA
MATTRAW HC	0540	INFRARED SPECTRUM OF IRIDIUM HEXAFLUORIDE.
	0891	INFRARED SPECTRUM OF PLUTONIUM HEXAFLUORIDE.
MAYER MG	1187	VIBRATIONAL SPECTRUM AND THERMODYNAMIC PROPERTIES OF URANIUM HEXAFLUORIDE GAS.
MAYLOTTE DH	1179	THEORETICAL CALCULATIONS OF THE ELECTRONIC STRUCTURE AND OPTICAL TRANSITIONS OF UR
MAZEEV MY	1161	CALCULATION OF THE DISSOCIATION ENERGY OF CALCIUM MONOFLUORIDE AND URANIUM MONOFLU
MCCORY LD	0045	INFRARED SPECTRA OF GASEOUS ALUMINUM TRIFLUORIDE, LITHIUM ALUMINUM TETRAFLUORIDE,
MCCREARY JR	0281	ENTROPIES AND ENTHALPIES OF SUBLIMATION OF CALCIUM AND CERIUM FLUORIDES. CORRELATI
	0456	ENTROPY AND ENTHALPY OF SUBLIMATION OF GADOLINIUM TRIFLUORIDE. ROLE OF CORRELATION
	0645	ENTROPIES AND ENTHALPIES OF SUBLIMATION OF NEODYMIUM AND TERBIUM TRIFLUORIDES.
	1353	ANALYSES OF THERMOCHEMICAL ERRORS AND SYSTEMATICS IN SUBLIMATION OF LANTHANIDE TRI
MCCULLOH KE	0378	PHOTOIONIZATION STUDY OF CHLORINE MONOFLUORIDE AND THE DISSOCIATION ENERGY OF FLUO
MCDIARMID R	0633	ASSIGNMENTS IN THE ULTRAVIOLET SPECTRA OF MOLYBDENUM HEXAFLUORIDE AND TUNGSTEN HEX
	0913	JAHN-TELLER EFFECTS IN THE 2E5/2G TO 2G3/2G TRANSITION OF RHENIUM HEXAFLUORIDE.
	0914	HIGHER ELECTRONIC STATES OF RHENIUM HEXAFLUORIDE.
MCDONALD GN	1251	VAPOR PRESSURE AND MELTING POINTS OF XENON DIFLUORIDE AND XENON TETRAFLUORIDE.
	1305	HEAT CAPACITY AND OTHER THERMODYNAMIC FUNCTIONS OF XENON HEXAFLUORIDE FROM 5 TO 35
MCDONALD JD	0441	MASS SPECTROMETRIC STUDIES AT HIGH TEMPERATURES. PART-9: SUBLIMATION PRESSURE OF C
	0442	MASS SPECTROMETRIC STUDIES AT HIGH TEMPERATURES. PART-9: SUBLIMATION PRESSURE OF C
	0947	SUBLIMATION PRESSURES OF SCANDIUM TRIFLUORIDE, YTTRIUM TRIFLUORIDE, AND LANTHANUM
	1382	SUBLIMATION PRESSURES OF REFRACTORY FLUORIDES.
MCDOWELL CA	0766	PHOTOELECTRON SPECTRA OF OXYGEN DIFLUORIDE AND OXYGEN DICHLORIDE.
	1076	PHOTOELECTRON SPECTRUM OF SULFUR HEXAFLUORIDE AT 584 ANGSTROMS.
MCDOWELL RS	0262	INFRARED SPECTRUM OF BROMINE PENTAFLUORIDE.
	0634	VIBRATIONAL SPECTRUM AND FORCE FIELD OF MOLYBDENUM HEXAFLUORIDE.
	0915	INFRARED BAND CONTOURS OF RHENIUM HEXAFLUORIDE.
	1063	OCTAHEDRAL FINE STRUCTURE SPLITTINGS IN NU-3 OF SULFUR HEXAFLUORIDE.
	1156	CORIOLIS CONSTANTS OF SPHERICAL TOP MOLECULES FROM LOW TEMPERATURE INFRARED STUDIE
	1196	VIBRATIONAL SPECTRUM AND FORCE FIELD OF URANIUM HEXAFLUORIDE.
	1487	CORIOLIS ZETA SUMS FOR SOME SPHERICAL TOP MOLECULES.
MCGEE JA	0565	MASS SPECTRUM AND MOLECULAR ENERGETICS OF KRYPTON DIFLUORIDE.
MCGURK J	0384	MOLECULAR MAGNETIC PROPERTIES OF CHLORINE MONOFLUORIDE.
MCKEAN DC	0232	ACCURATE FORCE CONSTANTS FROM HEAVY ISOTOPIC SUBSTITUTION. PART-2: BORON TRIFLUORI
	0320	ACCURATE FORCE CONSTANTS FROM HEAVY ISOTOPIC SUBSTITUTION. PART-1. CARBON TETRAFLU
	0697	INFRARED SPECTRUM OF NITROGEN TRIFLUORIDE AND FORCE FIELD.
MCKINLEY JD	0653	SPECTRA OF MATRIX ISOLATED NICKEL DIFLUORIDE AND NICKEL DICHLORIDE.
MCMATH HG	0394	THERMAL DISSOCIATION OF CHLORINE TRIFLUORIDE BEHIND INCIDENT SHOCK WAVES.
MECKE R	0190	FORCE CONSTANTS FOR MOLECULES WITH D3H SYMMETRY.
MEDINA A	1007	INTENSITIES OF THE INFRARED FUNDAMENTALS OF CARBON TETRAFLUORIDE, SILICON TETRAFLU
MEHTA ML	1443	MOLECULAR FORCE FIELDS OF XY6 MOLECULES.
MEINERT H	0263	SELF IONIZATION OF HALOGEN FLUORIDES. (BROMINE PENTAFLUORIDE)
	1241	INFRARED SPECTRA OF XENON DIFLUORIDE IN DIFFERENT SOLVENTS.
	1300	STRUCTURE OF THE IODINE HEXAFLUORIDE ANION. (XENON HEXAFLUORIDE)
MEISINGSETH E	0233	MEAN AMPLITUDES OF VIBRATION AND SHRINKAGE EFFECT OF BORON TRIHALIDES, CALCULATED
	0809	CALCULATION OF UREY-BRADLEY POTENTIAL CONSTANTS. (PHOSPHORUS TRIFLUORIDE, ARSENIC
	1393	CALCULATION OF UREY-BRADLEY POTENTIAL CONSTANTS. PLANAR XY3 MOLECULES.
	1394	VIBRATIONAL MEAN SQUARE AMPLITUDE MATRICES. (BASTIANSEN-MORINO SHRINKAGE EFFECT OF

MEISINGSETH E	1410	INFLUENCE OF ATOMIC MASSES ON THE CORIOLIS COUPLING COEFFICIENTS IN XY4 TETRAHEDRA
	1454	NORMAL COORDINATE ANALYSIS OF ROTATION VIBRATION OF OCTAHEDRAL XY6 AND Z6 MOLECULA
	1455	MEAN AMPLITUDES OF VIBRATION AND BASTIANSEN-MORINO SHRINKAGE EFFECTS IN SOME OCTAH
MELLISH CE	0860	DIRECTED VALENCY IN CERTAIN MOLECULES AND COMPLEX IONS. (PHOSPHORUS PENTAFLUORIDE,
MELTON CE	0344	CALCULATED IONIZATION POTENTIAL OF CHLORO- AND FLUOROMETHANES, TETRAFLUORO METHANE
MENZEL W	0345	VAPOR PRESSURE OF CARBON TETRAFLUORIDE AND NITROGEN TRIFLUORIDE AND THE TRIPLE POI
MENZINGER M	1227	PHOTOELECTRON SPECTRUM OF XENON DIFLUORIDE.
MERER AJ	0466	ROTATIONAL STRUCTURE OF THE A DOUBLET SIGMA(+), B DOUBLET SIGMA(+), AND A QUARTET
	0467	ROTATIONAL STRUCTURE IN SOME HIGHER EXCITED STATES OF THE GERMANIUM(I) FLUORIDE MO
	0476	NEW ELECTRONIC EMISSION SYSTEM OF GERMANIUM DIFLUORIDE.
	0971	A QUARTET SIGMA TO X DOUBLET PI ELECTRONIC TRANSITION OF SILICON MONOFLUORIDE.
	1133	ROTATIONAL ANALYSIS OF THE A DOUBLET SIGMA(+) TO X DOUBLET PI, B DOUBLET SIGMA(+)
MESSMER RP	1179	THEORETICAL CALCULATIONS OF THE ELECTRONIC STRUCTURE AND OPTICAL TRANSITIONS OF UR
METZ FI	0774	A REINVESTIGATION OF THE INFRARED SPECTRUM OF OXYGEN DIFLUORIDE.
	0775	FLUORINE NMR SPECTRA OF OXYGEN FLUORIDES. (OXYGEN DIFLUORIDE)
MEYER MG	1186	VIBRATIONAL SPECTRUM AND THERMODYNAMIC PROPERTIES OF URANIUM HEXAFLUORIDE GAS.
MEYER RT	0652	VAPOR PRESSURE OF NICKEL DIFLUORIDE.
MEZEI M	1395	CALCULATION OF FORCE CONSTANTS WITH THE MAXIMUM OVERLAP METHOD. MOLECULES WITH LON
	1416	CALCULATION OF FORCE CONSTANTS WITH THE MAXIMUM OVERLAP METHOD. TETRAHEDRAL MOLECU
MIKHAILOV VM	0182	ROTATIONAL SPECTRA OF NONPOLAR MOLECULES.
MIKULENOK VV	0669	THERMODYNAMIC PROPERTIES OF NIOBIUM AND TANTALUM PENTAFLUORIDES.
MILLE P	0602	STRUCTURE OF THE MAGNESIUM DIFLUORIDE MOLECULE.
MILLER FA	0825	INFRARED AND RAMAN SPECTRA AND STRUCTURE OF PHOSPHORUS TETRAFLUORIDE.
	0863	POTENTIAL FIELD AND MOLECULAR VIBRATIONS OF THE TRIGONAL BIPYRAMIDAL THE RAMAN SPE
MILLER HC	0710	LIQUID DENSITY, VAPOR PRESSURE AND CRITICAL TEMPERATURE AND PRESSURE OF NITROGEN T
MILLER TA	1020	GAS PHASE ELECTRON RESONANCE SPECTRA OF SULFUR MONOFLUORIDE, SELENIUM MONOFLUORIDE
	1021	GAS PHASE ELECTRON RESONANCE SPECTRA OF SULFUR MONOFLUORIDE AND SELENIUM MONOFLUOR
MILLIGAN DE	0653	SPECTRA OF MATRIX ISOLATED NICKEL DIFLUORIDE AND NICKEL DICHLORIDE.
	0691	MATRIX ISOLATION STUDY OF THE VACUUM ULTRAVIOLET PHOTOLYSIS OF NITROGEN TRIFLUORID
	0986	MATRIX ISOLATION STUDY OF THE VACUUM ULTRAVIOLET PHOTOLYSIS OF SILICON DIFLUORIDE
MILLS IM	0324	FORCE FIELDS OF CARBON TETRAFLUORIDE AND SILICON TETRAFLUORIDE.
	0743	SIGNS OF L DOUBLING CONSTANTS IN NITROGEN TRIFLUORIDE.
MILTON HT	1197	VAPOR PRESSURE AND CRITICAL CONSTANTS OF URANIUM HEXAFLUORIDE.
MIRRI AM	0717	GENERAL FORCE FIELD OF NITROGEN TRIFLUORIDE, PHOSPHORUS TRIFLUORIDE, AND ARSENIC T
	0718	MILLIMETER WAVE SPECTRUM AND CENTRIFUGAL DISTORTION CONSTANTS OF NITROGEN TRIFLUOR
MISHIRA RK	0168	VIBRATIONAL TRANSITION PROBABILITIES FOR THE A-X SYSTEM OF BORON MONOFLUORIDE.
MITCHELL KAR	1230	BONDING IN XENON DIFLUORIDE.
	1242	5D ORBITALS OF XENON IN ATOMIC VALENCE STATES AND IN THE MOLECULES XENON DIFLUORID
	1338	D ORBITALS IN THE NOBLE GAS DIHALIDES. (ARGON DIFLUORIDE, KRYPTON DIFLUORIDE, AND
MITCHELL SJ	0750	PREPARATION AND CRYSTAL STRUCTURE OF OSMIUM PENTAFLUORIDE.
MIURA M	0823	ESR SPECTRA OF PHOSPHORUS TETRAFLUORIDE RADICALS PRODUCED IN A SINGLE CRYSTAL OF P
MIZUSHIMA M	0385	SOME CORRECTIONS TO THE SECOND ORDER STARK EFFECT OF LINEAR MOLECULES. (CHLORINE M
MOCCIA R	1077	ELECTRONIC PROPERTIES OF SULFUR HEXAFLUORIDE. AB INITIO CALCULATION FOR THE GROUND
MODDEMAN WE	0319	HIGH RESOLUTION PHOTOELECTRON SPECTROSCOPY OF CARBON TETRAFLUORIDE AND SILICON TE
MODICA AP	0288	SHOCK TUBE DETERMINATION OF THE C2 (A TRIPLET PI TO X TRIPLET PI) AND CARBON MONOF
MOELLER MB	0167	FLUORESCENCE SPECTRA OF PHOSPHORUS NITRIDE AND BORON FLUORIDE.
MOFFITT W	0916	COLORS OF TRANSITION METAL HEXAFLUORIDES. (RHENIUM HEXAFLUORIDE, OSMIUM HEXAFLUORI
MOHAMED KA	0677	FRANCK-CONDON FACTORS AND R CENTROIDS FOR THE 6 SINGLET SIGMA- X TRIPLET SIGMA BAN
MOHAN H	0122	VIOLET AND ULTRAVIOLET SPECTRUM OF BARIUM MONOFLUORIDE MOLECULE.
	0123	THERMALLY EXCITED EMISSION SPECTRUM OF BARIUM MONOFLUORIDE.
MOHAN N	0234	UTILITY OF HEAVY ATOM ISOTOPIC SUBSTITUION DATA FOR THE CALCULATION OF FORCE CONST
	0729	GREEN'S FUNCTION ANALYSIS OF NITROGEN TRIFLUORIDE AND NITROGEN TRICHLORIDE.
	0730	NORMAL COORDINATE ANALYSIS OF SOME NITROGEN HALIDES. (NITROGEN TRIFLUORIDE)
	1093	GREEN'S FUNCTION ANALYSIS OF SULFUR HEXAFLUORIDE AND RELATED MOLECULES. (SELENIUM
MOHANTY BS	0346	FLUORINE RELAXATION BY NMR ABSORPTION IN GASEOUS CARBON TETRAFLUORIDE, SILICON TET
	0972	FRANCK-CONDON FACTORS AND R CENTROIDS OF SOME BAND SYSTEMS OF SILICON MONOFLUORIDE
MOHRY F	0345	VAPOR PRESSURE OF CARBON TETRAFLUORIDE AND NITROGEN TRIFLUORIDE AND THE TRIPLE POI
MONCSTORI B	0347	RAMAN SPECTRUM OF GASEOUS CARBON TETRAFLUORIDE.
MONTER B	1118	STARK EFFECT FOR THALLIUM MONOFLUORIDE AND POTASSIUM FLUORIDE.
MOORE DLG	0928	ELECTRONIC STATES OF GASEOUS SCANDIUM MONOFLUORIDE, YTTRIUM MONOFLUORIDE, AND LANT
MOORMAN J	1356	HIGH TEMPERATURE ENTHALPIES AND RELATED THERMODYNAMIC FUNCTIONS OF THE TRIFLUORIDE
MOOTZ A	1307	ELECTRICAL CONDUCTIVITY AND DIELECTRIC CONSTANT OF LIQUID XENON HEXAFLUORIDE.
MORCHER B	0992	THERMOCHEMISTRY OF SILICON DIFLUORIDE.
MORCILLO J	1007	INTENSITIES OF THE INFRARED FUNDAMENTALS OF CARBON TETRAFLUORIDE, SILICON TETRAFLU
MORINO Y	0221	MOLECULAR STRUCTURE AND FORCE CONSTANTS OF BORON TRIFLUORIDE AND BORON TRICHLORIDE
	0722	MICROWAVE SPECTRA OF NITROGEN TRIFLUORIDE IN THE EXCITED VIBRATIONAL STATES. (EQUI
	0723	MICROWAVE SPECTRA OF NITROGEN TRIFLUORIDE IN THE EXCITED VIBRATIONAL STATES. (ZETA
	0770	THIRD ORDER POTENTIAL CONSTANTS OF BENT XY2 MOLECULES. (OXYGEN DIFLUORIDE)
	0771	MICROWAVE SPECTRUM OF OXYGEN DIFLUORIDE IN VIBRATIONALLY EXCITED STATES, NU-1- 2 N
	0772	FERMI DIAD OF NU-1 AND 2NU-2 IN OXYGEN DIFLUORIDE.
	0805	MICROWAVE SPECTRUM OF PHOSPHORUS TRIFLUORIDE.
	0810	MOLECULAR STRUCTURE OF PHOSPHORUS TRIFLUORIDE STUDIED BY GAS ELECTRON DIFFRACTION.
	1383	ESTIMATION OF ANHARMONIC POTENTIAL CONSTANTS. PART-1. LINEAR XY2 MOLECULES.
	1398	INERTIA DEFECT AND PLANARITY OF FOUR ATOM MOLECULES. (XY3)
MORITANI T	0810	MOLECULAR STRUCTURE OF PHOSPHORUS TRIFLUORIDE STUDIED BY GAS ELECTRON DIFFRACTION.
MORRISON JD	1243	IONIZATION AND DISSOCIATION OF XENON DIFLUORIDE INDUCED BY PHOTON IMPACT.
MORSY TE	1089	THERMODYNAMIC FUNCTIONS FOR SULFUR HEXAFLUORIDE.
MORTON JB	1181	MEASUREMENT OF THE GAS DENSITY OF URANIUM HEXAFLUORIDE BY LASER RAMAN SCATTERING.
MORTON JR	0112	ESR SPECTRUM OF ARSENIC TETRAFLUORIDE.
	0271	ELECTRON SPIN RESONANCE SPECTRA OF HALOGEN HEXAFLUORIDES. (CHLORINE HEXAFLUORIDE,
	0413	EPR SPECTRUM OF THE RADICAL CHLORINE TETRAFLUORIDE.
	0546	ELECTRON SPIN RESONANCE SPECTRUM OF KRYPTON MONOFLUORIDE.
	0791	ESR SPECTRA OF PHOSPHORUS DIFLUORIDE AND SULFUR TRIFLUORIDE RADICALS.
	0959	PARAMAGNETIC HEXAFLUORIDE ANIONS OF GROUP-6. (SELENIUM PENTAFLUORIDE)
	1220	RADIATION DAMAGE IN XENON TETRAFLUORIDE. ELECTRON SPIN RESONANCE OF THE TRAPPED RA
	1222	ELECTRON SPIN RESONANCE SPECTRUM OF XENON MONOFLUORIDE IN GAMMA IRRADIATED XENON T
	1288	ELECTRON SPIN RESONANCE OF THE XENON FLUORIDE RADICAL.
MOSER CM	0164	CALCULATION OF RYDBERG LEVELS IN NITROGEN OXIDE AND BORON FLUORIDE.
	0932	ELECTRONIC STRUCTURE AND GROUND STATE OF SCANDIUM MONOFLUORIDE.
MOSKOWITZ JW	1188	CALCULATED ELECTRONIC STRUCTURE OF URANIUM HEXAFLUORIDE.

MOSKOWITZ JW   1224 SELF CONSISTENT FIELD STUDY OF THE SERIES XENON DIFLUORIDE, XENON TETRAFLUORIDE, X
MOSKVITINA EN   0230 STUDY OF THE ISOTOPE EFFECT AND A MORE PRECISE DETERMINATION OF THE FAR INFRARED S
MOULTON MC   0392 CHLORINE FLUORINE SYSTEM AT LOW TEMPERATURE. CHARACTERIZATION OF THE CHLORINE DIFL
                      0393 THE CHLORINE DIFLUORIDE FREE RADICAL.
MOULTON WG   1349 MAGNETIC RESONANCE AND SUSCEPTIBILITY OF RARE EARTH TRIFLUORIDES.
MOYNIHAN RE   0325 INFRARED BAND CONTOURS. PART-1: SPHERICAL TOP MOLECULES. (CARBON TETRAFLUORIDE)
MUELLER A   0085 MASS SPECTRUM OF ANTIMONY PENTAFLUORIDE.
                      0234 UTILITY OF HEAVY ATOM ISOTOPIC SUBSTITUION DATA FOR THE CALCULATION OF FORCE CONST
                      0348 COMPARISON OF CALCULATIONS OF FORCE CONSTANTS FOR BORON TETRAFLUORIDE(-), CARBON T
                      1396 MEAN AMPLITUDES OF VIBRATION IN SOME XY2Z MOLECULES OF C2V SYMMETRY.
                      1456 FORCE FIELD OF WEINSTOCK AND GOODMAN FOR OCTAHEDRAL HEXAFLUORIDES AND THE L MATRIX
                      1457 FORCE CONSTANTS FOR HEXAFLUORIDE MOLECULES.
MUENTER AA   0003 GAS PHASE STRUCTURES AND MASS SPECTRA OF BINARY PENTAFLUORIDES.
                      0349 MOLECULAR BEAM ELECTRIC DEFLECTION OF THE TETRAHALIDES CARBON TETRAFLUORIDE, CARBO
                      1213 POLAR STATES IN VANADIUM PENTAFLUORIDE.
                      1417 STARK ENERGIES FOR MOLECULES WITH VIBRATIONALLY INDUCED DIPOLE MOMENTS.
MUENTER JS   0375 HYPERFINE STRUCTURE CONSTANTS OF CHLORINE MONOFLUORIDE.
                      0529 POLAR DISTORTIONS IN RHENIUM HEPTAFLUORIDE AND IODINE HEPTAFLUORIDE.
                      1084 ELECTRIC DEFLECTION OF BINARY HEXAFLUORIDES. (SULFUR HEXAFLUORIDE, SELENIUM HEXAF
MUETTERTIES EL   0922 INTRAMOLECULAR LIGAND EXCHANGE IN SEVEN COORDINATE STRUCTURES. (RHENIUM HEPTAFLUOR
                      1048 STRUCTURE OF EXCHANGE PROCESSES IN SOME INORGANIC FLUORIDES BY NMR. (SULFUR TETRAF
                      1049 FLUORINE EXCHANGE IN SULFUR TETRAFLUORIDE.
                      1440 FURTHER EVIDENCE OF STEREOCHEMICAL NONRIGIDITY IN FIVE AND SEVEN COORDINATE STRUCT
MUIRHEAD JS   0705 LONE PAIR MODEL AND THE VIBRATIONAL FORCE CONSTANTS OF NITROGEN TRIFLUORIDE.
MULLER A   0235 CHEMICAL SHIFTS OF BORON, PHOSPHORUS, PHOSPHORYL, AND THIOPHOSPHORYL HALIDES.
                      0436 INFRARED SPECTRUM OF CHROMIUM HEXAFLUORIDE, MOLYBDENUM HEXAFLUORIDE, AND OSMIUM H
                      0735 FORCE FIELD AND MOLECULAR CONSTANTS OF NITROGEN TRIFLUORIDE.
                      0811 RAMAN AND INFRARED SPECTRA AND THERMODYNAMIC FUNCTIONS FOR VARIOUS PHOSPHORUS AND
                      0910 JAHN-TELLER VIBRATIONS OF RHENIUM HEXAFLUORIDE.
                      1216 MEAN AMPLITUDES OF VIBRATION AND THERMODYNAMIC FUNCTIONS FOR VANADIUM PENTAFLUORID
                      1265 GENERAL VALENCE FORCE FIELD FOR XENON TETRAFLUORIDE.
                      1390 MEAN AMPLITUDES OF VIBRATION FOR SOME PYRAMIDAL XY3 MOLECULES.
                      1399 CALCULATIONS OF MEAN AMPLITUDES OF VIBRATION AND UREY-BRADLEY FORCE FIELDS FOR PLA
                      1412 MOLECULES WITH D4H SYMMETRY. NEW METHOD FOR THE CALCULATION OF GVFF CONSTANTS.
                      1415 NORMAL COORDINATE TREATMENT OF XY4 TETRAHEDRAL MOLECULES AND IONS. (GVFF)
                      1418 NORMAL COORDINATE TREATMENT OF XY4 TETRAHEDRAL MOLECULES AND IONS. (UREY-BRADLEY A
                      1447 REVISED MEAN AMPLITUDES OF VIBRATION FOR SOME OCTAHEDRAL HEXAFLUORIDES.
                      1458 INFLUENCE OF ATOMIC MASS ON THE CORIOLIS CONSTANTS OF XY6 MOLECULES.
                      1482 FORMULAS FOR CORIOLIS CONSTANTS.
                      1488 THEORETICAL INTERPRETATION OF THE MASS INFLUENCE ON CORIOLIS COUPLING CONSTANTS IN
MULVIHILL M   0469 ENTHALPY OF FORMATION OF GERMANIUM TETRAFLUORIDE.
MURAD E   0129 MASS SPECTROMETRIC DETERMINATION OF THE DISSOCIATION ENERGY OF BERYLLIUM MONOFLUOR
                      0161 DISSOCIATION ENERGY OF BORON MONOFLUORIDE FROM MASS SPECTROMETRIC STUDIES.
MURATA Y   0221 MOLECULAR STRUCTURE AND FORCE CONSTANTS OF BORON TRIFLUORIDE AND BORON TRICHLORIDE
MURCHISON C   0561 STRUCTURE OF KRYPTON DIFLUORIDE.
MURRELL JN   0405 CONSTRUCTION OF HYBRID ORBITALS. (CHLORINE TRIFLUORIDE)
MURTY AAN   0814 CENTRIFUGAL DISTORTION CONSTANTS OF PHOSPHORUS TRIFLUORIDE.
MURTY PS   0144 NEW BAND SYSTEM OF BISMUTH FLUORIDE MOLECULE IN THE REGION 6200-7000 ANGSTROMS.
MUSHER JI   1027 MOLECULAR ORBITAL DESCRIPTION OF SULFUR DIFLUORIDE, SULFUR TETRAFLUORIDE AND SULFU
                      1085 MOLECULAR ORBITAL DESCRIPTION OF SULFUR COMPOUNDS OF VALENCES 2, 4, AND 6.
MUTHUSUBRAMANIAN P   0407 MOLECULAR FORCE FIELD FOR INTERHALOGEN COMPOUNDS. (CHLORINE TRIFLUORIDE AND BROMIN
                      0425 MOLECULAR FORCE FIELDS FOR INTERHALOGEN COMPOUNDS. (CHLORINE PENTAFLUORIDE AND BRO
                      0511 MOLECULAR FORCE FIELD FOR INTERHALOGEN COMPOUNDS. IODINE PENTAFLUORIDE, IODINE OXY
MYERS OE   0662 HEAT CAPACITY, ENTROPY, AND HEATS OF TRANSITION OF MOLYBDENUM HEXAFLUORIDE AND NIO
MYERS RJ   0686 INFRARED SPECTRUM AND THERMODYNAMIC FUNCTIONS OF THE NITROGEN DIFLUORIDE RADICAL.
                      0687 INFRARED SPECTRUM AND STRUCTURE OF THE NITROGEN DIFLUORIDE RADICAL.
NAGARAJAN G   0046 SPECTROSCOPIC STUDIES OF MOLECULAR CONSTANTS IN SOME POLYATOMIC MOLECULES.
                      0050 MEAN AMPLITUDES OF VIBRATION, CORIOLIS COUPLING CONSTANTS, SHRINKAGE EFFECT, AND T
                      0086 POTENTIAL FIELD AND FORCE CONSTANTS OF ANTIMONY PENTAFLUORIDE AND NIOBIUM PENTAFLU
                      0124 MEAN AMPLITUDES OF VIBRATION, BASTIANSEN-MORINO SHRINKAGE EFFECT, AND MOLECULAR PO
                      0236 THERMODYNAMIC PROPERTIES OF SOME BORON TRIHALIDES.
                      0350 POTENTIAL CONSTANTS AND THERMODYNAMIC FUNCTIONS OF CARBON TETRAFLUORIDE.
                      0487 MEAN AMPLITUDES OF VIBRATION OF SOME TETRAHEDRAL XY4 TYPE MOLECULES. PART-5: TETRA
                      0507 POTENTIAL CONSTANTS AND THERMODYNAMIC FUNCTIONS OF IODINE PENTAFLUORIDE.
                      0508 POTENTIAL CONSTANTS AND THERMODYNAMIC PROPERTIES OF IODINE PENTAFLUORIDE.
                      0509 POTENTIAL CONSTANTS AND THERMODYNAMIC FUNCTIONS FOR IODINE PENTAFLUORIDE AND IODIN
                      0532 POTENTIAL CONSTANTS OF IODINE HEPTAFLUORIDE.
                      0562 MEAN AMPLITUDES OF VIBRATION, BASTIANSEN-MORINO SHRINKAGE EFFECT AND THERMODYNAMIC
                      0619 MEAN AMPLITUDES OF VIBRATION FOR THE DIHALIDES OF MERCURY, BERYLLIUM, AND MAGNESIU
                      0635 POTENTIAL CONSTANTS FOR MOLYBDENUM HEXAFLUORIDE AND RHENIUM HEXAFLUORIDE.
                      0719 MEAN AMPLITUDES OF VIBRATION AND THERMODYNAMIC FUNCTIONS OF SOME GROUP-5 TRIHALIDE
                      0773 MEAN AMPLITUDES OF VIBRATION AND THERMODYNAMIC FUNCTIONS FOR SILICON DIFLUORIDE AN
                      0784 SPECTROSCOPIC STUDIES OF VIBRATIONAL CONSTANTS, STATISTCIAL THERMODYNAMICS AND QU
                      0864 MEAN AMPLITUDES OF VIBRATION FOR A TRIGONAL BIPYRAMIDAL MOLECULE WITH APPLICATION
                      0962 ROOT MEAN SQUARE AMPLITUDES IN SOME HEXAFLUORIDES OF OCTAHEDRAL SYMMETRY. (SELENIU
                      1008 MEAN AMPLITUDES OF VIBRATION OF SOME XY4 TETRAHEDRAL MOLECULES. (SILICON TETRAFLUO
                      1104 MEAN AMPLITUDES OF VIBRATION AND THERMODYNAMIC FUNCTIONS OF TECHNETIUM HEXAFLUORID
                      1216 MEAN AMPLITUDES OF VIBRATION AND THERMODYNAMIC FUNCTIONS FOR VANADIUM PENTAFLUORID
                      1244 MOLECULAR CONSTANTS AND THERMODYNAMIC FUNCTIONS OF SOME LINEAR SYMMETRICAL MOLECUL
                      1245 SPECTROSCOPIC STUDIES OF MOLECULAR CONSTANTS IN SOME POLYATOMIC MOLECULES. (XENON
                      1274 MEAN AMPLITUDES OF VIBRATION AND THERMODYNAMIC FUNCTIONS OF XENON TETRAFLUORIDE.
                      1372 MEAN AMPLITUDES OF VIBRATION AND BASTIANSEN-MORINO SHRINKAGE EFFECT IN SOME LINEAR
                      1384 MEAN AMPLITUDES OF VIBRATION FOR THE BORON DIFLUORIDE, CARBON DIFLUORIDE, NITROGEN
                      1385 POTENTIAL CONSTANTS AND CORIOLIS COUPLING CONSTANTS FOR SOME NONLINEAR SYMMETRICAL
                      1396 MEAN AMPLITUDES OF VIBRATION IN SOME XY2Z MOLECULES OF C2V SYMMETRY.
                      1397 MEAN AMPLITUDES OF VIBRATION AND BASTIANSEN-MORINO SHRINKAGE EFFECT IN SOME XY3 MO
                      1419 MEAN AMPLITUDES OF VIBRATION OF SOME XY4 TETRAHEDRAL MOLECULES.
                      1420 MEAN AMPLITUDES OF VIBRATION OF SOME TETRAHEDRAL XY4 TYPE MOLECULES. PART 3 AND 4.
                      1437 POTENTIAL CONSTANTS, MEAN AMPLITUDES OF VIBRATION, THERMODYNAMIC FUNCTIONS AND MOL
                      1438 THERMODYNAMIC FUNCTIONS OF SOME TRIGONAL BIPYRAMIDAL PENTAHALIDES.

NAGARAJAN G	1459	MEAN AMPLITUDES OF VIBRATION AND THERMODYNAMIC FUNCTIONS FOR SOME METAL HEXAFLUORI
	1460	MEAN AMPLITUDES OF THERMAL MOTION AND SHRINKAGES OF CHEMICAL BONDS: OCTAHEDRAL HEX
	1461	MEAN AMPLITUDES OF VIBRATION OF SOME XY6 MOLECULES.
	1462	POTENTIAL FIELD AND FORCE CONSTANTS FOR SOME XY6 MOLECULES.
	1463	STATISTICAL THERMODYNAMICS OF SOME HEXAFLUORIDES.
	1489	SHRINKAGES OF THE INTERNUCLEAR DISTANCES BY MOLECULAR VIBRATIONS.
NAIR KPR	0173	POTENTIAL ENERGY CURVES AND DISSOCIATION ENERGIES OF DIATOMIC BORON AND ALUMINUM H
	0458	POTENTIAL ENERGY CURVES AND DISSOCIATION ENERGIES OF DIATOMIC FLUORIDES AND CHLORI
NAKAGAWA I	0243	FORCE CONSTANTS OF CARBON TETRAFLUORIDE, SILICON TETRAFLUORIDE, BORON TRIFLUORIDE,
NAKAMOTO K	0838	UREY-BRADLEY FORCE CONSTANTS OF TRIGONAL BIPYRAMIDAL XY5 MOLECULES. (PHOSPHORUS PE
	1433	NORMAL COORDINATE ANALYSIS OF TRIGONAL BIPYRAMIDAL XY5 MOLECULES.
NAMPOORI VPN	0491	ROTATIONAL ANALYSIS OF THE C SINGLET PI- X SINGLET SIGMA(+) SYSTEM OF INDIUM MONOF
NARASIMHAM NA	0222	EMISSION SPECTRA OF CERTAIN POLYATOMIC FREE RADICALS OBTAINED IN ELECTRICAL DISCHA
	0559	ULTRAVIOLET EMISSION SPECTRA. (XENON DIFLUORIDE AND KRYPTON DIFLUORIDE)
NARAYANAN PS	0147	EMISSION SPECTRUM OF BISMUTH MONOFLUORIDE.
	0974	EMISSION SPECTRUM OF THE GAMMA SYSTEM OF SILICON MONOFLUORIDE.
NATKE CA	0667	ENTHALPIES OF FORMATION OF NIOBIUM PENTAFLUORIDE AND TANTALUM PENTAFLUORIDE.
NAUDE SM	0030	EMISSION SPECTRUM OF ALUMINUM MONOFLUORIDE IN THE VACUUM ULTRAVIOLET.
NAUMANN D	0500	VIBRATIONAL SPECTRUM OF IODINE TRIFLUORIDE.
NAUMOV VA	0460	ELECTRON DIFFRACTION ANALYSIS OF GALLIUM HALIDES. (GALLIUM TRIFLUORIDE)
	0644	ELECTRONOGRAPHIC INVESTIGATION OF THE STRUCTURE OF THE MOLECULES OF NEODYMIUM HALI
	1317	ELECTRON DIFFRACTION STUDY OF THE MOLECULAR STRUCTURE OF THE VAPOR PHASE HALIDES O
NEBGEN JW	0774	A REINVESTIGATION OF THE INFRARED SPECTRUM OF OXYGEN DIFLUORIDE.
	0775	FLUORINE NMR SPECTRA OF OXYGEN FLUORIDES. (OXYGEN DIFLUORIDE)
NELSON W	0178	ELECTRON SPIN RESONANCE AND STRUCTURE OF THE BORON DIFLUORIDE RADICAL.
	0792	ELECTRON SPIN RESONANCE OF TRAPPED PHOSPHORUS DIFLUORIDE AND PHOSPHORUS TETRAFLUOR
NENTWIG P	1195	ALPHA SPECTROMETRIC INVESTIGATION OF URANIUM HEXAFLUORIDE. PART-1.
NERESON NG	1063	OCTAHEDRAL FINE STRUCTURE SPLITTINGS IN NU-3 OF SULFUR HEXAFLUORIDE.
NESBET RK	0169	VALENCE EXCITED STATES OF NITROGEN, CARBON MONOXIDE, AND BORON FLUORIDE.
NESMEYANOV AN	0041	VAPOR PRESSURE OF ALUMINUM TRIFLUORIDE.
NEWMAN JB	0698	ABSORPTION SPECTRA OF MATRIX ISOLATED NITROGEN TRIFLUORIDE FROM 18 TO 50 GC.
NEWTON MD	0720	SELF CONSISTENT MOLECULAR ORBITAL METHODS. PART-5: AB INITIO CALCULATION OF EQUILI
NGAI LH	0351	A TRANSFERABLE UREY-BRADLEY FORCE FIELD AND THE ASSIGNMENTS OF SOME MIXED HALOMETH
NGU N	1015	MAGNETIC SUSCEPTIBILITY OF SILVER DIFLUORIDE.
NIBLER JW	0607	INFRARED AND RAMAN SPECTRA AND STRUCTURES OF MATRIX ISOLATED MAGNESIUM DIHALIDES:
	0895	LASER EXCITED RAMAN SPECTRA OF MATRIX ISOLATED PRASEODYMIUM TRIFLUORIDE.
	1123	INFRARED AND RAMAN SPECTRA AND STRUCTURE OF MATRIX ISOLATED THALLOUS HALIDE DIMERS
NICHOLSON AJC	1243	IONIZATION AND DISSOCIATION OF XENON DIFLUORIDE INDUCED BY PHOTON IMPACT.
NIECKE E	0235	CHEMICAL SHIFTS OF BORON, PHOSPHORUS, PHOSPHORYL, AND THIOPHOSPHORYL HALIDES.
	0811	RAMAN AND INFRARED SPECTRA AND THERMODYNAMIC FUNCTIONS FOR VARIOUS PHOSPHORUS AND
NIELSEN AH	0363	INFRARED SPECTRA OF CARBON TETRAFLUORIDE AND GERMANIUM TETRAFLUORIDE.
	0386	ANALYSIS OF THE INFRARED SPECTRUM OF CHLORINE MONOFLUORIDE.
	0627	MOLECULAR STRUCTURE OF MOLYBDENUM, TUNGSTEN, AND URANIUM HEXAFLUORIDES FROM INFRAR
	1005	INFRARED AND RAMAN SPECTRA OF SILICON TETRAFLUORIDE.
NIELSEN JR	1079	RAMAN SPECTRUM OF SULFUR HEXAFLUORIDE IN THE GASEOUS AND SOLID STATES.
NIELSEN U	1233	SPECTRA OF XENON DIFLUORIDE AND XENON TETRAFLUORIDE IN THE FAR ULTRAVIOLET REGION.
	1302	ELECTRONIC AND GEOMETRIC STRUCTURE OF THE FREE XENON HEXAFLUORIDE MOLECULE.
NIELSIN HH	1490	VIBRATION ROTATION ENERGIES OF MOLECULES.
NISAR M	0570	ELECTRONIC SPECTRA OF LANTHANUM FLUORIDE IN THE EXTREME ULTRAVIOLET.
	0582	ELECTRONIC SPECTRUM OF LANTHANUM TRIFLUORIDE.
	1017	FAR ULTRAVIOLET ELECTRONIC SPECTRA OF STRONTIUM DIFLUORIDE AND BARIUM DIFLUORIDE.
NISHIKIDA K	0270	ESR SPECTRUM AND STRUCTURE OF BROMINE HEXAFLUORIDE.
	0428	CHLORINE HEXAFLUORIDE RADICAL. PREPARATION, ELECTRON SPIN RESONANCE SPECTRUM, AND
NIWA K	0142	ENTHALPY OF BERYLLIUM FLUORIDE FROM 456 TO 1083 DEGREES K BY TRANSPOSED TEMPERATUR
NORDBERG R	0328	MOLECULAR SPECTROSCOPY BY MEANS OF ESCA. (CARBON TETRAFLUORIDE)
NORDINE PC	0387	DISSOCIATION ENERGY OF CHLORINE MONOFLUORIDE.
NORDLING C	0328	MOLECULAR SPECTROSCOPY BY MEANS OF ESCA. (CARBON TETRAFLUORIDE)
NORMAN N	1318	STRUCTURES OF YTTRIUM AND BISMUTH TRIFLUORIDES BY NEUTRON DIFFRACTION.
NORRIS CL	0384	MOLECULAR MAGNETIC PROPERTIES OF CHLORINE MONOFLUORIDE.
NORTHROP DA	1375	X-RAY DIFFRACTION STUDIES OF SOME TRANSITION METAL HEXAFLUORIDES. (SECOND ROW, THI
NOVICK JL	0783	VIRIAL THEOREM DECOMPOSITION OF MOLECULAR FORCE FIELDS.
NOVIKOV GI	0952	SATURATED VAPOR PRESSURE OF SCANDIUM TRIFLUORIDE, YTTRIUM TRIFLUORIDE, AND LANTHAN
	1357	SATURATED VAPOR PRESSURE OF RARE EARTH FLUORIDES.
NOWAK AV	1091	TEMPERATURE DEPENDENT ABSORPTION SPECTRUM OF THE NU-3 BAND OF SULFUR HEXAFLUORIDE
	1099	INFRARED DOUBLE RESONANCE IN SULFUR HEXAFLUORIDE.
NYHOLM RS	0617	INTERATOMIC BONDING IN CRYSTALLINE MANGANESE TRIFLUORIDE.
OAKLAND RL	0533	MOLECULAR ORBITAL TREATMENT OF IODINE HEPTAFLUORIDE.
OGIMACHI NN	0426	DENSITY, VAPOR PRESSURE, CRITICAL PROPERTIES, DIELECTRIC CONSTANT, AND SPECIFIC CO
OHNISHI K	0823	ESR SPECTRA OF PHOSPHORUS TETRAFLUORIDE RADICALS PRODUCED IN A SINGLE CRYSTAL OF P
OHRN Y	0672	CONFIGURATION INTERACTION STUDIES OF NITROGEN MONOFLUORIDE AND ITS POSITIVE ION.
OKA T	1398	INERTIA DEFECT AND PLANARITY OF FOUR ATOM MOLECULES. (XY3)
OLIVER GD	1197	VAPOR PRESSURE AND CRITICAL CONSTANTS OF URANIUM HEXAFLUORIDE.
OLMSTEAD M	0721	MOLECULAR ORBITAL STUDY OF NITROGEN TRIFLUORIDE, PHOSPHORUS TRIFLUORIDE, AND NITRO
ONAKA R	0170	STUDY OF THE A SINGLET PI- X SINGLET SIGMA(+) BANDS OF BORON-11 MONOFLUORIDE WITH
ORCHARD AF	1362	STUDY OF THE BONDING IN THE GROUP-4 TETRAHALIDES BY PHOTOELECTRON SPECTROSCOPY.
OREKHOV VT	0668	THERMODYNAMIC PROPERTIES OF NIOBIUM(V), TANTALUM(V), AND MOLYBDENUM(V) FLUORIDES.
ORGEL LE	1333	PAIRING ENERGIES IN THE ACTINIDE SERIES.
O'BRIEN TD	0781	PREPARATION AND PURIFICATION OF OXYGEN DIFLUORIDE AND DETERMINATION OF ITS VAPOR P
O'DONNELL TA	0002	FLUORIDES OF THE HEAVY TRANSITION METALS. (REVIEW)
	1243	IONIZATION AND DISSOCIATION OF XENON DIFLUORIDE INDUCED BY PHOTON IMPACT.
O'HARE PAG	0012	THERMODYNAMIC PROPERTIES OF SOME CHALCOGEN FLUORIDES.
	0093	THERMODYNAMIC PROPERTIES OF SOME ARSENIC FLUORIDES. (ARSENIC MONOFLUORIDE, ARSENIC
	0094	ARSENIC MONOFLUORIDE, 3 SIGMA. DISSOCIATION ENTHALPY, IONIZATION POTENTIAL, ELECTR
	0117	ENTHALPY OF FORMATION OF ARSENIC PENTAFLUORIDE.
	0290	MOLECULAR ORBITAL INVESTIGATION OF CARBON MONOFLUORIDE AND SILICON MONOFLUORIDE AN
	0468	THERMODYNAMIC PROPERTIES OF GASEOUS GERMANIUM, GERMANIUM TETRAFLUORIDE, GERMANIUM
	0469	ENTHALPY OF FORMATION OF GERMANIUM TETRAFLUORIDE.
	0678	QUANTUM CHEMICAL STUDY OF SOME PNICOGEN MONOFLUORIDES. (NITROGEN MONOFLUORIDE AND
	0757	OXYGEN MONOFLUORIDE, DOUBLET PI HARTREE-FOCK WAVE FUNCTION, BINDING ENERGY, IONIZA
	0788	THERMODYNAMIC PROPERTIES OF PHOSPHORUS(2), PHOSPHORUS(4), AND SOME PHOSPHORUS FLUO

O'HARE PAG   0865 FLUORINE BOMB CALORIMETRY. PART-18: STANDARD ENTHALPY OF FORMATION OF PHOSPHORUS P
              0955 DISSOCIATION ENTHALPY, SPECTROSCOPIC CONSTANTS, AND OTHER PROPERTIES OF SELENIUM M
              0973 PROPERTIES OF NITROGEN MONOFLUORIDE, SILICON MONOFLUORIDE, PHOSPHORUS MONOFLUORIDE
              1025 HARTREE-FOCK WAVE FUNCTIONS AND COMPUTED PROPERTIES FOR THE DOUBLET PI GROUND STAT
              1092 ENTHALPIES OF FORMATION OF SULFUR HEXAFLUORIDE, SELENIUM HEXAFLUORIDE, TELLURIUM H
              1157 ENTHALPY OF FORMATION OF TUNGSTEN HEXAFLUORIDE.
OSBORNE DW   0510 CALORIMETRIC STUDY OF IODINE PENTAFLUORIDE. HEAT CAPACITY BETWEEN 5 AND 350 DEGREE
              0636 HEAT CAPACITY AND OTHER THERMODYNAMIC PROPERTIES OF MOLYBDENUM HEXAFLUORIDE BETWEE
              0650 HEAT CAPACITY, ENTHALPY OF FUSION, AND THERMODYNAMIC PROPERTIES OF NEPTUNIUM HEXAF
              0885 EXPERIMENTAL HEAT CAPACITIES OF PLUTONIUM-242 TRIFLUORIDE AND PLUTONIUM-242 TETRAF
              0886 HEAT CAPACITY, ENTROPY, AND ENTHALPY OF PLUTONIUM-242 TRIFLUORIDE FROM 10 TO 350 D
              0889 HEAT CAPACITY, ENTROPY, ENTHALPY, AND GIBBS ENERGY OF PLUTONIUM-242 TETRAFLUORIDE
              1246 XENON DIFLUORIDE. HEAT CAPACITY FROM 5 TO 350 DEGREES K AND SOME DERIVED THERMODYN
              1275 XENON TETRAFLUORIDE. HEAT CAPACITY AND THERMODYNAMIC FUNCTIONS FROM 5 TO 350 DEGRE
              1301 PREPARATION AND SOME THERMAL PROPERTIES OF PURE XENON HEXAFLUORIDE.
              1305 HEAT CAPACITY AND OTHER THERMODYNAMIC FUNCTIONS OF XENON HEXAFLUORIDE FROM 5 TO 35
OSIPOVA LI   0075 THERMODYNAMIC FUNCTIONS OF ANTIMONY FLUORIDE, NIOBIUM BROMIDE, TANTALUM CHLORIDE,
OSMININ EN   0465 ANALYSIS OF THE ROTATIONAL STRUCTURE OF THE BANDS 0-0 AND 0-1 IN GERMANIUM MONOFLU
OTAKE M   0722 MICROWAVE SPECTRA OF NITROGEN TRIFLUORIDE IN THE EXCITED VIBRATIONAL STATES. (EQUI
              0723 MICROWAVE SPECTRA OF NITROGEN TRIFLUORIDE IN THE EXCITED VIBRATIONAL STATES. (ZETA
OTENG R   1139 ENVIRONMENT OF THE TIN ATOM IN ORTHORHOMBIC TIN DIFLUORIDE.
OUELLETTE TH   0626 VIBRATIONAL SPECTRA OF MOLYBDENUM PENTAFLUORIDE AND TUNGSTEN PENTAFLUORIDE.
OUELLETTE TJ   0625 VIBRATIONAL SPECTRA OF MOLYBDENUM AND TUNGSTEN PENTAFLUORIDES.
OUILLON R   1081 FORBIDDEN RAMAN BANDS OF SULFUR HEXAFLUORIDE. COLLISION-INDUCED RAMAN SCATTERING.
OVCHARENKO IE 0970 DETERMINATION OF THE MATRIX ELEMENT OF THE DIPOLE MOMENT OF AN A DOUBLET SIGMA TO
OVEREND J   0192 BORON TRIFLUORIDE. THE ANALYSIS OF 2NU-3 AND THE DETERMINATION OF THE 33 CORIOLIS
              0193 ROTATION VIBRATION BAND CONTOURS IN THE INFRARED SPECTRUM OF BORON TRIFLUORIDE.
              0207 NU-3 BANDS OF BORON TRIFLUORIDE.
              0208 NU-4 BANDS OF THE TEN AND ELEVEN ISOTOPES OF BORON TRIFLUORIDE AT HIGH RESOLUTION.
              0209 BORON TRIFLUORIDE. THE ANALYSIS OF NU-2 AND THE DETERMINATION OF THE MOLECULAR GEO
              0210 VIBRATIONAL ANHARMONICITY IN BORON TRIFLUORIDE.
              0338 INFRARED SPECTRA OF SIMPLE MOLECULES IN LIQUID ARGON SOLUTION. CARBON TETRAFLUORID
              0561 STRUCTURE OF KRYPTON DIFLUORIDE.
              0564 ABSENCE OF FERMI RESONANCE IN KRYPTON DIFLUORIDE.
              0745 VIBRATIONAL ANHARMONICITY IN THE PYRAMIDAL TRIFLUORIDES.
              0816 HIGH RESOLUTION SPECTRA OF ARSENIC TRIFLUORIDE AND PHOSPHORUS TRIFLUORIDE.
              1491 LEAST SQUARES ADJUSTMENT OF ANHARMONIC POTENTIAL CONSTANTS. APPLICATION TO CARBON
              1492 ADAPTATION OF THE UREY-BRADLEY FORCE FIELD TO AMMONIA, PHOSPHINE, STIBINE, AND ARS
              1493 VIBRATIONAL ANHARMONICITY IN THE METHYL HALIDES.
OZIER I   1287 MAGNETIC FIELD DEFLECTION OF XENON HEXAFLUORIDE.
OZIN GA   0475 MATRIX INFRARED AND LASER RAMAN SPECTRA, MOLECULAR STRUCTURES AND NORMAL COORDINAT
PACE EL   0361 THERMODYNAMIC PROPERTIES OF CARBON TETRAFLUORIDE FROM 12 DEGREES K TO ITS BOILING
              0724 INFRARED AND RAMAN SPECTRUM OF NITROGEN TRIFLUORIDE.
PACKER KH   0922 INTRAMOLECULAR LIGAND EXCHANGE IN SEVEN COORDINATE STRUCTURES. (RHENIUM HEPTAFLUOR
PAGLIA SRL   0725 VACUUM ULTRAVIOLET ABSORPTION SPECTRUM AND DIPOLE MOMENT OF NITROGEN TRIFLUORIDE.
PAHN DK   1484 USE OF A SIMPLE ELECTROSTATIC MODEL FOR CALCULATION OF ANGULAR FORCE CONSTANTS OF
PAINE RT   1196 VIBRATIONAL SPECTRUM AND FORCE FIELD OF URANIUM HEXAFLUORIDE.
PALMIERI P   0371 SCF- MO CALCULATIONS OF SOME MOLECULAR PROPERTIES OF THE ISOELECTRONIC SERIES CHLO
              1066 AB INITIO AND MZDO WAVE FUNCTIONS FOR SULFUR HEXAFLUORIDE.
PANKTRATOV AV 0696 THERMOCHEMICAL AND THERMODYNAMIC PROPERTIES OF INORGANIC NITROGEN FLUORIDES. (NITR
PANNETIER G   0089 ULTRAVIOLET SPECTRUM OF ARSENIC MONOFLUORIDE.
              0090 ANALYSIS OF THE ABSORPTION SPECTRUM AND OF TWO NEW SINGLET SYSTEMS OF ARSENIC MONO
              1142 EMISSION SPECTRUM OF TITANIUM MONOFLUORIDE.
PAPOUSEK D   0219 FORCE CONSTANTS AND THERMODYNAMIC CONSTANTS FOR THE BORON HALIDES.
PARISEAU MA   1491 LEAST SQUARES ADJUSTMENT OF ANHARMONIC POTENTIAL CONSTANTS. APPLICATION TO CARBON
              1492 ADAPTATION OF THE UREY-BRADLEY FORCE FIELD TO AMMONIA, PHOSPHINE, STIBINE, AND ARS
PARKINSON TF 0381 INFRARED AND RAMAN SPECTRA OF CHLORINE MONOFLUORIDE.
PARKS BH   1198 THERMODYNAMIC PROPERTIES OF GASEOUS URANIUM HEXAFLUORIDE.
PATEL MM   0057 ROTATIONAL ANALYSIS OF C3 SYSTEM OF THE ANTIMONY MONOFLUORIDE MOLECULE.
              0058 A NEW BAND SYSTEM OF THE ANTIMONY MONOFLUORIDE MOLECULE IN 4050-5450 ANGSTROM REGI
              0059 FINE STRUCTURE ANALYSIS OF THE C1 SYSTEM OF ANTIMONY MONOFLUORIDE.
              0060 EMISSION OF MOLECULAR ANTIMONY MONOFLUORIDE.
              0062 VIBRATIONAL ANALYSIS OF ULTRAVIOLET BANDS OF ANTIMONY MONOFLUORIDE.
              0147 EMISSION SPECTRUM OF BISMUTH MONOFLUORIDE.
              0491 ROTATIONAL ANALYSIS OF THE C SINGLET PI - X SINGLET SIGMA(+) SYSTEM OF INDIUM MONOF
PATHAK AN   0171 FRANCK-CONDON FACTORS AND R CENTROIDS FOR THE B TRIPLET SIGMA(+) TO A TRIPLET PI S
PATHAK CM   0121 GREEN BAND SYSTEM OF BARIUM MONOFLUORIDE.
PAULE RC   0045 INFRARED SPECTRA OF GASEOUS ALUMINUM TRIFLUORIDE, LITHIUM ALUMINUM TETRAFLUORIDE,
PAVLOV AA   0654 THERMODYNAMIC PROPERTIES OF NIOBIUM AND TANTALUM FLUORIDES AT HIGH TEMPERATURES. P
              0655 THERMODYNAMIC PROPERTIES OF NIOBIUM MONOFLUORIDE, NIOBIUM DIFLUORIDE, TANTALUM MON
              0656 THERMODYNAMIC PROPERTIES OF NIOBIUM TRIFLUORIDE, NIOBIUM TETRAFLUORIDE, TANTALUM T
              0657 THERMODYNAMIC PROPERTIES OF NIOBIUM AND TANTALUM FLUORIDES AT HIGH TEMPERATURES. P
              0665 THERMODYNAMIC PROPERTIES OF NIOBIUM AND TANTALUM FLUORIDES AT HIGH TEMPERATURES. P
PEACOCK CJ   0735 FORCE FIELD AND MOLECULAR CONSTANTS OF NITROGEN TRIFLUORIDE.
              1399 CALCULATIONS OF MEAN AMPLITUDES OF VIBRATION AND UREY-BRADLEY FORCE FIELDS FOR PLA
              1457 FORCE CONSTANTS FOR HEXAFLUORIDE MOLECULES.
PEACOCK RD   0013 TRANSITION METAL PENTAFLUORIDES AND RELATED COMPOUNDS.
              0749 PREPARATION AND CRYSTAL STRUCTURE OF OSMIUM PENTAFLUORIDE. TWO FLUORIDES OF OSMIUM
              0924 RUTHENIUM PENTAFLUORIDE AND RUTHENIUM OXYTETRAFLUORIDE. (VAPOR PRESSURES)
              0925 ELECTRONIC SPECTRA OF LIQUID RUTHENIUM AND MOLYBDENUM PENTAFLUORIDES.
PEARSON E   1299 FAR INFRARED AND MILLIMETER WAVE STUDIES OF XENON HEXAFLUORIDE.
PEARSON EF   0033 MILLIMETER AND SUBMILLETER WAVE SPECTRUM AND MOLECULAR CONSTANTS OF ALUMINUM MONOF
PEDERSEN L   0931 ROTATIONAL ANALYSIS OF THE SYSTEM B SINGLET PI TO SINGLET SIGMA(+) OF GASEOUS SCAN
PEEL JB   0835 VESCF-MO STUDIES OF PHOSPHORUS PENTAFLUORIDE, SULFUR TETRAFLUORIDE, AND CHLORINE T
              1029 NONEMPIRICAL VALENCE ELECTRON CALCULATIONS ON SMALL MOLECULES CONTAINING PHOSPHORU
              1373 VESCF-MO STUDIES OF MOLECULES CONTAINING ATOMS FROM THE SECOND ROW OF THE PERIODIC
              1474 UREY-BRADLEY POTENTIAL CONSTANTS FOR SECOND ROW FLUORIDES.
PEMSLER JP   0866 INFRARED SPECTRUM OF PHOSPHORUS PENTAFLUORIDE.
PEREZ M   1070 NU-3 AND NU-4 BANDS OF SULFUR HEXAFLUORIDE.
PERKINS HK   0543 ELECTRONIC STRUCTURE AND ANOMALOUS THERMAL EXPANSION ON IRON DIFLUORIDE AND VANADI

PERKINS PG	0184	ELECTRONIC STRUCTURE OF BORON TRIFLUORIDE.
	0185	GROUND STATE PROPERTIES OF THE GROUP-2 TRIHALIDES. (BORON TRIFLUORIDE)
PERNG CN	0116	PERPENDICULAR BAND CONTOURS AND VIBRATIONAL POTENTIAL FUNCTION OF THE E' VIBRATION
PERRIN F	0637	HIGH RESOLUTION NMR SPECTRUM OF LIQUID MOLYBDENUM HEXAFLUORIDE, TUNGSTEN HEXAFLUOR
	1015	MAGNETIC SUSCEPTIBILITY OF SILVER DIFLUORIDE.
	1189	VIBRATIONAL SPECTRUM AND STRUCTURE OF SOLID URANIUM HEXAFLUORIDE.
PERSON WB	0195	CNDO CALCULATION OF DIPOLE MOMENT DERIVATIVES AND INFRARED INTENSITIES OF BORON TR
	0352	A GEOMETRIC VISUALIZATION OF NORMAL COORDINATE TRANSFORMATIONS. APPLICATION TO THE
PERUMAL A	0087	VIBRATIONAL ASSIGNMENTS AND NORMAL COORDINATE ANALYSIS FOR ANTIMONY PENTAFLUORIDE.
PERVOV VS	1149	THERMODYNAMIC PROPERTIES OF TUNGSTEN PENTAFLUORIDE.
PETERSON SH	1295	STRUCTURE OF GASEOUS XENON HEXAFLUORIDE.
PETRONGOLO C	0762	DOUBLE ZETA SCF CALCULATIONS FOR NITRITE ION AND OXYGEN DIFLUORIDE.
	0763	CONFIGURATION INTERACTION CALCULATION FOR THE GROUND STATE OF OXYGEN DIFLUORIDE, N
	0776	MINIMAL BASIS SET SCF CALCULATIONS FOR THE GROUND STATE OF OZONE, NITRITE ION, NIT
PETROV KI	0073	RAMAN SPECTRUM OF ANTIMONY TRIFLUORIDE.
PETTY F	0047	HIGH TEMPERATURE NEGATIVE IONS. GASEOUS GROUP-3 FLUORIDES.
PETZEL VT	0452	PHASE INVESTIGATIONS AND VAPOR PRESSURE MEASUREMENTS ON EUROPIUM DIFLUORIDE.
	0950	THERMODYNAMIC PROPERTIES OF SCANDIUM TRIFLUORIDE.
PEYERIMHOFF SD	0764	GEOMETRY OF MOLECULES. (OXYGEN DIFLUORIDE)
PHILLIPS EW	1304	MULTIPLE SCATTERING X(ALPHA) CALCULATION OF THE ENERGY LEVEL STRUCTURE OF XENON HE
PHILLIPS WD	1048	STRUCTURE OF EXCHANGE PROCESSES IN SOME INORGANIC FLUORIDES BY NMR. (SULFUR TETRAF
	1049	FLUORINE EXCHANGE IN SULFUR TETRAFLUORIDE.
PHIPPS TE	0887	VAPOR PRESSURE OF PLUTONIUM HALIDES. (PLUTONIUM TRIFLUORIDE)
PICHAI R	1050	MOLECULAR FORCE FIELD FOR SULFUR TETRAFLUORIDE.
PIERCE L	0724	INFRARED AND RAMAN SPECTRUM OF NITROGEN TRIFLUORIDE.
	0777	SPIN ROTATIONAL HYPERFINE STRUCTURE IN THE MICROWAVE SPECTRUM OF OXYGEN DIFLUORIDE
	0778	CENTRIFUGAL DISTORTION EFFECTS IN ASYMMETRIC ROTOR MOLECULES. PART-1. QUADRATIC PO
	0779	MICROWAVE SPECTRUM, STRUCTURE, AND DIPOLE MOMENT OF OXYGEN DIFLUORIDE.
PIERCE SB	0516	NMR SPECTRA OF IODINE HEPTAFLUORIDE AND IODINE OXYGEN PENTAFLUORIDE.
PILIPOVICH D	0257	VIBRATIONAL SPECTRUM OF BROMINE TRIFLUORIDE.
	0534	MASS SPECTRA AND SUBLIMATION PRESSURES OF IODINE HEPTAFLUORIDE AND IODINE OXYGEN P
	1271	HEAT CAPACITY AND RELATED THERMODYNAMIC FUNCTIONS OF XENON TETRAFLUORIDE.
PILLAI MGK	0087	VIBRATIONAL ASSIGNMENTS AND NORMAL COORDINATE ANALYSIS FOR ANTIMONY PENTAFLUORIDE.
	0264	MOLECULAR FORCE FIELD FOR BROMINE PENTAFLUORIDE.
	0402	MOLECULAR FORCE FIELD FOR CHLORINE TRIFLUORIDE.
	1050	MOLECULAR FORCE FIELD FOR SULFUR TETRAFLUORIDE.
	1058	MOLECULAR VIBRATIONS OF SULFUR TETRAFLUORIDE.
	1400	MOLECULAR FORCE FIELD FOR SOME XY3 MOLECULES. (PHOSPHORUS TRIFLUORIDE AND ARSENIC
	1428	POTENTIAL CONSTANTS OF XY4 AND XY3Z TYPES OF MOLECULES AND RADICALS.
PILLAI PP	0264	MOLECULAR FORCE FIELD FOR BROMINE PENTAFLUORIDE.
	1400	MOLECULAR FORCE FIELD FOR SOME XY3 MOLECULES. (PHOSPHORUS TRIFLUORIDE AND ARSENIC
PIMENTEL GC	0310	INFRARED DETECTION OF GASEOUS CARBON TRIFLUORIDE RADICAL.
	0566	PREPARATION OF INERT GAS COMPOUNDS BY MATRIX ISOLATION: KRYPTON DIFLUORIDE.
	0567	KRYPTON DIFLUORIDE. PREPARATION BY THE MATRIX ISOLATION TECHNIQUE.
PINCELLI U	0196	BARRIER TO ELECTRON PASSAGE THROUGH ELECTRONEGATIVE ATOMS IN BORON TRIFLUORIDE.
	1066	AB INITIO AND MZDO WAVE FUNCTIONS FOR SULFUR HEXAFLUORIDE.
PIRKMAJER E	0883	SUSCEPTIBILITY MEASUREMENTS AND FLUORINE NMR OF PLATINUM HEXAFLUORIDE.
PISTORIUS CWFT	0353	HARMONIC POTENTIAL CONSTANTS OF OSMIUM TETROXIDE AND SOME TETRAHALIDE (CARBON TETR
	0354	FORCE CONSTANTS OF THE CARBON TETRAHALIDES. (CARBON TETRAFLUORIDE)
	0850	POTENTIAL FIELD AND FORCE CONSTANTS OF SOME TRIGONAL BIPYRAMIDAL PENTAHALIDES. (PH
	1401	EVALUATION OF FORCE CONSTANTS OF PLANE XY3 MOLECULES FROM VIBRATIONAL DATA.
	1421	CLASSICAL VIBRATION PROBLEM FOR THE PYRAMIDAL XY4 MOLECULE.
	1422	FORCE CONSTANTS OF TETRAHEDRAL MOLECULES.
	1464	POTENTIAL FIELD AND FORCE CONSTANTS OF OCTAHEDRAL MOLECULES.
PISTORIUS MC	0354	FORCE CONSTANTS OF THE CARBON TETRAHALIDES. (CARBON TETRAFLUORIDE)
PITZER KS	0830	POTENTIAL FUNCTION FOR THE NU-7 VIBRATION OF PHOSPHORUS PENTAFLUORIDE.
	1281	ELECTRIC FIELD DEFLECTION OF MOLECULES WITH LARGE AMPLITUDE MOTION. (XENON HEXAFLU
	1303	MOLECULAR STRUCTURE OF XENON HEXAFLUORIDE.
PLANCKAERT AA	0237	VACUUM ULTRAVIOLET ABSORPTION SPECTRA OF BORON TRIHALIDES.
PLANET WG	0866	INFRARED SPECTRUM OF PHOSPHORUS PENTAFLUORIDE.
PLATO V	0828	GILLESPIE-NYHOLM ASPECTS OF FORCE FIELDS. PART-1. POINTS-ON-A-SPHERE AND EXTENDED
PLOTNIKOVA AD	0826	CALCULATION OF FORCE CONSTANTS AND ROOT MEAN SQUARE VIBRATION AMPLITUDES OF PHOSPH
PLURIEN PL	1276	THERMOCHEMICAL STUDIES OF XENON TETRAFLUORIDE AND XENON HEXAFLUORIDE.
PLYLER EK	0343	INFRARED SPECTRUM OF CARBON TETRAFLUORIDE.
	0355	INFRARED SPECTRA OF HALOGEN-SUBSTITUTED METHANES. (CARBON TETRAFLUORIDE)
POCHAN JM	0744	ZEEMAN STUDIES INCLUDING THE MOLECULAR G VALUES, MAGNETIC SUSCEPTIBILITY, AND MOLE
	0780	MOLECULAR G VALUES, MAGNETIC SUSCEPTIBILITIES, MOLECULAR QUADRUPOLE MOMENTS AND SE
POLO SR	0748	INFRARED SPECTRA OF NITROGEN TRIFLUORIDE AND PHOSPHORUS TRIFLUORIDE.
POLYACHENOK OG	0052	THERMODYNAMICS OF THE VAPORIZATION OF ALUMINUM FLUORIDE AND CESIUM FLUORIDE.
	0577	ENTHALPY OF FORMATION OF LANTHANUM TRIFLUORIDE AND PRASEODYMIUM TRIFLUORIDE.
	0578	EXPERIMENTAL DETERMINATION OF THE ENTHALPIES OF FORMATION OF RARE EARTH FLUORIDES.
	0951	ENERGETICS AND STABILITY OF GAS PHASE SCANDIUM FLUORIDES.
PONOMAREV YI	0726	HARMONIC FREQUENCIES, POTENTIAL FUNCTION, CORIOLIS COUPLING CONSTANTS, CENTRIFUGAL
	0727	DETERMINATION OF THE STRUCTURE OF PYRAMIDAL MOLECULES OF THE XY3 TYPE BY SOLUTION
	1423	USE OF CORIOLIS INTERACTION CONSTANTS IN THE SOLUTION OF THE VIBRATIONAL PROBLEM F
POPLE JA	0707	MOLECULAR ORBITAL THEORY OF THE ELECTRONIC STRUCTURE OF ORGANIC COMPOUNDS. PART-3:
	0720	SELF CONSISTENT MOLECULAR ORBITAL METHODS. PART-5: AB INITIO CALCULATION OF EQUILI
POPOV M	1175	VAPOR PRESSURE OF URANIUM TETRAFLUORIDE.
POPPLEWELL RJL	0728	INFRARED SPECTRUM AND MOLECULAR CONSTANTS OF NITROGEN TRIFLUORIDE.
PORTER RF	0298	HEAT OF FORMATION OF CARBON DIFLUORIDE.
PORTER TL	0291	EMISSION SPECTRUM OF CARBON MONOFLUORIDE.
PORTIER J	0444	NEW FLUORIDES OF TRIVALENT COPPER.
PORTO SPS	0572	LATTICE VIBRATIONS AND STRUCTURE OF RARE EARTH HALIDES. (LANTHANUM TRIFLUORIDE)
	0615	RAMAN SPECTRA OF TITANIUM DIOXIDE, MAGNESIUM DIFLUORIDE, ZINC DIFLUORIDE, DIFLUORI
POTT CJ	0928	ELECTRONIC STATES OF GASEOUS CALCIUM MONOFLUORIDE, YTTRIUM MONOFLUORIDE, AND LANT
POTTER RL	0812	THERMODYNAMIC FUNCTIONS OF SOME PHOSPHORUS COMPOUNDS. (PHOSPHORUS TRIFLUORIDE)
POTTS AW	1361	PHOTOELECTRON SPECTRA OF THE HALIDES IN GROUP-3, GROUP-4, GROUP-5, AND GROUP-6.
POWELL FX	0307	MICROWAVE SPECTRUM OF CARBON DIFLUORIDE.
	1030	MICROWAVE SPECTRUM AND STRUCTURE OF SULFUR DIFLUORIDE.
	1031	CENTRIFUGAL DISTORTION EFFECTS IN SULFUR DIFLUORIDE. CALCULATION OF THE FORCE FIEL

POWELL JL	0964	SPECTRA OF SILICON MONOFLUORIDE, GERMANIUM MONOFLUORIDE, TIN MONOFLUORIDE, AND LEA
POWLING J	0761	VIBRATIONAL SPECTRA AND STRUCTURE OF INORGANIC MOLECULES. PART-1: INFRARED SPECTRU
POZHARSKAYA GV	0041	VAPOR PRESSURE OF ALUMINUM TRIFLUORIDE.
PRADAL F	0279	SPECTRA OF ELECTRONIC EXCITATIONS IN CALCIUM DIFLUORIDE, STRONTIUM DIFLUORIDE, AND
PREATON KF	0959	PARAMAGNETIC HEXAFLUORIDE ANIONS OF GROUP-6. (SELENIUM PENTAFLUORIDE)
PREMASWARUP D	0153	ROTATIONAL ANALYSIS OF BANDS OF THE C TRIPLET SIGMA, B TRIPLET SIGMA: A TRIPLET PI
PRESTON KF	0112	ESR SPECTRUM OF ARSENIC TETRAFLUORIDE.
	0271	ELECTRON SPIN RESONANCE SPECTRA OF HALOGEN HEXAFLUORIDES. (CHLORINE HEXAFLUORIDE,
	0413	EPR SPECTRUM OF THE RADICAL CHLORINE TETRAFLUORIDE.
	0791	ESR SPECTRA OF PHOSPHORUS DIFLUORIDE AND SULFUR TRIFLUORIDE RADICALS.
PREVOT F	0063	ROTATIONAL ANALYSIS OF THE ULTRAVIOLET BANDS OF THE ANTIMONY MONOFLUORIDE MOLECULE
	0071	BO(+)-X(21) AND BO(+)-X(10+) BANDS OF ANTIMONY MONOFLUORIDE.
	0585	B SINGLET SIGMA(+) TO X TRIPLET SIGMA(-) BAND SYSTEM IN LEAD MONOFLUORIDE.
PRICE WC	1361	PHOTOELECTRON SPECTRA OF THE HALIDES IN GROUP-3, GROUP-4, GROUP-5, AND GROUP-6.
PRIEST HF	1180	VAPOR PRESSURE OF URANIUM HEXAFLUORIDE IN EQUILIBRIUM WITH URANIUM PENTAFLUORIDE.
PRUSAKOV VN	0563	KRYPTON DIFLUORIDE. (PHYSICAL AND CHEMICAL PROPERTIES)
PULAY P	1036	VIBRATIONAL ASSIGNMENT OF SULFUR TETRAFLUORIDE.
	1395	CALCULATION OF FORCE CONSTANTS WITH THE MAXIMUM OVERLAP METHOD. MOLECULES WITH LON
	1416	CALCULATION OF FORCE CONSTANTS WITH THE MAXIMUM OVERLAP METHOD. TETRAHEDRAL MOLECU
PULLEN BP	0319	HIGH RESOLUTION PHOTOELECTRON SPECTROSCOPY OF CARBON TETRAFLUORIDE AND SILICON TE
PURCELL KF	0877	NORMAL COORDINATE ANALYSIS OF PHOSPHORUS PENTAFLUORIDE, PHOSPHORUS DIFLUORIDE DICH
PURETSKII AA	1064	SELECTIVE PHOTODISSOCIATION OF POLYATOMIC MOLECULES (SULFUR HEXAFLUORIDE) BY INFRA
PURUSHOTHAMAN C	0406	MEAN AMPLITUDES OF VIBRATION OF SELENIUM TRIOXIDE AND CHLORINE TRIFLUORIDE.
	0411	GENERALIZED MEAN AMPLITUDES AND CORIOLIS CONSTANTS IN DICHLORO BORANE, DIBROMO BOR
	0513	UREY-BRADLEY FORCE CONSTANTS, MEAN AMPLITUDES OF VIBRATION, SHRINKAGE EFFECT AND C
	1405	GENERALIZED MEAN SQUARE AMPLITUDES AND CORIOLIS CONSTANTS OF SOME PYRAMIDAL XY3 MO
PYSH ES	1247	FORBIDDEN ELECTRONIC TRANSITIONS IN XENON DIFLUORIDE AND XENON TETRAFLUORIDE.
QUAGLINO J	0426	DENSITY, VAPOR PRESSURE, CRITICAL PROPERTIES, DIELECTRIC CONSTANT, AND SPECIFIC CO
RACCAH PM	0454	MAGNETIC SUSCEPTIBILITY OF EUROPIUM TRIFLUORIDE.
RADHAKRISHNAN M	0596	UREY-BRADLEY FORCE FIELD. (LEAD TETRAFLUORIDE)
	1009	POTENTIAL CONSTANTS OF SOME XY4 TYPE OF SILANE COMPOUNDS. (SILICON TETRAFLUORIDE)
	1051	THERMODYNAMIC PROPERTIES OF SULFUR TETRAFLUORIDE AND SULFUR OXIDE TETRAFLUORIDE.
	1439	POTENTIAL CONSTANTS OF SOME XY5 MOLECULES.
RADTKE DD	1351	INFRARED SPECTRA AND GEOMETRIES OF RARE EARTH DIHALIDES.
RAFF L	0299	A THEORETICAL STUDY OF THE BOND INTERACTION FORCE CONSTANTS IN DIFLUORIDE MOLECULE
RAI DK	0018	FORCE CONSTANTS OF METAL HEXAFLUORIDES.
	0032	POTENTIAL ENERGY CURVES AND NATURE OF BINDING IN GROUP-3A MONOHALIDES.
	0061	DETERMINATION OF THE EFFECTIVE VIBRATIONAL TEMPERATURE IN THE A-2 SYSTEM OF ANTIMO
	0130	POTENTIAL CURVES FOR SOME DIATOMIC MOLECULES. (BERYLLIUM MONOFLUORIDE, SILICON MON
	0173	POTENTIAL ENERGY CURVES AND DISSOCIATION ENERGIES OF DIATOMIC BORON AND ALUMINUM H
	0458	POTENTIAL ENERGY CURVES AND DISSOCIATION ENERGIES OF DIATOMIC FLUORIDES AND CHLORI
	0589	ON THE DISSOCIATION OF LEAD MONOFLUORIDE AND BISMUTH MONOFLUORIDE.
	0641	VALENCE FORCE CONSTANTS OF THE HEXAFLUORIDES OF MOLYBDENUM, TECHNETIUM, AND RHENIU
	1136	ROTATIONAL ANALYSIS OF THE C DOUBLET DELTA 5/2- X DOUBLET PI 3/2 AND G DOUBLET DEL
	1141	IONICITY OF THE MX BOND IN THE TETRAHALIDES OF GROUP-4 ELEMENTS. (TIN TETRAFLUORID
	1425	APPLICATION OF UREY-BRADLEY FORCE FIELD TO SOME TETRAHEDRAL XY4 TYPE MOLECULES AND
	1466	MEAN SQUARE AMPLITUDES OF INTERATOMIC DISTANCES IN SOME OCTAHEDRAL HEXAFLUORIDE MO
	1468	EXTREMAL PROPERTIES OF FORCE CONSTANTS IN OCTAHEDRAL HEXAFLUORIDES.
RAI SB	1135	ROTATIONAL STRUCTURE AND ISOTOPIC SHIFT IN THE (1,0) BAND OF THE A DOUBLET SIGMA-
RAI SN	1141	IONICITY OF THE MX BOND IN THE TETRAHALIDES OF GROUP-4 ELEMENTS. (TIN TETRAFLUORID
	1402	INTRAMOLECULAR FORCE FIELDS IN GROUP-5 TRIHALIDES.
	1496	FORCE CONSTANTS OF MOLECULES WITH STRONGLY COUPLED VIBRATIONS.
RAJALAKSHMI KV	1406	UREY-BRADLEY FORCE FIELD AND THERMODYNAMIC PROPERTIES OF PYRAMIDAL XY3 MOLECULES.
	1407	MEAN AMPLITUDES OF VIBRATION FOR SOME PYRAMIDAL XY3 AND TETRAHEDRAL XY4 MOLECULES.
RAKOV EG	0669	THERMODYNAMIC PROPERTIES OF NIOBIUM AND TANTALUM PENTAFLUORIDES.
RAM RS	1136	ROTATIONAL ANALYSIS OF THE C DOUBLET DELTA 5/2- X DOUBLET PI 3/2 AND G DOUBLET DEL
RAMALHO DE AZEVEDO MM	1342	NOBLE GASES AND THE BINARY FLUORIDES OF XENON.
RAMASWAMY K	0087	VIBRATIONAL ASSIGNMENTS AND NORMAL COORDINATE ANALYSIS FOR ANTIMONY PENTAFLUORIDE.
	0238	UNIQUE FORCE FIELD FOR BORON TRIHALIDES.
	0407	MOLECULAR FORCE FIELD FOR INTERHALOGEN COMPOUNDS. (CHLORINE TRIFLUORIDE AND BROMIN
	0425	MOLECULAR FORCE FIELDS FOR INTERHALOGEN COMPOUNDS. (CHLORINE PENTAFLUORIDE AND BRO
	0511	MOLECULAR FORCE FIELD FOR INTERHALOGEN COMPOUNDS. IODINE PENTAFLUORIDE, IODINE OXY
	0729	GREEN'S FUNCTION ANALYSIS OF NITROGEN TRIFLUORIDE AND NITROGEN TRICHLORIDE.
	0730	NORMAL COORDINATE ANALYSIS OF SOME NITROGEN HALIDES. (NITROGEN TRIFLUORIDE)
	0867	MOLECULAR CONSTANTS OF PHOSPHORUS HALIDES. PART-4. (PHOSPHORUS PENTAFLUORIDE)
	0868	MOLECULAR CONSTANTS OF SOME PHOSPHORUS HALOGEN COMPOUNDS. (PHOSPHORUS PENTAFLUORID
	0958	VIBRATIONAL ASSIGNMENT AND NORMAL COORDINATE ANALYSIS OF SELENIUM TETRAFLUORIDE.
	1050	MOLECULAR FORCE FIELD FOR SULFUR TETRAFLUORIDE.
	1052	FORCE FIELD STUDY FOR SULFUR TETRAFLUORIDE.
	1093	GREEN'S FUNCTION ANALYSIS OF SULFUR HEXAFLUORIDE AND RELATED MOLECULES. (SELENIUM
	1403	UNIQUE FORCE FIELDS OF SOME XY3 TYPE MOLECULES.
	1424	MOLECULAR CONSTANTS OF SOME TETRAHEDRAL XY4 TYPE MOLECULES AND THEIR SUBSTITUTED D
	1465	MOLECULAR FORCE FIELD FOR SOME XY6 TYPE MOLECULES. (SELENIUM AND TELLURIUM HEXAFLU
RAMBIDI NG	0036	ELECTRON DIFFRACTION INVESTIGATION OF THE MOLECULAR STRUCTURES OF THE ALUMINUM HAL
	1170	ELECTRON DIFFRACTION STUDY OF URANIUM AND THORIUM TETRAHALIDES IN THE GAS PHASE. (
RAMSAY DA	0308	ELECTRONIC SPECTRA OF POLYATOMIC FREE RADICALS. (CARBON DIFLUORIDE)
RANGANATHAN V	1424	MOLECULAR CONSTANTS OF SOME TETRAHEDRAL XY4 TYPE MOLECULES AND THEIR SUBSTITUTED D
RANKIN WH	0800	IONIZATION BY ELECTRON IMPACT OF PHOSPHORUS TRIFLUORIDE AND DIFLUORO CYANO PHOSPHI
RAO BK	0813	MOLECULAR CONSTANTS OF PHOSPHORUS HALOGEN COMPOUNDS. (PHOSPHORUS TRIFLUORIDE)
	0867	MOLECULAR CONSTANTS OF PHOSPHORUS HALIDES. PART-4. (PHOSPHORUS PENTAFLUORIDE)
	0868	MOLECULAR CONSTANTS OF SOME PHOSPHORUS HALOGEN COMPOUNDS. (PHOSPHORUS PENTAFLUORID
RAO CGR	0814	CENTRIFUGAL DISTORTION CONSTANTS OF PHOSPHORUS TRIFLUORIDE.
RAO DR	0987	NEW ELECTRONIC EMISSION FROM SILICON DIFLUORIDE.
	0988	EMISSION SPECTRUM OF SILICON DIFLUORIDE. PART-1: THE BAND SYSTEM IN THE REGION 275
RAO DVK	0144	NEW BAND SYSTEM OF BISMUTH FLUORIDE MOLECULE IN THE REGION 6200-7000 ANGSTROMS.
RAO JVR	1125	BAND SPECTRA OF THALLIUM MONOIODIDE AND MONOFLUORIDE.
RAO KM	0145	VISIBLE EMISSION SPECTRUM OF BISMUTH MONOFLUORIDE.
	0587	ROTATIONAL ANALYSIS OF THE VISIBLE EMISSION BANDS OF LEAD MONOFLUORIDE.
RAO KN	0200	NU-3 OF BORON TRIFLUORIDE.
RAO PT	0064	EMISSION SPECTRA OF ANTIMONY MONOFLUORIDE AND BISMUTH MONOFLUORIDE.

RAO PT	0069	ROTATIONAL ANALYSIS OF THE C1 SYSTEM OF ANTIMONY MONOFLUORIDE.
	0144	NEW BAND SYSTEM OF BISMUTH FLUORIDE MOLECULE IN THE REGION 6200-7000 ANGSTROMS.
	0145	VISIBLE EMISSION SPECTRUM OF BISMUTH MONOFLUORIDE.
	0146	ROTATIONAL ANALYSIS OF THE VISIBLE BAND SYSTEM OF THE BISMUTH MONOFLUORIDE MOLECUL
	0586	ROTATIONAL ANALYSIS OF THE B-X SYSTEM OF LEAD MONOFLUORIDE IN ULTRAVIOLET.
	0587	ROTATIONAL ANALYSIS OF THE VISIBLE EMISSION BANDS OF LEAD MONOFLUORIDE.
	0610	COMPLEX BAND SYSTEM OF MANGANESE MONOFLUORIDE IN THE NEAR ULTRAVIOLET.
	0618	A NEW ELECTRONIC TRANSITION OF THE MERCURY MONOFLUORIDE MOLECULE.
	1125	BAND SPECTRA OF THALLIUM MONOIODIDE AND MONOFLUORIDE.
RAO TAP	0064	EMISSION SPECTRA OF ANTIMONY MONOFLUORIDE AND BISMUTH MONOFLUORIDE.
	0146	ROTATIONAL ANALYSIS OF THE VISIBLE BAND SYSTEM OF THE BISMUTH MONOFLUORIDE MOLECUL
RAO VK	0610	COMPLEX BAND SYSTEM OF MANGANESE MONOFLUORIDE IN THE NEAR ULTRAVIOLET.
RAO VM	0989	MICROWAVE SPECTRUM AND FORCE CONSTANTS OF SILICON DIFLUORIDE. CENTRIFUGAL DISTORTI
	0990	MICROWAVE SPECTRUM OF SILICON DIFLUORIDE.
RAO VRA	0429	BAND SPECTRUM OF CHROMIUM MONOFLUORIDE.
	1425	APPLICATION OF UREY-BRADLEY FORCE FIELD TO SOME TETRAHEDRAL XY4 TYPE MOLECULES AND
	1468	EXTREMAL PROPERTIES OF FORCE CONSTANTS IN OCTAHEDRAL HEXAFLUORIDES.
RAST HE	0453	OPTICAL ABSORPTION AND FLUORESCENCE SPECTRA OF EUROPIUM TRIFLUORIDE.
RATCLIFFE CT	0625	VIBRATIONAL SPECTRA OF MOLYBDENUM AND TUNGSTEN PENTAFLUORIDES.
	0626	VIBRATIONAL SPECTRA OF MOLYBDENUM PENTAFLUORIDE AND TUNGSTEN PENTAFLUORIDE.
RAYMOND JI	0755	MATRIX INFRARED SPECTRUM OF OXYGEN MONOFLUORIDE AND DETECTION OF LITHIUM OXYFLUORI
REDDY BR	0618	A NEW ELECTRONIC TRANSITION OF THE MERCURY MONOFLUORIDE MOLECULE.
REDDY SP	0610	COMPLEX BAND SYSTEM OF MANGANESE MONOFLUORIDE IN THE NEAR ULTRAVIOLET.
REDDY YP	0144	NEW BAND SYSTEM OF BISMUTH FLUORIDE MOLECULE IN THE REGION 6200-7000 ANGSTROMS.
	0586	ROTATIONAL ANALYSIS OF THE B-X SYSTEM OF LEAD MONOFLUORIDE IN ULTRAVIOLET.
REDINGTON RL	0113	INFRARED SPECTRA OF MATRIX ISOLATED ANTIMONY PENTAFLUORIDE AND ARSENIC PENTAFLUORI
	0239	INFRARED MATRIX ISOLATION STUDIES. (BORON TRIFLUORIDE, SULFUR TETRAFLUORIDE, ANTIM
	0399	INFRARED SPECTRA OF MATRIX ISOLATED CHLORINE TRIFLUORIDE, BROMINE TRIFLUORIDE, AND
	1053	FLUORINE EXCHANGE IN SULFUR TETRAFLUORIDE.
	1054	SULFUR TETRAFLUORIDE AND SULFUR OXIDE DIFLUORIDE. INFRARED EVIDENCE FOR DIMER FORM
REEVES LW	0517	FLUORINE NMR OF IODINE HEPTAFLUORIDE, RHENIUM HEPTAFLUORIDE, AND OXIDE PENTAFLUORI
REICH P	1300	STRUCTURE OF THE IODINE HEXAFLUORIDE ANION. (XENON HEXAFLUORIDE)
REICHMAN S	0210	VIBRATIONAL ANHARMONICITY IN BORON TRIFLUORIDE.
	0561	STRUCTURE OF KRYPTON DIFLUORIDE.
	0564	ABSENCE OF FERMI RESONANCE IN KRYPTON DIFLUORIDE.
	0731	HIGH RESOLUTION STUDY OF L-TYPE RESONANCE IN THE NU-4 BAND OF NITROGEN TRIFLUORIDE
	0732	1-TYPE RESONANCE IN AN OVERTONE BAND. THE 2NU-4 SPECTRUM OF NITROGEN TRIFLUORIDE.
	0815	NU-3 BAND OF PHOSPHORUS TRIFLUORIDE.
	0816	HIGH RESOLUTION SPECTRA OF ARSENIC TRIFLUORIDE AND PHOSPHORUS TRIFLUORIDE.
	0817	2NU-4 SPECTRUM OF PHOSPHORUS TRIFLUORIDE.
	0872	HIGH RESOLUTION INFRARED SPECTRA OF THE PARALLEL BANDS OF PHOSPHORUS PENTAFLUORIDE
	1248	GAS PHASE STRUCTURE OF XENON DIFLUORIDE.
	1493	VIBRATIONAL ANHARMONICITY IN THE METHYL HALIDES.
REIS A	0670	RAMAN SPECTRA OF LIQUID NIOBIUM PENTAFLUORIDE AND TANTALUM PENTAFLUORIDE.
REISFELD MJ	0897	PREPARATION, STRUCTURE AND SPECTRA OF PRASEODYMIUM TETRAFLUORIDE.
REYNOLDS DJ	0014	VIBRATIONAL SPECTRA OF INORGANIC FLUORIDES.
	0829	VIBRATIONAL SPECTRA OF MIXED PHOSPHORUS HALIDES. (PHOSPHORUS PENTAFLUORIDE)
RHEE KH	0825	INFRARED AND RAMAN SPECTRA AND STRUCTURE OF PHOSPHORUS TETRAFLUORIDE.
RICE S	1340	CHEMICAL PREDICTIONS BY MO THEORY. THE NOBLE GAS HALIDES.
RICE SA	1237	HEATS OF SUBLIMATION OF XENON DIFLUORIDE AND XENON TETRAFLUORIDE AND A CONJECTURE
	1238	THEORETICAL AND EXPERIMENTAL STUDIES OF THE ELECTRONIC STRUCTURE OF THE XENON FLUO
	1247	FORBIDDEN ELECTRONIC TRANSITIONS IN XENON DIFLUORIDE AND XENON TETRAFLUORIDE.
	1256	FAR ULTRAVIOLET SPECTROSCOPIC STUDY OF XENON DIFLUORIDE.
	1272	FAR ULTRAVIOLET SPECTROSCOPIC STUDY OF XENON TETRAFLUORIDE.
RICHARDS WG	0287	THEORETICAL STUDY OF THE SPECTROSCOPIC STATES OF THE CARBON MONOFLUORIDE MOLECULE.
	0937	REASSIGNMENT OF MOLECULAR ORBITAL CONFIGURATIONS OF THE ELECTRONIC STATES OF SCAND
RICHARDSON TJ	0485	PREPARATION AND CHARACTERIZATION OF GOLD PENTAFLUORIDE.
RIGINA IV	1330	NORMAL VIBRATION FREQUENCIES OF ZIRCONIUM TETRAFLUORIDE.
RIGNY P	0414	STUDY OF LIQUID CHLORINE FLUORIDES AND OXYFLUORIDES, CHLORINE PENTAFLUORIDE AND CH
	0637	HIGH RESOLUTION NMR SPECTRUM OF LIQUID MOLYBDENUM HEXAFLUORIDE, TUNGSTEN HEXAFLUOR
	0638	MOLECULAR MOTION AND FLUORINE RELAXATION IN THE LIQUIDS OF MOLYBDENUM HEXAFLUORIDE
	1189	VIBRATIONAL SPECTRUM AND STRUCTURE OF SOLID URANIUM HEXAFLUORIDE.
RINEHART EA	0689	A CAVITY SEARCH SPECTROMETER FOR FREE RADICAL MICROWAVE ROTATIONAL ABSORPTION STUD
	0690	MICROWAVE ROTATIONAL SPECTRUM OF THE NITROGEN DIFLUORIDE FREE RADICAL.
RIPD SM	0713	PHYSICAL CHEMICAL PROPERTIES OF INORGANIC NITROGEN FLUORIDES. (NITROGEN TRIFLUORID
RIPPON DM	0795	VAPOR PHASE RAMAN SPECTRA, RAMAN BAND CONTOUR ANALYSES, CORIOLIS CONSTANTS, FORCE
	1069	VAPOR PHASE RAMAN SPECTRA, RAMAN BAND CONTOUR ANALYSES, AND CORIOLIS CONSTANTS OF
ROACH AC	0904	JAHN-TELLER EFFECT IN RHENIUM HEXAFLUORIDE.
ROBBINS DJW	0488	ABSORPTION SPECTRUM OF GASEOUS HOLMIUM MONOFLUORIDE.
ROBERTO C	1144	SIMPLE THEORETICAL CALCULATIONS ON TITANIUM MONOFLUORIDE.
ROBERTS HL	1040	RAMAN AND INFRARED SPECTRUM OF SULFUR TETRAFLUORIDE.
ROBERTS JT	0880	ANALYSES OF THE VIBRATION ROTATION BANDS OF TRIFLUORO METHANE, PHOSPHORUS PENTAFLU
ROBERTS PJ	0341	PHOTOELECTRON SPECTRA OF HALIDES. PART-7: VARIABLE TEMPERATURE HELIUM-1 AND HELIUM
ROBIETTE AG	0265	GAS PHASE MOLECULAR STRUCTURES OF BROMINE PENTAFLUORIDE AND IODINE PENTAFLUORIDE F
	0421	MEAN AMPLITUDES OF VIBRATION FOR SOME PYRAMIDAL XY4Z MOLECULES. (IODINE PENTAFLUOR
ROBIN S	1017	FAR ULTRAVIOLET ELECTRONIC SPECTRA OF STRONTIUM DIFLUORIDE AND BARIUM DIFLUORIDE.
ROBINSON CP	1094	LASER ISOTOPE SEPARATION. (SULFUR HEXAFLUORIDE SPECTRUM)
ROBINSON DW	0172	THE ELECTRONIC SPECTRUM OF BORON FLUORIDE.
ROBINSON RJ	0768	MILLIMETER WAVE SPECTRUM AND STRUCTURE OF OXYGEN DIFLUORIDE.
ROCHESTER GD	0065	BAND SPECTRA OF BISMUTH MONOFLUORIDE AND ANTIMONY MONOFLUORIDE.
ROCHESTER L	0636	HEAT CAPACITY AND OTHER THERMODYNAMIC PROPERTIES OF MOLYBDENUM HEXAFLUORIDE BETWEE
RODE BM	0495	A SMALL GAUSSIAN BASIS SET FOR NONEMPIRICAL ALL-ELECTRON SCF CALCULATIONS IN IODIN
	0499	ELECTRONIC STRUCTURE AND CHEMISTRY OF IODINE COMPOUNDS. (FLUORIDES)
ROESCH N	1366	SCF- X(ALPHA) SCATTERED WAVE STUDIES ON BONDING AND IONIZATION POTENTIALS. PART-1:
ROESKY HW	0085	MASS SPECTRUM OF ANTIMONY PENTAFLUORIDE.
	0754	PREPARATION AND PROPERTIES (INFRARED SPECTRUM) OF OSMIUM HEPTAFLUORIDE.
ROFER-DEPOORTER CK	1191	ABSORPTION SPECTRUM OF URANIUM HEXAFLUORIDE FROM 2000 TO 4200 ANGSTROMS.
ROGERS HH	0426	DENSITY, VAPOR PRESSURE, CRITICAL PROPERTIES, DIELECTRIC CONSTANT, AND SPECIFIC CO
ROGERS MT	0108	ESR STUDY OF RADICALS IN GAMMA IRRADIATED POLYCRYSTALLINE ARSENIC TRIFLUORIDE AND
ROMANOV GV	0869	ELECTRON DIFFRACTION INVESTIGATIONS OF PHOSPHORUS PENTAFLUORIDE AND PHOSPHORUS PEN

ROMANOV GV	1217	ELECTRON DIFFRACTION STUDY OF THE MOLECULAR STRUCTURES OF VANADIUM PENTAFLUORIDE,
	1218	ELECTRON DIFFRACTION STUDY OF THE VANADIUM PENTAFLUORIDE MOLECULE IN THE VAPOR FOR
	1427	ESTIMATION OF CONFIGURATIONS AND INTERATOMIC DISTANCES IN GROUP-5 AND GROUP-6 TRAN
ROMASHKO BV	0240	THERMODYNAMIC PROPERTIES OF BORON HALIDES.
RON A	0275	VIBRATIONAL SPECTRA OF THE DIHALIDES OF MERCURY AND CADMIUM. (CADMIUM DIFLUORIDE A
	1323	VIBRATIONAL SPECTRA AND THERMODYNAMICS OF THE ZINC HALIDES. (ZINC DIFLUORIDE)
ROSE GVM	0930	ROTATIONAL ANALYSIS OF SOME SINGLET TRANSITIONS IN THE SPECTRUM OF GASEOUS SCANDIU
ROSE WB	0774	A REINVESTIGATION OF THE INFRARED SPECTRUM OF OXYGEN DIFLUORIDE.
	0775	FLUORINE NMR SPECTRA OF OXYGEN FLUORIDES. (OXYGEN DIFLUORIDE)
ROSEN A	1249	RELATIVISTIC MOLECULAR WAVE FUNCTIONS: XENON DIFLUORIDE.
ROSENBERG A	0356	FAR INFRARED ABSORPTION IN GASEOUS CARBON TETRAFLUORIDE.
	1095	FAR INFRARED SPECTRA OF GASEOUS AND LIQUID SULFUR HEXAFLUORIDE.
ROSS PA	0930	ROTATIONAL ANALYSIS OF SOME SINGLET TRANSITIONS IN THE SPECTRUM OF GASEOUS SCANDIU
ROSSI AR	0857	ELECTRONIC STRUCTURES OF PHOSPHORUS PENTAFLUORIDE AND TETRAFLUORO PHOSPHORANE.
ROTH A	0570	ELECTRONIC SPECTRA OF LANTHANUM FLUORIDE IN THE EXTREME ULTRAVIOLET.
	0582	ELECTRONIC SPECTRUM OF LANTHANUM TRIFLUORIDE.
ROTHENBERG S	0141	MOLECULAR PROPERTIES OF THE TRIATOMIC DIFLUORIDES BERYLLIUM DIFLUORIDE, BORON DIFL
ROUSSON R	0266	INFRARED AND RAMAN SPECTRA OF BROMINE PENTAFLUORIDE AND CHLORINE TRIFLUORIDE IN TH
ROZENBERG EL	0357	ELECTRONIC STRUCTURE OF THE CARBON TETRAFLUORIDE MOLECULE.
RUDZITIS E	0048	FLUORINE BOMB CALORIMETRY. PART-22: THE ENTHALPY OF FORMATION OF ALUMINUM TRIFLUOR
	0818	ENTHALPY OF FORMATION PHOSPHORUS TRIFLUORIDE.
	1130	ENTHALPY OF FORMATION OF THORIUM TETRAFLUORIDE BY FLUORINE BOMB CALORIMETRY.
	1319	ENTHALPY OF FORMATION OF YTTRIUM TRIFLUORIDE.
RUFF O	0049	VAPOR PRESSURE OF ZINC DIFLUORIDE, CADMIUM DIFLUORIDE, MAGNESIUM DIFLUORIDE, CALCI
	0639	PHYSICAL CONSTANTS (VAPOR PRESSURE) OF SILICON TETRAFLUORIDE, TUNGSTEN HEXAFLUORID
RUNDLE RE	1250	PROBABLE STRUCTURE OF XENON TETRAFLUORIDE AND XENON DIFLUORIDE.
RUOFF VA	0358	THE FORCE CONSTANTS OF VARIOUS ISOTOPES OF CARBON TETRAFLUORIDE, SILICON TETRAFLUO
	0733	ANHARMONIC POTENTIAL CONSTANTS FROM THE BAND CONTOURS OF HOT BANDS FOR A SYMMETRIC
	0735	FORCE FIELD AND MOLECULAR CONSTANTS OF NITROGEN TRIFLUORIDE.
	1096	FORCE CONSTANTS FOR SULFUR HEXAFLUORIDE.
	1126	INFRARED AND RAMAN SPECTRA AND THE STRUCTURE OF CRYSTALLINE THALLIUM MONOFLUORIDE.
RUSSEGGER P	0870	PSEUDOROTATION OF TRIGONAL BIPYRAMIDAL MOLECULES. BERRY ROTATION CONTRA TURNSTILE
RUSSELL DK	0954	ESR SPECTRUM OF BROMINE OXIDE, IODINE OXIDE AND SELENIUM MONOFLUORIDE IN J-5/2 ROT
	1019	ELECTRIC DIPOLE MOMENTS OF OPEN SHELL DIATOMIC MOLECULES. (SULFUR MONOFLUORIDE AND
RUSSELL JD	0758	EXTENDED HUCKEL THEORY AND THE SHAPE OF MOLECULES. (OXYGEN DIFLUORIDE)
RYABOV MA	0009	CHARACTERISTICS OF THE ELECTRONIC STRUCTURE OF FLUORIDES OF NONTRANSITION ELEMENTS
RYAN RR	1295	STRUCTURE OF GASEOUS XENON HEXAFLUORIDE.
RYBAKOV AG	0668	THERMODYNAMIC PROPERTIES OF NIOBIUM(V), TANTALUM(V), AND MOLYBDENUM(V) FLUORIDES.
RYMARCHUK YUA	0388	ABSORPTION SPECTRA OF CHLORINE FLUORIDE, CHLORINE TRIFLUORIDE, AND CHLORINE PENTAF
RYTTER E	1494	CLASSIFICATION OF NORMAL MODES OF VIBRATION. PART-2: INTERACTION TERMS IN ENERGY D
SABOL WW	0540	INFRARED SPECTRUM OF IRIDIUM HEXAFLUORIDE.
	0891	INFRARED SPECTRUM OF PLUTONIUM HEXAFLUORIDE.
SADOKHINA LA	0073	RAMAN SPECTRUM OF ANTIMONY TRIFLUORIDE.
SAFRONOV EK	0486	VAPOR PRESSURE AND HEAT OF SUBLIMATION OF ZIRCONIUM AND HAFNIUM TETRACHLORIDES.
SAHINI VE	0359	EVALUATION OF THE FORCE CONSTANTS OF SOME TETRAHEDRAL MOLECULES. (CARBON TETRAFLUO
SAITO S	0771	MICROWAVE SPECTRUM OF OXYGEN DIFLUORIDE IN VIBRATIONALLY EXCITED STATES, NU-1- 2 N
SANDHU JS	1076	PHOTOELECTRON SPECTRUM OF SULFUR HEXAFLUORIDE AT 584 ANGSTROMS.
SANDORFY C	0237	VACUUM ULTRAVIOLET ABSORPTION SPECTRA OF BORON TRIHALIDES.
SANKARANARAYANAN S	0147	EMISSION SPECTRUM OF BISMUTH MONOFLUORIDE.
	0974	EMISSION SPECTRUM OF THE GAMMA SYSTEM OF SILICON MONOFLUORIDE.
	0991	ULTRAVIOLET BAND SPECTRUM OF SILICON DIFLUORIDE.
SANYAL NK	0871	SPECTROSCOPIC STUDIES IN MEAN AMPLITUDES OF VIBRATION OF PENTAFLUORIDES OF ARSENIC
	1354	MEAN AMPLITUDES OF VIBRATION OF SOME LANTHANIDE TRIFLUORIDES.
SARASWATI V	1355	NMR IN RARE EARTH TRIFLUORIDES.
SATHIANANDAN K	0918	MOLECULAR CONSTANTS OF RHENIUM HEPTAFLUORIDE MOLECULE.
SAUNDERS VR	0802	AB INITIO CALCULATIONS OF THE BONDING IN PHOSPHINE, PHOSPHORUS TRIFLUORIDE AND TRI
SAUVAGEAU P	0237	VACUUM ULTRAVIOLET ABSORPTION SPECTRA OF BORON TRIHALIDES.
SAVOIE R	0327	VIBRATIONAL SPECTRA OF LIQUID AND CRYSTALLINE CARBON TETRAFLUORIDE.
	1001	INFRARED AND RAMAN SPECTRA OF LIQUID AND CRYSTALLINE SILICON TETRAFLUORIDE.
SAWODNY W	0734	AB INITIO STUDY OF THE FORCE CONSTANTS OF INORGANIC MOLECULES NITROGEN OXYFLUORIDE
	0735	FORCE FIELD AND MOLECULAR CONSTANTS OF NITROGEN TRIFLUORIDE.
	1035	RAMAN SPECTRUM AND FORCE CONSTANTS FOR SULFUR TETRAFLUORIDE.
	1036	VIBRATIONAL ASSIGNMENT OF SULFUR TETRAFLUORIDE.
SCHACK CJ	0534	MASS SPECTRA AND SUBLIMATION PRESSURES OF IODINE HEPTAFLUORIDE AND IODINE OXYGEN P
SCHAEFER HF	0141	MOLECULAR PROPERTIES OF THE TRIATOMIC DIFLUORIDES BERYLLIUM DIFLUORIDE, BORON DIFL
	0542	ELECTRONIC STRUCTURE OF IRON TRIFLUORIDE.
	0547	KRYPTON MONOFLUORIDE AND ITS POSITIVE ION. (SCF CALCULATIONS)
	0548	ELECTRONIC STRUCTURE AND PROPERTIES OF KRYPTON DIFLUORIDE.
	1221	PROBABLE NONEXISTENCE OF XENON MONOFLUORIDE AS A CHEMICALLY BOUND SPECIES IN THE G
SCHAFER VH	0992	THERMOCHEMISTRY OF SILICON DIFLUORIDE.
SCHATZ J	0732	1-TYPE RESONANCE IN AN OVERTONE BAND. THE 2NU-4 SPECTRUM OF NITROGEN TRIFLUORIDE.
	0872	HIGH RESOLUTION INFRARED SPECTRA OF THE PARALLEL BANDS OF PHOSPHORUS PENTAFLUORIDE
SCHATZ PN	0736	ABSOLUTE INFRARED INTENSITIES OF THE FUNDAMENTAL VIBRATIONS OF NITROGEN TRIFLUORID
	0737	POTENTIAL FUNCTION OF NITROGEN TRIFLUORIDE.
SCHERER V	0898	PROPERTIES OF PROMETHIUM COMPOUNDS. (PROMETHIUM TRIFLUORIDE)
SCHLECHT RG	1181	MEASUREMENT OF THE GAS DENSITY OF URANIUM HEXAFLUORIDE BY LASER RAMAN SCATTERING.
SCHMEISSER M	0500	VIBRATIONAL SPECTRUM OF IODINE TRIFLUORIDE.
SCHMULTZER R	0819	FLUORIDES OF PHOSPHORUS.
SCHNEIDER RT	1201	SPECIFIC HEAT RATIO OF URANIUM HEXAFLUORIDE MEASURED WITH A BALLISTIC PISTON COMPR
SCHNEPP O	0275	VIBRATIONAL SPECTRA OF THE DIHALIDES OF MERCURY AND CADMIUM. (CADMIUM DIFLUORIDE A
	1323	VIBRATIONAL SPECTRA AND THERMODYNAMICS OF THE ZINC HALIDES. (ZINC DIFLUORIDE)
SCHNIZLEIN JG	0781	PREPARATION AND PURIFICATION OF OXYGEN DIFLUORIDE AND DETERMINATION OF ITS VAPOR P
SCHOEN LH	0687	INFRARED SPECTRUM AND STRUCTURE OF THE NITROGEN DIFLUORIDE RADICAL.
SCHOMAKER V	0738	AN ELECTRON DIFFRACTION INVESTIGATION OF NITROGEN TRIFLUORIDE.
	1155	ELECTRON DIFFRACTION INVESTIGATION OF TUNGSTEN HEXAFLUORIDE, OSMIUM HEXAFLUORIDE,
SCHREINER F	0510	CALORIMETRIC STUDY OF IODINE PENTAFLUORIDE. HEAT CAPACITY BETWEEN 5 AND 350 DEGREE
	0552	INFRARED AND RAMAN SPECTRA OF KRYPTON DIFLUORIDE.
	0561	STRUCTURE OF KRYPTON DIFLUORIDE.
	0636	HEAT CAPACITY AND OTHER THERMODYNAMIC PROPERTIES OF MOLYBDENUM HEXAFLUORIDE BETWEE
	1248	GAS PHASE STRUCTURE OF XENON DIFLUORIDE.

SCHREINER F	1251	VAPOR PRESSURE AND MELTING POINTS OF XENON DIFLUORIDE AND XENON TETRAFLUORIDE.
	1275	XENON TETRAFLUORIDE. HEAT CAPACITY AND THERMODYNAMIC FUNCTIONS FROM 5 TO 350 DEGRE
	1301	PREPARATION AND SOME THERMAL PROPERTIES OF PURE XENON HEXAFLUORIDE.
	1305	HEAT CAPACITY AND OTHER THERMODYNAMIC FUNCTIONS OF XENON HEXAFLUORIDE FROM 5 TO 35
	1308	MAGNETIC SUSCEPTIBILITY OF XENON HEXAFLUORIDE.
SCHRODER J	1150	PREPARATION AND PROPERTIES (VAPOR PRESSURE) OF TUNGSTEN PENTAFLUORIDE.
	1151	ENTHALPY OF FORMATION OF TUNGSTEN PENTAFLUORIDE AND TUNGSTEN HEXAFLUORIDE.
SCHULZ DA	0282	VAPOR PRESSURE AND HEAT OF SUBLIMATION OF CALCIUM FLUORIDE.
SCHUMB WC	0506	VIBRATIONAL SPECTRA AND STRUCTURES OF IODINE PENTAFLUORIDE AND HEPTAFLUORIDE.
SCHWARTZ ME	0241	AB INITIO STUDIES OF THE ELECTRONIC STRUCTURES OF BORON TRIHYDRIDE, BORON DIHYDRID
SCHWARZ WHE	1233	SPECTRA OF XENON DIFLUORIDE AND XENON TETRAFLUORIDE IN THE FAR ULTRAVIOLET REGION.
	1302	ELECTRONIC AND GEOMETRIC STRUCTURE OF THE FREE XENON HEXAFLUORIDE MOLECULE.
	1306	CHEMICAL APPLICATIONS OF CORE ELECTRON EXCITATION SPECTROSCOPY. (XENON HEXAFLUORID
	1480	VIBRONIC AND SPIN ORBIT SPLITTING IN SPECTRA OF SYSTEMS EXHIBITING JAHN-TELLER EFF
SCHWEIGER JR	0846	PHOTOELECTRON SPECTRUM OF PHOSPHORUS PENTAFLUORIDE.
SCHWEITZER GK	0319	HIGH RESOLUTION PHOTOELECTRON SPECTROSCOPY OF CARBON TETRAFLUORIDE AND SILICON TE
SCOTT PR	0937	REASSIGNMENT OF MOLECULAR ORBITAL CONFIGURATIONS OF THE ELECTRONIC STATES OF SCAND
SCOTT RB	1166	LOW TEMPERATURE HEAT CAPACITIES, ENTHALPIES, AND ENTROPIES OF URANIUM TETRAFLUORID
SCROCCO E	0762	DOUBLE ZETA SCF CALCULATIONS FOR NITRITE ION AND OXYGEN DIFLUORIDE.
	0763	CONFIGURATION INTERACTION CALCULATION FOR THE GROUND STATE OF OXYGEN DIFLUORIDE, N
	0776	MINIMAL BASIS SET SCF CALCULATIONS FOR THE GROUND STATE OF OZONE, NITRITE ION, NIT
SCULLMAN R	0023	SINGLET ELECTRONIC STATES OF ALUMINUM MONOFLUORIDE.
SEARCY AW	0282	VAPOR PRESSURE AND HEAT OF SUBLIMATION OF CALCIUM FLUORIDE.
	0366	VAPOR PRESSURE AND HEAT OF SUBLIMATION OF CERIUM TRIFLUORIDE.
	0576	VAPOR PRESSURE, HEAT OF SUBLIMATION, AND EVAPORATION COEFFICIENT OF LANTHANUM TRIF
	0581	DEMONSTRATION OF THE EXISTENCE OF DILANTHANUM HEXAFLUORIDE GAS AND DETERMINATION O
	0896	VAPOR PRESSURE, HEAT OF SUBLIMATION, AND THE EVAPORATION COEFFICIENT OF PRASEODYMI
	1140	MELTING POINT, VAPOR PRESSURE, AND HEAT OF VAPORIZATION OF TIN DIFLUORIDE.
SEARS GW	0887	VAPOR PRESSURE OF PLUTONIUM HALIDES. (PLUTONIUM TRIFLUORIDE)
SEEL F	1044	FLUORINE-19 NMR INVESTIGATION OF THE ASSOCIATION OF SULFUR TETRAFLUORIDE.
	1055	PROBLEM OF FLUORINE EXCHANGE IN SULFUR TETRAFLUORIDE.
SEIBOLD EA	1363	ROOT MEAN SQUARE AMPLITUDES OF VIBRATION IN SOME GROUP-4 TETRAHALIDES.
SEIFERT RL	0887	VAPOR PRESSURE OF PLUTONIUM HALIDES. (PLUTONIUM TRIFLUORIDE)
SEIP HM	0640	STUDIES ON THE FAILURE OF THE FIRST BORN APPROXIMATION IN ELECTRON DIFFRACTION. (M
	1109	FAILURE OF THE FIRST BORN APPROXIMATION IN ELECTRON DIFFRACTION. (TELLURIUM HEXAFL
	1158	FAILURE OF THE FIRST BORN APPROXIMATION IN ELECTRON DIFFRACTION. (TUNGSTEN HEXAFLU
	1199	FAILURE OF THE FIRST BORN APPROXIMATION IN ELECTRON DIFFRACTION. (URANIUM HEXAFLUO
	1426	JAHN-TELLER EFFECT. EXPERIMENTAL AND THEORETICAL STUDIES ON VANADIUM TETRACHLORIDE
SEIP R	0640	STUDIES ON THE FAILURE OF THE FIRST BORN APPROXIMATION IN ELECTRON DIFFRACTION. (M
SELEZNEV VP	0669	THERMODYNAMIC PROPERTIES OF NIOBIUM AND TANTALUM PENTAFLUORIDES.
SELIG H	0408	INFRARED AND RAMAN SPECTRA OF CHLORINE TRIFLUORIDE AND BROMINE TRIFLUORIDE.
	0510	CALORIMETRIC STUDY OF IODINE PENTAFLUORIDE. HEAT CAPACITY BETWEEN 5 AND 350 DEGREE
	0512	RAMAN SPECTRUM OF IODINE PENTAFLUORIDE. EVIDENCE FOR POLYMERIZATION IN THE LIQUID
	0518	SPECTROSCOPIC STUDY OF IODINE HEPTAFLUORIDE AND IODINE HEXAFLUORIDE(X) IN ANHYDROU
	0522	VIBRATIONAL SPECTRA OF IODINE HEPTAFLUORIDE AND RHENIUM HEPTAFLUORIDE.
	0628	VIBRATIONAL SPECTRA OF MOLYBDENUM HEXAFLUORIDE AND TECHNETIUM HEXAFLUORIDE.
	0636	HEAT CAPACITY AND OTHER THERMODYNAMIC PROPERTIES OF MOLYBDENUM HEXAFLUORIDE BETWEE
	0670	RAMAN SPECTRA OF LIQUID NIOBIUM PENTAFLUORIDE AND TANTALUM PENTAFLUORIDE.
	0695	RAMAN SPECTRUM OF THE NITROGEN DIFLUORIDE RADICAL.
	0873	RAMAN SPECTRA OF ARSENIC PENTAFLUORIDE AND VANADIUM PENTAFLUORIDE AND THEIR FORCE
	0905	VIBRATIONAL SPECTRA OF RHENIUM HEXAFLUORIDE.
	0912	VAPOR PRESSURES AND OTHER PROPERTIES OF RHENIUM HEXAFLUORIDE AND RHENIUM HEPTAFLUO
	0917	MAGNETIC SUSCEPTIBILITY OF RHENIUM HEXAFLUORIDE.
	0919	VIBRATIONAL SPECTRA OF RHENIUM HEPTAFLUORIDE.
	0921	PREPARATION AND PROPERTIES OF RHENIUM HEPTAFLUORIDE.
	1072	RAMAN SPECTRA OF MOLYBDENUM HEXAFLUORIDE, TECHNETIUM HEXAFLUORIDE, RHENIUM HEXAFLU
	1105	VAPOR PRESSURE AND TRANSITION POINTS OF TECHNETIUM HEXAFLUORIDE.
	1211	VIBRATIONAL SPECTRA OF VANADIUM PENTAFLUORIDE.
	1263	XENON TETRAFLUORIDE. (INFRARED SPECTRUM)
	1307	ELECTRICAL CONDUCTIVITY AND DIELECTRIC CONSTANT OF LIQUID XENON HEXAFLUORIDE.
	1308	MAGNETIC SUSCEPTIBILITY OF XENON HEXAFLUORIDE.
	1378	RAMAN SPECTRA OF 5D TRANSITION METAL HEXAFLUORIDES IN THE VAPOR STATE. (THIRD ROW)
SEMENKOVICH SA	0031	ALUMINUM MONOFLUORIDE. (VAPOR PRESSURE)
SENSE KA	1331	VAPOR PRESSURE OF ZIRCONIUM FLUORIDE.
SEPPELT K	0525	VIBRATIONAL ASSIGNMENT FOR IODINE PENTAFLUORIDE, A NONRIGID D5H MOLECULE.
SERAFINI A	0820	BONDING IN PHOSPHORUS TRIFLUORIDE AND PHOSPHORUS OXYFLUORIDE. AN AB INITIO SCF STU
SESHADRI DN	0739	THERMODYNAMIC PROPERTIES OF NITROGEN TRIFLUORIDE.
SESHADRI KS	0609	GEOMETRY AND VIBRATIONAL SPECTRA OF THE ALKALINE EARTH DIHALIDES. PART-1: MAGNESIU
SESSA PA	0565	MASS SPECTRUM AND MOLECULAR ENERGETICS OF KRYPTON DIFLUORIDE.
SETTLE JL	1092	ENTHALPIES OF FORMATION OF SULFUR HEXAFLUORIDE, SELENIUM HEXAFLUORIDE, TELLURIUM H
	1147	HEATS OF FORMATION OF TITANIUM TETRAFLUORIDE AND HAFNIUM TETRAFLUORIDE.
SHAE RW	1229	ELECTRON DISTRIBUTION IN THE XENON DIFLUORIDE AND XENON OXIDE TETRAFLUORIDE BY ESCA
SHAFER MW	1206	SYNTHESIS AND CHARACTERIZATION OF VANADIUM DIFLUORIDE.
SHAH NP	0127	RYDBERG-KLEIN-REES FRANCK-CONDON FACTORS AND R CENTROIDS OF THE A-X BAND SYSTEM OF
SHAMIR J	0503	MOSSBAUER STUDY OF IODINE PENTAFLUORIDE AND IODINE HEPTAFLUORIDE.
	0740	RAMAN SPECTRUM OF GASEOUS NITROGEN TRIFLUORIDE.
SHANKER R	0066	ROTATIONAL STRUCTURE OF THE C3 SYSTEM OF ANTIMONY FLUORIDE MOLECULE.
	0067	ROTATIONAL ANALYSIS OF THE 2550-2750 ANGSTROM BAND SYSTEM AT ANTIMONY FLUORIDE MOL
	0068	ROTATIONAL ANALYSIS OF THE C2 SYSTEM OF ANTIMONY MONOFLUORIDE.
	0121	GREEN BAND SYSTEM OF BARIUM MONOFLUORIDE.
SHANMUGASUNDARAM G	0050	MEAN AMPLITUDES OF VIBRATION, CORIOLIS COUPLING CONSTANTS, SHRINKAGE EFFECT, AND T
SHARKOV VI	0826	CALCULATION OF FORCE CONSTANTS AND ROOT MEAN SQUARE VIBRATION AMPLITUDES OF PHOSPH
SHARMA D	1013	NEAR ULTRAVIOLET ABSORPTION SPECTRUM OF SILVER MONOFLUORIDE.
SHARP DWA	0007	FLUORIDES OF THE MAIN GROUP ELEMENTS.
	0625	VIBRATIONAL SPECTRA OF MOLYBDENUM AND TUNGSTEN PENTAFLUORIDES.
	0626	VIBRATIONAL SPECTRA OF MOLYBDENUM PENTAFLUORIDE AND TUNGSTEN PENTAFLUORIDE.
	0664	NIOBIUM-93 NUCLEAR QUADRUPOLE RESONANCE STUDIES OF NIOBIUM PENTAFLUORIDE AND ITS C
SHARPE AG	0015	TRANSITION METAL FLUORIDES AND THEIR COMPLEXES.
SHAW RW	0409	X-RAY PHOTOELECTRON SPECTROSCOPY OF CHLORINE TRIFLUORIDE, SULFUR TETRAFLUORIDE, AN
	0874	OBSERVATION BY ESCA OF INEQUIVALENT FLUORINES IN CHLORINE TRIFLUORIDE, SULFUR TETR

SHEARD JL   0781 PREPARATION AND PURIFICATION OF OXYGEN DIFLUORIDE AND DETERMINATION OF ITS VAPOR P
SHEEHAN DE   1271 HEAT CAPACITY AND RELATED THERMODYNAMIC FUNCTIONS OF XENON TETRAFLUORIDE.
SHEEHAN DF   0417 EQUILIBRIUM STUDIES OF CHLORINE PENTAFLUORIDE.
                             0534 MASS SPECTRA AND SUBLIMATION PRESSURES OF IODINE HEPTAFLUORIDE AND IODINE OXYGEN P
SHELDRICK GM   0242 FERMI RESONANCE IN BORON TRIFLUORIDE.
SHEN L   0579 NUCLEAR RELAXATION TIMES OF LANTHANUM TRIFLUORIDE.
SHENYAVSKAYA EA   0569 ELECTRONIC SPECTRUM OF LANTHANUM MONOFLUORIDE.
                             0580 BAND SYSTEM IN THE DISCHARGE SPECTRUM OF LANTHANUM TRIFLUORIDE VAPORS.
                             0934 ELECTRONIC SPECTRUM OF SCANDIUM MONOFLUORIDE.
                             0938 VIBRONIC SPECTRUM OF SCANDIUM MONOFLUORIDE, YTTRIUM MONOFLUORIDE, AND LANTHANUM MO
                             1315 VIBRONIC SPECTRA OF YTTRIUM MONOFLUORIDE.
SHERBORNE JE   0111 RAMAN SPECTRUM OF ARSENIC TRIFLUORIDE AND THE MOLECULAR CONSTANTS OF ARSENIC TRIFL
SHERIDAN J   0741 NUCLEAR QUADRUPOLE MOMENT OF NITROGEN-14 AND THE STRUCTURE OF NITROGEN TRIFLUORIDE
SHERMAN RJ   0634 VIBRATIONAL SPECTRUM AND FORCE FIELD OF MOLYBDENUM HEXAFLUORIDE.
SHIMANOUCHI T   0243 FORCE CONSTANTS OF CARBON TETRAFLUORIDE, SILICON TETRAFLUORIDE, BORON TRIFLUORIDE,
                             0360 UREY-BRADLEY FORCE FIELD FOR CARBON TETRAFLUORIDE.
                             1495 FORCE CONSTANTS OF SMALL MOLECULES. (UREY-BRADLEY)
SHIMIZU K   0244 EFFECTS OF THE LENNARD-JONES POTENTIALS ON THE UREY-BRADLEY FORCE CONSTANTS OF BOR
                             0246 UREY-BRADLEY FORCE CONSTANTS OF BORON HALIDES AND THEIR TRANSFERABILITY. (BORON TR
                             0247 NORMAL VIBRATIONS OF BORON TRIHALIDES. (BORON TRIFLUORIDE)
SHIMODA T   0326 THERMODYNAMIC PROPERTIES OF CARBON TETRAFLUORIDE FROM 4 DEGREES K TO ITS MELTING P
SHINODA T   1010 INTERMOLECULAR POTENTIAL FOR SILICON TETRAFLUORIDE.
SHIRK JS   0054 INFRARED SPECTRA OF ALUMINUM FLUORIDE IN SOLID ARGON.
                             1097 RAMAN SPECTRA OF MATRIX ISOLATED MOLECULES. (SULFUR HEXAFLUORIDE)
SHIRLEY DA   0683 MULTIPLET SPLITTING IN 1S HOLE STATES OF MOLECULES.
SHISHKOV YD   1328 VAPOR PRESSURE OF ZIRCONIUM TETRAFLUORIDE.
SHOJI H   0993 MICROWAVE SPECTRUM OF SILICON DIFLUORIDE IN THE EXCITED VIBRATIONAL STATES, EQUILI
SHOLTS VB   1208 MASS SPECTROMETRIC DETERMINATION OF THE VAPOR COMPOSITION AND VAPOR PRESSURE OF VA
SHORT LN   0482 RAMAN AND INFRARED SPECTRA OF GERMANIUM TETRAFLUORIDE.
SHURVELL HF   1098 RAMAN SPECTRUM OF SOLID SULFUR HEXAFLUORIDE.
SICHEL JM   0782 CNDO STUDY OF THE OZONIDE ION AND RELATED SPECIES. (OXYGEN DIFLUORIDE)
SICRE JE   0252 SPECTRUM OF BROMINE MONOFLUORIDE AND ITS DISSOCIATION ENERGY.
SIDOROV LN   0040 MASS SPECTROMETRIC INVESTIGATION OF THE THERMODYNAMIC PROPERTIES OF ALUMINUM TRIFL
                             1208 MASS SPECTROMETRIC DETERMINATION OF THE VAPOR COMPOSITION AND VAPOR PRESSURE OF VA
                             1326 VAPOR PRESSURE OF ZIRCONIUM TETRAFLUORIDE.
SIEBEN FJ   1151 ENTHALPY OF FORMATION OF TUNGSTEN PENTAFLUORIDE AND TUNGSTEN HEXAFLUORIDE.
SIEGBAHN K   0328 MOLECULAR SPECTROSCOPY BY MEANS OF ESCA. (CARBON TETRAFLUORIDE)
SIEGEL S   1375 X-RAY DIFFRACTION STUDIES OF SOME TRANSITION METAL HEXAFLUORIDES. (SECOND ROW, THI
SILLITOE JF   0301 WAVE FUNCTIONS FOR THE FOUR ELECTRON THREE CENTER BONDING OF FOUR AND EIGHT PI ELE
SILVERS SJ   0167 FLUORESCENCE SPECTRA OF PHOSPHORUS NITRIDE AND BORON FLUORIDE.
SIMONS G   0783 VIRIAL THEOREM DECOMPOSITION OF MOLECULAR FORCE FIELDS.
SIMPSON CJSM   1452 MOLECULAR FORCE FIELDS FOR SOME HEXAFLUORIDES.
SIMPSON OC   0887 VAPOR PRESSURE OF PLUTONIUM HALIDES. (PLUTONIUM TRIFLUORIDE)
SINANOGLU O   0174 METHODS FOR CORRELATING MOLECULES AND SOME OPTIMIZED VALENCE CONFIGURATION RESULTS
SINGH ID   0975 FRANCK-CONDON FACTORS AND R CENTROIDS FOR THE C-X SYSTEM OF SILICON MONOFLUORIDE M
SINGH IS   0066 ROTATIONAL STRUCTURE OF THE C3 SYSTEM OF ANTIMONY FLUORIDE MOLECULE.
                             0067 ROTATIONAL ANALYSIS OF THE 2550-2750 ANGSTROM BAND SYSTEM AT ANTIMONY FLUORIDE MOL
                             0068 ROTATIONAL ANALYSIS OF THE C2 SYSTEM OF ANTIMONY MONOFLUORIDE.
                             0590 ROTATIONAL ANALYSIS OF B-X2 SYSTEM OF LEAD-208 MONOFLUORIDE.
                             0591 ROTATIONAL ANALYSIS OF A-X1 BANDS OF LEAD MONOFLUORIDE.
                             0976 ROTATIONAL ANALYSIS OF THE BETA BAND SYSTEM OF SILICON MONOFLUORIDE MOLECULE.
SINGH J   0122 VIOLET AND ULTRAVIOLET SPECTRUM OF BARIUM MONOFLUORIDE MOLECULE.
                             0123 THERMALLY EXCITED EMISSION SPECTRUM OF BARIUM MONOFLUORIDE.
                             0173 POTENTIAL ENERGY CURVES AND DISSOCIATION ENERGIES OF DIATOMIC BORON AND ALUMINUM H
                             0458 POTENTIAL ENERGY CURVES AND DISSOCIATION ENERGIES OF DIATOMIC FLUORIDES AND CHLORI
                             0994 VIBRATIONAL TRANSITION PROBABILITIES AND R CENTROIDS FOR DIATOMIC FLUORIDES OF SIL
                             1135 ROTATIONAL STRUCTURE AND ISOTOPIC SHIFT IN THE (1,0) BAND OF THE A DOUBLET SIGMA-
                             1137 VARIATION OF ELECTRONIC TRANSITION MOMENT WITH THE INTERNUCLEAR SEPARATION IN SILI
SINGH M   0222 EMISSION SPECTRA OF CERTAIN POLYATOMIC FREE RADICALS OBTAINED IN ELECTRICAL DISCHA
                             0559 ULTRAVIOLET EMISSION SPECTRA. (XENON DIFLUORIDE AND KRYPTON DIFLUORIDE)
SINGH NI   0061 DETERMINATION OF THE EFFECTIVE VIBRATIONAL TEMPERATURE IN THE A-2 SYSTEM OF ANTIMO
SINGH ON   0590 ROTATIONAL ANALYSIS OF B-X2 SYSTEM OF LEAD-208 MONOFLUORIDE.
                             0590 ROTATIONAL ANALYSIS OF B-X2 SYSTEM OF LEAD-208 MONOFLUORIDE.
                             0591 ROTATIONAL ANALYSIS OF A-X1 BANDS OF LEAD MONOFLUORIDE.
                             0601 DISSOCIATION ENERGY OF MAGNESIUM MONOFLUORIDE.
                             0601 DISSOCIATION ENERGY OF MAGNESIUM MONOFLUORIDE.
                             0972 FRANCK-CONDON FACTORS AND R CENTROIDS OF SOME BAND SYSTEMS OF SILICON MONOFLUORIDE
                             0976 ROTATIONAL ANALYSIS OF THE BETA BAND SYSTEM OF SILICON MONOFLUORIDE MOLECULE.
                             1466 MEAN SQUARE AMPLITUDES OF INTERATOMIC DISTANCES IN SOME OCTAHEDRAL HEXAFLUORIDE MO
SINGH RB   0032 POTENTIAL ENERGY CURVES AND NATURE OF BINDING IN GROUP-3A MONOHALIDES.
                             0130 POTENTIAL CURVES FOR SOME DIATOMIC MOLECULES. (BERYLLIUM MONOFLUORIDE, SILICON MON
                             0589 ON THE DISSOCIATION OF LEAD MONOFLUORIDE AND BISMUTH MONOFLUORIDE.
                             0641 VALENCE FORCE CONSTANTS OF THE HEXAFLUORIDES OF MOLYBDENUM, TECHNETIUM, AND RHENIU
SINGH SP   0588 EMISSION SPECTRUM OF LEAD MONOFLUORIDE.
                             0597 THERMODYNAMIC FUNCTIONS OF LEAD TETRAHALIDES.
SINGH Z   0784 SPECTROSCOPIC STUDIES OF VIBRATIONAL CONSTANTS, STATISTCIAL THERMODYNAMICS AND QU
SINKE GC   0742 ENTHALPY OF DISSOCIATION OF NITROGEN TRIFLUORIDE.
SINYANSKII VF   0560 ABSORPTION SPECTRA OF CERTAIN FLUORIDES IN THE NEAR ULTRAVIOLET REGION. (KRYPTON D
SIU AKQ   0605 THEORETICAL STUDIES OF THE INTERACTION OF MAGNESIUM DIFLUORIDE WITH RARE GAS ATOMS
                             1360 NONEMPIRICAL LCAO MO SCF STUDIES OF THE GROUP-2A DIHALIDES BERYLLIUM DIFLUORIDE, M
SIVAJI C   0069 ROTATIONAL ANALYSIS OF THE C1 SYSTEM OF ANTIMONY MONOFLUORIDE.
SIVARAMAKRISHNAN TR   1052 FORCE FIELD STUDY FOR SULFUR TETRAFLUORIDE.
SKINNER HB   0581 DEMONSTRATION OF THE EXISTENCE OF DILANTHANUM HEXAFLUORIDE GAS AND DETERMINATION O
                             0896 VAPOR PRESSURE, HEAT OF SUBLIMATION, AND THE EVAPORATION COEFFICIENT OF PRASEODYMI
SKURATOV SM   0043 STANDARD ENTHALPY OF FORMATION OF ALUMINUM TRIFLUORIDE.
SLEGINA VI   0075 THERMODYNAMIC FUNCTIONS OF ANTIMONY FLUORIDE, NIOBIUM BROMIDE, TANTALUM CHLORIDE,
SLEIGHT TP   0925 ELECTRONIC SPECTRA OF LIQUID RUTHENIUM AND MOLYBDENUM PENTAFLUORIDES.
SLIVNIK J   0883 SUSCEPTIBILITY MEASUREMENTS AND FLUORINE NMR OF PLATINUM HEXAFLUORIDE.
SLOWINSKI EJ   0506 VIBRATIONAL SPECTRA AND STRUCTURES OF IODINE PENTAFLUORIDE AND HEPTAFLUORIDE.
SMARDZEWSKI RR   0255 MATRIX ISOLATION OF BROMINE MONOFLUORIDE.

SMIRNOV AD	0967	ELECTRONIC TRANSITION MOMENT AS A FUNCTION OF THE INTERNUCLEAR SEPARATION FOR THE
	0968	DISTRIBUTION OF INTENSITY IN THE ELECTRON EMISSION SPECTRUM OF SILICON MONOFLUORID
	0969	CORRELATION OF THE ELECTRONIC TRANSITION MOMENT WITH INTERNUCLEAR DISTANCE FOR BAN
	0977	CALCULATION OF THE FRANCK-CONDON FACTOR FOR MOLECULES OF SILICON FLUORIDE, GERMANI
SMIRNOV BM	0560	ABSORPTION SPECTRA OF CERTAIN FLUORIDES IN THE NEAR ULTRAVIOLET REGION. (KRYPTON D
SMITH DF	0410	MICROWAVE SPECTRUM AND STRUCTURE OF CHLORINE TRIFLUORIDE.
	0418	VIBRATIONAL SPECTRA AND VALENCE FORCE CONSTANTS OF SQUARE PYRAMIDAL MOLECULES. XEN
	0627	MOLECULAR STRUCTURE OF MOLYBDENUM, TUNGSTEN, AND URANIUM HEXAFLUORIDES FROM INFRAR
	0745	VIBRATIONAL ANHARMONICITY IN THE PYRAMIDAL TRIFLUORIDES.
	0817	2NU-4 SPECTRUM OF PHOSPHORUS TRIFLUORIDE.
	1252	XENON DIFLUORIDE. (FORCE CONSTANTS, SPECTRA)
SMITH DW	1155	ELECTRON DIFFRACTION INVESTIGATION OF TUNGSTEN HEXAFLUORIDE, OSMIUM HEXAFLUORIDE,
SMITH FJ	0022	SPECTROSCOPIC AND THERMODYNAMIC PROPERTIES OF ALUMINUM MONOFLUORIDE.
SMITH JH	0361	THERMODYNAMIC PROPERTIES OF CARBON TETRAFLUORIDE FROM 12 DEGREES K TO ITS BOILING
SMITH JJ	0691	MATRIX ISOLATION STUDY OF THE VACUUM ULTRAVIOLET PHOTOLYSIS OF NITROGEN TRIFLUORID
SMITH VH	1366	SCF- X(ALPHA) SCATTERED WAVE STUDIES ON BONDING AND IONIZATION POTENTIALS. PART-1:
SMOLYAR AE	0009	CHARACTERISTICS OF THE ELECTRONIC STRUCTURE OF FLUORIDES OF NONTRANSITION ELEMENTS
	0016	CALCULATIONS OF ELECTRON STRUCTURES OF COMPOUNDS OF HALOGENS AND INERT GASES BY TH
	1231	CALCULATIONS OF ELECTRONIC STRUCTURES OF COMPOUNDS OF HALOGENS AND INERT GASES BY
SMYRL N	0270	ESR SPECTRUM AND STRUCTURE OF BROMINE HEXAFLUORIDE.
	0428	CHLORINE HEXAFLUORIDE RADICAL. PREPARATION, ELECTRON SPIN RESONANCE SPECTRUM, AND
SMYTH CP	1200	DIPOLE MOMENT AND MOLECULAR STRUCTURE OF URANIUM HEXAFLUORIDE.
SNELSON A	0051	INFRARED SPECTRUM OF ALUMINUM TRIFLUORIDE, DIALUMINUM HEXAFLUORIDE, AND ALUMINUM M
	0125	INFRARED SPECTRA OF SOME ALKALINE EARTH HALIDES BY THE MATRIX ISOLATION TECHNIQUE.
SNIDER AM	0825	INFRARED AND RAMAN SPECTRA AND STRUCTURE OF PHOSPHORUS TETRAFLUORIDE.
SNYDER MJ	1331	VAPOR PRESSURE OF ZIRCONIUM FLUORIDE.
SOBHANADRI J	0107	CALCULATIONS ON IONIC CHARACTER AND HYBRILIZATION FROM DIPOLE MOMENTS FOR ARSENIC
SOGABE K	0823	ESR SPECTRA OF PHOSPHORUS TETRAFLUORIDE RADICALS PRODUCED IN A SINGLE CRYSTAL OF P
SOKOLOV VB	0563	KRYPTON DIFLUORIDE. (PHYSICAL AND CHEMICAL PROPERTIES)
SOMASUNDARAM V	1428	POTENTIAL CONSTANTS OF XY4 AND XY3Z TYPES OF MOLECULES AND RADICALS.
SONIN VI	0052	THERMODYNAMICS OF THE VAPORIZATION OF ALUMINUM FLUORIDE AND CESIUM FLUORIDE.
	0951	ENERGETICS AND STABILITY OF GAS PHASE SCANDIUM FLUORIDES.
SOOTS V	0274	RAMAN SPECTRA OF CADMIUM DIFLUORIDE AND LEAD DIFLUORIDE.
SORBE P	0444	NEW FLUORIDES OF TRIVALENT COPPER.
SORIANO J	0503	MOSSBAUER STUDY OF IODINE PENTAFLUORIDE AND IODINE HEPTAFLUORIDE.
SOROKIN YV	0240	THERMODYNAMIC PROPERTIES OF BORON HALIDES.
SPEDDING FH	1320	HIGH TEMPERATURE HEAT CONTENTS AND RELATED THERMODYNAMIC FUNCTIONS OF SEVEN RARE E
	1356	HIGH TEMPERATURE ENTHALPIES AND RELATED THERMODYNAMIC FUNCTIONS OF THE TRIFLUORIDE
SPIRIDONOV VP	0131	ELECTRON DIFFRACTION ANALYSIS OF THE MOLECULAR STRUCTURES OF GROUP-2 ELEMENTS. (BE
	0671	ELECTRON DIFFRACTION INVESTIGATION OF NIOBIUM PENTAFLUORIDE AND TANTALUM PENTAFLUO
	0869	ELECTRON DIFFRACTION INVESTIGATIONS OF PHOSPHORUS PENTAFLUORIDE AND PHOSPHORUS PEN
	1217	ELECTRON DIFFRACTION STUDY OF THE MOLECULAR STRUCTURES OF VANADIUM PENTAFLUORIDE,
	1218	ELECTRON DIFFRACTION STUDY OF THE VANADIUM PENTAFLUORIDE MOLECULE IN THE VAPOR FOR
	1332	ELECTRON DIFFRACTION INVESTIGATION OF THE STRUCTURE OF ZIRCONIUM TETRAFLUORIDE IN
	1427	ESTIMATION OF CONFIGURATIONS AND INTERATOMIC DISTANCES IN GROUP-5 AND GROUP-6 TRAN
SPIRO TG	0844	RAMAN INTENSITIES AND THE FORCE FIELD OF PHOSPHORUS PENTAFLUORIDE.
SPOHR R	1225	PHOTOIONIZATION MASS SPECTROMETRIC STUDY OF XENON DIFLUORIDE, XENON TETRAFLUORIDE,
SRIDHARAN T	1403	UNIQUE FORCE FIELDS OF SOME XY3 TYPE MOLECULES.
SRIVASTAVA MP	0591	ROTATIONAL ANALYSIS OF A-X1 BANDS OF LEAD MONOFLUORIDE.
SRIVASTAVA RD	0035	MASS SPECTROMETRIC DETERMINATION OF THE HEATS OF FORMATION OF THE GASEOUS MOLECULE
	0128	DISSOCIATION ENERGIES OF BERYLLIUM FLUORIDE AND BERYLLIUM CHLORIDE AND HEAT OF FOR
	0179	THERMODYNAMIC PROPERTIES OF THE BORON CHLORIDE FLUORIDE SYSTEM FROM MASS SPECTROME
SRIVASTAVA SC	0067	ROTATIONAL ANALYSIS OF THE 2550-2750 ANGSTROM BAND SYSTEM AT ANTIMONY FLUORIDE MOL
ST PETERS R	1179	THEORETICAL CALCULATIONS OF THE ELECTRONIC STRUCTURE AND OPTICAL TRANSITIONS OF UR
STAIR AT	1079	RAMAN SPECTRUM OF SULFUR HEXAFLUORIDE IN THE GASEOUS AND SOLID STATES.
STAMMREICH H	0389	HIGH RESOLUTION RAMAN SPECTROSCOPY IN THE RED AND NEAR INFRARED. (CHLORINE MONOFLU
STAUFFER JL	0037	SUBLIMATION OF ALUMINUM TRIFLUORIDE AND THE INFRARED SPECTRUM OF ALUMINUM TRIFLUOR
	0134	DETERMINATION OF THE GEOMETRY OF HIGH TEMPERATURE SPECIES BY ELECTRIC DEFLECTION A
	0526	MOLECULAR STRUCTURE OF XENON HEXAFLUORIDE AND IODINE HEPTAFLUORIDE.
	1368	GEOMETRY OF THE TRANSITION METAL DIHALIDES. THE DIFLUORIDES OF SOME FIRST ROW TRAN
STEELE RM	0620	PREPARATION AND CRYSTAL STRUCTURE OF MOLYBDENUM TRIFLUORIDE.
STEFFENS CC	0963	RAMAN SPECTRA AND MOLECULAR CONSTANTS OF THE HEXAFLUORIDES OF SULFUR, SELENIUM, AN
STEIGER RP	0047	HIGH TEMPERATURE NEGATIVE IONS. GASEOUS GROUP-3 FLUORIDES.
	0471	ENTHALPY OF SUBLIMATION OF GERMANIUM DIFLUORIDE AND THE THERMODYNAMICS OF SUBLIMAT
	0492	VAPORIZATION THERMODYNAMICS OF INDIUM TRIFLUORIDE.
STEIN L	0017	PHYSICAL AND CHEMICAL PROPERTIES OF HALOGEN FLUORIDES.
	0256	INFRARED STUDIES OF THE BROMINE FLUORIDES. (BROMINE MONOFLUORIDE, BROMINE TRIFLUOR
	1276	THERMOCHEMICAL STUDIES OF XENON TETRAFLUORIDE AND XENON HEXAFLUORIDE.
STEINDLER MJ	0892	ABSORPTION SPECTRUM OF PLUTONIUM HEXAFLUORIDE.
STEINFELD JI	1082	LOW TEMPERATURE ABSORPTION CONTOUR OF THE NU-3 BAND OF SULFUR HEXAFLUORIDE.
	1099	INFRARED DOUBLE RESONANCE IN SULFUR HEXAFLUORIDE.
STEINHARDT RG	0245	INFRARED ABSORPTION SPECTRUM OF LIQUID BORON TRIFLUORIDE.
STEPHAN G	0570	ELECTRONIC SPECTRA OF LANTHANUM FLUORIDE IN THE EXTREME ULTRAVIOLET.
	0582	ELECTRONIC SPECTRUM OF LANTHANUM TRIFLUORIDE.
STEPHENSON CV	0267	RAMAN SPECTRUM, STRUCTURE, FORCE CONSTANTS, AND THERMODYNAMIC PROPERTIES OF BROMIN
STERRITT DE	1201	SPECIFIC HEAT RATIO OF URANIUM HEXAFLUORIDE MEASURED WITH A BALLISTIC PISTON COMPR
STEVENSON PC	1186	VIBRATIONAL SPECTRUM AND THERMODYNAMIC PROPERTIES OF URANIUM HEXAFLUORIDE GAS.
	1187	VIBRATIONAL SPECTRUM AND THERMODYNAMIC PROPERTIES OF URANIUM HEXAFLUORIDE GAS.
STEVENSON PE	0459	MOLECULAR SCF CALCULATIONS FOR GALLIUM MONOFLUORIDE, GALLIUM TRIHYDRIDE, GERMANIUM
STEVIE FA	0485	PREPARATION AND CHARACTERIZATION OF GOLD PENTAFLUORIDE.
STOELEVIK R	1109	FAILURE OF THE FIRST BORN APPROXIMATION IN ELECTRON DIFFRACTION. (TELLURIUM HEXAFL
	1153	SPECTROSCOPIC STUDIES IN CONNECTION WITH ELECTRON DIFFRACTION INVESTIGATION OF SOM
STOLOV AL	0455	REFLECTION SPECTRA OF GADOLINIUM OXIDE AND GADOLINIUM TRIFLUORIDE POWDERS.
STONE AJ	0673	GAS PHASE ELECTRON PARAMAGNETIC RESONANCE SPECTRUM AND DIPOLE MOMENT OF NITROGEN M
STONE JMR	0709	AMBIGUITIES IN THE HARMONIC FORCE FIELDS OF XY3 MOLECULES. (NITROGEN TRIFLUORIDE,
	0743	SIGNS OF L DOUBLING CONSTANTS IN NITROGEN TRIFLUORIDE.
STONE RG	0744	ZEEMAN STUDIES INCLUDING THE MOLECULAR G VALUES, MAGNETIC SUSCEPTIBILITY, AND MOLE
	0780	MOLECULAR G VALUES, MAGNETIC SUSCEPTIBILITIES, MOLECULAR QUADRUPOLE MOMENTS AND SE
STREETS DG	1361	PHOTOELECTRON SPECTRA OF THE HALIDES IN GROUP-3, GROUP-4, GROUP-5, AND GROUP-6.
STRENG AG	0546	ELECTRON SPIN RESONANCE SPECTRUM OF KRYPTON MONOFLUORIDE.

STRENG AG	0785	OXYGEN FLUORIDES. (OXYGEN DIFLUORIDE)
STRICH A	0875	ELECTRONIC STRUCTURE OF PHOSPHORUS PENTAFLUORIDE AND POLYTOPAL REARRANGEMENT IN PH
STRIKER W	0390	O(+) TO SINGLET SIGMA(+) SYSTEMS OF BOTH ISOTOPES OF CHLORINE MONOFLUORIDE.
STROCK HB	0283	X-RAY STUDY OF ANHARMONIC VIBRATIONS IN CALCIUM DIFLUORIDE.
STUART MC	0214	HEATS OF FORMATION OF INORGANIC FLUORIDES ESPECIALLY THE ELEMENTS OF ATOMIC NUMBER
SUBHEDAR MK	1386	SIMPLE METHOD FOR EXACT SOLUTION OF 2 X 2 SECULAR EQUATION IN MOLECULAR VIBRATIONS
SUBRAMANIAN S	0108	ESR STUDY OF RADICALS IN GAMMA IRRADIATED POLYCRYSTALLINE ARSENIC TRIFLUORIDE AND
SUDARIKOV BN	0088	ASSOCIATION OF ANTIMONY PENTAFLUORIDE IN VAPOR.
SUKHOVERKHOV VF	0416	FLUORINE NMR OF CHLORINE PENTAFLUORIDE.
SUNDARAM S	0109	MEAN AMPLITUDES AND THERMODYNAMIC FUNCTIONS OF SOME TRIHALIDES. (ARSENIC TRIFLUORI
	1467	MEAN AMPLITUDES OF THERMAL VIBRATIONS AND SOME THERMODYNAMIC PROPERTIES OF METAL HEXAFL
	1470	EVALUATION OF FORCE CONSTANTS: MOLECULES OF THE TYPE XY6.
SUNDER WA	0003	GAS PHASE STRUCTURES AND MASS SPECTRA OF BINARY PENTAFLUORIDES.
	0367	ELECTRIC DEFLECTION AND THERMAL DECOMPOSITION STUDIES ON CERIUM TETRAFLUORIDE, TER
	0551	CRYSTAL STRUCTURE OF KRYPTON DIFLUORIDE AT -80C.
	1084	ELECTRIC DEFLECTION OF BINARY HEXAFLUORIDES. (SULFUR HEXAFLUORIDE, SELENIUM HEXAF
	1365	GAS PHASE STRUCTURES AND MASS SPECTRA OF BINARY PENTAFLUORIDES.
SUTTON DG	1099	INFRARED DOUBLE RESONANCE IN SULFUR HEXAFLUORIDE.
SUTTON LE	1041	INVESTIGATION BY ELECTRON DIFFRACTION OF THE MOLECULAR STRUCTURES OF SULFUR HEXAFL
SUTTON P	0174	METHODS FOR CORRELATING MOLECULES AND SOME OPTIMIZED VALENCE CONFIGURATION RESULTS
SUVOROV AL	0952	SATURATED VAPOR PRESSURE OF SCANDIUM TRIFLUORIDE, YTTRIUM TRIFLUORIDE, AND LANTHAN
	1357	SATURATED VAPOR PRESSURE OF RARE EARTH FLUORIDES.
SUZEAU P	0268	ROTATIONAL SPECTRA OF BROMINE PENTAFLUORIDE IN THE MILLIMETER WAVE LENGTH REGION.
	0423	MICROWAVE SPECTROSCOPY OF CHLORINE PENTAFLUORIDE AT 70 AND 140 GHZ
SUZUKI I	1491	LEAST SQUARES ADJUSTMENT OF ANHARMONIC POTENTIAL CONSTANTS. APPLICATION TO CARBON
SVEC HJ	1253	THERMOCHEMICAL PROPERTIES OF XENON DIFLUORIDE AND XENON TETRAFLUORIDE FROM MASS SP
SVERDLIN AS	1160	HEATS OF FORMATION OF GASEOUS URANIUM FLUORIDES. (URANIUM MONOFLUORIDE, URANIUM DI
	1329	FREQUENCIES OF ZIRCONIUM FLUORIDE (GAS) VIBRATIONS ACCORDING TO THERMODYNAMIC DATA
SVERDLOV LM	0362	THEORY OF INTENSITIES IN THE INFRARED SPECTRA OF THE MOLECULES CARBON TETRAFLUORID
TAIT JC	0959	PARAMAGNETIC HEXAFLUORIDE ANIONS OF GROUP-6. (SELENIUM PENTAFLUORIDE)
TAKEO H	0477	MICROWAVE SPECTRUM OF GERMANIUM DIFLUORIDE. QUADRUPOLE COUPLING AND CENTRIFUGAL DI
	0478	MICROWAVE SPECTRUM OF GERMANIUM DIFLUORIDE.
TAMRES M	0831	ORBITALS AND STRUCTURES OF PENTAFLUORIDES. (PHOSPHORUS PENTAFLUORIDE, ARSENIC PENT
TAMURA S	0142	ENTHALPY OF BERYLLIUM FLUORIDE FROM 456 TO 1083 DEGREES K BY TRANSPOSED TEMPERATUR
TANAKA T	0993	MICROWAVE SPECTRUM OF SILICON DIFLUORIDE IN THE EXCITED VIBRATIONAL STATES, EQUILI
TANDON SP	0646	ENERGY LEVELS AND SPECTROSCOPIC PARAMETERS OF NEODYMIUM TRIFLUORIDES.
TANNENBAUM IR	0890	PREPARATION AND PROPERTIES OF (VAPOR PRESSURE) OF PLUTONIUM HEXAFLUORIDE AND IDENT
TANNER KN	0642	RAMAN EFFECT AND ULTRAVIOLET ABSORPTION SPECTRA OF MOLYBDENUM AND TUNGSTEN HEXAFLU
TARASOV LK	0940	THERMODYNAMIC CHARACTERISTICS OF SCANDIUM DIHALIDES AND THE THE POSSIBILITY OF THE
TARASOV VI	1328	VAPOR PRESSURE OF ZIRCONIUM TETRAFLUORIDE.
TATAEVSKII VM	0230	STUDY OF THE ISOTOPE EFFECT AND A MORE PRECISE DETERMINATION OF THE FAR INFRARED S
	0292	NEW BANDS OF THE CARBON MONOFLUORIDE MOLECULE.
	0460	ELECTRON DIFFRACTION ANALYSIS OF GALLIUM HALIDES. (GALLIUM TRIFLUORIDE)
	0644	ELECTRONOGRAPHIC INVESTIGATION OF THE STRUCTURE OF THE MOLECULES OF NEODYMIUM HALI
	0693	CALCULATION OF FORCE CONSTANTS FOR NITROGEN DIFLUORIDE AND NITROGEN TRIFLUORIDE.
	0694	ELECTRONIC ABSORPTION SPECTRUM OF THE NITROGEN DIFLUORIDE RADICAL.
	1317	ELECTRON DIFFRACTION STUDY OF THE MOLECULAR STRUCTURE OF THE VAPOR PHASE HALIDES O
TAVARES Y	0389	HIGH RESOLUTION RAMAN SPECTROSCOPY IN THE RED AND NEAR INFRARED. (CHLORINE MONOFLU
TAYLOR JC	0443	STRUCTURES OF FLUORIDES. PART-6: PRECISE STRUCTURAL PARAMETERS IN COPPER DIFLUORID
	1202	STRUCTURES OF FLUORIDES. PART-10: NEUTRON POWDER DIFFRACTION PROFILE STUDIES OF UR
TEBBE FN	1440	FURTHER EVIDENCE OF STEREOCHEMICAL NONRIGIDITY IN FIVE AND SEVEN COORDINATE STRUCT
TEICHMAN S	0745	VIBRATIONAL ANHARMONICITY IN THE PYRAMIDAL TRIFLUORIDES.
TEIXEIRA-DIAS JJC	1404	CORRELATION OF INTERMOLECULAR EFFECTS IN THE VIBRATION SPECTRA OF BORON FLUORIDE W
TEMPLETON DH	1277	CRYSTAL AND MOLECULAR STRUCTURE OF XENON TETRAFLUORIDE.
TERAUDS K	1029	NONEMPIRICAL VALENCE ELECTRON CALCULATIONS ON SMALL MOLECULES CONTAINING PHOSPHORU
TETL'BAUM SD	1204	THERMODYNAMIC PROPERTIES OF URANIUM HEXAFLUORIDE.
THAKUR SN	0018	FORCE CONSTANTS OF METAL HEXAFLUORIDES.
	0032	POTENTIAL ENERGY CURVES AND NATURE OF BINDING IN GROUP-3A MONOHALIDES.
	0248	KINEMATICAL EVALUATION OF FORCE CONSTANTS. APPLICATION TO BORON TRIHALIDES.
	1100	FORCE CONSTANTS OF SULFUR HEXAFLUORIDE FROM ISOTOPIC SUBSTITUTION.
	1141	IONICITY OF THE MX BOND IN THE TETRAHALIDES OF GROUP-4 ELEMENTS. (TIN TETRAFLUORID
	1402	INTRAMOLECULAR FORCE FIELDS IN GROUP-5 TRIHALIDES.
	1468	EXTREMAL PROPERTIES OF FORCE CONSTANTS IN OCTAHEDRAL HEXAFLUORIDES.
	1496	FORCE CONSTANTS OF MOLECULES WITH STRONGLY COUPLED VIBRATIONS.
THANALAKSHMI R	1059	UREY-BRADLEY FORCE FIELD FOR SULFUR TETRAFLUORIDE.
	1387	UREY-BRADLEY FORCE FIELD AND THERMODYNAMIC PROPERTIES FOR BENT SYMMETRICAL XY2 MOL
	1407	MEAN AMPLITUDES OF VIBRATION FOR SOME PYRAMIDAL XY3 AND TETRAHEDRAL XY4 MOLECULES.
	1429	UREY-BRADLEY FORCE FIELD FOR XY4 TETRAHEDRAL MOLECULES.
	1498	POTENTIAL CONSTANTS OF POLYATOMIC MOLECULES. PART-1.
THEARD LP	0140	EFFUSION STUDIES, MASS SPECTRA, AND THERMODYNAMICS OF BERYLLIUM FLUORIDE VAPOR.
THIBAULT R	0343	INFRARED SPECTRUM OF CARBON TETRAFLUORIDE.
THOMAS EL	1453	RAMAN INTENSITIES FOR SPHERICALLY SYMMETRIC MODES OF SOME XY4 TETRAHEDRAL AND XY6
THOMAS PM	1111	ROTATIONAL ANALYSIS OF BANDS OF THE A TRIPLET PI ZERO(+), B TRIPLET PI(1) TO X SIN
THOMAS TD	0374	CHARGE DISTRIBUTION IN CHLORINE MONOFLUORIDE FROM CORE ELECTRON BINDING ENERGIES.
	0409	X-RAY PHOTOELECTRON SPECTROSCOPY OF CHLORINE TRIFLUORIDE, SULFUR TETRAFLUORIDE, AN
	0874	OBSERVATION BY ESCA OF INEQUIVALENT FLUORINES IN CHLORINE TRIFLUORIDE, SULFUR TETR
	1229	ELECTRON DISTRIBUTION IN THE XENON FLUORIDES AND XENON OXIDE TETRAFLUORIDE BY ESCA
THOMPSON HB	0515	STRUCTURE, PSEUDOROTATION, AND VIBRATIONAL MODE COUPLING IN IODINE HEPTAFLUORIDE.
	1280	MOLECULAR STRUCTURE OF XENON HEXAFLUORIDE.
THOMPSON HW	0728	INFRARED SPECTRUM AND MOLECULAR CONSTANTS OF NITROGEN TRIFLUORIDE.
	1479	FAR INFRARED SPECTRA OF PROLATE AND OBLATE SYMMETRIC TOPS.
THOMPSON JW	0186	INFRARED ABSORPTION SPECTRA OF SOME POLYATOMIC FLUORIDES. (SILICON TETRAFLUORIDE,
	0187	MOLECULAR STRUCTURE OF BORON TRIFLUORIDE.
THOMSON C	0180	AB INITIO INVESTIGATION OF THE GEOMETRY, BONDING, AND COUPLING CONSTANTS OF BORON
	0995	NONEMPIRICAL LCAO MO SCF CALCULATIONS OF THE ELECTRONIC STRUCTURE OF SILICON DIFLU
THORN RJ	0281	ENTROPIES AND ENTHALPIES OF SUBLIMATION OF CALCIUM AND CERIUM FLUORIDES. CORRELATI
	0456	ENTROPY AND ENTHALPY OF SUBLIMATION OF GADOLINIUM TRIFLUORIDE. ROLE OF CORRELATION
	0645	ENTROPIES AND ENTHALPIES OF SUBLIMATION OF NEODYMIUM AND TERBIUM TRIFLUORIDES.
	1353	ANALYSES OF THERMOCHEMICAL ERRORS AND SYSTEMATICS IN SUBLIMATION OF LANTHANIDE TRI
THRUSH BA	0293	PREDISSOCIATION IN THE ABSORPTION SPECTRA OF CARBON MONOFLUORIDE AND CARBON DIFLUO

THRUSH BA	0303	ABSORPTION SPECTRUM OF CARBON DIFLUORIDE AND ITS VIBRATIONAL ANALYSIS.
	0673	GAS PHASE ELECTRON PARAMAGNETIC RESONANCE SPECTRUM AND DIPOLE MOMENT OF NITROGEN M
THYAGARAJAN G	1386	SIMPLE METHOD FOR EXACT SOLUTION OF 2 X 2 SECULAR EQUATION IN MOLECULAR VIBRATIONS
THYNNE JCJ	0464	BOND DISSOCIATION ENERGIES, IONIZATION POTENTIALS AND ELECTRON AFFINITIES OF SOME
	0800	IONIZATION BY ELECTRON IMPACT OF PHOSPHORUS TRIFLUORIDE AND DIFLUORO CYANO PHOSPHI
TIEMANN E	0025	MICROWAVE ABSORPTION SPECTRA OF ALUMINUM MONOFLUORIDE, GALLIUM MONOFLUORIDE, INDIU
	0497	ROTATIONAL SPECTRUM OF IODINE FLUORIDE.
TIGELAAR HL	0254	MOLECULAR ZEEMAN EFFECT, MAGNETIC PROPERTIES, AND ELECTRIC QUADRUPOLE MOMENTS IN C
	0384	MOLECULAR MAGNETIC PROPERTIES OF CHLORINE MONOFLUORIDE.
TIMMS PL	0990	MICROWAVE SPECTRUM OF SILICON DIFLUORIDE.
TIMOSHININ VS	0074	CALCULATION OF THE FREQUENCIES OF THE NORMAL VIBRATIONS OF ANTIMONY TRIFLUORIDE.
	0479	FREQUENCIES OF THE DEFORMATION VIBRATIONS OF THE DIHALIDES OF GERMANIUM, TIN, AND
	0941	MOLECULAR CONSTANTS OF SCANDIUM, YTTRIUM, AND LANTHANUM HALIDES. (SCANDIUM DIFLUOR
TISCHER R	0026	ZEEMAN EFFECT IN THE MICROWAVE ROTATIONAL SPECTRUM OF THE ALUMINUM FLUORIDE MOLECU
	1120	HIGH TEMPERATURE MICROWAVE SPECTROMETER FOR MEASUREMENTS OF ZEEMAN EFFECT IN DIAMA
TOERRING T	0497	ROTATIONAL SPECTRUM OF IODINE FLUORIDE.
TOLLES WM	1056	STRUCTURE AND DIPOLE MOMENT OF SULFUR TETRAFLUORIDE.
TOMASI J	0762	DOUBLE ZETA SCF CALCULATIONS FOR NITRITE ION AND OXYGEN DIFLUORIDE.
	0763	CONFIGURATION INTERACTION CALCULATION FOR THE GROUND STATE OF OXYGEN DIFLUORIDE, N
	0776	MINIMAL BASIS SET SCF CALCULATIONS FOR THE GROUND STATE OF OZONE, NITRITE ION, NIT
TOMISAKA T	0606	FORCE FIELDS OF THE RUTILE COUNTERPARTS: TITANIUM DIOXIDE, MAGNESIUM FLUORIDE, ZIN
TONG DA	0664	NIOBIUM-93 NUCLEAR QUADRUPOLE RESONANCE STUDIES OF NIOBIUM PENTAFLUORIDE AND ITS C
TOOLE RC	0781	PREPARATION AND PURIFICATION OF OXYGEN DIFLUORIDE AND DETERMINATION OF ITS VAPOR P
TORRING T	0025	MICROWAVE ABSORPTION SPECTRA OF ALUMINUM MONOFLUORIDE, GALLIUM MONOFLUORIDE, INDIU
	0490	MICROWAVE ROTATIONAL SPECTRUM OF INDIUM MONOFLUORIDE.
TOSATTI E	0196	BARRIER TO ELECTRON PASSAGE THROUGH ELECTRONEGATIVE ATOMS IN BORON TRIFLUORIDE.
TOY MS	0259	ELECTRICAL CONDUCTIVITY OF SOLID CHLORINE TRIFLUORIDE AND BROMINE TRIFLUORIDE.
TOYUKI H	0246	UREY-BRADLEY FORCE CONSTANTS OF BORON HALIDES AND THEIR TRANSFERABILITY. (BORON TR
	0247	NORMAL VIBRATIONS OF BORON TRIHALIDES. (BORON TRIFLUORIDE)
TRAMBARULO R	0711	MICROWAVE SPECTROSCOPY IN THE REGION FROM TWO TO THREE MILLIMETERS, PART-2: (NITRO
TRICKEY SB	1304	MULTIPLE SCATTERING X(ALPHA) CALCULATION OF THE ENERGY LEVEL STRUCTURE OF XENON HE
TRIPATHI DN	0248	KINEMATICAL EVALUATION OF FORCE CONSTANTS. APPLICATION TO BORON TRIHALIDES.
TROMBETTI A	1022	SPECTRUM OF SULFUR MONOFLUORIDE.
TSAO P	1254	RAMAN SPECTRA FOR XENON DIFLUORIDE, XENON TETRAFLUORIDE, AND XENON OXIDE TETRAFLUO
TSYANGENKO MM	1068	INFRARED SPECTRA OF CRYOSYSTEMS. PART-2: SULFUR HEXAFLUORIDE.
TUMANOV YN	0004	REACTIVITY AND THERMAL STABILITY OF HEXAFLUORIDES.
	0422	THERMODYNAMIC PROPERTIES OF PENTAFLUORIDES AT HIGH TEMPERATURES. PART-1: CALCULATI
	0536	THERMODYNAMIC PROPERTIES OF PENTAFLUORIDES AT HIGH TEMPERATURES. PART-3: PENTAFLUO
	0622	EMPIRICAL METHOD FOR DETERMINING EFFECTIVE VIBRATIONAL AND ROTATIONAL CHARACTERIST
	0629	THERMODYNAMIC STABILITY OF HEXAFLUORIDES AT HIGH TEMPERATURES. PART-2: PLUTONIUM H
	0630	THERMAL STABILITY AND REACTIVITY OF D- AND F ELEMENT HEXAFLUORIDES.
	0647	THERMODYNAMIC PROPERTIES OF PENTAFLUORIDES AT HIGH TEMPERATURES. PART-4: PENTAFLUO
	0654	THERMODYNAMIC PROPERTIES OF NIOBIUM AND TANTALUM FLUORIDES AT HIGH TEMPERATURES. P
	0655	THERMODYNAMIC PROPERTIES OF NIOBIUM MONOFLUORIDE, NIOBIUM DIFLUORIDE, TANTALUM MON
	0656	THERMODYNAMIC PROPERTIES OF NIOBIUM TRIFLUORIDE, NIOBIUM TETRAFLUORIDE, TANTALUM T
	0657	THERMODYNAMIC PROPERTIES OF NIOBIUM AND TANTALUM FLUORIDES AT HIGH TEMPERATURES. P
	1177	THERMODYNAMIC PROPERTIES OF URANIUM PENTAFLUORIDE, NEPTUNIUM PENTAFLUORIDE, AND PL
	1203	THERMODYNAMIC STABILITY OF URANIUM HEXAFLUORIDE.
	1328	VAPOR PRESSURE OF ZIRCONIUM TETRAFLUORIDE.
	1379	THERMODYNAMIC PROPERTIES OF TRANSITION METAL PENTAFLUORIDES AT HIGH TEMPERATURE. (
	1434	THERMODYNAMIC PROPERTIES OF PENTAFLUORIDES AT HIGH TEMPERATURES. PART-2: PENTAFLUO
TUMANOV YUN	0665	THERMODYNAMIC PROPERTIES OF NIOBIUM AND TANTALUM FLUORIDES AT HIGH TEMPERATURES. P
TURKEVICH J	1186	VIBRATIONAL SPECTRUM AND THERMODYNAMIC PROPERTIES OF URANIUM HEXAFLUORIDE GAS.
	1187	VIBRATIONAL SPECTRUM AND THERMODYNAMIC PROPERTIES OF URANIUM HEXAFLUORIDE GAS.
TURNER DW	0980	PHOTOELECTRON SPECTRUM OF SILICON DIFLUORIDE.
	1362	STUDY OF THE BONDING IN THE GROUP-4 TETRAHALIDES BY PHOTOELECTRON SPECTROSCOPY.
TURNER JJ	0566	PREPARATION OF INERT GAS COMPOUNDS BY MATRIX ISOLATION: KRYPTON DIFLUORIDE.
	0567	KRYPTON DIFLUORIDE. PREPARATION BY THE MATRIX ISOLATION TECHNIQUE.
	0767	RAMAN SPECTRUM OF LIQUID OXYGEN DIFLUORIDE.
TVETEN A	0454	MAGNETIC SUSCEPTIBILITY OF EUROPIUM TRIFLUORIDE.
TYSON J	0873	RAMAN SPECTRA OF ARSENIC PENTAFLUORIDE AND VANADIUM PENTAFLUORIDE AND THEIR FORCE
UMEGAKI Y	0606	FORCE FIELDS OF THE RUTILE COUNTERPARTS: TITANIUM DIOXIDE, MAGNESIUM FLUORIDE, ZIN
UNLAND ML	0746	CHARACTERIZATION OF GROUND STATE WAVE FUNCTIONS BY MEASURED ELECTRONIC PROPERTIES
UPADHYA KN	1136	ROTATIONAL ANALYSIS OF THE C DOUBLET DELTA 5/2- X DOUBLET PI 3/2 AND G DOUBLET DEL
URCH DS	1309	THE STEREOCHEMICALLY INERT LONE PAIR. A SPECULATION ON THE BONDING IN ANTIMONY HEX
UREY HC	1497	VIBRATIONS OF PENTATOMIC TETRAHEDRAL MOLECULES.
UY OM	0035	MASS SPECTROMETRIC DETERMINATION OF THE HEATS OF FORMATION OF THE GASEOUS MOLECULE
UZIKOV AN	0294	ROTATIONAL CONSTANTS OF THE B STATE OF THE CARBON MONOFLUORIDE MOLECULE .
	0465	ANALYSIS OF THE ROTATIONAL STRUCTURE OF THE BANDS 0-0 AND 0-1 IN GERMANIUM MONOFLU
	0470	VIBRATIONAL ANALYSIS OF C-X AND C'-X' BAND SYSTEMS AND ENERGY OF DISSOCIATION IN G
	1138	ANALYSIS OF VIBRATIONAL AND ROTATIONAL STRUCTURE OF TIN MONOFLUORIDE IN THE 2500 A
VAISHNAVA PP	0646	ENERGY LEVELS AND SPECTROSCOPIC PARAMETERS OF NEODYMIUM TRIFLUORIDES.
VALERGA AJ	0296	THERMODYNAMIC, ELECTROCHEMICAL AND SYNTHETIC STUDIES OF THE GRAPHITE FLUORINE COM
VAN LEIRSBURG DA	1324	INTERACTION OF MATRIX ISOLATED NICKEL FLUORIDE AND NICKEL CHLORIDE WITH CARBON MON
VAN NIESSEN W	0175	DENSITY LOCALIZATION OF ATOMIC AND MOLECULAR ORBITALS. PART-3: HETERONUCLEAR DIATO
VANDERRYN J	0249	INFRARED SPECTRUM OF BORON TRIFLUORIDE.
VANDERVOET A	0475	MATRIX INFRARED AND LASER RAMAN SPECTRA, MOLECULAR STRUCTURES AND NORMAL COORDINAT
VANKA M	0893	PLUTONIUM HEXAFLUORIDE: ITS PREPARATION AND PROPERTIES.
VANWAZER JR	0857	ELECTRONIC STRUCTURES OF PHOSPHORUS PENTAFLUORIDE AND TETRAFLUORO PHOSPHORANE.
VASILE MJ	0003	GAS PHASE STRUCTURES AND MASS SPECTRA OF BINARY PENTAFLUORIDES.
	0019	APPLICATION OF A MOLECULAR BEAM SOURCE MASS SPECTROMETER TO THE STUDY OF REACTIVE
	0485	PREPARATION AND CHARACTERIZATION OF GOLD PENTAFLUORIDE.
	0876	ASSOCIATION OF GROUP-5 PENTAFLUORIDES IN THE GAS PHASE. (PHOSPHORUS PENTAFLUORIDE,
	1365	GAS PHASE STRUCTURES AND MASS SPECTRA OF BINARY PENTAFLUORIDES.
	1376	MOLECULAR BEAM MASS SPECTROMETRIC STUDY OF BINARY PENTAFLUORIDES. (ALSO ANTIMONY
	1377	APPLICATION OF A MOLECULAR BEAM SOURCE MASS SPECTROMETER TO THE STUDY OF REACTIVE
VASINI EJ	0392	CHLORINE FLUORINE SYSTEM AT LOW TEMPERATURE. CHARACTERIZATION OF THE CHLORINE DIFL
	0393	THE CHLORINE DIFLUORIDE FREE RADICAL.
VASUDEV R	0070	NEAR ULTRAVIOLET AND VISIBLE BANDS OF ANTIMONY MONOFLUORIDE.
VECHER AA	0432	THERMODYNAMIC PROPERTIES OF CHROMIUM DIFLUORIDE.

VECHER RA   0432 THERMODYNAMIC PROPERTIES OF CHROMIUM DIFLUORIDE.
VEILLARD A   0875 ELECTRONIC STRUCTURE OF PHOSPHORUS PENTAFLUORIDE AND POLYTOPAL REARRANGEMENT IN PH
VENKATESWARLU K   0411 GENERALIZED MEAN AMPLITUDES AND CORIOLIS CONSTANTS IN DICHLORO BORANE, DIBROMO BOR
                     0427 CORIOLIS COUPLING COEFFICIENTS OF BROMINE PENTAFLUORIDE AND CHLORINE PENTAFLUORIDE
                     0513 UREY-BRADLEY FORCE CONSTANTS, MEAN AMPLITUDES OF VIBRATION, SHRINKAGE EFFECT AND C
                     1057 MEAN AMPLITUDES OF VIBRATION: SULFUR TETRAFLUORIDE AND SULFUR PENTAFLUORO CHLORIDE
                     1058 MOLECULAR VIBRATIONS OF SULFUR TETRAFLUORIDE.
                     1059 UREY-BRADLEY FORCE FIELD FOR SULFUR TETRAFLUORIDE.
                     1278 FORCE FIELD, CORIOLIS COUPLING COEFFICIENTS, GENERALIZED MEAN SQUARE AMPLITUDES OF
                     1387 UREY-BRADLEY FORCE FIELD AND THERMODYNAMIC PROPERTIES FOR BENT SYMMETRICAL XY2 MOL
                     1405 GENERALIZED MEAN SQUARE AMPLITUDES AND CORIOLIS CONSTANTS OF SOME PYRAMIDAL XY3 MO
                     1406 UREY-BRADLEY FORCE FIELD AND THERMODYNAMIC PROPERTIES OF PYRAMIDAL XY3 MOLECULES.
                     1407 MEAN AMPLITUDES OF VIBRATION FOR SOME PYRAMIDAL XY3 AND TETRAHEDRAL XY4 MOLECULES.
                     1428 POTENTIAL CONSTANTS OF XY4 AND XY3Z TYPES OF MOLECULES AND RADICALS.
                     1429 UREY-BRADLEY FORCE FIELD FOR XY4 TETRAHEDRAL MOLECULES.
                     1469 UREY-BRADLEY FORCE FIELD OF SOME XY6 SYSTEMS.
                     1470 EVALUATION OF FORCE CONSTANTS: MOLECULES OF THE TYPE XY6.
                     1498 POTENTIAL CONSTANTS OF POLYATOMIC MOLECULES. PART-1.
VENKATESWARLU P   0309 EMISSION BANDS OF CARBON DIFLUORIDE.
                     0988 EMISSION SPECTRUM OF SILICON DIFLUORIDE. PART-1: THE BAND SYSTEM IN THE REGION 275
VERKHIVKER GP   1087 THERMODYNAMIC PROPERTIES OF SULFUR HEXAFLUORIDE.
                     1204 THERMODYNAMIC PROPERTIES OF URANIUM HEXAFLUORIDE.
VERMA RD   0978 A QUARTET SIGMA TO DOUBLET PI TRANSITION OF THE SILICON MONOFLUORIDE MOLECULE.
                     0999 ROTATIONAL ANALYSIS OF THE GAMMA BANDS OF SILICON MONOFLUORIDE.
VICKROY DG   0392 CHLORINE FLUORIDE SYSTEM AT LOW TEMPERATURE. CHARACTERIZATION OF THE CHLORINE DIFL
                     0393 THE CHLORINE DIFLUORIDE FREE RADICAL.
VIJAYARAGHAVAN R   0927 ANOMALOUS NMR SHIFT IN SAMARIUM TRIFLUORIDE.
                     1355 NMR IN RARE EARTH TRIFLUORIDES.
VINOGRADOV AS   1101 X-RAY SPECTRA OF SULFUR AND FLUORINE IN SULFUR HEXAFLUORIDE.
VINOGRADOVA VN   1430 PROGRESSIVE STIFFNESS METHOD APPLIED TO XY4 MOLECULES IN THE LIGHT OF THE THEORY O
                     1441 APPLICATION OF THE METHOD OF PROGRESSIVE RIGIDITY TO OCTAHEDRAL MX6 MOLECULES.
                     1448 CALCULATION OF THE FORCE CONSTANTS OF HEXAFLUORIDES.
VIRLET J   0638 MOLECULAR MOTION AND FLUORINE RELAXATION IN THE LIQUIDS OF MOLYBDENUM HEXAFLUORIDE
VISWANATH DS   0739 THERMODYNAMIC PROPERTIES OF NITROGEN TRIFLUORIDE.
VIZI B   1219 FORCE CONSTANTS, MEAN AMPLITUDES OF VIBRATION, SHRINKAGE EFFECTS, CORIOLIS CONSTAN
                     1369 MEAN AMPLITUDES AND SHRINKAGE EFFECTS OF VIBRATION FOR SOME LINEAR SYMMETRICAL DIF
VOLAVSEK B   1310 MAGNETIC SUSCEPTIBILITY OF XENON HEXAFLUORIDE.
VOLKOV VM   1162 STRUCTURE OF MOLECULAR URANIUM DIFLUORIDE AND URANIUM TRIFLUORIDE.
VON NIESSEN W   1090 ELECTRONIC STRUCTURE OF MOLECULES BY A MANY BODY APPROACH. PART-6: THE ASSIGNMENT
VOORN PCVD   0877 NORMAL COORDINATE ANALYSIS OF PHOSPHORUS PENTAFLUORIDE, PHOSPHORUS DIFLUORIDE DICH
VROOM DA   1076 PHOTOELECTRON SPECTRUM OF SULFUR HEXAFLUORIDE AT 584 ANGSTROMS.
WABER JT   0881 MOLECULAR ORBITALS OF PLATINUM HEXAFLUORIDE AND ELEMENT 110 HEXAFLUORIDE CALCULATE
WAHL AC   0094 ARSENIC MONOFLUORIDE, 3 SIGMA. DISSOCIATION ENTHALPY, IONIZATION POTENTIAL, ELECTR
                     0174 METHODS FOR CORRELATING MOLECULES AND SOME OPTIMIZED VALENCE CONFIGURATION RESULTS
                     0290 MOLECULAR ORBITAL INVESTIGATION OF CARBON MONOFLUORIDE AND SILICON MONOFLUORIDE AN
                     0678 QUANTUM CHEMICAL STUDY OF SOME PNICOGEN MONOFLUORIDES. (NITROGEN MONOFLUORIDE AND
                     0757 OXYGEN MONOFLUORIDE, DOUBLET PI HARTREE-FOCK WAVE FUNCTION, BINDING ENERGY, IONIZA
                     1025 HARTREE-FOCK WAVE FUNCTIONS AND COMPUTED PROPERTIES FOR THE DOUBLET PI GROUND STAT
WALKER MA   0391 METHOD OF DETERMINING SATURATED LIQUID AND SATURATED VAPOR ENTROPY. (CHLORINE MONO
WALSH AD   0806 ABSORPTION SPECTRA OF THE HYDRIDES, DEUTERIDES, AND HALIDES OF GROUP-5 ELEMENTS. (
WALTER JA   0378 PHOTOIONIZATION STUDY OF CHLORINE MONOFLUORIDE AND THE DISSOCIATION ENERGY OF FLUO
WALTER TA   1112 PHOTOIONIZATION OF HIGH TEMPERATURE VAPORS. PART-4: THALLIUM MONOFLUORIDE, THALLIU
WANCZEK KP   1026 ESTIMATION OF INFRARED FREQUENCIES OF SULFUR DIFLUORIDE.
WANG DKW   0071 BO(+)-X(21) AND BO(+)-X(10+) BANDS OF ANTIMONY MONOFLUORIDE.
WANG JL   0480 ENTHALPY OF FORMATION OF GERMANIUM TRIFLUORIDE.
                     0998 EMISSION SPECTRUM OF SILICON TRIFLUORIDE.
WANG JLF   0047 HIGH TEMPERATURE NEGATIVE IONS. GASEOUS GROUP-3 FLUORIDES.
WANG SY   1311 CRYSTAL FIELD MODEL STUDY OF XENON HEXAFLUORIDE. PART-1: ENERGY LEVELS AND MOLECUL
WANNOUS G   0599 ROTATIONAL ANALYSIS OF SELECTED BANDS FROM THE ELECTRONIC SPECTRUM OF THE LUTETIUM
WARD JC   1164 PREPARATION AND PROPERTIES OF URANIUM TRIFLUORIDE. (CRYSTAL STRUCTURE)
WARSOP PA   0806 ABSORPTION SPECTRA OF THE HYDRIDES, DEUTERIDES, AND HALIDES OF GROUP-5 ELEMENTS. (
WATERS JH   1343 BOND ENERGIES AND IONIC CHARACTER OF INERT GAS HALIDES. HELIUM-2 PHOTOELECTRON SPE
WATSON JKG   0709 AMBIGUITIES IN THE HARMONIC FORCE FIELDS OF XY3 MOLECULES. (NITROGEN TRIFLUORIDE,
WAUGH JS   1037 NMR SPECTRUM AND STRUCTURE OF SULFUR TETRAFLUORIDE.
WAZER JRV   0746 CHARACTERIZATION OF GROUND STATE WAVE FUNCTIONS BY MEASURED ELECTRONIC PROPERTIES.
WEATHERLY TL   0101 CENTRIFUGAL STRETCHING OF TRIFLUORO CHLORO METHANE AND SOME OTHER SYMMETRIC TOPS.
WEAVER EE   0882 PREPARATION AND SOME PROPERTIES OF PLATINUM HEXAFLUORIDE.
                     1255 THE XENON- FLUORINE SYSTEM.
                     1312 XENON HEXAFLUORIDE. (SPECTRUM)
                     1336 VAPOR PRESSURES AND THERMODYNAMIC CALCULATIONS FOR SOME ACTINIDE FLUORIDES
WEDLEIN J   1126 INFRARED AND RAMAN SPECTRA AND THE STRUCTURE OF CRYSTALLINE THALLIUM MONOFLUORIDE.
WEI MSS   0793 ELECTRON SPIN RESONANCE STUDIES OF PHOSPHORUS DIFLUORIDE, PHOSPHORUS DICHLORIDE, A
WEIGEL F   0898 PROPERTIES OF PROMETHIUM COMPOUNDS. (PROMETHIUM TRIFLUORIDE)
WEINER RS   0747 CALCULATION OF PERPENDICULAR INFRARED BAND CONTOURS: APPLICATION TO THE NU-4 BAND
WEINSTOCK B   0020 SOME PROPERTIES OF THE HEXAFLUORIDE MOLECULES. (REVIEW)
                     0021 VIBRATIONAL PROPERTIES OF HEXAFLUORIDE MOLECULES.
                     0396 VIBRATIONAL SPECTRA AND THERMODYNAMIC PROPERTIES OF CHLORINE TRIFLUORIDE AND BROMI
                     0539 SEARCH FOR A JAHN-TELLER EFFECT IN IRIDIUM HEXAFLUORIDE.
                     0649 INFRARED SPECTRA OF NEPTUNIUM HEXAFLUORIDE AND PLUTONIUM HEXAFLUORIDE.
                     0650 HEAT CAPACITY, ENTHALPY OF FUSION, AND THERMODYNAMIC PROPERTIES OF NEPTUNIUM HEXAF
                     0752 VIBRATIONAL SPECTRA OF OSMIUM HEXAFLUORIDE AND PLATINUM HEXAFLUORIDE.
                     0753 OSMIUM HEXAFLUORIDE AND ITS IDENTITY WITH THE PREVIOUSLY REPORTED OCTAFLUORIDE.
                     0882 PREPARATION AND SOME PROPERTIES OF PLATINUM HEXAFLUORIDE.
                     0894 PROPERTIES OF PLUTONIUM HEXAFLUORIDE. (MOLECULAR STRUCTURE, ABSORPTION SPECTRUM)
                     0902 NEAR INFRARED SYSTEM OF RHENIUM HEXAFLUORIDE.
                     0916 COLORS OF TRANSITION METAL HEXAFLUORIDES. (RHENIUM HEXAFLUORIDE, OSMIUM HEXAFLUORI
                     0926 JAHN-TELLER EFFECT IN THE E(8) VIBRATIONAL MODE OF HEXAFLUORIDE MOLECULES. THE INF
                     1155 ELECTRON DIFFRACTION INVESTIGATION OF TUNGSTEN HEXAFLUORIDE, OSMIUM HEXAFLUORIDE,
                     1159 HEAT CAPACITIES AND ELECTRONIC SPECTRA OF THE PLATINUM GROUP METAL HEXAFLUORIDE MO
                     1190 RAMAN SPECTRUM OF URANIUM HEXAFLUORIDE.

WEINSTOCK B	1205	SOME RECENT STUDIES WITH HEXAFLUORIDES. (URANIUM HEXAFLUORIDE, PLUTONIUM HEXAFLUOR
	1255	THE XENON- FLUORINE SYSTEM.
	1312	XENON HEXAFLUORIDE. (SPECTRUM)
	1336	VAPOR PRESSURES AND THERMODYNAMIC CALCULATIONS FOR SOME ACTINIDE FLUORIDES
	1471	JAHN-TELLER EFFECT IN VIBRATIONAL SPECTRA OF HEXAFLUORIDES.
WEISS H	0847	RAMAN AND INFRARED SPECTRUM OF TRIFLUORO PHOSPHORANE. (PHOSPHORUS PENTAFLUORIDE)
WEISS S	1182	INFRARED SPECTRUM OF URANIUM HEXAFLUORIDE.
WELLS EJ	0517	FLUORINE NMR OF IODINE HEPTAFLUORIDE, RHENIUM HEPTAFLUORIDE, AND OXIDE PENTAFLUORI
WENDLING EJL	0535	FORCE CONSTANTS AND MEAN VIBRATION AMPLITUDES OF IODINE HEPTAFLUORIDE CALCULATED B
	0878	GENERALIZED VALENCE FORCE FIELD FORCE CONSTANTS AND MEAN SQUARE AMPLITUDES OF VIBR
	1431	NEW METHOD FOR RESOLVING WILSON'S SECULAR EQUATIONS. APPLIED TO TETRAHEDRAL MOLECU
	1472	FORCE CONSTANTS OF OCTAHEDRAL MOLECULES CALCULATED BY THE LOGARITHMIC STEP METHOD.
WENSKY DA	0821	SELF CONSISTENT CHARGE AND CONFIGURATION MOLECULAR ORBITAL CALCULATIONS ON PHOSPHO
WENTINK T	0295	OSCILLATOR STRENGTHS OF CARBON MONOFLUORIDE AND COMMENTS ON HEATS OF FORMATION OF
WENTZ RE	0880	ANALYSES OF THE VIBRATION ROTATION BANDS OF TRIFLUORO METHANE, PHOSPHORUS PENTAFLU
WERTHER HU	0754	PREPARATION AND PROPERTIES (INFRARED SPECTRUM) OF OSMIUM HEPTAFLUORIDE.
WESLEY RD	0583	GEOMETRY AND INFRARED SPECTRA OF MATRIX ISOLATED RARE EARTH HALIDES. (ALSO LANTHAN
	1316	INFRARED SPECTRA AND GEOMETRIES OF MATRIX ISOLATED YTTRIUM DIFLUORIDE AND YTTRIUM
	1351	INFRARED SPECTRA AND GEOMETRIES OF RARE EARTH DIHALIDES.
WESTLEY F	1358	BIBLIOGRAPHY ON SPECTROSCOPY OF FLUORIDES AND OXIDES BELONGING TO THE LANTHANIDE S
WESTRUM EF	1124	HEAT CAPACITY AND THERMODYNAMIC PROPERTIES OF THALLIUM FLUORIDE FROM 5 TO 445 DEGR
	1159	HEAT CAPACITIES AND ELECTRONIC SPECTRA OF THE PLATINUM GROUP METAL HEXAFLUORIDE MO
WESTWOOD NPC	0996	PHOTOELECTRON SPECTRUM OF SILICON DIFLUORIDE.
WEVER A	0347	RAMAN SPECTRUM OF GASEOUS CARBON TETRAFLUORIDE.
WHALEN JJ	0698	ABSORPTION SPECTRA OF MATRIX ISOLATED NITROGEN TRIFLUORIDE FROM 18 TO 50 GC.
WHANGBO MH	1366	SCF- X(ALPHA) SCATTERED WAVE STUDIES ON BONDING AND IONIZATION POTENTIALS. PART-1:
WHARTON L	0126	GEOMETRY OF THE ALKALINE EARTH DIHALIDES. (BARIUM DIFLUORIDE, STRONTIUM DIFLUORIDE
WHITE D	0609	GEOMETRY AND VIBRATIONAL SPECTRA OF THE ALKALINE EARTH DIHALIDES. PART-1: MAGNESIU
WHITEWAY SG	0273	MELTING POINTS OF INORGANIC FLUORIDES. (CADMIUM DIFLUORIDE, MAGNESIUM DIFLUORIDE,
WHITTLE E	0311	INFRARED SPECTRUM OF THE FREE RADICAL CARBON TRIFLUORIDE ISOLATED IN INERT MATRICE
WHITTLE MJ	0260	K-TYPE DOUBLING IN THE MILLIMETER WAVE SPECTRUM OF BROMINE PENTAFLUORIDE.
	0269	MICROWAVE SPECTRUM OF BROMINE PENTAFLUORIDE.
	0502	MICROWAVE SPECTRUM OF IODINE PENTAFLUORIDE.
WIEDER GM	1075	INFRARED INTENSITIES IN CRYSTALLINE SULFUR HEXAFLUORIDE.
WIELAND K	0298	HEAT OF FORMATION OF CARBON DIFLUORIDE.
WILKINSON MK	0620	PREPARATION AND CRYSTAL STRUCTURE OF MOLYBDENUM TRIFLUORIDE.
WILLETT RD	1313	BONDING IN XENON HEXAFLUORIDE.
WILLIAMS F	0270	ESR SPECTRUM AND STRUCTURE OF BROMINE HEXAFLUORIDE.
	0428	CHLORINE HEXAFLUORIDE RADICAL. PREPARATION, ELECTRON SPIN RESONANCE SPECTRUM, AND
WILLIAMS GR	0176	AB INITIO CALCULATIONS ON SOME SMALL RADICALS BY THE UNRESTRICTED HARTREE-FOCK MET
	0682	ELECTRONIC STRUCTURE OF THE NITROGEN DIFLUORIDE RADICAL.
WILLIAMS Q	0101	CENTRIFUGAL STRETCHING OF TRIFLUORO CHLORO METHANE AND SOME OTHER SYMMETRIC TOPS.
WILLIAMSON SM	1266	HEAT OF FORMATION OF XENON TETRAFLUORIDE.
	1267	XENON TETRAFLUORIDE: HEAT OF FORMATION.
	1277	CRYSTAL AND MOLECULAR STRUCTURE OF XENON TETRAFLUORIDE.
WILSON EB	1024	MICROWAVE AND MASS SPECTRA OF SULFUR MONOFLUORIDE.
WILSON EG	1237	HEATS OF SUBLIMATION OF XENON DIFLUORIDE AND XENON TETRAFLUORIDE AND A CONJECTURE
	1238	THEORETICAL AND EXPERIMENTAL STUDIES OF THE ELECTRONIC STRUCTURE OF THE XENON FLUO
	1256	FAR ULTRAVIOLET SPECTROSCOPIC STUDY OF XENON DIFLUORIDE.
	1272	FAR ULTRAVIOLET SPECTROSCOPIC STUDY OF XENON TETRAFLUORIDE.
WILSON MK	0748	INFRARED SPECTRA OF NITROGEN TRIFLUORIDE AND PHOSPHORUS TRIFLUORIDE.
WILSON PW	0443	STRUCTURES OF FLUORIDES. PART-6: PRECISE STRUCTURAL PARAMETERS IN COPPER DIFLUORID
	0472	ENTHALPY OF FORMATION OF GERMANIUM DIFLUORIDE.
	0478	MICROWAVE SPECTRUM OF GERMANIUM DIFLUORIDE.
	0484	A STUDY OF THE RAMAN AND BROAD LINE NMR SPECTRA OF GERMANIUM TETRAFLUORIDE.
	1194	ELECTRONIC ABSORPTION SPECTRUM OF GASEOUS URANIUM HEXACHLORIDE AND URANIUM HEXAFLU
	1202	STRUCTURES OF FLUORIDES. PART-10: NEUTRON POWDER DIFFRACTION PROFILE STUDIES OF UR
WILT PM	0861	E' FORCE FIELD OF PHOSPHORUS PENTAFLUORIDE.
	0880	ANALYSES OF THE VIBRATION ROTATION BANDS OF TRIFLUORO METHANE, PHOSPHORUS PENTAFLU
WINFIELD JM	0664	NIOBIUM-93 NUCLEAR QUADRUPOLE RESONANCE STUDIES OF NIOBIUM PENTAFLUORIDE AND ITS C
WINTERNITZ J	0153	ROTATIONAL ANALYSIS OF BANDS OF THE C TRIPLET SIGMA, B TRIPLET SIGMA: A TRIPLET PI
WIRICK MP	0584	NEAR ULTRAVIOLET OPTICAL CONSTANTS OF LANTHANUM TRIFLUORIDE.
WIRSAM B	0997	ELECTRONIC SPECTRUM OF SILICON DIFLUORIDE. THEORETICAL STUDY.
WISE SS	0250	FLUORINE BOMB CALORIMETRY. PART-3: THE HEAT OF FORMATION OF BORON TRIFLUORIDE.
WITHERS GL	1115	ENTHALPY OF FORMATION AND THE DISSOCIATION ENERGY OF THALLIUM MONOFLUORIDE
WITKE K	1300	STRUCTURE OF THE IODINE HEXAFLUORIDE ANION. (XENON HEXAFLUORIDE)
WITT JD	0879	RAMAN SPECTRA OF GASES. PART-10: RAMAN BAND CONTOUR OF THE NU-7 (E') FUNDAMENTAL O
WITT WP	0053	HEAT OF SUBLIMATION OF ALUMINUM TRIFLUORIDE AND HEAT OF FORMATION OF ALUMINUM MONO
WOLGA GJ	1074	OPTICAL SATURATION OF A SINGLE VIBRATION- ROTATION TRANSITION IN THE NU-3 FUNDAMEN
WOLTZ PJH	0363	INFRARED SPECTRA OF CARBON TETRAFLUORIDE AND GERMANIUM TETRAFLUORIDE.
	1005	INFRARED AND RAMAN SPECTRA OF SILICON TETRAFLUORIDE.
WOOD JL	0296	THERMODYNAMIC, ELECTROCHEMICAL AND SYNTHETIC STUDIES OF THE GRAPHITE FLUORINE COM
WOODS HP	0685	ABSORPTION SPECTRUM OF NITROGEN DIFLUORIDE.
WOODWARD LA	0482	RAMAN AND INFRARED SPECTRA OF GERMANIUM TETRAFLUORIDE.
	1040	RAMAN AND INFRARED SPECTRUM OF SULFUR TETRAFLUORIDE.
WOOLF AA	0110	STANDARD HEAT OF FORMATION OF ARSENIC TRIFLUORIDE.
WU E	1492	ADAPTATION OF THE UREY-BRADLEY FORCE FIELD TO AMMONIA, PHOSPHINE, STIBINE, AND ARS
WUSE FC	0033	MILLIMETER AND SUBMILLETER WAVE SPECTRUM AND MOLECULAR CONSTANTS OF ALUMINUM MONOF
WYATT R	0880	ANALYSES OF THE VIBRATION ROTATION BANDS OF TRIFLUORO METHANE, PHOSPHORUS PENTAFLU
XUEN VD	1414	ESTIMATION OF THE FORCE CONSTANTS AND INTERATOMIC DISTANCES OF HALIDE COMPLEX MOLE
YAKEL HL	0620	PREPARATION AND CRYSTAL STRUCTURE OF MOLYBDENUM TRIFLUORIDE.
YAKOVLEV GN	0056	INVESTIGATION OF THE CHEMISTRY OF AMERICIUM. (AMERICIUM TETRAFLUORIDE)
YAMADA S	1344	MOLECULAR STRUCTURES OF RARE GAS COMPOUNDS.
YAMAMOTO S	0772	FERMI DIAD OF NU-1 AND 2NU-2 IN OXYGEN DIFLUORIDE.
YANG YS	0054	INFRARED SPECTRA OF ALUMINUM FLUORIDE IN SOLID ARGON.
YARUNIN VS	0498	SPIN ORBITAL INTERACTION AND RELATIVE INTENSITIES OF FORBIDDEN TRANSITIONS IN IODI
YAWS CL	1062	PHYSICAL AND THERMODYNAMIC PROPERTIES OF SULFUR HEXAFLUORIDE. (REVIEW)
YEE KK	0092	ROTATIONAL ANALYSIS OF FOUR SINGLET SYSTEMS OF ARSENIC MONOFLUORIDE.
	0095	VIBRATIONAL ANALYSIS OF THE SINGLET SYSTEMS OF ARSENIC MONOFLUORIDE.
YERANOS W	1257	A THEORETICAL EVALUATION OF THE ROOT MEAN SQUARE AMPLITUDES OF VIBRATION IN XENON

YOKOKAWA T   0142   ENTHALPY OF BERYLLIUM FLUORIDE FROM 456 TO 1083 DEGREES K BY TRANSPOSED TEMPERATUR
YOST DM   0111   RAMAN SPECTRUM OF ARSENIC TRIFLUORIDE AND THE MOLECULAR CONSTANTS OF ARSENIC TRIFL
                      0251   RAMAN SPECTRUM OF BORON TRIFLUORIDE GAS.
                      0364   RAMAN SPECTRA OF CARBON AND SILICON TETRAFLUORIDES.
                      0822   RAMAN SPECTRA AND MOLECULAR CONSTANTS OF PHOSPHORUS TRIFLUORIDE AND PHOSPHINE.
                      0963   RAMAN SPECTRA AND MOLECULAR CONSTANTS OF THE HEXAFLUORIDES OF SULFUR, SELENIUM, AN
YURKOV GN   0789   THERMODYNAMIC FUNCTIONS OF DIATOMIC GASES WITH MOLECULES IN TRIPLET SIGMA STATES.
YUSHIN AS   0075   THERMODYNAMIC FUNCTIONS OF ANTIMONY FLUORIDE, NIOBIUM BROMIDE, TANTALUM CHLORIDE,
ZACHARIASEN WH   0447   EVIDENCE FOR CURIUM TETRAFLUORIDE.
ZAITSEV AA   0574   VALENCE VIBRATION FREQUENCIES OF TRIHALIDE MOLECULES OF LANTHANUM HALIDES.
ZALKIN A   1277   CRYSTAL AND MOLECULAR STRUCTURE OF XENON TETRAFLUORIDE.
ZASORIN EZ   0036   ELECTRON DIFFRACTION INVESTIGATION OF THE MOLECULAR STRUCTURES OF THE ALUMINUM HAL
ZAUCER M   0181   ESR PARAMETERS OF THE BORON DIFLUORIDE RADICAL.
ZAVALISHIN NI   0149   THERMODYNAMIC FUNCTIONS OF BISMUTH TRIHALIDE VAPORS.
ZEEMAN PB   0153   ROTATIONAL ANALYSIS OF BANDS OF THE C TRIPLET SIGMA, B TRIPLET SIGMA: A TRIPLET PI
                      0457   ROTATIONAL ANALYSIS OF BANDS OF THE A TRIPLET PI ZERO(+) B TRIPLET PI(1) TO X SING
                      1111   ROTATIONAL ANALYSIS OF BANDS OF THE A TRIPLET PI ZERO(+), B TRIPLET PI(1) TO X SIN
ZEMSKII GT   0088   ASSOCIATION OF ANTIMONY PENTAFLUORIDE IN VAPOR.
ZHEGUL'SKAYA NA   0040   MASS SPECTROMETRIC INVESTIGATION OF THE THERMODYNAMIC PROPERTIES OF ALUMINUM TRIFL
ZHIGULA LA   0313   INFRARED SPECTRA OF A CRYOSYSTEM. CARBON TETRAFLUORIDE.
ZILBERMAN TB   0432   THERMODYNAMIC PROPERTIES OF CHROMIUM DIFLUORIDE.
ZIMKINA TM   1033   SOFT X-RAY SPECTROSCOPY ON SULFUR TETRAFLUORIDE AND SULFUR HEXAFLUORIDE.
                      1065   ELECTRONIC STRUCTURE OF SULFUR HEXAFLUORIDE STUDIED BY ULTRALONG WAVE X-RAY SPECTR
                      1101   X-RAY SPECTRA OF SULFUR AND FLUORINE IN SULFUR HEXAFLUORIDE.
                      1102   ULTRASOFT X-RAY ABSORPTION SPECTRUM OF SULFUR HEXAFLUORIDE.
ZMBOV KF   0434   SUBLIMATION PRESSURES OF CHROMIUM TRIFLUORIDE, MANGANESE TRIFLUORIDE, AND IRON TRI
                      0448   MASS SPECTROMETRIC STUDIES AT HIGH TEMPERATURES. PART-17: SUBLIMATION AND VAPOR PR
                      0449   STABILITIES OF DYSPROSIUM TRIFLUORIDE, HOLMIUM TRIFLUORIDE, AND ERBIUM TRIFLUORIDE
                      0450   MASS SPECTROMETRIC STUDIES AT HIGH TEMPERATURES. PART-13: STABILITIES OF SAMARIUM,
                      0592   MASS SPECTROMETRIC STUDIES AT HIGH TEMPERATURES. (TIN DIFLUORIDE, LEAD DIFLUORIDE,
                      0643   SUBLIMATION PRESSURE OF NEODYMIUM TRIFLUORIDE AND THE STABILITIES OF GASEOUS NEODY
                      0939   MASS SPECTROMETRIC STUDIES AT HIGH TEMPERATURES. PART-18: THE STABILITIES OF THE M
                      0947   SUBLIMATION PRESSURES OF SCANDIUM TRIFLUORIDE, YTTRIUM TRIFLUORIDE, AND LANTHANUM
                      0953   MASS SPECTROMETRIC STUDIES OF SCANDIUM TRIFLUORIDE, YTTRIUM TRIFLUORIDE, LANTHANUM
                      1014   MASS SPECTROMETRIC STUDIES AT HIGH TEMPERATURES. PART-14: VAPOR PRESSURE AND DISSO
                      1103   STABILITIES OF TANTALUM PENTAFLUORIDE AND TANTALUM OXIDE TRIFLUORIDE.
                      1128   HEATS OF FORMATION GASEOUS THORIUM TRIFLUORIDE AND THORIUM DIFLUORIDE FROM MASS SP
                      1132   SUBLIMATION PRESSURES AND HEATS OF SUBLIMATION OF THULIUM TRIFLUORIDE, YTTERBIUM T
                      1145   MASS SPECTROMETRIC STUDIES AT HIGH TEMPERATURES. PART-16: SUBLIMATION PRESSURES FO
                      1359   MASS SPECTROMETRIC STUDIES OF SCANDIUM, YTTRIUM, LANTHANUM, AND RARE EARTH FLUORID
                      1382   SUBLIMATION PRESSURES OF REFRACTORY FLUORIDES.
ZOBEL CR   0365   VACUUM ULTRAVIOLET ABSORPTION SPECTRA OF SOME HALOGEN DERIVATIVES OF METHANE. (CAR
ZORN C   1227   PHOTOELECTRON SPECTRUM OF XENON DIFLUORIDE.
ZORN JC   0489   HYPERFINE STRUCTURE OF INDIUM MONOFLUORIDE.
ZUBOVA NV   1175   VAPOR PRESSURE OF URANIUM TETRAFLUORIDE.
ZUPAN J   0883   SUSCEPTIBILITY MEASUREMENTS AND FLUORINE NMR OF PLATINUM HEXAFLUORIDE.
ZWOLENIK JJ   0293   PREDISSOCIATION IN THE ABSORPTION SPECTRA OF CARBON MONOFLUORIDE AND CARBON DIFLUO
ZYUBIN AS   0009   CHARACTERISTICS OF THE ELECTRONIC STRUCTURE OF FLUORIDES OF NONTRANSITION ELEMENTS
                      1231   CALCULATIONS OF ELECTRONIC STRUCTURES OF COMPOUNDS OF HALOGENS AND INERT GASES BY
ZYUBINA TS   0009   CHARACTERISTICS OF THE ELECTRONIC STRUCTURE OF FLUORIDES OF NONTRANSITION ELEMENTS